Key Stage 3 Mathematics:

Revision and Practice

M Bindley   C Oliver   A Ledsham   R Elwin

OXFORD

UNIVERSITY PRESS

9112000026017l

OXFORD
UNIVERSITY PRESS

Great Clarendon Street, Oxford OX2 6DP

Oxford University Press in a department of the University of Oxford. It furthers the University's objective of excellence in research, scholarship, and education by publishing worldwide in

Oxford New York

Auckland Cape Town Dar es Salaam Hong Kong Karachi
Kuala Lumpur Madrid Melbourne Mexico City Nairobi
New Delhi Shanghai Taipei Toronto

With offices in

Argentina Austria Brazil Chile Czech Republic France
Greece Guatemala Hungary Italy Japan Poland Portugal
Singapore South Korea Switzerland Thailand Turkey
Ukraine Vietnam

Oxford is a registered trade mark of Oxford University Press
In the UK and in certain other countries

© Oxford University Press
First published 1998
20
ISBN 978 - 0199145607

A CIP catalogue record for this book is available from the British Library.

Typeset by Tech-Set Limited,Gateshead, Tyne and Wear
Illustrations by Tech-Set Limited and Oxford Illustrators

Printed and bound in China by Printplus Ltd.

Contents

Introduction

Key Stage 3 Mathematics: Revision and Practice has been created from the *Curriculum Mathematics Practice* series, also published by Oxford University Press. The present book has been designed to cover all the mathematics required for National Curriculum Levels 4 to 7. Because the vast majority of students are entered for Key Stage 3 tests at these levels, the book will be an excellent resource for Years 7, 8 and 9.

Each teaching unit begins with an exact statement of the National Curriculum skill to be covered. The required knowledge and skill examples are then clearly presented. They are followed by a vast range of carefully constructed and graded skill-practice exercises. The skill statements at the start of each unit are drawn directly from the Level Descriptions in the National Curriculum. It will therefore be easy for a mathematics department to use them as the basis of a simple and effective recording system.

The teaching units are organised into Levels within Attainment Targets. Thus, all of Number and Algebra Level 4 is presented first, then Number and Algebra Level 5 and so on. The only exception to this is transformation geometry, which is specified in the Programme of Study but which does not have a full set of corresponding Attainment Targets. The author has included this as a Level 5/6 Shape, space and measures unit. The order of presentation is not intended as a teaching sequence. Because each unit starts with a clear statement of Attainment Target skills, it will be easy for teachers to plan a coherent mathematical progression to suit the levels they wish to cover with their particular students.

A range of 100 actual SAT questions is presented at the end of the book, collected into three sections: Number and Algebra, Shape, space and measures and Handling data. These questions have been selected to illustrate the style of SAT questions over the past five years, which differ in several respects from questions in other types of examination. These will provide an excellent source of revision material in the final weeks before the examination. Some questions have had to be amended slightly to suit the book format and the available space, but their content and style remain essentially intact.

This book has the same objective as the *Curriculum Mathematics Practice* series. That is, to enable students 'to gain confidence in their abilities and master the fundamental processes so necessary for future success'.

The *Answer Book* provides solutions for all exercises with numerical or symbolic answers. Solutions are not provided for some questions leading only to an illustration (e.g. a graph).

Mark Bindley
January 1998

Unit 1

Number and Algebra   Level 4

Skill   You use your understanding of place value to multiply and divide whole numbers by 10 or 100.

Skill Example

Give the largest and smallest numbers that can be made using *all* the following digits.

a   3, 8, 6 and 1   **b**   9, 0, 1 and 9

a   Largest is 8631; smallest is 1368.
b   Largest is 9910; smallest is 1099.

Skill Practice One

Give the largest and smallest numbers that can be made using *all* the following digits.

1   3, 5 and 2   **2**   4, 9 and 3
3   6, 8 and 1   **4**   5, 3 and 5
5   7, 6 and 0   **6**   4, 3, 6 and 5
7   4, 1, 9 and 7   **8**   2, 8, 1 and 5
9   5, 1, 1 and 7   **10**   6, 3, 3 and 6
11   3, 5, 0 and 2   **12**   9, 7, 0 and 0

Knowledge

To multiply a number by 10, move each figure one place to the left and put a nought in the empty units place.

So   **a** $3 \times 10 = 3$ tens and 0 units $= 30$

        b $406 \times 10 = 4060$

Th	H	T	U		Th	H	T	U
	4	0	6		4	0	6	0

Skill Practice Two

Multiply each of these numbers by 10.

1 4	**2** 7	**3** 15	**4** 19
5 10	**6** 24	**7** 43	**8** 52
9 69	**10** 20	**11** 50	**12** 126
13 155	**14** 317	**15** 632	**16** 510
17 850	**18** 700	**19** 2317	**20** 3547
21 4620	**22** 7050	**23** 5300	**24** 8000
25 9901			

Knowledge

To divide a number by 10, move each figure one place to the right. The units figure becomes the remainder.

So   **a** $50 \div 10 = 5$ with a remainder of 0
            i.e. 5

        b $632 \div 10 = 63$ with a remainder of 2
            i.e. 63 r 2

Skill Practice Three

Divide each of these numbers by 10.

1 30	**2** 50	**3** 70	**4** 65
5 54	**6** 73	**7** 120	**8** 160
9 370	**10** 820	**11** 200	**12** 900
13 445	**14** 638	**15** 307	**16** 402
17 1320	**18** 4780	**19** 2300	**20** 5200
21 7000	**22** 5176	**23** 3078	**24** 4204
25 9001			

Knowledge

To multiply a number by 100, move each figure two places to the left.

So   **a** $3 \times 100 = 300$

      b $406 \times 100 = 40\,600$

Skill Practice Four

Multiply each of these numbers by 100.

1 3	**2** 9	**3** 13	**4** 16
5 10	**6** 25	**7** 51	**8** 78
9 30	**10** 60	**11** 124	**12** 237
13 519	**14** 708	**15** 630	**16** 810
17 400	**18** 900	**19** 100	**20** 999

Knowledge

To divide a number by 100, move each figure two places to the right.

So   **a** $500 \div 100 = 5$

      b $7263 \div 100 = 72$ r 63

Skill Practice Five

Divide each of these numbers by 100.

1 200	**2** 600	**3** 800	**4** 850
5 853	**6** 764	**7** 704	**8** 502
9 3200	**10** 9700	**11** 9710	**12** 4680
13 4683	**14** 5162	**15** 9375	**16** 4118
17 1384	**18** 3725	**19** 3705	**20** 5107
21 5007	**22** 2008	**23** 2030	**24** 8070

Skill Practice Six

1 Windows for this new building cost £54 each. What is the cost of providing windows for the front of the building?

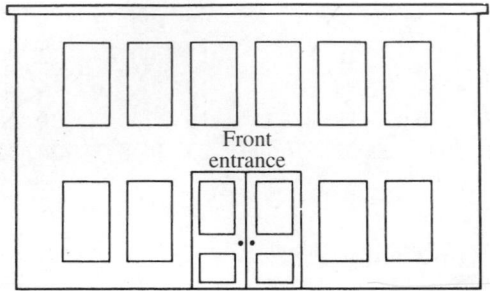

2 What is the height of the landing above the floor in **a** centimetres **b** metres?
(1 metre = 100 centimetres)

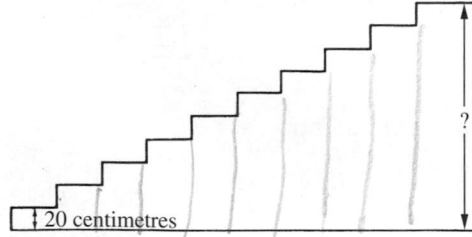

3 There are 100 kerb blocks between one end of Park Avenue and the other.
What is the length of the avenue in
a centimetres **b** metres?
(1 metre = 100 centimetres)

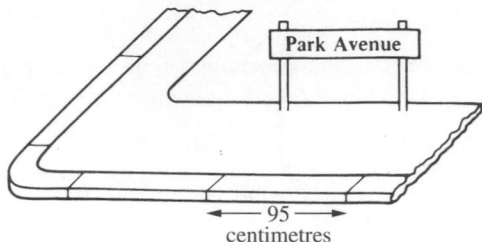

4 A teacher has 125 sheets of paper to share out amongst 10 pupils.
a How many sheets does each pupil receive?
b How many sheets does the teacher have left over?

5 A coal merchant has only 1318 bags of coal in his stock which he has to share out between 100 customers.
a How many bags can he supply to each customer?
b How many bags will he have left over?

6 What is the weight of 20 bars of chocolate like this one?

Unit 2

8 **a** $34 + 25$ **b** $28 + 31$
   c $49 + 11$ **d** $17 + 42$

9 **a** $43 + 35$ **b** $59 + 19$
   c $54 + 25$ **d** $26 + 52$

10 **a** $81 + 17$ **b** $75 + 22$
   c $18 + 79$ **d** $51 + 46$

11 **a** $62 + 55$ **b** $66 + 51$
   c $28 + 89$ **d** $19 + 99$

12 **a** $131 + 26$ **b** $140 + 17$
   c $45 + 122$ **d** $89 + 68$

Skill Example

Add each of the following 'in your head' to find the 'odd answer out'.

a $16 + 22$ **b** $27 + 11$
c $22 + 19$ **d** $19 + 19$

This is one way to think the problems through:

a $16 + 22 = 10 + 6 + 20 + 2 = 30 + 8 = 38$

b $27 + 11 = 20 + 7 + 10 + 1 = 30 + 8 = 38$

c $22 + 19 = 20 + 2 + 10 + 9 = 30 + 11 = 41$

d $19 + 19 = 10 + 9 + 10 + 9 = 20 + 18 = 38$

So **c** has the 'odd answer out' because its answer is 41.

Skill Example

Add each of the following to find the 'odd answer out'.

a $6 + 8 + 4 + 2$ **b** $7 + 2 + 9 + 3$
c $6 + 4 + 5 + 6$ **d** $8 + 1 + 5 + 7$

a	6	**b**	7	**c**	6	**d**	8
	8		2		4		1
	4		9		5		5
	$+\,2$		$+\,3$		$+\,6$		$+\,7$
	20		21		21		21

So **a** has the 'odd answer out' because its answer is 20.

Skill Practice One

For each question, add each pair of numbers 'in your head' to find the 'odd answer out'.

1 **a** $12 + 7$ **b** $11 + 8$
   c $13 + 7$ **d** $14 + 5$

2 **a** $14 + 14$ **b** $13 + 15$
   c $19 + 9$ **d** $14 + 15$

3 **a** $18 + 8$ **b** $16 + 12$
   c $17 + 9$ **d** $13 + 13$

4 **a** $21 + 12$ **b** $25 + 7$
   c $14 + 18$ **d** $15 + 17$

5 **a** $25 + 18$ **b** $27 + 17$
   c $20 + 23$ **d** $22 + 21$

6 **a** $51 + 12$ **b** $45 + 18$
   c $49 + 11$ **d** $48 + 15$

7 **a** $23 + 27$ **b** $34 + 17$
   c $37 + 13$ **d** $34 + 16$

Skill Practice Two

For each question, add each set of numbers to find the 'odd answer out'.

1	**a**	2	**b**	7	**c**	8	**d**	4
		5		2		1		3
		3		3		6		7
		$+\,8$		$+\,6$		$+\,4$		$+\,4$

2	**a**	4	**b**	6	**c**	8	**d**	5
		3		2		1		7
		8		9		7		3
		$+\,7$		$+\,5$		$+\,6$		$+\,8$

3	**a**	7	**b**	6	**c**	8	**d**	6
		8		8		4		9
		9		5		5		4
		$+\,3$		$+\,7$		$+\,9$		$+\,7$

4	**a**	8	**b**	5	**c**	9	**d**	7
		7		9		2		5
		6		7		8		9
		$+\,3$		$+\,4$		$+\,5$		$+\,3$

4

5 a $9 + 6 + 7 + 7$ **b** $8 + 5 + 6 + 9$
  c $7 + 9 + 4 + 8$ **d** $8 + 8 + 5 + 7$

6 a $8 + 9 + 7 + 6$ **b** $9 + 7 + 9 + 5$
  c $9 + 8 + 8 + 5$ **d** $7 + 7 + 8 + 9$

7

a		b		c		d	
	16		17		15		18
	15		12		13		11
	13		12		13		12
	+ 11		+ 13		+ 14		+ 14

8

a		b		c		d	
	19		17		18		16
	13		14		15		15
	15		15		14		17
	+ 12		+ 12		+ 11		+ 10

9 a $15 + 13 + 17 + 11$ **b** $14 + 12 + 16 + 15$
  c $16 + 14 + 14 + 13$ **d** $13 + 12 + 19 + 13$

10 a $18 + 13 + 22 + 24$ **b** $17 + 15 + 11 + 34$
   c $13 + 27 + 11 + 26$ **d** $32 + 14 + 15 + 15$

11

a		b		c		d	
	15		13		9		3
	6		8		11		8
	4		5		6		12
	+ 2		+ 2		+ 1		+ 4

12

a		b		c		d	
	21		14		12		3
	16		22		9		11
	13		7		24		13
	+ 6		+ 12		+ 10		+ 28

13 a $11 + 8 + 13 + 9$ **b** $15 + 4 + 17 + 5$
   c $9 + 15 + 14 + 4$ **d** $7 + 14 + 16 + 4$

14 a $24 + 7 + 18 + 5$ **b** $15 + 8 + 26 + 4$
   c $6 + 28 + 16 + 3$ **d** $4 + 19 + 25 + 5$

15 a $32 + 6 + 19 + 8$ **b** $7 + 26 + 28 + 4$
   c $11 + 8 + 37 + 9$ **d** $5 + 29 + 22 + 8$

Skill Example

Add each of the following to find the 'odd answer out'.

a $201 + 73 + 429$ **b** $300 + 16 + 397$
c $199 + 187 + 317$

a		b		c	
	201		300		199
	73		16		187
	+ 429		+ 397		+ 317
	703		713		703

So **b** has the 'odd answer out' because its answer is 713.

Skill Practice Three

For each question, add each set of numbers to find the 'odd answer out'.

1

a		b		c	
	234		127		15
	116		47		256
	+ 28		+ 205		+ 107

2

a		b		c	
	215		124		173
	43		231		64
	+ 174		+ 77		+ 205

3 a $167 + 45 + 352$
  b $241 + 187 + 36$
  c $154 + 82 + 328$

4 a $326 + 145 + 44$
  b $275 + 36 + 214$
  c $33 + 255 + 227$

5

a		b		c	
	232		42		61
	20		251		43
	+ 72		+ 31		+ 230

6 a $46 + 213 + 36$
  b $230 + 39 + 27$
  c $64 + 209 + 23$

7

a		b		c	
	223		42		36
	18		217		24
	32		25		212
	+ 12		+ 11		+ 13

8 a $34 + 13 + 32 + 147$
  b $15 + 34 + 114 + 53$
  c $25 + 132 + 33 + 26$

9 a $275 + 41 + 16 + 52$
  b $57 + 240 + 34 + 43$
  c $63 + 24 + 245 + 42$

10 a $55 + 20 + 31 + 211$
   b $43 + 21 + 241 + 12$
   c $31 + 222 + 24 + 50$

11 a $260 + 113 + 32 + 52$
   b $14 + 151 + 231 + 62$
   c $173 + 12 + 30 + 242$

12 a $57 + 12 + 303 + 211$
   b $144 + 401 + 26 + 13$
   c $22 + 214 + 333 + 15$

Skill Example

Subtract each of the following 'in your head' to find the 'odd answer out'.

a $26 - 12$ **b** $88 - 74$
c $50 - 36$ **d** $43 - 28$

This is one way to think the problems through:

a $(20 - 10)$ $(6 - 2)$
     10 $+$ 4 $= 14$

b $(80 - 70)$ $(8 - 4)$
     10 $+$ 4 $= 14$

c $(50 - 30)$ $(0 - 6)$
     20 $-$ 6 $= 14$

d $(40 - 20)$ $(3 - 8)$
     20 $-$ 5 $= 15$

So **d** has the 'odd answer out' because its answer is 15.

Skill Practice Four

For each question, subtract each pair of numbers 'in your head' to find the 'odd answer out'.

1 a $58 - 16$ **2 a** $97 - 43$ **3 a** $68 - 12$
   b $84 - 41$ **b** $74 - 20$ **b** $97 - 51$
   c $75 - 33$ **c** $89 - 35$ **c** $86 - 40$
   d $97 - 55$ **d** $66 - 11$ **d** $79 - 33$

4 a $96 - 33$ **5 a** $78 - 26$ **6 a** $96 - 51$
   b $75 - 11$ **b** $94 - 43$ **b** $49 - 4$
   c $87 - 24$ **c** $59 - 7$ **c** $75 - 40$
   d $69 - 6$ **d** $66 - 14$ **d** $57 - 12$

7 a $39 - 3$ **8 a** $79 - 35$ **9 a** $72 - 37$
   b $67 - 31$ **b** $54 - 20$ **b** $94 - 59$
   c $46 - 10$ **c** $86 - 52$ **c** $63 - 28$
   d $98 - 52$ **d** $97 - 63$ **d** $81 - 45$

10 a $75 - 28$ **11 a** $44 - 28$ **12 a** $62 - 34$
   b $91 - 44$ **b** $63 - 46$ **b** $45 - 16$
   c $63 - 15$ **c** $71 - 55$ **c** $86 - 57$
   d $86 - 39$ **d** $55 - 39$ **d** $74 - 45$

Skill Example

Subtract to find the difference between each pair of numbers and so find the 'odd answer out'.

a 676 and 394
b 309 and 27
c 584 and 292

a $676 - 394 = 282$ **b** $309 - 27 = 282$ **c** $584 - 292 = 292$

So **c** has the 'odd answer out' because its answer is 292.

Skill Practice Five

For each question, subtract to find the difference between each pairs of numbers and so find the 'odd answer out'.

1 a 797 and 352   **2 a** 586 and 214
   b 856 and 421   **b** 794 and 432
   c 669 and 234   **c** 678 and 306

3 a 897 and 384   **4 a** 895 and 432
   b 759 and 236   **b** 668 and 205
   c 983 and 460   **c** 789 and 327

5 a 397 and 74   **6 a** 729 and 243
   b 365 and 52   **b** 857 and 381
   c 386 and 63   **c** 648 and 162

7 a 615 and 253   **8 a** 381 and 257
   b 538 and 186   **b** 550 and 426
   c 807 and 445   **c** 272 and 138

9 a 422 and 143   **10 a** 614 and 31
   b 614 and 325   **b** 638 and 54
   c 541 and 262   **c** 606 and 23

Skill Example

In a competition, five balls are rolled into numbered slots. If four balls have been rolled as shown, what is:

a the total score

b the score that the fifth ball must make to win (i) a 'free go' (ii) a prize?

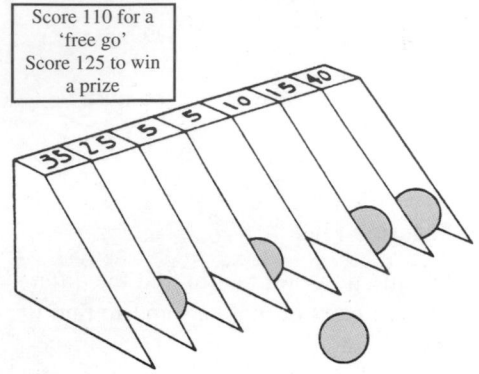

Score 110 for a 'free go'
Score 125 to win a prize

a Total score = 25 + 5 + 15 + 40 = 85

b (i) To win a 'free go' the fifth ball must score

$$110 - 85 = 25$$

(ii) To win a prize the fifth ball must score

$$125 - 85 = 40$$

Always use commonsense to check your answers.

Suppose you had set out the calculation for part **a** like this:

$$
\begin{array}{r}
25 \\
5 \quad \longleftarrow \text{Mistake} \\
15 \\
+\,40 \\
\hline
130 \\
\hline
\end{array}
$$

The size of the numbers should tell you that your answer cannot possibly be 130.

Skill Practice Six

1 The milkman uses a crate containing 20 bottles to deliver milk to Short Street. How many bottles will he have left when he gets to the end of the street?

> **MILK ORDERS FOR SHORT STREET**
>
> | | |
> |---|---|
> | Mrs. Jones, No. 1 | 4 Bottles |
> | Mr. Adams, No. 2 | 2 Bottles |
> | Mrs. Butler, No. 3 | 1 Bottle |
> | Miss. Haynes, No. 4 | None today |
> | Mrs. Ashurst, No. 5 | 3 Bottles |
> | Mr. Biggs, No. 6 | 1 Bottle |

2 Natalie arrived at the railway station 15 minutes before her train was due, but the train arrived 27 minutes late.
How many minutes did she have to wait?
How many minutes less than one hour did she have to wait?

3 A man bought a coffee and a slice of cake at the snack bar.
If he gave a £5 note, what change should he have received?

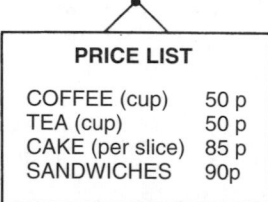

PRICE LIST	
COFFEE (cup)	50 p
TEA (cup)	50 p
CAKE (per slice)	85 p
SANDWICHES	90p

4 A school relay team complete a 4 by 100 metres run in 51 seconds.
If the first, second and third runners complete their sprints in 12, 14 and 13 seconds respectively, what is the last runner's time?

5 A footballer scores an equalizing goal after 57 minutes. If the match is scheduled to last 90 minutes, how many minutes are left after he scores?

However, the match ends in a draw, and 30 minutes of extra time are played.

How long after his goal does the match finally end?

6

At Willow Bank School all competitors have to enter for their events at least 14 days before Sports Day.

What is the latest date for entries?

If Sports Day is 13 days before the end of the term, on what date does the school close for the summer holiday?

7 At Willow Bank School there are 455 girls and 428 boys.

How many pupils arc there altogether?

How many more girls are there than boys?

8 My car has 23 litres of petrol in its tank. I use up 8 litres in driving to the seaside where I fill up the tank.

If the tank holds 32 litres, how many litres do I put in?

9 Peter and William compete in an archery contest. Look at their scores below.

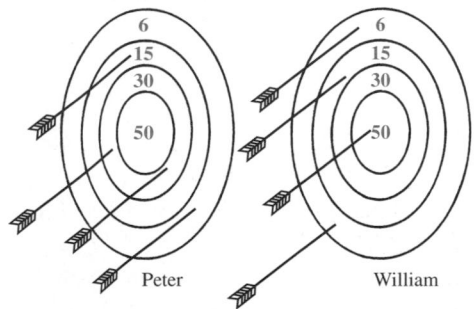

Who has won the contest and by how many points?

10 Which is the shorter way from Liverpool to Manchester and by how many kilometres?

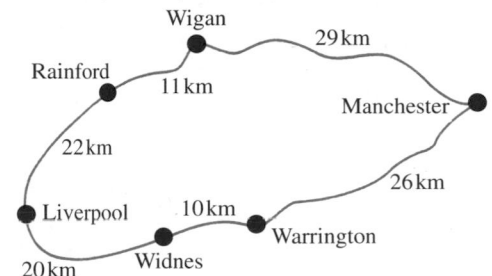

11 At West Hill School, the library books are classed as fiction, non-fiction or technical. The number of books borrowed by four pupils during a certain week are shown below.

	Jandeep	Peter	Michael	Tanya
Fiction	1	0	0	2
Non-fiction	2	0	1	1
Technical	2	1	1	0

a How many technical books did Michael borrow?

b How many fiction books did Jandeep borrow?

c How many fiction books were borrowed?

d How many books did Peter borrow?

e Who borrowed no technical books?

f How many books were borrowed?

12 All long-distance trains from Newcastle upon Tyne station use platform 8, platform 9 or platform 10. The table shows the numbers of trains to the various destinations.

	London	Scotland	Birmingham	Liverpool
Platform 8	4	4	5	2
Platform 9	6	2	3	1
Platform 10	6	4	1	3

a How many trains for Birmingham leave from platform 9?
b Which platform has only one train leaving for Liverpool?
c How many of the trains leave from platform 10?
d How many of the trains leave for Scotland?
e Which platform has the most long-distance trains leaving?
f How many long-distance trains leave Newcastle upon Tyne station?

13 The table below shows how many vehicles used a ferry on each day of an Easter holiday period.

	Cars	Vans	Motorcycles
Friday	50	20	20
Saturday	40	15	15
Sunday	60	15	25
Monday	50	10	30
Tuesday	30	5	15

a How many vans used the ferry on Monday?
b On which day were the greatest number of motor cycles carried?
c On which day were the greatest number of vehicles carried?
d On which day were the least number of vehicles carried?
e How many cars were carried during the holiday?
f How many vehicles were carried during the holiday?

14 At West Hill School the pupils are given their examination results by being told that they have either a pass, a credit or a failure. The table below shows the results of four pupils.

	Kelly	Theo	Luke	Shona
Passes	5	3	4	5
Credits	1	4	1	3
Failures	2	1	3	0

a Which pupil did not fail any examination?
b Which pupil gained the most credits?
c In how many subjects did the pupils take examinations?
d How many pass grades were obtained altogether?
e A pass is worth 3 points, a credit is worth 7 points and a failure is worth 0 points. How many points did Luke score?
f Which pupil scored the most points?

Knowledge

These are important number facts.

	1	2	3	4	5	6	7	8	9	10
1	1	2	3	4	5	6	7	8	9	10
2	2	4	6	8	10	12	14	16	18	20
3	3	6	9	12	15	18	21	24	27	30
4	4	8	12	16	20	24	28	32	36	40
5	5	10	15	20	25	30	35	40	45	50
6	6	12	18	24	30	36	42	48	54	60
7	7	14	21	28	35	42	49	56	63	70
8	8	16	24	32	40	48	56	64	72	80
9	9	18	27	36	45	54	63	72	81	90
10	10	20	30	40	50	60	70	80	90	100

Skill Example

Multiply each of the following to find the 'odd answer out'.

a 6×6 b 5×7 c 12×3 d 4×9

a 6×6 b 5×7 c 12×3 d 4×9
  $= 36$ $= 35$ $= 36$ $= 36$

So b has the 'odd answer out' because its answer is 35.

Skill Practice Seven

For each question, multiply each pair of numbers to find the 'odd answer out'.

1 a 8×3 b 5×5 c 6×4 d 2×12

2 a 5×4 3 a 7×2 4 a 4×5
   b 3×7 b 3×4 b 9×2
   c 10×2 c 2×6 c 3×6

5 a 6×8 6 a 4×10 7 a 3×5
   b 12×4 b 7×6 b 2×8
   c 7×7 c 8×5 c 4×4

8 a 10×7 9 a 3×10 10 a 12×5
   b 8×9 b 6×5 b 7×9
   c 6×12 c 8×4 c 6×10

Skill Example

Multiply each of the following to find the 'odd answer out'.

a 49×6 b 42×7 c 38×8

a   49   b   42   c   38
  $\times 6$ $\times 7$ $\times 8$
  294   294   304

So c has the 'odd answer out' because its answer is 304.

Skill Practice Eight

For each question, multiply each pair of numbers to find the 'odd answer out'.

1 a 24×4 2 a 57×4 3 a 15×7
   b 16×6 b 76×3 b 27×5
   c 14×7 c 34×7 c 45×3

4 a 35×9 5 a 22×7 6 a 197×4
   b 67×5 b 16×9 b 266×3
   c 45×7 c 18×8 c 114×7

7 a 119×8 8 a 153×5 9 a 124×7
   b 314×3 b 85×9 b 207×4
   c 136×7 c 105×7 c 138×6

Skill Example

Multiply each of the following:

a 32×30 b 27×600

a $32 \times 30 = 320 \times 3$ b $27 \times 600 = 2700 \times 6$
  $= 960$ $= 16\,200$

Skill Practice Nine

1 36×20 2 14×60

3 16×70 4 135×30

5 216×40 6 142×60

7 156×50 8 80×60

9 400×20 10 28×200

11 32×300 12 148×200

13 232×400 14 104×700

15 105×800 16 160×600

10

Skill Example

Find the product of:

a 43 and 27   **b** 104 and 16

You can use a calculator, or find the answers like this:

a	43		**b**	104	
	×27			×16	
	860	(43 × 20)		624	(104 × 6)
	301	(43 × 7)		1040	(104 × 10)
	1161			1664	

Skill Practice Ten

For each question, multiply each pair of numbers to find the 'odd answer out'.

1 a 21 × 16
  b 19 × 18
  c 24 × 14

2 a 34 × 18
  b 32 × 19
  c 51 × 12

3 a 62 × 11
  b 48 × 14
  c 42 × 16

4 a 36 × 15
  b 35 × 16
  c 40 × 14

5 a 28 × 24
  b 32 × 21
  c 31 × 22

6 a 32 × 27
  b 31 × 28
  c 36 × 24

7 a 46 × 33
  b 42 × 36
  c 54 × 28

8 a 42 × 24
  b 39 × 26
  c 36 × 28

9 a 135 × 14
  b 124 × 15
  c 105 × 18

10 a 114 × 17
  b 102 × 19
  c 137 × 14

Skill Example

A market gardener planted 27 rows of potatoes and put 42 potatoes in each row.

How many potatoes did she plant altogether?

Number of potatoes planted = 27 × 42

27	
×42	
1080	(27 × 40)
54	(27 × 2)
1134	

The market gardener planted 1134 potatoes.

Skill Practice Eleven

1 How high is the wall if each brick is 8 cm in height?

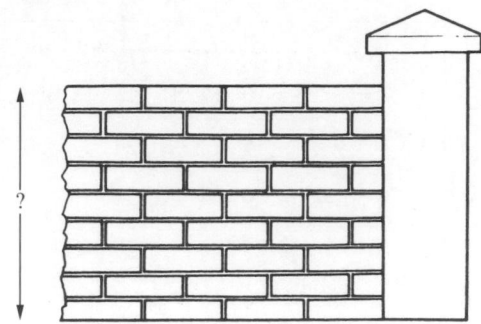

2 The distance from Dover to Calais is 32 kilometres.
  a What is the distance from Newhaven to Dieppe if it is 3 times as far?
  b What is the distance from Weymouth to St Hélier if it is 4 times as far?

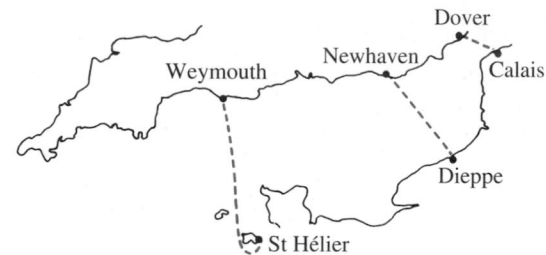

3 Walton Hill, near Halesowen, is 316 metres above sea level. Helvellyn in Cumbria is three times this height.
What is the height of Helvellyn?

4 What is the length of this viaduct?

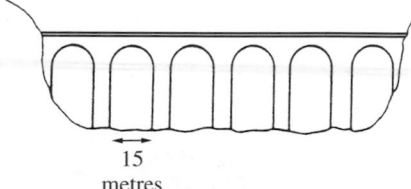

15
metres

5 How many small panes are there in the windows at the front of this school building?

LEA ROAD SCHOOL

6 A railway locomotive is pulling a train of 12 coaches, each of which has 8 wheels.
If the locomotive also has 8 wheels, how many wheels are rolling altogether?

7 A van delivers 5 cases of lemonade cans to a shop. If each case contains 24 cans, how many cans are delivered?

8 In a new office building 35 doors are required. If each door is fastened by means of 3 hinges, how many hinges are needed?
What is the total number of screws that are required if each hinge has 6 screw-holes?

9 During a certain week 26 lorry loads of stone were removed from a quarry.
If each lorry carried 15 tonnes, how many tonnes of stone were removed?

10 The house illustrated is the first of 25 similar ones in a terrace.
What is the length of the terrace?

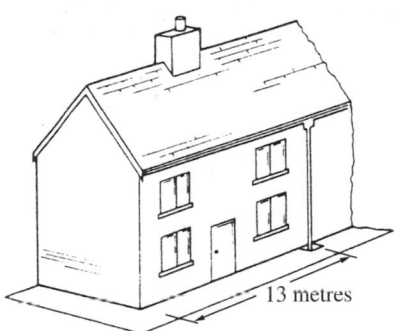

13 metres

Skill Example

a $72 \div 8$ **b** $72 \div 4$

a
$$8\overline{)72} \quad \frac{9}{}$$
$$\underline{72}$$

b
$$4\overline{)72} \quad \frac{18}{}$$
$$\underline{4} \quad (4 \times 1)$$
$$32$$
$$\underline{32} \quad (4 \times 8)$$

Skill Practice Twelve

For each question, divide each pair of numbers to find the 'odd answer out'.

1	**a** $42 \div 7$	**2**	**a** $18 \div 3$	**3**	**a** $63 \div 9$
	b $56 \div 8$		**b** $54 \div 9$		**b** $48 \div 6$
	c $30 \div 5$		**c** $36 \div 6$		**c** $64 \div 8$
	d $24 \div 4$		**d** $35 \div 7$		**d** $40 \div 5$
4	**a** $42 \div 3$	**5**	**a** $78 \div 6$	**6**	**a** $85 \div 5$
	b $98 \div 7$		**b** $56 \div 4$		**b** $51 \div 3$
	c $84 \div 6$		**c** $65 \div 5$		**c** $72 \div 4$
	d $52 \div 4$		**d** $91 \div 7$		**d** $34 \div 2$
7	**a** $76 \div 4$	**8**	**a** $36 \div 3$	**9**	**a** $38 \div 2$
	b $36 \div 2$		**b** $84 \div 7$		**b** $95 \div 5$
	c $54 \div 3$		**c** $96 \div 8$		**c** $64 \div 4$
	d $90 \div 5$		**d** $70 \div 5$		**d** $57 \div 3$

Skill Example

Divide each of the following to find the 'odd answer out'.

a $224 \div 7$ **b** $256 \div 8$ **c** $165 \div 5$

a
$$7\overline{)224} \quad \frac{32}{}$$
$$\underline{21} \quad (7 \times 3)$$
$$14$$
$$\underline{14} \quad (7 \times 2)$$

b
$$8\overline{)256} \quad \frac{32}{}$$
$$\underline{24} \quad (3 \times 8)$$
$$16$$
$$\underline{16} \quad (2 \times 8)$$

c
$$5\overline{)165} \quad \frac{33}{}$$
$$\underline{15} \quad (5 \times 3)$$
$$15$$
$$\underline{15} \quad (5 \times 3)$$

So **c** has the 'odd answer out' because its answer is 33.

12

Skill Practice Thirteen

For each question, divide each pair of numbers to find the 'odd answer out'.

1	a	$195 \div 5$	2	a	$288 \div 6$	3	a	$366 \div 6$
	b	$117 \div 3$		b	$376 \div 8$		b	$248 \div 4$
	c	$259 \div 7$		c	$336 \div 7$		c	$549 \div 9$
	d	$312 \div 8$		d	$240 \div 5$		d	$305 \div 5$

4	a	$369 \div 3$	5	a	$847 \div 7$	6	a	$952 \div 8$
	b	$791 \div 7$		b	$524 \div 4$		b	$645 \div 5$
	c	$565 \div 5$		c	$655 \div 5$		c	$714 \div 6$

7	a	$696 \div 6$	8	a	$545 \div 5$	9	a	$640 \div 4$
	b	$742 \div 7$		b	$981 \div 9$		b	$540 \div 3$
	c	$954 \div 9$		c	$476 \div 4$		c	$800 \div 5$

Skill Practice Fourteen

1 The bottle contains 620 millilitres of lemonade and its contents exactly fill all of the glasses shown.
What is the capacity of each of the glasses?

2 A poultry farmer has 396 eggs which are to be placed in cartons containing 6 eggs each.
How many cartons will he need?

3 Jack's birthday is 133 days after Aisha's.
How many weeks are there between their birthdays?

4 A multistorey block has 104 flats altogether and there are 8 flats on each floor.
How many storeys does the building have?

5 A pipeline is to be laid from a village to a nearby reservoir 960 metres away.
If the pipes are each of length 15 metres, how many will be needed?

6 In this gate the spaces between the wooden palings are the same width as the palings.
What is the width of each paling?

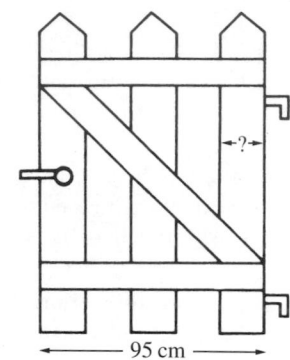

7 The distance by rail from London to Carmarthen is 357 kilometres.
If the stations shown are equally spaced, what is the distance from Swindon to Cardiff?

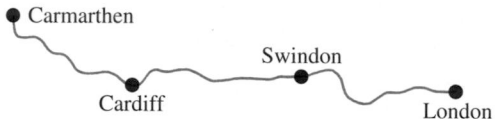

8 A van weighs 2000 kg when unloaded and 2360 kg when loaded with 8 bags of coal.
What is the weight of each bag?

Unit 3

Skill Example

To add decimals, put the decimal points underneath each other to make sure that each figure is in its proper place.

Add each of the following to find the 'odd answer out'.

a $0.4 + 0.5 + 0.6$ **b** $0.2 + 0.7 + 0.6$
c $0.9 + 0.7 + 0.9$ **d** $0.8 + 0.2 + 0.5$

a	0.4	**b**	0.2	**c**	0.9	**d**	0.8
	0.5		0.7		0.7		0.2
	+0.6		+0.6		+0.9		+0.5
	1.5		1.5		2.5		1.5

So **c** is the 'odd answer out'.

Skill Practice One

For each question, add each set of numbers to find the 'odd answer out'.

1 **a** $0.2 + 0.7 + 0.4$ **2** **a** $0.6 + 0.8 + 0.4$
  b $0.6 + 0.1 + 0.8$ **b** $0.5 + 0.9 + 0.2$
  c $0.3 + 0.2 + 0.8$ **c** $0.4 + 0.5 + 0.7$
  d $0.7 + 0.3 + 0.3$ **d** $0.3 + 0.8 + 0.5$

3 **a** $0.9 + 0.8 + 0.6$ **4** **a** $0.9 + 0.7 + 0.8$
  b $0.6 + 0.5 + 1.2$ **b** $0.2 + 0.5 + 1.5$
  c $1.2 + 0.7 + 0.4$ **c** $1.3 + 0.4 + 0.7$
  d $0.6 + 0.8 + 0.7$ **d** $1.1 + 1.2 + 0.1$

5 **a** $1.3 + 0.5 + 0.8$ **6** **a** $1.4 + 1.5 + 0.3$
  b $0.4 + 1.7 + 0.5$ **b** $0.2 + 1.7 + 1.5$
  c $1.2 + 1.4 + 0.2$ **c** $1.7 + 0.9 + 0.6$
  d $0.9 + 0.9 + 0.8$ **d** $0.5 + 1.8 + 0.9$

7 **a** $1.5 + 1.3 + 1.8$ **8** **a** $2.8 + 1.9 + 0.8$
  b $2.8 + 0.6 + 1.4$ **b** $2.5 + 2.8 + 0.4$
  c $3.8 + 0.7 + 0.3$ **c** $1.5 + 0.9 + 3.3$
  d $2.2 + 2.5 + 0.1$ **d** $2.2 + 1.4 + 2.1$

Skill Example

Add each of the following to find the 'odd answer out'.

a $3.46 + 1.2 + 5.36$
b $3.34 + 4 + 2.78$ (Hint: write 4 as 4.0)
c $2.14 + 5.08 + 2.9$

a	3.46	**b**	3.34	**c**	2.14
	1.2		4.0		5.08
	+5.36		+2.78		+2.9
	10.02		10.12		10.12

So **a** is the 'odd answer out'.

Skill Practice Two

For each question, add each set of numbers to find the 'odd answer out'.

1 **a** $2.13 + 4.37 + 3.12$ **2** **a** $2.28 + 2.16 + 3.4$
  b $2.65 + 4.21 + 2.56$ **b** $4.29 + 1.3 + 2.15$
  c $5.81 + 2.37 + 1.44$ **c** $3.9 + 2.31 + 1.53$

3 **a** $2.63 + 1.4 + 2.5$ **4** **a** $5.82 + 3.31 + 1.33$
  b $2.1 + 3.38 + 1.15$ **b** $4.92 + 3.14 + 2.4$
  c $2.89 + 0.24 + 3.5$ **c** $6.56 + 1.7 + 2.3$

5 **a** $5.49 + 2.8 + 2.46$ **6** **a** $3.53 + 4.14 + 1.39$
  b $4.93 + 3.09 + 2.75$ **b** $2.21 + 5.12 + 1.75$
  c $5.97 + 1.78 + 3$ **c** $5.46 + 3.02 + 0.6$

7 **a** $8.43 + 4.07 + 2.82$ **8** **a** $7.62 + 3.24 + 1.64$
  b $5.78 + 3.44 + 6.2$ **b** $5.19 + 3.06 + 4.25$
  c $2.09 + 4.33 + 9$ **c** $3.97 + 5.57 + 2.76$

9 **a** $5.65 + 0.81 + 2.44$
  b $4.73 + 1.02 + 2.34$
  c $2.53 + 5.2 + 1.17$

10 **a** $5.41 + 4.32 + 1.27$
  b $4.6 + 5.16 + 2.24$
  c $3.92 + 4.6 + 3.48$

11 **a** $6.5 + 9.84 + 3.66$
  b $9.57 + 4.43 + 7$
  c $8.91 + 5.03 + 6.06$

12 **a** $12.48 + 11.37 + 1.15$
  b $11.01 + 10.65 + 2.34$
  c $12.08 + 11.3 + 1.62$

Skill Example

To subtract decimals, make sure that the figures are in their right places by putting the decimal points underneath each other.

Subtract each of the following to find the 'odd answer out'.

a 2.6 − 1.4   b 2.4 − 1.2
c 3.7 − 2.5   d 9.6 − 8.5

a	2.6	b	2.4	c	3.7	d	9.6
	− 1.4		− 1.2		− 2.5		− 8.5
	1.2		1.2		1.2		1.1

So **d** is the 'odd answer out'.

Skill Example

Subtract each of the following to find the 'odd answer out'.

a 4 − 2.6   (Write 4 as 4.0)
b 5.2 − 2.8
c 9 − 6.6   (Write 9 as 9.0)
d 6.6 − 4.2

a	4.0	b	5.2	c	9.0	d	6.6
	− 2.6		− 2.8		− 6.6		− 4.2
	1.4		2.4		2.4		2.4

So **a** is the 'odd answer out'.

Skill Practice Three

For each question, subtract each pair of numbers to find the 'odd answer out'.

1 a 3.9 − 2.3   2 a 5.9 − 3.6
   b 6.6 − 5.2   b 4.8 − 2.3
   c 5.5 − 4.1   c 7.5 − 5.2
   d 2.8 − 1.4   d 6.4 − 4.1

3 a 5.6 − 3.4   4 a 4.8 − 1.5
   b 8.9 − 6.5   b 7.6 − 4.3
   c 9.3 − 7.1   c 6.5 − 3.2
   d 3.5 − 1.3   d 9.7 − 6.6

5 a 7.8 − 3.3   6 a 9.9 − 6.5
   b 5.9 − 1.4   b 5.6 − 2.2
   c 8.7 − 4.1   c 8.4 − 5
   d 6.5 − 2   d 6.8 − 3.3

7 a 6.9 − 1.8   8 a 8.5 − 2.2
   b 8.7 − 3.5   b 7.6 − 1.4
   c 9.6 − 4.4   c 9.3 − 3.1
   d 7.8 − 2.6   d 6.9 − 0.7

9 a 6.9 − 1.4   10 a 9.6 − 3.2
   b 7.8 − 2.3   b 7.8 − 1.3
   c 9.7 − 4.1   c 6.9 − 0.5
   d 8.5 − 3   d 8.5 − 2.1

Skill Practice Four

For each question, subtract each pair of numbers to find the 'odd answer out'.

1 a 4.4 − 1.8   2 a 7.4 − 5.6
   b 6.5 − 3.9   b 8.5 − 6.8
   c 7.1 − 4.5   c 3.2 − 1.5
   d 5.2 − 2.7   d 5.6 − 3.9

3 a 5.2 − 2.3   4 a 6.2 − 2.6
   b 7.5 − 4.6   b 5.5 − 1.8
   c 6.6 − 3.8   c 8.3 − 4.7
   d 8.3 − 5.4   d 7.1 − 3.5

5 a 8.2 − 3.4   6 a 8.3 − 2.7
   b 6.5 − 1.7   b 7.2 − 1.5
   c 7.4 − 2.6   c 9.5 − 3.8
   d 9.7 − 4.8   d 6.6 − 0.9

7 a 9.3 − 4.5   8 a 5.3 − 2.8
   b 6.6 − 1.8   b 6.4 − 3.9
   c 8.4 − 3.7   c 4.1 − 1.7
   d 5.2 − 0.4   d 7 − 4.5

9 a 8.5 − 4.9   10 a 7.5 − 1.8
   b 5.4 − 1.8   b 9.3 − 3.7
   c 4.3 − 0.7   c 8.1 − 2.5
   d 9 − 5.6   d 6 − 0.4

Skill Example

Subtract to find the difference between each pair of numbers and so find the 'odd answer out'.

a 7.5 and 1.26   (Write 7.5 as 7.50)
b 34 and 27.66   (Write 34 as 34.00)
c 10.63 and 4.39

a	7.50	**b**	34.00	**c**	10.63
	− 1.26		− 27.66		− 4.39
	6.24		6.34		6.24

So **b** is the 'odd answer out'.

Skill Practice Five

For each question, subtract to find the difference between each pair of numbers and so find the 'odd answer out'.

1 a 8.59 and 3.25
  b 6.62 and 1.38
  c 9.23 and 3.89

2 a 9.39 and 4.83
  b 5.73 and 1.27
  c 8.14 and 3.68

3 a 8.32 and 2.87
  b 9.04 and 3.49
  c 7.07 and 1.52

4 a 6.58 and 0.23
  b 8.36 and 2.01
  c 9.27 and 3.02

5 a 5.49 and 0.86
  b 5.17 and 0.64
  c 9.62 and 5.09

6 a 7.95 and 4.5
  b 5.35 and 1.8
  c 9.15 and 5.6

7 a 7.27 and 2.9
  b 4.87 and 0.4
  c 5.17 and 0.8

8 a 9.7 and 4.28
  b 6.8 and 1.19
  c 8.2 and 2.78

9 a 5.2 and 1.66
  b 8 and 4.46
  c 6.7 and 3.06

10 a 6.8 and 2.08
  b 4.81 and 0.09
  c 4.6 and 0.08

Skill Example

Find the 'odd answer out'.

a £1.63 + £2.07 + £3.49
b £10.93 − £2.74
c £20 − £12.81

a	£	**b**	£	**c**	£
	1.63		10.93		20.00
	2.07		− 2.74		− 12.81
	+ 3.49		8.19		7.19
	7.19				

So **b** is the 'odd answer out'.

Skill Practice Six

For each question, find the 'odd answer out'.

1 a £12.52 + £21.14 + £11.34
  b £18.26 + £12.41 + £13.33
  c £16.34 + £17.20 + £11.46

2 a £12.10 + £10.75 + £2.15
  b £11.38 + £11.15 + £2.47
  c £13.80 + £2.76 + £7.44

3 a £56.89 − £13.65
  b £61.92 − £19.58
  c £59.11 − £16.77

4 a £31.70 − £13.54
  b £34 − £15.84
  c £30.08 − £13.90

5 a £12.68 − £8.12
  b £11.29 − £6.83
  c £10.25 − £5.79

6 a £15.80 + £10.30 + £6.26
  b £19.05 + £7.30 + £6.01
  c £51.82 − £18.56

7 a £10.52 + £3.15 + £1.76
  b £11 + £2.69 + £0.84
  c £22.71 − £7.28

8 a £3.53 + £4.48 + £1.74
  b £24.78 − £15.13
  c £22.84 − £13.19

9 a £4 + £0.60 + £3.85
  b £9.19 − £0.64
  c £9.20 − £0.75

10 a £20 + £16.26 + £3.79
  b £52.40 − £12.35
  c £52 − £1.96

Skill Practice Seven

1 Find the length of the spade.

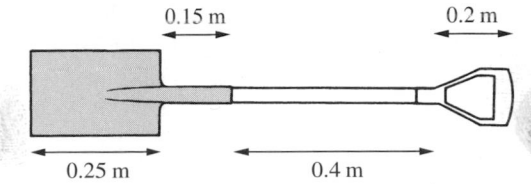

0.15 m   0.2 m

0.25 m   0.4 m

2 Find the height of the clock tower.

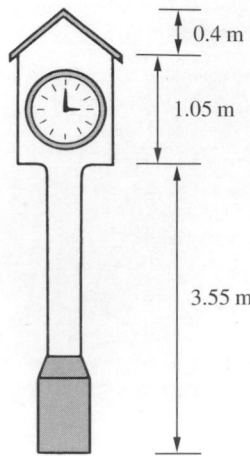

0.4 m

1.05 m

3.55 m

3

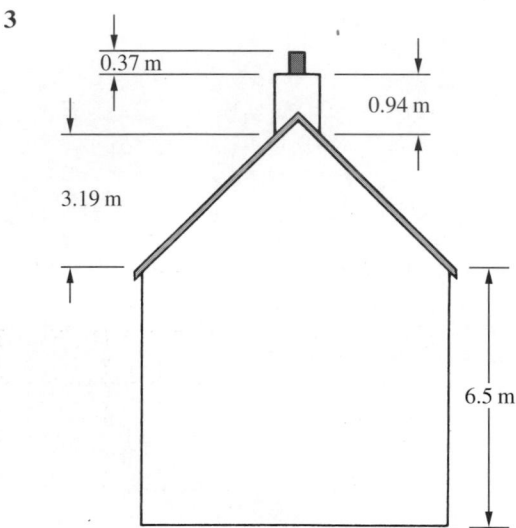

0.37 m

0.94 m

3.19 m

6.5 m

Find the height of the chimney top above the ground.

4 Mrs Johnson has a packet of currants which weighs 2 kg.
If she uses 0.25 kg for making a cake, what weight of currants will be left in the packet?

5 I go to the grocer's shop and buy the following.

Half a kilogram of cheese	£2.05
Half a kilogram of butter	£1.20
One loaf of bread	£0.78
Six eggs	£0.82

If I pay with a £5 note, how much change should I receive?

6 I have to make a picture-frame like the one illustrated.

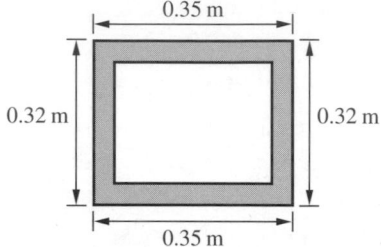

0.35 m

0.32 m

0.32 m

0.35 m

I have a piece of wood which is 1 metre in length.
Can I use this piece of wood for making three sides of the picture-frame?
Which three sides can I make?

7 Peter cycles from York to Scarborough. The distance on the tripmeter fitted to his bicycle shows the following figures before and after the journey:

At York	3212.4 km
At Scarborough	3280.2 km

At a later, date his father makes the same journey by car. The tripmeter on the car shows the following figures:

At York	37 986.7 km
At Scarborough	38 054.5 km

Are the two tripmeters accurate?
If so, what is the distance from York to Scarborough?

Unit 4

Number and Algebra   Level 4

Skill   You recognise approximate proportions of a whole and use simple fractions and percentages to describe these.

Knowledge

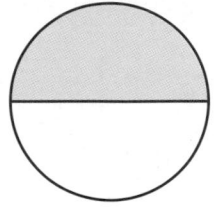 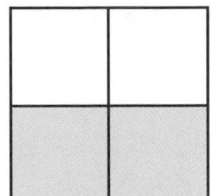

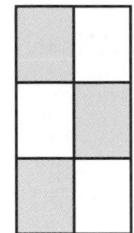 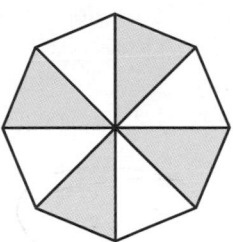

All of the above shapes have one half shaded.

$$\tfrac{1}{2} = \tfrac{2}{4} = \tfrac{3}{6} = \tfrac{4}{8}$$

Fractions having the same value are called **equivalent fractions**. Equivalent fractions can be found by two methods.

- Multiplying the top and the bottom by the same number. For example:

  a $\dfrac{2}{3} = \dfrac{2 \times 2}{3 \times 2} = \dfrac{4}{6}$

  b $\dfrac{3}{5} = \dfrac{3 \times 6}{5 \times 6} = \dfrac{18}{30}$

- Dividing the top and the bottom by the same number. For example:

  a $\dfrac{4}{6} = \dfrac{4 \div 2}{6 \div 2} = \dfrac{2}{3}$

  b $\dfrac{18}{30} = \dfrac{18 \div 2}{30 \div 2} = \dfrac{9}{15} = \dfrac{9 \div 3}{15 \div 3} = \dfrac{3}{5}$

Skill Example

Find the simplest form of the fraction shaded in each of these drawings.

a

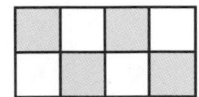

Number of parts shaded = 4
Number of equal parts = 8
Therefore, fraction $= \tfrac{4}{8} = \tfrac{1}{2}$

b

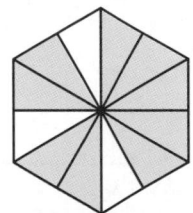

Number of parts shaded = 9
Number of equal parts = 12
Therefore, fraction $= \tfrac{9}{12} = \tfrac{3}{4}$

Skill Practice One

Find the simplest form of the fraction shaded in each of these drawings.

1 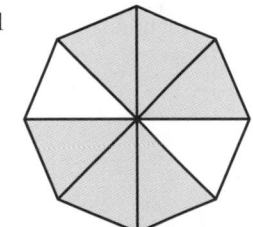   **2**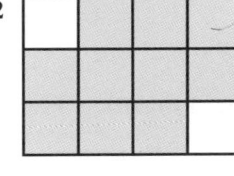

3    **4**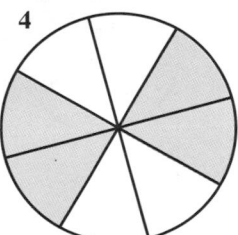

5 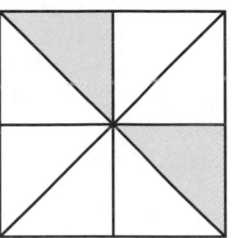   **6**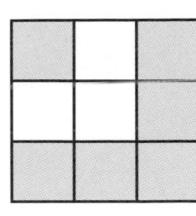

7 8
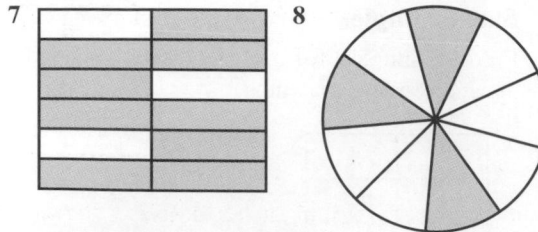

Skill Example

Which of the three diagrams has a different fraction shaded?

a   b   c

a 1 out of 3 parts is shaded.
  Therefore, fraction = $\frac{1}{3}$

b 2 out of 6 parts are shaded.
  Therefore, fraction = $\frac{2}{6} = \frac{1}{3}$

c 1 out of 4 parts is shaded.
  Therefore, fraction = $\frac{1}{4}$

Therefore, **c** has a different fraction shaded.

Skill Practice Two

For each question, state which of the three diagrams has a different fraction shaded.

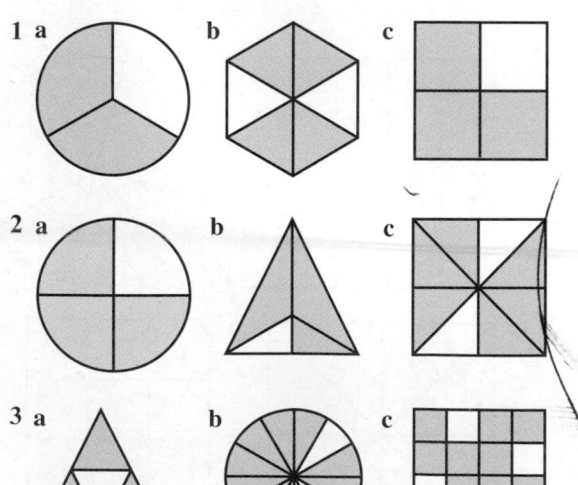

4 a   b   c

5 a   b   c

6 a   b   c

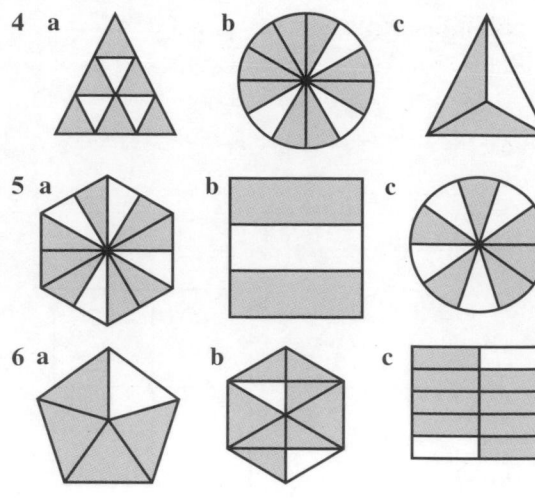

Skill Example

Find which fraction is different from the other three.

a $\frac{4}{6}$ **b** $\frac{10}{15}$ **c** $\frac{10}{18}$ **d** $\frac{14}{21}$

Changing each fraction to its simplest form gives:

a $\frac{4}{6} = \frac{4 \div 2}{6 \div 2} = \frac{2}{3}$ **b** $\frac{10}{15} = \frac{10 \div 5}{15 \div 5} = \frac{2}{3}$

c $\frac{10}{18} = \frac{10 \div 2}{18 \div 2} = \frac{5}{9}$ **d** $\frac{14}{21} = \frac{14 \div 7}{21 \div 7} = \frac{2}{3}$

Therefore, **c** is different from the other three because it is not equivalent to $\frac{2}{3}$.

Skill Practice Three

For each question, find which fraction is different from the others.

1	**a** $\frac{12}{16}$	**b** $\frac{18}{24}$	**c** $\frac{25}{30}$	**d** $\frac{27}{36}$
2	**a** $\frac{16}{24}$	**b** $\frac{18}{30}$	**c** $\frac{12}{18}$	**d** $\frac{30}{45}$
3	**a** $\frac{10}{12}$	**b** $\frac{40}{48}$	**c** $\frac{15}{24}$	**d** $\frac{21}{24}$
4	**a** $\frac{12}{30}$	**b** $\frac{10}{25}$	**c** $\frac{4}{10}$	**d** $\frac{15}{40}$
5	**a** $\frac{12}{36}$	**b** $\frac{8}{32}$	**c** $\frac{7}{28}$	**d** $\frac{4}{16}$
6	**a** $\frac{3}{15}$	**b** $\frac{7}{35}$	**c** $\frac{6}{30}$	**d** $\frac{5}{20}$
7	**a** $\frac{4}{24}$	**b** $\frac{6}{30}$	**c** $\frac{7}{42}$	**d** $\frac{3}{18}$
8	**a** $\frac{6}{18}$	**b** $\frac{15}{45}$	**c** $\frac{9}{36}$	**d** $\frac{4}{12}$
9	**a** $\frac{24}{30}$	**b** $\frac{20}{32}$	**c** $\frac{30}{48}$	**d** $\frac{10}{16}$
10	**a** $\frac{15}{36}$	**b** $\frac{6}{16}$	**c** $\frac{18}{48}$	**d** $\frac{12}{32}$
11	**a** $\frac{14}{16}$	**b** $\frac{40}{48}$	**c** $\frac{35}{40}$	**d** $\frac{21}{24}$
12	**a** $\frac{20}{36}$	**b** $\frac{28}{48}$	**c** $\frac{30}{54}$	**d** $\frac{10}{18}$

Skill Example

Copy the diagrams and shade in $\frac{2}{3}$ of the pattern on each.

a 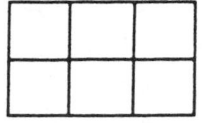   b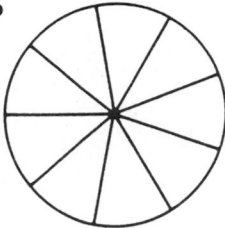

a There are 6 equal parts, so 4 parts have to be shaded because $\frac{4}{6} = \frac{2}{3}$.

b There are 9 equal parts, so 6 parts have to be shaded because $\frac{6}{9} = \frac{2}{3}$.

Skill Practice Four

1 Copy the diagrams and shade in $\frac{1}{2}$ the pattern on each.

a 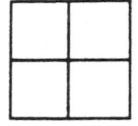   b

2 Copy the diagram and shade in $\frac{1}{5}$ of the pattern.

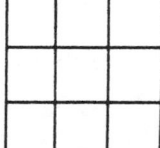

3 Copy the diagram and shade in $\frac{1}{3}$ of the pattern.

4 Copy the diagram and shade in $\frac{2}{3}$ of the pattern.

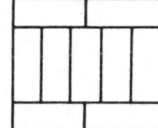

5 Copy the diagram and shade in $\frac{3}{4}$ of the pattern.

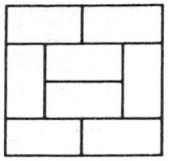

6 Copy the diagram and shade in $\frac{3}{5}$ of the pattern.

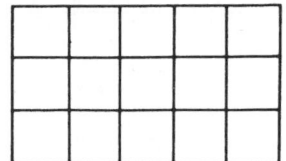

7 Copy the diagram and shade in $\frac{3}{8}$ of the pattern.

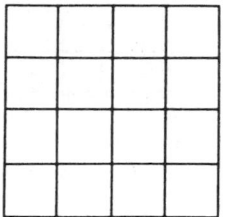

8 Copy the diagram and shade in $\frac{5}{8}$ of the pattern.

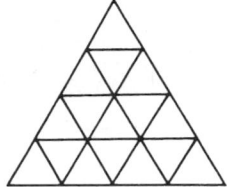

Skill Practice Five

Copy and complete by filling in the missing numbers.

1 $\frac{1}{3} = \frac{}{6}$ 2 $\frac{3}{4} = \frac{}{8}$ 3 $\frac{1}{2} = \frac{}{8}$ 4 $\frac{2}{5} = \frac{}{15}$

5 $\frac{1}{5} = \frac{}{20}$ 6 $\frac{3}{8} = \frac{}{32}$ 7 $\frac{1}{6} = \frac{}{30}$ 8 $\frac{5}{8} = \frac{}{40}$

9 $\frac{3}{5} = \frac{}{30}$ 10 $\frac{2}{3} = \frac{}{18}$ 11 $\frac{1}{4} = \frac{}{36}$ 12 $\frac{1}{5} = \frac{}{60}$

13 $\frac{1}{8} = \frac{2}{}$ 14 $\frac{7}{10} = \frac{14}{}$ 15 $\frac{1}{12} = \frac{3}{}$ 16 $\frac{4}{5} = \frac{12}{}$

17 $\frac{1}{9} = \frac{4}{}$ 18 $\frac{5}{7} = \frac{20}{}$ 19 $\frac{3}{8} = \frac{15}{}$ 20 $\frac{5}{6} = \frac{25}{}$

21 $\frac{3}{4} = \frac{18}{}$ 22 $\frac{2}{5} = \frac{14}{}$ 23 $\frac{4}{5} = \frac{32}{}$ 24 $\frac{2}{3} = \frac{24}{}$

25 $\frac{1}{4} = \frac{15}{}$ 26 $\frac{4}{6} = \frac{}{3}$ 27 $\frac{10}{12} = \frac{}{6}$ 28 $\frac{9}{15} = \frac{}{5}$

29 $\frac{21}{36} = \frac{}{12}$ 30 $\frac{16}{20} = \frac{}{5}$ 31 $\frac{32}{36} = \frac{}{9}$ 32 $\frac{5}{40} = \frac{}{8}$

33 $\frac{15}{35} = \frac{}{7}$ 34 $\frac{24}{30} = \frac{}{5}$ 35 $\frac{6}{24} = \frac{}{4}$ 36 $\frac{24}{32} = \frac{}{4}$

Knowledge

A percentage indicates the number of **hundredths** in a share of a given quantity. The special sign '%' is used for a percentage.

This diagram shows percentages expressed as fractions.

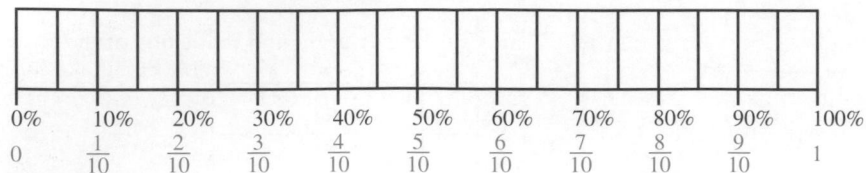

For example, if Surbajit was absent on 15% of the days during a school year, this means he was absent for:

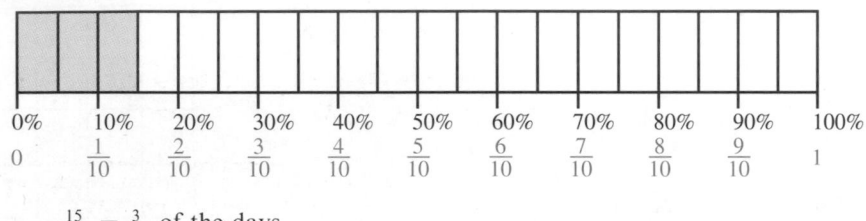

$\frac{15}{100} = \frac{3}{20}$ of the days

Skill Example

Write each of the following as a fraction in its simplest form.

a 57%   **b** 18%   **c** 35%

a $57\% = \frac{57}{100}$ **b** $18\% = \frac{18}{100} = \frac{9}{50}$

c $35\% = \frac{35}{100} = \frac{7}{20}$

Skill Example

Write each of the following as a percentage.

a $\frac{11}{100}$ **b** $\frac{9}{10}$ **c** $\frac{3}{8}$

a $11 \div 100 \times 100\% = 11\%$

b $9 \div 10 \times 100\% = 90\%$

c $3 \div 8 \times 100\% = 37.5\%$

Skill Practice Six

Write each as a fraction in its simplest form.

1 63%	**2** 29%	**3** 13%	**4** 43%	**5** 77%
6 9%	**7** 22%	**8** 46%	**9** 82%	**10** 14%
11 34%	**12** 98%	**13** 58%	**14** 55%	**15** 85%
16 15%	**17** 5%	**18** 48%	**19** 64%	**20** 72%

Knowledge

Changing a fraction to a percentage is best done with a calculator.

You divide the top number by the bottom number and then multiply by 100 %.

Skill Practice Seven

Write each of the following as a percentage.

1 $\frac{31}{100}$	**2** $\frac{27}{100}$	**3** $\frac{87}{100}$	**4** $\frac{99}{100}$
5 $\frac{3}{100}$	**6** $\frac{1}{100}$	**7** $\frac{9}{50}$	**8** $\frac{21}{50}$
9 $\frac{19}{50}$	**10** $\frac{27}{50}$	**11** $\frac{9}{20}$	**12** $\frac{13}{20}$
13 $\frac{19}{20}$	**14** $\frac{4}{25}$	**15** $\frac{6}{25}$	**16** $\frac{1}{25}$
17 $\frac{8}{25}$	**18** $\frac{11}{25}$	**19** $\frac{13}{25}$	**20** $\frac{14}{25}$

Unit 5

Number and Algebra   Level 4

Skill   You explore and describe number patterns, and relationships including multiple, factor and square.

Knowledge

An **even number** can be divided exactly by 2. The sequence of even numbers starts:

2, 4, 6, 8, . . .

An **odd number** has a remainder of 1 when divided by 2. The sequence of odd numbers starts:

1, 3, 5, 7, . . .

Skill Practice One

1 Six children have each thrown two dice. Look at their scores below.

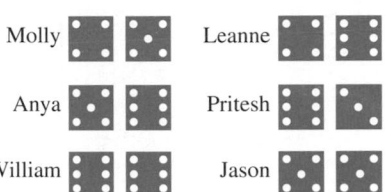

Molly   Leanne

Anya   Pritesh

William   Jason

a Whose scores are even numbers?
b Whose scores are odd numbers?

2 List the next five even numbers starting from:
   a 4   b 16   c 32   d 88   e 100

3 List the next five odd numbers starting from:
   a 5   b 13   c 31   d 69   e 117

Knowledge

A **square number** is any number that can be represented by counters arranged in a square.

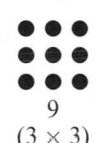

1   4   9
(1 × 1)   (2 × 2)   (3 × 3)

A square number is found by multiplying any number by itself.

A **rectangular number** is any number that can be represented by counters arranged in a rectangle.

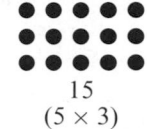

6   8   15
(3 × 2)   (4 × 2)   (5 × 3)

A **triangular** number is any number that can be represented by counters arranged in a triangle.

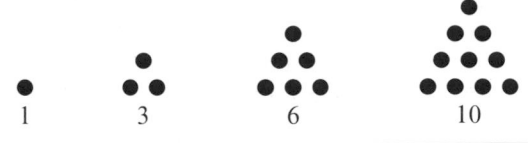

1   3   6   10

Skill Practice Two

1 Look at the numbers on the signposts A, B and C.

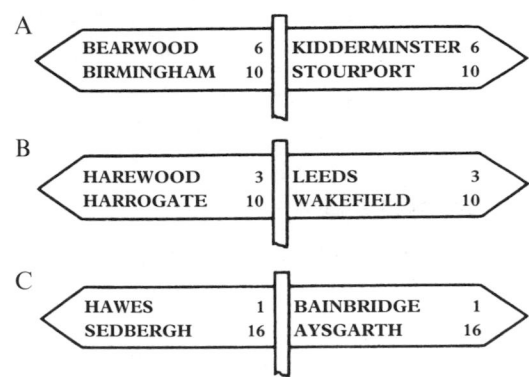

A
BEARWOOD   6   KIDDERMINSTER 6
BIRMINGHAM   10   STOURPORT   10

B
HAREWOOD   3   LEEDS   3
HARROGATE   10   WAKEFIELD   10

C
HAWES   1   BAINBRIDGE   1
SEDBERGH   16   AYSGARTH   16

a On which signpost are *all* the numbers square?
b On which signpost are *all* the numbers rectangular?
c On which signpost are *all* the numbers triangular?

2 State whether the following numbers are square (S), rectangular (R), triangular (T) or none of these (N). For example, 15 is (R,T), 17 is (N).
   a 12   b 14   c 16   d 21   e 7
   f 18   g 19   h 25   i 28   j 36

3 Can you find the following?
   a Any odd numbers which are also square numbers.
   b Any numbers which are both rectangular and odd.
   c Any even numbers which are not rectangular.
   Give examples if you can.

3 5711

4 Find the sum of:
   a the first three odd numbers
   b the first four odd numbers
   c the first five odd numbers
   d the first ten odd numbers
   e the first hundred odd numbers.

5 Make a list of the first six triangular numbers. Add together:
   a the first and second
   b the second and third
   c the third and fourth
   d the fourth and fifth
   e the fifth and sixth.
   f The numbers in the above answers are all of the same kind. What are they?

Knowledge

A **prime number** is one that cannot be divided by any number apart from 1 and itself.

The first five prime numbers are 2, 3, 5, 7 and 11.

Skill Practice Three

1 Look at the four buses **A**, **B**, **C** and **D**.

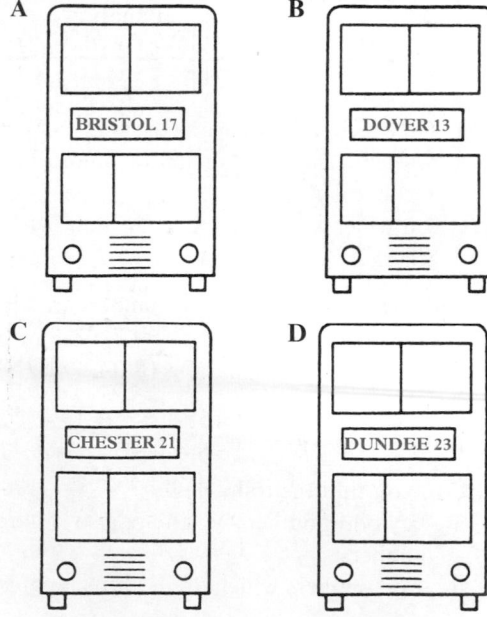

One of the buses is not displaying a prime number. Which one is it?

2 List all the prime numbers between 1 and 30.

3 Which of the following are prime numbers?
   a 31   b 33   c 37   d 35   e 49   f 43
   g 51   h 41   i 53   j 57   k 63   l 61

Knowledge

A **sequence** is a set of numbers such that each number is related to the next in the same way.

a 1, 4, 7, 10, . . . is a sequence in which each term is three more than the one before it.
b 1, 4, 16, 64, . . . is a sequence in which each term is four times the one before it.

Skill Practice Four

1 Find the heights of the next two arches of the viaduct below.

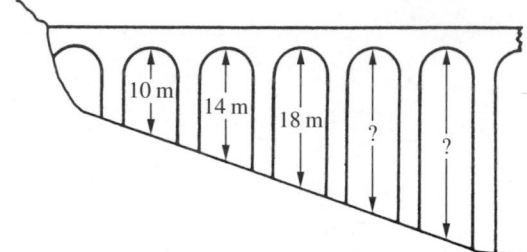

2 Find the next five terms of these sequences and state the rule in each case.
   a 2, 4, 6, 8, . . .   b 3, 6, 9, 12, . . .
   c 12, 24, 36, 48, . . .   d 50, 45, 40, 35, . . .
   e 66, 60, 54, 48, . . .   f 56, 49, 42, 35, . . .
   g 8, 13, 18, 23, . . .   h 5, 11, 17, 23, . . .
   i 40, 36, 32, 28, . . .   j 90, 82, 74, 66, . . .

3 What is the distance to Stratford if the number is the next one in the sequence?

HAGLEY	1
CLENT	3
BROMSGROVE	9
STRATFORD	?

4 Find the next three terms of these sequences and state the rule in each case.
   a 1, 2, 4, 8, . . .   b 1, 4, 16, 64, . . .
   c 2, 20, 200, . . .   d 32, 16, 8, . . .
   e 243, 81, 27, . . .   f 3125, 625, 125, . . .

5 In each of the following, state which term is the odd one out.

a 2, 4, 6, 9, 10   b 3, 5, 7, 9, 10
c 4, 8, 12, 14, 20   d 100, 90, 85, 70, 60
e 90, 81, 72, 64, 54   f 55, 49, 42, 35, 28
g 1, 3, 4, 8, 16   h 1, 4, 6, 16, 25
i 100, 81, 64, 50, 36   j 2, 3, 5, 7, 11, 15, 17
k 1, 3, 6, 10, 16, 21   l 6, 8, 9, 12, 14, 15

Skill Example

List the first four multiples of 8.

$1 \times 8 = 8$ $3 \times 8 = 24$
$2 \times 8 = 16$ $4 \times 8 = 32$

So the first four multiples of 8 are 8, 16, 24, 32.

Skill Practice Five

List the first four multiples of:

1 2	**2** 5	**3** 3	**4** 6
5 10	**6** 7	**7** 9	**8** 11
9 12	**10** 20	**11** 50	**12** 40
13 60	**14** 25	**15** 15	

Skill Example

14, 28, 49, 64, 70

Which of the above numbers is *not* a multiple of 7?

$14 \div 7 = 2$ $64 \div 7 = 9 \text{ r } 1$
$28 \div 7 = 4$ $70 \div 7 = 10$
$49 \div 7 = 7$

So 64 is not a multiple of 7.

Skill Practice Six

1 9, 13, 18, 21, 27.
Which one of these numbers is not a multiple of 3?

2 30, 36, 40, 48, 54.
Which one of these numbers is not a multiple of 6?

3 40, 48, 56, 62, 80.
Which one of these numbers is not a multiple of 8?

4 20, 28, 32, 38, 44.
Which one of these numbers is not a multiple of 4?

5 45, 54, 63, 74, 81.
Which one of these numbers is not a multiple of 9?

6 66, 88, 99, 112, 121.
Which one of these numbers is not a multiple of 11?

7 34, 48, 60, 72, 96.
Which one of these numbers is not a multiple of 12?

8 80, 100, 120, 140, 150.
Which one of these numbers is not a multiple of 20?

9 30, 45, 60, 70, 90.
Which one of these numbers is not a multiple of 15?

10 100, 125, 150, 185, 200.
Which one of these numbers is not a multiple of 25?

Skill Example

Find all the factors of 24.

$24 \div 1 = 24$	So 1 and 24 are factors because $1 \times 24 = 24$.
$24 \div 2 = 12$	So 2 and 12 are factors because $2 \times 12 = 24$.
$24 \div 3 = 8$	So 3 and 8 are factors because $3 \times 8 = 24$.
$24 \div 4 = 6$	So 4 and 6 are factors because $4 \times 6 = 24$.
$24 \div 5 = 4 \text{ r } 4$	So 5 is *not* a factor.
$24 \div 6 = 4$	So 6 and 4 are factors, but these have already been found.

Therefore, all the factors have now been found. All the factors of 24 are 1, 2, 3, 4, 6, 8, 12, and 24.

Skill Practice Seven

Find all the factors of:

1	18	**2**	20	**3**	12	**4**	10
5	8	**6**	14	**7**	22	**8**	15
9	21	**10**	27	**11**	35	**12**	26
13	28	**14**	32	**15**	30	**16**	40
17	36	**18**	9	**19**	25	**20**	16

Knowledge

Two special factors of 10 are 2 and 5 because they are both prime numbers and their product is 10.

2 and 5 are called the **prime factors** of 10.

The prime numbers which are factors of 12 are 2 and 3.

12 can be written as a multiplication of prime factors like this: $12 = 2 \times 2 \times 3$.

Skill Example

Write as a multiplication of prime factors:

a 42   **b** 100

a 2)42 $42 = 2 \times 3 \times 7$
     3)21
     7)7
       1

b 2)100 $100 = 2 \times 2 \times 5 \times 5$
     2)50
     5)25
     5 5
        1

Skill Practice Eight

Write as a multiplication of prime factors:

1	30	**2**	66	**3**	70	**4**	78
5	110	**6**	130	**7**	154	**8**	210
9	84	**10**	140	**11**	132	**12**	88
13	104	**14**	56	**15**	40	**16**	120
17	72	**18**	168	**19**	80	**20**	48
21	112	**22**	180	**23**	108	**24**	162

Unit 6

Number and Algebra   Level 4

Skill   You have begun to use simple formulae expressed in words.

Skill Example

a An electrician charges a call out fee of £25 which is added to the cost of the work he does. His total charge can be calculated from the formula:

Total charge = Cost of repairs + £25

What is the total charge when the cost of the repairs is £33?

Total charge = £33 + £25 = £58

b A taxi driver charges a fixed price of 80 p plus 90 p for each mile of the journey. The total fare can be calculated from the formula:

Total fare = Number of miles × 90 p + 80 p

What is the total fare for a journey of 7 miles?

Total fare = 7 × 90 + 80 = 710 p = £7.10

Skill Practice One

1 A mail-order shoe company charges £2.50 postage and packaging with every order. The total cost of a pair of shoes can be calculated from the formula:

Total cost = Price of shoes + £2.50

What is the total cost of ordering a pair of shoes priced at:

a £27   b £33.50
c £21.28   d £39.99?

2 A baker works out the number of bread rolls needed to fill bags, each containing a dozen rolls, with this formula:

Number of rolls needed = Number of bags × 12

How many rolls are needed to fill:

a 5 bags   b 3 bags
c 100 bags   d 83 bags?

3 A shopkeeper starts the week with 200 cans of Cola in stock. During the week, she works out how many cans she has left using this formula:

Number left = 200 − Number sold

How many cans has she left when the number sold is:

a 15   b 28   c 105   d 125?

4 A syndicate of 40 teachers has agreed to share any lottery wins using the formula

One share = Amount won ÷ 40

How much will one share be worth if the syndicate wins:

a £40   b £4000
c £103 440   d £3 075 800?

5 A cinema sells ice-cream cones and tubs. A profit of 15 p is made on each cone sold and a profit of 18 p is made on each tub sold. This formula is used to calculate the total profit:

Total profit = Number of cones sold × 15 p
               + Number of tubs sold × 18 p

Calculate the total profit on a night when the cinema sells:

a 100 cones and 85 tubs
b 250 cones and 160 tubs
c 321 cones and 286 tubs
d 583 cones and 411 tubs

6 A teacher is organising a trip to a theme park. Entry will cost £6.99 per person and a coach will cost £350 to hire for the trip. She uses two formulae to work out the costs:

Total cost = £350 + £6.99
             × Number going on trip

Cost per person = Total cost
                  ÷ Number going on trip

Work out the total cost and the cost per person if the number going on the trip is:

a 25   b 28   c 35   d 51

7 An infant is a child less than 12 months old. An infant's dose of a medicine can be calculated from the formula:

Infant's dose = Adult's dose × Age of infant in months ÷ 150

Calculate the infant's dose of a medicine for which the adult's dose is 125 millilitres, when the infant has an age of:

a 3 months   **b** 9 months
c 8 months   **d** 10 months

8 The depth, in metres, of a well can be found by dropping in a stone and measuring the time, in seconds, it takes before you hear a splash.
The depth is found from this formula:

Depth = 5 × Time × Time

Find the depth of a well when you hear a splash after:

a 1 second   **b** 2 seconds
c 3 seconds   **d** 2.5 seconds

Unit 7

Number and Algebra   Level 4

Skill   You use and interpret coordinates in the first quadrant.

Knowledge

The position of a point on a graph is fixed by referring to its **coordinates**. These are the two distances of the point, measured along the axes, from a fixed point, O, called the **origin**.

The horizontal axis from O is called the x-axis. The vertical axis from O is called the y-axis.

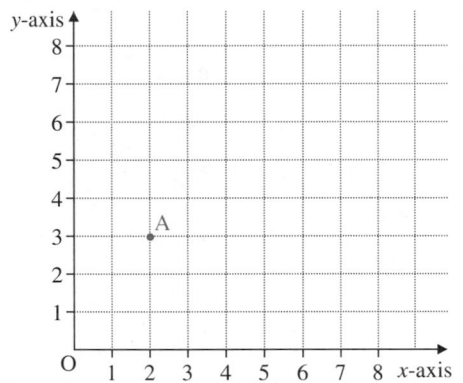

So, the position of any point given by the coordinates (x, y) is x units to the 'east' of the origin and y units to the 'north' of the origin.

On the graph shown above, the coordinates $(2, 3)$ gives the position of the point A.

Skill Example

Give the coordinates necessary to fix the position of each point shown in the diagram.

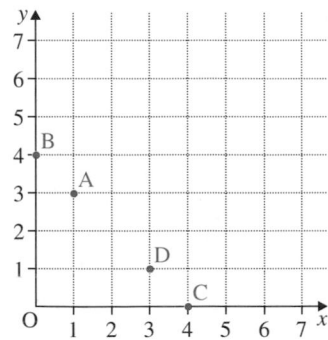

A is the point (1, 3)   B is the point (0, 4)
C is the point (4, 0)   D is the point (3, 1)

Skill Practice One

Give the coordinates necessary to fix the position of each point shown in the diagram below.

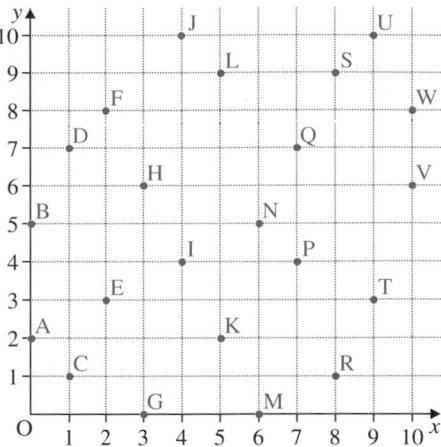

Skill Example

On graph paper, with the x-axis numbered from 0 to 10 and the y-axis numbered from 0 to 8, plot the positions of the following points.

(1, 4), (1, 6), (3, 8), (5, 6), (5, 4), (7, 4), (7, 6), (9, 7), (10, 6), (9, 6), (9, 4), (8, 3), (8, 0), (7, 2), (5, 0), (5, 1), (3, 3), (3, 2), (1, 4)

Join each point to the next with a straight line. Then suggest a name for the picture you have drawn.

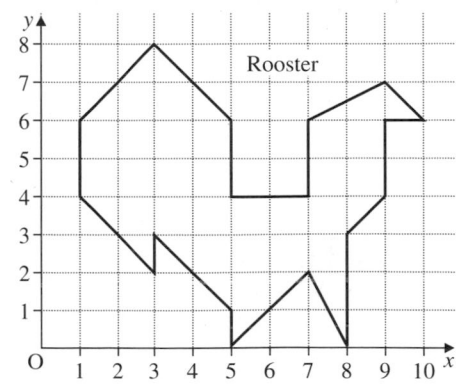

28

Skill Practice Two

On graph paper, with the x-axis numbered from 0 to 10 and the y-axis numbered from 0 to 8, plot the positions of the points in each question.
Join each point to the next with a straight line.
Then suggest a name for the picture you have drawn.

1 (1, 2), (0, 4), (2, 4), (2, 5), (3, 5), (3, 6), (4, 6), (4, 5), (5, 5), (5, 6), (6, 6), (6, 5), (7, 5), (7, 4), (10, 4), (8, 2), (1, 2)

2 (5, 1), (5, 3), (0, 5), (0, 6), (5, 4), (5, 5), (10, 5), (10, 1), (5, 1)

3 (2, 0), (3, 1), (4, 1), (4, 4), (2, 4), (3, 8), (6, 8), (7, 4), (5, 4), (5, 1), (6, 1), (7, 0), (2, 0)

4 (2, 1), (3, 2), (4, 2), (4, 5), (0, 5), (0, 6), (4, 7), (4, 8), (5, 8), (5, 7), (9, 6), (9, 5), (5, 5), (5, 2), (6, 2), (7, 1), (5, 1), (5, 0), (4, 0), (4, 1), (2, 1)

5 (1, 4), (0, 4), (0, 7), (1, 7), (1, 6), (7, 6), (7, 7), (10, 7), (10, 4), (7, 4), (7, 5), (1, 5), (1, 4)

6 (1, 2), (1, 4), (0, 4), (0, 5), (7, 5), (8, 7), (9, 7), (10, 5), (10, 4), (9, 2), (8, 2), (7, 4), (3, 4), (3, 2), (1, 2)

7 (2, 3), (1, 3), (0, 4), (1, 4), (1, 5), (0, 5), (1, 6), (2, 6), (3, 5), (7, 5), (8, 6), (9, 6), (10, 5), (8, 5), (8, 4), (10, 4), (9, 3), (8, 3), (7, 4), (3, 4), (2, 3)

8 (5, 8), (8, 6), (10, 6), (10, 5), (8, 5), (5, 3), (10, 3), (10, 2), (5, 2), (8, 0), (6, 0), (3, 2), (1, 2), (1, 3), (3, 3), (6, 5), (1, 5), (1, 6), (6, 6), (3, 8), (5, 8)

Knowledge

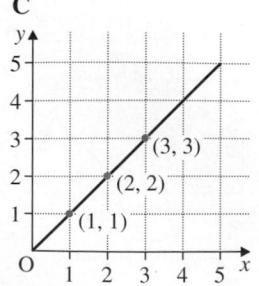

In graph A, all the coordinates which give points on the line have an x-value of 2. The line is the graph of $x = 2$.

In graph B, all the coordinates which give points on the line have a y-value of 3. The line is the graph of $y = 3$.

In graph C, all the coordinates which give points on the line have their x-value equal to their y-value. The line is the graph of $y = x$.

Skill Example

Give the equation of the line for the graphs **a** and **b**.

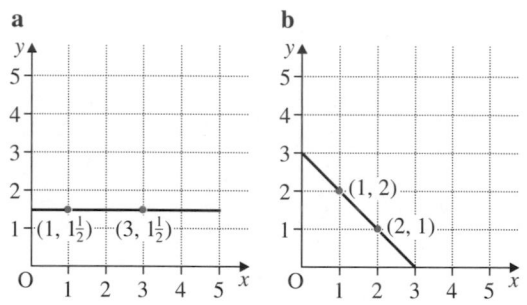

a As all the points on the line have a y-value of $1\frac{1}{2}$, the graph is $y = 1\frac{1}{2}$.

b As all the points on the line are such that the x-value added to the y-value equals 3, the graph is $x + y = 3$.

Skill Practice Three

Give the equation of the line for each of the following graphs.

1

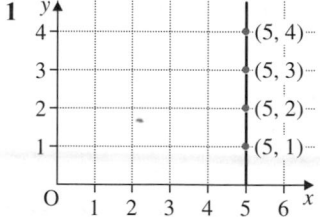

2

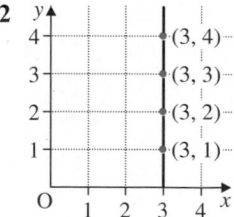

3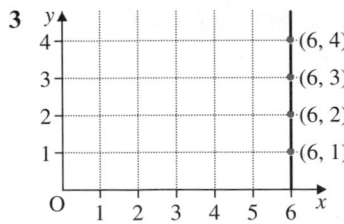

4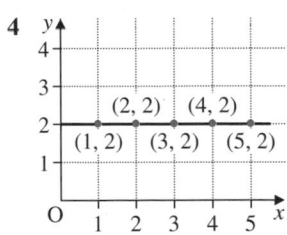

5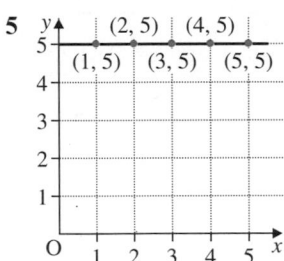

6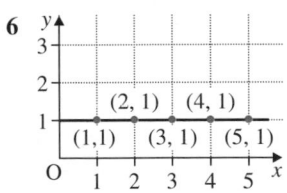

7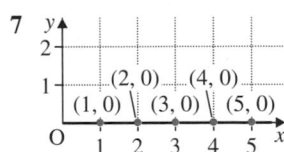

8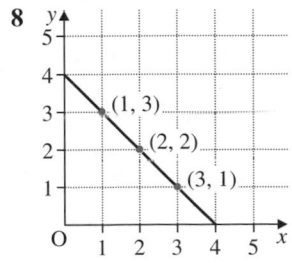

9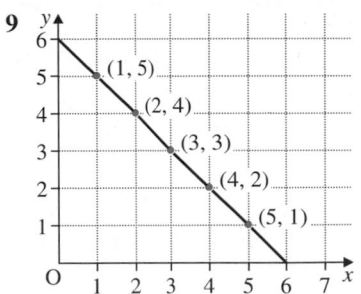

10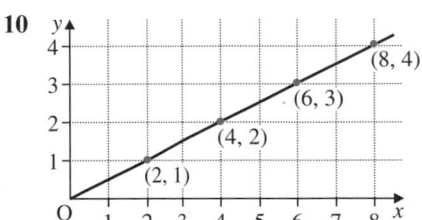

Draw the graphs of the following equations.

11 $x = 4$ **12** $x = 1$ **13** $x = 7$ **14** $x = 0$

15 $y = 4$ **16** $y = 6$ **17** $y = 3$ **18** $x + y = 5$

19 $x + y = 2$ **20** $y = 3x$

Skill Example

The table shows the pairs of numbers whose sum is 12.

1st number	0	1	2	3	4	5	6
2nd number	12	11	10	9	8	7	6

1st number	7	8	9	10	11	12
2nd number	5	4	3	2	1	0

Plot a graph of these numbers and join up the points with a suitable line.

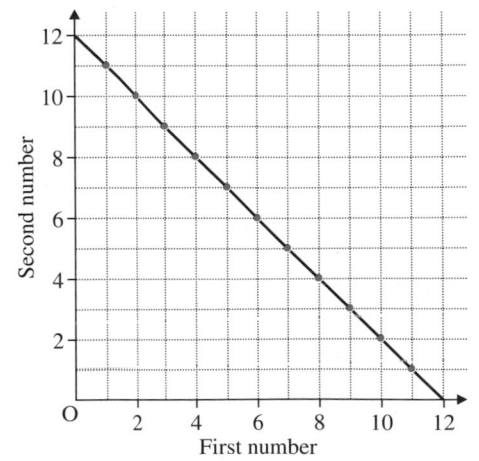

The points can be joined by a straight line.

Skill Example

The table shows the pairs of numbers whose product is 12.

1st number	1	2	3	4	6	12
2nd number	12	6	4	3	2	1

Plot a graph of these numbers and join up the points with a suitable line.

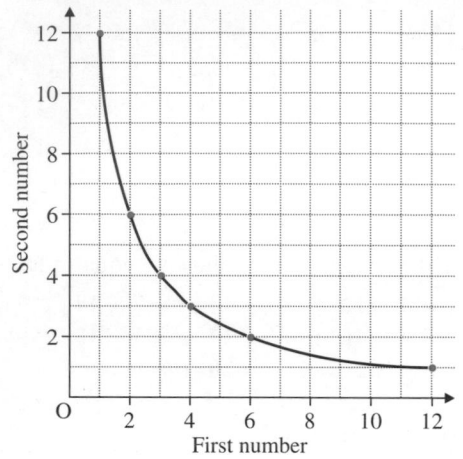

The points on this graph have to be joined by a *smooth* curve.

Skill Practice Four

1 The table below shows the pairs of numbers whose sum is 10.

1st number	0	1	2	3	4	5	6	7	8	9	10
2nd number	10	9	8	7	6	5	4	3	2	1	0

Using a scale of 1 cm to 1 unit on both axes, plot this graph.
Find from your graph what must be added to each of the following to make 10:
a 3.6 **b** 8.2

2 The following table shows pairs of angles that add to 180°.

angle 1	0°	30°	60°	90°	120°	150°	180°
angle 2	180°	150°	120°	90°	60°	30°	0°

Using a scale of 1 cm to 10° on both axes, plot this graph.
Find from your graph the angle to add to each of the following to make 180°:
a 24° **b** 72° **c** 144° **d** 168°

3 The following table shows the change out of £1 for certain prices.

price (p)	20	40	60	80
change (p)	80	60	40	20

Using a scale of 1 cm to 10 p on both axes, plot the graph.
From the graph find the change out of £1 for the following prices:
a 16 p **b** 42 p **c** 54 p **d** 76 p

4 The table below shows the pairs of numbers whose product is 18.

1st number	1	2	3	6	9	18
2nd number	18	9	6	3	2	1

Using a scale of 1 cm to 1 unit on both axes, plot this graph.
Find from your graph by how much the following must be multiplied to make 18:
a 3.6 **b** 15

5 The following table shows certain pairs of numbers whose product is 180.

1st number	2	3	5	6	10	18	30	36	60	90
2nd number	90	60	36	30	18	10	6	5	3	2

Using a scale of 1 cm to 5 units on both axes, plot this graph.
Find from your graph what the second number is when the first number is:
a 4 **b** 9 **c** 12

6 The following table shows the lengths and widths of certain rectangles which all have an area equal to 96 cm².

Width (cm)	2	4	8	12	24	48
Length (cm)	48	24	12	8	4	2

Using a scale of 1 cm to 5 cm on both axes, plot this graph.
Find from your graph the length of a rectangle of this same area when the width is equal to:
a 3 cm **b** 6 cm

Unit 8

Number and Algebra   Level 5

Skills   You use your understanding of place value to multiply and divide whole numbers and decimals by 10, 100 and 1000.

Knowledge

To multiply a number by 1000, move each figure three places to the left.

So   **a** $3 \times 1000 = 3000$
     b $406 \times 1000 = 406\,000$

Skill Practice One

Multiply each of these numbers by 1000.

1	8	**2**	27	**3**	70	**4**	103
5	200	**6**	570	**7**	507	**8**	999
9	623	**10**	620	**11**	603	**12**	2418

Knowledge

To divide a number by 1000, move each figure three places to the right.

So   **a** $5000 \div 1000 = 5$
     b $7263 \div 1000 = 7 \text{ r } 263 \text{ or } 7.263$

Skill Practice Two

Divide each of these numbers by 1000.

1	6000	**2**	2000	**3**	2004	**4**	5070
5	7050	**6**	7500	**7**	6814	**8**	68 140
9	681 400	**10**	6 814 000	**11**	2308	**12**	8320

Knowledge

Reminder

To multiply a number by 10, move each figure one place to the left.

$$3 \times 10 = 30$$

To divide a number by 10, move each figure one place to the right.

$$50 \div 10 = 5$$

To multiply a number by 100, move each figure two places to the left.

$$3 \times 100 = 300$$

To divide a number by 100, move each figure two places to the right.

$$500 \div 100 = 5$$

Skill Practice Three

Copy the following and fill in the missing numbers.

1	$26 \times 10 =$	**2**	$\times 10 = 260$
3	$26 \times \;\; = 260$	**4**	$260 \div 10 =$
5	$\div 10 = 26$	**6**	$260 \div \;\; = 26$
7	$70 \times 10 =$	**8**	$\times 10 = 700$
9	$70 \times \;\; = 70$	**10**	$700 \div 10 =$
11	$\div 10 = 70$	**12**	$700 \div \;\; = 70$
13	$145 \times 10 =$	**14**	$\times 10 = 1450$
15	$145 \times \;\; = 1450$	**16**	$1450 \div 10 =$
17	$\div 10 = 145$	**18**	$1450 \div \;\; = 145$
19	$240 \times 10 =$	**20**	$\times 10 = 2400$
21	$240 \times \;\; = 2400$	**22**	$2400 \div 10 =$
23	$\div 10 = 240$	**24**	$2400 \div \;\; = 240$
25	$400 \times 10 =$	**26**	$\times 10 = 4000$
27	$400 \times \;\; = 4000$	**28**	$4000 \div 10 =$
29	$\div 10 = 400$	**30**	$4000 \div \;\; = 4$
31	$13 \times 100 =$	**32**	$\times 100 = 1300$
33	$13 \times \;\; = 13000$	**34**	$1300 \div 100 =$
35	$\div 100 = 13$	**36**	$1300 \div \;\; = 13$
37	$50 \times 1000 =$	**38**	$\times 100 = 5000$
39	$50 \times \;\; = 5000$	**40**	$5000 \div 100 =$
41	$\div 100 = 50$	**42**	$5000 \div \;\; = 5$

Knowledge

To multiply a decimal by 10, move each figure one place to the left.

To multiply a decimal by 100, move each figure two places to the left.

Skill Example

Work out the following.

a 4.6×10 **b** 1.1×100 **c** 0.23×10
d 0.05×10 **e** 0.004×100

a $4.6 \times 10 = 46.0$ **b** $1.1 \times 100 = 110$

c $0.23 \times 10 = 2.3$ **d** $0.05 \times 10 = 0.5$

e $0.004 \times 100 = 0.4$

32

Skill Practice Four

Multiply each of the following by 10.

1 5.36	**2** 4.05	**3** 5.4
4 9.1	**5** 0.57	**6** 0.3
7 0.1	**8** 0.07	**9** 0.641
10 0.402	**11** 0.004	**12** 0.054

Multiply each of the following by 100.

13 0.453	**14** 0.121	**15** 0.906
16 0.038	**17** 0.002	**18** 0.92
19 0.04	**20** 0.3	**21** 0.1
22 3.94	**23** 2.03	**24** 5.2

Knowledge

To divide a decimal by 10, move each figure one place to the right.

To divide a decimal by 100, move each figure two places to the right.

Skill Example

Work out the following.

a $6.4 \div 10$	**b** $6 \div 10$	
c $7.3 \div 100$	**d** $0.4 \div 100$	

a $6.4 \div 10 = 0.64$

b $6 \div 10 = 0.6$

c $7.3 \div 100 = 0.073$

d $0.4 \div 100 = 0.004$

Skill Practice Five

Divide each of the following by 10.

1 3.2	**2** 4.35	**3** 2.16
4 5	**5** 2	**6** 67.1
7 50.6	**8** 75	**9** 0.19
10 0.55	**11** 0.7	**12** 0.05

Divide each of the following by 100.

13 35.2	**14** 40.5	**15** 37
16 50	**17** 236	**18** 9.95
19 1.08	**20** 3.6	**21** 0.27
22 0.32	**23** 0.08	**24** 0.8

Skill Practice Six

Copy the following and fill in the missing numbers.

1 $2.6 \times 10 =$	**2** $\quad \times 10 = 26$
3 $2.6 \times \quad = 26$	**4** $3.25 \times 10 =$
5 $\quad \times 10 = 32.5$	**6** $3.25 \times \quad = 32.5$
7 $5.42 \times 100 =$	**8** $\quad \times 100 = 542$
9 $5.42 \times \quad = 542$	**10** $4.8 \times 100 =$
11 $\quad \times 100 = 480$	**12** $4.8 \times \quad = 480$
13 $18.6 \div 10 =$	**14** $\quad \div 10 = 1.86$
15 $18.6 \div \quad = 1.86$	**16** $3.41 \div 10 =$
17 $\quad \div 10 = 0.341$	**18** $3.41 \div \quad = 0.341$
19 $61.5 \div 100 =$	**20** $\quad \div 100 = 0.615$

Knowledge

To multiply a decimal by 1000, move each figure three places to the left.

To divide a decimal by 1000, move each figure three places to the right.

Skill Example

Work out the following.

a 3.4×1000	**b** 78.1×1000	
c 213.0456×1000	**d** $3.4 \div 1000$	
e $78 \div 1000$	**f** $213.0456 \div 1000$	

a $3.4 \times 1000 = 3400$

b $78.1 \times 1000 = 78\,100$

c $213.0456 \times 1000 = 213\,045.6$

d $3.4 \div 1000 = 0.0034$

e $78 \div 1000 = 0.078$

f $213.0456 \div 1000 = 0.213\,045\,6$

Skill Practice Seven

Multiply each of the following by 1000.

1 2.45	**2** 13
3 1.6	**4** 5.9
5 3.66	**6** 0.056
7 0.1	**8** 123
9 12.3	**10** 1.23
11 683.34	**12** 0.001 23
13 6833.4	**14** 68 334
15 0.000 001	**16** 45.99

Divide each of the following by 1000.

17	67.8	**18**	6.78	**19**	0.0678
20	4000	**21**	40	**22**	4
23	0.04	**24**	0.004	**25**	5555.5
26	555.55	**27**	5.5555	**28**	0.555 55
29	60.04	**30**	12 789	**31**	7
32	0.0007				

Skill Example

a Write 345 as another number multiplied by 1000.

b Write 345 as another number divided by 1000.

a $345 = 0.345 \times 1000$

b $345 = 345\,000 \div 1000$

Skill Practice Eight

Write each of these numbers as another number multiplied by 1000.

1	2000	**2**	20
3	0.2	**4**	0.002
5	756	**6**	34.26
7	5000	**8**	8904.5
9	0.855	**10**	0.017

Write each of these numbers as another number divided by 1000.

11	5000	**12**	50
13	0.5	**14**	0.005
15	23	**16**	4.007
17	1.24	**18**	0.001
19	1000	**20**	3.142

Unit 9

Number and Algebra   Level 5

Skills You understand and use an appropriate non-calculator method for solving problems that involve multiplying and dividing any three-digit by any two-digit number.

You check your solutions by applying inverse operations or estimating using approximations.

Knowledge

Suppose you are asked the question:

What is 79×53?

You can find an answer in three ways:
- With a calculator:

     4187

- With pen and paper:

  $$\begin{array}{r} 79 \\ \times\,53 \\ \hline 3950 \\ 237 \\ \hline 4187 \end{array}$$

  3950 (79×50)
  237 (79×3)
  4187 (79×53)

- With an estimated approximate answer:

  79×53 is approximately 80×50

  So 79×53 is approximately 4000

To estimate approximate answers:
- Leave numbers between 1 and 10 unchanged, round numbers between 10 and 99 to the nearest 10.
- Round numbers between 100 and 999 to the nearest 100.

Skill Example

Estimate approximate answers for:

a 7×19 **b** 23×38 **c** 467×72

a 7×19 is approximately $7 \times 20 = 140$

b 23×38 is approximately $20 \times 40 = 800$

c 467×72 is approximately $500 \times 70 = 35\,000$

Skill Practice One

Estimate an approximate answer for each of the following.
Find an accurate answer using either pen and paper or a calculator.

1 18×6	**2** 34×9	**3** 2×92
4 8×76	**5** 12×52	**6** 7×59
7 57×9	**8** 14×25	**9** 54×43
10 96×69	**11** 37×82	**12** 296×24
13 372×51	**14** 649×77	**15** 45×712
16 $35 \times 41 \times 29$		

Skill Example

When you find answers using a calculator, you can check them using **inverse operations**:

Multiply with a calculator and check the answer with an inverse operation:

a 79×8 **b** 127×65

a $79 \times 8 = 632$
  Check: $632 \div 8 = 79$

b $127 \times 65 = 8255$
  Check: $8255 \div 65 = 127$

Find an approximate answer and an accurate answer for 897×45.
Check your accurate answer with an inverse operation.

897×45 is approximately $900 \times 50 = 45\,000$

  $897 \times 45 = 40\,365$

Check: $40\,365 \div 45 = 897$

Skill Practice Two

Find an approximate answer and an accurate answer for each of the following.
Check your accurate answer with an inverse operation.

1 5×63	**2** 85×4	**3** 65×6
4 8×79	**5** 27×72	**6** 83×38
7 44×27	**8** 39×17	**9** 45×31
10 405×72	**11** 831×16	
12 550×13	**13** 92×896	
14 65×624	**15** 83×167	

Skill Example

Work out the following divisions:

a $576 \div 18$ **b** $920 \div 23$ **c** $653 \div 31$

You can use a calculator or find the answers like this:

a
$$
\begin{array}{r}
32 \\
18\overline{)576} \\
54 \quad (18 \times 3) \\
\hline
36 \\
36 \quad (18 \times 2) \\
\hline
\end{array}
$$

b
$$
\begin{array}{r}
40 \\
23\overline{)920} \\
92 \quad (23 \times 4) \\
\hline
00 \\
00 \quad (23 \times 0) \\
\hline
\end{array}
$$

c
$$
\begin{array}{r}
21 \\
31\overline{)653} \\
62 \quad (31 \times 2) \\
\hline
33 \\
31 \quad (31 \times 1) \\
\hline
2 \quad \text{remainder}
\end{array}
$$

Therefore, $576 \div 18 = 32$
               $920 \div 23 = 40$
               $653 \div 31 = 21 \text{ r } 2$

Skill Practice Three

Divide the following to find the 'odd answer out'.

1 a $182 \div 13$ **2 a** $224 \div 16$
   b $221 \div 17$ **b** $266 \div 19$
   c $234 \div 18$ **c** $210 \div 14$

3 a $176 \div 11$ **4 a** $255 \div 17$
   b $204 \div 12$ **b** $240 \div 15$
   c $208 \div 13$ **c** $270 \div 18$

5 a $299 \div 13$ **6 a** $378 \div 18$
   b $345 \div 15$ **b** $374 \div 17$
   c $384 \div 16$ **c** $418 \div 19$

7 a $294 \div 21$ **8 a** $504 \div 21$
   b $336 \div 24$ **b** $506 \div 22$
   c $276 \div 23$ **c** $600 \div 25$

9 a $792 \div 24$ **10 a** $748 \div 22$
   b $672 \div 21$ **b** $864 \div 27$
   c $832 \div 26$ **c** $782 \div 23$

Find the answer and remainder for each of the following.

11 $159 \div 12$ **12** $310 \div 20$
13 $360 \div 25$ **14** $204 \div 15$
15 $184 \div 16$ **16** $222 \div 18$

Skill Example

Suppose you are asked the question:

'What is $985 \div 47$?'

You can find an answer in three ways:

- With a calculator:

 20.957447

     $47 \times 20 = 940$

     $985 - 940 = 45$

     $985 \div 47 = 20 \text{ r } 45$

- With an estimated approximate answer:

     $985 \div 47$ is approximately $1000 \div 50$

     $1000 \div 50 = 100 \div 5 = 20$

     $985 \div 47$ is approximately 20

- With pen and paper:

$$
\begin{array}{r}
20 \\
47\overline{)985} \\
94 \\
\hline
45 \\
00 \\
\hline
45 \quad \text{remainder}
\end{array}
$$

$985 \div 47 = 20 \text{ r } 45$

Skill Example

Estimate approximate answers for each of the following:

a $74 \div 34$ **b** $684 \div 57$ **c** $627 \div 23$

a $74 \div 34$ is approximately $70 \div 30$
           $= 7 \div 3$ which is approximately 2

b $684 \div 57$ is approximately $700 \div 60$
           $= 70 \div 6$ which is approximately 12

c $627 \div 23$ is approximately $600 \div 20$
           $= 60 \div 2 = 30$

Skill Practice Four

Estimate approximate answers for each of the following.
Then find an accurate answer using either pen and paper or a calculator.

1 96 ÷ 16	**2** 95 ÷ 19
3 192 ÷ 24	**4** 715 ÷ 13
5 275 ÷ 25	**6** 504 ÷ 56
7 858 ÷ 22	**8** 858 ÷ 39
9 817 ÷ 19	**10** 279 ÷ 31
11 335 ÷ 67	**12** 538 ÷ 24
13 438 ÷ 38	**14** 632 ÷ 33
15 868 ÷ 37	**16** 377 ÷ 29
17 390 ÷ 42	**18** 890 ÷ 19
19 661 ÷ 42	**20** 427 ÷ 26

Skill Example

When you find answers with a calculator, you can check them using **inverse operations**.

Divide with a calculator and check each answer with an inverse operation:

a 79 ÷ 8 **b** 127 ÷ 65

a 79 ÷ 8 = 9.875

Check: 9.875 × 8 = 79

b 127 ÷ 65 = 1.953 846 2

Check: 1.953 846 2 × 65 = 127

Skill Example

Find an approximate answer and an accurate answer for 897 ÷ 45.
Check your accurate answer with an inverse operation.

897 ÷ 45 is approximately 900 ÷ 50

$$= 90 ÷ 5 = 18$$

897 ÷ 45 = 19.933 333

Check: 19.93 333 × 45 = 897

Skill Practice Five

Find an approximate answer and an accurate answer for each of the following.
Check your accurate answer with an inverse operation.

1 923 ÷ 34	**2** 798 ÷ 37
3 302 ÷ 21	**4** 259 ÷ 13
5 721 ÷ 22	**6** 587 ÷ 33
7 602 ÷ 39	**8** 941 ÷ 19
9 826 ÷ 13	**10** 99 ÷ 21
11 998 ÷ 46	**12** 957 ÷ 36
13 985 ÷ 101	**14** 587 ÷ 195
15 783 ÷ 351	**16** 938 ÷ 384
17 597 ÷ 14	**18** 651 ÷ 15
19 921 ÷ 35	**20** 276 ÷ 56

Skill Example

How many jars of instant coffee, each containing 58 grams, can be filled from a carton containing 1000 grams?
How much coffee is left?

Number of jars = 1000 ÷ 58

```
        17
    58)1000
        58    (58 × 1)
       420
       406    (58 × 7)
        14    remainder
```

Therefore, 17 jars can be filled and 14 grams are left.

Skill Practice Six

1 An egg carton, which can hold 12 eggs, weighs 20 g when empty and 440 g when full.
What is the weight of each egg?

2 A warehouse manager orders 2400 tins of baked beans. They arrive on a lorry which is carrying 50 boxes of tins.
How many tins are there in each box?

3 The track through West Hill Tunnel is laid
   with steel rails each of length 20 metres.
   How many rails are there between one end of
   the tunnel and the other?

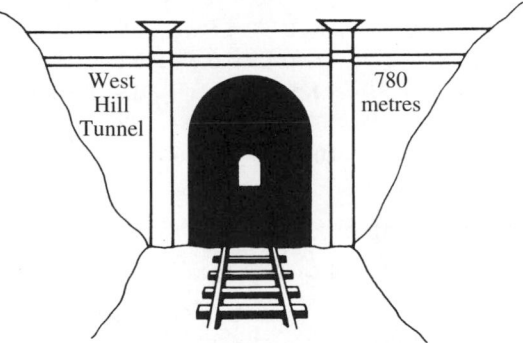

4 I drive my car from London to Birmingham
   and use 14 litres of petrol in doing so.
   Given the distance is 196 kilometres, how far
   can my car travel on 1 litre of petrol?

5 A library shelf is 120 cm long and 24 books of
   equal thickness fit exactly on it.
   What is the thickness of each book?

6 Mrs Williams has made 1000 ml of jam and
   she wishes to put it into small jars which each
   have a capacity of 60 ml.
   How many full jars will there be?
   How many millilitres will there be in the one
   jar which will not be full?

7 A train is to move 300 tonnes of coal from a
   colliery to a power station.
   Given the wagons can carry 18 tonnes each,
   how many full wagons will there be?
   How many tonnes will there be in the one
   wagon which will not be full?

Unit 10

Skill Practice One

Revise your knowledge of decimal addition and subtraction by trying these.
For each question, find the 'odd answer out'.

1 a $31.75 + 20.34 + 12.53$
   b $27.88 + 12.23 + 24.41$
   c $39.16 + 10.24 + 15.12$

6 a $43.98 - 29.13$
   b $32.92 - 18.17$
   c $51.04 - 36.29$

2 a $15.24 + 12.32 + 4.75$
   b $19.43 + 8.57 + 4.41$
   c $8.86 + 6.42 + 17.03$

7 a $19.94 - 4.48$
   b $18.25 - 2.79$
   c $23.82 - 8.46$

3 a $23.8 + 18.58 + 12.27$
   b $16.6 + 24.64 + 13.51$
   c $21.9 + 22.3 + 10.45$

8 a $18.54 - 15.3$
   b $14.7 - 11.56$
   c $18 - 14.76$

4 a $16.38 + 5.46 + 6.5$
   b $17.6 + 6.6 + 4.14$
   c $9.74 + 9.5 + 9$

9 a $7.66 - 4.3$
   b $12.96 - 9.7$
   c $10.16 - 6.9$

5 a $15.63 + 9.85 + 0.28$
   b $18.3 + 7.19 + 0.37$
   c $21.38 + 3.9 + 0.58$

10 a $14.6 - 8.18$
    b $9.4 - 2.88$
    c $12 - 5.58$

Skill Example

Find the value of each of the following

a 0.3×6 **b** 0.4×0.2

A calculator can be used, or the calculations can be completed in the following way.

a $0.3 \times 6 = 3 \div 10 \times 6$
$= 3 \times 6 \div 10$
$= 18 \div 10 = 1.8$

b $0.4 \times 0.2 = 4 \div 10 \times 2 \div 10$
$= 4 \times 2 \div 100$
$= 8 \div 100 = 0.08$

Skill Practice Two

Find the value of each of the following.

1 0.2×7 **2** 0.4×8 **3** 1.6×8
4 2.3×5 **5** 0.45×3 **6** 0.24×7
7 0.07×9 **8** 0.08×5 **9** 0.8×0.3
10 0.7×0.6 **11** 0.15×0.6 **12** 0.24×0.8
13 0.04×0.2 **14** 0.02×0.3 **15** 1.3×0.5
16 1.6×0.4 **17** 4.8×0.4 **18** 3.6×0.5
19 1.92×0.2 **20** 1.04×0.6

Skill Example

Find the value of each of the following.

a 2.6×1.3 **b** 6.8×0.21

A calculator can be used or the calculations can be completed in the following way.

a 2.6×1.3
$= 26 \div 10 \times 13 \div 10$
$= 26 \times 13 \div 100$
$= 338 \div 100 = 3.38$

$$\begin{array}{r} 26 \\ \times 13 \\ \hline 260 \\ 78 \\ \hline 338 \end{array}$$

b 6.8×0.21
$= 68 \div 10 \times 21 \div 100$
$= 68 \times 21 \div 1000$
$= 1428 \div 1000 = 1.428$

$$\begin{array}{r} 68 \\ \times 21 \\ \hline 1360 \\ 68 \\ \hline 1428 \end{array}$$

Skill Practice Three

Find the value of each of the following.

1 2.3×1.4 **2** 3.2×1.6 **3** 4.3×1.5
4 2.1×1.8 **5** 3.9×1.2 **6** 2.5×2.1
7 2.4×1.5 **8** 3.5×2.4 **9** 8.2×1.6
10 9.8×1.2 **11** 9.3×1.5 **12** 5.3×2.4
13 4.8×2.1 **14** 8.8×1.5 **15** 6.4×2.5
16 3.2×0.14 **17** 4.5×0.13 **18** 2.4×0.18
19 6.3×0.12 **20** 2.8×0.21

Skill Example

Find the value of each of the following.

a $12.5 \div 5$ **b** $4.9 \div 7$ **c** $4 \div 5$

A calculator can be used or the calculations can be completed in the following way.

a
$$\begin{array}{r} 2.5 \\ 5\overline{)12.5} \\ \underline{10} \quad (5 \times 2) \\ 2.5 \\ \underline{2.5} \quad (5 \times 0.5) \end{array}$$

b $4.9 \div 7$

$$\begin{array}{r} 0.7 \\ 7\overline{)4.9} \\ \underline{4.9} \quad (7 \times 0.7) \end{array}$$

So $4.9 \div 7 = 0.7$

c $4 \div 5$

We must write 4 as 4.0.

$$\begin{array}{r} 0.8 \\ 5\overline{)4.0} \\ \underline{4.0} \quad (5 \times 0.8) \end{array}$$

So $4 \div 5 = 0.8$

Skill Example

Find the value of each of the following.

a $1.25 \div 0.5$ **b** $0.036 \div 0.6$

A calculator can be used or the calculations can be completed in the following way.

a $1.25 \div 0.5 = \dfrac{1.25}{0.5}$

Multiply the numerator and the denominator by 10 to make the denominator a whole number.

$$\frac{1.25}{0.5} = \frac{1.25 \times 10}{0.5 \times 10}$$

$$= \frac{12.5}{5}$$

$$= 2.5$$

$$\begin{array}{r} 2.5 \\ 5\overline{)12.5} \\ \underline{10} \\ 2.5 \\ \underline{2.5} \end{array}$$

b $0.036 \div 0.06 = \dfrac{0.036}{0.06}$

Multiply the numerator and the denominator by 100 to make the denominator a whole number.

$$\frac{0.036}{0.06} = \frac{0.036 \times 100}{0.06 \times 100}$$

$$= \frac{3.6}{6}$$

$$= 0.6$$

$$\begin{array}{r} 0.6 \\ 6\overline{)3.6} \\ \underline{3.6} \end{array}$$

Skill Practice Four

Find the value of each of the following.

1 $13.6 \div 4$		**2** $11.4 \div 3$	
3 $14.5 \div 5$		**4** $15.6 \div 6$	
5 $27.2 \div 8$		**6** $26.4 \div 6$	
7 $37.1 \div 7$		**8** $40.5 \div 9$	
9 $4.8 \div 3$		**10** $3.4 \div 2$	
11 $5.2 \div 4$		**12** $8.7 \div 3$	
13 $9.6 \div 4$		**14** $4.5 \div 5$	
15 $3.5 \div 7$		**16** $3.2 \div 4$	
17 $5.6 \div 8$		**18** $3.6 \div 6$	
19 $2.4 \div 3$		**20** $8.1 \div 9$	
21 $3 \div 5$		**22** $4 \div 8$	
23 $6 \div 4$		**24** $8 \div 5$	

Skill Practice Five

Find the value of each of the following.

1 $5.28 \div 0.3$		**2** $7.25 \div 0.5$	
3 $8.54 \div 0.7$		**4** $6.36 \div 0.6$	
5 $8.05 \div 0.5$		**6** $7.2 \div 0.4$	
7 $7.8 \div 0.3$		**8** $8.7 \div 0.6$	
9 $9.6 \div 0.5$		**10** $1.34 \div 0.2$	
11 $5.16 \div 0.6$		**12** $5.76 \div 0.8$	
13 $3.44 \div 0.4$		**14** $0.78 \div 0.6$	
15 $0.98 \div 0.7$		**16** $0.9 \div 0.5$	
17 $0.6 \div 0.4$		**18** $0.85 \div 0.05$	
19 $0.76 \div 0.02$		**20** $1.36 \div 0.04$	
21 $2.52 \div 0.09$		**22** $2.8 \div 0.08$	
23 $1.7 \div 0.05$		**24** $0.6 \div 0.03$	

Skill Example

The price of printed cotton material is £5.24 per metre.

What is the price of a dress length of this material, 2.5 m in length?

A calculator can be used or the calculation can be completed in the following way.

Price = £5.24 × 2.5

= £524 ÷ 100 × 25 ÷ 10

= £524 × 25 ÷ 1000

= £13 100 ÷ 1000 = £13.10

	524
×	25
	10 480
	2 620
	13 100

Skill Practice Six

1 Find the cost of carpeting a corridor of length 3.6 m if the price of the carpet is £7.50 per metre.

2 The weight of 1 metre of electric wiring is 0.06 kg. What is the weight of this wiring on a 50 metre reel?

3 The length and height of a section of fencing is shown below.

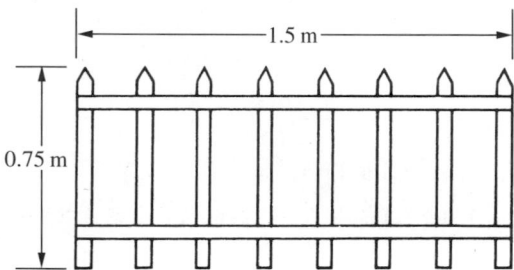

   a Find the total length of wood required.
   b The price of the wood is £1.20 per metre. What is the total cost of the wood?

4 Peter weighs 31.6 kg.
Find the weights of the following members of his family.
   a Peter's father who is 3 times heavier than Peter.
   b Peter's mother who is 28.8 kg lighter than his father.
   c His baby sister Jodie who weighs $\frac{1}{4}$ as much as his mother.

5 Wendy's folder contains 120 sheets of paper, each of weight 0.72 g.
If the empty folder weighs 113.6 g, what is the total weight of her folder and the paper?

6 A large Thermos flask holds sufficient coffee to fill exactly 2 large cups of capacity 0.24 litres and 4 smaller ones of capacity 0.13 litres.
What is the capacity of the flask?

7 What is the length of each sausage in the string of sausages shown below?

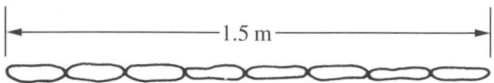

8 A pile of exercise books is 12 cm high.
If each book is 0.8 cm thick, how many are there in the pile?

9 How many pies can be made from 0.5 kg of flour if each pie requires 0.02 kg of the flour?

10 A bottle contains 0.9 litres of lemonade.
How many glasses, each of capacity 0.15 litres, can be filled from it?

11 A milkman is carrying a crate which contains 12 bottles and weighs 11.5 kg.
If the empty crate weighs 0.7 kg, what is the weight of each bottle of milk?

12 From the dimensions of the garage shown in the diagram below, find each of the following.
   a The overall width of the garage.
   b The height of the garage door.

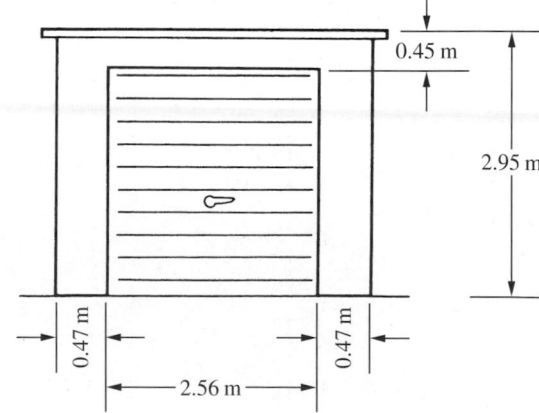

Unit 11

Skill Example

Find:

a $\frac{1}{2}$ of 144   **b** $\frac{1}{5}$ of 34   **c** $\frac{3}{4}$ of 68   **d** $\frac{5}{8}$ of 68

a $\frac{1}{2}$ of $144 = 144 \div 2 = 72$

b $\frac{1}{5}$ of $34 = 34 \div 5 = 6.8$

c $\frac{3}{4}$ of $68 = (68 \div 4) \times 3 = 17 \times 3 = 51$

d $\frac{5}{8}$ of $68 = (68 \div 8) \times 5 = 8.5 \times 5 = 42.5$

Skill Practice One

Find:

1 $\frac{1}{2}$ of 46	**2** $\frac{1}{3}$ of 99	**3** $\frac{1}{4}$ of 64
4 $\frac{1}{5}$ of 1000	**5** $\frac{1}{9}$ of 72	**6** $\frac{2}{3}$ of 330
7 $\frac{3}{4}$ of 224	**8** $\frac{2}{5}$ of 115	**9** $\frac{5}{6}$ of 30
10 $\frac{1}{7}$ of 700	**11** $\frac{1}{8}$ of 800	**12** $\frac{3}{8}$ of 800
13 $\frac{2}{9}$ of 900	**14** $\frac{6}{7}$ of 350	**15** $\frac{3}{5}$ of 80
16 $\frac{3}{5}$ of 8	**17** $\frac{7}{8}$ of 16	**18** $\frac{7}{8}$ of 8
19 $\frac{7}{8}$ of 4	**20** $\frac{8}{9}$ of 90	

Skill Example

There are 24 000 spectators at a football match and police estimate that $\frac{7}{8}$ of them support the home team.

How many home supporters does this statement represent?

$\frac{7}{8}$ of $24\,000 = (24\,000 \div 8) \times 7 = 21\,000$

Therefore, there are 21 000 home supporters.

Skill Practice Two

1 There are 144 apples in a box.
Calculate how many apples each statement represents.

   a $\frac{1}{2}$ of the apples are red.

   b $\frac{3}{4}$ of the apples are French.

   c $\frac{5}{12}$ of the apples weigh more than 110 grams.

   d $\frac{5}{36}$ of the apples are bad.

   e $\frac{1}{6}$ of the apples are in top layer.

2 There are 260 girls and 240 boys in a school.
Calculate how many pupils each statement represents.

   a $\frac{2}{5}$ of the girls come to school by bike.

   b $\frac{3}{10}$ of the boys stay to school dinner.

   c $\frac{7}{10}$ of the girls wear earrings.

   d All the boys and $\frac{1}{5}$ of the girls wear ties.

   e $\frac{3}{10}$ of the girls and $\frac{4}{10}$ of the boys play for a school team.

   f $\frac{81}{100}$ of the pupils live within 2 miles of the school.

   g $\frac{11}{100}$ of the pupils wear glasses.

   h $\frac{33}{100}$ of the pupils are left handed.

   i $\frac{19}{100}$ of the pupils are in Year 7.

   j $\frac{21}{100}$ of the pupils are in Year 10.

3 A farmer has a flock of 300 sheep.
Calculate how many sheep each statement represents.

   a $\frac{2}{25}$ of the sheep are black.

   b $\frac{5}{12}$ of the sheep are lambs.

   c $\frac{7}{30}$ of the sheep are in the North Field.

   d $\frac{3}{20}$ of the sheep are in the West Field.

   e $\frac{7}{10}$ of the sheep have been dipped.

   f $\frac{31}{100}$ of the sheep are over 2 years old.

   g $\frac{83}{100}$ of the sheep are ewes (females).

   h The vet saw $\frac{7}{100}$ of the sheep last month.

   i $\frac{19}{100}$ of the sheep will be sold next week.

   j $\frac{41}{100}$ of the sheep have been shorn.

4 There are 240 marks available in a mathematics examination with three papers: Paper 1, Paper 2 and Paper 3.
Calculate how many marks each statement represents.

  a Amy scored $\frac{3}{8}$ of the available marks.

  b Janice scored $\frac{5}{6}$ of the available marks.

  c Siloben scored $\frac{29}{30}$ of the available marks.

  d $\frac{25}{100}$ of the marks are for algebra questions.

  e $\frac{20}{100}$ of the marks are for geometry questions.

  f $\frac{40}{100}$ of the marks are for arithmetic questions.

  g $\frac{30}{100}$ of the marks are for Paper 1.

  h $\frac{55}{100}$ of the marks are for Paper 2.

  i $\frac{15}{100}$ of the marks are for Paper 3.

  j $\frac{95}{100}$ or more of the marks are needed for a pass with distinction.

Knowledge

A percentage of a quantity can be found by changing the percentage into a fraction and calculating this fraction of the quantity.

Skill Example

Find:

a 15% of £500   **b** 66% of 250 g

a $15\% = \frac{15}{100}$

  $\frac{15}{100}$ of £500 $= (£500 \div 100) \times 15 = £75$

b $66\% = \frac{66}{100}$

  $\frac{66}{100}$ of 250 g $= (250\,g \div 100) \times 66 = 165\,g$

Skill Practice Three

Find:

1 8% of £500   **2** 5% of £900
3 6% of 400 g   **4** 3% of £1200
5 20% of £60   **6** 60% of £90
7 50% of 30 cm   **8** 40% of £120
9 80% of £110   **10** 40% of £25

11 60% of £15   **12** 20% of 75 cm
13 80% of 35 cm   **14** 25% of £96
15 75% of £28   **16** 75% of £64
17 35% of £60   **18** 45% of 80 cm
19 15% of 120 g   **20** 40% of £1.50
21 60% of £1.20   **22** 80% of 1 m 10 cm
23 20% of £3.50   **24** 60% of £2.50

Skill Example

In a school of 1400 pupils, 45% of them are boys.
Find the number of boys in the school

Number of boys $= 45\%$ of 1400

  $45\% = \frac{45}{100}$

  $\frac{45}{100}$ of 1400 $= (1400 \div 100) \times 15 = 630$

Skill Practice Four

1 A football club has 25 players, but only 60% of them have ever played for the first team.
Find the number who have played for the first team.

2 It takes me 45 minutes to get to school and I spend 80% of that time travelling on the bus. How long does my bus journey last?

3 A room has an area of 30 m^2 and a carpet covers 90% of this area.
Find the area of the carpet.

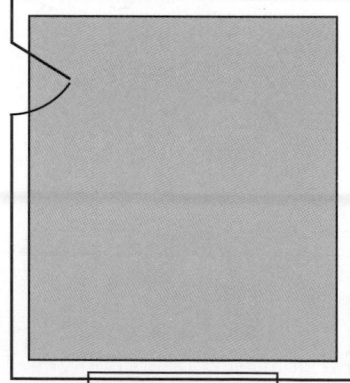

4 A farmer has 40 sheep and 35% of them are black.
Find the number of black sheep.

5 A car's petrol tank can hold 36 litres.
How many litres are there in it if it is 75% full?

6 There are 32 boys in class 5A and one day
$12\frac{1}{2}$% of them are absent.
Find the number who are absent.

7 At 3 pm a newspaper seller is given 350 papers
and by 5 pm she has sold 40% of them.
Find the number that she has sold by 5 pm.

8 A match box had 50 matches inside when it
was bought, but only 70% of them are left.
How many matches have been used?

9 There are 20 boys in class 1B and they have
three sports options to choose from.
If 25% choose athletics, 35% choose
swimming and 40% choose cricket, find the
number who choose each of the three sports.

10 At Northgate School there are 750 pupils. The
percentage absent on each day of a certain
week is shown below.

Monday	8%	Thursday	2%
Tuesday	10%	Friday	4%
Wednesday	6%		

Find the number absent on each day.

Unit 12

Number and Algebra   Level 5

Skill   You order, add and subtract negative numbers in context.

8 −90 °C, −85 °C, −92 °C, −91 °C, −80 °C
9 −11 °C, 8 °C, −3 °C, −2 °C, 0 °C, 5 °C, 4 °C, −8 °C, −1 °C, 12 °C
10 −6 °C, 3 °C, −2 °C, 6 °C, −1 °C, 0 °C, −3 °C, 2 °C, −4 °C, 4 °C

Knowledge

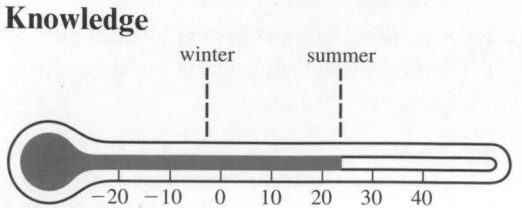

On the Celsius temperature scale, the freezing point of water is 0 degrees, written as 0 °C.

The temperature on a summer's day might be 23 degrees, written as +23 °C.

The temperature on a frosty day in winter might be 2 degrees below zero, written as −2 °C.

Normally, the temperature is above 0 °C, so only those temperatures below the freezing point need a sign.

Thus, body temperature is 37 °C; the temperature inside a freezer is −3 °C. This is a **negative** temperature, read as minus 3 °C.

Skill Example

These are the lowest temperatures over five days in December.
Arrange them in ascending order,.

   −3 °C   0 °C   −5 °C   2 °C   −1 °C

The order is:

   −5 °C   −3 °C   −1 °C   0 °C   2 °C

Skill Practice One

Arrange each set of temperatures in ascending order.

1 5 °C, 4 °C, −7 °C, −2 °C, 1 °C
2 −3 °C, 3 °C, −2 °C, 2 °C, 1 °C
3 2 °C, 0 °C, −1 °C, 1 °C, −2 °C
4 −10 °C, 5 °C, −5 °C, 10 °C, 0 °C
5 8 °C, 0 °C, −4 °C, 4 °C, −8 °C
6 17 °C, 63 °C, −15 °C, −21 °C, −33 °C
7 −8 °C, −6 °C, −9 °C, −11 °C, −7 °C

Skill Example

a The temperature is 10 °C.

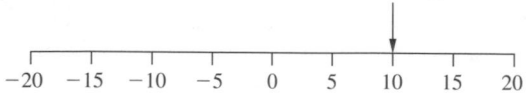

Find the new temperature when:

(i) the temperature rises by 5 degrees
(ii) the temperature falls by 5 degrees.

(i) New temperature = 10 + 5 = 15 °C
(ii) New temperature = 10 − 5 = 5 °C

b The temperature is 0 °C.

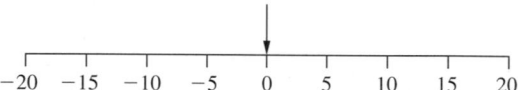

Find the new temperature when:

(i) the temperature rises by 5 degrees
(ii) the temperature falls by 5 degrees.

(i) New temperature = 0 + 5 = 5 °C
(ii) New temperature = 0 − 5 = −5 °C

c The temperature is −5 °C.

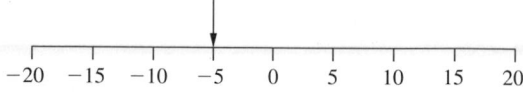

Find the new temperature when:

(i) the temperature rises by 5 degrees
(ii) the temperature falls by 5 degrees.

(i) New temperature = −5 + 5 = 0 °C
(ii) New temperature = −5 − 5 = −10 °C

Skill Practice Two

Copy and complete the table.

	Temperature	Change °C	New temperature
1	10 °C	rises 5	
2	13 °C	+6	
3	11 °C	+4	
4	15 °C	falls 3	
5	18 °C	−5	
6	17 °C	−6	
7	9 °C		11 °C
8	14 °C		10 °C
9	12 °C		5 °C
10		+5	13 °C
11		−6	5 °C
12	0 °C	falls 6	
13	5 °C	falls 7	
14	3 °C	−4	
15	4 °C		−2 °C
16	0 °C		−5 °C
17		−3	−1 °C
18		−8	−3 °C
19	−3 °C	rises 7	
20	−5 °C	+8	
21	−4 °C		5 °C
22	−3 °C		3 °C
23		+8	6 °C
24		+9	8 °C
25	−2 °C	falls 5	
26	−5 °C	−4	
27	−4 °C		−6 °C
28	−1 °C		−7 °C
29		−5	−8 °C
30		−3	−7 °C

Knowledge

The amount of money that is in your bank account is called the **balance**.

Taking money out of your account is called **making a withdrawal**.

Paying money into your account is called **making a deposit**.

Writing a cheque to pay for something is equivalent to making a withdrawal.

If you withdraw more money than is in your account, your account is **overdrawn** and you owe the bank money.

Skill Example

a Ms Jones has a balance of £230 in her bank account.
If she writes a cheque for £350, how much is she overdrawn?

£350 is £120 more than £230, so Ms Jones has −£120 in her account. She is £120 *overdrawn*.

b Mr Smith was £455 overdrawn but then made a deposit of £600.
What is his new balance?

Mr Smith's balance before he paid in the £600 was −£455. £600 is £145 more than £455, so Mr Smith's new balance is £145.

Skill Practice Three

Copy and complete this table.

	Balance	Change	New balance
1	£340	deposit £400	
2	£235	deposit £1675	
3	£389	withdraw £350	
4	£1745	withdraw £705	
5	£3450	cheque £2775	
6	£209	cheque £209	
7	−£400	deposit £500	
8	−£783	deposit £800	
9	−£1000	deposit £1200	
10	−£275	deposit £645	
11	£500	withdraw £650	
12	£456	withdraw £457	
13	£371	withdraw £567	
14	−£300	deposit £200	
15	−£150	deposit £80	
16	−£321	deposit £212	
17	−£765	deposit £89	
18	−£200	withdraw £200	
19	−£47	cheque £56	
20	−£913	cheque £274	

46

Knowledge

Most recent types of calculator can deal with negative numbers. Those which can have a key like this:

The key is used to enter the sign of a number. It is used after the number. So, to enter −6, you press these keys:

The display will show

-6

To enter −8 + −14, you press

Skill Example

Copy this table and complete it using a calculator.

First number	Operation sign	Second number	Answer
8	+	14	
−8	+	14	
8	+	−14	
−8	+	−14	
8	−	14	
−8	−	14	
8	−	−14	
−8	−	−14	

This is the completed table.

First number	Operation sign	Second number	Answer
8	+	14	22
−8	+	14	6
8	+	−14	−6
−8	+	−14	−22
8	−	14	−6
−8	−	14	−22
8	−	−14	22
−8	−	−14	6

Skill Practice Four

Copy this table and complete it using a calculator.

	First number	Operation sign	Second number	Answer
1	2	+	3	
2	−2	+	3	
3	2	+	−3	
4	−2	+	−3	
5	2	−	3	
6	−2	−	3	
7	2	−	−3	
8	−2	−	−3	
9	5	+	4	
10	−5	+	4	
11	5	+	−4	
12	−5	+	−4	
13	5	−	4	
14	−5	−	4	
15	5	−	−4	
16	−5	−	−4	
17	9	+	7	
18	−9	+	7	
19	9	+	−7	
20	−9	+	−7	
21	9	−	7	
22	−9	−	7	
23	9	−	−7	
24	−9	−	−7	
25	7	+	8	
26	−7	+	8	
27	7	+	−8	
28	−7	+	−8	
29	7	−	8	
30	−7	−	8	
31	7	−	−8	
32	−7	−	−8	
33	10	+	12	
34	−10	+	12	
35	10	+	−12	
36	−10	+	−12	
37	10	−	12	
38	−10	−	12	
39	10	−	−12	
40	−10	−	−12	

Skill Example

Eight addition and subtraction problems can be created using only two numbers and both signs.

For example, starting with 8 and 5:

$$8 + 5 =$$
$$-8 + 5 =$$
$$8 + -5 =$$
$$-8 + -5 =$$
$$8 - 5 =$$
$$-8 - 5 =$$
$$8 - -5 =$$
$$-8 - -5 =$$

The answers to the first six problems are straightforward.

The answers to the last two can be found by using this rule:

> Subtracting a negative number is the same as adding the same postive number.

So,

$$8 + 5 = 13$$
$$-8 + 5 = -3$$
$$8 + -5 = 3$$
$$-8 + -5 = -13$$
$$8 - 5 = 3$$
$$-8 - 5 = -13$$
$$8 - -5 = 8 + 5 = 13$$
$$-8 - -5 = -8 + 5 = -3$$

Skill Practice Five

Copy and complete each of the following *without* using a calculator.

1	$6 + 5$	**2**	$-6 + 5$
3	$6 + -5$	**4**	$-6 + -5$
5	$6 - 5$	**6**	$-6 - 5$
7	$6 - -5$	**8**	$-6 - -5$
9	$3 + 9$	**10**	$-3 + 9$
11	$3 + -9$	**12**	$-3 + -9$
13	$3 - 9$	**14**	$-3 - 9$
15	$3 - -9$	**16**	$-3 - -9$
17	$4 + 7$	**18**	$-4 + 7$
19	$4 + -7$	**20**	$-4 + -7$
21	$4 - 7$	**22**	$-4 - 7$
23	$4 - -7$	**24**	$-4 - -7$
25	$11 + 10$	**26**	$-11 + 10$
27	$11 + -10$	**28**	$-11 + -10$
29	$11 - 10$	**30**	$-11 - 10$
31	$11 - -10$	**32**	$-11 - -10$
33	$13 + 7$	**34**	$-13 + 7$
35	$13 + -7$	**36**	$-13 + -7$
37	$13 - 7$	**38**	$-13 - 7$
39	$13 - -7$	**40**	$-13 - -7$

Skill Practice Six

Create and solve eight different addition or subtraction problems from each of these pairs of numbers.

1 3 and 4
2 5 and 9
3 6 and 7
4 10 and 5
5 17 and 3

Unit 13

Skill Example

Wendy has m pencils.

Write each of these people's number of pencils using m.

a Susan has 4 more pencils than Wendy.
b John has 5 less pencils than Susan.

a Susan has $m + 4$ pencils.
b John has $m + 4 - 5 = m - 1$ pencils.

Skill Practice One

1 Julie plants x trees in her garden.
Write the number of trees each of these people plant, using x.
  a Lyndsey, who plants 6 more trees than Julie.
  b Alan, who plants 8 less trees than Julie.
  c Kathleen, who plants 5 more trees than Julie.
  d Christopher, who plants 8 less trees than Kathleen.

2 Andrew buys 'pick and-mix' and puts n sweets in his bag.
Write the number of sweets each of these people put in their bags, using n.
  a Leanne, who puts 7 less sweets than Andrew in her bag.
  b Emma, who puts 9 more sweets than Leanne in her bag.
  c Amy, who puts 12 more sweets than Andrew in her bag.
  d Douglas, who puts 8 less sweets than Amy in his bag.

3 Matthew was absent from school on s days last year.
Write the number of days each of these people were absent, using s.
  a Nathan, who was absent for 4 more days than Matthew.
  b Carmen, who was absent for 5 more days than Nathan.
  c Priya, who was absent for 6 days less than Matthew.
  d Adam, who was absent for 2 days less than Priya.

4 Ronnie owns y computer games.
Write the number of computer games each of these people own, using y.
  a Nicky, who owns 9 more games than Ronnie.
  b Jenna, who owns 12 less games than Nicky.
  c Kirsty, who owns 14 less games than Ronnie.
  d Michelle, who owns 20 more games than Kirsty.

Skill Example

In a bottle of medicine there are 30 doses. How many doses are there in:

a 2 bottles   **b** 5 bottles
c x bottles   **d** $3a$ bottles?

In 1 bottle of medicine there are 30 doses.

a In 2 bottles of medicine there are
$30 \times 2 = 60$ doses

b In 5 bottles of medicine there are
$30 \times 5 = 150$ doses

c In x bottles of medicine there are
$30 \times x = 30x$ doses

d In $3a$ bottles of medicine there are
$30 \times 3a = 30 \times 3 \times a = 90a$ doses

Skill Practice Two

1 In a box of chalks there are 12 sticks.
How many sticks of chalk are there in:
  a 2 boxes   **b** 5 boxes
  c x boxes   **d** $2x$ boxes?

2 In a book of stamps there are 10 stamps altogether.
How many stamps are there in:
  a 3 books   **b** 6 books
  c x books   **d** $3x$ books?

3 In a matchbox there are 50 matches.
How many matches are there in:
a 2 boxes   **b** 3 boxes
c x boxes   **d** $4x$ boxes?

4 A wrapped loaf of bread has 20 slices.
How many slices are there in:
a 3 loaves   **b** 5 loaves
c y loaves   **d** $10y$ loaves?

5 A writing pad contains 40 sheets of paper.
How many sheets are there in:
a 2 pads   **b** 5 pads
c y pads   **d** $5y$ pads?

6 A bottle contains 100 tablets.
How many tablets are there in:
a 3 bottles   **b** 6 bottles
c x bottles   **d** $4y$ bottles?

7 In a box of chocolates there are 25 chocolates altogether.
How many chocolates are there in:
a 2 boxes   **b** 3 boxes
c x boxes   **d** $2y$ boxes?

8 A milkman uses crates which each hold 15 bottles.
How many bottles can be placed in:
a 2 crates   **b** 4 crates
c p crates   **d** $4q$ crates?

Skill Example

Simplify:

a $5x + 3x$ **b** $4a - 3a$ **c** $8b + b - 9b$

a $5x + 3x = 8x$

b $4a - 3a = a$ (Note: this is not written $1a$)

c $8b + b - 9b = 0$ (Note: this is not written $0b$)

Skill Practice Three

Collect the terms in each question to give a single term.

1 $a + a$ **2** $c + c + c + c$
3 $l + 2l$ **4** $n + 5n$
5 $6q + q$ **6** $2t + 3t$
7 $4v + 6v$ **8** $8y + 3y$
9 $4b + 10b + b$ **10** $4m + m + 7m$
11 $2l + 3l + 7l$ **12** $6q + 7q + 7q$
13 $7b - 4b$ **14** $10l - 4l$
15 $4n - n$ **16** $12q - q$
17 $8t - 7t$ **18** $2v - v$
19 $6b + 2b - 3b$ **20** $7l + 4l - 5l$
21 $5n + 2n - 7n$ **22** $9q - 4q - 3q$
23 $7s - 2s - s$ **24** $8u - 5u - 2u$
25 $7x - 5x - x$

Skill Example

Remember: expressions such as $3x + 2y$ cannot be written as a single term, because the terms are not alike.

Collect the like terms in the expression $3x + 4y + 2x$.

$$3x + 4y + 2x = 3x + 2x + 4y$$
$$= 5x + 4y$$

Skill Practice Four

Collect the like terms in each expression.

1 $2x + 5y + 4x$ **2** $3u + 7v + 5u$
3 $4a + 2b + 5a$ **4** $6x + 4y + x$
5 $5u + 4v - 2u$ **6** $7x + 9y - 5x$
7 $8p + 9q - p$ **8** $4l + 7m - 3l$
9 $2m + 6n + 5m + 3n$ **10** $3p + 7q + 2p + 5q$
11 $2u + 5v + u + 3v$ **12** $2l + 7m + 3l - 4m$
13 $8b + 9c + 4b - 2c$ **14** $5q + 9r + 2q - r$
15 $11a + 12b + a - b$ **16** $7x + 5y - 4x + 2y$
17 $9a + 3b - 5a + 6b$ **18** $8m + 5n - 7m + 3n$
19 $9p + 7q - 4p - 3q$ **20** $10u + 8v - 4u - 3v$

Skill Example

Find a formula for the perimeter p of each shape.

a

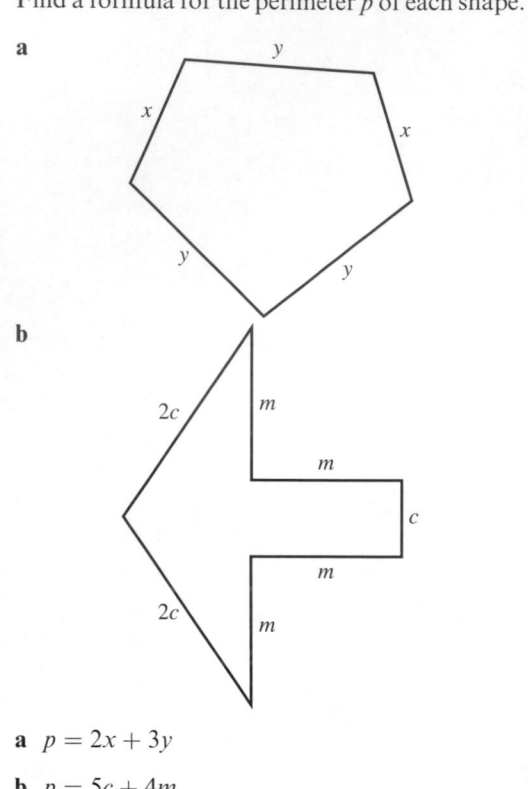

b

a $p = 2x + 3y$

b $p = 5c + 4m$

Skill Practice Five

Find a formula for the perimeter p of each shape.

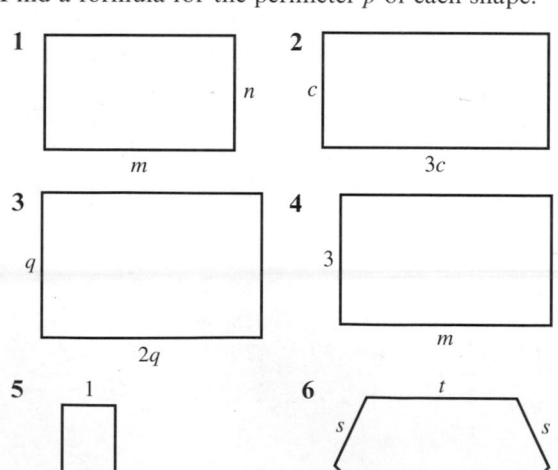

7

8

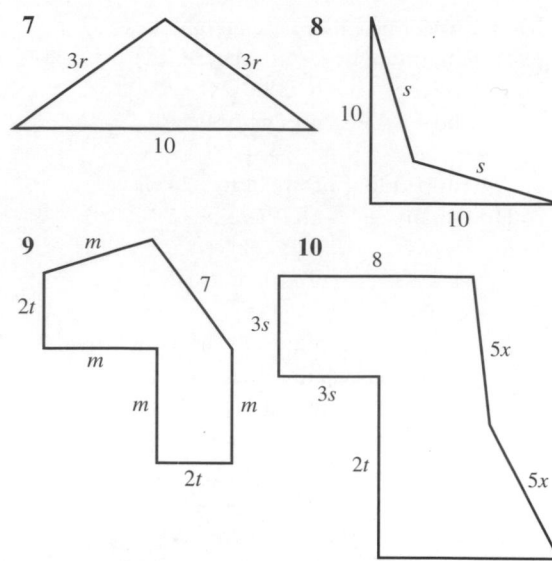

9

10

Skill Example

These boxes contain cans.
Find a formula for the weight w of one can from each box.

a

b

a $w = q \div 100$

Usually written

$$w = \frac{q}{100}$$

b $w = \frac{25m}{50} = \frac{m}{2}$

Skill Practice Six

Find a formula for the weight w of one can from each box.

1

2

3

4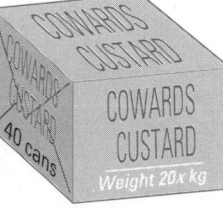

5

6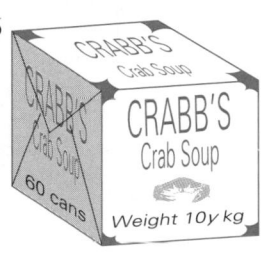

7

8

9

10

Knowledge

2^3 is read as 'two cubed'

$$2^3 = 2 \times 2 \times 2 = 8$$

5^2 is read as 'five squared'

$$5^2 = 5 \times 5 = 25$$

a^3 is read as 'a cubed'

$$a^3 = a \times a \times a$$

x^2 is read as 'x squared'

$$x^2 = x \times x$$

In the term 2^3, the figure 3 is called the **index** (plural 'indices').

Skill Example

Find the value of each of the following.

a 4^3 **b** 7^2 **c** $2^3 \times 3^2$

a $4^3 = 4 \times 4 \times 4$
$\quad = 16 \times 4 = 64$

b $7^2 = 7 \times 7 = 49$

c $2^3 \times 3^2 = 2 \times 2 \times 2 \times 3 \times 3$
$\quad\quad\quad\quad = 8 \times 9 = 72$

Skill Practice Seven

Find the value of each of the following.

1 2^2	**2** 3^2	**3** 4^2
4 6^2	**5** 10^2	**6** 3^3
7 6^3	**8** 10^3	**9** 5^3
10 1^3	**11** $2^2 \times 2^2$	**12** $2^2 \times 2^3$
13 $2^3 \times 2^3$	**14** $2^3 \times 2$	**15** $2^2 \times 3^2$
16 $2^2 \times 5^2$	**17** $3^2 \times 3^2$	**18** $3^3 \times 3$
19 $3^3 \times 5$	**20** $3^2 \times 5^2$	**21** $3^2 \times 10^2$
22 $10^2 \times 10$	**23** $5^2 \times 5$	**24** $10^2 \times 10^2$
25 $5^2 \times 10^2$		

Skill Example

Express each of the following in index form.

a $4 \times 4 \times 4$ **b** $x \times x \times x$ **c** $3 \times 3 \times 5 \times 5$

a $4 \times 4 \times 4 = 4^3$

b $x \times x \times x = x^3$

c $3 \times 3 \times 5 \times 5 = 3^2 \times 5^2$

Skill Practice Eight

Express each of the following in index form.

1	8×8	2	7×7
3	9×9	4	$6 \times 6 \times 6$
5	$8 \times 8 \times 8$	6	$10 \times 10 \times 10$
7	$12 \times 12 \times 12$	8	$a \times a$
9	$p \times p$	10	$t \times t$
11	$b \times b \times b$	12	$m \times m \times m$
13	$z \times z \times z$	14	$2 \times 2 \times 4 \times 4$
15	$3 \times 3 \times 5 \times 5$	16	$6 \times 6 \times 10 \times 10$
17	$2 \times 2 \times 2 \times 7 \times 7$	18	$4 \times 4 \times 4 \times 9 \times 9$
19	$5 \times 5 \times 5 \times 6 \times 6$	20	$2 \times 2 \times 5 \times 5 \times 5$
21	$8 \times 8 \times 9 \times 9 \times 9$	22	$x \times x \times x \times y \times y$
23	$m \times m \times n \times n$	24	$u \times u \times v \times v$
25	$a \times a \times a \times b \times b$	26	$y \times y \times y \times z \times z$
27	$u \times u \times u \times v \times v$	28	$m \times m \times n \times n \times n$
29	$p \times p \times q \times q \times q$	30	$c \times c \times d \times d \times d$

Skill Example

Simplify the following.

a $x^2 \times x$ **b** $2a \times 3a$ **c** $4b \times 3b^2$

a $x^2 \times x = x \times x \times x = x^3$

b $2a \times 3a = 2 \times a \times 3 \times a$
$= 2 \times 3 \times a \times a$
$= 6 \times a \times a$
$= 6 \times a^2 = 6a^2$

c $4b \times 3b^2 = 4 \times b \times 3 \times b \times b$
$= 4 \times 3 \times b \times b \times b$
$= 12 \times b \times b \times b$
$= 12 \times b^3 = 12b^3$

Skill Practice Nine

Simplify the following.

1	$x^2 \times x$	2	$y^2 \times y$	3	$a \times a^2$
4	$b \times b^2$	5	$3p \times 2p$	6	$5q \times 4q$
7	$3r \times 3r$	8	$6s \times s$	9	$4x^2 \times x$
10	$7y^2 \times y$	11	$a^2 \times 3a$	12	$b^2 \times 9b$
13	$5m^2 \times 3m$	14	$6n^2 \times 2n$	15	$4t^2 \times 4t$
16	$2u^2 \times 3u$	17	$3v^2 \times 8v$	18	$4z \times 5z^2$
19	$3a \times 4a^2$	20	$2b \times 7b^2$	21	$6c \times 4c^2$
22	$(3p)^2$	23	$(4q)^2$	24	$(2x)^2$
25	$(10y)^2$				

Skill Example

Find the area, A, of each of these rectangles.

a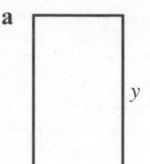

b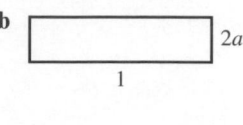

c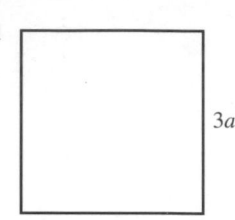

d

a $A = x \times y$
   $A = xy$

b $A = 3 \times 2a$
   $A = 3 \times 2 \times a$
   $A = 6a$

c $A = 2a \times b$
   $A = 2 \times a \times b$
   $A = 2ab$

d $A = 3a \times 3a$
   $A = 3 \times a \times 3 \times a$
   $A = 3 \times 3 \times a \times a$
   $A = 9a^2$

Skill Practice Ten

Find the area of each of these rectangles.

1   2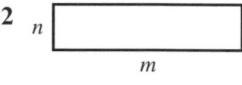

3   4

5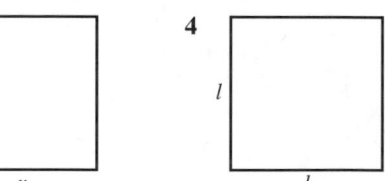

6

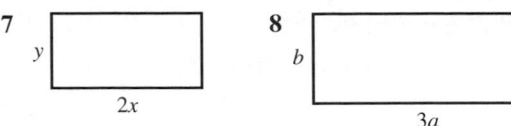

7 y, 2x

8 b, 3a

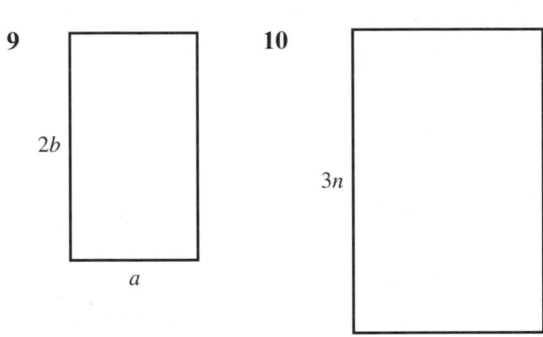

9 2b, a

10 3n, m

Skill Example

If $a = 4$, $b = 3$, $c = 2$, find the value of each of the following.

a $a + b$ **b** $b - c$ **c** $2a$ **d** $3c + a$

a $a + b = 4 + 3 = 7$

b $b - c = 3 - 2 = 1$

c $2a = 2 \times a = 2 \times 4 = 8$

d $3c + a = (3 \times c) + a$
$= (3 \times 2) + 4$
$= 6 + 4$
$= 10$

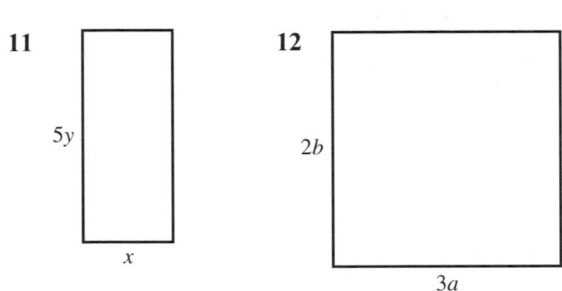

11 5y, x

12 2b, 3a

Skill Practice Eleven

Find the value of each of the following if $a = 4$, $b = 3$ and $c = 2$.

1 $a + b$ **2** $2a$ **3** $2a + b$ **4** $a + c$
5 $2b$ **6** $2a + c$ **7** $b + c$ **8** $2c$
9 $2b + c$ **10** $b + a$ **11** $3a$ **12** $2b + a$
13 $c + b$ **14** $4b$ **15** $a + 3c$ **16** $c + a$
17 $3c$ **18** $b + 4a$ **19** $a - b$ **20** $4c$

Find the value of each of the following if $x = 6$, $y = 5$ and $z = 2$.

21 $x + y$ **22** $2x$ **23** $3x + y$
24 $x - y$ **25** $3y$ **26** $2y + z$
27 $y + z$ **28** $4z$ **29** $4z + x$
30 $y - z$ **31** $3x$ **32** $3z + y$
33 $x + y + z$ **34** $5y$ **35** $5y + z$
36 $y + z - x$ **37** $5z$ **38** $4z - y$
39 $x + z - y$ **40** $5x$

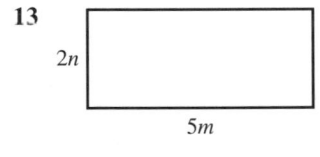

13 2n, 5m

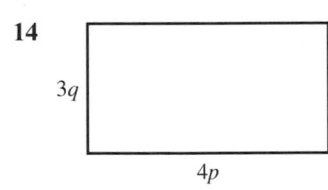

14 3q, 4p

Skill Example

If $a = 5$, $b = 4$, $c = 2$, $d = 0$, find the value of each of the following.

a $3a$ **b** ab **c** bcd **d** $2bc$

a $3a = 3 \times a = 3 \times 5 = 15$

b $ab = a \times b = 5 \times 4 = 20$

c $bcd = b \times c \times d$ **d** $2bc = 2 \times b \times c$
$= 4 \times 2 \times 0$ $= 2 \times 4 \times 2$
$= 8 \times 0$ $= 8 \times 2$
$= 0$ $= 16$

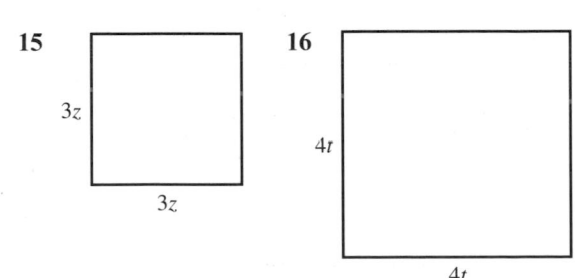

15 3z, 3z

16 4t, 4t

Skill Practice Twelve

Find the value of each of the following if $a = 4$, $b = 2$ and $c = 3$.

1 $3a$	**2** $7b$	**3** $5c$	**4** ab
5 $2ab$	**6** $3ab$	**7** bc	**8** $3bc$
9 $5bc$	**10** ac	**11** $4ac$	**12** abc

Find the value of each of the following if $p = 6$, $q = 4$, $r = 3$ and $s = 1$.

13 $3p$	**14** $6s$	**15** pq	**16** $5pq$
17 qr	**18** $2qr$	**19** pr	**20** $5pr$
21 rs	**22** $3rs$	**23** $4ps$	**24** $3qs$
25 pqr	**26** pqs	**27** prs	**28** qrs

Find the value of each of the following if $r = 6$, $s = 3$, $t = 2$ and $v = 0$.

29 $5r$	**30** $7v$	**31** rs	**32** $2rs$
33 tv	**34** $5tv$	**35** $2st$	**36** ts
37 rt	**38** $4rt$	**39** $3rv$	**40** rst
41 tsr	**42** stv	**43** tvr	**44** vrs

Skill Example

If $a = 3$, $b = 2$, find the value of each of the following.

a a^2 **b** b^2 **c** a^3 **d** $2b^3$ **e** ab^2

a $a^2 = a \times a = 3 \times 3 = 9$

b $b^2 = b \times b = 2 \times 2 = 4$

c $a^3 = a \times a \times a$
$= 3 \times 3 \times 3$
$= 9 \times 3$
$= 27$

d $2b^3 = 2 \times b^3$
$= 2 \times b \times b \times b$
$= 2 \times 2 \times 2 \times 2$
$= 4 \times 2 \times 2$
$= 8 \times 2$
$= 16$

e $ab^2 = a \times b^2$
$= a \times b \times b$
$= 3 \times 2 \times 2$
$= 6 \times 2$
$= 12$

Skill Practice Thirteen

Find the value of each of the following if $a = 2$ and $b = 3$.

1 a^2	**2** b^2	**3** $2a^2$	**4** $2b^2$	**5** $4a^2$
6 $3b^2$	**7** a^3	**8** b^3	**9** ab^2	**10** a^2b

Find the value of each of the following if $x = 4$ and $y = 1$.

11 x^2	**12** y^2	**13** $3x^2$	**14** $4y^2$	**15** $5x^2$
16 $10y^2$	**17** x^3	**18** y^3	**19** xy^2	**20** x^2y

Skill Example

A newspaper charges for advertisements using the rule:

Cost is $80\,\text{p} + 90\,\text{p}$ per line

We can write this as the formula

$$C = 90L + 80$$

where C is the cost (in pence) and L is the number of lines.

Find the cost of this advertisement.

> **COSTA BLANCA** Complex with pool, tennis, 2 beautiful villas slps 6 each. Tel 0354 840337.

There are 3 lines in the advertisement. So,

$$C = (3 \times 90) + 80$$
$$C = 350$$

The cost is £3.50.

Skill Practice Fourteen

1 Find the cost of each of these advertisements, using the formula given in the Skill Example.

a
> **CHATEAUX**. Region of Loire. Designer's cottages with pools. Medieval town, walk shops. Golf/wine. 0579 796 515.

b
> **COTE D'AZUR**. Grasse very lovely stone mas in lge olive grove: 4–5 bedrms/4 bathrms; seaviews; Sat. TV: 50ft pool & poolhse; games rm; total peace & privacy. £950–£2950 pw. Tel 02034 98936885 fax 3904.

c

> **COTE D'AZUR**. Lovely old Mod Frmhse in 2 acre terraced olive grove, 20 mins coast nr Grasse 4 dbl bed, 2 bath, pool, views. 8th Jul-16th Sep £800-1200 pw inc Maid Service. 0324 773534.

d

> **FRANCE** Dordogne. Delightful converted barn in tranquil, listed hill village. Slps 8, 4 dble bdrms, 2 bthrms, lge pool, (safely fenced in), dishwasher and all mod cons. Phone us in Belgium 020 233 5336 57 33 and we'll call straight back, or fax 020 233 5336 07 67.

e

> **FREE DRINKS**, all Sports, Full Board at this House Party sporting holiday in the Dordogne. Tennis, Clay Pigeon Shooting, Badminton and Snooker etc. £270 pp pw. Nothing else to pay. Sorry most weeks no children. Phone Mandy 00425 621544 or John 0300 35 38 52 09.

2 The formula for the weight (W) in kg of a certain piglet is:

$$W = 1.8 + 0.5D$$

where D is the number of days since the piglet was born.

a Find the weight of the piglet after
(i) 3 days   (ii) 5 days   (iii) 25 days
b How much did the piglet weigh when it was born?

3 The formula $S = 3L - 25$ links foot length (L) measured in inches with shoe size (S).

a What shoe sizes fit feet of the following lengths?
(i) 10 inches (ii) 11 inches (iii) $9\frac{1}{3}$ inches
(iv) 12 inches (v) $10\frac{2}{3}$ inches
b John Thrupp of Stratford-upon-Avon has the biggest feet in England. They are $15\frac{1}{3}$ inches long.
What size of shoe does he wear?
c The largest known feet in the world are those of Muhammad Adlam Channa of Pakistan. He wears a size 22 shoe.
How long are his feet?

d If they were made, how long would these shoes be?
(i) size 0   (ii) size -1 (iii) size -10

4 The formula to convert a temperature measured in degrees Celsius (C) into a temperature measured in degrees Fahrenheit (F) is:

$$F = 9C/5 + 32$$

a Convert these Celsius temperatures into Fahrenheit.
(i) $20\,°C$ (ii) $45\,°C$ (iii) $62\,°C$ (iv) $100\,°C$
b Copy and complete this conversion table for Celsius to Fahrenheit.

°C	0	20	40	60	80	100	120	140	160	180	200	220
°F	32											

c A simple approximate formula to convert a temperature measured in degrees Celsius (C) into a temperature measured in degrees Fahrenheit (F) is:

$$F = 2C + 30$$

(i) Make a conversion table using this formula.
(ii) Does the accurate formula ever agree with the approximate formula?
(iii) Do you think the approximate formula is accurate enough to use to convert cooking temperatures from Celsius to Fahrenheit?

5 The depth (D) of a well is linked to the time (T) it takes a stone to drop to the bottom by the formula $D = 5T^2$, where D is measured in metres and T is measured in seconds.

a How deep is the well if a stone drops to the bottom in:
(i) 1 second   (ii) 1.5 seconds
(iii) 2.5 seconds   (iv) 3.5 seconds?
b How long will a stone take to drop in a well with a depth of:
(i) 45 metres   (ii) 20 metres
(iii) 51.2 metres   (iv) 28.8 metres?

6 In Britain, stopping distances for cars are calculated using the formula

$$D = \frac{(S^2 + 20S)}{60}$$

where S is the speed of the car in miles per hour and D is the stopping distance in metres.

a Copy and complete this table for speed and stopping distance.

Speed (mph)	0	10	20	30	40	50	60	70
Stopping distance (metres)								

b In the Netherlands, the simpler rule

$$D = \frac{4S}{5}$$

is used, where S is the speed of the car in miles per hour and D is the stopping distance in metres.

Make a table for this formula.

c Which rule do you think is safer? Give your reasons.

Unit 14

Skill You order and approximate decimals when solving numerical problems and equations such as $x^2 = 20$, using trial-and-improvement methods.

Skill Example

Arrange the following numbers in order of size, starting with the smallest.

a 4.04, 40.4, 0.404, 404

The order is: 0.404, 4.04, 40.4, 404

b 6, 6.006, 6.6, 6.016, 6.06

The order is: 6, 6.006, 6.016, 6.06, 6.6

Skill Practice One

Arrange the numbers in each question in order of size, starting with the smallest.

 1 5.02, 5.22, 5, 5.2
 2 4.33, 4, 4.03, 4.3
 3 7.04, 7.4, 7.004, 7.44, 7.044, 7
 4 8.005, 8.55, 8.5, 8.055, 8, 8.05
 5 3.11, 3, 3.1, 3.01, 3.011, 3.001
 6 5.03, 5.33, 0.53, 5.3
 7 7.2, 7.22, 7.02, 0.72
 8 6.05, 0.65, 6.55, 6.5
 9 3.024, 0.324, 3.24, 32.4, 3.204
10 5.061, 56.1, 5.601, 0.561, 5.61
11 4.803, 0.483, 4.083, 4.83, 48.3, 48.03
12 2.705, 27.5, 2.75, 0.275, 2.075, 27.05

Skill Example

Decimal quantities are often approximated.

Give £16.35

a to the nearest ten pence
b to the nearest pound
c to the nearest ten pounds.

a As 35 p is halfway between 30 p and 40 p, it is rounded up, so
  £16.35 = £16.40 to the nearest ten pence.

b £16.35 is nearer to £16 than it is to £17, so
  £16.35 = £16 to the nearest pound

c £16.35 is nearer to £20 than it is to £10, so
  £16.35 = £20 to the nearest ten pounds

Skill Practice Two

Express each of the following:

a to the nearest ten pence
b to the nearest pound
c to the nearest ten pounds.

 1 £17.87   2 £36.94   3 £57.12
 4 £42.43   5 £63.28   6 £24.76
 7 £38.19   8 £14.53   9 £52.98
10 £49.76   11 £26.07   12 £30.81

Knowledge

A decimal number can be 'rounded off' (approximated) as required. For example:

  4.62 = 4.6 *to the nearest tenth.*

This is usually written

  4.62 = 4.6   correct to 1 decimal place
      or   correct to 1 dp or (to 1 dp)

Skill Example

a Give 3.74 correct to 1 decimal place.
b Give 4.25 correct to 1 decimal place.
c Give 1.98 correct to 1 decimal place.
d Give 0.486 correct to 2 decimal places.

a 3.74 = 3.7 correct to 1 dp
b 4.25 = 4.3 correct to 1 dp
c 1.98 = 2.0 correct to 1 dp. Note that the answer is not 2 but 2.0 because it is required correct to one decimal place.
d 0.486 = 0.49 correct to 2 dp

Skill Practice Three

For questions **1** to **12**, express each term correct to 1 decimal place.

 1 1.37   2 4.59   3 5.63   4 3.41
 5 6.28   6 8.76   7 7.85   8 8.07
 9 0.92   10 5.14   11 2.03   12 4.98

For questions **13** to **28**, express each term correct to 2 decimal places.

13 1.543   14 3.954   15 2.617   16 6.579
17 4.285   18 0.892   19 5.971   20 1.658
21 6.116   22 4.509   23 3.402   24 2.008
25 1.397   26 5.698   27 4.996   28 1.999

Skill Example

Find the value of the following, correct to 2 decimal places.

a $12.7 \div 3.4$ **b** $321 \div 28$

c The cost of 6.78 metres of cloth at £12.34 a metre.

Using an eight-digit display calculator gives:

a $12.7 \div 3.4 = 3.735\,294\,1$
$\qquad\qquad = 3.74$ (to 2 dp)

b $321 \div 28 = 11.464\,285$
$\qquad\qquad = 11.46$ (to 2 dp)

c $6.78 \times £12.34 = £83.6652$
$\qquad\qquad\quad = £83.67$ (to 2 dp)

Skill Practice Four

For questions **1** to **10**, find each value correct to 1 decimal place.

1 $12.3 \div 7$	**2** $0.56 \div 0.25$	**3** $327 \div 11$
4 $524 \div 0.89$	**5** $0.7 \div 0.03$	**6** $5.33 \div 0.9$
7 $67.4 \div 0.72$	**8** $1.102 \div 7$	**9** $58.91 \div 2.7$
10 $6.0222 \div 8$		

For questions **11** to **20**, find each value correct to 2 decimal places.

11 $329 \div 465$	**12** $0.65 \div 0.89$
13 $12.47 \div 7$	**14** $517 \div 35$
15 $623 \div 0.815$	**16** $4589 \div 12$
17 $0.983 \div 0.451$	**18** $0.0045 \div 0.5$
19 $10.45 \div 1.47$	**20** $3.142 \div 17$

Skill Example

A certain type of cloth costs £3.45 a metre. Find the cost of:

a 9 metres   **b** 7.9 metres

a $9 \times £3.45 = £31.05$

b $7.9 \times £3.45 = £27.255$
But £27.255 is not a real amount of money, so we must round off the answer to £27.26.

Skill Practice Five

1 Look at the prices of these cheeses.

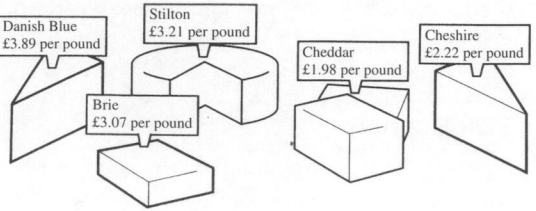

Find the cost of the following.

a 2.7 pounds of Stilton.
b 0.9 pounds of Cheddar.
c 1.2 pounds of Cheshire.
d 1.3 pounds of Brie.
e 0.6 pounds of Danish Blue.

2 A bus company calculates the cost of its fares by charging 12 pence for each mile of the journey. For a return ticket the cost is calculated by adding the two single fares together and then multiplying by 0.9.
Find the cost for these journeys of the following length.

a 8 miles
b 8.9 miles
c 5.6 miles
d Return ticket for a 15.4 mile journey.
e Return ticket for a 22.9 mile journey.

3 Many people form syndicates for the football pools or the National Lottery.
In each case, find the value of each share of the following wins.

a £345 000 shared by 7 people.
b £800 000 shared by 11 people.
c £1 234 555 shared by 3 people.
d £2 768 345 shared by 8 people.
e £517 shared by 35 people.

4 This is part of the menu in a pizzeria.

Pizzas	Large	Medium	Small
Cheese	£4.56	£3.21	£2.85
Mushroom	£4.86	£3.51	£3.15
Chilli	£4.96	£3.61	£3.25
Drinks			
Beer	£1.42		
Wine	£1.13 (per glass)		
Soft drink	£0.96		

If five friends agree to share the bill equally, how much does each pay if they order:

a Two large cheese pizzas, one small chilli pizza, two beers and three glasses of wine?

b One medium mushroom pizza, three medium chilli pizzas and five soft drinks?

c Four large cheese pizzas, three beers and three soft drinks?

Skill Example

Diana wants to solve the equation $x^2 = 20$. The square root key on her calculator is broken, so she draws up a table like this:

x	x^2
1	1
2	4
3	9
4	16
5	25

The table tells her that the required value of x must be between 4 and 5, so she draws up a second table like this:

x	x^2
4.1	16.81
4.2	17.64
4.3	18.49
4.4	19.36
4.5	20.25

The second table tells her that the required value of x must be between 4.4 and 4.5, so she draws up a third table like this:

x	x^2
4.41	19.4481
4.42	19.5364
4.43	19.6249
4.44	19.7136
4.45	19.8025
4.46	19.8916
4.47	19.9809
4.48	20.0704

The third table tells her that the required value of x must be between 4.47 and 4.48.

Diana stops at this point, because she knows that the required value of x, correct to one decimal place, is 4.5.

Skill Practice Six

1 Use a calculator (but *not* the square root key) to solve these equations correct to one decimal place.

a $x^2 = 10$ **b** $x^2 = 15$ **c** $x^2 = 43$

2 Kelly has a loop of string 36 cm long. She wants to use it to make a rectangle with an area of 58 cm^2. The sides of the rectangle are x cm and $(18 - x)$ cm.

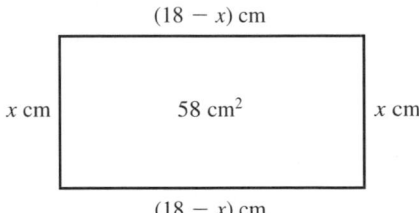

This table shows Kelly that the required value of x is between 4 and 5.

x	$(18 - x)$	Area
1	17	17
2	16	32
3	15	45
4	14	56
5	13	65

Make two more tables to find the value of x correct to one decimal place.

3 Kevin wants to find a value of x which makes $3x^2$ equal to $8 - x$. He constructs this table, which shows him the required value of x is between 1 and 2.

x	$3x^2$	$8 - x$	Difference
1	3	7	-4
2	12	6	6

a Explain why the table shows that the required value of x is between 1 and 2.

b Copy and complete this table.

x	$3x^2$	$8 - x$	Difference
1.1			
1.2			
1.3			
1.4			
1.5			

c Between which two one-decimal place numbers does the required value of x lie?

4 The stopping distance of a car can be calculated using this formula:

$$D = \frac{(S^2 + 20S)}{60}$$

where S is the speed of the car in miles per hour and D is the stopping distance in metres.

After an accident, skid marks 90 m long were measured on the road. This table shows that the car which made the marks was travelling at between 60 and 70 miles per hour.

Speed (miles/h)	50	60	70
Stopping distance (metres)	58.3	80	105

Find the speed of the car correct to the nearest 1 mile per hour.

Unit 15

Number and Algebra   Level 6

Skill   You understand and use the equivalence between fractions, decimals and percentages.

You are aware of which number to consider as 100 per cent, or a whole, in problems involving comparisons, and use this to evaluate one number as a fraction or percentage of another.

Knowledge

A percentage indicates the number of **hundredths** in a share of a given quantity. The special sign '%' is used for a percentage.

This diagram shows the relationship between percentages, fractions and decimals.

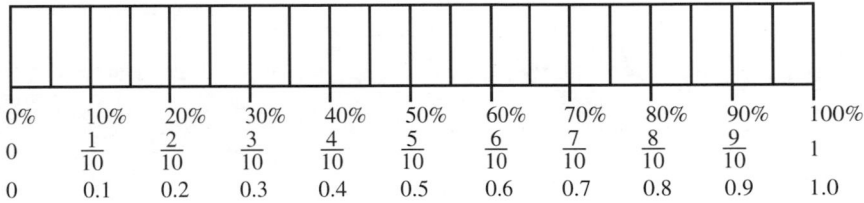

For example, if Surbajit was absent on 15% of the days during a school year, this means he was absent for:

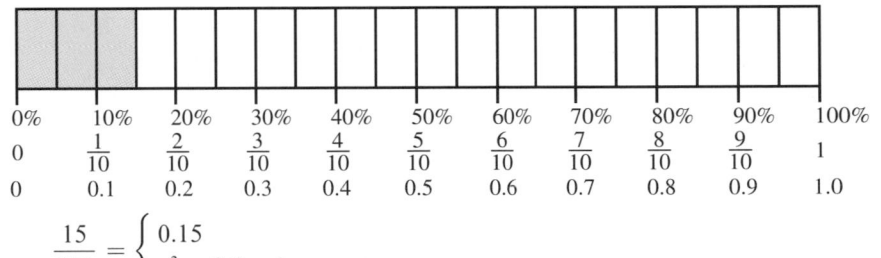

$$\frac{15}{100} = \begin{cases} 0.15 \\ \frac{3}{20} \text{ of the days} \end{cases}$$

Skill Example

Write each of the following as a decimal and a percentage.

a $\frac{1}{4}$ **b** $\frac{3}{50}$ **c** $\frac{5}{8}$

a $\frac{1}{4} = 1 \div 4 = 0.25$
$= 0.25 \times 100\% = 25\%$

b $\frac{3}{50} = 3 \div 50 = 0.06$
$= 0.06 \times 100\% = 6\%$

c $\frac{5}{8} = 5 \div 8 = 0.625$
$= 0.625 \times 100\% = 62.5\%$

Skill Practice One

Write each of the following as a decimal and a percentage.

1 $\frac{3}{10}$ **2** $\frac{9}{10}$ **3** $\frac{1}{10}$

4 $\frac{3}{5}$ **5** $\frac{1}{5}$ **6** $\frac{7}{50}$

7 $\frac{11}{50}$ **8** $\frac{17}{50}$ **9** $\frac{29}{50}$

10 $\frac{3}{20}$ **11** $\frac{7}{20}$ **12** $\frac{1}{20}$

13 $\frac{11}{20}$ **14** $\frac{9}{25}$ **15** $\frac{12}{25}$

Skill Example

Write each of the following as a fraction and a decimal.

$$57\% = \tfrac{57}{100} = \begin{cases} 0.57 \text{ (as a decimal)} \\ \tfrac{57}{100} \text{ (as a fraction)} \end{cases}$$

$$18\% = \tfrac{18}{100} = \begin{cases} 0.18 \text{ (as a decimal)} \\ \tfrac{9}{50} \text{ (as a fraction)} \end{cases}$$

$$35\% = \tfrac{35}{100} = \begin{cases} 0.35 \text{ (as a decimal)} \\ \tfrac{7}{20} \text{ (as a fraction)} \end{cases}$$

Skill Practice Two

Write each of the following as a fraction and a decimal.

1 3%	**2** 11%	**3** 39%	**4** 53%
5 81%	**6** 42%	**7** 38%	**8** 86%
9 45%	**10** 65%	**11** 32%	**12** 56%
13 4%	**14** 90%	**15** 20%	

Skill Example

Write each of the following as a fraction and a percentage.

a 0.32   **b** 0.07   **c** 0.375

a $0.32 = \tfrac{32}{100} = \tfrac{8}{25}$

$0.32 = 0.32 \times 100\% = 32\%$

b $0.07 = \tfrac{7}{100}$

$0.07 = 0.07 \times 100\% = 7\%$

c $0.375 = \tfrac{375}{1000} = \tfrac{3}{8}$

$0.375 = 0.375 \times 100\% = 37.5\%$

Skill Practice Three

Write each of the following as a fraction and a percentage.

1 0.15	**2** 0.29	**3** 0.48	**4** 0.53
5 0.76	**6** 0.93	**7** 0.9	**8** 0.7
9 0.4	**10** 0.2	**11** 0.09	**12** 0.06
13 0.04	**14** 0.625	**15** 0.575	

Knowledge

One quantity can be written as a percentage of another quantity, provided each quantity is written in the **same unit**.

First, write one quantity as a fraction of the other. Then change this fraction to a percentage.

Skill Example

Find:

a 10 cm as a percentage of 50 cm
b £1.50 as a percentage of £5
c 250 g as a percentage of 2 kg.

a 10 as a fraction of $50 = \dfrac{10}{50}$

$10 \div 50 = 0.2$

$0.2 \times 100\% = 20\%$

b £1.5 as a fraction of $£5.00 = \dfrac{1.5}{5.00}$

$1.5 \div 5.00 = 0.3$

$0.3 \times 100\% = 30\%$

c 250 g as a fraction of $2\,\text{kg} = \dfrac{250}{2000}$
(kg changed to g)

$250 \div 2000 = 0.125$

$0.125 \times 100\% = 12.5\%$

Skill Practice Four

Find:

1 £72 as a percentage of £800
2 £42 as a percentage of £600
3 56 g as a percentage of 700 g
4 £66 as a percentage of £1100
5 £35 as a percentage of £50
6 £27 as a percentage of £90
7 £54 as a percentage of £60
8 32 cm as a percentage of 80 cm
9 24 cm as a percentage of 30 cm
10 £45 as a percentage of £75
11 £18 as a percentage of £45
12 £100 as a percentage of £125

13 £21 as a percentage of £105

14 £28 as a percentage of £80

15 66 g as a percentage of 120 g

16 £24 as a percentage of £160

17 27 cm as a percentage of 60 cm

18 54 cm as a percentage of 72 cm

19 £27 as a percentage of £108

20 £1.20 as a percentage of £1.50

Skill Practice Five

1 There are 20 eggs in a fridge and 6 of them are brown.
Find the percentage which are brown.

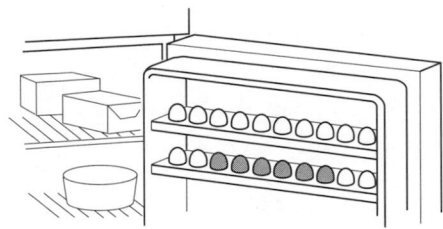

2 A passage has an area of 25 m² and there is a carpet on its floor which has an area of 20 m².
What percentage of the floor's area is covered by the carpet?

3 An examination is marked out of 120 and one girl gets 84 marks.
Find her mark as a percentage.

4 Sixty pupils are entered for an examination and 45 of them pass.
Find the percentage who pass.

5 A factory employs 160 workers and 72 of them travel to work by bus.
Find the percentage who use the bus.

6 There are 24 girls in class 2B and one day 18 of them are present.
Find the percentage who are present.

7 At a football match a programme seller is supplied with 1200 programmes and he sells 1080.
What percentage of them does he sell?

8 10 m of wood is bought to make the window frame illustrated.
What percentage of the wood is used?

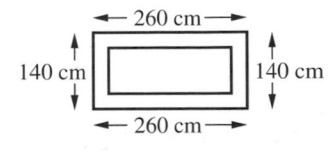

9 There are 30 girls in class 4C and they have four sports options to choose from.
If 12 choose tennis, 9 choose swimming, 3 choose rounders and 6 choose athletics, find the percentage who choose each sport.

10 At Manor Grange School there are 450 pupils. The number absent on each day of a certain week is shown below.

Monday	27	Thursday	36
Tuesday	18	Friday	54
Wednesday	45		

Find the percentage who are absent each day.

Unit 16

Knowledge

A **ratio** is used when we wish to compare two or more quantities.

'Filla' is a powder that is mixed with water to form a paste.

The proportions used are 3 parts of powder to 1 part of water. We say that the ratio is 'three to one' and usually write it as 3 to 1 or 3 : 1.

Ratios can be simplified like fractions by dividing by a common factor.

Skill Example

Give each ratio in its simplest form.

a 6 : 8   **b** £1.50 : £5.00

a 6 and 8 have a common factor of 2, so dividing through by 2 gives:

$$6 : 8 = 3 : 4$$

b Changing to pence and then dividing through by the common factor of 50 give:

$$£1.50 : £5.00 = 150 : 500$$
$$= 3 : 10$$

Skill Practice One

Give each ratio in its simplest form.

1	10 : 12	**2**	10 : 16
3	9 : 12	**4**	15 : 24

5	3 : 9	**6**	16 : 20
7	10 : 15	**8**	10 : 25
9	5 : 25	**10**	30 : 36
11	24 : 40	**12**	18 : 27
13	£1.50 : £2.00	**14**	£1.60 : £4.00
15	£1.20 : £6.00	**16**	£1.50 : £9.00
17	£0.90 : £3.00	**18**	1 m 60 cm : 2 m
19	4 m 50 cm : 6 m	**20**	1 m 50 cm : 4 m
21	60 cm : 2 m	**22**	1 cm 8 mm : 2 cm
23	7 cm 5 mm : 10 cm	**24**	1 cm 5 mm : 6 cm
25	6 mm : 3 cm		

Skill Example

In a 50-seater bus, there are 16 seats for smokers. Find the ratio of seats for smokers to those for non-smokers.

No. of seats for smokers $= 16$

So, no. of seats for non-smokers $= 50 - 16 = 34$

Therefore, the ratio is $16 : 34 = 8 : 17$

Skill Practice Two

1 On a supermarket shelf, there are 6 bags of plain flour and 9 bags of self-raising flour. Find the ratio of plain flour to self-raising flour.

2 Anne weighs 28 kg and Jandeep weighs 40 kg. Find the ratio of Anne's weight to Jandeep's.

3 At a party, 24 children asked for tea and 30 asked for lemonade.
Find the ratio of those who had tea to those who had lemonade.

4 Jack has 21 marbles and Tom has 28.
Find the ratio of the number of Jack's marbles to those of Tom.

5 Kirsty has picked 27 flowers and Melanie has picked 45.
Find the ratio of the number that Kirsty has picked to those of Melanie.

6 Liam is 96 cm tall and Robert is 120 cm tall. Find the ratio of Liam's height to Robert's.

7 In a class of 30 pupils there are 12 boys. Find the ratio of boys to girls.

8 In the bread shop there are 20 loaves on the shelf; 5 of them are brown and the rest are white loaves. Find the ratio of brown loaves to white loaves.

9 A farmer has a flock of 60 sheep; 12 of them are black and the rest are white.
Find the ratio of black sheep to white sheep.

10 A railway carriage has 56 seats and 24 of them are for first-class passengers.
Find the ratio of first-class seats to second-class seats.

Skill Example

Share:

a £45 in the ratio 2 : 1
b 90 litres in the ratio 4 : 5 : 6

a Total number of shares = 2 + 1 = 3

Therefore, one share = £45 ÷ 3 = £15

Therefore, value of first share = £15 × 2
= £30

and value of second share = £15 × 1
= £15

Therefore, the two amounts are £30 and £15.

Check that these two shares total £45.

b Total number of shares = 4 + 5 + 6 = 15

Therefore, one share = (90 ÷ 15) litres
= 6 litres

Therefore, the first share = (6 × 4) litres
= 24 litres

second share = (6 × 5) litres
= 30 litres

and third share = (6 × 6) litres
= 36 litres

Therefore, the three amounts are 24, 30 and 36 litres.

Check that these three shares total 90 litres.

Skill Practice Three

1 Share £48 in the ratio 2 : 1.

2 Share £60 in the ratio 3 : 1.

3 Share £80 in the ratio 4 : 1.

4 Share £91 in the ratio 6 : 1.

5 Share £70 in the ratio 3 : 2.

6 Share £120 in the ratio 5 : 3.

7 Share 112 ml of milk between the cat and her kitten in the ratio 4 : 3.

8 Share 162 ml of milk between the cat and the dog in the ratio 4 : 5.

9 Share 200 g of sweets between Aisha and Nicola in the ratio 3 : 5.

10 Share 120 g of cereal between Waseem and David in the ratio 3 : 7.

11 Share £108 in the ratio 3 : 2 : 1.

12 Share 315 g of flour between Mrs Smith, Mrs Johnson and Mrs Bates in the ratio 4 : 2 : 1.

13 Share 450 kg of soil between three gardeners in the ratio 4 : 5 : 6.

14 Share 1500 ml of paraffin between Mr Brown, Mr Jones and Mr Patel in the ratio 2 : 3 : 5.

15 A bottle containing 560 ml of lemonade exactly fills three glasses belonging to Jill, Jane and Paul. If the capacities of the glasses are in the ratio of 3 : 5 : 6 respectively, how much lemonade does each child receive?

Skill Example

A sum of money is shared in the ratio 2 : 3.
If the smaller share is 50 p, what is the larger share?

    50 p is equal to 2 shares
so   25 p is equal to 1 share

Therefore, the larger share = 25 p × 3 = 75 p

Check that the two shares are in the ratio 2 : 3.

Skill Practice Four

1 A sum of money is shared in the ratio 2 : 3.
If the smaller share is 30 p, what is the larger share?

2 A sum of money is shared in the ratio 2 : 5.
If the smaller share is 16 p, what is the larger share?

3 A sum of money is shared in the ratio 4 : 5.
If the smaller share is 36 p, what is the larger share?
How much money was shared out?

4 A sum of money is shared in the ratio 3 : 7.
If the smaller share is £12, what is the larger share?
How much money was shared out?

5 A sum of money is shared in the ratio 5 : 8.
If the smaller share is £20, what is the larger share?
How much money was shared out?

6 A sum of money is shared out in the ratio 2 : 3 : 5.
If the smallest share is £10, what are the other two shares?

7 A sum of money is shared out in the ratio 3 : 5 : 7.
If the smallest share is £9, what are the other two shares?

8 A sum of money is shared out in the ratio 5 : 6 : 9.
If the smallest share is 25 p, what are the other two shares?
How much money is shared out altogether?

9 At a bread shop the prices of a white and a brown loaf are in the ratio of 5 : 6.
If a white loaf costs 30 p, what is the price of a brown loaf?

10 The heights of two sisters Lynn and Sophie are in the ratio of 4 : 5.
If Lynn is 120 cm tall, how tall is Sophie?

11 The weights of two brothers Martin and Richard are in the ratio of 3 : 4.
If Martin's weight is 45 kg, how much does Richard weigh?

12 A Thermos flask can exactly fill two cups whose capacities are in the ratio of 3 : 5.
If the smaller one has a capacity of 150 ml, what is the capacity of the larger one?
What is the capacity of the flask?

13 In class 3A the ratio of boys to girls is 6 : 7.
If there are 12 boys in the class, find:
a the number of girls in the class and
b the number of pupils in the class altogether.

14 A long, thin piece of wood is cut into two pieces, the ratio of whose lengths is 9 : 11.
If the shorter piece is 45 cm long, what is the length of the longer piece?
What was the length of the original piece?

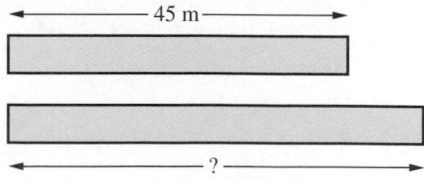

15 Some sweets are shared between Tom, Luke and Kim in the ratio $2:3:4$.
    If Tom has 100 g, what weight has:
    a Luke
    b Kim?
    What total weight of sweets is shared out?

Skill Example

a If 3 kg of apples cost 99 p, what would 5 kg cost?

b If 12 eggs cost 60 p, how many eggs could be bought for 45 p?

a If 3 kg of apples cost 99 p,
    then 1 kg of apples cost $99 \div 3 = 33$ p
    Therefore, 5 kg cost, $33 \times 5 = 165$ p $= £1.65$

b If 12 eggs cost 60 p,
    then 1 egg costs $60 \div 12 = 5$ p
    Therefore, the number of eggs for 45 p is $45 \div 5 = 9$

Skill Practice Five

1 If 5 kg of potatoes cost 90 p, what is the cost of 3 kg?

2 If 6 litres of paraffin cost 96 p, what is the cost of 4 litres?

3 If 20 postcards cost 180 p, what is the cost of 50?

4 If it takes me 45 minutes to walk 5 km, how long will it take me to walk:
   a 9 km   **b** 4 km?

5 If 12 m² of carpet cost £60, find the cost of:
   a 5 m²   **b** 8 m²

6 If 5 kg of bananas cost £2.40, find the cost of 2 kg.

7 If 4 kg of tomatoes cost £3.60, find the cost of 3 kg.

8 If 6 kg of apples cost £1.92, find the cost of:
   a 4 kg   **b** 5 kg

9 If 5 kg of pears cost £1.80, find the cost of:
   a 3 kg   **b** 4 kg

10 If 5 m of curtain track cost £12, find the cost of:
    a 3 m   **b** 8 m

11 If 5 m of dress fabric cost £14, find the cost of:
    a 4 m   **b** 12 m

12 If 10 m of a certain kind of electric cable cost £5, find the cost of:
    a 3 m   **b** 7 m

13 If 10 m² of vinyl flooring cost £12, find the cost of:
    a 3 m²   **b** 8 m²   **c** 12 m²

14 If 10 tonnes of garden soil cost £75, find the cost of:
    a 3 tonnes   **b** 4 tonnes   **c** 12 tonnes

15 If 20 litres of petrol cost £10, find the cost of:
    a 8 litres   **b** 12 litres   **c** 30 litres

16 If 8 oranges cost 96 p, how many can be bought for 60 p?

17 If 5 grapefruits cost 80 p, how many can be bought for 48 p?

18 If 6 eggs cost 36 p, how many can be bought for 90 p?

19 If 5 bread buns cost 30 p, how many can be bought for:
    a 18 p   **b** 48 p?

20 If 5 doughnuts cost 60 p, how many can be bought for:
    a 24 p   **b** 84 p?

Unit 17

Knowledge

The sequence 1, 2, 3, 4, 5, 6, 7, 8, ... is the most simple of all. This sequence is called the **natural number sequence**.

The terms of other sequences can often be formed from the sequence of natural numbers.

Skill Example

a Describe how the sequence of even numbers is produced from the sequence of natural numbers.

b Describe how the sequence of odd numbers is produced from the sequence of natural numbers.

a Compare the two sequences:

1	2	3	4	5	6
↓	↓	↓	↓	↓	↓
2	4	6	8	10	12

We see that the even numbers are produced by multiplying the natural numbers by 2.

b Compare the two sequences:

1	2	3	4	5	6
↓	↓	↓	↓	↓	↓
1	3	5	7	9	11

We see that the odd numbers are produced by multiplying the natural numbers by 2 and then subtracting 1.

Skill Example

Write down the first 5 terms of the sequence produced by multiplying the natural numbers by 8 and subtracting 5.

Applying these operations to the natural numbers gives:

1	2	3	4	5
↓	↓	↓	↓	↓
8 − 5	16 − 5	24 − 5	32 − 5	40 − 5
↓	↓	↓	↓	↓
3	11	19	27	35

Skill Practice One

Write down the first 5 terms of the sequence produced by changing the sequence of natural numbers in each of the following ways.

1 Multiplying by 7.
2 Adding 11.
3 Subtracting 5.
4 Dividing by 2.
5 Multiplying by 2 and adding 11.
6 Adding 11 and multiplying by 2.
7 Multiplying by 7 and subtracting 1.
8 Subtracting 1 and multiplying by 7.
9 Dividing by 2 and adding 5.
10 Adding 5 and dividing by 2.

Skill Example

Describe the way in which this sequence is produced from the natural numbers:

    6   10   14   18   ...

Number the terms in the sequence and find the difference between each term and the next.

1st	2nd	3rd	4th
↓	↓	↓	↓
6	10	14	18

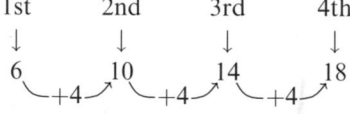

Because the difference is 4, this sequence will be based on the 4 times table.

1st	2nd	3rd	4th	
↓	↓	↓	↓	
4	8	12	16	← 4 times table
↓	↓	↓	↓	
6	10	14	18	

The sequence is produced by multiplying the natural numbers by 4, then adding 2.

Skill Practice Two

Describe how each sequence is produced from the sequence of natural numbers and give the next two terms.

1 3, 6, 9, 12, . . .   **2** 2, 5, 8, 11, . . .
3 −5, −4, −3, −2, . . .   **4** 4, 8, 12, 16, . . .
5 5, 9, 13, 17, . . .   **6** 6, 10, 14, 18, . . .
7 3, 7, 11, 15, . . .   **8** 2, 6, 10, 14, . . .
9 5, 10, 15, 20, . . .   **10** 6, 11, 16, 21, . . .
11 7, 12, 17, 22, . . .   **12** 8, 13, 18, 23, . . .
13 4, 9, 14, 19, . . .   **14** 0, 5, 10, 15, . . .
15 −1, 4, 9, 14, . . .   **16** 5, 11, 17, 23, . . .
17 7, 13, 19, 25, . . .   **18** 13, 23, 33, 43, . . .
19 7, 17, 27, 37, . . .   **20** 0.1, 0.2, 0.3, 0.4, . . .

Skill Example

Write down the first 5 terms of the sequence produced by the rule:

nth term of the sequence $= 7 \times n + 10$

The terms are:

1st	2nd	3rd	4th	5th
↓	↓	↓	↓	↓
17	24	31	38	45

Skill Practice Three

Write down the first 5 terms of the sequences produced by these rules.

1 nth term $= 5 \times n + 4$
2 nth term $= 5 \times n - 4$
3 nth term $= 4 \times n + 5$
4 nth term $= 5 \times n - 5$
5 nth term $= 5 \times n \div 4$
6 nth term $= 3 \times n + 8$
7 nth term $= 2 \times n + 7$
8 nth term $= 2 \times n - 7$
9 nth term $= 7 \times n - 2$
10 nth term $= 6 \times n + 11$

Skill Example

Find a rule for the nth term of this sequence:

−3   2   7   12   17

Number the sequence and find the difference between each term and the next.

1st	2nd	3rd	4th	5th
↓	↓	↓	↓	↓
−3	2	7	12	17

+5   +5   +5   +5

Because the difference is 5, this sequence is based on the 5 times table.

1st	2nd	3rd	4th	5th	
↓	↓	↓	↓	↓	
5	10	15	20	25	← 5 times table
↓	↓	↓	↓	↓	
−3	2	7	12	17	

The rule for the nth term is:

nth term $= 5 \times n - 8$

Skill Practice Four

Write a rule for the nth term of each of the following sequences.

1 3, 6, 9, 12, . . .
2 4, 8, 12, 16, . . .
3 5, 10, 15, 20, . . .
4 6, 12, 18, 24, . . .
5 9, 18, 27, 36, . . .
6 9, 10, 11, 12, . . .
7 7, 8, 9, 10, . . .
8 101, 102, 103, 104, 105, . . .
9 −5, −4, −3, −2, . . .
10 −10, −9, −8, −7, . . .
11 3, 5, 7, 9, . . .
12 6, 9, 12, 15, . . .
13 11, 21, 31, 41, . . .
14 20, 30, 40, 50, . . .
15 4, 6, 8, 10, . . .

70

Skill Example

A gardener plants borders with roses and marigolds using the following pattern.

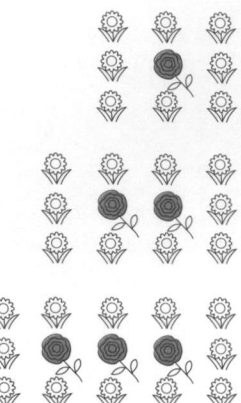

Find a rule for the number of marigolds that will be planted with *n* roses.
How many marigolds will be planted with 10 roses?

The number of roses is the sequence of natural numbers. So, arranging the number of marigolds as the terms of a sequence gives:

```
Roses        1        2        3
             ↓        ↓        ↓
Marigolds    8       10       12
               ⌣+2⌣    ⌣+2⌣
```

This sequence is based on the 2 times table.

```
Roses        1        2        3
             ↓        ↓        ↓
             2        4        6  ← 2 times table
             ↓        ↓        ↓
Marigolds    8       10       12
```

The rule is:

Number of marigolds $= 2 \times$ Number of roses $+ 6$

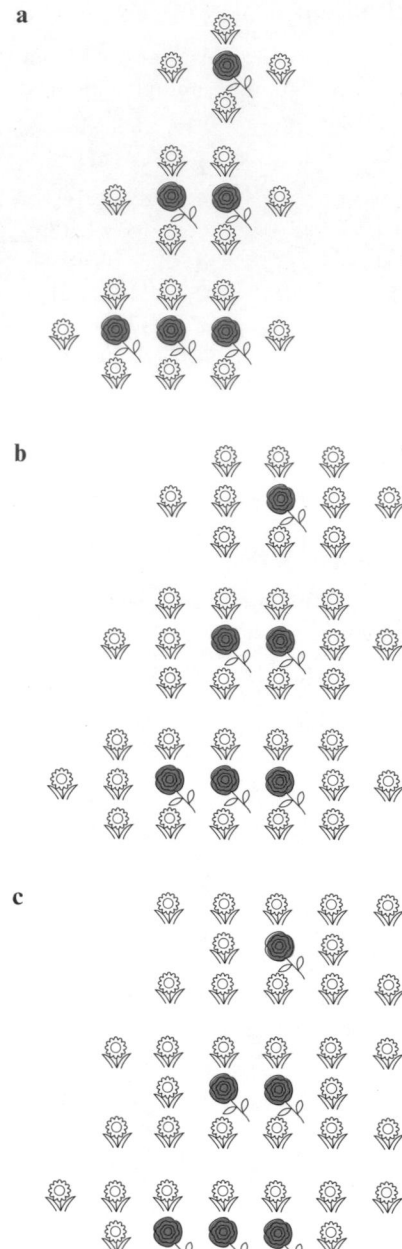

a

b

c

Skill Practice Five

1 A gardener plants borders with roses and marigolds using the following patterns.
   In each case, find a rule for the number of marigolds that will be planted with *n* roses.
   How many marigolds will be planted with 10 roses?

2 Jo has £20 in her piggy bank.
   In each case, find a rule for the amount of money she will have in the piggy bank after *n* weeks if she saves:
   a £3 a week   b £1 a week
   c £2 a week   d £5 a week
   e £10 a week

3 Caroline has won a prize of 1000 tins of dog food.

In each case, find a formula for the number of tins she will have left after *n* weeks if her dog eats:

a 5 tins a week   **b** 3 tins a week

c 7 tins a week   **d** 10 tins a week

e 14 tins a week

4 An author has signed a contract to write a book of 400 pages.

In each case, find a formula for the number of pages left to write after *n* days if the author writes:

a 10 pages a day   **b** 20 pages a day

c 5 pages a day   **d** 17 pages a day

e 25 pages a day

5 Find a rule for the number of tiles in the *n*th pattern of each of these sequences.

a

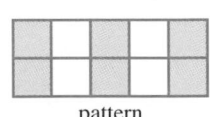

pattern number 1   pattern number 2   pattern number 3

b

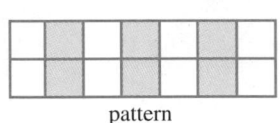

pattern number 1   pattern number 2

pattern number 3

c

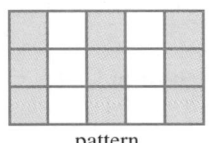

pattern number 1   pattern number 2

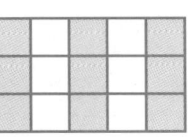

pattern number 3

Unit 18

Number and Algebra   Level 6
Skill   You formulate and solve linear equations with whole-number coefficients.

Skill Example

Solve the following equations.

a $4x = 20$ **b** $3b = -9$

a $4x = 20$

   Divide both sides by 4:

$$\frac{4x}{4} = \frac{20}{4}$$

$$\Rightarrow \quad x = 5$$

b $3b = -9$

   Divide both sides by 3:

$$\frac{3b}{3} = \frac{-9}{3}$$

$$\Rightarrow \quad b = -3$$

Skill Practice One

Solve the following equations.

1 $3x = 12$ **2** $5x = 30$ **3** $4x = 28$
4 $6y = 48$ **5** $2y = 30$ **6** $3z = 48$
7 $5a = 90$ **8** $4b = 64$ **9** $6c = 84$
10 $7d = 105$ **11** $3x = -54$ **12** $5x = -65$
13 $4y = -64$ **14** $8z = -104$ **15** $7t = -112$
16 $9a = -126$ **17** $6b = -108$ **18** $5c = -120$
19 $6m = -150$ **20** $8n = -160$

Skill Example

Solve the following equations.

a $x + 6 = 9$ **b** $3x + 5 = 11$ **c** $4a - 10 = 22$

a $x + 6 = 9$
    $\Rightarrow \quad x = 3$ (Subtract 6 from both sides)

b $3x + 5 = 11$
    $\Rightarrow \quad 3x = 6$ (Subtract 5 from both sides)
    $\Rightarrow \quad x = 2$ (Divide both sides by 3)

c $4a - 10 = 22$
    $\Rightarrow \quad 4a = 32$ (Add 10 to both sides)
    $\Rightarrow \quad a = 8$ (Divide both sides by 4)

Skill Practice Two

Solve the following equations.

1 $x + 5 = 7$ **2** $y + 4 = 8$
3 $z + 2 = 10$ **4** $t + 5 = 11$
5 $a + 9 = 16$ **6** $x - 3 = 4$
7 $y - 4 = 5$ **8** $z - 5 = 3$
9 $a - 8 = 4$ **10** $b - 8 = 10$
11 $3x + 4 = 16$ **12** $4y + 3 = 19$
13 $8z + 9 = 25$ **14** $7b + 8 = 36$
15 $9m + 5 = 41$ **16** $6x + 7 = 13$
17 $8t + 9 = 41$ **18** $4a + 11 = 19$
19 $7c + 11 = 60$ **20** $12p + 13 = 85$
21 $3x - 11 = 10$ **22** $6y - 5 = 1$
23 $6z - 7 = 11$ **24** $3a - 5 = 16$
25 $4c - 5 = 19$ **26** $5m - 7 = 18$
27 $9p - 8 = 28$ **28** $8x - 12 = 20$
29 $6y - 15 = 21$ **30** $12z - 14 = 22$

Skill Example

Solve the following equations.

a $6x - 3x + 2x = 10$ **b** $5a + 5 + 3a = 21$

a $6x - 3x + 2x = 10$
   $\Rightarrow \quad 3x + 2x = 10$
     $\Rightarrow \quad 5x = 10$ (Collect like terms)
      $\Rightarrow \quad x = 2$ (Divide both sides by 5)

b $5a + 5 + 3a = 21$
   $\Rightarrow \quad 8a + 5 = 21$ (Collect like terms)
     $\Rightarrow \quad 8a = 16$ (Subtract 5 from both sides)
      $\Rightarrow \quad a = 2$ (Divide both sides by 8)

Skill Practice Three

Solve the following equations.

1 $6x + 3x + 2x = 33$ **2** $5y + 2y + y = 56$
3 $3p + 4p - 2p = 20$ **4** $5q + 3q - 6q = 16$
5 $6r + 4r - r = 36$ **6** $8s + 2s - 9s = 12$
7 $8a - 3a + 2a = 21$ **8** $9b - 6b + 4b = 35$
9 $12c - 5c + 2c = 54$ **10** $5d - d + 4d = 40$
11 $9m - 3m - 2m = 8$ **12** $12n - 2n - 7n = 15$
13 $11u - 4u - u = 30$ **14** $15v - 5v - 9v = 7$
15 $4a + 7 + 2a = 25$ **16** $6b + 5 + 3b = 50$
17 $3c + 9 + 5c = 17$ **18** $7d + 15 + d = 55$
19 $3p - 8 + 2p = 22$ **20** $5q - 3 + 3q = 29$

Skill Example

a Solve $3a = 8 - a$.

$$3a = 8 - a$$

Add a to both sides $\quad 3a + a = 8 - a + a$

Collect like terms $\quad\quad 4a = 8$

Divide both sides by 4 $\quad a = 2$

b Solve $3x + 5 = 2x$.

$$3x + 5 = 2x$$

Take $2x$ from both sides

$$3x - 2x + 5 = 2x - 2x$$

Collect like terms $\quad\quad x + 5 = 0$

Take 5 from both sides $\quad x = -5$

Skill Practice Four

Solve:

1 $3x = 15 - 2x$ **2** $5x = 16 - 3x$

3 $4y = 36 - 5y$ **4** $6z = 35 - z$

5 $3t = 4 - t$ **6** $5a = 8 + 3a$

7 $8b = 18 + 5b$ **8** $9c = 15 + 4c$

9 $6d = 2 + 5d$ **10** $7e = 30 + e$

11 $9m = 7m + 10$ **12** $6n = 2n + 28$

13 $12p = 10p + 4$ **14** $9q = 8q + 6$

15 $5r = r + 32$ **16** $6x = 2x - 8$

Skill Example

a Solve $3a + 5 = a + 9$.

$$3a + 5 = a + 9$$

Take a from both sides $\quad 2a + 5 = 9$

Take 5 from both sides $\quad 2a = 4$

Divide both sides by 2 $\quad a = 2$

b Solve $2b - 4 = 8 - b$.

$$2b - 4 = 8 - b$$

Add b to both sides $\quad 3b - 4 = 8$

Add 4 to both sides $\quad 3b = 12$

Divide both sides by 3 $\quad b = 4$

Skill Practice Five

Solve:

1 $6a + 2 = 2a + 10$ **2** $9b + 3 = 6b + 18$

3 $12c + 9 = 7c + 14$ **4** $11d + 9 = 4d + 30$

5 $5e + 8 = 4e + 15$ **6** $7f + 8 = f + 20$

7 $8p - 7 = 6p + 3$ **8** $9q - 8 = 3q + 16$

9 $11r - 12 = 8r + 6$ **10** $20s - 3 = 11s + 6$

11 $15t - 2 = 14t + 5$ **12** $5u - 12 = u + 20$

13 $5x - 9 = 2x - 3$ **14** $7y - 10 = 5y - 2$

15 $9z - 14 = 6z - 5$ **16** $15t - 56 = 7t - 16$

17 $10u - 20 = 9u - 11$ **18** $9v - 40 = v - 24$

19 $4m - 5 = 7 - 2m$ **20** $5n - 8 = 32 - 3n$

Skill Example

Sometimes brackets have to be removed before an equation can be solved.

a Solve $3(x + 2) = 18 - x$.

$$3(x + 2) = 18 - x$$

Remove the bracket $\quad 3x + 6 = 18 - x$

Add x to and subtract 6

from both sides $\quad 3x + x = 18 - 6$

Collect like terms $\quad 4x = 12$

Divide both sides by 4 $\quad x = 3$

b Solve $4(a - 2) = 2(a + 5)$.

$$4(a - 2) = 2(a + 5)$$

Remove the brackets $\quad 4a - 8 = 2a + 10$

Add 8 to and subtract

$2a$ from both sides $\quad 4a - 2a = 10 + 8$

Collect like terms $\quad 2a = 18$

Divide both sides by 2 $\quad a = 9$

Skill Practice Six

Solve:

1 $5(x + 2) = 2x + 16$ **2** $7(x + 3) = 5x + 29$

3 $4(y + 1) = y + 13$ **4** $8(z - 2) = 3z + 9$

5 $6(t - 3) = 2t + 10$ **6** $9(u - 1) = 8u + 3$

7 $5(v - 4) = 2v - 5$ **8** $7(m - 2) = 5m - 4$

9 $4(n - 5) = n - 2$ **10** $3(a - 2) = 9 - 2a$

11 $4(b - 5) = 8 - 3b$ **12** $6(c - 3) = 4 - 5c$

13 $2(p + 3) = 21 - 3p$ **14** $5(q + 1) = 12 - 2q$

15 $4(r + 2) = 33 - r$ **16** $8(x + 2) = 3(x + 7)$

17 $7(y + 2) = 2(y + 12)$ **18** $9(z + 1) = 5(z + 5)$

19 $5(t - 2) = 2(t + 4)$ **20** $7(u - 3) = 3(u + 5)$

Unit 19

Number and Algebra   Level 6

Skill   You represent mappings expressed algebraically, interpreting general features and using graphical representation in four quadrants where appropriate.

Knowledge

The axes of a graph may be drawn in four directions from the origin, O, to form four **quadrants**.

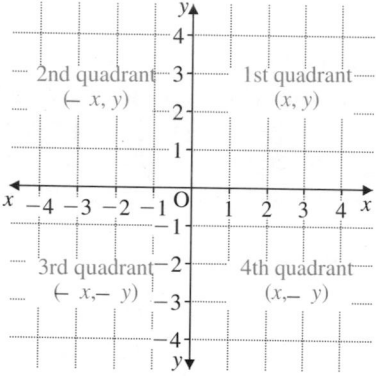

Distances to the right of the origin are positive.
Distances to the left of the origin are negative.
Distances upwards from the origin are positive.
Distances downwards from the origin are negative.

So both positive and negative values of x and y may be plotted. The ordered pair (x, y) which gives the position of any point is called its **cartesian coordinates**.

Skill Example

Write down the coordinates of the points shown on the following graph.

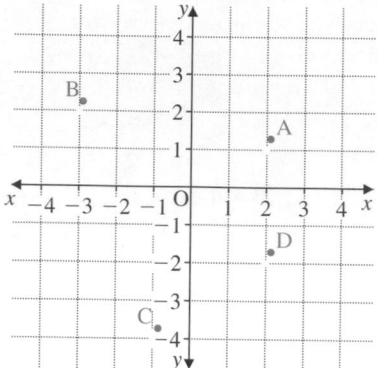

A is the point $(2, 1)$ B is the point $(-3, 2)$
C is the point $(-1, -4)$ D is the point $(2, -2)$

Skill Practice One

Write down the coordinates of each point shown on the graph.

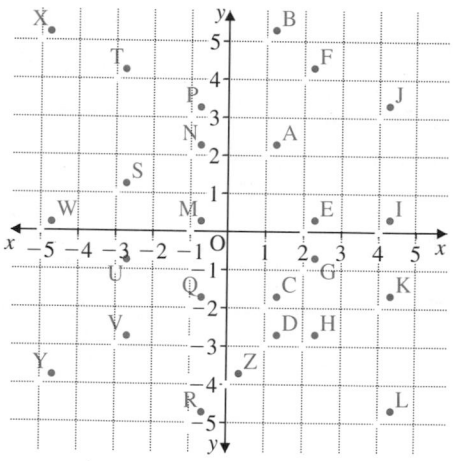

Skill Example

Plot the following points and join each point to the next with a straight line.
Suggest a name for the picture you have drawn.

(2, −1), (2, −5), (3, −4), (5, −4), (4, −5), (−3, −5), (−1, −4), (−1, −2), (−3, 0), (−2, 1), (−4, 3), (−4, 6), (−2, 5), (0, 6), (0, 3), (−1, 2), (2, −1).

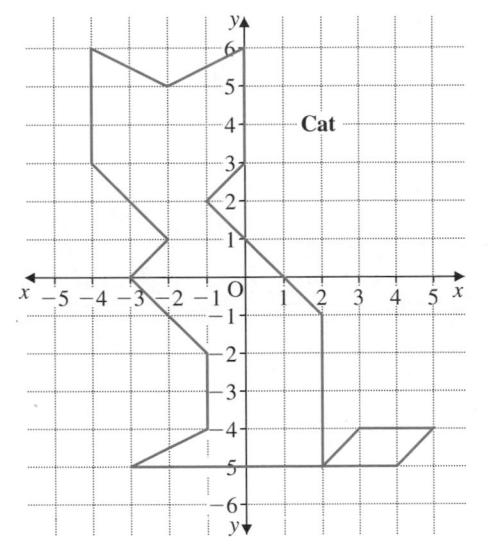

Skill Practice Two

On graph paper, draw an x-axis numbered from −5 to 5 and a y-axis numbered from −10 to 10. Plot the positions of the points in each question on your axes.

Join each point to the next with a straight line. Then suggest a name for each picture you have drawn.

1 (1, 10), (1, 5), (2, 4), (2, −5), (1, −6), (0, −6), (−1, −5), (−1, 4), (0, 5), (0, 10), (1, 10).

2 (0, 10), (4, −1), (1, −1), (1, −2), (2, −2), (1, −6), (−1, −6), (−2, −2), (−1, −2), (−1, −1), (−4, −1), (0, 10).

3 (3, 10), (4, 9), (4, 7), (3, 6), (5, 6), (5, −6), (3, −6), (4, −7), (4, −9), (3, −10), (2, −9), (2, −7), (3, −6), (1, −6), (1, 6), (3, 6), (2, 7), (2, 9), (3, 10).

4 (0, 9), (3, 4), (1, 4), (1, −4), (3, −6), (3, −7), (1, −5), (1, −6), (3, −8), (3, −9), (1, −7), (1, −9), (−1, −9), (−1, −7), (−3, −9), (−3, −8), (−1, −6), (−1, −5), (−3, −7), (−3, −6), (−1, −4), (−1, 4), (−3, 4), (0, 9).

5 (1, 10), (1, −1), (2, 9), (3, 10), (4, 9), (4, −7), (3, −10), (3, −5), (4, −5), (3, −5), (3, 9), (2, −1), (2, −3), (−1, −3), (1, −3), (1, −7), (−1, −7), (1, −7), (0, −10), (−1, −7), (−1, −3), (−2, −3), (−2, −1), (1, −1), (−1, −1), (−1, 10), (1, 10).

Skill Example

The graph below shows the line $y = 2x + 1$.

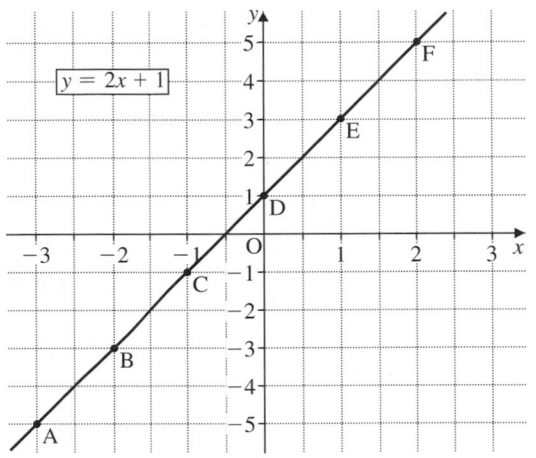

The table is for the coordinates of all the lettered points which lie on the line $y = 2x + 1$.

	A	B	C	D	E	F
x	−3			0		
y	−5			1		

Copy and complete the table by filling in the missing values of x and y. The points A and D are already done for you. The coordinates of A are (−3, −5), and those of D are (0, 1).

From the graph:
point B is (−2, −3)
point C is (−1, −1)
point E is (1, 3)
point F is (2, 5)

The completed table then looks like this:

	A	B	C	D	E	F
x	−3	−2	−1	0	1	2
y	−5	−3	−1	1	3	5

Skill Practice Three

For each graph, copy and complete the table of values, giving the coordinates of the lettered points which lie on the line.

1 $y = 2x - 3$

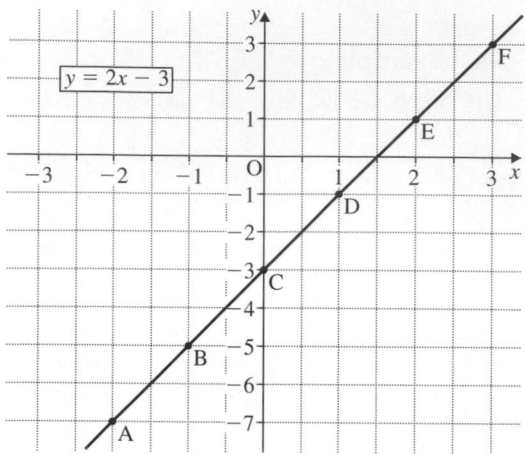

	A	B	C	D	E	F
x	−2		0	1		
y	−7		−3	−1		

2 $y = 2x + 4$

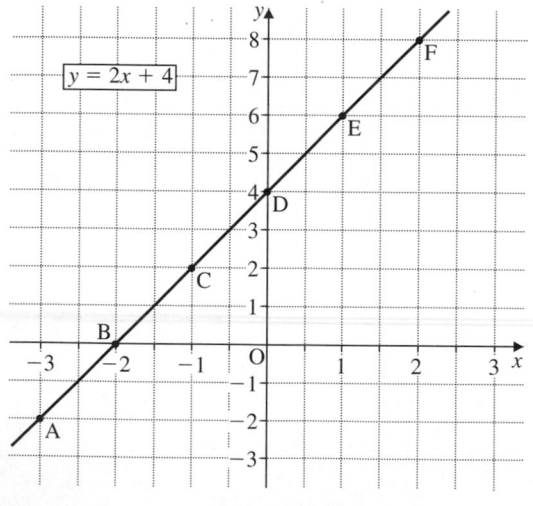

	A	B	C	D	E	F
x		−2	−1	0		
y		0	2	4		

3 $y = 2x - 4$

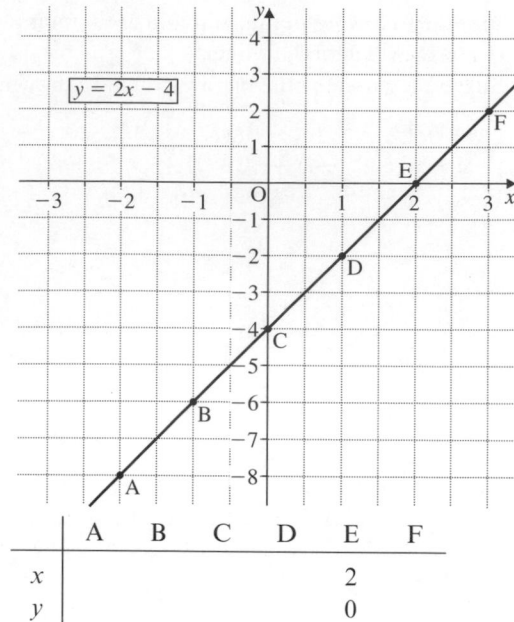

	A	B	C	D	E	F
x					2	
y					0	

4 $y = 3x + 1$

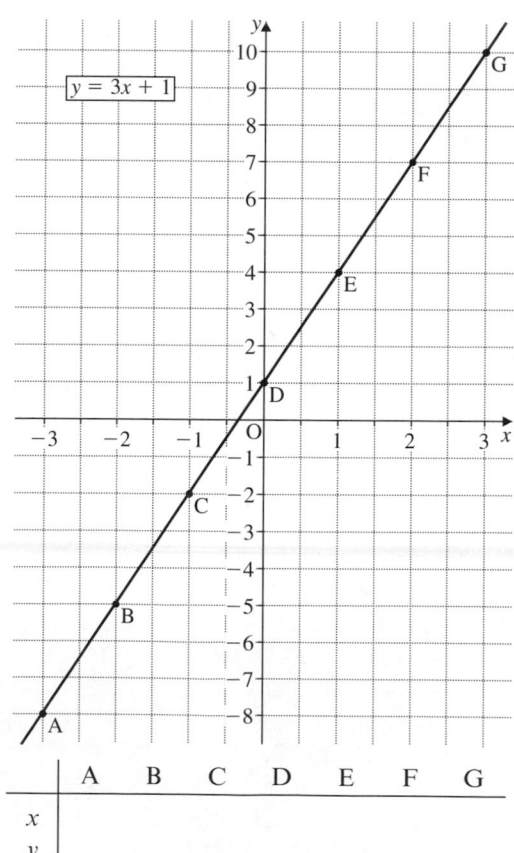

	A	B	C	D	E	F	G
x							
y							

Skill Example

The table of values gives the coordinates of points that lie on the line $y = 1 - x$.

x	-3	-2	-1	0	1	2	3
y	4	3	2	1	0	-1	-2

Plot the line $y = 1 - x$ from these values.

Your completed graph should look like this.

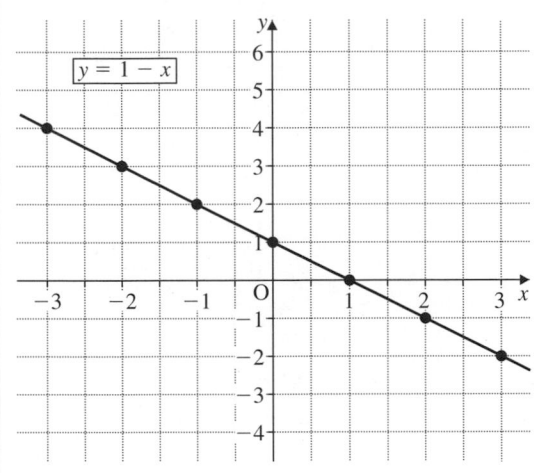

Skill Practice Four

For each question, plot on squared paper the graph of the equation from the table of values given.

1 $y = x + 4$

x	-3	-2	-1	0	1	2	3
y	1	2	3	4	5	6	7

2 $y = x + 3$

x	-3	-2	-1	0	1	2	3
y	0	1	2	3	4	5	6

3 $y = x - 2$

x	-3	-2	-1	0	1	2	3
y	-5	-4	-3	-2	-1	0	1

4 $y = x - 4$

x	-3	-2	-1	0	1	2	3
y	-7	-6	-5	-4	-3	-2	-1

5 $y = 4 - x$

x	-3	-2	-1	0	1	2	3
y	7	6	5	4	3	2	1

6 $y = 2 - x$

x	-3	-2	-1	0	1	2	3
y	5	4	3	2	1	0	-1

Skill Example

A table of values must first be made out before the graph of an equation can be plotted.

Make out a table of values for the equation $y = 2x + 1$ from $x = -2$ to $x = 2$.

Values of x	-2	-1	0	1	2
$2x$	-4	-2	0	2	4
$+1$	$+1$	$+1$	$+1$	$+1$	$+1$
Values of y	-3	-1	$+1$	$+3$	$+5$

The first value of y is found as follows.

When $x = -2$: $2x = 2 \times -2 = -4$

$\Rightarrow \quad y = 2x + 1 = -4 + 1 = -3$

Skill Practice Five

Copy and complete each table of values.

1 $y = 2x + 5$

x	-2	-1	0	1	2
$2x$					
$+5$					
y					

2 $y = 2x + 8$

x	-2	-1	0	1	2
$2x$					
$+8$					
y					

3 $y = 4x + 3$

x	-2	-1	0	1	2
$4x$					
$+3$					
y					

4 $y = 4x + 1$

x	-2	-1	0	1	2
$4x$					
$+1$					
y					

5 $y = 4x - 1$

x	-2	-1	0	1	2
$4x$					
-1					
y					

Skill Example

Draw the graph of $y = 3x - 2$ for values of x from -2 to $+2$.

Use a scale of 1 cm to 1 unit on the x-axis and a scale of 1 cm to 2 units on the y-axis.

Step 1

Produce a table of values as follows.

x	-2	-1	0	1	2
$3x$	-6	-3	0	3	6
-2	-2	-2	-2	-2	-2
y	-8	-5	-2	1	4

Step 2

Draw the axes, plot the points and then join up the points on the graph with a straight line.

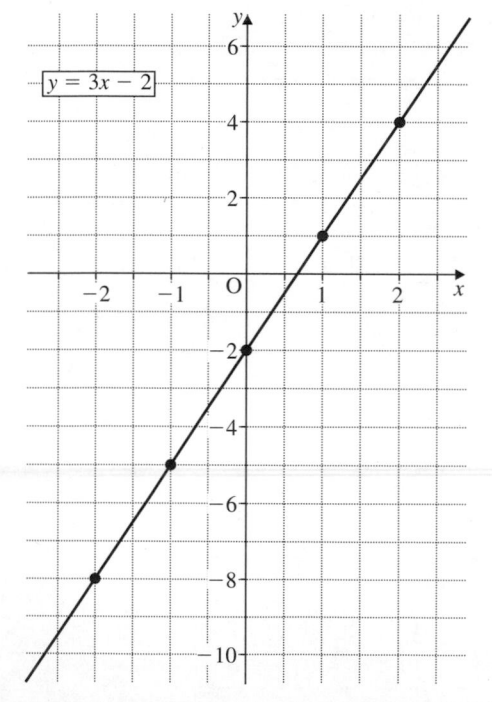

$y = 3x - 2$

Skill Practice Six

For questions **1** to **5**, copy and complete each table, then plot the graph of the equation given. Use a scale of 1 cm to 1 unit on the x-axis and a scale of 1 cm to 2 units on the y-axis.

1 $y = 2x - 2$

x	-2	-1	0	1	2
$2x$					
-2					
y					

2 $y = 3x - 1$

x	-2	-1	0	1	2
$3x$					
-1					
y					

3 $y = x - 1$

x	-2	-1	0	1	2
x					
-1					
y					

4 $y = x + 2$

x	-2	-1	0	1	2
x					
$+2$					
y					

5 $y = 2x + 3$

x	-2	-1	0	1	2
$2x$					
$+3$					
y					

For questions **6** to **10**, draw up a table of values from $x = -2$ to $x = +2$ for each equation. Then plot the graph of the equation. Use a scale of 1 cm to 1 unit on the x-axis and a scale of 1 cm to 2 units on the y-axis.

6 $y = 3x + 2$ **7** $y = 3x + 4$ **8** $y = 2x + 2$
9 $y = 2x$ **10** $y = 2x - 1$

Skill Example

Plot the graph of each of these equations on the same set of axes.

$$2x + 3y = 12$$
$$3y - 2x = -6$$

Graphs of such equations are usually drawn by completing a table with first x and then y taken to be zero, as shown below.

x	0	
y		0

Using such a table, we obtain

$2x + 3y = 12$

x	0	6
y	4	0

$3y - 2x = -6$

x	0	3
y	-2	0

Then, drawing both lines on the same axes we get the graphs shown on the right.

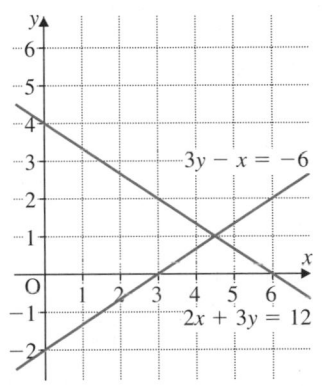

Skill Practice Seven

Plot the graph of each of these equations.

1 $2x + 4y = 8$ **2** $y - x = -2$ **3** $3y - x = 6$
4 $4y - x = 2$ **5** $2x + 4y = 8$ **6** $x + y = -2$
7 $x + y = 7$ **8** $2x + 3y = 12$ **9** $y - x = 2$
10 $2y - x = 2$

Knowledge

The **intercept** of a graph is the value of y at the point where the graph crosses the y-axis.

The **gradient** (or **slope**) of a graph can be measured between any two selected points on the graph:

$$\text{Gradient} = \frac{\text{Change in } y}{\text{Change in } x}$$

Skill Example

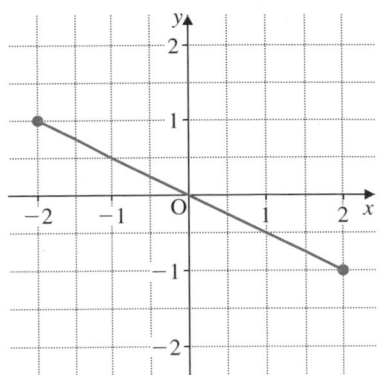

a Find the intercept of the graph.
b Find the gradient of the graph using the two points marked with dots.

a The intercept $= 0$

b Between points $(-2, 1)$ and $(2, -1)$:

$$\text{Gradient} = \frac{\text{Change in } y}{\text{Change in } x} = \frac{-1 - 1}{2 - -2}$$

$$= \frac{-2}{4} = -\frac{1}{2}$$

Finding the gradient of this graph involves calculations with negative numbers. It may be easier to read the changes in x and y directly from the graph like this:

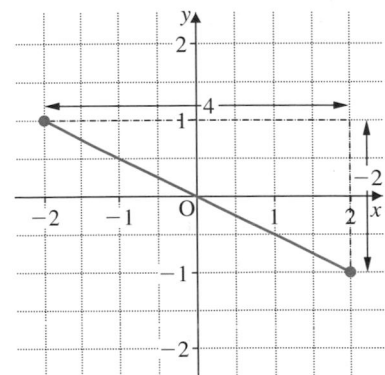

$$\text{Gradient} = \frac{-2}{4} = -\frac{1}{2}$$

Skill Practice Eight

In questions **1** to **10**, for each question find:

a the intercept of the graph
b the gradient of the graph using the two points marked with dots.

1

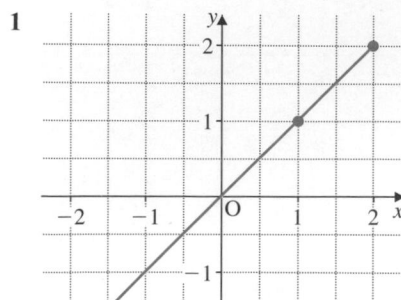

3

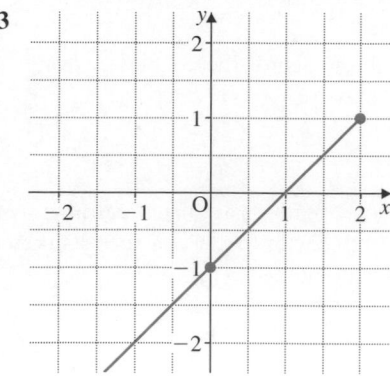

2

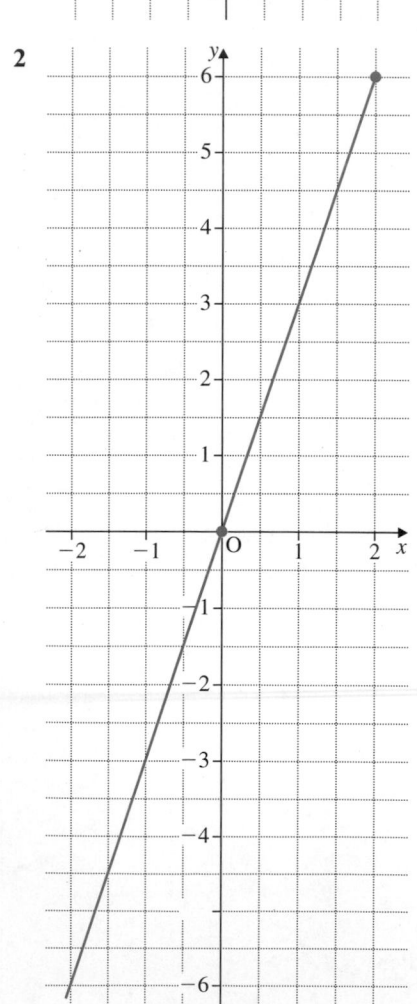

4

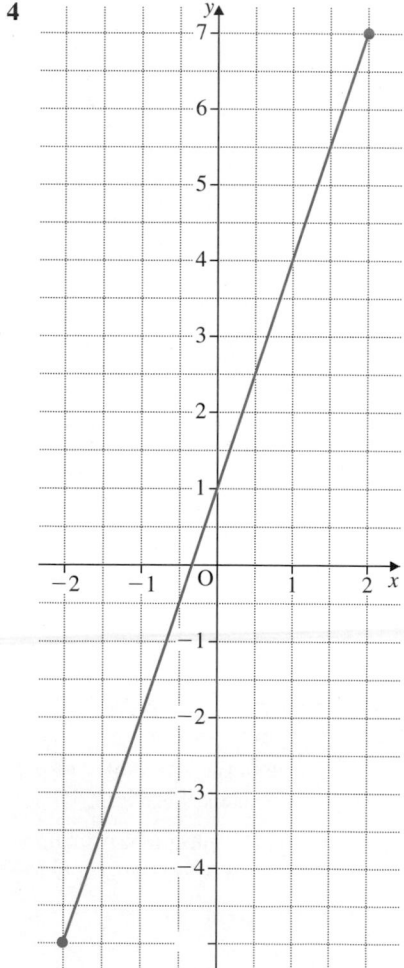

5

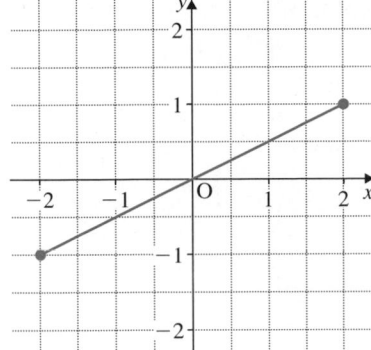

6

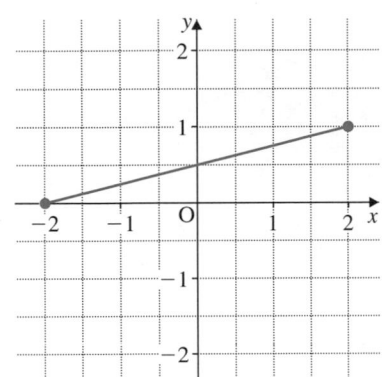

7

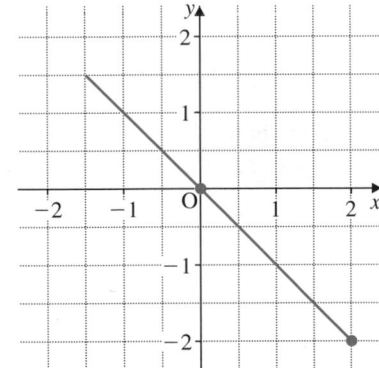

8

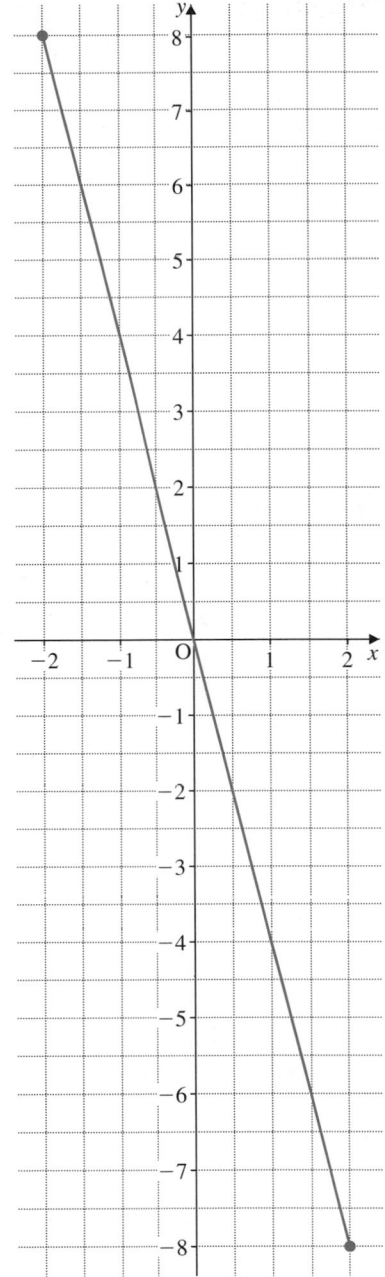

9

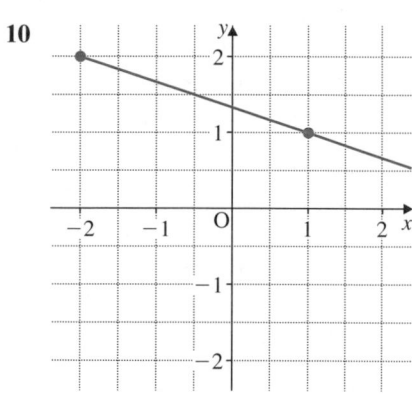

10

Skill Example

a Draw the graph of $y = 2x - 1$ for values of x from -2 to $+2$.

b Find the intercept of $y = 2x - 1$.

c Find the gradient of $y = 2x - 1$.

a

x	-2	-1	0	1	2
$2x$	-4	-2	0	2	4
-1	-1	-1	-1	-1	-1
y	-5	-3	-1	1	3

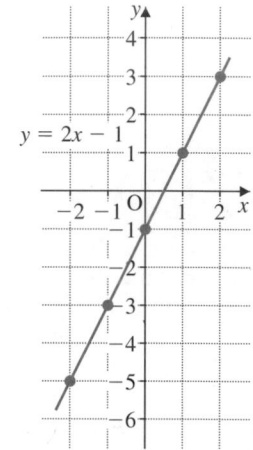

$y = 2x - 1$

b The graph crosses the y-axis at $(0, -1)$
Therefore, the intercept $= -1$

c Points $(1, 1)$ and $(2, 3)$ are selected to avoid working with negative numbers.
Between points $(1, 1)$ and $(2, 3)$:

$$\text{Gradient} = \frac{\text{Change in } y}{\text{Change in } x} = \frac{3 - 1}{2 - 1} = \frac{2}{1}$$

Skill Practice Nine

In questions **1** to **16**, for each equation:

a draw up a table of values from $x = -2$ to $x = +2$

b draw suitable axes and plot the graph of the equation

c find the intercept of the graph

d find the gradient of the graph.

1 $y = x + 6$	**2** $y = x - 4$
3 $y = 2x - 4$	**4** $y = 3x + 4$
5 $y = 4x + 3$	**6** $y = 4x - 2$
7 $y = 5x - 5$	**8** $y = 6 - 2x$

9 $y = 6 - 3x$

10 $y = -4x - 1$

11 $y = -5x$

12 $y = \dfrac{x}{2} + 1$

13 $y = \dfrac{x}{2} + 4$

14 $y = \dfrac{x}{2} - 3$

15 $y = \dfrac{x}{4}$

16 $y = \dfrac{x}{4} - 2$

Knowledge

When the equation of a straight line graph is written in the form

$$y = mx + c$$

the value of m is the **gradient** of the graph and the value of c is the **intercept** of the graph.

Skill Example

a Write down the gradient and intercept of the graph with equation $y = 5x + 2$.

b What is the equation of the graph with a gradient of 4 and an intercept of -7?

a Gradient $= 5$ Intercept $= 2$

b $y = 4x - 7$

Skill Practice Ten

In questions **1** to **12**, write down the gradient and intercept of the graph of each equation.

1 $y = 4x + 5$ **2** $y = 4x - 5$

3 $y = 5x + 4$ **4** $y = 5x - 4$

5 $y = 4 - 5x$ **6** $y = 5 - 4x$

7 $y = -4x - 5$ **8** $y = -4 - 5x$

9 $y = 11x$ **10** $y = -13x$

11 $y = \dfrac{x}{3}$ **12** $y = \dfrac{-x}{5}$

In questions **13** to **24**, write down the equation of each graph with the given gradient and intercept.

13 Gradient $= 3$ intercept $= 2$

14 Gradient $= 3$ intercept $= -2$

15 Gradient $= 2$ intercept $= 3$

16 Gradient $= 2$ intercept $= -3$

17 Gradient $= -3$ intercept $= 2$

18 Gradient $= -3$ intercept $= -2$

19 Gradient $= -2$ intercept $= 3$

20 Gradient $= -2$ intercept $= -3$

21 Gradient $= 7$ intercept $= 0$

22 Gradient $= -4$ intercept $= 0$

23 Gradient $= \frac{1}{4}$ intercept $= 0$

24 Gradient $= \frac{5}{4}$ intercept $= 0$

Unit 20

Knowledge

The first non-zero digit of a number is called its **first significant figure**.

For example, the first significant figure of each of these numbers is underlined.

$\underline{2}793$ $6\underline{5}.27$ $0.\underline{7}85$ $0.0\underline{9}$

Skill Practice One

Underline the first significant figure of each of these numbers.

1 34	**2** 506	**3** 6.7
4 0.55	**5** 6.09	**6** 0.05
7 0.001 09	**8** 1000.6	**9** 1999
10 0.004		

Knowledge

Numbers can be approximated and rounded to a number of **significant figures** as well as to a number of decimal places.

When rounding to a number of significant figures, you always start with the first significant figure.

You then count out the number of significant figures you require and apply normal rules for rounding.

The abbreviation 'sf' can be written instead of 'significant figure'.

Skill Example

a Give 37.5 correct to 2 significant figures.
b Give 3.75 correct to 2 significant figures.
c Give 0.342 correct to 2 significant figures.
d Give 0.003 42 correct to 2 significant figures.

a $37.5 = 38$ correct to 2 sf

b $3.75 = 3.8$ correct to 2 sf

c $0.342 = 0.34$ correct to 2 sf

d $0.003\,42 = 0.0034$ correct to 2 sf

Skill Practice Two

Give each number correct to 2 significant figures.

1 67.8	**2** 78.3	**3** 75.7
4 39.9	**5** 92.7	**6** 12.5
7 67.6	**8** 3.05	**9** 4.51
10 4.15	**11** 5.41	**12** 1.45
13 5.99	**14** 0.567	**15** 0.862
16 0.682	**17** 0.0457	**18** 0.0317
19 0.0713	**20** 0.0731	**21** 34.2
22 77.7	**23** 0.004 27	**24** 17.855

Skill Example

a Give 425.7 correct to 3 significant figures.
b Give 3.75 correct to 1 significant figure.
c Give 0.3429 correct to 3 significant figures.
d Give 0.003 42 correct to 1 significant figure.

a $425.7 = 426$ correct to 3 sf

b $3.75 = 4$ correct to 1 sf

c $0.3429 = 0.343$ correct to 3 sf

d $0.003\,42 = 0.003$ correct to 1 sf

Skill Practice Three

In questions **1** to **10**, give each number correct to 3 significant figures.

1 375.3	**2** 373.5	**3** 735.3
4 7.893	**5** 6.625	**6** 6.126
7 9.015	**8** 0.036 21	**9** 0.045 83
10 0.002 199		

In questions **11** to **20**, give each number correct to 1 significant figure.

11	5.56	**12**	7.37	**13**	6.72
14	6.27	**15**	2.67	**16**	2.51
17	2.99	**18**	0.051	**19**	0.054
20	0.0059				

Skill Example

a Give 347.9 correct to 1 significant figure.
b Give 375.4 correct to 2 significant figures.
c Give 375 200 correct to 3 significant figures.

a 347.9 = 300 correct to 1 sf. Note that the answer is 300, not 3, which would be a very poor approximation for 347.9.

b 375.4 = 380 correct to 2 sf

c 375 200 = 375 000 correct to 3 sf

Skill Practice Four

1 Give each number correct to 3 sf.

a	2345	**b**	2354	**c**	2453
d	2435	**e**	2543	**f**	2534
g	3245	**h**	3254	**i**	3452
j	3425				

2 Give each number in question **1** correct to 2 sf.

3 Give each number in question **1** correct to 1 sf.

Skill Example

Estimate the value of $(3.4 \times 27.8) \div 7.3$.

Rounding each number to 1 significant figure gives the estimate:

$$(3 \times 30) \div 7 = 90 \div 7$$
$$= 10 \text{ to } 1 \text{ sf}$$

The calculated answer is 13.0 correct to 1 dp.

Skill Practice Five

In questions **1** to **15**, estimate the value of each calculation, giving your answer correct to 1 sf. Use a calculator to find an answer correct to 1 dp and write this after your estimate.

1	3.45×7.27	**2**	89.7×1.237
3	92.6×3.69	**4**	653×2.99
5	0.685×111	**6**	613×9.73
7	759×76.2	**8**	345×30.1
9	2345×2.45	**10**	73.9×4.01
11	$(23.6 \times 2.89) \div 5.1$	**12**	$(98.1 \times 4.09) \div 7.62$
13	$(201 \times 3.55) \div 378$	**14**	$(75.2 \times 99.9) \div 6753$
15	$(23.4 \times 78.9) \div 32.3$		

16 There are 650 children in a school who each spend £6.25 a week on school meals.
  a Estimate correct to 1 sf the total amount spent each week on school meals.
  b Calculate correct to 1 dp the total amount spent each week on school meals.

17 A carpet costs £7.89 a square metre.
  a Estimate correct to 1 sf the area of a piece of carpet 2.5 metres by 4.3 metres.
  b Estimate correct to 1 sf the price of a piece of carpet 2.5 metres by 4.3 metres.
  c Calculate correct to 1 dp the price of a piece of carpet 2.5 metres by 4.3 metres.

18 A car averages 14.7 kilometres per litre of petrol. It uses petrol which costs 54.9 p per litre.
  a Estimate correct to 1 sf the quantity of petrol used during a journey of 293.8 km.
  b Estimate correct to the nearest 10 p the cost of the petrol used during a journey of 293.8 km.
  c Calculate correct to 1 dp the cost of the petrol used during a journey of 293.8 km.

Skill Example

A shop is selling cheese for £4.80 per kilogram. Find the cost of these weights of cheese.

a 1.5 kg   **b** 150 g

a Cost = 15×4.8 = £7.20

b 150 g = 0.15 kg
Cost = 0.15×4.8 = £0.72 = 72 p

Notice that when you multiply a quantity by a number less than one, the answer is *less* than the original quantity.

86

Skill Practice Six

1 Look at the prices of these cheeses.

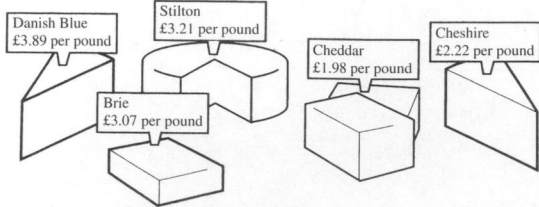

Danish Blue £3.89 per pound
Stilton £3.21 per pound
Brie £3.07 per pound
Cheddar £1.98 per pound
Cheshire £2.22 per pound

Find, to the nearest penny, the cost of each of the following amounts.

a 0.5 pounds of Stilton and 0.7 pounds of Danish Blue.

b 1.1 pounds of Cheddar and 1.3 pounds of Cheshire.

c 1.5 pounds of Brie and 1.4 pounds each of Stilton and Cheddar.

d 0.2 pounds of Danish Blue, 0.3 pounds of Cheddar and 0.4 pounds of Brie.

e 0.5 pounds of each cheese.

2 One litre of Cola costs 85 p. Find, to the nearest penny, the cost of the following quantities of Cola.

a 3 litres   **b** 2.5 litres
c 0.5 litres   **d** 330 ml

3 A bus company calculates the cost of its fares by charging 12 pence for each mile of the journey. For a return ticket the cost is calculated by adding the two single fares together and then multiplying by 0.9.
Find the cost of each of the following.

a Journey of 15.9 miles.
b Journey of 23.2 miles.
c Return ticket for a 6.7 mile journey.
d Return ticket for an 8.5 mile journey.
e Return ticket for a 9.3 mile journey.

4 A large orange contains 0.016 grams of vitamin C.
Find the weight of vitamin C in:

a 4 oranges   **b** 2 oranges
c half an orange   **d** a quarter of an orange

5 It costs a manufacturer £1.20 to produce a kilogram of savoury snack biscuits.
Find the cost, to the nearest penny, of the biscuits in packs which weigh:

a 44 g   **b** 56 g   **c** 100 g   **d** 150 g

Skill Example

A human white blood cell has a diameter of 0.001 243 cm.

How many white cells would it take to make a line 2 cm long?

Number of cells $= 2 \div 0.001\,243$

$$= 1609 \quad \text{(to nearest whole number)}$$

Notice that when you divide a quantity by a number less than one, the answer is *more* than the original quantity.

Skill Practice Seven

1 The diameter of a human red blood cell is 0.000 714 cm.
How many would it take to make a line:

a 5 cm long   **b** 2 cm long
c 1 cm long   **d** 0.5 cm long
e 0.1 cm long?

2 A large orange contains 0.016 grams of vitamin C.
How many oranges would you need to eat to get:

a 1 gram of vitamin C
b 5 grams of vitamin C
c 3 grams of vitamin C
d 0.5 grams of vitamin C
e 0.32 grams of vitamin C?

3 A bag of crisps weighs 0.044 kg.
How many bags must you buy to obtain the following weights of crisps?

a 0.44 kg   **b** 0.528 kg
c 1.276 kg   **d** 1.012 kg

4 A bus company calculates the cost of its fares by charging 12 pence for each mile of the journey. For a return ticket the cost is calculated by adding the two single fares together and then multiplying by 0.9.
Find the length of the journey for which a return tickets costs:.

a £4.32   **b** £5.40   **c** £3.24   **d** £7.56

Knowledge

All recent scientific calculators have these keys:

$1/x$	works out $1 \div x$
$\sqrt{x}$	works out the square root of x
x^2	works out the square of x

where x represents any number you enter.

Skill Example

Calculate:

a $\frac{1}{16}$ **b** $\sqrt{362}$ **c** $43^2 + 12^2$

a Pressing

$\boxed{1}\ \boxed{6}\ \boxed{1/x}$

gives 0.0625

b Pressing

$\boxed{3}\ \boxed{6}\ \boxed{2}\ \boxed{\sqrt{x}}$

gives 19.03 (correct to 2 dp)

c Pressing

$\boxed{4}\ \boxed{3}\ \boxed{x^2}\ \boxed{+}\ \boxed{1}\ \boxed{2}\ \boxed{x^2}\ \boxed{=}$

gives 1993

Skill Practice Eight

1 Calculate:

   a $\frac{1}{20}$ **b** $\frac{1}{40}$ **c** $\frac{1}{50}$

   d $\frac{1}{121}$ **e** $\frac{1}{2.3 + 4.8}$

2 Calculate:

   a $\sqrt{2209}$ **b** $\sqrt{262.44}$ **c** $\sqrt{0.36}$

   d $\sqrt{0.7225}$ **e** $\sqrt{11.3 + 3.14}$

3 Calculate:

   a 72^2 **b** $(3.7 + 5.1)^2$ **c** $3.7^2 + 5.1^2$

   d $(2.9 - 1.5)^2$ **e** $2.9^2 - 1.5^2$

Skill Example

Calculate:

a $\frac{1}{3 + \sqrt{7}}$ **b** $\sqrt{\frac{1}{3.14 \times 5^2}}$ **c** $\frac{5}{2 + \sqrt{5}}$

a Pressing

$\boxed{3}\ \boxed{+}\ \boxed{7}\ \boxed{\sqrt{}}\ \boxed{=}\ \boxed{1/x}$

gives 0.18 (correct to 2 dp)

b Pressing

$\boxed{3}\ \boxed{\cdot}\ \boxed{1}\ \boxed{4}\ \boxed{\times}\ \boxed{5}\ \boxed{x^2}\ \boxed{=}\ \boxed{1/x}\ \boxed{\sqrt{}}$

gives 0.11 (correct to 2 dp)

c Pressing

$\boxed{2}\ \boxed{+}\ \boxed{5}\ \boxed{\sqrt{}}\ \boxed{=}\ \boxed{1/x}\ \boxed{\times}\ \boxed{5}\ \boxed{=}$

gives 1.18 (correct to 2 dp)

Skill Practice Nine

Calculate:

1 $\frac{1}{3 + \sqrt{2}}$ **2** $\frac{1}{3^2 + 4^2}$ **3** $\frac{1}{\sqrt{3^2 + 4^2}}$

4 $\sqrt{13^2 - 12^2}$ **5** $\frac{1}{\sqrt{2} + \sqrt{3}}$ **6** $\sqrt{\frac{5 + 1.8^2}{3}}$

7 $\frac{3}{\sqrt{5 + 1.8^2}}$ **8** $\frac{5}{5 + \sqrt{5}}$ **9** $\frac{10}{3.14 \times \sqrt{10}}$

10 $\frac{\sqrt{8}}{2 \times \sqrt{2}}$

Knowledge

All recent scientific calculators have a key marked

$\boxed{\text{M}}$ or $\boxed{\text{Min}}$ or $\boxed{\text{STO}}$

to store a number in the memory.

Another key marked

$\boxed{\text{MR}}$ or $\boxed{\text{RCL}}$

recalls the number from the memory.

You use these keys to avoid writing down the result of each stage of a complicated calculation.

Skill Example

Calculate:

$$\frac{3.14 + 6.78}{2.39 \times 0.65}$$

Write down only your final answer.

Pressing

$$\boxed{2}\ \boxed{.}\ \boxed{3}\ \boxed{9}\ \boxed{\times}\ \boxed{0}\ \boxed{.}\ \boxed{6}\ \boxed{5}\ \boxed{=}\ \boxed{\text{STO}}$$

$$\boxed{3}\ \boxed{.}\ \boxed{1}\ \boxed{4}\ \boxed{+}\ \boxed{6}\ \boxed{.}\ \boxed{7}\ \boxed{8}\ \boxed{=}\ \boxed{\div}\ \boxed{\text{RCL}}\ \boxed{=}$$

gives 6.39 (correct to 2 dp)

Skill Practice Ten

Calculate each of the following, writing down only your final answer.

1 $\dfrac{23 + 97}{84 - 33}$

2 $\dfrac{5.3 \times 1.5}{1.5 + 4.9}$

3 $\dfrac{2.6 \times 8.1}{3.4 \times 5.8}$

4 $\dfrac{5 \times 87}{23 \div 0.9}$

5 $\dfrac{\sqrt{7} + 7}{\sqrt{7} \times (1 + \sqrt{7})}$

6 $\dfrac{4 \times 36}{3.14 \times \sqrt{6^2 + 6^2}}$

7 $\dfrac{(2.15^2 + 1.6^2)}{\sqrt{2.15^2 + 1.6^2}}$

8 $\dfrac{\left(1 + \dfrac{1}{\sqrt{8}}\right)^2}{2 + \sqrt{2}}$

9 $\sqrt{\dfrac{5.1^2 - 3.2^2}{2 \times 7.9}}$

10 $2.8 + \sqrt{\dfrac{2.8^2 - (4 \times 1.2 \times 1.3)}{2 \times 1.2}}$

Unit 21

Skill Example

$$35\% = \tfrac{35}{100} = \begin{cases} 0.35 \text{ (as a decimal)} \\ \tfrac{7}{20} \text{ (as a fraction)} \end{cases}$$

Skill Practice One

Write the following as fractions and decimals.

1 18%	**2** 26%	**3** 37%	**4** 41%
5 67%	**6** 84%	**7** 80%	**8** 60%
9 30%	**10** 10%		

Skill Example

a Find the interest charged if £500 is borrowed for 1 year at 16% pa (per annum).

b Find the interest charged if £250 is borrowed for 3 years at 10% pa.

a Interest = 16% of £500 = (£500 ÷ 100) × 16
$$= £80$$

b Interest charged for 1 year = 10% of £250
$$= (£250 ÷ 100) × 10$$
$$= £25$$

So, interest charged for 3 years = £25 × 3
$$= £75$$

Skill Practice Two

1 A man borrows £4000 to buy a car at an interest rate of 10% pa. If it is 3 years before he has repaid the debt, how much interest will be charged?

2 A man borrows £800 to buy a motor cycle at an interest rate of 12% pa.
If it is 3 years before he has repaid the debt, how much interest will be charged?

3 A woman borrows £200 to buy a colour television at an interest rate of 15% pa.
If it is 2 years before she has repaid the debt, how much interest will be charged?

4 A woman borrows £150 to buy a washing machine at an interest rate of 10% pa.
If it is 2 years before she has repaid the debt, how much interest will be charged?

5 Saleem's mother has to borrow £80 at an interest rate of 5% pa in order to buy a new bicycle.
If it is one year before she has repaid the debt, how much interest will be charged?

6 Kelly's father has to borrow £40 at an interest rate of 15% pa in order to buy a new vacuum cleaner.
If it is one year before he has repaid the debt, how much interest will be charged?

7 Mrs Green has to borrow £400 in order to buy a new three-piece suite. There are two ways in which she can borrow the money:
a at an interest rate of 20% pa for 2 years
b at an interest rate of 15% pa for 3 years.
Which option involves paying less interest and by how much?

8 Mr Bates has to borrow £600 in order to buy new kitchen units. There are two ways in which he can borrow the money:
a at an interest rate of 5% pa for 5 years
b at an interest rate of 8% pa for 3 years.
Which option involves paying less interest and by how much?

9 Natalie has to borrow £150 in order to buy a new electric cooker. There are two ways in which she can borrow the money:
a at an interest rate of 10% pa for 3 years
b at an interest rate of 8% pa for 4 years.
Which option involves paying less interest and by how much?

10 Peter has to borrow £240 in order to buy a new moped. There are two ways in which he can borrow the money:
a at an interest rate of 10% pa for 2 years
b at an interest rate of $7\tfrac{1}{2}$% pa for 3 years.
Which option involves paying less interest and by how much?

Skill Example

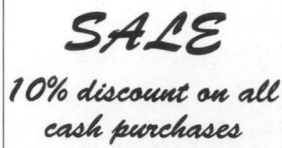

SALE

10% discount on all cash purchases

There are two different methods used to calculate such a percentage reduction.

Method 1
Find the value of the discount and the amount actually paid if a discount of 10% is offered on a marked price of £64.

Value of discount $= 10\%$ of £64

$$= \frac{10}{100} \text{ of } £64$$

$$= (£64 \div 100) \times 10$$

$$= £6.40$$

Amount paid $= £64 - £6.40 = £57.60$

Method 2
Find the amount actually paid and the value of the discount if a discount of 10% is offered on a marked price of £64.

A discount of 10% means you pay 90% of the marked price.

Amount paid $= 90\%$ of £64

$$= \frac{90}{100} \text{ of } £64$$

$$= (£64 \div 100) \times 90$$

$$= £57.60$$

Value of discount $= £64 - £57.60 = £6.40$

Skill Practice Three

Use Method 1 to find the value of the discount and the amount actually paid for the following.

Item	Marked price	Discount offered
1 Radio	£50	10%
2 Kettle	£20	5%
3 Cassette player	£80	25%
4 Refrigerator	£90	20%
5 Bicycle	£60	15%

6 Washing machine	£180	10%
7 Exercise bike	£140	5%
8 Colour television	£500	25%
9 Electric cooker	£135	20%
10 Tumble dryer	£160	15%

Skill Practice Four

Use Method 2 to find the amount actually paid and the value of the discount for the following.

Item	Marked price	Discount offered
1 Vacuum cleaner	£45	10%
2 Fan heater	£32	20%
3 Electric toaster	£42	25%
4 Telephone	£50	5%
5 Electric fire	£30	15%
6 Cassette recorder	£125	10%
7 Book cabinet	£108	20%
8 Armchair	£110	25%
9 Kitchen units	£210	5%
10 Bed	£150	15%

Skill Example

Many items in shops are priced with VAT already added, but sometimes VAT has to be added to the marked price.

There are two different methods used to calculate such a percentage increase.

Method 1
Find the amount of VAT to be added and the amount a purchaser will pay for this computer.

£1200.00 +VAT

BUY NOW PAY LATER

VAT $= 17.5\%$ of £1200

$$= \frac{17.5}{100} \text{ of } £1200$$

$$= (£200 \div 100) \times 17.5$$

$$= £210$$

Amount paid $= £1200 + £210 = £1410$

Method 2

£340.00 +VAT

INTEREST FREE CREDIT

An increase of 17.5% means you pay 117.5% of the marked price.

Amount paid = 117.5% of £340

$$= \frac{117.5}{100} \text{ of } £340$$

$$= (£340 \div 100) \times 117.5$$

$$= £399.50$$

VAT = £399.50 − £340 = £59.50

Skill Practice Five

Use Method 1 to find the VAT and the amount paid for each of the following.

Item	Price before VAT
1 Washing machine	£200
2 Television	£500
3 Car	£6000
4 Motor cycle	£1200
5 Cassette player	£80
6 Exercise bike	£140
7 Electric cooker	£180
8 Radio	£50
9 Refrigerator	£90
10 CD player	£110

Skill Practice Six

Use Method 2 to find the amount paid and the VAT for each of the following.

Item	Price before VAT
1 Typewriter	£210
2 Electric fire	£42
3 Electricity bill	£68
4 Telephone bill	£48
5 Vacuum cleaner repair	£14.80
6 Television repair	£25.40
7 Gas bill	£82.60
8 Pair of shoes	£28.50
9 Pair of trousers	£23.20
10 Watch	£35.60

Knowledge

If an article is bought for £50 and sold for £60, the difference (£60 − £50) is called the **profit**.

Profit = £60 − £50 = £10

Profit = Selling price − Buying price

If an article is bought for £50 and sold for £30, the difference (£50 − £30) is called the **loss**.

Loss = £50 − £30 = £20

Loss = Buying price − Selling price

Skill Example

Find the profit or loss in each of these cases.

a Buying price = £10, selling price = £14.50
b Buying price = £14.50, selling price = £12.75

a The selling price is greater than the buying price, so profit = £14.50 − £10.00 = £4.50
b The buying price is greater than the selling price, so loss = £14.50 − £12.75 = £1.75

Skill Practice Seven

For questions 1 to 10, find the profit or loss.

	Buying price	Selling price
1	£15.00	£18.50
2	£13.00	£19.50
3	£9.00	£13.50
4	£18.00	£15.00
5	£10.00	£6.50
6	£14.50	£17.00
7	£16.50	£20.00
8	£8.50	£16.00
9	£20.50	£17.00
10	£22.50	£18.00

11 A student buys a new book for £5.25 and then sells it for £3.50.
Find the loss that he makes.

12 A woman buys a new bicycle for £80.50 and sells it later for £58.
Find the loss that she makes.

13 A man buys a new refrigerator for £120 and sells it later for £85.
Find the loss that he makes.

14 A car dealer buys a second-hand car for £2150 and then sells it for £2400.
Find the profit that he makes.

15 A woman buys a house for £18 750 and sells it later for £22 500.
Find the profit that she makes.

Skill Example

A greengrocer buys a crate containing 200 oranges for £20 and then sells the oranges in packs of five for 60 p a pack. Calculate his profit or loss on the crate.

Crate contains $\dfrac{200}{5} = 40$ packs

So, selling price of the crate $= 40 \times 60\,\text{p}$
$$= 2400\,\text{p}$$
$$= \text{£}24.00$$

But buying price of the crate $= \text{£}20.00$

Selling price is greater than buying price.

So, profit $= \text{£}24.00 - \text{£}20.00 = \text{£}4.00$

Skill Practice Eight

For each question, find the total profit or loss.

		Buying price	Selling price
1	Bars of chocolate	£1.00 for 10	12 p each
2	Pencils	80 p for 10	10 p each
3	Pens	£4.00 for 10	46 p each
4	Grapefruits	£20.00 for 100	18 p each
5	Oranges	£5.00 for 50	8 p each
6	Writing pads	£20.00 for 50	38 p each
7	Packets of crisps	£2.50 for 25	11 p each
8	Cans of lemonade	£10.00 for 50	25 p each
9	Blocks of ice cream	£30.00 for 50	69 p each
10	Large envelopes	£3.00 for 25	9 p each

Skill Example

a An electric clock is bought for £10 and then sold for £13.
What is the percentage profit?

$$\text{Profit} = \text{£}13 - \text{£}10 = \text{£}3$$

Therefore, the profit as a fraction of the buying price $= \dfrac{3}{10} = 0.3$

So, percentage profit $= 0.3 \times 100\%$
$$= 30\%$$

b A calculator is bought for £15 and then sold for £12.
What is the percentage loss?

$$\text{Loss} = \text{£}15 - \text{£}12 = \text{£}3$$

Therefore, the loss as a fraction of the buying price $= \dfrac{3}{15} = 0.2$

So, percentage loss $= 0.2 \times 100\%$
$$= 20\%$$

Skill Practice Nine

For questions **1** to **5**, find the percentage profit.

	Buying price	Selling price
1	£25	£30
2	£16	£20
3	£120	£126
4	£50	£65
5	£200	£230

For questions **5** to **10**, find the percentage loss.

	Buying price	Selling price
6	£24	£18
7	£40	£32
8	£20	£14
9	£160	£152
10	£120	£102

Skill Example

A newsagent buys 20 magazines for £12.00 and sells them for 75 p each.
Find the percentage profit that he makes on each magazine.

Buying price of one magazine = £12.00 ÷ 20
= 60 p

Therefore, the profit made on one magazine
= 75 p − 60 p
= 15 p

So, the profit as a fraction of the buying price
$$= \frac{15}{60} = 0.25$$

and percentage profit = 0.25 × 100%
= 25%

Skill Practice Ten

For each question, find the percentage profit or loss.

		Buying price	Selling price
1	Pencils	£6.00 for 20	27 p each
2	Iced buns	£5.00 for 20	20 p each
3	Tubes of sweets	£5.00 for 25	21 p each
4	Exercise books	£40.00 for 50	68 p each
5	Bars of chocolate	£12.00 for 20	57 p each
6	Cans of cola	£15.00 for 50	33 p each
7	Packets of crisps	£1.20 for 5	30 p each
8	Sandwiches	£4.80 for 6	84 p each
9	Apples	£3.60 for 12	24 p each
10	Large nails	£1.92 for 24	6 p each

Unit 22

Number and Algebra   Level 7

Skill   You find and describe in symbols the next term or nth term of a sequence where the rule is quadratic.

Knowledge

We can use mathematical shorthand to describe sequences.

We use an italic capital letter followed by a suffix to represent each term. For example, if T is used to represent the terms of the sequence, 2, 4, 6, 8, ..., then

$$T_1 = 2 \quad T_2 = 4 \quad T_3 = 6 \quad T_4 = 8$$

A general term of the sequence is represented by T_n, where n can be any whole number. The term after T_n will be T_{n+1}, the term after that T_{n+2} and so on.

The rule for the sequence can also be written in mathematical shorthand. For this sequence the rule is:

$$T_{n+1} = T_n + 2$$

This states that, to find the term after T_n, add 2 to T_n.

Skill Example

Write the rule for each sequence, using mathematical shorthand.

a   1, 5, 9, 13, ...

Each term is formed by adding 4 to the one before it, so the rule is:

$$T_{n+1} = T_n + 4$$

b   1, 5, 25, 125, ...

Each term is formed by multiplying the one before it by 5, so the rule is:

$$T_{n+1} = 5T_n$$

c   1, 3, 7, 15, ...

Each term is formed by multiplying the one before it by 2, and then adding 1, so the rule is:

$$T_{n+1} = 2T_n + 1$$

Skill Practice One

Write the rule for each sequence, using mathematical shorthand.

1   1, 8, 15, 22, ...
2   2, 4, 8, 16, ...
3   3, 7, 11, 15, ...
4   5, 15, 45, 135, ...
5   17, 19, 21, 23, ...
6   1, 7, 49, 343, ...
7   12, 10, 8, 6, ...
8   100, 50, 25, 12.5, ...
9   200, 190, 180, 170, ...
10   1000, 100, 10, 1, ...
11   1, 2, 4, 8, ...
12   1, 3, 7, 15, ...
13   1, 5, 13, 29, ...
14   1, 6, 16, 36, ...
15   1, 3, 9, 27, ...
16   1, 4, 13, 40, ...

Skill Example

Write the first 5 terms of the sequence for which $T_1 = 1$ and $T_{n+1} = 2T_n + 5$

Starting with 1 and applying the rule, we have

$$1, 7, 19, 43, 91, ...$$

Skill Practice Two

Write down the first five terms of each of the following sequences.

1	$T_1 = 1$	$T_{n+1} = T_n + 13$
2	$T_1 = 1$	$T_{n+1} = T_n + 4$
3	$T_1 = 1$	$T_{n+1} = 6T_n$
4	$T_1 = 1$	$T_{n+1} = 8T_n$
5	$T_1 = 20$	$T_{n+1} = T_n - 5$
6	$T_1 = 12$	$T_{n+1} = T_n - 1$
7	$T_1 = 1000$	$T_{n+1} = T_n \div 10$
8	$T_1 = 512$	$T_{n+1} = T_n \div 2$
9	$T_1 = 1$	$T_{n+1} = 2T_n + 6$
10	$T_1 = 1$	$T_{n+1} = 6T_n + 2$
11	$T_1 = 1$	$T_{n+1} = 5T_n - 3$
12	$T_1 = 1$	$T_{n+1} = 7T_n - 1$

Skill Example

a The sequence of square numbers is:

$$
\begin{array}{ccccc}
T_1 & T_2 & T_3 & T_4 & T_5 \quad \ldots \\
\downarrow & \downarrow & \downarrow & \downarrow & \downarrow \\
1 & 4 & 9 & 16 & 25 \quad \ldots
\end{array}
$$

We can see that the formula for the nth term of this sequence is:

$$T_n = n^2$$

b Write down the first 5 terms of the sequence produced by the formula $T_n = n^2 + n$.

Applying the formula, we have:

$$
\begin{aligned}
T_1 &= 1 \times 1 + 1 = 2 \\
T_2 &= 2 \times 2 + 2 = 6 \\
T_3 &= 3 \times 3 + 3 = 12 \\
T_4 &= 4 \times 4 + 4 = 20 \\
T_5 &= 5 \times 5 + 5 = 30
\end{aligned}
$$

Skill Practice Three

Write down the first 5 terms of the sequence produced by each formula.

1 $T_n = 2n^2$
2 $T_n = 3n^2$
3 $T_n = 5n^2$
4 $T_n = n^2 + 1$
5 $T_n = n^2 - 1$
6 $T_n = n^2 + 2$
7 $T_n = n^2 - 2$
8 $T_n = n^2 - n$
9 $T_n = 2n^2 + 1$
10 $T_n = 2n^2 + n$
11 $T_n = n^2 + n + 1$
12 $T_n = n^2 + n + 5$

Skill Practice Four

1 The triangular numbers are:

$$
\begin{array}{ccccc}
T_1 & T_2 & T_3 & T_4 & T_5 \quad \ldots \\
\downarrow & \downarrow & \downarrow & \downarrow & \downarrow \\
1 & 3 & 6 & 10 & 15 \quad \ldots
\end{array}
$$

When two identical triangular numbers are placed together, they form a rectangle:

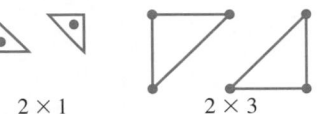

2×1 2×3

a Draw the rectangles formed by the next 5 triangular numbers.
b How big would the rectangle formed from the 12th triangular number be?
c How big would the rectangle formed from the 100th triangular number be?
d How big would the rectangle formed from the nth triangular number be?
e Write down a formula for the nth triangular number.

2 When groups of French people meet, usually everyone shakes hands with everyone else. Three French friends meet. Each person shakes hands with the other two.

a How many handshakes will there be in total?
b How many handshakes will there be in total if 4 friends meet?
c How many handshakes will there be in total if 5 friends meet?
d How many handshakes will there be in total if n friends meet?

3 There are 22 teams in the football's Premier League. Each team plays all the other teams both 'at home' and 'away'.
a How many games are played each season?
b How many games would be played in a league with n teams?

4 A 3 cm cube of wood is painted red and then cut into individual 1 cm cubes.

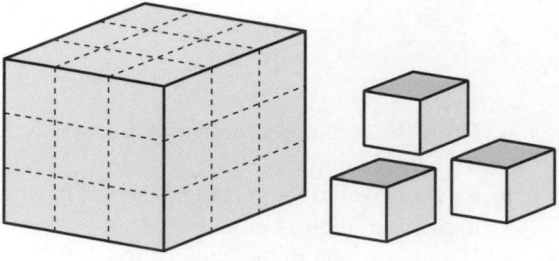

a How many of the 1 cm cubes will have no painted faces?

b How many of the 1 cm cubes will have one painted face?

c How many of the 1 cm cubes will have two painted faces?

d How many of the 1 cm cubes will have three painted three faces?

e How many of each type of 1 cm cube would there be if you cut up a 5 cm cube?

f Find formulae for the number of each type of 1 cm cube if you cut up an n cm cube.

Unit 23

Skill   You use algebraic and graphical methods to solve simultaneous linear equations in two variables.

Knowledge

Consider

$$x + y = 7$$

There are many values of x and y which will satisfy this equation. For example:

$$x = 6 \text{ and } y = 1$$
$$x = 4 \text{ and } y = 3 \text{ and so on} \ldots$$

Now consider

$$x - y = 1$$

There are many values of x and y which will satisfy this second equation. For example:

$$x = 6 \text{ and } y = 5$$
$$x = 2 \text{ and } y = 1 \text{ and so on} \ldots$$

However, if both $x + y = 7$
and $x - y = 1$ at the same time, there is only one value of x and one value of y that will together satisfy both equations.

These values are $x = 4$ and $y = 3$, because

$$4 + 3 = 7$$
and $$4 - 3 = 1$$

Such pairs of equations are called **simultaneous equations**.
When they are very simple, they can be solved by trial and improvement.

Skill Example

Solve this pair of simultaneous equations:

$$x + y = 6$$
$$x - y = 2$$

By inspection, the solution is $x = 4$ and $y = 2$, because $4 + 2 = 6$
and $4 - 2 = 2$

Skill Practice One

Solve these simultaneous equations by inspection.

1	$x + y = 8$	**2**	$u + v = 8$
	$x - y = 2$		$u - v = 6$
3	$r + s = 9$	**4**	$x + y = 7$
	$r - s = 3$		$x - y = 3$
5	$x + y = 5$	**6**	$m + n = 4$
	$x - y = 3$		$m - n = 2$
7	$b + c = 3$	**8**	$p + q = 10$
	$b - c = 1$		$p - q = 2$
9	$x + y = 12$	**10**	$x + y = 20$
	$x - y = 6$		$x - y = 10$

Knowledge

When both the x-terms in a pair of simultaneous equations are the same, the equations can be solved by **elimination**.

When the signs of the equal terms are different, *add* the two equations to eliminate x.

This method can also be used when the y-terms in the equation are the same.

Skill Example

Solve this pair of simultaneous equations:

$$x + y = 7 \qquad \text{[i]}$$
$$2x - y = 5 \qquad \text{[ii]}$$

The y-terms in both equations are the same, but the signs are different. So add the two equations to eliminate y:

$$x + y = 7 \qquad \text{[i]}$$
$$2x - y = 5 \qquad \text{[ii]}$$

$$x + 2x + y - y = 7 + 5$$

$$\Rightarrow \quad 3x = 12$$
$$\Rightarrow \quad x = 4$$

Then put this value of x in equation [i], which gives

$$\Rightarrow \quad 4 + y = 7$$
$$\Rightarrow \quad y = 3$$

The solution is $x = 4$ and $y = 3$.

Skill Example

Solve for x and y:

$$3x + 4y = 11 \qquad \text{[i]}$$
$$-3x + 2y = 1 \qquad \text{[ii]}$$

Add [i] and [ii] to eliminate x:

$$3x - 3x + 4y + 2y = 1 + 11$$
$$\Rightarrow \quad 6y = 12$$
$$\Rightarrow \quad y = 2$$

Then put this value of y in equation [i]:

$$3x + (4 \times 2) = 11$$
$$3x + 8 = 11$$
$$\Rightarrow \quad 3x = 3 \quad (\text{because } 8 + 3 = 11)$$
$$\Rightarrow \quad x = 1$$

The solution is $x = 1$ and $y = 2$.

Skill Practice Two

Solve:

1 $3x + y = 10$
$x - y = 2$

2 $5x + y = 22$
$x - y = 2$

3 $2p + q = 8$
$p - q = 1$

4 $4m + n = 17$
$m - n = 3$

5 $5x + y = 11$
$3x - y = 5$

6 $3x + y = 15$
$2x - y = 5$

7 $4a + b = 14$
$3a - b = 7$

8 $u + v = 9$
$4u - v = 1$

9 $3x + 2y = 19$
$x - 2y = 1$

10 $5x + 2y = 32$
$x - 2y = 4$

11 $4b + 3c = 38$
$b - 3c = 2$

12 $2q + 3r = 29$
$q - 3r = 1$

13 $5x + 2y = 19$
$3x - 2y = 5$

14 $7x + 2y = 22$
$5x - 2y = 2$

15 $4d + 3e = 15$
$2d - 3e = 3$

Skill Example

Solve for x and y:

$$2x + y = 7 \qquad \text{[i]}$$
$$x + y = 4 \qquad \text{[ii]}$$

Here the x-terms are the same and the signs are the same. So subtract one equation from the other to eliminate y:

$$2x + y = 7 \qquad \text{[i]}$$
$$x + y = 4 \qquad \text{[ii]}$$
$$2x - x + y - y = 7 - 4$$
$$\Rightarrow \quad x = 3$$

Then put this value of x in equation [ii]:

$$3 + y = 4$$
$$\Rightarrow \quad y = 1$$

The solution is $x = 3$ and $y = 1$.

Skill Example

Solve for x and y:

$$2x - 2y = 6 \qquad \text{[i]}$$
$$x - 2y = 1 \qquad \text{[ii]}$$

Subtract equation [ii] from equation [i] to eliminate y:

$$2x - x - 2y - (-2y) = 6 - 1$$

Remember that $-(-2y)$ is the same as $+2y$:

$$\Rightarrow \quad x = 5$$

Then put this value of x in equation [i]:

$$(2 \times 5) - 2y = 6$$
$$10 - 2y = 6$$
$$\Rightarrow \quad 2y = 4 \quad (\text{because } 10 - 4 = 6)$$
$$\Rightarrow \quad y = 2$$

The solution is $x = 5$ and $y = 2$.

Skill Practice Three

Solve:

1 $3x + y = 11$
$x + y = 5$

2 $4p + q = 17$
$p + q = 8$

3 $5x + y = 22$
$2x + y = 10$

4 $8u + v = 28$
$3u + v = 13$

5 $3x + 2y = 18$
$x + 2y = 14$

6 $4r + 3s = 27$
$r + 3s = 18$

7 $5x + 2y = 26$
$3x + 2y = 18$

8 $9s + 3t = 30$
$2s + 3t = 9$

9 $4x - y = 18$
$x - y = 3$

10 $2p - q = 7$
$p - q = 1$

11 $5a - b = 13$
$3a - b = 7$

12 $7x - 2y = 19$
$x - 2y = 1$

13 $4c - 3d = 31$
$c - 3d = 1$

14 $5x - 2y = 16$
$3x - 2y = 8$

15 $7q - 3r = 19$
$4q - 3r = 7$

Skill Example

Solve for x and y:

$$2x + 3y = 13 \qquad \text{[i]}$$
$$x + 2y = 8 \qquad \text{[ii]}$$

Here neither the x-terms nor the y-terms are the same. So form equation [iii] by multiplying equation [ii] by 2 to obtain two equations with x-terms that are the same.

$$2x + 3y = 13 \qquad \text{[i]}$$
$$2x + 4y = 16 \qquad \text{[iii]}$$

Subtract [i] from [iii] to eliminate x:

$$2x + 4y = 16 \qquad \text{[iii]}$$
$$2x + 3y = 13 \qquad \text{[i]}$$
$$2x - 2x + 4y - 3y = 16 - 13$$
$$\Rightarrow \quad y = 3$$

Then put this value of y in equation [i]:

$$2x + 9 = 13$$
$$\Rightarrow \quad 2x = 4$$
$$\Rightarrow \quad x = 2$$

The solution is $x = 2$ and $y = 3$.

Skill Example

Solve for x and y:

$$2x - 3y = 14 \qquad \text{[i]}$$
$$3x - 2y = 16 \qquad \text{[ii]}$$

Form equations [iii] and [iv] by multiplying equation [i] by 3 and equation [ii] by 2:

$$6x - 9y = 42 \qquad \text{[iii]}$$
$$6x - 4y = 32 \qquad \text{[iv]}$$

Subtract (iv) from (iii) to eliminate x:

$$6x - 6x - 9y - (-4y) = 42 - 32$$
$$\Rightarrow \quad -5y = 10$$
$$\Rightarrow \quad y = -2$$

Then put this value of y in equation [i]:

$$2x + 6 = 14$$
$$\Rightarrow \quad 2x = 8$$
$$\Rightarrow \quad x = 4$$

The solution is $x = 4$ and $y = -2$.

Skill Practice Four

Solve:

1 $\quad 3x + 2y = 12$
$\quad\quad 2x - y = 1$

2 $\quad 4u - 3v = 13$
$\quad\quad u + v = 5$

3 $\quad 3m + 2n = 10$
$\quad\quad 5m - 4n = -20$

4 $\quad 5x + y = 12$
$\quad\quad x - 2y = -2$

5 $\quad 3s + 2t = 27$
$\quad\quad 5s - 6t = 17$

6 $\quad 3x + 2y = 10$
$\quad\quad 4x + 4y = 12$

7 $\quad 2x + y = 3$
$\quad\quad 5x + 3y = 8$

8 $\quad 8m = 14 - 2n$
$\quad\quad 4m = 3n + 11$

9 $\quad 7s + 3n = 22$
$\quad\quad s + 2n = 0$

10 $\quad 8v = 4m - 4$
$\quad\quad 2v = 3m - 7$

11 $\quad 5v + 3m = 4$
$\quad\quad v - 4m = 10$

12 $\quad m + 5n = 35$
$\quad\quad 3m + n = 35$

13 $\quad 8y = 5x + 19$
$\quad\quad 7y = x + 20$

14 $\quad p - 3t = 3$
$\quad\quad 5p + 2t = 83$

15 $\quad m + 4n = 9$
$\quad\quad 6m - 3n = -27$

16 $\quad 16q + 5p = -3$
$\quad\quad 4q - 2p = 22$

Skill Practice Five

Solve:

1 $\quad 2x - 3y = -1$
$\quad\quad 3x + 5y = 8$

2 $\quad 4x + 2y = 7$
$\quad\quad 3x - 3y = 3$

3 $\quad 4x + 2y = 26$
$\quad\quad 3x + 3y = 20$

4 $\quad 4m - 2n = -2$
$\quad\quad 3m + 3n = -6$

5 $\quad 5e + 3f = 51$
$\quad\quad 2e - 7f = 4$

6 $\quad 2u + 2v = 12$
$\quad\quad 3u - 3v = 0$

7 $\quad 5s + 2t = 33$
$\quad\quad 4s + 5t = 40$

8 $\quad 3m + 6n = 6$
$\quad\quad 2m - 5n = 13$

9 $\quad 3x + 4y = 25$
$\quad\quad 5x + 6y = 39$

10 $\quad 8x - 3y = 19$
$\quad\quad 5x + 4y = 6$

11 $\quad 5p + 3q = 44$
$\quad\quad 7p - 8q = 25$

12 $\quad 2r + 3s = 15$
$\quad\quad 5r + 5s = 25$

13 $\quad 2m + 3n = 51$
$\quad\quad 3m - 2n = 5$

14 $\quad 4p - 2q = 2$
$\quad\quad 3p + 5q = 21$

15 $\quad 4s + 5t = 1$
$\quad\quad 5s + 6t = 2$

16 $\quad 3m - 4n = -11$
$\quad\quad 4m - 3n = -10$

Skill Example

Graphs offer an alternative method of solving simultaneous equations.

Solve graphically:

$$2x + 3y = 12$$
$$3y - 2x = -6$$

Graphs of such equations are usually drawn by completing a table with first x and then y taken to be zero, as shown below.

x	0	
y		0

Using such a table, we obtain:

$2x + 3y = 12$

x	0	6
y	4	0

$3y - 2x = -6$

x	0	3
y	-2	0

Then, drawing both lines on the same axes, we get:

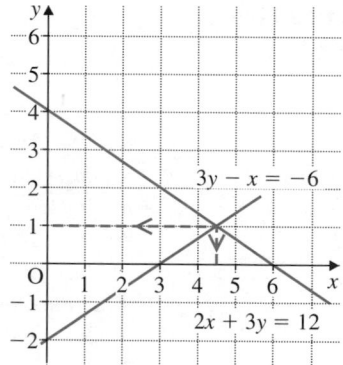

The point where the two lines cross gives the solution: $x = 4.5$ and $y = 1$.

Skill Practice Six

Solve graphically:

1 $y = 3 - x$
  $2x + 4y = 8$

2 $y = -x$
  $y - x = -2$

3 $3y - x = 6$
  $3y - 2x = 3$

4 $y = 1$
  $4y - x = 2$

5 $2x + 4y = 8$
  $y = 1 - x$

6 $y = x$
  $x + y = -2$

7 $x = y$
  $x + y = 7$

8 $2x + 3y = 12$
  $x + y = 5$

9 $y = 2x + 1$
  $y - x = 2$

10 $6x + 3y = 18$
  $2y - x = 2$

Unit 24

Knowledge

$$3x + 5 > 14$$

A relationship like this is called a **linear inequality**.

It means that three times the number x, plus 5, always gives an answer greater than 14.

We solve the inequality by simplifying it until we have a statement about x itself.

We use rules like those used to solve equations.

$$3x + 5 > 14$$

Subtract 5 from both sides:

$$\Rightarrow \quad 3x > 9$$

Divide both sides by 3:

$$\Rightarrow \quad \frac{3x}{3} > \frac{9}{3}$$

$$\Rightarrow \quad x > 3$$

So, our original relationship $3x + 5 > 14$ can be simplified to $x > 3$, or, in words, x is any number greater than 3.

The solution of an inequality can be represented on the number line. In this case, it would be:

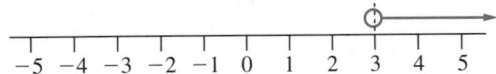

The open circle shows that the number 3 *is not part* of the solution.

Knowledge

Reminder

$<$ means less than

$\leqslant$ means less than or equal to

$>$ means greater than

$\geqslant$ means greater than or equal to

Skill Example

Solve the inequality $4q + 5 \leqslant 1$ and illustrate the solution on the number line.

$$4q + 5 \leqslant 1$$

Subtract 5 from both sides:

$$\Rightarrow \quad 4q \leqslant -4$$

Divide both sides by 4:

$$\Rightarrow \quad q \leqslant -1$$

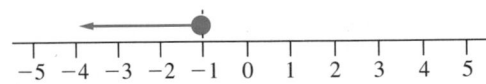

Here, the solid circle shows that the number -1 *is part* of the solution.

Skill Practice One

Solve the following inequalities and illustrate the solutions on number lines.

1	$x + 5 \leqslant 8$	**2**	$m + 2 > 4$
3	$a - 1 < 0$	**4**	$y - 2 \geqslant 1$
5	$2x \leqslant 8$	**6**	$3p > 15$
7	$2t + 1 < 5$	**8**	$3d - 5 \geqslant 7$
9	$x + 4 \leqslant 1$	**10**	$f + 4 > 1$
11	$2z < -4$	**12**	$3e \leqslant -9$
13	$10q > -20$	**14**	$w - 4 > -4$
15	$e - 7 \geqslant -7$	**16**	$r - 11 \geqslant -11$
17	$t + 3 < 3$	**18**	$2y + 3 > 1$
19	$u + 6 \leqslant 2$	**20**	$3p - 2 > -8$
21	$5a + 7 < 2$	**22**	$3s + 20 \geqslant 5$
23	$2d - 1 \leqslant -7$	**24**	$2f + 5 > -15$

Skill Example

You can multiply or divide both sides of an inequality by any number. But if that number is negative, you must *reverse the direction of the inequality sign*, as this example shows.

Solve the inequality $6 - 4m < 19$ and illustrate the solution on the number line.

$$6 - 4m < 18$$

Subtract 6 from both sides:

$$\Rightarrow \quad -4m < 12$$

Divide both sides by -4 and reverse the sign:

$$\Rightarrow \quad m > -3$$

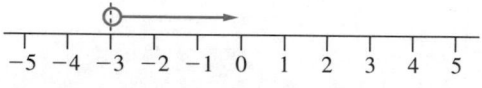

Skill Practice Two

Solve the following inequalities and illustrate the solutions on number lines.

1	$6 - x > 8$	**2**	$8 - x < 6$
3	$10 - x > 7$	**4**	$7 - x < 7$
5	$5 - x > 4$	**6**	$4 - x < 5$
7	$3 - x > 9$	**8**	$9 - x < 3$
9	$-2x > 4$	**10**	$-3x < 15$
11	$-2x > -4$	**12**	$-3x < -15$
13	$-6x < 12$	**14**	$-6x < -12$
15	$-8x < 24$	**16**	$-8x > -24$
17	$8 - 2x < 10$	**18**	$10 - 2x > 8$
19	$5 - 3x < 20$	**20**	$20 - 3x > 5$

Unit 25

Shape, space and measures   Level 4

Skill   You make 3-D mathematical models by linking given faces or edges.

Knowledge

A **net** is a shape which folds to make a hollow 3-D shape.

For example, this shape is a net for a cube.

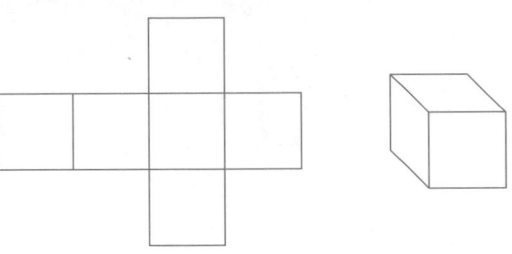

Skill Practice One

Copy each of these shapes on squared paper, then cut them out and fold them. Which of the shapes are nets for a cube?

1

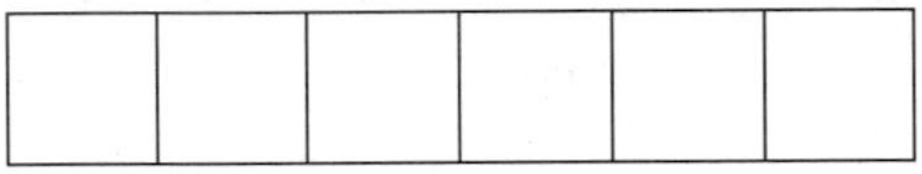

2

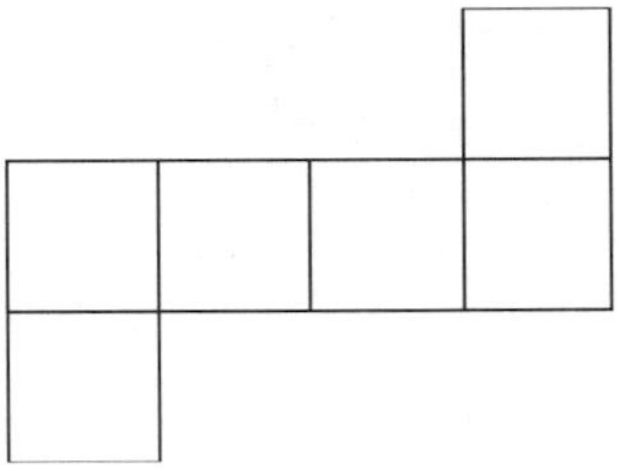

3

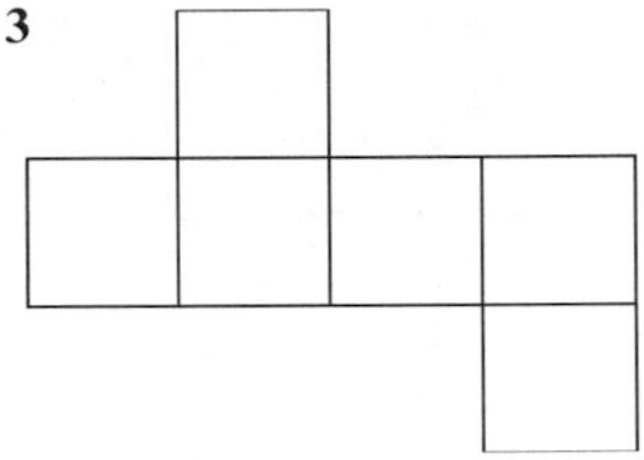

4

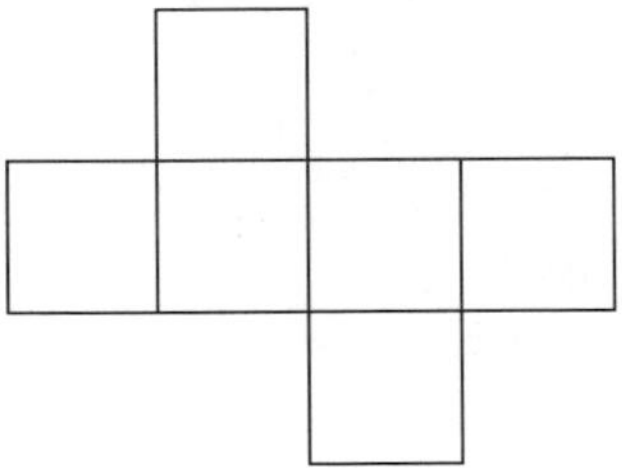

5

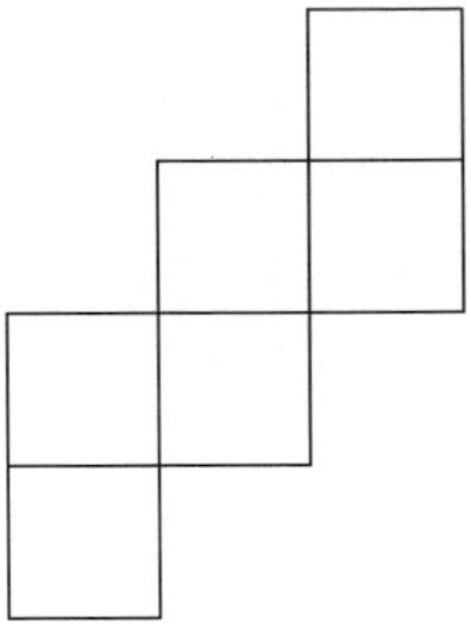

6

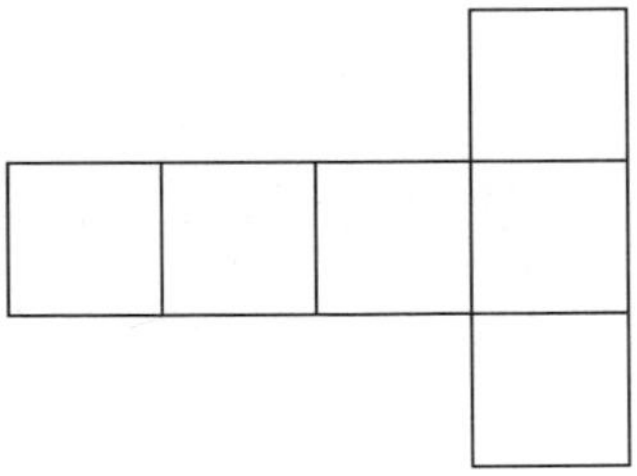

7

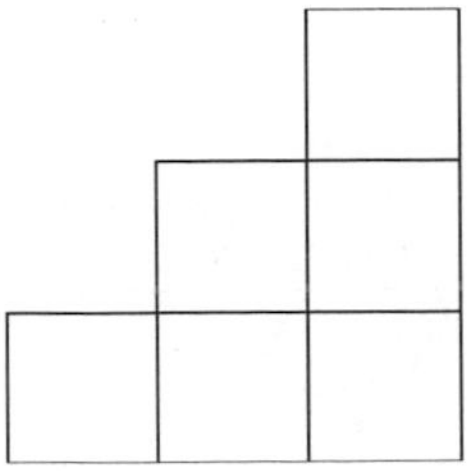

8

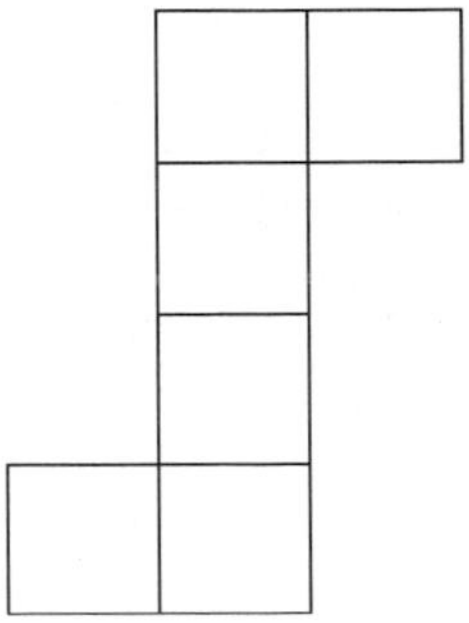

9

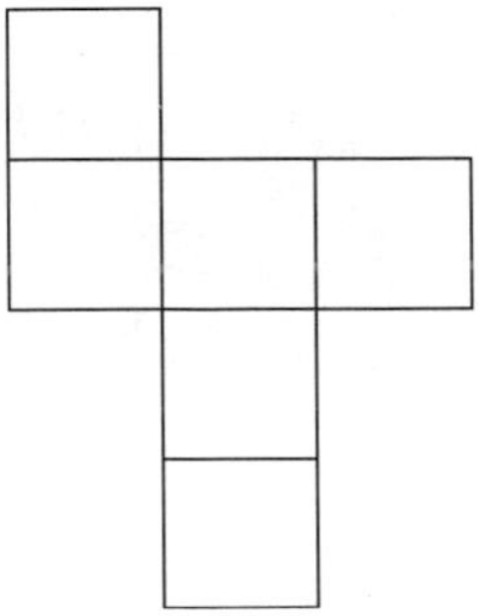

Skill Practice Two

This is the net (without glue flaps) for a box designed to hold 12 stock cubes.

1 How big are the stock cubes?

2 On squared paper, draw nets for two different shaped boxes which could each hold 12 stock cubes.

3 On squared paper, draw nets for two different shaped boxes which could each hold 16 stock cubes.

Skill Practice Three

This is the net for a regular tetrahedron (triangle-based pyramid).

It is most easily drawn on **isometric** or triangular graph paper.

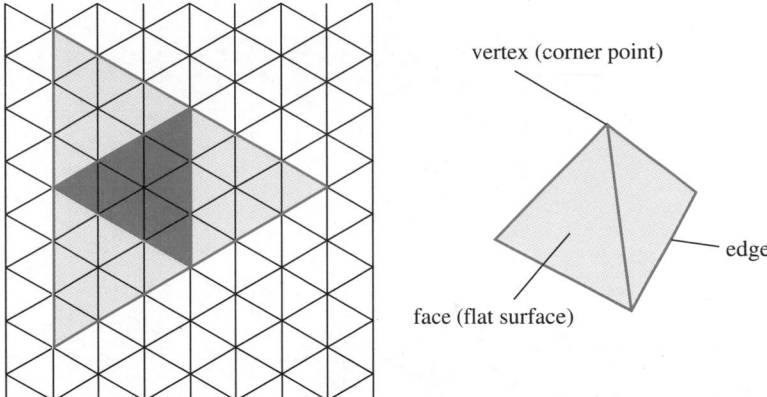

1 Use isometric graph paper to make a regular tetrahedron with edges 6 cm long. You will need to add glue flaps to the net so that you can assemble it.
2 How many faces does the tetrahedron have?
3 How many vertices does the tetrahedron have?
4 How many edges does the tetrahedron have?

Skill Practice Four

This is the net for a regular octahedron.

It is most easily drawn on isometric graph paper.

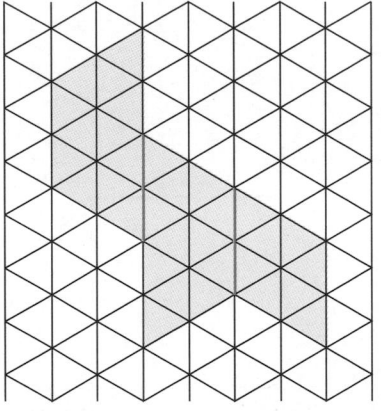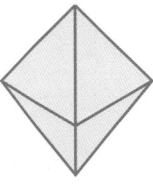

1 Use isometric graph paper to make a regular octahedron with edges 6 cm long. You will need to add glue flaps to the net so that you can assemble it.
2 How many faces does the octahedron have?
3 How many vertices does the octahedron have?
4 How many edges does the octahedron have?

Skill Practice Five

This is the net for a triangular prism.

It is most easily drawn on isometric graph paper.

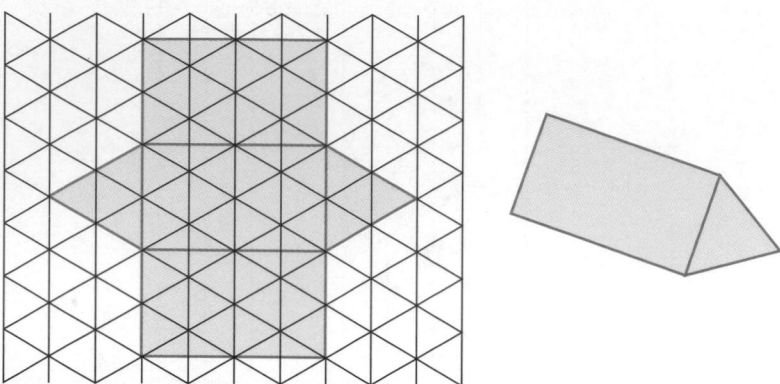

1 Use isometric graph paper to make a triangular prism with rectangular faces 4 cm wide and 10 cm long. You will need to add glue flaps to the net so that you can assemble it.
2 How many faces does the triangular prism have?
3 How many vertices does the triangular prism have?
4 How many edges does the triangular prism have?

Skill Practice Six

The diagrams below show you how to make a puzzle cube by fitting together three pyramids. The diagrams have been drawn at half size. Copy them at full size on squared paper.

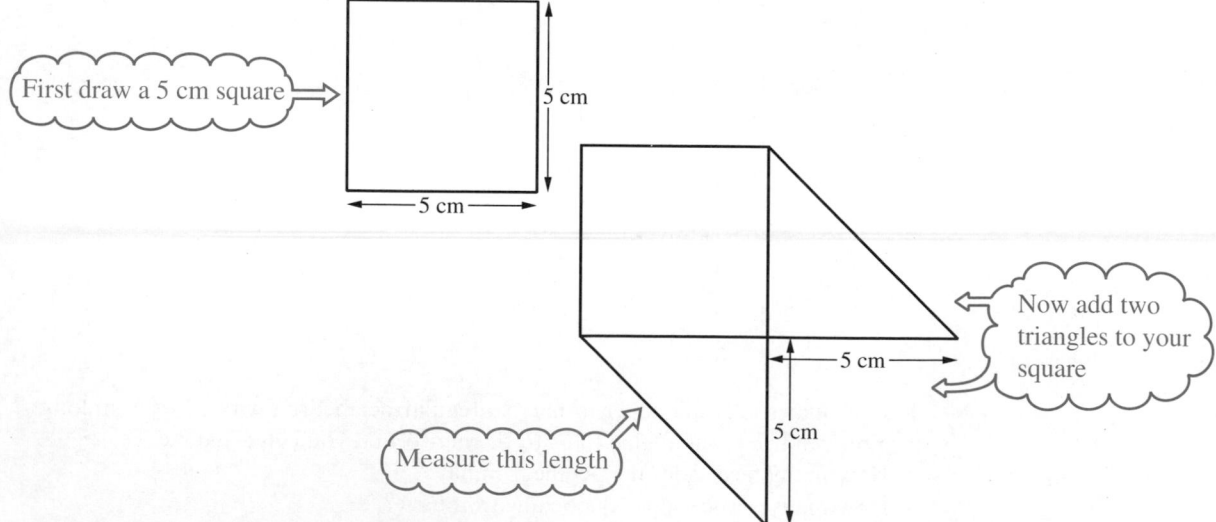

Your net will assemble into a square-based pyramid which leans to one side. Make three of these pyramids, or work with two friends and make one pyramid each.

The three pyramids fit together to make the puzzle cube.

Experiment with your pyramids until you can fit them together easily.

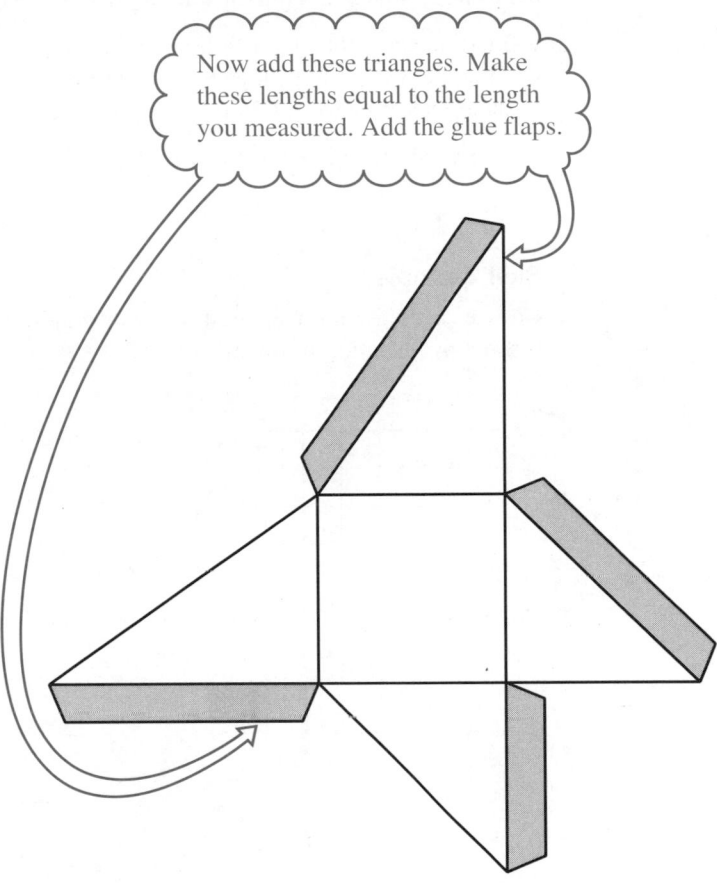

Now add these triangles. Make these lengths equal to the length you measured. Add the glue flaps.

Unit 26

Knowledge

To **reflect** a shape in a mirror line, we take these two steps:

- First, imagine the mirror line to be line of symmetry.
- Then draw a second shape (called the **image**) which makes the completed diagram symmetrical about the mirror line.

The original shape is known as the **object**.

Skill Example

Copy each diagram on squared paper or graph paper. Then draw the image formed by reflection in the mirror line.

a

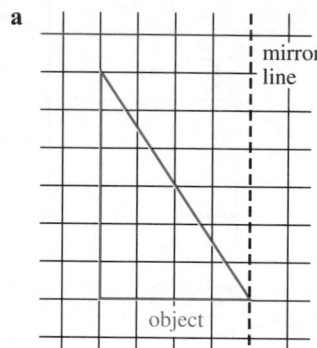
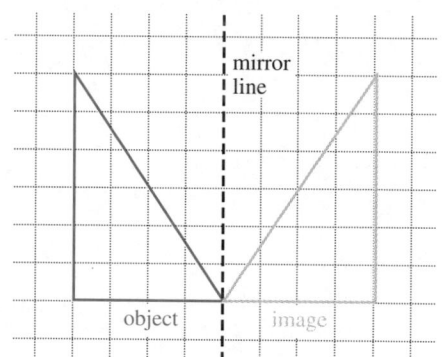

b

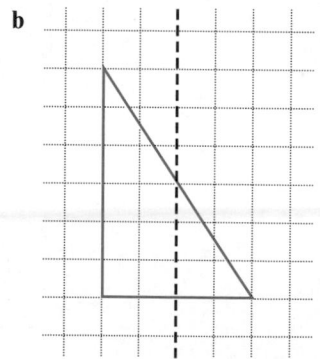
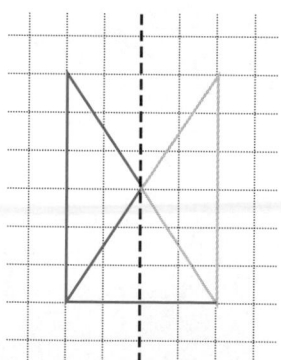

c

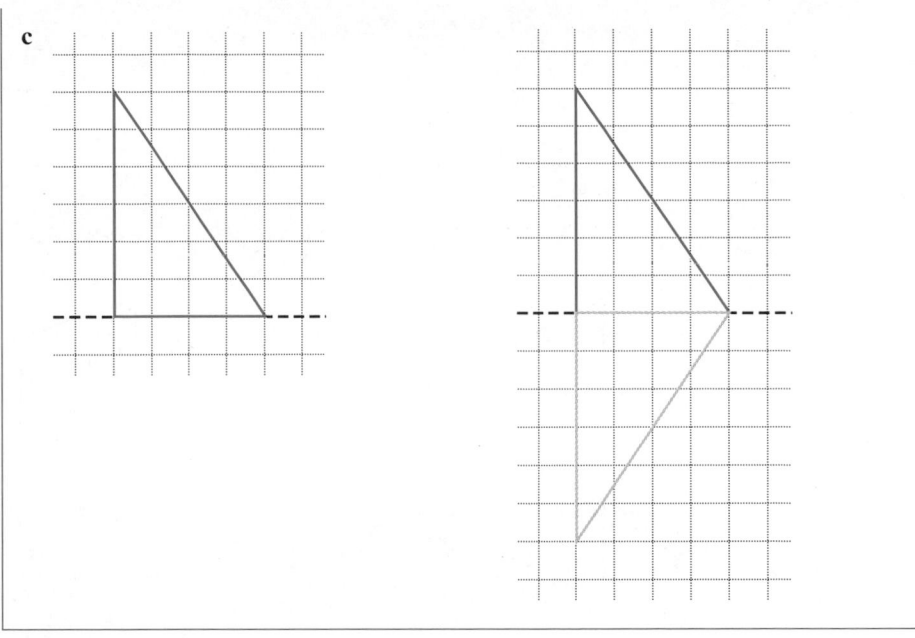

Skill Practice One

Copy each of the following on squared paper or graph paper. Then draw the image formed by reflection in the mirror line.

1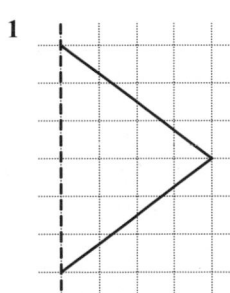

2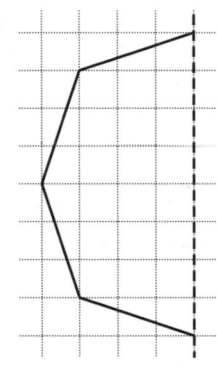

3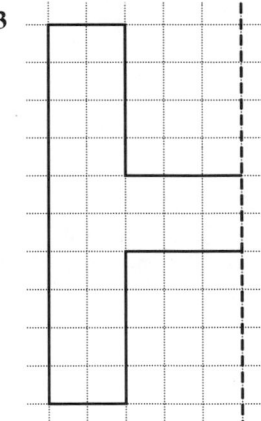

4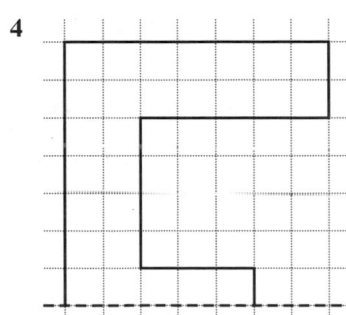

5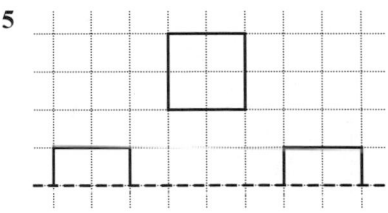

6

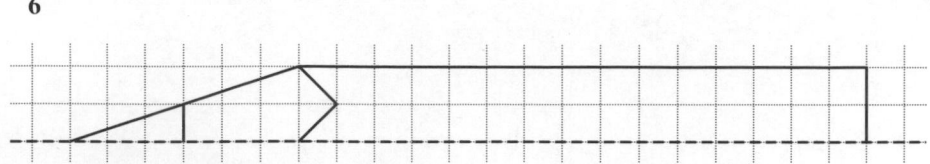

7

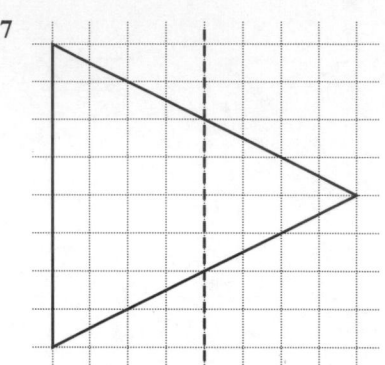

8

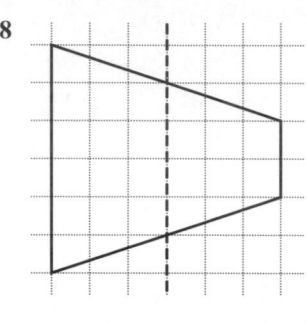

9

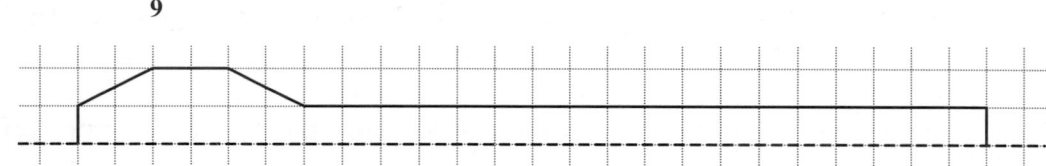

Skill Example

Copy each of the following on squared paper or graph paper and then draw the image which is formed by reflection in the mirror line.

a

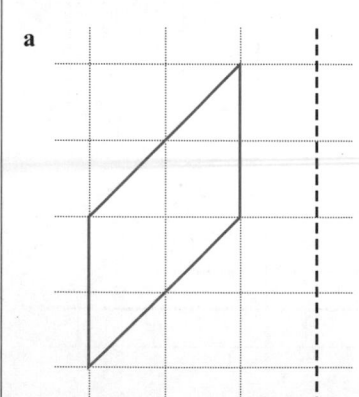

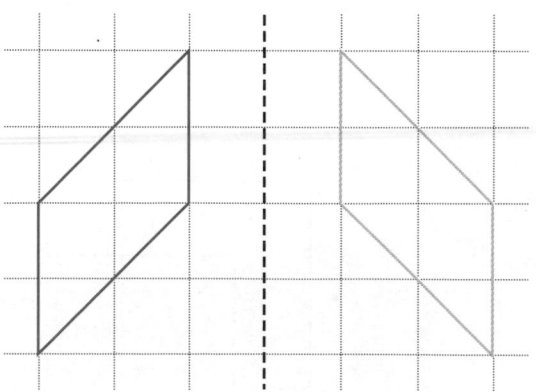

b

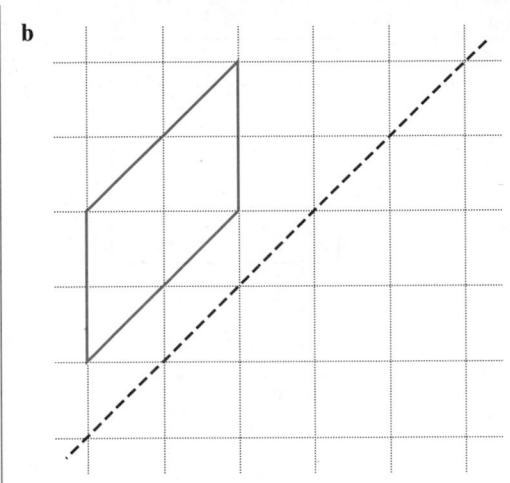 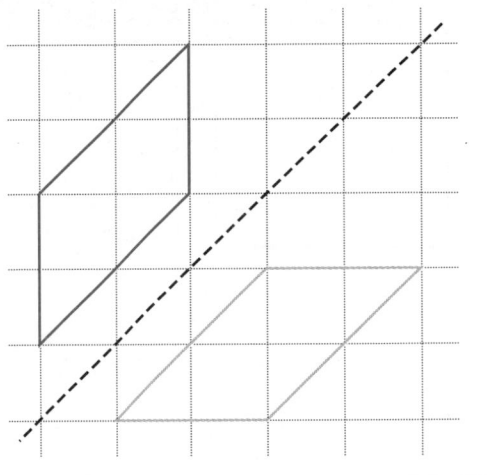

Skill Practice Two

Copy each of the following on squared paper or graph paper and then draw the image which is formed by reflection in the mirror line.

1

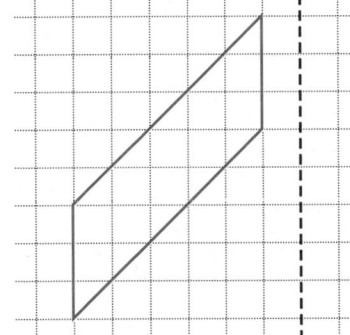

2

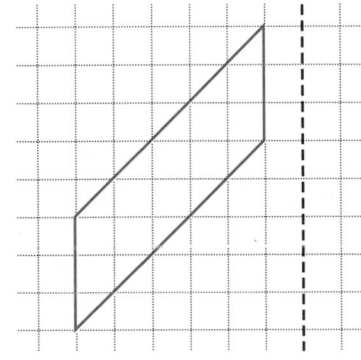

3

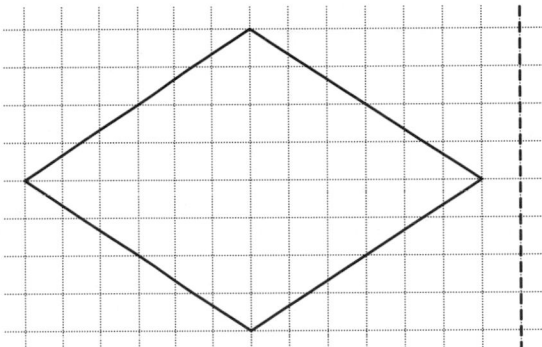

4

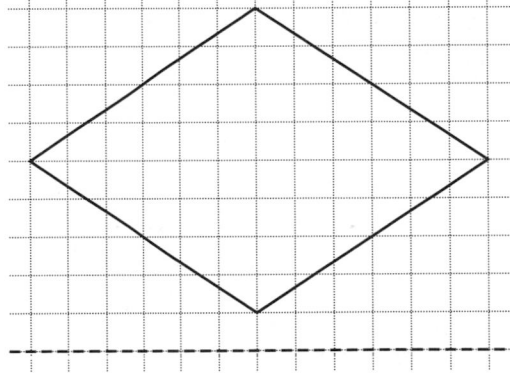

112

5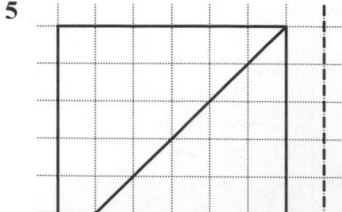

6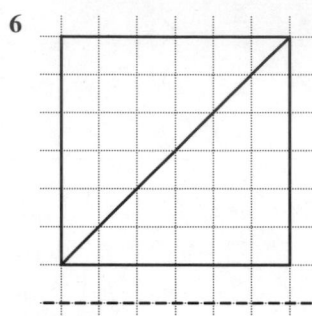

7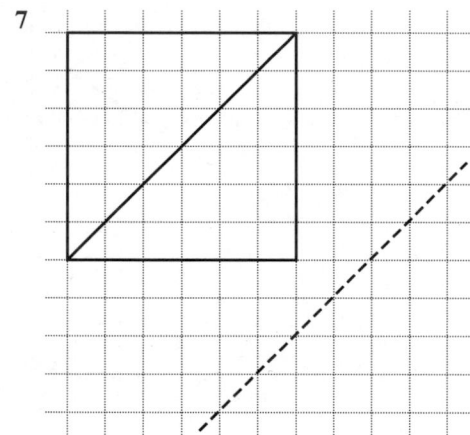

8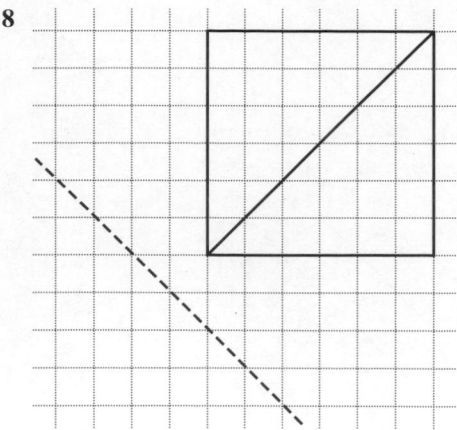

9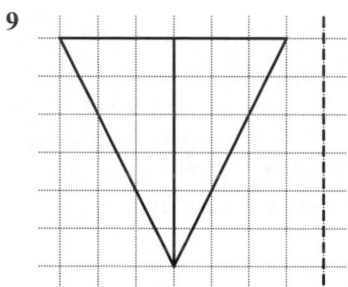

10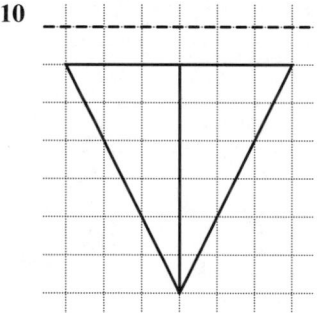

Unit 27

Shape, space and measures   Level 4

Skills   You draw common 2-D shapes in different orientations on grids.

You identify orders of rotational symmetry.

You identify congruent shapes.

Knowledge

The change in position of an object is known as a **transformation**.

We have just looked at reflection, in which an object is transformed to its mirror image.

Rotation is another kind of transformation.

The object is rotated about a fixed point (the centre of rotation) to form the image.

Clockwise rotations are *negative*.
Anticlockwise rotations are *positive*.

Skill Example

Copy the shape.
Then draw the image formed by rotating the object:

a   +90° about the dot
b   −90° about the dot

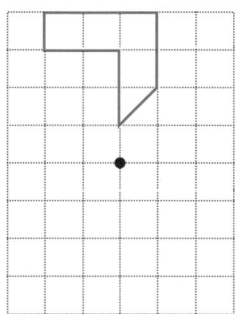

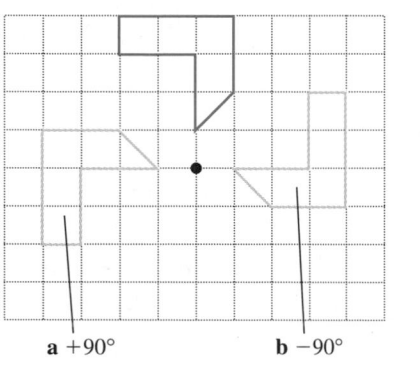

a +90°   **b** −90°

Skill Practice One

Copy each diagram on squared paper.
Then draw the image formed by rotating the object through the given angle about the dot.

1 +90°

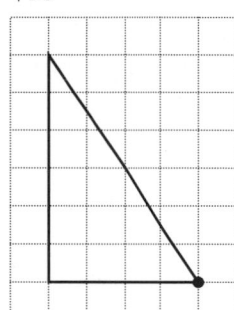

2 −90°

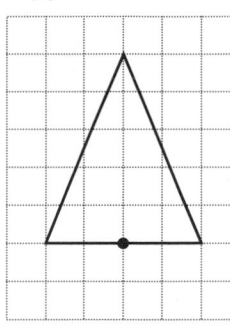

3 180°

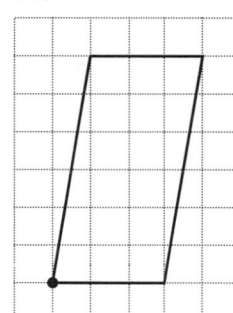

4 +270°

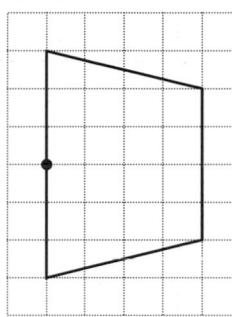

5 −270°

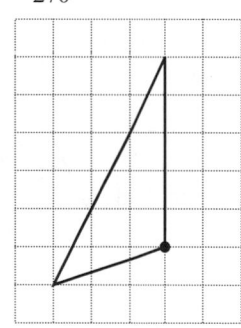

6 +90°

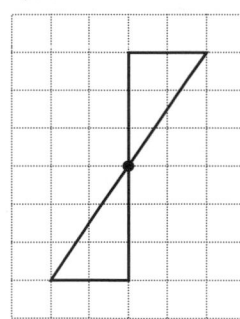

7 −90°

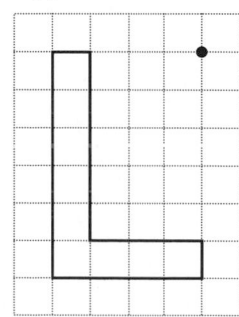

8 180°

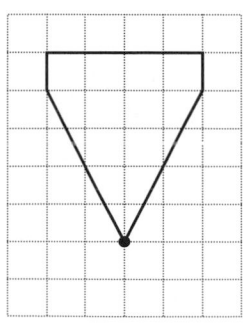

114

9 +270°

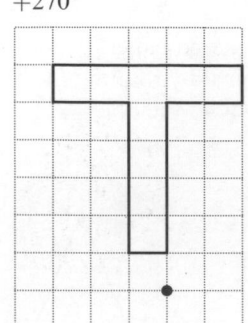

10 −270°

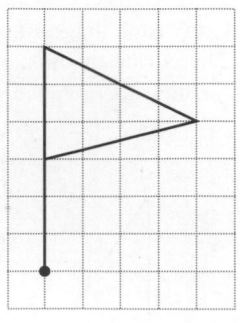

17 −270°

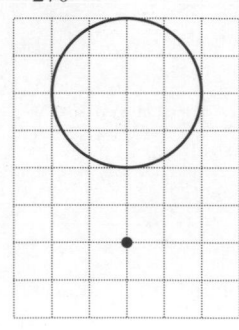

18 +270°

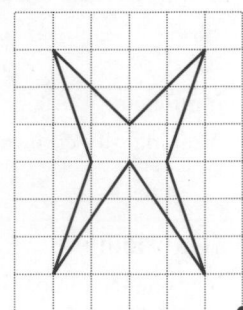

11 180°

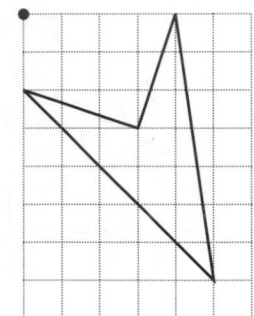

12 180°

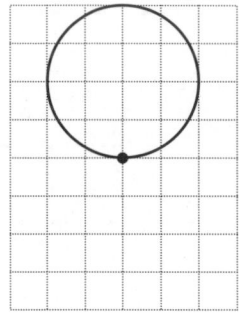

19 +90°

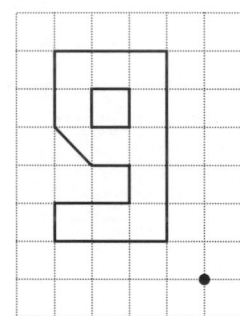

20 −90°

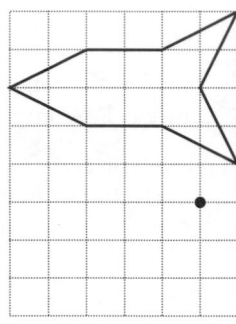

13 +90°

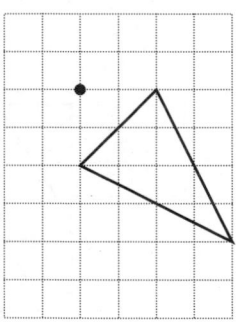

14 −90°

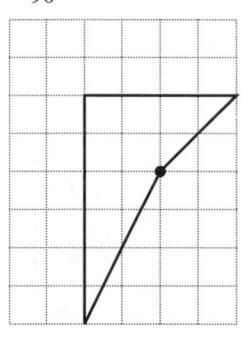

Knowledge

Some shapes fit into the same position more than once when rotated through 360°, as in the case of the shape shown on the right.

When this happens, the shape is said to have **rotational symmetry**.

The **order of rotational symmetry** is the number of times the shape fits into the same position.

In this case, the shape has an order of rotational symmetry of 3.

Shapes with no rotational symmetry fit only once and have an order of rotational symmetry of 1.

15 +270°

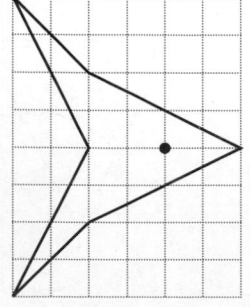

16 180°

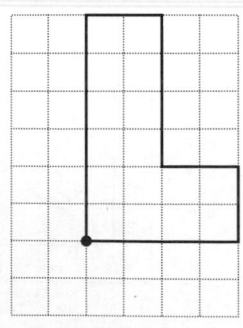

Skill Example

Which shape is different from the others, i.e. has no rotational symmetry?
State the order of rotational symmetry for each shape.

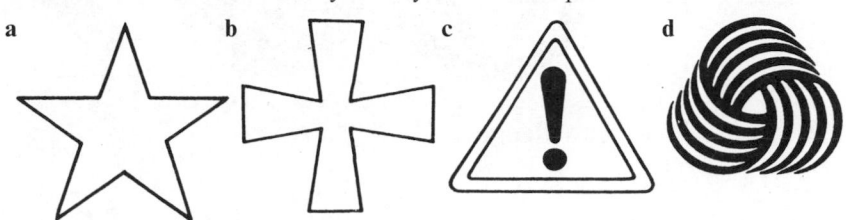

The shape that is different is **c**, because it has no rotational symmetry.
The orders of rotational symmetry are

| **a** 5 | **b** 4 | **c** 1 | **d** 3 |

Skill Practice Two

For each of the following questions, find which shape is different from the others,
i.e. has no rotational symmetry.
State the order of rotational symmetry for each shape.

6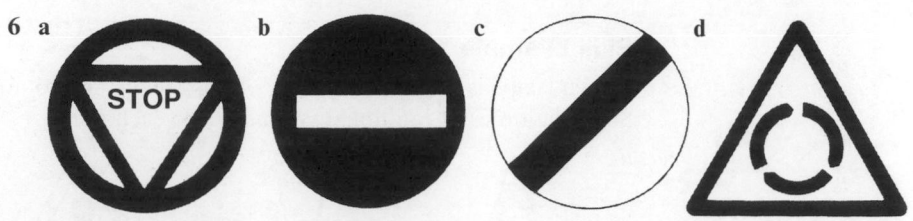

Knowledge

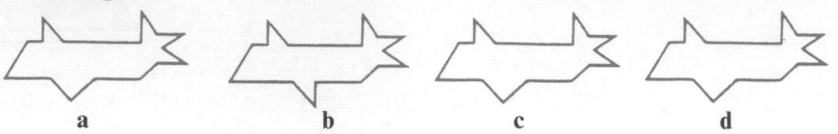

If you look at the above drawings carefully, you should see that three of them, **a**, **c** and **d**, are exactly the same.

Shapes that are identical in every possible way – sides, angles and area – are said to be **congruent**.

Skill Example

Pick out the congruent shapes from the following.

a, **b** and **d** are the congruent shapes.

Skill Practice Three

Pick out the congruent shapes from the following.

1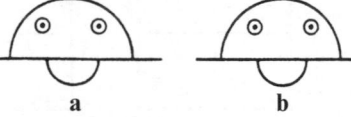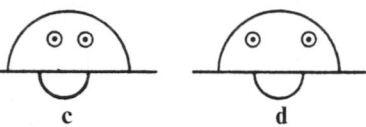

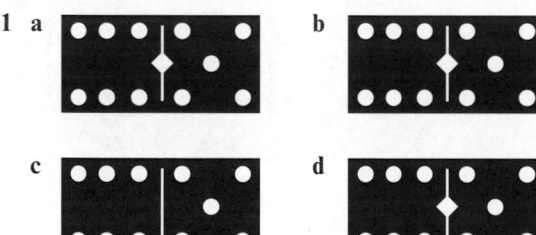

2

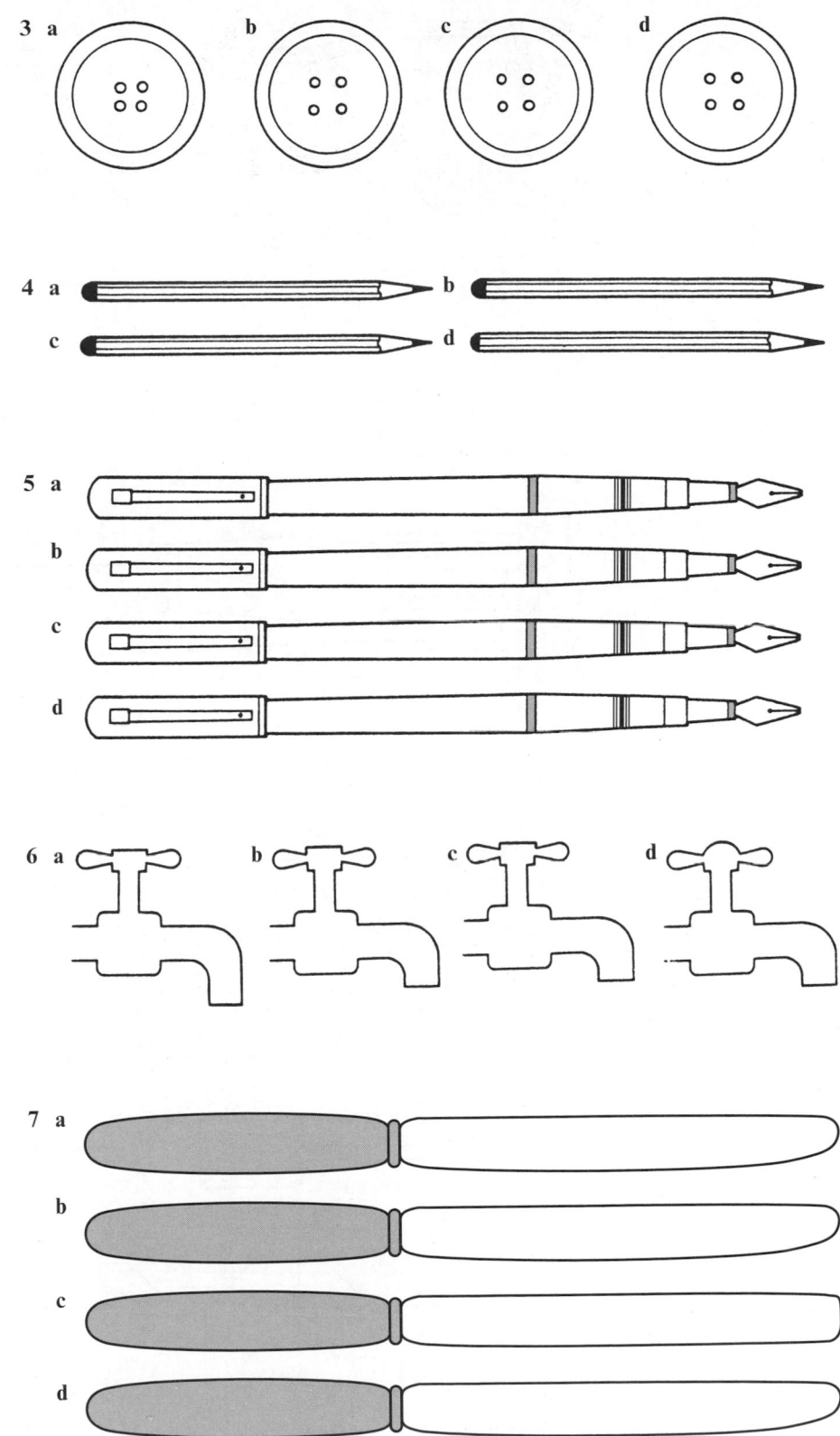

3 a   b   c   d

4 a   b   c   d

5 a   b   c   d

6 a   b   c   d

7 a   b   c   d

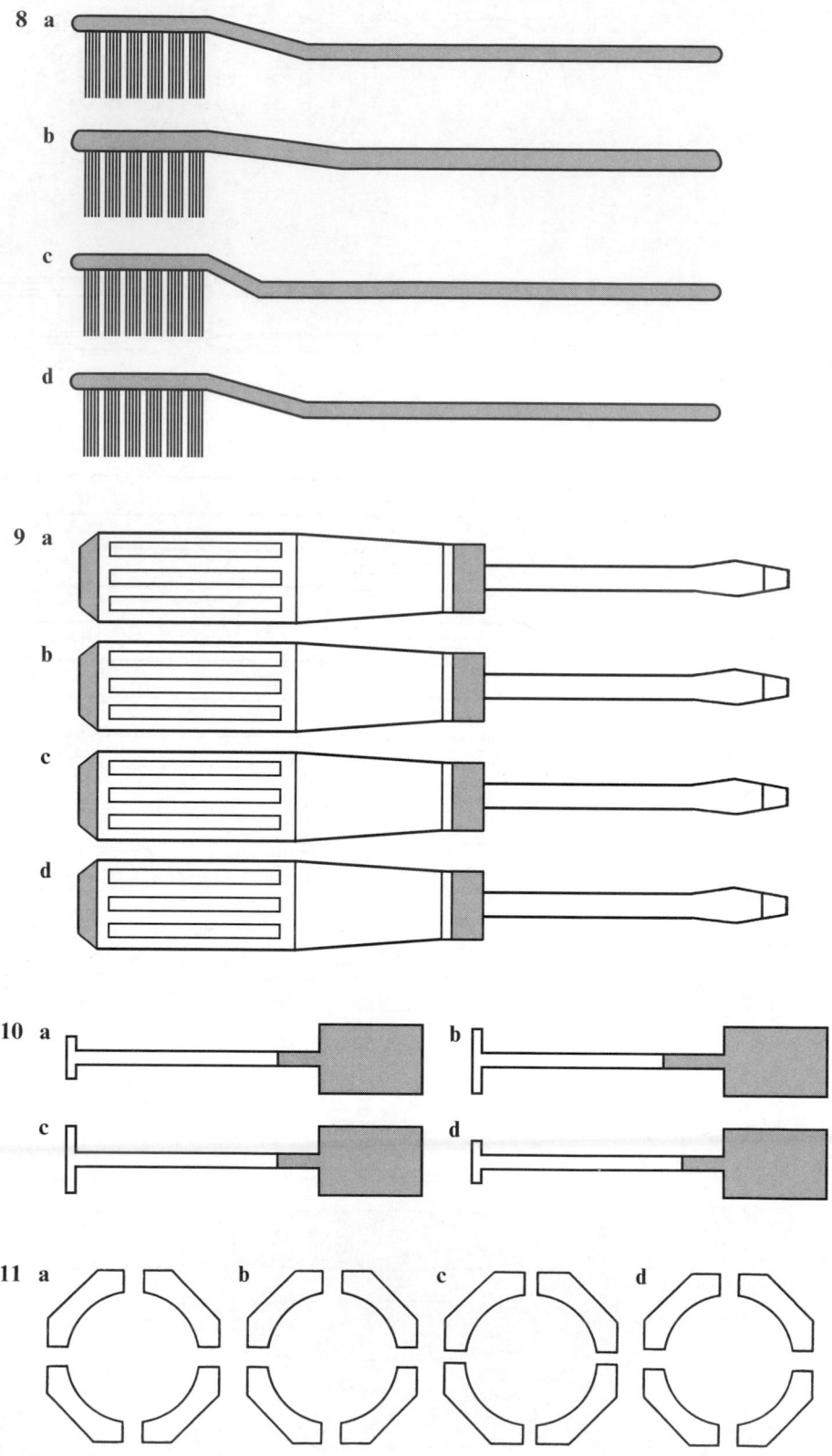

12

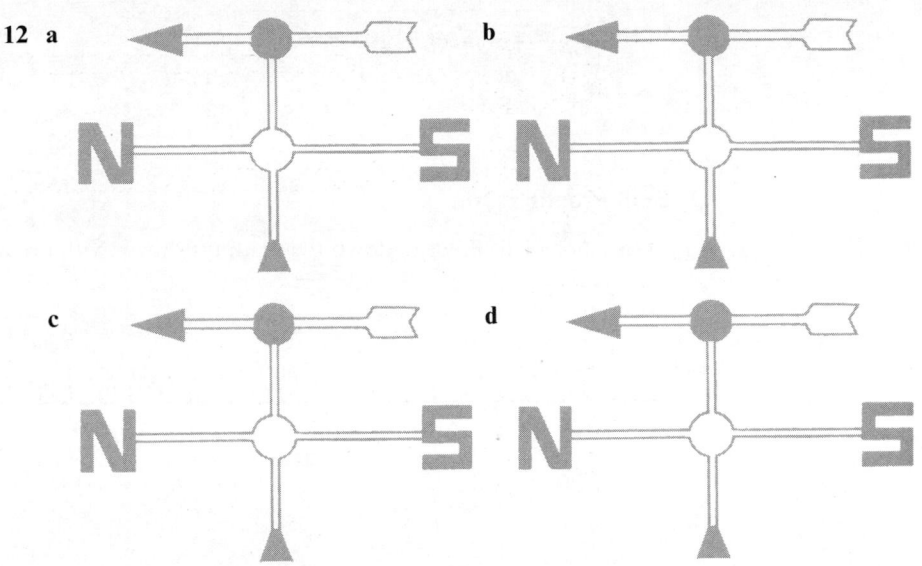

a   b

c   d

Unit 28

Skill Practice One

In questions **1** to **8**, write down the readings shown with arrows on each scale.

1

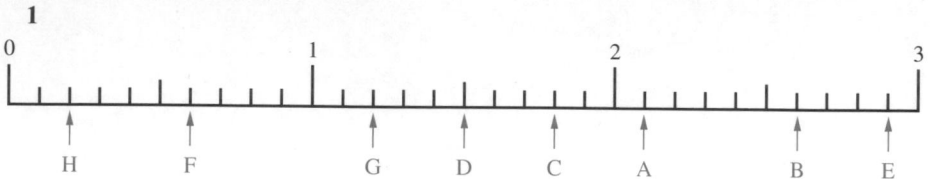

2

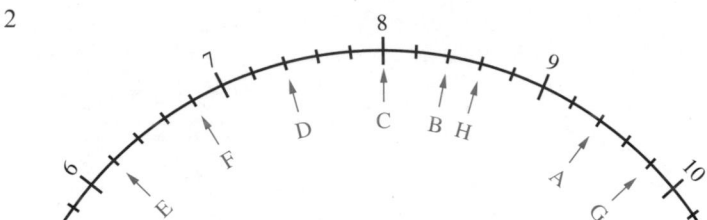

3

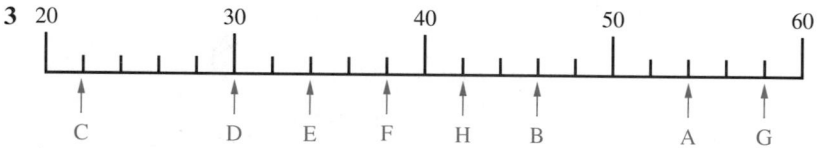

4

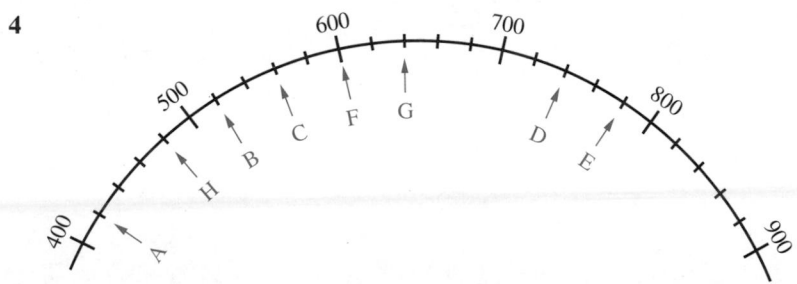

5

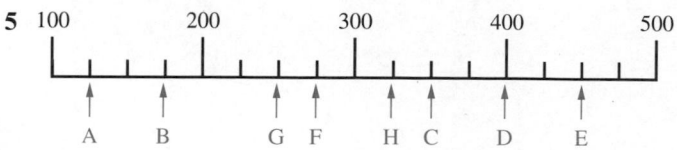

6

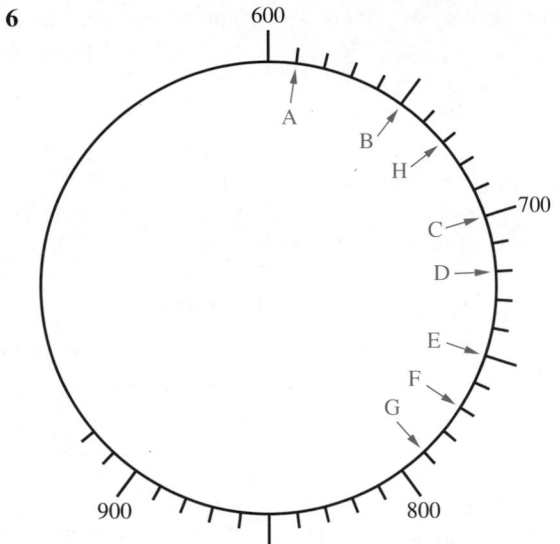

7

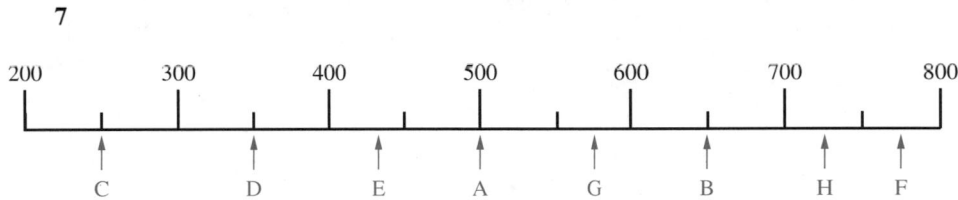

8

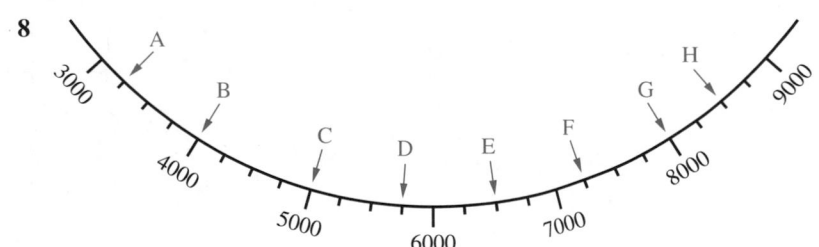

Unit 29

Skill   You find perimeters of simple shapes, find areas by counting squares, and find volumes by counting cubes.

Knowledge

The **perimeter** of a shape is the total distance round the outside of the shape.

The **area** of a shape is the size of the surface covered by the shape.

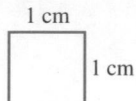

This square has a perimeter of 4 cm and an area of 1 square centimetre (written 1 cm^2).

Each side of the square is 1 cm long.

The diagonal of the square is approximately 1.4 cm long.

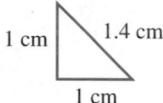

This half square has a perimeter of 3.4 cm (1 cm + 1 cm + 1.4 cm) and an area of $\frac{1}{2}$ cm^2.

Skill Example

Find the perimeter and area of each of the following shapes.
Each square has an area of 1 cm^2.

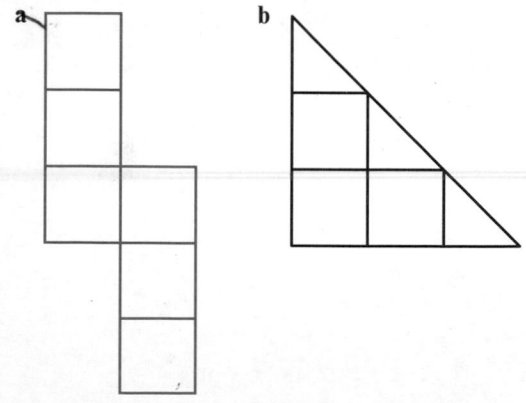

a   This figure has a perimeter of 14 cm and an area of 6 cm^2.

b   This triangle has a perimeter of 6 whole sides and 3 diagonals. So, its perimeter is:

$$6 + 1.4 + 1.4 + 1.4 = 10.2 \text{ cm}$$

It has an area of 3 whole squares and 3 half squares. So, its area is:

$$3 + 1\tfrac{1}{2} = 4\tfrac{1}{2} \text{ cm}^2$$

Skill Practice One

Find the perimeter and area of each of the following shapes.
Each square has an area of 1 cm^2.

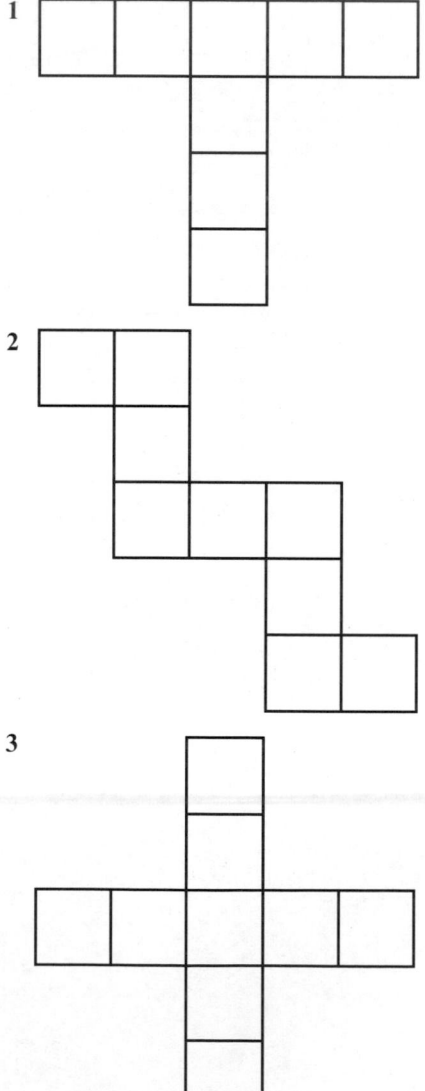

4

5

6

7

8

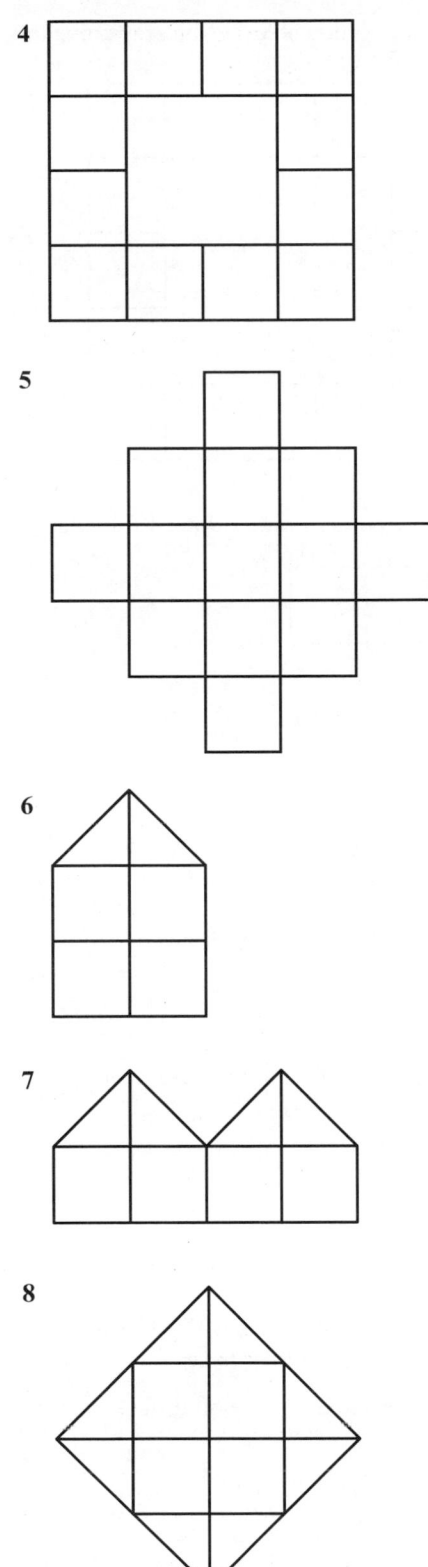

9

10

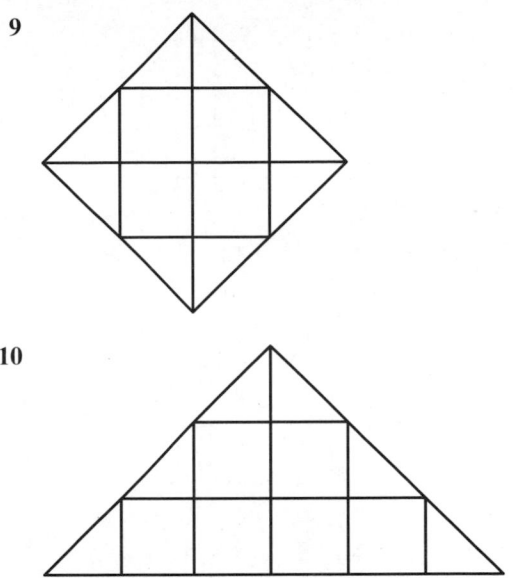

Skill Practice Two

Each of the following words has been drawn on a centimetre grid. (The diagrams have been reduced.)

Find the perimeter and area of each letter.

1

2

3

124

4

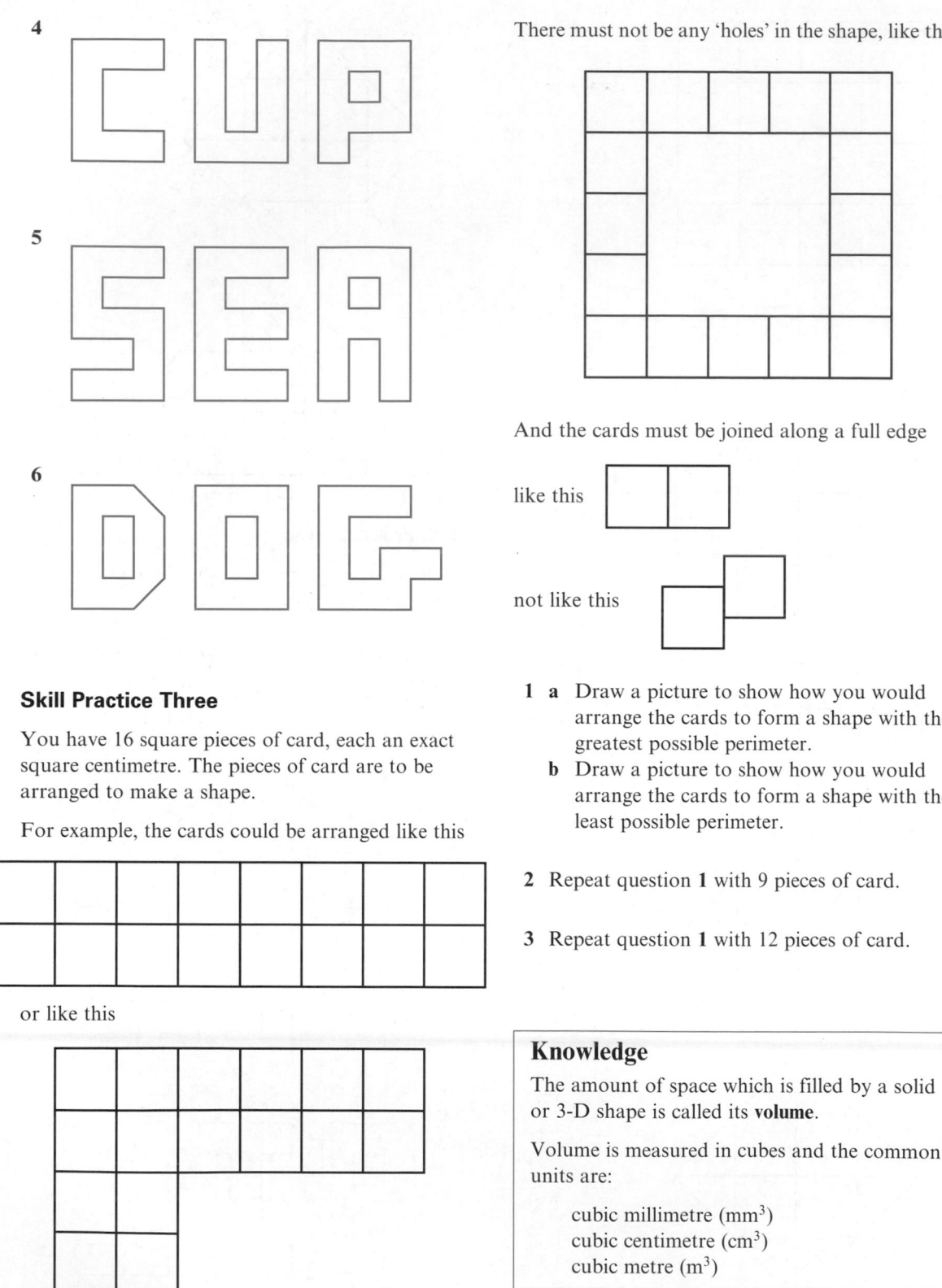

5

6

Skill Practice Three

You have 16 square pieces of card, each an exact square centimetre. The pieces of card are to be arranged to make a shape.

For example, the cards could be arranged like this

or like this

There must not be any 'holes' in the shape, like this

And the cards must be joined along a full edge

like this

not like this

1 a Draw a picture to show how you would arrange the cards to form a shape with the greatest possible perimeter.
  b Draw a picture to show how you would arrange the cards to form a shape with the least possible perimeter.

2 Repeat question **1** with 9 pieces of card.

3 Repeat question **1** with 12 pieces of card.

Knowledge

The amount of space which is filled by a solid or 3-D shape is called its **volume**.

Volume is measured in cubes and the common units are:

    cubic millimetre (mm³)
    cubic centimetre (cm³)
    cubic metre (m³)

Skill Example

Each cube has a volume of $1\,cm^3$. Find the volume of each of the following shapes.

a

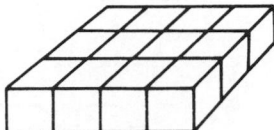

b

c

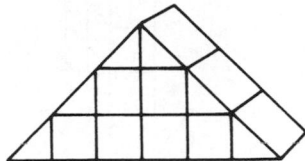

a The shape is made up of 12 cubes, so its volume is $12\,cm^3$.

b The shape is made up of 15 cubes, so its volume is $15\,cm^3$.

c The shape is made up of 6 cubes and 6 half cubes, so its volume is:

$$6 + (\tfrac{1}{2} \times 6) = 6 + 3 = 9\,cm^3$$

Skill Practice Four

For each of the following questions, find which shape has a different volume from the other two.

1 a

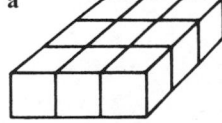

b

c

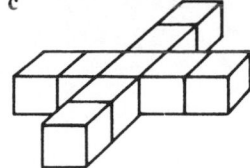

2 a

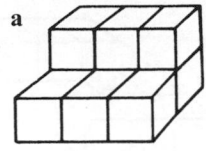

b

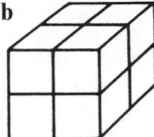

c

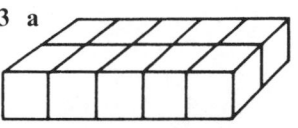

3 a

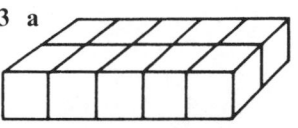

b

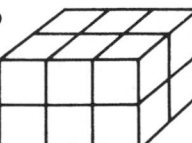

c

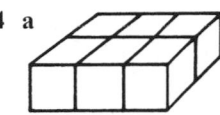

4 a

b

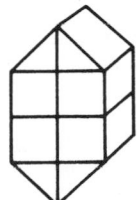

c

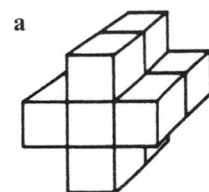

5 a

b

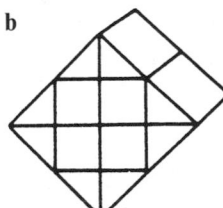

c

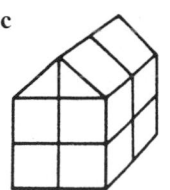

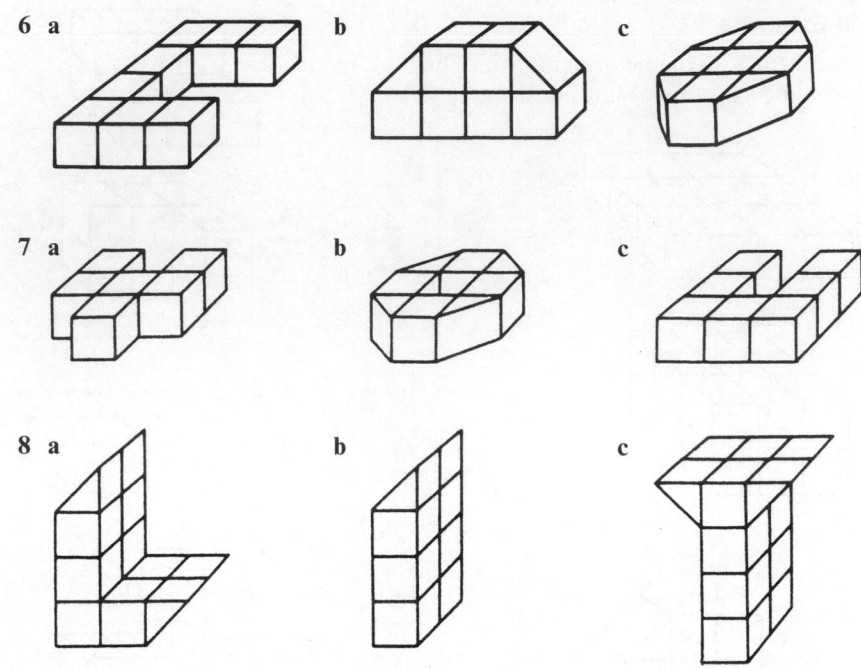

6 a   b   c

7 a   b   c

8 a   b   c

Unit 30

Shape, space and measures Level 5

Skill   When constructing models and when drawing or using shapes, you measure and draw angles to the nearest degree, and use language associated with angle.

Skill Example

a Measure the size of PQ̂R.

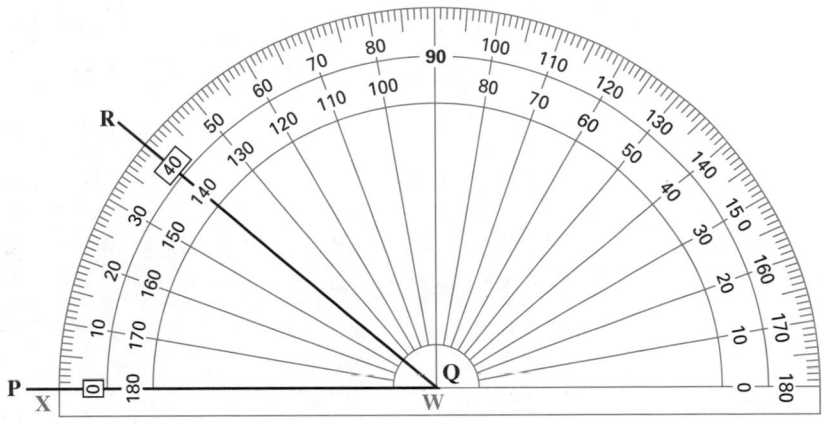

1 Place the protractor on the angle so that W (the centre) is on Q and WX (the base line) lies on PQ, as shown.
2 Where the line QR cuts the scale, read off the angle on the scale, starting from 0° at P.
3 PQ̂R = 40°.

b Measure the size of LM̂N.

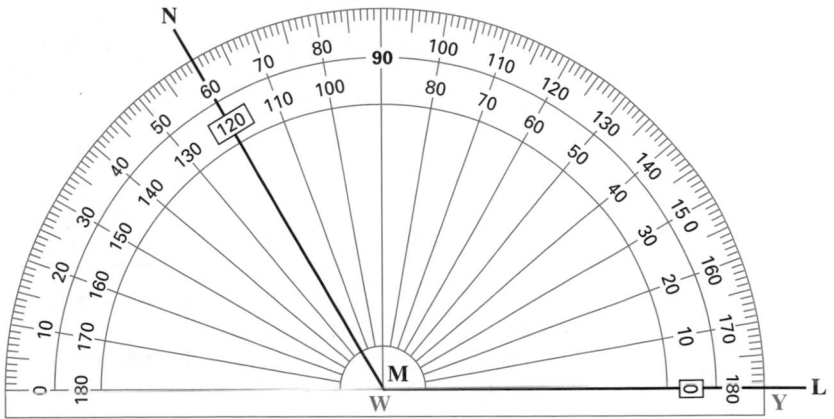

Place W on M and WY on ML.

From the inner scale, LM̂N = 120°.

Skill Practice One

For each question, look at the diagram, then copy and complete the details.

1

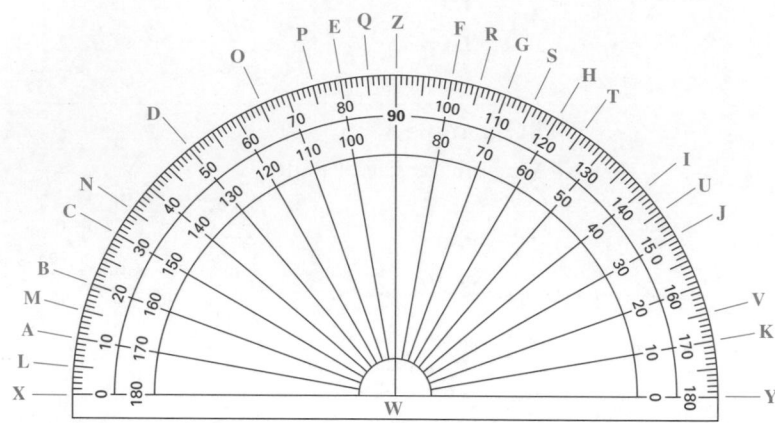

a $X\widehat{W}A = 10°$ $X\widehat{W}B = 20°$ $X\widehat{W}C =$ $X\widehat{W}D =$ $X\widehat{W}E =$
b $X\widehat{W}Z =$ $X\widehat{W}F = 100°$ $X\widehat{W}G = 110°$ $X\widehat{W}H =$ $X\widehat{W}I =$
c $X\widehat{W}J =$ $X\widehat{W}K =$ $X\widehat{W}Y =$ $X\widehat{W}L = 5°$ $X\widehat{W}M = 15°$
d $X\widehat{W}N =$ $X\widehat{W}O =$ $X\widehat{W}P =$ $X\widehat{W}Q =$ $X\widehat{W}R = 105°$
e $X\widehat{W}S = 115°$ $X\widehat{W}T =$ $X\widehat{W}U =$ $X\widehat{W}V =$ $Y\widehat{W}K = 10°$
f $Y\widehat{W}J = 30°$ $Y\widehat{W}I =$ $Y\widehat{W}H =$ $Y\widehat{W}G =$ $Y\widehat{W}F =$
g $Y\widehat{W}Z =$ $Y\widehat{W}E = 100°$ $Y\widehat{W}D =$ $Y\widehat{W}C =$ $Y\widehat{W}B =$
h $Y\widehat{W}A =$ $Y\widehat{W}X =$ $Y\widehat{W}V = 15°$ $Y\widehat{W}U = 35°$ $Y\widehat{W}T =$
i $Y\widehat{W}S =$ $Y\widehat{W}R =$ $Y\widehat{W}Q = 95°$ $Y\widehat{W}P = 105°$ $Y\widehat{W}O = 115°$
j $Y\widehat{W}N =$ $Y\widehat{W}M =$ $Y\widehat{W}L =$

2

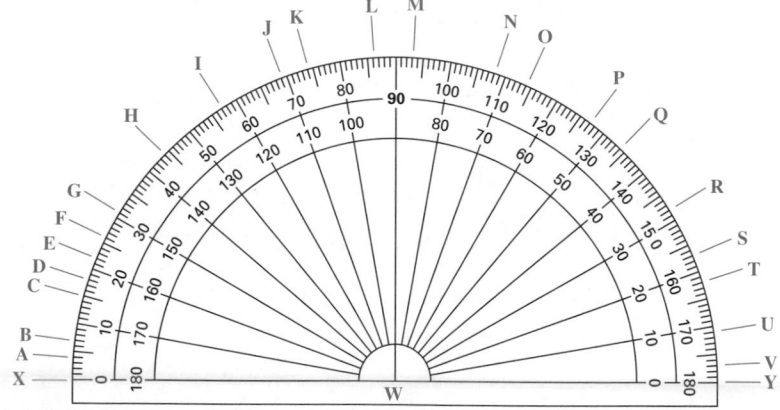

a $X\widehat{W}A = 4°$ $X\widehat{W}B = 7°$ $X\widehat{W}C = 13°$ $X\widehat{W}D =$ $X\widehat{W}E = 22°$
b $X\widehat{W}F =$ $X\widehat{W}G =$ $X\widehat{W}H =$ $X\widehat{W}I =$ $X\widehat{W}J =$
c $X\widehat{W}K =$ $X\widehat{W}L =$ $X\widehat{W}M = 93°$ $X\widehat{W}N = 108°$ $X\widehat{W}O =$
d $X\widehat{W}P =$ $X\widehat{W}Q =$ $X\widehat{W}R =$ $X\widehat{W}S =$ $X\widehat{W}T =$
e $X\widehat{W}U =$ $X\widehat{W}V =$ $Y\widehat{W}V = 3°$ $Y\widehat{W}U =$ $Y\widehat{W}T = 18°$
f $Y\widehat{W}S = 23°$ $Y\widehat{W}R =$ $Y\widehat{W}Q =$ $Y\widehat{W}P =$ $Y\widehat{W}O =$
g $Y\widehat{W}N =$ $Y\widehat{W}M =$ $Y\widehat{W}L = 94°$ $Y\widehat{W}K = 106°$ $Y\widehat{W}J =$
h $Y\widehat{W}I =$ $Y\widehat{W}H =$ $Y\widehat{W}G =$ $Y\widehat{W}F =$ $Y\widehat{W}E =$
i $Y\widehat{W}D =$ $Y\widehat{W}C =$ $Y\widehat{W}B =$ $Y\widehat{W}A =$

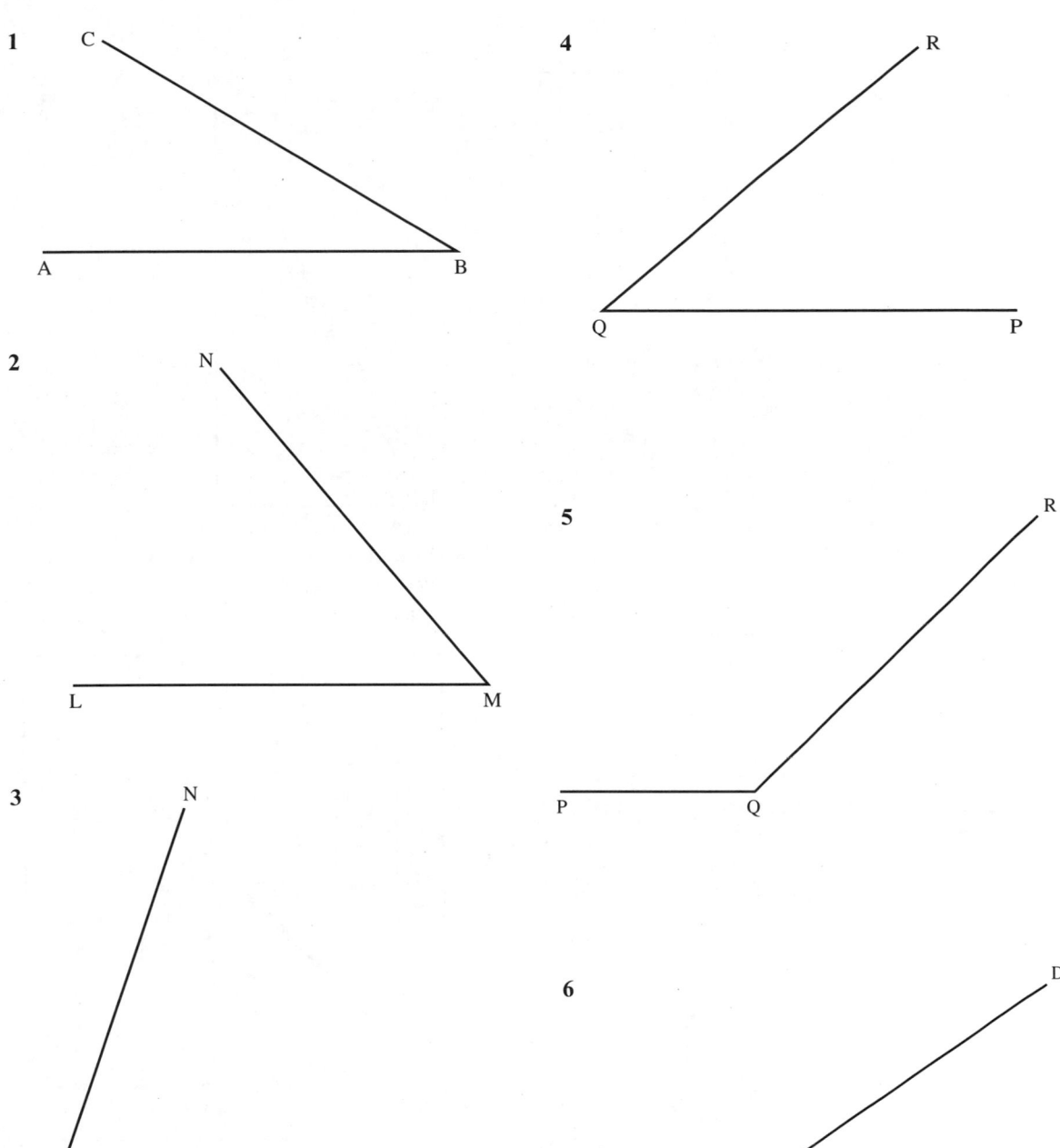

Skill Practice Two

With a protractor, measure each of the following angles.

1   C ... B   A

2   N   L   M

3   N   M   L

4   R   Q   P

5   R   P   Q

6   D   B   C

7

8

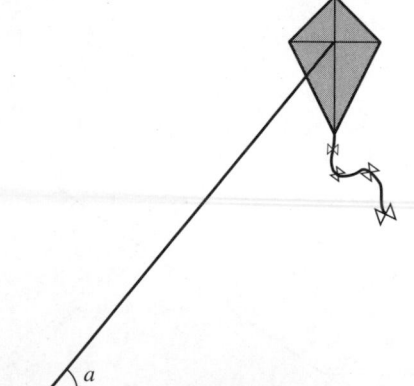

9 Measure the angle *a* that the kite string makes with the ground.

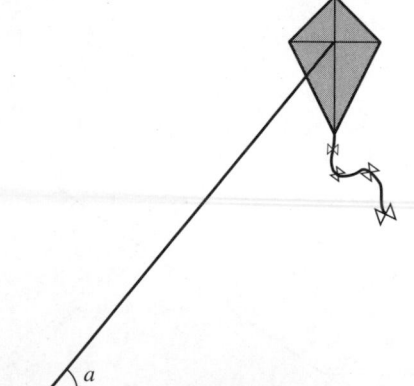

10 Measure the angle *a* that the ladder makes with the ground, and the angle *b* that the ladder makes with the wall.

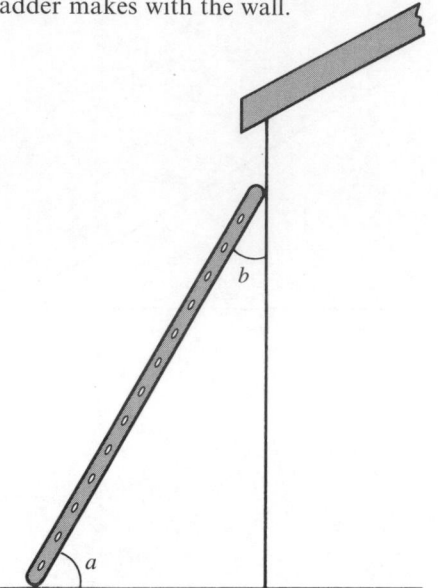

11 Measure the angle *a* that the diagonal strut makes with the horizontal and the angle *b* that the diagonal strut makes with the vertical.

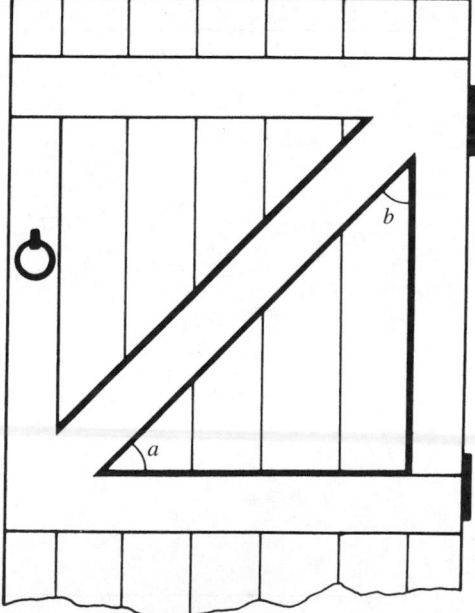

12 Measure the angle *a* between the two parts of the folding ruler.

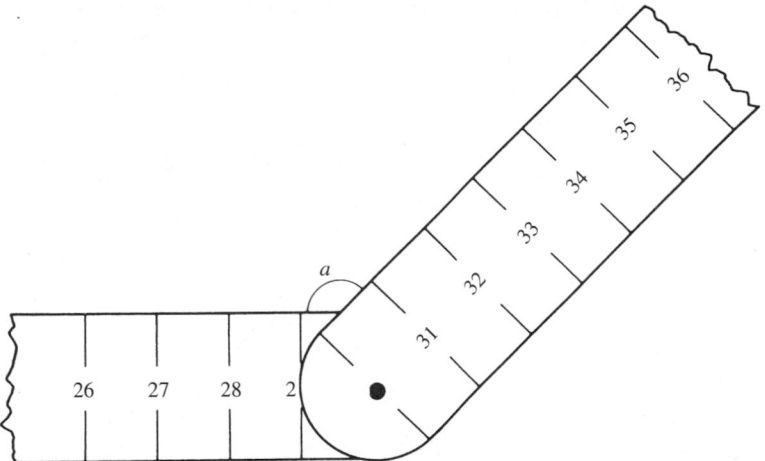

13 Measure the two angles *a* and *b* on the blade of this modelling knife.

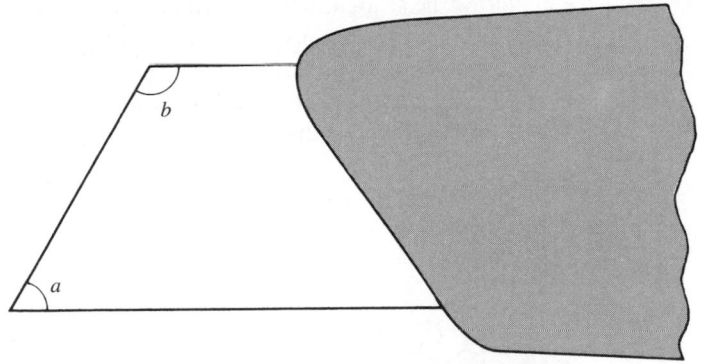

14 Measure the angle *a* between the wall and the slope of the roof, and the angle *b* between the two sloping parts of the roof.

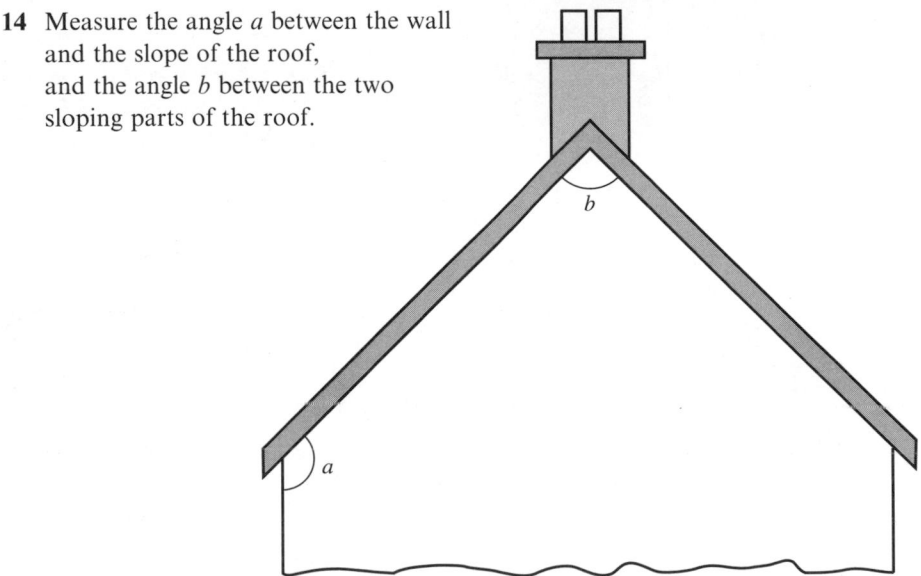

Skill Example

Using a protractor, draw:

a $\widehat{ABC} = 50°$ **b** $\widehat{XYZ} = 170°$

a

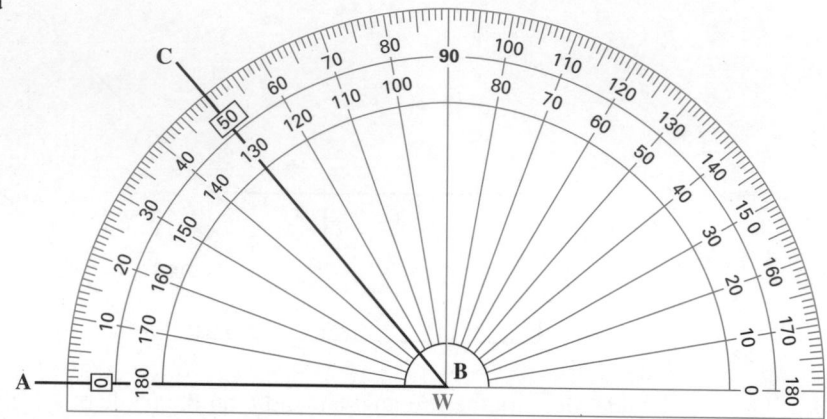

1 Draw the straight line AB. Make it 6 cm long.
2 Place the protractor on the paper with its base line on the line AB and its centre point W on B, as shown.
3 Mark the point C at 50° on the scale, starting from 0° at A.
4 Join BC, and the angle 50° is now complete.

b

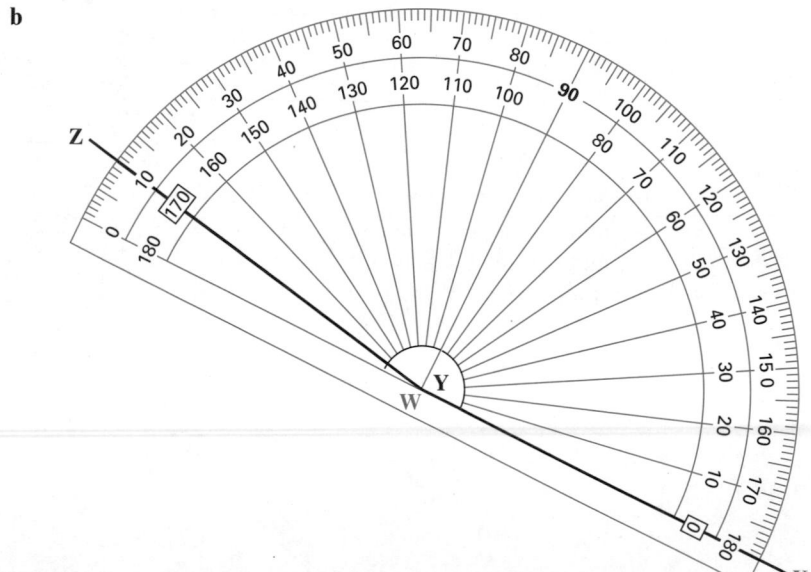

1 Draw the straight line XY. Make it 6 cm long.
2 Place the protractor on the paper with its base line on the line XY and its centre point W on Y, as shown.
3 Mark the point Z at 170° on the scale, starting from 0° at X.
4 Join YZ and the angle of 170° is now complete.

Skill Practice Three

For each question, draw a line AB 6 cm long.
Then draw the angle.

1	$\widehat{ABC} = 60°$	**2**	$\widehat{ABC} = 45°$
3	$\widehat{ABC} = 35°$	**4**	$\widehat{BAC} = 80°$
5	$\widehat{BAC} = 55°$	**6**	$\widehat{ABC} = 120°$
7	$\widehat{ABC} = 155°$	**8**	$\widehat{ABC} = 115°$
9	$\widehat{BAC} = 110°$	**10**	$\widehat{BAC} = 145°$

Skill Practice Four

For each question, draw the triangle ABC from
the details given.
Then measure $\widehat{C}$ with a protractor.

1	AB = 5 cm	$\widehat{A} = 40°$	$\widehat{B} = 60°$
2	AB = 5 cm	$\widehat{A} = 30°$	$\widehat{B} = 80°$
3	AB = 5 cm	$\widehat{A} = 50°$	$\widehat{B} = 40°$
4	AB = 5 cm	$\widehat{A} = 60°$	$\widehat{B} = 90°$
5	AB = 5 cm	$\widehat{A} = 50°$	$\widehat{B} = 50°$
6	AB = 5 cm	$\widehat{A} = 60°$	$\widehat{B} = 60°$
7	AB = 6 cm	$\widehat{A} = 30°$	$\widehat{B} = 40°$
8	AB = 6 cm	$\widehat{A} = 40°$	$\widehat{B} = 40°$
9	AB = 6 cm	$\widehat{A} = 50°$	$\widehat{B} = 100°$
10	AB = 6 cm	$\widehat{A} = 20°$	$\widehat{B} = 110°$

Skill Example

An angle can be described with either three
letters or one letter.

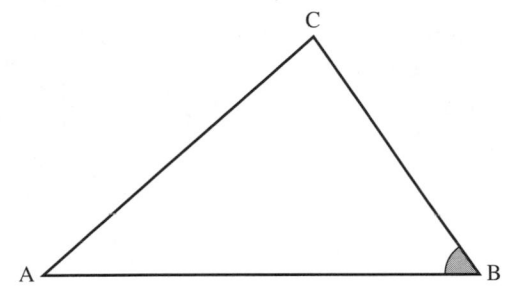

In the diagram, the shaded angle can be
described as $\widehat{ABC}$ or as $\widehat{B}$.

Draw a triangle ABC with AB = 5 cm,
$\widehat{A} = 50°$ and $\widehat{B} = 70°$.
Measure the third angle with a protractor.

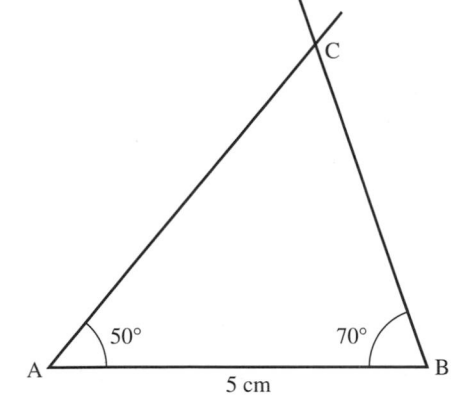

5 cm

By measurement, $\widehat{C} = 60°$.

Knowledge

A bearing (or direction) is measured **clockwise**
from north, and is *always* given as a three-digit
number (to avoid errors).

For example:

a

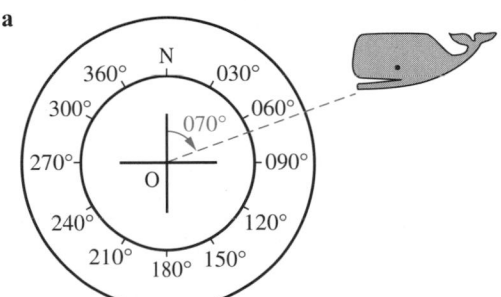

The whale has a bearing of 070° from O.

b

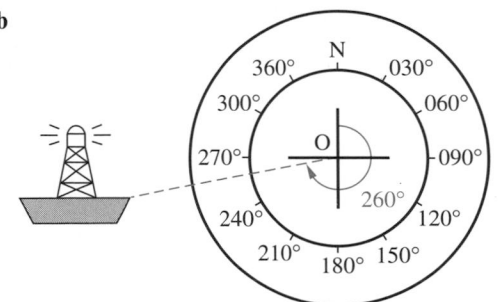

The lightship has a bearing of 260° from O.

Skill Example

a Give the bearing of B from A.

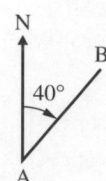

The bearing of B from A is 040°.

b Give the bearing of Q from P.

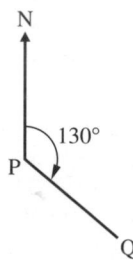

The bearing of Q from P is 130°.

c Give the bearing of Y from X.

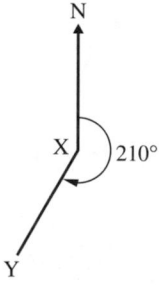

The bearing of Y from X is 210°.

d Give the bearing of C from B.

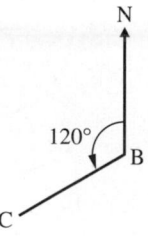

The bearing of C from B is

$$360° - 120° = 240°$$

Skill Practice Five

Remember that bearings are always measured *clockwise* from north.

Give the bearings of the following.

1 B from A

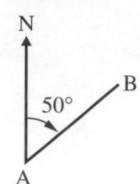

2 Q from P

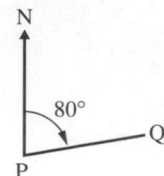

3 Y from X

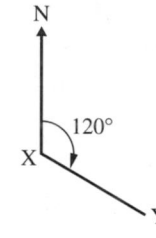

4 C from B

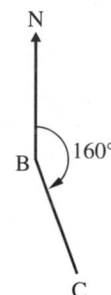

5 R from Q

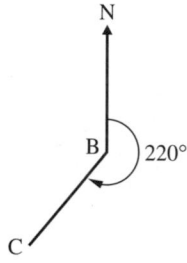

6 Z from Y

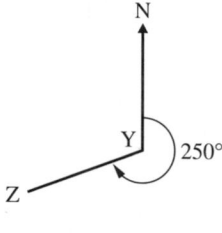

7 D from C

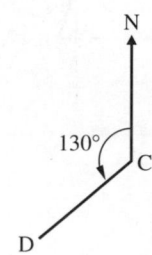

8 V from U

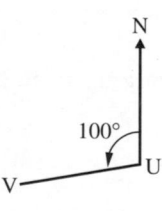

9 B from A

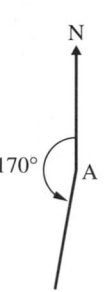

170° A

B

10 Q from P

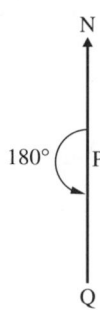

180° P

Q

11 Y from X

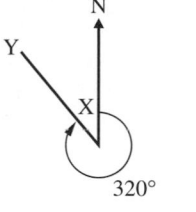

320°

12 C from B

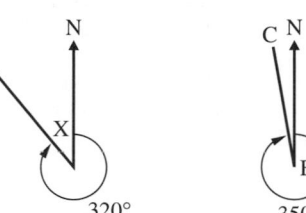

350°

13 R from Q

R

30°

Q

14 Z from Y

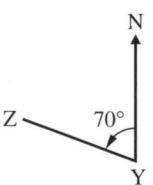

Z 70°

Y

15 V from U

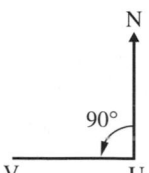

90°

V U

16 B from A

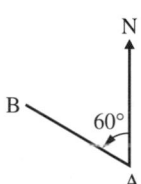

B 60°

A

Skill Example

B is on a bearing of 065° from A and 3 km from A. Use a scale of 1 cm to 1 km to show the relative positions of A and B.

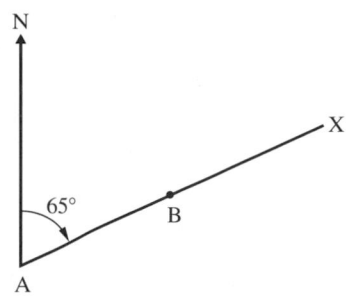

1 Draw a vertical line through the point A to fix north, N.
2 With your protractor, draw N$\hat{A}$X at 65° in the clockwise direction from the line AN.
3 Mark off point B 3 cm along the line AX.

Skill Practice Six

For each question, draw a diagram to show the relative positions of the points. Use a scale of 1 cm to 1 km.

1 B is on a bearing of 030° from A and is 3 km from A.

2 Q is on a bearing of 070° from P and is 3 km from P.

3 Y is on a bearing of 045° from X and is 4 km from X.

4 C is on a bearing of 150° from B and is 4 km from B.

5 R is on a bearing of 135° from Q and is 5 km from Q.

6 Z is on a bearing of 200° from Y and is 5 km from Y.

7 D is on a bearing of 225° from C and is 3 km from C.

8 V is on a bearing of 300° from U and is 3 km from U.

9 B is on a bearing of 340° from A and is 4 km from A.

10 Q is on a bearing of 315° from P and is 4 km from P.

Knowledge

You can plot the course followed by a ship or an aircraft if its distance and bearing at each change of direction are known. This may be done by drawing the distances to a suitable scale.

Skill Example

The first hole on a golf course is 280 m from the tee on a bearing of 090°.
On his first shot, a golfer drives the ball a distance of 160 m on a bearing of 060°.
Make a scale drawing to find the distance and bearing on which he should aim his next shot at the hole. Use a scale of 1 cm to 40 m.

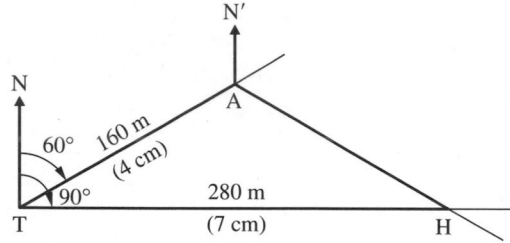

1 Draw a vertical line through T to fix north.
2 With your protractor, draw NT̂H at 90° in the clockwise direction from NT.
   Mark off point H 7 cm along this line TH: the distance from tee to hole is:
   280 ÷ 40 = 7 cm
3 Again with your protractor, draw NT̂A at 60° in the clockwise direction from NT.
   Mark off the point A 4 cm along this line TA: the ball travels 160 ÷ 40 = 4 cm on the first shot.
4 Draw a vertical line N′A through A to represent north.
   Then join the line AH.
5 Measure the length of AH and the size of the clockwise angle N′ÂH. AH = 4 cm, so the distance is 4 × 40 = 160 m.
   N′ÂH = 120°, so the bearing is 120°.

Skill Practice Seven

1 On the sports field, Jill threw the discus a distance of 35 m on a bearing of 060°. Belinda threw the discus a distance of 40 m on a bearing of 090°.

This is a sketch of their throws.

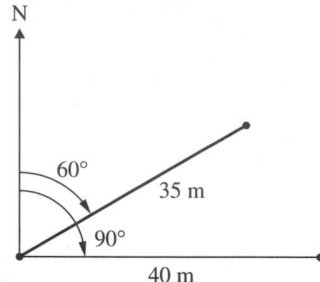

Make a scale drawing to find the distance and bearing of Belinda's discus from Jill's. Use a scale of 1 cm to 5 m.

2 There are three jetties A, B and C on a boating lake. B is 50 m due north of A. C is 50 m due east of A.

This is a sketch of the jetties' relative positions.

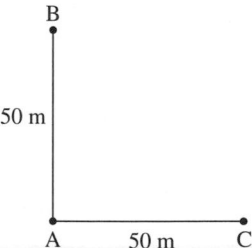

Use a scale drawing of 1 cm to 5 m to find the distance and bearing of C from B.

3 Sheffield is 100 km from Liverpool on a bearing of 090°. Stoke-on-Trent is 70 km from Liverpool on a bearing of 135°.

This is a sketch map.

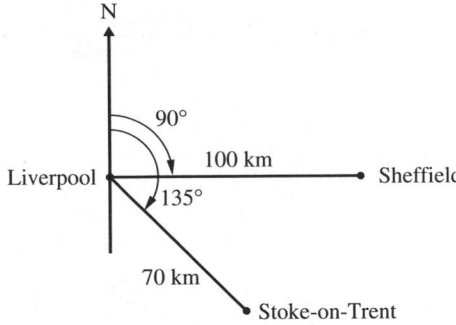

By drawing a map to a scale of 1 cm to 10 km, find the distance and bearing of Sheffield from Stoke-on-Trent.

4 In a cricket match the bowler is 20 m from the batsman on a bearing of 180°. The batsman hits the ball to a fielder 17.5 m away on a bearing of 150°. The fielder catches the ball and throws it back to the bowler.

This is a sketch plan.

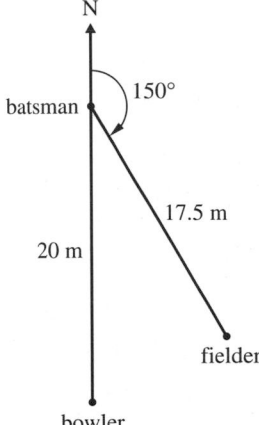

Find from a scale drawing the distance and bearing for the fielder to throw the ball back. Use a scale of 1 cm to 2 m.

Unit 31

Skill Example

Describe the symmetry of this regular shape.

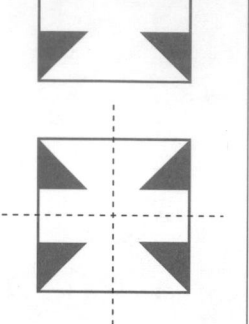

It has two lines of reflective symmetry and rotational symmetry of order 2.

Skill Practice One

Describe the symmetry of each shape.

1

2

3

4

5

6

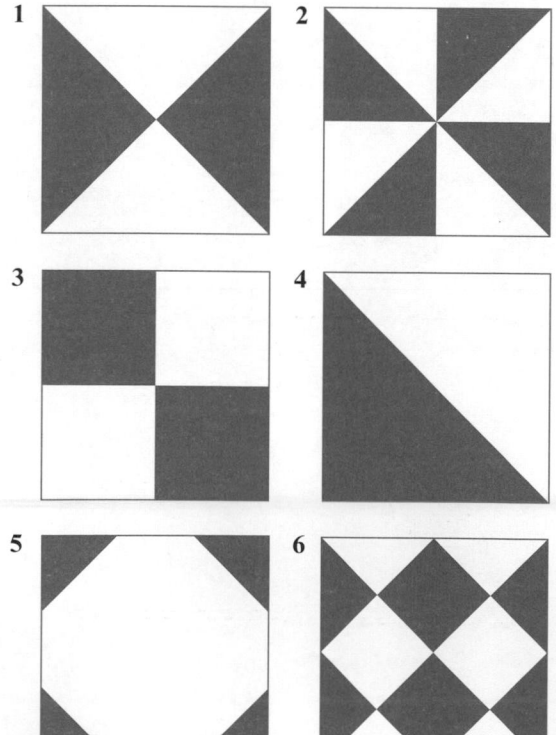

7

8

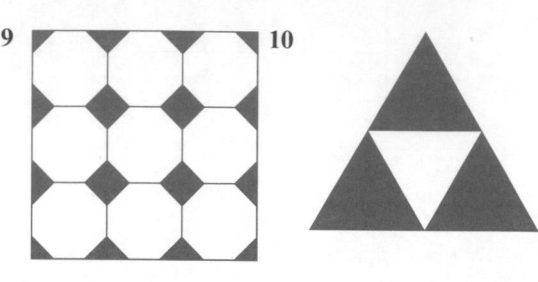

9

10

11

12

Skill Practice Two

1 Draw three shapes of your own whose order of rotational symmetry are:

   a 2   **b** 3   **c** 4

   How many lines of symmetry does each of your shapes have?

2 Draw three shapes of your own with the following number of lines of symmetry:

   a 2   **b** 3   **c** 4

   What is the order of rotational symmetry of each of your shapes?

Unit 32

Shape, space and measures   Level 5

Skills   You know the rough metric equivalents of imperial units still in daily use and convert one metric unit to another.

You make sensible estimates of a range of measures in relation to everyday situations.

Knowledge

Measure the length of the pencil.

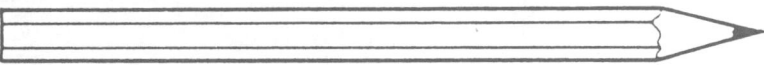

The length of an object like this is measured either in centimetres (cm) or in millimetres (mm).

The pencil is 10 cm long, or 100 mm long, because

$$10\,mm = 1\,cm$$

Larger distances are measured in metres (m):

$$100\,cm = 1\,m$$

or in kilometres (km):

$$1000\,m = 1\,km$$

These are the metric units of length in daily use.

Skill Practice One

In questions **1** to **7**, measure the length of the object shown, using the given units.

1 In centimetres.

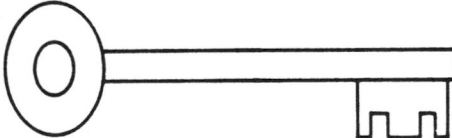

2 In centimetres.

3 In centimetres.

4 In cm and mm.

5 In mm.

6 In mm.

7 In mm.

8 Look at the car and trailer below.
Find the length of:

   a the car only   **b** the trailer only   **c** the car and trailer coupled together

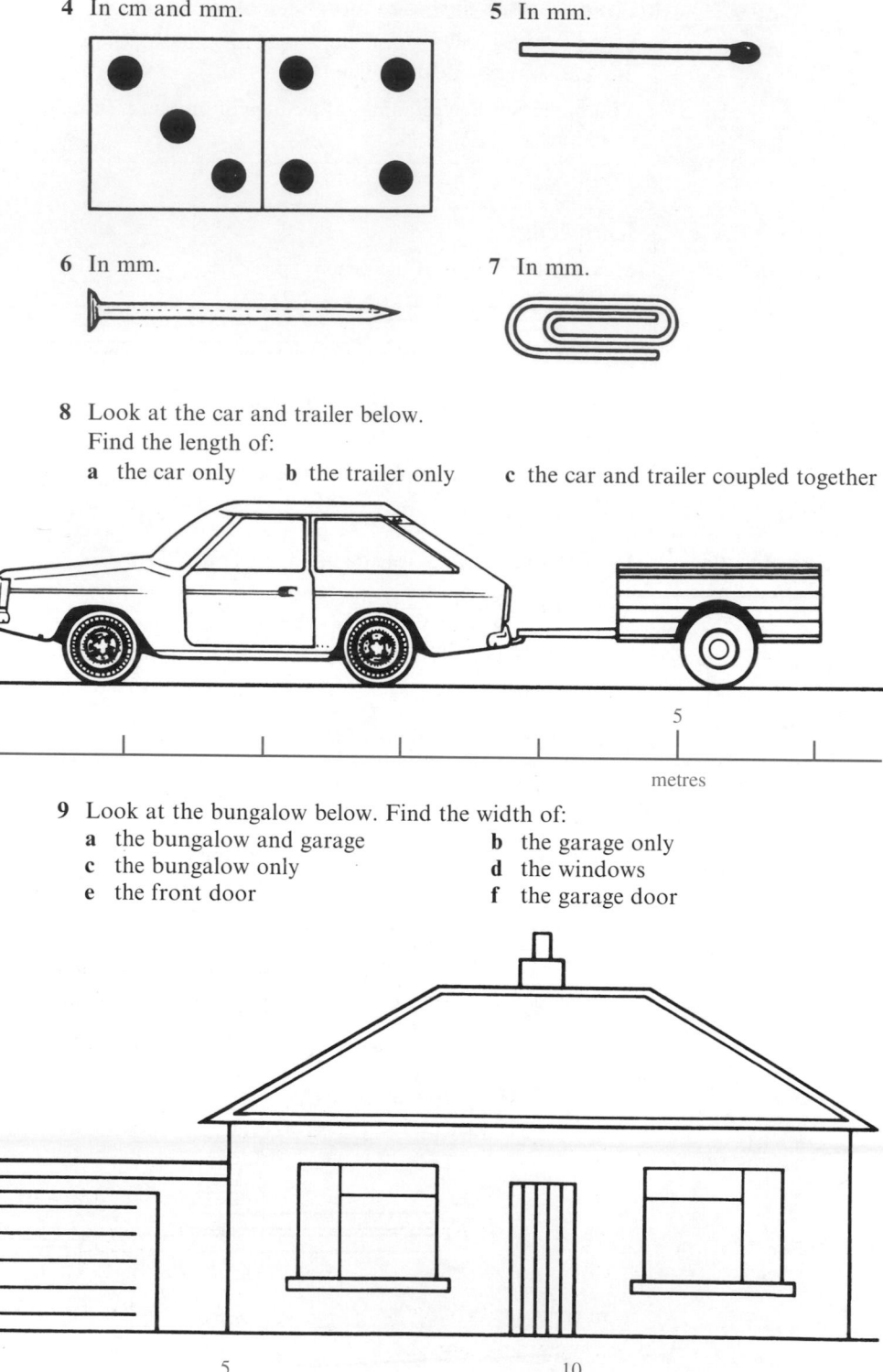

0   5

metres

9 Look at the bungalow below. Find the width of:

   a the bungalow and garage   **b** the garage only
   c the bungalow only   **d** the windows
   e the front door   **f** the garage door

0   5   10

metres

10 Look at the map below.
Find the distance from:

    a Littlehampton to Worthing
    b Worthing to Brighton
    c Littlehampton to Brighton

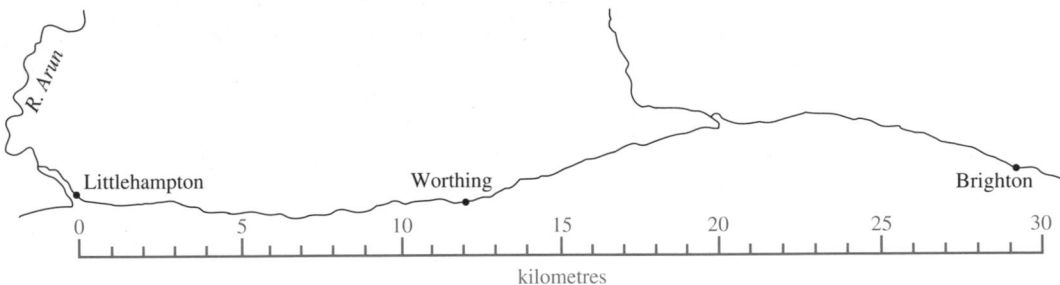

kilometres

Skill Example

a Change 160 millimetres (mm) to centimetres (cm).
b Change 4 kilometres (km) to metres (m).
c Change 275 centimetres (cm) to metres and centimetres.

a $160 \text{ mm} = (160 \div 10) \text{ cm} = 16 \text{ cm}$
b $4 \text{ km} \quad = (4 \times 1000) \text{ m} \ = 4000 \text{ m}$
c $275 \text{ cm} \ = (275 \div 100) \text{ m}$
$\qquad\qquad = 2.75 \text{ m or } 2 \text{ m } 75 \text{ cm}$

Skill Practice Two

Change each of the following measurements into the units given.

  1 7 cm to mm   **2** 28 cm to mm

  3 190 mm to cm   **4** 60 mm to cm

  5 8 m to cm   **6** 236 m to cm

  7 500 m to cm   **8** 7200 cm to m

  9 65 000 cm to m   **10** 4000 cm to m

11 79 km to m   **12** 50 km to m

13 100 km to m   **14** 215 000 m to km

15 460 000 m to km   **16** 215 cm to m and cm

17 304 cm to m and cm **18** 4326 m to km and m

19 3400 m to km and m **20** 54 mm to cm and mm

21 8 cm 6 mm to mm   **22** 1 m 45 cm to cm

23 5 m 7 cm to cm   **24** 2 km 356 m to km

25 4 km 80 m to m

142

Knowledge

The metric units of weight most commonly used are the gram, the kilogram and the tonne.

>   1000 grams (g) = 1 kilogram (kg)
>
>   1000 kilograms (kg) = 1 tonne (t)

The gram is a very small weight.

A sugar lump weighs about 5 g.

An apple weighs about 100 g.

A full one-pint bottle of milk weighs about 1 kg.

A plastic two-gallon bucket full of water weighs about 10 kg.

An electric cooker weighs about 50 kg.

This weight is too heavy for most people to lift.

An ordinary saloon car when empty weighs about 1.25 t.

Skill Practice Three

Look at the examples illustrated on the left, then give the most sensible unit for measuring the weight of each of the following.

1 A cotton reel
2 A sack of potatoes
3 A ballpoint pen
4 A lorry
5 A light bulb
6 A television set
7 A large bunch of bananas
8 An aircraft
9 A full bottle of ink
10 A bicycle

Skill Example

Change the following as indicated.

a 3000 g to kg   **b** 6 t to kg
c 2 kg 625 g to g

a 3000 g = 3000 ÷ 1000 kg = 3 kg

b 6 t = 6 × 1000 kg = 6000 kg

c 2 kg 625 g = (2 × 1000) + 625 g
             = 2625 g

Skill Practice Four

Change each of the following measurements into the units given.

1 5000 g to kg
2 32 000 g to kg
3 8 kg to g
4 41 kg to g
5 7000 kg to t
6 96 000 kg to t
7 9 t, to kg
8 80 t, to kg
9 2520 g to kg and g
10 8075 g to kg and g
11 4372 kg to t and kg
12 5004 kg to t and kg
13 3 kg 450 g to g
14 5 kg 32 g to g
15 6 t 321 kg to kg
16 2 t 9 kg to kg

Knowledge

The metric units of capacity most commonly used are the millilitre, and the litre.

The millilitre is equivalent to the volume of a cube 1 cm by 1 cm by 1 cm, called a cubic centimetre.

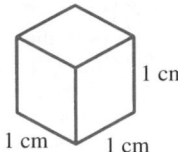

1000 millilitres (ml) = 1 litre (l)

Small quantities, such as a carton of milk or a dose of medicine, are measured in millilitres (ml). A full teaspoon, for example, contains approximately 5 ml.

An Oxo cube has a volume of 8 ml.

A can of Cola has a volume of 330 ml.

Bottles of lemonade come in 1 litre, 2 litre and 3 litre sizes.

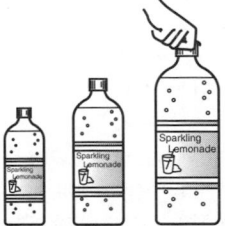

Skill Practice Five

Give the most sensible unit for measuring the capacity of each of the following.

1 A fountain pen ink cartridge
2 A swimming pool
3 A fridge
4 A bottle of perfume
5 A can of paint
6 A car's petrol tank
7 A hypodermic syringe
8 A hot air balloon
9 A tea cup
10 A garden watering can

Skill Example

Change the following measurements into the units given.

a 4500 ml to litres   **b** 3.45 litres to ml

a $4500 \div 1000 = 4.5$ litres

b $3.45 \times 1000 = 3450$ ml

Skill Practice Six

Change the following measurements into the units given.

1 5000 ml to litres
2 3200 ml to litres
3 6780 ml to litres
4 5555 ml to litres
5 345 ml to litres
6 8 litres to ml
7 17 litres to ml
8 2.3 litres to ml
9 3.45 litres to ml
10 0.934 litres to ml

Skill Practice Seven

1 There are 35 English books in the pile on teacher's desk.
   If each book is 8 mm thick, what is the height of the pile:
   a in millimetres   **b** in centimetres?

2 What is the distance between the wall and the left-hand gatepost shown in the picture below:
   a in centimetres   **b** in metres?

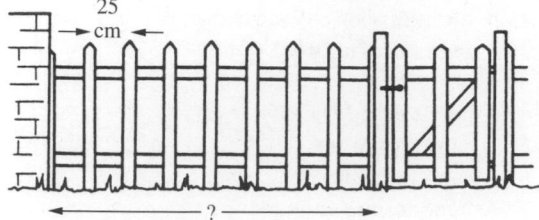

3 The picture shows a path made from paving stones that runs alongside a garage.

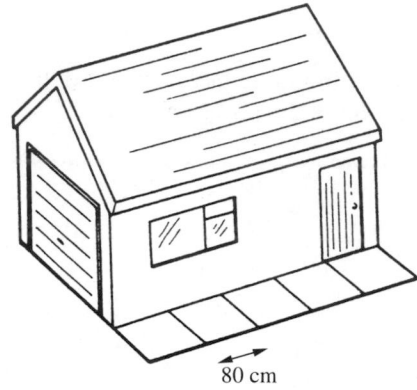

What is the length of the garage:
a in centimetres   **b** in metres?

4 Larch Avenue is 874 m long. Beech Avenue is 345 m long. Elm Avenue is 781 m long.

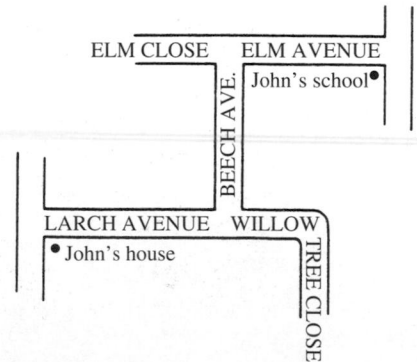

How far does John walk to school:
a in metres   **b** in kilometres?

5 Mrs Patel buys the following at the supermarket:
A packet of butter which weighs 250 g
A piece of cheese which weighs 450 g
A jar of jam which weighs 520 g
A packet of soap powder weighing 810 g
A bunch of bananas which weighs 690 g
If her empty shopping bag weighs 280 g, what is the total weight that she carries home:
a in grams   **b** in kilograms?

6 A cardboard case contains 24 cans of lemonade, each of weight 325 g.
If the empty case weighs 200 g, what is the weight of the full case:
a in grams   **b** in kilograms?

7 A man who weighs 80 kg loads 12 crates, each of weight 55 kg, on to a trolley of weight 260 kg. He then pushes the full trolley into a lift cage where there is a notice as follows:

Load not to exceed 1 tonne

Is it safe to start the lift?

8 Twelve coloured pencils, each of weight 8 g, are contained in a cardboard packet of weight 29 g.
   a What is the weight of the full packet?
   b How many of the same packets would together weigh 1 kg?

9 A cardboard box of weight 34 g contains six golf balls.
If the box and the balls weigh 250 g altogether, what is the weight of one golf ball?

10 A chocolate Easter egg contains 25 chocolate drops. The chocolate shell weighing 113 g is wrapped in a decorative pack that weighs 12 g. If the total weight is 250 g, what is the weight of each chocolate drop?

11 Mrs Jones buys a bag of flour containing 400 g. If she uses 150 g for making some scones and 165 g for making a pie, how much flour will she have left over?

12 The total weight of a van and its load is 5 tonnes. If the van carries 10 crates, each having a weight of 270 kg, find the weight of the van when empty.

13 To make a fruit punch for a party, Tammy mixes three 1 litre boxes of orange juice with two 1 litre boxes of pineapple juice and ten 500 ml bottles of lemonade.

What is the total volume of the fruit punch?

14 A pack of 12 cans of drink contains a total volume of 3.96 litres.
  a What volume of drink is contained in each can?
  b What volume of drink would be contained in a pack of 6 of the cans?

15 A large coffee urn contains 15 litres of tea and is used to fill cups with 125 ml of tea in each.
  a How many cups can be filled from a full urn of tea?
  b If 40 cups have been filled from a full urn, what volume of tea is left in the urn?

Knowledge

Britain started using metric measurements during the 1970s. Until that time, length, weight and capacity were measured using **imperial** units.

The imperial table for length is:　　12 inches = 1 foot

3 feet = 1 yard

1760 yards = 1 mile

One inch is approximately 2.5 centimetres.

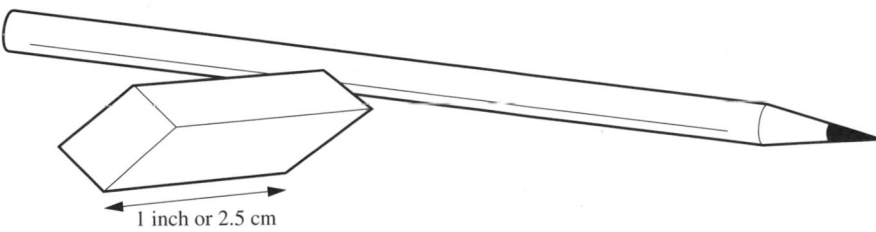

1 inch or 2.5 cm

One yard is approximately 0.9 metres.

2 metres

2 yards

One kilometre is approximately $\frac{5}{8}$ of a mile. This is more easily remembered as:

8 kilometres = 5 miles

Skill Example

a This pencil is approximately 6 inches long.

What is the length of the pencil in centimetres?

b A family see this sign while driving in France.

How far is it in miles to Paris?

a The pencil is approximately $6 \times 2.5 = 15$ cm long

b It is $160 \div 8 \times 5 = 100$ miles to Paris

Skill Practice Eight

In questions **1** to **6**, find the approximate length of each object in centimetres.

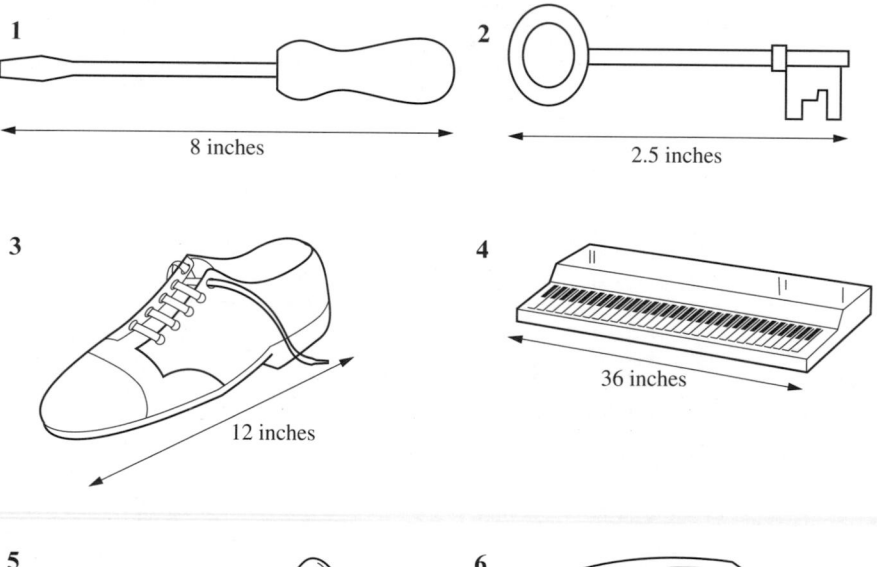

1

8 inches

2

2.5 inches

3

12 inches

4

36 inches

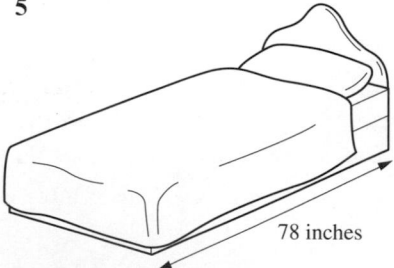

5

78 inches

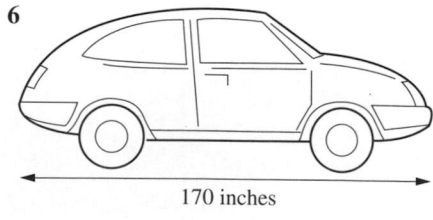

6

170 inches

In questions **7** to **12**, convert the distance shown on each sign into miles (to the nearest mile).

7

8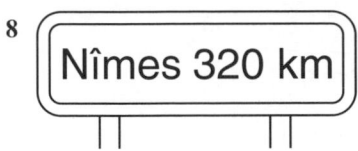

9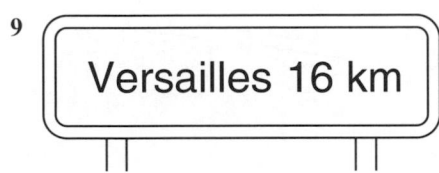

10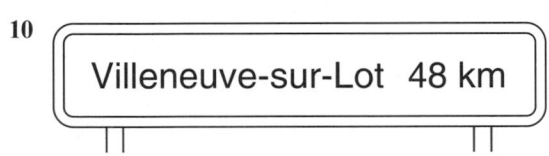

11

12

Knowledge

The imperial table for weight is:

$$16 \text{ ounces} = 1 \text{ pound}$$
$$14 \text{ pounds} = 1 \text{ stone}$$
$$8 \text{ stones} = 1 \text{ hundredweight}$$
$$20 \text{ hundredweight} = 1 \text{ ton}$$

One kilogram is approximately 2.2 pounds.

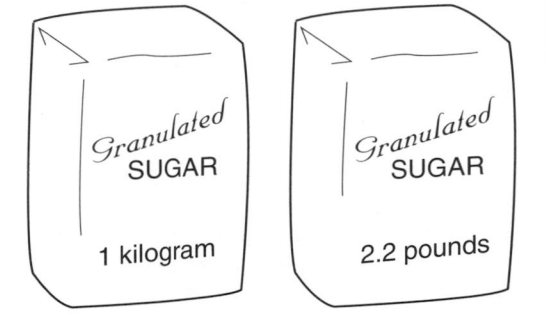

Knowledge

The imperial table for capacity is:

$$8 \text{ pints} = 1 \text{ gallon}$$

One litre is approximately 1.75 pints.

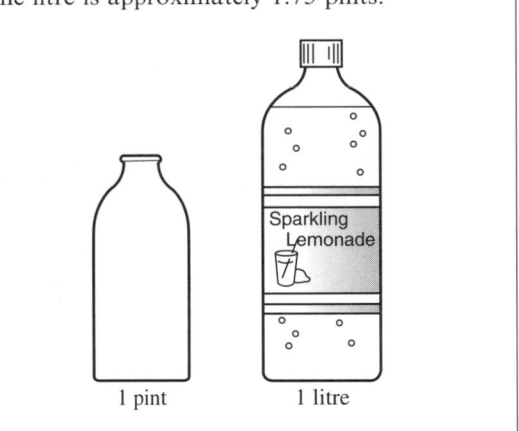

Skill Example

a Mr Wilson has just bought an 11 pound packet of grass seed.
How many kilograms of grass seed has Mr Wilson bought?

To convert pounds to kg, divide by 2.2:

$$11 \div 2.2 = 5$$

Mr Wilson has bought 5 kg of grass seed.

b Mrs Wilson has just put 40 litres of petrol in her car.
How many gallons of petrol has Mrs Wilson put in her car?

First, convert 40 litres to pints:

$$40 \times 1.75 = 70$$

Then convert 70 pints to gallons:

$$70 \div 8 = 8.75$$

So Mrs Wilson has put 8.75 gallons of petrol in her car.

Skill Practice Nine

Copy and complete each weight label.

1

Weight
154 pounds
___ kilograms

2

Weight
330 pounds
___ kilograms

3

Weight
1430 pounds
_____ kilograms

4
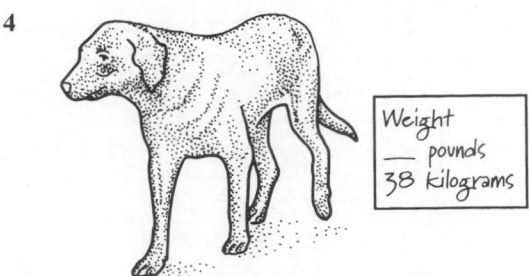

Weight
___ pounds
38 kilograms

5

Weight
___ pounds
4 kilograms

6
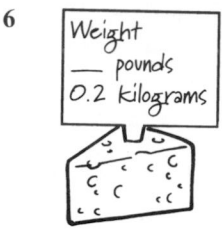

Weight
___ pounds
0.2 kilograms

How many pints of petrol are there in each can?

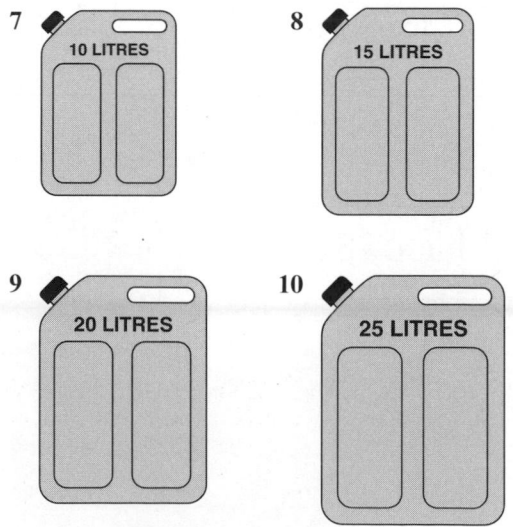

7 10 LITRES

8 15 LITRES

9 20 LITRES

10 25 LITRES

How many gallons of petrol are in each can in questions **7** to **10**?

Unit 33

Shape, space and measures   Level 6

Skills   You recognise and visualise the transformations of translation, reflection and rotation and their combinations in two dimensions.

You understand the notations used to describe them.

Knowledge

We have already looked at two different types of **transformation**: reflection and rotation.

Translation is another kind of transformation.

This diagram shows a shaded triangle moved to eight new positions

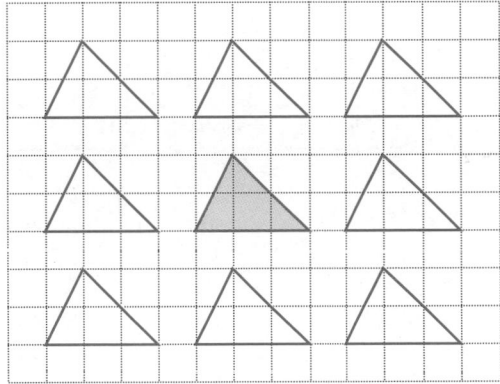

The translation that moves the shaded triangle to each new position is defined by two numbers in a bracket.

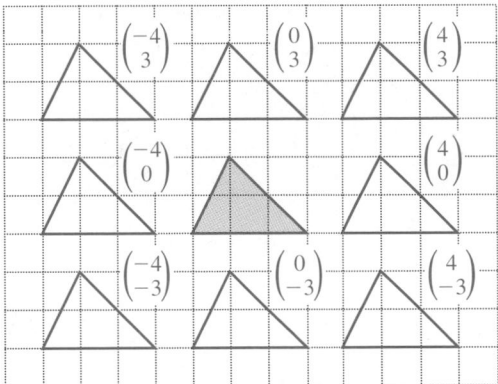

These numbers are respectively the horizontal and vertical distances moved by each point on the shape.

$$\begin{pmatrix} \overset{-}{\leftarrow} \text{ Horizontal movement } \overset{+}{\rightarrow} \\[2pt] \uparrow+ \\ \text{Vertical movement} \\ \downarrow- \end{pmatrix}$$

Skill Practice One

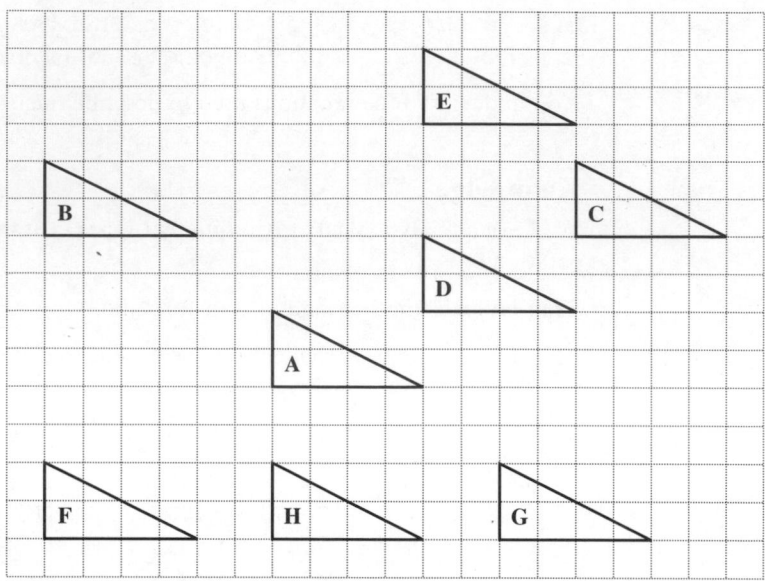

Describe, in each case, the translation which moves the first triangle on to the second triangle.

1 A to B	**2** A to C	**3** A to D	**4** A to E	**5** A to F
6 A to G	**7** A to H	**8** B to A	**9** B to C	**10** B to D
11 B to E	**12** B to F	**13** B to G	**14** B to H	**15** C to A
16 C to B	**17** C to D	**18** C to E	**19** C to F	**20** C to G

Skill Example

The triangle P has vertices (corner points) at (2, 7), (3, 4) and 6, 5).
Show on one graph the triangle P and its images Q, R and S after these translations:

$$\begin{pmatrix} 3 \\ -4 \end{pmatrix} \qquad \begin{pmatrix} 2 \\ 3 \end{pmatrix} \qquad \begin{pmatrix} -2 \\ -3 \end{pmatrix}$$

Skill Practice Two

1 The triangle P has vertices at (2, 7), (3, 4) and (6, 5).
Show on one graph the triangle P and its images Q, R and S after the translations

$$\begin{pmatrix} 3 \\ 3 \end{pmatrix} \quad \begin{pmatrix} 0 \\ -4 \end{pmatrix} \quad \begin{pmatrix} -2 \\ 3 \end{pmatrix}$$

2 The triangle W has vertices at (0, 0), (0, 2) and (4, 2).
Show on one graph the triangle W and its images X, Y and Z after the translations

$$\begin{pmatrix} 6 \\ 6 \end{pmatrix} \quad \begin{pmatrix} 0 \\ 6 \end{pmatrix} \quad \begin{pmatrix} 6 \\ 0 \end{pmatrix}$$

3 The triangle A has vertices at (7, 8), (10, 8) and (9, 10).
Show on one graph the triangle A and its images B, C and D after the translations

$$\begin{pmatrix} -6 \\ -6 \end{pmatrix} \quad \begin{pmatrix} 0 \\ -6 \end{pmatrix} \quad \begin{pmatrix} -6 \\ 0 \end{pmatrix}$$

4 The triangle J has vertices (7, 8), (10, 8) and (9, 10).
Show on one graph the triangle A and its images K, L and M after the translations

$$\begin{pmatrix} 4 \\ 2 \end{pmatrix} \quad \begin{pmatrix} 4 \\ -2 \end{pmatrix} \quad \begin{pmatrix} 0 \\ 4 \end{pmatrix}$$

5 The rectangle E has vertices at (1, 1), (1, 2), (3, 1) and (3, 2).
Show on one graph the rectangle E and its images F, G and H after the translations

$$\begin{pmatrix} -1 \\ -1 \end{pmatrix} \quad \begin{pmatrix} 7 \\ 7 \end{pmatrix} \quad \begin{pmatrix} 7 \\ -1 \end{pmatrix}$$

6 The hexagon S has vertices at (3, 5), (3, 6), (4, 7), (5, 6), (5, 5) and (4, 4).
Show on one graph the hexagon S and its images T, U and V after the translations

$$\begin{pmatrix} 0 \\ -4 \end{pmatrix} \quad \begin{pmatrix} 5 \\ 3 \end{pmatrix} \quad \begin{pmatrix} -3 \\ 1 \end{pmatrix}$$

Skill Practice Three

In questions **1** to **4**, copy each diagram on squared paper, then draw the image formed by:

a reflection in the *x*-axis
b reflection in the *y*-axis.

1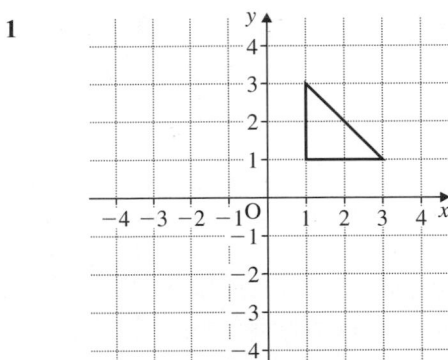

2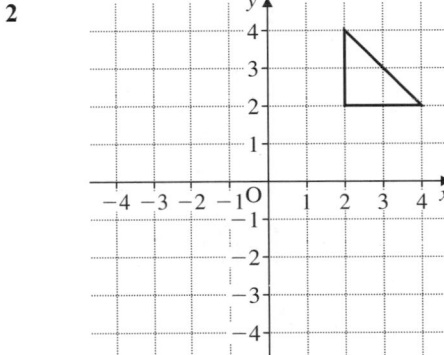

3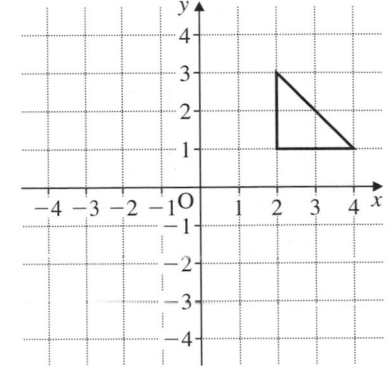

4

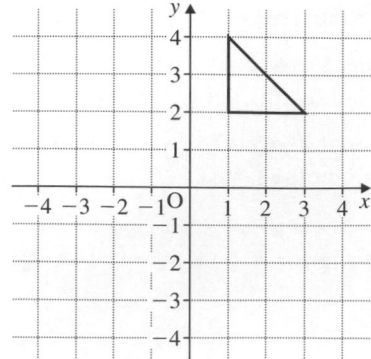

7

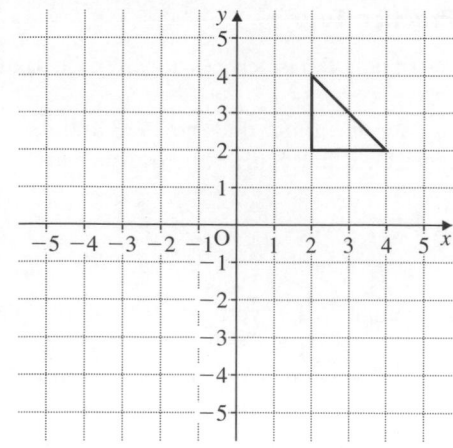

In questions **5** to **7**, copy each diagram on squared paper, then draw the image formed by rotation about (0, 0):

a through 180°
b through 90° anticlockwise
c through 90° clockwise.

In questions **8** to **12**, write a translation to describe the transformation of each triangle P to its image Q.

5

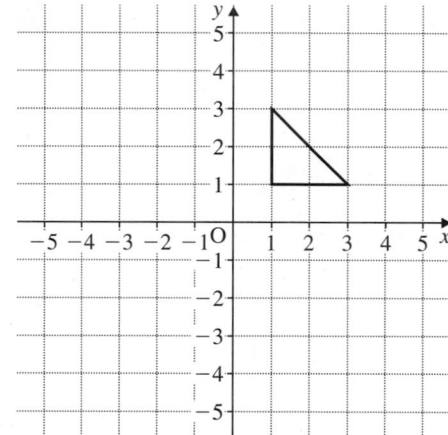

6

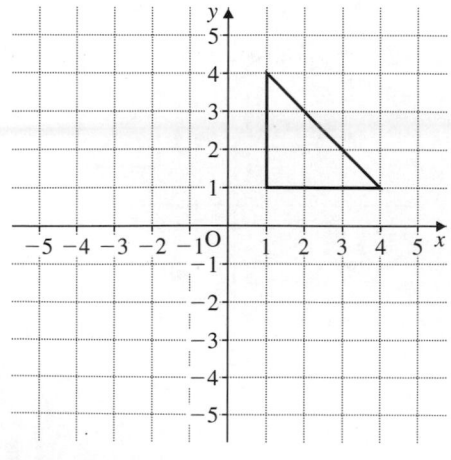

8

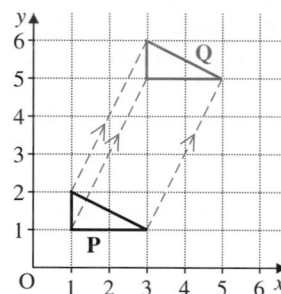

9

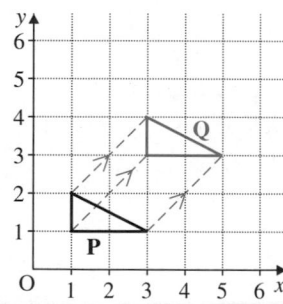

10

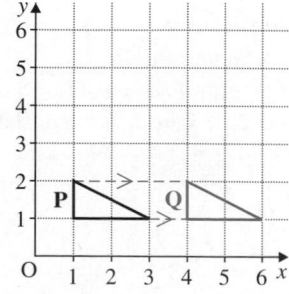

11

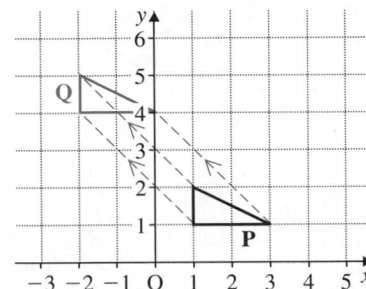

12

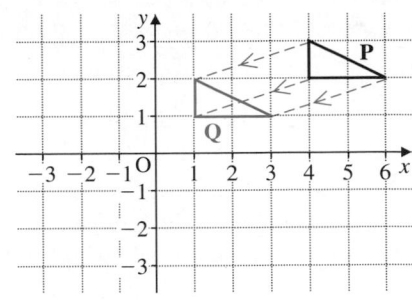

Skill Example

Draw a graph with x-values from 0 to 8 and y-values from 0 to 6. On your graph, show the triangle P with vertices at (3, 1), (3, 2) and (6, 1).

Show the new position of P after the triangle has been first reflected in the line $x = 3$ and then translated $\begin{pmatrix} 5 \\ 2 \end{pmatrix}$.

This is the completed diagram.

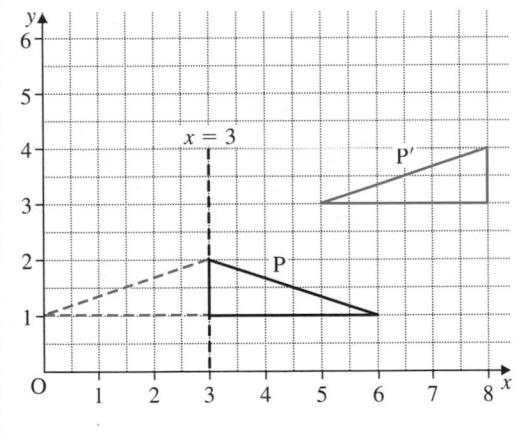

Skill Practice Four

For each question, draw a graph with x and y-values from 0 to 8. On your graph, show the triangle P with vertices at (3, 1), (3, 2) and (6, 1). Show the new position of P after each pair of transformations.

1 Reflection in $y = 2$, followed by the translation $\begin{pmatrix} -2 \\ 3 \end{pmatrix}$.

2 Rotation of $+90°$ about (3, 1), followed by the translation $\begin{pmatrix} 2 \\ 2 \end{pmatrix}$.

3 Translation $\begin{pmatrix} 2 \\ 2 \end{pmatrix}$, followed by rotation of $+90°$ about (3, 1).

4 Reflection in $y = 3$, followed by rotation of $-90°$ about (6, 5).

5 Rotation of $180°$ about (3, 1), followed by reflection in $x = 4$.

Unit 34

Skill Example

Isometric or triangular grids can help you to draw two-dimensional (2-D) drawings of three-dimensional (3-D) objects.

Draw this model made from cubes on isometric paper.

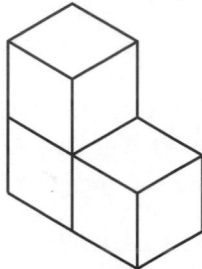

This is the drawing.

Skill Practice One

Make each of these objects from cubes.
Then draw the objects on isometric paper.

1

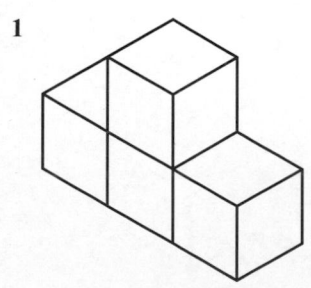

2

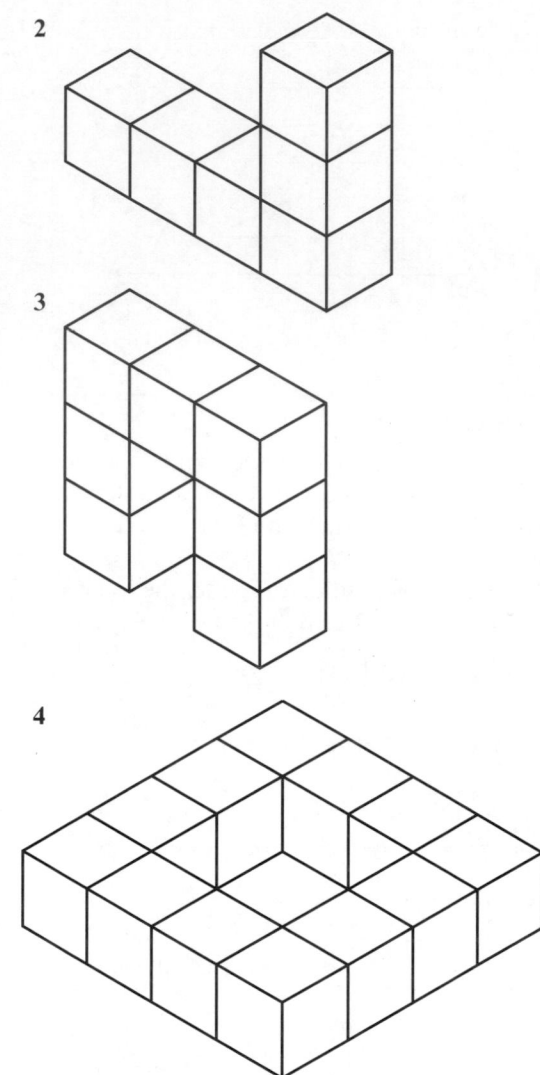

3

4

Skill Example

Lenny cut some letters out of a block of wood and painted them. He put the letter L down on his desk when the paint was still wet and it left an outline like this.

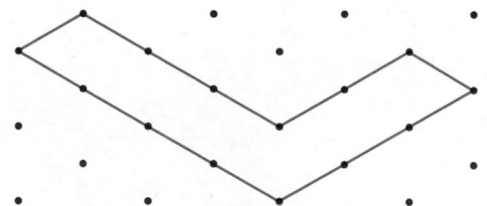

Draw the letter on isometric paper.

The drawing looks like this.

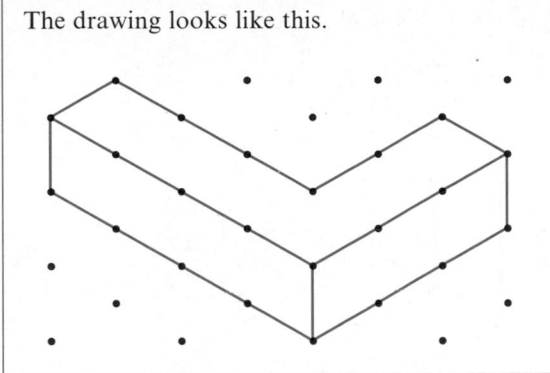

4

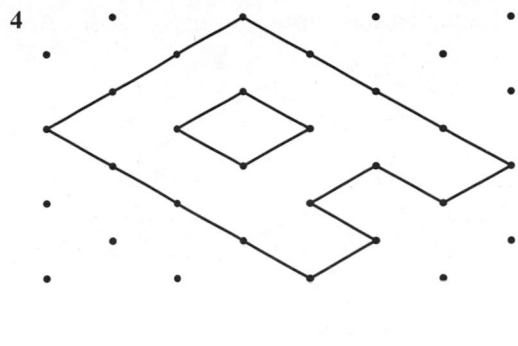

Skill Practice Two

Use isometric paper to draw the letters which left outlines like these on Lenny's desk.

1

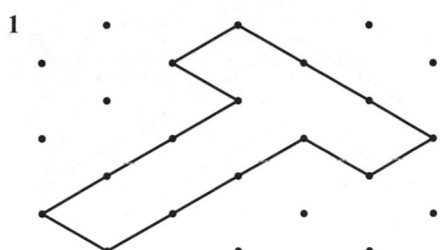

5

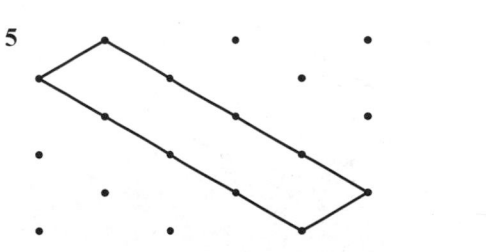

2

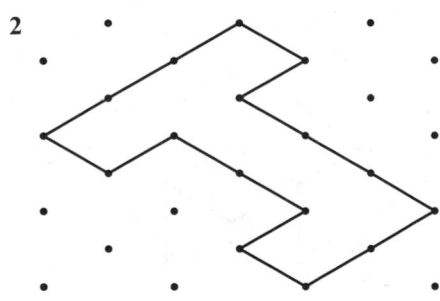

6

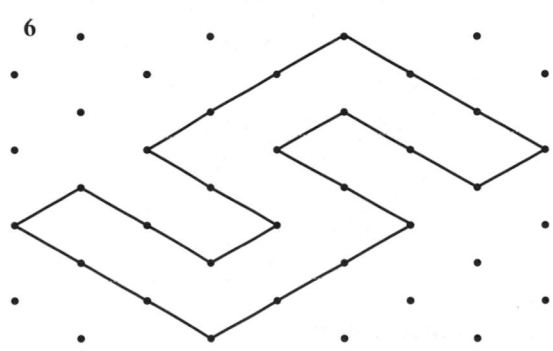

3

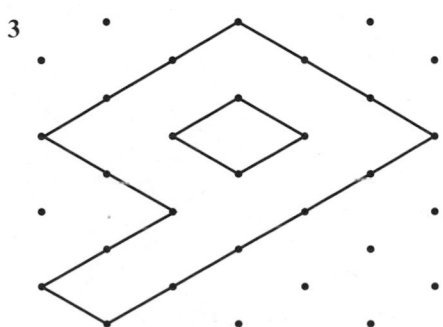

Skill Example

This solid

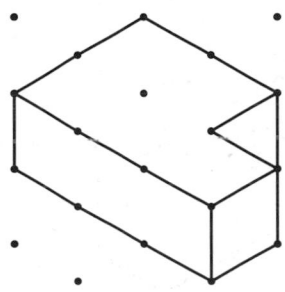

156

is resting on this base.

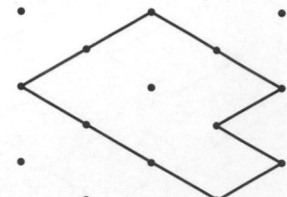

Draw the solid resting on this base.

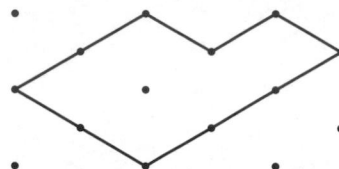

The drawing looks like this.

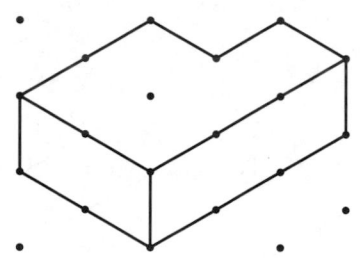

Skill Practice Three

1 Draw the solid resting on this base.

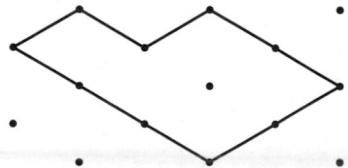

2 Draw the solid resting on this base.

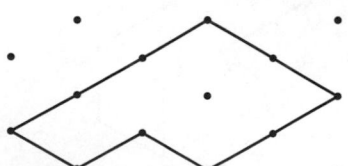

3 Draw this solid

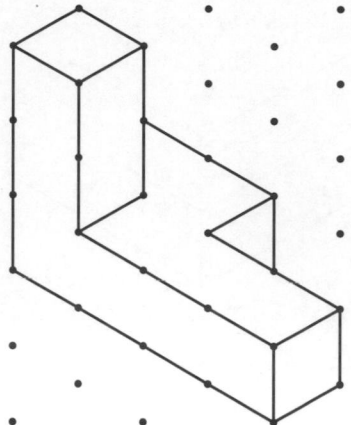

resting on this base.

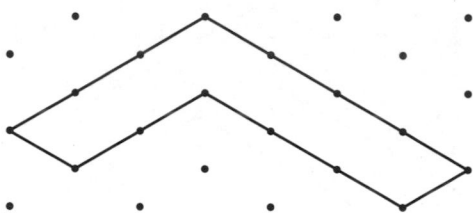

4 Draw this solid

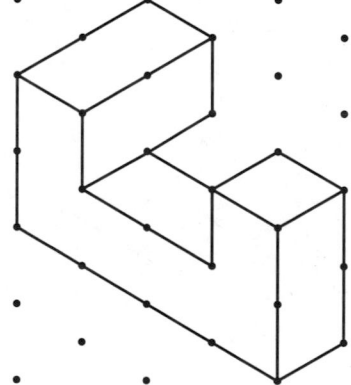

resting on this base.

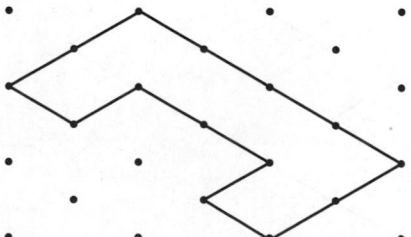

Skill Example

Here is one of Lenny's letters lying on a desk. Three small creatures are looking at it.

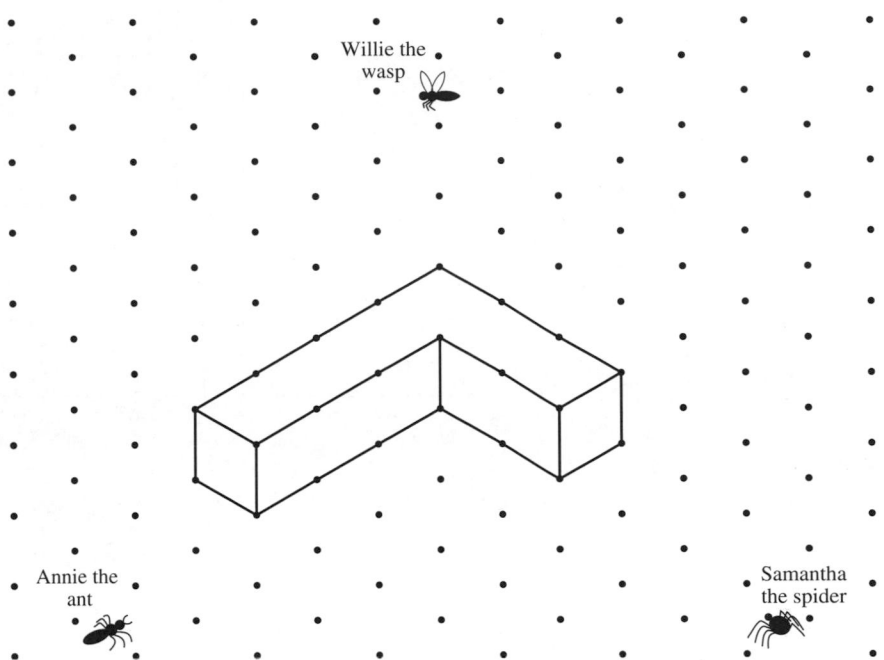

Here are three different views of the letter.

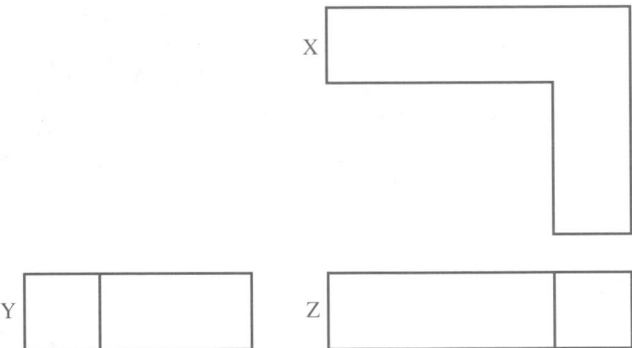

a Which is Annie's view of the letter?
b Which is Samantha's view of the letter?
c Which is Willie's view of the letter?

a Y is Annie's view of the letter.
b Z is Samantha's view of the letter.
c X is Willie's view of the letter.

Skill Practice Four

Draw Willie's, Annie's and Samantha's view of each of these letters. Make each drawing like this.

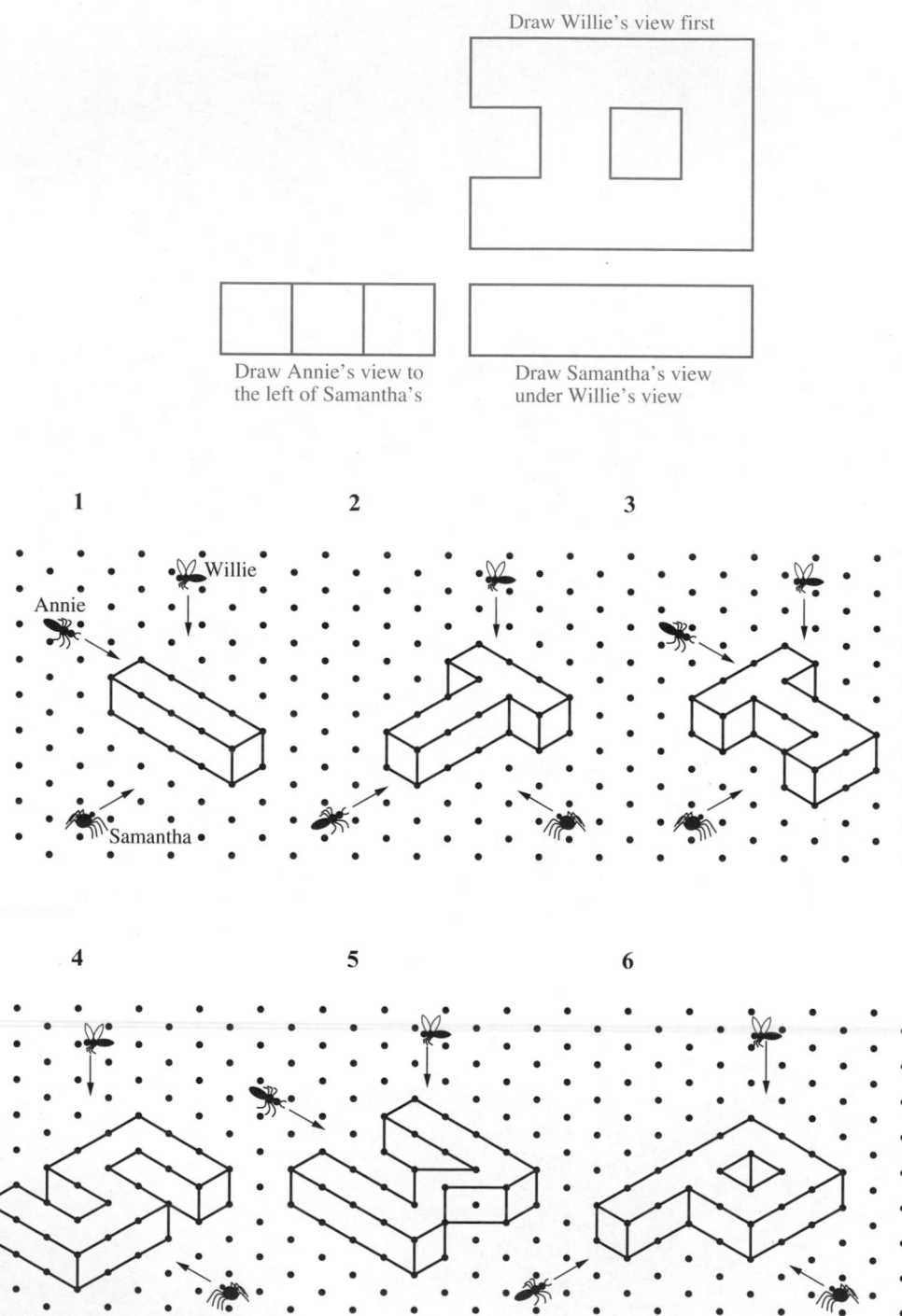

Draw Willie's view first

Draw Annie's view to the left of Samantha's

Draw Samantha's view under Willie's view

1   2   3

Willie

Annie

Samantha

4   5   6

Knowledge

The view seen from directly above an object is called the **plan**.

The view seen from the front of an object is called the **front elevation**.

The view seen from the side of an object is called the **side elevation**.

Skill Example

Draw the plan, the front elevation and the side elevation of this object.

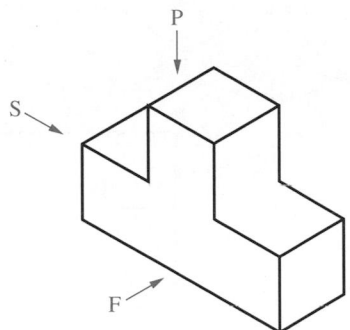

These are the required views:

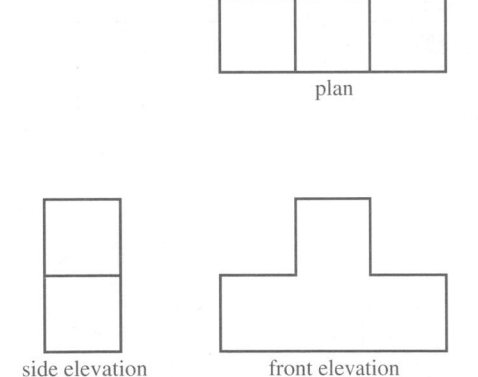

plan

side elevation   front elevation

Skill Practice Five

Draw the plan (P), the front elevation (F) and the side elevation (S) of each object.

1

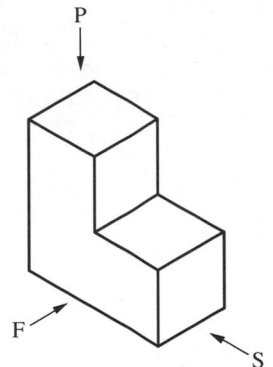

2

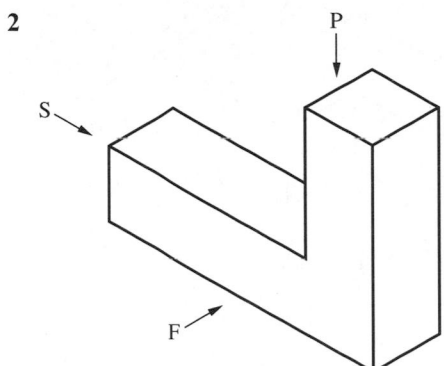

3

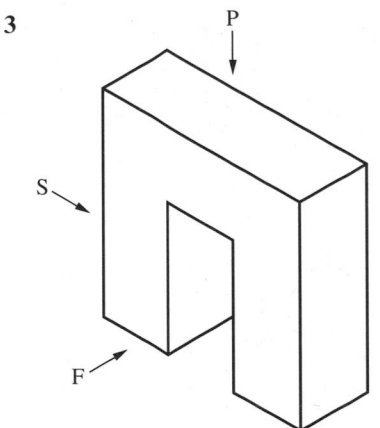

160

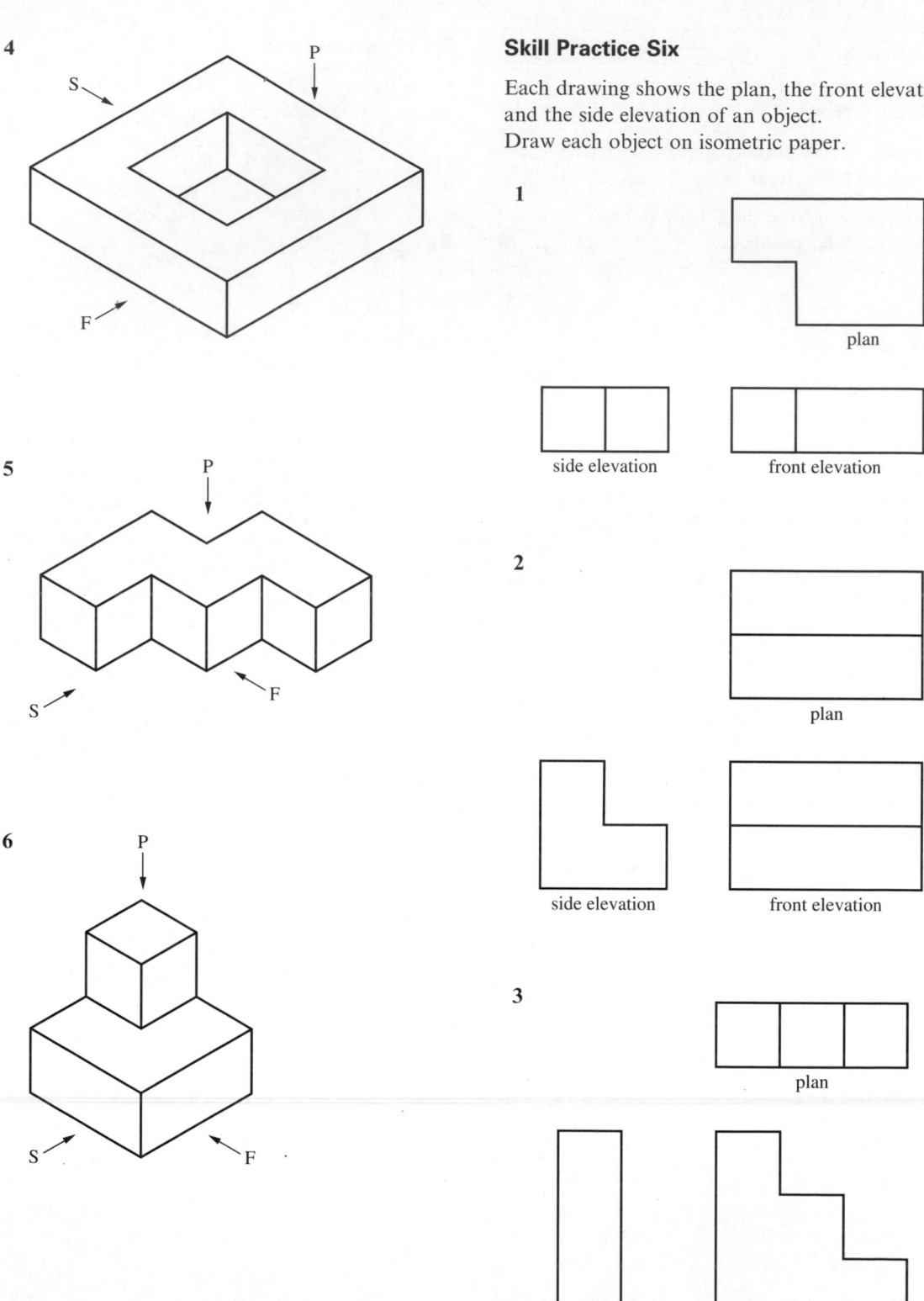

4

5

6

Skill Practice Six

Each drawing shows the plan, the front elevation
and the side elevation of an object.
Draw each object on isometric paper.

1

plan

side elevation   front elevation

2

plan

side elevation   front elevation

3

plan

side elevation   front elevation

4

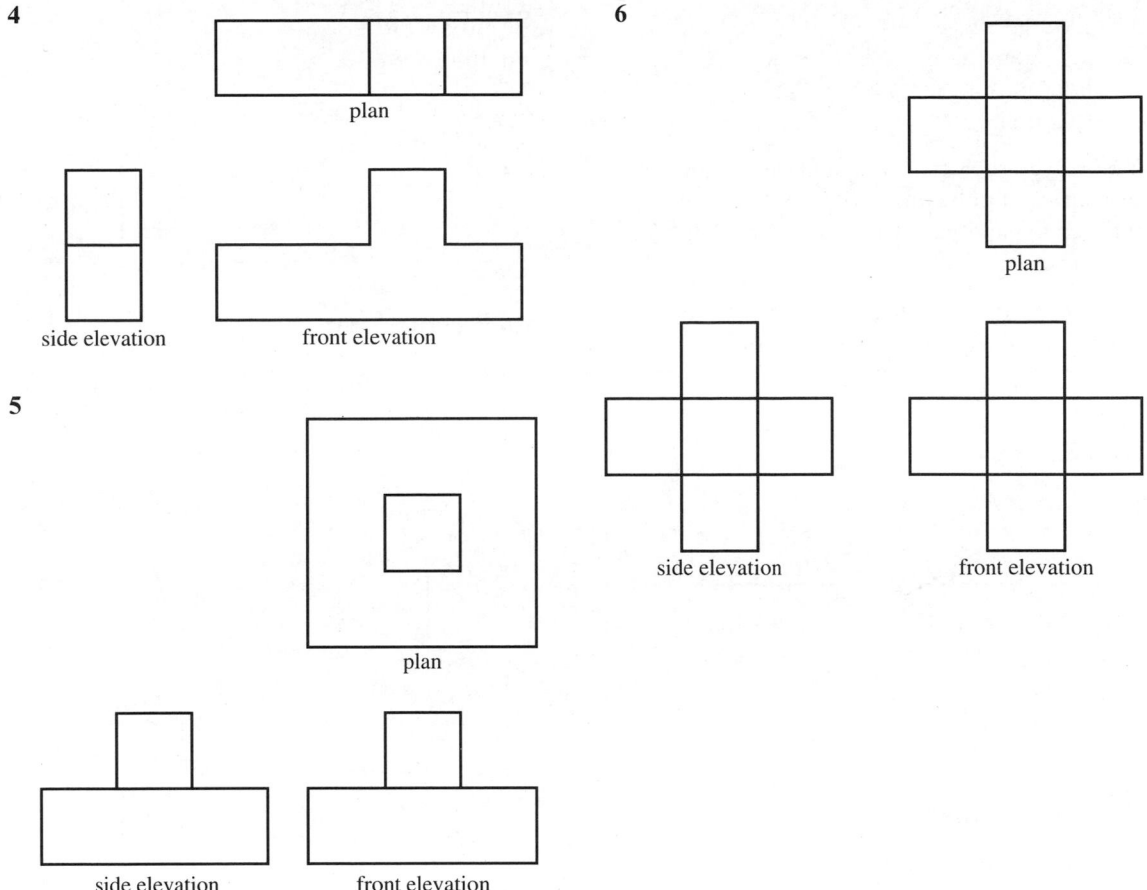

plan

side elevation

front elevation

5

plan

side elevation

front elevation

6

plan

side elevation

front elevation

Unit 35

Shape, space and measures   Level 6

Skills   You know and use the properties of quadrilaterals in classifying different types of quadrilateral.

You solve problems using angle and symmetry properties of polygons and properties of intersecting and parallel lines, and explain these properties.

You devise instructions for a computer to generate and transform shapes and paths.

Skill Example

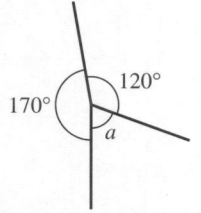

Find the size of a.

$a = 360° - 170° - 120°$
$= 190° - 120°$
$= 70°$

Skill Practice Two

For each question, find the size of the lettered angle.

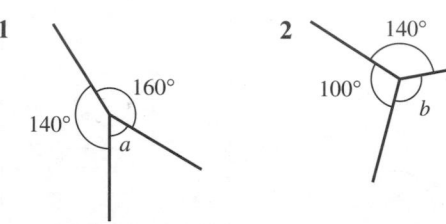
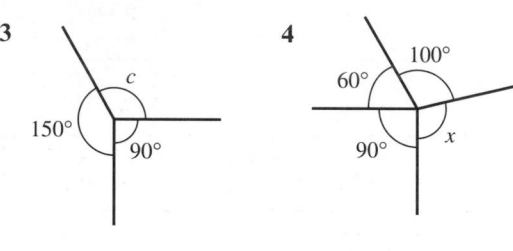
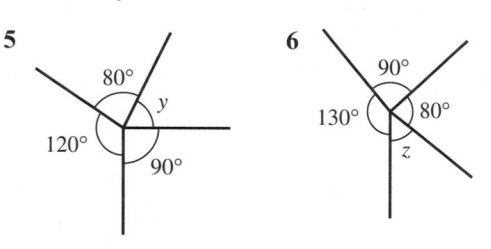

Skill Example

a

115° x

b

y
40°   50°

Find the size of x.
$x = 180° - 115°$
$\quad = 65°$

Find the size of y.
$y = 180° - 40° - 50°$
$\quad = 140° - 50°$
$\quad = 90°$

Skill Practice One

For each question, find the size of the lettered angle.

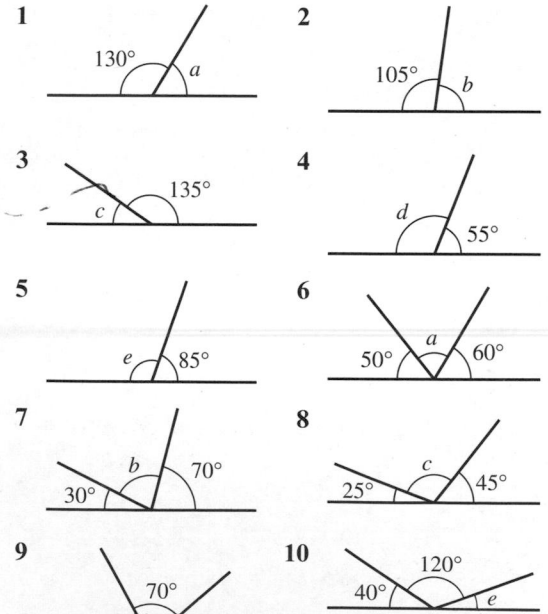

Skill Example

Find the size of angles a, b, c formed by the two straight lines in the diagram.

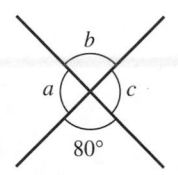

a Because a and 80° form a straight line,
$a = 180° - 80° = 100°$

b Because b is opposite to 80°, $b = 80°$

c Because c is opposite to a, $c = 100°$

Skill Practice Three

Find the lettered angles formed by the straight lines in each diagram.

1
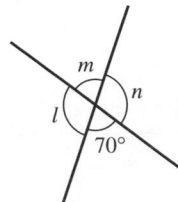

2
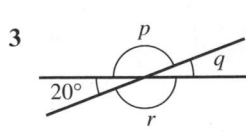

3
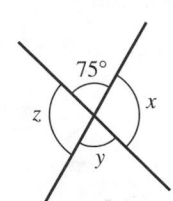

4
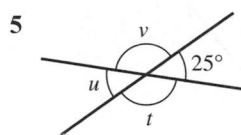

5
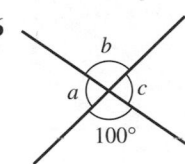

6
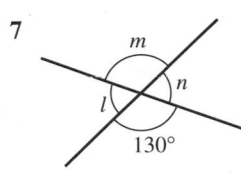

7
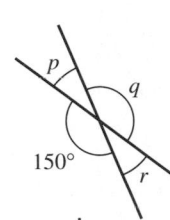

8
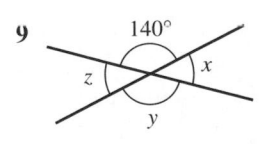

9
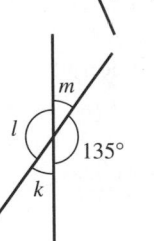

10
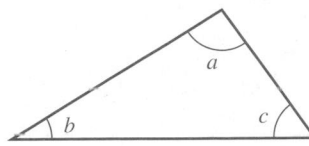

There are three main types of triangle.

- Acute-angled triangle: all its angles are less than 90°.

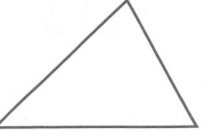

- Right-angled triangle: its largest angle equals 90°

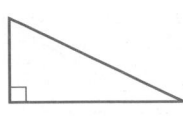

- Obtuse-angled triangle: its largest angle is greater than 90°.

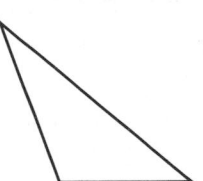

Skill Example

Find the unknown angle in the following triangles and state the types of triangle.

a

The unknown angle is $180° - 40° - 30°$
$$= 140° - 30°$$
$$= 110°$$

The triangle is obtuse-angled.

b
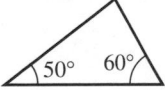

The unknown angle is $180° - 50° - 60°$
$$= 130° - 60°$$
$$= 70°$$

The triangle is acute-angled.

Knowledge

A triangle has three sides that form three angles.

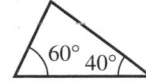

The sum of these three angles is always 180°:

$$a + b + c = 180°$$

Skill Practice Four

Find the unknown angle in each triangle and state the type of triangle.

1
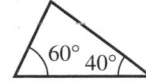

2
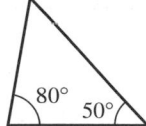

164

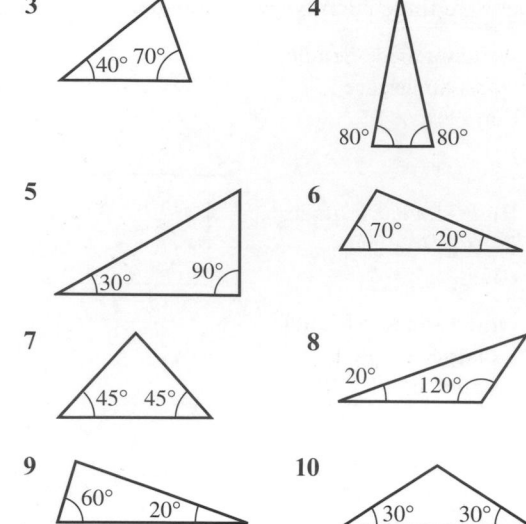

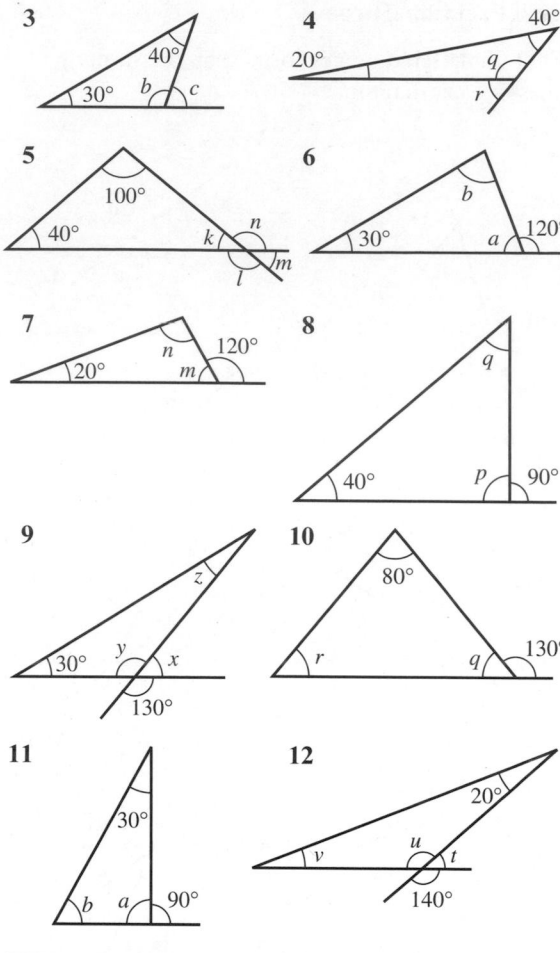

Skill Example

To find some unknown angles, you will need to use not only the angle properties of triangles but also those of straight lines.

Find a and b in the diagram.

The sum of the angles in a triangle is 180°. So,

$a = 180° - 80° - 30°$

$= 100° - 30°$

$= 70°$

a and b are on a straight line. So,

$b = 180° - a$

$= 180° - 70°$

$= 110°$

Knowledge

Lines in the same plane which do not meet, however far they are extended, are called **parallel lines**.

To show that lines are parallel, arrowheads are placed on them.

transversal

A straight line that cuts parallel lines is called a **transversal**.

Skill Practice Five

Find the lettered angles in each of the following.

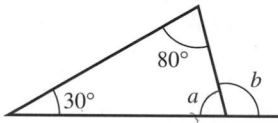

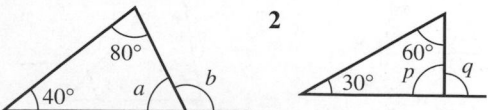

Adjacent angles

The sum of two adjacent angles is 180°.

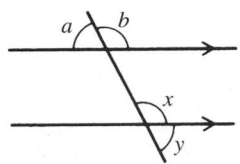

$a + b = 180°$ because they are adjacent angles.

$x + y = 180°$ because they are also adjacent.

Vertically opposite angles

Vertically opposite angles are equal.

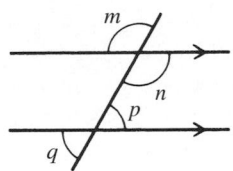

$m = n$ because they are vertically opposite angles.

$p = q$ because they are also vertically opposite angles.

Alternate angles

Alternate angles are equal.

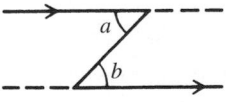 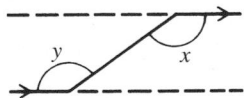

$a = b$ because they are alternate angles.

$x = y$ because they are alternate angles.

Skill Example

a Find a, giving reasons.

As a and 80° are alternate angles,
$a = 80°$

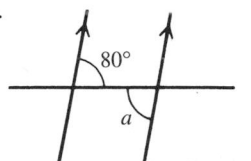

b Find x and y, giving reasons.

As x and 110° are adjacent angles,

$x = 180° - 110°$
$\quad = 70°$

As x and y are alternate angles,

$x = y = 70°$

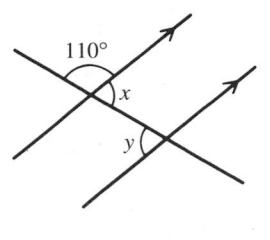

Skill Practice Six

Find each angle that is marked with a letter, giving reasons.

1
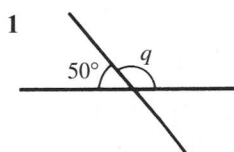

2
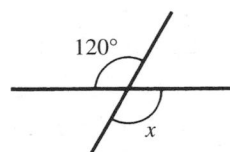

3
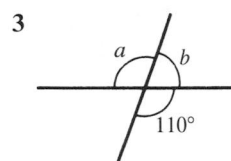

4
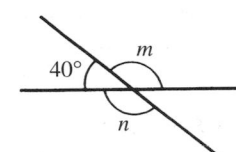

5
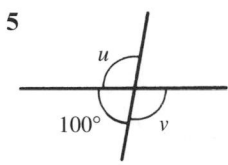

6
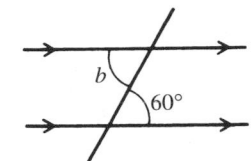

7
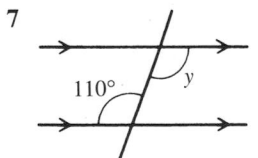

8
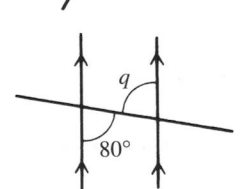

9
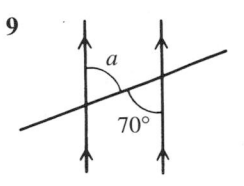

10
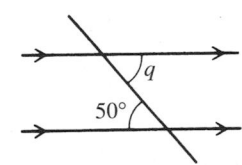

11
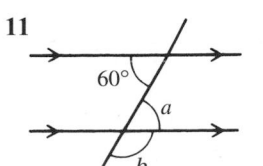

12
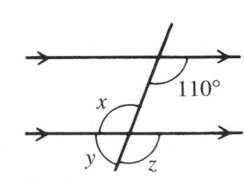

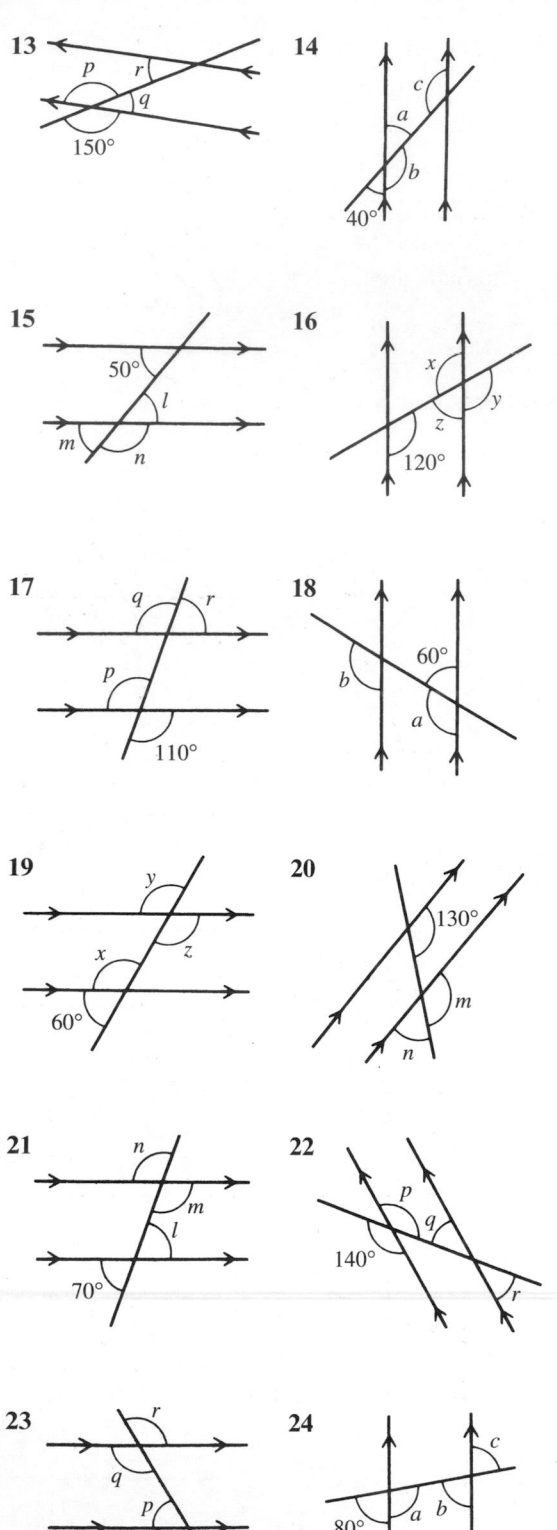

Knowledge

You are already familiar with acute-angled triangles, right-angled triangles and obtuse-angled triangles.

Here are three special types of triangle that you also need to know.

Equilateral triangle

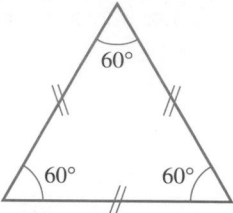

The sides are all equal in length.
The interior angles are all equal to 60°.
The equilateral triangle has three lines of symmetry, and also rotational symmetry.

Isosceles triangle

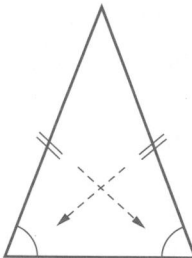

Two of the sides are equal in length.
The angles opposite the equal sides are equal.
The isosceles triangle has one line of symmetry, but no rotational symmetry.

Scalene triangle

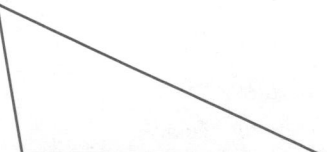

The sides are all of different lengths.
The angles are all of different size.
The scalene triangle has no symmetry.

Skill Example

Find the lettered angles in the following triangles.

a

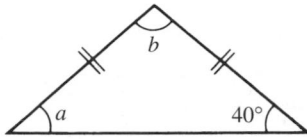

As the triangle is isosceles, $a = 40°$.

The sum of the angles in a triangle is $180°$. So,

$$a + 40° + b = 180°$$
$$40° + 40° + b = 180°$$
$$\Rightarrow \quad b = 100°$$

b

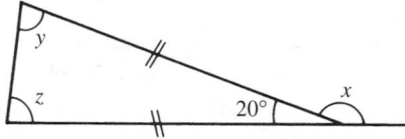

x and $20°$ are adjacent angles on a straight line. So,

$$x + 20° = 180°$$
$$\Rightarrow \quad x = 160°$$

The triangle is isosceles. So,

$$y = \frac{180° - 20°}{2} = \frac{160°}{2} = 80°$$

As $y = z$, then $z = 80°$.

Skill Practice Seven

Find the lettered angles in each of the following.

1

2
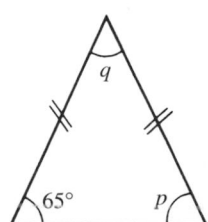

3
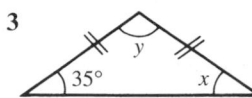

4
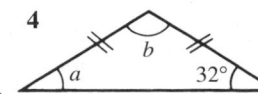

5
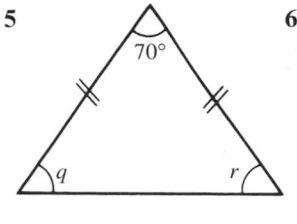

6
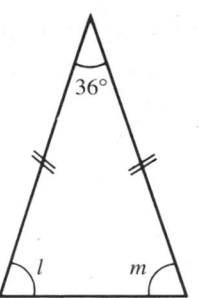

7
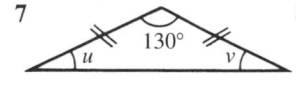

8
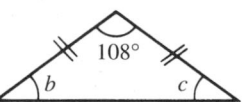

9
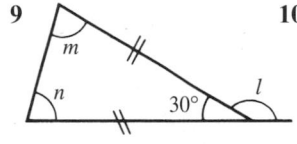

10
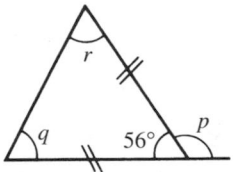

11
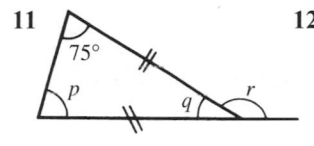

12
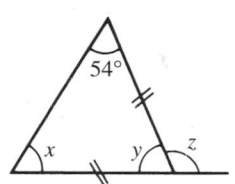

Skill Example

Find a, b and c.

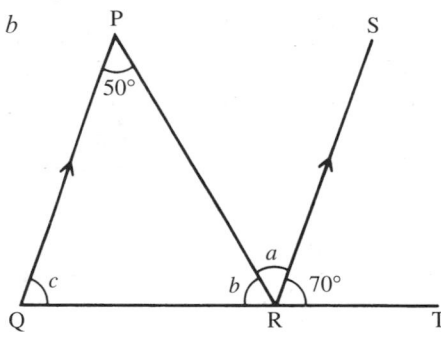

PQ is parallel to SR, so $a = 50°$ because a and $50°$ are alternate angles.

QRT is a straight line. So,

$$b + a + 70° = 180°$$
$$b + 50° + 70° = 180°$$
$$\Rightarrow \quad b = 60°$$

PQR is a triangle. So,

$$b + 50° + c = 180°$$
$$60° + 50° + c = 180°$$
$$\Rightarrow \quad c = 70°$$

Skill Practice Eight

Find the lettered angles in each of the following.

1

2

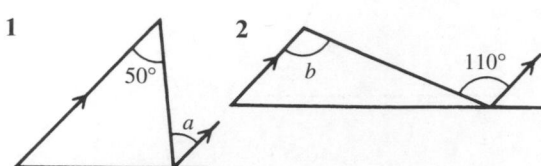

11

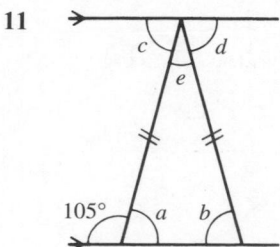

3

4

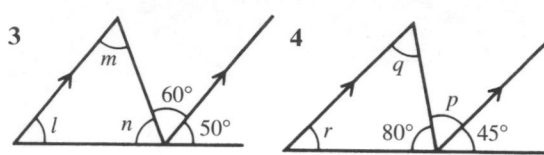

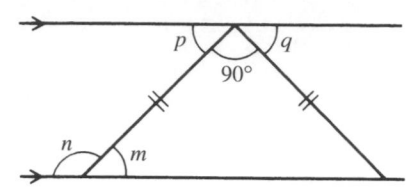

12

5

6

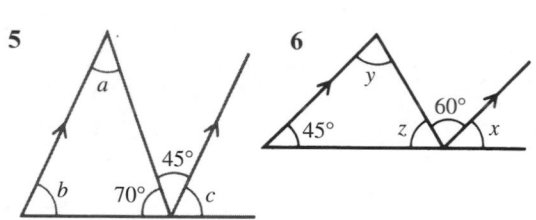

13

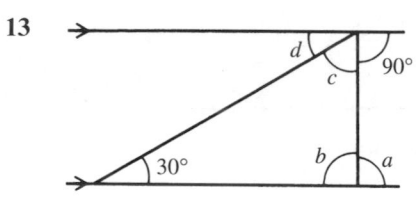

7

8

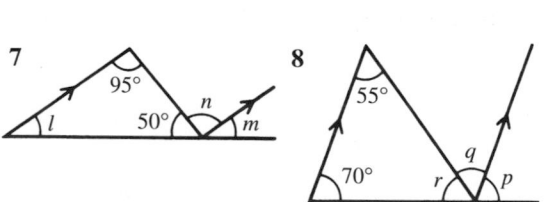

14

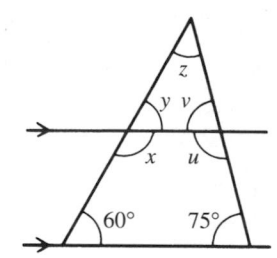

9

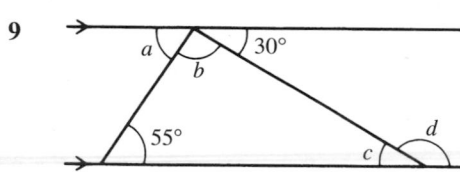

Knowledge

A **quadrilateral** is a closed figure with four sides and four angles. The sum of these four angles is 360°.

10

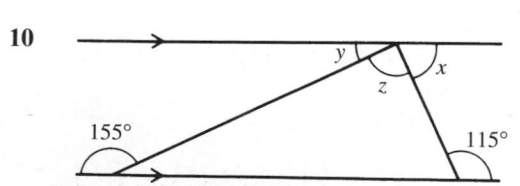

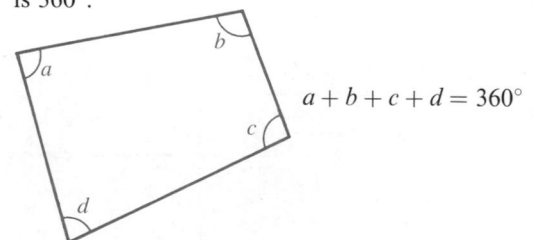

$$a + b + c + d = 360°$$

Skill Practice Nine

Copy these special quadrilaterals carefully. You
may find it easier to trace them.

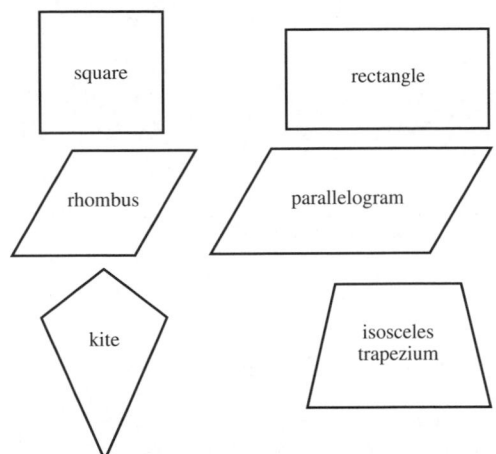

Then answer the following questions for each of
the six shapes.

1 Are the sides all the same length?
2 Are the opposite sides equal?
3 Are the adjacent sides equal?
4 Are the opposite sides parallel?
5 Are the angles all right angles?
6 Are the opposite angles the same size?
7 Are the diagonals the same length?
8 Do the diagonals cut each other at right angles?
9 Is the shape symmetrical?
   If so, state
   a the number of lines of symmetry
   b the order of rotational symmetry (if any).

Skill Practice Ten

1 Look at the two right-angled triangles in the
   illustration.

Draw a diagram to show how the pair can be
arranged to form:
a a rectangle
b a parallelogram
c a kite.

2 Look at the two isosceles triangles in the
   illustration.

Draw a diagram to show how the pair can be
arranged to form:
a a parallelogram
b a rhombus
c a kite.

3 Look at the four right-angled triangles in the
   illustration.

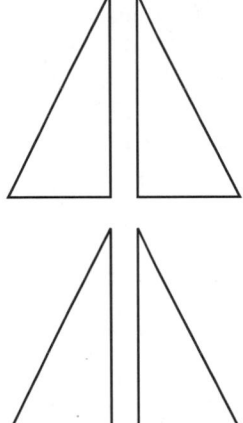

Draw a diagram to show how they can be
arranged to form:
a a rectangle
b a rhombus
c a parallelogram
d an isosceles trapezium.

4 Draw diagrams to show how:
   a two equilateral triangles can form a
      rhombus
   b three equilateral triangles can form an
      isosceles trapezium
   c four equilateral triangles can form a
      parallelogram
   d four right-angled isosceles triangles can
      form a square.

Knowledge

Look at this diagram of a regular octagon.

The octagon has rotational symmetry of order 8. This means that it will fit back into its original position 8 times as it is rotated through 360°.

Therefore, its external angles must total 360°.

Answer the following questions.

a What is the size of one of the exterior angles?
b What is the sum of an exterior angle and an interior angle?
c What is the size of one of the interior angles?

a The size of one exterior angle is 360° ÷ 8 = 45°.
b Each pair of interior and exterior angles forms a straight line, so the sum of any pair is 180°.
c The size of one interior angle = 180° − 45° = 135°.

Skill Practice Eleven

Look at these five regular polygons.

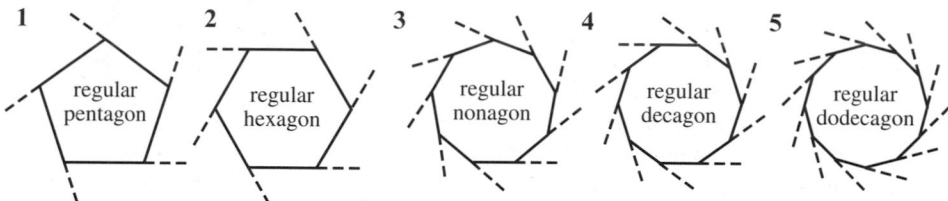

1 regular pentagon **2** regular hexagon **3** regular nonagon **4** regular decagon **5** regular dodecagon

Then answer the following questions for each polygon.

a How many equal exterior angles does the polygon have?
b What is the sum of the exterior angles?
c What is the size of one of the equal exterior angles?
d What is the sum of an exterior angle and an interior angle?
e What is the size of one of the equal interior angles?

6 Look at this star pattern.
  a Name the polygon ABCDEF.
  b Find the sizes of the following angles:
    (i) $F\widehat{A}B$ (ii) $P\widehat{A}B$
    (iii) the vertex angle $A\widehat{P}B$
  c What is the sum of all the vertex angles in the pattern?

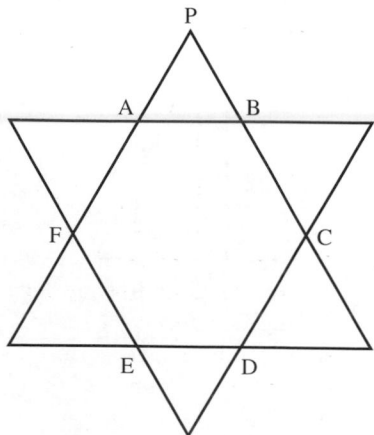

7 Look at this star pattern.

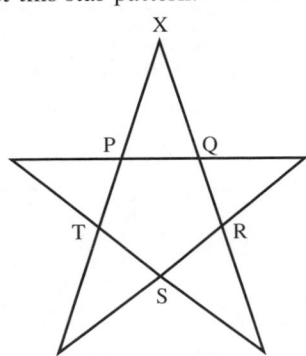

a Name the polygon PQRST.
b Find the sizes of the following angles:
  (i) TP̂Q
  (ii) XP̂Q
  (iii) the vertex angle PX̂Q
c What is the sum of all the vertex angles in the pattern?

8 The diagram illustrates the top of a fence stake. Find the size of angles *a* and *b*.

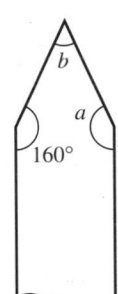

9 The head of the arrow is an isosceles triangle and its two flights are parallelograms.

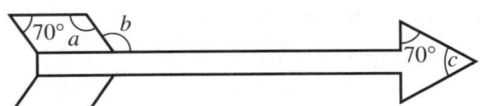

Find the size of the angles *a*, *b* and *c*.

Skill Example

Write down a set of instructions to produce this path.

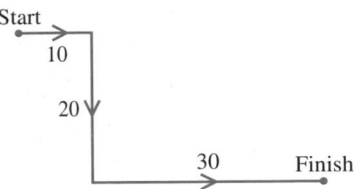

The instructions are:

> Forward 10, Right 90, Forward 20, Left 90, Forward 30

The instructions can be shortened to:

> FD10, RT 90, FD 20, LT 90, FD 30

Skill Practice Twelve

1 Write down a set of instructions to produce each of these paths.

a

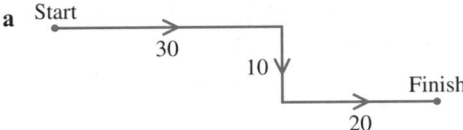

b

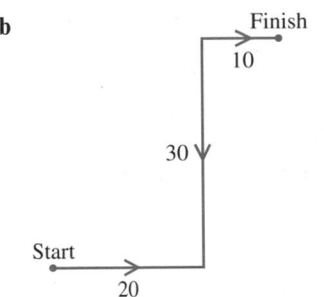

c

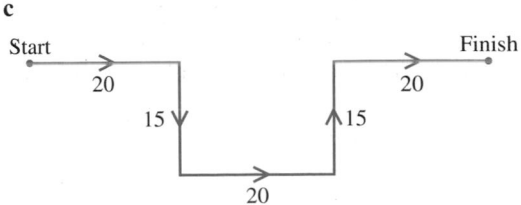

172

d

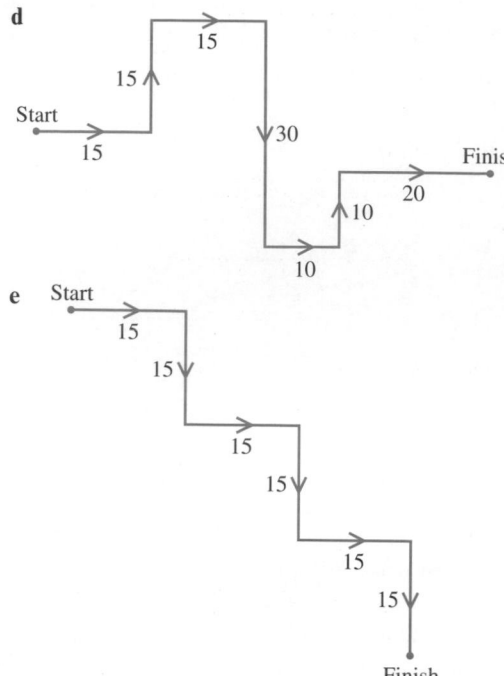

e Start

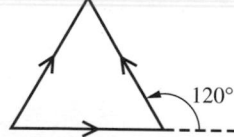

2 Draw each of the paths produced from these sets of instructions

a FD20, RT90, FD30, LT90, FD10

b FD10, LT90, FD10, RT90, FD10, LT90, FD10, RT90, FD10

c FD30, RT90, FD30, RT90, FD30, RT90, FD30

d FD30, LT90, FD60, LT90, FD30, LT90, FD60

e FD15, LT90, FD15, LT90, FD15, RT90, FD15, RT90, FD15

Skill Example

Write down a set of instructions to produce this equilateral triangle with a side length of 50 units.

The exterior angle of an equilateral triangle is 120°.
So, the set of instructions is:

FD50, LT120°, FD50, LT120°, FD50

Skill Practice Thirteen

Write down a set of instructions to produce each of these regular polygons with a side length of 50 units.

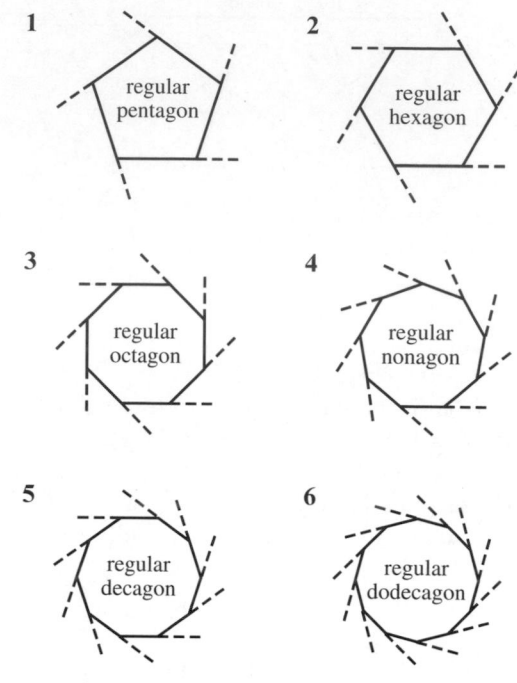

Skill Example

Write down a repeat instruction to produce an equilateral triangle with a side length of 50 units.

The instruction is:

REPEAT 3[FD50, LT120°]

Skill Practice Fourteen

Write down a repeat instruction to produce each of the regular polygons in Skill Practice Thirteen.

Unit 36

Shape, space and measures   Level 6

Skill   You understand and use appropriate
formulae for finding the areas of plane
rectilinear figures.

Skill Example

Find the area and the perimeter of each of
these rectangles.

a 9 mm [rectangle] 12 mm

b 25 mm [rectangle] 6 cm

c 1.2 m [square] 1.2 m

a Area $= 12\,\text{mm} \times 9\,\text{mm}$
   $= 108\,\text{mm}^2$
Perimeter $= 12\,\text{mm} + 9\,\text{mm} + 12\,\text{mm} + 9\,\text{mm}$
   $= 42\,\text{mm}$

b Area $= 25\,\text{mm} \times 6\,\text{cm}$
   $= 2.5\,\text{cm} \times 6\,\text{cm}$
   $= 15\,\text{cm}^2$
Perimeter $= 2.5\,\text{cm} + 6\,\text{cm} + 2.5\,\text{cm} + 6\,\text{cm}$
   $= 17\,\text{cm}$

c Area $= 1.2\,\text{m} \times 1.2\,\text{m}$
   $= 1.44\,\text{m}^2$
Perimeter $= (1.2 + 1.2 + 1.2 + 1.2)\,\text{m}$
   $= 4.8\,\text{m}$

Skill Practice One

For each question find which rectangle has a
different area from the other two.

1 a
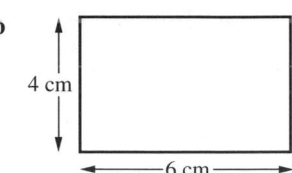
3 cm, 8 cm

b
4 cm, 6 cm

c
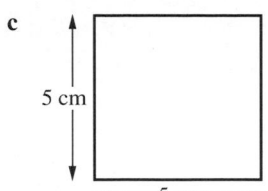
5 cm, 5 cm

2 a
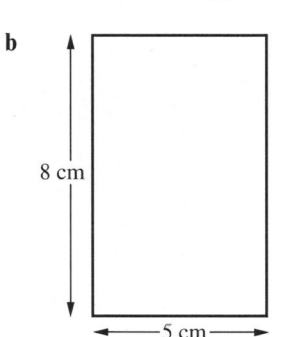
6 cm, 7 cm

b
8 cm, 5 cm

c
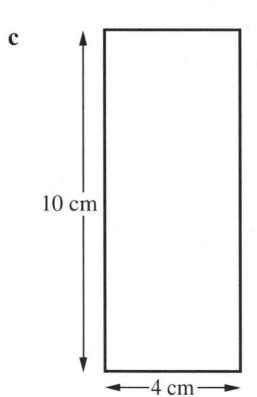
10 cm, 4 cm

3 a
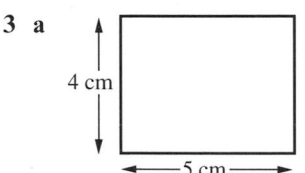
4 cm, 5 cm

b
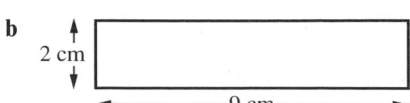
2 cm, 9 cm

174

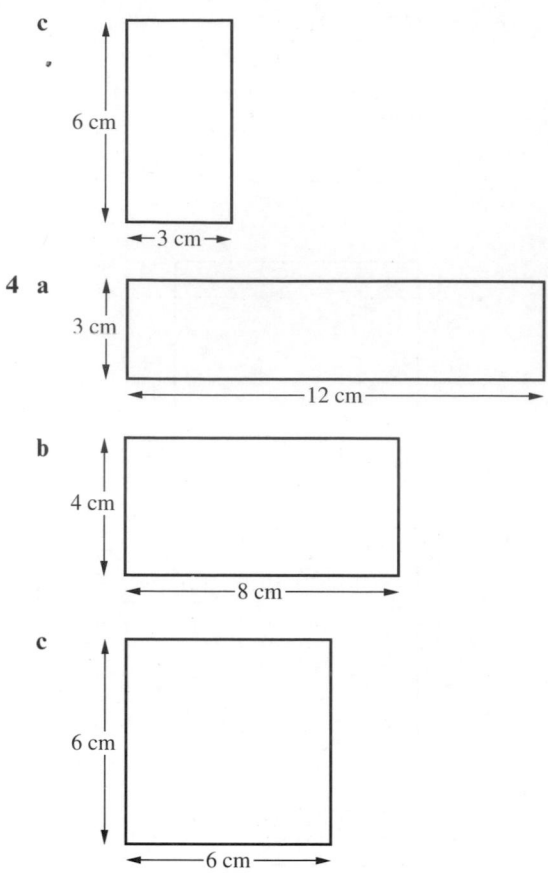

c

6 cm

←3 cm→

4 a

3 cm

←————12 cm————→

b

4 cm

←———8 cm———→

c

6 cm

←———6 cm———→

Skill Practice Two

For each question in Skill Practice One, find which rectangle has a different perimeter from the other two.

Skill Practice Three

Copy and complete the following table.

	Length	Width	Area	Perimeter
1	9 cm	4 cm		
2	7 cm	8 cm		
3	6 cm	9 cm		
4	5 cm	7 cm		
5	8 cm		48 cm^2	
6	50 mm		2000 mm^2	
7		5 cm	45 cm^2	
8		4 m	60 m^2	
9	6 cm			16 cm
10	7 cm			22 cm
11		4 m		18 m
12		5 m		26 m

Skill Example

Find **a** the area, **b** the perimeter of the shape illustrated.

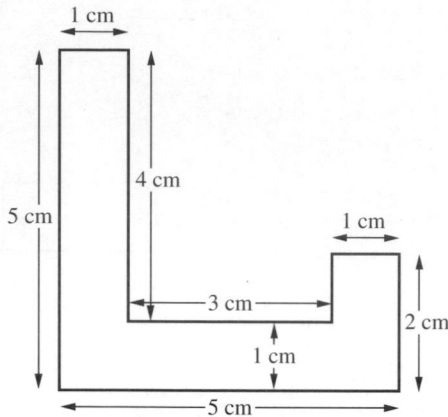

The area is found by dividing the shape into three rectangles, as shown below, and then adding together the three separate areas.

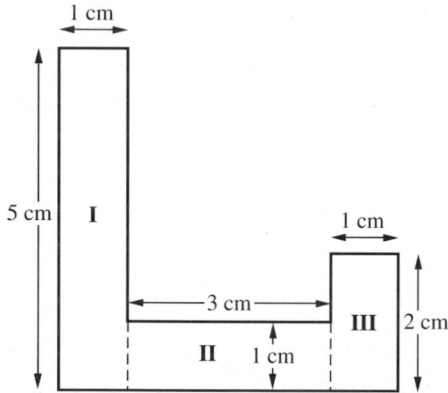

Area of rectangle I = $5 \times 1 = 5 \, \text{cm}^2$

Area of rectangle II = $3 \times 1 = 3 \, \text{cm}^2$

Area of rectangle III = $2 \times 1 = 2 \, \text{cm}^2$

Therefore, area of the shape is:

$$5 \, \text{cm}^2 + 3 \, \text{cm}^2 + 2 \, \text{cm}^2 = 10 \, \text{cm}^2$$

Perimeter of the shape is:

$$5 + 5 + 2 + 1 + 1 + 3 + 4 + 1 = 22 \, \text{cm}$$

Skill Practice Four

Find **a** the area, **b** the perimeter of each of the
following shapes.

1

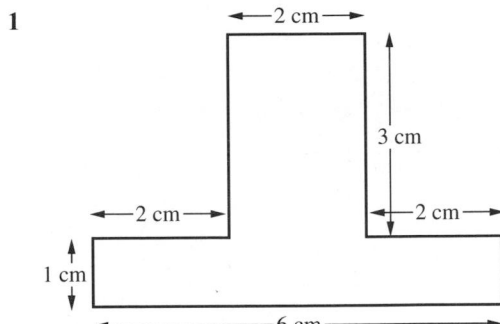

2

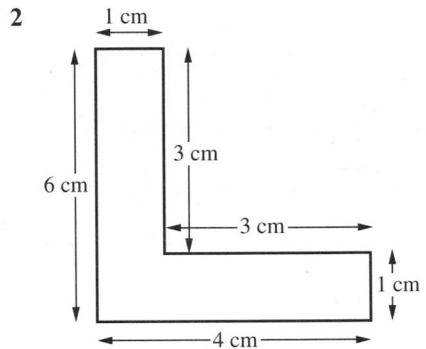

3

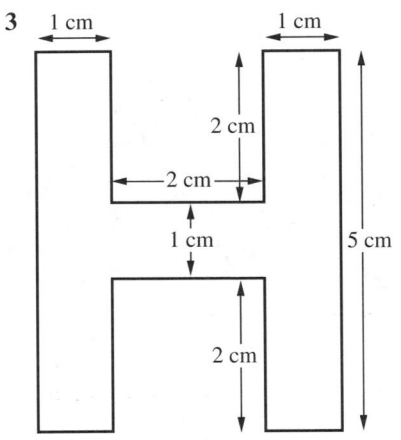

4

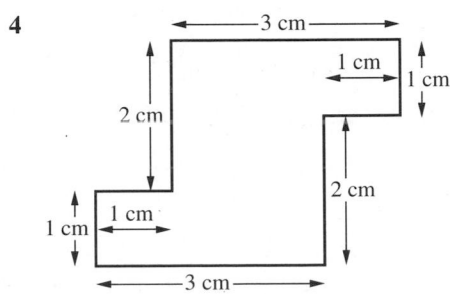

5

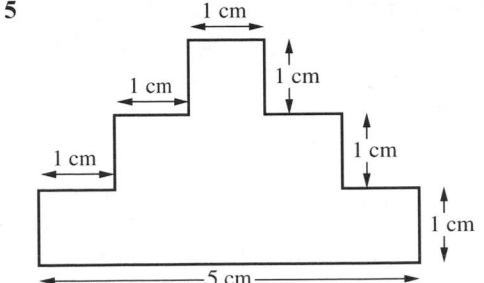

Knowledge

Area of a triangle
$$= \tfrac{1}{2} \times \text{Length of base} \times \text{Height}$$

Skill Example

Find the area of each of these triangles.

a

7 cm

8 cm

b

5.5 cm

4 cm

c

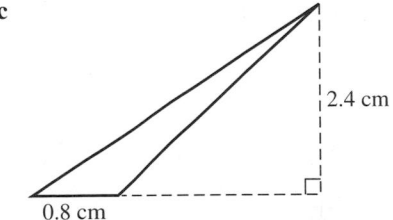

2.4 cm

0.8 cm

a Area $= \tfrac{1}{2} \times 8\,\text{cm} \times 7\,\text{cm} = 4 \times 7 = 28\,\text{cm}^2$

b Area $= \tfrac{1}{2} \times 4\,\text{cm} \times 5.5\,\text{cm} = 2 \times 5.5$
$= 11.0\,\text{cm}^2$

c Area $= \tfrac{1}{2} \times 0.8\,\text{cm} \times 2.4\,\text{cm} = 0.4 \times 2.4$
$= 0.96\,\text{cm}^2$

176

Skill Practice Five

Find the area of each of the following triangles.

1

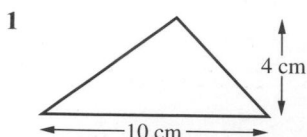

2

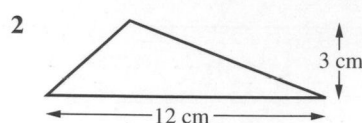

3

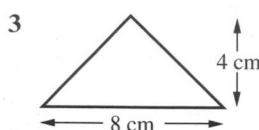

4

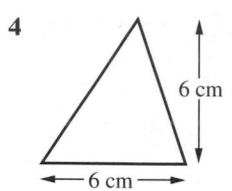

5

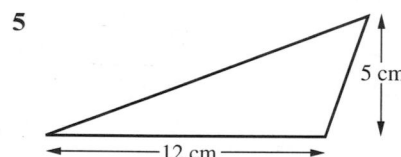

6

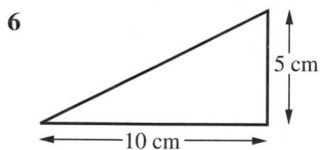

7

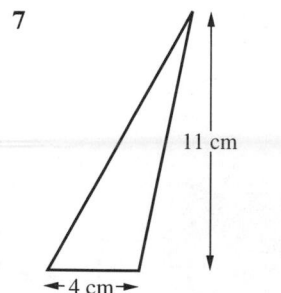

8

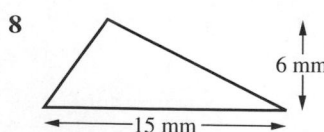

9

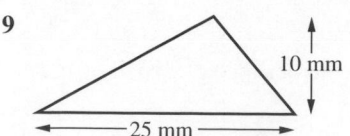

10

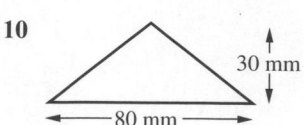

11

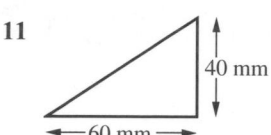

12

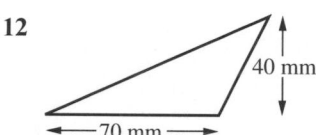

Skill Example

Find the area of the kite ABCD.

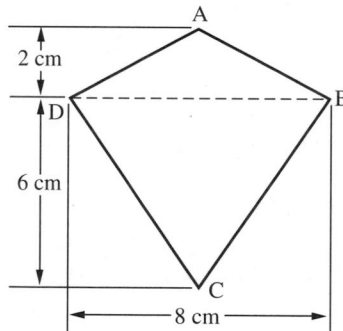

The line BD splits the kite into two triangles, ABD and BCD.

Area of ABD $= \frac{1}{2}$ (Base $\times$ Height)

$= \frac{1}{2} (8 \times 2) = 8\,\text{cm}^2$

Area of BCD $= \frac{1}{2}$ (Base $\times$ Height)

$= \frac{1}{2} (8 \times 6) = 24\,\text{cm}^2$

So, area of kite $= 8\,\text{cm}^2 + 24\,\text{cm}^2$

$= 32\,\text{cm}^2$

Skill Example

Find the area of the trapezium PQRS.

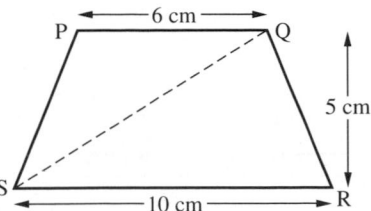

The line QS splits the trapezium into two triangles, PQS and QRS.

Area of PQS $= \frac{1}{2}$ (Base × Height)

$\qquad = \frac{1}{2}(6 \times 5) = 15\,\text{cm}^2$

Area of QRS $= \frac{1}{2}$ (Base × Height)

$\qquad = \frac{1}{2}(10 \times 5) = 25\,\text{cm}^2$

So, area of trapezium $= 15\,\text{cm}^2 + 25\,\text{cm}^2$

$\qquad\qquad\qquad\qquad = 40\,\text{cm}^2$

Skill Practice Six

Find the area of each of the following shapes.

1

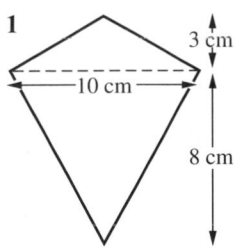

2

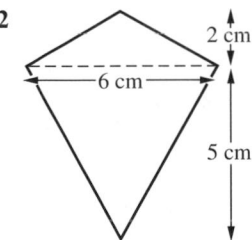

3

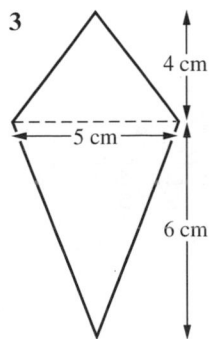

4

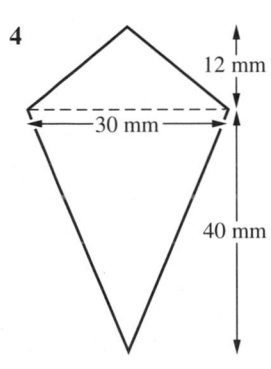

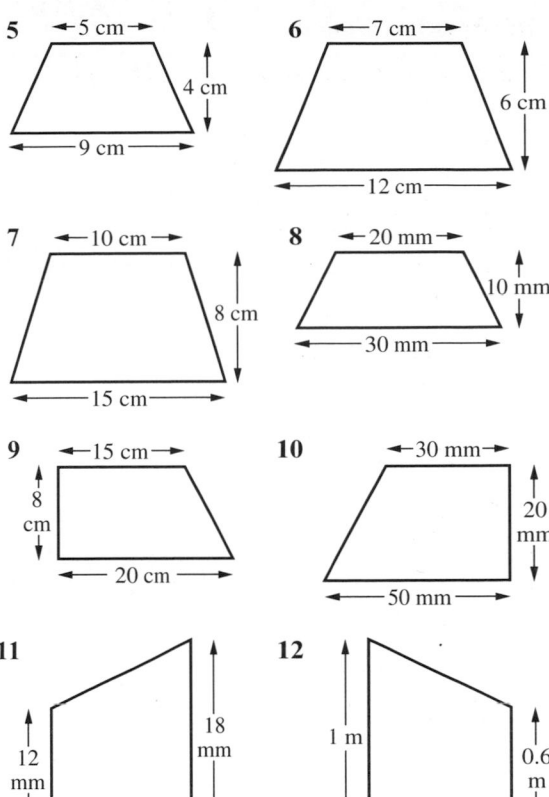

5 5 cm · 4 cm · 9 cm

6 7 cm · 6 cm · 12 cm

7 10 cm · 8 cm · 15 cm

8 20 mm · 10 mm · 30 mm

9 15 cm · 8 cm · 20 cm

10 30 mm · 20 mm · 50 mm

11 18 mm · 12 mm · 15 mm

12 1 m · 0.6 m · 0.75 m

Skill Example

Find the area of each of the following parallelograms.

a

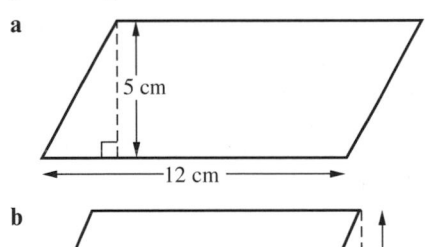

b

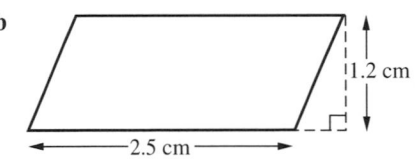

a Area = Base × Height

$\qquad = 12 \times 5$

$\qquad = 60\,\text{cm}^2$

b Area = Base × Height

$\qquad = 2.5 \times 1.2$

$\qquad = 3.00 \text{ or } 3\,\text{cm}^2$

Skill Practice Seven

Find the area of each of the following parallelograms.

1

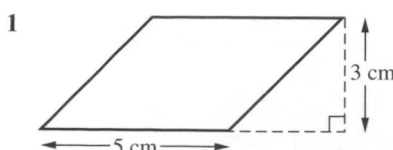

2

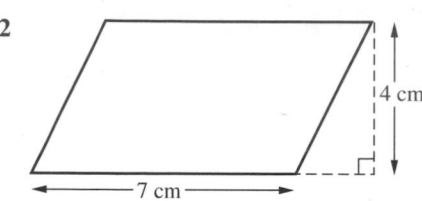

3

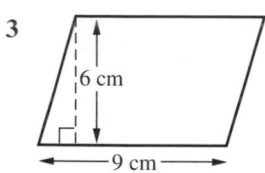

4

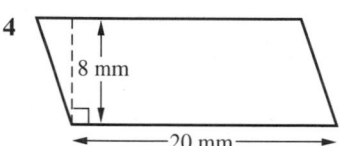

5

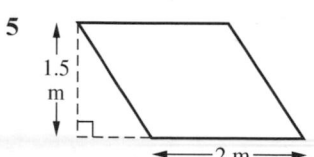

Skill Example

The areas of some shapes may be found more easily by subtraction.

A rectangular lawn measures 8 m by 5 m. It is surrounded by a path of width $\frac{1}{2}$ m.
Find the area of the path, shown shaded.

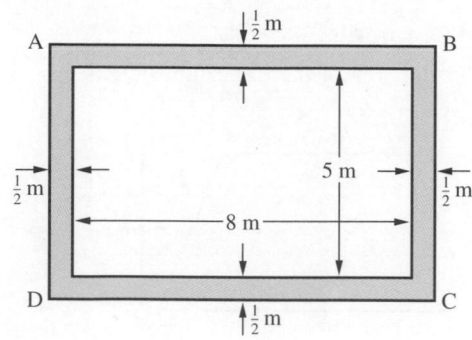

$AB = \frac{1}{2}\,m + 8\,m + \frac{1}{2}\,m = 9\,m$

$BC = \frac{1}{2}\,m + 5\,m + \frac{1}{2}\,m = 6\,m$

Therefore, area $ABCD = 9\,m \times 6\,m = 54\,m^2$

Area of lawn $= 8\,m \times 5\,m = 40\,m^2$

So, area of path $= 54\,m^2 - 40\,m^2 = 14\,m^2$

Skill Practice Eight

Find the area of the shaded part of each shape.

1

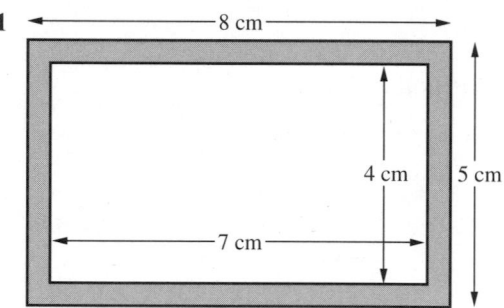

2

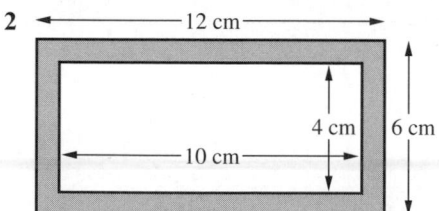

3

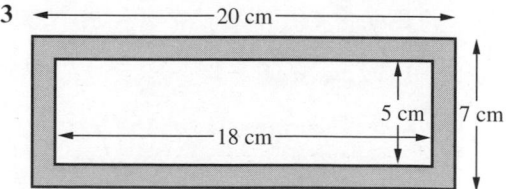

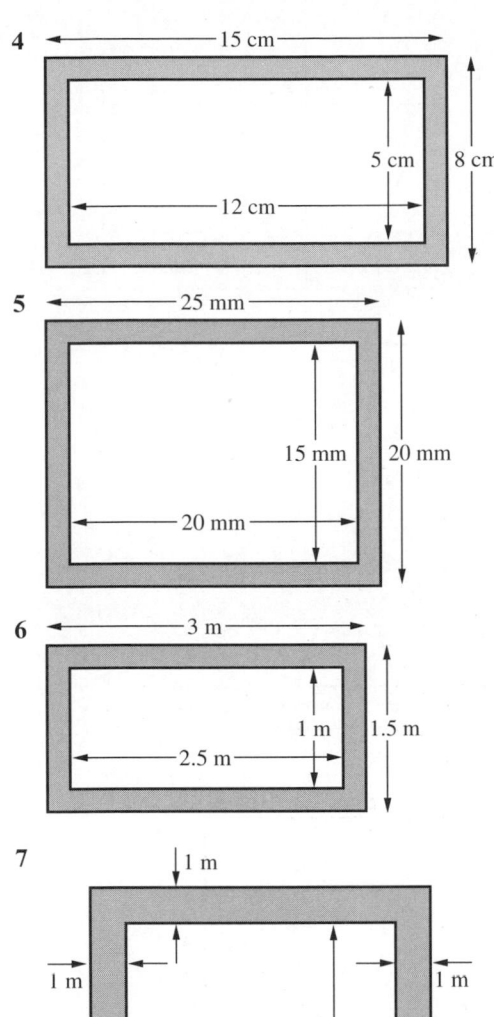

4 15 cm · 8 cm · 5 cm · 12 cm

5 25 mm · 20 mm · 15 mm · 20 mm

6 3 m · 1.5 m · 1 m · 2.5 m

7 1 m · 1 m · 1 m · 5 m · 6 m · 1 m

8 $\frac{1}{2}$ cm · $\frac{1}{2}$ cm · $\frac{1}{2}$ cm · 3 cm · 4 cm · $\frac{1}{2}$ cm

Skill Example

Find the area of the shaded part.

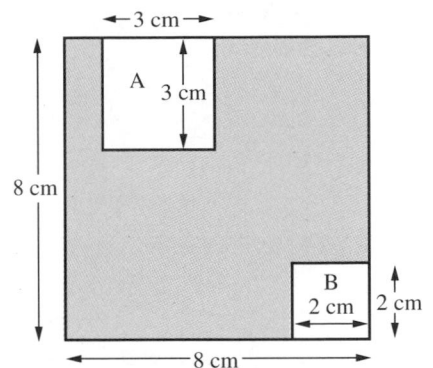

3 cm · A · 3 cm · 8 cm · B · 2 cm · 2 cm · 8 cm

Area of whole shape $= 8 \times 8 = 64\,\text{cm}^2$

Area of A $= 3 \times 3 = 9\,\text{cm}^2$

Area of B $= 2 \times 2 = 4\,\text{cm}^2$

So, shaded area $= 64 - 9 - 4 = 51\,\text{cm}^2$

Skill Practice Nine

Find the area of the shaded part in each shape.

1

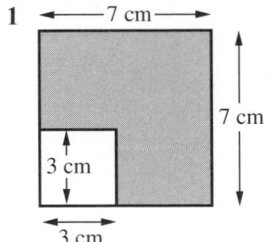

7 cm · 7 cm · 3 cm · 3 cm

2

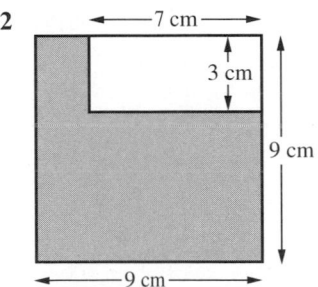

7 cm · 3 cm · 9 cm · 9 cm

3

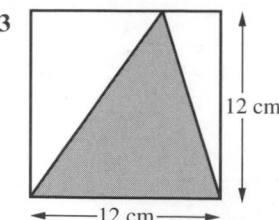

12 cm

12 cm

4

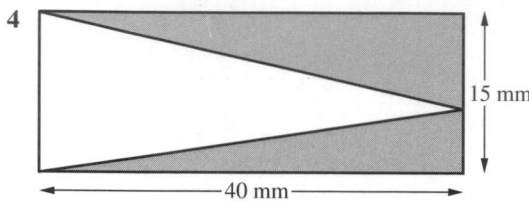

15 mm

40 mm

5

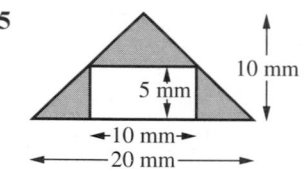

5 mm

10 mm

10 mm

20 mm

6

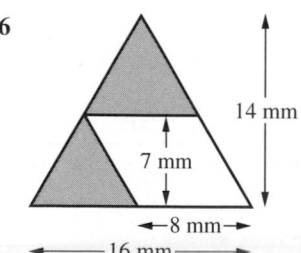

14 mm

7 mm

8 mm

16 mm

7

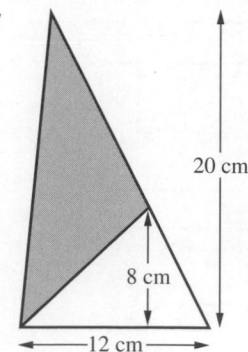

20 cm

8 cm

12 cm

8

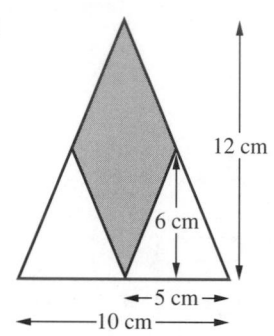

12 cm

6 cm

5 cm

10 cm

9

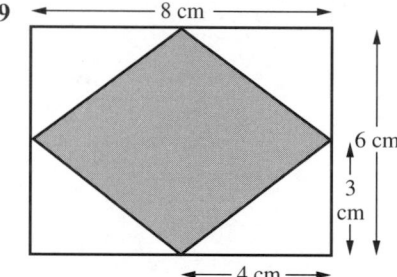

8 cm

6 cm

3 cm

4 cm

10

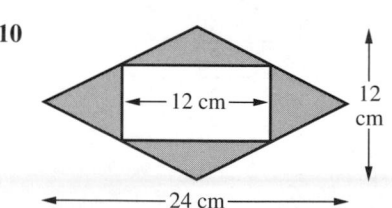

12 cm

12 cm

24 cm

Skill Example

One wall of a living room is to be covered with wallpaper costing £3.50 per m^2.
The dimensions are shown in the diagram.
Find the cost of the wallpaper.

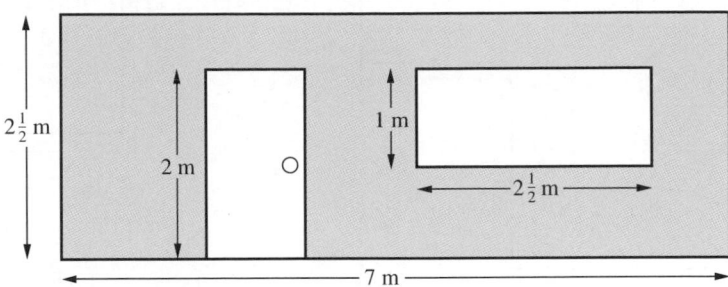

Area of wall = $7\,\text{m} \times 2\frac{1}{2}\,\text{m} = 17\frac{1}{2}\,\text{m}^2$

Area of door = $2\,\text{m} \times 1\,\text{m} = 2\,\text{m}^2$

Area of window = $2\frac{1}{2}\,\text{m} \times 1\,\text{m} = 2\frac{1}{2}\,\text{m}^2$

So, area to be covered with wallpaper

$$= 17\frac{1}{2}\,\text{m}^2 - 2\,\text{m}^2 - 2\frac{1}{2}\,\text{m}^2 = 13\,\text{m}^2$$

Therefore, cost = £3.50 × 13 = £45.50

Skill Practice Ten

1 A bedroom wall has one window.
All the dimensions are shown in
the diagram.
Find the area of wallpaper
required to cover this wall.

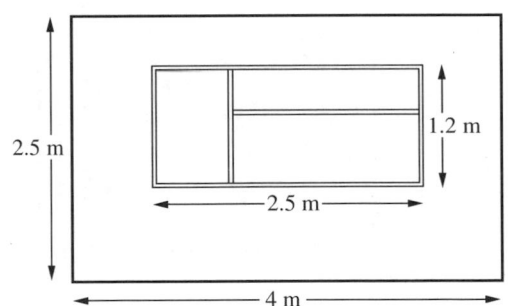

2 The diagram shows the plan of
a loft which is to be insulated.
Find the area of insulating
material required.

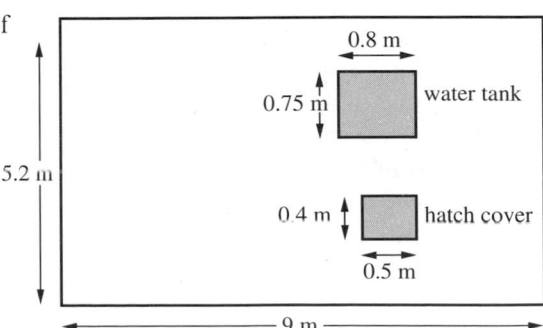

3 The diagram shows the dimensions of a yard.

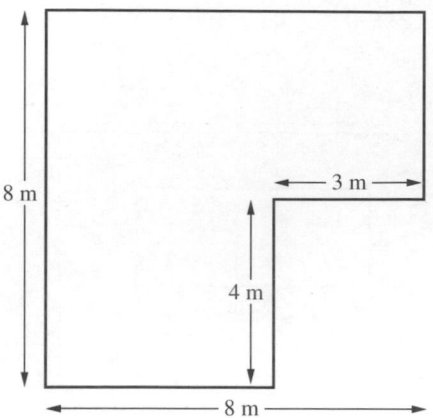

Find:

a the area of the yard

b the number of paving slabs 0.8 m × 0.5 m required to cover the yard.

4 A square lawn is bordered by a shingle path.

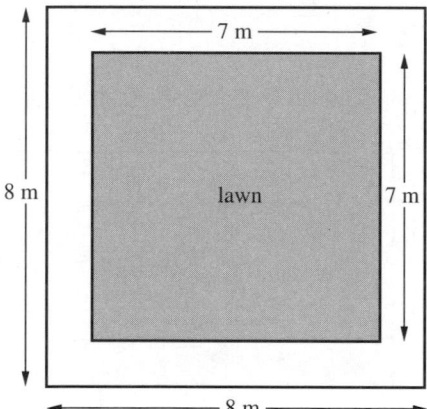

Find:

a the area of the path

b the number of bags of shingle required to cover the path if it is supplied in 50 kg bags and 1 m² of path requires 40 kg of shingle.

5 The ends of a rabbit hutch are to be made from two pieces of wood measuring 50 cm × 50 cm. The dimensions are shown in the diagram (top right).

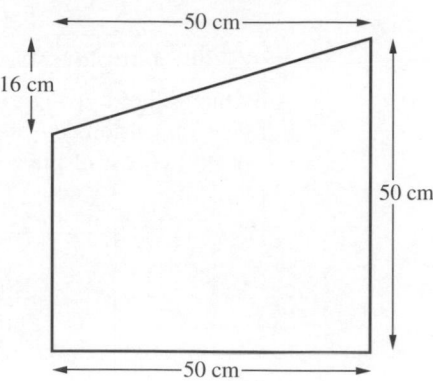

Find the area of each of the ends.

6 An arrow for a signpost is to be cut from a sheet of metal measuring 20 cm × 20 cm.

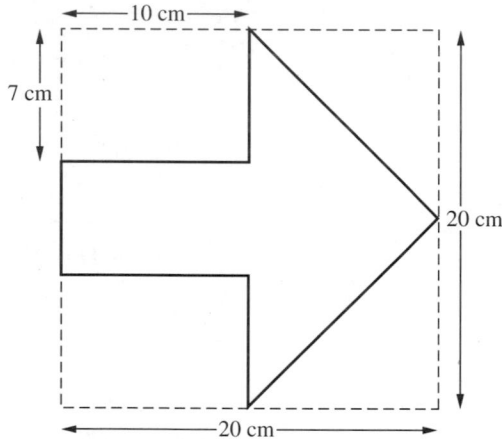

The dimensions are shown in the diagram. Find the area of the arrow.

7 The diagram shows a piece of metal which is to be used for making a saw blade. Find its area.

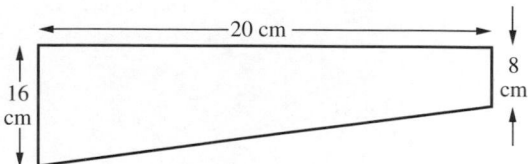

8 The diagram shows the end of a terrace of houses.
Find the area of this end wall.

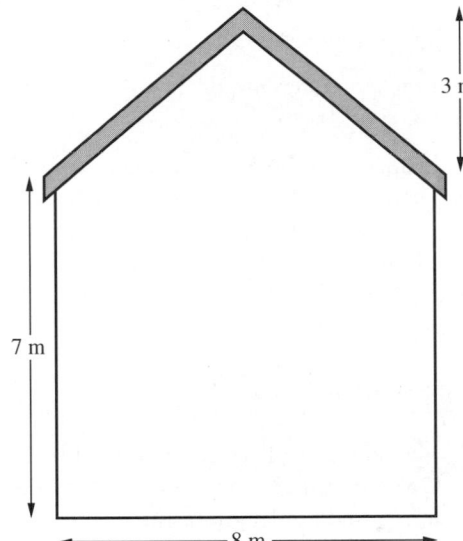

9 Find the area of this spinner dice.

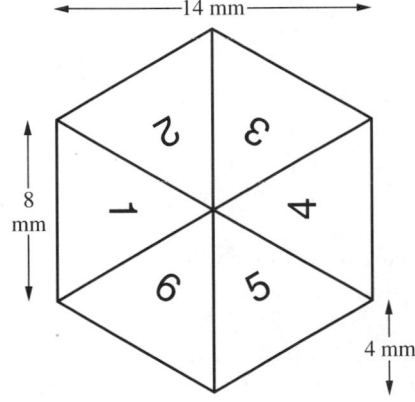

10 The diagram below shows a piece of wood which has been cut for making the deck of a toy boat.
Find its area.

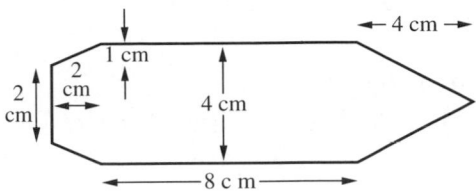

11 The diagram shows a joiner's tri-square.
Find:
a the area of the metal blade
b the area of the wooden handle
c the area of the blade which is enclosed by the handle in both mm² and cm².

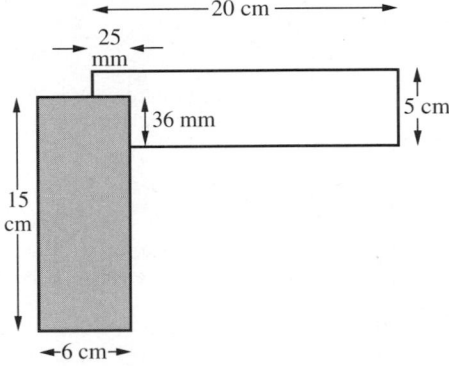

Unit 37

Shape, space and measures   Level 6

Skill   You understand and use appropriate formulae for finding the circumferences and areas of circles.

Knowledge

The diagram shows the names of the parts of a circle.

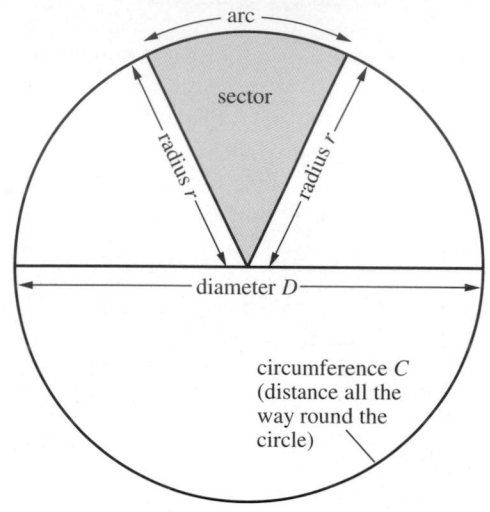

The ratio $\dfrac{C}{D}$ is the same for all circles. It is represented by the Greek letter π (pi).

$$\pi = \frac{C}{D} \approx 3.14$$

The symbol $\approx$ means 'is approximately equal to'.

Because $\dfrac{C}{D} = \pi$

then $C = \pi D = 2\pi r$

where r is the radius of the circle.

Skill Example

Find the circumference of each of the following circles (take $\pi = 3.14$).

a Diameter $= 6$ cm   **b** Radius $= 7$ cm

a Circumference $= \pi D$
$$= 3.14 \times 6 = 18.84 \text{ cm}$$

b Circumference $= \pi D$
$$= 2\pi r = 2 \times 3.14 \times 7$$
$$= 43.96 \text{ cm}$$

Skill Practice One

Find the circumference of the following circles (take $\pi = 3.14$).

1 Diameter $= 21$ cm   **2** Diameter $= 35$ cm
3 Diameter $= 49$ cm   **4** Diameter $= 63$ cm
5 Diameter $= 77$ mm   **6** Diameter $= 140$ mm
7 Radius $= 35$ cm   **8** Radius $= 14$ cm
9 Radius $= 21$ cm   **10** Radius $= 28$ cm
11 Radius $= 42$ mm   **12** Radius $= 105$ mm

In questions **13** to **16**, take $\pi = 3.14$.

13 The diameter of a reel of wire is 14 cm.
Given that the wire is wound round the reel 50 times, find the length of the wire:
a in centimetres   **b** in metres

14 Ranjit's bicycle has a front wheel of diameter 50 cm.
Find the distance that he has cycled after the front wheel has turned round 500 times.
Give your answer in metres.

15 Josh has a marble of diameter 1.4 cm. He rolls it towards another marble which is 66 cm away.
How many revolutions will his marble make before hitting the second one?

16 Jolene has a hoop of diameter 1.05 m.
How many revolutions will it make as she rolls it along her garden path which is 16.5 m in length?

Knowledge

The area of a circle is given by:

$$\text{Area} = \pi r^2$$

Skill Example

Find the area of each of these circles
(take $\pi = 3.14$).

a Radius $= 10$ cm

b Diameter $= 14$ m

a Area $= \pi r^2 = 3.14 \times 10 \times 10 = 31.4 \times 10$
$$= 314 \text{ cm}^2$$

b Radius $= \frac{1}{2} \times$ Diameter $= \frac{1}{2} \times 14 = 7$ m

Area $= \pi r^2 = 3.14 \times 7 \times 7$
$$= 153.86 \text{ m}^2$$

Skill Practice Two

Find the area of each of the following circles
(take $\pi = 3.14$).

1 Radius $= 2$ cm   **2** Radius $= 4$ cm
3 Diameter $= 6$ cm   **4** Diameter $= 10$ cm
5 Radius $= 0.1$ m   **6** Radius $= 70$ mm
7 Radius $= 3\frac{1}{2}$ cm **8** Radius $= 14$ cm
9 Diameter $= 42$ cm   **10** Diameter $= 1.4$ m
11 Radius $= 1\frac{2}{5}$ m

In questions **12** to **16**, take $\pi = 3.14$.

12 A push button switch which operates a door
bell has a radius of 5 mm.
Find its surface area.

13 A dartboard has a radius of 28 cm.
Find its surface area.

14 The diagram shows the dimensions of a metal
washer.
Find its area.

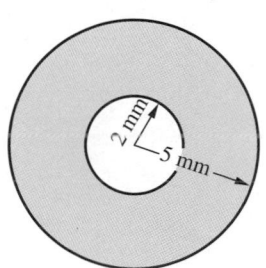

15 The diagram shows a 'No Entry' sign which
has a radius of 35 cm.
If the white rectangle measures 60 cm by
10 cm, find the area of the sign that is painted
red.

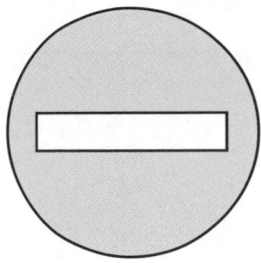

16 Find the shaded area of each of the two
diagrams below.

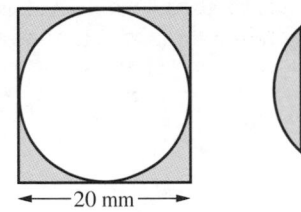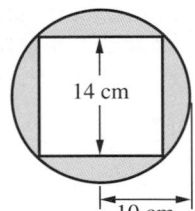

Skill Example

A running track has two 'straights', each 112 m
long and placed 56 m apart. They are joined by
semicircles, as shown below.

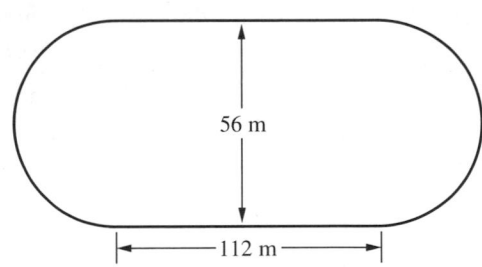

Find the length of one lap round the track.
Assume $\pi = 3.14$.

Total distance $=$ (Length of straights)
$\qquad\qquad\quad +$ (Circumference of circle)
$\qquad\quad = (112 \times 2)\,\text{m} + (3.14 \times 56)\,\text{m}$
$\qquad\quad = 224\,\text{m} + 175.84\,\text{m} = 399.84\,\text{m}$

Skill Practice Three

For all questions, assume $\pi = 3.14$.

1 A running track has two 'straights', each 68 m long and 84 m apart, which are joined by semi-circles as shown.

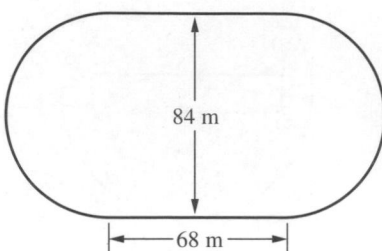

Find the length of one lap round the track.

2 The diagram shows the dimensions of a dodg'em car circuit at a fairground.

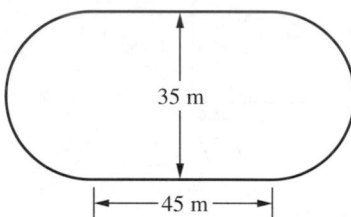

Find the perimeter of the circuit, given that both curved ends are semicircles.

3 The diagram shows the dimensions of a model railway track which consists of two straight sections and two semicircular sections.

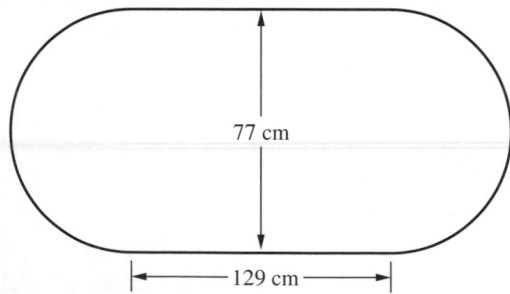

Find the length of one lap around the track.

4 From the dimensions given in the illustration, find the perimeter of the rim of the bath. Each curved end is a semicircle.

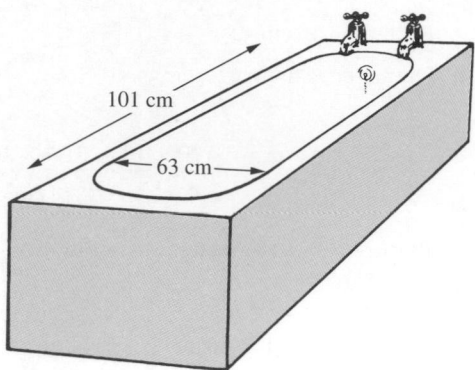

5 The illustration shows a ticket office at an underground station. Find its perimeter, given that each end is a semicircle.

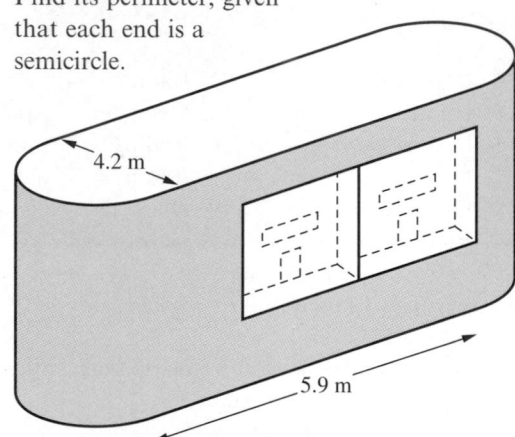

6 From the dimensions given in the illustration, find the perimeter of the swimming pool. Each end is a semicircle.

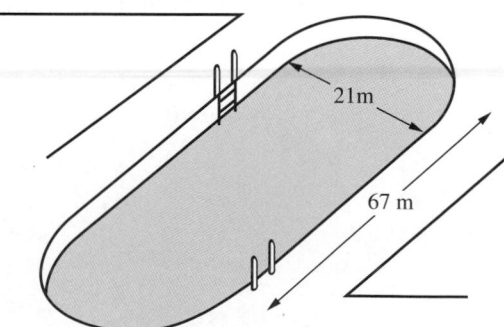

7 Find the area of each mirror shown below.

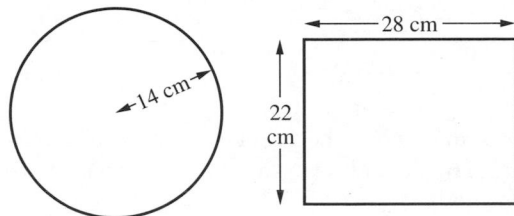

What do you notice about the two areas?

8 Find the area of the circular piece of glass covering the face of this clock.

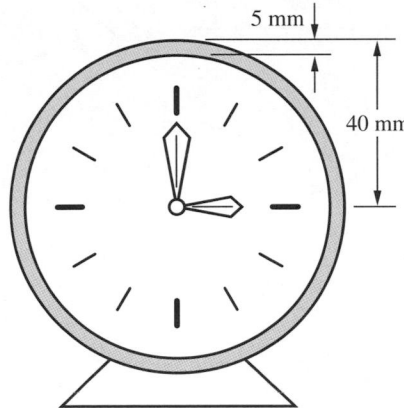

9 Mr Briggs lives at 10 Park Street. He cuts two metal figures for his door number from a rectangular metal plate, to the sizes shown.

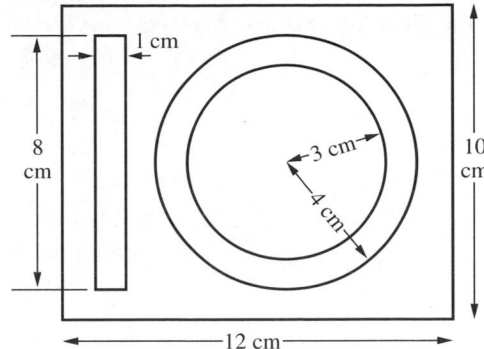

a Find the total area of metal that forms the numbers.

b Find the area of metal that is wasted when he has cut out the numbers from the metal plate.

10 Find the shaded area of each shape.

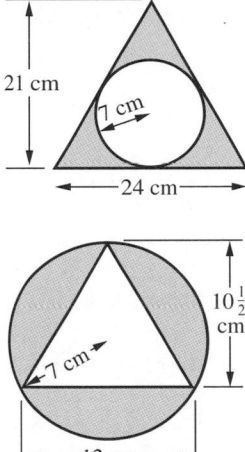

Unit 38

Shape, space and measures   Level 6

Skill   You understand and use appropriate formulae for finding the volume of cuboids.

Knowledge

Instead of counting the number of unit cubes, the volume of a cuboid can be found by multiplying the length (L) by the width (W) by the height (H), provided each one is measured in the same units.

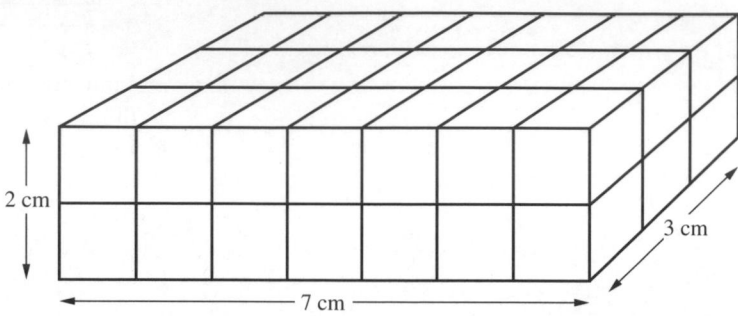

For the cuboid above:

$$\text{Volume } (V) = L \times W \times H$$
$$= 7\,\text{cm} \times 3\,\text{cm} \times 2\,\text{cm}$$
$$= 42\,\text{cm}^3$$

Skill Example

Find the volume of the following cuboids.

a Length = 10 cm   **b** Length = 30 mm
   Width = 10 cm   Width = 30 mm
   Height = 10 cm   Height = 4 cm

c Length = 4 m
   Width = 2.5 m
   Height = 1.5 m

a Volume = $L \times W \times H$
          = 10 cm × 10 cm × 10 cm
          = 1000 cm^3

b Volume = $L \times W \times H$
          = 30 mm × 30 mm × 40 mm
          = 3 cm × 3 cm × 4 cm
          = 36 cm^3

c Volume = $L \times W \times H$ = 4 m × 2.5 m × 1.5 m
                   = 10 × 1.5 m^3
                   = 15 m^3

Skill Practice One

For each question, find which cuboid has a different volume from the other two.

1 a
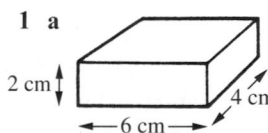

  b
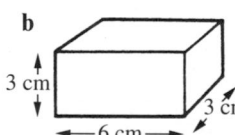

  c
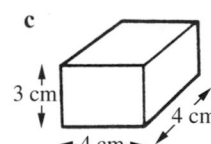

2 a
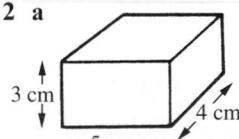

  b
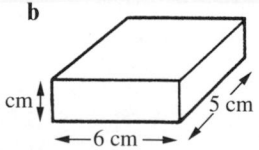

  c
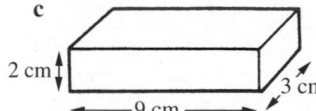

	Length	Width	Height
3 a	12 cm	3 cm	2 cm
b	8 cm	2 cm	4 cm
c	4 cm	4 cm	4 cm
4 a	6 cm	6 cm	2 cm
b	8 cm	2 cm	5 cm
c	3 cm	3 cm	8 cm
5 a	7 cm	2 cm	6 cm
b	10 cm	4 cm	2 cm
c	4 cm	4 cm	5 cm
6 a	3 cm	3 cm	10 cm
b	9 cm	5 cm	2 cm
c	7 cm	4 cm	3 cm

Skill Example

A cuboid has a volume of $72\,\text{cm}^3$. If its length is 6 cm, and its width is 4 cm, find its height.

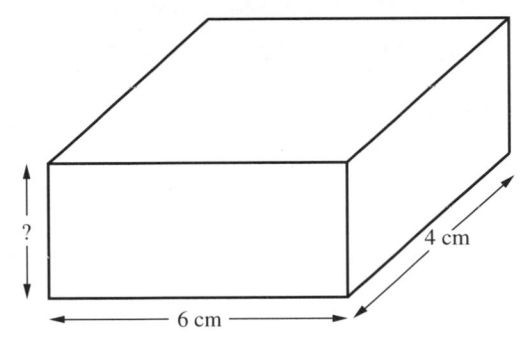

$$\text{Volume} = L \times W \times H$$

$$72 = 4 \times 6 \times H$$

$$72 = 24 \times H$$

$$\Rightarrow \quad H = \frac{72}{24} = 3\,\text{cm}$$

Therefore, the height of the cuboid is 3 cm.

Skill Practice Two

Copy and complete the following table.

	Length	Width	Height	Volume
1	5 cm	2 cm		$30\,\text{cm}^3$
2	8 cm	2 cm		$32\,\text{cm}^3$
3	5 cm		2 cm	$40\,\text{cm}^3$
4	10 cm		2 cm	$60\,\text{cm}^3$
5		2 cm	3 cm	$48\,\text{cm}^3$
6		5 cm	2 cm	$70\,\text{cm}^3$

Knowledge

An oil storage tank, as illustrated, measures 2 m by $1\frac{1}{2}$ m by 1 m, and so has a volume of

$$2 \times 1\tfrac{1}{2} \times 1 = 3\,\text{m}^3$$

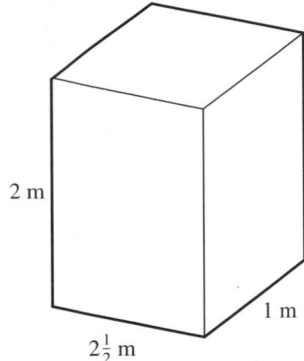

Its capacity is therefore 3000 litres of oil, because

$$1\,\text{m}^3 = 1000 \text{ litres}$$

The **capacity** of the tank is the amount of liquid (in this case oil) which it contains when full.

Skill Example

Write as litres:

a $2500\,\text{cm}^3$ **b** $3.75\,\text{m}^3$ **c** 25 ml

a $2500\,\text{cm}^3 = 2500 \div 1000$ litres $= 2.5$ litres
b $3.75\,\text{m}^3 = 3.75 \times 1000$ litres $= 3750$ litres
c 25 ml $= 25 \div 1000$ litres $= 0.025$ litres

Skill Practice Three

Write as litres:

1	$4000\,\text{cm}^3$	**2**	3000 ml	**3**	$12\,000\,\text{cm}^3$
4	$4500\,\text{cm}^3$	**5**	$6300\,\text{cm}^3$	**6**	$5\,\text{m}^3$
7	$2\,\text{m}^3$	**8**	$10\,\text{m}^3$	**9**	$4.2\,\text{m}^3$
10	$1.25\,\text{m}^3$	**11**	750 ml	**12**	600 ml
13	65 ml	**14**	$15\,\text{cm}^3$	**15**	$80\,\text{cm}^3$

Skill Example

Find the capacity in litres of each of the following tanks.

a $L = 30\,\text{cm}$ **b** $L = 3\,\text{m}$ **c** $L = 4\,\text{m}$
    $W = 10\,\text{cm}$ $W = 4\,\text{m}$ $W = 50\,\text{cm}$
    $H = 20\,\text{cm}$ $H = 2\,\text{m}$ $H = 20\,\text{cm}$

190

a $V = 30\,\text{cm} \times 10\,\text{cm} \times 20\,\text{cm} = 6000\,\text{cm}^3$
Therefore, capacity $= 6000 \div 1000$ litres
$= 6$ litres

b $V = 3\,\text{m} \times 4\,\text{m} \times 2\,\text{m} = 24\,\text{m}^3$
Therefore, capacity $= 24 \times 1000$ litres
$= 24\,000$ litres

c Volume (in m^3) $= 4 \times 0.5 \times 0.2 = 0.4\,\text{m}^3$
Therefore, capacity $= 0.4 \times 1000$ litres
$= 400$ litres

Skill Practice Four

Find the capacity in litres of each of the following tanks.

1 $L = 40\,\text{cm}$
$W = 20\,\text{cm}$
$H = 10\,\text{cm}$

2 $L = 50\,\text{cm}$
$W = 30\,\text{cm}$
$H = 20\,\text{cm}$

3 $L = 50\,\text{cm}$
$W = 30\,\text{cm}$
$H = 30\,\text{cm}$

4 $L = 4\,\text{m}$
$W = 3\,\text{m}$
$H = 1\,\text{m}$

5 $L = 4\,\text{m}$
$W = 2\,\text{m}$
$H = 2\,\text{m}$

6 $L = 2\,\text{m}$
$W = 50\,\text{m}$
$H = 40\,\text{m}$

7 $L = 2\,\text{m}$
$W = 60\,\text{cm}$
$H = 1\,\text{m}$

8 $L = 2\,\text{m}$
$W = 1.5\,\text{m}$
$H = 60\,\text{cm}$

Skill Example

How many 5 ml spoonfuls of medicine can be taken from a bottle with a capacity of 150 ml? If you have to take 3 spoonfuls a day, how long will the bottle last?

Number of spoonfuls in the bottle
$$= 150 \div 5 = 30$$

Number of days
$$= 30 \div 3 = 10$$

So the bottle will last 10 days.

Skill Practice Five

1 The water tank in a house has a square base measuring 40 cm by 40 cm.
If it is filled with water to a depth of 50 cm, how many litres does it contain?

2 A rectangular coffee urn, which is 40 cm high, has a base which measures 30 cm by 25 cm. How many litres of coffee does it contain when full?
How many cups, each of capacity 200 ml, can be filled from the urn when full?

3 At a café, orange squash is served from a plastic container with a square base which measures 40 cm by 40 cm.
If the orange squash is poured in to a depth of 15 cm, how many litres are in the container? How many glasses, each of capacity 250 ml, can now be filled?

4 A paraffin can has a rectangular base measuring 30 cm by 25 cm, and it is filled to a depth of 60 cm.
How many times can the tank of a heater be filled from this quantity of paraffin if the tank measures 25 cm by 20 cm by 10 cm?

5 A small oil can has dimensions 10 cm by 7.5 cm by 4 cm.
a Find its volume in cm^3.
b Find its capacity in millilitres.
c Find its capacity in litres.

6 A car has a petrol tank whose dimensions are 80 cm by 25 cm by 20 cm.
How far can the car be driven on a full tank if it consumes one litre of petrol for every 12 km travelled?

7 A woman is driving a car which suddenly runs out of petrol. In the boot of the car is a full can of petrol which measures 25 cm by 15 cm by 8 cm.
If the woman is 50 km from home and her car travels 17 km on every litre of petrol, has she enough to get home?

8 The picture shows the dimensions of a paddling pool.

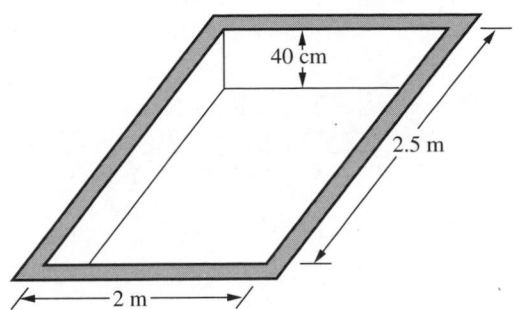

   40 cm

   2.5 m

   2 m

a Find the volume of the paddling pool in cubic metres.

b How many litres of water are required to fill the pool?

9 The dimensions of a cattle trough are shown below.

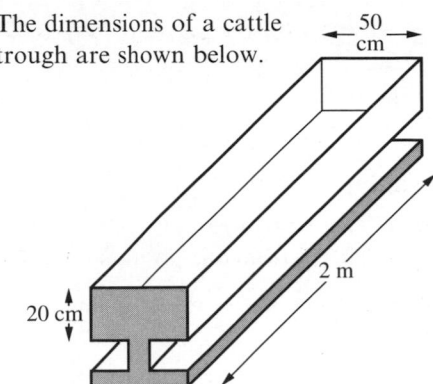

   50 cm

   2 m

   20 cm

a Find the volume of the cattle trough in cubic metres.

b What is the capacity of the trough in litres?

c The trough is filled using a bucket of capacity 25 litres.
How many full buckets are required?

Unit 39

Shape, space and measures   Level 6
Skills   You enlarge shapes by a positive whole-number scale factor.

Knowledge

Reminder There are three ways to transform a point (or a shape) from one position to another.

Translation The point is moved through a set distance.

Reflection The point is moved to the same distance on the other side of a mirror line.

Rotation The point is turned through a set angle about the centre of rotation.

There is also a fourth transformation, **enlargement**, whereby an object is changed into another object (its *image*) which has the *same* shape but a *different* size. The image is said to be **similar** to the object.

Skill Example

Complete an enlargement of this triangle with centre of enlargement X and scale factor 2.

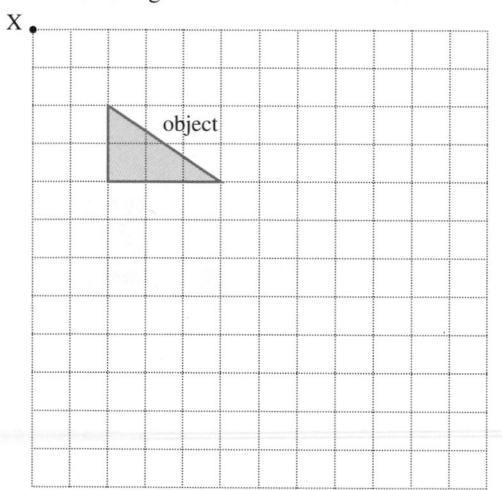

First, draw lines from X through all three vertices of the object.

Then measure the distance of each vertex of the object from X.

Since the scale factor is 2, each vertex of the image has to be twice as far from X as the corresponding vertex of the object.

Mark the vertices of the image and join them with a ruler.

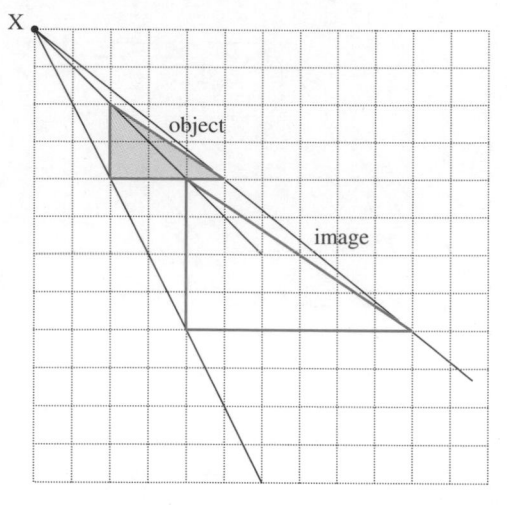

Skill Example

Complete each of the following enlargements.
Centre T, scale factor 3.
Centre M, scale factor 4.

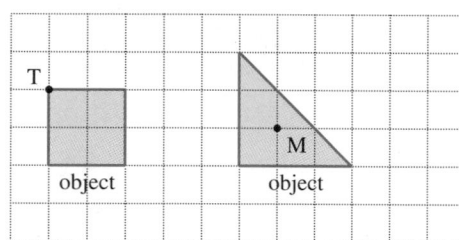

These are the completed enlargements.

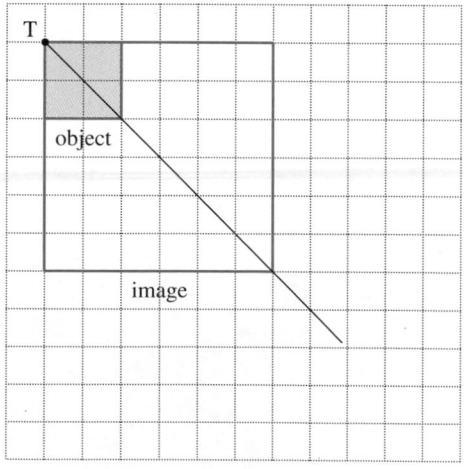

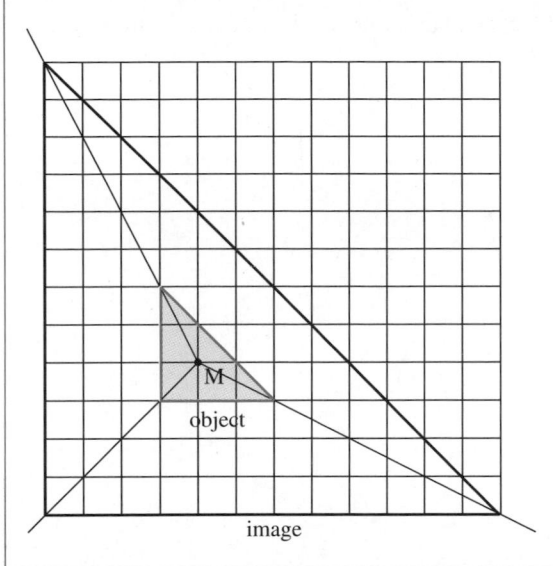

object

image

Skill Practice One

In questions **1** to **16**, copy each diagram on squared paper and complete the enlargement about the centre marked.

1 Scale factor 2.

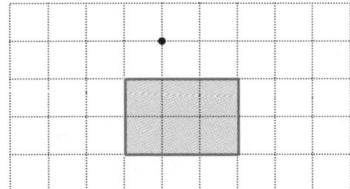

2 Scale factor 3.

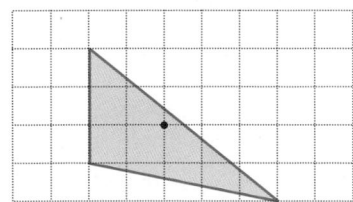

3 Scale factor 2.

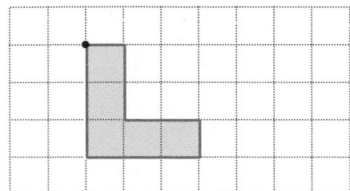

4 Scale factor 3.

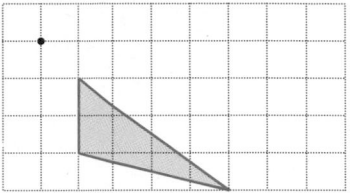

5 Scale factor 4.

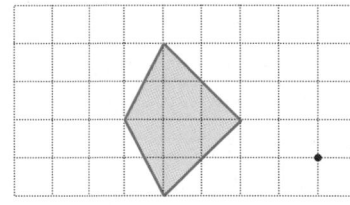

6 Scale factor 3.

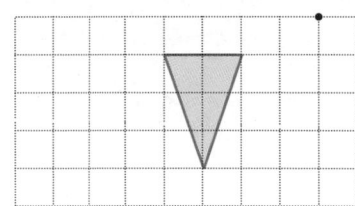

7 Scale factor 2.

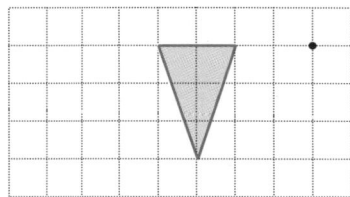

8 Scale factor 2.

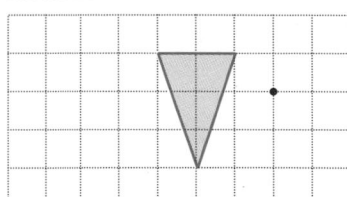

9 Scale factor 2.

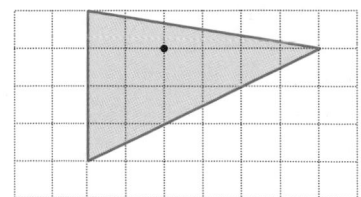

10 Scale factor 2.

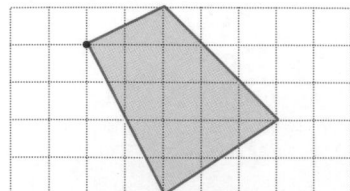

11 Scale factor 4.

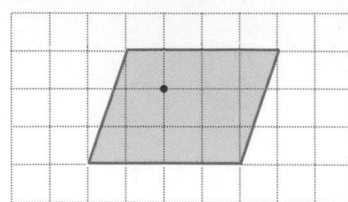

12 Scale factor 2.

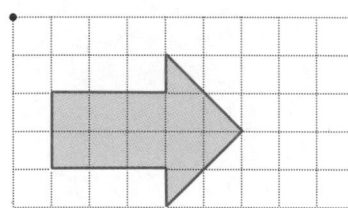

13 Scale factor 2.

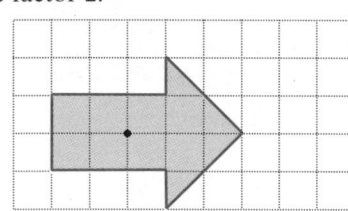

14 Scale factor 2.

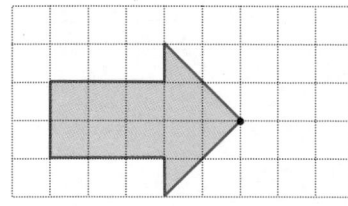

15 Scale factor 3.

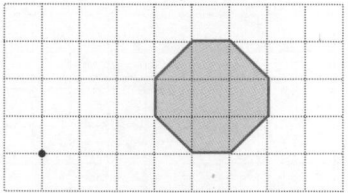

16 Scale factor 4.

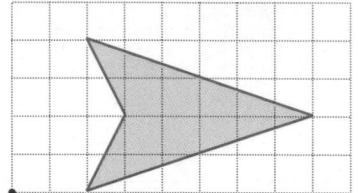

17 Draw a graph with x and y-axes from 0 to 12.
Plot on the graph the object with vertices at (1, 1), (1, 2) and (2, 1).
Draw in the images of this object after enlargements with centre (0, 0) and scale factors of:

  a 2   **b** 4   **c** 6

18 Draw a graph with x and y-axes from −12 to 12.
Plot on the graph the object with vertices at (0, 2), (−2, 2), (−1, −2) and (1, −2).
Draw in the images of this object after enlargements with centre (0, 0) and scale factors of:

  a 2   **b** 3
  c 4   **d** 5

Unit 40

Skill You understand and apply Pythagoras' theorem when solving problems in two dimensions.

Knowledge

$$7^2 = 7 \times 7 = 49$$

or in words

seven squared equals forty-nine

49 is the square of 7; and 7 is the square root of 49 or

$$\sqrt{49} = 7$$

The table on the right gives the squares of all numbers between 1 and 100.

Skill Example

From the table, find the value of 36^2.

Find 36 in the column headed n and read off the square in the column headed n^2 on its right.

$$36^2 = 1296$$

Skill Practice One

From the table of squares, find the value of each of the following.

1 20^2	**2** 13^2	**3** 14^2	**4** 18^2
5 22^2	**6** 26^2	**7** 27^2	**8** 29^2
9 31^2	**10** 40^2	**11** 60^2	**12** 70^2
13 35^2	**14** 65^2	**15** 75^2	**16** 100^2

Skill Example

From the table of squares, find the value of $\sqrt{5329}$.

Find 5329 in the column headed n^2 and read off the square root in the column headed n on its left.

$$\sqrt{5329} = 73$$

Skill Practice Two

From the table of squares, find the value of each of the following.

1 $\sqrt{900}$	**2** $\sqrt{225}$	**3** $\sqrt{625}$	**4** $\sqrt{256}$
5 $\sqrt{289}$	**6** $\sqrt{361}$	**7** $\sqrt{441}$	**8** $\sqrt{529}$

9 $\sqrt{576}$	**10** $\sqrt{784}$	**11** $\sqrt{2500}$	**12** $\sqrt{6400}$
13 $\sqrt{8100}$	**14** $\sqrt{2025}$	**15** $\sqrt{3025}$	**16** $\sqrt{7225}$

Table of squares

n	n^2	n	n^2
1	1	51	2601
2	4	52	2704
3	9	53	2809
4	16	54	2916
5	25	55	3025
6	36	56	3136
7	49	57	3249
8	64	58	3364
9	81	59	3481
10	100	60	3600
11	121	61	3721
12	144	62	3844
13	169	63	3969
14	196	64	4096
15	225	65	4225
16	256	66	4356
17	289	67	4489
18	324	68	4624
19	361	69	4761
20	400	70	4900
21	441	71	5041
22	484	72	5184
23	529	73	5329
24	576	74	5476
25	625	75	5625
26	676	76	5776
27	729	77	5929
28	784	78	6084
29	841	79	6241
30	900	80	6400
31	961	81	6561
32	1024	82	6724
33	1089	83	6889
34	1156	84	7056
35	1225	85	7225
36	1296	86	7396
37	1369	87	7569
38	1444	88	7744
39	1521	89	7921
40	1600	90	8100
41	1681	91	8281
42	1764	92	8464
43	1849	93	8649
44	1936	94	8836
45	2025	95	9025
46	2116	96	9216
47	2209	97	9409
48	2304	98	9604
49	2401	99	9801
50	2500	100	10 000

196

Skill Example

Use the table of squares to evaluate the following.

a $\sqrt{48^2 + 36^2}$ **b** $\sqrt{85^2 - 40^2}$

a $\sqrt{48^2 + 36^2} = \sqrt{2304 + 1296}$
$= \sqrt{3600}$
$= 60$

b $\sqrt{85^2 - 40^2} = \sqrt{7225 - 1600}$
$= \sqrt{5625}$
$= 75$

Skill Practice Three

Use the table of squares to evaluate each of the following.

1 $\sqrt{15^2 + 20^2}$ 2 $\sqrt{10^2 + 24^2}$
3 $\sqrt{16^2 + 30^2}$ 4 $\sqrt{14^2 + 48^2}$
5 $\sqrt{20^2 + 21^2}$ 6 $\sqrt{28^2 + 45^2}$
7 $\sqrt{9^2 + 40^2}$ 8 $\sqrt{11^2 + 60^2}$
9 $\sqrt{12^2 + 35^2}$ 10 $\sqrt{60^2 + 80^2}$
11 $\sqrt{17^2 - 15^2}$ 12 $\sqrt{20^2 - 16^2}$

Skill Example

A square garden has an area of $1024 \, \text{m}^2$.

area = 1024 m²

What length of fencing is required to surround it?

If the side of the square is n, then its area is
$$n \times n = n^2$$

For the garden, $n^2 = 1024 \, \text{m}^2$

So, from the table of squares, $n = 32 \, \text{m}$ and the length of fencing required is
$$32 \times 4 = 128 \, \text{m}$$

Skill Practice Four

1 A square lawn has an area of $324 \, \text{m}^2$.

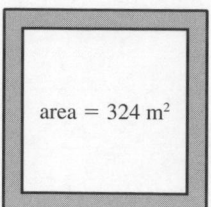

area = 324 m²

A border path is to be made around it. What will the length of this path be?

2 A square picture has an area of $676 \, \text{cm}^2$.

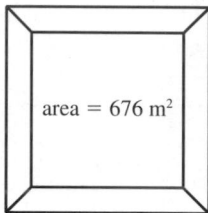

area = 676 m²

Find the length of wood required to make its frame.

3 A square tray has a base area of $1089 \, \text{cm}^2$. Find the length of wood required for its sides.

4 A single-pane square window has an area of $5625 \, \text{cm}^2$. Find the length of wood required to make its frame.

5 A school playground is square in shape and has an area of $2401 \, \text{m}^2$. A boy cycles around it and finds that the tripmeter on his bicycle records $0.2 \, \text{km}$. Is this figure accurate? If not, what is the error?

Skill Example

A farmer has $220 \, \text{m}$ of fencing. What is the area of the largest square enclosure that he can make?

Length all the way around = $220 \, \text{m}$
Length of one side = $220 \div 4 = 55 \, \text{m}$

So, from the table of squares, the largest area is
$$55^2 = 3025 \, \text{m}^2$$

Skill Practice Five

1 An inspection cover in the road fits into a
square hole which is lined with metal of total
length 244 cm.

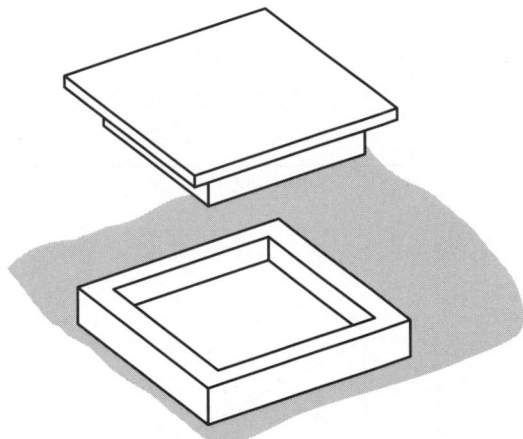

Find the area of the cover.

2 A farmer drives his tractor around the edge of
a square field and finds the distance to be
exactly 0.30 km.
a Find the side length of the field in metres.
b Find the area of the field in square metres.

3 A square room requires 22.5 m of skirting
board.
If there are two doors into the room, each of
width 75 cm, find:
a the perimeter of the room
b the side length of the room
c the area of the room.

4 Look at the chessboard below.

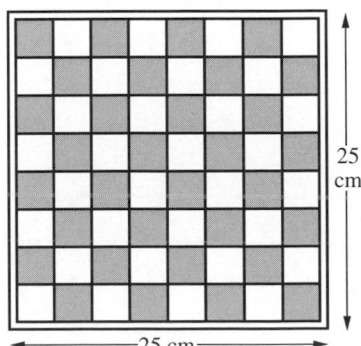

25 cm
25 cm

Find:
a the outer perimeter
b the perimeter of the playing surface if it is
4 cm less than the outer perimeter
c the side length of the playing surface
d the side length of each square
e the area of each square
f the area of the whole playing surface.

5 The diagram shows a refrigerator shelf which
is square in shape.

If 480 cm of plastic-covered wire was used to
make the shelf, find:
a the side length of the shelf
b the area of the shelf.

6 The diagram shows the wire guard of a gas
fire.

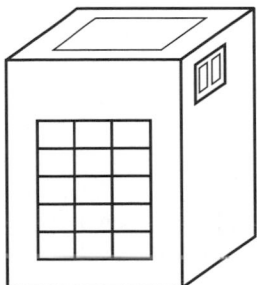

If the guard is square in shape and made from
450 cm of wire, find its area.

Knowledge

Pythagoras' theorem

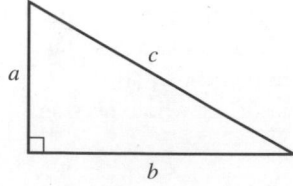

For this right-angled triangle:

$$a^2 + b^2 = c^2$$

This relationship is true for *all* right-angled triangles.

The theorem can be used to find the unknown length of a side in any right-angled triangle.

The longest side of any right-angled triangle is called the **hypotenuse**.

Skill Example

In this right-angled triangle, find the length of side c.

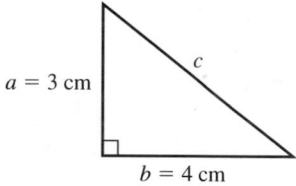

The triangle is right-angled. So,

$$a^2 + b^2 = c^2$$
$$3^2 + 4^2 = c^2$$
$$9 + 16 = c^2$$
$$\Rightarrow \quad 25 = c^2$$
$$\Rightarrow \quad c = \sqrt{25} = 5\,\text{cm}$$

Skill Practice Six

In questions **1** to **5**, find the length of c.

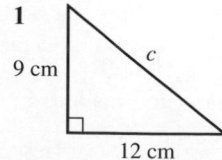

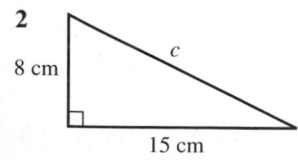

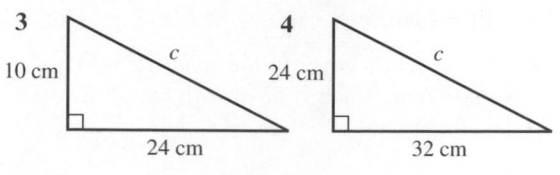

In questions **6** to **10** find the length of AB.

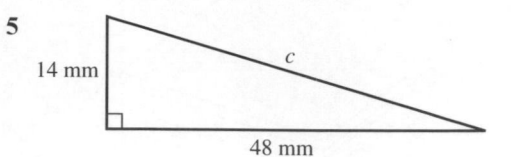

6 AC = 60 cm, BC = 80 cm
7 AC = 54 cm, BC = 72 cm
8 AC = 45 mm, BC = 60 mm
9 AC = 60 mm, BC = 25 mm
10 AC = 27 mm, BC = 36 mm

Knowledge

Most calculators have a square-root key, marked with the square-root sign.

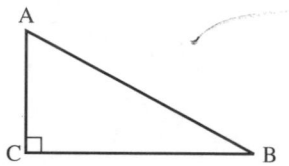

Skill Example

A telegraph pole, 15 m high, is secured with four wires from the top of the pole to points on the ground 10 m from the base of the pole.

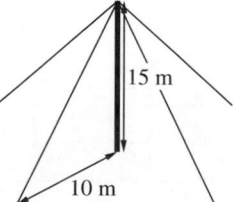

Find the length of each wire.

Each wire forms a right-angled triangle with the pole and the ground. Each wire is the hypotenuse of its triangle. So,

$$a^2 + b^2 = c^2$$

$$15^2 + 10^2 = c^2$$

$$225 + 100 = c^2$$

$$\Rightarrow \quad c^2 = 325$$

To find the value of c using a calculator, enter 325 and then press the square-root key.

$$c = \sqrt{325} = 18.0\,\text{m} \quad (\text{to 1 dp})$$

Skill Practice Seven

1 A tent pole, 2 m tall, is secured with wires fixed to pegs 1.6 m from the base of the pole.

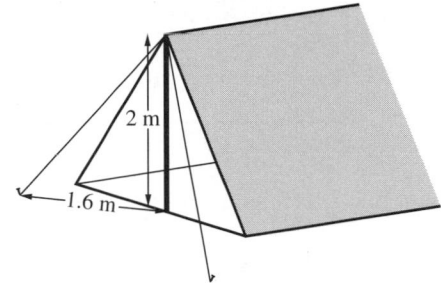

Find the length of the wires.

2 A diagonal brace is fitted to a square wire frame with sides 1.2 m long.

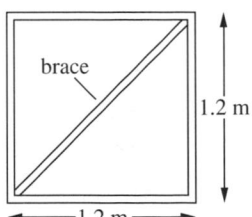

Find the length of the brace.

3 A ship leaves harbour and sales 40 km east and then 25 km north.

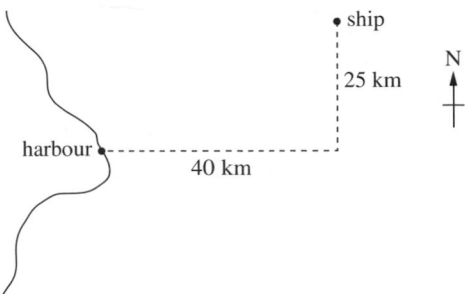

How far is it from the harbour to the ship?

4 A path runs diagonally across a rectangular playing field which is 500 m by 600 m.

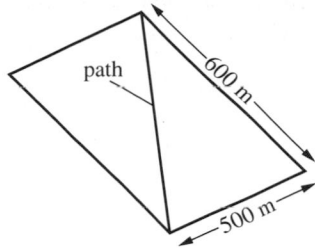

How long is the path?

5 A spider is sitting at the bottom left-hand corner of a wall which is 4 m long and 2.1 m high. A fly is sitting in the top right-hand corner.

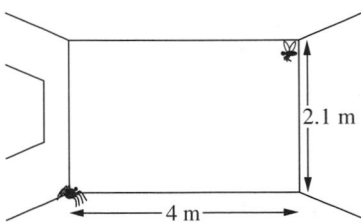

What is the shortest distance between the spider and the fly?

6 Judy wants to build a wheelchair ramp for a step 25 cm high. The ramp will start 1.5 m from the step.

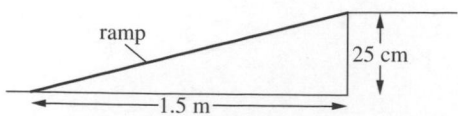

How long will the ramp be?

7 Can you fit a walking stick 1.1 m long in a suitcase which is 1 m by 0.5 m?

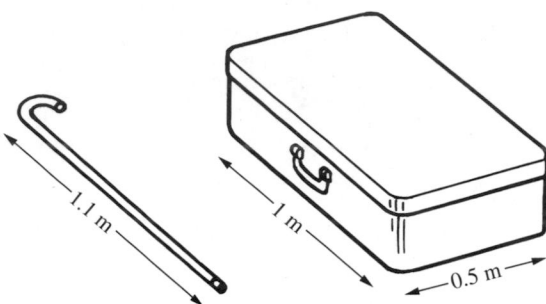

8 A football pitch is 100 m long and 73 m wide.

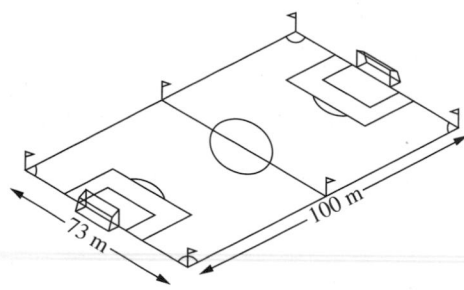

What is the greatest possible distance there can be between two players on this pitch?

Skill Example

Find the unknown length of this triangle.

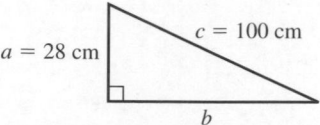

The triangle is right-angled. Therefore,

$$a^2 + b^2 = c^2$$
$$28^2 + b^2 = 100^2$$
$$784 + b^2 = 10\,000$$
$$\Rightarrow \quad b^2 = 9216$$
$$\Rightarrow \quad b = \sqrt{9216} = 96 \text{ cm}$$

Skill Practice Eight

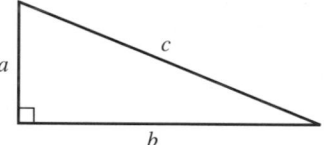

In questions **1** to **5**, find the length of a.

1 $c = 25$ cm, $b = 24$ cm

2 $c = 52$ cm, $b = 48$ cm

3 $c = 58$ cm, $b = 42$ cm

4 $c = 82$ mm, $b = 18$ mm

5 $c = 65$ mm, $b = 63$ mm

In questions **6** to **10**, find the length of b.

6 $c = 25$ cm, $a = 15$ cm

7 $c = 39$ cm, $a = 36$ cm

8 $c = 30$ cm, $a = 24$ cm

9 $c = 85$ mm, $a = 75$ mm

10 $c = 78$ mm, $a = 72$ mm

11 The diagram shows a ladder resting against a vertical wall.

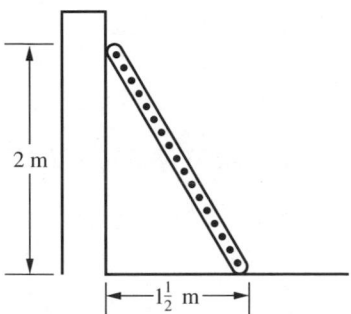

2 m

$1\frac{1}{2}$ m

Find the length of the ladder.

12 On a boating lake, Pritesh and Liam row due west from A to B. They then row due north from B to C, and finally back to A.

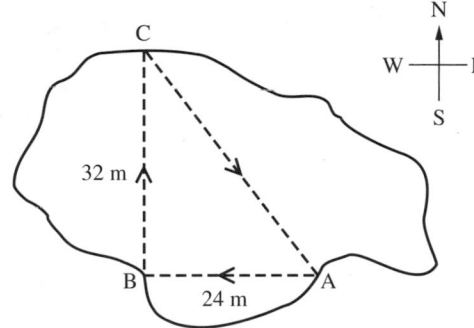

N

W — E

S

C

32 m

B

24 m

A

How far altogether do they row?

13 Find the length of the diagonal of this rectangle.

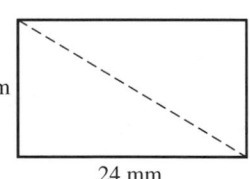

18 mm

24 mm

14 Find the length of one of the equal sides of the isosceles triangle below.

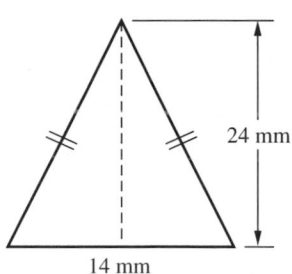

24 mm

14 mm

15 The rhombus illustrated has diagonals of lengths 12 cm and 16 cm.
Find the length of its sides.

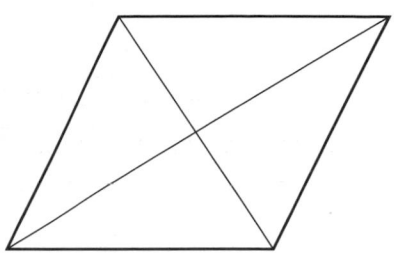

16 A ship has travelled 15 km from a harbour and is now 8 km west of its starting point.

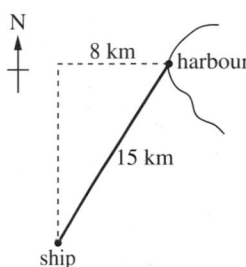

N

8 km

harbour

15 km

ship

How far is the ship to the south of its starting point?

17 A ladder 2.5 m long is resting against a wall with its foot 0.7 m from the base of the wall.

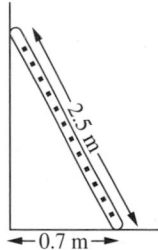

2.5 m

0.7 m

How far up the wall does the ladder reach?

18 Wendy is flying a kite on a 55-metre string. Her hand is 1 m from the ground. Simon is standing 15 m away from Wendy. How high is the kite when it is directly over Simon's head?

55 m

Wendy

Simon

15 m

Unit 41

Knowledge

A **prism** is any solid which has a uniform cross-section throughout its length.

A right prism has its end (or base) perpendicular to its other faces.

A 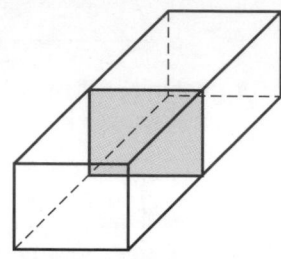   B 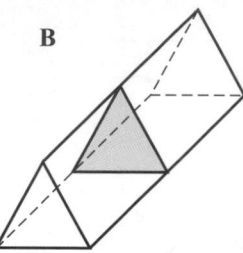   C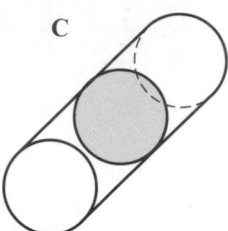

A prism takes its name from its cross-section. Thus B is a **triangular** prism.

A prism with rectangular cross-section, such as A, is a **cuboid**.

A prism with a circular cross-section, such as C, is a **cylinder**.

The volume V of any prism of uniform cross-section is given by:

$$V = A \times l$$

where A is the area of the cross-section and l is the length of the prism.

Skill Example

Find the volume of each of the following solids (take $\pi = 3.14$ when required).

a 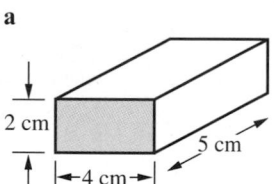   b 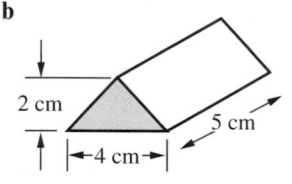   c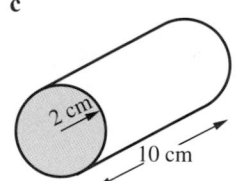

a Area of cross-section $= 4 \times 2 = 8\,\text{cm}^2$
  Length of solid $= 5\,\text{cm}$
  So, volume of solid $= 8 \times 5 = 40\,\text{cm}^3$

b Area of cross-section $= \frac{1}{2} \times 4 \times 2 = 4\,\text{cm}^2$
  Length of solid $= 5\,\text{cm}$
  So, volume of solid $= 4 \times 5 = 20\,\text{cm}^3$

c Area of cross-section $= \pi \times 2 \times 2 = 3.14 \times 2 \times 2 = 12.56\,\text{cm}^2$
  Length of solid $= 10\,\text{cm}$
  So, volume of solid $= 12.56 \times 10 = 125.6\,\text{cm}^3$

Skill Practice One

In questions **1** to **4**, find the volume of each cuboid.

1

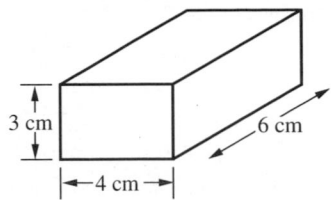

Width = 4 cm; height = 3 cm;
length = 6 cm

2

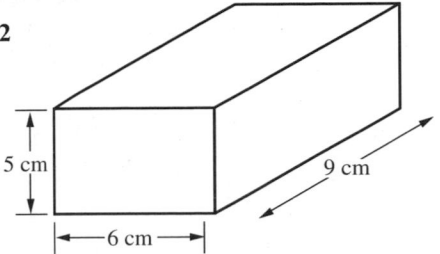

Width = 6 cm; height = 5 cm;
length = 9 cm

3 Width = 5 mm; height = 8 mm;
length = 12 mm

4 Width = 2 m; height = 1.5 m;
length = 4 m

In questions **5** to **10**, find the volume of each triangular prism.

5

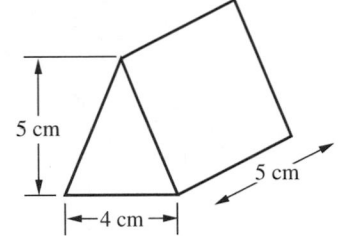

Width = 4 cm; height = 5 cm;
length = 5 cm

6

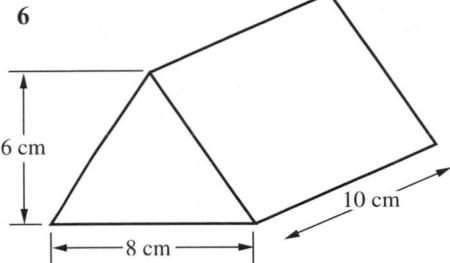

Width = 8 cm; height = 6 cm;
length = 10 cm

7 Width = 6 cm; height = 5 cm;
length = 7 cm

8 Width = 4 cm; height = 6 cm;
length = 8 cm

9 Width = 3 cm; height = 8 cm;
length = 12 cm

10 Width = 10 mm; height = 6 mm;
length = 12 mm

In questions **11** to **16**, find the volume of each cylinder (take $\pi = 3.14$).

11

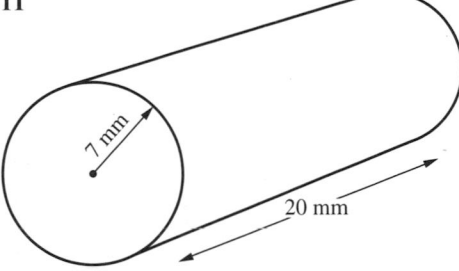

Radius = 7 mm; length = 20 mm

12

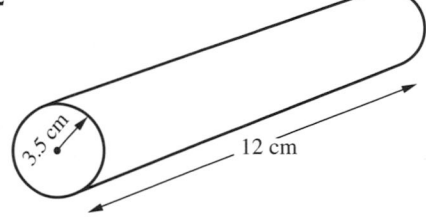

Radius = 3.5 cm; length = 12 cm

13 Radius = 5 cm; length = 8 cm

14 Radius = 4 cm; length = 5 cm

15 Radius = 3 cm; length = 7 cm

16 Radius = 10 mm; length = 35 mm

Skill Example

The solid illustrated has a uniform cross-section, and the dimensions shown.
Find its volume.

Area of cross section

$$= (3 \times 1) + (1 \times 2) + (2 \times 1)$$
$$= 3 + 2 + 2 = 7\,\text{cm}^2$$

Length $= 5\,\text{cm}$

So, volume $= 7 \times 5 = 35\,\text{cm}^3$

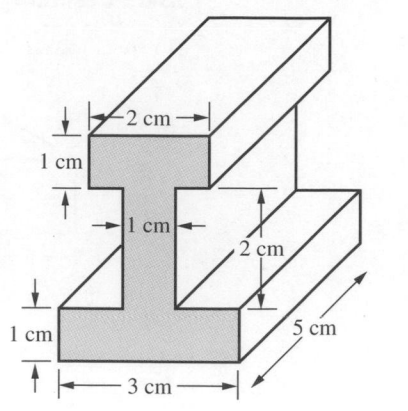

Skill Practice Two

Each of the solids illustrated has a uniform cross-section.
Find the volume of each solid from the dimensions given.

1

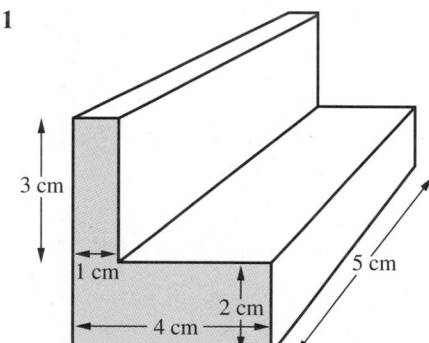

2

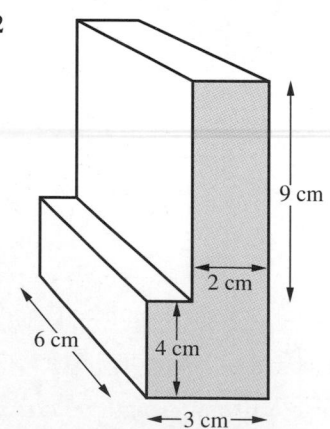

3

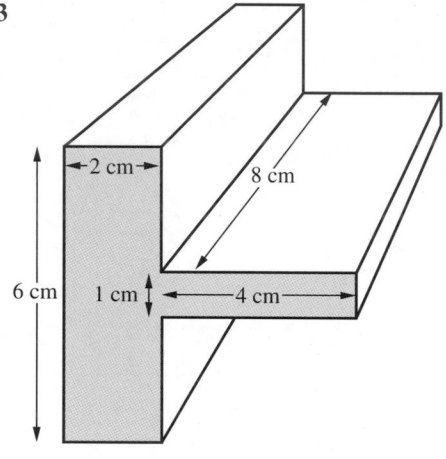

4

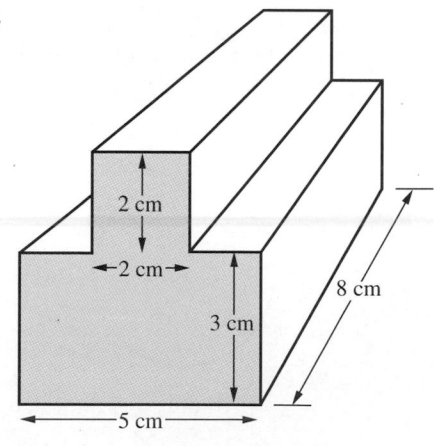

5

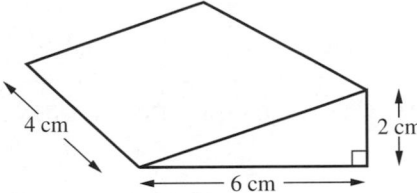

6 This wedge is used as a door-stop. Its dimensions are shown below.
Find its volume.

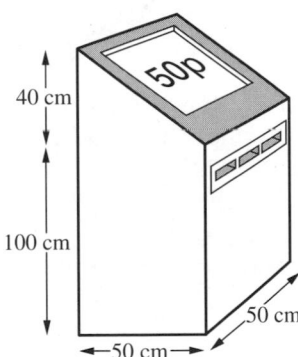

7 The picture below shows the dimensions of a ticket machine at an underground station.
Find the volume of the machine.

40 cm
100 cm
50 cm
50 cm
50p

8 Find the volume of this lean-to.

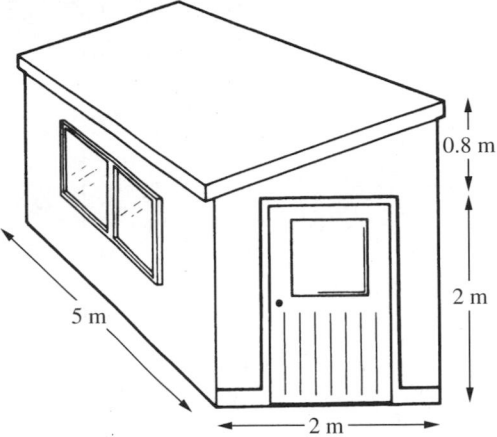

9 The diagram shows the dimensions of the roof of a house.
Find the volume enclosed by the loft.

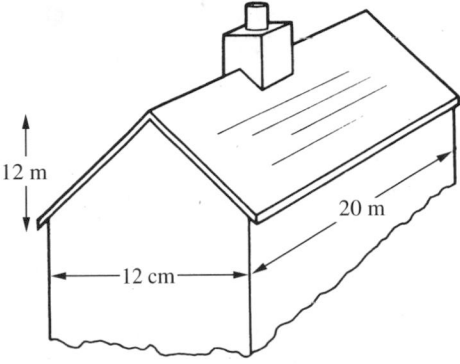

12 m
12 cm
20 m

10 Find the volume of the garage illustrated below.

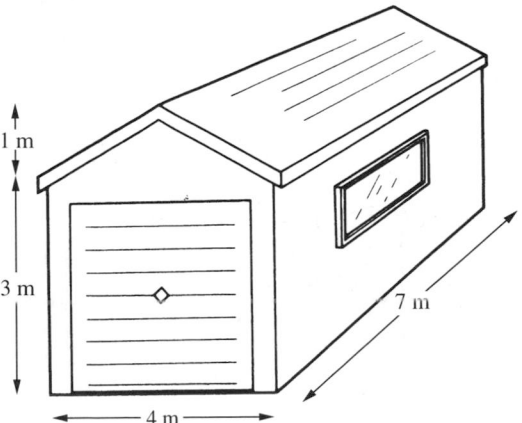

1 m
3 m
4 m
7 m

Skill Example

Find the capacity in litres of each of the tanks shown. (1 m³ = 1000 litres; take π = 3.14)

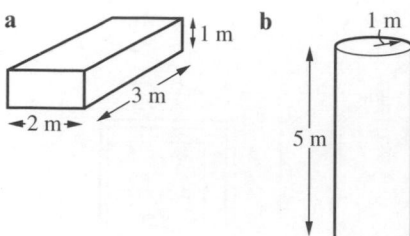

a Area of cross-section = 2 × 1 = 2 m²
Length of tank = 3 m
Therefore, volume = 2 × 3 = 6 m³
So, capacity = 6 × 1000
= 6000 litres

b Area of cross-section = π × 1 × 1 = 3.14 m²
Length (height) of tank = 5 m
Therefore, volume = 3.14 × 5 = 15.7 m³
So, capacity = 15.7 × 1000
= 15 700 litres

Skill Practice Three

Find the capacity in litres of each tank.
(1 m³ = 1000 litres; 1000 cm³ = 1 litre; take π = 3.14)

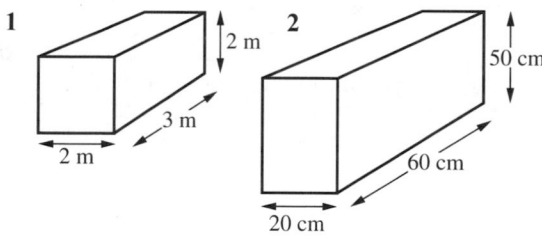

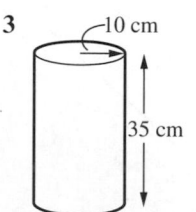

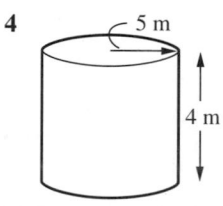

5 A house is heated by an oil-fired boiler. The supply tank has a rectangular base measuring 120 cm by 70 cm.
  a The depth of oil in the tank when full is 100 cm. How many litres of oil does the tank contain?
  b The burner consumes 12 litres of oil per day.
  For how many weeks will this oil supply last?

6 In a café, cold milk is served from a machine. The milk is contained in a plastic tank measuring 30 cm by 30 cm by 30 cm.
  a How many litres of milk does the tank contain when full?
  b The milk is served in plastic cups each containing $\frac{1}{5}$ litre.
  How many cups can be served from the full tank?

7 An inflatable paddling pool has an inside radius of 70 cm.

  a How many litres of water does the pool contain if the depth of water is 20 cm? (Take π = 3.14)
  b The pool is filled to this depth using a bucket of capacity 28 litres.
  How many buckets full of water are required?

8 A cylindrical coffee urn has a radius of 20 cm.
  a If the depth of coffee in the urn is 35 cm, how many litres does it contain? (Take π = 3.14)
  b How many cups of coffee, each of capacity $\frac{11}{50}$ of a litre, can be served?

Unit 42

Shape, space and measures   Level 7

Skills   You enlarge shapes by a fractional scale factor.

You develop an understanding of scale, including using and interpreting maps and drawings.

Knowledge

When the scale factor for an enlargement is a fraction, the distances of points on the image from the centre of enlargement will be reduced and the image will be *smaller* than the object.

Skill Example

Complete the following enlargements.

Centre T, scale factor $\frac{1}{3}$.

Centre M, scale factor $\frac{1}{2}$.

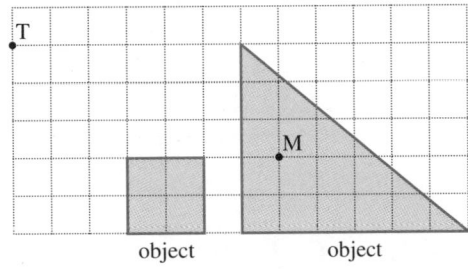

These are the completed enlargements.

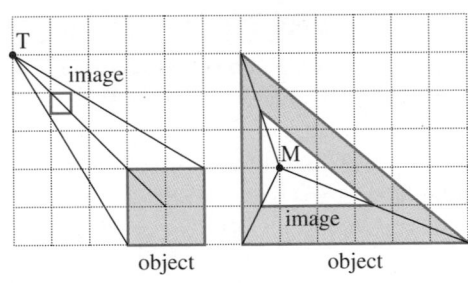

Skill Practice One

In questions **1** to **16**, copy each diagram on squared paper and complete the enlargement about the centre marked.

1 Scale factor $\frac{1}{2}$.

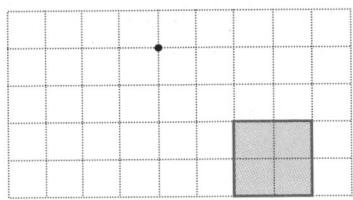

2 Scale factor $\frac{1}{2}$.

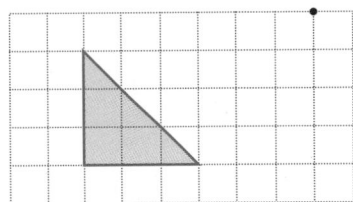

3 Scale factor $\frac{1}{3}$.

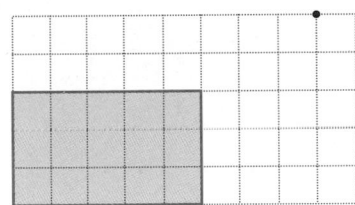

4 Scale factor $\frac{1}{4}$.

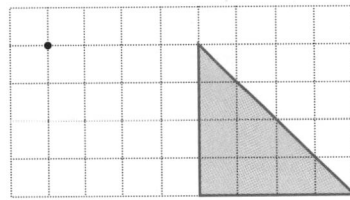

5 Scale factor $\frac{1}{2}$.

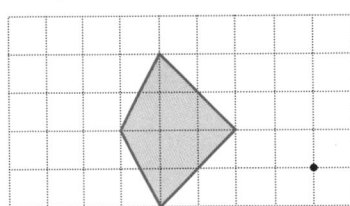

6 Scale factor $\frac{1}{2}$.

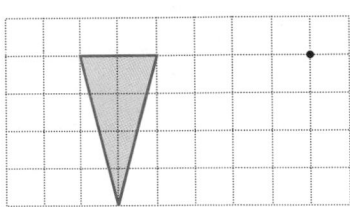

7 Scale factor $\frac{1}{2}$.

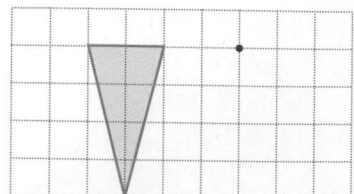

8 Scale factor $\frac{1}{2}$.

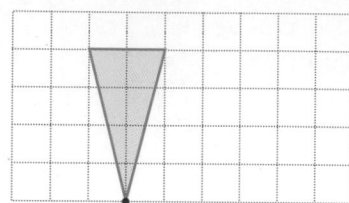

9 Scale factor $\frac{1}{2}$.

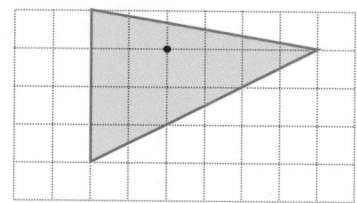

10 Scale factor $\frac{1}{4}$.

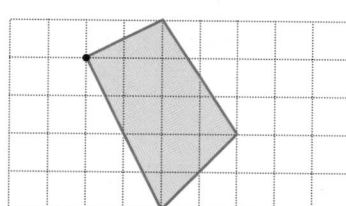

11 Scale factor $\frac{1}{2}$.

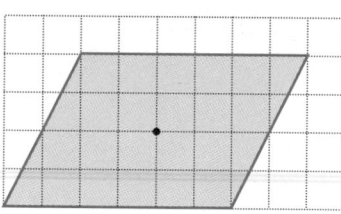

12 Scale factor $\frac{1}{3}$.

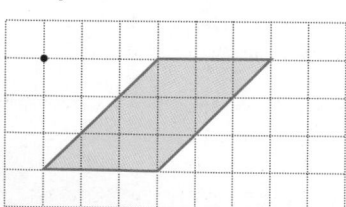

13 Scale factor $\frac{1}{2}$.

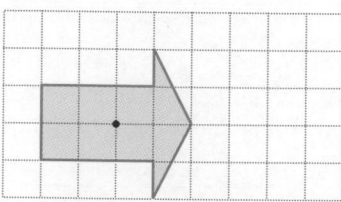

14 Scale factor $\frac{1}{4}$.

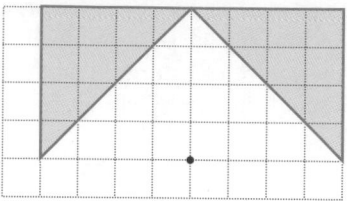

15 Scale factor $\frac{1}{3}$.

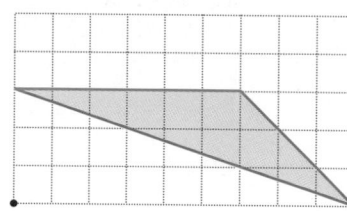

16 Scale factor $\frac{1}{2}$.

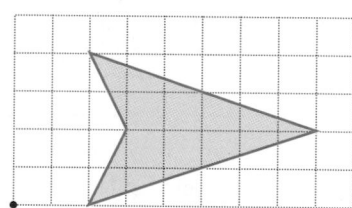

17 Draw a graph with x and y-axes from 0 to 12.
Plot on the graph the object with corner
points at (6, 12), (12, 6) and (12, 12).
Draw in the images of this object after
enlargements with centre (0, 0) and scale
factors of:

  a $\frac{1}{2}$ **b** $\frac{1}{3}$ **c** $\frac{1}{4}$ **d** $\frac{1}{6}$

18 Draw a graph with x and y-axes from -12 to 12.
Plot on the graph the object with corner points
at $(-12, 12)$, $(6, 12)$, $(12, -12)$ and $(-6, -12)$.
Draw in the images of this object after
enlargements with centre (0, 0) and scale
factors of:

  a $\frac{1}{2}$ **b** $\frac{1}{3}$ **c** $\frac{1}{4}$ **d** $\frac{1}{6}$

209

Knowledge

This is a map of North Lancashire and the Lake District.

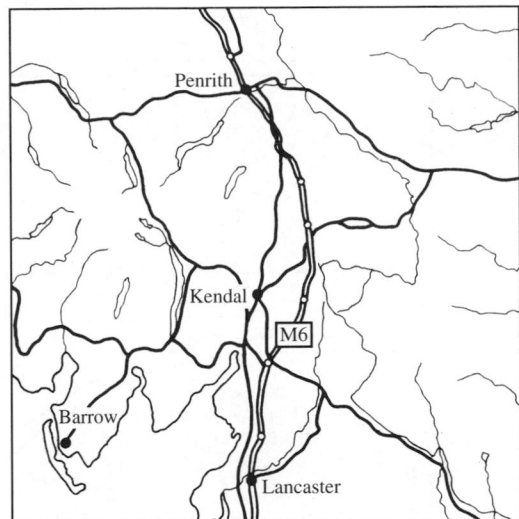

The map is drawn to a scale of 1 cm to 10 km. This means that 1 cm on the map represents 10 km on the land.

The scale of the map can be written

    1 : 1 000 000

because 1 cm represents 10 km

$$= (10 \times 1000)\,\text{m}$$
$$= (10 \times 1000 \times 100)\,\text{cm}$$
$$= 1\,000\,000\,\text{cm}$$

Skill Example

Write the following map scales in the form $1 : n$.
a 1 cm represents 10 m
b 1 cm represents 0.5 km

a 1 cm represents 10 m
$$= (10 \times 100)\,\text{cm}$$
$$= 1000\,\text{cm}$$
Therefore, the scale is 1 : 1000.

b 1 cm represents 0.5 km
$$= (0.5 \times 1000)\,\text{m}$$
$$= 500\,\text{m}$$
$$= (500 \times 100)\,\text{cm}$$
$$= 50\,000\,\text{cm}$$
Therefore, the scale is 1 : 50 000.

Skill Practice Two

Write each of the following scales in the form $1 : n$.

1 1 cm represents 1 m
2 1 cm represents 5 m
3 1 cm represents 2 m
4 1 cm represents 20 m
5 1 cm represents 50 m
6 1 cm represents 25 m
7 1 cm represents 100 m
8 1 cm represents 0.5 m
9 1 cm represents 0.2 m
10 1 cm represents 0.25 m
11 1 cm represents 1 km
12 1 cm represents 5 km
13 1 cm represents 2 km
14 1 cm represents 0.2 km
15 1 cm represents 0.25 km

Skill Example

The plan of a house is drawn to a scale of 1 : 20.

a Find the actual width of the house if the width on the plan is 50 cm.
b Find the actual height of a room if the plan height is 12.5 cm.

a Actual width $= 50\,\text{cm} \times 20$
$$= 1000\,\text{cm}$$
$$= 10\,\text{m}$$

b Actual height $= 12.5\,\text{cm} \times 20$
$$= 250\,\text{cm}$$
$$= 2.5\,\text{m}$$

Skill Practice Three

1 On a drawing, a table top measures 20 cm by 15 cm.
If the drawing is made to a scale of 1 : 6, find the dimensions of the actual table.

2 On the plan of a house, a door measures 4 cm by 10 cm.
If the plan is drawn to a scale of 1 : 20, find the actual dimensions of the door.

3 On the plan of a house, a window measures 10 cm by 5 cm.
If the plan is made to a scale of 1 : 25, find the actual dimensions of the window.

4 A model of an aeroplane has a length of 45 cm and is built to a scale of 1 : 40.
Find the length of the real aeroplane.

5 On a plan, a garden measures 15 cm by 12 cm. If the plan is drawn to a scale of 1 : 100, find the dimensions of the real garden.

6 On a model railway a coach is 20 cm long, 8 cm high and 4 cm wide.
If it is built to a scale of 1 : 50, find the dimensions of its real counterpart.

7 A dolls' house is a 1 : 15 scale model of a real house.
If the dolls' house is 80 cm long, 60 cm wide and 50 cm high, find the dimensions of the real house.

8 A model ship is 40 cm long and is built to a scale of 1 : 150.
Find the length of its real counterpart.

9 The diagram shows Highpark Road as it appears on a street plan.

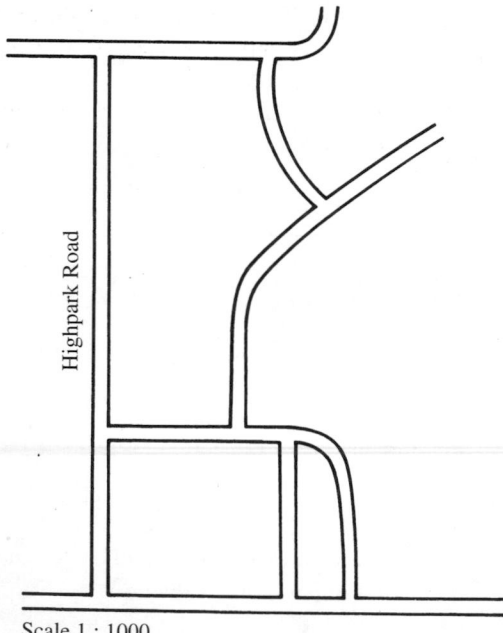

Scale 1 : 1000

Measure the length of Highpark Road on the diagram and then find its real length.

10 The actual length of Hazel Avenue is 180 m. The street plan below shows the surrounding roads.

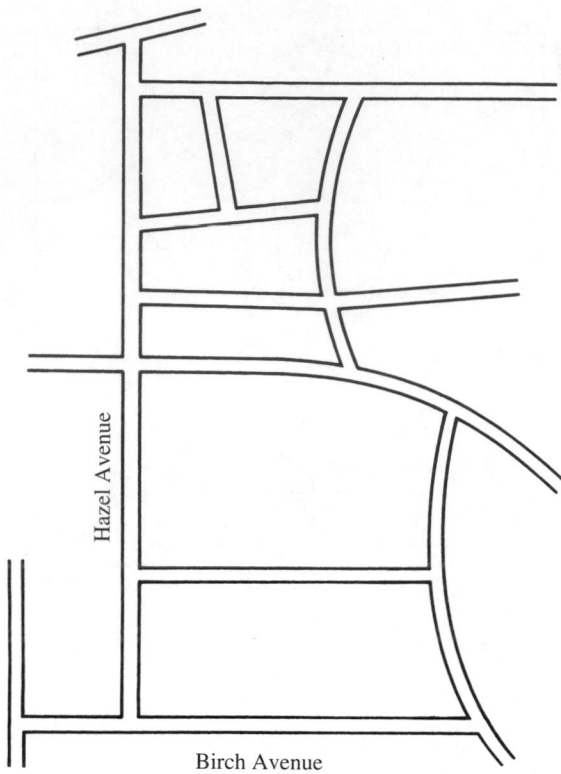

a Measure the length of Hazel Avenue on the street plan and hence find the scale of the plan.
b Measure the length of Birch Avenue and find its real length from the scale.

Skill Example

Some children draw a plan of their classroom using a scale of 1 : 25. If the classroom measures 8 m by 5.5 m, find the dimensions of the classroom on the plan.

$$8 \, m = (8 \times 100) \, cm = 800 \, cm$$

An actual distance of 25 cm is equal to 1 cm on the plan.

That is, $25 \, cm \equiv 1 \, cm$

Therefore, $800 \, cm \equiv \dfrac{800}{25} \, cm = 32 \, cm$ on plan

$$5.5 \, m = (5.5 \times 100) \, cm = 550 \, cm$$

Therefore, $550 \, cm \equiv \dfrac{550}{25} \, cm = 22 \, cm$ on plan

Skill Practice Four

1 A tray measures 60 cm by 30 cm.
   Find its dimensions on a drawing of scale 1 : 3.

2 A wall cupboard has dimensions 100 cm by 60 cm.
   Find its dimensions on a drawing of scale 1 : 5.

3 The miniature railway on Romney Marsh is built to a scale of 1 : 4.
   Given that the standard gauge of British mainline tracks is 144 cm, find the gauge of the Romney Marsh Railway.

4 A car has a length of 3.6 m.
   If a model is made of it to a scale of 1 : 60, find the length of the model.

5 A school yard has dimensions of 40 m by 24 m.
   Some children draw a plan of the yard to a scale of 1 : 200.
   Find the dimensions of the yard on their plan.

6 The porch on the front of a house is 2.4 m high and 1.5 m wide.
   What will be the dimensions of this porch on a plan of the house which is drawn to a scale of 1 : 30?

7 A garden shed is 4 m long, 3 m wide and 2 m high.
   What will be its dimensions on a plan which is drawn to a scale of 1 : 20?

8 A church has a tower of height 60 m.
   What will be the height of the tower on a plan drawn to a scale of 1 : 250?

9 Northmead Grove is a straight road of length 40 m.
   How long will it be on a street plan drawn to a scale of 1 : 500?

10 Primrose Lane is a straight road of length 60 m.
    If it is 5 cm long on a street plan, what is the scale of the plan?
    If Bluebell Way is 84 m long, how long will it be on the same plan?

Unit 43

Knowledge

The **locus** of a point is the path it traces out when moving according to a rule.

For example, a conker is swung on the end of a 50 cm string.

The conker is moving according to the rule that its distance from the centre of swing is always 50 cm.

The locus traced out by the conker is a circle with a radius of 50 cm.

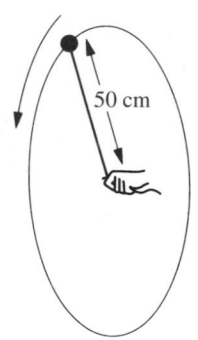

Skill Example

2 metres

The diagram shows the plan view of a goat tethered by a 4 m rope to a ring in the wall of a barn in a field.

Copy the diagram and mark on it the locus of the goat's movements as it grazes.

The goat can move in a semicircle of 4 m radius until the rope meets the corner of the barn. The rope will then be shortened to 2 m and the goat can move in a quarter circle of 2 m radius. Drawing these segments of a circle on the diagram produces this locus.

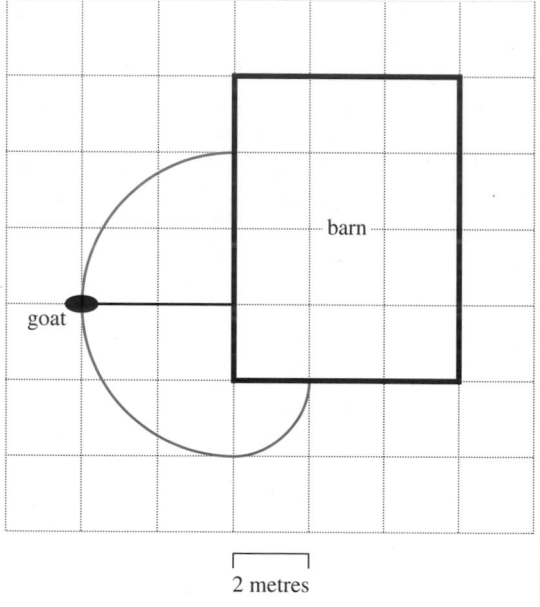

2 metres

Skill Practice One

Each diagram shows a plan view of a guard-dog tethered to a ring in the wall of a house.
Copy each diagram and mark on the boundary of the area that the dog can guard effectively.

1

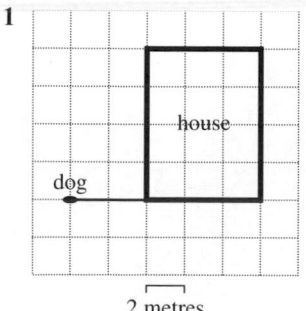

2 metres

2

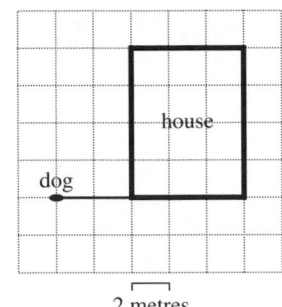

2 metres

3

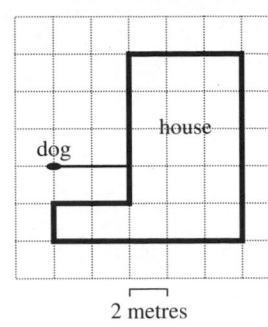

2 metres

4

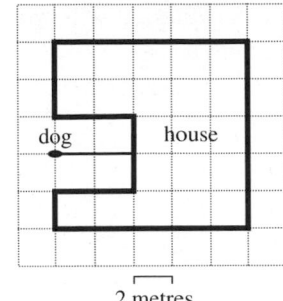

2 metres

5

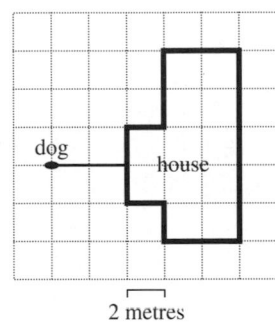

2 metres

6

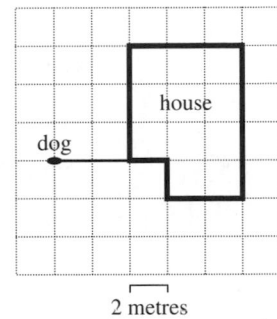

2 metres

Knowledge

The word 'bisect' means to cut exactly in half.

These drawings show how to bisect angles of 60° and 90°. The same construction can be used to bisect *any* other angle.

30°: half of 60°

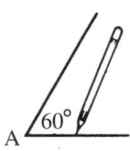 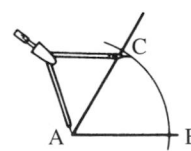 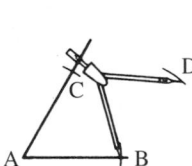 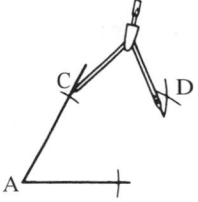 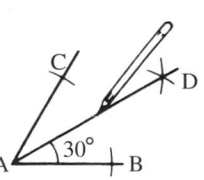

45°: half of 90°

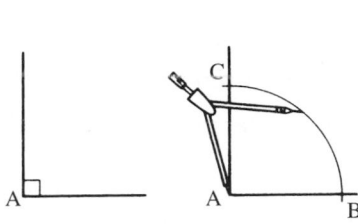 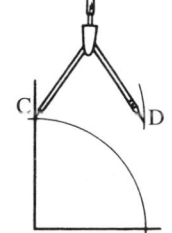 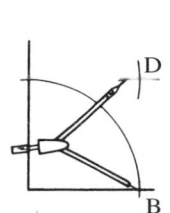 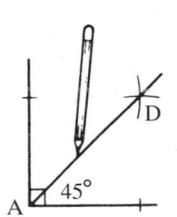

These drawings show how to bisect a line AB.

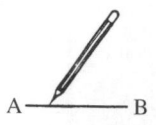 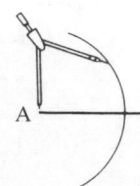 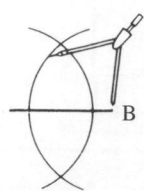 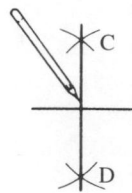 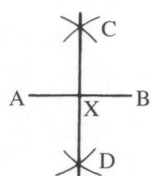

The line CD bisects line AB at right angles at point X.

Skill Practice Two

1 Draw lines of the following lengths.
   a 8 cm   **b** 12 cm   **c** 6.4 cm
   d 10.2 cm   **e** 9.6 cm   **f** 8.5 cm
   Bisect each line and check your accuracy by measuring both parts.

2 With your protractor draw angles of:
   a 40°   **b** 64°   **c** 88°
   d 160°   **e** 106°   **f** 170°
   Bisect each angle and check your accuracy by measuring each part.

3 Construct a triangle ABC in which
   AB = 10 cm, $B\widehat{A}C = 60°$ and $A\widehat{B}C = 45°$.
   Bisect each side and mark where these bisectors meet with the letter X.
   Draw a circle of radius XA with its centre at X.
   What do you notice about this circle?

4 Reconstruct the triangle referred to in question **3**.
   Bisect each angle and mark where these bisectors meet with the letter Y.
   Draw a circle of radius 2.4 cm with its centre at Y.
   What do you notice about this circle?

5 Construct a triangle PQR in which
   PQ = PR = 7 cm, $Q\widehat{P}R = 90°$ and $P\widehat{Q}R = 45°$.
   Bisect each side and mark where these bisectors meet with the letter Z.
   What do you notice about the position of the point Z?

Skill Example

A soldier wants to crawl between two guard-posts, A and B, so that he is always as far away as possible from each post.

soldier
•

post A •   • post B

Copy the diagram and mark on it the locus of the soldier's movement.

The soldier needs to maintain an equal distance between himself and the two posts. Therefore, he must crawl along the perpendicular bisector of AB.

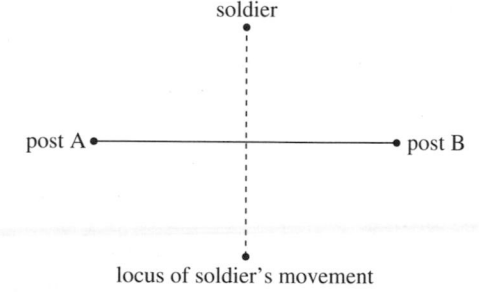

Skill Practice Three

1 A soldier realises he has crawled into the gap between two electrified fences which meet at an angle of 60°. He wishes to crawl away and keep as far away as possible from both fences.

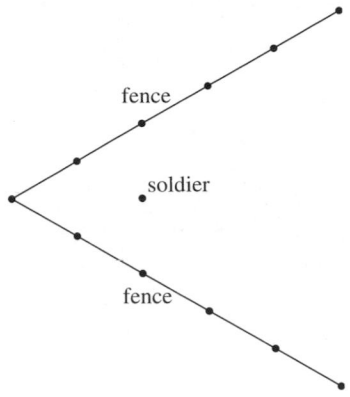

Copy the diagram and mark on it the locus of the soldier's movement.

2 A donkey is tethered by a rope to a ring which can move freely along a wire stretched between two short posts 5 metres apart.

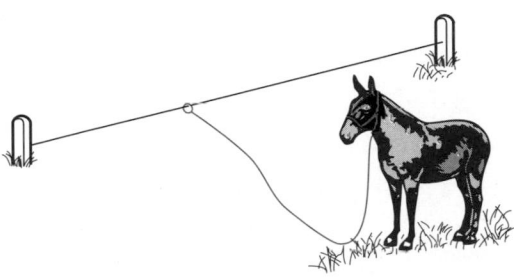

The rope is long enough to allow the donkey to move 3 m away from the wire.
This is a plan view.

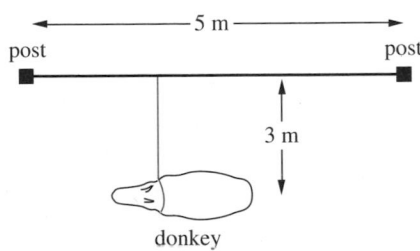

Copy the diagram using a scale of 2 cm to 1 m and mark on it the locus of the donkey's movements when the rope is fully extended.

3 A boat's captain sees two rocks sticking out of the sea. The rocks are 0.5 km apart.
This is the plan view.

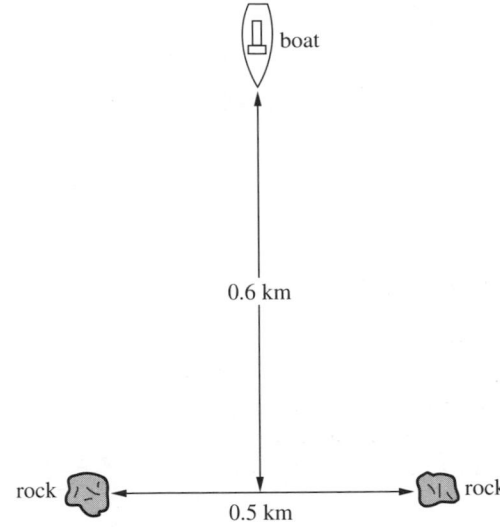

Copy the diagram using a scale of 20 cm to 1 km and mark on it the boat's course if the captain decides to sail between the rocks but as far away from each rock as possible.

4 Two rows of houses in a close are at an angle of 40° to each other.
This is the plan view.

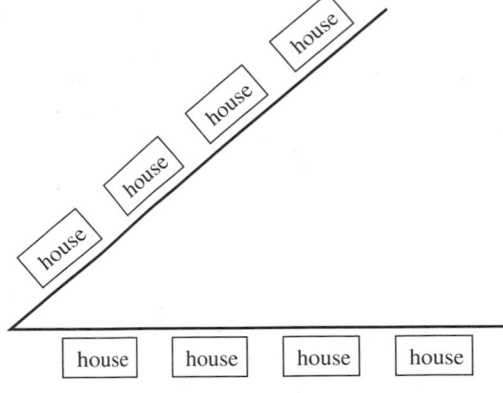

A cable engineer wishes to lay a cable down the middle of the close so that the distance from the cable to each pair of houses on opposite sides of the close is equal.
Copy the diagram and mark on it the position of the cable.

5 A pirate knows that treasure was buried at an equal distance from two rivers which meet at an angle of 45°.

He also knows that the treasure was buried at an equal distance from two trees.

Copy the plan view (right) using a suitable scale, and mark on it the location of the treasure.

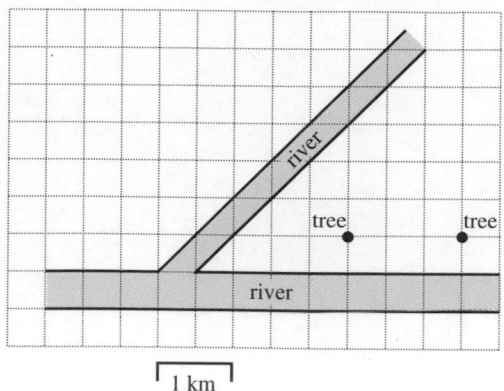

1 km

Unit 44

Knowledge

A car travels on a motorway at a constant speed. The journey of 270 km is completed in 3 hours.

So, in 1 hour the car travels a distance of

$$\frac{270}{3} = 90\,km$$

We say that the **average speed** of the car is 90 kilometres per hour (90 km/h).

$$\text{Average speed} \atop \text{(km/h)} = \frac{\text{Distance travelled (km)}}{\text{Time taken (h)}}$$

Skill Example

Write each as a decimal of one hour.

a 15 minutes   **b** 50 minutes   **c** 35 minutes

a $15\,min = \dfrac{15}{60}\,h = 0.25\,h$

b $50\,min = \dfrac{50}{60}\,h \approx 0.83\,h$

c $35\,min = \dfrac{35}{60}\,h \approx 0.58\,h$

Skill Practice One

Write each of the following as a decimal of one hour.

1	20 minutes	**2**	40 minutes	**3**	30 minutes
4	10 minutes	**5**	12 minutes	**6**	48 minutes
7	24 minutes	**8**	36 minutes	**9**	6 minutes
10	18 minutes	**11**	54 minutes	**12**	5 minutes

Skill Example

Find the average speed, in km/h, for a journey of:

a 135 km in 3 hours   **b** 150 km in $2\frac{1}{2}$ hours

a Average speed $= \dfrac{\text{Distance}}{\text{Time}} = \dfrac{135}{3} = 45\ km/h$

b Average speed $= \dfrac{\text{Distance}}{\text{Time}} = \dfrac{150}{2\frac{1}{2}} = 150 \div 2.5$

$$= 60\,km/h$$

Skill Practice Two

Find the average speed for each of the following journeys.

	Distance covered	Time taken		Distance covered	Time taken
1	100 km	2 h	**2**	180 km	3 h
3	220 km	4 h	**4**	420 km	5 h
5	378 km	6 h	**6**	656 km	8 h
7	536 km	4 h	**8**	708 km	6 h
9	515 km	5 h	**10**	630 km	6 h
11	100 km	$2\frac{1}{2}$ h	**12**	120 km	$1\frac{1}{2}$ h

Skill Example

Find the average speed, in km/h, for a journey of:

a 30 km in 20 min   **b** 165 km in 2 h 45 min

a Average speed $= \dfrac{\text{Distance (km)}}{\text{Time (h)}}$

$$= 30 \div (20 \div 60)$$

Using a calculator gives:

$$20 \div 60 = 0.333\,333\,3$$

$$30 \div (20 \div 60) = 90\,km/h$$

b Average speed $= \dfrac{\text{Distance (km)}}{\text{Time (h)}}$

$$= \frac{165}{2\frac{45}{60}}$$

$$= \frac{165}{2.75}$$

$$= 165 \div 2.75$$

$$= 60\,km/h$$

Skill Practice Three

For questions **1** to **8**, find the average speed for each journey.

	Distance covered	Time taken		Distance covered	Time taken
1	150 km	1 h 30 min	**2**	120 km	2 h 30 min
3	90 km	1 h 15 min	**4**	105 km	1 h 45 min
5	189 km	2 h 15 min	**6**	275 km	2 h 45 min
7	100 km	1 h 20 min	**8**	160 km	1 h 40 min

9 A train travels a distance of 630 km from London to Edinburgh in 4 h 30 min. Find the average speed.

10 A bus travels a distance of 98 km from Newcastle to Berwick in 2 h 20 min. Find the average speed.

11 A woman drives her car a distance of 13 km from Birmingham to Halesowen in 20 min. Find her average speed.

12 A bus travels a distance of 24 km from Leeds to Harrogate in 45 min. Find the average speed.

Skill Example

Find the distance travelled when a car travels for:

a 4 hours at an average speed of 45 km/h

b $2\frac{1}{2}$ hours at an average speed of 60 km/h

c 20 minutes at an average speed of 75 km/h

a Distance $= 45 \times 4 = 180$ km

b Distance $= 60 \times 2.5 = 150$ km

c Distance $= 75 \times \dfrac{20}{60} = \dfrac{75 \times 20}{60} = 25$ km

---

Skill Practice Four

For questions **1** to **4**, find the distance covered for each journey.

	Average speed	*Time of journey*		*Average speed*	*Time of journey*
1	80 km/h	3 hours	**2**	90 km/h	4 hours
3	75 km/h	6 hours	**4**	45 km/h	8 hours

5 A car travelling at an average speed of 75 km/h takes 4 hours to travel from London to Manchester. Find the distance between the two cities.

6 A car travelling at an average speed of 66 km/h takes 3 h 30 min to travel from Newcastle upon Tyne to Glasgow. Find the distance between the two cities.

7 A bus travelling at an average speed of 28 km/h takes 2 h 15 min to travel from Birmingham to Shrewsbury. Find the distance between the two bus stations.

8 A ship sailing at an average speed of 27 km/h takes 1 h 20 min to cross the English Channel from Dover to Calais. Find the distance between the two ports.

9 A car travelling at an average speed of 54 km/h takes 20 minutes to travel from Brighton to Worthing. Find the distance between the two resorts.

10 An aeroplane travelling at an average speed of 480 km/h takes 45 minutes to fly from Birmingham to Dublin. Find the distance between the two airports.

Skill Example

Find the time taken to travel a distance of:

a 750 km at an average speed of 75 km/h
b 150 km at an average speed of 60 km/h
c 60 km at an average speed of 80 km/h

a Time taken $= 750 \div 75 = \dfrac{750}{75}$ h $= 10$ h

b Time taken $= 150 \div 60 = \dfrac{150}{60}$ h $= 2.5$ h

c Time taken $= 60 \div 80 = \dfrac{60}{80}$ h $= 0.75$ h

$= 45$ min

Skill Practice Five

For questions **1** to **4**, find the time taken for each journey.

	Distance covered	Average speed		Distance covered	Average speed
1	150 km	50 km/h	**2**	320 km	80 km/h
3	240 km	30 km/h	**4**	200 km	40 km/h

5 Find the time taken by a train which travels a distance of 480 km from London to Penzance at an average speed of 96 km/h.

6 Find the time taken by a car which travels a distance of 432 km from London to Newcastle upon Tyne at an average speed of 72 km/h.

7 Find the time taken by a car which travels a distance of 630 km from Bristol to Glasgow at an average speed of 70 km/h.

8 Find the time taken by a train which travels a distance of 297 km from London to Leeds at an average speed of 132 km/h.

9 Find the time taken by an aeroplane which flies a distance of 125 km from Manchester to Birmingham at an average speed of 500 km/h.

10 Find the time taken by an aeroplane which flies a distance of 320 km from Leeds to Belfast at an average speed of 480 km/h.

Unit 45

Shape, space and measures   Level 7

Skill   You appreciate the continuous nature of measurement and recognise that a measurement given to the nearest whole number may be inaccurate by up to one half in either direction.

Knowledge

A length measured correct to a given unit will have a real length that lies between two limits.

For example, if a length is measured as 20 cm, correct to the nearest centimetre, the actual length can be anywhere between 19.5 cm and 20.5 cm.

Strictly speaking, the length could be anything up to 20.5 cm but not exactly 20.5 cm, since this would round up to 21 cm.

This means the upper limit could be 20.49 cm or 20.499 cm or 20.4999 cm and so on.

For practical reasons, the upper limit is therefore taken as 20.5 cm, even though it cannot have this value.

Skill Example

Between what limits will a length lie which is measured as 25 m correct to the nearest metre?

The length will be between 24.5 m and 25.5 m.

Skill Practice One

State the limits between which each measurement must lie.

1 A weight of 2 kg correct to the nearest kilogram.

2 A length of 100 m correct to the nearest metre.

3 A length of 100 cm correct to the nearest centimetre.

4 A time of 65 minutes correct to the nearest minute.

5 An area of 60 cm^2 correct to the nearest square centimetre.

6 A length of 56 mm correct to the nearest millimetre.

7 A weight of 55 g correct to the nearest gram.

8 A weight of 500 g correct to the nearest gram.

9 A time of 3 minutes correct to the nearest minute.

10 An area of 25 m^2 correct to the nearest square metre.

Skill Example

A square carpet tile is manufactured so that its side length is 30 cm, correct to the nearest centimetre.

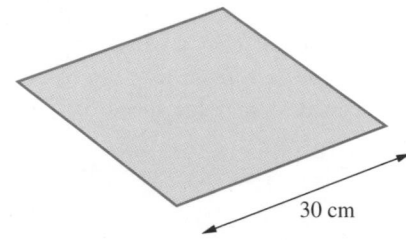

30 cm

a Within what limits will the side length of such a tile lie?
b What is the least area such a tile can have?
c What is the greatest area such a tile can have?

a The side length of a tile will be between 29.5 cm and 30.5 cm.

b The least area a tile can have is
$$29.5 \times 29.5 = 870.25 \text{ cm}^2$$

c The greatest area a tile can have is
$$30.5 \times 30.5 = 930.25 \text{ cm}^2$$

Skill Practice Two

1 A trap-door 50 cm wide by 90 cm long (correct to the nearest cm) is cut from a sheet of wood.

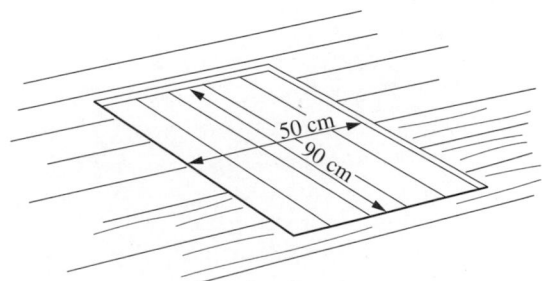

50 cm
90 cm

a What is the least possible width for the trap-door?

b What is the greatest possible width for the trap-door?

c What is the least possible length for the trap-door?

d What is the greatest possible length for the trap-door?

e What is the least possible area for the trap-door?

f What is the greatest possible area for the trap-door?

2 A car completes a 240 mile journey (correct to the nearest mile) at an average speed of 60 miles per hour, correct to the nearest mile per hour.

a What is the least possible value for the average speed of the car?

b What is the greatest possible value for the average speed of the car?

c What is the least possible value for the length of the journey?

d What is the greatest possible value for the length of the journey?

e What is the least possible time the journey will take?

f What is the greatest possible time the journey will take?

3 A length of cloth is 25 m long (to the nearest metre). It is to be cut into pieces of length 80 cm (to the nearest centimetre).

a What is the least possible value for the length of the cloth in metres?

b What is the greatest possible value for the length of the cloth in metres?

c What is the least possible value for the length of the cloth in centimetres?

d What is the greatest possible value for the length of the cloth in centimetres?

e What is the least possible value for the length of each piece in centimetres?

f What is the greatest possible value for the length of each piece in centimetres?

g What is the least number of pieces which can be cut from the cloth?

h What is the greatest number of pieces which can be cut from the cloth?

4 A window has 8 separate panes of glass, each 200 mm wide by 300 mm high (to the nearest millimetre).

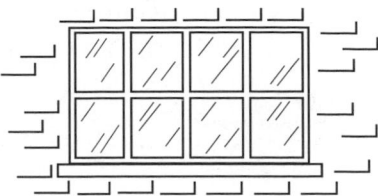

a What is the least possible width for one pane of glass?

b What is the greatest possible width for one pane of glass?

c What is the least possible area of glass in the window?

d What is the greatest possible area of glass in the window?

5 A square garden has side 55 m long (correct to the nearest metre). It is to be covered with grass seed which comes in boxes that cover 64 m² (correct to the nearest square metre).

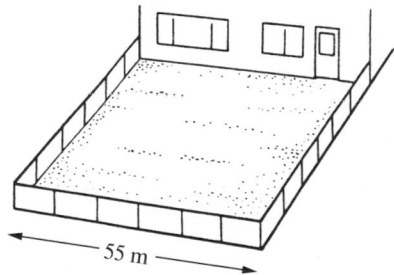

55 m

a What is the least possible value for the side length of the garden?

b What is the greatest possible value for the side length of the garden?

c What is the least possible value for the area of the garden?

d What is the greatest possible value for the area of the garden?

e What is the least number of boxes of seed that will be required?

f What is the greatest number of boxes of seed that will be required?

Unit 46

Handling data   Level 4

Skill   You collect discrete data and record them using a frequency table.

Knowledge

The following list gives the numbers of goals scored by 30 footballers during one season.

20	14	19	14	12	17	13	16	14	17
18	13	16	13	21	16	13	15	13	15
15	12	16	13	18	20	16	15	18	14

This information can be illustrated on a bar chart. The first step is to draw up a **tally chart** from the data to find the frequency of each number of goals scored.

Goals scored	Tally	Frequency
12	//	2
13	//// /	6
14	////	4
15	////	4
16	////	5
17	//	2
18	///	3
19	/	1
20	//	2
21	/	1
	Total	30

Here is the bar chart for this data.

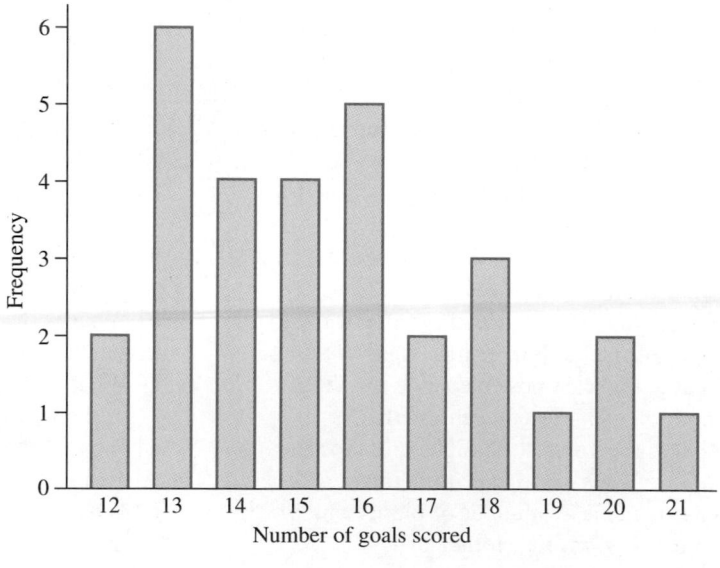

Skill Practice

For each set of data **1** to **6**, first draw up a tally chart. Then display the
information on a bar chart.

1 The marks out of 10 obtained by each pupil in Year 11 in a science test:

3	6	5	7	1	8	6	7	8	4
7	8	3	0	9	7	2	3	5	6
5	9	7	8	10	4	1	6	7	6

2 The marks obtained by each pupil in class 5A in a French test:

2	6	4	6	5	7	2	3	8	5
5	2	7	1	3	8	5	9	4	7
3	7	1	8	9	0	7	6	2	8

3 The number of bottles of milk delivered each day to each house in Park Close:

3	1	3	2	4	2	1
2	4	3	5	2	1	6
4	5	2	3	1	2	

4 The outside temperature, in degrees Celsius, on each of the 30 days in April:

5	4	5	8	10	7	6	9	6	2
2	1	0	0	3	4	5	4	6	8
10	10	8	7	5	6	9	7	4	5

5 The number of cars sold by dealer on each of the 30 days in June:

2	4	1	0	3	1	2	0	1	4
5	3	1	2	0	3	4	1	0	2
0	1	4	3	1	2	0	3	5	2

6 The number of pupils in each of the 25 classes at Ash Green School:

26	28	27	28	25	29	28	25	29
27	28	25	29	27	26	29	28	26
29	25	29	28	28	27	25		

Unit 47

Knowledge

In any set of data, the most frequently occurring item is called the **mode** or **modal average**.

Skill Example

a During one week, the marks in Kate's English exercise book were:

5, 7, 7, 9, 7

What was the modal mark?

The modal mark was 7 because this mark occurred more frequently than any other.

b The shoe sizes of a class of 20 pupils are:

3, 5, 4, 5, 3, 3, 4, 4, 5, 7, 7, 4, 4, 4, 6, 7, 6, 7, 3, 5

Find the mode.

Construct a tally chart, as below.

Shoe size	Tally	Frequency
3	////	4
→4	////// /	6←
5	////	4
6	//	2
7	////	4
	Total = 20	

The most popular size is 4. This is the size that occurs the most frequently.

So, 4 is the mode.

Skill Practice One

1 Twelve school pupils apply to go on an outward bound course. Their ages are:

16, 14, 15, 16, 15, 17, 15, 16, 15, 17, 16, 15

Find the mode of their ages.

2 The list below shows how many pupils in class 2B were absent on each day of a three-week period.

Mon	0	Mon	2	Mon	0
Tue	1	Tue	1	Tue	0
Wed	2	Wed	1	Wed	1
Thur	2	Thur	0	Thur	2
Fri	1	Fri	1	Fri	2

Find the modal number of absentees.

3 A cricketer plays for one club for 15 seasons. The list below shows how many centuries he scored in each of the seasons.

4, 3, 0, 5, 4, 5, 2, 4, 1, 2, 4, 1, 0, 2, 0

Find his modal number of centuries.

4 The list below shows how many suits were sold at a tailor's shop on each day of a three-week period.

Mon	2	Mon	1	Mon	4
Tue	1	Tue	4	Tue	1
Wed	4	Wed	3	Wed	5
Thur	2	Thur	2	Thur	3
Fri	3	Fri	5	Fri	2
Sat	5	Sat	2	Sat	4

Find the modal number of suits sold.

5 A dice is thrown 20 times, giving the following scores:

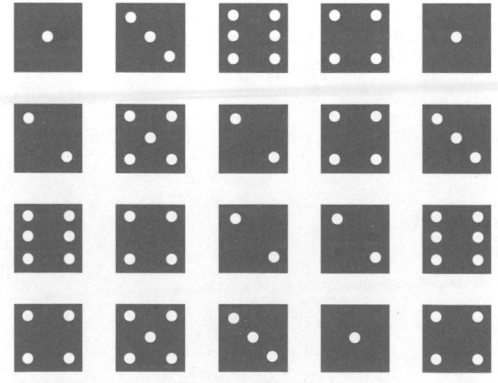

Find the modal score.

6 The list below shows how many lunches a cafe served on each day of a three-week period.

Mon	5	Mon	4	Mon	5
Tue	7	Tue	7	Tue	5
Wed	6	Wed	5	Wed	8
Thur	5	Thur	6	Thur	7
Fri	6	Fri	5	Fri	6
Sat	4	Sat	6	Sat	7

Find the modal number of lunches served.

7 A footballer made 20 appearances during a certain season and scored the following numbers of goals in the matches.

0, 1, 2, 1, 0, 2, 3, 0, 1, 2,
1, 0, 3, 0, 1, 2, 1, 3, 3, 2

Find the modal number of goals that he scored.

8 The list below shows how many pupils there are in each of the 20 classes at Westmead School.

26, 25, 24, 22, 24, 23, 26, 25, 23, 25,
26, 24, 25, 25, 22, 23, 23, 25, 24, 23

Find the modal number of pupils per class.

9 A tennis club has 25 members and the list below shows their ages.

16, 18, 15, 18, 16, 15, 14, 16, 17, 15, 17, 16, 17, 14, 18, 16, 18, 15, 18, 17, 17, 16, 17, 18, 17

Find the modal age of the members.

10 During a certain season a football team played in 30 matches and scored the following numbers of goals.

1, 0, 6, 2, 7, 2, 2, 3, 4, 3, 0, 6, 3, 2, 0,
1, 4, 3, 1, 5, 3, 0, 2, 1, 4, 0, 2, 6, 1, 5

Find the modal number of goals per match.

Knowledge

If a number of items are arranged in order of size, the middle one of the items is called the **median**.

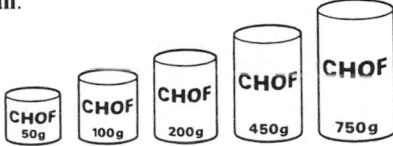

Look at these cans of dog food. The median size of cans of 'Chof' is 200 g.

Skill Example

Two dice are thrown seven times. The results are as shown.

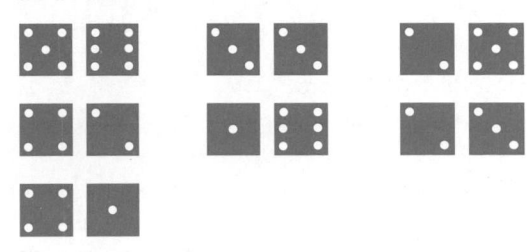

What is the median score?

The total scores are 11, 6, 7, 6, 7, 5, 5

When arranged in ascending order, the list becomes

5, 5, 6, <u>6</u>, 7, 7, 11

The middle figure is 6, so the median score is 6.

Knowledge

When there is an even number of items in a list, there will be no exact middle item. In this case, the median is found by taking the average of the two items on each side of the exact middle position.

Skill Example

The rainfall, in mm, in a town is recorded each month. These are the results after one year.

20, 12, 15, 21, 12, 10, 5, 6, 12, 18, 21, 20

What is the median rainfall?

When arranged in ascending order, the list becomes:

5, 6, 10, 12, 12, <u>12</u>, <u>15</u>, 18, 20, 20, 21, 21

Median $= \dfrac{(12 + 15)}{2} = 13.5\,\text{mm}$

Skill Practice Two

1 One week, Peter goes to school by bus and finds that the bus arrives late by the following numbers of minutes:

Mon	5	Wed	9	Fri	2
Tue	7	Thur	1		

Find the median number of minutes late.

2 One week, a travelling saleswoman calls at a garage every day and buys the following quantities of petrol:

Mon	15 litres	Thur	24 litres
Tue	21 litres	Fri	12 litres
Wed	16 litres		

Find the median quantity that she buys.

3 A coach driver has to make five journeys from a railway station to a nearby holiday camp. The numbers of passengers that he carries are:

36, 40, 33, 31, 25

Find the median number that he carries.

4 Mrs Andrews buys half a kilogram of tomatoes on six different occasions. The prices that she pays are:

59 p, 62 p, 57 p, 61 p, 55 p, 58 p

Find the median price that she pays.

5 A rugby team wins a trophy after playing in a five-round contest. They win their matches by the following scores.

First round	24 10	Semi-final	27 14
Second round	9 6	Final	19 4
Third round	16 12		

Find:
a the median number of points that they score
b the median number of points that are scored against them.

6 The midday temperatures for a certain week during July were as follows.

Mon	22 °C	Thur	21 °C	Sat	26 °C
Tue	20 °C	Fri	23 °C	Sun	24 °C
Wed	19 °C		

Find the median temperature.

7 The list below shows how far I travel in my car on each day of a certain week.

Mon 42 km   Thur 61 km   Sat 35 km
Tue 53 km   Fri 94 km   Sun 90 km
Wed 47 km

Find the median distance that I travel per day.

8 A train from Birmingham to Manchester makes six stops on the way. The list below shows the number of passengers in the train for each of the seven stages of the journey.

Birmingham to Wolverhampton	240
Wolverhampton to Stafford	200
Stafford to Stoke-on-Trent	230
Stoke-on-Trent to Congleton	190
Congleton to Macclesfield	180
Macclesfield to Stockport	210
Stockport to Manchester	170

Find the median number of passengers in the train.

9 A policeman who works at night found that in one year he required ten batteries for his bicycle headlamp. The times that they lasted were:

42, 39, 36, 47, 51, 40,
44, 43, 33 and 32 days

Find the median time for which a battery lasted.

10 A cricket team were all out for 250 runs. The eleven players scored:

65, 43, 35, 8, 21, 0,
6, 17, 32, 13 and 10 runs

Find their median score.

Unit 48

Handling data   Level 4

Skill   You group data, where appropriate, in equal class intervals, represent collected data in frequency diagrams and interpret such diagrams.

Skill Example

One hundred pupils took a mathematics test marked out of 20. These are their scores.

1	1	1	2	2	2	2	3	3	3
3	3	4	4	4	4	5	5	5	5
5	5	6	6	6	6	7	7	7	7
7	7	7	8	8	8	8	9	9	9
9	9	10	10	11	11	11	11	12	12
12	12	12	13	13	13	13	13	13	14
14	14	14	14	14	14	14	14	15	15
15	15	15	15	15	15	15	16	16	16
16	16	16	16	17	17	17	17	17	18
18	18	18	19	19	19	19	20	20	20

This data could be illustrated with a bar chart like this:

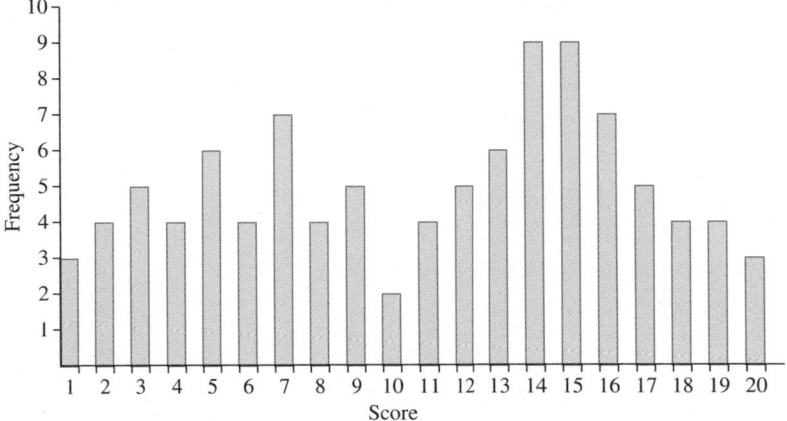

This bar chart has a large number of bars and takes quite a time to construct. It may be preferable to **group** the data into a table like this:

Score	1–4	5–8	9–12	13–16	17–20
Frequency	16	21	16	31	16

This is a bar chart for the **grouped** scores:

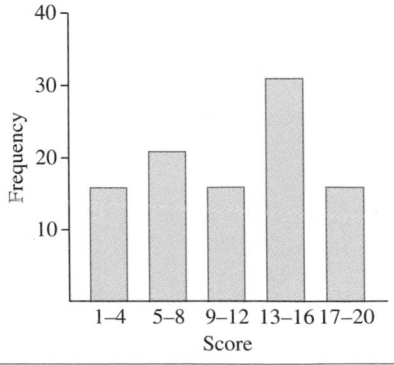

Skill Practice One

1 Make a table for the 100 marks given in the Skill Example using groups 1–2, 3–4, 5–6, 7–8, 9–10 and so on.
Draw a bar chart to illustrate your table.

2 Make a table for the 100 marks given in the Skill Example using groups 1–5, 6–10 and so on.
Draw a bar chart to illustrate your table.

3 Two darts players, Bill and Terri, each threw a set of three darts 50 times. The 50 scores for each player are shown in this table.

| Score | Frequency | |
	Bill	Terri
1–20	4	1
21–40	8	12
41–60	8	14
61–80	9	12
81–100	8	1
101–120	5	5
121–140	4	3
141–160	3	2
161–180	1	0

Draw two bar charts to illustrate this data.

4 The weekly incomes of two groups of 50 people are shown in this table.

| Income (£) | Frequency | |
	Group 1	Group 2
0–50	2	0
51–100	3	0
101–150	7	2
151–200	10	8
201–250	18	13
251–300	7	17
301–350	2	7
351–400	1	3

Draw two bar charts to illustrate this data.

5 Two groups of 80 students are compared on an English test, marked out of 100. The results are shown below.

Group 1

15	12	17	64	56	67	51	89
45	56	67	34	92	28	31	32
36	74	47	83	87	98	89	21
11	23	33	34	56	35	47	68
82	42	34	45	31	47	33	56
39	38	47	36	46	37	65	70
20	32	51	62	52	53	54	67
72	81	77	83	85	37	36	36
29	80	40	53	33	33	34	51
30	63	66	55	44	33	71	70

Group 2

21	56	57	62	73	82	83	90
34	57	83	66	74	74	57	60
34	56	71	79	73	74	56	63
64	81	88	92	99	34	35	63
22	82	20	90	65	56	72	77
75	23	34	43	56	65	67	76
80	81	45	59	92	85	75	73
62	64	68	66	74	73	77	78
23	39	47	74	77	82	94	90
38	49	60	55	55	73	77	82

a Collect this data into a table with these headings:

| Mark | Frequency | |
	Group 1	Group 2
1–10		
11–20		
etc.		

b Draw two bar charts to illustrate this data.

6 This graph shows percentages of the population of Great Britain who say their diet is mainly vegetarian.

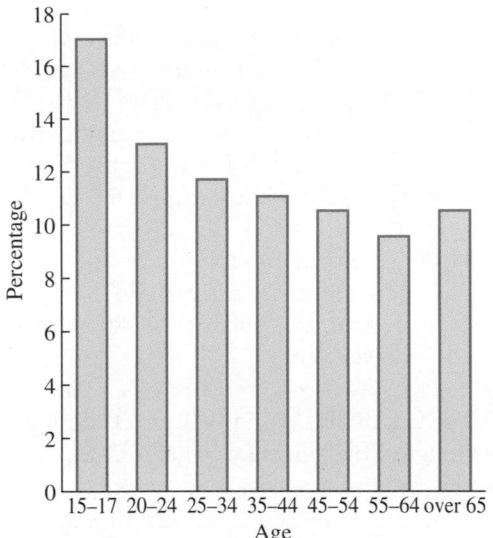

a Estimate the percentage of people in each age range who say their diet is mainly vegetarian.

b Write some comments on the graph.

7 This graph shows the age distribution for the projected UK female population for the year 2031.

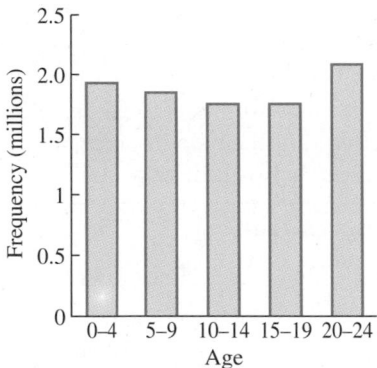

Estimate the number of females who will be in each age range.

Unit 49

Skill Example

The table below shows the real distance, in kilometres, and the map distance, in centimetres, between three towns.

Real distance (km)	5	7.5	12.5
Map distance (cm)	2	3	5

Using a scale of 1 cm to represent 2 km on the horizontal axis and a scale of 1 cm to represent 1 cm on the vertical axis, draw a suitable graph.

From your graph, estimate:

a the distance on the map between two towns which are 9 km apart

b the real distance between two towns when the map distance is 2.5 cm.

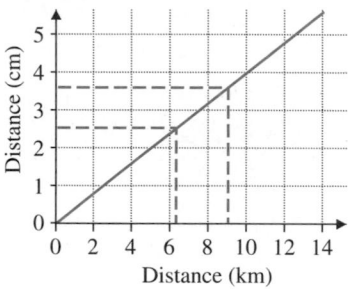

a Map distance is 3.6 cm.

b Real distance is 6.3 km.

Skill Practice One

1 A flight of stairs reaches a height of 3 metres above the lower floor. The table shows the height of certain stairs above the floor.

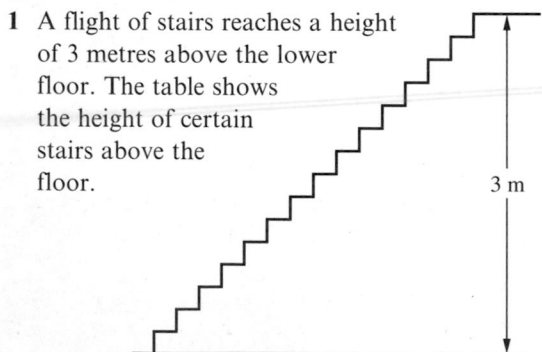

3 m

No. of stairs	2	4	5	10	13	15
Height above floor (cm)	40	80	100	200	260	300

Draw a graph of the figures in the table using a horizontal scale of 1 cm to 1 stair and a vertical scale of 1 cm to 20 cm.

a Find from your graph the height above the floor of (i) the third stair, (ii) the seventh stair, (iii) the ninth stair, (iv) the eleventh stair.

b Find from your graph which stair I am standing on if my feet are (i) 120 cm, (ii) 160 cm, (iii) 240 cm, (iv) 280 cm above the lower floor.

2 The length of a spring is 10 cm. The spring is stretched by hanging weights from its end. The new lengths are given in the table.

Weight (g)	0	10	15	30	45	60
Length (cm)	10	12	13	16	19	22

Draw a graph of this information using a horizontal scale of 1 cm to 5 g and a vertical scale of 1 cm to 1 cm.

a Find from your graph the length of the spring when it supports a weight of (i) 5 g, (ii) 25 g, (iii) 40 g, (iv) 55 g.

b Find from your graph the weight required to stretch the spring to a length of (i) 14 cm, (ii) 17 cm, (iii) 20 cm.

3 The table shows the number of sheets of writing paper in a pile of a given height.

No. of sheets	100	250	350	500
Height of pile (mm)	8	20	28	40

Draw a graph of this information using a horizontal scale of 1 cm to 50 sheets and a vertical scale of 1 cm to 2 mm.

a What is the height of a pile containing (i) 50 sheets, (ii) 150 sheets, (iii) 400 sheets?

b Find the number of sheets of paper in a pile of height (i) 16 mm, (ii) 24 mm, (iii) 36 mm.

4 The table gives the weight of each pile of writing paper listed in question **3**.

No. of sheets	100	250	350	500
Weight of pile (g)	400	1000	1400	2000

Display this information on a graph using a horizontal scale of 1 cm to 50 sheets and a vertical scale of 1 cm to 100 g.
  a Use your graph to find the weight of a pile of (i) 50 sheets, (ii) 200 sheets, (iii) 300 sheets.
  b What is the number of sheets in a pile weighing (i) 600 g, (ii) 1600 g, (iii) 1800 g?

5 The table shows the number of lumps of sugar of equal size contained in three different weights of packet.

Weight of packet (g)	200	320	400	600
No. of lumps	60	96	120	180

Draw a graph to illustrate this information using a horizontal scale of 1 cm to 50 g and a vertical scale of 1 cm to 10 sugar lumps.
  a How many sugar lumps could be contained in a packet weighing (i) 500 g, (ii) 560 g?
  b Find the weight of a packet containing 144 lumps of sugar.

6 A girl produced this table of money equivalents for her Belgian pen friend.

British pounds	1	3	4.50	7	8
Belgian francs	60	180	270	420	480

Plot a graph of this information using a horizontal scale of 2 cm to £1 and a vertical scale of 2 cm to 100 Belgian francs.
  a From your graph, find the value in Belgian francs of (i) 50 p, (ii) £2, (iii) £3.50, (iv) £7.50.
  b What is the value in British pounds of (i) 150 Belgian francs, (ii) 240 Belgian francs, (iii) 390 Belgian francs?

7 The cost of a certain kind of curtain track is given in the table.

Length of window (cm)	80	100	140	200	250
Cost of curtain track	48 p	60 p	84 p	£1.20	£1.50

Show this information on a graph using a horizontal scale of 1 cm to 10 cm and a vertical scale of 1 cm to 10 p.
What is the cost of curtain track for a window of length (i) 150 cm, (ii) 180 cm, (iii) 240 cm?

8 Six children are standing together in the sunshine. For some of the children the heights and shadow lengths are given in the table.

Name	Aisha	Scott	Polly	Sanjay
Height (cm)	100	120	132	160
Shadow length (cm)	150	180	198	240

Using a scale of 1 cm to 10 cm on both axis, plot a graph of the above figures. (Use the horizontal axis for the heights.)
Find from your graph:
  a the length of Daniel's shadow, who is 140 cm tall
  b the length of Kelly's shadow, who is 108 cm tall.

9 Eight children have their photographs taken together. For some of the children, their real heights and their heights on the snapshot are given in the table.

Name	Annie	Robert	Jason	Chantelle
Real height (cm)	110	120	140	180
Height on snapshot (mm)	55	60	70	90

Using a scale of 1 cm to 10 cm on the horizontal axis and a scale of 1 cm to 5 mm on the vertical axis, draw a graph of the above figures.
Find the following from your graph
  a The height on the snapshot of (i) Melanie, who is 130 cm tall, and (ii) Samita, who is 160 cm tall.
  b The real height of (i) Jodie, who is 75 mm tall on the snapshot, and (ii) Liam, who is 85 mm tall on the snapshot.

10 The girls in class 4A have either auburn, blonde or dark hair. Details are given in the table.

Colour of hair	Auburn	Blonde	Dark
Number of girls	2	6	12
Percentage of girls	10%	30%	60%

Using a scale of 1 cm to 1 girl on the horizontal axis and a scale of 1 cm to 5% on the vertical axis, draw a graph of the above figures.
Find the answers to the following from your graph.

a If 8 girls are hockey players, what percentage of the class are hockey players?
b If 11 girls are netball players, what percentage of the class are netball players?
c If 14 girls have blue eyes, what percentage of the class have blue eyes?
d One day, 15% of the girls arrive late. How many girls arrive late?
e 25% of the girls cycle to school. What number of girls cycle to school?
f On a certain day, 80% of the girls are present. What number of girls are present?

Unit 50

Knowledge

Millions of things may or may not happen tomorrow.

Each may be classified as one of the following:

> Certain to happen
> Likely to happen
> Not likely to happen
> Impossible to happen

Skill Example

Classify each of the following.

a The sun will rise tomorrow.
b Tomorrow, the day will be 25 hours long.
c You will see a car tomorrow.
d You will see a space ship tomorrow.

a Certain
b Impossible
c Likely to happen
d Not likely to happen

Skill Practice One

Classify each of the following events as one of the following:

> certain, impossible, likely to happen tomorrow, not likely to happen tomorrow

1 You will win £1 000 000.
2 You will buy some sweets.
3 You will eat chips.
4 Your hair will grow.
5 You will meet Queen Victoria.
6 You will meet the prime minister.
7 You will watch television.
8 You will play football.
9 You will play netball.
10 You will wear shoes.
11 You will wear a bowler hat.
12 You will work hard at school.
13 You will become headteacher of your school.
14 You will talk to a friend.
15 You will talk to somebody from Iceland.

Knowledge

Often, there are several possible outcomes in a given situation. For example, each day somebody might decide whether to wear red or blue socks.

If the choice is made at **random**, it is done without deliberately trying to select any particular colour.

If each colour is **equally likely** to be selected, the choice is said to be **fair**.

Skill Example

Ali wants to decide whether to wear red or blue socks.

Are the following fair or unfair ways of deciding?

a Look out of the window. If it is raining, wear red socks. If it is not, wear blue socks.
b Put two socks in a bag, one of each colour. Choose one sock from the bag without looking and wear that colour.
c Select a word at random from the newspaper. If the word contains the letter 'z', wear red socks. If not, wear blue.

a This is not a fair way of deciding, because it is not raining for much more of the time than it is raining, even in Britain.

b This is a fair way of deciding, because he is just as likely to pick a red sock as a blue one.

c This is not a fair way of deciding, because there are far more words which do not contain a 'z' than words which do.

Skill Practice Two

1 Bill and Wendy both hate doing the washing up. They consider various ways to decide who will wash up on a particular day.
Are the following fair or unfair ways of deciding?
   a Toss a coin. If it's heads, Bill washes up.
   b If the day of the week contains the letter 'n', Bill washes up.
   c If the month of the year contains the letter 'r', Wendy washes up.
   d If there are leaves on the pear tree in the garden, Bill washes up.
   e Both cut a deck of cards. If she gets the highest card, Wendy washes up.
   f The tallest person does the washing up.
   g The person with the longest name does the washing up.
   h Both guess what colour the next car to drive up the street will be. The first person to be right does not have to wash up.

2 Salima and Lucy are on a train. They are trying to decide who will have the last sweet.
Are the following fair or unfair ways of deciding?
   a The person who bought the sweets eats the last one.
   b Lucy puts the sweet in one hand behind her back. Salima picks a hand. If the sweet is in it she eats it.
   c They look out of the window. The first person to spot a horse in a field eats the sweet.
   d They ask each other general knowledge questions. The first person who gets a question wrong loses and the other person eats the sweet.
   e The person who has the biggest shoe size eats the sweet.
   f They wait for the ticket collector to come. The person he speaks to first eats the sweet.
   g They put their names on two pieces of paper, put them in a bag and then pick one out without looking.
   h They check their pockets. The person with the largest number of 10 p coins eats the sweet.

3 John, Nilesh and Gavin all want to watch different TV programmes.
Are the following fair or unfair ways of deciding who chooses the programme?

   a They all roll two dice. The person with the highest score chooses the programme they watch.
   b On the word 'go', they all hold up either an odd or an even number of fingers. They repeat this until there is an 'odd one out'. He chooses the programme they watch.
   c They all go for a race around the block. The first one back chooses the programme they watch.
   d They go out into the garden. The first one to find a worm chooses the programme they watch.
   e They all play a computer game. The person with the highest score chooses the programme they watch.
   f They ask somebody else to think of a number between 1 and 10. They take turns to guess the number and the first person to guess correctly chooses the programme they watch.

4 Each day, a teacher has to choose two pupils from her class of 30 to pick up litter.
Are the following fair or unfair ways of choosing the pupils?
   a The teacher picks the first two pupils on the class register.
   b The teacher writes all the pupils' names on slips of paper, puts them in a box and then chooses two at random.
   c The teacher picks the two pupils who came bottom in the last maths test.
   d The teacher closes her eyes, waves a pin over the class register, then sticks it in twice to pick two names.
   e The teacher picks the last two pupils that she saw dropping litter.
   f The teacher gives all the pupils a number from 1 to 30. She then rolls five dice, adds the scores and selects the pupil with that number. She repeats this to select the other pupil.
   g The teacher obtains 28 normal drinking straws and two straws with a drop of red paint on one end. She puts them in a cup so that it is impossible to see which ones have red paint on them. Each pupil then picks a straw and the ones who pick the straws with red paint pick up litter.
   h The teacher picks the last two pupils to arrive in the classroom.

5 Daniel and Tom are lost. They have come to a crossroads and don't know whether to turn left, turn right or go straight on.

Are the following fair or unfair ways of deciding?

a Throw a pointed stick in the air and then go in whichever direction it points when it lands.

b Toss a coin first to decide between left and right. Then, having selected left or right, toss a coin again to decide between this direction and straight on.

c Daniel shuts his eyes and points with one arm. Tom spins him round and round. They go in whichever direction Daniel is pointing when he stops spinning.

d Go in the direction which will require the smallest turn of the handlebars.

e Write 'left', 'straight on' and 'right' on three pieces of paper. Put them in a pocket and select one at random.

f Each choose a direction and call it out on the word 'now'. Repeat this until two different directions are selected. Go in the direction that is not selected.

g Toss a coin to decide who will choose. Whoever wins selects a direction.

Unit 51

Handling data   Level 5

Skills   You understand and use the mean of discrete data.

You compare two simple distributions, using the range and one of the measures of average.

Knowledge

The **average height** of this group of children is found by adding together all their individual heights and dividing by the number of children in the group. There are 5 children, so divide by 5.

Their **average weight** can be found in the same way. Find the sum of all their weights, and then divide by 5.

$$\text{Average} = \frac{\text{Sum of all the items}}{\text{Total number of items}}$$

Skill Example

Find the average of:

a 6, 9 and 18   **b** £1.25, £2.75, £3.50 and £4.50

a $\text{Average} = \dfrac{\text{Sum of all the items}}{\text{Total number of items}}$

$= \dfrac{6 + 9 + 18}{3}$

$= \dfrac{33}{3} = 11$

b $\text{Average} = \dfrac{\text{Sum of all the items}}{\text{Total number of items}}$

$= \dfrac{£1.25 + £2.75 + £3.50 + £4.50}{4}$

$= \dfrac{£12}{4} = £3$

Skill Practice One

1 Over a six-week period a man's weekly wages were £173, £174, £181, £177, £171 and £174. Find his average wage.

2 Over a six-week period a newspaper girl's wages were £8.16, £7.75, £7.94, £8.10, £7.80 and £8.25.
Find her average wage.

3 Mrs Robinson buys a kilogram of tomatoes on six different days. The prices that she pays are £1.15, £1.06, £1.04, 97 p, 93 p and 85 p. Find the average price.

4 A newspaper seller takes the following amounts of money during one week:
Mon £31.40   Tue £30.50   Wed £30.20
Thur £33.10   Fri £32.60   Sat £40.20
Find his average daily takings.

5 The heights of four shrubs in a garden are 1 m 45 cm, 2 m 23 cm, 1 m 97 cm and 2 m 35 cm.
Find the average height.

6 The head-to-tail lengths of four dogs are 1 m 13 cm, 1 m 8 cm, 95 cm and 84 cm.
Find the average length.

7 Over a four-week period a woman's wages were £265.46, £275.38, £281.05 and £266.11. Find her average wage.

8 The ages of eight girls are 16, 11, 13, 10, 11, 12, 16 and 15 years.
Find their average age.

9 The ages of eight boys are 13, 14, 11, 13, 10, 8, 9 and 10 years.
Find their average age.

10 In eight innings a batsman makes the following scores: 35, 14, 15, 0, 37, 10, 6 and 3 runs.
Find his average score.

11 There are twelve pupils in class 1A. Their marks out of 50 in an English test are 45, 46, 46, 47, 48, 49, 44, 47, 33, 35, 38 and 50.
Find the average mark.

12 The weights of four cats are 9 kg 150 g, 8 kg 320 g, 8 kg 860 g, and 9 kg 670 g.
Find their average weight.

13 The head-to-tail lengths of five white mice are 11 cm 7 mm, 12 cm 5 mm, 11 cm 9 mm, 13 cm 6 mm and 10 cm 3 mm.
Find their average length.

14 The heights of five canaries are 10 cm 1 mm, 10 cm 9 mm, 9 cm 7 mm, 9 cm 5 mm and 9 cm 8 mm.
Find their average height.

Knowledge

The average you have been calculating in Skill Practice One is usually known as the **mean**, to distinguish it from two other types of average widely used in statistics: the **mode** and the **median**.

You have already met the mode and the median on pages 225 and 226 respectively.

Skill Example

In ten boxes of matches, the number of matches was:

37, 45, 41, 41, 42, 44, 39, 40, 41, 40

Find the mean number of matches per box.

Total number of matches =

$$37 + 45 + 41 + 41 + 42$$
$$+ 44 + 39 + 40 + 41 + 40 = 410$$

Total number of boxes = 10

So, mean number per box = $\frac{410}{10} = 41$

Skill Practice Two

1 Six boys have each bought 100 g of assorted sweets. The actual numbers of sweets that they have are:

19, 22, 20, 23, 20 and 22

Find the mean number of sweets in a bag.

2 A theatre showed the same play for six nights of the same week. The attendances were:

Mon 105   Wed 115   Fri 120
Tue 108   Thur 103   Sat 109

Find the mean attendance figure.

3 A hockey club won a trophy after a six-round contest. The scores in the matches were:

1st round   5 3   2nd round   6 2
3rd round   4 3   4th round   2 0
Semi-final   4 2   Final   4 2

Find:
a the mean number of goals that they scored
b the mean number of goals that were scored against them.

4 Mrs Charlton had the following numbers of bottles of milk delivered during the course of a certain week.

Mon 4   Wed 4   Fri 2   Sun 1
Tue 4   Thur 4   Sat 2

Find the mean number of bottles that she had per day.

5 The ages of the seven girls in a netball team are:

16, 19, 21, 18, 17, 17 and 18

Find their mean age.

6 A taxi driver answers eight calls. The number of passengers that he carries on each trip are:

5, 4, 6, 3, 2, 6, 1 and 5

Find the mean number of passengers per trip.

7 Eight similar bottles of pain-killing tablets have the following contents:

59, 62, 61, 57, 64, 63, 60 and 62 tablets

Find the mean number of tablets per bottle.

8 A primary school has the following numbers of boys and girls in each of its four classes.

Class	Boys	Girls
1	12	13
2	11	8
3	14	12
4	15	11

Find:
a the mean number of boys per class
b the mean number of girls per class.

9 The list below shows how many hours a bricklayer worked on each day of a two-week period.

Mon	8	Mon	8
Tue	8	Tue	11
Wed	12	Wed	12
Thur	12	Thur	10
Fri	11	Fri	8

Find the mean number of hours per day that he worked.

10 Ten packets of paper clips had the following contents:

101, 102, 104, 108, 104,
102, 103, 107, 109, 110

Find the mean number of paper clips per packet.

11 Mr Chavda bought the following numbers of loaves of bread per week over a period of three months:

4, 3, 2, 2, 5, 4, 3, 2, 4, 1, 4, 2

Find the mean number of loaves per week that he bought.

12 A chicken laid the following numbers of eggs per week over a three-month period:

5, 2, 0, 1, 3, 2, 3, 1, 4, 0, 1, 2

Find the mean number of eggs laid per week.

13 Five train journeys from Buxton to Manchester took:

57 min, 58 min, 55 min, 1 h 6 min and 1 h 4 min

Find the mean journey time.

14 Five bus journeys from Ludlow to Birmingham took:

1 h 58 min, 1 h 56 min, 1 h 59 min,
1 h 57 min and 2 h 10 min

Find the mean journey time.

15 A football cup-tie had to be played three times because two matches ended as draws. The three attendance figures were:

52 804, 39 007, 45 760

Find the mean number of spectators per match.

Skill Practice Three

For each question, find which of the mean, the median or the mode is different from the other two.

1 Tom travels to school by bus every day. One week he had to wait at the bus stop for the following times.

Monday	5 minutes	Thursday	5 minutes
Tuesday	9 minutes	Friday	4 minutes
Wednesday	7 minutes		

2 Jamila has her English book marked every day.
One week her marks out of ten were as follows.

Monday	7	Thursday	6
Tuesday	4	Friday	9
Wednesday	9		

3 A small hotel serves lunches daily. One week the numbers served on each day were as follows.

Monday	33	Friday	36
Tuesday	37	Saturday	34
Wednesday	15	Sunday	31
Thursday	31		

4 One week, a bricklayer worked on all seven days. The numbers of hours that he worked on each day were as follows.

Monday	10	Friday	9
Tuesday	9	Saturday	13
Wednesday	12	Sunday	9
Thursday	8		

5 Kerry's examination marks at the end of one term were as follows.

Mathematics	61	French	64
English	63	Science	70
History	40	Technology	68
Geography	61		

6 Mrs Jones buys a pack of tomatoes at a supermarket. The weights of each of the tomatoes in the pack are:

72 g, 65 g, 70 g, 39 g, 71 g, 73 g and 65 g

7 At a fairground, William has five shots with a rifle. The picture below shows his scores.

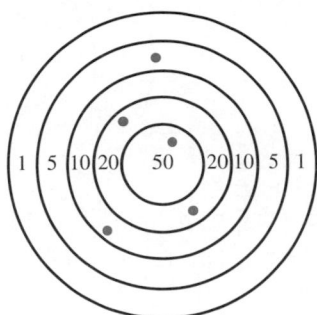

8 Anne throws a dice nine times. These are her scores:

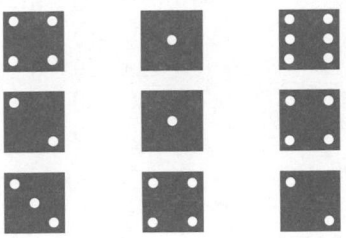

Knowledge

The **range** of a group of values is found by subtracting the smallest value from the largest value.

Skill Example

Find the range of:

5, 6, 7, 5, 3, 3, 3, 5, 7, 8, 6, 2, 10, 5, 2, 7, 12, 9, 10, 5

Largest value = 12

Smallest value = 2

Range = 12 − 2 = 10

Skill Practice Four

Find the range of each group of values.

1 1, 2, 4, 8, 9, 4, 5, 7
2 23, 56, 29, 20, 19, 54, 53, 20
3 17, 17, 17, 18, 19, 20, 15, 29
4 0, 8, 9, 0, 2, 2, 3, 4, 5, 3, 6, 7, 2
5 100, 96, 95, 92, 93, 92, 91

6 62, 62, 62, 63, 65, 66, 66, 68, 66, 65, 65, 65
7 8, 8, 8, 8, 7, 7, 9, 9, 9, 6, 6, 6
8 50, 50, 50, 50, 50, 51, 51, 51, 51, 51, 51
9 0.5, 0.6, 0.7, 0.5, 0.6, 0.7, 0.8, 0.9, 0.5
10 0, 2, −2, 3, −4, 4, 3, −3, −2, 0, 4

Skill Example

A group of values or scores is called a **distribution**.

For example, Jake and Sally each take end-of-year examinations in 10 subjects. These are the distributions of their marks:

Jake's: 52, 64, 55, 61, 67, 55, 51, 60, 57, 62

Sally's: 85, 93, 72, 95, 23, 88, 17, 89, 92, 87

Two distributions can be compared using their mean and their range. So,

Mean of Jake's marks = 58.4
Mean of Sally's marks = 74.1

Range of Jake's marks = 16
Range of Sally's marks = 78

We compare the distributions and conclude that:
- Sally's mean mark is much higher than Jake's mean mark. She has done better overall.
- The range of Jake's marks is much less than the range of Sally's marks. This shows that he has been more consistent, having done roughly equally well in all his subjects.
The range of Sally's marks show that she has done extremely well in some subjects but very badly in others.

Skill Example

These are the distributions of scores when two darts players each threw his set of three darts 12 times.

Eric's scores	98	100	100	100
	100	104	105	106
	120	160	160	160
Jocky's scores	5	5	41	60
	65	66	90	100
	100	180	180	180

a Find the median and range for each player's distribution.

b Compare the two distributions.

a Median of Eric's marks = 104.5
Median of Jocky's marks = 78

Range of Eric's marks = 62
Range of Jocky's marks = 175

b Eric has a higher median score than Jocky. On this evidence, he is the more accurate player. Jocky got the maximum score (180) three times, but the range of his distribution compared with Eric's shows that he is less consistent than Eric. His scores are much more widely spread than Eric's.

Skill Practice Five

1 The contents of 20 boxes of Striko matches and 20 boxes of Katchwell matches are counted.

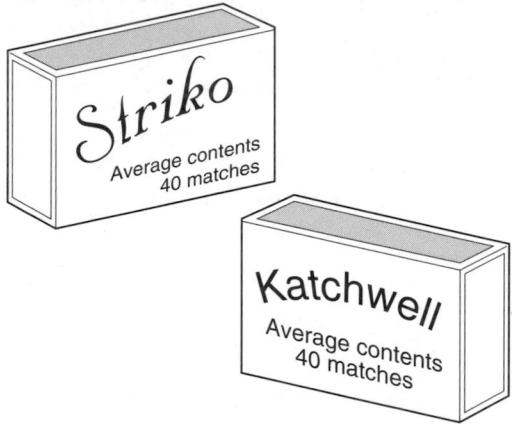

These are the distributions of the number of matches per box.

Striko	37	38	38	38
	39	39	40	40
	40	40	40	40
	41	41	41	41
	42	42	43	44

Katchwell	35	36	36	37
	38	38	39	39
	40	40	40	40
	40	41	43	45
	45	47	48	51

a Find the mean and the range for each distribution.
b Compare the two distributions.

2 Amarjit and Martha each fired 50 shots at a target.

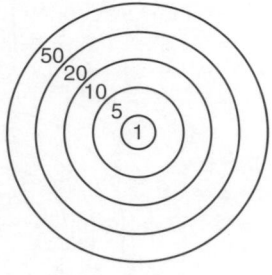

These are the distributions of their scores.

Amarjit	1	1	1	1	1
	1	1	1	1	1
	1	1	5	5	5
	5	5	5	5	5
	10	10	10	10	10
	10	10	10	10	10
	10	10	10	10	10
	20	20	20	20	20
	20	20	20	50	50
	50	50	50	50	50

Martha	5	5	5	5	5
	5	5	5	5	5
	5	5	5	5	5
	5	5	5	5	5
	5	5	10	10	10
	10	10	10	10	10
	10	10	10	10	10
	10	10	10	10	10
	10	10	20	20	20
	20	20	20	20	20

a Find the mean and the range for each distribution.
b Compare the two distributions.

3 The captain of a cricket team is considering the distributions of batting scores for two of his players, Bill and Ted.
These are the distributions of their scores.

Bill	0	0	0	0	0
	1	3	29	33	42
	50	52	67	84	100
	101	120	147	153	184

Ted	0	27	35	42	44
	45	46	48	48	52
	58	62	66	71	74
	77	82	83	84	91

a Find the mean and range for each distribution.
b Compare the two distributions.

4 The cost in pence of 1 litre of petrol was checked at 50 petrol station in London and 50 petrol stations in Newcastle upon Tyne. These are the distributions of prices.

London	46.9	47.5	49.5	49.5	49.5
	49.5	49.5	49.9	49.9	50.9
	50.9	50.9	50.9	50.9	50.9
	52.9	52.9	52.9	52.9	52.9
	52.9	52.9	52.9	52.9	52.9
	52.9	52.9	52.9	52.9	52.9
	52.9	52.9	52.9	52.9	52.9
	52.9	52.9	52.9	52.9	52.9
	55.9	55.9	55.9	55.9	55.9
	55.9	55.9	57.9	59.9	59.9

Newcastle	47.9	47.9	47.9	48.9	48.9
	49.5	49.5	49.5	49.5	49.5
	49.5	49.5	49.9	49.9	49.9
	49.9	49.9	49.9	49.9	49.9
	49.9	49.9	49.9	49.9	49.9
	50.9	50.9	50.9	50.9	50.9
	50.9	50.9	50.9	50.9	50.9
	50.9	50.9	50.9	50.9	50.9
	51.9	51.9	51.9	51.9	51.9
	51.9	52.9	52.9	52.9	54.9

a Find the median and the range for each distribution.

b Compare the two distributions.

5 Two ovens, the TooCool and the BurnAll, were tested by a consumer magazine.
In one test, the ovens were set to 200 °C and the actual temperature inside each oven was measured every minute for one hour.

These are the distributions of actual temperatures.

TooCool	194	194	194	194	195
	195	195	195	195	195
	195	195	195	195	195
	196	196	196	196	196
	196	196	197	197	197
	197	197	197	197	197
	197	197	197	197	198
	198	198	198	198	198
	198	198	198	198	198
	198	198	198	198	198
	198	199	199	199	199
	199	199	199	199	199

BurnAll	198	198	198	198	198
	198	198	198	198	198
	198	198	198	198	198
	199	199	199	199	199
	199	199	199	199	200
	200	200	200	200	200
	200	200	200	200	201
	201	201	202	202	202
	202	202	202	202	202
	202	202	203	203	203
	203	203	203	203	204
	205	205	206	207	208

a Find the median and the range for each distribution.

b Compare the two distributions.

Knowledge

One hundred pupils take a mathematics test marked out of 10. These are their scores.

0	0	1	1	1	1	1	2	2	2
2	2	2	2	2	2	3	3	3	3
3	3	3	3	3	3	3	3	3	4
4	4	4	4	4	4	4	4	4	5
5	5	5	5	5	5	5	5	5	5
5	5	5	6	6	6	6	6	6	6
6	6	6	6	6	6	6	6	6	7
7	7	7	7	7	7	7	7	8	8
8	8	8	8	8	8	9	9	9	9
9	9	9	9	10	10	10	10	10	10

A distribution of scores like this can be displayed in a **frequency table**.

Score	0	1	2	3	4	5	6	7	8	9	10
Frequency	2	5	9	13	10	14	16	9	8	8	6

The **mode**, the **median** and the **range** can be found from the frequency table *without* using the original list of scores.

The mode, the most frequent score, can be seen from the table to be 6.

The median is the middle score, which in a list of 100 scores will be between the 50th and 51st scores. To find the median, we add **cumulative frequencies** to the above table.

Score	0	1	2	3	4	5	6	7	8	9	10
Frequency	2	5	9	13	10	14	16	9	8	8	6
Cumulative frequency	2	7	16	29	39	53	69	78	86	94	100

The cumulative frequencies show that there were 39 scores of 4 or less and 53 scores of 5 or less. The 50th and 51st scores must therefore both have been 5. Therefore, the median is 5.

The range is the greatest score (10) minus the least score (0). So, the range is 10.

Skill Example

This table shows the number of children per family in 100 families.

Number of children	0	1	2	3	4	5	6
Frequency	12	14	37	18	9	6	4

a Find the mode of the distribution.
b Find the median of the distribution.
c Find the range of the distribution.

a The mode is 2 children per family.
b Adding cumulative frequencies to the table gives:

Number of children	0	1	2	3	4	5	6
Frequency	12	14	37	18	9	6	4
Cumulative frequency	12	26	63	81	90	96	100

In a distribution of 100 values, the median will be between the 50th and 51st values.

The cumulative frequencies show that there were 26 families with 1 child or fewer and 63 families with 2 children or fewer. The 50th and 51st families must both have had 2 children.

Therefore, the median = 2 children per family

c The range of the distribution $= 6 - 0 = 6$

Skill Practice Six

Find the mode, the median and the range of each distribution.

1 The times in which a 400-metre runner completed 20 races:

Time (seconds)	55	56	57	58	59
Frequency (no. of races)	1	4	7	5	3

2 The weights of each of the 30 children in a class:

Weight (kg)	40	41	42	43	44	45	46
Frequency	2	3	5	8	7	4	1

3 The number of passengers a taxi driver carried after answering each of the 30 calls he received on a certain day:

Number of passengers	1	2	3	4	5	6
Frequency (no. of calls)	5	6	7	5	4	3

4 The number of wickets a bowler took in each match of a twenty-match season:

Number of wickets	0	1	2	3	4	5	6
Frequency (no. of matches)	1	2	3	5	6	3	0

5 The number of centuries a batsman scored in each season of his playing career:

Number of centuries	0	1	2	3	4	5	6
Frequency (no. of seasons)	0	3	4	5	2	1	0

Knowledge

The mean can also be found directly from a frequency table.

This table shows the scores of a group of pupils in a mathematics test.

Score (S)	0	1	2	3	4	5	6	7	8	9	10
Frequency (F)	2	5	9	13	10	14	16	9	8	8	6

To find the mean, first add to the table a row giving each score multiplied by its frequency.

Score (S)	0	1	2	3	4	5	6	7	8	9	10
Frequency (F)	2	5	9	13	10	14	16	9	8	8	6
$S \times F$	0	5	18	39	40	70	96	63	64	72	60

Adding together all the frequencies gives the total number of scores.

Adding together all the products in the score × frequency row gives the total of all the scores.

												Total
Score (S)	0	1	2	3	4	5	6	7	8	9	10	
Frequency (F)	2	5	9	13	10	14	16	9	8	8	6	100
S × F	0	5	18	39	40	70	96	63	64	72	60	527

The mean is equal to the total of all the scores divided by the total number of scores.

Therefore, the mean $= 527 \div 100 = 5.27$

Skill Example

This table shows the number of children per family in 100 families.

Number of children (N)	0	1	2	3	4	5	6
Frequency (F)	12	14	37	18	9	6	4

Find the mean number of children.

Add to the table the row of values of N × F.

								Total
Number of children (N)	0	1	2	3	4	5	6	
Frequency (F)	12	14	37	18	9	6	4	100
N × F	0	14	74	54	36	30	24	232

Therefore, the mean $= 232 \div 100 = 2.32$

Skill Practice Seven

Find the mean of each distribution.

1 The table below shows how many goals a hockey player scored in each of the matches that she played during a certain season.

Number of goals	0	1	2	3	4	5
Frequency (no. of matches)	8	13	10	6	2	1

2 The table below shows the finishing positions over 20 seasons for a rugby club which plays in a ten-club league.

Position	1st	2nd	3rd	4th	5th	6th	7th	8th	9th
Frequency (no. of seasons)	0	1	1	2	5	7	3	1	0

3 The table below shows how many school lunches were taken by the children in class 2B on each day of a 60-day term.

Number of lunches taken	1	2	3	4	5	6	7	8	9	10	11
Frequency (no. of days)	0	1	4	5	7	8	10	11	8	3	3

4 Kelly travelled to school by bus on each day of a 60-day term. The table below shows the times that each bus journey took.

Time (minutes)	30	31	32	33	34	35
Frequency (no. of journeys)	5	8	10	16	14	7

Skill Practice Eight

1 A small hotel has accommodation for ten overnight guests and is open during June, July, August and September. The table below shows how many guests stayed on each night of an opening season.

Number of guests	0	1	2	3	4	5	6	7	8	9	10
Frequency (no. of nights)	5	10	12	15	20	16	15	10	8	6	5

Find the mode, the median and the range of the distribution.

2 The data below shows how many international caps a footballer earned in each season of his international career.

1967 1	1971 3	1975 3	1979 2	1968 2
1972 5	1976 4	1980 2	1969 3	1973 6
1977 3	1981 3	1970 4	1974 4	1978 5

a Copy and complete the tally chart below.

Number of caps per season	Tally	Frequency
1	/	1
2		
3		
4		
5		
6		

b Copy the frequency distribution table below and use your tally chart to complete it.

No. of caps per season	1	2	3	4	5	6
Frequency						

c Find the mode, the median, the mean and the range of the distribution.

3 The table below shows how many bottles of milk a milkman delivered each day to each of the 40 houses in South Street.

Number 1	2	Number 15	3	Number 29	6
Number 2	3	Number 16	3	Number 30	1
Number 3	4	Number 17	2	Number 31	3
Number 4	1	Number 18	1	Number 32	4
Number 5	5	Number 19	3	Number 33	6
Number 6	3	Number 20	2	Number 34	2
Number 7	2	Number 21	2	Number 35	4
Number 8	3	Number 22	1	Number 36	5
Number 9	4	Number 23	3	Number 37	3
Number 10	2	Number 24	3	Number 38	1
Number 11	3	Number 25	4	Number 39	4
Number 12	1	Number 26	2	Number 40	2
Number 13	4	Number 27	4		
Number 14	2	Number 28	5		

a Copy and complete the tally chart below.

Number of bottles per house	Tally	Frequency
1	$\cancel{||||}$ $|$	6
2		
3		
4		
5		
6		

b Copy the frequency distribution table below and use your tally chart to complete it.

No. of bottles per house	1	2	3	4	5	6
Frequency						

c Find the mode, the median, the mean and the range of the distribution.

Skill Example

The numbers of children in 80 households in two different areas of a town were recorded.

Number of children	0	1	2	3	4	5	6
Frequency (area X)	12	13	28	22	3	1	1

Number of children	0	1	2	3	4	5	6
Frequency (area Y)	42	23	15	0	0	0	0

a Find the mode, the median, the mean and the range of each distribution.
b Draw bar charts to illustrate each distribution.
c Compare and comment on the two distributions.

a For area X:

$$\text{Mode} = 2 \text{ children}$$

$$\text{Median} = 2 \text{ children}$$

$$\text{Mean} = \frac{158}{80} = 2.0 \text{ children (to 1 dp)}$$

$$\text{Range} = 6 - 0 = 6 \text{ children}$$

For area Y:

$$\text{Mode} = 1 \text{ child}$$

$$\text{Median} = 0 \text{ children}$$

$$\text{Mean} = \frac{53}{80} = 1.0 \text{ child (to 1 dp)}$$

$$\text{Range} = 2 - 0 = 2 \text{ children}$$

b

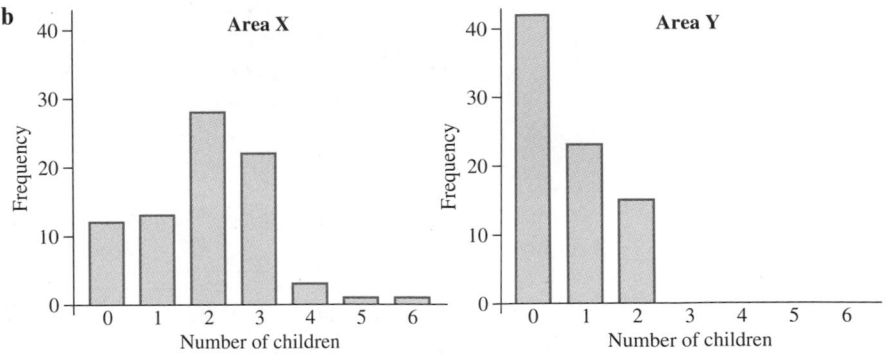

c All the averages indicate that there are far fewer children living in households in area Y than in area X.

The spread of children per household is smaller in area Y with a range of 2 to 0 children, whereas area X has a range of 6 to 0 children.

The bar charts illustrate these differences.

Skill Practice Nine

1 These tables show the number of tests taken by the students at two driving schools before they passed.

Number of tests	1	2	3	4	5
Frequency (school M)	28	12	5	3	2

Number of tests	1	2	3	4	5
Frequency (school N)	45	30	21	4	0

a Find the mode, the median, the mean and the range of each distribution.
b Draw bar charts to illustrate each distribution.
c Compare and comment on the two distributions.

2 These tables show the number of visits per year made to their doctor by two groups of patients.

Number of visits	0	1	2	3	4	5	6	7	8	9	10
Frequency (group A)	15	20	25	15	10	6	5	3	1	0	0

Number of visits	0	1	2	3	4	5	6	7	8	9	10
Frequency (group B)	2	6	10	12	20	30	50	40	20	6	4

a Find the mode, the median, the mean and the range of each distribution.
b Draw bar charts to illustrate each distribution.
c Compare and comment on the two distributions.

3 These tables show the number of matches per box in samples of two different types of match.

Number of matches per box	37	38	39	40	41	42	43	44	45
Frequency (Striko)	8	18	23	31	27	11	13	4	6

Number of matches per box	37	38	39	40	41	42	43	44	45
Frequency (Katchwell)	0	27	45	39	21	0	0	0	0

a Find the mode, the median, the mean and the range of each distribution.
b Draw bar charts to illustrate each distribution.
c Compare and comment on the two distributions.

Unit 52

Handling data   Level 5

Skill   You interpret graphs and diagrams, including pie charts, and draw conclusions.

Skill Example

From this graph, estimate:
a the temperature at 11 am and at 3 pm
b the two-hour period during which the temperature rises the most rapidly and by how many degrees.

From the graph:
a The temperature is 5.5 °C at 11 am and 12 °C at 3 pm.
b The temperature rises the most rapidly between 10 am and 12 noon. During this time, it rises by 7 °C.

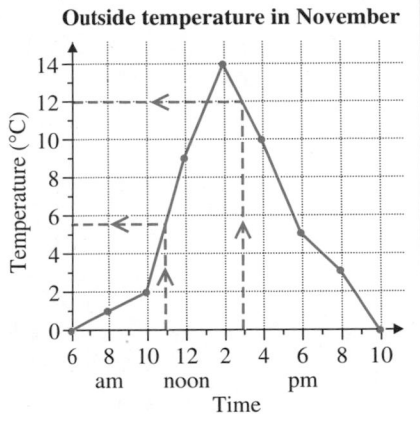

Outside temperature in November

Skill Practice One

1 Mrs Jackson spent all day knitting a scarf. She started at 8 am and finished at 5 pm. The final length of the scarf was one metre. The graph below shows her progress.

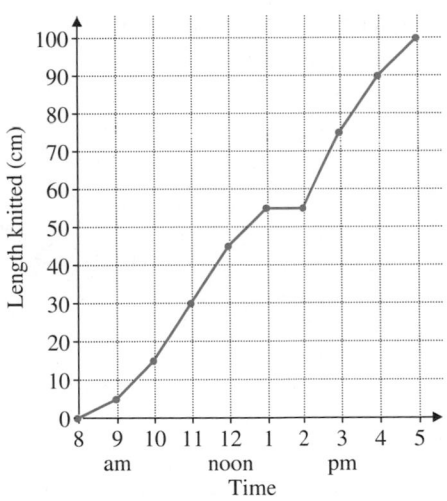

a What length had she knitted by the following times?
   (i) 9.30 am   (ii) 12.30 pm
   (ii) 2.30 pm   (iv) 4.30 pm

b During which one-hour period did she knit the greatest length? What length was this?

c During which one-hour period did she knit the shortest length? What length was this?

d Between what times did Mrs Jackson take a lunch break?

2 A crowd of 10 000 people attended a football match which started at 3 pm. The graph below shows the number of people who had entered the ground by various times.

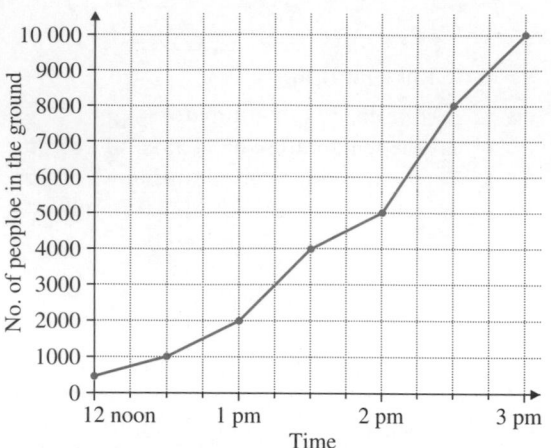

a How many people had entered the ground by the following times?
(i)   1.15 pm   (ii)   2.15 pm
(iii) 12.45 pm   (iv)   1.45 pm
(v)   2.45 pm

b During which half-hour interval did the most people enter the ground? How many people was this?

3 A group of hikers walked 20 km between 10 am and 2.30 pm. The graph below shows how far they had walked by certain times.

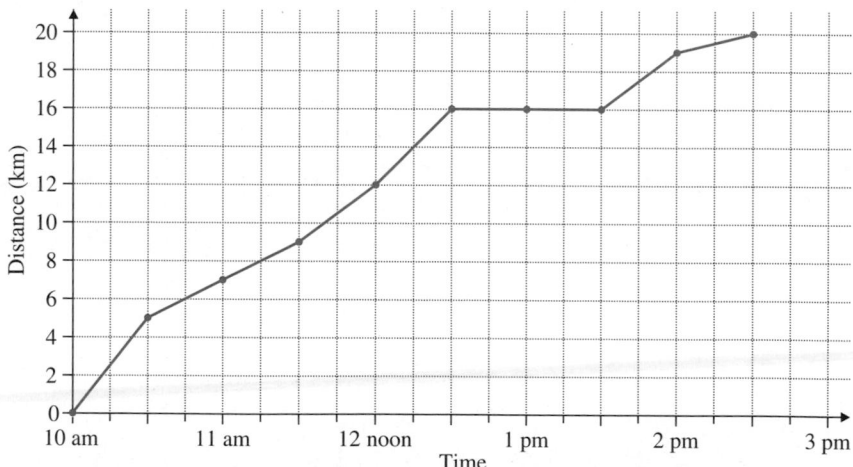

a What distance had the hikers walked by the following times?
(i)   11.15 am   (ii)   12.15 pm
(iii) 10.45 am   (iv)   12.45 pm

b During which half-hour interval did they walk the furthest? What distance was this?

c Between which times did they stop for lunch?

4 On a long journey by car, I recorded the number of litres of petrol in the tank every 100 km. The graph shows this information.

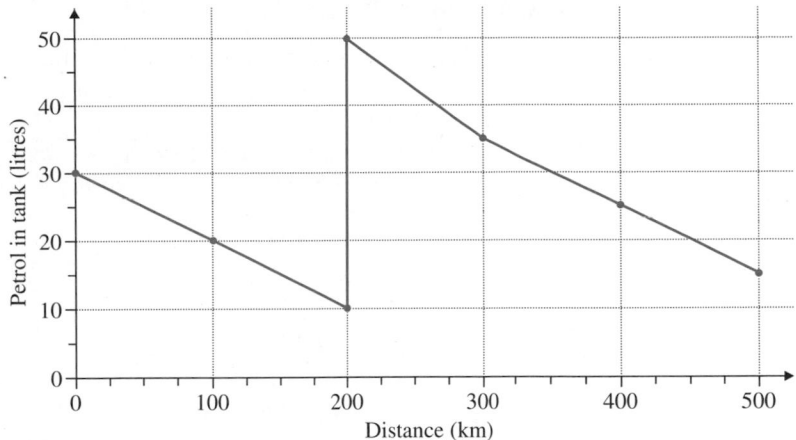

a How many litres of petrol were in the tank after:
(i) 350 km   (ii) 450 km
(iii) 50 km   (iv) 150 km?

b Over which 100-km stretch did I use the most petrol? How many litres was this?

c How much petrol did I use over the whole 500-km journey?

5 A boy ran an 800-metre race. The time he took to run a given distance in the race is shown in the graph below.

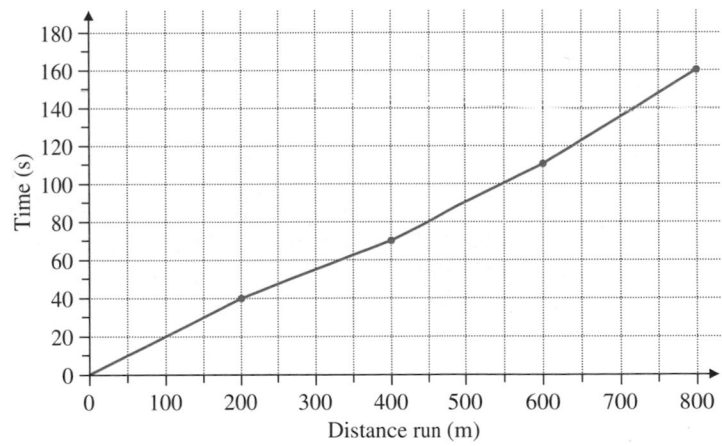

a How many seconds did he take from the start to run these distances?
(i) 100 m   (ii) 500 m
(iii) 550 m   (iv) 450 m
(v) 150 m   (vi) 50 m

b Over which of the 200-metre stretches did he run the quickest? How many seconds did it take him?

c Over which of the 200-metre stretches did he run the slowest? How many seconds did it take him?

Skill Example

The graph shows the journey of a car travelling at a steady speed of 40 km/h from Brighton to Portsmouth.

a Name the place where the car is:
  (i) after 30 min
  (ii) after 1 h 15 min

b Find the time when the car is:
  (i) at Chichester
  (ii) at Havant
  (iii) at Portsmouth

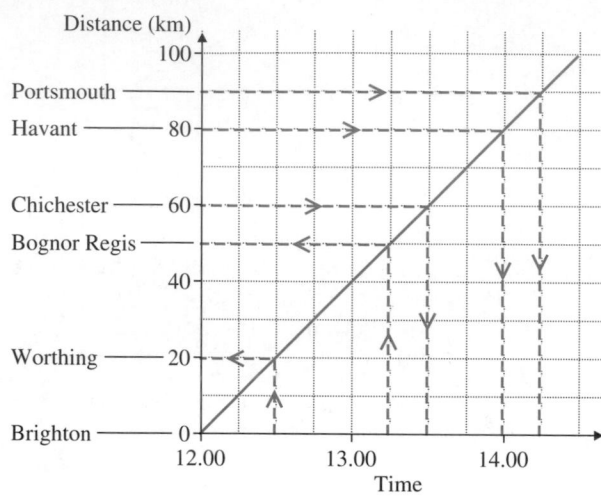

a (i) Worthing
   (ii) Bognor Regis

b (i) 13.30
   (ii) 14.00
   (iii) 14.15

Skill Practice Two

1 The graph shows the journey of a train travelling at a steady speed of 120 km/h from London (Paddington) to Cardiff.

  a Name the place where the train is:
    (i) after 30 min
    (ii) after 45 min
    (iii) after 1 h 30 min
    (iv) after 1 h 45 min

  b Find the time when the train is:
    (i) at Slough
    (ii) at Swindon
    (iii) at Cardiff

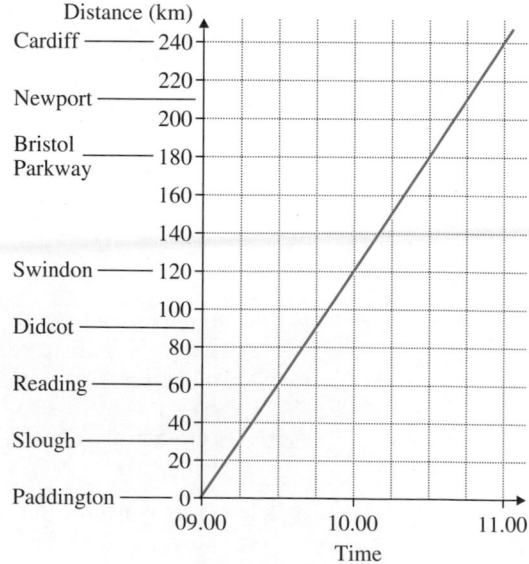

2 The graph shows the journey of a car travelling at a steady speed of 80 km/h from Doncaster to Durham.

   a Name the place where the car was:

     (i)   after 1 h

     (ii) after 1 h 15 min

     (iii) after 2 h

   b Find the time when the car was:

     (i)   at Bramham Moor

     (ii) at Scotch Corner

     (iii) at Aycliffe

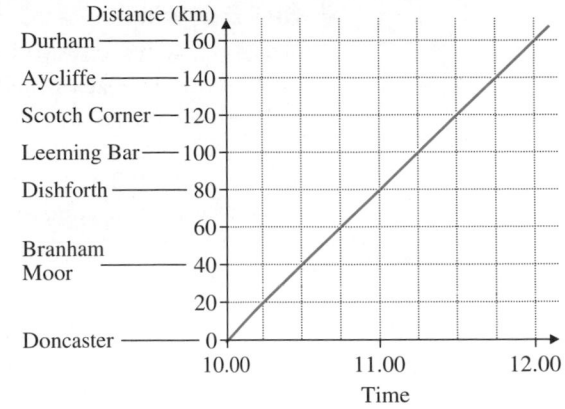

3 The graph shows the details of a hike through Cannock Chase by a group of school children who walked at a steady speed of 4 km/h.

   a Name the place where the group was:

     (i)   after 15 min

     (ii) after 1 h 15 min

     (iii) after 2 h 30 min

   b Find the time when the group was:

     (i)   at The Old Pump House

     (ii) at Oat Hill

     (iii) at Shugborough Park Gates

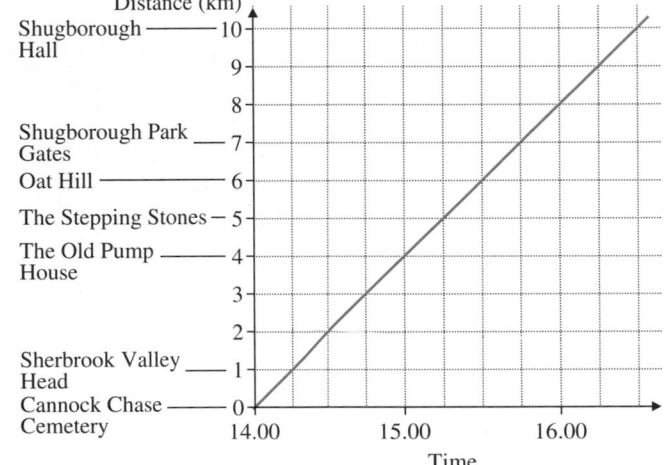

4 The graph shows the journey of a cyclist from Manchester to Liverpool.

He cycled at a steady speed of 24 km/h.

   a Name the place where the cyclist was:

     (i)   after 1 h 30 min

     (ii) after 1 h 45 min

     (iii) after 2 h 15 min

   b Find the time when the cyclist reached:

     (i)   Irlam

     (ii) Warrington

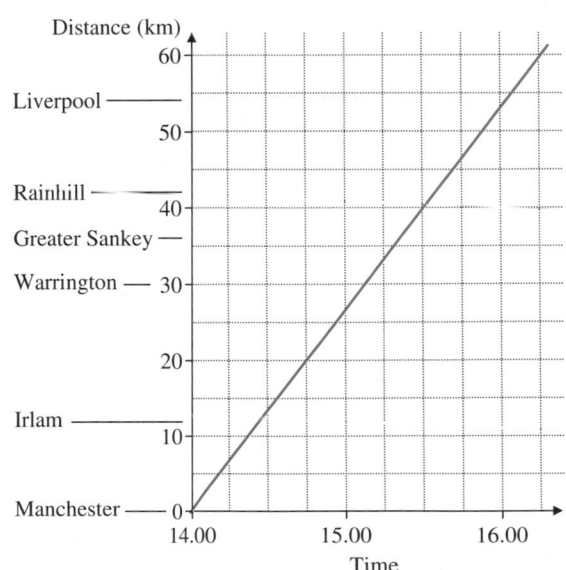

Skill Example

The graph shows a cyclist's journey from Scunthorpe to Cleethorpes and back.
a When did he reach Brigg?
b When did he leave Brigg?
c How far is Cleethorpes from Brigg?
d How long did he stop in Cleethorpes?
e What was his average speed on the return journey?

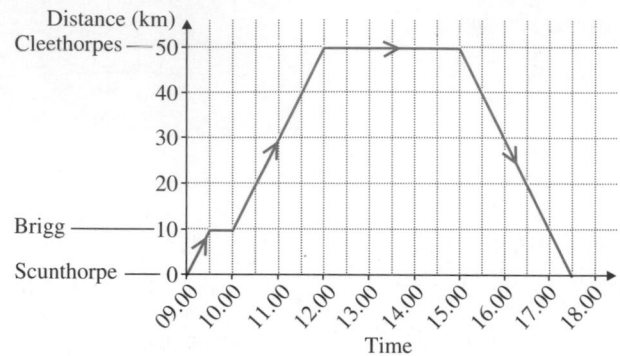

a He reached Brigg at 09.30.
b He left Brigg at 10.00.
c Distance from Brigg to Cleethorpes is $(50 - 10)\,\text{km} = 40\,\text{km}$.
d He arrived in Cleethorpes at 12.00; he left at 15.00.
   Therefore, time spent in Cleethorpes is 3 hours.
e Return journey took $2\tfrac{1}{2}\,\text{h}$, a distance of 50 km. Therefore,

$$\text{Average speed} = \frac{\text{Distance}}{\text{Time}} = \frac{50}{2\tfrac{1}{2}} = 20\,\text{km/h}$$

Skill Practice Three

1 The graph shows the return journey of
   a lorry driver from Birmingham to Bristol.
   a When did he reach Cheltenham?
   b When did he leave Cheltenham?
   c What is the distance from Cheltenham
      to Bristol?
   d How long did he stop in Bristol?
   e What was his average speed on the
      return journey?

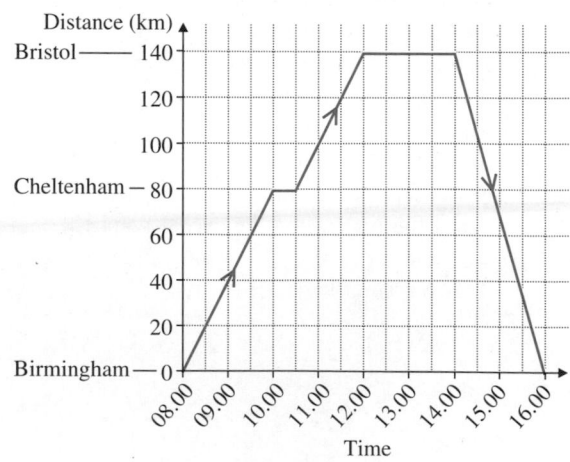

2 The graph shows a day's walk by a group of hikers from Grasmere to the top of Helvellyn and back.
   a When did they reach Grisedale Tarn?
   b When did they set off again from Grisedale Tarn?
   c How long did they stop at the top of Helvellyn?
   d What was their average speed on the return journey?

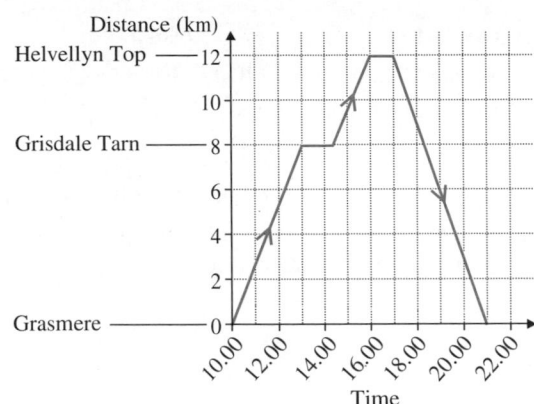

3 The graph shows a man's car journey from Hull to Newcastle upon Tyne and back.
   a When did he reach York?
   b When did he leave York?
   c How far is it from York to Newcastle?
   d What was his average speed on the return journey?

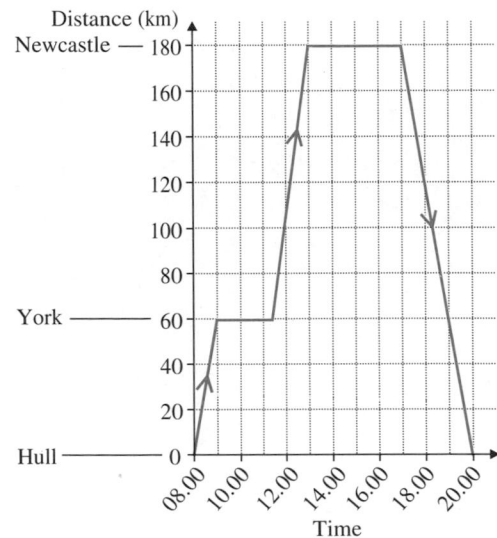

4 The return journey by train from Liverpool to York is shown in the graph. It involved changing trains at Leeds on the outward journey.
   a When did the train arrive at Leeds?
   b When did the second train leave Leeds?
   c How long was the stay in York?
   d Find the average speed of the return journey.

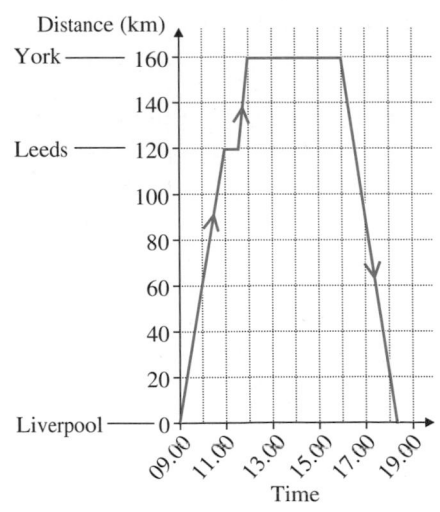

256

Knowledge

A **pie chart** shows how a distribution is divided among different categories.

The frequency of each category is represented by a sector ('slice') of a circle (the 'pie').

The diagram below shows the principle of the pie chart: a circle divided into percentage sectors.

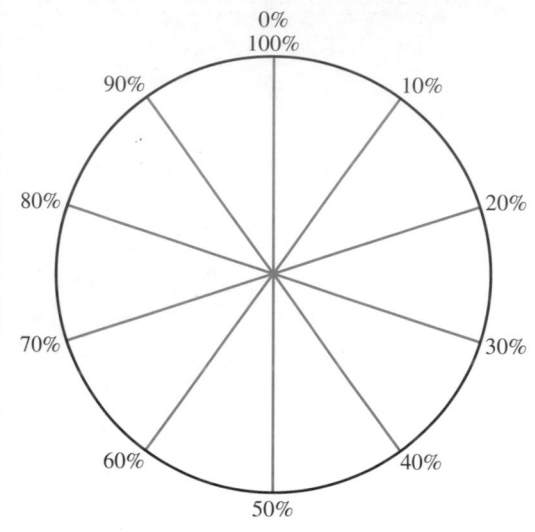

Skill Example

Estimate the percentage represented by each part of this pie chart.

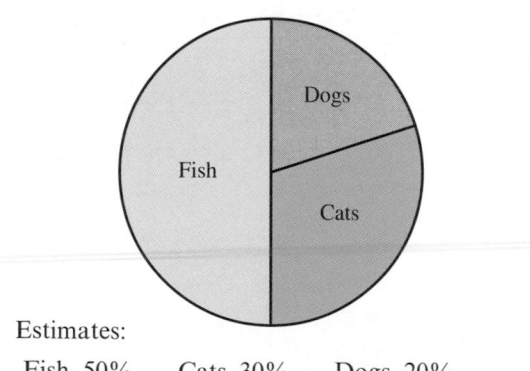

Estimates:

Fish 50%,   Cats 30%,   Dogs 20%

Skill Practice Four

Estimate the percentage represented by each part of these pie charts.

1

2

3

4

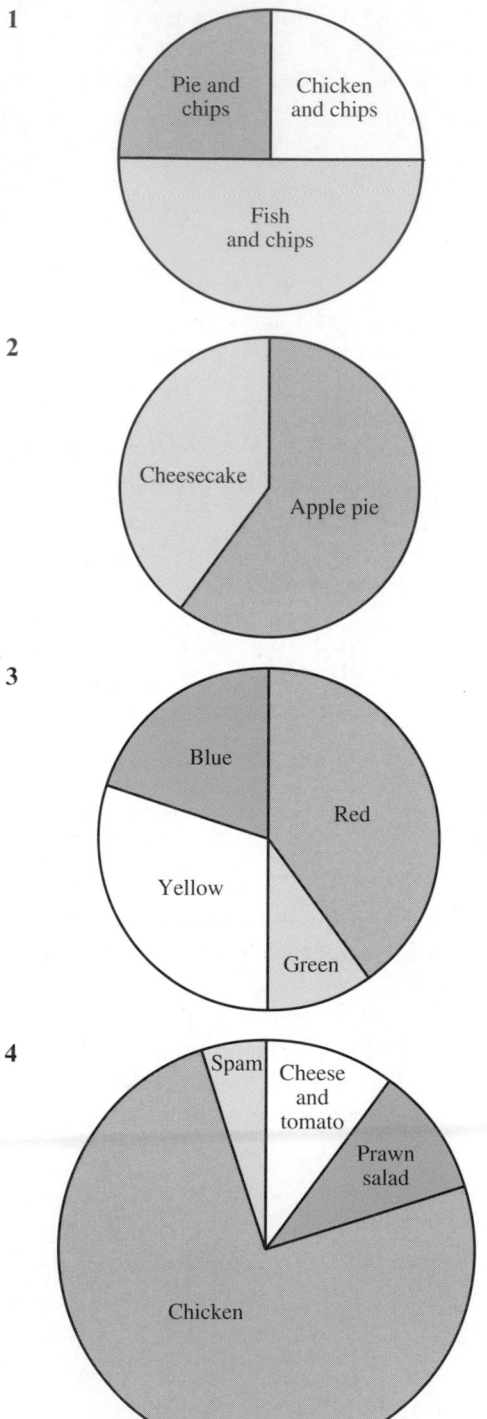

5

Pie chart with segments: Monday, Tuesday, Wednesday, Thursday, Friday

6

Pie chart with segments: Strawberry, Raspberry, Banana

7

Pie chart with segments: Italian, English, German, French

8

Pie chart with segments: Pizza, Cheese and onion, Roast chicken, Salt and vinegar, Plain

Skill Practice Five

1 These charts show information about the decrease in meat eating.

Percentage of 11–17 year olds who avoid red meat, 1996
Source: SMRC Survey of 979 7–19 year olds

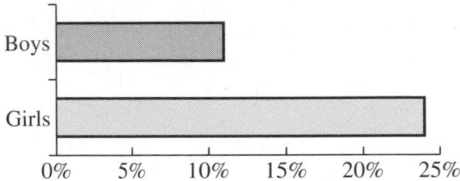

Percentage of adults who ate less meat than 5 years ago, 1995
Del Monte Survey of 1000 adults

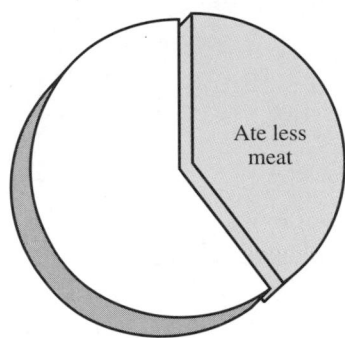

Estimate:
a The percentage of 11–17 year old boys who avoided red meat in 1996.
b The percentage of 11–17 year old girls who avoided red meat in 1996.
c The percentage of adults who ate less meat in 1995 than in 1990.
d Sketch a pie chart to show the percentage of 11–17 year old girls who avoided red meat in 1996.

2 These charts show information about the types of fuel used to generate electricity in the UK.

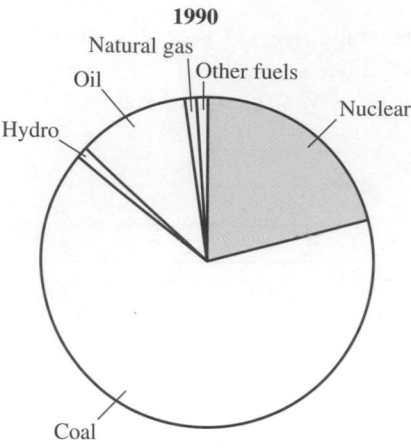

1990

Natural gas
Oil
Other fuels
Hydro
Nuclear
Coal

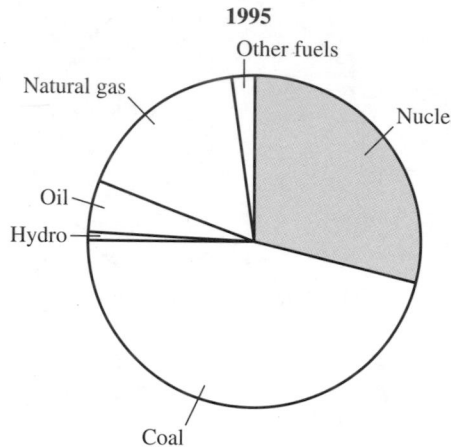

1995

Other fuels
Natural gas
Nuclear
Oil
Hydro
Coal

a Estimate the percentage of electricity that was generated in 1990 by:
(i) natural gas   (ii) coal

b Estimate the percentage of electricity that was generated in 1995 by:
(i) natural gas   (ii) coal

c Comment on the changes in the fuels used to generate electricity between 1990 and 1995.

3 These charts show the responses of young people in a survey of their eating habits.

Healthy eating claims of young people, England, 1995
Source: Young People in 1995

When choosing what to eat, do you consider your health?

Girls aged 14–15

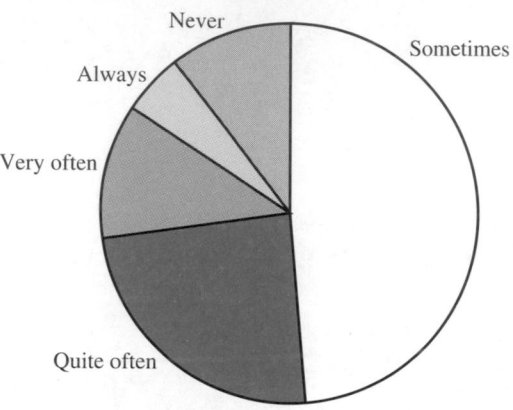

Never
Always
Sometimes
Very often
Quite often

Boys aged 14–15

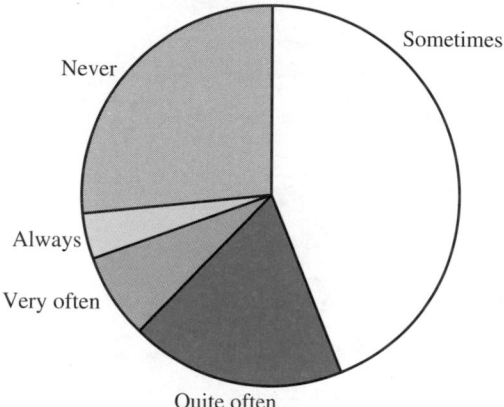

Never
Sometimes
Always
Very often
Quite often

a Estimate the percentages of girls who replied 'Always' and 'Never'.

b Estimate the percentages of boys who replied 'Always' and 'Never'.

c If 800 girls answered the survey, how many said they sometimes considered their health when choosing what to eat?

d If 1000 boys answered the survey, how many said they quite often considered their health when choosing what to eat?

e Comment on the differences between boys and girls as shown in the charts.

4 This chart shows how the money from the
sales of National Lottery tickets is divided.

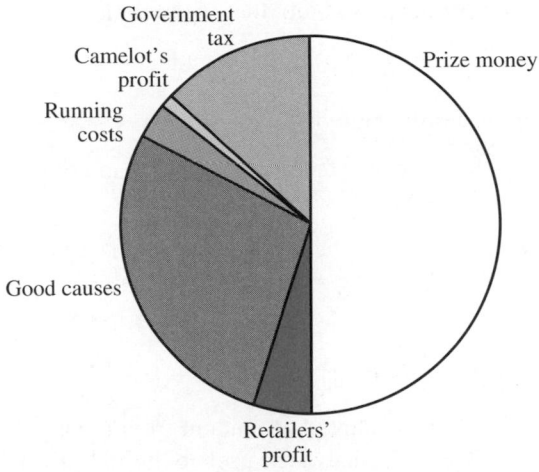

a Estimate the percentage of the money from
sales which is used for prizes.

b Estimate the percentage of the money from
sales which goes to good causes.

c Average weekly sales of lottery tickets
amount to £70 million. Calculate how
much of this money goes to each of the
six categories shown in the pie chart.

Unit 53

Knowledge

If we are certain that selection is fair and all outcomes are equally likely, we can calculate the **probability** that a particular outcome will happen.

An umpire tosses a coin to decide whether the red team or the blue team will bully off. The red team's captain calls 'heads'.

What is the probability that the red team will bully off?

There are two possible outcomes: 'heads' or 'tails'. So, the probability that the red team will bully off is $\frac{1}{2}$.

Skill Example

A card is picked from a normal pack (without jokers).
What is the probability that it is a red queen?

There are 52 cards in a pack of which two are red queens. So the probability is

$$\frac{2}{52} = \frac{1}{26}$$

Skill Example

A dice is rolled.
What is the probability that the score is:

a a 4   **b** a factor of 6   **c** a factor of 5?

a The probability is $\frac{1}{6}$

b There are four factors of 6 (1, 2, 3 and 6), so the probability is $\frac{4}{6} = \frac{2}{3}$

c There are two factors of 5 (1 and 5), so the probability is $\frac{2}{6} = \frac{1}{3}$

Skill Practice One

1 If a letter is chosen at random from the word SUCCESS, what is the probability that it will be:
   a the letter S   **b** the letter C?

2 If a letter is chosen at random from the word PEPPER, what is the probability that it will be:
   a the letter P   **b** the letter E?

3 If a letter is chosen at random from the name GEORGE, what is the probability that it will be:
   a the letter E   **b** the letter G
   c a vowel   **d** a consonant?

4 If a letter is chosen at random from the word NEEDLEWORK, what is the probability that it will be:
   a the letter E   **b** a vowel
   c a consonant?

5 On a supermarket shelf there are 16 bags of sugar, 12 of which contain white sugar and 4 of which contain brown sugar.
   If a bag is taken at random, what is the probability that it will contain:
   a white sugar   **b** brown sugar?

6 In class 3A there are 12 boys and 8 girls.
   If the pupils leave their classroom and walk to the assembly hall in any random order, what is the probability that the first pupil to enter the hall will be:
   a a boy   **b** a girl?

7 A farmer has 25 white sheep and 5 black sheep.
If they are rounded up for shearing in any random order, what is the probability that the first one to be sheared will be:
a white   **b** black?

8 A box of sweets contains 15 chocolates, 9 toffees and 6 nougats. If a sweet is taken from the box at random, what is the probability that it will be:
a a chocolate   **b** a toffee   **c** a nougat?

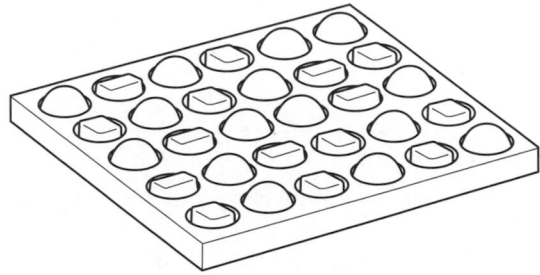

Knowledge

Probabilities can also be written as **decimals** or **percentages**.

Remember that any fraction can be changed into a decimal by dividing the top number by the bottom number.

Remember also that any decimal can be changed to a percentage by multiplying by 100.

Skill Example

If a letter is chosen at random from the word SELECTED, what is the probability that it will be:

a the letter S   **b** the letter E
c neither an S nor an E?

Give each answer as a fraction, a decimal and a percentage.

a $\frac{1}{8} = 0.125 = 12.5\%$

b $\frac{3}{8} = 0.375 = 37.5\%$

c $\frac{4}{8} = \frac{1}{2} = 0.5 = 50\%$

Skill Practice Two

In questions **1** to **10**, give each answer as a fraction, a decimal and a percentage.

1 In class 2B there are 18 girls with dark hair, 10 girls with fair hair and 2 girls with red hair. If the teacher asks one girl at random to give out some books, what is the probability that she will have:
a dark hair   **b** fair hair   **c** red hair?

2 A £1 cash bag contains six 10 p coins, four 5 p coins, six 2 p coins and eight 1 p coins.
If a coin is removed from the bag, what is the probability that it will be:
a a 10 p coin   **b** a 5 p coin
c a 2 p coin   **d** a 1 p coin
e a silver coin   **f** a copper coin?

3 On a supermarket shelf there are 8 packets of ready salted crisps, 5 packets of cheese and onion crisps, 3 packets of salt and vinegar crisps and 4 packets of prawn cocktail crisps.
If a bag is removed from the shelf at random, what is the probability that it will contain:
a ready salted crisps
b cheese and onion crisps
c salt and vinegar crisps
d prawn cocktail crisps
e any kind of flavoured crisps?

4 If a dice is thrown, what is the probability that the score will be:
a a six
b an odd number
c an even number
d a multiple of 3
e a prime number
f a square number
g a triangular number?

5 Twelve counters numbered 1, 2, 3, 4, 5, 6, 7, 8, 9, 10, 11 and 12 are placed in a bag.
If a counter is removed from the bag, what is the probability that the number on it will be:
a a prime number
b a square number
c a triangular number
d a multiple of 3
e a multiple of 5?

262

6 Certain geometrical shapes are drawn on cards as shown and the eight cards are then placed in a bag.

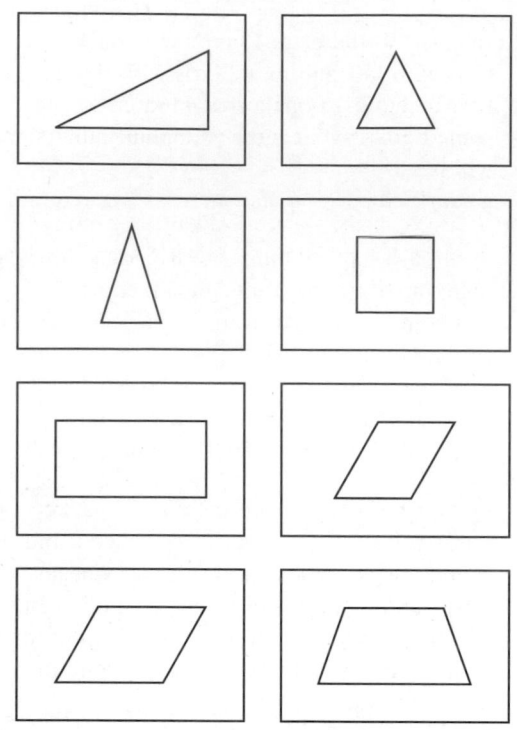

If a card is then removed from the bag, what is the probability that the figure on it has:

a three sides

b four sides

c four equal sides

d all its sides equal

e two pairs of equal sides

f one pair of equal sides?

Knowledge

If a bag contains 7 yellow counters, the probability of picking out a yellow counter is $\frac{7}{7}$ or 1.

Thus the probability of a **certainty** is **1**.

The probability of picking out a red counter is $\frac{0}{7} = 0$.

Thus the probability of an **impossibility** is **0**.

All other probabilities must be somewhere between 0 and 1. That is, somewhere between impossible and certain.

Skill Example

A case contains 3 pens and 5 pencils.
If one is selected at random, what is the probability that it is a pen?
Give your answer as a fraction and illustrate its position on the number line between 0 and 1.

The probability is $\frac{3}{8}$.

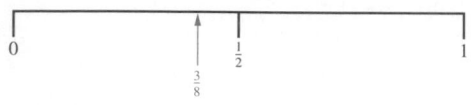

Skill Example

A bag contains 3 strawberry and 2 lemon flavoured sweets.
If one is selected at random, what is the probability that it is strawberry flavoured?
Give your answer as a decimal and illustrate its position on the number line between 0 and 1.

The probability is $\frac{3}{5} = 0.6$

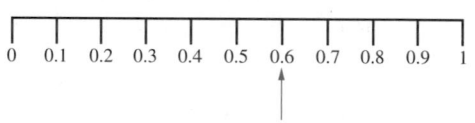

Skill Example

In a packet, there are 3 chocolate biscuits and 7 plain biscuits.
If one is selected at random, what is the probability that it is plain?
Give your answer as a percentage and illustrate its position on the number line between 0% and 100%.

The probability is $\frac{7}{10} = 70\%$

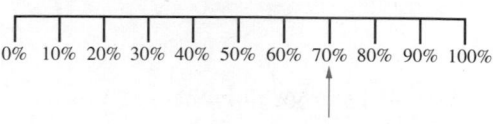

Skill Practice Three

In questions **1** to **3**, give your answer as a fraction and illustrate its position on the number line between 0 and 1.

1 If a season of the year is picked at random, what is the probability that it is Spring?

2 If one bead is picked from a bag containing 2 red, 3 blue and 3 green beads, what is the probability that it is red?

3 If a letter is chosen at random from the word SUCCESSFUL, what is the probability that it is the letter S?

In questions **4** to **6**, give your answer as a decimal and illustrate its position on the number line between 0 and 1.

4 There are 1000 possible combinations on a three-digit cycle lock.
If you try 100 different combinations, what is the probability that you will find the correct combination to open the lock?

5 In a raffle, 500 tickets are sold.
If Mr Chaudray has bought 25 tickets, what is the probability that he wins the first prize?

6 There are 14 girls and 11 boys in class 7E.
If one pupil is selected at random, what is the probability that it is a girl?

In questions **7** to **10**, give your answer as a percentage and illustrate its position on a line marked from 0 to 100%.

7 A dice is made from a regular tetrahedron, with faces numbered from 1 to 4. The score is taken as the number on the bottom face.

What is the probability that when the dice is rolled the score will be a prime number?

8 A restaurant menu offers a choice of one vegetable from peas, carrots, parsnips, broccoli and sweet corn.
What is the probability that a vegetable picked at random is green?

9 If one letter is picked at random from the word DISASTER, what is the probability that it is a vowel?

10 If a day of the week is selected at random, what is the probability that it is not Saturday or Sunday?

Skill Example

A glass tumbler, an egg, a knife and a calculator are all dropped on the floor.

a Estimate the probability that each will break and show your answers on the number line.

b How could you check your estimates?

a We can estimate, using common sense, that the egg is almost certain to break, the glass will probably break, the calculator will probably not break and the knife will almost certainly not break.

We can show these estimates on the number line.

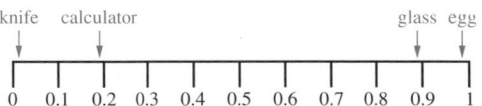

b To check these estimates, we could conduct an experiment, perhaps dropping 100 eggs, 100 glasses, 100 calculators and 100 knives to see how many broke.

Knowledge

There are three ways to establish a probability:

- By logical argument, when we are certain that all outcomes are equally likely.
- By using existing data, for example, school registers or weather records.
- By conducting a survey or experiment.

Skill Example

a What is the probability of scoring 1 when a dice is rolled?

b Estimate the probability that a Year 11 pupil picked at random in your school will have had 100% attendance since Year 7. How could you check your estimate?

c Estimate the probability that when a slice of buttered bread is dropped, it lands butter side up.
How could you check your estimate?

d Estimate the probability that a pupil selected at random from your school will like chips. How would you check your estimate?

a $\frac{1}{6}$ (by logical argument based on equally likely outcomes).

b This probability is likely to be very low, since few pupils attend until Year 11 with no absences.
A sensible estimate might be $\frac{1}{100}$.
The estimate could be checked by using existing data, in this case the school registers.

c This probability is likely to be close to $\frac{1}{2}$.
The estimate could be checked by an experiment, perhaps dropping a buttered slice of bread 100 times.

d This probability is likely to be very high, because chips are a very popular food.
A sensible estimate might be $\frac{19}{20}$ or even $\frac{99}{100}$.
The estimate could be checked by conducting a survey of perhaps 100 pupils selected at random.

Skill Practice Four

1 A drawing pin is dropped on a table.
  a Illustrate on the number line your estimate of the probability that it will land point up.
  b How could you check your estimate?

2 This chair falls off a table where it has been placed while the floor is swept.

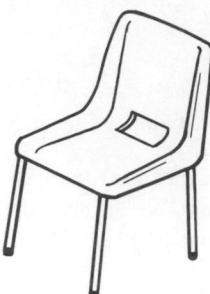

  a Illustrate on the number line your estimate of the probability that it will *not* land on all four legs.
  b Illustrate on the same line your estimate of the probability that it will land balanced on one leg.
  c How could you check your estimates?

3 Chantal has planned a week's holiday in England in July.
  a Illustrate on the number line your estimate of the probability that it will rain on at least one day during her holiday.
  b How could you check your estimate?

4 A motor insurance agent is considering offering a one-year policy to an 18-year-old driver.
  a Illustrate on the number line your estimate of the probability that the driver will have an accident during the year.
  b How could you check your estimate?

5 A pupil is selected at random from your school.
  a Illustrate on the number line your estimate of the probability that the pupil will be a Manchester United supporter.
  b How could you check you estimate?

6 A pupil is selected at random from your school.
  a Illustrate on the number line your estimate of the probability that the pupil will eat school dinners.
  b How could you check your estimate?

Unit 54

Handling data   Level 6

Skills   You collect and record continuous data, choosing appropriate equal class intervals over a sensible range to create frequency tables.

You construct and interpret frequency diagrams.

Knowledge

Discrete data is obtained by **counting**.

For example, we could dig up the soil in a metre square and count how many earth worms we found.

Our result would be a **whole number**. We would not find 27.35 or 19.7 earth worms.

Continuous data is obtained by **measuring**.

For example, we could measure the length of every earth worm we found.

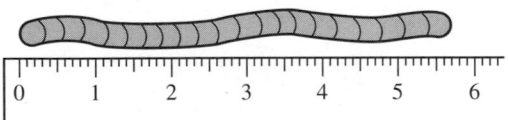

Our results would be measurements such as 2.7 cm or 4.8 cm.

To record continuous data, we must decide on the **accuracy** we need to use.

For example, we might decide to measure each earth worm's length to the nearest centimetre. In this case, our recording table might look like this:

Length (cm)	1.5–2.5	2.5–3.5	3.5–4.5	4.5–5.5	5.5–6.5
Frequency	0	4	7	8	1

We could then illustrate our data with the type of frequency diagram shown below, called a **histogram**.

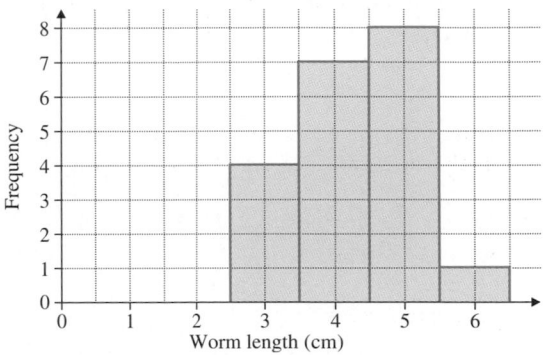

The **frequency polygon**, shown below, is an alternative to the histogram. To draw a frequency polygon, you plot points at the **middle** of each interval and join them with straight lines.

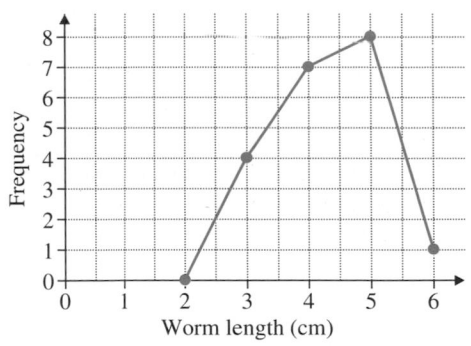

Skill Practice One

1 This histogram shows the times taken by a group of 100 runners to complete a race.

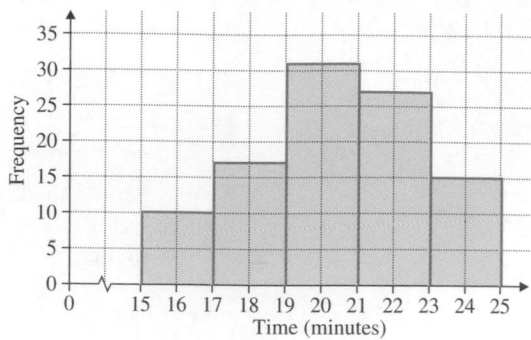

a How many runners took from 17 minutes to 19 minutes to complete the race?
b How many runners took 21 minutes or longer to finish the race?
c How many runners finished the race?
d The results were recorded in a grouped frequency table like this:

Time (min)	15–17	17–19
Frequency	10	17

Copy and complete the table.
e In which class interval would a runner who was timed at 17 minutes exactly be recorded?
f How many runners took less than 19 minutes to finish the race?
g Draw a frequency polygon to illustrate the data.

2 This histogram shows the heights of a group of children, measured to the nearest centimetre.

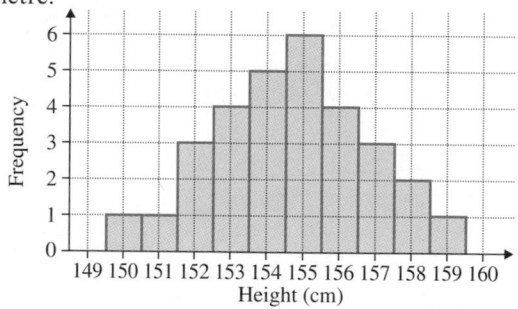

a How many children were 154 cm tall (to the nearest centimetre)?
b How many children were less than 153 cm tall (to the nearest centimetre)?
c How many children were more than 155 cm tall (to the nearest centimetre)?
d The first bar of the histogram represents children who are from 149.5 cm to 150.5 cm tall.
What does the fifth bar represent?
e How many children were there in the group?
f The results were recorded in a grouped frequency table like this:

Height (cm)	148.5–149.5	149.5–150.5
Frequency	0	1

Copy and complete the table.
g Draw a frequency polygon to illustrate the data.

3 This histogram shows the weights (to the nearest kilogram) of a group of piglets.

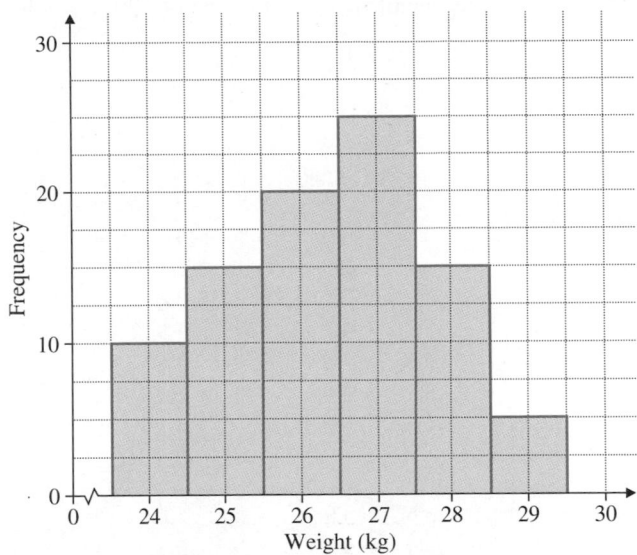

a The results were recorded in a grouped frequency table like this:

Weight (kg)	23.5–24.5	24.5–25.5
Frequency	10	15

Copy and complete the table.
b How many piglets were there in the group?
c How many piglets weighed more than 26 kg (to the nearest kg)?
d How many piglets weighed 26 kg or less (to the nearest kg)?
e Can you tell from the histogram the weight of the heaviest piglet?
f In which range of weights were the most piglets recorded?
g Draw a frequency polygon to illustrate the data.

4 Sixty members of a slimming club are weighed before and after a diet. The results are shown in this table.

Weight (kg)	Frequency	
	Before	After
75–80	3	5
80–85	1	5
85–90	2	6
90–95	3	9
95–100	6	8
100–105	17	14
105–110	15	8
110–115	13	5

a Draw two histograms to illustrate this data, drawing columns for 75–80, 80–85, 85–90 and so on.
b Draw two frequency polygons to illustrate this data.

5 Two sets of 50 seedlings are grown under different conditions. After four weeks, the heights (in centimetres) of all the seedlings are measured.

The results are shown in these lists

Set 1

2.4	5.3	3.4	2.7	2.9
2.1	2.8	3.5	3.4	4.1
2.7	2.6	2.5	2.9	3.7
2.8	3.6	3.8	4.3	4.2
4.7	5.0	3.0	3.2	2.9
4.3	2.8	3.3	3.5	3.4
3.7	4.3	5.4	1.9	2.7
2.6	3.5	2.7	3.6	2.8
3.7	2.9	3.8	2.0	4.2
4.0	5.0	3.9	4.1	6.7

Set 2

3.6	3.9	4.5	5.6	6.3
6.4	6.1	5.2	6.2	7.0
6.0	5.5	2.5	3.5	6.8
5.6	6.7	5.7	6.8	5.8
4.9	4.9	4.0	4.1	6.2
6.7	6.5	5.6	7.1	7.5
6.1	3.4	3.1	5.6	6.0
5.3	5.3	5.2	4.9	4.7
4.3	6.1	7.1	6.6	5.7
6.7	7.6	4.0	3.9	3.8

a Collect this data into a table with these headings

Height (to nearest cm)	Frequency	
	Set 1	*Set 2*
1		
2		
etc.		

b Draw two histograms to illustrate this data, drawing columns centred on 1 cm, 2 cm, 3 cm and so on.

c Draw two frequency polygons to illustrate this data.

Unit 55

Knowledge

Pie charts are usually constructed
using a **pie-chart scale**.

This is a circular disc divided
into percentage sectors.

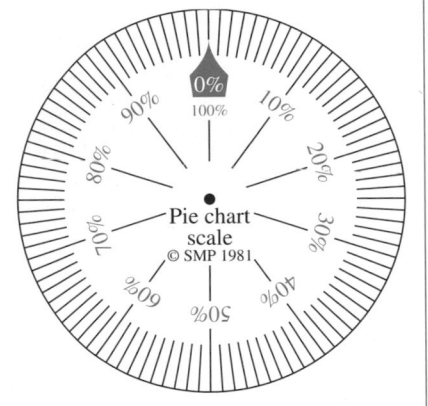

Skill Example

This table shows sources of government revenue.

Income tax	22%
Social security contributions	15%
VAT	15%
Business rates and local tax	8%
Excise duties	10%
Corporation tax	7%
Borrowing	12%
Other	11%

Illustrate the table with a pie chart.

A circle of suitable size is drawn, and the percentage sectors are marked off with
the pie-chart scale.

This is the completed pie chart.

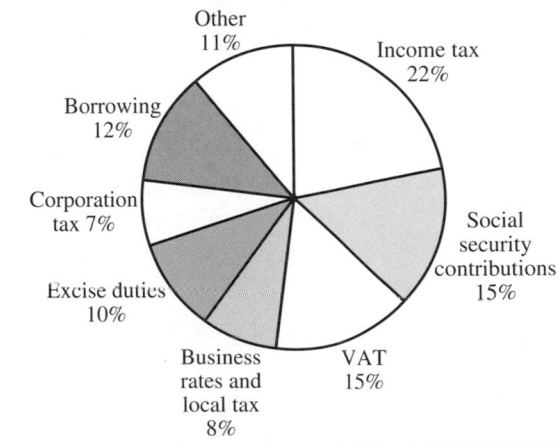

Skill Practice One

1 In a *Radio Times* survey, 1000 people were asked the question:

'Should zoos be abolished?'

These are the results:

Don't know	3%
Yes	27%
No	70%

Illustrate the results with a pie chart.

2 A second question in the *Radio Times* survey asked:

'What is the most important function of a zoo?'

These are the results:

Conservation of endangered species	53%
Education about how animals live	28%
Entertainment	11%
Research into animal behaviour and physiology	7%
None of these is important	1%

Illustrate the results with a pie chart.

3 Global warming is believed to be caused by the discharge into the atmosphere of four main gases produced by industry and other human activity. Their concentrations are estimated to be:

CFCs (chlorofluorocarbons)	13%
N_2O (nitrous oxide)	5%
CH_4 (methane)	10%
CO_2 (carbon dioxide)	72%

Illustrate this data with a pie chart.

4 A survey found these were the main methods used by men and women to search for a job.

	Men	Women
Study situations vacant in newspapers	29%	43%
Visit Job Centre	30%	22%
Ask friends, relatives, colleagues	14%	9%
Apply direct to employers	11%	9%
Answer advertisements in newspapers	9%	11%
Other	7%	6%

Illustrate this data with two pie charts.

Skill Example

A survey of Year 10 boys found the following numbers were involved at least once a week in these sports.

	Numbers involved
Cycling	5157
Football	5090
Golf	2545
Athletics	2525
Swimming	2264

Illustrate these results with a pie chart.

First a percentage column must be added to the table:

	Numbers involved	% Involved
Cycling	5157	29%
Football	5090	29%
Golf	2545	14%
Athletics	2525	14%
Swimming	2264	13%
Total	17 581	99%

For example, the percentage for cycling is calculated like this:

$$\text{Percentage} = \frac{5157}{17\,581} \times 100 = 29.33\ldots = 29\% \text{ (correct to nearest whole number)}$$

Notice that the total percentage is 99%. This is because the decimals have been corrected to the nearest whole number. The small gap will not be obvious in the final pie chart.

This is the completed pie chart.

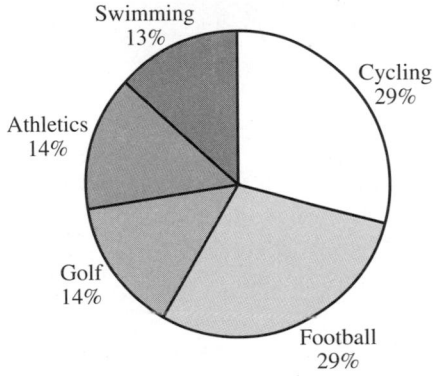

Skill Practice Two

1 A survey of Year 10 girls found the following numbers were involved at least once a week in these activities.

Aerobics	1320
Swimming	1185
Riding a bike	1170
Dancing	1025
Netball	900

Illustrate these results with a pie chart.

2 A large group of Year 7 pupils were asked the question:

'How long did you spend watching television after school yesterday?'

These were the results:

	Boys	Girls
None at all	369	314
Up to 1 hour	694	730
Up to 2 hours	570	566
Up to 3 hours	226	254
More than 3 hours	273	273

Illustrate these results with two pie charts.

3 Attendances at the five most popular tourist attractions in England in 1994 were:

	Attendance (millions)
Blackpool Pleasure Beach	7.2
British Museum	5.9
National Gallery	4.3
Palace Pier, Brighton	3.5
Alton Towers	3.0

Illustrate this data with a pie chart.

4 A survey discovered the following information about the weekly spending of different types of household.

Goods and services	Two adults with children	%	Retired occupants	%
Food	£68.54	18.2	£36.30	25.1
Housing	£67.04	17.8	£21.12	14.6
Fuel and power	£15.06	4.0	£11.28	7.8
Alcohol and tobacco	£18.83	5.0	£8.68	
Clothing and footwear	£24.48	6.5	£5.93	4.1
Household goods	£29.00		£14.03	9.7
Transport and vehicles	£54.99	14.6	£14.46	
Leisure	£58.37	15.5	£19.67	13.6
Other	£39.92		£13.16	9.1
Total	£376.23		£144.63	

Copy and complete the table and illustrate this data with two pie charts.

Unit 56

Handling data   Level 6

Skill   You draw conclusions from scatter diagrams, and have a basic understanding of correlation.

Knowledge

When two variables are said to be **correlated**, this means that they are linked in some way.

If, as the value of one variable increases, the value of the other variable also tends to increase, the variables are **positively correlated**.

If, as the value of one variable increases, the value of the other variable tends to decrease, the variables are **negatively correlated**.

For example, we might say: 'Tall people tend to be heavier than short people'. This statement is equivalent to saying: 'In humans, height and weight are positively correlated'.

Or we might say: 'As cars age, their value falls'.
This statement is equivalent to saying: 'In cars, age and value are negatively correlated'.

A **scatter diagram** can be used to test whether two variables are correlated.

Skill Example

These are the test scores for 10 students in mathematics and science.

Student	A	B	C	D	E	F	G	H	I	J
Maths score	9	3	6	7	4	5	1	6	8	9
Science score	8	4	5	6	5	4	2	7	8	10

a Draw a scatter diagram.
b Comment on any correlation indicated in the scatter diagram.

a This is the scatter diagram, with each student's pair of scores represented by a point. For example, Student D scored 7 and 6, which are represented by the point (7, 6).

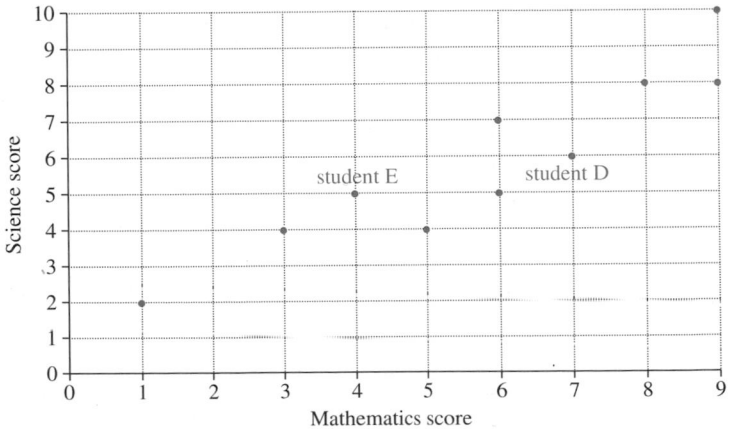

b The scatter diagram indicates there is a positive correlation between mathematics and science marks. This means that students who score a high mark for mathematics tend also to score a high mark for science. Students who score a low mark for mathematics tend also to score a low mark for science.

Skill Example

These are the ages of 10 children and their times to run 100 m.

Child	A	B	C	D	E	F	G	H	I	J
Age	8	4	11	8	6	7	5	4	9	10
Time (seconds)	17	23	14	15	20	20	22	26	15	16

a Draw a scatter diagram.
b Comment on any correlation indicated in the scatter diagram.

a

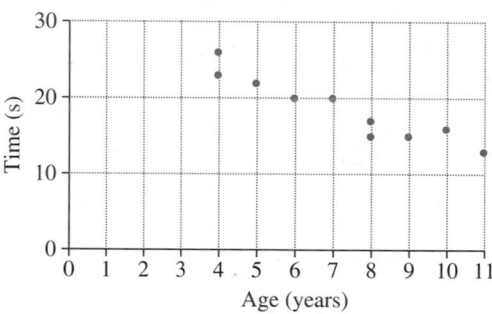

b The scatter diagram indicates there is a negative correlation between age and the time taken to run 100 m. This means that as the age *increases* the time taken to run 100 m tends to *decrease*.

Knowledge

When the points in a scatter diagram are distributed at random, this indicates there is no correlation between the variables.

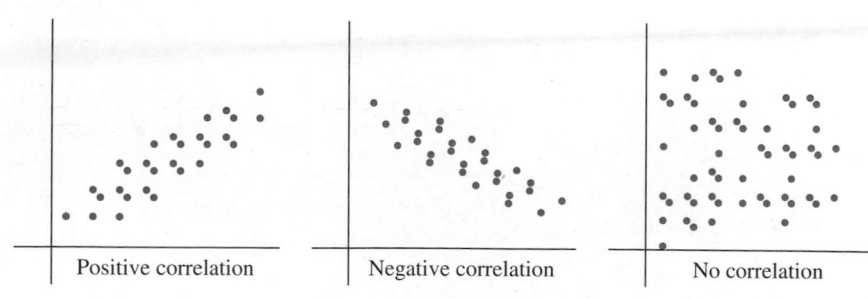

Positive correlation   Negative correlation   No correlation

Skill Practice One

1 This table records the year group of 20 students and the number of days they were absent from school in the year.

Student	A	B	C	D	E	F	G	H	I	J
Year	7	8	7	9	11	10	7	8	10	11
Days absent	0	5	3	5	15	15	5	0	5	20

Student	K	L	M	N	O	P	Q	R	S	T
Year	7	8	9	9	9	11	11	10	10	8
Days absent	10	7	8	10	12	10	12	10	13	10

   a Draw a scatter diagram.
   b Comment on any correlation indicated in the scatter diagram.

2 This table records the temperature each day in June and the number of ice-creams a shopkeeper sold.

Date	1st	2nd	3rd	4th	5th	6th	7th	8th	9th	10th
Temperature	20	19	19	18	16	14	15	17	16	14
Number sold	21	20	22	17	14	10	16	18	15	13

Date	11th	12th	13th	14th	15th	16th	17th	18th	19th	20th
Temperature	16	18	20	21	22	24	27	29	27	26
Number sold	14	18	22	14	14	27	27	32	18	25

Date	21st	22nd	23rd	24th	25th	26th	27th	28th	29th	30th
Temperature	24	26	22	21	17	18	19	16	12	9
Number sold	26	27	22	22	16	16	19	12	11	12

   a Draw a scatter diagram.
   b Comment on any correlation indicated in the scatter diagram.

3 The length of hair (measured in inches) of 20 girls and their shoe sizes are recorded in this table.

Girl	A	B	C	D	E	F	G	H	I	J
Hair length	5	6	12	8	8	9	12	6	7	8
Shoe size	4	3	4	5	7	2	5	4	5	3

Girl	K	L	M	N	O	P	Q	R	S	T
Hair length	15	6	7	5	2	9	10	11	7	12
Shoe size	4	8	5	7	7	5	3	4	7	3

   a Draw a scatter diagram.
   b Comment on any correlation indicated in the scatter diagram.

4 A security firm experiments with the number of guards it deploys each night on a large industrial estate. This table records the number of guards deployed during 20 different months and the number of reported thefts.

Month		1	2	3	4	5	6	7	8	9	10
Number of guards		4	3	2	4	5	7	6	8	10	12
Number of thefts		15	20	23	17	13	12	10	9	7	5

Month		11	12	13	14	15	16	17	18	19	20
Number of guards		15	14	12	10	8	7	6	4	4	8
Number of thefts		3	4	6	8	10	11	12	16	13	8

a Draw a scatter diagram.
b Comment on any correlation indicated in the scatter diagram.

5 This table records the ages of 20 nurses and their heights (in centimetres).

Nurse	A	B	C	D	E	F	G	H	I	J
Age	23	32	41	25	26	45	33	49	38	20
Height	155	160	160	165	165	175	165	160	165	150

Nurse	K	L	M	N	O	P	Q	R	S	T
Age	36	36	30	43	50	28	35	46	40	21
Height	160	155	170	155	155	160	150	150	165	160

a Draw a scatter diagram.
Start your horizontal axis at 20 and use a scale of 2 cm to 5 years.
Start your vertical axis at 140 cm and use a scale of 2 cm to 10 cm (of height).
b Comment on any correlation indicated in the scatter diagram.

6 This table records the heights (in centimetres) of a group of 20 adults and their weights (in kilograms).

Adult	A	B	C	D	E	F	G	H	I	J
Height	160	170	155	175	165	185	175	160	180	170
Weight	43	60	45	65	54	70	60	50	73	57

Adult	K	L	M	N	O	P	Q	R	S	T
Height	155	165	180	170	180	155	165	175	185	160
Weight	40	51	69	52	63	38	45	63	81	45

a Draw a scatter diagram.
Start your horizontal axis at 140 cm and use a scale of 2 cm to 10 cm (of height).
Start your vertical axis at 20 kg and use a scale of 2 cm to 10 kg.
b Comment on any correlation indicated in the scatter diagram.

Unit 57

Handling data   Level 6

Skills   When dealing with a combination of two experiments, you identify all the outcomes, using diagrammatic, tabular or other forms of communication.

In solving problems, you use your knowledge that the total probability of all the mutually exclusive outcomes of an experiment is 1.

Skill Example

Two coins are tossed.

a What is the probability that the result is two heads?

b If the two coins are tossed 1000 times, how many of the results would you predict will be two heads?

When we are dealing with two events, a table like a graph helps to identify all the possible outcomes. This is a table for two coins.

Second coin		
T	HT	TT
H	HH	TH
	H	T

First coin

a The table shows us there are 4 possible results:

Head Head   Head Tail
Tail Head   Tail Tail

Therefore, the probability of the result Head Head is $\frac{1}{4}$ or 0.25.

b In 1000 throws, we would predict
$$1000 \times 0.25 = 250 \text{ heads}$$

Skill Example

Two fair spinners are made, one numbered 1, 2, 3 and the other numbered 1, 2, 3, 4.

 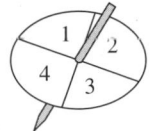

A score is obtained by spinning both spinners and adding their scores.

a Draw a table identifying all the possible results when the two spinners are spun.

b What is the probability that the total score is greater than 5?

c What is the probability of a score of 3?

d If the spinners are each spun 100 times, how many of the results would you predict will be scores of 3?

a This table identifies all the possible results.

Second spinner			
4	5	6	7
3	4	5	6
2	3	4	5
1	2	3	4
	1	2	3

First spinner

b The probability that the score is greater than 5 is $\frac{3}{12}$ or 0.25.

c The probability that the score is 3 is $\frac{2}{12}$ or 0.17 (to 2 decimal places).

d In 100 spins, we would predict
$$100 \times 0.17 = 17 \text{ scores of } 3$$

Skill Practice One

1 If two coins are tossed 2000 times, how many of the results would you predict will be
a two heads   **b** two tails
c one head and one tail?

2 Two fair spinners are made, one numbered 1, 2, 3 and the other numbered 1, 2, 3, 4, 5.

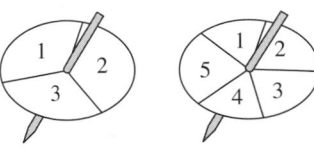

A score is obtained by spinning both spinners and adding their scores.

a Draw a table identifying all the possible results when the two spinners are spun.

b What is the probability that the score is greater than 5?

c What is the probability that the score is 5?

3 Two fair spinners are made, one numbered 1, 3, 5 and the other numbered 2, 4, 6.

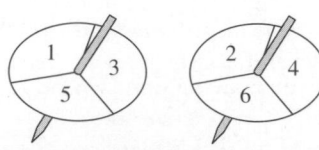

A score is obtained by spinning both spinners and adding their scores.
a Draw a table identifying all the possible results when the two spinners are spun.
b What is the probability that the score is greater than 5?
c What is the probability that the score is 5?

4 Two fair spinners are made, one numbered 1, 2, 3 and the other numbered 1, 2, 3, 4.

 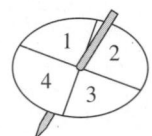

A score is obtained by spinning both spinners and multiplying their scores.
a Draw a table identifying all the possible results when the two spinners are spun.
b What is the probability that the score is greater than 4?
c What is the probability that the score is an even number?
d If the two spinners are spun 1200 times, how many of the results would you predict will be:
(i) 2   (ii) 9?

5 Two fair spinners are made, one numbered 1, 2, 3 and the other numbered 1, 2, 3, 4, 5.

A score is obtained by spinning both spinners and multiplying the two scores.
a Draw a table identifying all the possible results when the two spinners are spun.
b What is the probability that the score is:
(i) 1 (ii) 2 (iii) 3 (iv) 4 (v) 5
(vi) 6 (vii) 8 (viii) 9 (ix) 10 (x) 15?

c If the two spinners are spun 1000 times, how many of the results would you predict will be:
(i) 2   (ii) 15?

6 Two fair dice, each numbered from 1 to 6, are rolled. A score is obtained by adding the scores on the two dice.

a Draw a table identifying all the possible results when the two dice are rolled.
b What is the probability that the score is
(i) 1   (ii) 2   (iii) 3   (iv) 4
(v) 5   (vi) 6   (vii) 7   (viii) 8
(ix) 9   (x) 10   (xi) 11   (xii) 12
(xiii) an odd number
(xiv) a factor of 12
(xv) greater than 5?
c If the two dice are rolled 3600 times, how many of the results would you predict will be:
(i) 2   (ii) 7?

7 A pack of cards is cut twice and the suits (hearts, diamonds, spades or clubs) are noted.

a Draw a table identifying all the possible results when the two suits are noted.
b What is the probability that the suits noted will be:
(i)   two hearts
(ii) one heart and one club
(iii) one black and one red card
(iv) a pair which contains one diamond
(v)   a pair which contains at least one heart?
c If the experiment is repeated 1600 times, how many of the results would you predict will not contain a club?

8 Two special dice are made, one numbered 0, 0, 1, 1, 2, 2 and the other numbered 0, 0, 0, 1, 2, 3. William rolls the first dice and Mary rolls the second dice.

 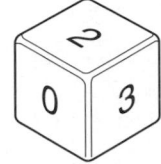

a Draw a table identifying all the possible results when William and Mary roll the dice. Use the code W when William wins, M when Mary wins and D when it is a draw.

b They roll their dice 360 times. How many times do you predict that William will beat Mary?

Knowledge

A **tree diagram** is an alternative way of identifying all the outcomes when we are dealing with two events.

When two coins are tossed, the possible results can be illustrated on this tree diagram, as shown below.

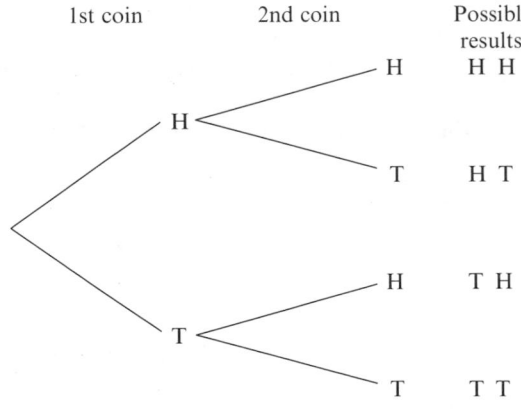

Questions on probability can be answered from this diagram.

For example, the probability of 2 heads $= \frac{1}{4}$ or 0.25.

Skill Example

A box contains a red pen, a blue pen and a green pen. A pen is taken out, used, put back and a second pen taken out.
Draw a tree diagram to show all the possible results and from the diagram find:

a the probability that both pens are red

b the probability that both pens are different colours.

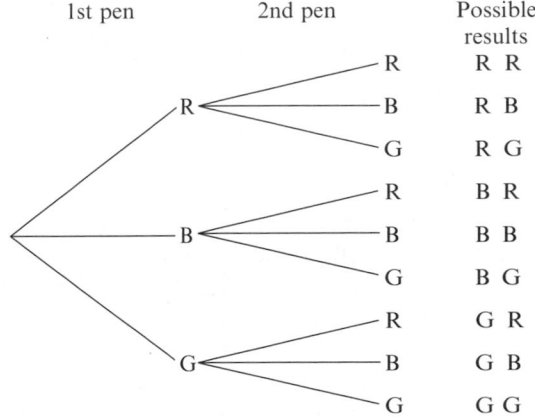

a Total number of possibilities = 9
Total number when both are red = 1
So, the probability that both pens are red is $\frac{1}{9}$ or 0.11 (to 2 dp).

b Total number of possibilities = 9
Total number involving different colours = 6
So, the probability that the two pens are of different colour is $\frac{6}{9}$ or 0.67 (to 2 dp).

Skill Practice Two

1 In my wallet there is a £5 note, a £10 note and a £20 note.
I remove a note, put it back and then remove a note again. Draw a tree diagram to show all the possible results.
From the diagram find:

a the probability that both notes are the same

b the probability that the two notes are different

c the probability that both are £5 notes

d the probability that both are £10 notes

e the probability that both are £20 notes.

2 In my wallet there is a 10 p stamp, a 12 p stamp and a 20 p stamp.

I remove a stamp, put it back and then remove a stamp again. Draw a tree diagram to show all the possible results.

From the diagram find:

a the probability that both stamps are the same

b the probability that the two stamps are different

c the probability that both stamps are worth more than 10 p.

3 In my pocket I have a 2 p coin, a 5 p coin and a 10 p coin.

I remove a coin, put it back and then remove a coin again. Draw a tree diagram to show all the possible results.

From the diagram find:

a the probability that both coins are the same

b the probability that both coins are silver

c the probability that one coin is silver and the other is copper

d the probability that both coins are copper.

4 Three counters marked 1, 2 and 3 are placed in a bag.

A counter is removed from the bag, replaced and then a counter is removed again. Draw a tree diagram to show all the possible results.

From the diagram find:

a the probability that both counters are the same

b the probability that both counters are marked with odd numbers

c the probability that both counters are marked with even numbers.

5 Three counters marked 1, 4 and 6 are placed in a bag.

A counter is removed from the bag, replaced and then a counter is removed again. Draw a tree diagram to show all the possible results.

From the diagram find:

a the probability that both counters are marked with odd numbers

b the probability that both counters are marked with even numbers

c the probability that both counters are marked with square numbers

d the probability that both counters are marked with triangular numbers.

6 Three counters marked A, B and C are placed in a bag.

A counter is removed from the bag, replaced and then a counter is removed again. Draw a tree diagram to show all the possible results.

From the diagram find:

a the probability that both counters are marked with consonant letters

b the probability that both counters are marked with vowel letters

c the probability that one counter is marked with a consonant and the other with a vowel.

7 The three dominoes illustrated are placed face-down on a table.

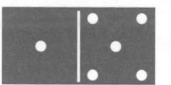  

One domino is picked up, replaced and then a domino is picked up again. Draw a tree diagram to show all the possible results.

From the diagram find:

a the probability that both dominoes have the same number of dots

b the probability that both dominoes have six dots

c the probability that both dominoes have nine dots.

8 Three playing cards, the three of diamonds, the six of hearts and the six of clubs, are placed face-down on a table.

One card is picked up, replaced and then a card is picked up again. Draw a tree diagram to show all the possible results.

From the diagram find:

a the probability that both cards are of the same number

b the probability that both cards are of a red suit

c the probability that both cards have a triangular number printed on them.

9 Three playing cards, the four of hearts, the six of clubs and the nine of spades, are placed face-down on a table.
One card is picked up, replaced and then a card is picked up again. Draw a tree diagram to show all the possible results.

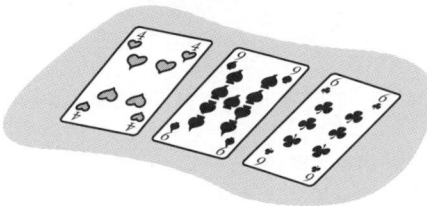

From the diagram find:
a the probability that both cards are of a black suit
b the probability that both cards have square numbers printed on them.

10 Four counters marked 1, 2, 3, and 4 are placed in a bag.
A counter is removed from the bag, replaced and then a counter is removed again. Draw a tree diagram to show all the possible results.
From the diagram:
a the probability that both counters are marked with odd numbers
b the probability that both counters are marked with even numbers
c the probability that both counters show numbers which are multiples of 2
d the probability that both counters show numbers which are factors of 6
e the probability that both counters show numbers which are factors of 12.

Knowledge

Events are said to be **mutually exclusive** if they cannot happen at the same time.

In such cases, the probabilities of the events happening must add up to 1.

Skill Example

a The probability that it will rain tomorrow is 0.3.
What is the probability that it will not rain tomorrow?

Raining and not raining are mutually exclusive events, so the probability that it does not rain is

$$1 - 0.3 = 0.7$$

b There are three kinds of sweet in a bag: mints, toffees and lime creams. If a sweet is picked at random, the probability that it is a mint is 0.37 and the probability that it is a toffee is 0.51.
What is the probability that it is a lime cream?

Selecting the three different sweets are mutually exclusive events, so the probability that the sweet is a lime cream is

$$1 - (0.37 + 0.51) = 1 - 0.88 = 0.12$$

Skill Practice Three

1 The probability that Peter is late for school is 0.06.
What is the probability that Peter is not late for school?

2 The probability that a weekday selected at random is a Wednesday is 0.2.
What is the probability that a weekday selected at random is not a Wednesday?

3 The probability that a light bulb fails in the first 1000 hours of use is 0.15.
What is the probability that it does not fail in the first 1000 hours of use?

4 A bag contains red and white beads. The probability that a bead selected at random is red is 0.67.
What is the probability that a bead selected at random is white?

5   A bag contains red and white beads. The
    probability that a bead selected at random is
    white is 0.25.
    a   What is the probability that a bead selected
        at random is red?
    b   If there are 4 beads in the bag, how many
        are red?
    c   If there are 28 beads in the bag, how many
        are white?
    d   Explain why there cannot be 10 beads in
        the bag.

6   A bag contains red and white beads. The
    probability that a bead selected at random is
    white is 0.375.
    a   What is the probability that a bead selected
        at random is red?
    b   If there are 8 beads in the bag, how many
        are red?
    c   If there are 64 beads in the bag, how many
        are white?
    d   Explain why there cannot be 28 beads in
        the bag.

7   A bag contains red, white and blue beads. The
    probability that a bead selected at random is
    red is 0.1 and the probability that a bead
    selected at random is white is 0.3.
    a   What is the probability that a bead selected
        at random is blue?
    b   If there are 10 beads in the bag, how many
        are red?
    c   If there are 80 beads in the bag, how many
        are blue?
    d   Explain why there cannot be 12 beads in
        the bag.

8   A bag contains red, white and blue beads. The
    probability that a bead selected at random is
    red is 0.125 and the probability that a bead
    selected at random is white is 0.2.

    a   What is the probability that a bead selected
        at random is blue?
    b   If there are 40 beads in the bag, how many
        are red?
    c   If there are 120 beads in the bag, how
        many are blue?
    d   Explain why there cannot be 12 beads in
        the bag.

9   The probability that it rains on a June day in
    Gedney is 0.13.
    a   What is the probability that it does not rain
        on a June day in Gedney?
    b   On how many days in Gedney would you
        predict it will rain next June?
    c   On how many days in Gedney would you
        predict it will not rain next June?

10 In a restaurant, all meals are offered with
    either chips or baked potatoes. The
    probability that a person chooses chips with
    their meal is 0.75.
    a   What is the probability that a person
        chooses baked potatoes with their meal?
    b   If the restaurant serves 120 meals in one day,
        estimate how many are served with chips.
    c   If the restaurant serves 2260 meals in one
        month, estimate how many are served with
        baked potatoes.

11 The approximate probability of getting
    6 numbers correct and winning the Jackpot in
    the National Lottery is 0.000 000 071.
    Mr Smith plays 5 lines each week.
    a   What is the probability that Mr Smith wins
        the Jackpot in any given week?
    b   What is the probability that Mr Smith does
        not win the Jackpot in any given week?

Unit 58

Handling data   Level 7

Skill   You specify hypotheses and test them by designing and using appropriate methods that take account of bias.

Knowledge

A **hypothesis** is a statement which can be tested statistically to see whether it is true or false.

For example:

At least 75% of the meals in this school are served with chips.

To test a hypothesis, a survey is carried out. For example, the above hypothesis could be tested by counting the number of meals served with chips, and the number served without chips.

A simple **data collection sheet** like this could be used to make the counts.

Type of meal	Tally	Number
With chips	卌 ‖	117
Without chips	卌 卌 卌 卌 卌 卌 卌 卌 卌 卌 ‖‖	48

When testing a hypothesis, **bias** must be avoided. This survey could be biased in several ways:

- If only Year 7 students are counted. They might eat less or more chips than other year groups, and so the count would not be representative of the whole school.
- If the count is made on only one day. The menu might change from day to day, and thus affect the choices made.
- If food is served from several counters and the survey is made at only one counter. The results of counting at, say, the Salad Bar would be different from the results of counting at, say, the Fast Food Bar.

To complete this survey without introducing bias, it would be necessary to count every year group and every counter on several different days.

Skill Example

A restaurant owner has noticed that there is no take-out pizza service in a certain small town. So, she wants to test the hypothesis: *There is a demand for a take-out pizza service*. Design a question or series of questions that could be used to test this hypothesis. How would you avoid bias when conducting a survey using your question(s)? A possible question is:

If this town had a take-out pizza service, how often would you buy pizzas?

Frequently (at least once a week) ☐

Sometimes (at least once a month) ☐

Infrequently (at least once a year) ☐

Never ☐

Bias could be avoided by ensuring that a **representative sample** of the community is questioned. This would involve:

- Not conducting your survey only during the day. Otherwise, you will miss people who are at work.
- Not conducting your survey in only one place. (A wealthy or deprived area would not produce results representative of the whole town.)
- Not restricting your survey to one age group.

Skill Practice One

Design a question or series of questions that could be used to test each of the following hypotheses. Explain how you would avoid bias when conducting a survey using your question(s).

1 'The most popular flavour of crisps in my school is cheese and onion.'

2 'Most pupils in my school get at least £2.50 a week pocket money.'

3 'In my school, most pupils and their parents would support the idea to abandon school uniform.'

4 'Soaps are the most popular television programmes among Year 7 pupils, but documentaries are more popular with sixth formers.'

5 'The most popular night for young people to go out is Friday, but older people prefer Saturday.'

Unit 59

Skills You determine the modal class and estimate the mean, median and range of sets of grouped data, selecting the statistic most appropriate to your line of enquiry.

You use measures of average and range, with associated frequency polygons, as appropriate, to compare distributions and make inferences.

Knowledge

The median and the range can be found from a grouped-frequency table without using the original list of values.

This grouped-frequency table shows the results of 100 pupils in a mathematics test.

Score	Frequency
1–5	22
6–10	22
11–15	33
16–20	23

The mode becomes a **modal class**, simply defined as the class interval which has the highest frequency.

The modal class for this distribution is 11–15.

The range is still defined as the greatest value minus the least value.

The range for this distribution is $20 - 1 = 19$.

To find the median, we add cumulative frequencies to the table.

Score	Frequency	Cumulative frequency
1–5	22	22
6–10	22	44
11–15	33	77
16–20	23	100

The cumulative frequencies show that the 50th score is in the 11–15 class.

The cumulative frequencies also show that the 50th score is probably much closer to 11 than 15.

An estimate for the median is 12.

Skill Example

This table shows the lengths of two groups of 20 worms, measured to the nearest centimetre.

Group 1		Group 2	
Length	Frequency	Length	Frequency
1.5–2.5	0	1.5–2.5	3
2.5–3.5	4	2.5–3.5	7
3.5–4.5	7	3.5–4.5	8
4.5–5.5	8	4.5–5.5	2
5.5–6.5	1	5.5–6.5	0

a Find the modal class and range of each distribution.
b Estimate the median of each distribution.
c Draw a frequency polygon to illustrate each distribution.
d Compare the two distributions.

a For Group 1: the modal class is $4.5 - 5.5\,\text{cm}$
    the range is $6.5 - 2.5 = 4\,\text{cm}$
  For Group 2: the modal class is $3.5 - 4.5\,\text{cm}$
    the range is $5.5 - 1.5 = 4\,\text{cm}$

b Adding cumulative frequencies to the table gives:

Group 1		
Length	Frequency	Cumulative frequency
1.5–2.5	0	0
2.5–3.5	4	4
3.5–4.5	7	11
4.5–5.5	8	19
5.5–6.5	1	20

Group 2		
Length	Frequency	Cumulative frequency
1.5–2.5	3	3
2.5–3.5	7	10
3.5–4.5	8	18
4.5–5.5	2	20
5.5–6.5	0	20

For Group 1, the 10th length is towards the end of the 3.5–4.5 class. The median is estimated to be 4.4.

For Group 2, the 10th length is at the very end of the 2.5–3.5 class. The median is estimated to be 3.5.

c

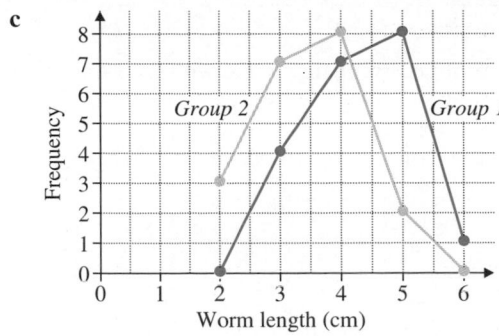

Worm length (cm)

d The distributions are similar in terms of the spread of values, both having a range of 4 cm. The median for Group 2 is lower than that for Group 1, which indicates that the worms in Group 2 are 'on average' shorter than those in Group 1.

Skill Practice One

1 Two darts players, Bill and Terri, each throws his set of three darts 50 times.
The 50 scores for each player are shown in this table.

Score	Frequency	
	Bill	Terri
1–20	4	1
21–40	8	12
41–60	8	14
61–80	9	12
81–100	8	1
101–120	5	5
121–140	4	3
141–160	3	2
161–180	1	0

a Find the modal class and the range of each distribution.
b Estimate the median of each distribution.
c Draw a frequency polygon to illustrate each distribution.
d Compare the distributions.

2 Sixty rugby players are weighed before and after playing a six-day tournament in very hot conditions.
The results are shown in this table.

Weight (kg)	Frequency	
	Before	After
75–80	3	5
80–85	1	5
85–90	2	6
90–95	3	9
95–100	6	8
100–105	17	14
105–110	15	8
110–115	13	5

a Find the modal class and the range of each distribution.
b Estimate the median of each distribution.
c Draw a frequency polygon to illustrate each distribution.
d Compare the distributions.

3 The weekly incomes of two groups of 50 people are shown in this table.

Income (£)	Frequency	
	Group 1	Group 2
1–50	2	0
51–100	3	0
101–150	7	2
151–200	10	8
201–250	18	13
251–300	7	17
301–350	2	7
351–400	1	3

a Find the modal class and the range of each distribution.
b Estimate the median of each distribution.
c Draw a frequency polygon to illustrate each distribution.
d Compare the distributions.

Knowledge

The mean can be estimated from a grouped-frequency table without using the original list of values.

This grouped-frequency table shows the scores of 100 pupils in a mathematics test.

Score	1–5	6–10	11–15	16–20
Frequency	22	22	33	23

To estimate the mean, use the middle value of each class interval.

Score	1–5	6–10	11–15	16–20
Mid-interval value (m)	3	8	13	18
Frequency (f)	22	22	33	23

Then use the same method that is used to find the mean in a simple frequency table.

Score	1–5	6–10	11–15	16–20	Total
Mid-interval value (m)	3	8	13	18	
Frequency (f)	22	22	33	23	100
$m \times f$	66	176	429	414	1085

Estimated mean = $1085 \div 100 = 10.85$

Skill Example

This table shows the lengths of a sample of 20 worms.
Estimate the mean length.

Length	1.5–2.5	2.5–3.5	3.5–4.5	4.5–5.5	5.5–6.5
Frequency	0	4	7	8	1

Add the mid-interval values (m) and the $m \times f$ values to the table.

Length	1.5–2.5	2.5–3.5	3.5–4.5	4.5–5.5	5.5–6.5	Total
Mid-interval (m)	1	2	3	4	5	
Frequency (f)	0	4	7	8	1	20
$m \times f$	0	8	21	32	5	66

Estimated mean = $66 \div 20 = 3.3 \, \text{cm}$

Skill Practice Two

1 Two darts players, Bill and Terri, each throws his set of three darts 50 times.
The 50 scores for each player are shown in this table.

Score	Frequency	
	Bill	Terri
1–20	4	1
21–40	8	12
41–60	8	14
61–80	9	12
81–100	8	1
101–120	5	5
121–140	4	3
141–160	3	2
161–180	1	0

Estimate the mean of each distribution, using the mid-interval values 10.5, 30.5 and so on.

2 Sixty rugby players are weighed before and after playing a six-day tournament in very hot conditions.
The results are shown in this table.

Weight (kg)	Frequency	
	Before	After
75–80	3	5
80–85	1	5
85–90	2	6
90–95	3	9
95–100	6	8
100–105	17	14
105–110	15	8
110–115	13	5

Estimate the mean of each distribution, using the mid-interval values 77.5, 82.5, 87.5, 92.5 and so on.

3 The weekly incomes of two groups of 50 people are shown in this table.

Income (£)	Frequency	
	Group 1	Group 2
1–50	2	0
51–100	3	0
101–150	7	2
151–200	10	8
201–250	18	13
251–300	7	17
301–350	2	7
351–400	1	3

Estimate the mean of each distribution, using the mid-interval values 25.5, 75.5, 125.5 and so on.

Skill Practice Three

1 This table shows the frequency of car journeys made by two social workers in one month.

Length (miles)	Worker 1	Worker 2
1–5	12	36
6–10	16	24
11–15	32	20
16–20	18	10
21–25	12	7
26–30	8	3
31–35	2	0

a Illustrate each distribution with a frequency polygon.
b Find the modal class and range of each distribution.
c Estimate the median of each distribution.
d Estimate the mean of each distribution.
e Compare the two distributions.

2 Two groups of pupils were given a mathematics examination.
Their results are shown in this table.

Mark	Group 1	Group 2
1–10	3	5
11–20	4	8
21–30	7	12
31–40	10	18
41–50	16	24
51–60	20	13
61–70	16	10
71–80	10	9
81–90	8	1
91–100	6	0

a Illustrate each distribution with a frequency polygon.
b Find the modal class and range of each distribution.
c Estimate the median of each distribution.
d Estimate the mean of each distribution.
e Compare the two distributions.

3 The lengths of two samples of leaves are measured (to the nearest centimetre).
The results are shown in this table.

Length	Sample 1	Sample 2
2	8	7
3	18	13
4	25	20
5	15	25
6	8	15
7	4	0
8	2	0

a Illustrate each distribution with a frequency polygon.
b Find the modal class and range of each distribution.
c Estimate the median of each distribution.
d Estimate the mean of each distribution.
e Compare the two distributions.

Unit 60

Handling data   Level 7

Skill   You draw a line of best fit on a scatter diagram, by inspection.

Knowledge

When two variables are perfectly correlated, all the points in their scatter diagram lie on the same line.

This scatter diagram shows the correlation between the diameter and circumference of 20 circles.

Because circumference and diameter are perfectly correlated, all the points lie on the same straight line.

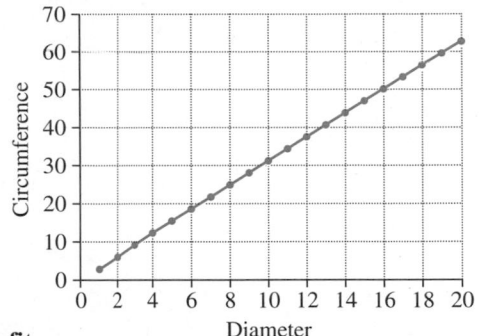

When a scatter diagram indicates that two variables are highly correlated, we can draw a **line of best fit**.
This is the straight line which goes through the point representing the mean of each set of data, passing close to as many points as possible.

Skill Example

This table shows the ages of 10 children and their times to run 100 m.

Child	A	B	C	D	E	F	G	H	I	J
Age	8	4	11	8	6	7	5	4	9	10
Time (seconds)	17	23	14	15	20	20	22	26	15	16

a Draw a scatter diagram and add the line of best fit.
b Estimate the time a child of 8 years and 6 months would take to run 100 m.

a We draw the scatter diagram and then, using a transparent ruler, judge by eye the position of the line of best fit, balancing points above the line with points below the line.

Here is the scatter diagram with the line of best fit added.

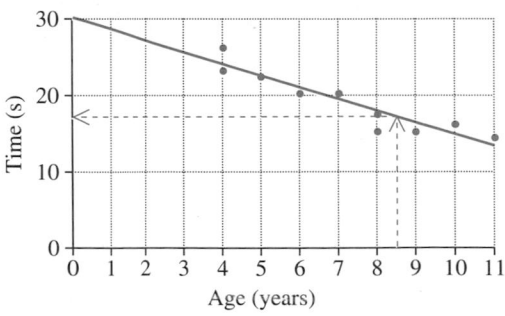

b We draw a line up from the age axis to the line of best fit and then across to the time axis. By measuring the intercept, we obtain the estimate that a child of 8 years 6 months will run 100 m in about 17 seconds.

Skill Practice One

1 This scatter graph shows the correlation between the engine capacity (in cubic centimetres) and the urban fuel consumption (in miles per gallon) of a group of cars.
A line of best fit has been added.

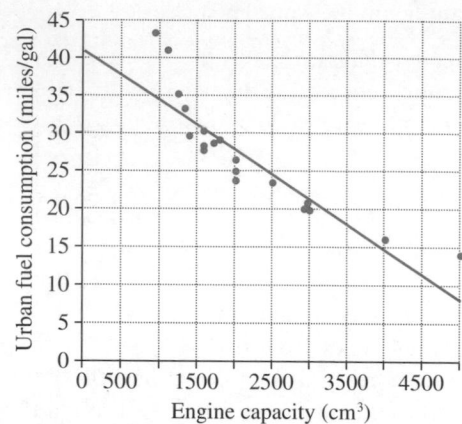

a Comment on any correlation indicated in the scatter diagram.
b Estimate the urban fuel consumption of cars with engine capacities of:
(i) 2000 cm^3 (ii) 3500 cm^3 (iii) 4500 cm^3 (iv) 1000 cm^3
c Estimate the engine capacity of a car with an urban fuel consumption of:
(i) 30 miles/gal   (ii) 10 miles/gal   (iii) 25 miles/gal

2 This scatter graph shows the correlation between the engine capacity (in cubic centimetres) and the top speed (in miles per hour) of a group of cars.
A line of best fit has been added.

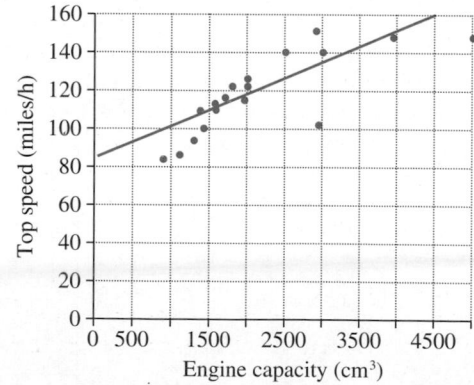

a Comment on any correlation indicated in the scatter diagram.
b Estimate the top speed of cars with engine capacities of:
(i) 2000 cm^3 (ii) 3500 cm^3 (iii) 4500 cm^3 (iv) 1000 cm^3
c Estimate the engine capacity of a car with a top speed of:
(i) 100 miles/h   (ii) 130 miles/h   (iii) 90 miles/h   (iv) 110 miles/h

3 A science student measures the current (in amps) that flows through a circuit as the voltage is increased from 0 to 100 volts.

Voltage	0	10	20	30	40	50	60	70	80	90	100
Current	0.0	1.1	1.8	3.2	3.7	5.3	6.2	6.8	8.1	9.3	10.0

a Draw a scatter diagram on 2 mm graph paper.
Plot the voltage on the horizontal axis, using a scale of 2 cm to 20 volts.
Plot the current on the vertical axis, using a scale of 2 cm to 1 amp.
b Comment on any correlation indicated in the scatter diagram.
c Add a line of best fit the scatter diagram.
d Estimate the current that will flow through the circuit if the voltage is:
(i) 25 volts   (ii) 55 volts   (iii) 75 volts   (iv) 15 volts
e Estimate the voltage if the current which flows through the circuit is:
(i) 0.5 amps   (ii) 3.5 amps   (iii) 6.5 amps   (iv) 8.7 amps

4 A group of students have holiday jobs in a factory. They complete a survey linking the number of hours they worked each week with the number of hours they spent watching television.

Hours worked	15	15	20	20	20	25	25	30	30	30
Hours watching TV	30	34	35	30	32	29	25	28	26	25

Hours worked	30	35	35	35	35	35	40	40	45	50
Hours watching TV	22	18	20	22	15	21	20	18	15	12

a Draw a scatter diagram on 2 mm graph paper.
Plot the hours worked on the horizontal axis, using a scale of 2 cm to 5 hours.
Plot the hours watching television on the vertical axis, using a scale of 2 cm to 5 hours.
b Comment on any correlation indicated in the scatter diagram.
c Add a line of best fit the scatter diagram.
d Estimate the time spent watching television by a student who works:
(i) 22 hours   (ii) 27 hours   (iii) 36 hours   (iv) 43 hours
e Estimate the hours worked by a student who watches television for:
(i) 22 hours   (ii) 27 hours

5 A horticultural scientist measures the water used on 10 trial plots of seedlings and the average growth of the seedlings. These are her results.

Water used (1000 litres)	1	1	2	2	3	3	4	4	5	5
Average growth (cm)	3.2	2.7	3.8	3.5	5.2	4.5	6.7	5.0	6.0	6.9

a Draw a scatter diagram on 2 mm graph paper.
Plot the water used on the horizontal axis, using a scale of 2 cm to 1000 litres.
Plot the average growth on the vertical axis, using a scale of 2 cm to 1 cm (of growth).
b Comment on any correlation indicated in the scatter diagram.
c Add a line of best fit to the scatter diagram.
d Estimate the average growth of a plot of seedlings watered with:
(i) 1500 litres   (ii) 3500 litres   (iii) 6000 litres   (iv) 500 litres
e Estimate the water used on a plot of seedlings with an average growth of:
(i) 2 cm   (ii) 6.5 cm   (iii) 4.5 cm   (iv) 6 cm

6 A science student is investigating the correlation between the length of a pendulum and the time it takes to complete one oscillation (one to and fro swing). She suspects there is a correlation between the square root of the pendulum length and the time.
This is her table of results.

Length (cm)	10	20	30	40	50	60	70	80	90	100
Square root of length (1 dp)	3.2	4.5	5.5	6.3	7.0	7.7	8.4	8.9	9.5	10
Time (s)	0.5	0.9	1.1	1.2	1.4	1.5	1.7	1.7	1.8	2.0

a Draw a scatter diagram on 2 mm graph paper.
Plot the square root of the length on the horizontal axis, using a scale of 2 cm to 1 unit.
Plot the time on the vertical axis, using a scale of 4 cm to 1 second.
b Comment on any correlation indicated in the scatter diagram.
c Add a line of best fit to the scatter diagram.
d Estimate the time of a single swing for a pendulum with a length of:
(i) 36 cm (ii) 49 cm (iii) 64 cm (iv) 81 cm
e Estimate the length of a pendulum with a time for one swing of:
(i) 1 second (ii) 1.6 seconds.

Unit 61

Skill   You understand relative frequency as a measure of probability and use this to compare outcomes of experiments.

Knowledge

Reminder

The probability of an event happening is:

$$\frac{\text{Number of outcomes which contain the event}}{\text{Total number of outcomes}}$$

A probability can be written as a fraction, a decimal or a percentage. Decimal answers are easier to illustrate and compare on the **probability scale** from 0 (impossible) to 1 (certain).

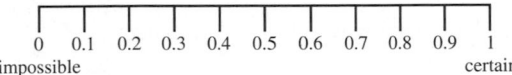

When decimal answers are very long, they should be rounded to a sensible number of decimal places.

Skill Example

In a class of 24 girls, 12 wear dresses, 8 wear skirts and blouses and the rest wear jeans.

a Find the probability that the first girl to go into the classroom will be wearing
   (i) a skirt and blouse   (ii) jeans

b Illustrate your answers on the probability scale.

a (i) Number of outcomes which contain the event = 8
      Total number of outcomes = 24
      So, the probability = $\frac{8}{24}$ or 0.33 (to 2 dp)

   (ii) Number of outcomes which contain the event = 4
      Total number of outcomes = 24
      So, the probability = $\frac{4}{24}$ or 0.17 (to 2 dp)

b
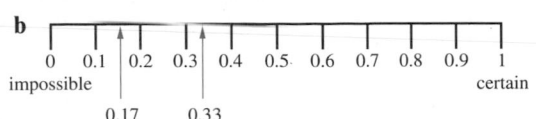

Skill Practice One

In each question, give your answers as fractions and decimals (correct to a sensible number of places).
Illustrate all the answers to each question on a single probability scale.

1 There are 20 boys in class 5B. Twelve of them have dark hair, 6 of them have blonde hair and 2 of them have ginger hair.
   Find the probability that the first boy to go into the classroom will have:
   a dark hair
   b blonde hair
   c ginger hair.

2 There are 60 passengers on a bus. Thirty are sitting downstairs, 24 are sitting upstairs and 6 are standing.
   If the bus stops and one passenger gets off, find the probability that this passenger was:
   a sitting downstairs
   b sitting upstairs
   c standing.

3 In a £1 cash bag there are 8 one penny coins, 6 two pence coins, 4 five pence coins and 6 ten pence coins.
   If a coin is removed from the bag, find the probability that it will be:
   a a one penny coin
   b a two pence coin
   c a five pence coin
   d a ten pence coin
   e a copper coin
   f a silver coin.

4 The dominoes illustrated below are shuffled and placed face-down on a table.

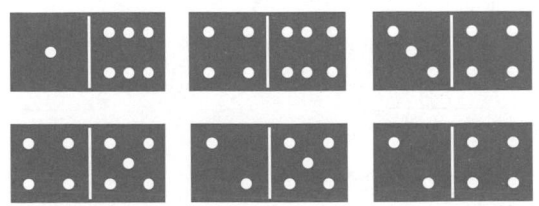

   If one is then picked up, find the probability that it will have:
   a 7 dots
   b any odd number of dots
   c any even number of dots.

5 The names of the four seasons of the year are written on cards.

If any one card is chosen at random, find the probability that:

a the first letter on the card is S

b the last letter on the card is R

c there are six letters on the card.

Knowledge

Reminder

There are three ways to establish a probability:

- By logical argument, when we are certain that all outcomes are equally likely.
- By using existing data, for example, school registers or weather records.
- By conducting a survey or experiment.

The last method uses **relative frequency** as an estimate of probability.

If the experiment or survey is repeated only a few times, the estimate will not be very accurate. To obtain an accurate estimate, the experiment must be repeated a large number of times.

Skill Example

Two pupils each select 100 seeds at random from a bag containing a very large number of red and white seeds. Each seed is returned to the bag before another is selected.

These are their results:

Carl	red	57
	white	43
Sherene	red	47
	white	53

a Write down the relative frequencies of red or white seeds for each pupil's results.

b Write down the relative frequencies of red or white seeds for the combined results.

c Which relative frequencies are likely to be the most accurate estimate of the probabilities that a selected seed is red or white?

d If there are 10 000 seeds in the bag, estimate the number of red seeds.

a The relative frequencies are:

Carl	red	$\frac{57}{100} = 0.57$
	white	$\frac{43}{100} = 0.43$
Sherene	red	$\frac{47}{100} = 0.47$
	white	$\frac{53}{100} = 0.53$

b The combined relative frequencies are:

red $\frac{104}{200} = 0.52$

white $\frac{96}{200} = 0.48$

c The combined results are likely to be the most accurate.

d The best estimate will be

$$0.52 \times 10\,000 = 52\,000$$

Skill Practice Two

1 Five pupils each toss two coins 20 times. These are their results.

Nicola	two heads	6
	two tails	4
	head and tail	10
Edward	two heads	4
	two tails	9
	head and tail	7
Wayne	two heads	5
	two tails	4
	head and tail	11
Clare	two tails	2
	head and tail	12
Madrina	two heads	2
	two tails	5
	head and tail	13

a Write down the relative frequencies of two heads, two tails, or a head and a tail for each pupil's results.

b Write down the relative frequencies of two heads, two tails, or a head and a tail for the combined results.

c Which relative frequencies are likely to be the most accurate estimate of the probabilities that two tossed coins land as two heads, two tails or as a head and a tail?

2 Four research students each open 25 pea pods and count the number of peas inside each pod. These are their results.

Student A

Number of peas	3	4	5	6	7	8
Frequency	3	5	11	3	2	1

Student B

Number of peas	3	4	5	6	7	8
Frequency	4	5	6	7	3	0

Student C

Number of peas	3	4	5	6	7	8
Frequency	1	8	5	6	0	5

Student D

Number of peas	3	4	5	6	7	8
Frequency	2	4	7	8	4	0

a Write down the relative frequencies of 3, 4, 5, 6, 7 or 8 peas per pod for each student's results.

b Write down the relative frequencies of 3, 4, 5, 6, 7 or 8 peas per pod for the combined results.

c Which relative frequencies are likely to be the most accurate estimate of the probabilities that a pod contains 3, 4, 5, 6, 7 or 8 peas?

d In a sample of 1000 pods, how many would you expect to contain 5 peas?

e In a sample of 2000 pods, how many would you expect to contain fewer than 5 peas?

3 Five trading standards officers each open 20 boxes of Striko Matches and count the contents.

These are their results.

Officer A

Number of matches	38	39	40	41	42
Number of boxes	2	6	7	5	0

Officer B

Number of matches	38	39	40	41	42
Number of boxes	1	5	8	5	1

Officer C

Number of matches	38	39	40	41	42
Number of boxes	0	4	12	4	0

Officer D

Number of matches	38	39	40	41	42
Number of boxes	3	4	5	7	1

Officer E

Number of matches	38	39	40	41	42
Number of boxes	1	8	10	0	1

a Write down the relative frequencies of 38, 39, 40, 41 or 42 matches per box for each officer's results.

b Write down the relative frequencies of 38, 39, 40, 41 or 42 matches per box for the combined results.

c Which relative frequencies are likely to be the most accurate estimate of the probabilities that a box contains 38, 39, 40, 41 or 42 matches?

d In a survey of 1000 boxes, how many would you expect to contain 40 matches?

e In a survey of 2000 boxes, how many would you expect to contain more than 40 matches?

4 In an experiment, 4 groups of 50 people are asked to select their favourite colour from red, blue, green and yellow.
These are the results.

Group A

Colour	R	B	G	Y
Number	21	12	7	10

Group B

Colour	R	B	G	Y
Number	17	18	5	10

Group C

Colour	R	B	G	Y
Number	28	13	7	7

Group D

Colour	R	B	G	Y
Number	20	16	11	3

a Write down the relative frequencies of red, blue, green and yellow for each group's results.

b Write down the relative frequencies of red, blue, green and yellow for the combined results.

c Which relative frequencies are likely to be the most accurate estimate of the probabilities that a person selects red, blue, green or yellow as their favourite colour?

d If 1500 people are asked to select their favourite colour from red, blue, green and yellow, how many would you expect to select red?

e If 5000 people are asked to select their favourite colour from red, blue, green and yellow, how many would you expect to select a colour other than red?

Standard Assessment Tests

Number and Algebra

1a Claire puts a **2** digit whole number into her calculator.

She **multiplies** the number by **10**.

Write down **one** other digit which you **know** must be on the calculator display.

b Claire starts again with the **same** 2 digit whole number.

This time she **multiplies** it by 100.

Write down all the digits that must be on the calculator display.

2 Kate is adding up the number of pupils in Year 9 and the number of pupils in Year 10.

| Year 9 | 127 pupils |
| Year 10 | 154 pupils |

Kate says that there is a total of 2711 pupils in Years 9 and 10.

a How can you tell that Kate's total is wrong **before** you work out the correct total?

This is Kate's work:

$$\begin{array}{r} 127 \\ + \ 154 \\ \hline = 2711 \end{array}$$

b What do you think Kate did wrong?

3 Chris goes shopping. He buys some fruit and vegetables.

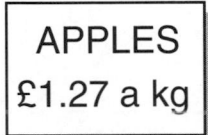

APPLES
£1.27 a kg

The shopkeeper says: 'The apples weigh 3 kg.'

a How much does Chris pay for the apples?

b How much do the potatoes weigh?

The shopkeeper says:
'The carrots weigh 2.8 kg.'

c How much does Chris pay for the carrots?

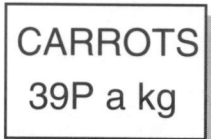

4 Stamps are **19 p** each.

Gwyn wants to buy **9** stamps.

He know that he will have to pay **less than £2**.

a Show how you can tell that he will have to pay less than £2 **without** working out the exact answer.

b Gwyn buys **9** stamps at **19 p** each.

Work out exactly how much he must pay. Do not use a calculator.

5 A class is planning a trip to a funfair.

The pupils have found out the prices at these two funfairs:

The teacher says: 'There will be time for 8 rides.'

a How much money do you need to get into Milltown Funfair and have **8** rides?

b How much money do you need to get into Seaview Funfair and have **8** rides?

Ben says: 'I've only got £5 to get in and pay for the rides.'

c How many rides would Ben get at each funfair?

6   Simon is growing vegetables in three vegetables patches.

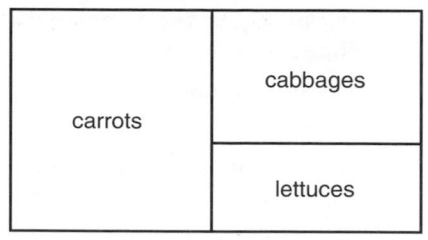

a   About **50%** of this vegetable patch is for **carrots**.

Copy the sentences below.

Fill in each gap with a **percentage**.

About ...... % of the patch is for **cabbages**.

About ...... % of the patch is for **lettuces**.

b   About $\frac{1}{8}$ of this vegetable patch is for **beetroot**.

Copy the sentences below.

Fill in each gap with a **percentage**.

About ...... % of the patch is for **broad beans**.

About ...... % of the patch is for **peas**.

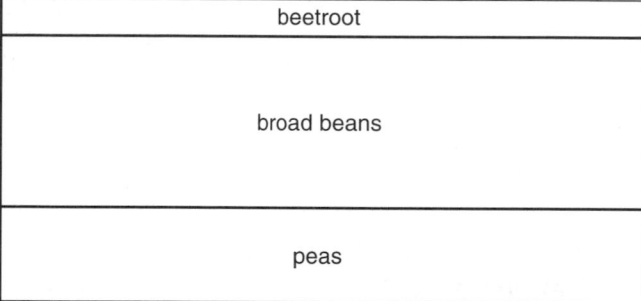

c   About $\frac{4}{5}$ of this vegetable patch is for **potatoes**.

Copy the diagram and draw a **straight line** to show how much of the patch is for potatoes. **Shade in** the area for potatoes.

The rest of the patch is for **turnips**.

About what fraction of the patch is for **turnips**?

7   There are **50** children altogether in a playgroup.

a   **How many** of the children are **girls**?

What **percentage** of the children are girls?

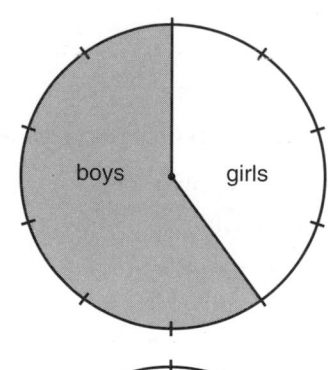

b   **25** of the children are **4 years old**.

20 of the children are **3 years old**.

5 of the children are **2 years old**.

Show this information on a copy of the diagram.

Label each part clearly.

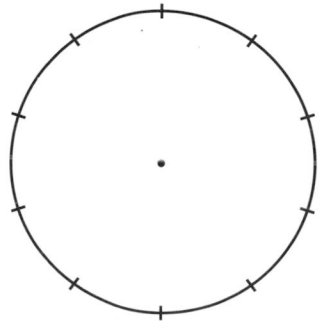

8 Some pupils are climbing up the ropes in the gym.

These are their positions after climbing for a few seconds.

a Dylan is about $\frac{1}{2}$ of the way **up** the rope.

Fill each gap with a fraction.

Lena is about ...... of the way up the rope.

John is about ...... of the way up the rope.

b Dylan is about 50% of the way **up** the rope.

Fill each gap with a percentage.

Mindu is about ...... % of the way up the rope.

Mary is about ...... % of the way up the rope.

Lena   John   Dylan   Mindu   Mary

c Anna is also climbing a rope.

She has climbed $\frac{2}{5}$ of the way up the rope.

Draw a rope and put a ✗ on to show Anna's position.

9a Look at this part of a number line.

What are the 2 missing numbers?

$$^{-}7 \quad ...... \quad 1 \quad 5 \quad 9 \quad ...... \quad 17$$

Finish this sentence:

The numbers on this number line go **up** in steps of ......

b This is a **different** number line.

What are the 3 missing numbers?

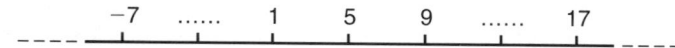

$$7.5 \quad 7.6 \quad 7.7 \quad 7.8 \quad ...... \quad ...... \quad ......$$

Finish this sentence:

The numbers on this number line go **up** in steps of ......

10a Here is a number chain:

$$2 \rightarrow 4 \rightarrow 6 \rightarrow 8 \rightarrow 10 \rightarrow 12 \rightarrow$$

The rule is: **add on 2 each time**

A different number chain is:

$$2 \rightarrow 4 \rightarrow 8 \rightarrow 16 \rightarrow 32 \rightarrow 64 \rightarrow$$

What could the rule be?

b

Some number chains start like this:

$1 \rightarrow 5 \rightarrow$

Show three **different** ways to continue this number chain.

For each chain, write down the **next three** numbers. Then write down the rule you are using.

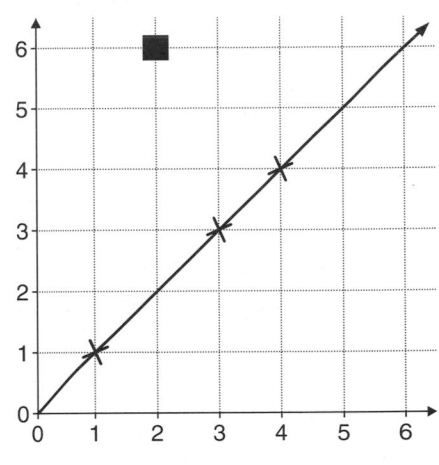

11a

Three points on this line are marked with X.

Their coordinates are:

(1, 1), (3, 3) and (4, 4).

Look at the **numbers** in the coordinates of each point.

What do you notice?

b

The point $\left(?, 14\frac{1}{2}\right)$ is **on** the line.

Fill in its missing coordinate.

c

The point ■ is **above** the line.

Four points are at (10, 10), (10, 12), (12, 10) and (12, 12).

Which **one** of these points is **above** the line?

Explain why.

d

The point (?, 15) is **above** the line.

Fill in a possible coordinate for the point.

e

Look at triangles A and B.

	Triangle A	Triangle B
Coordinates of ●	(4, 3)	(3, 4)
Coordinates of X	(2, 1)	(1, 2)
Coordinates of ■	(6, 2)	(2, 6)

Triangle A was reflected onto triangle B.

What happened to the **numbers** in the coordinates of each corner?

f

Elen wants to reflect the point (20, 13) in the mirror line.

What point will (20, 13) go to?

12

Andy has these number cards:

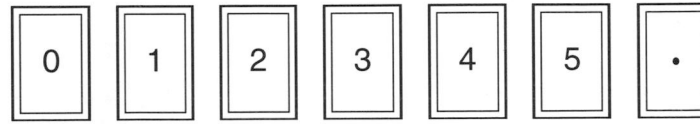

He made the number 42.5 with four of his cards.

a

Use some of Andy's cards to show the number **10 times** as big as 42.5.

b

Use some of Andy's cards to show the number **100 times** as big as 42.5.

13 Steve needs to know the answer to 214 × 92.

He works it out as 2354 but he thinks that he may have made a mistake.

a Make a **rough estimate** of 214 × 92 and show the numbers that you use.

b Compare your estimate with Steve's answer.

Do you think he made a mistake in his calculation? Say why.

c Now work out the exact answer to 214 × 92 without using a calculator.

14 Do not use a calculator to answer this question.

Gwen makes kites to sell. She sells the kites for **£4.75** each.

a Gwen sells **26** kites.

How much does she get for the **26** kites?

b Gwen has a box of **250** staples. She uses **16** staples to make each kite.

How many **complete** kites can she make using the **250** staples?

15a Bethan's class is collecting money for an old people's home. They want to get £1000 for a TV and video.

So far, they have got 40% of their target total.

Show this on a copy of the diagram.

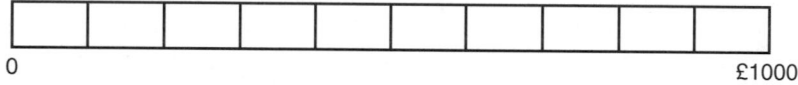

0   £1000

b Peter's class is also collecting money. They want to get £500.

Peter marks this diagram to show how much they have got so far.

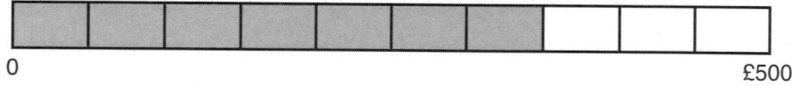

0   £500

What percentage of their target total have they got so far?

c How much money has Bethan's class collected?

Show how you work this out.

d How much money has Peter's class collected?

Show how you work this out.

e What percentage of £500 is the same as 40% of £1000?

Your answers to the other parts may help you.

16 Two shops usually sell the same *Go-Fast* trainers at the same price.

Today there is a sale at both shops.

Each shop has a different offer on *Go-Fast* trainers.

Jane's Sports Shop	Brian's Sports Shop

GO-FAST TRAINERS **30% Off**	**GO-FAST** TRAINERS $\frac{1}{3}$ **Off**

Which shop sells the cheaper *Go-Fast* trainers today?

Show your working.

17 The price of a computer has been reduced by 20% in a sale. It now costs £80.

Terri thinks that it cost £96 before the sale.

a Say why Terri is wrong.

b What was the correct price of the computer before the sale?

18 Emlyn is doing a project on world population.

He has found some data about the population of the regions of the world in 1950 and 1990.

Regions of the world	Population in 1950 (in millions)	Population in 1990 (in millions)
Africa	222	642
Asia	1558	3402
Europe	393	498
Latin America	166	448
North America	166	276
Oceania	13	26
World	2518	5292

a In **1950**, what percentage of the world's population lived in **Asia**?

Show each step in your working.

b In **1990**, for every person who lived in **North America** how many people lived in **Asia**?

Show your working.

c   For every person who lived in **Africa** in **1950**, how many people lived in **Africa** in **1990**?

Show your working.

d   Emlyn thinks that from **1950** to **1990** the population of **Oceania** went up by **100%**.

Is Emlyn right? Explain your answer.

19   The cost of an old toy vehicle depends on its condition and on whether it is in its original box.

Condition	Value
excellent, and in its box	100%
good, and in its box	85%
poor, and in its box	50%
excellent, but not in its box	65%
good, but not in its box	32%
poor, but not in its box	15%

A mail van in excellent condition, and in its box, costs **£125**.

a   How much is a mail van in **good** condition, and in its box?

b   How much is a mail van in **good** condition, **but not in its box**?

c   A petrol tanker in excellent condition, and in its box, costs £152.

Another petrol tanker should be sold for £98.80.

Using the chart above, what is its condition and does it have its box?

20   A clothes shop had a closing down sale.

The sale started on Tuesday and finished on Saturday.

For each day of the sale, prices were reduced by 15% of the prices on the day before.

A shirt had a price of £19.95 on Monday.

Kevin bought it on Wednesday.

How much did he pay?

Show your working.

21 Look at these number cards:

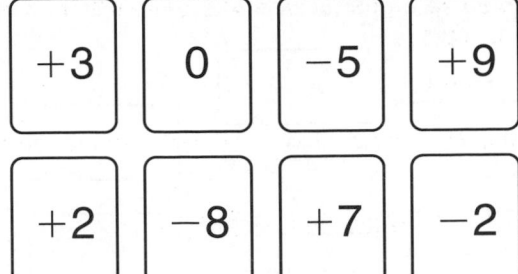

a Choose a card to give the answer 4.

$$\boxed{+3} + \boxed{-5} + \boxed{} = 4$$

b Choose a card to give the **lowest** possible answer, and work out the answer.

$$\boxed{-2} + \boxed{} = \dots\dots\dots\dots\dots\dots$$

c Choose a card to give the **lowest** possible answer, and work out the answer.

$$\boxed{-2} - \boxed{} = \dots\dots\dots\dots\dots\dots$$

d Now choose a card to give the **highest** possible answer, and work out the answer.

$$\boxed{-2} - \boxed{} = \dots\dots\dots\dots\dots\dots$$

22 Jo is planting a small orchard. She plants **cherry** trees, **plum** trees, **apple** trees and **pear** trees.
n stands for the number of **cherry** trees Jo plants.

a Jo plants the same number of **plum** trees as **cherry** trees.
How many **plum** trees does she plant?

b Jo plants **twice** as many **apple** trees as **cherry** trees.
How many **apple** trees does she plant?

c Jo plants **7 more pear** trees than **cherry** trees.
How many **pear** trees does she plant?

d How many trees does Jo plant **altogether**?
Write your answer as simply as possible.

23 Here are some algebra cards:

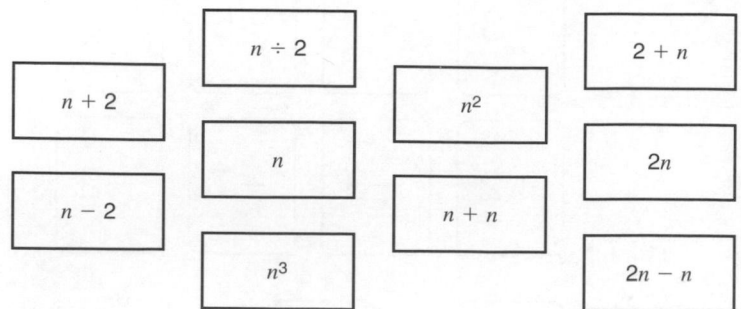

a One of the cards will always give the same answer as $\frac{n}{2}$

Which card is it?

b One of the cards will always give the same answer as $n \times n$

Which cards is it?

c **Two** of the cards will always give the same answer as $2 \times n$

Which cards are they?

d Write a **new** card which will always give the same answer as

$3n + 2n$

24 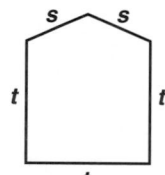 The perimeter of this shape is $3t + 2s$.

$$p = 3t + 2s$$

Write an expression for the perimeters of each of these shapes.
Write each expression in its simplest form.

a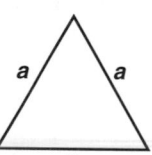

b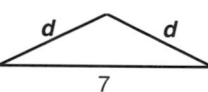

c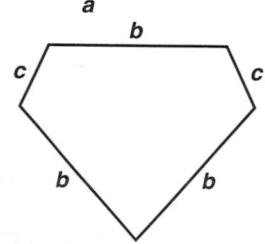

d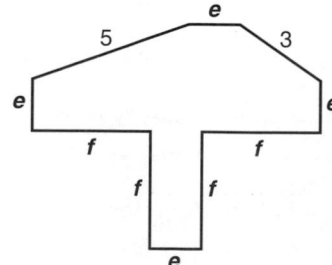

25 In these walls, each brick is made by **adding** the **two** bricks underneath it.

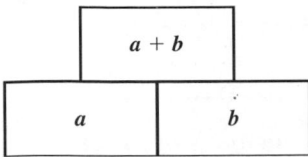

a Write an expression for the top brick in this wall.

Write your expression as simply as possible.

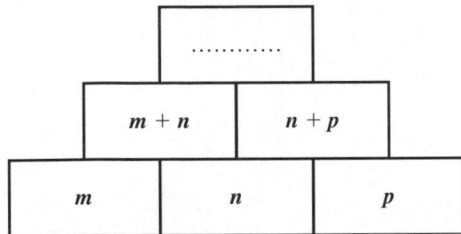

b Fill in the missing expressions on these walls.

Write your expression as simply as possible.

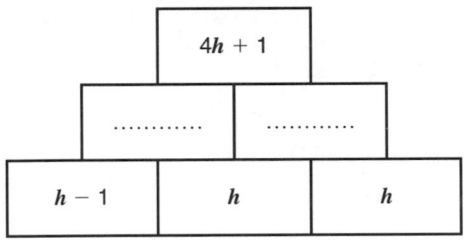

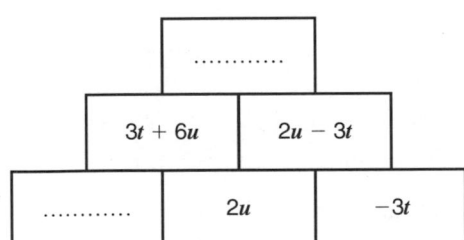

c In the wall below, *h*, *j* and *k* can be any whole numbers.

Explain why the top brick of the wall must **always** be an **even** number.

You can fill in the missing expressions if you want to.

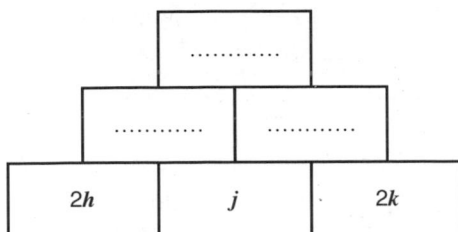

26 This is a function machine.

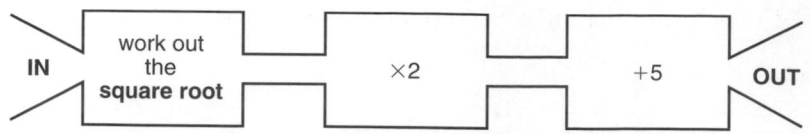

a Copy this diagram and fill in the spaces to find the numbers which come out.

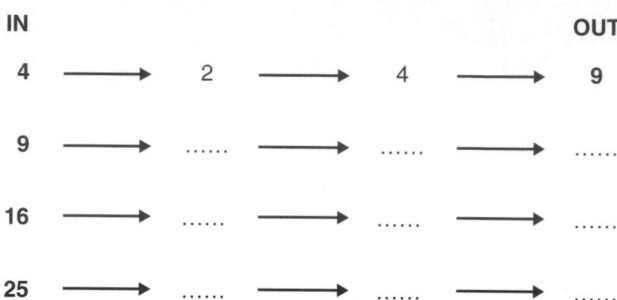

b **Two** numbers in the OUT column are **prime numbers**.

Write these numbers.

27 Ice creams are sold as cones or tubs at the Beach Kiosk.

A cone costs 60 pence. A tub costs 40pence.

The income (F) in pence of the beach Kiosk can be calculated from the equation

$$F = 60x + 40y$$

where x is the number of cones sold and y is the number of tubs sold.

a On 1 June 1997, $x = 65$ and $y = 80$.

Work out the income. Show your working.

b On 2 June 1997, $F = 4800$ and $x = 50$.

Work out how many tubs were sold. Show your working.

c During the first week of summer 1997, 950 ice creams were sold. 437 of them were tubs.

What percentage of the ice creams sold were tubs?

d **Estimate** the total income in pounds for the summer of 1998 using the information in the box.

Write down the number you will use instead of 14 723.

Write down the value you will use for the cost of an ice cream.

Write down your estimate of the total income for the summer of 1998.

> 14 723 ice creams were sold in the summer of 1997.
>
> Roughly the same number of ice creams is likely to be sold in the summer of 1998.
>
> The ratio of cones to tubs sold is likely to be about 1 : 1
>
> The cost of a cone is to stay at 60 pence.
>
> The cost of a tub is to stay at 40 pence.

28 Karen tries to find the square root of 5.

She can use her calculator but she **must not** use the $\sqrt{}$ key.

> *1st try* $2 \times 2 = 4$
> *2nd try* $2.4 \times 2.4 = 5.76$
> *3rd try* $2.2 \times 2.2 = 4.84$

This is called 'trial and improvement.'

a Continue Karen's trials.

Use at least 4 more sensible trials.

Try to get as close to 5 as you can.

b Solve this equation:

$$x^2 + 4 = 9$$

You can use what you did in part **a** to help you.

29 Lucy is investigating areas and perimeters of shapes.

She makes a **square** with a perimeter of 24 cm.

a Calculate the area of her square.

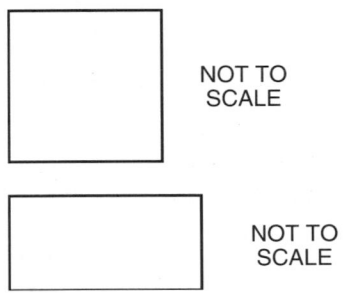

NOT TO SCALE

Lucy makes a rectangle with a perimeter of 24 cm.

The **length** is **twice** the **width**

b Calculate the area of her rectangle.

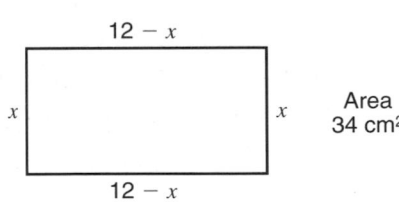

NOT TO SCALE

c Lucy makes a different rectangle with an area of 34 cm².

The sides of Lucy's rectangle are x cm and $(12 - x)$ cm.

She wants to find a value of x so that $x(12 - x) = 34$.

Between which **one decimal place** numbers does x lie?

You may use a table like this:

$12 - x$

x x Area 34 cm²

$12 - x$

x	$12 - x$	Area
3	9	27

30 Rosemary drew these squares using a computer:

The sides of the squares are in the ratio $1:3:6:9$.

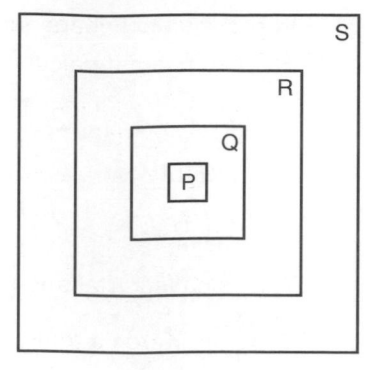

a The side of square Q is 12 mm.

Use the ratios above to find the sides of squares P, R and S.

b Write this ratio in its **simplest form**:

Side of square R : Side of square S

c Write the ratio

Perimeter of square R : perimeter of square S

d Compare the ratios in parts **b** and **c**.

What do you notice about the ratios?

31a One morning last summer, Ravi carried out a survey of the birds in the school garden.

He saw 5 pigeons, 20 crows, 25 seagulls and 45 sparrows.

Copy and complete the line below to show the ratios.

Pigeons : Crows : Seagulls : Sparrows

1 : ...... : ...... : ......

b What percentage of all the birds Ravi saw were sparrows?

c One morning this spring Ravi carried out a second survey. This time he saw:

the same number of pigeons

25% fewer crows

60% more seagulls

two thirds of the number of sparrows

Copy and complete the line below to show the ratios for the second survey

Pigeons : Crows : Seagulls : Sparrows

1 : ...... : ...... : ......

32 This is a **square tile**.

The edge of the tile is n centimetres long.

The **perimeter** of the tile is **4n** centimetres.

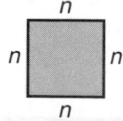

This **T-shape** is made with **6 square tiles**.

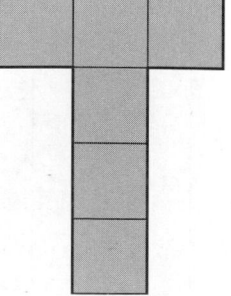

a Write an **expression** for the **perimeter** of the **T-shape**.

The expression should be a **number multiplied by** n.

b The perimeter of the T-shape is **28** centimetres.

Use your expression from part **a** to write an **equation** involving n.

Solve your equation to find the **value** of n.

33 On a farm many years ago, the water tanks were filled using a bucket from a well.

a The table shows the numbers of buckets, of different capacities, needed to fill a tank of capacity 2400 pints.

Copy and complete the table.

Capacity of bucket (pints)	8	10	12	15	16		
Number of buckets			200		150	100	80

b Write an equation using symbols to connect **T**, the capacity of the tank, **B**, the capacity of a bucket, and **N**, the number of buckets.

c Now tanks are filled through a hosepipe connected to a tap. The rate of flow through the hosepipe can be varied.

A tank of capacity **4000** litres fills at a rate of **12.5** litres per minute.

How long in hours and minutes does it take to fill the tank?

Show your working.

d Another tank took **5 hours** to fill at **a different rate** of flow.

How long would it have taken to fill this tank if this rate of flow had been increased by **100%**?

34

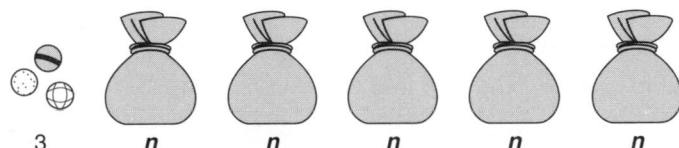

$3 \quad n \quad n \quad n \quad n \quad n$

A teacher has 5 bags of marbles and 3 extra marbles. Each bag has n marbles inside.

She asks: 'How many marbles are there altogether?'

The pupils say:

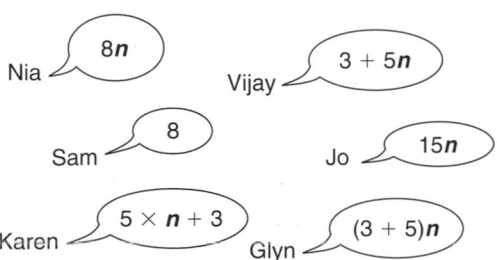

Nia $8n$

Vijay $3 + 5n$

Sam 8

Jo $15n$

Karen $5 \times n + 3$

Glyn $(3 + 5)n$

There are 88 marbles altogether.

Use one pupil's right answer to write down an equation.

Solve the equation to find n, the number of marbles in a bag.

35 Four people each have a card.

a Work out the value for *b* which will make Ann and Rob's cards worth the **same** as each other.

b There are two people whose cards will **never** be worth the same as each other.

Who are they? Explain your answer.

36 Four people play a game with counters.

Each person starts with one or more bags of counters.

Each bag has *m* counters in it.

The table shows what happened during the game. Copy the table.

a Write an expression in the table to show what **Cal** and **Fiona** had at the end of the game.

Write each expression **as simply as possible**.

	Start	During game	End of game
Lisa	3 bags	lost 5 counters	$3m - 5$
Ben	2 bags	won 3 counters	$2m + 3$
Cal	1 bag	lost 2 counters	
Fiona	4 bags	won 6 counters, and lost 2 counters	

b At the end of the game, **Lisa** and **Ben** had the **same** number of counters.

Write an **equation** to show this.

c Solve the equation to find *m*, the number of counters in each bag at the start of the game.

37 Lucy was investigating straight **lines** and their **equations**.

She drew these lines.

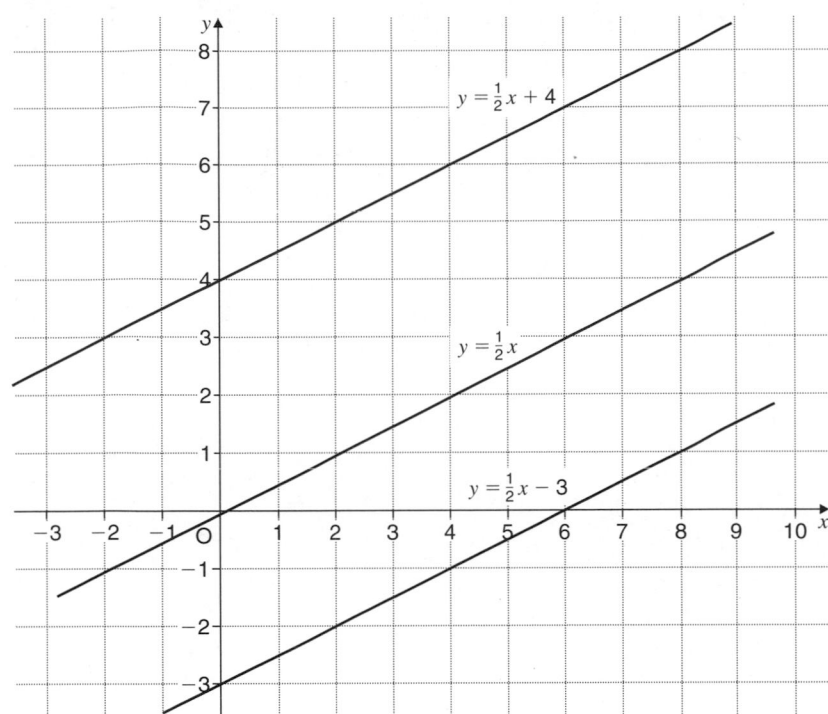

a $y = \frac{1}{2}x$ is in each equation.

Write one fact this tells you about all the **lines**.

b   The lines cross the y-axis at $(0, -3)$, $(0, 0)$ and $(0, 4)$.

Which part of each **equation** helps you to see where the line crosses the y-axis?

c   Lucy decided to investigate more lines. She needed longer axes.

Where will the line $y = \frac{1}{2}x - 20$ cross the y-axis?

d   Write down the equation of another line on the graph which is parallel to $y = \frac{1}{2}x$.

38a Draw the graph of the straight line $y = 2x$.

Label your line $y = 2x$.

b   Write the equation of another straight line which goes through the point $(0, 0)$.

c   The straight line with the equation $y = x - 1$ goes through the point $(4, 3)$.

Draw the graph of the straight line $y = x - 1$.

Label your line $y = x - 1$.

d   Write the equation of the straight line which goes through the point $(0, -1)$ and is **parallel** to the straight line $y = 3x$.

39 Look at this octagon:

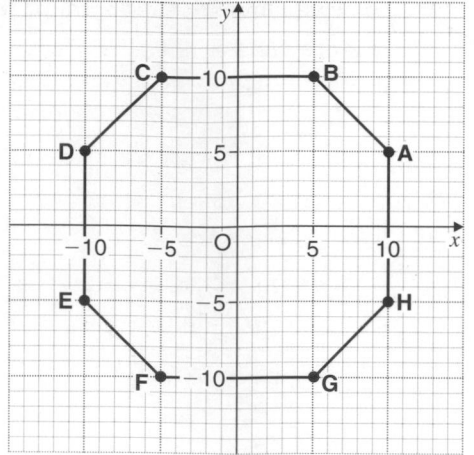

a The line through A and H has the equation $x = 10$.
What is the equation of the line through **F** and **G**?

b Fill in the gaps below:

$x + y = 15$ is the equation of the line
through ...... and ......

c The octagon has four lines of symmetry.

One of the lines of symmetry has the
equation $y = x$.

On a copy of the diagram, draw and **label**
the line $y = x$.

d The octagon has three **other** lines of symmetry.

Write the equation of **one** of these three **other**
lines of symmetry.

e The line through D and B has the equation $3y = x + 25$.

The line through G and H has the equation $x = y + 15$.

Solve the simultaneous equations

$$3y = x + 25$$
$$x = y + 15$$

Show your working.

f Complete this sentence:

The line through D and B meets the line through G and H
at (......, ...)

40 David is studying blood cells through a microscope.

The diameter of a red cell is 0.000 714 cm and the diameter of a white
cell is 0.001 243 cm.

a Use a calculator to work out the difference between the diameter of a
red cell and the diameter of a white cell.

Give your answer in **millimetres**.

David wants to explain how small the cells are.

He calculates how many white cells would fit across a full stop which
has a diameter of 0.65 mm.

b How many whole white cells would fit across the full stop?

41 Bill, Ravi and Eric are three divers in a competition.

Each type of dive has a **dive rating**.

Easy dives have a **low** rating. **Hard** dives have a **high** rating.

Every dive is marked by five judges who each give a **mark out of 10**.

How to calculate the score for a dive:

1 Look at all five marks. Remove the highest and the lowest marks.
2 Add together the middle three marks to give a total.
3 Multiply this total by the dive rating.

a Bill does a dive with a dive rating of 3.34.

The judges give the marks   7.0   7.5   8.0   8.0   8.5

What is Bill's score?

b Ravi scored 82.68 on his first dive.

The dive had a dive rating of 3.18

What was the **total** of the middle three marks given by the judges?

c Eric is getting ready to take his final dive.

He needs to score at least 102.69 to win the competition.

Eric decides to do a dive with a dive rating of 3.26

Explain why Eric has made a poor decision.

Show your working.

42 For each of these cards, n can be any positive number.

The **answers** given by the cards are all positive numbers.

$$\boxed{n^2} \quad \boxed{0.8n} \quad \boxed{\sqrt{n}} \quad \boxed{\dfrac{n}{0.8}} \quad \boxed{\dfrac{1}{n}}$$

a Which card will **always** give an answer **less than** n?

b When n **is 1**, which cards will give the answer **1**?

b When n **is 4**, which cards will give an answer **less than 4**?

43 Rectangles with length and width in the special

ratio $1 : \dfrac{2}{1 + \sqrt{5}}$ are called golden rectangles.

Some artists use them because the proportions look attractive.

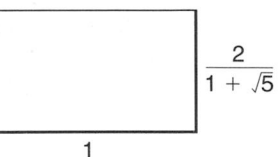

a Work out $\dfrac{2}{1 + \sqrt{5}}$.

Write the answer using all the decimal places your calculator shows.

b Ramesh used his calculator efficiently to work out $\dfrac{2}{1 + \sqrt{5}}$.

He did **not** have to write down any numbers and then key them back into his calculator.

Show the steps he might have used with his calculator.

c   Four faces of this cuboid are golden rectangles.
The volume of the cuboid is

$$\left(\frac{2}{1+\sqrt{5}}\right) \times \left(\frac{2}{1+\sqrt{5}}\right) \times 1$$

Use a **short** and **accurate** method on your calculator to calculate this.

Write the volume using all the decimal places your calculator shows.

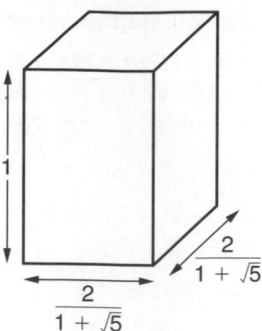

d   Show the steps you used to make your method **short** and **accurate**.

44   Bryn wants to use the formulae

$$P = s + t + \frac{5\sqrt{s^2 + t^2}}{3} \quad \text{and} \quad A = \tfrac{1}{2}st + \frac{(s^2 + t^2)}{9}$$

to work out the perimeter (P) and area (A) of shapes like this:

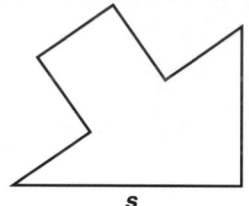

For this shape, Bryn substitutes $s = 4.5$ and $t = 6$ into the formulae:

a   Work out the values of:

$$4.5 + 6 + \frac{5 \times \sqrt{4.5^2 + 6^2}}{3} \quad \text{and} \quad \tfrac{1}{2} \times 4.5 \times 6 + \frac{(4.5^2 + 6^2)}{9}$$

For this shape, Bryn substitutes $s = 1.7$ and $t = 0.9$ into the formulae:

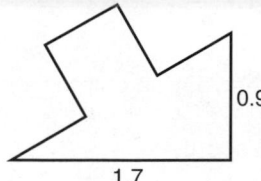

a   Work out the values of:

$$1.7 + 0.9 + \frac{5 \times \sqrt{1.7^2 + 0.9^2}}{3} \quad \text{and} \quad \tfrac{1}{2} \times 1.7 \times 0.9 + \frac{(1.7^2 + 0.9^2)}{9}$$

45 This is a series of patterns with grey and black tiles:

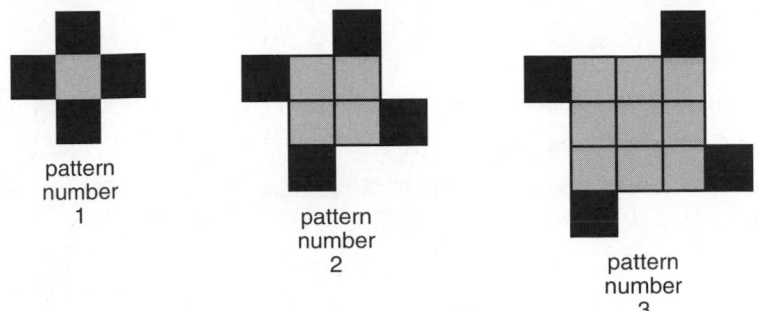

pattern
number
1

pattern
number
2

pattern
number
3

a How many grey tiles and black tiles will there be in pattern number 8?

b How many grey tiles and black tiles will there be in pattern number 16?

c How many grey tiles and black tiles will there be in pattern number P?

d T = total number of grey tiles and black tiles in a pattern

P = pattern number

Use symbols to write an equation connecting T and P.

46 Class 9H were playing a number game.

Elin said:

'Multiplying my number by 4 and then subtracting 5 gives the same answer as multiplying my number by 2 and then adding 1.'

a Lena called Elin's number x and formed an equation:

$4x - 5 = 2x + 1$

Solve this equation and write down the **value** of x.

Show your working.

Aled said:

'Multiplying my number by 2 and then adding 5 gives the same answer as subtracting my number from 23.'

b Call Aled's number y and form an equation:

Work out the **value** of Aled's number.

c Lena thought of two numbers, which she called a and b.

She wrote down this information about them in the form of equations:

$a + 3b = 25$

$2a + b = 15$

Work out the values of a and b.

Show your working.

Shape, space and measures

1   These patterns come from Egypt.

The first pattern looks the same after **part** of a turn.

It will look the same in **4** different positions.

In how many positions will each of these patterns look the same?

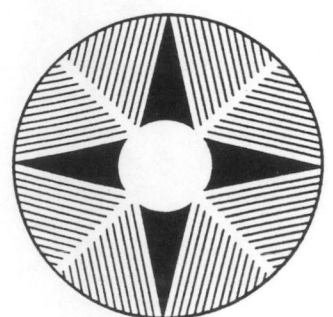

a

b

c

d

2   Nina is making Rangoli patterns.

To make a pattern, she draws some
lines on a grid.

Then she reflects them in a mirror line.

For example:

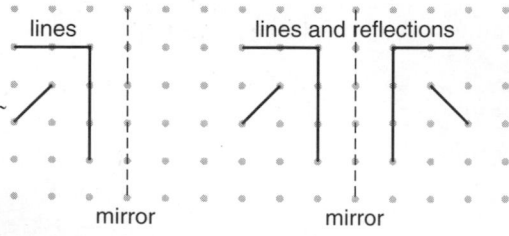

Reflect each group of lines in its mirror line to make a pattern.
You may use a real mirror or tracing paper to help you.

a

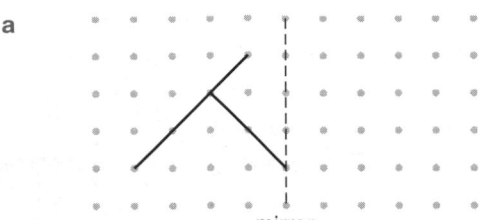

mirror

b

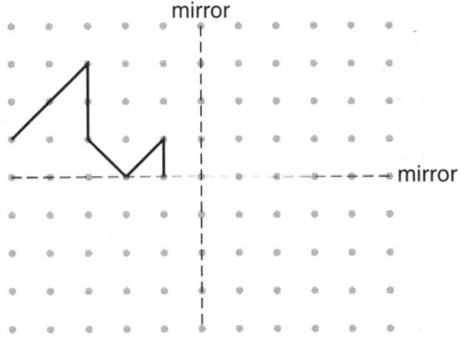

mirror

c   Now use two mirror lines to make a pattern.

First reflect the group of lines in one mirror line to make a pattern.
Then reflect the whole pattern in the other mirror line.

mirror

mirror

3a The scale shows how long
    Laura was when she was born.

    How long was Laura?

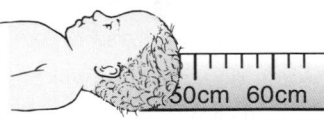

b   When Laura was one month old,
    she was put on the scales.

    What mass do the scales show?

c   Now Laura is older.

    She is **1.03 m** tall.

    Write Laura's height in centimetres.

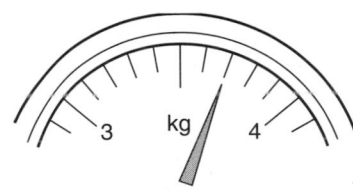

4 Pepe is working out the size of the classroom.
He counts his paces along each side.

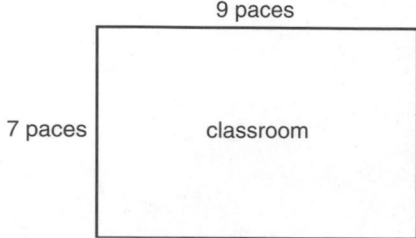

9 paces

7 paces   classroom

a How far is it all round the classroom?
He measures **one** pace. It is 85 cm.

b Work out the length and width of the classroom in centimetres.
Do not use a calculator.

5a Carl is putting packs of biscuits into a box.
He starts to put in the bottom layer.
The box holds **5 packs across** and is
4 packs wide.
How many packs will fit altogether
on the bottom layer?

b The box holds **6 layers**.
How many packs will fit in the box when
it is **full**?

c Aziz is putting packs of tea into a box.
The box holds **5 packs across** and is
6 packs wide.
The box holds **3 layers**.
How many packs of tea will fit in the
box when it is **full**?

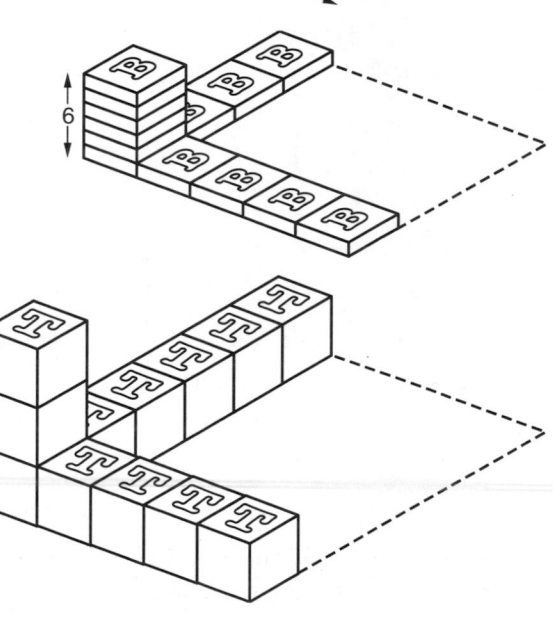

d Fill in the gaps below to show one way of filling a **different** box with **24**
packs in **2** layers.

Total: 24 packs   2 layers

...... packs across   ...... packs wide

6   Darren has a large empty matchbox.

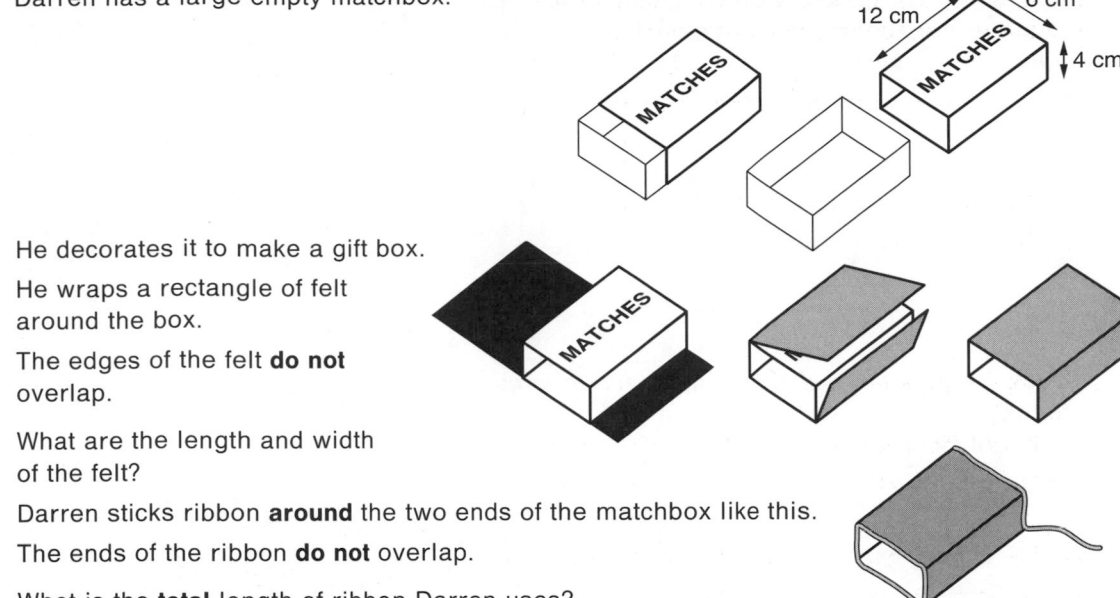

12 cm   6 cm
4 cm

He decorates it to make a gift box.

He wraps a rectangle of felt around the box.

The edges of the felt **do not** overlap.

a   What are the length and width of the felt?

Darren sticks ribbon **around** the two ends of the matchbox like this.

The ends of the ribbon **do not** overlap.

b   What is the **total** length of ribbon Darren uses?

7   Some of these nets can be folded to make cuboids.

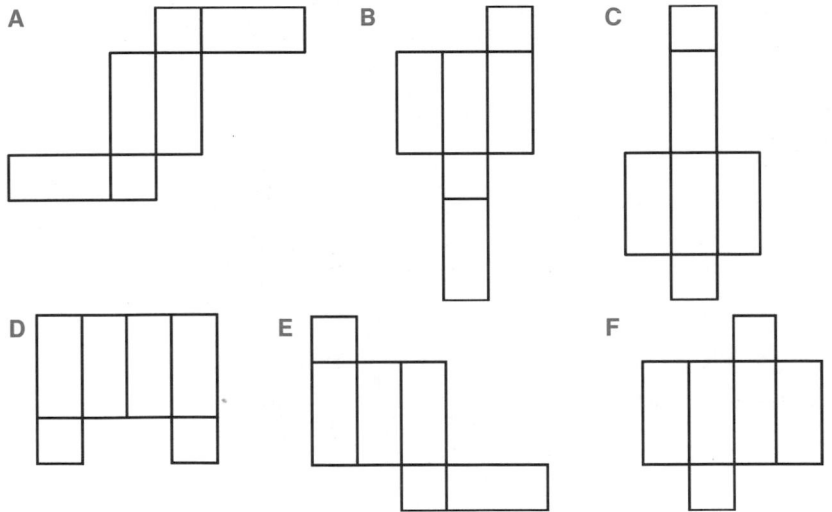

A   B   C

D   E   F

a   Which nets **can** be folded to make cuboids?

b   Choose **two** nets which **cannot** be folded to make cuboids.
    Explain why **one** of the nets cannot be folded to make a cuboid.
    You can write your explanation, or show it on a diagram.
    Say which net you have chosen.

c   Explain why the **other** net cannot be folded to make a cuboid.
    You can write your explanation, or show it on a diagram.
    Say which net you have chosen.

8a Copy the diagram and shade in **2 more squares** to make a shape which has the dashed line as a **line of symmetry**.

You may use a mirror or tracing paper to help you.

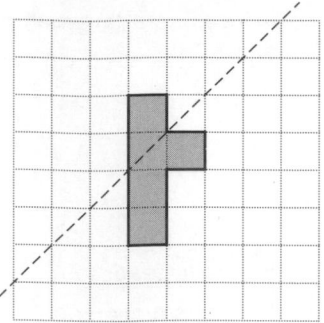

b Copy the diagram and shade in **2 more squares** to make a shape which has the dashed line as a **line of symmetry**.

You may use a mirror or tracing paper to help you.

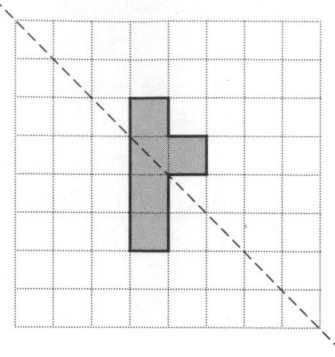

9 Each measurement below can be given in different units.

Fill in the 2 units which would be used.

Choose your units from this list: **feet   kilometres   metres   miles**

a The height of a girl is 1.53 .... or 5 ....

b The distance from Cardiff to Birmingham is 196 .... or 122 ....

10 The same measurement is sometimes given in different units.

Write the correct units below.

Choose your units from this list: **grams   litres   pints   pounds**

a Volume of milk: 2 ....   or   1.13 ....

b Weight of jam: 1 ....   or   454 ....

11 This is the sign in a lift.

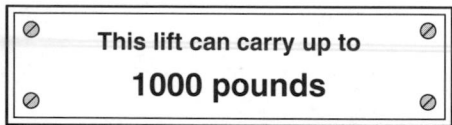

This lift can carry up to
1000 pounds

Six people want to use the lift. They weigh:

65 kg   85 kg   114 kg   72 kg   93 kg   79 kg

One kilogram is just over two pounds.

Can they all go in the lift at the same time?

Show how you work this out.

12 Julie has written a computer program to transform pictures of tiles.

There are **only two instructions** in her program:

Reflect vertical   or   **Rotate 90° clockwise**

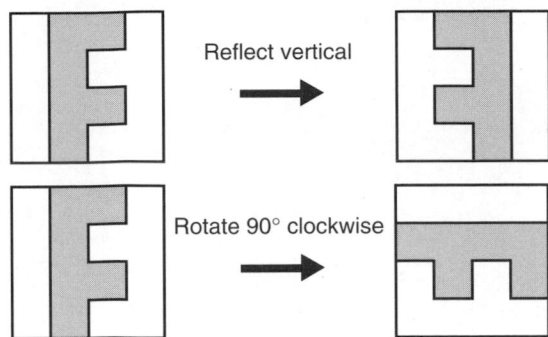

a Julie wants to transform the first pattern to the second pattern.

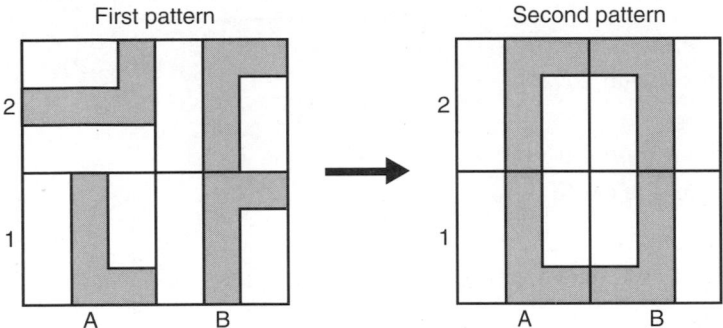

Copy and complete the instructions to transform tiles B1 and B2.

You must use only **Reflect vertical** or **Rotate 90° clockwise**.

    A1 *Tile is in the correct position.*

    A2 *Reflect vertical, and then rotate 90° clockwise.*

    B1 *Rotate 90° clockwise, and then . . . .*

    B2 *. . . . .*

b Paul starts with the first pattern that was on the screen.

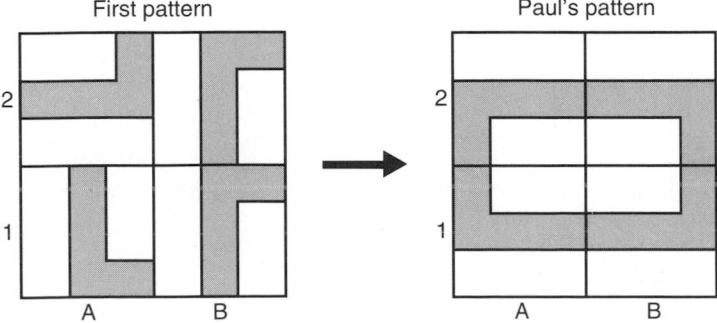

Complete the instructions for the transformations of tiles A2, B1 and B2 to make Paul's pattern.

You must use only **Reflect vertical** or **Rotate 90° clockwise**.

A1 *Reflect vertical, and then rotate 90° clockwise.*

A2 *Rotate 90° clockwise, and then* . . . .

B1 . . . .

B2 . . . .

13   Julie wants to make a card with a picture of a boat.

She makes a rough sketch of the boat.

It is made out of a triangle and a trapezium.

Make an accurate full-sized drawing of the **triangle** for the sail and the **trapezium** for the boat.

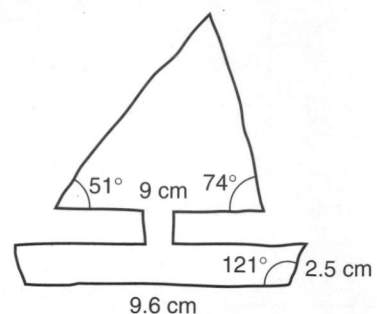

14   Ursula made a solid letter U with seven plastic cubes.

She drew this picture of the U on a grid.

She did not draw any edge she could not see.

a   Ursula then turned the U so that it was upside down.

Draw on a grid what the solid looks like.

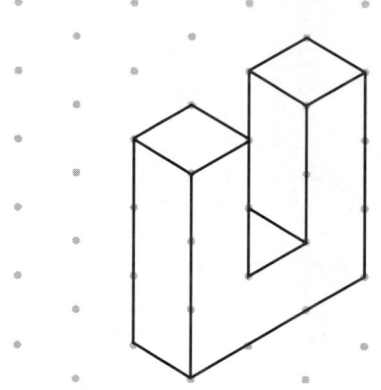

b   Ursula wants to draw solid letters in a row so that they make her initials UIT.

Complete UIT on a grid.

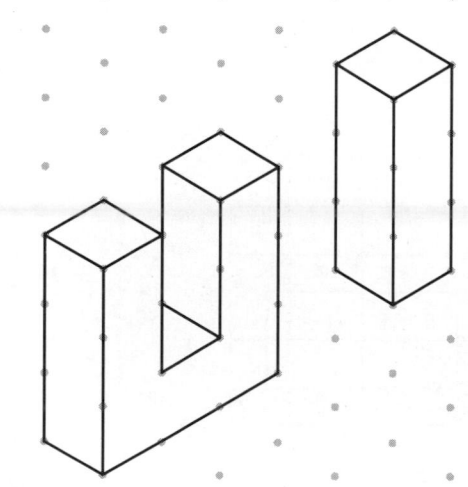

15 Kay is drawing shapes on her computer.

a She wants to draw this triangle.
She needs to know angles a, b and c.

Calculate angles a, b and c.

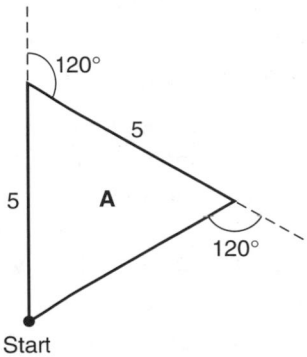

Not to scale

b Kay draws a rhombus:

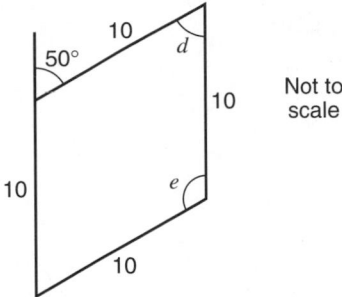

Not to scale

Calculate angles d and e.

c Kay types the instructions to draw a regular pentagon:

Repeat 5 [forward 10, left turn 72]

Complete the instructions to draw a regular hexagon:

Repeat 6 [forward 10, left turn . . . .]

16 Shape A is an equilateral triangle.
The instructions to draw shape A are:

Forward 5

Turn right 120°

Forward 5

Turn right 120°

Forward 5

a Write instructions to draw a triangle that has sides double those of shape A.

Shape B is a parallelogram.

b Complete the instructions to draw shape B:

Forward 8

. . . . . .
. . . . . .
. . . . . .
. . . . . .
. . . . . .
. . . . . .

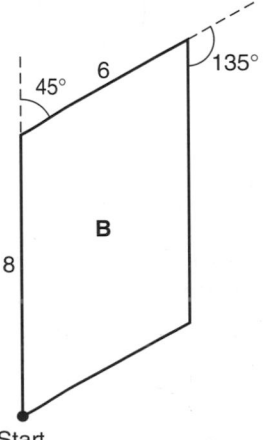

17 Wyn and Jay are using their wheelchairs to measure distances.

a The large wheel on **Wyn's** wheelchair has a **diameter** of **60 cm**.

Wyn pushes the wheel round **exactly** once.

Calculate how far Wyn has moved.

Show your working.

b The large wheel on **Jay's** wheelchair has a **diameter** of **52 cm**.

Jay moves her wheelchair forward **950 cm**.

Calculate how many times the large wheel goes round.

Show your working.

18 In a competition all boats have sails the same size.

a Work out the area of the sail.

Show your working.
Some boats have advertisements on their sails.

The competition has this rule:

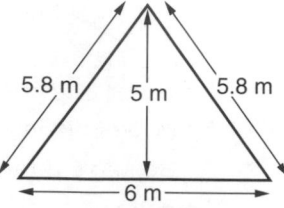

> Advertisements must cover **less** than 20% of the sail.

b The advertisement on one sail covers exactly 20% of the sail.

Work out the area of the advertisement.

Show your working.

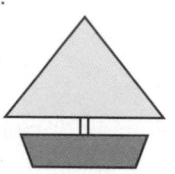

c Work out the area of this advertisement to see if it would be allowed on the sail.

Show your working.

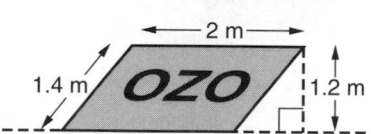

d Work out the area of this advertisement.

You will need to remember a formula to help you answer this.

Use 3.14 or your calculator key for π.

Show your working.

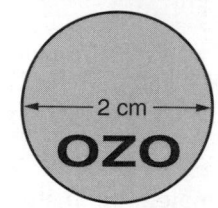

19a What is the volume of this **standard size** box of salt?

b What is the volume of a **special offer** box of salt, which is **20% bigger?**

The **standard size** box contains enough salt to fill up **10** salt pots.

c How many salt pots may be filled up from the **special offer** box of salt?

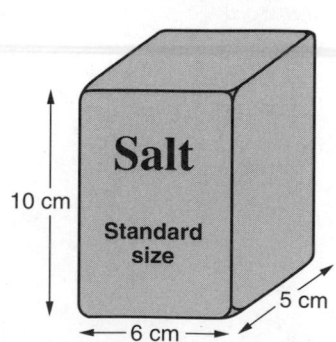

20 A children's paddling pool is in the shape of a cuboid.

It is 14.6 m long and 6.7 m wide.

It has been emptied for cleaning.

20 m³ of water is poured into the pool.

a How deep is the water?

Show your working.

b Explain why your answer to part **a** is a **sensible** depth of water for the pool.

21 A boat sails from the harbour to a buoy.

The buoy is 6 km to the east and 4 km to the north of the harbour.

Calculate the shortest distance between the buoy and the harbour.

Give your answer to 1 decimal place.

Show your working.

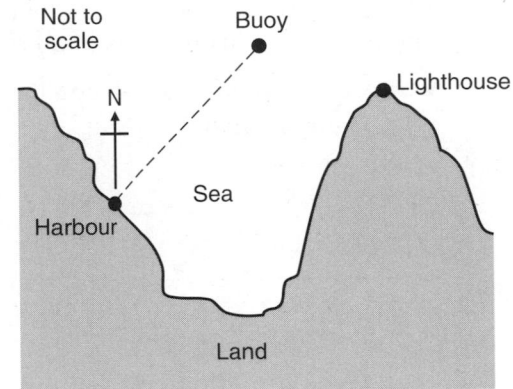

22 Calculate the area of this triangle.

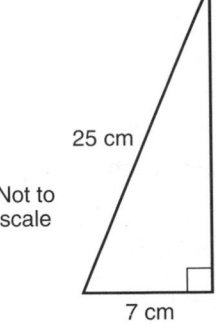

25 cm

Not to scale

7 cm

23 TJ's Cat Food is sold in tins shaped like this.

Each tin has an internal height of 5 cm.

a The area of the lid of the tin is 35 cm².

Work out the volume of cat food that the tin contains.

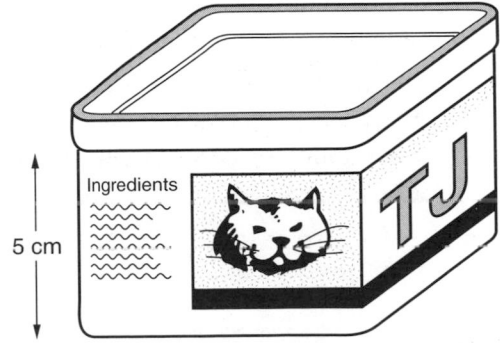

b The label that goes around the tin overlaps by 1 cm.

The area of the label is 134 cm².

Work out the distance around the tin.

Show your working.

TJ's Cat Food plans to use tins that are the shape of cylinders.
The internal measurements of a tin are shown.

c Work out the volume of cat food that the tin contains.

Show your working.

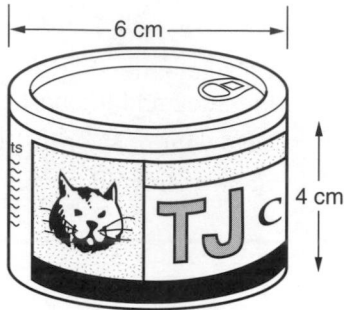

24 The California Beach leisure complex has a swimming pool.

One whole side of the pool is a viewing window.

You can see the swimmers under the water.

The diagram opposite is a sketch of the window. It is not drawn to scale.

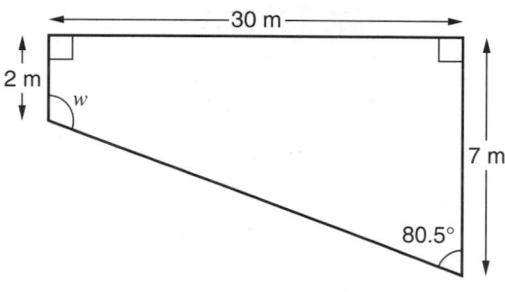

a Work out the angle (*w*) in the window.

b What facts about angles did you use?

c Work out the area of the window.

Show your working.

d The length of the pool is 30 metres and the width of the pool is 12 metres.

Work out the **volume** of water you would need to fill the pool right up to the top.

Show your working.

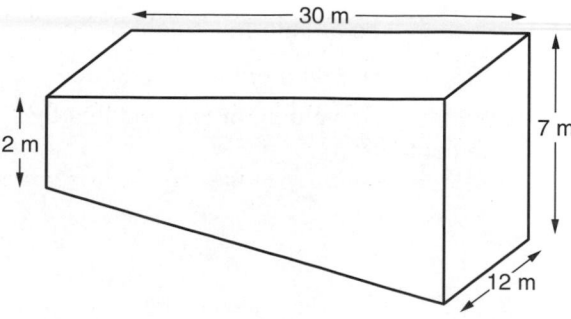

25 Two goats, A and B, are tied to opposite sides of a shed.

They both have chains 5 m long.

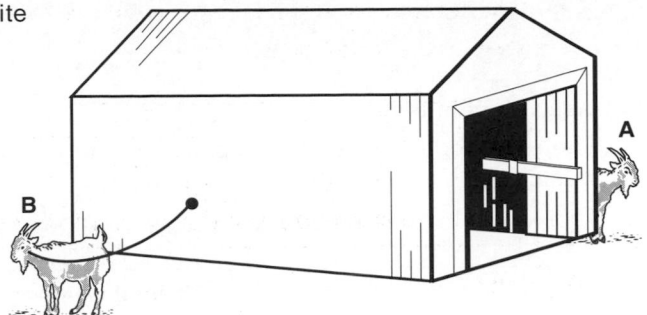

a Copy this plan and sketch the area that **goat A** can reach.

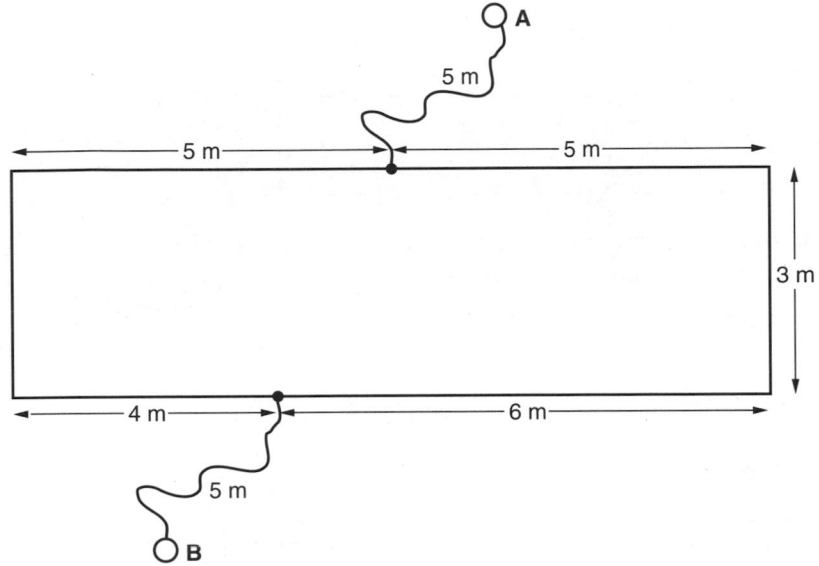

b On the same plan, sketch the area that **goat B** can reach.

26 At an athletics meeting, the discus throws are measured to the nearest centimetre.

a Viv's best throw was measured as 35.42 m.

Could Viv's throw actually have been more than 35.42 m?

Explain your answer.

b Chris won the hurdles race in a time of 14.6 seconds, measured to the nearest tenth of a second.

Between what two values does Chris's time actually lie?

Handling data

1a Lisa works in a shoe shop.

She recorded the size of each pair of trainers that she sold during a week.

This is what she wrote down.

	Sizes of trainers sold						
Monday	7	7	5	6			
Tuesday	6	4	4	8			
Wednesday	5	8	6	7	5		
Thursday	7	4	5				
Friday	7	4	9	5	7	8	
Saturday	6	5	7	6	9	4	7

Using a **tallying method** to make a table showing how many pairs of trainers of each size were sold during the whole week.

b Which size of trainer did Lisa sell the **most** of?

c Lisa said: 'Most of the trainers sold were bigger than size 6.'

How can you tell from **your** table that Lisa is **wrong**?

2a Joe has these cards:

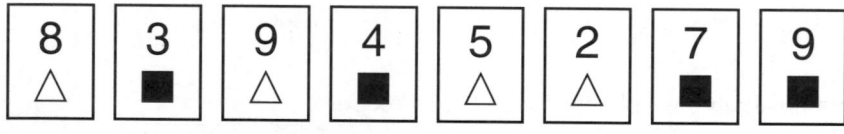

Sara takes a card without looking.

Joe says: 'On Sara's card, ■ is more likely than △.'
Explain why Joe is **wrong**.

Here are some words and phrases:

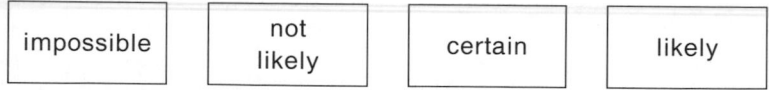

Choose a word or phrase to fill in the gaps below.

It is . . . . that the number on Sara's card will be **smaller than 10**.

It is . . . . that the number on Sara's card will be an **odd number**.

b Joe still has these cards:

| 8 △ | 3 ■ | 9 △ | 4 ■ | 5 △ | 2 △ | 7 ■ | 9 ■ |

He mixes them up and puts them face down on the table.

Then he turns the first card over, like this:

| 5 △ | | | | | | |

Joe is going to turn the next card over.

Complete this sentence:

On the next card, ... is **less likely** than ....

The number on the next card could be higher than 5 or lower than 5.

Which is **more likely**?

---

3 Kim has 5 dogs.

Their ages are: 3 years   3 years   4 years   5 years   10 years

The **mean** age of the dogs is 5.

The **median** age of the dogs is 4.

The **mode** is 3.

Arlene also has 5 dogs.

The mean, mode and median ages of her dogs are the **same** as Kim's dogs.

Arlene's dogs are **not all** the same ages as Kim's dogs.

What ages could they be?

4a James has these four number cards: 1, 8, 5 and 2.

The **mean** is 4.

James takes another card.

The mean of the **five** cards is still 4.

What number is on his new card?

| 1 | 8 | 5 | 2 |

b Tara has these four number cards: 10, 3, 2 and 5.

She takes another card.

The mean goes **up** by **2**.

What number is on her new card?

Show your working.

| 10 | 3 | 2 | 5 |

c   Ali has six cards: 10, 10, 10, 10, ? and ?

The **mean** of the six cards is **10**.

The **range** of the six cards is **4**.

What are the numbers on the other two cards?

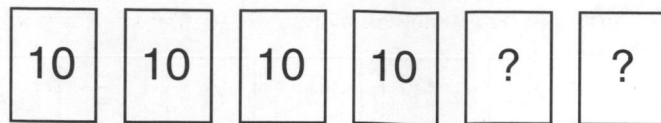

5   Nikos and Sharon are planning to go on a cycling holiday in October. They are trying to choose where to go. They have the following information.

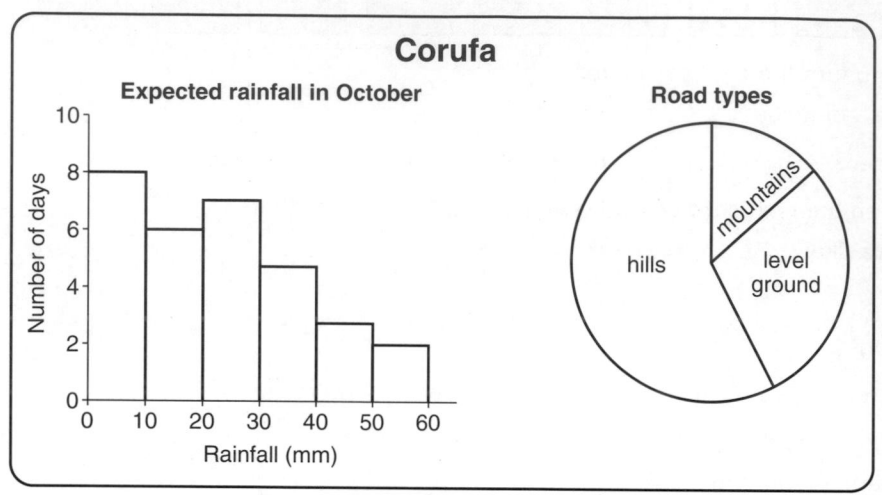

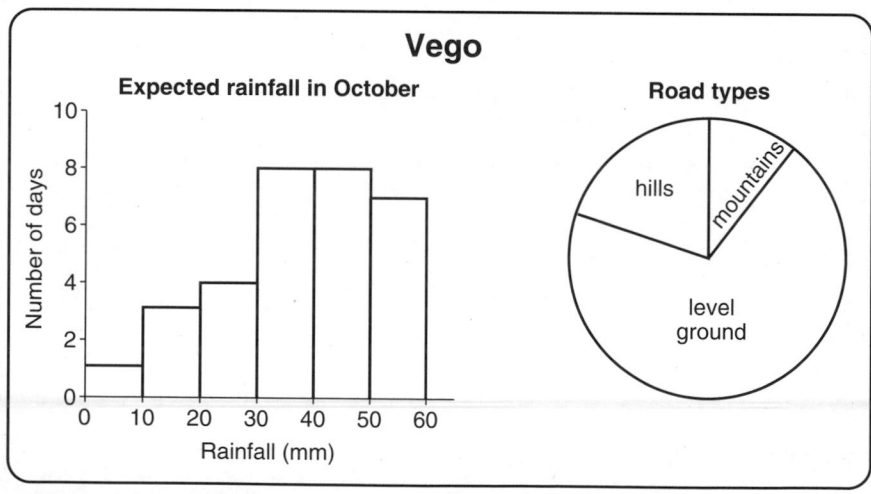

Choose which place you would most like to go to for a cycling holiday.

Use the 2 frequency diagrams and the 2 pie charts to explain your choice.

It doesn't matter which place you choose to go to, but you must write something about the rainfall **and** the road types.

6 The two frequency diagrams below show the amount of ran that fell in two different months.

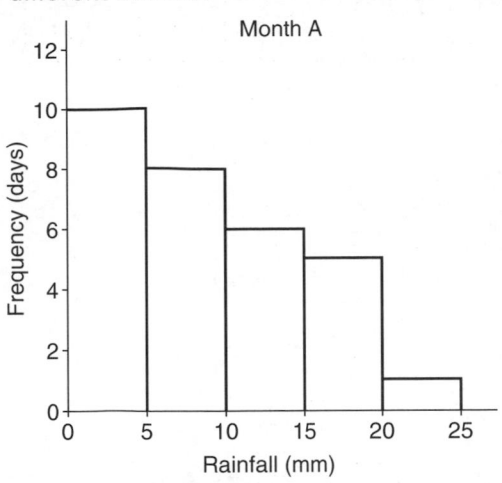

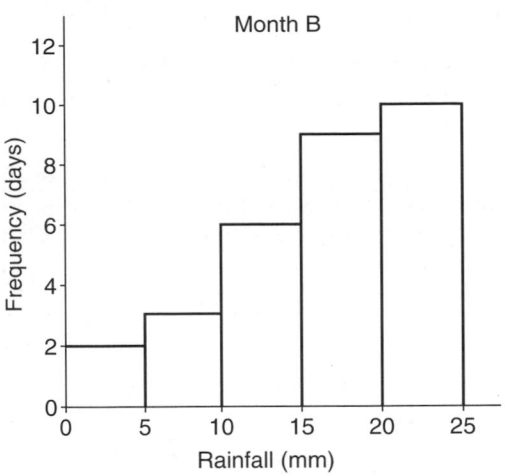

a Kath said: 'There were 30 days in month A.'

Explain how you know she was **right**.

b Carl asked 5 friends: 'How much rain fell during month A?'

They said:

Jon   Dipta   Ian   Nerys   Sharon

5 mm   25 mm   30 mm   75 mm   250 mm

Only **one** friend could have been right.
You can tell who it is **without** trying to work out the **total** rainfall.

Which **one** of Carl's friends could have been right?

Explain how you know.

c Sudi said: 'The diagram for month B shows that it rained more at the end of the month.'

Sudi is wrong.

Explain why the diagram does **not** show this.

7 A child is having a bath.

The simplified graph shows the depth of water in the bath.

a From A to B **both taps** are turned **full on**.

What might be happening at point B?

b Which part of the graph shows the child getting into the bath?

c Which part of the graph shows the child getting out of the bath?

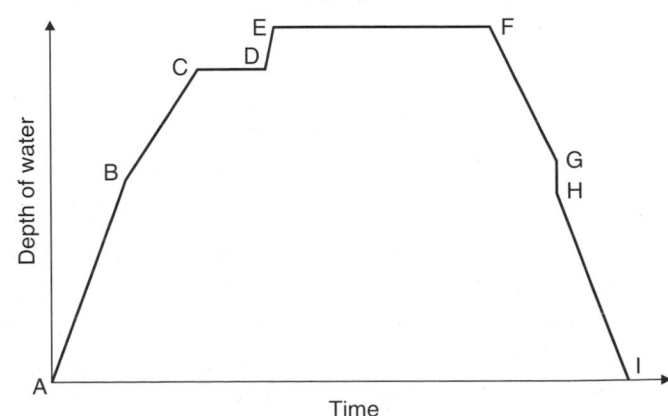

8   The *Highway Code* states the minimum distance there should be between cars. There are different distances for bad weather and good weather.

The graph below shows this.

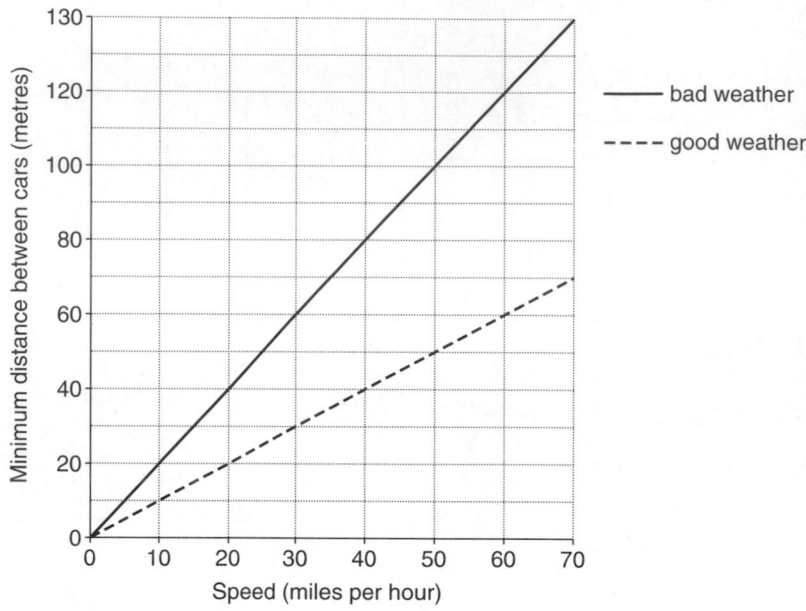

a   The weather is **bad**.

A car is travelling at **40 miles per hour**.

What is the minimum distance it should be from the car in front?

b   The weather is **good**.

A car is travelling at **55 miles per hour**.

What is the minimum distance it should be from the car in front?

c   Mr Evans is driving **30 metres** behind another car.

The weather is **bad**.

What is the maximum speed at which Mr Evans should be driving?

d   Mrs Singh is driving at **50 miles per hour** in **good** weather.

She is the minimum distance from the car in front.

It begins to rain heavily.

Both cars slow down to **30 miles per hour**.

Use the graph to work out how much Mrs Singh must increase her distance from the car in front.

Show your working.

9 A machine sells sweets in **five** different colours: red, green, orange, yellow, purple.

You **cannot choose** which colour you get.

There are the **same number** of each colour in the machine.

Ken and Colin want to buy a sweet each.

Ken says: 'I don't like yellow or orange ones.'

Colin says: 'I like all of them.'

a What is the **probability** that **Ken** will get a sweet that he **likes**?

b What is the **probability** that **Colin** will get a sweet that he **likes**?

c Draw an arrow on a probability scale to show the probability that **Ken** will get a sweet that he likes.

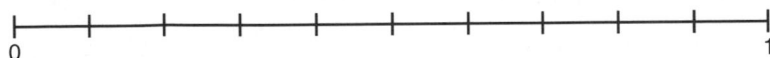

d Draw an arrow on a probability scale to show the probability that **Colin** will get a sweet that he likes.

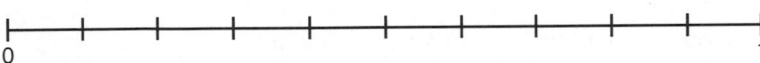

e Mandy buys one sweet.

The arrow on this scale shows the probability that Mandy gets a sweet that she likes.

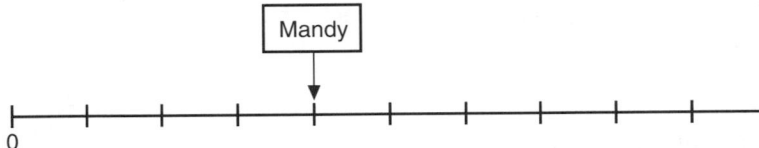

Write a sentence that **could** describe which sweets Mandy likes.

10 There are different ways of estimating probabilities.

For example:

Method A   Use equally likely outcomes

Method B   Look back at data

Method C   Survey or experiment to collect data

Look at the following situations. Say whether you would use Method A, Method B or Method C to estimate the probability.

a The probability that a cassette tape will last exactly 42 minutes.

b The probability that a car in London will be broken into on a Saturday night.

c The probability that a person chosen at random from a sports club is left-handed.

d The probability that a girl's name will be picked at random if 50 girls' names and 50 boys' names are put in a box.

e The probability that the hottest day in Britain next year will be in July.

11a Brenda wants to spin this spinner.

What colour is she **most** likely to get?

Explain why.

What colour is she **least** likely to get?

Explain why.

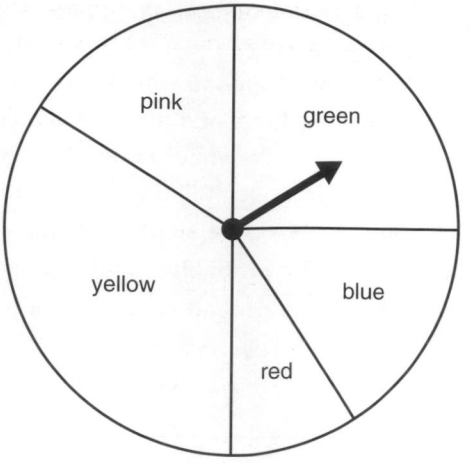

b Brenda says: 'I've got an equal chance of getting pink or blue.'

Is Brenda right or wrong? Give a reason for your answer.

c Brenda then says: 'My chance of getting green is about $\frac{1}{2}$.'

She is wrong.

Make a better estimate of the chance that Brenda will get green.

d Brenda also says: 'The probability that I will get yellow is $\frac{1}{5}$ because there are 5 colours.'

She is wrong. Explain why.

Make a better estimate of the probability that Brenda will get yellow.

e Write down a rough estimate of the probability of getting each colour.

12 The table shows some information about pupils in a school.

	Left-handed	Right-handed
Girls	32	180
Boys	28	168

There are **408 pupils** in the school.

a What **percentage** of the pupils are **boys**?

Show your working.

b What is the **ratio** of **left-handed** pupils to **right-handed** pupils?

Write your ratio in the form 1 : ......

Show your working.

c One pupil is chosen at random from the whole school.

What is the **probability** that the pupil chosen is a **girl** who is **right-handed**?

13 A doctor measures the pulse rate of 400 athletes and 400 company
    directors. Her results are shown in the tables below.

Pulse rate of athletes (beats/min)		
Class interval		Frequency
at least	below	
55	57	70
57	59	100
59	61	120
61	63	100
63	65	10

Pulse rate of company directors (beats/min)		
Class interval		Frequency
at least	below	
61	63	10
63	65	20
65	67	40
67	69	50
69	71	70
71	73	70
73	75	60
75	77	40
77	79	30
79	81	10

a   The frequency polygon below shows the pulse rates for company
    directors. Copy the diagram and on it draw a frequency polygon for
    athletes.

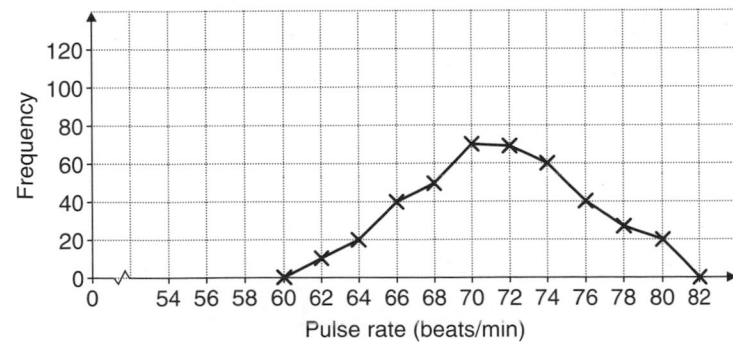

b   Using the two frequency polygons describe the two main differences
    between the data for athletes and company directors.

14a At a sports centre, people take part in one of five different sports. This table shows the percentage of people who played badminton, football and squash on Friday.

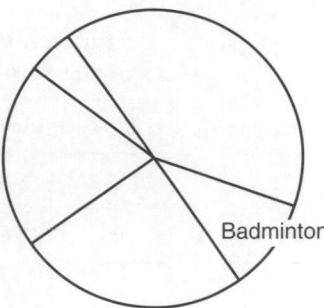

Friday	
Badminton	10%
Football	40%
Squash	5%
Swimming	?
Tennis	?

Copy the pie chart and label the correct two sections of the pie chart **football** and **squash**.

Badminton has been labelled for you.

b On Friday, **more** people went swimming than played tennis.

Use the chart to estimate the percentage of people who went swimming.

Use the chart to estimate the percentage of people who played tennis.

Make sure you have accounted for all the people.

c Altogether **260** people played the different sports on Friday.

Complete this table to show how many people played badminton, football and squash on Friday.

Friday Sport	Percentage	Number of people
Badminton	10%	26
Football	40%	. . . . . .
Squash	5%	. . . . . .

d Altogether **260** people played the different sports on **Friday**
700 people played the different sports on **Saturday**

40% of the people played football on **Friday**
but only **20%** of the people played football on **Saturday**

Mike said: '40% is more than 20%, so more people played football on Friday.'

Explain why Mike is **wrong**.

15 There are **24 pupils** in Jim's class.

He did a survey of how the pupils in his class travelled to school.

He started to draw a pie chart to show his results. Copy the pie chart.

a **4** pupils travelled to school by **train**.

Show this on Jim's pie chart as accurately as you can.

Label this part **train**. Label the remaining part **car**.

Jim's class (24 pupils)

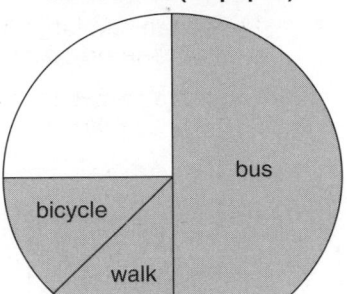

b There are **36 pupils** in Sara's class.

She did the same survey and drew a pie chart to show her results.

15 pupils travelled by **bus** and **6** pupils **walked**.

On a copy of Sara's pie chart, write how many pupils travelled to school by **train**, **car** and **bicycle**

Sara's class (36 pupils)

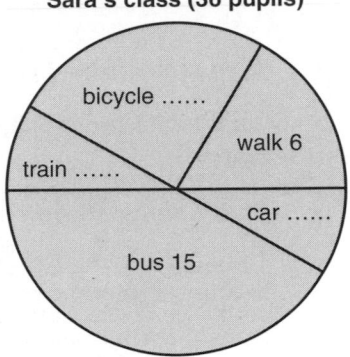

c Jim says:

'**15** pupils in Sara's class travelled by **bus**. Only **12** pupils in my class travelled by **bus**. Sara's pie chart shows **fewer** people travelling by bus than mine does. So Sara's chart must be wrong.'

Explain why Jim is **wrong**.

16 Alun has these two spinners.

Alun spins both spinners and then adds up the numbers to get a total.

He makes a list of all the possible totals.

Write down Alun's complete list.

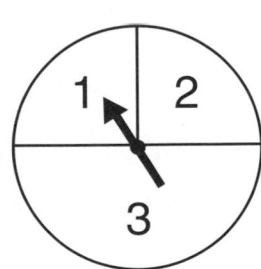

17 Les, Tom, Nia and Ann are in a singing competition.

To decide the order in which they will sing, all four names are put into a bag.

Each name is taken out of the bag, one at a time, without looking.

a Write down **all** the possible orders with **Tom** singing **second**.

b In a different competition there are 8 singers.

The probability that Tom sings second is $\frac{1}{8}$.

Work out the probability that Tom does **not** sing second.

18   Karen and Huw each have three cards, numbered 2, 3 and 4.

They each take any **one** of their own cards. Then they **add** together the numbers on the two cards.

The table shows all possible answers.

Karen

+	2	3	4
2	4	5	6
3	5	6	7
4	6	7	8

(Huw labels the rows)

a   What is the **probability** that their answer is an **even** number?

b   What is the **probability** that their answer is a number that is **greater than 6**?

c   Both Karen and Huw still have their three cards, numbered 2, 3 and 4.

They each take any one of their own cards. Then they **multiply** together the numbers on the two cards.

Draw a table to show all possible answers.

Use your table to fill in the gaps below:

The probability that their answer is a number that is less than
...... is $\frac{8}{9}$.

The probability that their answer is a number that is less than
...... is **zero**.

19a   Tracy has a bag of black beads and white beads.

The **probability** of picking a **black bead** from Tracy's bag is $\frac{7}{13}$.

What is the probability of picking a white bead from Tracy's bag?

b   How many black beads and how many white beads could be in Tracy's bag?

c   Peter has a different bag of black beads and white beads.

Peter has **more beads in total** than Tracy.

The probability of picking a black bead from peter's bag is **also** $\frac{7}{13}$.

How many black beads and how many white beads could be in Peter's bag?

20 The Stealers and the Allstars are two hockey teams.

They have played each other 20 times over the past two years.

The Stealers have won 30% of these matches.

The Allstars have won 50 of the matches.

Kath says:

'The probability that the Stealers will win the next match is 0.3
The probability that the Allstars will win the next match is 0.5.'

Use Kath's probabilities to answer these questions.

a What is the probability that the next match will be a draw?

b What is the probability that the next match will **not** be a draw?

21 Tony, Ben, Sally and Chitra want to find out what people think and do about animal welfare.

They are writing a questionnaire. Here are some questions they suggest:

Tony	Are you a member of an animal welfare organisation?	Yes/No
Ben	Are animals important?	Yes/No
Chitra	Don't you agree that experimenting on live animals is very, very cruel?	Yes/No
Sally	Do you buy products that have been tested on animals?	Yes/No/ Don't know

a Choose **two** questions which **you** think should **not** be used.

Whose questions are they?

b Explain why you think these two questions should not be used.

c Write an extra question **you** would use.

People must be able to answer you question with yes or no.

22 Some pupils wanted to find out if people liked a new biscuit.

They decided to do a survey and wrote a questionnaire.

a One question was:

How old are you (in years)?

☐   ☐   ☐   ☐   ☐

20 or younger   20 to 30   30 to 40   40 to 50   50 or older

Mary said: 'The labels for the middle three boxes need changing.'

Explain why Mary was **right**.

b A different question was:

How much do you usually spend on biscuits each week?

☐ a lot   ☐ a little   ☐ nothing   ☐ don't know

Mary said: 'Some of these labels need changing too.'

Write new labels for any boxes that need changing.

You may change as many labels as you want to.

The pupils decide to give their questionnaire to 50 people.

Jon said: 'Let's ask 50 pupils in our school.'

c Give **one disadvantage** of Jon's suggestion.

d Give **one advantage** of Jon's suggestion.

23 A company makes breakfast cereal containing nuts and raisins.

They counted the number of nuts and raisins in 100 small packets.

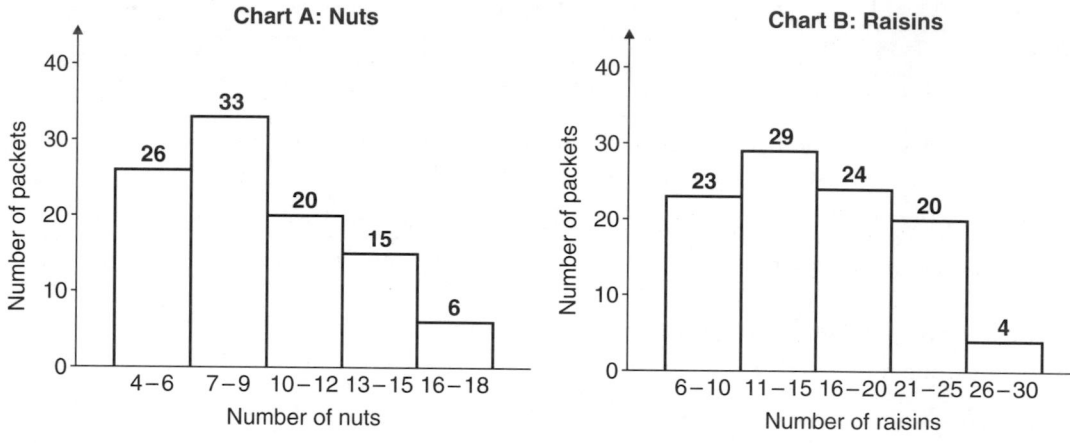

a Calculate an estimate of the **mean** number of **nuts** in a packet.

Show your working.

b Calculate an estimate of the **number** of packets that contain **24 or more raisins**.

Number of nuts	Mid-point of bar (x)	Number of packets (f)	fx
4–6	5	26	130
7–9	8	33	
10–12	11	20	
13–15	14	15	
16–18	17	6	
		100	

c Which of the two charts shows the **greater range**?

Explain your answer.

d A packet is chosen at random.

Calculate the probability that it contains **9 nuts or fewer**.

24   A school has 5 year groups.

80 pupils from the school took part in a sponsored swim.

Lara and Jack drew these graphs.

Lara's graph: Jack's graph:

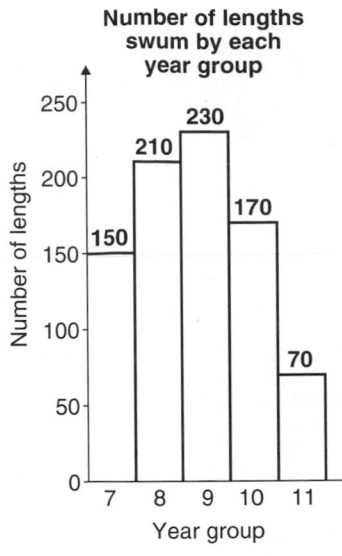
Number of lengths swum by each year group

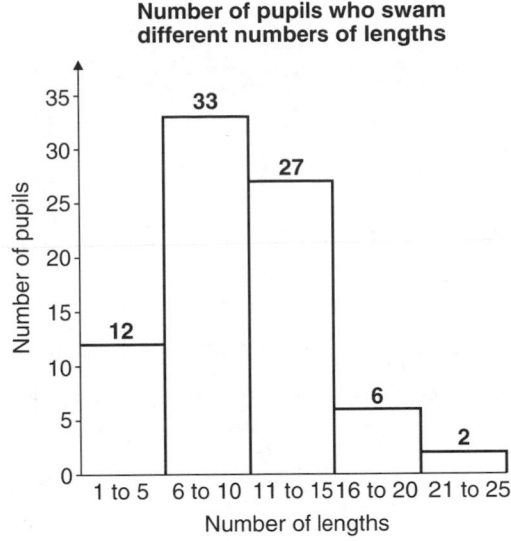
Number of pupils who swam different numbers of lengths

a   Look at **Lara's** graph. Is it possible to tell if **Year 10** had **fewer** pupils taking part in the swim than **Year 7**?

Explain your answer.

b   Use **Lara's graph** to work out the mean number of lengths swum by each of the 80 pupils.

Show your working.

c   Use **Jack's graph** to work out the mean number of lengths swum by each of the 80 pupils.

Show your working.

d   Explain why the means calculated from Lara's graph and Jack's graph are different.

25   A company has 10 pizza shops. All the shops sell the same types of pizza at the same prices.

Mr Bal asked:

'Is there a **relationship** between the **number of pizzas sold** and the **number of people who live within 3 miles** of a shop?'

Mrs Evans asked:

'Is there a **relationship** between the **number of pizzas sold** and the **floor area** of a shop?'

They looked at these graphs.

Graph 1

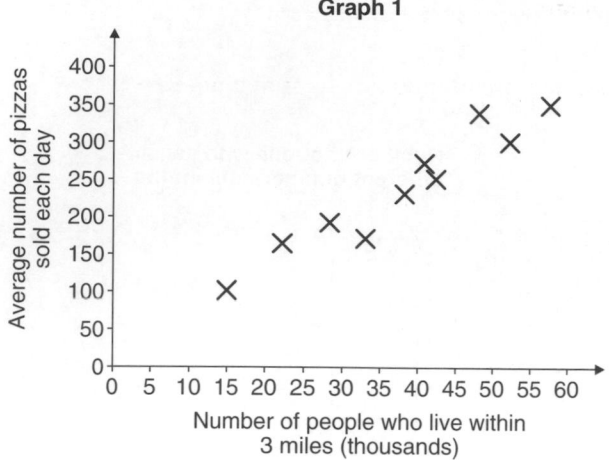

Graph 2

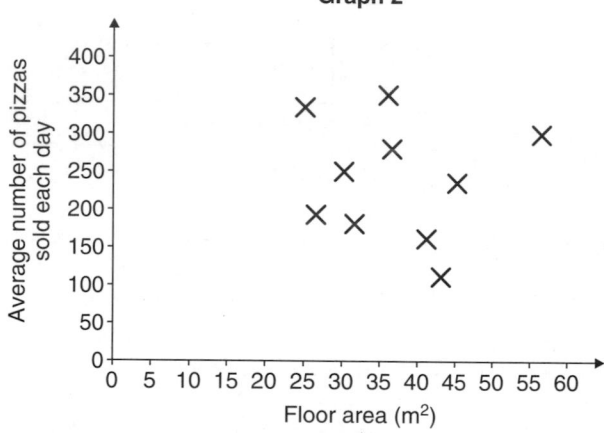

a   What does **graph 1** show about the relationship between the average number of pizzas sold each day the number of people who live within 3 miles of a shop?

b   What does **graph 2** show about the relationship between the average number of pizzas sold each day and the floor area of a shop?

c   The company opens a new shop. It has a floor area of 42 m^2 and 35 thousand people live within 3 miles.

Use **one** of the graphs to estimate the average number of pizzas this shop is likely to sell each day.

Say which graph you used **and** explain how you made your estimate.

d   The company plans to open four other new shops.

	Shop A	Shop B	Shop C	Shop D
Floor area (m^2)	32	48	42	30
No. of people within 3 miles (thousands)	55	29	21	31

Which two of these four shops are the most likely to have similar sales of pizzas?

26 The scatter diagram shows the total amounts of sunshine and rainfall for 12 seaside towns during one summer.

Each town has been given a letter.

The dashed lines drawn go through the **mean** amounts of sunshine and rainfall.

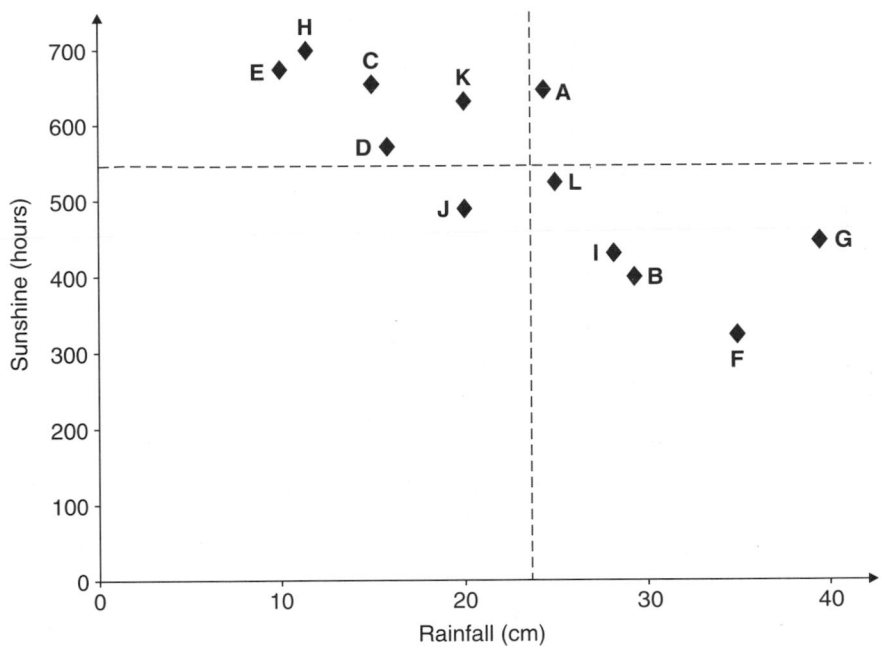

a   Which town's rainfall was closest to the mean?

b   Copy the scatter diagram. On it draw a **line of best fit**.

Use your line to find an estimate of the hours of sunshine for a seaside town that had 30 cm of rain.

27 Barry is doing an experiment.

He drops 20 matchsticks at random onto a grid of parallel lines.

Barry does the experiment 10 times and records his results. He wants to work out an estimate of probability.

Number of the 20 matchsticks that have fallen across a line									
5	7	6	4	6	8	5	3	5	7

a   Use Barry's data to work out the probability that a **single matchstick** when dropped will fall across one of the lines.

b   Barry continues the experiment until he has dropped the 20 matchsticks 60 times.

About how many matchsticks **in total** would you expect to fall across one of the lines?

28   As part of a biology project, Dan and Ama are counting the number of peas in a sample of pea pods.

The table shows their results for the first 50 pods:

Number of peas in a pod	Number of pods
3	2
4	7
5	14
6	12
7	10
8	5

Dan and Ama correctly worked out the mode as 5 and the median as 6.

a   Work out the mean number of peas in a pod in their sample.
Show your working.

b   Work out the number of peas in 200 pods.
You should first decide whether to use the mode, median or mean.
Show your method.

c   About how many pods out of 200 would you expect to have 3 or 4 peas?

d   Dan takes another pod at random from the sample, opens it, and counts the number of peas.
Work out the probability that the pod contains **more** than 6 peas.

Acknowledgement

Oxford University Press wishes to thank the Qualifications and Curriculum Authority for granting permission to use the following questions from past papers for the Schools Curriculum and Assessment Authority (SCAA) Key Stage 3 Mathematics Tests.

	Year	Tier	Paper	Question
Number and Algebra				
Q1	1995	4–6	P2	3
Q2	1993	3–6	P1	9
Q3	1992	3–6	T2	7
Q4	1995	4–6	P2	5
Q5	1994	3–5	P1	6
Q6	1996	4–6	P1	2
Q7	1997	4–6	P2	6
Q8	1995	4–6	P1	2
Q9	1997	4–6	P1	2
Q10	1997	4–6	P2	3
Q11	1994	3–5	P2	6
Q12	1996	5–7	P1	1
Q13	1993	3–6	P3	15
Q14	1996	5–7	P1	6
Q15	1992	5–8	T1	8
Q16	1992	5–8	T2	4
Q17	1993	3–6	P2	18
Q18	1996	5–7	P2	6
Q19	1997	5–7	P2	4
Q20	1996	5–7	P2	13
Q21	1997	5–7	P1	4
Q22	1995	5–7	P2	2
Q23	1997	5–7	P1	3
Q24	1995	5–7	P1	5
Q25	1997	5–7	P2	7
Q26	1994	3–5	P2	11
Q27	1995	5–7	P1	7
Q28	1993	3–6	P1	15
Q29	1996	5–7	P2	10
Q30	1994	6–8	P1	2
Q31	1995	5–7	P2	6
Q32	1996	5–7	P2	5
Q33	1997	5–7	P1	12
Q34	1994	6–8	P1	1
Q35	1994	6–8	P2	4
Q36	1996	5–7	P1	5
Q37	1994	6–8	P2	5
Q38	1996	5–7	P2	9
Q39	1997	5–7	P1	9
Q40	1995	5–7	P1	15
Q41	1996	5–7	P1	9
Q42	1997	5–7	P1	11
Q43	1994	6–8	P1	8
Q44	1995	5–7	P2	12
Q45	1996	5–7	P2	7
Q46	1996	5–7	P1	11
Shape, space and measures				
Q1	1995	4–6	P2	2
Q2	1995	4–6	P1	1
Q3	1997	4–6	P2	5
Q4	1992	3–6	T2	5

	Year	Tier	Paper	Question
Q5	1997	4–6	P2	2
Q6	1993	3–6	P1	16
Q7	1995	5–7	P2	1
Q8	1996	5–7	P2	9
Q9	1993	3–6	P1	12
Q10	1993	3–6	P3	14
Q11	1992	5–8	T1	1
Q12	1996	5–7	P1	8
Q13	1995	5–7	P1	4
Q14	1995	5–7	P1	8
Q15	1997	5–7	P1	7
Q16	1995	5–7	P2	5
Q17	1996	5–7	P2	8
Q18	1994	3–5	P2	9
Q19	1996	5–7	P2	4
Q20	1992	5–8	T2	10
Q21	1996	5–7	P1	15
Q22	1997	5–7	P2	9
Q23	1995	5–7	P2	9
Q24	1992	5–8	T3	6
Q25	1993	7–10	P3	3
Q26	1995	5–7	P1	11
Handling data				
Q1	1995	4–6	P1	3
Q2	1997	4–6	P2	4
Q3	1993	7–10	P3	4
Q4	1997	5–7	P1	8
Q5	1992	5–8	T1	4
Q6	1994	3–5	P1	11
Q7	1997	5–7	P1	6
Q8	1997	5–7	P2	2
Q9	1996	5–7	P2	2
Q10	1992	5–8	T3	2
Q11	1994	3–5	P2	5
Q12	1997	5–7	P2	8
Q13	1992	5–8	T1	10
Q14	1995	5–7	P1	3
Q15	1996	5–7	P1	4
Q16	1995	5–7	P1	6
Q17	1996	5–7	P1	7
Q18	1997	5–7	P1	2
Q19	1994	6–8	P1	4
Q20	1993	3–6	P1	18
Q21	1993	3–6	P3	10
Q22	1997	5 7	P2	5
Q23	1997	5–7	P2	11
Q24	1996	5–7	P2	11
Q25	1996	5–7	P1	10
Q26	1997	5–7	P1	10
Q27	1996	5–7	P2	14
Q28	1995	5–7	P2	11

Index

e Power of Practice

With your purchase of a new copy of this textbook you received a Student Access Kit for MyFinanceLab for *Corporate Finance and Investment, sixth edition*. Follow the instructions on the card to register successfully and start making the most of the resources.

Don't throw it away!

The Power of Practice:

MyFinanceLab is an online study and testing resource that puts you in control of your study, providing extensive practice exactly where and when you need it.

MyFinanceLab gives you unrivalled resources:

- Sample tests for each chapter allowing you to see how much you have learned and where you still need practice
- A personalised study plan, which constantly adapts to your strengths and weaknesses taking you to exercises you can practice over and again with different variables every time
- Guided solutions which break the problem into its component steps and guide you through with hints
- Audio animations to guide you step by step through the key concepts in corporate finance.
- Podcasts analysing topical issues and recent news items in finance.
- Video clips of CEOs and Financial Directors from large companies explaining how they make financial decisions for each part of the book.
- E text
- Online glossary which defines key terms and provides examples

See Using MyFinanceLab on page xxii for more details.

To activate your registration go to **www.myfinancelab.com/register** and follow the instructions on screen to register as a new user.

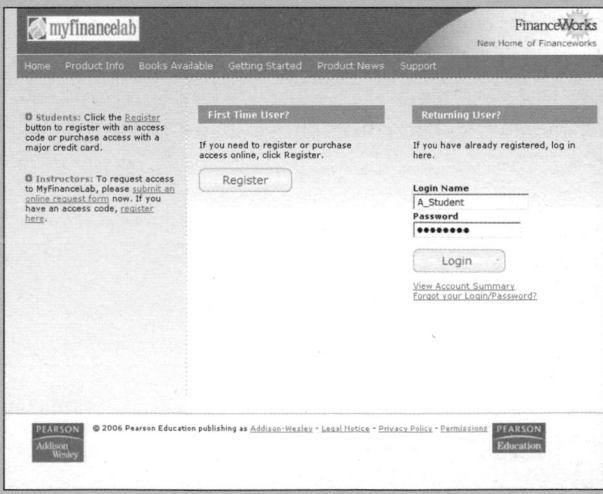

CORPORATE FINANCE AND INVESTMENT

DECISIONS & STRATEGIES

Sixth Edition

Richard Pike and Bill Neale

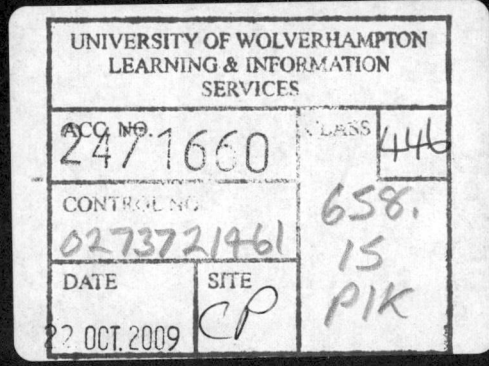

FT Prentice Hall
FINANCIAL TIMES

An imprint of **Pearson Education**

Harlow, England • London • New York • Boston • San Francisco • Toronto
Sydney • Tokyo • Singapore • Hong Kong • Seoul • Taipei • New Delhi
Cape Town • Madrid • Mexico City • Amsterdam • Munich • Paris • Milan

To our wives, Carol and Jean

Pearson Education Limited

Edinburgh Gate
Harlow
Essex CM20 2JE

and Associated Companies throughout the world

Visit us on the World Wide Web at:
www.pearsoned.com

First published 1993
Sixth edition published 2009

ISBN 978-0-273-71550-4

British Library Cataloguing-in-Publication Data
A catalogue record for this book is available from the British Library

Library of Congress Cataloging-in-Publication Data
Pike, Richard (Richard H.)
  Corporate finance and investment : decisions & strategies/Richard Pike and Bill Neale. — 6th ed.
     p. cm.
  ISBN 978-0-273-71550-4 (pbk.)
  1. Corporations—Great Britain—Finance. 2. Investments—Great Britain.
  3. Corporations—Europe—Finance. I. Neale, Bill. II. Title.
  HG4135.P35 2009
  658.150941—dc22

                                                                    2008039373

10 9 8 7 6 5 4 3 2 1
13 12 11 10 09

Typeset in 9.5/12 Palatino by 73.
Printed and bound by Rotolito Lombarda, Italy.

The publisher's policy is to use paper manufactured from sustainable forests.

Brief Contents

Contents

List of figures and tables

List of figures

List of tables

Preface

Not all textbooks survive to a sixth edition. As one of the lucky survivors, we wish to preface this edition with yet another 'thank you' – thank you to the lecturers who have recommended our book and also to the students who have purchased and used it. Hopefully, you have all obtained good value from it.

We first began work on this project around 1990, almost two decades ago. Over this period, there have been many changes in the financial arena. For example, a radical downshift in inflationary expectations, increasing integration of world financial markets, powered by the ongoing revolution in communications, the end of the 'Japanese Miracle', and the introduction of the euro. We have seen several financial meltdowns – at the national level, the 'Asian Crisis', Argentina, and at the micro-level, the 'dotcom' boom and bust, the crisis in corporate governance and the 'credit-crunch' starting in 2007.

It is not surprising that financial issues increasingly dominate the news bulletins, emphasising the need for both students of business and also business practitioners to have at least a working knowledge of finance. Yet academic courses are becoming increasingly fragmented, for example, with the move to semesterisation. At the same time, within academic courses, the emphasis now placed on formal mathematical and statistical training, and even economics, is also being reduced.

These considerations reinforce our view that finance should be about developing, explaining and, above all, *applying* key concepts and techniques to a broad range of contemporary management and business policy concerns and challenges. It is becoming more appropriate, certainly at the undergraduate level, to demonstrate the role finance has to play in explaining and shaping business development rather than concentrating on rigorous, quantitative aspects.

The focus of the sixth edition, as in previous ones, is distinctly corporate, examining financial issues from a managerial standpoint. To simplify greatly, we have tried, wherever possible, to present the reader with the question 'OK, but how does this help the managerial decision-maker?' and also to provide a few answers, or at least pointers.

Some might say we should include chapters on other financial issues deemed to have a degree of importance equivalent to those covered here. Yet we believe, as ever, that there is a trade-off between comprehensiveness and manageability. This edition is directed at those issues, which in our experience are regarded as the central issues in finance.

■ Distinctive features

The sixth edition retains a set of distinctive features, including the following:

- *A strategic focus*. Students often regard financial management as a subject quite distinct from management and business policy. We attempt to relate the subject to these matters, emphasizing the integration of the finance function within the context of managerial decision-making and corporate planning, and to the wider external environment.
- *A practical approach*. Financial theory increasingly dominates some texts. Theory has its place, and this text covers an appreciable amount; however, we seek to blend theory and practice: to ask why they sometimes differ, and to assess the role of less-sophisticated financial approaches. In other words, we do not elevate theory above common sense and intuition.

- *A clear and accessible style.* Personal experience and feedback suggests that much of our target readership prefers a more descriptive, rather than heavily mathematical, approach but appreciates worked examples and illustrations. There is a place for formulae, proofs and quantitative analysis; however, where possible, an alternative narrative explanation is provided. Appendices are often used to deal with rather more complex mathematical aspects.
- *An international perspective.* Although emanating from the UK, our text uses, where appropriate, examples drawn from other regions and countries, especially mainland Europe and the USA.

■ Teaching and learning features

A range of teaching and learning features is provided, including the following:

- *Mini case studies.* Topical cameos, applying financial management principles to well-known companies, are presented at the start of chapters and elsewhere within the text.
- *Learning objectives.* Specified at the outset of each chapter, these highlight what the reader should achieve in terms of concepts, terminology and skills.
- *Worked examples.* Integrated throughout the text to illustrate the key principles.
- *Extracts from the press.* Each chapter includes at least one article mainly from either the *Financial Times* and the *Economist* focusing on one of the key issues addressed in the chapter.
- *Key revision points.* Provided at the end of each chapter to summarise the main concepts covered.
- *Annotated further reading.* At the end of each chapter, a number of key books and articles are suggested to offer additional perspectives and enable subjects to be studied in more depth. Full details of all books and articles are given in the References at the end of the book.
- A quick reference *glossary* of simple definitions.

■ Assessment features

Flexible study and assessment is facilitated by a variety of activities:

- *Self-assessment activities (SAAs).* These include both short questions and simple numerical exercises designed to reinforce a point made in the text or to encourage the reader to pursue a particular line of thought. Questions are inserted in the text at appropriate points and the answers are packaged together at the end of the book.
- *Questions.* These test a mix of numerical, analytical and descriptive skills, offering a spread of difficulty. A selection of solutions is also provided in Appendix B at the end of the text, making these suitable for self-assessment, tutorial or examination purposes.
- *Practical assignments.* These provide the opportunity to look beyond the confines of the text to consider the application of concepts to a company or organisation, or to published financial reports and data, and are suitable where group or individually assessed coursework is set.

■ Readership

The text has proved successful both for newcomers to finance and also for students with a prior knowledge of the subject. It is particularly relevant to undergraduate, MBA and other postgraduate and post-experience courses in corporate finance or financial management. Students seeking a professionally accredited qualification will also find it especially relevant to the financial management papers of the Association of Chartered Certified Accountants, Institute of Chartered Secretaries and Administrators, Certified Diploma in Finance and Accounting, Chartered Institute of Management Accountants and the Institute of Chartered Accountants in England and Wales.

■ Changes to the sixth edition

As with previous editions, our revisions are based on extensive market research including reviewers' questionnaires and direct feedback from adopters and users. Feedback, while always interesting and helpful, was sometimes contradictory. Some wished for a more comprehensive, and sometimes more rigorous treatment, while others expressed concern that we might lean too far in the direction of strategy. Hopefully, we have achieved a balance between academic rigour and practical application.

In preparing this edition, we have battled with two opposing forces. We wanted to avoid expanding the text to an unmanageable size, yet we have been aware of several gaps in our coverage in previous editions, and the need for 'infill'.

The main changes to this edition in structure and in content are summarised below.

■ Structural and other changes

The main structural change is the removal of Chapter 4 from Part I, where we thought it was rather prematurely placed, to now appear in revised form as Chapter 12, with a new title. The material on valuation of shares has been moved to Chapter 3, with the dividend valuation model, including constant and growing perpetuities, now used as an application of discounting formulae.

The new Chapter 12 now includes the material on shareholder value analysis, formerly in the old Chapter 11, but now presented as a further application of DCF approaches to valuation. It provides the culmination of Part III, and thus benefits from a greater 'feed-in' regarding the appropriate required rate of return. We decided that it would be most appropriate to present shareholder value analysis (SVA) in an all-equity context in order to defer issues of debt financing, e.g. ungearing of Betas, until Chapters 18 and 19. Chapter 10 now contains a greater input on factor models and Arbitrage Pricing Theory. Chapter 13 has greater focus on derivatives and interest rate management.

Chapter 16, on sources of long-term finance, now includes a section on Islamic finance, or, more accurately, on Islamic bonds (*sukuk*). Our judgement is that there is little or no difference between equity financing under orthodox and Islamic approaches. We appreciate that there are more types of Islamic financing instruments that can be used, especially short-term ones, but for space reasons we have concentrated on the long-term debt form. Chapter 22 now includes a section on private equity under the restructuring heading, to augment the material on this topic already in Chapter 16. The concluding chapter, Chapter 23, now has a section which reviews the main developments in corporate finance. This offers a useful revision of the main concepts within the book, before going on to consider behavioural finance.

Every chapter now contains a detailed worked example, often taken from a recent CIMA Paper 9 exam. More extensive guides to further reading appear at the end of chapters with a sharper focus on key empirical references.

■ Structure and outline

An outline of the text is given below; however, a further description of the purpose and content of each section is given in the introduction to each.

Part I considers the underlying framework for corporate financing and investment decisions; key aspects of this part are the financial objectives of business, the financial environment within which firms operate, the time-value of money and the concept of value.

Part II addresses investment decisions and strategies within firms. Emphasis is placed on evaluation procedures, including treatments of taxation, inflation and capital rationing. Because, in practice, investment decision-making often bears little relationship to the theoretical approaches outlined in some texts, we persist in our attempt to promote an understanding of the practical evaluation of investment decisions by firms.

The importance of value, risk and the expected rate of return are examined in Part III, with six chapters devoted to this theme. The first two chapters consider the investment project in isolation, including the rapidly developing and exciting field of options analysis. Other chapters view risk and return more from a shareholder perspective. Fundamental to this section are the rate of return on investment required by shareholders and the valuation of the enterprise.

Part IV discusses the short-term financing decisions and policies for acquiring assets. It covers treasury and working capital management.

Part V addresses long-term strategic financing and policy issues. What are the main sources of finance? How much should a company pay in dividends? How much should it borrow? The culminating chapter focuses on corporate restructuring with particular reference to acquisitions.

Part VI examines international financial management issues. It explains the operation of the foreign currency markets and how firms can hedge against adverse foreign exchange movements, and sets out the principles underpinning firms' evaluation of foreign investment decisions. A concluding chapter reviews developments in corporate finance, with specific focus on market efficiency and behavioural finance.

Guided tour

SETTING THE SCENE

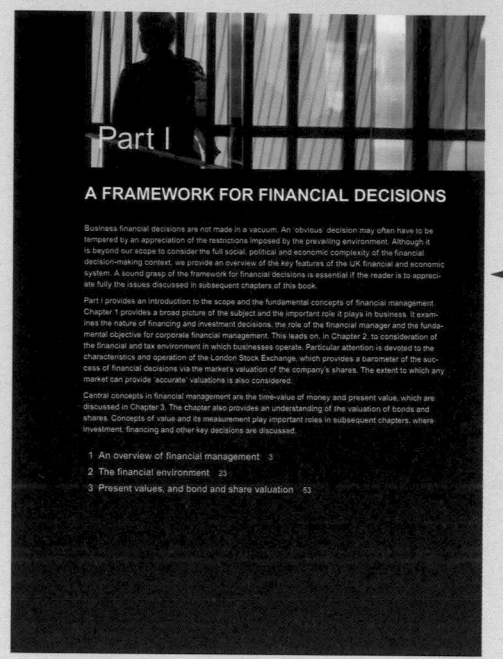

Part introductions divide the book into six parts introducing the upcoming chapter topics and help you navigate your way through the book.

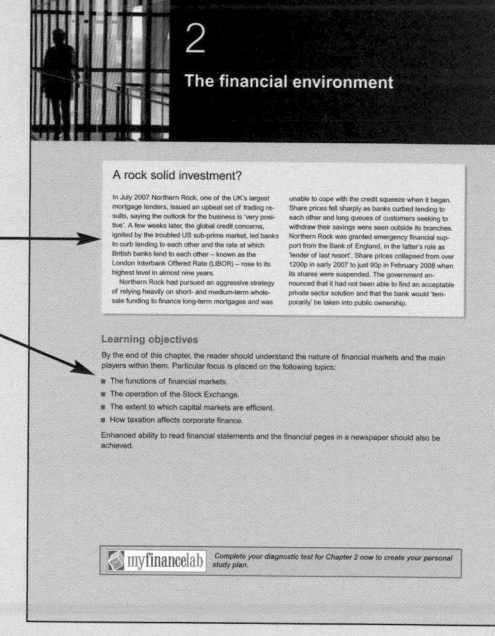

Topical cameos at the beginning of the chapter set the scene applying financial management principles to well-known companies.

Learning objectives enable you to see exactly where the chapter is going.

AIDING YOUR UNDERSTANDING

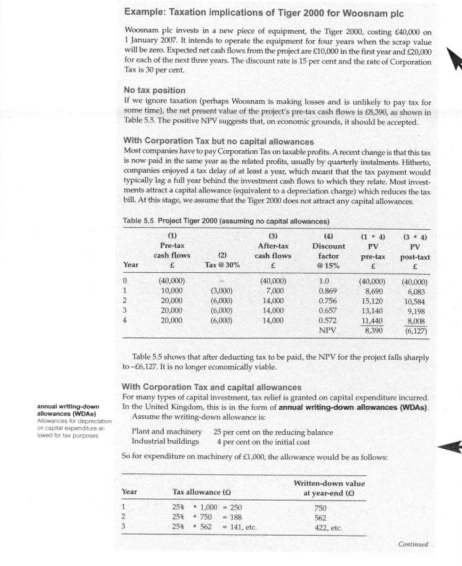

Examples break the concepts down into more manageable steps and illustrate key principles.

Key terms are defined in the margin when they first appear and, collected together in a **glossary** at the end of the book, provide simple definitions to help you remember and understand financial terminology.

AIDING YOUR UNDERSTANDING continued

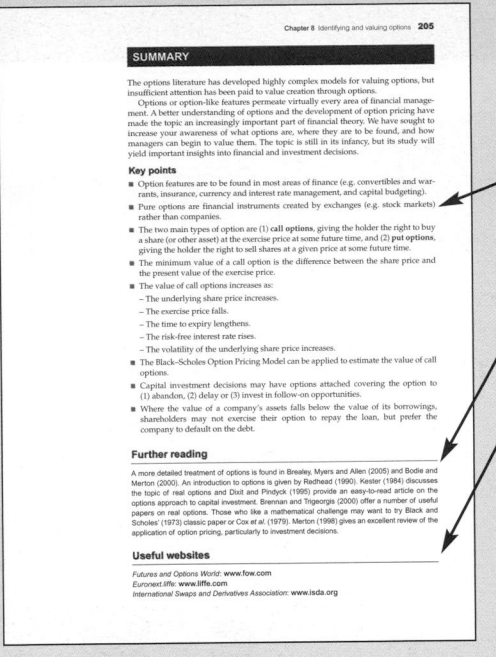

Key points provided at the end of each chapter summarise the main concepts covered.

Further reading at the end of each chapter offer suggestions for additional perspectives to study the topic in more depth.

Useful websites direct you to regularly updated sites keeping you up-to-date with recent news events and in touch with real data.

TEST YOUR LEARNING

Topical articles from a wide range of newspapers, including the *Financial Times,* use real companies to illustrate the principles of financial management in action and test your understanding of corporate finance in the real world.

Self-assessment activities encourage self-learning by reinforcing points made in the chapter. Answers can be found in Appendix A at the back of the book.

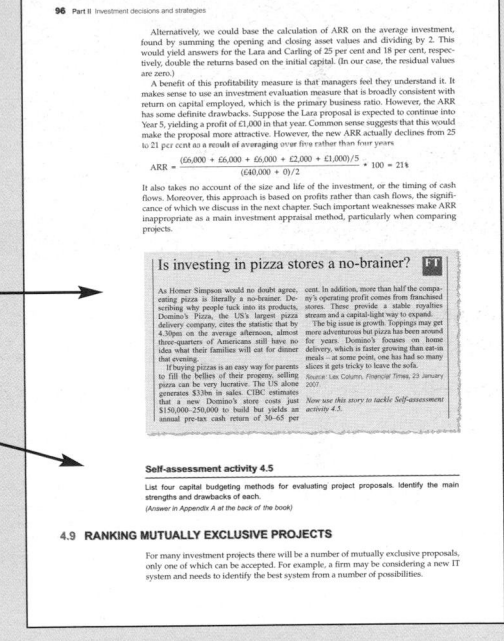

Questions at the end of each chapter, ranging in difficulty, mix numerical, analytical and descriptive problems to test your knowledge. Many of the questions are taken from the CIMA and ACCA examination papers. Selected answers can be found in Appendix B in the back of the book.

Questions with the **question mark** icon have a corresponding question for you to practice in your online learning resource **MyFinanceLab**.

Practical assignments ask you to apply concepts learned from your reading of the text to the real world using companies, organizations or published financial reports.

Packaged with every new copy of the sixth edition of *Corporate Finance and Investment*, **MyFinanceLab** puts you in control of your study. To register as a new user go to **www.myfinancelab.com/pikeneale** and follow the instructions on-screen using the code in your student access kit.

By using MyFinanceLab you test your understanding and practise what you have learned.

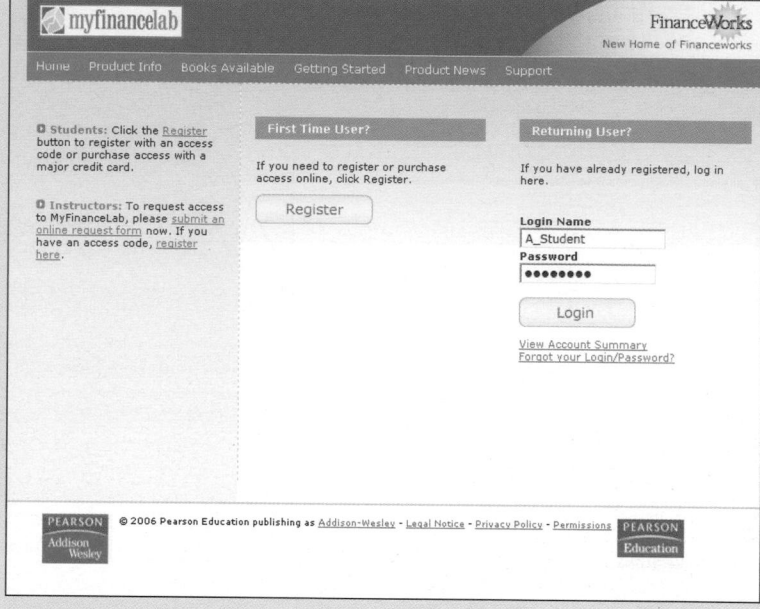

Sample tests (two for each chapter) enable you to test your understanding and identify the areas in which you need to do further work.

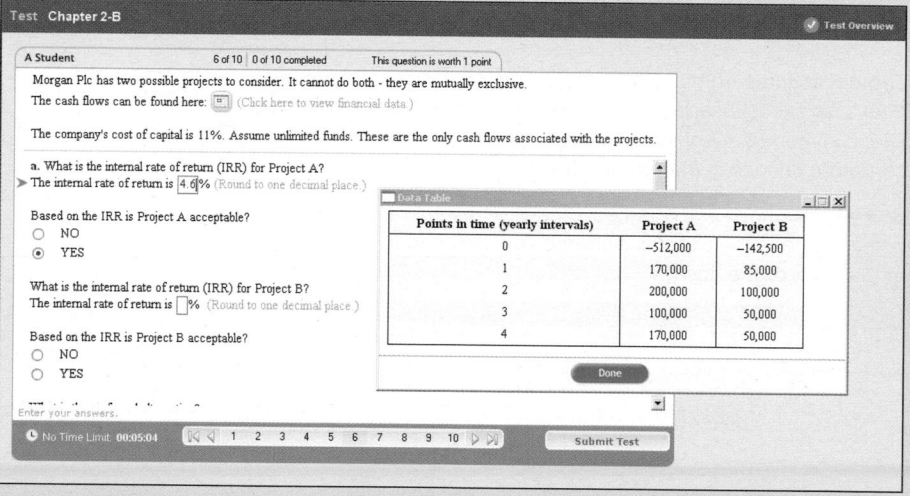

 *Complete your diagnostic test for Chapter X now to create your personal study plan.*

When you see this icon in the text at the beginning of a chapter complete your **sample test (a)** in MyFinanceLab to create your personal study plan for the chapter.

 *Now retake your diagnostic test for Chapter X to check your progress and update your personal study plan.*

When you see this icon in the text at the end of a chapter go back to MyFinanceLab and take your **sample test (b)** to see how much you have improved.

MyFinanceLab creates a personal **study plan** for you based on your performances in tests. The study plan diagnoses areas that need more practice and consists of a series of additional exercises with detailed step-by-step guided solutions and additional study tools to help you complete the exercises.

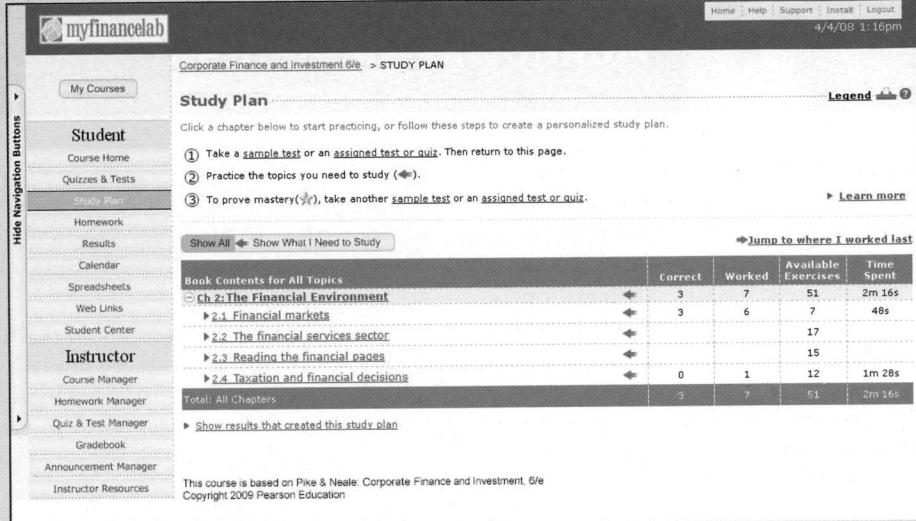

From the study plan exercises you can link out to the step-by-step guided solutions to help you complete the exercise.

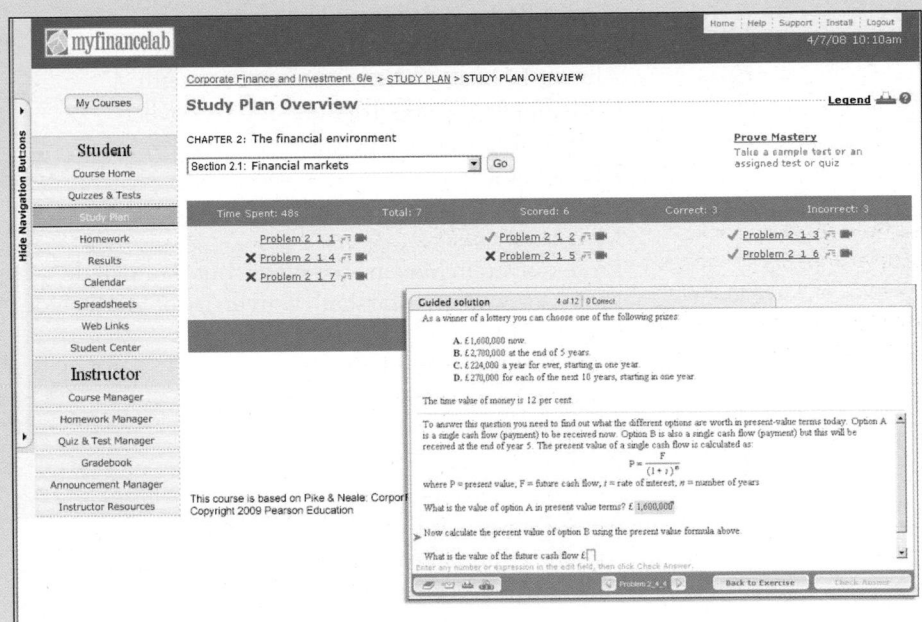

Additional resources such as **podcasts**, **videos**, **audio animations** of key concepts in corporate finance and an **electronic version of your text book** are also on hand to help you.

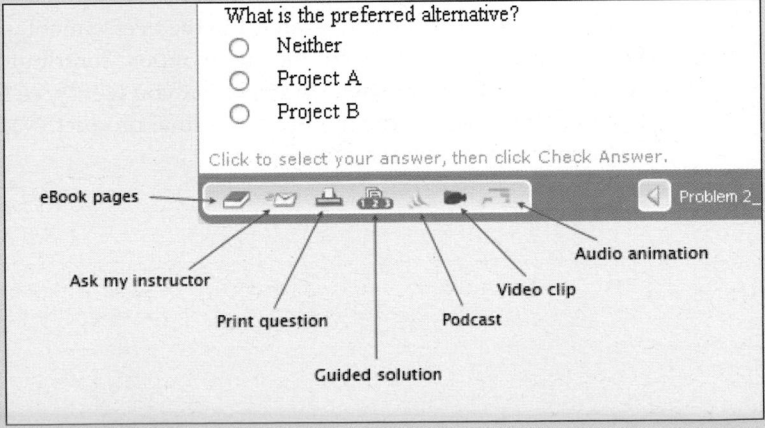

Acknowledgements

All textbooks include 'acknowledgements' but, on reflection, this seems too weak a word to use when assistance has so often been so freely given. *Roget's Thesaurus* offers as a synonym, 'the act of admitting to something', suggesting rather grudging recognition!

Our recognition of the wide range of people and organizations is anything but grudging. We extend our warm appreciation of the helpful comments provided by you over the years, and also for consent to use your material.

To the ever-lengthening roll of honour, we wish to add the following names and organizations, whom we sincerely hope will be happy to be associated with our efforts:

John Ward – Dealogic
Andrew Carr – CIMA
Janine of the Barclays Capital Equity-Gilt Study Team
Patrick Barber – Bradford University
Kathy Grieve – Birmingham City University
Ahmed El-Masry – University of Plymouth
Christopher Brown – JP Morgan Cazenove
Patrick McColgan – Aberdeeen University
Himanshu Dubey
Andrew Barfield
Maxim Kakareka
Professor Colin Mason – University of Strathclyde
Professor Andrew Marshall, University of Strathclyde
Sue Lane
Peter Blankenhorn – E.On AG
Andrew Naughton-Doe – Corus UK plc
Pat Rowham – LBS
Peter Aubusson – DS Smith plc
Sue Cox – BAA plc
ASJR Ramsay – International Power plc
"Sarah" at British Airways plc
Jane Lanyon – Thorntons plc
Ian Lomas – DTI
Ian Patterson – HM Customs & Excise

As ever, we apologise for any omissions.

Finally, we are especially grateful to the ever-patient, ever-tolerant editorial staff at Pearson Education, and to the anonymous contributors to the market research conducted by the publisher. We hope that you will agree that your comments have led to an improvement in the quality of the final product. Naturally, as ever, we claim sole responsibility for any remaining errors.

Richard Pike, University of Bradford
Bill Neale, Bournemouth University

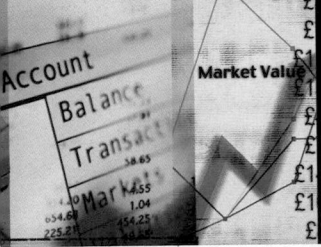

Publisher's acknowledgements

We are grateful to the Financial Times Limited for permission to reprint the following material:

Chapter 2 Inion plans £30m public offering, © *Financial Times*, 10 November 2004; Chapter 2 We are all chartists now, © *Financial Times*, 8 March 2007; Chapter 3 Back to the future, © *Financial Times*, 23 December 2004; Chapter 4 Pizza's attractions, © *Financial Times*, 3 January 2007; Chapter 4 Internal rate of return, © *Financial Times*, 1 June 2005; Chapter 7 Space tourist insurers eye up the final frontier, © *Financial Times*, 8 March 2007; Chapter 9 Metro and the weather, © *Financial Times*, 29 October 2004; Chapter 11 Counting the cost, © *Financial Times*, 24 March 2003; Chapter 12 Capturing the indefinable value of a brand, © *Financial Times*, 9 February 2005; Chapter 12 Heidelberg/Hanson, © *Financial Times*, 16 May 2007; Chapter 12 Ericsson's cash flow, © *Financial Times*, 16 January 2007; Chapter 16 Hargreaves Lansdown soars on debut, © *Financial Times*, 16 May 2007; Chapter 16 Aim flotation for London's Capital Pub Company, © *Financial Times*, 17 May 2007; Chapter 16 Investcorp puts Welcome Break up for sale with £500m price tag, © *Financial Times*, 20 October 2007; Chapter 16 Why even discounted rights issues need underwriting, © *Financial Times*, 25 April 2008; Chapter 16 GT achieves a first with $200m issue, © *Financial Times*, 18 December 2007; Chapter 16 Islamic bonds hit by growing religious concerns, © *Financial Times*, 7 February 2008; Chapter 17 Charter set for dividend pay-out, © *Financial Times*, 13 September 2007; Chapter 17 GSK doubles buy-back pledge, © *Financial Times*, 26 July 2007; Chapter 17 Chevron to spend $15bn on buying back shares, © *Financial Times*, 27 September 2007; Chapter 17 Ill-judged buy-backs are collectively destroying billions in shareholder value, 16 December 2007; Chapter 17 Dividend rise marks shift in investor rewards, © *Financial Times*, 6 February 2008; Chapter 18 Premier Foods, © *Financial Times*, 4 March 2008; Chapter 18 UK utilities, © *Financial Times*, 19 October 2007; Chapter 18 Kwik Save poised to enter administration, © *Financial Times*, 6 July 2007; Chapter 18 Rights issue to cut SMG debt by £91m, © *Financial Times*, 7 November 2007; Chapter 20 Screen Saver, © *Financial Times*, 16 May 2007; Chapter 20 TV producer Shed buys Twenty Twenty, © *Financial Times*, 20 September 2007; Chapter 20 JCB digs up its second takeover, © *Financial Times*, 20 July 2005; Chapter 20 BAT Tekel deal to draw market share from Turkey's Marlboro men, © *Financial Times*, 23 February 2008; Chapter 20 Eddie Stobart drives on to LSE, © *Financial Times*, 16 August 2007; Chapter 20 Fortis sets out rights issue for ABN bid, © *Financial Times*, 22 September 2007; Chapter 20 Pfizer's strategy a sign that bigger is rarely better, © *Financial Times*, 1 October 2007; Chapter 20 Staples offers ?2,5bn for Dutch rival, © *Financial Times*, 20 February 2008; Chapter 20 Japan hit by new 'poison pill' concern, 25 February 2008; Chapter 20 Smooth integration of GB helps EasyJet soar, © *Financial Times*, 20 February 2008; Chapter 20 Acquisitions in US 'disastrous' for British companies, © *Financial Times*, 11 October 2004; Chapter 20 Ebay writes down Skype value by $1.4bn, © *Financial Times*, 2 October 2007; Chapter 20 Smiths Group, © *Financial Times*, 14 September 2007; Chapter 20 Branson lets go of record store chain, © *Financial Times*, 17 September 2007; Chapter 20 Safestore to unlock value with a £449m flotation, © *Financial Times*, 10 March 2007; Chapter 20 Holidaybreak acquires PGL, © *Financial Times*, 19 May 2007; Chapter 20 Morocco factory key to car alliance's schemes, © *Financial Times*, 3 September 2007; Chapter 20 Private equity and the secret recipe of success, © *Financial Times*, 28 March 2007; Chapter 21 Exporters curse dollar's drag on profits, © *Financial Times*, 13 May 2007; Chapter 21 Yen rise to have 'big impact' on Toyota, © *Financial Times*,

8 March 2008; Chapter 21 BMW bets on rebound for falling US dollar, © *Financial Times*, 18 March 2004; Chapter 21 BMW steers a tricky course, © *Financial Times*, 30 October 2007; Chapter 22 Mallya buys W&M and eyes listing, © *Financial Times*, 17 May 2007; Chapter 22 Peugeot to build plant in Russia, © *Financial Times*, 30 January 2008; Chapter 22 Yen's rise puts brakes on Japan carmakers, © *Financial Times*, 8 February 2008; Chapter 22 Corruption conundrum, © *Financial Times*, 10 January 2008; Chapter 22 Tesco launches first dollar bond, © *Financial Times*, 30 October 2007; Chapter 23 Costain announces its first dividend in 17 years, © *Financial Times*, 13 March 2008; Chapter 23 The time has come for the CAPM to RIP, © *Financial Times*, 10 February 2007; Chapter 23 Investors show various traits of behaviour, © *Financial Times*, 27 March 2004.

We are grateful to the following for permission to reproduce copyright material:

Chapter 1 and Chapter 18 extracts from *Tomkins plc Annual Report 2006*; Figure 10.1, Tables 12.1, 12.3 and 18.1 from *DS Smith Plc Annual Report 2007*; Figure 3.3 from Yield-Curve © YieldCurve.publishing 2003, 2008; Table 6.1 from Strategic Capital Investment Decision-making: A Role for Emergent Analysis Tools? A Study of Practice in Large UK Manufacturing Companies in *British Accounting Review*, 38(12), 149–173, Elsevier (June 2006); Table 10.3 from Risk Measurement Service, London Business School, October–December 2007, *The London Business School*; Chapter 4 Question 6 from *CIMA Financial Strategy November 2006*, CIMA; Chapter 7 New risks put scenario planning in favour from *Financial Times Limited*, 19 August 2003, © Awi Federgruen and Garrett Van Ryzin; Chapter 15 Worked Example 15.11 from *CIMA Strategic Financial Management exam paper 2007*, CIMA; Chapter 16 Worked Example 16.6 from *CIMA paper 9, November 2006*, CIMA; Table 17.1 from *Scottish and Southern Energy plc.*; Chapter 18 extract from *E.on Annual Report 2006*; Chapter 21 'Carry on Speculating' from *The Economist*, 24th February 2007; Chapter 21 'The Big Mac Index' from *The Economist*, © The Economist Newspaper Limited, London 7th July 2007; Chapter 22 'Ikea shelves Thai store plans after new curbs on foreign ownership', *The Daily Telegraph*, 29th September 2007; Chapter 22 Worked Example 22.17 from *CIMA May 2006 P9 Financial Strategy exam paper*, CIMA; Table 22.5 Reproduced by permission of *Compass Group PLC Annual Report 2006*; Chapter 23 Watch the herd, but don't join it from *The Financial Times Limited*, 3 July 2004, © Brian Bloch.

In some instances we have been unable to trace the owners of copyright material, and we would appreciate any information that would enable us to do so.

Part I

A FRAMEWORK FOR FINANCIAL DECISIONS

Business financial decisions are not made in a vacuum. An 'obvious' decision may often have to be tempered by an appreciation of the restrictions imposed by the prevailing environment. Although it is beyond our scope to consider the full social, political and economic complexity of the financial decision-making context, we provide an overview of the key features of the UK financial and economic system. A sound grasp of the framework for financial decisions is essential if the reader is to appreciate fully the issues discussed in subsequent chapters of this book.

Part I provides an introduction to the scope and the fundamental concepts of financial management. Chapter 1 provides a broad picture of the subject and the important role it plays in business. It examines the nature of financing and investment decisions, the role of the financial manager and the fundamental objective for corporate financial management. This leads on, in Chapter 2, to consideration of the financial and tax environment in which businesses operate. Particular attention is devoted to the characteristics and operation of the London Stock Exchange, which provides a barometer of the success of financial decisions via the market's valuation of the company's shares. The extent to which any market can provide 'accurate' valuations is also considered.

Central concepts in financial management are the time-value of money and present value, which are discussed in Chapter 3. The chapter also provides an understanding of the valuation of bonds and shares. Concepts of value and its measurement play important roles in subsequent chapters, where investment, financing and other key decisions are discussed.

1

An overview of financial management

Working for shareholders

Tomkins plc, the global engineering and manufacturing group, has enjoyed one of the fastest growth rates over the past 30 years. In its *2006 Annual Report* it states its primary objective as:

> the creation of shareholder value by achieving long-term sustainable growth in the economic value of Tomkins

Source: 2006 Annual Report, www.tomkins.co.uk.

Cadbury plc, the world's largest confectionery company, has a similar goal:

> Our objective is to consistently deliver superior shareholder returns. We are committed to this objective although we recognise that the company does not operate in isolation. We have clear obligations to consumers, customers and suppliers, to our colleagues and to the society, communities and natural environment in which we operate.

Source: 2006 Annual Report, www.cadburyschw.com.

Learning objectives

By the end of this chapter, you should understand the following:

■ What corporate finance and investment decisions involve.

■ How financial management has evolved.

■ The finance function and how it relates to its wider environment and to strategic planning.

■ The central role of cash in business.

■ The goal of shareholder wealth creation and how investors can encourage managers to adopt this goal.

 *Complete your diagnostic test for Chapter 1 now to create your personal study plan.*

1.1 INTRODUCTION

The objectives of Tomkins, summarised at the start, suggest that its management has a clear idea of its purpose and key objectives. Its mission is to deliver economic value to its shareholders in the form of dividend and capital growth. An organisation such as Tomkins, with a broad range of products, understands the importance of meeting the requirements of its existing and potential customers. But it also recognises that the most important 'customers' are the **shareholders** – the owners of the business. Its objectives, strategies and decisions are all directed towards creating value for them.

One of the challenges in any business is to make investments that consistently yield rates of return to shareholders in excess of the cost of financing those projects and better than the competition. This book centres on that very issue: *how can firms create value through sound investment decisions and financial strategies?*

This chapter provides a broad picture of financial management and the fundamental role it plays in achieving financial objectives and operating successful businesses. First, we consider where financial management fits into the strategic planning process for a new business. This leads to an outline of the finance function and the role of the financial manager, and what objectives he or she may follow. Central to the subject is the nature of these financial objectives and how they affect shareholders' interests. Finally, we introduce the underlying principles of finance, which are developed in later chapters.

■ Starting a business: Brownbake Ltd

Ken Brown, a recent business graduate, decides to set up his own small bakery business. He recognises that a clear business strategy is required, giving a broad thrust to be adopted in achieving his objectives. The main issues are market identification, competitor analysis and business formation. He identifies a suitable market with room for a new entrant and develops a range of bakery products that are expected to stand up well, in terms of price and quality, against the existing competition.

Brown and his wife become the directors of a newly-formed limited company, Brownbake Ltd. This form of organisation has a number of advantages not found in a sole proprietorship or partnership:

- *Limited liability.* The financial liability of the owners is limited to the amount they have paid in. Should the company become insolvent, those with outstanding claims on the company cannot compel the owners to pay in further capital.
- *Transferability of ownership.* It is generally easier to sell shares in a company, particularly if it is listed on a stock market, than to sell all or part of a partnership or sole proprietorship.
- *Permanence.* A company has a legal identity quite separate from its owners. Its existence is unaffected by the sale of shares or death of a shareholder.
- *Access to markets.* The above benefits, together with the fact that companies enable large numbers of shareholders to participate, mean that companies can enjoy financial economies of scale, giving rise to greater choice and lower costs of financing the business.

Brown should have a clear idea of why the business exists and its financial and other objectives. He must now concentrate on how the business strategy is to be implemented. This requires careful planning of the decisions to be taken and their effect on the business. Planning requires answers to some important questions. What resources are required? Does the business require premises, equipment, vehicles and material to produce and deliver the product?

The key to industrial capitalism: limited liability

Shares or 'equities' were first issued in the 16th century, by Europe's new joint-stock companies, led by the Muscovy Company, set up in London in 1553, to trade with Russia. (Bonds, from the French government, made their debut in 1555.) Equity's popularity waxed and waned over the next 300 years or so, soaring with the South Sea and Mississippi bubbles, then slumping after both burst in 1720. But share-owning was mainly a gamble for the wealthy few, though by the early 19th century, in London, Amsterdam and New York, trading had moved from the coffee houses into specialised exchanges. Yet the key to the future was already there. In 1811, from America, came the first limited-liability law. In 1854, Britain, the world's leading economic power, introduced similar legislation.

The concept of limited liability, whereby the shareholders are not liable, in the last resort, for the debts of their company, can be traced back to the Romans. But it was rarely used, most often being granted only as a special favour to friends by those in power.

Before limited liability, shareholders risked going bust, even into a debtors' prison maybe, if their company did. Few would buy shares in a firm unless they knew its managers well and could monitor their activities, especially their borrowing, closely. Now, quite passive investors could afford to risk capital – but only what they chose – with entrepreneurs. This unlocked vast sums previously put in safe investments; it also freed new companies from the burden of fixed-interest debt. The way was open to finance the mounting capital needs of the new railways and factories that were to transform the world.

Source: Based on *The Economist*, 31 December 1999.

Once these issues have been addressed, an important further question is: how will such plans be funded? However sympathetic his bank manager, Brown will probably need to find other investors to carry a large part of the business risk. Eventually, these operating plans must be translated into financial plans, giving a clear indication of the investment required and the intended sources of finance. Brown will also need to establish an appropriate finance and accounting function (even if he does it himself), to keep himself informed of financial progress in achieving plans and ensure that there is always sufficient cash to pay the bills and to implement plans. Such issues are the principal concern of financial management, which applies equally to small businesses, like Brownbake Ltd, and large multinational corporations, like Tomkins plc.

1.2 THE FINANCE FUNCTION

In a well-organised business, each section should arrange its activities to maximise its contribution towards the attainment of corporate goals. The finance function is very sharply focused, its activities being specific to the financial aspects of management decisions. Figure 1.1 illustrates how the accounting and finance functions may be structured in a large company. This book focuses primarily on the roles of finance director and treasurer.

It is the task of those within the finance function to plan, raise and use funds in an efficient manner to achieve corporate financial objectives. Two central activities are as follows:

1 Providing the link between the business and the wider financial environment.
2 Investment and financial analysis and decision-making.

■ Link with financial environment

The finance function provides the link between the firm and the financial markets in which funds are raised and the company's shares and other financial instruments are traded. The financial manager, whether a corporate treasurer in a multinational

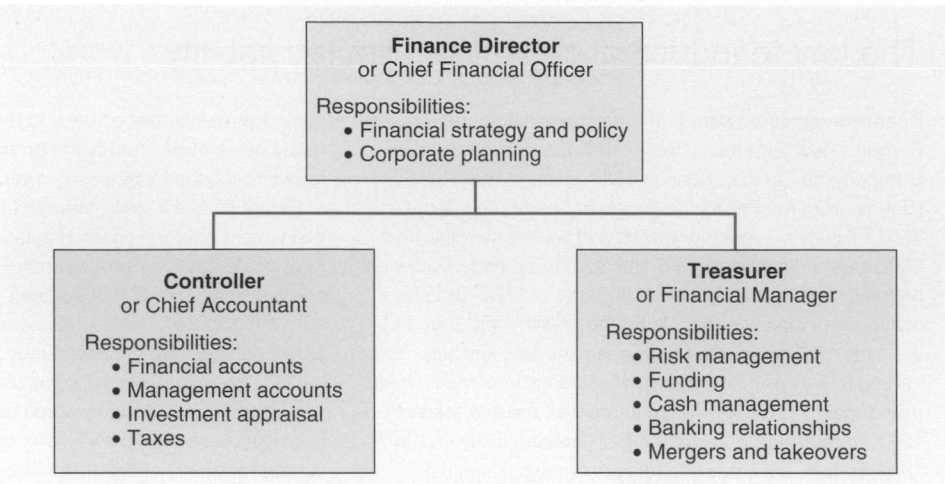

Figure 1.1 The finance function in a large organisation

company or the sole trader of a small business, acts as the vital link between financial markets and the firm. Corporate finance is therefore as much about understanding financial markets as it is about good financial management within the business. We examine financial markets in Chapter 2.

1.3 INVESTMENT AND FINANCIAL DECISIONS

Financial management is primarily concerned with investment and financing decisions and the interactions between them. These two broad areas lie at the heart of financial management theory and practice. Let us first be clear what we mean by these decisions.

The *investment decision*, sometimes referred to as the capital budgeting decision, is the decision to acquire assets. Most of these assets will be *real assets* employed within the business to produce goods or services to satisfy consumer demand. Real assets may be tangible (e.g. land and buildings, plant and equipment, and stocks) or intangible (e.g. patents, trademarks and 'know-how'). Sometimes a firm may invest in *financial assets* outside the business, in the form of short-term securities and deposits.

The basic problems relating to investments are as follows:

1 How much should the firm invest?
2 In which projects should the firm invest (fixed or current, tangible or intangible, real or financial)? Investment need not be purely internal. Acquisitions represent a form of external investment.

The *financing decision* addresses the problems of how much capital should be raised to fund the firm's operations (both existing and proposed), and what the best mix of financing is. In the same way that a firm can hold financial assets (e.g. investing in shares of other companies or lending to banks), it can also sell claims on its own real assets, by issuing shares, raising loans, undertaking lease obligations etc. A financial security, such as a share, gives the holder a claim on the future profits in the form of a dividend, while a bond (or loan) gives the holder a claim in the form of interest payable. Financing and investment decisions are therefore closely related.

Self-assessment activity 1.1

Take a look at the balance sheet of Brownbake Ltd.

Assets employed	£
Machinery and equipment	15,000
Vehicles	8,000
Patents	12,000
Stocks	10,000
Debtors	3,000
Cash and bank deposit	4,000
	52,000

Liabilities and shareholders' funds	
Trade creditors	12,000
Loans	8,000
Shareholders' equity	32,000
	52,000

Identify the tangible real assets, intangible assets and financial assets. Who has financial claims on these assets?

(Answer in Appendix A at the back of the book)

1.4 CASH – THE LIFEBLOOD OF THE BUSINESS

Central to the whole of finance is the generation and management of cash. Figure 1.2 illustrates the flow of cash for a typical manufacturing business. Rather like the bloodstream in a living body, cash is viewed as the 'lifeblood' of the business, flowing to all essential parts of the corporate body. If at any point the cash fails to flow properly, a 'clot' occurs that can damage the business and, if not addressed in time, can prove fatal!

Good cash management therefore lies at the heart of a healthy business. Let us now consider the major sources and uses of cash for a typical business.

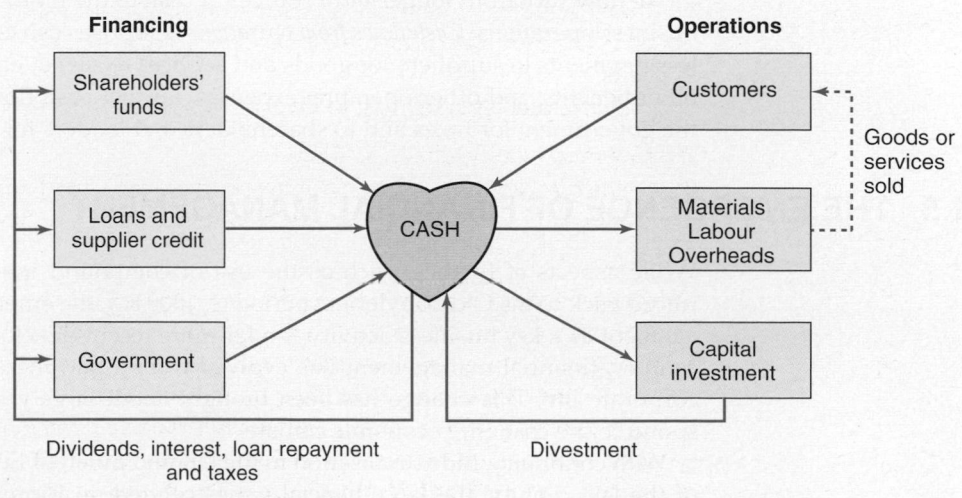

Figure 1.2 Cash – the lifeblood of the business

■ Sources and uses of cash

Shareholders' funds

**shareholders' funds/
equity capital**
Money invested by share-
holders and profits retained
in the company

The largest proportion of long-term finance is usually provided by shareholders and is termed **shareholders' funds or equity capital**. By purchasing a portion of, or shares in, a company, almost anyone can become a shareholder with some degree of control over a company.

Ordinary share capital is the main source of new money from shareholders. They are entitled both to participate in the business through voting in general meetings and to receive dividends out of profits. As owners of the business, the ordinary sharehold-ers bear the greatest risk, but enjoy the main fruits of success in the form of dividends and share price growth.

Retained profits

For an established business, the majority of equity funds will normally be internally generated from successful trading. Any profits remaining after deducting operating costs, interest payments, taxation and dividends are reinvested in the business (i.e. ploughed back) and regarded as part of the equity capital. As the business reinvests its cash surpluses, it grows and creates value for its owners. The purpose of the business is to do just that – create value for the owners.

Loan capital

debt finance/loan capital
Capital raised with an
obligation to pay interest
and repay principal

Money lent to a business by third parties is termed **debt finance or loan capital**. Most companies borrow money on a long-term basis by issuing loan stocks (or debentures). The terms of the loan will specify the amount of the loan, rate of interest and date of payment, redemption date, and method of repayment. Loan stock carries a lower risk than equity capital and, hence, offers a lower return.

The finance manager will monitor the long-term financial structure by examining the relationship between loan capital, where interest and loan repayments are contrac-tually obligatory, and ordinary share capital, where dividend payment is at the discre-tion of directors. This relationship is termed **gearing** (known in the USA as leverage).

gearing
Proportion of the total capital
that is borrowed

Government

Governments and the European Union (EU) provide various financial incentives and grants to the business community. A major cash outflow for successful businesses will be taxation.

We now turn from longer-term sources of cash to the more regular cash flows from business operations. *Cash flows from operations* comprise cash collected from customers less payments to suppliers for goods and services received, employees for wages and other benefits, and other operating expenses. Further cash flows include payments to the government for taxes and to shareholders and lenders for dividends and interest.

1.5 THE EMERGENCE OF FINANCIAL MANAGEMENT

While aspects of finance, such as the use of compound interest in trading, can be traced back to the Old Babylonian period (*c.* 1800 BC), the emergence of financial man-agement as a key business activity is a far more recent development. During the 20th century, financial management has evolved from a peripheral to a central aspect of corporate life. This change has been brought about largely through the need to re-spond to the changing economic climate.

With continuing industrialisation in the UK and much of Europe in the first quarter of the last century, the key financial issues centred on forming new businesses and raising capital for expansion. Legal and descriptive consideration was given to the types of security issued, company formations and mergers.

As the focus of business activity moved from growth to survival during the depression of the 1930s, finance evolved by focusing more on business liquidity, reorganisation and insolvency.

Successive Companies Acts, Accounting Standards and corporate governance mechanisms have been designed to increase investors' confidence in published financial statements and financial markets. However, the US accounting scandals in 2002, involving such giants as Enron and Worldcom, have dented this confidence.

In 2007, the mortgage crisis resulting from banks expanding their lending to sub-prime (i.e. riskier) borrowers developed into a world-wide banking crisis with a number of well-established banks going out of business, or being acquired or nationalised. By September 2008, the banking crisis had developed to such an extent that the Federal Reserve was forced to put forward a $700bn financial bail-out plan aimed at regaining global financial stability and investor confidence.

Recent years have seen the emergence of financial management as a major contributor to the analysis of investment and financing decisions. The subject continues to respond to external economic and technical developments:

1 Successive waves of merger activity over the past forty years have increased our understanding of valuation and takeover tactics. With governments committed to freedom of markets and financial liberalisation, acquisitions, mega-mergers and management buy-outs have become a regular part of business life.
2 Technological progress in communications and the liberalisation of markets have led to the globalisation of business. The single European market has created a major financial market with generally unrestricted capital movement. Modern computer technology not only makes globalisation of finance possible, but also brings complex financial calculations and financial databases within easy reach of every manager.
3 Complexities in taxation and the enormous growth in new financial instruments for raising money and managing risk have made some aspects of financial management highly specialised. The collapse in 1995 of Barings, the highly respected merchant bank, resulted from a lack of internal controls in the complex derivatives market.
4 Deregulation in the City is an attempt to make financial markets more efficient and competitive. The full adoption of the euro in 2002 for most European countries has reduced the risk and cost of doing business between such nations.
5 The requirement that from 2006 all EU-listed companies must use International Financial Reporting Standards (IFRS) for their accounts, gives greater transparency in company performance in Europe.
6 Greater awareness of the need to view all decision-making within a strategic framework is moving the focus away from purely technical to more strategic issues. For example, a good deal of corporate restructuring has taken place, breaking down large organisations into smaller, more strategically compatible businesses.

1.6 THE FINANCE DEPARTMENT IN THE FIRM

The organisational structure for the finance department will vary with company size and other factors. The board of directors is appointed by the shareholders of the company. Virtually all business organisations of any size are limited liability companies, thereby reducing the risk borne by shareholders and, for companies whose shares are listed on a stock exchange, giving investors a ready market for disposal of their holdings or further investment.

The financial manager can help in the attainment of corporate objectives in the following ways:

1 *Strategic investment and financing decisions.* The financial manager must raise the finance to fund growth and assist in the appraisal of key capital projects.
2 *Dealing with the capital markets.* The financial manager, as the intermediary between the markets and the company, must develop good links with the company's bankers and other major financiers, and be aware of the appropriate sources of finance for corporate requirements.
3 *Managing exposure to risk.* The finance manager should ensure that exposure to adverse movements in interest and exchange rates is adequately managed. Various techniques for hedging (a term for reducing exposure to risk) are available.
4 *Forecasting, coordination and control.* Virtually all important business decisions have financial implications. The financial manager should assist in and, where appropriate, coordinate and control activities that have a significant impact on cash flow.

Self-assessment activity 1.2

What are the financial manager's primary tasks?

(Answer in Appendix A at the back of the book)

1.7 THE FINANCIAL OBJECTIVE

For any company, there are likely to be a number of corporate goals, some of which may, on occasions, conflict. In finance, we assume that the objective of the firm is to *maximise shareholder value*. Put simply, this means that managers should create as much wealth as possible for the shareholders. Given this objective, any financing or investment decision expected to improve the value of the shareholders' stake in the firm is acceptable. You may be wondering why shareholder wealth maximisation is preferred to profit maximisation. Quite apart from the problems associated with profit measurement, it ignores the *timing* and *risks* of the profit flows. As will be seen later, value is heavily dependent on when costs and benefits arise and the uncertainty surrounding them.

The Quaker Oats Company was one of the first firms to adopt this goal:

> Our objective is to maximise value for shareholders over the long term . . . Ultimately, our goal is the goal of all professional investors – to maximise value by generating the highest cash flow possible.

However, many practising managers might take a different view of the goal of their firm. In recent years, a wide variety of goals have been suggested, from the traditional goal of profit maximisation to goals relating to sales, employee welfare, manager satisfaction, survival and the good of society. It has also been questioned whether management attempts to maximise, by seeking optimal solutions, or to seek merely satisfactory solutions.

Managers often seem to pursue a sales maximisation goal subject to a minimum profit constraint. As long as a company matches the average rate of return for the industry sector, the shareholders are likely to be content to stay with their investment. Thus, once this level is attained, managers will be tempted to pursue other goals. As sales levels are frequently employed as a basis for managerial salaries and status, managers may adopt goals that maximise sales subject to a minimum profit constraint.

earnings per share
Profit available for distribution to shareholders divided by the number of shares issued

A popular performance target is **earnings per share** (EPS). It focuses on the shareholder, rather than the company's performance, by calculating the earnings (i.e. profits after tax) attributable to each equity share.

Other subsidiary targets may be employed, often more in the form of a constraint ensuring that management does not threaten corporate survival in its pursuit of shareholder goals. Examples of such secondary goals which are sometimes employed include targets for:

1 *Profit retention.* For example, 'distributable profits must always be, say, at least three times greater than dividends'.
2 *Borrowing levels.* For example, 'long-term borrowing should not exceed 50 per cent of total capital employed'.
3 *Profitability.* For example, 'return on capital employed should be at least 18 per cent'.
4 *Non-financial goals.* These take a variety of forms but basically recognise that shareholders are not the only group interested in the company's success. Other stakeholders include trade creditors, banks, employees, the government and management. Each stakeholder group will measure corporate performance in a slightly different way. It is therefore to be expected that the targets and constraints discussed above will, from time to time, conflict with the overriding goal of shareholder value, and management must seek to manage these conflicts.

The financial manager has the specific task of advising management on the financial implications of the firm's plans and activities. The shareholder wealth objective should underlie all such advice, although the chief executive may sometimes allow non-financial considerations to take precedence over financial ones. It is not possible to translate this objective directly to the public sector or not-for-profit organisations. However, in seeking to create wealth in such organisations, the 'value for money' goal perhaps comes close.

Self-assessment activity 1.3

The past ten years have seen a much greater emphasis on investor-related goals, such as earnings per share and shareholder wealth. Why do you think this has arisen?

(Answer in Appendix A at the back of the book)

1.8 THE AGENCY PROBLEM

Potential conflict arises where ownership is separated from management. The ownership of most larger companies is widely spread, while the day-to-day control of the business rests in the hands of a few managers who usually own a relatively small proportion of the total shares issued. This can give rise to what is termed *managerialism* – self-serving behaviour by managers at the shareholders' expense. Examples of managerialism include pursuing more perquisites (splendid offices and company cars, etc.) and adopting low-risk survival strategies and 'satisficing' behaviour. This conflict has been explored by Jensen and Meckling (1976), who developed a theory of the firm under agency arrangements. Managers are, in effect, agents for the shareholders and are required to act in their best interests. However, they have operational control of the business and the shareholders receive little information on whether the managers are acting in their best interests.

A company can be viewed as simply a set of contracts, the most important of which is the contract between the firm and its shareholders. This contract describes the **principal–agent** relationship, where the shareholders are the principals and the management team the agents. An efficient agency contract allows full delegation of decision-making authority over use of invested capital to management without the risk of that authority being abused. However, left to themselves, managers cannot be expected to act in the shareholders' best interests, but require appropriate incentives

principal–agent
The agent, such as board of directors, is expected to act in the best interests of the principal (e.g. the shareholder)

and controls to do so. **Agency costs** are the difference between the return expected from an efficient agency contract and the actual return, given that managers may act more in their own interests than the interests of shareholders.

Self-assessment activity 1.4

Identify some potential agency problems that may arise between shareholders and managers.
(Answer in Appendix A at the back of the book)

1.9   MANAGING THE AGENCY PROBLEM

To attempt to deal with such agency problems, various incentives and controls have been recommended, all of which incur costs. Incentives frequently take the form of bonuses tied to profits (profit-related pay) and share options as part of a remuneration package scheme.

Managerial incentives: Blanco plc

Relating managers' compensation to achievement of owner-oriented targets is an obvious way to bring the interests of managers and shareholders closer together. A group of major institutional shareholders of Blanco plc has expressed concern to the chief executive that management decisions do not appear to be fully in line with shareholder requirements. They suggest that a new remuneration package is introduced to help solve the problem. Such packages have increasingly been introduced to encourage managers to take decisions that are consistent with the objectives of the shareholders.

The main factors to be considered by Blanco plc might include the following:

1 Linking management compensation to changes in shareholder wealth, where possible reflecting managers' contributions.
2 Rewarding managerial efficiency, not managerial luck.
3 Matching the time horizon for managers' decisions to that of shareholders. Many managers seek to maximise short-term profits rather than long-term shareholder wealth.
4 Making the scheme easy to monitor, inexpensive to operate, clearly defined and incapable of managerial manipulation. Poorly devised schemes have sometimes 'backfired', giving senior managers huge bonuses.

Two performance-based incentive schemes that Blanco plc might consider are rewarding managers with shares or with share options.

1 *Long-term incentive plans (LTIPs).* Such schemes typically incentivise performance over a period of three or more years, with the manager receiving the award at the end of the period. Shares are allotted to managers on attaining performance targets. Commonly employed performance measures are growth in earnings per share, return on equity and return on assets. Managers are allocated a certain number of shares to be received on attaining prescribed targets. While this incentive scheme offers managers greater control, the performance measures may not be entirely consistent with shareholder goals. For example, adoption of return on assets as a measure, which is based on book values, can inhibit investment in wealth-creating projects with heavy depreciation charges in early years.
2 *Executive share option schemes.* These are long-term compensation arrangements that permit managers to buy shares at a given price (generally today's) at some future date (generally 3–10 years). Subject to certain provisos and tax rules, a share option scheme usually entitles managers to acquire a fixed number of shares over a fixed period of time for a fixed price. The shares need not be paid for until the option is exercised – normally 3–10 years after the granting of the option. For example, a manager may be granted 20,000 share options. She can purchase these shares at any time over the next three years at £1 a share. If she decides to exercise her option when the share price has risen to £4, she would have gained £60,000 (i.e. buying 20,000 shares at £1, now worth £80,000).

Share options only have value when the actual share price exceeds the option price; managers are thereby encouraged to pursue policies that enhance long-term wealth-creation. Most large UK companies now operate share option schemes, which are spreading to managers well below board level. The figure is far higher for companies recently coming to the stock market: virtually all of them have executive share option schemes, and many of these operate an all-employee scheme. However, a major problem with these approaches is that general stock market movements, due mainly to macroeconomic events, are sometimes so large as to dwarf the efforts of managers. No matter how hard a management team seeks to make wealth-creating decisions, the effects on share price in a given year may be undetectable if general market movements are downward. A good incentive scheme gives managers a large degree of control over achieving targets. Chief executives in a number of large companies have recently come under fire for their 'outrageously high' pay resulting from such schemes.

Executive compensation schemes, such as those outlined above, are imperfect, but useful, mechanisms for retaining able managers and encouraging them to pursue goals that promote shareholder value.

Another way of attempting to minimise the agency problem is by setting up and monitoring managers' behaviour. Examples of these include:

1 audited accounts of the company;
2 management audits and additional reporting requirements; and
3 restrictive covenants imposed by lenders, such as ceilings on the dividend payable on the maximum borrowings.

To what extent does the agency problem invalidate the goal of maximising the value of the firm? In an efficient, highly competitive stock market, the share price is a 'fair' reflection of investors' perceptions of the company's expected future performance. So agency problems in a large publicly quoted company will, before long, be reflected in a lower than expected share price. This could lead to an *internal* response – the shareholders replacing the board of directors with others more committed to their goals – or an *external* response – the company being acquired by a better-performing company where shareholder interests are pursued more vigorously.

1.10 SOCIAL RESPONSIBILITY AND SHAREHOLDER WEALTH

Is the shareholder wealth maximisation objective consistent with concern for social responsibility? In most cases it is. As far back as 1776, Adam Smith recognised that, in a market-based economy, the wider needs of society are met by individuals pursuing their own interests: 'It is not from the benevolence of the butcher, the brewer, or the baker, that we expect our dinner, but from their regard to their own interest.' The needs of customers and the goals of businesses are matched by the 'invisible hand' of the free market mechanism.

Of course, the market mechanism cannot differentiate between 'right' and 'wrong'. Addictive drugs and other socially undesirable products will be made available as long as customers are willing to pay for them. Legislation may work, but often it simply creates illegal markets in which prices are much higher than before legislation. Other products have side-effects adversely affecting individuals other than the consumers, e.g. passive smoking and car exhaust emissions.

There will always be individuals in business seeking short-term gains from unethical activities. But, for the vast majority of firms, such activity is counterproductive in the longer term. Shareholder wealth rests on companies building long-term relationships with suppliers, customers and employees, and promoting a reputation for honesty, financial integrity and corporate social responsibility. After all, a major company's most important asset is its good name.

Not all large businesses are dominated by shareholder wealth goals. The John Lewis Partnership, which operates department stores and Waitrose supermarkets, is a partnership with its staff electing half the board. The Partnership's ultimate aim, as described in its constitution, 'shall be the happiness in every way of all its members'.

The Partnership rule book makes it clear, however, that pursuit of happiness shall not be at the expense of business efficiency. Its constitution requires it to take account of its suppliers, customers and local community.

Stakeholder theory asserts that managers should make decisions that take into account the interests of all the firm's stakeholders. This will include shareholders, employees, suppliers, customers, local communities, the government and environment. It is undoubtedly true that management should consider all stakeholders in its decision-making, but where interests conflict it becomes a highly complex task to maximise multiple objectives. However, as we saw with Cadbury plc at the start of the chapter, the company recognised its obligations to all its stakeholders, but had a single-valued objective function based on delivering superior shareholder returns. This shareholder focus that also recognises the needs of other stakeholders is sometimes termed 'enlightened shareholder value'.

Environmental concerns have in recent years become an important consideration for the boards of large companies, including the source of supplies, such as timber and paper from 'managed forests'. Investors are also becoming more socially aware and many are channelling their funds into companies that employ environmentally and socially responsible practices.

1.11 THE CORPORATE GOVERNANCE DEBATE

In recent years, there has been considerable concern in the UK about standards of corporate governance, the system by which companies are directed and controlled. While, in company law, directors are obliged to act in the best interests of shareholders, there have been many instances of boardroom behaviour difficult to reconcile with this ideal.

There have been numerous examples of spectacular collapses of companies, often the result of excessive debt financing in order to finance ill-advised takeovers, and sometimes laced with fraud. Many companies have been criticised for the generosity with which they reward their leading executives. The procedures for remunerating executives have been less than transparent, and many compensation schemes involve payment by results in one direction alone. Many chief executives have been criticised for receiving pay increases several times greater than the increases awarded to less exalted staff.

In the train of these corporate collapses and scandals, a number of committees have reported on the accountability of the board of directors to their stakeholders and risk management procedures, brought together as the 'Combined Code'.

The **Combined Code on Corporate Governance,** introduced in 2003, applies to all listed companies. Its main requirements for financial management are summarised below.

1 *Directors and the board*
   - There should be a clear division of responsibilities between the running of the board (chairman) and the executive responsibility for the running of the business (chief executive).
   - The board should include a balance of executive and independent non-executive directors.
   - It should be supplied in a timely manner with information in a form and quality appropriate to enable it to discharge its duties.
2 *Directors' remuneration*
   - Levels of remuneration should be sufficient to attract, retain and motivate directors, but should not be more than is necessary for the purpose.
   - No director should be involved in deciding his/her remuneration.
   - The performance-related elements of remuneration should form a significant proportion of the total remuneration package of executive directors and be designed to align their interests with those of shareholders.

3 *Accountability and audit*

- The board should present a balanced and understandable assessment of the company's position and prospects.
- The directors should report that the business is a going concern, with supporting assumptions or qualifications as necessary.
- The board should maintain a sound system of internal control to safeguard shareholders' investment and the company's assets.
- The board should establish an audit committee to monitor the integrity of financial statements.

4 *Relations with shareholders*

- The board should maintain a satisfactory dialogue with shareholders and keep in touch with shareholder opinion in whatever ways are practical and efficient.
- The board should use the AGM to communicate to investors and encourage participation.

Corporate governance is an important issue throughout the world and most countries have developed a code or recommendations. (A website for the relevant country codes is given at the end of this chapter.) In the US, for example, the Sarbanes-Oxley Act of 2002 is intended to protect investors by improving the accuracy and reliability of corporate reporting.

A manager's real responsibility

Businesses fail. As Joseph Schumpeter, the great Austrian economist, pointed out almost a century ago, such 'creative destruction' lies at the heart of the market economy's dynamism. Coming at the end of an era of rapid growth, swift technological change and widespread euphoria, a big corporate failure, such as Enron's, cannot be that surprising. There could be many more. Yet the Enron case also sheds intriguing light on conflicts of interest inherent in corporate capitalism.

The corporation is a wonderful institution. But it contains inherent drawbacks, at the core of which are conflicts of interest. Control over the company's resources is vested in the hands of top managers who may rationally pursue their interests at the expense of all others. Economists call this the 'principal–agent' problem. In the modern economy, where shares are held by fund managers, there is not just one set of principal–agent relations but a long chain of them.

The principal–agent problem is exacerbated by two others: asymmetric information and obstacles to collective action. Corporate managers know more about what is going on in the business than anybody else and have an interest in keeping at least some of this information to themselves. Equally, dispersed shareholders have a weak incentive to act, because they would share the gains with others but bear much of the cost themselves.

The upshot is the chronic vulnerability of the corporation to managerial incompetence, self-seeking, deceit or downright malfeasance. In practice, there are five (interconnected) ways of reducing these risks. The first is market discipline, since failure will ultimately find managers out. The second is internal checks, with independent directors or requirements for voting by institutional shareholders. The third is regulation covering the composition of boards, structure of businesses and reporting requirements. The fourth is transparency,

including accounting standards and independent audits. The last is simply values of honest dealing.

Economists are very uncomfortable with the notion of morality. Yet it seems to have rather a clear meaning in the business context. It consists of acting honestly even when the opposite may be to one's advantage. Such morality is essential for all trustee relationships. Without it, costs of supervision and control become exorbitant. At the limit, a range of transactions and long-term relationships becomes impossible and society remains impoverished. Corporate managers are trustees. So are fund managers. The more they view themselves (and are viewed) as such, the less they are likely to exploit opportunities created by the conflicts of interest within the business. What has all this to do with Enron? The answer is that the checks failed. The conflicts of interest of those responsible for transparency (the auditors) were huge and rules governing accounting proved inadequate. Because information was insufficient, the company was able to pursue its bets well beyond a sensible limit. The vast personal wealth available to top management also created big incentives for such behaviour.

None of this is unique to Enron. In what will surely come to be called the US bubble era, top managers were allowed to do many things that made little sense for anybody but themselves. Lavish share options that failed to align their interests with those of shareholders were just one example. The response will be to tighten up on regulation. Some of this is necessary, particularly over the role of auditors and the probity of accounts. Yet care must be taken. Any system guaranteed to prevent bankruptcies would damage the risk-taking essential to economic dynamism.

Source: Based on Martin Woolf, *Financial Times*, 30 January 2002, p. 19.

The main reservations centre on the issues of compliance and enforcement. These changes in the rules and responsibilities of directors and auditors are non-statutory. The Stock Exchange may not withdraw the listings of companies that fail to comply, although it hopes that any adverse publicity will whip offenders into line. This lack of 'teeth' has raised suspicions that determined wrongdoers can still exert their influence on weak boards of directors, to the detriment of the relatively ill-informed private investor in particular.

1.12 THE RISK DIMENSION

Some financial decisions incur very little risk (e.g. investing in government stocks, since the interest is known); others may carry far more risk (e.g. investing in shares). Risk and expected return tend to be related: the greater the perceived risk, the greater the return required by investors. This is seen in Figure 1.3.

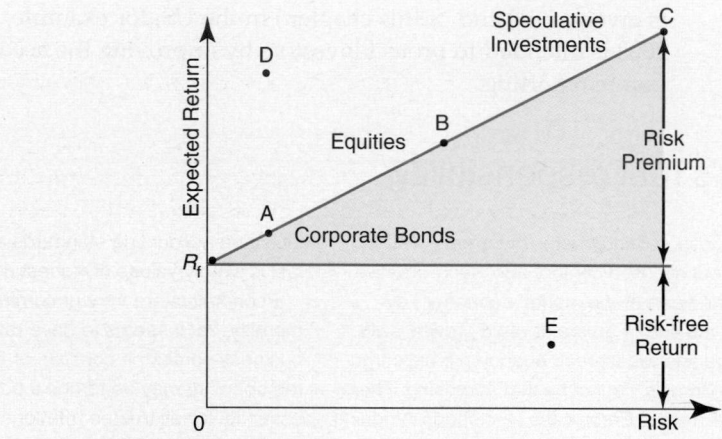

Figure 1.3 The risk–return trade-off

When the finance manager of a company seeks to raise funds, potential investors take a view on the risk related to the intended use of the funds. This can best be measured in terms of a **risk premium** above the risk-free rate (R_f) obtainable from, say, government stocks to compensate investors for taking risk. The capital market offers a host of investment opportunities for private and corporate investors, but in all cases there exists a clear relationship between the perceived degree of risk involved and the expected return. For example, R_f in Figure 1.3 represents the return on three-month Treasury Bills; point A represents a long-term fixed interest corporate bond; point B, a portfolio of ordinary shares in major listed companies; and point C, a more speculative investment, such as non-quoted shares. Studies indicate that the long-term average return on an investment portfolio consisting of the market index (e.g. the FTSE-100) is up to 6 percentage points higher than that from holding risk-free government securities.

One task of the financial manager is to raise funds in the capital markets at a cost consistent with the perceived risk, and to invest such funds in wealth-creating opportunities in the business. Here it is quite possible – because of a firm's competitive advantage, or possession of superior brand names – to make highly profitable capital projects with relatively little risk (see D in diagram). It is also possible to find the reverse, such as project E. If the goal is to deliver cash flows to shareholders at rates above their cost of capital, managers should seek to invest in projects, such as D, that offer returns better than those obtainable on the capital market for the same degree of risk (A in the diagram).

1.13 THE STRATEGIC DIMENSION

To enhance shareholder value, managers could adopt a wide range of strategies. Strategic management may be defined as a systematic approach to positioning the business in relation to its environment to ensure continued success and offer security from surprises. No approach can guarantee continuous success and total security, but an integrated approach to strategy formulation, involving all levels of management, can go some way.

Strategy can be developed at three levels:

1 *Corporate strategy* is concerned with the broad issues, such as the types of business the company should be in. Strategic finance has an important role to play here. For example, the decision to enter or exit from a business – whether through corporate acquisitions, organic growth, divestment or buy-outs – requires sound financial analysis. Similarly, the appropriate capital structure and dividend policy form part of strategic development at the corporate level.

2 *Business or competitive strategy* is concerned with how strategic business units compete in particular markets. Business strategies are formulated which influence the allocation of resources to these units. This allocation may be based on the attractiveness of the markets in which business units operate and the firm's competitive strengths.

3 *Operational strategy* is concerned with how functional levels contribute to corporate and business strategies. For example, the finance function may formulate strategies to achieve a new dividend policy identified at the corporate strategy level. Similarly, a foreign currency exposure strategy may be developed to reduce the risk of loss through currency movements. A typical strategic planning process is shown in Figure 1.4.

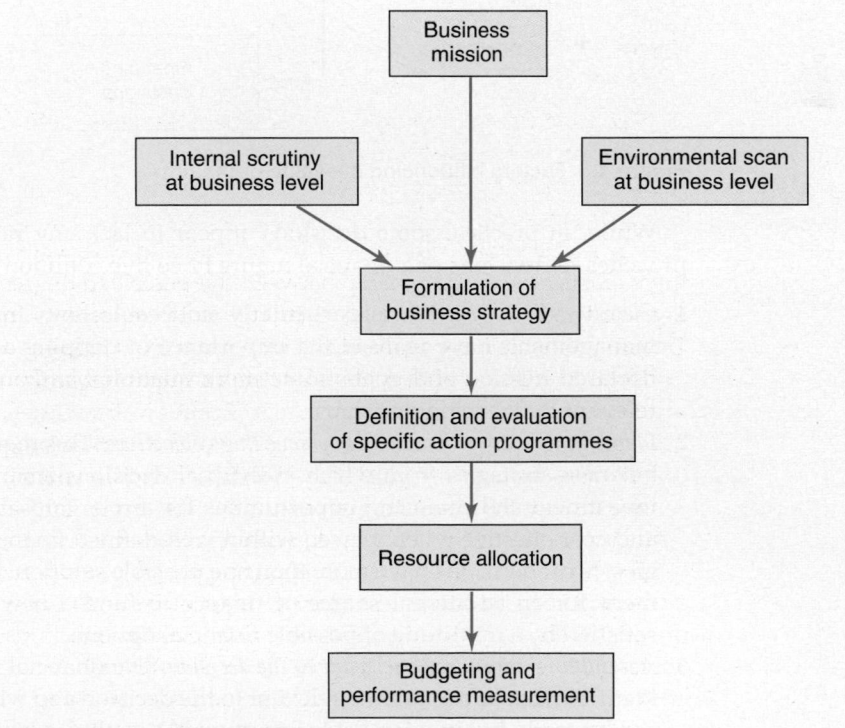

Figure 1.4 Main elements in strategic planning

■ Strategic planning and value creation

The importance of competitive forces in determining shareholder wealth cannot be overestimated. They largely determine the price at which goods and services can be sold, the quantities sold, the cost of production, the level of required investment and the risks inherent in the business.

However, individual companies can develop strategies leading to long-term financial performance well above the industry average. Figure 1.5 illustrates the main factors influencing the value of the firm. In any industry, all firms will be subject to much the same underlying economic conditions. Rates of inflation, interest and taxation and competitive forces in the industry will affect all businesses, although not necessarily to the same degree. The firm will develop corporate, business and operating strategies to exploit economic opportunities and to create sustainable competitive advantage. We are mainly concerned with those strategies affecting investment, financing and dividends. Operating and investment decisions create cash flows for the business, while financing decisions influence the cost of capital. The value of the firm depends upon the cash flows generated from business operations – their size, timing and riskiness – and the firm's cost of capital. Depending on the success of the firm's strategies and decisions, the value of the firm will increase or shrink.

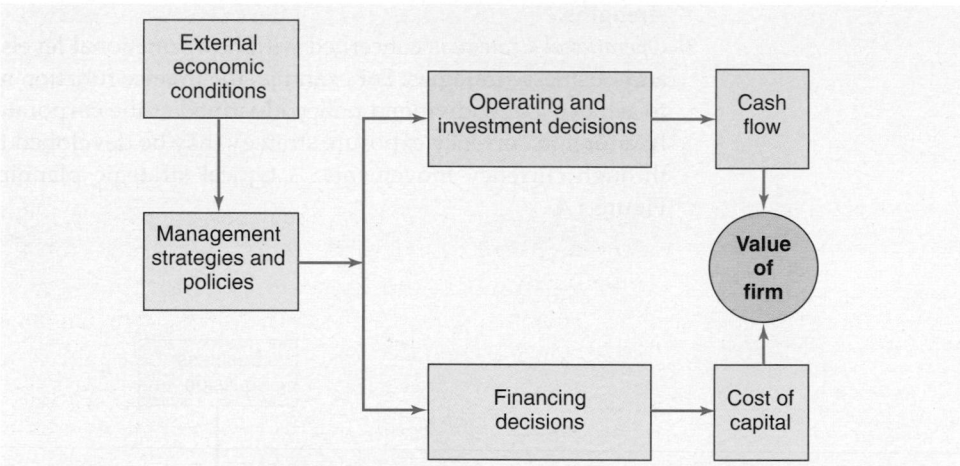

Figure 1.5 Factors influencing the value of the firm

While, in practice, some decisions appear to lack any rational process, most approaches to decisions of a financial nature have five common elements:

1 *Clearly-defined goals.* It is particularly noticeable how, in recent years, corporate managements have realised the importance of defining and communicating their declared mission and goals, some more quantifiable than others, and some more relevant to financing decisions.

2 *Identifying courses of action to achieve these objectives.* This requires the development of business strategies from which individual decisions emanate. The search for new investment and financing opportunities for any organisation is far better focused and cost-effective when viewed within well-defined financial objectives and strategies. Most decisions have more than one possible solution. For instance, the requirement for an additional source of finance to fund a new product launch can be satisfied by a multitude of possible financial options.

3 *Assembling information relevant to the decision.* The financial manager must be able to identify what information is relevant to the decision and what is not. Data gathering can be costly, but good, reliable information greatly facilitates decision analysis and confidence in the decision outcome.

4　*Evaluation.* Analysing and interpreting assembled information lies at the heart of financial analysis. A large part of this book is devoted to techniques of appraising financial decisions.

5　*Monitoring the effects of the decision taken.* However sophisticated a firm's financial planning system, there is no real substitute for experience. Feedback on the performance of past decisions provides vital information on the reliability of data gathered, the efficacy of the method employed in decision appraisal and the judgement of decision-makers.

Throughout this book, we shall attempt to allow for practical, real-world considerations when considering appropriate financial policy decisions. However, we hope that a clearer understanding of the concepts, together with an awareness of the degree of realism in their underlying assumptions, will enable the reader to make sound and successful investment and financial decisions in practice.

SUMMARY

This chapter has provided an overview of strategic financial management and the critical role it plays in corporate survival and success. We have examined how financial management has evolved over the years, its main functions and objectives.

Key points

- It is the task of the financial manager to plan, raise and use funds in an efficient manner to achieve corporate financial objectives. This implies (1) involvement in investment and financing decisions, (2) dealing with the financial markets, and (3) forecasting, coordinating and controlling cash flows.

- Cash is the lifeblood of any business. Financial management is concerned with cash generation and control.

- Financial management evolved during the last century, largely in response to economic and other external events (e.g. inflation and technological developments), making globalisation of finance a reality and the need to concentrate on more strategic issues essential.

- The distinction should be drawn between accounting – the mere provision of relevant financial information for internal and external users – and financial management – the utilisation of financial and other data to assist financial decision-making.

- In finance, we assume that the primary corporate goal is to maximise value for the shareholders.

- The agency problem – managers pursuing actions not totally consistent with shareholders' interests – can be reduced both by managerial incentive schemes and also by closer monitoring of their actions.

- Investors require compensation for taking risks in the form of enhanced potential returns.

- Most of the assumptions underlying pure finance theory are not particularly realistic. In practice, market and other imperfections must also be considered in practical financial decision-making.

- Financial management has an essential role in strategic development and implementation at strategic, business and operational levels. Competitive forces, together with business strategy, influence the value drivers that impact on shareholder value.

Further reading

Students should get into the habit of reading the *Financial Times* and relevant pages of *The Economist* and *Investors Chronicle*.

Jensen (2001) and Wallace (2003) provide a useful discussion of the relevance of the value maximisation and stakeholder goals as the corporate objective function for the firm. Miller (2000) offers a fascinating history of finance.

For a fuller discussion on managerial compensation, see Lambert and Larcker (1985). Jensen and Meckling (1976) and Fama (1980) provide the best articles on agency costs while Brickley *et al.* (2003) give a useful insight into organisational ethics and social responsibility. Grinyer (1986) provides an alternative to the shareholder wealth goal while Doyle (1994) argues for a 'stakeholder' approach to goal-setting. On the other hand, Koller *et al.* (2005) argue that shareholder wealth creation is good for all stakeholders, productivity and employment. Details on these and other references are provided at the end of the book.

Useful websites

Financial Times: **www.FT.com**
Guardian: **www.guardian.co.uk/money**
The Economist: **www.economist.com**
Corporate governance codes in other countries: **www.ecgi.org/codes/all_codes**
Companies House: **www.companieshouse.gov.uk**

QUESTIONS

 **myfinancelab** | *Questions with an icon are also available for practice in myfinancelab with additional supporting resources.*

Questions with a **coloured number** have solutions in Appendix B on page 729.

1 Why is the goal of maximising owners' wealth helpful in analysing capital investment decisions? What other goals should also be considered?

2 Go4it plc is a young dynamic company which became listed on the stock market three years ago. Its management is very keen to do all it can to maximise shareholder value and, for this reason, has been advised to pursue the goal of maximising earnings per share. Do you agree?

(Solution on companion website **www.booksites.net/pikeneale**)

3 (a) 'Managers and owners of businesses may not have the same objectives.' Explain this statement, illustrating your answer with examples of possible conflicts of interest.
(b) In what respects can it be argued that companies need to exercise corporate social responsibility?
(c) Explain the meaning of the term 'Value for Money' in relation to the management of publicly owned services/utilities.

(ACCA)

4 Discuss the importance and limitations of ESOPs (executive share option plans) to the achievement of goal congruence within an organisation.

(ACCA)

5 (a) A group of major shareholders of Zedo plc wishes to introduce a new remuneration scheme for the company's senior management. Explain why such schemes might be important to the shareholders. What factors should shareholders consider when devising such schemes?
(b) Eventually a short-list of three possible schemes is agreed. All pay the same basic salary plus:
(i) A bonus based upon at least a minimum pre-tax profit being achieved.
(ii) A bonus based upon turnover growth.
(iii) A share option scheme.
Briefly discuss the advantages and disadvantages of each of these three schemes.

(ACCA)

6 The primary financial objective of companies is usually said to be the maximisation of shareholders' wealth. Discuss whether this objective is realistic in a world where corporate ownership and control are often separate, and environmental and social factors are increasingly affecting business decisions.

7 The main principles of financial management may be applied to most organisations. However, the role of the financial manager may be affected by the type of organisation in which he or she works.

Required
Describe the key characteristics of the financial management function and the role of the financial manager in each of the following types of organisation.
(a) Quoted high-growth company
(b) Quoted low-growth company
(c) Unquoted company aiming for a stock exchange listing
(d) Small family-owned business
(e) Non-profit-making organisation, for example a charity
(f) Public sector, for example a government department

(CIMA)

8 (a) The Cleevemoor Water Authority was privatised in 2000, to become Northern Water plc (NW). Apart from political considerations, a major motive for the privatisation was to allow access for NW to private sector supplies of finance. During the 1980s, central government controls on capital expenditure had resulted in relatively low levels of investment, so that considerable investment was required to enable the company to meet more stringent water quality regulations. When privatised, it was valued by the merchant bankers advising on the issue at £100 million and was floated in the form of 100 million ordinary shares (par value 50p), sold fully paid for £1 each. The shares reached a premium of 60 per cent on the first day of stock market trading.

Required

In what ways might you expect the objectives of an organisation like Cleevemoor/NW to alter following transfer from public to private ownership?

(b) Selected biannual data from NW's accounts are provided below relating to its first six years of operation as a private sector concern. Also shown, for comparison, are the pro forma data as included in the privatisation documents. The pro forma accounts are notional accounts prepared to show the operating and financial performance of the company in its last year under public ownership as if it had applied private sector accounting conventions. They also incorporate a dividend payment based on the dividend policy declared in the prospectus.

 The activities of privatised utilities are scrutinised by a regulatory body which restricts the extent to which prices can be increased. The demand for water in the area served by NW has risen over time at a steady 2 per cent per annum, largely reflecting demographic trends.

Required

Using the data provided, assess the extent to which NW has met the interests of the following groups of stakeholders in its first six years as a privatised enterprise.

Key financial and operating data for year ending 31 December (£m)

	2000 (pro forma)	2002 (actual)	2004 (actual)	2006 (actual)
Turnover	450	480	540	620
Operating profit	26	35	55	75
Taxation	5	6	8	10
Profit after tax	21	29	47	65
Dividends	7	10	15	20
Total assets	100	119	151	191
Capital expenditure	20	30	60	75
Wage bill	100	98	90	86
Directors' emoluments	0.8	2.0	2.3	3.0
Employees (number)	12,000	11,800	10,500	10,000
P:E ratio (average)	–	7.0	8.0	7.5
Retail Price Index	100	102	105	109

If relevant, suggest what other data would be helpful in forming a more balanced view.
(i) shareholders
(ii) consumers
(iii) the workforce
(iv) the government, through NW's contribution to the achievement of macroeconomic policies of price stability and economic growth.

(ACCA)

Practical assignment

Examine the annual report for a well-known company, particularly the chairman's statement. Are the corporate goals clearly specified? What specific references are made to financial management? What does it say about corporate governance and risk management?

 **myfinancelab** | *Now retake your diagnostic test for Chapter 1 to check your progress and update your study plan.*

2

The financial environment

A rock solid investment?

In July 2007 Northern Rock, one of the UK's largest mortgage lenders, issued an upbeat set of trading results, saying the outlook for the business is 'very positive'. A few weeks later, the global credit concerns, ignited by the troubled US sub-prime market, led banks to curb lending to each other and the rate at which British banks lend to each other – known as the London Interbank Offered Rate (LIBOR) – rose to its highest level in almost nine years.

Northern Rock had pursued an aggressive strategy of relying heavily on short- and medium-term wholesale funding to finance long-term mortgages and was unable to cope with the credit squeeze when it began. Share prices fell sharply as banks curbed lending to each other and long queues of customers seeking to withdraw their savings were seen outside its branches. Northern Rock was granted emergency financial support from the Bank of England, in the latter's role as 'lender of last resort'. Share prices collapsed from over 1200p in early 2007 to just 90p in February 2008 when its shares were suspended. The government announced that it had not been able to find an acceptable private sector solution and that the bank would 'temporarily' be taken into public ownership.

Learning objectives

By the end of this chapter, the reader should understand the nature of financial markets and the main players within them. Particular focus is placed on the following topics:

■ The functions of financial markets.

■ The operation of the Stock Exchange.

■ The extent to which capital markets are efficient.

■ How taxation affects corporate finance.

Enhanced ability to read financial statements and the financial pages in a newspaper should also be achieved.

myfinancelab *Complete your diagnostic test for Chapter 2 now to create your personal study plan.*

2.1 INTRODUCTION

The corporate financial manager, whether in Northern Rock or any company, has the important task of ensuring that there are sufficient funds available to meet all the likely needs of the business. To do this properly, he or she requires a clear grasp both of the future financial requirements of the business and of the workings of the financial markets. This chapter provides an overview of these markets, and the major institutions within them, paying particular attention to the Stock Exchange.

2.2 FINANCIAL MARKETS

financial market
Any market in which financial assets and liabilities are traded

A **financial market** is any mechanism for trading financial assets or securities. A **security** is a legal contract giving the right to receive future benefits under a stated set of conditions. Examples of financial securities range from the mortgage on a house or lease on a car to securities that are traded on financial markets, termed marketable securities.

Frequently, there is no physical marketplace, traders conducting transactions via computer. London is widely regarded as the leading European financial centre and is the largest by volume of dealing. Figure 2.1 shows the main financial markets, which are further explained below.

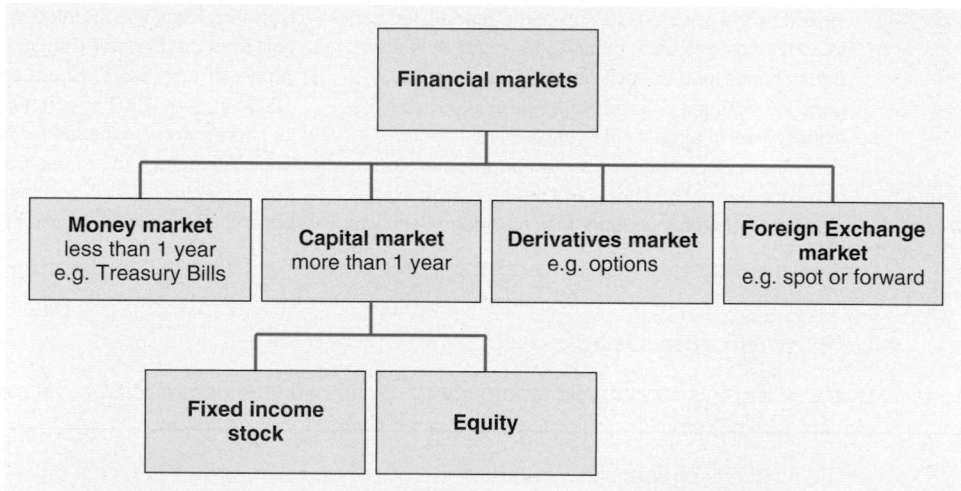

Figure 2.1 Financial markets

money market
The market for short-term money, broadly speaking for repayment within about a year

1 The **money market** channels wholesale funds, usually for less than one year, from lenders to borrowers. The market is largely dominated by the major banks and other financial institutions, but local government and large companies also use it for short-term lending and borrowing purposes. The official market is where approved institutions deal in financial instruments with the Central Bank. Other money markets include the inter-bank market, where banks lend short-term funds to each other, and the Euro-currency market where banks lend and borrow in foreign currencies.

securities/capital market
The market for long-term finance

2 The **securities** or **capital market** deals with long-dated securities such as shares and loan stock. The London Stock Exchange is the best-known institution in the UK capital market, but there are other important markets, such as the bond market (for long-dated government and corporate borrowing) and the Eurobond market.

spot
The spot, or cash market, is where transactions are settled immediately

3 The foreign exchange market is a market for buying and selling one currency against another. Deals are either on a **spot** basis (for immediate delivery) or on a **forward** basis (for future delivery).

forward
The forward market is where contracts are made for future settlement at a price specified now

4 The London International Financial Futures and Options Exchange (LIFFE) **www.liffe.com** provides various means of hedging (i.e. protecting) or speculating against

derivatives
Securities that are traded separately from the assets from which they are derived

future
A tradable contract to buy or sell a specified amount of an asset at a specified price at a specified future date

option
The right but not the obligation to buy or sell a particular asset

movements in shares, currencies and interest rates. These are called **derivatives** because they are derived from the underlying security. A **future** is an agreement to buy or sell an asset (e.g. foreign currency, shares etc.) at an agreed price at some future date. An **option** is the right, but not the obligation, to buy or sell such assets at an agreed price at, or within, an agreed time period.

The financial markets provide mechanisms through which the corporate financial manager has access to a wide range of sources of finance and instruments.

Capital markets function in two important ways:

1 *Primary market* – providing new capital for business and other activities, usually in the form of share issues to new or existing shareholders (equity), or loans.
2 *Secondary market* – trading existing securities, thus enabling share or bond holders to dispose of their holdings when they wish. An active secondary market is a necessary condition for an effective primary market, as no investor wants to feel 'locked in' to an investment that cannot be realised when desired.

Imagine what business life would be like if these capital markets were not available to companies. New businesses could start up only if the owners had sufficient personal wealth to fund the initial capital investment; existing businesses could develop only through re-investing profits generated; and investors could not easily dispose of their shareholdings. In many parts of the world where financial markets are embryonic or even non-existent, this is exactly what does happen. *The development of a strong and healthy economy rests very largely on efficient, well-developed financial markets.*

Financial markets promote savings and investment by providing mechanisms whereby the financial requirements of lenders (suppliers of funds) and borrowers (users of funds) can be met. Figure 2.2 shows in simple terms how businesses finance their operations.

financial intermediaries
Institutions that channel funds from savers and depositors with cash surpluses to people and organisations with cash shortages

Financial institutions (e.g. pension funds, insurance companies, banks, building societies, unit trusts and specialist investment institutions) act as **financial intermediaries**, collecting funds from savers to lend to their corporate and other customers through the money and capital markets, or directly through loans, leasing and other forms of financing.

Businesses are major users of these funds. The financial manager raises cash by selling claims to the company's existing or future assets in financial markets (e.g. by issuing shares, debentures or Bills of Exchange) or borrowing from financial institutions. The cash is then used to acquire fixed and current assets. If those investments are successful, they will generate positive cash flows from business operations. This cash surplus is used to service existing financial obligations in the form of dividends, interest etc., and to make repayments. Any residue is re-invested in the business to replace existing assets or to expand operations.

We focus in this chapter on the financial institutions and financial markets shown in Figure 2.2.

■ Financial institutions provide essential services

The needs of lenders and borrowers rarely match. Hence, there is an important role for financial intermediaries, such as banks, if the financial markets are to operate efficiently. Financial intermediaries perform the following functions:

1 *Re-packaging, or pooling, finance*: gathering small amounts of savings from a large number of individuals and repackaging them into larger bundles for lending to businesses. The banks have an important role here.
2 *Risk reduction*: placing small sums from numerous individuals in large, well-diversified investment portfolios, such as unit trusts.

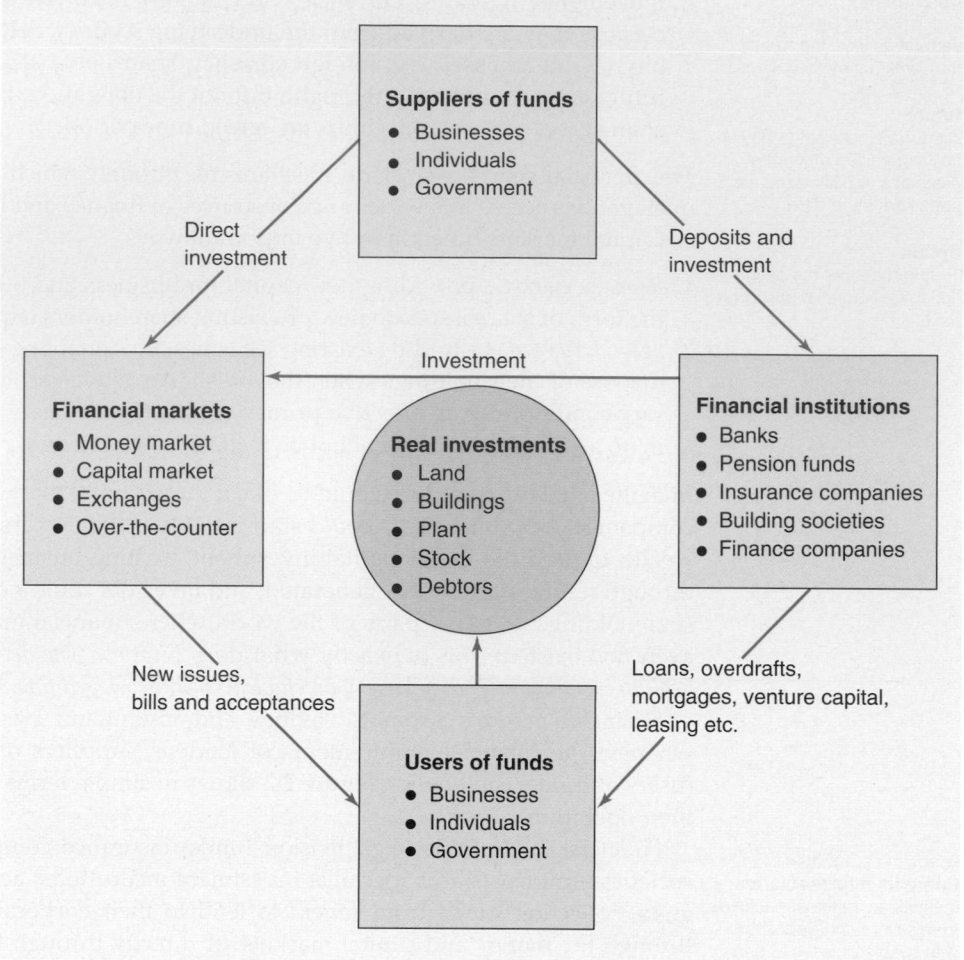

Figure 2.2 Financial markets, institutions, suppliers and users

3 *Liquidity transformation*: bringing together short-term savers and long-term borrowers (e.g. building societies and banks). Borrowing 'short' and lending 'long' is acceptable only where relatively few savers will want to withdraw funds at any given time. The financial difficulties faced by Northern Rock in the UK in 2007 shows that this is not always the case.

4 *Cost reduction*: minimising transaction costs by providing convenient and relatively inexpensive services for linking small savers to larger borrowers.

5 *Financial advice*: providing advisory and other services for both lender and borrower.

2.3 THE FINANCIAL SERVICES SECTOR

The financial services sector can be divided into three groups: institutions engaged in (1) deposit-taking, (2) contractual savings and (3) other investment funds.

■ Deposit-taking institutions

clearing banks
Banks (mainly the High Street banks) that are members of the Central Clearing House that arranges the mutual off-setting of cheques drawn on different banks

Clearing banks have three important roles: they manage nationwide networks of High Street branches and online facilities; they operate a national payments system by clearing cheques and by receiving and paying out notes and coins; and they accept deposits in varying amounts from a wide range of customers. Hence, these operations

retail banking
Retail banks accept deposits from the general public who can draw on these accounts by cheque (or ATM), and lend to other people and organisations seeking funds

are often called **retail banking**. As well as being the dominant force in retail banking, the clearing banks have diversified into wholesale banking and are continuing to expand their international activities. (Useful websites: **www.bba.org.uk**, **www.bcsb.co.uk**)

The balance sheet of any clearing bank reveals that the main sterling assets are advances to the private sector, other banks, the public sector in the form of Treasury Bills and government stock, local authorities and private households. Nowadays, the main instruments of lending by retail banks are overdrafts, term loans and mortgages.

In 2004 an international standard, called Basel II, was launched with regulations about how much capital banks need to put aside to guard against the types of financial and operational risks banks face. These rules mean that the greater the risk to which the bank is exposed, the greater the amount of capital the bank needs to hold to safeguard its solvency and overall economic stability.

Wholesale banks

accepting houses
Accepting houses are specialist institutions that discount or 'accept' Bills of Exchange, especially short-term government securities (see Chapter 15)

discount houses
Discount houses bid for issues of short-term government securities at a discount and either hold them to maturity or sell them on in the money market

merchant banks
Merchant banks are wholesale banks that arrange specialist financial services like mergers and acquisition funding, finance of international trade fund management

Wholesale banking (or merchant banking) developed out of the need to finance the enormous growth in world trade in the 19th century. **Accepting houses** were formed whose main business was to accept Bills of Exchange (promising to pay a sum of money at some future date) from less well-known traders, and from **discount houses** which provided cash by discounting such bills. **Merchant banks** nowadays concentrate on dealing with institutional investors, large corporations and governments. They have three major activities, frequently organised into separate divisions: corporate finance, mergers and acquisitions, and fund management.

Merchant banks' activities include *giving financial advice to companies and arranging finance* through syndicated loans and new security issues. Merchant banks are also members of the Issuing Houses Association, an organisation responsible for the *flotation of shares* on the Stock Exchange. This involves advising a company on the correct mix of financial instruments to be issued and on drawing up a prospectus and underwriting the issue. They also play a leading role in the *development of new financial products*, such as swaps, options and other derivative products, that have become very widely traded in recent years.

Another area of activity for wholesale banks is advising companies on corporate *mergers, acquisitions and restructuring*. This involves both assisting in the negotiation of a 'friendly' merger of two independent companies and developing strategies for 'unfriendly' takeovers, or acting as an adviser for a company defending against an unwanted bidder.

Finally, merchant banks fulfil a major role as *managers of the investment portfolios* of some pension funds, insurance companies, investment and unit trusts, and various charities. Whether in arranging finance, advising on takeover bids or managing the funds of institutional investors, merchant banks exert considerable influence on both corporate finance and the capital market.

The growth of *overseas banking* has been closely linked to the development of Eurocurrency markets and to the growth of multinational companies. Over 300 foreign banks operate in London. A substantial amount of their business consists of providing finance to branches or subsidiaries of foreign companies.

building societies
Financial institutions whose main function is to accept deposits from customers and lend for house purchases

Building societies (**www.bsa.org.uk**) are a form of savings bank specialising in the provision of finance for house purchases in the private sector. As a result of deregulation of the financial services industry, building societies now offer an almost complete set of private banking services, and the distinction between them and the traditional banks is increasingly blurred. Indeed, many societies have given up their mutual status to become public limited companies.

Self-assessment activity 2.1

What are financial intermediaries and what economic services do they perform?

(Answer in Appendix A at the back of the book)

■ Institutions engaged in contractual savings

pension funds
Financial institutions that manage the pension schemes of large firms and other organisations

Pension funds accumulate funds to meet the future pension liabilities of a particular organisation to its employees. Funds are normally built up from contributions paid by the employer and employees. They can be divided into **self-administered schemes**, where the funds are invested directly in the financial markets, and **insured schemes**, where the funds are invested by, and the risk is covered by, a life assurance company. Pension schemes have enormous and rapidly growing funds available for investment in the securities markets. Pension funds enjoy major tax advantages. Subject to certain restrictions, individuals enjoy tax relief on their subscriptions to a fund. In turn, the fund's income and capital gains are tax-free. Together with insurance companies, pension funds comprise the major purchasers of company securities.

self-administered schemes
A pension fund that invests clients' contributions directly into the stock market and other investments

Insurance companies' activities (www.abi.org.uk) can be divided into long-term and general insurance. Long-term insurance business consists mainly of *life assurance* and *pension provision*. Policyholders pay premiums to the companies and are guaranteed either a lump sum in the event of death, or a regular annual income for some defined period. With a guaranteed premium inflow and predictable aggregate future payments, there is no great need for liquidity, so life assurance funds are able to invest heavily in long-term assets, such as ordinary shares.

insured schemes
A pension fund that uses an insurance company to invest contributions and to insure against actuarial risks (e.g. members living longer than expected)

General insurance business (e.g. fire, accident, motor, marine and other insurance) consists of contracts to cover losses within a specified period, normally 12 months. As liquidity is important here, a greater proportion of funds is invested in short-term assets, although a considerable proportion of such funds is invested in securities and property.

insurance companies
Financial institutions that guarantee to protect clients against specified risks, including death, and general risks in return for the payment of an annual premium

The investment strategy of both pension fund managers and insurance companies tends to be long-term. They invest in **portfolios** of company shares and government stocks, direct loans and mortgages.

portfolios
Combinations of securities of various kinds invested in a diversified fund

■ Other investment funds

As we shall see in Section 2.4, private investors, independently managing their own investment portfolios, are a dying breed. Increasingly, they are being replaced by financial institutions that manage widely-diversified portfolios of securities, such as unit trusts and investment trusts (www.investmentfunds.org.uk). These pool the funds of large numbers of investors, enabling them to achieve a degree of diversification not otherwise attainable owing to the prohibitive transactions costs and time required for active portfolio management. However, there are important differences between these institutions.

Investment trusts

Investment trusts are limited companies, whose shares are usually quoted on the Stock Exchange, and set up specifically to invest in securities. The company's share price depends on the value of the securities held in the trust, but also on supply and demand. As a result, these shares often sell at values different from their net asset values, usually at a discount.

They are traditionally 'closed-ended' in the sense that the company's articles restrict the number of shares, and hence the amount of share capital, that can be issued. However, several open-ended investment trust companies (OEICS) have now been launched. To realise their holdings, shareholders can sell their shares on the stock market.

Unit trusts

Unit trusts are investment syndicates, established by trust deed and regulated by trust law. Investors' funds are pooled into a portfolio of investments, each investor being allocated tranches or 'units' according to the amount of the funds they subscribe. They are mainly operated by banks and insurance companies, which appoint managers whose conduct is supervised by a set of trustees.

Unit prices are fixed by the managers, but reflect the value of the underlying securities. Prices reflect the costs of buying and selling, via an initial charge. Managers also apply annual charges, usually about 1 per cent of the value of the fund. Unit-holders can realise their holdings only by selling units back to the trust managers.

They are 'open-ended' in the sense that the size of the fund is not restricted and the managers can advertise for funds.

Hedge funds

A hedge fund is a fund used by wealthy individuals and institutions. They are exempt from many of the rules and regulations governing other institutions, which allows them to accomplish aggressive investing strategies that are unavailable to others. These include:

- *Arbitrage* – simultaneous buying and selling of a financial instrument in different markets to profit from differences in prices.
- *Short-selling* – selling a security one does not own with the expectation of buying it in the future at a lower cost.
- *Leverage* – borrowing money for investment purposes.
- *Hedging* – using derivative instruments to offset potential losses on the underlying security.

Hedge fund managers receive their remuneration in the form of a management fee plus incentive fees tied to the fund's performance.

Disintermediation and securitisation

disintermediation
Business-to-business lending that eliminates the banking intermediary

securitisation
The capitalisation of a future stream of income into a single capital value that is sold on the capital market for immediate cash

While financial intermediaries play a vital role in the financial markets, **disintermediation** is an important new development. This is the process whereby companies borrow and lend funds directly between themselves without recourse to banks and other institutions. Allied to this is the process of **securitisation**, the development of new financial instruments to meet ever-changing corporate needs (i.e. financial engineering). Some assets generate predictable cash returns and offer security. Debt can be issued to the market on the basis of the returns and suitable security. Securitisation usually also involves a credit rating agency assessing the issue and giving it a credit rating. Securitisation can also be used to create value through 'unbundling' traditional financial processes. For example, a conventional loan has many elements, such as loan origination, credit status evaluation, financing and collection of interest and principal. Rather than arranging the whole process through a single intermediary, such as a bank, the process can be 'unbundled' and handled by separate institutions, which may lower the cost of the loan.

Securitising the Beatles

Chrysalis, the media group, has completed a complex cross-border securitisation deal to unlock £60 million over 15 years against the future value of its music publishing catalogue which includes artists ranging from Blondie, the Beatles and Jethro Tull to David Gray and Moloko.

Music publishing is a separate business from recorded music, comprising the rights to the written composition of a song, performance rights such as radio airplay, a share of CD sales and synchronisation rights from use in advertisements or films. Chrysalis's revenues from its catalogue were £8 million in 2000.

The Chrysalis securitisation deal took 18 months to structure because of the complexity in bringing together publishing rights in the UK, US, Germany, Sweden and Holland under their different tax regimes.

Chrysalis follows in the footsteps of the singer-songwriter David Bowie, who recently raised $55 million via a bond issue against his share of the publishing rights to his compositions.

Source: Based on *Financial Times*, 2 March 2001.

Securitisation and disintermediation have permitted larger companies to create alternative, more flexible forms of finance. This, in turn, has forced banks to become more competitive in the services offered to larger companies. Recent more exotic forms of securitisation include pubs, gate receipts from a football club, future income from a pop star's recordings, and even the football World Cup competition for 2006.

2.4 THE LONDON STOCK EXCHANGE (LSE)

The capital market is the market where long-term securities are issued and traded. The London Stock Exchange is the principal trading market for long-dated securities in the UK (**www.londonstockexchange.com**).

A stock exchange has two principal economic functions: to enable companies to *raise new capital* (the primary market), and to *facilitate the trading of existing shares* (the secondary market) through the negotiation of a price at which title to ownership of a company is transferred between investors.

■ A brief history of the London Stock Exchange

The world's first joint-stock company – the Muscovy Company – was founded in London in 1553. With the growth in such companies, there arose the need for shareholders to be able to sell their holdings, leading to a growth in brokers acting as intermediaries for investors. In 1760, after being ejected from the Royal Exchange for rowdiness, a group of 150 brokers formed a club at Jonathan's Coffee House to buy and sell shares. By 1773, the club was renamed the Stock Exchange.

The Exchange developed rapidly, playing a major role in financing UK companies during the Industrial Revolution. New technology began to have an impact in 1872, when the Exchange Telegraph tickertape service was introduced.

For over a century, the Exchange continued to expand and become more efficient, but fundamental changes did not occur until 27 October 1986 – 'Big Bang' – the most important of which were:

1 All firms became brokers/dealers able to operate in a dual capacity – either buying securities from, or selling them to, clients without the need to deal through a third party. Firms could also register as market-makers committed to making firm bid (buying) and offer (selling) prices at all times.
2 Ownership of member firms by an outside corporation was permitted, enabling member firms to build a large capital base to compete with competition from overseas.
3 Minimum scales of commission were abolished to improve competitiveness.
4 Trading moved from being conducted face-to-face on a single market floor to being performed via computer and telephone from separate dealing rooms. Computer-based systems were introduced to display share price information, such as SEAQ (Stock Exchange Automated Quotations).

In 2007 the Exchange launched a new trading system with greater capacity and speed of trading. Today, the London Stock Exchange is viewed as one of the leading and most competitive places to do business in the world, second only to New York in total market value terms.

The LSE has two tiers. The bigger market is the Main List, providing a quotation for nearly 3,000 companies. To obtain a full listing, companies have to satisfy rigorous criteria laid down in the Stock Exchange's 'Listing Rules' (or 'Yellow Book'). These relate to size of issued capital, financial record, trading history and acceptability of board members. These details are set out in a document called the company's 'listing particulars'.

The second tier is the Alternative Investment Market (AIM). It attempts to minimise the cost of entry and membership by keeping the rules and application process as simple as possible. A nominated adviser firm (typically a stockbroker or bank) both introduces the new company to the market and acts as a mentor, ensuring that it complies with market rules. Although the majority of companies are capitalised at between £2 million and £20 million, it also includes start-up operations at one end and companies capitalised at over £200 million at the other. However, the requirement to observe existing obligations in relation to publication of price-sensitive information and annual and interim accounts remains. The AIM is unlikely to appeal to private investors unless they are prepared to invest in relatively high-risk businesses. In 2008 AIM had over 1,700 listed companies and provides a relatively low-cost and flexible route to obtaining a listing.

While the vast majority of share trading takes place through the Stock Exchange, it is not the only trading arena. For some years, there has been a small, but active over-the-counter (OTC) market, where organisations trade their shares, usually on a 'matched bargain' basis, via an intermediary.

Self-assessment activity 2.2

What type of company would be most likely to trade on:

(a) the main securities market?
(b) the Alternative Investment Market?
(c) The over-the-counter market?

(Answer in Appendix A at the back of the book)

■ Regulation of the market

Investor confidence in the workings of the stock market is paramount if it is to operate effectively. Even in deregulated markets, there is still a requirement to provide strong safeguards against unfair or incompetent trading and to ensure that the market operates as intended. The mechanism for regulating the whole UK financial system was established by the Financial Services Act 1986 (FSA86), which provided a structure based on 'self-regulation within a statutory framework'.

In 1997, statutory powers were vested in a supervisory body, the Financial Services Authority (FSA), responsible to the Treasury. Its objectives are to sustain confidence in the UK's financial services industry and monitor, detect and prevent financial crime (**www.fsa.gov.uk**). This involves the regulation of the financial markets, investment managers and investment advisers.

The FSA also takes on additional responsibilities for monitoring the money markets, building societies and the insurance market. The hope is that, by having a single regulator covering all financial markets, there will be greater efficiency, lower costs, clearer accountability and a single point of service for customer enquiries and complaints.

In an attempt to enhance London's reputation for clean and fair markets, the FSA has introduced new powers, effective from 2000, to deal with insider dealing and attempts to distort prices. It is a criminal offence to undertake 'investment business' without due authorisation. A Recognised Investment Exchange (RIE), of which the London Stock Exchange is one, may also receive authorisation. Recognition exempts an exchange (but not its members) from needing authorisation for any activity constituting investment business.

The Stock Exchange discharges its responsibilities by:

■ vetting new applicants for membership
■ monitoring members' compliance with its rules

- providing services to aid trading and settlement of members' business
- supervising settlement activity and management of settlement risk
- investigating suspected abuse of its markets.

Market abuse includes three strands:

(a) Market distortion – acting in such a manner as to force up a company's share price.

(b) Misuse of information – e.g. buying or selling shares on the basis of privileged information.

(c) Creating false information – e.g. putting false information on to a website.

Other bodies also keep a watchful eye on the workings of capital markets. These include the Bank of England (**www.bankofengland.co.uk**), the Competition Commission (CC), the Panel on Takeovers and Mergers, the Office of Fair Trading, the press and various government departments.

Self-assessment activity 2.3

To what extent does an effective primary capital market depend on a healthy secondary market?

(Answer in Appendix A at the back of the book)

■ Share ownership in the UK

Back in 1963, over half (54 per cent) of all UK equities were held by private individuals. This proportion had dropped to 13 per cent by the end of 2006 (**www.statistics.gov.uk**). Today, share ownership is dominated by financial institutions (the pension funds, insurance groups and investment and unit trusts). Together, including both UK and foreign institutions, they own around 80 per cent of the value of UK traded companies. These impersonal bodies, acting for millions of pensioners and employees, policyholders and small investors, have vast power to influence the market and the companies they invest in. Institutional investors employ a variety of investment strategies, from passive index-tracking funds, which seek to reflect movements in the stock market, to actively managed funds.

Institutional investors have important responsibilities, and this can create a dilemma: on the one hand, they are expected to speak out against corporate management policies and decisions that are deemed unacceptable environmentally, ethically or economically. But public opposition to the management could well adversely affect share price. Institutions therefore have a conflict between their responsibilities as major shareholders and their investment role as managers seeking to outperform the markets.

A further indication of changing patterns of share ownership is the proportion of the adult population that holds shares. Successive governments have promoted a 'share-owning democracy', particularly through privatisation programmes. However, individuals tend to hold small, undiversified portfolios, which exposes them to a greater degree of risk than from investing in a diversified investment portfolio.

■ Towards a European stock market?

The European Union is meant to be about removing barriers and providing easier access to capital markets. Until recently, this was still a pipe dream, with some 30 stock exchanges within the EU, most of which had different regulations. With the introduction of a single currency, there will undoubtedly be strong pressure towards a single capital market. But does this mean a single European stock exchange, with one set of rules for share listing and trading?

Euronext was formed in 2000 as a result of the merger of the Amsterdam, Brussels and Paris stock exchanges. As the first pan-European stock exchange, it has undertaken further mergers with other smaller exchanges in Europe.

Inion plans £30m public offering

Inion, a Finnish medical devices company, is planning to raise £30m in an initial public offering on the London Stock Exchange.

The indicative price range for the flotation has been set at between 113p and 136p a share, giving a market capitalisation of between £80m and £90m.

The company, which makes biodegradable polymer implants, was set up in 1999 by senior researchers from Bionics, a Nasdaq-listed Finnish implants company.

Auvo Kaikkonen, chief executive, said listing alongside other medical technology companies in London would ensure better liquidity than floating on the Helsinki Stock Exchange, which is dominated by Nokia. 'We wanted a market where there was an experienced analyst and investor community,' he said.

Inion's products include biodegradable screws, plates and meshes to stabilise broken and damaged bones while they heal.

Inion incurred a pre-tax loss of approximately €3m (£2.1m) on revenues of €2.4m in the first half of 2004. Mr Kaikkonen said it would break even when revenues reached €20m.

Source: Financial Times, 10 November 2004.

2.5 ARE FINANCIAL MARKETS EFFICIENT?

If financial managers are to achieve corporate goals, they require well-developed financial markets where transfers of wealth from savers to borrowers are efficient in both pricing and operational cost.

allocative efficiency
The most efficient way that a society can allocate its overall stock of resources

operating/technical efficiency
The most cost-effective way of producing an item, or organising a process

social efficiency
The extent to which a socio-economic system accords with prevailing social and ethical standards

pricing/information efficiency
The extent to which available information is impounded into the current set of share prices

fair game
A competitive process in which all participants have equal access to information and therefore similar chances of success

Efficiency can mean many things. The *economist* talks about **allocative efficiency** – the extent to which resources are allocated to the most productive uses, thus satisfying society's needs to the maximum. The *engineer* talks about **operating** or **technical efficiency** – the extent to which a mechanism performs to maximum capability. The *sociologist* and *the political scientist* talk about **social efficiency** – the extent to which a mechanism conforms to accepted social and political values. The most important concept of efficiency for our purposes is **pricing** or **information efficiency**. This refers to the extent to which available information is built into the structure of share prices. If information relevant for assessing a company's future earnings prospects (including both past information and relevant information relating to future expected events) is widely and cheaply available, then this will be impounded into share prices by an efficient market. As a result, the market should allow all participants to compete on an equal basis in a so-called **fair game**.

We often hear of the shares of a particular company being 'under-valued' or 'over-valued', the implication being that the stock market pricing mechanism has got it wrong and that analysts know better. *In an efficient stock market, current market prices fully reflect available information* and it is impossible to outperform the market consistently, except by luck.

Consider any major European stock market. On any given trading day, there are hundreds of analysts – representing the powerful financial institutions which dominate the market – closely tracking the daily performance of the share price of, say, Taylor Wimpey, the construction company. They each receive at the same time new information from the company – a major order, a labour dispute or a revised profits forecast. This information is rapidly evaluated and reflected in the share price by their decisions to buy or sell Taylor Wimpey shares. *The measure of efficiency is seen in the extent and speed with which the market reflects new information in the share price.*

arbitrage
The process whereby astute entrepreneurs identify and exploit opportunities to make profits by trading on differentials in price of the same item as between two locations or markets

The law of one price suggests that equivalent securities must be traded at the same price (excluding differences in transaction costs). If this is not the case, **arbitrage** opportunities arise whereby a trader can buy a security at a lower price and simultaneously sell it at a higher price, thereby making a profit without incurring any risk. In an efficient market, arbitrage activity will continue until the price differential is eliminated.

■ The efficient markets hypothesis (EMH)

Information can be classified as historical, current or forecast. Only current or historical information is certain in its effect on price. The more information that is available, the better the situation. Informed decisions are more likely to be correct, although the use of inside information to benefit from investment decisions (insider dealing) is illegal in the UK.

Company information is available both within and without the organisation. Those within the organisation will obviously be better informed about the state of the business. They have access to sensitive information about future investment projects, contracts under negotiation, forthcoming managerial changes, etc. The additional knowledge will vary according to a person's level of responsibility and place in the organisational hierarchy.

Outsider investors fall into two categories: individual investors and the institutions. Of these two groups, the institutions are the better informed, as they have greater access to senior management, and may be represented on the board of directors.

Different amounts of financial information are available to different groups of people. There is unequal access to the information, called 'information asymmetry', which may affect a company's share price. If you are one of the well-informed, this gives you the opportunity to keep one step ahead of the market. Otherwise, you may lose out. The share price reflects who knows what about the company. You should note, however, that in the UK, share dealings by company directors are tightly circumscribed; for example, they can only buy and sell at specific times, and details of all such trades must be publicly disclosed.

Market efficiency evolved from the notion of perfect competition, which assumes free and instantly available information, rational investors and no taxes or transaction costs. Of course, such conditions do not exist in capital markets, so just how do we assess their level of efficiency? Market efficiency, as reflected by the efficient markets hypothesis (EMH), may exist at three levels:

weak form
A weak-form efficient share market does not allow investors to look back at past share price movements and identify clear, repetitive patterns

1 The **weak form** of the EMH states that current share prices fully reflect *all information contained in past price movements*. If this level of efficiency holds, there is no value in trying to predict future price movements by analysing trends in past price movements. Efficient stock market prices will fluctuate more or less randomly, any departure from randomness being too expensive to determine. Share prices are said to follow a **random walk**.

semi-strong form
A semi-strong efficient share market incorporates newly released information accurately and quickly into the structure of share prices

2 The **semi-strong form** of the EMH states that current market prices reflect not only all past price movements, but *all publicly available information*. In other words, there is no benefit in analysing existing information, such as that given in published accounts, dividend and profits announcements, appointment of a new chief executive or product breakthroughs, after the information has been released. The stock market has already captured this information in the current share price.

strong form
In a strong-form efficient share market, all information including inside information is built into share prices

3 The **strong form** of the EMH states that current market prices reflect *all relevant information* – even if privately held. The market price reflects the 'true' or intrinsic value of the share based on the underlying future cash flows. The implications of such a level of market efficiency are clear: no one can consistently beat the market and earn abnormal returns. Few would go so far as to argue that stock markets are

efficient at this level, although investors are often good at predicting what is happening inside companies before the information is officially released.

You will have noticed that as the EMH strengthens, the opportunities for profitable speculation reduce. Competition between well-informed investors drives share prices to reflect their intrinsic values.

■ The EMH and fundamental and technical analysis

intrinsic worth
The inherent or fundamental value of a company and its shares

fundamental analysis
Analysis of the fundamental determinants of company financial health and future performance prospects, such as endowment of resources, quality of management, product innovation record, etc.

technical analysis
The detailed scrutiny of past time series of share price movements attempting to identify repetitive patterns

chartists
Analysts who use technical analysis

Investment analysts who seek to determine the **intrinsic worth** of a share based on underlying information undertake **fundamental analysis**. The EMH implies that fundamental analysis will not identify under-priced shares unless the analyst can respond more quickly to new information than other investors, or has inside information.

Another approach is **technical analysis**, its advocates being labelled **chartists** because of their reliance upon graphs and charts of price movements. Chartists are not interested in estimating the intrinsic value of shares, preferring to develop trading rules based on patterns in share price movement over time, or 'breakout' points of change. Charts are used to predict 'floors' and 'ceilings', marking the end of a share price trend. Figure 2.3 shows how charts are used to detect patterns of 'resistance' (for shares on the way up) and 'support' (for shares on the way down). This approach can often prove to be a 'self-fulfilling prophecy'. In the short term, if analysts predict that share prices will rise, investors will start to buy, thus creating a bull market and resulting in upward pressure on prices.

Even in its weak form, the EMH questions the value of technical analysis; future price changes cannot be predicted from past price changes. However, the fact that many analysts, using fundamental or technical analysis, make a comfortable living from their investment advice suggests that many investors find comfort in the advice given.

Considerable empirical tests on market efficiency have been conducted over many years. In the USA and the UK, until the 1987 stock market crash, the evidence broadly supported the semi-strong form of efficiency. More specifically, it suggests the following:

1 *There is little benefit in attempting to forecast future share price movements by analysing past price movements.* As the EMH seems to hold in its weak form, the value of charts must be questioned.

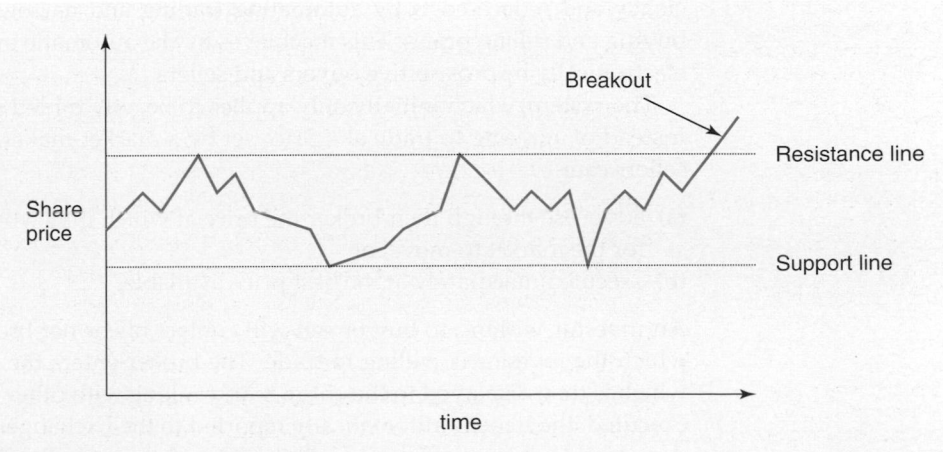

Figure 2.3 Chart showing breakout beyond resistance line

2 For quoted companies that are regularly traded on the stock market, analysts are *unlikely to find significantly over- or under-valued shares through studying publicly held information.* Studies indicate (e.g. Ball and Brown, 1968) that most of the information content contained in annual reports and profit announcements is reflected in share prices anything up to a year before release of the information, as investors make judgements based on press releases and other information during the year. However, analysts with specialist knowledge, paying careful attention to smaller, less well-traded shares, may be more successful. Equally, analysts able to respond to new information slightly ahead of the market may make further gains. The semi-strong form of the EMH seems to hold fairly well for most quoted shares.

3 The strong form of the EMH does not hold, so superior returns can be achieved by those with 'inside knowledge'. However, it is the duty of directors to act in the shareholders' best interests, and it is a criminal offence to engage in insider trading for personal gain. The fact that cases of insider trading have led to the conviction of senior executives shows that market prices do not fully reflect unpublished information.

Recent governments have encouraged greater market efficiency in several ways:

- Stock market deregulation and computerised dealing have enabled speedier adjustment of share prices in response to global information.
- Mergers and takeovers have been encouraged as ways of improving managerial efficiency. Poorly-performing companies experience depressed share prices and become candidates for acquisition.
- Governments have seen privatisation of public utilities as a means of subjecting previously publicly-owned organisations to market pressures.

How people trade in London

The Big Bang in 1986 gave the London Stock Exchange a huge advantage over most of its competitors. The result was strong growth in trading activity and international participation. But Big Bang was only a partial revolution – automating the distribution of price information, but stopping short of automating the trading function itself.

Since 1986, global equity markets have become increasingly complex, with investors constantly looking for greater choice and lower costs. The London Stock Exchange made various attempts to retain its reputation as one of the most efficient stock markets. It took a major step by moving from a quote-based trading system, under which share dealing is conducted by telephone, to order-driven trading, termed SETS – the Stock Exchange Electronic Trading Service. The aim was to improve efficiency and reduce costs by automating trading and narrowing the spread between buying and selling prices. This it achieves by the automatic matching of orders placed electronically by prospective buyers and sellers.

The system, which initially only applies to heavily traded shares, works as follows. Instead of agreeing to trade at a price set by a market-maker, prospective buyers and sellers can:

(a) advertise through their broker the price at which they would like to deal, and wait for the market to move, or
(b) execute immediately at the best price available.

An investor wishing to buy or sell will contact his or her broker and agree a price at which the investor is willing to trade. The broker enters the order in the order book, which is then displayed to the entire market along with other orders. Once the order is executed, the trade is automatically reported to the Exchange. Time will tell whether it does lead to greater efficiency, but it is hoped that it will offer users more attractive, transparent and flexible trading opportunities.

We are all chartists now

Chartists (or 'technicians' in the US), study the movements of markets, rather than the underlying fundamentals of securities, when predicting price moves. Their methods can sound like voodoo, but at times of high confusion, charts offer the only way to navigate. Only once the markets find a new level, can fundamentals reassert themselves.

Trading chatter is now almost exclusively about technical patterns. And they suggest more turbulence is to come. In stocks, the charts show that corrections during a bull market follow a 'two-wave' pattern. After a first downdraft, there is a recovery, or relief rally (charmingly known as a 'dead cat bounce' in London), followed by a second or 'capitulation' downdraft. This pattern was seen in March and then April of 2005, and again in May and then June of last year.

Another popular tool of chartists is the moving average – an average of an index over the past 30, 50 or 200 days. If a price drops below a moving average, this can be an indication that the market has lost confidence. Last Tuesday's huge afternoon sell-off in New York came in large part because the main indices fell through their 50-day moving averages, for the first time since last July.

Whether charts really have predictive power is for the moment not the question. Enough people act on the assumption that they do to ensure that they have an effect. In the long run, fundamentals matter – economies tend to grow, and stocks grow with them. In the long run, we need not worry too much about the charts. But traders go along with Keynes: in the long run we are all dead.

Source: John Authers, *Financial Times*, 8 March 2007, p. 19.

■ Implications of market efficiency for corporate managers

In quoted companies, managers and investors are directly linked through stock market prices, corporate actions being rapidly reflected in share prices. This indicates the following:

1 Investors are not easily fooled by glossy financial reports or 'creative accounting' techniques, which boost corporate reported earnings but not underlying cash flows.
2 Corporate management should endeavour to make decisions that maximise shareholder wealth.
3 The timing of new issues of securities is not critical. Market prices are a 'fair' reflection of the information available and accurately reflect the degree of risk in shares.
4 Where corporate managers possess information not yet released to the market, there is an opportunity for influencing prices. For example, a company may retain information so that, in the event of an unwelcome takeover bid, it can offer positive signals.

We return to the issue of market efficiency in discussing behavioural finance in Chapter 23.

Self-assessment activity 2.4

Consider why a dealing rule like 'Always buy in early December' should be doomed to failure. This rule is designed to exploit the so-called 'end-of-year effect' claiming that share prices 'always' rise at the end of the year.

(Answer in Appendix A at the back of the book)

Self-assessment activity 2.5

Share prices of takeover targets invariably rise before the formal announcement of a takeover bid. What does this suggest for the EMH?

(Answer in Appendix A at the back of the book)

2.6 READING THE FINANCIAL PAGES

Corporate finance is changing so quickly that it is essential for students of finance to read the financial pages in newspapers on a regular basis. In this section, we explain the main information contained in the Share Service pages of the *Financial Times*, and other newspapers.

■ The FT-SE Index

Every day, shares move up or down with the release of information from within the firm, such as a revised profits forecast, or from an external source, such as the latest government statistics on inflation or unemployment. To indicate how the whole share market has performed, a share index is used, the most common being the **FT-SE 100** – familiarly known as 'Footsie'. This index is based on the share prices of the 100 most valuable UK quoted companies (sometimes termed 'blue chips'), mostly those with capitalisations above £3 billion, with each company weighted in proportion to its total market value. All the world's major stock markets have similar indices (for example, the Nikkei Index in Japan, the Dow Jones Index in the USA and the CAC-40 in France).

Every share index is constructed on a base date and base value. The FT-SE 100 started with a base value of 1,000 at the end of 1983. By February 2008, the index stood at around 6,029. Despite the collapse in world markets in 2000, and the subsequent slow recovery in confidence (punctuated by a fresh collapse at the time of the Iraq war), this still represented an annual compound growth rate of about 8 per cent, well above both the rate of inflation and the yield on low-risk investments over the same period. Moreover, it includes only capital appreciation – inclusion of dividend income would raise this percentage to about a 12 per cent return.

The FT-SE Actuaries Share Indices reveal share movements by sector. Their total gives the All-Share Index, representing the more frequently traded quoted companies, and between 98 and 99 per cent of market capitalisation.

Other FT indices

In recent years, the Stock Exchange has introduced several new indices:

- FT-SE Actuaries 350 provides the benchmark for investors who wish to focus on the more actively traded large and medium-sized UK companies, and covers 95 per cent of trading by value.
- FT-SE Small Cap offers investors a daily measure of the performance of smaller companies.
- FT-SE Fledgling covers smaller companies taken from the main listing and the Alternative Investment Market (AIM).

Using the published information

Financial managers and investors need to study the performance of the shares of their company, both against the appropriate sector as a whole and also against competitors within that sector. Two performance statistics that are most commonly reported are the **dividend yield** and **price:earnings ratio**.

Dividend yield

This is the gross, or pre-tax, dividends of companies and whole sectors in the last year as a percentage of their market value. Generally, sectors with low dividend yields are those with companies where the market expects high growth. Often we observe that the dividend yield for leading shares, and also on the overall index, is well below the return investors could currently earn on a safe investment in Treasury Bills. This is because shareholders are looking to a capital gain on top of the dividend yield to recompense them for the higher risks involved.

Price:earnings (P:E) ratio

The P:E ratio is a much-used performance indicator. It is the share price divided by the most recently reported earnings, or profit, per share. So for the sector, it is the total market value of the companies represented divided by total sector earnings. The P:E ratio is a measure of the market's confidence in a particular company or industry. A high P:E usually indicates that investors have confidence that profits will grow strongly in future, perhaps after a short-term setback, although irregular events like a rumoured takeover bid will raise the P:E ratio if they lead to a higher share price.

Let us now turn to the performance of individual companies. Table 2.1 is an extract from the London Share Service pages in the *Financial Times*, giving the food and drug retailing sector. Different information is provided on Mondays than on other days of the week. We will focus on a major supermarket chain, Tesco Group. Bold names indicate members of the FTSE 100 index.

Transactions and prices of stocks are published continuously through SEAQ (the Stock Exchange Automatic Quotation). Quoted prices assume that shareholders buying a share are entitled to any forthcoming dividend (cum div) unless this is expressly precluded. The symbol 'xd' (ex div, i.e. excluding dividends) would mean that new investors are too late to qualify for it. The share price will accordingly be lower to reflect the forgone dividend.

Tesco's closing share price of $320\frac{1}{4}$p is up $\frac{1}{2}$p from the previous day's trading with over 14 million shares traded in the day. The current price is just below its highest over the past year. Every Monday, the *Financial Times* publishes the dividend cover and market capitalisation value (i.e. number of issued shares times current share price).

price:earnings
The ratio of price per ordinary share to earnings (i.e. profit after tax) per share (EPS)

The **yield** of 2.2 per cent is the dividend yield, i.e. the dividend expressed as a proportion of the current share price. The **price:earnings (P:E) ratio** of 19.4 suggests that it would take over 20 years for investors to get their money back in profit terms. Why should anyone be willing to wait that long? Remember that the calculation compares the last reported earnings per share with the current share price. Investors expect the payback period to be far quicker than 20 years because they anticipate strong earnings growth for Tesco.

Table 2.1 Share price information for the food retail sector

	Price	Chng	52 week high	52 week low	Yld	P/E	Vol '000s
Big Food	94	$-\frac{1}{4}$	182	80	3.1	11.7	13,104
CaffeNro	127	......	$132\frac{1}{2}$	48	–	28.3	5
Dairy Fm	$131\frac{1}{4}$	$+1\frac{1}{4}$	154	$88\frac{1}{2}$	2.2	26.6	5
Greggs	3750	+50	3750	3075	2.3	14.5	32
Morrison	$215\frac{1}{2}$	$-1\frac{1}{2}$	256	$171\frac{1}{4}$	1.5	32.3	19,842
Sainsbry	$270\frac{1}{4}$	−2	$311\frac{1}{2}$	242	5.0	–	9,912
Somerfld	$154\frac{3}{4}$	$-\frac{3}{4}$	$172\frac{1}{4}$	$125\frac{1}{2}$	1.4	26.4	1,236
Tesco	$320\frac{1}{4}$	$+\frac{1}{2}$	$321\frac{3}{4}$	237	2.2	19.4	14,082
Thorntns	$153\frac{1}{2}$	$+\frac{1}{2}$	168	135	4.4	22.8	82
WhitrdCh	$189\frac{1}{2}$	+1	220	161	1.8	17.9	22

Source: Financial Times, 11 January 2005.

2.7   TAXATION AND FINANCIAL DECISIONS

Few financial decisions are immune from taxation considerations. *Corporate and personal taxation affects both the cash flows received by companies and the dividend income received by shareholders.* Consequently, financial managers need to understand the tax consequences of investment and financing decisions. Taxation may be important in three key areas of financial management:

1 *Raising finance.* There are clear tax benefits in raising finance by issuing debt rather than capital. Interest on borrowings attracts tax relief, thereby reducing the company's tax bill, while a dividend payment on equity capital does not attract tax relief. The tax system is thereby biased in favour of debt finance.

2 *Investment in fixed assets.* Spending on certain types of fixed asset attracts a form of tax relief termed **capital allowances**. This is intended to stimulate certain types of investment, such as in industrial plant and machinery. The taxation implications of an investment decision can be very important. We discuss capital allowances and tax implications for investment decisions in Chapter 5.

3 *Paying dividends.* Until 1973, in the UK, company profits were effectively taxed twice – first on the profits achieved and then again on those profits paid to shareholders in the form of dividends. Such a 'classical' tax system (which still exists in certain countries) is clearly biased in favour of retaining profits rather than paying out large dividends. The UK taxation system is more neutral, the same tax bill being paid (for companies making profits) regardless of the dividend policy.

Finally, the corporate financial manager should understand not only how taxation affects the company, but also how it affects the company's shareholders (**www.inlandrevenue.gov.uk**). For example, some financial institutions (e.g. pension funds) pay no tax; some shareholders pay tax at 20 per cent, while others pay higher-rate income tax at 40 per cent. Some may prefer capital gains to dividends.

Self-assessment activity 2.6

Explain why it is important to consider the tax implications of financial and investment decisions.

(Answer in Appendix A at the back of the book)

SUMMARY

This chapter has introduced readers to the financial and tax environment within which financial and investment decisions take place.

Key points

■ Financial markets consist of numerous specialist markets where financial transactions occur (e.g. the money market, capital market, foreign exchange market, derivatives markets).

■ Financial institutions (e.g. banks, building societies, pension funds) provide a vital service by acting as financial intermediaries between savers and borrowers.

■ Securitisation and disintermediation have permitted larger companies to create alternative, more flexible forms of finance.

■ The London Stock Exchange operates two tiers: the Main List for larger established companies, and the Alternative Investment Market which mainly caters for very young companies.

- An efficient capital market is one where investors are rational and share prices reflect all available information. The efficient markets hypothesis has been examined in its various forms (weak, semi-strong and strong). In all but the strong form, it seems to hold up reasonably well, but it is increasingly unable to explain 'special' circumstances.

- Taxation can play a key role in financial management, particularly in raising finance, investing in fixed assets and paying dividends.

Further reading

Brett (2003) provides a clear explanation of how to read the financial pages in the press. Clear and more extensive introductions to capital markets are found in Foley (1991), Weston and Copeland (1992), O'Shea (1986), Redhead (1990), and Levinson (2002).

Two classic review articles on market efficiency were written by Fama (1970 and 1992), while Rappaport (1987) examines the implications for managers. Tests of capital market efficiency are found in Ross, Westerfield and Jordan (2008), Copeland and Weston (2004) and Keane (1983), while some exceptions to efficiency are found in the June 1977 special issue of *Journal of Financial Economics.* Peters (1991, 1993) applies Chaos Theory to stock markets. Discussion on short-termism in the City is found in Marsh (1990) and Ball (1991). *Mastering Finance* (1997) offers useful articles on securitisation, financial intermediaries, the role of financial markets, market efficiency and short-termism.

Useful websites

www.moneyfactor.co.uk
www.moneysupermarket.com
www.moneyextra.com
www.find.co.uk
www.moneynet.co.uk
www.ftourmoney.co.uk

Appendix
FINANCIAL STATEMENT ANALYSIS

balance sheet/position statement
A financial statement that lists the assets held by a business at a point in time and explains how they have been financed (i.e. by owners' capital and by third-party liabilities)

profit and loss account/ income statement
A financial statement that details for a specific time period the amount of revenue earned by a firm, the costs it has incurred, the resulting profit and how it has been distributed ('appropriated')

Most readers will previously have undertaken a module in accounting and be familiar with financial statements. This appendix provides a summary of the key elements in analysing financial statements and the main ratios involved in interpreting accounts.

Investors, whether shareholders or bank managers, ask three basic questions when they examine the accounts of a business:

- *Position* – what is the current financial position, or state of affairs, of the business? This question is addressed by examining the **balance sheet**, sometimes referred to as the **position statement**.

- *Performance* – how well has the business performed over the period of time we are interested in, for example, the past year? This question is addressed by looking at the **profit and loss account**, otherwise termed the **income statement**.

■ *Prospects* – what are the likely prospects of the business for which we are considering investment? A bank manager would probably request a cash flow forecast, showing the expected cash receipts and payments for the coming year. However, published accounts are historical documents and the shareholder will have to settle for the **cash flow statement** for the past year. Clues as to the expected future prospects may be found in the Chairman's Statement frequently published with the accounts.

cash flow statement
A financial statement that explains the reasons for cash inflows and outflow of a business, and highlights the resulting change in cash position

We will examine the three financial statements, drawing on the abridged accounts of a fictitious company called *Foto-U*, a business specialising in offering instant photographs through photo booths in public places throughout Europe.

■ The balance sheet

Imagine it is possible to take a financial snapshot of *Foto-U* on 31 March 2010, the end of its trading year. What we would see are the very things we find in the balance sheet. Looking at *Foto-U*'s balance sheet in Table 2.2, we see three main categories–assets, creditors (or liabilities), and capital and reserves. This statement demonstrates the 'accounting equation': the money invested in the business by shareholders and creditors is represented by the assets in which they have been invested.

fixed assets
Assets that remain in the balance sheet for more than one accounting period (i.e. they are fixed in the balance sheet)

Where the cash came from $\quad=\quad$ *Where the cash went*
$\quad$ *(sources of funds)* $\qquad\qquad$ *(uses of funds)*

Shareholders' funds of £78 m + Creditors £60 m = Assets £138 m

current assets
Assets that will leave the balance sheet in the next accounting period

The more permanent assets (typically those with a life beyond a year) are termed **fixed assets** while the less permanent are termed **current assets**. For *Foto-U*, **intangible** fixed assets refer to patents and goodwill, the latter arising from acquiring another company and paying more for it than the balance sheet value of its underlying assets. **Tangible** fixed assets include land and buildings, photo booths, plant and machinery, vehicles, and fixtures and fittings. Their values are not stated at what they could be sold for, but at their **net book value** – what they originally cost less an estimate of the extent to which they have depreciated in value with use or age.

intangible
Intangible assets cannot be seen or touched, e.g. the image and good reputation of a firm

Current assets represent the less permanent items (typically less than a year) the business owns at the balance sheet date. Our financial snapshot for *Foto-U* captures four items – stocks, debtors, investments and cash. Unlike fixed assets, these items are continuously changing (or 'turning over') as trading takes place. Trade creditors and bank overdraft, where the amount has to be settled within one year, are deducted from the current assets to give the **net current assets** figure, commonly termed **working capital**. This is the amount of money likely to be turned into cash over the coming weeks. Creditors to be paid after more than a year are typically in the form of medium/long-term loans. Finally, **shareholders' funds** represent the capital originally paid in by shareholders plus any reserves created since then. The most common reserve will be the profit retained in the business rather than paid to the shareholders as dividends.

tangible
Tangible assets can quite literally be seen and touched, e.g. machinery and buildings

net book value
The original cost of buying an asset less accumulated depreciation charges to date

net current assets
Current assets less current liabilities

shareholders' funds
The value of the owners' stake in the business – identically equal to net assets, or equity

■ Does the balance sheet show the worth of the business?

Although the shareholders' funds for *Foto-U* of £78 million is the difference between what it *owns*, in the form of various assets, and what it *owes* to third parties, it would not be correct to say that this is what their investment is worth. The market value for the company is based on what investors are willing to pay for it. But the assets and liabilities are valued according to **Generally Accepted Accounting Principles (GAAP)**. We cannot explore them all here, but one principle is that assets are usually valued at their historical cost less a provision for such things as depreciation, in the case of fixed assets, and bad or doubtful debts, in the case of debtors. The key difference is that

Table 2.2 *Foto-U* plc

Balance Sheet as at 31 March 2010

	2010 £m	2009 £m
Fixed assets		
Intangible assets	15	10
Tangible assets	117	92
	132	102
Current assets		
Stocks	25	24
Debtors	29	25
Investments and short-term deposits	3	–
Cash at bank and in hand	9	5
	66	54
Creditors: amounts falling due within one year	(60)	(62)
Net current assets (liabilities)	6	(8)
Total assets less current liabilities	138	94
Creditors*: amounts falling due after more than one year	(60)	(23)
Net assets	78	71
Capital and reserves		
Called-up share capital	2	2
Reserves	76	69
Shareholders' funds	78	71
*The creditors figures include trade creditors	40	42

Profit and Loss Account for the year ended 31 March 2010

Turnover – continuing operations	**200**	**190**
Cost of sales (including depreciation of £27 m)	(157)	(160)
Gross profit	43	30
Administration expenses	(21)	(20)
Operating profit (Earnings before interest and taxes)	22	10
Interest payable	(2)	(2)
Profit before taxation	20	8
Tax on profit	(6)	(4)
Profit after tax attributable to shareholders	14	4
Dividends	(7)	(3)
Retained profit for year	7	1

Cash Flow Statement for the year ended 31 March 2010

Net cash inflow from operations	42	46
Servicing of finance	(2)	(2)
Taxation	(6)	(4)
Capital expenditure	(57)	(33)
Dividends paid	(7)	(3)
Financing	37	2
Increase in cash in the year	7	6

Other data: *Foto-U* has 200 million shares in issue.
Share price at 31 March 2010 is 120p (50p for 2009).

book values, based on GAAP, are backward-looking, while market values are forward-looking, based on expected future profits and cash flows.

To get some idea of the difference between the market and book values of the shareholders' funds we can look at the share price listed on the Stock Exchange on the balance sheet date. For *Foto-U* the share price at the balance sheet date was 120p. There are 200 million issued shares so the **market capitalisation** is:

(200 million shares $\times$ 120p a share) = £240 million

Comparing the market value with the book value for shareholders' equity, we find a ratio of approximately 3:1 (£240 m/£78 m). We should not be surprised to find that the market value is so much higher. Successful businesses are much more than a collection of assets less liabilities. They include creative people, successful trading strategies, profitable brands and much more. Generally, we can say that the greater the market-to-book value ratio, the more successful the business.

■ The profit and loss account

profit after tax (PAT)
Profit available to pay dividends to shareholders after tax has been paid

dividend
A periodic payment to a firm's owners – usually once or twice a year – made out of profits after tax

retained profit
Profit that remains for reinvestment in the business after a dividend is paid out

To gain an impression of how well *Foto-U* has performed over the past year we need to turn to the profit and loss account or income statement. This shows the sales income less the costs of trading. Shareholders are primarily interested in the **profit after tax (PAT)** available for distribution to them in the form of dividends. *Foto-U* has made a PAT of £14 million of which half has or will be paid to shareholders in the form of **dividend**, the remainder being **retained profit**, reinvested in the business, hopefully to earn a higher profit next year.

Investors also want to know how much profit (or earnings) has been made from its trading, before the cost of financing is deducted. **Earnings before interest and taxes (EBIT)** for *Foto-U* is:

EBIT = total revenues − operating costs (including depreciation)

= £200 m − £178 m

= £22 m

operating profit
Revenues less total operating costs, both variable and fixed – as distinct from financial costs such as interest payments

This is also termed **operating profit**.

Profit is not the sole consideration for investors. They are perhaps more interested in how much cash has been created through successful trading. This can be estimated by adding back the depreciation (a non-cash cost) previously deducted in calculating EBIT. This is termed **earnings before interest, taxes, depreciation and amortisation** (amortisation is just a fancy name for depreciating intangible assets) or **EBITDA.** For our company, this is:

EBITDA = EBIT + Depreciation

= £22 m + £27 m

= £49 m

■ The cash flow statement

The third and final financial statement in a published set of accounts is the cash flow statement. This statement is valuable because it reveals the main sources of cash and how it has been applied. For *Foto-U*, the main two sources of cash during the year are additional finance raised from new loans and net cash from operations (basically the EBITDA referred to above plus a few other adjustments for non-cash items). The main applications of this cash generated from trading are investment in capital expenditure, and dividends. The final line on this statement shows that, during the year, cash and cash equivalent has increased by £7 million.

■ A financial health check using ratios

Accountants and bank managers have formulated dozens of financial ratios to help diagnose the financial health of the business, its position, performance and prospects. We shall restrict our focus to those key financial ratios that every finance manager and investor should be acquainted with. These are summarised in Table 2.3 and discussed briefly below.

Table 2.3 *Foto-U* key ratios

Ratio	Form	2010	2009
Profitability			
Gross profit margin	%	21.5	15.8
Net profit margin	%	11.0	5.3
Return on capital employed (ROCE)	%	15.9	10.6
Activity ratios			
Net asset turnover	times	1.4	2.0
Debtors	days	53	48
Stock	days	58	54
Supplier credit period	days	93	96
Liquidity and financing ratios			
Current ratio	times	1.1	0.9
Quick (acid test) ratio	times	0.7	0.5
Gearing	%	43.5	24.5
Interest cover	times	11	5
Investor ratios			
Return on shareholders' funds	%	17.9	5.6
Dividend per share	pence	3.5	1.5
Earnings per share	pence	7	2
Dividend cover	times	2	1.3
Price:earnings	times	17.1	25
Dividend yield	%	2.9	3

Profitability ratios

To assess the performance of *Foto-U*, we study a number of profitability ratios.

Profit margin

This ratio shows how much profit is generated from every £ of sales. It can be considered in the form of a percentage at both the gross and net profit levels.

Gross profit margin

$$\frac{\text{Gross profit}}{\text{Sales}} \times 100 = \frac{43}{200} \times 100$$

$$= 21.5\% \quad (15.8\% \text{ last year})$$

Net profit margin

$$\frac{\text{EBIT}}{\text{Sales}} \times 100 = \frac{22}{200} \times 100$$

$$= 11\% \quad (5.3\% \text{ last year})$$

(EBIT is earnings before interest and tax, i.e. operating profit.)

Return on capital employed (ROCE)

This ratio, also termed the primary ratio, examines the rate of profit the business makes on the long-term capital invested in it. *Foto-U* has shareholders' funds of £78 million and long-term creditors of £60 million giving long-term capital of £138 million. This is represented by the total assets less current liabilities figure on the balance sheet.

$$\text{ROCE} = \frac{\text{EBIT}}{\text{Long-term capital}} \times 100 = \frac{22}{138} \times 100$$

$$= 15.9\% \quad (10.6\% \text{ last year})$$

Activity ratios

Here we examine how efficiently *Foto-U* manages its assets in terms of the level of sales obtained from the assets invested.

Asset turnover

$$\frac{\text{Sales}}{\text{Total assets} - \text{current liabilities}} = \frac{200}{138}$$

$$= 1.45 \text{ times} \quad (2 \text{ times last year})$$

This can also be expressed in terms of each type of asset, but here, we usually express it in terms of days. For example, the average number of days it takes for debtors to pay is given by debtor days.

Debtor days

$$\frac{\text{Debtors}}{\text{Credit sales}} \times 365 = \frac{29}{200} \times 365$$

$$= 53 \text{ days} \quad (48 \text{ days last year})$$

Note also that we have used the asset figure at the year-end. A more accurate picture is given by finding the average asset value based on the values at the start and end of the year.

Similar calculations can be made for stock and creditors, but with one important difference. Stock and trade creditors are valued in the balance sheet at original cost so instead of using sales, we use cost of sales, i.e., what it cost the firm to build these stocks.

Stockholding period

$$\frac{\text{Stock}}{\text{Cost of sales}} \times 365 = \frac{25}{157} \times 365$$

$$= 58 \text{ days} \quad (54 \text{ days last year})$$

Supplier credit days

$$\frac{\text{Trade creditors}}{\text{Cost of sales}} \times 365 = \frac{40}{157} \times 365$$

$$= 93 \text{ days} \quad (96 \text{ days last year})$$

It is preferable to use purchases rather than cost of sales, although this figure is not always available.

Liquidity and financing ratios

To assess whether the company is able to meet its financial obligations as they fall due, we need to compare short-term assets with short-term creditors. Two such ratios are commonly employed.

Current ratio

$$\frac{\text{Current assets}}{\text{Current liabilities}} = \frac{66}{60}$$

$$= 1.1 \text{ times} \quad (0.9 \text{ times last year})$$

Quick assets

For most firms, it is not easy to convert stock into cash with any great speed. The quick assets (or acid-test ratio) is a more prudent liquidity ratio which excludes stock entirely.

$$\frac{\text{Current assets} - \text{stock}}{\text{Current liabilities}} = \frac{66 - 25}{60}$$

$$= 0.7 \text{ times} \quad (0.5 \text{ times last year})$$

As a general rule of thumb, we would typically expect the current ratio to be 2 and the quick assets to match creditors (i.e. a quick ratio of 1). However, this guide may differ from industry to industry depending on the trade credit periods granted to customers and claimed from suppliers.

Gearing

A rather different question asks how the capital employed in the business is financed. The gearing ratio shows the proportion of capital employed funded by long-term borrowings.

$$\frac{\text{Long-term borrowings}}{\text{Debt} + \text{Equity capital}} \times 100 = \frac{60}{138} \times 100$$

$$= 43.5\% \quad (24.5\% \text{ last year})$$

An equally acceptable way of expressing the gearing ratio is by the Debt/Equity ratio.

$$\frac{\text{Long-term borrowings}}{\text{Shareholders' funds}} = \frac{60}{78} = 0.77{:}1$$

Interest cover

Another way of considering gearing is to look to the profit and loss account by assessing the degree of profits cover the firm has to meet its interest payments.

$$\frac{\text{Earnings before interest and taxes}}{\text{Interest payable}} = \frac{22}{2}$$

$$= 11 \text{ times} \quad (5 \text{ times last year})$$

An interest cover of 11 times is very safe. But were it to fall to, say, below three or four, concern may arise that taxation and dividends cannot be paid.

Investor ratios

Shareholders are more interested in the return they obtain on *their* investment rather than the return the company makes on the total business.

Return on shareholders' funds (return on equity)

This indicates how profitable the company has been for its shareholders.

$$\frac{\text{Earnings after tax and preference dividends}}{\text{Shareholders' funds}} \times 100 = \frac{14}{78} \times 100$$

$$= 17.9\% \quad (5.6\% \text{ last year})$$

Shareholders will also be interested in the earnings per share (what dividend could be paid) and dividend per share (what dividend is paid) for the year.

Earnings per share (EPS)

$$\frac{\text{Earnings after tax and preference dividends}}{\text{Number of ordinary shares in issue}} = \frac{14}{200}$$

$$= 7 \text{ pence per share}$$

$$(2 \text{ pence last year})$$

In practice, the EPS calculation is usually more complex than this, but the notes to the accounts will explain the calculation.

Dividend per share (DPS)

$$\frac{\text{Total ordinary dividend}}{\text{Number of ordinary shares in issue}} = \frac{7}{200}$$

$$= 3.5 \text{ pence per share} \quad (1.5 \text{ pence last year})$$

Dividend cover

This links the DPS and the EPS to indicate how many times the dividend *could* be paid, and, hence, how safe it is, in terms of exposure to a fall in EPS.

$$\frac{\text{Earnings per share}}{\text{Dividend per share}} = \frac{7}{3.5}$$

$$= 2 \text{ times} \quad (1.3 \text{ times last year})$$

The final two ratios relate earnings and dividends to stock market performance as reflected in the current share price. If the current share price for *Foto-U* is 120p, we can calculate the price:earnings ratio and dividend yield.

Price:earnings ratio (P:E)

$$\frac{\text{Current share price}}{\text{Earnings per share}} = \frac{120}{7}$$

$$= 17.1 \text{ times} \quad (25 \text{ times last year})$$

The share price, of course, is based on investors' expectations of *future* profits. A high P:E ratio indicates that investors expect future profits to grow – the higher the P:E, the greater the profit growth expectation.

Dividend yield

$$\frac{\text{Dividend per share (p)}}{\text{Share price (p)}} \times 100 = \frac{3.5}{120} \times 100$$

$$= 2.9\% \quad (3\% \text{ last year})$$

In the UK, income tax at 10% is deducted at source, so the calculation should therefore be based on the gross dividend.

■ Interpretation of the accounts and ratios

The financial manager or investor needs to put together all the clues suggested by ratio analysis and reading the accounts to gain insights into the financial position, performance and prospects of the company. This will probably involve looking at the trend of financial indicators, not simply comparison with the previous year, together with comparison with industry and competitor data. It certainly requires a reasonable grasp of the business, its objectives and strategies. Table 2.4 offers a brief report to senior management of *Foto-U* by the finance manager on the company's published accounts.

Table 2.4 *Foto-U* annual corporate performance report

To: Senior Management of *Foto-U*
From: Finance Manager

Subject: **Annual corporate performance**
30 April 2010

I have reviewed the published accounts for the past year to establish how successful *Foto-U* was in financial terms.

Profitability. The Return on Capital Employed has improved over the year from 10.6% to 16%. This is a significant improvement and well above the risk-free return we would expect from investing in say building society deposits, but we need also to compare the return against that achieved by our competitors. ROCE is a combination of two subsidiary ratios – net profit margin and asset turnover:

	ROCE	=	**Net profit margin**	×	**Asset turnover**
2009	10.6%	=	5.3%	×	2 times
2010	15.9%	=	11%	×	1.45 times

Both the gross and net profit margins have improved significantly as a result of the £10 million growth in sales over the year without any increase in costs. However, this growth has come at the expense of a poorer utilisation of our assets, as reflected in the significant decline in asset turnover. This is mainly attributable to a major capital expenditure programme during the year, the benefits of which will not be fully experienced for at least another year. A further factor is the increase in working capital. Last year, we actually managed to have negative working capital (i.e. our trade creditors and overdraft financed more than our current assets). This year, there has been a slight deterioration in all elements of working capital:

- We take five more days to collect cash from customers
- Stockholding period has increased by four days
- We pay suppliers a little quicker.

Liquidity. Our current and quick asset ratios are both well below the typical level for the industry of 1.8 and 1.0 respectively. However, this is largely due to the fact that our suppliers have been willing to grant us extended credit periods of about three months. Realistically, we cannot expect this to continue. Were they to demand payment within say, 45 days, it is difficult to see where we would be able to find the cash. It is not good financial management for us to rely on the generous credit of suppliers over whom we have no control, and we need to address this issue urgently. Linked to this, we have just raised a large medium-term loan in order to fund our capital expenditure in the coming year. Our gearing ratio has now nearly doubled and we will have to find cash both for additional interest payments and, eventually, the loan repayments. Unless the new investment very rapidly produces higher profits and cashflow, I am concerned that we could be in serious financial difficulty, despite the strong level of profits. Perhaps it is time to consider asking shareholders to invest more capital in the business, or to reduce dividend payments.

Investment attractiveness. The company's share price has progressed from 50 pence to 120 pence over the year. No doubt this is due to the growth in sales, profits and dividends in the year. Many of the investment performance indicators have improved, particularly earnings per share and return on shareholders' funds, the latter looking much healthier at nearly 18%. However, the price:earnings ratio has slipped a little, suggesting that investors do not expect the company's profits and share price to continue to grow at quite the same rate as this year.

In summary, *Foto-U* has improved its performance over the past year, but there remain concerns regarding its liquidity. Management is urged to give urgent attention to this matter.

QUESTIONS

 **myfinancelab** *Questions with an icon are also available for practice in myfinancelab with additional supporting resources.*

Questions with a **coloured number** have solutions in Appendix B on page 729.

1 When a company seeks a listing for its shares on a stock exchange, it usually recruits the assistance of a merchant bank.

(a) Explain the role of a merchant bank in a listing operation with respect to the various matters on which its advice will be sought by a company.

(b) Identify the conflicts which might arise if the merchant bank were part of a group providing a wide range of financial services.

(CIMA)

2 (a) Briefly outline the major functions performed by the capital market and explain the importance of each function for corporate financial management. How does the existence of a well-functioning capital market assist the financial management function?

(b) Describe the efficient markets hypothesis and explain the differences between the three forms of the hypothesis which have been distinguished.

(c) Company A has 2 million shares in issue and company B 6 million. On day 1 the market value per share is £2 for A and £3 for B. On day 2, the management of B decides, at a private meeting, to make a cash takeover bid for A at a price of £3.00 per share. The takeover will produce large operating savings with a value of £3.2 million. On day 4, B publicly announces an unconditional offer to purchase all shares of A at a price of £3.00 per share with settlement on day 15. Details of the large savings are not announced and are not public knowledge. On day 10, B announces details of the savings which will be derived from the takeover.

Required

Ignoring tax and the time-value of money between days 1 and 15, and assuming the details given are the only factors having an impact on the share prices of A and B, determine the day 2, day 4 and day 10 share prices of A and B if the market is:

1 semi-strong form efficient, and
2 strong form efficient

in each of the following separate circumstances:

(i) the purchase consideration is cash as specified above, and
(ii) the purchase consideration, decided upon on day 2 and publicly announced on day 4, is one newly issued share of B for each share of A.

(ACCA)

? **3** You are an accountant with a practice that includes a large proportion of individual clients, who often ask for information about traded investments. You have extracted the following data from a leading financial newspaper.

(i) Stock	**Price**	**P:E ratio**	**Dividend yield (% gross)**
Buntam plc	160p	20	5
Zellus plc	270p	15	3.33

(ii) Earnings and dividend data for Crazy Games plc are given below:

	1993	1994	1995	1996	1997
EPS	5p	6p	7p	10p	12p
Div. per share (gross)	3p	3p	3.5p	5p	5.5p

The estimated before tax return on equity required by investors in Crazy Games plc is 20%.

Required

Draft a report for circulation to your private clients which explains:

(a) the factors to be taken into account (including risks and returns) when considering the purchase of different types of traded investments.

(b) the role of financial intermediaries, and their usefulness to the private investor.

(c) the meaning and the relevance to the investor of each of the following:

(i) Gross dividend (pence per share)

(ii) EPS

(iii) Dividend cover

Your answer should include calculation of, and comment upon, the gross dividends, EPS and dividend cover for Buntam plc and Zellus plc, based on the information given above.

(ACCA)

4 Beta plc has been trading for twelve years and during this period has achieved a good profit record. To date, the company has not been listed on a recognised stock exchange. However, Beta plc has recently appointed a new chairman and managing director who are considering whether or not the company should obtain a full Stock Exchange listing.

Required

(a) What are the advantages and disadvantages which may accrue to the company and its shareholders, of obtaining a full stock exchange listing?

(b) What factors should be taken into account when attempting to set an issue price for new equity shares in the company, assuming it is to be floated on a stock exchange?

(Certified Diploma)

5 Collingham plc produces electronic measuring instruments for medical research. It has recorded strong and consistent growth during the past 10 years since its present team of managers bought it out from a large multinational corporation. They are now contemplating obtaining a stock market listing.

Collingham's accounting statements for the last financial year are summarised below. Fixed assets, including freehold land and premises, are shown at historic cost net of depreciation. The debenture is redeemable in two years although early redemption without penalty is permissible.

Profit and Loss Account for the year ended 31 December 1994 (£m)

Turnover	80.0
Cost of sales	(70.0)
Operating profit	10.0
Interest charges	(3.0)
Pre-tax profit	7.0
Corporation tax (after capital allowances)	(1.0)
Profits attributable to ordinary shareholders	6.0
Dividends	(0.5)
Retained earnings	5.5

Balance Sheet as at 31 December 1994 (£m)

Assets employed		
Fixed: Land and premises	10.0	
Machinery	20.0	30.0
Current: Stocks	10.0	
Debtors	10.0	
Cash	3.0	23.0
Current liabilities: Trade creditors	(15.0)	
Bank overdraft	(5.0)	(20.0)
Net current assets		3.0
Total assets less current liabilities		33.0
14% Debentures		(5.0)
Net assets		28.0
Financed by:		
Issued share capital (par value 50p):		
Voting shares		2.0
Non-voting 'A' shares		2.0
Profit and Loss Account		24.0
Shareholders' funds		28.0

The following information is also available regarding key financial indicators for Collingham's industry.

Return on (long-term) capital employed	22% (pre-tax)
Return on equity	14% (post-tax)
Operating profit margin	10%
Current ratio	1.8:1
Acid test	1.1:1
Gearing (total debt/equity)	18%
Interest cover	5.2
Dividend cover	2.6
P:E ratio	13:1

Required

(a) Briefly explain why companies like Collingham seek stock market listings.

(b) Discuss the performance and financial health of Collingham in relation to that of the industry as a whole.

(c) In what ways would you advise Collingham:
   (i) to restructure its balance sheet prior to flotation?
   (ii) to change its financial policy following flotation?

Practical assignment

Select two companies from one sector in the *Financial Times* share information service. Analyse the share price and other data provided and compare this with the FT All-Share Index data for the sector. Suggest why the P:E ratios for the companies differ.

 *Now retake your diagnostic test for Chapter 2 to check your progress and update your study plan.*

3

Present values, and bond and share valuation

An investment parable

A man, going off to another country, called together his servants and loaned them money to invest for him while he was gone. He gave £500 to one, £200 to another and £100 to the last – dividing it in proportion to their abilities – and then left on his trip. The man who received the £500 began immediately to buy and sell with it and soon earned another £500. The man with £200 went right to work, too, and earned another £200. But the man who received the £100 dug a hole in the ground and hid the money for safe keeping.

After a long time their master returned from his trip and called them to him to account for his money. The man to whom he had entrusted the £500 brought him £1,000. His master praised him for good work. 'You have been faithful in handling this small amount,' he told him, 'so now I will give you many more responsibilities.' Next came the man who had received £200, with the report, 'Sir, you gave me £200 to use, and I have doubled it.' 'Good work', his master said. 'You have been faithful over this small amount, so now I will give you much more.'

Then the man with the £100 came and said, 'Sir, I knew you were a hard man, and I was afraid you would rob me of what I earned, so I hid your money in the earth and here it is!'

But his master replied, 'You lazy rogue! Since you knew I would demand your profit, you should at least have put my money into the bank so I could have some interest.'

Source: Matthew, Chapter 25, *Living Bible.*

Learning objectives

Having completed this chapter, you should have a sound grasp of the time-value of money and discounted cash flow concepts. In particular, you should understand the following:

- The time-value of money.
- The financial arithmetic underlying compound interest and discounting.
- Present value formulae for single amounts, annuities and perpetuities.
- The valuation of bonds and shares.

Skills developed in discounted cash flow analysis, using both formulae and tables, will help enormously in subsequent chapters.

 **myfinancelab** *Complete your diagnostic test for Chapter 3 now to create your personal study plan.*

3.1 INTRODUCTION

The introductory investment parable, taken from business life in 1st century Palestine, is equally appropriate to present times. Managers are expected to make sound long-term decisions and to manage resources in the best interests of the owners. To do otherwise is to risk the wrath of an unmerciful stock market! Rather like the lazy servant in the parable, Eurotunnel put the £10 billion entrusted to it by shareholders and bankers into a 'hole in the ground' stretching from Dover to Calais. From an investment perspective they would have done better letting it earn interest in a bank.

To assess whether investment ideas are wealth-creating, we need to have a clear understanding of cash flow and the time-value of money. Capital investment decisions, security and bond value analyses, financial structure decisions, lease vs. buy decisions and the tricky question of the required rate of return can be addressed only when you understand exactly what the old expression 'time is money' really means.

In this chapter we will consider the measurement of wealth and the fundamental role it plays in the decision-making process; the time-value of money, which underlies the discounted cash flow concept; and the net present value approach for analysing investment decisions.

3.2 MEASURING WEALTH

'Cash is King' seems to be the message for businesses today. Spectacular business collapses in recent years demonstrate that reliance on sales, profits or earnings per share as measures of performance can be dangerous.

The chairman of a fast-growing company that went out of business stated in the annual report: 'Last year, we delivered a 425% increase in turnover from £19.9 million to £109.8 million.' But when the firm was placed into the hands of the receiver the following year, it was not the lack of sales or even profits that put it there. It was the lack of cash. Businesses go 'bust' because they run out of the cash required to fulfil their financial obligations. Of course, there are always reasons why this happens – recession, an over-ambitious investment programme, rapid growth without adequate long-term finance – but basically corporate survival and success come down to cash flow and value creation.

Boo.com, the internet fashion retailer, thought it had a promising future at the start of 2000. It had raised $135 million to set up the new business and invest in marketing to break into the competitive fashion retail sector. But less than six months later, it had virtually run out of cash and was forced into liquidation.

Recall from Chapter 1 that the assumed objective of the firm is to create as much wealth as possible for its shareholders. A successful business is one that creates value for its owners. Wealth is created when the market value of the outputs exceeds the market value of the inputs, i.e. the benefits are greater than the costs. Expressed mathematically:

$$V_j = B_j - C_j$$

The value (V_j) created by decision j is the difference between the benefits (B_j) and the costs (C_j) attributable to the decision. This leads to an obvious decision rule: accept only those investment or financing proposals that enhance the wealth of shareholders, i.e. accept if $B_j - C_j > 0$.

time-value of money
Money received in the future is usually worth less than today because it could be invested to earn interest over this period

Nothing could be simpler in concept – the problems emerge only when we probe more deeply into how the benefits and costs are measured and evaluated. One obvious problem is that benefits and costs usually occur at different times and over a number of years. This leads us to consider the **time-value of money**.

Boo.com collapses as investors refuse funds

Boo.com, the online sportswear retailer, became Europe's first big internet casualty when the refusal of its backers to continue funding its heavy losses forced it into liquidation. The company – one of the highest profile internet retailers in Europe – appointed KPMG as liquidator, having spent all but $500,000 of the $135 million it had raised.

Boo's founders, including former model Kajsa Leander and Ernst Malmsten, chief executive, own about 40 per cent of the equity. Ernst Malmsten said: 'We have been too visionary. We wanted everything to be perfect, and we have not had control of costs. My mistake has been not to have a counterpart who was a strong financial controller.'

After a high-profile launch, the company was dogged by technical problems that delayed the site going live by five months. Boo needed $430 million to implement an emergency restructuring plan that would have seen redundancies among the 300-strong workforce and closure of some overseas offices. But investors were not prepared to back the plan with more money.

Source: Based on Financial Times, 18 May 2000.

3.3 TIME-VALUE OF MONEY

An important principle in financial management is that the value of money depends on *when* the cash flow occurs – £100 *now* is worth more than £100 at some *future* time. There are a number of reasons for this:

1 *Risk*. One hundred pounds now is certain, whereas £100 receivable next year is less certain. This 'bird-in-the-hand' principle affects many aspects of financial management that will be covered in later chapters.
2 *Inflation*. Under inflationary conditions, the value of money, in terms of its purchasing power over goods and services, declines.
3 *Personal consumption preference*. Most of us have a strong preference for immediate rather than delayed consumption.

More fundamental than any of the above, however, is the time-value of money. Money – like any other desirable commodity – has a price. If you own money, you can 'rent' it to someone else, say a banker, and earn interest. A business which carries unnecessarily high cash balances incurs an *opportunity cost* – the lost opportunity to earn money by investing it to earn a higher return. The overall investor's return, which reflects the time-value of money, therefore comprises:

(a) the risk-free rate of return rewarding investors for forgoing immediate consumption, plus
(b) compensation for risk and loss of purchasing power.

Self-assessment activity 3.1

Imagine you went to your bank manager asking for a £50,000 loan, for five years, to start up a burger bar under a McDonald's franchise. Which of the considerations in the previous paragraph would the bank manager consider?

(Answer in Appendix A at the back of the book)

Before proceeding further, we need to understand the essential financial arithmetic for the time-value of money. This will stand readers in good stead not only in analysing capital and financial investments in the remainder of this book, but also in handling their personal finances. For example, it will provide a better understanding of how interest is calculated for credit cards, bank loans, repayment mortgages and hire purchase arrangements.

3.4 FINANCIAL ARITHMETIC FOR CAPITAL GROWTH

■ Simple and compound interest

compound interest
Interest paid on the sum
which accumulates, i.e. the
principal plus interest

The future value (FV) of a sum of money invested at a given annual rate of interest will depend on whether the interest is paid only on the original investment (simple interest), or whether it is calculated on the original investment plus accrued interest (**compound interest**). Suppose you win £1,000 on the National Lottery and decide to invest it at 10 per cent for five years, simple interest. The future value will be the original £1,000 capital plus five years' interest of £100 a year, giving a total future value of £1,500.

With compound interest, the interest is paid on the original capital plus accrued interest, as shown in Table 3.1. The process of compounding provides a convenient way of adjusting for the time-value of money. An investment made now in the capital market of V_0 gives rise to a cash flow of $V_0(1 + i)^2$ after two years, and so on. In general, the future value of V_0 invested today at a compound rate of interest of i per cent for n years will be:

$$FV_{(i,n)} = V_0(1 + i)^n$$

where $FV_{(i,n)}$ is the future value at time n, V_0 is the original sum invested, sometimes termed the principal (note that the o subscript refers to the time period, i.e. today), and i is the annual rate of interest.

Using this formula in the above example we obtain the same future value as in Table 3.1.

$$FV_5 = £1,000(1 + 0.10)^5 = £1,610$$

Note that the effect of compound interest yields a higher value than simple interest, which yielded only £1,500.

Table 3.1 Compound interest on £1,000 over five years (at 10%)

Year	Starting balance £	+	Interest £	=	Closing balance £
1	1000		100		1100
2	1100		110		1210
3	1210		121		1331
4	1331		133		1464
5	1464		146		1610

■ More frequent compounding and annual percentage rates

annual percentage rate
The true annual interest rate
charged by the lender which
takes account of the timing
of interest and principal
payments

Unless otherwise stated, it is assumed that compounding or discounting is an annual process; cash payments of benefits arise either at the start or the end of the year. Frequently, however, the contractual payment period is less than one year. Building societies and government bonds pay interest semi-annually or quarterly. Interest charged on credit cards is applied monthly. To compare the true costs or benefits of such financial contracts, it is necessary to determine the **annual percentage rate** (APR), or effective annual interest rate, taking into account any costs such as one-off fees. In the US and the UK, lenders are required to disclose the APR before the loan is finalised. Taking compounding to its limits, we can adopt a continuous discounting approach.*

Examples of more frequent compound interest

Returning to our earlier example of £1,000 invested for five years at 10 per cent compound interest, we now assume 5 per cent payable every six months.

After the first six months, the interest is £50, which is reinvested to give interest for the second half year of (£1,050 × 5%) = £52. The end-of-year value is therefore (£1,050 + £52) = £1,102. We can still use the compound interest formula, but with i as the six-monthly interest rate and n the six-monthly, rather than annual, interval:

$$\text{After 1 year, } FV_1 = £1,000(1 + 0.05)^2$$
$$= £1,102$$
$$\text{After 5 years, } FV_5 = £1,000(1 + 0.05)^{10}$$
$$= £1,629$$

Note that this value is higher than the £1,610 value based on the earlier annual interval calculation. In converting the annual compounding formula to another interest payment frequency, the trick is simply to divide the annual rate of interest (i) and multiply the time (n) by the number of payments each year.

If, in the above example, interest is calculated at weekly intervals over five years, the future value will be:

$$FV_5 = £1,000\left(1 + \frac{0.10}{52}\right)^{52(5)} = £1,648$$

We calculate below the APRs based on a range of interest payment frequencies for a 22 per cent per annum loan. By charging compound interest on a daily basis, the effective annual rate is 24.6 per cent, some 2.6 per cent higher than on an annual basis.

Annually	$(1 + 0.22)$	$- 1$	$= 0.22$ or 22%
Semi-annually	$\left(1 + \frac{0.22}{2}\right)^2$	$- 1$	$= 0.232$ or 23.2%
Monthly	$\left(1 + \frac{0.22}{12}\right)^{12}$	$- 1$	$= 0.244$ or 24.4%
Daily	$\left(1 + \frac{0.22}{365}\right)^{365}$	$- 1$	$= 0.246$ or 24.6%

3.5 PRESENT VALUE

present value
The current worth of future cash flows

discounting
The process of reducing cash flows to present values

An alternative way of assessing the worth of an investment is to invert the compounding process to give the **present value** of the future cash flows. This process is called **discounting**.

The time-value of money principle argues that, given the choice of £100 now or the same amount in one year's time, it is always preferable to take the £100 now because it

---

*When the number of compounding periods each year approaches infinity, the future value is found by:

$$FV_n = V_o e^{in}$$

where i is the annual interest rate, n is the number of years and e is the value of the exponential function. Using a scientific calculator, this is shown as 2.71828 (to five decimal places).

Using the same example as before:

$$FV_4 = V_o e^{in} = £1,000\ e^{(0.1)5}$$

$$= £1,648.72 \text{ (slightly more than compounding on a weekly basis)}$$

could be invested over the next year at, say, a 10 per cent interest rate to produce £110 at the end of one year. If 10 per cent is the best available annual rate of interest, then one would be indifferent to (i.e. attach equal value to) receiving £100 now or £110 in one year's time. Expressed another way, the *present value* of £110 received one year hence is £100.

We obtained the present value (PV) simply by dividing the future cash flow by 1 plus the rate of interest, i, i.e.

$$PV = \frac{£110}{(1 + 0.10)} = \frac{£110}{(1.1)} = £100$$

Discounting is the process of adjusting future cash flows to their present values. It is, in effect, compounding in reverse.

Recall that earlier we specified the future value as:

$$FV_n = V_0(1 + i)^n$$

Dividing both sides by $(1 + i)^n$ we find the present value:

$$V_o = \frac{FV_n}{(1 + i)^n}$$

which can be read as the present value of future cash flow FV receivable in n years' time given a rate of interest i. This is the process of discounting future sums to their present values.

Let us apply the present value formula to compute the present value of £133 receivable three years hence, discounted at 10 per cent:

$$PV_{(10\%,\,3\,\text{yrs})} = \frac{£133}{(1 + 0.10)^3} = \frac{£133}{1.33} = £100$$

The message is: do not pay more than £100 today for an investment offering a certain return of £133 after three years, assuming a 10 per cent market rate of interest.

Calculator tip

Your calculator should have a power function key, usually x^y. Try the following steps for the previous example.

Input	1.1
Press	x^y function key
Input	3
Press	=
Display	1.331
Press	1/x
Multiply	133
Press	=
Answer	99.9

Self-assessment activity 3.2

Calculate the present value of £623 receivable in eight years' time plus £1,092 receivable eight years after that, assuming an interest rate of 7 per cent.

(Answer in Appendix A at the back of the book)

■ Discount tables

Much of the tedium of using formulae and power functions can be eased by using discount tables or computer-based spreadsheet packages. In the previous example, the discount factor for £1 for a 10 per cent discount rate in three years' time is:

$$\frac{1}{(1.10)^3} = \frac{1}{1.33} = 0.751$$

This can be found in Appendix C by locating the 10 per cent column and the 3-year row. We call this the present value interest factor (PVIF) and express it as $PVIF_{(10\%,\,3\,yrs)}$ or $PVIF_{(10,3)}$.

Multiplying the cash flow of £133 by the discount factor yields the same result as before:

$$PV = £133 \times 0.751 = £100 \text{ (subject to rounding)}$$

annuity
A constant annual cash flow for a prescribed period of time

With a constant annual cash flow, termed an **annuity**, we can shorten the discounting operation. Appendix D provides the present value interest factor for an annuity (PVIFA). Thus, if £133 is to be received in each of the next three years, the present value is:

$$PV = £133 \times PVIFA_{(10\%,\,3\,yrs)}$$
$$= £133 \times 2.4868 = £331$$

It is standard practice to write interest factors as: Interest factor(rate, period).

Examples:
$PVIF_{(8,10)}$ is the present value interest factor at 8 per cent for ten years.
$PVIFA_{(10,4)}$ is the present value interest factor for an annuity at 10 per cent for four years.

Example of present values: Soldem Pathetic FC Ltd

Soldem Pathetic Football Club has recently been bought up by a wealthy businessman who intends to return the club to its former glory days. He also wants to pay a good dividend to the shareholders of the newly-formed quoted company by making sound investments in quality players. One such player the manager would dearly like in his squad is Bryan Riggs, currently on the market for around £9 million. The chairman reckons that, quite apart from the extra income at the turnstiles from buying him, he could be sold for £11 million by the end of the year, given the way transfer prices are moving. Should he bid for Riggs?

Assuming a 10 per cent rate of interest as the reward that the other shareholders demand for accepting the delayed payoff, the present value (PV) of £11 million receivable one year hence is:

$$PV = \text{discount factor} \times \text{future cash flow} = \frac{1}{1.10} \times £11 \text{ million}$$
$$= £10 \text{ million}$$

How much better off will the club be if it buys Riggs? The answer is, in present value terms:

$$£10 \text{ million} - £9 \text{ million} = £1 \text{ million}$$

net present value
The present value of the future net benefits less the initial cost

We call this the **net present value** (NPV). The decision to buy the player makes economic sense; it promises to create wealth for the club and its shareholders, even excluding the likely additional gate receipts. Of course, Riggs could break a leg in the very first game for his new club and never play again. In such an unfortunate situation, the club would achieve a negative NPV of £9 million, the initial cost. Alternatively, he could be insured against such injury, in which case there would be premiums to pay, resulting in a lower net present value.

Continued

Another way of looking at this issue is to ask whether the investment offers a return greater than could have been achieved by investing in financial, rather than human, assets. The return over one year from acquiring Riggs' services is:

$$\text{Return} = \frac{\text{Profit}}{\text{Investment}} = \frac{£11\text{ m} - £9\text{ m}}{£9\text{ m}} \times 100 = 22.2\%.$$

If the available rate of interest is 10 per cent, the investment in Riggs is a considerably more rewarding prospect.

In the highly simplified example above, we assumed that the future value was certain and the interest rate was known. Of course, a spectrum of interest rates is listed in the financial press. This variety of rates arises predominantly because of uncertainty surrounding the future and imperfections in the capital market. To simplify our understanding of the time-value of money concept, let us 'assume away' these realities. The lender knows with certainty the future returns arising from the proposal for which finance is sought, and can borrow or lend on a perfect capital market. The latter assumes the following:

1 Relevant information is freely available to all participants in the market.
2 No transaction costs or taxes are involved in using the capital market.
3 No participant (borrower or lender) can influence the market price for funds by the scale of its activities.
4 All participants can lend and borrow at the same rate of interest.

Under such conditions, the corporate treasurer of a major company like Shell can raise funds no more cheaply than the chairman of Soldem Pathetic. A single market rate of interest prevails. Borrowers and lenders will base time-related decisions on this unique market rate of interest. The impact of uncertainty will be discussed in later chapters; for now, these simplistic assumptions will help us to grasp the basics of financial arithmetic.

■ The effect of discounting

Figure 3.1 shows how the discounting process affects present values at different rates of interest between 0 and 20 per cent. The value of £1 decreases very significantly as

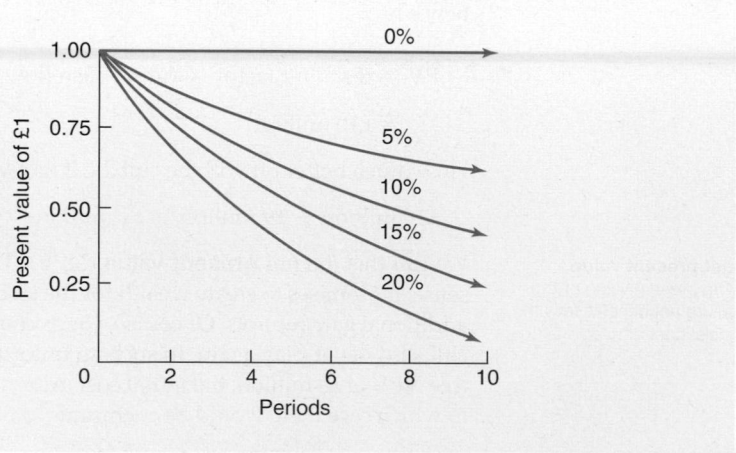

Figure 3.1 The relationship between present value of £1 and interest over time

the rate and period increase. Indeed, after 10 years, for an interest rate of 20 per cent, the present value of a cash flow is only a small fraction of its nominal value.

Table 3.2 summarises the discount factors for three rates of interest. It is useful to develop a 'feel' for how money changes with time for these rates of interest. The 15 per cent discount rate is particularly useful, because investment surveys (e.g. Pike 1988) suggest that this is a popular discount rate for evaluating capital projects. It also happens to be easy to remember: every five years the discounted value halves. Thus, with a 15 per cent discount rate, after five years the value of £1 is 50p, after 10 years 25p, etc.

Table 3.2 Present value of a single future sum

Year	10%	15%	20%
0	£1.00	£1.00	£1.00
5	0.60	0.50	0.40
10	0.40	0.25	0.16
15	0.24	0.12	0.06
20	0.15	0.06	0.03
25	0.09	0.03	0.01

Self-assessment activity 3.3

Your company is just about to sign a deal to purchase a fleet of lorries for £1 million. The payment terms are £500,000 down payment and £500,000 at the end of five years. No one present has a calculator or discount tables to hand. If the cost of capital for the company is 15 per cent, what is the present value cost of the purchase?

(Answer in Appendix A at the back of the book)

3.6 PRESENT VALUE ARITHMETIC

We have seen that the present value of a future cash flow is found by multiplying the cash flow by the present value interest factor. The present value concept is not difficult to apply in practice. This section explains the various present value formulae, and illustrates how they can be applied to investment and financing problems. Throughout, we shall use the symbol X to denote annual cash flow in pounds and i to denote the interest, or discount, rate (expressed as a percentage). Recall that PVIF is the present value interest factor and PVIFA is the PVIF for an annuity.

■ Present value

We know that the present value of X receivable in n years is calculated from the expression:

$$PV_{(i,n)} = \frac{X_n}{(1 + i)^n}$$
$$= X \text{ times } PVIF_{(i,n)}$$

Example

Calculate the present value of £1,000 receivable in 10 years' time, assuming a discount rate of 14 per cent:

$$PVIF_{(14\%, \, 10 \, yrs)} = \frac{1}{(1.14)^{10}} = 0.26974$$

Alternatively, the table in Appendix C provides PVIF of 0.26974 for $n = 10$ and $i = 14$ per cent:

PV = £1,000 × 0.26974 = £269.74

The present value of £1,000 receivable ten years hence, discounted at 14 per cent, is thus £269.74.

Self-assessment activity 3.4

Calculate the present value of £1,000 receivable 12 years hence, assuming the discount rate is 12 per cent.

(Answer in Appendix A at the back of the book)

Example: Pay cash up front or by instalments?

Mary has agreed to purchase a new car for £18,500. She is considering whether to pay this amount in full now or by instalments involving £9,000 now and payments of £5,000 at the start of each of the next two years.

Her first thought is that by paying through instalments she pays £19,000 which is £500 more than the single payment option. She then recalls that the time-value of money principle argues that all future cash flows should be converted to present values to make a valid comparison. She estimates that the rate she could earn on her savings is 6%. The calculations are:

		Present value
Down payment	£9,000	£9,000
Second payment	£5,000/1.06	£4,717
Third payment	$£5,000/(1.06)^2$	£4,450
Total present value		£18,167

Mary decides that it is better to go for the deferred payment package as it will cost her £18,167 in present value terms which is £333 cheaper than outright purchase. (The reader may wish to refer to Appendix C to check the present value calculations.)

In practice, most managers will use spreadsheets to do present value calculations, particularly when they involve multiple cash flows. We illustrate this using Microsoft Excel™ below.

1	A	B	C	D
2	Year	Cash flow	Present value	Formula in column C
3	0	−£9,000	−£9,000	=PV(B9,A3,0,B3)
4	1	−£5,000	−£4,717	=PV(B9,A4,0,B4)
5	2	−£5,000	−£4,450	=PV(B9,A5,0,B5)
6				
7	Total present value		−£18,167	=SUM(C3:C5)
8				
9	Discount rate	0.06		

■ Valuing perpetuities

perpetuity
A constant annual cash flow for an infinite period of time

Frequently, an investment pays a fixed sum each year for a specified number of years. A series of annual receipts or payments is termed an annuity. The simplest form of annuity is the infinite series or **perpetuity**. For example, certain government stocks offer a fixed annual income, but there is no obligation to repay the capital. The present value of such stocks (called irredeemables) is found by dividing the annual sum received by the annual rate of interest:

$$\text{PV perpetuity} = \frac{X}{i}$$

Example

Uncle George wishes to leave you in his will an annual sum of £10,000 a year, starting next year. Assuming an interest rate of 10 per cent, how much of his estate must be set aside for this purpose? The answer is:

$$\text{PV perpetuity} = \frac{£10,000}{0.10} = £100,000$$

Suppose that your benevolent uncle now wishes to compensate for inflation, estimated to be at 5 per cent per annum. The formula can be adjusted to allow for growth at the rate of g per cent p.a. in the annual amount. (The derivation of the present value of a growing perpetuity is found in Appendix II at the end of the chapter.)

$$\text{PV} = \frac{X}{i - g}$$

As long as the growth rate is less than the interest rate, we can compute the present value required:

$$\text{PV} = \frac{£10,000}{0.10 - 0.05} = £200,000$$

This formula plays a key part in analysing financial decisions and will be developed further, below, when we consider the valuation of assets, shares and companies.

■ Valuing annuities

An annuity is an investment paying a fixed sum each year for a specified period of time. Examples of annuities are many credit agreements and house mortgages.

The life of an annuity is less than that of a perpetuity, so its value will also be somewhat less. In fact, the formula for calculating the present value of an annuity of £A is found by calculating the present value of a perpetuity and deducting the present value of that element falling beyond the end of the annuity period. This gives the somewhat complicated formula (see Appendix II at the end of the chapter for the derivation) for the present value of an annuity (PVA):

$$\text{PVA}_{(i,n)} = A\left(\frac{1}{i} - \frac{1}{i(1 + i)^n}\right)$$
$$= A \times \text{PVIFA}_{(i,n)}$$

In words, the present value of an annuity for n years at i per cent is the annual sum multiplied by the appropriate present value interest factor for an annuity.

Suppose an annuity of £1,000 is issued for 20 years at 10 per cent. Using the table in Appendix D, we find the present value as follows:

$$\text{PVA}_{(10\%,\ 20\ \text{yrs})} = £1,000 \times \text{PVIFA}_{(10,20)}$$
$$= £1,000 \times 8.5136 = £8,513.60$$

Self-assessment activity 3.5

Calculate the present value of £250 receivable annually for 21 years plus £1,200 receivable after 22 years, assuming an interest rate of 11 per cent.

(Answer in Appendix A at the back of the book)

■ Calculating interest rates

Sometimes, the present values and future cash flows are known, but the rate of interest is not given. A credit company may offer to lend you £1,000 today on condition that you repay £1,643 at the end of three years. To find the compound rate of interest on the loan, we solve the present value formula for i:

$$PV_{(i,n)} = PVIF_{(i,n)} \times FV$$

Rearranging the formula,

$$PVIF_{(i,3)} = \frac{PV}{FV} = \frac{£1,000}{£1,643} = 0.60864$$

Turning to the tables in Appendix C and looking for 0.6086 under the year-3 column, we find the rate of interest is 18 per cent. As we shall see in Chapter 4, this calculation is fundamental to investment and finance decisions and is termed the **internal rate of return**.

internal rate of return
The rate of return that equates the present value of future cash flows with initial investment outlay

Alternatively, it is also possible to solve the present value formula for i:

$$PV = \frac{FV}{(1 + i)^n}$$

$$(1 + i)^n = FV/PV$$

$$i = (FV/PV)^{1/n} - 1$$

In the above example:

$$i = (1,643/1,000)^{1/3} - 1 = 0.18 \text{ or } 18\%$$

Who wants to be a millionaire?

An advertisement in the financial press read: 'How to become a millionaire? Invest £9,138 in the M&G Recovery unit trust in 1969 and wait for 25 years.' So, for those of us who missed out on this investment, let us grudgingly calculate its annual return:

$$i = (FV/PV)^{1/n} - 1$$
$$= (£1 \text{ million}/£9,138)^{1/25} - 1$$
$$= 20.66\%$$

By investing in a unit trust earning an annual rate of return of around 21 per cent, £9,138 turns you into a millionaire in 25 years' time. All you have to do is find an investment giving 21 per cent for 25 years!

3.7 VALUING BONDS

DCF analysis
The process of analysing financial instruments and decisions by discounting cash flows to present values

Now that we have explored the essential financial arithmetic of discounting, we can apply it to **discounted cash flow (DCF) analysis** in the analysis and valuation of financial instruments and investment projects. This chapter will cover the valuation of shares and Chapter 4 the valuation of capital investment projects. We now turn our

bond
A debt obligation with a maturity of more than a year

attention to the valuation of fixed income securities, better known as **bonds**. When a company wants to make long-term investments it may look to raising a long-term loan to finance it. One way of doing this is by issuing corporate bonds, promising investors that it will make a series of fixed interest payments and then repay the initial loan. A bond is a long-term (more than 1 year) loan which promises to pay interest and repay the loan in accordance with agreed terms. Governments, local authorities, companies and other organisations frequently seek to raise funds by issuing fixed interest bonds, offering a specific payment schedule for interest and repayment of **principal**. The return offered to the investor will depend on the creditworthiness of the issuer. For example, a UK government bond is seen as less risky than an unsecured corporate bond where the risk of default (the inability to meet its payment obligations) is higher. Accordingly, the return required for the corporate bond would typically be higher.

principal
The principal or face value or par value is the amount of the debt excluding interest

Once issued, bonds are traded in the bond markets. Although a bond has a par, or nominal, value – typically £100 – its actual value will vary according to the cash flows it pays (interest and repayments) and the prevailing rate of interest for this type of bond. The fair price is the present value of the future interest and repayments.

$$V_o = \text{PV (interest payments)} + \text{PV (redemption value)}$$

Example: Bondo Ltd

coupon rate
The nominal annual rate of interest expressed as a percentage of the principal value

Bondo Ltd issues a two-year bond with a 10 per cent **coupon rate** and interest payable annually. The bond is priced at its face value of £100:

$$£100 = \frac{£10}{1.10} + \frac{£10 + £100}{(1.10)^2}$$

The bond value above includes the present value of the first year's interest plus the present value of the two elements of the Year 2 cash flow (i.e. interest and redemption value).

Bond prices are subject to interest rate risk, increasing when interest rates fall and dropping when market interest rates rise. Typically, the longer the term of the bond, the greater the exposure to interest rate risk.

discount
The amount below the face value of a financial instrument at which it sells

Assume that the market interest rate unexpectedly rises to 12 per cent. The bond is now priced in the market at a **discount** at the lower value of £96.62, reflecting the fact that the 10 per cent interest rate is now less attractive to investors:

$$£96.62 = \frac{£10}{1.12} + \frac{£10 + £100}{(1.12)^2}$$

Assume now that the market interest rate falls to 8 per cent. The bond would now be viewed as more attractive and lead it to be priced at a **premium**:

premium
The amount above the face value of a financial instrument at which it sells

$$£103.57 = \frac{£10}{1.08} + \frac{£10 + £100}{(1.08)^2}$$

From the above example we may conclude that bonds will sell:

- at a discount where the coupon rate is below the market interest rate, and
- at a premium where the coupon rate is above the market interest rate.

yield to maturity
The interest rate at which the present value of the future cash flows equals the current market price

In the above example, the market interest was known. It may be that we know the bond prices and wish to calculate the **yield to maturity**. This measures the average rate of return to an investor who holds the bond until maturity. Here we use the same formula but the unknown is the interest rate:

$$£103.57 = \frac{£10}{1 + i} + \frac{£10 + £100}{(1 + i)^2}$$

Thus, in the above where the market price is £103.57 we solve the equation (using a computer or trial and error) to find that 8 per cent is the yield to maturity. The bond has a 10 per cent coupon and is priced at £103.57 to yield 8 per cent.

Example: Valuing a bond in Millie Meter plc

Some time ago you purchased an 8 per cent bond in the fashion chain Millie Meter. Today, it has a par value of £100 and two years to maturity. Interest is payable half-yearly. What is it worth?

Assuming the current comparable rate of interest is 8 per cent, the value should equal the par value of £100.

$$V_o = \frac{4}{(1.04)} + \frac{4}{(1.04)^2} + \frac{4}{(1.04)^3} + \frac{4}{(1.04)^4} + \frac{100}{(1.04)^4} = £100$$

Notice that because payments are made half-yearly, both the interest and discount rate are half the annual figures.

In reality, the required rate of return demanded by investors may be different from the original coupon rate. Let us say it is 10 per cent. As this is higher than the coupon rate, the bond value for Millie Meter will fall *below* its par value:

$$V_o = \frac{4}{(1.05)} + \frac{4}{(1.05)^2} + \frac{4}{(1.05)^3} + \frac{4}{(1.05)^4} + \frac{100}{(1.05)^4} = £96.45$$

This example shows that an investor would have to pay £96.45 for a bond offering a 4 per cent coupon rate (i.e. based on the par value of £100) plus the redemption value in two years' time, assuming that the market rate of interest for this security is 10 per cent.

For actively traded bonds there is little need to value them in this way because, if the bond market is efficient, it is already done for you. All you need do is to look at the latest quoted price. However, the required rate of return is less easy to obtain. Who says, in the above example, that 10 per cent is the return expected by the market for this type of bond? The answer is simple. If we know the current bond price, we put this in the above equation to find that discount rate which equates price with the discounted future cash flows – 10 per cent in the previous example.

Back to the future

'Tis the season to be jolly but there's always someone to cry 'Humbug'. According to Guy Monson from Saracen Investment Fund in London, things are pretty much as they were back in 1843 when Charles Dickens gave the world Ebenezer Scrooge, miser extraordinaire, in his novel *A Christmas Carol*.

Interest rates, government bond yields and inflation are all within a whisker of where they stood 141 years ago. There's also much living beyond one's means: that exercised Scrooge then and worries analysts now.

If that wasn't enough, some things have actually got worse since the days of poverty that Dickens so savagely chronicled. Back then, income tax stood at just 5 per cent.

As old Ebenezer so charmingly put it: 'Every idiot who goes around with Merry Christmas on his lips should be buried with his own pudding.'

Source: Financial Times, 23 December 2004, p. 12.

■ Factors affecting interest rates

It is common in financial management to talk about 'the interest rate ruling in the money market'. However, it is important to realise that there is never a single prevailing rate. At any time, there is a spectrum of interest rates on offer – along this spectrum the rates depend on the identity of the borrower, e.g. firm or government, and hence the degree of risk faced by the lender, the amount lent or borrowed and the period over which the loan is made available. The last of these aspects is referred to as the **term structure of interest rates**. This shows how the yields offered for loans of different maturities vary as the term of the loan increases. We discuss this, together with the yield curve, in Appendix I to this chapter.

term structure of interest rates
Pattern of interest rates on bonds of the same risk with different lengths of time to maturity

3.8 VALUING SHARES: THE DIVIDEND VALUATION MODEL

Bond valuation is relatively straightforward because the cash flows and life of the bond are known in advance. When we consider valuing shares we realise that the share may exist for as long as the company exists, and the cash flows to the shareholder are far from certain. The main cash flow arising to a shareholder will be the dividend payment, but this can only be paid if the company has built up sufficient profits, and the dividend policy pursued by companies varies. Shareholders attach value to shares because they expect to receive a stream of dividends and hope to make an eventual capital gain. Although shareholders are legally entitled to the earnings of a company, in the case of a company with a dispersed ownership body, their influence on the dividend payout is limited by their ability to exert their voting power on the directors. Other things being equal, shareholders prefer higher to lower dividends, but issues such as capital investment strategy and taxation may cloud the relationship between dividend policy and share value. With this reservation in mind, we now develop the **dividend valuation model (DVM)**. *This is appropriate for valuing part shares of companies rather than whole enterprises.* This is because minority shareholders have little or no control over dividend policy and thus it is reasonable to project past dividend policy, especially as companies and their owners are known to prefer a steadily rising dividend pattern rather than more erratic payouts. Conversely, if control changes hands, the new owner can appropriate the earnings as it chooses.

■ Valuing the dividend stream

The DVM states that the value of a share now, P_0, is the sum of the stream of future discounted dividends plus the value of the share as and when sold, in some future year, n:

$$P_o = \frac{D_1}{(1 + k_e)} + \frac{D_2}{(1 + k_e)^2} + \frac{D_3}{(1 + k_e)^3} + \cdots + \frac{D_n}{(1 + k_e)^n} + \frac{P_n}{(1 + k_e)^n}$$

However, since the new purchaser will, in turn, value the stream of dividends after year n, we can infer that the value of the share at any time may be found by valuing all future expected dividend payments over the lifetime of the firm.

Zero growth

If the lifespan is assumed infinite and the annual dividend is constant, we have:

$$P_o = \sum_{t=1}^{\infty} \frac{D_t}{(1 + k_e)^t} = \frac{D_1}{k_e}, \quad \text{where} \quad D_1 = D_2 = D_3 \text{ etc.}$$

This is another application of valuing a perpetuity.

For example, the shares of Nogrow Ltd, whose owners require a return of 15 per cent, and which is expected to pay a constant annual dividend of 30p per share through time would be valued thus:

$$P_o = \frac{30p}{0.15} = £2.00 \text{ per share}$$

In reality, the assumptions underlying this basic model are suspect. The annual dividend is unlikely to remain unchanged indefinitely, and it is difficult to forecast a varying stream of future dividend flows. To a degree, the forecasting problem is moderated by the effect of applying a risk-adjusted discount rate because more distant dividends are more heavily discounted. For example, discounting at 20 per cent, the present value of a dividend of £1 in 15 years' time is only 6p, while £1 received in 20 years adds only 3p to the value of a share. In other words, for a plausible cost of equity, we lose little by assuming a time-horizon of, say, 15 years. Even so, reliable valuations still require estimates of dividends over the intervening years, and by the same token, any errors will have a magnified effect during this period.

■ Allowing for future dividend growth

Dividends fluctuate over time, largely because of variations in the company's fortunes, although most firms attempt to grow dividends more or less in line with the company's longer-term earnings growth rate. For reasons explained in Chapter 17, financial managers attempt to 'smooth' the stream of dividends. For companies operating in mature industries, the growth rate will roughly correspond to the underlying growth rate of the whole economy. For companies operating in activities with attractive growth opportunities, dividends are likely to grow at a faster rate, at least over the medium term.

■ Allowing for dividend growth: the DGM

The constant dividend valuation model can be extended to cover constant growth thus becoming the dividend growth model (DGM). This states that the value of a share is the sum of all discounted dividends, growing at the annual rate g:

$$P_o = \frac{D_o(1+g)}{(1+k_e)} + \frac{D_o(1+g)^2}{(1+k_e)^2} + \frac{D_o(1+g)^3}{(1+k_e)^3} + \cdots + \frac{D_o(1+g)^n}{(1+k_e)^n}$$

If D_o is this year's recently paid dividend,* $D_o(1+g)$ is the dividend to be paid in one year's time (D_1), and so on.

Such a series growing to infinity has a present value of:

$$P_o = \frac{D_o(1+g)}{(k_e - g)} = \frac{D_1}{(k_e - g)}$$

The growth version of the model is often used in practice by security analysts (it is popularly known as 'the dividend discount model'), at least as a reference point, but it makes some key assumptions. Dividend growth is assumed to result from earnings growth, generated solely by new investment that is financed by retained earnings. Such investment is, of course, worthwhile only if the anticipated rate of return, R, is in excess of the cost of equity, k_e. Furthermore, it is assumed that the company will retain a constant fraction of earnings and invest these in a continuous stream of projects all offering a return of R. It also breaks down if g exceeds k_e.

---

*If the dividend has recently been paid, i.e. the next dividend will be paid in, say, a year's time, the shares are said to be 'ex-dividend'. They trade without entitlement to a dividend for some considerable time.

Example: Growmore Ltd

Growmore Ltd has just paid a dividend of 6p per share. The dividend grows at a steady rate of 5 per cent per year and the cost of equity is 12 per cent. Using the dividend growth model, the price per share is:

$$P_0 = D_1/(k_e - g)$$
$$= 6p \times 1.05/(0.12 - 0.05)$$
$$= 6.3p/0.07$$
$$= 90p$$

Where did Growmore's dividend growth rate, g, come from? It is a compound of the proportion of profits retained in the company and the return it expects to make on those reinvested profits. If we term the retention ratio b, and return on invested capital R, we can say:

$$g = (b \times R)$$

If Growmore regulary reinvested 40 per cent of its earnings and expected to get a 15 per cent return on the reinvested earning, the dividend growth rate would be 6 per cent:

$$g = (b \times R)$$
$$= 0.40 \times 0.15$$
$$= 0.06 \text{ or } 6 \text{ per cent}$$

An alternative approach to estimating the dividend growth rate is to determine the historical rate of growth in dividend over a reasonable period of time.

In Chapter 17, we examine more fully the issues of whether and how a change in dividend policy can be expected to alter share value. For the moment, we are mainly concerned with the mechanics of the DGM and rely simply on the assumption that any retained earnings are used for worthwhile investment. If this applies, the value of the equity will be higher with retentions-plus-reinvestment than if the investment opportunities were neglected, i.e. the decision to retain earnings benefits shareholders because of company access to projects that offer returns higher than the owners could otherwise obtain.

Self-assessment activity 3.6

XYZ plc currently earns 16p per share. It retains 75 per cent of its profits to reinvest at an average return of 18 per cent. Its shareholders require a return of 15 per cent. What is the ex-dividend value of XYZ's shares? What happens to this value if investors suddenly become more risk-averse by seeking a return of 20 per cent?

(Answer in Appendix A at the back of the book)

3.9 PROBLEMS WITH THE DIVIDEND GROWTH MODEL

The Dividend Growth Model, while possessing some convenient properties, has some major limitations.

■ What if the company pays no dividend?

The company may be faced with highly attractive investment opportunities that cannot be financed in other ways. According to the model, such a company would have no value at all! Total retention is fairly common, either because the company has suffered an actual or expected earnings collapse, or because, as in some European economies (e.g. Switzerland), the expressed policy of some firms is to pay no dividends at all. The cash-rich American computer software firm Microsoft paid its first dividend only in 2003, while two other computer firms, Dell and Apple, have yet to pay dividends at all. Yet we observe that shares in such companies do not have zero values. Indeed, nothing could be further from the truth.

In the case of Dell, $100 invested in its initial public offering in June 1988, would have been worth about $28,000 by January 2008 following 100 per cent profits retention, and seven stock splits. Apple's history is more chequered. It managed to survive the major strategic blunder of omitting to license out the Macintosh operating system to other manufacturers. Having gone public in 1980 at an issue price of $22, its share price plummeted to $7 in 1998, soaring to nearly $70 in the dotcom bubble before receding to $15 in 2003. However, this firm is enjoying a 'second bite at the cherry' with the spectacular success of the iPod digital music player. Its product, iTunes, registered its 200 millionth download in December 2004, just ten months after launch, making Apple the world leader in legally downloaded music. During 2004, its shares rose from $20 to $65, including a 20 per cent jump in November on the announcement of its first quarter 2004 results.

By July 2007, the share price was nudging $139 on the continuing success of iTunes, but also following the spectacular launch of the 'Jesus phone', the iPhone. It was reported that queues formed several days in advance at many of its 164 shops in America and the number of iPhones sold over the launch weekend approached a million.

A distressed company like Apple, in its 'dog days', would have a positive value so long as its management were thought capable of staging a corporate recovery, i.e. the market is valuing more distant dividends on hopes of a turnaround in earnings. If recovery is thought unlikely, the company is valued at its break-up value.

For inveterate non-dividend payers, the market is implicitly valuing the liquidating dividend when the company is ultimately wound up. Until this happens, the company is adding to its reserves as it reinvests, and continually enhancing its assets, its earning power and its value. In effect, the market is valuing the stream of future earnings that are legally the property of the shareholders.

■ Will there always be enough worthwhile projects in the future?

The DGM implies an ongoing supply of attractive projects to match the earnings available for retention. It is most unlikely that there will always be sufficient attractive projects available, each offering a constant rate of return, R, sufficient to absorb a given fraction, b, of earnings in each future year. While a handful of firms do have very lengthy lifespans, corporate history typically parallels the marketing concept of the product life cycle – introduction, (rapid) growth, maturity, decline and death – with paucity of investment opportunities a very common reason for corporate demise. It is thus rather hopeful to value a firm over a perpetual lifespan. However, remember that the discounting process compresses most of the value into a relatively short lifespan.

■ What if the growth rate exceeds the discount rate?

The arithmetic of the model shows that if $g > k_e$, the denominator becomes negative and value is infinite. Again, this appears nonsensical, but, in reality, many companies do experience periods of very rapid growth. Usually, however, company growth settles down to a less dramatic pace after the most attractive projects are exploited, once the firm's markets mature and competition emerges. There are two ways of redeeming the model in these cases. First, we may regard g as a long-term average or 'normal' growth rate. This is not totally satisfactory, as rapid growth often occurs early in the life cycle and the value computed would thus understate the worth of near-in-time dividends. Alternatively, we could segment the company's lifespan into periods of varying growth and value these separately. For example, if we expect fast growth in the first five years and slower growth thereafter, the expression for value is:

P_o = [Present value of dividends during year 1–5]

   + [Present value of all further dividends]

Note that the second term is a perpetuity beginning in year 6, but we have to find its present value. Hence it is discounted down to year zero as in the following expression:

$$P_o = \frac{D_0(1 + g_f)}{(1 + k_e)} + \frac{D_0(1 + g_f)^2}{(1 + k_e)^2} + \cdots + \frac{D_0(1 + g_f)^5}{(1 + k_e)^5} + \left(\frac{D_5(1 + g_s)}{(k_e - g_s)} \times \frac{1}{(1 + k_e)^5} \right)$$

$$= \sum_{t=1}^{5} \frac{D_0(1 + g_f)}{(1 + k_e)^t} + \sum_{t=6}^{\infty} \frac{D_5(1 + g_s)}{(1 + k_e)^t}$$

where g_f is the rate of fast growth during years 1–5 and g_s is the rate of slower growth beginning in year 6 (i.e. from the end of year 5).

The DGM may be used to examine the impact of changes in dividend policy, i.e. changes in b. Detailed analysis of this issue is deferred to Chapter 17.

Example: The case of unequal growth rates

Consider the case of dividend growth of 25 per cent for years 1–5 and 7 per cent thereafter. Assuming shareholders require a return of 10 per cent, and that the dividend in year zero is 10p, the value of the share is calculated as follows:

	For years 1–5		
Year	Dividend (p)	Discount factor at 10%	PV (p)
1	10(1.25)　　 = 12.5	0.909	11.4
2	$12.50(1.25)^2$ = 15.6	0.826	12.9
3	etc.　　　 = 19.5	0.751	14.6
4	= 24.4	0.683	16.7
5	= 30.5	0.621	18.9
			Total 74.5

For later years, we anticipate a perpetual stream growing from the year 5 value at 7 per cent p.a. The present value of this stream as at the end of year 5 is:

$$\frac{D_6}{k_e - g_s} = \frac{D_5(1 + 7\%)}{(10\% - 7\%)} = \frac{30.5p(1.07)}{0.03} = \frac{32.64p}{0.03} = £10.88$$

This figure, representing the PV of all dividends following year 5, is now converted into a year zero present value:

PV = £10.88 (PVIF$_{10,5}$) = (£10.88 × 0.621) = £6.76

Adding in the PV of the dividends for the first five years, the PV of the share right now is:

PV = (£0.745 + £6.76) = £7.51

However, we may note here that valuation of the dividend stream implies a known dividend policy. Because dividends are not controlled by shareholders, but by the firm's

Continued

directors, the DGM is more applicable to the valuation of small investment stakes in companies than to the valuation of whole companies, as in takeover situations. When company control changes hands, control of dividend policy is also transferred. It seems particularly unrealistic, therefore, to assume an unchanged dividend policy when valuing a company for takeover. However, the growth formula can be used to value the earnings stream, i.e. by assuming all earnings are paid as dividend as, in effect, they would be if the enterprise became a 100%-owned subsidiary of an acquiring firm.

SUMMARY

We have examined the meaning of wealth and its fundamental importance in financial management. For most investments, there is a time-lag between the initial investment outlay and the receipt of benefits. Consideration therefore must be given to both the timing and size of the costs and benefits. Whenever there is an alternative opportunity to use funds committed to a project (e.g. to invest in the capital market), cash today is worth more than cash received tomorrow. These concepts were then applied to valuing bonds and shares.

Key points

■ Money, like any other scarce resource, has a cost. We allow for the time-value of money by discounting. The higher the interest cost for a future cash flow, the lower its present value.

■ Discount tables take away much of the tedium of discounting – but computer spreadsheets eliminate it altogether.

■ Standard discount factors are:

  PVIF = the present value interest factor,

  PVIFA = the present value interest factor for an annuity.

Conventional shorthand is:

  Interest factor (rate of interest, number of years)

e.g. $PVIFA_{(10,3)}$ reads 'the present value interest factor for an annuity at 10 per cent for three years'.

■ Bonds are valued by discounting the interest payments and final repayment by the market interest rate for comparable bonds. The yield to maturity is the interest rate that equates the present value of bond payments to the bond price.

■ Shares are more difficult to value because the future dividends are difficult to forecast. The dividend growth model offers a valuation approach where the dividend growth rate is constant.

■ The value of a share can be found by discounting all future expected dividend payments.

■ The retention of earnings for worthwhile investment enhances future earnings, dividends and, therefore, the current share price.

■ The Dividend Valuation Model must be treated with caution. It embodies many critical assumptions.

■ The term structure of interest rates shows how yields on bonds vary as the durations of loans increase.

Further reading

Early writers on discounted cash flow include Fisher (1930) and Dean (1951). Copeland, Koller and Murrin (2000) discuss a range of valuation issues. Ross, Westerfield and Jordan (2008) have good chapters on bond and share valuation.

Useful websites

Discounted cash flow: www.investopedia.com
Annual percentage rate: www.moneyextra.com
www.investinginbonds.com
www.YieldCurve.com

Appendix I
THE TERM STRUCTURE OF INTEREST RATES AND THE YIELD CURVE

We saw in Section 3.7 that the interest rate depends on a number of factors, one of which is duration of the investment or loan. This relationship is called the term structure of interest rates. It shows how the yields offered for loans of different maturities vary as the term of the loan increases.

Relating this to bonds issued by the state, or government stock, the term structure shows the rate of return expected, or yield, by today's purchaser of stock who plans to hold to **maturity**, or redemption, i.e. when the stock will be repaid, or redeemed, by the government. It also shows how the yield varies for different lengths of time to maturity. In graphical terms, it is shown by a relationship called the **yield curve**.

yield curve
A graph depicting the relationship between interest rates and length of time to maturity

Normally, we find that yields to maturity increase as the term increases. In other words, rates of interest on 'longs' are higher than on 'shorts', as Figure 3.2 shows. Notice that the relevant yield is the gross redemption yield, which includes both interest payments and any capital gain or loss at redemption.

By tradition, short-dated stocks, with up to five years to maturity are called *shorts*, *mediums* have between five and 15 years before repayment and *longs* will be paid beyond 15 years. Notice that *longs* include a number of irredeemables or perpetuities which quite literally will never be repaid but will attract interest forever. These are also called **undated** stocks. Figure 3.3 presents the actual yield curves for UK Gilt and US Treasuries at 3 September 2007. Here we see that US Treasuries follow the normal curve while UK Gilts follow an inverted yield curve. For example, the 3-month yield for Gilts is 5.82% compared with a 30-year yield of only 4.50%.

■ Explaining the shape of the yield curve

Three theories have been proposed to explain the shape of the yield curve – the expectations theory, the liquidity preference theory and the market segmentation theory. These are not mutually exclusive explanations – the influences incorporated in each theory all tend to operate at any one time but with different degrees of pressure. Sometimes, investors' expectations (e.g. about future inflation) are predominant, while, at other times, investors' desire for liquidity may govern the shape of the curve.

Expectations theory

This theory asserts that investors' expectations about future interest rates exert the dominant influence. When the curve rises with years to maturity, this suggests that

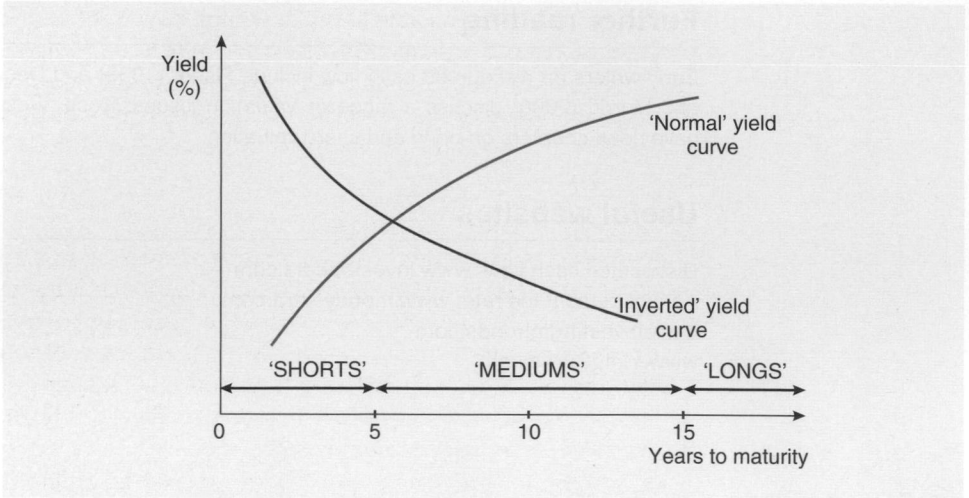

Figure 3.2 The term structure of interest rates

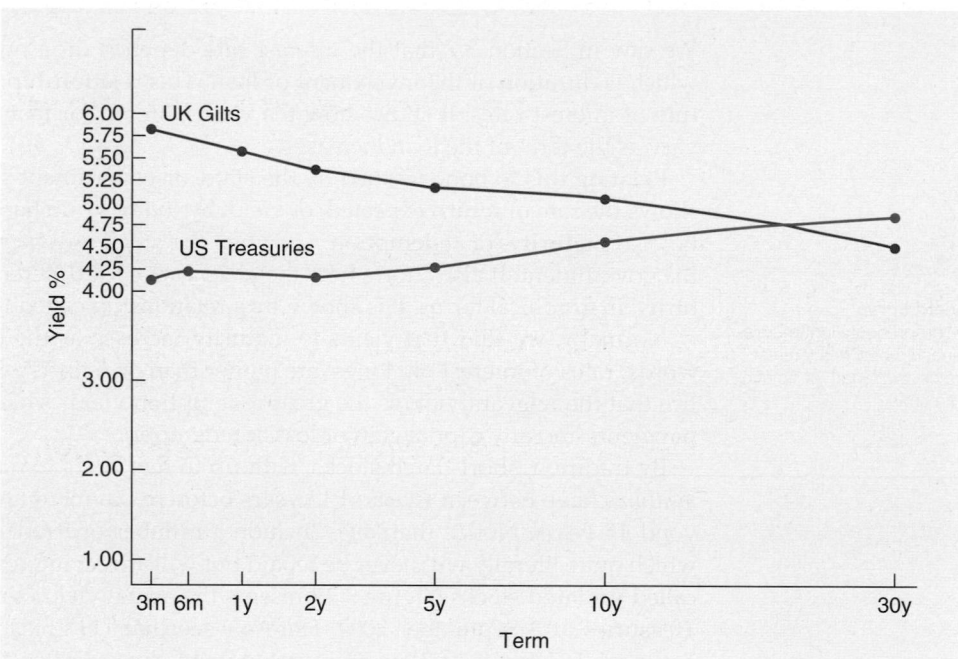

Figure 3.3 Comparison of Yield Curves for UK Gilts and US Treasuries 3 September 2007
Source: YieldCurve.com

people expect interest rates to rise in the future. This is reflected in the relative demand for short-dated and long-dated securities – investors expect to be able to earn higher rates in the future so they defer buying long-dated stocks, preferring to invest in shorts. This pushes up the price on shorts, and thus lowers the yields on them, and conversely, for longer-dated stock.

Liquidity preference theory
Most investors, being risk-averse, prefer to hold cash rather than securities – cash is effectively free of risk (although banks do go bust!), while even the shortest-dated government stocks carry a degree of risk. Here, by risk, we mean not the risk of default,

but the risk of not being able to find a willing buyer of the stock at an acceptable price, i.e. liquidity risk. Consequently, investors need to be compensated for having to wait for the return of their money. Preference for liquidity now, and risk avoidance, thus explains the shape of the yield curve. The longer the time to maturity, the greater the risk of illiquidity and the higher the compensation required.

Market segmentation theory

In developed markets, there is a wide range of investors with different needs and time-horizons who, therefore, focus on different segments of the yield curve. For example, some financial institutions, such as banks, are anxious to protect their ability to allow investors to withdraw their deposits freely – for them, shorts are very attractive as they need liquidity. Conversely, pension funds have far longer-term liabilities and wish to match the maturity stream of their assets to these quite predictable liabilities. For them, longs are more suitable.

According to this view, the 'short' market is quite distinct from the 'long' market and the two ends could behave quite differently under similar conditions. For example, if the government is expected to be a net repayer of its debt in the future, this suggests a shortage of longs. This is likely to increase the demand for those stocks presently available and thus reduce their yields. This would explain the case of the 'inverted' i.e. downward-sloping, yield curve, shown by the red line in Figure 3.2.

In Chapter 16, we will examine how firms can use the information contained in the yield curve for their financial planning.

Appendix II
PRESENT VALUE FORMULAE

■ Formula for the present value of a perpetuity

This formula derives from the present value formula:

$$PV = \frac{X}{1 + i} + \frac{X}{(1 + i)^2} + \frac{X}{(1 + i)^3} + \cdots$$

Let $X/(1 + i) = a$ and $1/(1 + i) = b$. We now have:

(i) $PV = a(1 + b + b^2 + \cdots)$

Multiplying both sides by b gives us:

(ii) $PVb = a(b + b^2 + b^3 + \cdots)$

Subtracting (ii) from (i) we have:

$PV(1 - b) = a$

Substituting for a and b,

$$PV\left(1 - \frac{1}{1 + i}\right) = \frac{X}{1 + i}$$

Multiplying both sides by $(1 + i)$ and rearranging, we have:

$$PV = \frac{X}{i}$$

■ Formula for the present value of a growing perpetuity

In the formula above, we obtained:

$$PV (1 - b) = a$$

Redefining $b = (1 + g)/(1 + i)$ and keeping $a = X/(1 + i)$:

$$PV\left(1 - \frac{1 + g}{1 + i}\right) = \frac{X}{1 + i}$$

Multiplying both sides by $(1 + i)$ and rearranging, we have:

$$PV = \frac{X}{i - g}$$

■ The present value of annuities

The above perpetuities were special cases of the annuity formula. To find the present value of an annuity, we can first use the perpetuity formula and deduct from it the years outside the annuity period. For example, if an annuity of £100 is issued for 20 years at 10 per cent, we would find the present value of a perpetuity of £100 using the formula:

$$PV = \frac{X}{i} = \frac{100}{0.10} = £1,000$$

Next, find the present value of a perpetuity for the same amount, starting at year 20, using the formula:

$$PV = \frac{X}{i(1 + i)^t} = \frac{£100}{0.10(1 + 0.10)^{20}} = £148.64$$

The difference will be:

$$PV \text{ of annuity} = \frac{X}{i} - \frac{X}{i(1 + i)^t}$$

$$= £1,000 - £148.64 = £851.36$$

The present value of an annuity of £100 for 20 years discounted at 10 per cent is £851.36.

The formula may be simplified to:

$$PV \text{ of annuity} = X\left(\frac{1}{i} - \frac{1}{i(1 + i)^t}\right)$$

Appendix III
THE P:E RATIO AND THE CONSTANT DIVIDEND VALUATION MODEL

If we examine the P:E ratio more closely, we find it has a close affinity with the growth version of the DVM. The P:E ratio is defined as price per share (PPS) divided by earnings per share (EPS). In its reciprocal form, it measures the **earnings yield** of the firm's shares:

earnings yield
The earnings per share (EPS)
divided by market share price

$$\frac{1}{P:E} = \frac{EPS}{PPS} = \frac{\text{Earnings}}{\text{Company value}} = \frac{E}{V}$$

This equals the dividend yield plus retained earnings (bE) per share. As in the DGM, the growth version of the DVM, we define the fraction of earnings retained as b. We can then write:

$$\frac{E}{V} = \frac{D}{V} + \frac{bE}{V}$$

The ratio E/V is the overall rate of return *currently* achieved. If this equals R, the rate of return on reinvested funds, then bE/V is equivalent to the growth rate g in the DGM. In other words, the earnings yield, E/V, comprises the dividend yield plus the growth rate or 'capital gains yield' for a company retaining a constant fraction of earnings and investing at the rate R. The two approaches thus look very similar. However, this apparent similarity should not be overemphasised for three important reasons:

1   The earnings yield is expressed in terms of the current earnings, whereas the DGM deals with the *prospective* dividend yield and growth rate, i.e. the former is historic in its focus, while the latter is forward-looking.
2   The DGM relies on discounting cash returns, while the earnings figure is based on accounting principles. It does not follow that cash flows will coincide with accounting profit, not least due to depreciation adjustments.
3   For the equivalence to hold, the current rate of return, E/V, would have to equal the rate of return expected on future investments.

Despite these qualifications, it is still common to find the earnings yield presented as the rate of return required by shareholders, and hence the cut-off rate for new investment projects. Unfortunately, this confuses a historical accounting measure with a forward-looking concept.

QUESTIONS

 **myfinancelab** | *Questions with an icon are also available for practice in myfinancelab with additional supporting resources.*

Questions with a coloured number have solutions in Appendix B on page 730.

1 Explain the difference between accounting profit and cash flow.

2 Calculate the present value of a ten-year annuity of £100, assuming an interest rate of 20 per cent.

3 A firm is considering the purchase of a machine which will cost £20,000. It is estimated that annual savings of £5,000 will result from the machine's installation, that the life of the machine will be five years, and that its residual value will be £1,000. Assuming the required rate of return to be 10 per cent, what action would you recommend?

4 Brymo Ltd issued bonds two years ago that pay interest on an annual basis at 8%. The bonds are due for repayment in two years' time. They will be redeemed at £110 per £100 nominal value. A yield of 10% is required by investors for such bonds. What is the expected market value?

5 The gross yield to redemption on government stocks (gilts) are as follows:

Treasury 8.5% 2000	7.00%
Exchequer 10.5% 2005	6.70%
Treasury 8% 2015	6.53%

(a) Examine the shape of the yield curve for gilts, based upon the information above, which you should use to construct the curve.
(b) Explain the meaning of the term 'gilts' and the relevance of yield curves to the private investor.

? **6** Calculate the net present value of projects A and B, assuming discount rates of 0 per cent, 10 per cent and 20 per cent.

	A (£)	B (£)
Initial outlay	1,200	1,200
Cash receipts:		
Year 1	1,000	100
Year 2	500	600
Year 3	100	1,100

Which is the superior project at each discount rate? Why do they not all produce the same answer?

? **7** Brosnan plc generates cash flows of £5 million p.a. after allowing for tax and depreciation, which is used for reinvestment. It has issued 10 million shares. Shareholders require a 12 per cent return.

Required
Value each share:
(i) assuming all cash flows are distributed as dividend.
(ii) assuming 50 per cent of cash flows are retained, with a return on retained earnings of 15 per cent.
(iii) as for (ii), but assuming 10 per cent return on reinvestment.
(iv) assuming that cash flows grow at 7.5 per cent for each of the first three future years, then at 5 per cent thereafter.
Note: assume all cash flows are perpetuities.

8 Insert the missing values in the following table:

	P_o	D_o	D_1	g	b	R	k_e
(i)	£8.44	£0.35	?	8.5%	0.5	17%	13.0%
(ii)	£4.98	£0.20	£0.219	?	0.6	16%	14.0%
(iii)	?	£0.10	£0.108	8.0%	0.4	20%	15.0%
(iv)	£2.75	?	£0.220	10.0%	0.5	20%	18.0%
(v)	£10.20	£0.60	£0.610	2.0%	?	10%	8.0%
(vi)	£0.60	£0.05	£0.054	8.0%	0.8	20%	?
(vii)	£1.47	£0.12	£0.133	10.5%	0.7	?	19.5%

Note: answers may have some minor rounding errors.

9 Leyburn plc currently generates profits before tax of £10 million, and proposes to pay a dividend of £4 million out of cash holdings to its shareholders. The rate of corporation tax is 30 per cent. Recent dividend growth has averaged 8 per cent p.a. It is considering retaining an extra £1 million in order to finance new strategic investment. This switch in dividend policy will be permanent, as management believe that there will be a stream of highly attractive investments available over the next few years, all offering returns of around 20 per cent after tax. Leyburn's shares are currently valued 'cum-dividend'. Shareholders require a return of 14 per cent. Leyburn is wholly equity-financed.

Required

(a) Value the equity of Leyburn assuming no change in retention policy.

(b) What is the impact on the value of equity of adopting the higher level of retentions? (Assume the new payout ratio will persist into the future.)

Practical assignment

List three decisions in a business with which you are familiar where cash flows arise over a lengthy time period and where discounted cash flow (DCF) may be beneficial. To what extent is DCF applied (formally or intuitively)? What are the dangers of ignoring the time-value of money in these particular cases?

 **myfinancelab** | *Now retake your diagnostic test for Chapter 3 to check your progress and update your study plan.*

Part II

INVESTMENT DECISIONS AND STRATEGIES

Chapters 4 to 6 examine in depth the investment decision and how it is evaluated. The concepts of time-value of money and present value are extensively applied. The available methods for assisting the financial manager to evaluate investment proposals are examined in Chapter 4, both when capital is freely available and when it is in short supply. Methods of appraisal that do not utilise discounting procedures are also examined.

In Chapter 5, investment appraisal procedures are applied to practical situations, incorporating the impact of both taxation and inflation. Consideration is given to identifying the relevant information for project evaluation, particularly for replacement decisions.

Chapter 6 sets the whole project appraisal system in a strategic perspective and explores the wider aspects of the investment appraisal system within companies. It dispels the notion that investment analysis hinges solely on methods of appraisal, and it reveals how companies approach their project evaluations in practice.

4

Investment appraisal methods

Cigarettes can damage your wealth

Cigarette companies have for years looked for the Holy Grail of a smokeless cigarette. R. J. Reynolds Tobacco, US maker of Camel and other cigarette brands, launched a smokeless cigarette called Premier. It spent $800 million developing and marketing the new brand, which had vast wealth-creating potential and was socially more acceptable to passive smokers.

After test marketing it for several months, the company finally recognised that it had created one of the biggest new product flops on record. Smokers complained about the taste, which some said left a charcoal flavour in the mouth. With 400 brands of cigarette in the USA, launching a new product is costly and risky. But the idea of a smokeless cigarette was still seen by the company as worth pursuing and it began trials on a new smokeless cigarette brand, Eclipse, that heats, rather than burns, tobacco. Since the earlier flop, however, the market has changed, with passive smoking becoming a bigger issue. Time will tell whether this brand generates a positive net present value. However, a spokesperson for the American Health Foundation said, 'The best cigarette is no cigarette.'

Learning objectives

Having read this chapter, you should have a good grasp of the investment appraisal techniques commonly employed in business, and have developed skills in applying them. Particular attention will be devoted to the following:

■ The net present value approach and why it is consistent with shareholder goals.

■ The three discounted cash flow approaches – net present value, internal rate of return and profitability index.

■ The underlying strengths and limitations of the above methods.

■ How net present value and internal rate of return methods can be reconciled when they conflict.

■ Non-discounting methods.

■ Analysing investments when capital availability is an important constraint.

 *Complete your diagnostic test for Chapter 4 now to create your personal study plan.*

4.1 INTRODUCTION

Every day managers and investors make long-term investment decisions. How do they go about this? A major US company explains how it employs the net present value (NPV) approach in assessing capital projects:

> We measure all potential projects by their cash flow merit. We then discount projected cash flows back to present value in order to compare the initial investment cost with a project's future returns to determine if it will add incremental value after compensating for a given level of risk.

There are, however, a number of alternative techniques to the NPV method. The aim of this chapter is to present the main methods of investment appraisal and to consider their strengths and limitations. In a later chapter, we consider their practical application in business, large and small.

4.2 CASH FLOW ANALYSIS

The investment decision is the decision to commit the firm's financial and other resources to a particular course of action. Confusingly, the same term is often applied to both real investment, such as buildings and equipment, and financial investment, such as investment in shares and other securities. While the principles underlying investment analysis are basically the same for both types of investment, it is helpful for us to concentrate here on the former category, usually referred to as capital investment. Our particular emphasis on strategic capital projects concentrates on the allocation of a firm's long-term capital resources.

Self-assessment activity 4.1

Investment projects do not only include investment in plant and equipment or buildings. Think of some other types of capital projects.

(Answer in Appendix A at the back of the book)

■ Cash flow matters more than profit

Managers in business usually view profit as the best measure of performance. It might, therefore, be assumed that capital project appraisal should seek to assess whether the investment is expected to be 'profitable'. Indeed, many firms do use such an approach.

There are, however, many problems with the profit measure for assessing future investment performance. Profit is based on accounting concepts of income and expenses relating to a particular accounting period, based on the *matching principle*. This means that income receivable and expenses payable, but not yet received or paid, along with depreciation charges, form part of the profit calculation.

Consider the case of the Oval Furniture Company with expected annual sales from its new factory of £400,000 and profits of £60,000. In order to stimulate demand, customers are offered two years' credit. While this decision has no impact on the reported profit, it certainly affects the cash position – no cash flow being received for two years. Cash flow analysis considers all the cash inflows and outflows resulting from the investment decision. Non-cash flows, such as depreciation charges and other accounting policy adjustments, are not relevant to the decision. We seek to estimate the stream of cash flows arising from a particular course of action and their timing.

■ Timing of cash flows

Project cash flows will usually arrive throughout the year. For example, if we acquire a machine with a four-year life on 1 January 2010, the subsequent cash flows related to it may involve the monthly payment purchases and expenses and daily receipt of cash from customers throughout each year. Strictly speaking, these cash flows should be identified on a monthly, even daily, basis and discounted using appropriate discount factors.

In practice, to facilitate the use of annual discount tables, cash flows arising during the year are treated as occurring at the year end. Thus, while the initial outlay is assumed to occur at the start of the project (frequently termed Year 0), subsequent cash flows are deemed to arrive later than they actually arise. This has the effect of producing an NPV slightly lower than the true NPV, assuming that subsequent cash flows are positive.

Decision-making can be viewed as an *incremental* activity. Businesses generally operate as going concerns with fairly clear strategies and well-established management processes. Decisions are part of a sequence of actions seeking to move the organisation from its current to its intended position. The same idea is apparent in analysing projects – the decision-maker must assess how the business changes as a direct result of selecting the project. Every project can be either accepted or rejected, and it is the difference between these two alternatives in any time period, t, expressed in cash flow terms (CF_t), that is taken into the appraisal.

Incremental analysis

Project CF_t = CF_t for firm *with* project − CF_t for firm *without* project.

4.3 NET PRESENT VALUE

We have assumed that the paramount objective of the firm is to create as much wealth as possible for its owners through the efficient use of existing and future resources. To create wealth, the present value of all future cash inflows must exceed the present value of all anticipated cash outflows. Quite simply, *an investment with a positive net present value increases the owners' wealth*. The elements of investment appraisal are shown in Figure 4.1.

Most decisions involve both costs and benefits. Usually, the initial expenditure incurred on an investment undertaken is clear-cut: it is what we pay for it. This includes the cash paid to the supplier of the asset plus any other costs involved in making the project operational. The problems start in measuring the worth of the investment project. What an asset is worth may have little to do with what it cost or what value is placed on it in the firm's balance sheet. A machine standing in the firm's books at £20,000 may be worth far more if it is essential to the manufacture of a highly profitable product, or far less than this if rendered obsolete through the advent of new technology. To measure its worth, we need to consider the *value of the current and future benefits less costs* arising from the investment. Wherever possible, these benefits should be expressed in terms of *cash flows*. Sometimes (as will be discussed later) it is impossible to quantify benefits so conveniently. Typically, investment decisions involve an initial capital expenditure followed by a stream of cash receipts and disbursements in subsequent periods. The net present value (NPV) method is applied to evaluate the desirability of investment opportunities. NPV is defined as:

$$\text{NPV} = \frac{X_1}{(1 + k)} + \frac{X_2}{(1 + k)^2} + \frac{X_3}{(1 + k)^3} + \cdots + \frac{X_n}{(1 + k)^n} - I$$

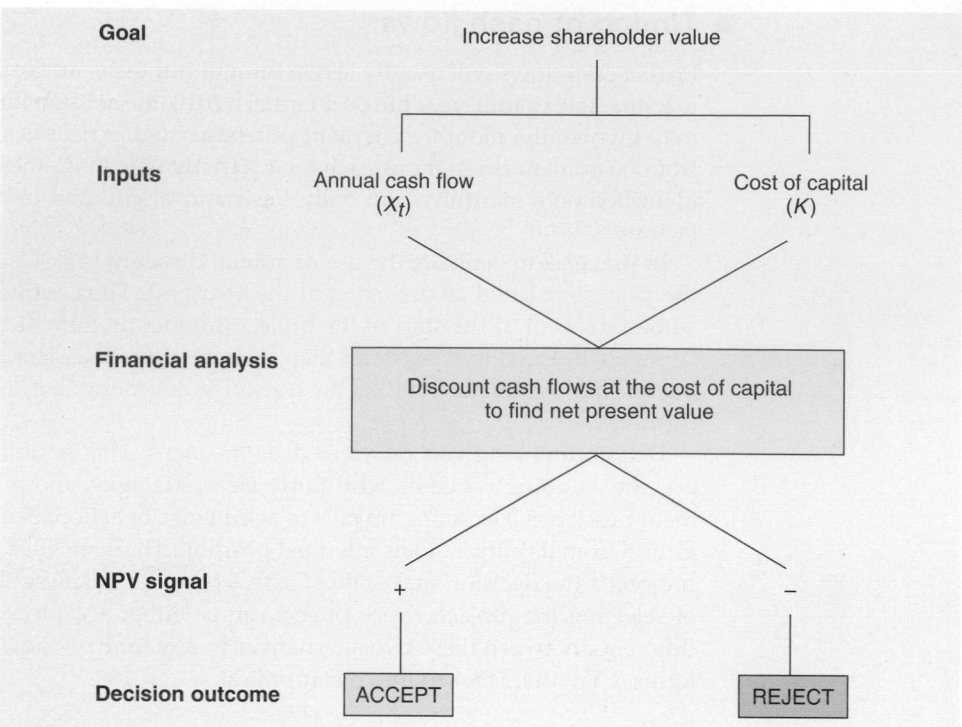

Figure 4.1 Investment appraisal elements

which may be summarised as:

$$\text{NPV} = \sum_{t=1}^{n} \frac{X_t}{(1 + k)^t} - I$$

where X_t is the net cash flow arising at the end of year t, I is the initial cost of the investment, n is the project's life, and k the minimum required rate of return on the investment (or discount rate). (The Greek letter Σ, or sigma, denotes the sum of all values in a particular series.)

A project's net present value (NPV) is determined by summing the net annual cash flows, discounted at a rate that reflects the cost of an investment of equivalent risk on the capital market, and deducting the initial outlay.

Self-assessment activity 4.2

Define the main elements in the capital investment decision.

(Answer in Appendix A at the back of the book)

The net present value rule

Wealth is maximised by accepting all projects that offer positive net present values when discounted at the required rate of return for each investment.

Most of the main elements in the NPV formula are largely externally determined. For example, in the case of investment in a new piece of manufacturing equipment, management has relatively little influence over the price paid, the life expectancy or the discount rate. These elements are determined, respectively, by the price of capital goods, the rate of new technological development and the returns required by the capital market. Management's main opportunity for wealth creation lies in its ability to implement and manage the project so as to generate positive net cash flows over the project's economic life.

An NPV example: Gazza Ltd

The management of Gazza Ltd is currently evaluating an investment in hair dye products costing £10,000. Anticipated net cash inflows are £6,000 received at the end of year 1 and a further £6,000 at the end of year 2. Assuming a discount rate of 10 per cent, calculate the project's net present value.

We can compute the NPV for Gazza using three different approaches, all of which will be employed in later chapters.

1 *Formula approach*

$$\text{NPV} = \frac{£6,000}{1.1} + \frac{£6,000}{(1.1)^2} - £10,000$$

$$= £5,454 + £4,959 - £10,000$$
$$= £413$$

2 *Present value tables* (using Appendix C)

Year	Cash flow £		Discount factor at 10%		Present value £
1	6,000	×	0.90909	=	5,454
2	6,000	×	0.82645	=	4,959
			1.73554		10,413
	Less initial cost				(10,000)
			NPV		413

3 *Present value annuity tables* (using Appendix D)

$$\text{NPV} = (£6,000 \times \text{PVIFA}_{(10,2)}) - £10,000$$
$$= (£6,000 \times 1.7355) - £10,000$$
$$= £413$$

This approach is appropriate only when annual cash flows are constant. Notice that the present value interest factor for an annuity at 10 per cent for two years (taken from Appendix D) is simply the cumulative total of the individual factors in the previous approach.

How would the net present value differ if the perceived project risk were greater? The risk-averse management of Gazza would probably require a higher return from the project, reflected in a higher discount rate. Let us repeat the exercise using 13 per cent (average risk) and 16 per cent (high risk).

Using 13 per cent:

$$\text{NPV} = (£6,000 \times \text{PVIFA}_{(13,2)}) - £10,000$$
$$= (£6,000 \times 1.6681) - £10,000$$
$$= £8 \text{ (i.e. approximately zero)}$$

Using 16 per cent:

$$\text{NPV} = (£6,000 \times 1.6052) - £10,000$$
$$= (369)$$

Looking at the net present values, what interpretation can be made? With a 10 per cent discount rate, the project offers a *positive* NPV of £413. If the projected cash flows are generally expected to be achieved, the market value of the firm should rise by £413. Hence, the project should be accepted. On the other hand, if the project is classified as high risk, the cash inflows are discounted at a rate of 16 per cent and the NPV is estimated at −£369. Its acceptance would reduce the firm's market value by £369. Hence, the project should not be accepted. Clearly, it would not be wise to exchange £10,000 today for future cash flows having a present value of less than this amount.

Continued

If the project is classified as having average risk, the discount rate used is 13 per cent, yielding an NPV of £8. The project is just acceptable; it yields 13 per cent, which is the required rate of return. We can draw two important conclusions:

1 Project acceptability depends upon cash flows and risk.
2 The higher the risk of a given set of expected cash flows (and the higher the discount rate), the lower will be its present value. In other words, the value of a given expected cash flow *decreases* as its risk *increases*.

■ Why NPV makes sense

The main rationale for the net present value approach may be summarised as follows:

1 Managers are assumed to act in the best interests of the owners or shareholders, even if agency costs – in the form of incentives or controls – have to be incurred. They seek to increase shareholders' wealth by maximising cash flows through time. The market rate of exchange between current and future wealth is reflected in the current rate of interest.
2 Managers should undertake all projects up to the point at which the marginal return on the investment is equal to the rate of interest on equivalent financial investments in the capital market. This is exactly the same as the net present value rule: accept all investments offering positive net present values when discounted at the equivalent market rate of interest. The result is an increase in the market value of the firm and thus in the market value of the shareholders' stake in the firm.
3 Management need not concern itself with shareholders' particular time patterns of consumption or risk preferences. In well-functioning capital markets, shareholders can borrow or lend funds to achieve their personal requirements. Furthermore, by carefully combining risky and safe investments, they can achieve the desired risk characteristics for those consumption requirements.

Mini case

How NPV is used in debt relief to the poorest nations

The International Monetary Fund (IMF) and World Bank have designed a framework to provide special assistance for heavily indebted poor countries. It entails coordinated action by the international financial community, including banks and multinational companies, to reduce and reschedule the debt burden to levels that countries can service through exports and aid.

Net present value is central to the calculation of the sustainable debt level. The face value of debt stock is not a good measure of a country's debt burden if a significant part of it is contracted on concessional terms, for example with an interest rate below the prevailing market rate. The net present value of debt is used to find the sum of all future debt-service obligations (interest and principal) on existing debt, discounted at the market interest rate. Whenever the interest rate on the loan is lower than the market rate, the resulting NPV of debt is smaller than its face value, with the difference reflecting the grant element.

Question

Explain to a government official from one of the world's poorest countries why the NPV approach is an appropriate method for calculating the sustainable debt level.

Self-assessment activity 4.3

Why should managers seek to maximise net present value? Is business not about maximising profit?

(Answer in Appendix A at the back of the book)

4.4 INVESTMENT TECHNIQUES – NET PRESENT VALUE

Discounted cash flow (DCF) analysis is a family of techniques, of which the NPV method is just one variant. Two other DCF methods are the internal rate of return (IRR) and the profitability index (PI) approaches. Many managers prefer to use non-discounting approaches such as the payback and return on capital methods; others use both approaches. The following example illustrates the various approaches to investment appraisal.

Example: Appraising the Lara and Carling projects

Sportsman plc is a manufacturer of sports equipment. The firm is considering whether to invest in one of two automated processes, the Lara or the Carling, both of which give rise to staffing and other cost savings over the existing process. The relevant data relating to each are given below:

	Lara (£)	Carling (£)
Investment outlay (payable immediately)	(40,000)	(50,000)
Year 1 Annual cost savings	16,000	17,000
2 Annual cost savings	16,000	17,000
3 Annual cost savings	16,000	17,000
4 Annual cost savings	12,000	17,000

The required return is 14 per cent p.a.

The investment outlays are obviously additional cash outflows, while the annual cost savings are cash flow benefits because total annual expenditures are reduced as a result of the investment.

Should the company invest in either of the two proposals and if so, which is preferable?

The NPV solution

The net present value for the Lara machine is found by multiplying the annual cash flows by the present value interest factor (PVIF) at 14 per cent (using the tables) and finding the total, as shown in Table 4.1. An immediate cash outlay (treated as Year 0) is not discounted as it is already expressed in present value terms. The same factors could be applied to

Table 4.1 Net present value calculations

Year		Cash flow (£)	PVIF at factor 14%	Present value (£)
Lara proposal				
0	Outlay	(40,000)	1	(40,000)
1	Cost savings	16,000	0.87719	14,035
2	Cost savings	16,000	0.76947	12,312
3	Cost savings	16,000	0.67497	10,800
4	Cost savings	12,000	0.59208	7,105
	Net present value at 14%			4,252
Carling proposal				
Cost savings	£17,000 × PVIFA$_{(14\%,4\text{ yrs})}$		2.9137	49,533
Outlay				(50,000)
	Net present value at 14%			(467)

Continued

evaluate the Carling proposal. However, as the annual savings are constant, it is far simpler to use the present value interest factor for an annuity (PVIFA) at 14 per cent for four years.

Comparison of the two proposals reveals the following:

1 The Lara machine offers a positive NPV of £4,252, and would increase shareholder wealth.
2 The Carling machine offers a negative NPV of £467 and would reduce value.
3 Given that the proposals are mutually exclusive (i.e. only one is required), the Lara proposal should be accepted.

What does an expected NPV of £4,252 from the Lara proposal really mean? The project's future cash flows are sufficient for the firm to pay all costs associated with financing the project and to provide an adequate return to shareholders. From the shareholders' viewpoint, it means that the firm could borrow £44,252 (the cost plus the NPV) to purchase the machine and pay out a dividend today of £4,252, and still have sufficient funds from the project to pay off the interest at 14 per cent p.a. and annual repayments (see Table 4.2).

In practice, it is unlikely that the lender will agree to a repayment schedule that exactly matches the expected annual cash flows of the project. It is also somewhat imprudent to pay as a dividend the whole of the expected NPV before the project commences! However, in theory at least, the proposal creates wealth of £4,252 and the shareholders are that much better off than they were prior to the decision. Note that we assume that borrowing and lending rates of interest are the same. We discuss in later chapters how the discount rate is estimated; suffice it to say that it is the required rate of return that investors can expect on comparable alternative investment in the marketplace.

Table 4.2 Why NPV makes sense for shareholders

			£
Year 0	Borrow: machine	£40,000	
	Pay NPV as dividend	£4,252	44,252
1	Interest: £44,252 at 14%		6,195
			50,447
	Less: repayment		(16,000)
	(through annual savings)		34,447
2	Interest: £34,447 at 14%		4,822
			39,269
	Less: repayment		(16,000)
			23,269
3	Interest: £23,269 at 14%		3,257
			26,526
	Less: repayment		(16,000)
			10,526
4	Interest: £10,526 at 14%		1,474
			12,000
	Less: repayment		(12,000)
			—

4.5 INTERNAL RATE OF RETURN

IRR or DCF yield
The rate of return that equates the present value of future cash flows with the initial investment outlay

Managers frequently ask: 'What rate of return am I getting on my investment?' To calculate the correct return, or yield, requires us to find the rate that equates the present value of future benefits to the initial cash outlay. We call this the **internal rate of return (IRR)**, or **DCF yield**.

The IRR is that discount rate, r, which, when applied to project cash flows (X_t), produces a net present value of zero. It is found by solving the equation for r:

$$\sum_{t=0}^{n} \frac{X_t}{(1 + r)^t} = 0$$

Where the IRR exceeds the required rate of return ($r > k$) the project should be accepted.

Suppose a savings scheme offers a plan whereby, for an initial investment of £100, you would receive £112 at the year end. The IRR is thus 12 per cent:

$$£100(1 + r) = £112$$
$$r = 12\%$$

If another scheme offered a single payment of £148 in three years' time, from an initial investment of £100, the IRR is found by solving:

$$£100(1 + r)^3 = £148$$

or

$$\frac{1}{(1 + r)^3} = \frac{£100}{£148} = 0.6757$$

Turning to the present value interest factor (PVIF) table (Appendix C) for three years, and looking for the rate that comes closest to 0.6757, we find that the IRR for the investment is approximately 14 per cent. The same approach is used to find the IRR for capital investment, but here the annual cash flows may differ. We find the IRR by solving for the rate of return at which the present value of the cash inflows equals the present value of the cash outflows. That is, we have to solve for

$$I_o = \frac{X_1}{1 + r} + \frac{X_2}{(1 + r)^2} + \cdots + \frac{X_n}{(1 + r)^n}$$

This is the same as finding the rate of return that produces an NPV of zero.

The IRR Solution

In our earlier example, the Lara produced an NPV of £4,252 at 14 per cent. Given a 'normal' pattern of cash flows, i.e. an outlay followed by cash inflows, we can see that as the discount rate increases, the NPV falls. Trial and error will give us the discount rate that yields a zero NPV.

Trying 18 per cent, as shown in Table 4.3, gives a positive NPV of £976. Trying 20 per cent gives a *negative* NPV of £510. Clearly the IRR giving a zero NPV falls between 18 and 20 per cent, probably closer to 20 per cent. Using linear interpolation, we estimate the IRR by applying the formula:

$$\text{IRR} = r_1 + \left(\frac{N_1}{N_1 + N_2} \times (r_2 - r_1) \right)$$

where r_1 is the rate of interest and N_1 the NPV for the first guess, and r_2 and N_2, the NPV for the second guess. Applying the formula:

$$\text{IRR} = 18\% + \left(\frac{£976}{£976 + £510} \times 2\% \right) = 19.31\%$$

Note that the calculation includes the class interval, in this case $(20\% - 18\%) = 2\%$.

Continued

Table 4.3 IRR calculations for Lara proposal

Year	Cash flow (£)	PVIF at 18%	PV (£)	PVIF at 20%	PV (£)
0	(40,000)	1.0	(40,000)	1.0	(40,000)
1	16,000	0.84746	13,559	0.83333	13,333
2	16,000	0.71818	11,490	0.69444	11,111
3	16,000	0.60863	9,738	0.57870	9,259
4	12,000	0.51579	6,189	0.48225	5,787
NPV			976		(510)

$$\text{IRR} = 18\% + \left(\frac{976}{976 + 510} \times 2\% \right) = 19.31\%$$

In the Lara example, the NPV at various rates of interest is shown in Figure 4.2. The graph shows a clearer relationship between IRR and NPV. We also have an idea of the break-even rate of interest – or IRR – at around 19–20 per cent, as calculated earlier. The IRR of 19.31 per cent is well above the required rate of 14 per cent and the project is, therefore, wealth-creating.

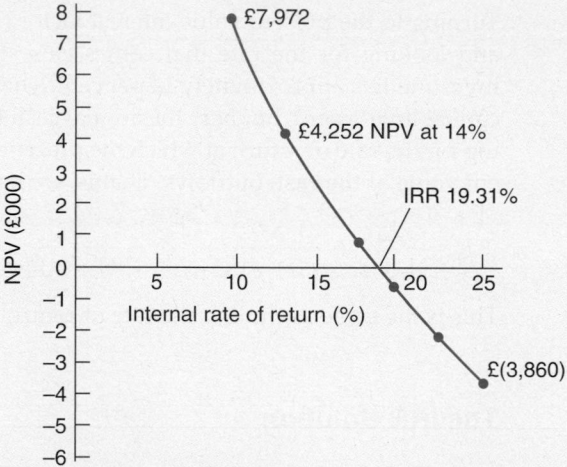

Figure 4.2 Lara proposal: NPV–IRR graph

Most managers have access to computer spreadsheets that solve the equation in a fraction of a second and avoid tedious manual effort. However, our analysis explains the logic behind the computer calculation.

For the Carling proposal, the IRR calculation is much more straightforward as the annual cash flows are constant.

$$£17,000 \times \text{PVIFA}_{(r,4\text{yrs})} = £50,000$$

$$\text{PVIFA}_{(r,4\text{yrs})} = \frac{£50,000}{£17,000} = 2.9411$$

Referring to annuity tables (Appendix D), we find that for four years at 13 per cent, the factor is 2.9745, and at 14 per cent it is 2.9137. The IRR is therefore between 13 and 14 per cent. This return falls just below the 14 per cent requirement, making it an uneconomic proposal.

Prince makes 500% profit on Canary Wharf

Prince Al Waleed Bin Talal Bin Abdul Aziz, the Saudi prince said to be the richest businessman outside the US, yesterday revealed that he had realised 500 per cent profit by selling most of his stake in Canary Wharf.

The Prince was one of a group of investors who funded Canary Wharf chairman Paul Reichmann to buy back the 85-acre estate in London's docklands from its bankers in 1995 for £800 million.

When Canary Wharf emerged from administration in 1993, it had attracted interest from few potential tenants and most investors gave it little chance of success.

On Tuesday the Prince completed the sale of two-thirds of his stake, raising £122 million. The Prince calculates that the internal rate of return, over the five years of the investment, has been a healthy 47.7 per cent per year. He will retain the remaining third of his investment. 'He likes it,' a spokesman said. Asked how the money will be reinvested, a spokesman said, 'Very wisely.'

Source: Based on Norma Cohen, *Financial Times*, 18 January 2001.

4.6 PROFITABILITY INDEX

Another method for evaluating capital projects is the **profitability index (PI)**, sometimes called the benefit–cost ratio.

The profitability index

The profitability index is the ratio of the present value of project benefits to the present value of initial costs. The decision rule is that projects with a PI greater than 1.0 are acceptable.

Example: The PI decision rule

Referring back to the present values calculated in Table 4.1, we can find for the Lara proposal:

$$PI = \frac{PV \text{ benefits}}{PV \text{ outlay}} = \frac{£44,252}{£40,000} = 1.1063$$

while for the Carling proposal:

$$PI = \frac{£49,533}{£50,000} = 0.9906$$

From this we see that the Lara is acceptable on financial grounds as the PI exceeds 1. The higher the PI, the more attractive the project.

For *independent* projects, the PI gives the same advice as NPV and IRR methods, although there are important reservations when projects are 'mutually exclusive' (see Section 4.9).

The PI can also be expressed as the net present value per £1 invested, i.e.

$$PI = \frac{NPV}{PV \text{ of outlays}}$$

If NPV per £1 invested exceeds zero, then the project should be accepted.

Self-assessment activity 4.4

What are the three main DCF methods and how do you know when to accept a capital project with each?

(Answer in Appendix A at the back of the book)

4.7 PAYBACK PERIOD

Over the years, managers have come to rely upon a number of rule-of-thumb approaches to analyse investments. Two of the most popular methods are the payback period and the accounting rate of return.

payback period
Period of time a project's annual net cash flows take to match the initial cost outlay

The **payback period** (PB) is the period of time taken for the future net cash inflows to match the initial cash outlay.

Table 4.4 gives the cumulative cash flows for the two projects in our earlier example. After two years, the cumulative cash flow for Lara has reduced to −$8,000; but by the end of the third year it has improved to +$8,000. The project therefore breaks even, or pays back, in two and a half years. Similarly, the Carling pays back in 2.9 years. Many companies set payback requirements for capital projects. For example, if all projects are required to pay back within three years, both the Lara and Carling are acceptable.

Table 4.4 Payback period calculation

		Lara cash flow		Carling cash flow	
Year		Annual	Cumulative	Annual	Cumulative
0	Cost	(40,000)	(40,000)	(50,000)	(50,000)
1	Cost savings	16,000	(24,000)	17,000	(33,000)
2	Cost savings	16,000	(8,000)	17,000	(16,000)
3	Cost savings	16,000	8,000	17,000	1,000
4	Cost savings	12,000	20,000	17,000	18,000

$$\text{Payback: Lara} \quad 2 + \frac{8,000}{16,000} \text{ years} = 2.5 \text{ years}$$

$$\text{Carling} \quad 2 + \frac{16,000}{17,000} \text{ years} = 2.9 \text{ years}$$

discounted payback
Period of time the present value of a project's annual net cash flows take to match the initial cost outlay

A number of modifications to simple payback are possible. **Discounted payback** addresses the problem of comparing cash flows in different time periods. It calculates how quickly discounted cash flows recoup the initial investment. Referring back to the NPV calculation for the Lara, the discounted payback period at 14 per cent interest is approximately three and a half years (see below). The cumulative present values recoup the initial outlay only in the final year.

Year	Present value @14%	Cumulative PV
0	(40,000)	(40,000)
1	14,035	(25,965)
2	12,312	(13,653)
3	10,800	(2,853)
		Payback period 3.5 years
4	7,105	4,252
NPV	4,252	

A fuller discussion of the popularity of the payback period will be given in Chapter 5. However, we should note that this approach has some serious problems as a measure of investment worth:

1 The time-value of money is ignored (except in the case of discounted payback).
2 Cash flows arising after the payback period are ignored.
3 The payback period criterion that firms stipulate for assessing projects has little the-oretical basis. How do firms justify setting, say, a two-year payback requirement?

4.8 ACCOUNTING RATE OF RETURN

return on capital employed
Operating profit expressed as a percentage of capital employed

A key ratio in analysing accounts is the **return on capital employed**, or ROCE. This is calculated as:

$$\frac{\text{Profit before interest and tax}}{\text{Capital employed}} \times 100$$

This indicates a company's efficiency in generating profits from its asset base. All new investment should at least match existing assets in terms of its earning power. How-ever, the annual ROCE on a project will change each year. Typically, it is less profitable in the early years but improves over time as the project's sales build up and as the book value of the asset (i.e. cost less depreciation) declines.

accounting rate of return
Return on investment over the whole life of a project

The **accounting rate of return** (ARR) seeks to provide a measure of project prof-itability over the entire asset life. It compares the average profit of the project with the book value of the asset acquired. The ARR can be calculated on the *original* capital invested or on the *average* amount invested over the life of the asset.

Accounting rate of return

$$\text{ARR (total investment)} = \frac{\text{Average annual profit}}{\text{Initial capital invested}} \times 100$$

$$\text{ARR (average investment)} = \frac{\text{Average annual profit}}{\text{Average capital invested}} \times 100$$

Returning to our example, suppose the depreciation policy is to depreciate assets over their useful lives on a straight-line basis. The annual depreciation for the Lara will be £10,000 (i.e. £40,000 over four years) and for the Carling, £12,500. The annual profit from the proposals will be the annual cash saving less the annual depreciation. The ARRs based on initial capital invested for the two proposals are shown in Table 4.5.

Table 4.5 Calculation of the ARR on initial capital invested

	Year					
	1	**2**	**3**	**4**	**Average**	**ARR**
Project						
Lara						
Cash flow (£)	16,000	16,000	16,000	12,000	–	
Depreciation* (£)	(10,000)	(10,000)	(10,000)	(10,000)	–	
Accounting profit (£)	6,000	6,000	6,000	2,000	5,000	5,000/40,000 = $12\frac{1}{2}$%
Carling						
Cash flow (£)	17,000	17,000	17,000	17,000	–	
Depreciation* (£)	(12,500)	(12,500)	(12,500)	(12,500)	–	
Accounting profit (£)	4,500	4,500	4,500	4,500	4,500	4,500/50,000 = 9%

*Straight-line depreciation is used in each case.

Alternatively, we could base the calculation of ARR on the average investment, found by summing the opening and closing asset values and dividing by 2. This would yield answers for the Lara and Carling of 25 per cent and 18 per cent, respectively, double the returns based on the initial capital. (In our case, the residual values are zero.)

A benefit of this profitability measure is that managers feel they understand it. It makes sense to use an investment evaluation measure that is broadly consistent with return on capital employed, which is the primary business ratio. However, the ARR has some definite drawbacks. Suppose the Lara proposal is expected to continue into Year 5, yielding a profit of £1,000 in that year. Common sense suggests that this would make the proposal more attractive. However, the new ARR actually declines from 25 to 21 per cent as a result of averaging over five rather than four years.

$$ARR = \frac{(£6,000 + £6,000 + £6,000 + £2,000 + £1,000)/5}{(£40,000 + 0)/2} \times 100 = 21\%$$

It also takes no account of the size and life of the investment, or the timing of cash flows. Moreover, this approach is based on profits rather than cash flows, the significance of which we discuss in the next chapter. Such important weaknesses make ARR inappropriate as a main investment appraisal method, particularly when comparing projects.

Is investing in pizza stores a no-brainer?

As Homer Simpson would no doubt agree, eating pizza is literally a no-brainer. Describing why people tuck into its products, Domino's Pizza, the US's largest pizza delivery company, cites the statistic that by 4.30pm on the average afternoon, almost three-quarters of Americans still have no idea what their families will eat for dinner that evening.

If buying pizzas is an easy way for parents to fill the bellies of their progeny, selling pizza can be very lucrative. The US alone generates $33bn in sales. CIBC estimates that a new Domino's store costs just $150,000–250,000 to build but yields an annual pre-tax cash return of 30–65 per cent. In addition, more than half the company's operating profit comes from franchised stores. These provide a stable royalties stream and a capital-light way to expand.

The big issue is growth. Toppings may get more adventurous but pizza has been around for years. Domino's focuses on home delivery, which is faster growing than eat-in meals – at some point, one has had so many slices it gets tricky to leave the sofa.

Source: Lex Column, *Financial Times*, 3 January 2007.

Now use this story to tackle Self-assessment activity 4.5.

Self-assessment activity 4.5

List four capital budgeting methods for evaluating project proposals. Identify the main strengths and drawbacks of each.

(Answer in Appendix A at the back of the book)

4.9 RANKING MUTUALLY EXCLUSIVE PROJECTS

For many investment projects there will be a number of mutually exclusive proposals, only one of which can be accepted. For example, a firm may be considering a new IT system and needs to identify the best system from a number of possibilities.

Example: Ranking project performance

The manufacturers of the Lara also make the Bruno – a larger, more powerful, but more erratic model – offering a further 50 per cent in cost savings each year, but costing a further 50 per cent to purchase. The NPV will be 50 per cent greater than for the Lara, but the other measures of performance – based on ratios or percentages – will be the same, as shown in Table 4.6.

Table 4.6 Comparison of various appraisal methods

	Lara	Bruno	Carling
Net present value (£)	4,252	6,378	(467)
Internal rate of return (%)	19.3	19.3	13.5
Profitability index	1.1	1.1	0.99
Payback period (years)	2.5	2.5	2.9
Accounting rate of return (%)	25.0	25.0	18.0

In ranking mutually exclusive capital projects, we can reject the Carling for having a negative NPV and performance indicators that are consistently inferior to the alternatives. While the Bruno and Lara are, pound-for-pound, identical, the Bruno creates £2,126 additional wealth and is preferred.

Under the conditions typically found in business, no single method is ideal, which is why three or four different measures are often calculated. The ready availability of spreadsheet packages with graphics facilities makes this a straightforward and inexpensive procedure. Investment appraisal techniques are tools to assist managers in assessing the worth of a given project.

■ NPV or IRR?

In many cases, the choice of DCF method has no effect on the investment advice, and it is simply a matter of personal preference. In certain circumstances, however, the choice does matter. We shall consider three such situations:

1 Mutually exclusive projects.
2 Variable discount rates.
3 Unconventional cash flows.

■ Mutually exclusive projects

The decision to accept or reject a project cannot always be separated from other investment projects. For example, a company may have a spare plot of land that could be used to build a warehouse or a sports centre. In such cases, the problem is to evaluate mutually exclusive alternatives.

The earlier worked examples comparing the Lara, Carling and Bruno proposals are mutually exclusive. Recall that, while the Lara and Bruno offered the same IRR, the latter offered a much higher NPV because it was on a larger scale. The weakness of IRR is that it ignores the scale of the project. It implies that firms would prefer to make, say, a 60 per cent IRR on an investment of £1,000 than a 30 per cent return on a £1 million project. Clearly, project scale should be taken into consideration, which is why we recommend the NPV method when assessing mutually exclusive projects of different size or duration.

■ Variable discount rates

It is common to discount cash flows at a constant rate of return throughout a project's life. But this may not always be appropriate. The required rate of return is linked to underlying interest rates and cash flow uncertainties, both of which can change over time.

This presents little difficulty in the case of NPV: different discount rates can be set for each period. The IRR method, however, is compared against a single required rate of return and cannot handle variable rates.

■ Unconventional cash flows

There are three basic cash flow profiles:

Type	Cash flow pattern	Example
Conventional	Outlay followed by inflows ($-+++$)	Capital project
Reverse	Inflow followed by outflow ($+---$)	Loan
Unconventional	More than one change of sign ($-+-+$)	Two-stage development project

For a reverse cash flow pattern, such as a loan where cash is received and interest paid in subsequent periods, the IRR can be usefully applied. But in interpreting the result, remember that the lower the rate of return the better, so the decision rule is to accept the loan proposal if the IRR is below the required rate of return.

Unconventional cash flow patterns create particular difficulty for the IRR approach. Consider the following project cash flows and NPV calculation at 10 per cent required rate of return.

		£	PVIF at 10%	PV (£000)
Initial outlay	0	−100,000	1.00	−100
Year	1	+360,000	0.909	327
	2	−432,000	0.826	−357
	3	+173,000	0.751	130
		NPV		0

With an NPV of zero, the IRR is, by definition, 10 per cent. But at certain other rates, such as 20 per cent and 30 per cent, the NPV is still zero!

Multiple solutions may occur where there are multiple changes of sign. In our example there are three changes in sign – from negative cash flow at the start to positive in Year 1, negative in Year 2 and positive in Year 3. While a conventional project has only one IRR, unconventional projects may have as many IRRs as there are changes in the cash flow sign.

Self-assessment activity 4.6

Why do problems arise in evaluating mutually exclusive projects? What approach would you recommend in such circumstances?

(Answer in Appendix A at the back of the book)

To summarise, the use of NPV and IRR is a matter of personal preference in most instances. But where the evaluation is for mutually exclusive projects, where the discount

Table 4.7 Comparison of mutually exclusive projects

Proposal	Cash flows (£)					Undiscounted cash flow	IRR	NPV at 10%
	Year 0	Year 1	Year 2	Year 3	Year 4			
X	−18,896	8,000	8,000	8,000	8,000	13,104	25%	6,463
Y	−18,896	0	4,000	8,000	26,164	19,268	22%	8,290

rate is not constant throughout the project's life, or where an unconventional cash flow pattern is suspected, we recommend use of the net present value approach. To underline the superiority of NPV we need to examine the respective reinvestment assumptions of the two methods.

The NPV method assumes that all cash flows can be reinvested at the firm's cost of capital. This is entirely sensible, since the discount rate is an opportunity cost of capital that should reflect the alternative use of funds. The IRR method assumes that a project's annual cash flows can be reinvested at the project's internal rate of return. Thus, a project offering a 30 per cent IRR, given a 12 per cent cost of capital, assumes that interim cash flows are compounded forward at the project's rate of return (30 per cent) rather than at the cost of capital (12 per cent). In effect, therefore, the IRR method includes a bonus of the assumed benefits accruing from the reinvestment of interim cash flows at rates of interest in excess of the cost of capital. This is a serious error for projects with IRRs well above the cost of capital.

Consider the mutually exclusive investment proposals given in Table 4.7: X and Y each cost £18,896. Project rankings reveal that X has the higher internal rate of return but the lower net present value. Figure 4.3 shows how this apparent anomaly occurs. (Strictly speaking, the graphs should be curvilinear.)

While Project Y has the higher NPV when discounted at 10 per cent, it has the lower IRR, the two projects intersecting in the graph at around 17 per cent. Wherever there is a sizeable difference between the project IRR and the discount rate, this problem becomes a distinct possibility.

Harry Potter and the global sales hopes of Coca-Cola

When the long-awaited *Harry Potter* movie opened one of the biggest stars was not even seen on film. As millions enjoy *Harry Potter and the Philosopher's Stone*, Coca-Cola is assuming the role of exclusive marketing partner.

Never has so much been poured into one movie by one company. Since lengthy negotiations with Warner Bros Pictures for exclusivity last year, the beverage group has sunk $150 million into a global marketing programme usually preserved for world sporting events such as the Olympics.

In many ways, *Harry Potter* is able to do what Coca-Cola has been attempting for many years – to reach out to a younger audience while not alienating adults. That is crucial as Coca-Cola reinvents itself as an all-beverage company, offering from fun juice drinks to gourmet coffees. But *Harry*

Potter also serves another purpose: instantly elevating the Coke brand by its sheer popularity worldwide, something its own advertising campaigns have failed to do. Such a powerful platform seems to justify spending nearly 10 per cent of the group's global marketing budget on *Harry Potter*.

The biggest critics Coke has to worry about are its shareholders. Its share price has been relatively flat since the announcement of the *Harry Potter* campaign. 'Investors are simply looking for Coke to meet volume goals. That would be enough,' says Ms Levy, a spokesperson for the firm. 'If this can help re-establish the brand in the hearts of consumers, then putting 10 per cent of the budget into *Harry Potter* won't be a bad investment.'

Source: Based on *Financial Times*, 15 November 2001.

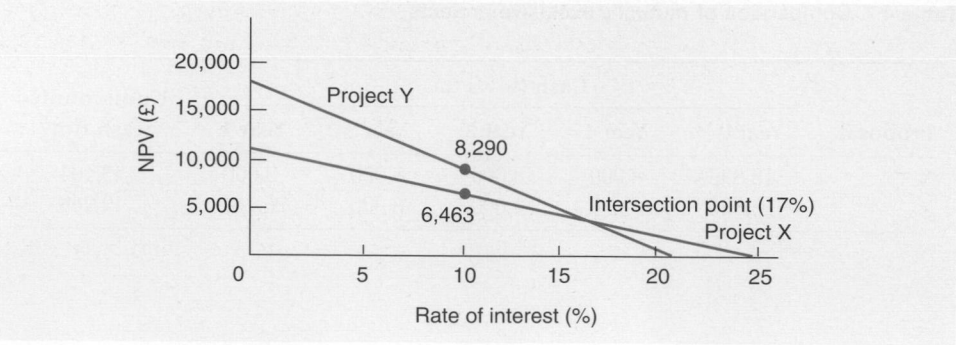

Figure 4.3 NPV and IRR compared

Self-assessment activity 4.7

Take another look at the graphs in Figure 4.3. How would you explain to a manager that Project X, with the higher IRR, is actually less attractive than Project Y?

(Answer in Appendix A at the back of the book)

4.10 INVESTMENT EVALUATION AND CAPITAL RATIONING

We have seen that, under the somewhat limiting assumptions specified, the wealth of a firm's shareholders is maximised if the firm accepts all investment proposals that have positive net present values. Alternatively, the NPV decision rule may be restated as: accept investments that offer rates of return in excess of their opportunity costs of capital. The opportunity cost of capital is the return shareholders could obtain for the same level of risk by investing their capital elsewhere. Implicit in the NPV decision rule is the notion that capital is always available at some cost to finance investment opportunities.

In this section, we relax another assumption of perfect capital markets to include the situation where firms are restricted from undertaking all the investments offering positive net present values. Although individual projects cannot be accepted/rejected on the basis of the NPV rule, the essential problem remains: namely, to determine the package of investment projects that offers the highest total net present value to the shareholders.

◼ The nature of constraints on investment

In imperfect markets, the capital budgeting problem may involve the allocation of scarce resources among competing, economically desirable projects, not all of which can be undertaken. This **capital rationing** applies equally to non-capital, as well as capital, constraints. For example, the resource constraint may be the availability of skilled labour, management time or working capital requirements. Investment constraints may even arise from the insistence that top management appraise and approve all capital projects, thus creating a backlog of investment proposals.

capital rationing
The process of allocating capital to projects where there is insufficient capital to fund all value-creating proposals

◼ Hard and soft rationing

Capital rationing may arise either because a firm cannot obtain funds at market rates of return, or because of internally-imposed financial constraints by management. Externally-imposed constraints are referred to as hard rationing and internally-imposed constraints as soft rationing.

The Wilson Committee (1980) found no evidence of any general shortage of finance for industry at prevailing rates of interest and levels of demand. A survey of managers (Pike 1983) found that:

1 The problem of low investment essentially derives not from a shortage of finance but from an inadequate demand for funds.
2 Capital constraints, where they exist, tend to be internally imposed rather than externally imposed by the capital market.
3 Capital constraints are more acutely experienced by smaller, less profitable and higher-risk firms.

■ Soft rationing

Why should the internal management of a company wish to impose a capital expenditure constraint that may actually result in the sacrifice of wealth-creating projects? Soft rationing may arise because of the following:

1 Management sets maximum limits on borrowing and is unable or unwilling to raise additional equity capital in the short term. Investment is restricted to internally generated funds.
2 Management pursues a policy of stable growth rather than a fluctuating growth pattern with its attendant problems.
3 Management imposes divisional ceilings by way of annual capital budgets.
4 Management is highly risk-averse and operates a rationing process to select only highly profitable projects, hoping to reduce the number of project failures.

The capital budget forms an essential element of the company's complex planning and control process. It may sometimes be expedient for capital expenditure to be restricted – in the short term – to permit the proper planning and control of the organisation. Divisional investment ceilings also provide a simple, if somewhat crude, method of dealing with biased cash flow forecasts. Where, for example, a division is in the habit of creating numbers to justify the projects it wishes to implement, the institution of capital budget ceilings forces divisional management to set its own priorities and to select those offering highest returns.

It is clear that capital rationing can be explained, in part, by imperfections in both the capital and labour markets, and agency costs arising from the separation of ownership from management. Of particular relevance are the problems of **information asymmetry** and transaction costs.

Information asymmetry

Shareholders and other investors in a business do not possess all the information available to management. Nor do they always have the necessary expertise to appreciate fully the information they do receive. Capital rationing may arise because senior managers, convinced that their set of investment proposals is wealth-creating, cannot convince a more sceptical group of potential investors who have far less information on which to make an assessment and who may be influenced by the company's recent performance record.

Transaction costs

The issuing and other costs associated with raising long-term capital do not vary in direct proportion to the amount raised. Corporate treasurers in large organisations will not want to go to the capital market each year for relatively small sums of money if the costs can be significantly reduced by raising much larger sums at less frequent intervals. Capital rationing is therefore a distinct possibility in the intervening years, although this usually means delaying the start date for investments rather than outright rejection.

■ One-period capital rationing

The simplest form of capital rationing arises when financial limits are imposed for a single period. For that period of time, the amount of funds available becomes the limiting factor.

Example: Mervtech plc

The manufacturing division of Mervtech plc has been set an upper limit on capital spending for the coming year of £20 million. It is not normal practice for the group to set investment ceilings, and it is anticipated that the capital constraint will not extend into future years. Assuming a cost of capital of 10 per cent, which of the investment opportunities set out in Table 4.8 should divisional management select?

Table 4.8 Investment opportunities for Mervtech plc

Project	Initial cost (£m)	Cash flows (£m) Year 1	Cash flows (£m) Year 2	Present value at 10% (£m)	NPV at 10% (£m)
A	−15	+17	+17	30	15
B	−5	+5	+10	13	8
C	−12	+12	+12	21	9
D	−8	+12	+11	20	12
E	−20	+10	+10	17	−3

In the absence of any financial constraint, projects A–D, each with positive net present values, would be selected. Once this information has been communicated to investors, the total stock market value would, in theory at least, increase by £44 million – the sum of their net present values.

However, a financial constraint may prevent the selection of all profitable projects. If so, it becomes necessary to select the investment package that offers the highest net present value within the £20 million expenditure limit. A simple method of selecting projects under these circumstances is the profitability index. Recall that this measure is defined as:

Profitability index = Present value/Investment outlay

Project selection is made on the basis of the highest ratio of present value to investment outlay. This method is valuable under conditions of capital rationing because it focuses attention on the net present value of each project relative to the scarce resource required to undertake it. Appraising projects according to the NPV per £1 of investment outlay can give different rankings from those obtained from application of the NPV rule. For example, while in the absence of capital rationing, project A ranks highest (using the NPV rule), project B ranks highest when funds are limited, as shown in Table 4.9. Assuming project independence and infinite divisibility, divisional management will obtain the maximum net

Table 4.9 NPV vs. PI for Mervtech plc

Project	Profitability index	Outlay (£m)	Outlay (£m)	NPV (£m)
B	2.6	5 accept	5	8
D	2.5	8 accept	8	12
A	2.0	15 accept 7/15	7	7
C	1.7	12 reject	–	–
E	0.8	20 reject	–	–
		60	20	27

present value from its £20 million investment expenditure permitted by accepting projects B and D in total and £7 million or 7/15 of project A.

However, the profitability index rarely offers optimal solutions in practice. First, few investment projects possess the attribute of divisibility. Where it is possible for projects to be scaled down to meet expenditure limits, this is frequently at the expense of profitability. Let us suppose that projects are not capable of division. How would this affect the selection problem? The best combination of projects now becomes A and B, giving a total net present value of £23 million. Project D, which ranked above A using the profitability index, is now excluded. Even more fundamental than this, however, is the limitation that the profitability index is appropriate only when capital rationing is restricted to a single period. This is not usually the case. Firms experiencing either hard or soft capital rationing tend to experience it over a number of periods.

In summary, the profitability index provides a convenient method of selecting projects under conditions of capital rationing when investment projects are divisible and independent, and when only one period is subject to a resource constraint. Where, as is more commonly the case, these assumptions do not hold, investment selections should be made after examining the total net present values of all the feasible alternative combinations of investment opportunities falling within the capital outlay constraints.

Self-assessment activity 4.8

What do you understand by 'soft' and 'hard' forms of capital rationing? Give two approaches available to resolve capital rationing problems.

(Answer in Appendix A at the back of the book)

■ Multi-period capital rationing

Many business problems have similar characteristics to those exhibited in the capital rationing problem, namely:

1 Scarce resources have to be allocated between competing alternatives.
2 An overriding objective that the decision-maker is seeking to attain.
3 Constraints, in one form or another, imposed on the decision-maker.

As the number of alternatives and constraints increases, so the decision-making process becomes more complex. In such cases, mathematical programming models are particularly valuable in the evaluation of decision alternatives, for two reasons:

1 They provide descriptive representations of real problems using mathematical equations. Because they capture the critical elements and relationships existing in the real system, they provide insights about a problem without having to experiment directly on the actual system.
2 They provide optimal solutions – that is, the best solution for a given problem representation.

A mathematical programming approach to solving more complex capital rationing problems is provided in Appendix II to this chapter.

SUMMARY

We have examined a number of commonly employed investment appraisal techniques and asked the question: to what extent do they assist managers in making wealth-creating decisions? The primary methods advocated involve discounting the incremental cash flows resulting from the investment decision, although non-discounting techniques are useful secondary methods for evaluating capital projects.

Key points

- The net present value (NPV) method discounts project cash flows at the firm's required return and then sums the cash flows. The decision rule is: accept all projects whose NPV is positive.

- The internal rate of return (IRR) is that discount rate which, when applied to project cash flows, produces a zero NPV. Projects with IRRs above the required return are acceptable.

- The profitability index (PI) is the ratio of the present value of project benefits to the present value of investment costs. The decision rule is to accept projects with a PI greater than 1.

- The NPV, IRR and PI methods give the same investment advice for independent projects. But where projects are mutually exclusive, differences can arise in rankings.

- The NPV approach is viewed as more sound than the IRR method because it assumes reinvestment at the required return rather than the project's IRR.

- The modified IRR (MIRR) is that rate of return which, when the initial outlay is compared with the terminal value of the project's cash flows reinvested at the cost of capital, gives an NPV of zero. This method provides a rate of return consistent with the NPV approach. This is discussed in Appendix I.

- Payback is a useful method, but ignores cash flows beyond the payback period. Simple payback also ignores the time-value of money.

- Accounting rate of return (ARR) compares the average profit of the project against the book value of the asset acquired. Its main merit is that, as a measure of profitability, it can be related to the accounts of the business. However, it takes no account of the timing of cash flows or of the size and life of the investment.

- Capital rationing, where it exists, tends to be of the 'softer' form where management voluntarily imposes investment ceilings in the short term.

- Single-period capital rationing is resolved by ranking projects according to their profitability index. More complex multi-period capital rationing problems demand a mathematical programming approach.

Further reading

Most good finance texts cover the topic of investment appraisal well, including Ross, Westerfield and Jordan (2008), and Brealey, Myers and Allen (2005). These texts also address the capital rationing problem. More detailed treatment of capital rationing is found in Pike (1983), Elton (1970), Lorie and Savage (1955) and Weingartner (1977). For a fuller discussion on the modified IRR, see McDaniel *et al.* (1988).

Appendix I
MODIFIED IRR

Most managers prefer the IRR to the NPV method. The modified IRR seeks to adjust the IRR so that it has the same reinvestment assumption as the NPV approach.

■ The modified internal rate of return (MIRR)

MIRR is that rate of return which, when the initial outlay is compared with the terminal value of the project's net cash flows reinvested at the cost of capital, gives an NPV of zero.

This involves a two-stage process:

1 Calculate the terminal value of the project by compounding forward all interim cash flows at the cost of capital to the end of the project.
2 Find the rate of interest that equates the terminal value with the initial cost.

Example: Lara revisited

We established earlier that the Lara proposal offered an NPV of £4,252 (Table 4.1) and an IRR of approximately 19 per cent (Table 4.3). Table 4.10 shows that by compounding the interim cash flows at 14 per cent to the end of Year 4, the project offers a terminal value of £74,738. To find the Year 4 present value factor that comes closest to equating the terminal value with the initial outlay, we divide the initial outlay by the terminal value and look up the interest rate that gives this factor in Year 4 (see Appendix B). For the Lara project, the MIRR is approximately 17 per cent, a good 2 per cent below the IRR figure. For more profitable projects, the deviation would be greater.

Table 4.10 Modified IRR for Lara

Year	Cash flow (£)	Future value factor @ 14%	Terminal value (£)
(a) Find the terminal value			
1	16,000	$(1.14)^3$	23,704
2	16,000	$(1.14)^2$	20,794
3	16,000	1.14	18,240
4	12,000	1.00	12,000
			74,738

(b) Find the rate of interest (denoted by x) which equates the terminal value with initial cost:

$$\text{PVIF}_{(x\%,4\text{ yrs})} = £40,000/£74,738 = 0.535$$

Using tables (Appendix C) for four years we find that 17 per cent gives a PVIF of 0.534.

	£
To check: £74,738 × 0.534 =	39,910
less initial investment	(40,000)
NPV	(90) i.e. close to zero

The modified IRR is approximately 17 per cent compared with the IRR of 19.3 per cent.

You might like to check this out using an Excel spreadsheet. In cells A1 to A5 type in −40,000, 16,000, 16,000, 16,000, 12,000. In cell A6 click on the *fx* icon and select Financial/ MIRR. In the box enter for Values, A1:A5, and .14 for both Finance rate and Reinvestment rate.

Self-assessment activity 4.9

Describe how the modified IRR is calculated. What advantages does the MIRR have over the IRR in assessing capital investment decisions?

(Answer in Appendix A at the back of the book)

Dangers with internal rate of return

Imagine an investment that requires an upfront payment of $5,000 and then produces positive cash flows of $3,000 in years two to 10. At a 10 per cent discount rate, this project will have a positive net present value of just over $11,160 and an internal rate of return of 59 per cent.

Now assume that the investor can sell out at fair value in year three, receiving $14,600 (the NPV of the cash flows in years four to 10). The project's overall NPV remains exactly the same. But the IRR more than doubles to 120 per cent as a decade worth of cash flows has been concentrated into three years – even though no extra value has been created.

This example is hardly news to cognoscenti. But all too often, it creeps into practice: IRR is the private equity industry's main yardstick for judging performance, raising funds and rewarding managers. Indeed, one of the sector's selling points is continued promises of superior – 25 per cent or so – returns. Yet as the example demonstrates, such IRR-based numbers can be artificially boosted by extracting cash early – through sales (including to rival firms), listings or recapitalisations. Indeed, the theoretical investor could accept as little as $5,000 in year three, thereby destroying considerable shareholder value,

and still trumpet an IRR of more than 59 per cent.

In fact, extracting cash early is justified only if it can be reinvested at the original IRR or a higher one. A more realistic assumption would be to assume reinvestment at the cost of capital. Because it is intuitive and relatively easy to calculate, IRR will remain a popular metric for appraising investments. But it should not be used in isolation.

Source: Lex Column, *Financial Times*, 1 June 2005.

Appendix II
MULTI-PERIOD CAPITAL RATIONING AND MATHEMATICAL PROGRAMMING

Where an overriding financial objective exists (such as maximising shareholder wealth) and financial constraints are expected to operate over a number of years, the allocation of capital resources to investment projects is best solved by the mathematical programming approach.

Many programming techniques have been developed. We shall concentrate on the most common technique: linear programming. The assumptions and limitations underlying the LP approach will be discussed in a subsequent section. Problem-solving using the LP approach involves four basic steps:

1 *Formulate the problem.* This requires specification of the objective function, input parameters, decision variables and all relevant constraints. Take a firm that produces two products, A and B, with contributions per unit of £5 and £10 respectively. The firm wishes to determine the product mix that will maximise its total contribution. The objective function may be expressed as follows:

maximise contribution: £5A + £10B

A and B are the decision variables representing the number of units of products A and B that should be produced. The input values £5 and £10 specify the unit contribution values for products A and B respectively. Constraint equations may also be determined to describe any limitations on resources, whether imposed by managerial policies or the external environment.

2 *Solve the LP problem.* Simple problems can be solved using either a graphical approach or the simplex method. More complex problems require a computer-based solution algorithm.

3 *Interpret the optimal solution.* Examine the effect on the total value of the objective function if a binding constraint were marginally slackened or tightened.

4 *Conduct sensitivity analysis.* Assess, for each input parameter, the range of values for which the optimal solution remains valid.

These four stages in the LP process are illustrated in the following example.

Example: Multi-period capital rationing in Flintoff plc

Flintoff's five-year planning exercise shows that the cost of its six major projects, forming the basis of the firm's investment programme, exceeds the planned finance available. Flintoff is already highly geared and control is in the hands of a few shareholders who are reluctant to introduce more equity funds. Accordingly, the main source of funds is through cash generated from existing operations, estimated to be £300,000 p.a. over the next five years. The six projects are independent and cannot be delayed or brought forward. Each project has a similar risk complexion to that of the existing business. If necessary, projects are capable of division but no more than one of each is required. The planned investment schedule and associated cash flows are given in Table 4.11.

Table 4.11 Flintoff plc: planned investment schedule (£000)

	Project outlays						Total	Available
Year	A	B	C	D	E	F	outlay	capital
0	−200	–	−220	−110	−24	–	−554	300
1	−220	−220	−100	−150	−48	–	−718	300
2		−66	−50			−500	−616	300
3					−200		−200	300
NPV	130	184	35	42	186	280		
Total NPV = £857,000								

The six projects, if implemented, are forecast to produce a total NPV of £857,000. However, the annual capital constraint of £300,000 means that for the next three years the required investment expenditure exceeds available investment finance, i.e. there is a capital rationing problem.

The solution sequence is as follows.

1 Specify the problem

The objective function seeks to maximise the NPV from the given set of projects available.

Max NPV:$130A + 184B + 35C + 42D + 186E + 280F$

However, given the capital expenditure constraints over the coming years, we must express for each year the capital required for each project and the maximum capital available each year (i.e. £300,000):

Year 0 $200A + 220C + 110D + 24E \leq 300$
Year 1 $220A + 220B + 100C + 150D + 48E \leq 300$
Year 2 $66B + 50C + 500F \leq 300$
Year 3 $200E \leq 300$

In addition to the capital constraints, we need to define the bounds for each variable. As no more than one of each project is required and projects are divisible, we can specify the bounds as:

$A, B, C, D, E, F \geq 0 \leq 1$

Continued

This linear programming formulation tells us to find the mix of projects producing the highest total net present value, given the constraint that only £300,000 can be spent in any year and that not more than one of each project is permitted.

2 Solve the problem

Using a linear program on the computer gives the solution in Table 4.12. Flintoff plc should accept investment proposals B and E in full plus 14.5 per cent of project A and 46.8 per cent of project F. This will produce the highest possible total net present value available, £520,000. This is significantly less than the £857,000 total NPV if no constraints are imposed.

Table 4.12 Projects accepted based on LP solution

| Project | Proportion accepted | NPV (£000) | Capital outlay (£000) | | | |
			Year 0	Year 1	Year 2	Year 3
A	0.145	19	−29	32	–	–
B	1	184		220	66	
C	0	–				
D	0	–				
E	1	186	−29	−48	–	−200
F	0.468	131	–	–	−234	–
		520	−58	−300	−300	−200

3 Interpret the optimal solution

Table 4.12 shows that only in Years 1 and 2 is the full £300,000 utilised. These years then impose binding constraints – their existence limits the company's freedom to pursue its objective of NPV maximisation because it restricts the investment finance available to the firm in those years. Conversely, Years 0 and 3 are non-binding: they do not constrain the firm in its efforts to achieve its objective. Hence while there is no additional opportunity cost (besides that already incorporated in the discount rate), for non-binding periods there is an additional opportunity cost attached to the use of investment finance in the two years where constraints are binding. These additional opportunity costs are termed shadow prices (or dual values). Shadow prices show how much the decision-maker would be willing to pay to acquire one additional unit of each constrained resource. In this case, computer analysis reveals that the shadow prices are:

Year	Shadow price (£)	Constraint
0	0	non-binding
1	0.59	binding
2	0.56	binding
3	0	non-binding

A £1 increase (reduction) in capital spending in Year 1 would produce an increase (reduction) in total NPV of £0.59. Similarly, for Year 2 a £1 change in investment expenditure would result in a £0.56 change in total NPV. Because the capital constraints in Years 0 and 3 are non-binding, their shadow prices are zero and a marginal change in capital spending in those years will have no impact on the NPV objective function. Shadow prices, while of value in indicating the additional opportunity cost, can be used only within a specific range. In addition, it is desirable to ascertain the effect of changes in input parameters on the optimal solution. These issues require some form of sensitivity analysis.

4 Perform sensitivity analysis

The computer output provides two additional pieces of information. First, it tells the decision-maker the maximum variation for each binding constraint. In our example, the shadow price for the Year 1 constraint has a range of −36 to +188. In other words, the shadow price of £0.59 would hold up to an increase in capital expenditure for that year of £188,000, or a reduction of £36,000.

The program also indicates the margin of error permitted for input parameters before the optimal solution differs. In our example, the actual NPV for the optimal investment mix could fall as indicated below and still not change the optimal solution:

Project	Maximum permitted fall in NPV (£000)
A	−68
B	−17
E	−158
F	−280

This facility is particularly appropriate as a means of assessing the margin of error permitted for risky projects under conditions of capital rationing.

■ LP assumptions

In order to assess the value of the basic linear programming approach, we must consider the assumptions underlying its application. These are as follows:

1 All input parameters to the LP model are certain.
2 There is a single objective to be optimised.
3 The objective function and all constraint equations are linear.
4 Decision variables are continuous (i.e. divisible).
5 There is independence among decision variables and resources available.

Most, if not all, of these limiting assumptions can be relaxed by using more complex mathematical programming. For example, uncertainty, multiple objectives and non-linearity can be better addressed by other approaches such as stochastic LP, goal programming and quadratic programming respectively.

For most businesses, the LP assumption that projects are divisible is unrealistic. Even if a project could be operated on a reduced scale, it is unlikely that the NPV would reduce *pro rata* because many of the fixed costs would remain while the benefits of sale would be reduced. *Integer programming* is more appropriate when projects are non-divisible. This is a special case of linear programming where variables can take only the values 0 (reject the project) or 1 (accept it *in toto*).

Applying integer programming to Flintoff plc requires only one change in the problem specification. The bounds become:

$A, B, C, D, E, F, = 0$ or 1

The solution, provided by an integer programming computer application, shows that only two projects should be accepted: projects B and E. These offer a combined NPV of £370,000, which is the best available given the capital constraints. This is well under half the £857,000 total NPV achievable in the absence of capital rationing (see Table 4.11). Were the shareholders of Flintoff plc aware of these lost wealth-creating opportunities, they might well be concerned and ask the chairman whether the capital constraints were really as fixed as they appeared to be!

QUESTIONS

 **myfinancelab** *Questions with an icon are also available for practice in myfinancelab with additional supporting resources.*

Questions with a **coloured number** have solutions in Appendix B on page 731.

1 The directors of Yorkshire Autopoints are considering the acquisition of an automatic car-washing installation. The initial cost and setting-up expenses will amount to about £140,000. Its estimated life is about seven years, and estimated annual accounting profit is as follows:

Year	1	2	3	4	5	6	7
Operations cash flow (£)	30,000	50,000	60,000	60,000	30,000	20,000	20,000
Depreciation (£)	20,000	20,000	20,000	20,000	20,000	20,000	20,000
Accounting profit (£)	10,000	30,000	40,000	40,000	10,000	–	–

At the end of its seven-year life, the installation will yield only a few pounds in scrap value. The company classifies its projects as follows:

Required rate of return

Low risk	20 per cent
Average risk	30 per cent
High risk	40 per cent

Car-washing projects are estimated to be of average risk.

(a) Should the car-wash be installed?
(b) List some of the popular errors made in assessing capital projects.

2 Microtic Ltd, a manufacturer of watches, is considering the selection of one from two mutually exclusive investment projects, each with an estimated five-year life. Project A costs £1,616,000 and is forecast to generate annual cash flows of £500,000. Its estimated residual value after five years is £301,000. Project B, costing £556,000 and with a scrap value of £56,000, should generate annual cash flows of £200,000. The company operates a straight-line depreciation policy and discounts cash flows at 15 per cent p.a.

Microtic Ltd uses four investment appraisal techniques: payback period, net present value, internal rate of return and accounting rate of return (i.e. average accounting profit to initial book value of investment).

Make the appropriate calculations and give reasons for your investment advice.

3 Mace Ltd is planning its capital budget for 19_7 and 19_8. The company's directors have reduced their initial list of projects to five, the expected cash flows of which are set out below:

Project	19_7	19_8	19_9	19_0	NPV
1	−60,000	+30,000	+25,000	+25,000	+1,600
2	−30,000	−20,000	+25,000	+45,000	+1,300
3	−40,000	−50,000	+60,000	+70,000	+8,300
4	0	−80,000	+45,000	+55,000	+900
5	−50,000	+10,000	+30,000	+40,000	+7,900

None of the five projects can be delayed and all are divisible. Cash flows arise on the first day of the year. The minimum return required by shareholders of Mace Ltd is 10 per cent p.a. Which projects should Mace Ltd accept if the capital available for investment is limited to £100,000 on 1 January 19_7, but readily available at 10 per cent p.a. on 1 January 19_8 and subsequently?

4 The directors of Mylo Ltd are currently considering two mutually exclusive investment projects. Both projects are concerned with the purchase of new plant. The following data are available for each project:

	Project	
	1 (£)	2 (£)
Cost (immediate outlay)	100,000	60,000
Expected annual net profit (loss)		
Year 1	29,000	18,000
2	(1,000)	(2,000)
3	2,000	4,000
Estimated residual value	7,000	6,000

The company has an estimated cost of capital of 10 per cent and employs the straight-line method of depreciation for all fixed assets when calculating net profit. Neither project would increase the working capital of the company. The company has sufficient funds to meet all capital expenditure requirements.

Required

(a) Calculate for each project:

    (i) the net present value

    (ii) the approximate internal rate of return

    (iii) the profitability index

    (iv) the payback period

(b) State which, if any, of the two investment projects the directors of Mylo Ltd should accept, and why.

(c) State, in general terms, which method of investment appraisal you consider to be most appropriate for evaluating investment projects and why.

(Certified Diploma)

5 Mr Cowdrey runs a manufacturing business. He is considering whether to accept one of two mutually exclusive investment projects and, if so, which one to accept. Each project involves an immediate cash outlay of £100,000. Mr Cowdrey estimates that the net cash inflows from each project will be as follows:

Net cash inflow at end of:	Project A (£)	Project B (£)
Year 1	60,000	10,000
Year 2	40,000	20,000
Year 3	30,000	110,000

Mr Cowdrey does not expect capital or any other resource to be in short supply during the next three years.

Required

(a) Prepare a graph to show the functional relationship between net present value and the discount rate for the two projects (label the vertical axis 'net present value' and the horizontal axis 'discount rate').

(b) Use the graph to estimate the internal rate of return of each project.

(c) On the basis of the information given, advise Mr Cowdrey which project to accept if his cost of capital is (i) 6 per cent; (ii) 12 per cent.

(d) Describe briefly any additional information you think would be useful to Mr Cowdrey in choosing between the two projects.

(e) Discuss the relative merits of net present value and internal rate of return as methods of investment appraisal.

Ignore taxation.

(ICAEW)

? **6** The directors of XYZ plc wish to expand the company's operations. However, they are not prepared to borrow at the present time to finance capital investment. The directors have therefore decided to use the company's cash resources for the expansion programme.

Three possible investment opportunities have been identified. Only £400,000 is available in cash and the directors intend to limit the capital expenditure over the next 12 months to this amount. The projects are not divisible (i.e. cannot be scaled down) and none of them can be postponed. The following cash flows do not allow for inflation, which is expected to be 10 per cent per annum constant for the foreseeable future.

Expected net cash flows (including residual values)

Project	Initial investment £	Year 1 £	Year 2 £	Year 3 £
A	−350,000	95,000	110,000	200,000
B	−105,000	45,000	45,000	45,000
C	−35,000	−40,000	−25,000	125,000

The company's shareholders currently require a return of 15 per cent nominal on their investment. Ignore taxation.

Required

(a) (i) Calculate the expected net present value and profitability indexes of the three projects; and

(ii) comment on which project(s) should be chosen for the investment, assuming the company can invest surplus cash in the money market at 10 per cent. (Note: you should assume that the decision not to borrow, thereby limiting investment expenditure, is in the best interests of its shareholders.)

(b) Discuss whether the company's decision not to borrow, thereby limiting investment expenditure, is in the best interests of its shareholders.

(CIMA)

7 Raiders Ltd is a private limited company financed entirely by ordinary shares. Its effective cost of capital, net of tax, is 10 per cent p.a. The directors are considering the company's capital investment programme for the next two years, and have reduced their initial list of projects to four. Details of the projects' cash flows (net of tax) are as follows (in £000):

Project	Immediately	After 1 year	After 2 years	After 3 years	NPV (at 10%)	IRR (to nearest 1%)
A	−400	+50	+300	+350	+157.0	26%
B	−300	−200	+400	+400	+150.0	25%
C	−300	+150	+150	+150	+73.5	23%
D	0	−300	+250	+300	+159.5	50%

None of the projects can be delayed. All projects are divisible; outlays may be reduced by any proportion and net inflows will then be reduced in the same proportion. No project can be undertaken more than once. Raiders Ltd is able to invest surplus funds in a bank deposit account yielding a return of 7 per cent p.a., net of tax.

Required

(a) Prepare calculations showing which projects Raiders Ltd should undertake if capital for immediate investment is limited to £500,000, but is expected to be available without limit at a cost of 10 per cent p.a. thereafter.

(b) Provide a mathematical programming formulation to assist the directors of Raiders Ltd in choosing investment projects if capital available immediately is limited to £500,000, capital available after one year is limited to £300,000, and capital is available thereafter without limit at a cost of 10 per cent p.a.

(c) Outline the limitations of the formulation you have provided in (b).

(d) Comment briefly on the view that in practice capital is rarely limited absolutely, provided that the borrower is willing to pay a sufficiently high price, and in consequence a technique for selecting investment projects that assumes that capital is limited absolutely is of no use.

(ICAEW)

Practical assignment

Either drawing on your own experience, or by asking someone you know in management, find out the primary investment appraisal techniques employed in an organisation. How well does the appraisal system appear to operate?

 **myfinancelab** *Now retake your diagnostic test for Chapter 4 to check your progress and update your study plan.*

5

Project appraisal – applications

To boldly go into space-age investment

There is a danger that investment analysis can become bogged down in unnecessary detail. So how does an entrepreneur like Sir Richard Branson, inventor of the Virgin brand, make investment decisions?

Like many other top managers, Branson places more reliance on experience and 'hunch' in decision-making than detailed financial analyses. However, he also works through the risks involved and whether they can be managed. In his autobiography, Branson claims to make up his mind about whether a business proposal excites him within about thirty seconds of looking at it. He relies more on gut instinct than researching huge quantities of statistics. The idea of operating a Virgin airline grabbed his imagination, but he had to work out in his own mind what the potential risks were.

In 2004 Branson took a giant step in announcing that he would boldly go into the space age through his new company – Virgin Galactic – the world's first commercial space business. A fleet of five Virgin spacecraft will carry 3,000 passengers into space between 2007 and 2012. The company expects to spend $100 million over five years to develop the spacecraft and the first Virgin astronauts will have to pay £105,000 for a flight.

Branson also argues that fun is at the core of the way he likes to do business and the main secret of Virgin's success. He observes that the idea of business being fun and creative goes right against the grain of convention, and it's certainly not how they teach it at some of those business schools, where, as he puts it, 'business means hard grind and lots of discounted cash flows and net present values!' (Branson, 1998, p. 490).

Learning objectives

Having read this chapter, you should be well equipped to handle most capital investment decision problems found either on examination papers or in business. Skills should develop in the following areas:

- Identifying the relevant information in investment analysis.
- Evaluating replacement and other investment decisions.
- Handling inflation.
- Assessing the effects of taxation on investment decisions.
- Investment appraisal practices, strengths and limitations.
- Identifying the appropriate discount rate.

 *Complete your diagnostic test for Chapter 5 now to create your personal study plan.*

5.1 INTRODUCTION

In the previous chapter, we examined a variety of approaches to assessing investment projects. The focus was almost exclusively on the appropriate appraisal method. But even the best appraisal method is of little use unless we can first identify the relevant information.

Investment decisions, particularly larger ones with strategic implications, are not usually made on 'the spur of the moment'. The whole process, from the initial idea through to project authorisation, usually takes many months, or even years. A vital part of this process is gathering information to identify the *incremental cash flows* pertaining to the investment decision. In this chapter, we consider the principles underlying economic feasibility analysis and apply them to particular situations. We pay particular attention to the treatment of inflation and taxation in project evaluation.

■ Need for relevant information

In financial management, as with all areas of management, an effective manager needs to identify the right information for decision-making. In the case of capital investment decisions, committing a substantial proportion of the firm's funds to non-routine, largely irreversible actions can be risky and demands a careful examination of all the relevant information available.

Information on the likely costs and benefits of an investment proposal, its expected economic life, appropriate inflation rates and discount rates should be gathered to provide a clearer picture of the project's economic feasibility. Frequently, we find that the reliability of the information source varies. For example, a demand forecast from a marketing executive with a track record of making wildly inaccurate forecasts will be viewed differently from an official quote for the cost of a machine. The accounting system and formal reports provide a part of the relevant information, the remainder coming through informal channels, frequently more qualitative than quantitative in nature.

In identifying and analysing information, managers should remember that effective information should, wherever possible, be relevant, reliable, timely, accurate and cost-efficient.

5.2 INCREMENTAL CASH FLOW ANALYSIS

We stressed in the previous chapter that the financial input into any investment decision analysis should be based on the incremental cash flows arising as a consequence of the decision. These can be found by calculating the differences between the forecast cash flows from going ahead with the project and the forecast cash flows from not accepting the project.

This is not always easy. To illustrate this point, we consider how investment analysis handles opportunity costs, sunk costs, associated cash flows, working capital changes, interest costs and fixed overheads.

Self-assessment activity 5.1

What do you understand by the term 'incremental cash flow'?

(Answer in Appendix A at the back of the book)

■ Remember opportunity costs

Capital projects frequently give rise to opportunity costs. For example, a company has developed a patent to produce a new type of lawnmower. If it makes the product, the expected NPV is £70,000. However, this ignores the alternative course of action: to sell the patent to another company for £90,000. This opportunity cost is a fundamental element in the investment decision to manufacture the product and should be deducted from the £70,000, giving a negative NPV of £20,000. Alternatively, the sale of the patent is viewed as a mutually exclusive proposal which ranks above the proposal to manufacture the product.

Opportunity cost example: Belfry plc

We often see opportunity costs in replacement decisions. In Belfry plc, an existing machine can be replaced by an improved model costing £50,000, which generates cash savings of £20,000 each year for five years, after which it will have a £5,000 scrap value. The equipment manufacturers are prepared to give an allowance on the existing machine of £15,000, making a net initial cash outlay of £35,000. But in pursuing this course of action, we terminate the existing machine's life, preventing it from yielding £3,000 scrap value in three years. The prospective scrap value denied is the opportunity cost of replacing the existing machine. The cash flows associated with the replacement decision are therefore:

Year 0	Net cost	(£35,000)
Years 1–5	Annual cash savings	(£20,000)
Year 3	Opportunity cost (scrap value forgone on old machine)	(£3,000)
Year 5	Scrap value on new machine	£5,000

■ Ignore sunk costs

By definition, any costs incurred or revenues received prior to a decision are not relevant cash flows; they are sunk costs. This does not necessarily imply that previously incurred costs did not produce relevant information. For example, externally conducted feasibility studies are often undertaken to provide important technical, marketing and cost data prior to a major new investment. However, the costs of the study are excluded from project analysis. We are concerned with future cash flows arising as a consequence of the particular course of action.

■ Look for associated cash flows

Investment in capital projects may have company-wide cash flow implications. Those involved in forecasting cash flows may not realise how the project affects other parts of the business – senior management should therefore carefully consider whether there are any additional cash flows associated with the investment decision. The decision to produce and launch a new product may influence the demand for other products within the product range. Similarly, the decision to invest in a new manufacturing plant in Eastern Europe, or to take over existing facilities, may have an adverse effect on the company's exports to such countries.

Self-assessment activity 5.2

Waxo plc has developed a new wonder earache drug. The management is currently putting together an investment proposal to produce and sell the drug, but is not sure whether to include the following:

1 The original cost of developing the drug.
2 Production of the new product will have an adverse effect on the sale of related products in another division of Waxo.
3 Instead of producing the drug internally, the patent could be sold for £10 million.

How would you advise Waxo on the relevant costs?

(Answer in Appendix A at the back of the book)

■ Include working capital changes

It is easy to forget that the total investment for capital projects can be considerably more than the fixed asset outlay. Normally, a capital project gives rise to increased stocks and debtors to support the increase in sales. This increase in working capital forms part of the investment outlay and should be included in project appraisal. If the project takes a number of years to reach its full capacity, there will probably be additional working capital requirements in the early years, especially for new products where the seller may have to tempt purchasers by offering more than usually generous credit terms. The investment decision implies that the firm ties up fixed and working capital for the life of the project. At the end of the project, whatever is realised is returned to the firm. For fixed assets, this will be scrap or residual value – usually considerably less than the original cost, except in the case of land and some premises. For working capital, the whole figure – less the value of damaged stock and bad debts – is treated as a cash inflow in the final year, because the finance tied up in working capital can now be released for other purposes.

Occasionally, the introduction of new equipment or technology reduces stock requirements. Here the stock reduction is a positive cash flow in the start year; but an equivalent negative outflow at the end of the project should be included only if it is assumed that the firm will revert to the previous stock levels. A more realistic assumption may be that any replacement would at least maintain existing stock levels, in which case no cash flow for stock in the final year is necessary.

■ Separate investment and financing decisions

Capital projects must be financed. Commonly, this involves borrowing, which requires a series of cash outflows in the form of interest payments. These interest charges should not be included in the cash flows because they relate to the financing rather than the investment decision. Were interest payments to be deducted from the cash flows, it would amount to double-counting, since the discounting process already considers the cost of capital in the form of the discount rate. To include interest charges as a cash outflow could therefore result in seriously understating the true NPV.

Some companies include interest on short-term loans (such as for financing seasonal fluctuations in working capital) in the project cash flows. If so, it is important that both the timing of the receipt and the repayment of the loan are also included. For example, the NPV on a 15 per cent one-year loan of £100,000, assuming a 15 per cent discount rate, must be zero: £100,000 cash received today less the present value of interest and loan repaid after a year (i.e. £115,000/1.15).

■ Fixed overheads can be tricky

Only additional fixed overheads incurred as a result of the capital project should be included in the analysis. In the short term, there will often be sufficient factory space to house new equipment without incurring additional overheads, but ultimately some additional fixed costs (for rent, heating and lighting, etc.) will be incurred. Most factories operate an accounting system whereby all costs, including fixed overheads, are charged on some agreed basis to cost centres. Investment in a new process or machine frequently attracts a share of these overheads. While this may be appropriate for accounting purposes, only *incremental* fixed overheads incurred by the decision should be included in the project analysis.

Self-assessment activity 5.3

Rick Faldo – the marketing manager of a manufacturer of golf equipment – has recently submitted a proposal for the production of a range of clubs for beginners. He has just received the following response from the managing director:

To: Rick Faldo

From: Sid Torrance

I have examined your proposal for the new 'Clubs for Beginners' range which you say promises a three-year payback and a 30 per cent DCF return. Some hope! You seem to have forgotten the following relevant points.

1 We have a policy that all investment is subject to a depreciation charge of 25 per cent on the reducing balance.
2 The accounting department will need to recover factory fixed overheads on the new machine.
3 We need to charge against the project the £8,000 marketing research conducted to assess the size of the market for the new range.
4 I'd have to pay 15 per cent to finance the project.

Projects like this I can do without!

How would you reply to this e-mail if you were Faldo?

(Answer in Appendix A at the back of the book)

5.3 REPLACEMENT DECISIONS

The decision to replace an existing machine which has yet to reach the end of its useful life is often necessary because of developments in technology and generous trade-in values offered by manufacturers. In analysing replacement decisions, we assess the additional costs and benefits arising from the replacement, rather than the attractiveness of the new machine in isolation.

Example of replacement analysis: Sevvie plc

Sevvie plc manufactures components for the car industry. It is considering automating its line for producing crankshaft bearings. The automated equipment will cost £750,000. It will replace equipment with a residual value of £80,000 and a written-down book value of £200,000. It is anticipated that the existing machine has a further five years to run, after which its scrap value would be £5,000.

At present, the line has a capacity of 1.25 million units per annum but, typically, it has only been run at 80 per cent of capacity because of the lack of demand for its output. The new line has a capacity of 1.4 million units per annum. Its life is expected to be five years and its scrap value at that time £105,000. The main benefits of the new proposal are a reduction in staffing levels and an improvement in price due to its superior quality.

The accountant has prepared the cost estimates shown in Table 5.1 based on output of 1 million units p.a. Fixed overheads include depreciation on the old machine of £40,000 p.a. and £130,000 for the new machine. It is considered that for the company overall, other fixed overheads are unlikely to change.

The introduction of the new machine will enable the average level of stocks held to be reduced by £160,000. After five years, the machine will probably be replaced by a similar one.

Table 5.1 Profitability of Sevvie's project

	Old line (per unit) (p)	New line (per unit) (p)
Selling price	150	155
Materials	(40)	(36)
Labour	(22)	(15)
Variable overheads	(14)	(14)
Fixed overheads	(34)	(40)
Profit per unit	40	50

Continued

The company uses a 10 per cent discount rate. We shall ignore taxation.

The solution is given in Table 5.2. Several comments are worthy of note:

1 It has been assumed that no benefits can be obtained from the additional capacity due to the sales constraints. In reality, it would be useful to explore whether – for example, by investing in advertising – demand could be increased.

2 Fixed costs are not relevant. Depreciation is not a cash flow, and we are told that other fixed costs will not alter with the decision. The incremental cash flow per unit is therefore 16p, giving £160,000 (i.e. 1 million units at 16p) additional cash each year on the expected sales.

3 In addition to the scrap values of £80,000 in Year 0 and £105,000 in Year 5 on the old and new machines respectively, there is a £5,000 opportunity cost in Year 5. This is the scrap value no longer available as a consequence of the replacement decision.

4 Working capital will be reduced by £160,000 for the period of the project and it therefore appears as a benefit in Year 0.

5 The book value of the existing machine represents the undepreciated element of the original cost, a sunk cost which is not relevant to the decision. The book value of assets, however, may be important in practice, as it can sometimes mean a heavy accounting loss in the year of acquisition. In this case, the loss would be £120,000 (i.e. book value of £200,000 less £80,000 residual value). This is not a cash flow, but, in practice, it may still be regarded as undesirable to depress reported profit figures in this way. This, of course, raises issues of market efficiency – will the market see through the accounting adjustment?

The replacement decision is a wealth-creating opportunity offering an NPV of £157,000, although the cumulative present value calculation in Table 5.2 shows that the project does not come into surplus, in net present value terms, until the final year.

Table 5.2 Sevvie plc solution

	Old line (pence per unit)	New line (pence per unit)
Selling price	150	155
Less:		
Materials	(40)	(36)
Labour	(22)	(15)
Variable overheads	(14)	(14)
Variable costs	(76)	(65)
Cash contribution	74	90
Incremental cash flow per unit (90–74)		16p
Total incremental cash flow on 1 million unit sales		£160,000

Year (£000)	0	1	2	3	4	5
Cost savings		160	160	160	160	160
New machine	(750)					105
Scrap old machine	80					(5)
Working capital reduction	160					
Net cash flow	(510)	160	160	160	160	260
Net present value						
10% discount factor	1.000	0.909	0.826	0.751	0.683	0.621
Present value	(510)	145	132	120	109	161
Cumulative present value	(510)	(365)	(233)	(113)	(4)	157
NPV	157					

5.4 INFLATION CANNOT BE IGNORED

Inflation can have a major impact on the ultimate success or failure of capital projects. In considering how it should be treated in discounted cash flow analysis, two problems arise: first, how does inflation affect the estimated cash flows from the project; and second, how does it affect the discount rate?

For example, a machine costs £18,000 and is projected to produce, in current prices, cash flows of £6,000, £10,000 and £7,000 respectively over the next three years. The expected rate of inflation is 6 per cent and the firm's cost of capital is 16.6 per cent.

We can adopt one of two approaches:

1 Forecast cash flows in *money* terms and discount at the nominal or *money* cost of capital including inflation (i.e. 16.6 per cent), or
2 Forecast cash flows in *constant* (i.e. current) money terms and discount at the *real* cost of capital.

'Money terms' here means the actual price levels that are forecast to obtain at the date of each cash flow; 'constant terms' means the price level prevailing today; and 'real cost of capital' means the net of inflation cost.

In Table 5.3 cash flows expressed at constant prices are converted to actual money cash flows by compounding at $(1 + I)$, where I is the inflation rate. These cash flows are then discounted in the normal manner at the money discount factor (the reason for such an awkward rate will become apparent later) to give a positive NPV of £977. Had we not adjusted cash flows for inflation, the NPV would have been incorrectly expressed as a negative value.

Table 5.3 The money terms approach

Year	Cash flow current prices (£)	Actual money prices (£)	Discount factor @ 16.6%	Present value (£)
0	$(18,000) \times 1.0$	(18,000)	1	(18,000)
1	$6,000 \times 1.06$	6,360	$\dfrac{1}{1.166}$	5,454
2	$10,000 \times (1.06)^2$	11,236	$\dfrac{1}{(1.166)^2}$	8,264
3	$7,000 \times (1.06)^3$	8,337	$\dfrac{1}{(1.166)^3}$	5,259
			NPV	977

In this example, we undertook both compounding and discounting. The process could be simplified by multiplying the two elements. For example, in Year 2 we could multiply $(1.06)^2$ by $1/(1.166)^2$ to obtain a net of inflation discount factor of 0.8264 which, when multiplied by the cash flow in current prices, gives a present value of £8,264, as stated above. This gives rise to the formula for the real cost of capital, denoted by P.

Calculating the real cost of capital

$$(1 + P) = \frac{1 + M}{1 + I}$$

or

$$P = \frac{1 + M}{1 + I} - 1$$

where M is the money cost of capital, I is the inflation rate and P is the real cost of capital. In our example, this gives us a real cost of capital of:

$$P = \frac{1.166}{1.06} - 1 = 0.10, \text{ i.e. a rate of 10 per cent.}$$

Applying the real cost of capital gives the same NPV as before, as shown in Table 5.4.

While the latter approach may be simpler, it is not without difficulties. In business, the use of a single indicator of the rate of inflation, such as the Retail Price Index, may be inappropriate. Selling prices, wage rates, material costs and overheads rarely change at exactly the same rate each year. Rent may be fixed for a five-year period; selling prices may be held for more than a year. Furthermore, when taxation is introduced into the analysis, we find that tax relief on capital investment is not subject to inflation. Such complexities lead us to recommend that both cash flows and discount rates should include inflation.

Table 5.4 The real terms approach

Year	Cash flow current prices (£)	Real discount rate @ 10%	PV (£)
0	(18,000)	1	(18,000)
1	6,000	$\dfrac{1}{1.1}$	5,454
2	10,000	$\dfrac{1}{(1.1)^2}$	8,264
3	7,000	$\dfrac{1}{(1.1)^3}$	5,259
		NPV	977

Self-assessment activity 5.4

What is the impact of firms not adjusting their investment 'hurdle' rates for changing levels of inflation? How would you advise a company which employs a 20 per cent discount rate which was based on a calculation made when inflation was twice the current level?

(Answer in Appendix A at the back of the book)

5.5 TAXATION IS A CASH FLOW

In Chapter 2, we introduced the subject of taxation and its broad implications for financial management. In this section, we examine in greater depth the taxation considerations for capital investment projects.

Recall that in the UK, Corporation Tax is assessed by the Inland Revenue on the profits of the company after certain adjustments. While it is not calculated on a project basis by the Inland Revenue, the actual tax bill will increase with every new project offering additional profits and reduce with every project offering losses. Corporation Tax is charged on the profits, gains and income of an accounting period, usually the period for which accounts are made up annually. In arriving at taxable profits, a deduction is made for capital allowances on certain types of capital investment. Following the principle outlined earlier of identifying the incremental cash flow, we need to ask: by how much will the Corporation Tax bill for the company change each year as a result of the decision? To answer this, we must consider the tax charged on project operating profits and the tax relief obtained on the capital investment outlay.

Example: Taxation implications of Tiger 2000 for Woosnam plc

Woosnam plc invests in a new piece of equipment, the Tiger 2000, costing £40,000 on 1 January 2007. It intends to operate the equipment for four years when the scrap value will be zero. Expected net cash flows from the project are £10,000 in the first year and £20,000 for each of the next three years. The discount rate is 15 per cent and the rate of Corporation Tax is 30 per cent.

No tax position

If we ignore taxation (perhaps Woosnam is making losses and is unlikely to pay tax for some time), the net present value of the project's pre-tax cash flows is £8,390, as shown in Table 5.5. The positive NPV suggests that, on economic grounds, it should be accepted.

With Corporation Tax but no capital allowances

Most companies have to pay Corporation Tax on taxable profits. A recent change is that this tax is now paid in the same year as the related profits, usually by quarterly instalments. Hitherto, companies enjoyed a tax delay of at least a year, which meant that the tax payment would typically lag a full year behind the investment cash flows to which they relate. Most investments attract a capital allowance (equivalent to a depreciation charge) which reduces the tax bill. At this stage, we assume that the Tiger 2000 does not attract any capital allowances.

Table 5.5 Project Tiger 2000 (assuming no capital allowances)

Year	(1) Pre-tax cash flows £	(2) Tax @ 30%	(3) After-tax cash flows £	(4) Discount factor @ 15%	(1 × 4) PV pre-tax £	(3 × 4) PV post-taxt £
0	(40,000)	–	(40,000)	1.0	(40,000)	(40,000)
1	10,000	(3,000)	7,000	0.869	8,690	6,083
2	20,000	(6,000)	14,000	0.756	15,120	10,584
3	20,000	(6,000)	14,000	0.657	13,140	9,198
4	20,000	(6,000)	14,000	0.572	11,440	8,008
				NPV	8,390	(6,127)

Table 5.5 shows that after deducting tax to be paid, the NPV for the project falls sharply to –£6,127. It is no longer economically viable.

With Corporation Tax and capital allowances

annual writing-down allowances (WDAs)
Allowances for depreciation on capital expenditure allowed for tax purposes

For many types of capital investment, tax relief is granted on capital expenditure incurred. In the United Kingdom, this is in the form of **annual writing-down allowances (WDAs)**. Assume the writing-down allowance is:

Plant and machinery   25 per cent on the reducing balance
Industrial buildings   4 per cent on the initial cost

So for expenditure on machinery of £1,000, the allowance would be as follows:

Year	Tax allowance (£)	Written-down value at year-end (£)
1	25% × 1,000 = 250	750
2	25% × 750 = 188	562
3	25% × 562 = 141, etc.	422, etc.

Continued

Table 5.6 Woosnam plc – Tiger 2000 tax reliefs

End of accounting year	Tax written-down value £	Writing-down allowance £	30% tax relief £
1– Initial outlay	40,000		
WDA at 25%	10,000	10,000	3,000
	30,000		
2– WDA at 25%	7,500	7,500	2,250
	22,500		
3– WDA at 25%	5,625	5,625	1,688
	16,875		
4– Sale proceeds	–		
Balancing allowance	16,875	16,875	5,062
		40,000	12,000

Clearly, the tax allowances diminish over time. Companies are allowed to write assets down for tax purposes to their disposal value. Any discrepancy between written-down value (WDV) and disposal value may trigger a tax liability (balancing charge) or qualify for tax relief (balancing allowance). In the above example, disposal of the asset for £500 after three years would mean that the capital allowances have been over-generous to the extent of £78 (i.e. disposal value of £500 – WDV of £422). This *balancing charge* of £78 would then be subject to Corporation Tax. Disposal for, say, £300 would qualify the company for a *balancing allowance* of £122 (i.e. £422 – £300), a loss that would be set against the taxable profits.

Let us return to the Woosnam plc example, this time assuming that the Tiger 2000 attracts a 25 per cent writing-down allowance. Table 5.6 calculates the WDAs. Tax is payable in the same year as the investment cash flows to which they relate.

The difference between what the investment finally sold for (in this case zero) and the balance at the start of the year is a balancing allowance, which is treated in the same way as the writing-down allowance. (We will not introduce further complications such as the election to pool plant and machinery in this book.) A useful check is to see that the total WDA (column 3) equals the initial investment, and the tax benefit (column 4) on this total corresponds, in this case £40,000 at 30% = £12,000.

These cash flows can then be added to the earlier example, as in Table 5.7, showing that the investment offers a positive NPV of £2,185 after tax.

Table 5.7 Woosnam plc – Tiger 2000 with tax relief

Year	Pre-tax cash flows £	Tax at 30%	Tax relief on WDA £	Net cash flows £	Discount factor 15%	PV £
0	(40,000)	–		(40,000)	1.000	(40,000)
1	10,000	(3,000)	3,000	10,000	0.869	8,690
2	20,000	(6,000)	2,250	16,250	0.756	12,285
3	20,000	(6,000)	1,688	15,688	0.657	10,307
4	20,000	(6,000)	5,062	19,062	0.572	10,903
						2,185

How would the after-tax NPV differ were Woosnam plc a small or medium-size company? Such firms currently have a further tax incentive to invest by attracting a 40 per cent initial allowance, rather than 25 per cent. The effect is to reduce the tax bill in the early years, deferring it to later years. Because later cash flows are less valuable, this means that the NPV will increase.

Taxation therefore affects cash flows from investments. It is payable on taxable profits arising from the investment decision after deduction of capital allowances. However, as the FT article below makes clear, many of the most profitable firms pay little or no tax because of the tax allowances on interest and investment.

One-third of biggest businesses pays no tax

Almost a third of the UK's 700 biggest businesses paid no corporation tax in the 2005–06 financial year while another 30 per cent paid less than £10m each, an official study has found. Of the tax paid by these businesses, two-thirds came from just three industries – banking, insurance, and oil and gas – while the alcohol, tobacco, car and real estate sectors contributed only a few hundred million pounds.

This analysis by the spending watchdog could fuel the debate surrounding the UK's generous tax treatment of interest costs, which allow highly geared (borrowed) companies to cut their bills. This aspect of the tax regime has come under scrutiny when used by private equity-backed companies but is also used widely by multinationals with intra-group loans, which allow them to concentrate debt in countries where they can make the most use of the tax relief on interest costs.

Explanations for low tax bills of many large companies include low profitability, the availability of tax losses from previous years, high financing costs and the impact of capital investment and pension fund contributions.

Source: Vanessa Houlder, *Financial Times*, 28 August 2007.

Self-assessment activity 5.5

Your boss says: 'We only assess capital projects before tax. Every firm has to pay tax, so we can ignore it.' Do you agree?

(Answer in Appendix A at the back of the book)

5.6 USE OF DCF TECHNIQUES

It is a common misconception that the discounted cash flow approach is a relatively recent phenomenon. Historical records reveal an understanding of compound interest (upon which discounted cash flow techniques are based) as far back as the Old Babylonian period (*c.* 1800–1600 BC) in Mesopotamia. The earliest manuscripts setting out compound interest tables date back to the 14th century, while the first recorded reference to the net present value rule is found in a book by Stevin published in 1582.

In these early days, the application of discounted cash flow methods was restricted to financial investments such as loans and life assurance, where either the cash flows were known or their probabilities could be determined based on actuarial evidence. Only in the 19th century, with the Industrial Revolution well established, did the scale of capital investments lead some engineering economists to begin to apply discounted cash flow concepts to capital assets. However, in practice, these concepts were largely ignored until the early 1950s in the USA and the early 1960s in the UK.

Surveys between 1981 and 2003 provide a clearer picture of the changing trends in the practices of larger firms in the UK. Table 5.8 shows that, while all firms surveyed conduct financial appraisals on capital projects, the choice of method varies considerably, and most firms employ a combination of appraisal techniques.

The use of the NPV approach in large firms increased from 39 per cent in 1981 to 99 per cent by 2003. The IRR approach has also increased significantly over the same period. However, the use of the average accounting rate of return has not changed much over time.

Table 5.8 Capital investment evaluation methods in large UK firms

Firms using:	1981 (%)	1992 (%)	2003 (%)
Payback	81	94	96
Average accounting rate of return	49	50	60
Internal rate of return	57	81	89
Net present value	39	74	99

Sources: Pike (1988, 1996), Alkaraan and Northcott (2006).

The payback method has always been popular with smaller firms but the table shows that virtually all larger firms use it within their evaluation process. It is clear that firms do not normally rely on any single appraisal measure, but prefer to employ a combination of simple and more sophisticated techniques. DCF methods therefore complement, rather than substitute for, traditional approaches.

■ Dangers with DCF

While we have argued that DCF analysis offers a conceptually sound approach for appraising capital projects, a word of caution is appropriate.

From the emphasis devoted by most textbooks to advanced capital budgeting methods, one might be forgiven for assuming that successful investment is exclusively attributable to the correct evaluation method. However, DCF methods often create an illusion of exactness that the underlying assumptions do not warrant. As top management places more weight on the quantifiable element, there is a danger that the unquantifiable aspects of the decision, which frequently have a critical bearing on a project's success or failure, will be devalued. The human element is particularly important with regard to the project sponsor. The margin between a project's success or failure often hinges on the enthusiasm and commitment of the person sponsoring and implementing it.

Managers cannot afford to treat investment decisions in a vacuum, ignoring the complexities of the business environment. Any attempt to incorporate such complexities, however, will at best consist of abstractions from reality relying on generalised and simplified assumptions concerning business relationships and environments. A fundamental assumption underlying DCF methods is that decision-makers pursue the primary goal of maximising shareholders' wealth. For many firms, this may not be the case.

Common errors in applying DCF

- Discount rates are calculated on a pre-tax basis, while operating cash flows are calculated after tax.

- Discount rates are increased to compensate for non-economic statutory and welfare investments.

- Including interest charges in cash flows.

- Cash flows are specified in today's money (excluding inflation), while hurdle rates are based on the money cost of capital (including inflation).

- Managerial aversion to uncertainty frequently results in conservative project life and terminal value assumptions.

- Use of a uniform cut-off rate instead of a rate reflecting individual project risk. This often leads to rejection of low-risk/low-return replacement projects.

- Failure to include scrap values.

- Neglect of working capital movements.

Critical errors may often be seen in the way DCF theory is applied by managers. Usually, these errors are biased against investment. For example, many firms do not adjust their operating cash flows for inflation, but discount them at the money cost of capital, rather than the real rate of return before inflation. The effect is that cash flows in later years (typically the strong positive cash flows) are unduly deflated by the high discount factor, giving a lower net present value than should be the case.

Perhaps even more important, DCF methods ignore the value of investment options. This key topic is the subject of Chapter 8.

5.7 TRADITIONAL APPRAISAL METHODS

Managers have developed and come to rely upon simple rule-of-thumb approaches to analysing investment worth. Two of the most popular traditional methods are the *payback period* and the *accounting rate of return*, both of which were described in earlier chapters. Our present concern is to ask whether they have a valuable role to play in the modern capital budgeting process. Do they offer anything to the decision-maker that cannot be found in the DCF approaches?

■ Accounting rate of return (ARR)

We discussed the basic application of the ARR approach in Chapter 4. Table 5.8 reports that a little over half the companies surveyed employ the accounting rate of return approach in assessing investment decisions. This is not altogether surprising, given that the rate of return on capital is a very important financial goal in practice.

The ARR can be criticised on at least two counts: it uses accounting profits rather than cash flows, and it ignores the time-value of money. Nevertheless there has been a certain amount of support for the ARR in the literature. The absence of ARR leads to an inconsistency between the methods commonly used to *report* a firm's results and the techniques most frequently employed to *appraise* investment decisions. This is most acutely experienced where the divisional manager of an investment centre is expected to use a DCF approach in reaching investment decisions, while his or her short-term performance is being judged on a return on investment basis. Little wonder, then, that the divisional manager generally shows a marked reluctance to enter into any profitable long-term investment decisions that produce low returns in the early years.

A common assumption among managers is that the accounting rate of return and the internal rate of return produce much the same solutions. But while there is a relationship between a project's discounted return and the ARR, the relationship is not simple. Consider an investment costing £10,000 and generating an annual stream of net cash flows of £3,000. Assuming straight-line depreciation, the relationship between the internal rate of return and the accounting rate of return calculated on both the total investment and the average investment is as shown in Table 5.9.

Table 5.9 Relationship between ARR and IRR

Project duration (years)	5	10	20	25
IRR (%)	15.2	27.3	29.8	30
ARR on total investment (%)	10	20	25	27.5
Deviation from IRR	−5.2	−7.3	−4.8	−2.5
ARR on average investment (%)	20	40	50	55
Deviation from IRR	+4.8	+12.7	+20.2	+25

From this example we can see that the accounting rate of return on *total* investment consistently *understates*, and the accounting rate of return on *average* investment *overstates*, the internal rate of return. The case for retaining the accounting rate of return is, therefore, valid only when applied as a secondary criterion to highlight the likely impact on the organisation's profitability upon which the divisional manager is judged.

Residual income approach: Pluto Electronics

residual income
Operating profit less the charge for capital

While the average accounting return can be a misleading decision indicator for capital projects, it is possible to employ a profit-based approach that is in line with net present value. This involves calculating the **residual income** (RI), the profit less a cost of capital charge based on the book value of the assets employed.

Pluto Electronics has acquired the rights to manufacture a product for three years and has set up a new division to do so. The investment outlay is £60 million and annual cash flows are forecast to be £30 million. The company operates a straight-line depreciation policy and has a cost of capital of 10 per cent.

We can calculate the NPV (£m) as:

$$\text{NPV} = -60 + (30 \times \text{PVIFA}_{10,3}) = -60 + (30 \times 2.4868) = £14.6 \text{ m}$$

The same answer is given by calculating the residual income for each year and discounting at the cost of capital, as shown below. The annual profit is £10 million (i.e. £30 million cash flow less £20 million depreciation).

	£m	PV @ 10% (£m)
Yr 1 profit	10	
Investment outlay: £60 m		
10% capital charge on investment	(6)	
Residual income	4	× 0.909 = 3.636
Yr 2 profit	10	
Book value of assets: £40 m		
10% capital charge	(4)	
	6	× 0.826 = 4.956
Yr 3 profit	10	
Book value of assets: £20 m		
10% capital charge	(2)	
	8	× 0.751 = 6.008
Net present value		14.600

■ Payback period

Most finance texts have condemned the use of the payback period as potentially misleading in reaching investment decisions. However, Table 5.8 shows that it continues to flourish, being employed by most firms surveyed. Typically, the payback period required by firms is within 2–4 years (Arnold and Hatzopoulos 2000). Why is payback so popular? Does it possess certain qualities not so apparent in more sophisticated approaches?

The two main objections to payback (PB)

1 It ignores all cash flows beyond the payback period.

2 It does not consider the profile of the project's cash flows within the payback period.

Although such theoretical shortcomings could fundamentally alter a project's ranking and selection, the payback criterion possesses a number of merits.

1 PB estimates DCF return

The payback period provides a crude measure of investment profitability. When the annual cash receipts from a project are uniform, the payback reciprocal is the internal rate of return for a project of infinite life, or a good approximation to this rate for long-lived projects.

$$\text{IRR} = \frac{1}{\text{payback period}}$$

In the case of *very* long-lived projects where the cash inflows are, on average, spread evenly over the life of the project, the payback reciprocal is a reasonable proxy for the internal rate of return. For example, a project offering permanent cash savings and giving a four-year payback period with relatively stable annual cash returns will have approximately a 25 per cent internal rate of return (i.e. the reciprocal of payback period). However, if the project life is only ten years, the IRR would fall to 21 per cent – some four percentage points below the payback reciprocal. In fact, the payback reciprocal consistently *overstates* the true rate of return for finite project lives.

2 PB considers uncertainty

Whereas more sophisticated techniques attempt to model the uncertainty surrounding project returns, payback assumes that risk is time-related; the longer the period, the greater the chance of failure. General economic uncertainty makes the task of forecasting cash flows extremely difficult; but for the most part, cash flows are correlated over time. If the operating returns are below the expected level in the early years, they will probably also be below plan in the later years.

Discounted cash flow, as practised in most firms, ignores this increase in uncertainty over time. Early cash flows, therefore, have an important information content on the degree of accuracy of subsequent cash flows. By concentrating on the early cash flows, the payback approach analyses the data where managers have greater confidence. If such evaluation provides a different signal from DCF methods, it highlights the need for a more careful consideration of the project's risk characteristics.

3 PB as a screening device

Payback provides a relatively efficient method for ranking projects when constraints prevail. The most obvious constraint is the time that managers can devote to initial product screening. Only a handful of the investment ideas may stand up to serious and thorough financial investigation. Payback period serves as a simple, first-level screening device which, in the case of marginal projects, tends to operate in their favour and permits them to go forward for more thorough investigation.

Many firms also resort to payback period when experiencing liquidity constraints. Such a policy may make sense when funds are constrained and better investment ideas are in the pipeline. The attractiveness of investment proposals considered during the interim period will be a function more of their ability to pay back rapidly than of their overall profitability. This does not necessarily lead to optimal solutions.

4 PB assists communication

Managers feel more comfortable with payback period than with DCF. In the first place, it is simple to calculate and understand. The non-quantitative manager is reluctant to rely on the recommendations of 'sophisticated' models when he or she lacks both the time and expertise to verify such outcomes. Confidence in and commitment to a proposal depend to some degree on how thoroughly the evaluation model is comprehended. The payback method offers a convenient shorthand for the desirability of each investment that is understandable at all levels of the organisation: namely, how quickly will the project recover its initial outlay? Some firms use a project classification system in which the payback period indicates how rapidly proposals should be processed and put into operation.

Ultimately, it is the manager – not the method – who makes investment decisions and is appraised on their outcome. Payback period is particularly attractive to managers not only because it is convenient to calculate and communicate, but also because it signals good investment decisions at the earliest opportunity.

While the payback concept may lack the refinements of its more sophisticated evaluation counterparts, it possesses many endearing qualities that make it irresistible to most managers; hence its resilience.

Self-assessment activity 5.6

The following reasons for using payback were given by finance executives from three different companies:

'We use payback in support of other methods. It is not a sufficiently reliable tool to be used in isolation.'

'When liquidity is under pressure, payback is particularly relevant.'

'Payback helps to give some idea of the riskiness of the project – a long time to get one's money back is obviously more risky than a short time.'

To what extent do you agree with these views?

■ The appropriate discount rate

So far the examples used have simply stated the project discount rate based on the cost of capital, the rate of return required by investors. We discuss in some depth the appropriate discount rate in later chapters. Here, we outline one approach, the weighted average cost of capital (WACC).

This measures the rate of return that the firm must achieve in order to satisfy all of the people who invest in it. All of these investors incur an opportunity cost when placing their money in the hands of the firm's managers. This is the rate of return they could have achieved on the next best alternative investment.

Example: WACC

Wacky Ideas PLC Ltd produces novelty toys. It currently finances its business one-third through loans and two-thirds through equity and reserves. Looking ahead it does not expect to change this funding mix. The accountant estimates that the cost of equity is 12 per cent while the after-tax cost of borrowing is lower at 9 per cent. Given this information we can calculate the average cost of capital for the company, duly weighted according to the proportion of capital represented by equity and borrowings respectively. For Wacky this is:

Source of capital	Proportion		Cost of capital		Weighted Cost
Equity	67%	×	12%	=	8%
Loans	33%	×	9%	=	3%
WACC					11%

The weighted average cost of capital (WACC) approach multiplies the cost of each source of capital by the proportion of the total capital it represents. The results are summed to provide a WACC estimate of 11 per cent in Wacky's case. If we assume that each new investment project receives a slice of the total capital in the same 2:1 equity:borrowing proportion, and that the project has the same level of risk as the typical investments in the firm, we can apply a discount rate of 11 per cent in calculating the project's net present value. We leave the issue of determining the cost of capital for each source of finance to a later chapter.

Self-assessment activity 5.7

Major plc has 20 million £0.50 ordinary shares and irredeemable loan capital with a nominal value of £40 million in issue. The ordinary shares have a current market value of £2.40 per share and the loan capital is quoted at £80 per £100 nominal value. The cost of ordinary shares is estimated at 11% and the cost of loan capital is calculated to be 8%. The rate of corporation tax is 25%. What is the weighted average cost of capital for the company?

(Answer in Appendix A at the back of the book)

SUMMARY

One of the most difficult aspects of capital budgeting is identifying and gathering the relevant information for analysis. This chapter has examined the incremental cash flow approach to project analysis. Specific attention has been paid to the replacement decision and to the impact of inflation and taxation on investment decisions.

Key points

- Include only future, incremental cash flows relating to the investment decision and its consequences. This implies the following:

  1 Only additional fixed overheads are included.

  2 Depreciation (a non-cash item) is excluded.

  3 Sunk (or past) costs are not relevant.

  4 Interest charges are financing (not investment) cash flows and are therefore excluded from the cash flow profile.

  5 Opportunity costs (e.g. the opportunity to rent or sell premises if the proposal is not acceptable) are included.

- Replacement decision analysis examines the change in cash flows resulting from the decision to replace an existing asset with a new asset.

- Inflation can have important effects on project analysis. Two approaches are possible: (1) specify all cash flows at 'money-of-the-day' (i.e. including inflation) prices and discount at the money cost of capital, or (2) specify cash flows at today's prices and discount at the real (i.e. net of inflation) cost of capital. We recommend the former in most cases.

- Taxation is for most organisations a cash flow. Tax is calculated by deducting any cash benefits from tax relief on the initial capital expenditure from tax payable on additional cash flows. Care should be taken in estimating the timing of tax cash flows.

- In practice, most firms, particularly larger companies, employ a combination of DCF and traditional appraisal methods.

- One way of estimating the discount rate to be used is to calculate the firm's weighted average cost of capital.

Further reading

Most finance texts are not particularly strong on the applied aspects of capital budgeting. Pohlman *et al.* (1988) describe cash flow estimation practices in large firms. Levy and Sarnat (1994) has useful chapters, but the US tax system is employed. For a discussion on the investment appraisal criteria under low inflation, see the Bank of England (1994). Pointon (1980) examines the effect of capital allowances on investment, while Hodgkinson (1989) surveys tax treatment in corporate investment appraisal.

Studies on the investment practices of UK firms are well worth reading. See, for example, Pike (1982, 1988, 1996), McIntyre and Coulthurst (1985), Mills (1988), Pike and Wolfe (1988), Arnold and Hatzopoulos (2000), and Alkaraan and Northcott (2006). Useful references on the capital budgeting process are Cooper (1975), Pinches (1982) and Neale and Holmes (1991). Tomkins (1991) and Butler *et al.* (1993) explore the strategic and organisational aspects in greater depth.

Appendix
THE PROBLEM OF UNEQUAL LIVES: ALLIS PLC

Comparing mutually exclusive projects – such as retaining the old asset or replacing it with a new one – frequently involves the problem of assessing projects with different economic lives.

Allis plc is seeking to modernise and speed up its production process. Two proposals have been suggested to achieve this: the purchase of a number of forklift trucks and the acquisition of a conveyor system. The accountant has produced cost savings figures for the two proposals using a 10 per cent discount rate, shown in Table 5.10.

At first sight, the more expensive conveyor system appears more wealth-creating. But it is not appropriate to compare projects with different lives without making some adjustment. Two approaches can be employed for this: the replacement chain approach and the equivalent annual annuity approach.

replacement chain approach
The process of comparing like-for-like replacement decisions for mutually exclusive projects with different lives over a common time period

The **replacement chain approach** recognises that while, for convenience, we usually consider only the time-horizon of the proposal, most investments form part of a replacement chain over a much longer time period. We therefore need to compare mutually exclusive projects over a common period. In the example, this period is six years, two forklift truck proposals (one following the other) being equivalent to one conveyor system proposal. Assuming the cash flows for the original forklift trucks also apply to their replacements in Year 4 (a pretty big assumption, given inflation, improvements in technology, etc.), the replacement will produce a further NPV of £5,010 at the start of Year 4.

To convert this to the present value (i.e. Year 0) we must discount this figure to the present using the discount factor for 10 per cent for a cash flow three years hence:

$$PV = £5,010 \times PVIF_{(10,3)} = £5,010 \times 0.7513 = £3,764$$

The NPV for the forklift truck proposal, assuming like-for-like replacement after three years, is therefore £5,010 + £3,764 = £8,774. This is well in excess of the NPV of the conveyor system proposal of £6,538 over the same time period.

Table 5.10 Allis plc cash flows for two projects

Year	Forklift trucks	Conveyor system
0	(30,000)	(66,000)
1	10,000	12,000
2	15,000	20,000
3	18,000	20,000
4		18,000
5		15,000
6		15,000
NPV at 10%	5,010	6,538

Allis plc NPV comparison

Year		Forklift trucks £	Conveyor system £
1–3		5,010	
4–6	£5,010 × 0.7513	3,764	
1–6		8,774	6,538

equivalent annual annuity (EAA)
The constant annual cash flow offering the same present value as the project's net present value

A second approach, the **equivalent annual annuity (EAA)** approach, is easier than its name suggests. It seeks to determine the constant annual cash flow that offers the same present value as the project's NPV. This is found by dividing the project's NPV by the relevant annuity discount factor (i.e. 10 per cent over three years):

$$EAA = \frac{NPV}{PVIFA_{(10,3)}}$$

For the forklift proposal:

$$EAA = \frac{£5,010}{2.4869} = £2,015$$

For the conveyor system proposal:

$$EAA = \frac{NPV}{PVIFA_{(10,6)}} = \frac{£6,538}{4.3553} = £1,501$$

The forklift proposal offers the higher equivalent annual annuity and is to be preferred. Assuming continuous replacement at the end of their project lives, the NPVs for the projects over an infinite time-horizon are found by dividing the EAA by the discount rate:

NPV forklift truck = £2,015/0.10 = £20,150

NPV conveyor = £1,501/0.10 = £15,010

QUESTIONS

 **myfinancelab** *Questions with an icon are also available for practice in myfinancelab with additional supporting resources.*

Questions with a coloured number have solutions in Appendix B on page 732.

1 Most capital budgeting textbooks strongly recommend NPV, but most firms prefer IRR. Explain.

2 A project costing £20,000 offers an annual cash flow of £5,000 over its life.

(a) Calculate the internal rate of return using the payback reciprocal assuming an infinite life.
(b) Use tables to test your answer assuming the project life is (i) 20 years, (ii) eight years.
(c) What conclusions can be drawn as to the suitability of the payback reciprocal in measuring investment profitability?

? 3 Your firm uses the IRR method and asks you to evaluate the following mutually exclusive projects:

	Year				
Cash flows (£)	0	1	2	3	4
Proposal L	−47,232	20,000	20,000	20,000	20,000
Proposal M	−47,232	0	10,000	20,000	65,350

Using the appropriate IRR method, evaluate these proposals assuming a required rate of return of 10 per cent. Compare your answer with the net present value method.

? 4 State two ways in which inflation can be handled in investment analysis. Which way would you recommend and why?

5 Bramhope Manufacturing Co. Ltd has found that, after only two years of using a machine for a semi-automatic process, a more advanced model has arrived on the market. This advanced model will not only produce the current volume of the company's product more efficiently, but allow increased output of the product. The existing machine had cost £32,000 and was being depreciated straight-line over a ten-year period, at the end of which it would be scrapped. The market value of this machine is currently £15,000 and there is a prospective purchaser interested in acquiring it.

The advanced model now available costs £123,500 fully installed. Because of its more complex mechanism, the advanced model is expected to have a useful life of only eight years. A scrap value of £20,500 is considered reasonable.

A comparison of the existing and advanced model now available shows the following:

	Existing machine	Advanced model
Capacity p.a.	200,000 units	230,000 units
	£	£
Selling price per unit	0.95	0.95
Production costs per unit		
Labour	0.12	0.08
Materials	0.48	0.46
Fixed overheads (allocation of portion of company's fixed overheads)	0.25	0.16

The sales director is of the opinion that additional output could be sold at 95p per unit.

If the advanced model were to be run at the old production level of 200,000 units per annum, the operators would be freed for a proportionate period of time for reassignment to the other operations of the company.

The sales director has suggested that the advanced model should be purchased by the company to replace the existing machine.

The required return is 15 per cent.

(i) You are required to calculate:

    (a) payback period
    (b) the net present value
    (c) the internal rate of return (to the nearest per cent)

(ii) What recommendation would you make to the sales director? What other considerations are relevant?

6 Argon Mining plc is investigating the possibility of purchasing an open-cast coal mine in South Wales at a cost of £2.5 million which the British Government is selling as part of its privatisation programme. The company's surveyors have spent the last three months examining the potential of the mine and have incurred costs to date of £0.2 million. The surveyors have prepared a report which states that the company will require equipment and vehicles costing £12.5 million in order to operate the mine and that these assets can be sold for £2.5 million in four years time when the coal reserves of the mine are exhausted.

The assistant to the Chief Financial Officer of the company has prepared the following projected profit and loss accounts for each year of the life of the mine.

Projected Profit and Loss Accounts (£m)

	Year			
	1	2	3	4
Sales	9.4	9.8	8.5	6.3
less Wages and salaries	(2.3)	(2.5)	(2.6)	(1.8)
Selling and distribution costs	(1.3)	(1.2)	(1.5)	(0.6)
Materials and consumables	(0.3)	(0.4)	(0.4)	(0.2)
Depreciation and equipment	(2.5)	(2.5)	(2.5)	(2.5)
Head office expenses	(0.6)	(0.6)	(0.6)	(0.6)
Survey costs	(0.4)			
Interest charges	(1.2)	(1.2)	(1.2)	(1.2)
Net profit (loss)	0.8	1.4	(0.3)	(0.6)

In his report to the Chief Financial Officer, the assistant recommends that the company should not proceed with the acquisition of the mine as the profitability of the proposal is poor.

The following additional information is available:

(i) The project will require an investment of £0.5 million of working capital from the beginning of the project until the end of the useful life of the mine.

(ii) The wages and salaries expenses include £0.5 million of working capital in Year 1 for staff who are already employed by the company but who would be without productive work until Year 2 if the project does not proceed. However, the company has no intention of dismissing these staff. After Year 1, these staff will be employed on another project of the company.

(iii) One-third of the head office expenses consists of amounts directly incurred in managing the new project and two-thirds represents an apportionment of other head office expenses to the project to ensure that it bears a fair share of these expenses.

(iv) The survey costs include those costs already incurred to date, and which are to be written off in the first year of the project, as well as costs to be incurred in the first year if the project is accepted.

(v) The interest charges relate to finance required to purchase the equipment and vehicles necessary to carry out the project.

(vi) After the mine has been exhausted, the company will be required to clean up the site and to make good the damage to the environment resulting from its mining operations. The company will incur costs of £0.4 million in Year 5 in order to do this.

The company has a cost of capital of 12 per cent.
Ignore taxation.

Required

(a) Using what you consider to be the most appropriate investment appraisal method, prepare calculations which will help the company to decide whether or not to proceed with the project.
(b) State, giving reasons, whether you think the project should go ahead.
(c) Explain why you consider the investment appraisal method selected in (a) above to be most appropriate for evaluating investment projects.

7 Consolidated Oilfields plc is interested in exploring for oil near the west coast of Australia. The Australian government is prepared to grant an exploration licence to the company for a five-year period for a fee of £300,000 p.a. The option to acquire the rights must be taken immediately, otherwise another oil company will be granted the rights. However, Consolidated Oilfields is not in a position to commence operations immediately, and exploration of the oilfield will not start until the beginning of the second year. In order to carry out the exploration work, the company will require equipment costing £10,400,000, which will be made by a specialist engineering company. Half of the equipment cost will be payable immediately and half will be paid when the equipment has been built and tested to the satisfaction of Consolidated Oilfields. It is estimated that the second instalment will be paid at the end of the first year. The company commissioned a geological survey of the area and the results suggest that the oilfield will produce relatively small amounts of high-quality crude oil. The survey cost £250,000 and is now due for payment.

The assistant to the project accountant has produced the following projected profit and loss accounts for the project for Years 2–5 when the oilfield is operational.

	£000	£000	£000	£000	£000	£000	£000	£000
		2		**3**		**4**		**5**
Sales		7,400		8,300		9,800		5,800
Less expenses								
Wages and salaries	550		580		620		520	
Materials and consumables	340		360		410		370	
Licence fee	600		300		300		300	
Overheads	220		220		220		220	
Depreciation	2,100		2,100		2,100		2,100	
Survey cost written off	250		–		–		–	
Interest charges	650		650		650		650	
		4,710		4,210		4,300		3,160
Profit		2,690		4,090		5,500		2,640

The following additional information is available:

1 The licence fee charge appearing in the accounts in Year 2 includes a write-off for all the annual fee payable in Year 1. The licence fee is paid to the Australian government at the end of each year.
2 The overheads contain an annual charge of £120,000, which represents an apportionment of head office costs. This is based on a standard calculation to ensure that all projects bear a fair share of the central administrative costs of the business. The remainder of the overheads relate directly to the project.

3 The survey costs written off relate to the geological survey already undertaken and due for payment immediately.
4 The new equipment costing £10,400,000 will be sold at the end of the licence period for £2,000,000.
5 The project will require a specialised cutting tool for a brief period at the end of Year 2, which is currently being used by the company in another project. The manager of the other project has estimated that he will have to hire machinery at a cost of £150,000 for the period the cutting tool is on loan.
6 The project will require an investment of £650,000 working capital from the end of the first year to the end of the licence period.

The company has a cost of capital of 10 per cent.
Ignore taxation.

Required
(a) Prepare calculations that will help the company to evaluate further the profitability of the proposed project.
(b) State, with reasons, whether you would recommend that the project be undertaken.
(c) Explain how inflation can pose problems when appraising capital expenditure proposals, and how these problems may be dealt with.

(Certified Diploma)

8 You are the chief accountant of Deighton plc, which manufactures a wide range of building and plumbing fittings. It has recently taken over a smaller unquoted competitor, Linton Ltd. Deighton is currently checking through various documents at Linton's head office, including a number of investment appraisals. One of these, a recently rejected application involving an outlay on equipment of £900,000, is reproduced below. It was rejected because it failed to offer Linton's target return on investment of 25 per cent (average profit-to-initial investment outlay). Closer inspection reveals several errors in the appraisal.

Evaluation of profitability of proposed project NT17
(all values in current year prices)

Item (£000)	0	1	2	3	4
Sales		1,400	1,600	1,800	1,000
Materials		(400)	(450)	(500)	(250)
Direct labour		(400)	(450)	(500)	(250)
Overheads		(100)	(100)	(100)	(100)
Interest		(120)	(120)	(120)	(120)
Depreciation		(225)	(225)	(225)	(225)
Profit pre-tax		155	255	355	55
Tax at 33%		(51)	(84)	(117)	(18)
Post-tax profit		104	171	238	37
Outlay					
Stock	(100)				
Equipment	(900)				
Market research	(200)				
	(1,200)				

$$\text{Rate of return} = \frac{\text{Average profit}}{\text{Investment}} = \frac{£138}{£1,200} = 11.5\%$$

You discover the following further details:

1 Linton's policy was to finance both working capital and fixed investment by a bank overdraft. A 12 per cent interest rate applied at the time of the evaluation.
2 A 25 per cent writing-down allowance (WDA) on a reducing balance basis is offered for new investment. Linton's profits are sufficient to utilise fully this allowance throughout the project.
3 Corporation tax is paid a year in arrears.
4 Of the overhead charge, about half reflects absorption of existing overhead costs.

5 The market research was actually undertaken to investigate two proposals, the other project also having been rejected. The total bill for all this research has already been paid.

6 Deighton itself requires a nominal return on new projects of 20 per cent after taxes, is currently ungeared and has no plans to use any debt finance in the future.

Required

Write a report to the finance director in which you:

(a) Identify the mistakes made in Linton's evaluation.

(b) Restate the investment appraisal in terms of the post-tax net present value to Deighton, recommending whether the project should be undertaken or not.

(ACCA)

9 (a) Explain how inflation affects the rate of return required on an investment project, and the distinction between a real and a nominal (or 'money terms') approach to the evaluation of an investment project under inflation.

(b) Howden plc is contemplating investment in an additional production line to produce its range of compact discs. A market research study, undertaken by a well-known firm of consultants, has revealed scope to sell an additional output of 400,000 units p.a. The study cost £0.1 million, but the account has not yet been settled.

The price and cost structure of a typical disc (net of royalties) is as follows:

	£	£
Price per unit		12.00
Costs per unit of output		
Material cost per unit	1.50	
Direct labour cost per unit	0.50	
Variable overhead cost per unit	0.50	
Fixed overhead cost per unit	1.50	
		(4.00)
Profit		8.00

The fixed overhead represents an apportionment of central administrative and marketing costs. These are expected to rise in total by £500,000 p.a. as a result of undertaking this project. The production line is expected to operate for five years and require a total cash outlay of £11 million, including £0.5 million of materials stocks. The equipment will have a residual value of £2 million. Because the company is moving towards a JIT stock management policy, it is expected that this project will involve steadily reducing working capital needs, expected to decline at about 3 per cent p.a. by volume. The production line will be accommodated in a presently empty building for which an offer of £2 million has recently been received from another company. If the building is retained, it is expected that property price inflation will increase its value to £3 million after five years.

While the precise rates of price and cost inflation are uncertain, economists in Howden's corporate planning department make the following forecasts for the average annual rates of inflation relevant to the project:

Retail Price Index	6% p.a.
Disc prices	5% p.a.
Material prices	3% p.a.
Direct labour wage rates	7% p.a.
Variable overhead costs	7% p.a.
Other overhead costs	5% p.a.

Note: You may ignore taxes and capital allowances in this question.

Required

(a) Given that Howden's shareholders require a real return of 8.5 per cent for projects of this degree of risk, assess the financial viability of this proposal.

(b) Briefly discuss how inflation may complicate the analysis of business financial decisions.

(ACCA)

Practical assignment: Engineering Products case study

The following case study brings together many of the issues raised in Part 2 of this book on the analysis of strategic investment decisions. In answering certain parts the student should also read Chapters 6 and 7.

Roger Davis, the newly appointed financial analyst of the Steel Tube division of Engineering Products plc, shut his office door and walked over to his desk. He had just 24 hours to re-examine the accountant's profit projections and come up with a recommendation on the proposed new computer numerically controlled (CNC) milling machine.

At the meeting he had just left, the managing director made it quite clear: 'If the project can't pay for itself in the first three years, it's not worth bothering with.' Davis was unhappy with the accountant's analysis which showed that the project was a loss maker. But as the MD said, 'Unless you can convince me by this time tomorrow that spending £240,000 on this capital project makes economic sense, you can forget the whole idea.'

His first task was to re-examine the accountant's profitability forecast (Table 5.11) in the light of the following facts that emerged from the meeting:

1 Given the rapid developments in the market, it was unrealistic to assume that the product had more than a four-year life. The machinery would have no other use and could not raise more than £20,000 in scrap metal at the end of the project.
2 The opening stock in Year 1 would be acquired at the same time as the machine. All other stock movement would occur at the year ends.
3 This type of machine was depreciated over six years on a straight-line basis.
4 Within the 'other production expenses' were apportioned fixed overheads equal to 20 per cent of labour costs. As far as could be seen, none of these overheads were incurred as a result of the proposal.
5 The administration charge was an apportionment of central fixed overheads.

Table 5.11 Profit projection for CNC milling machine (£000)

	Year			
	1	2	3	4
Sales	400	600	800	600
Less costs				
Materials				
Opening stock	40	80	80	60
Purchases	260	300	360	240
Closing stock	(80)	(80)	(60)	–
Cost of sales	220	300	380	300
Labour	80	120	120	80
Other production expenses	80	90	92	100
Depreciation	40	40	40	40
Administrative overhead	54	76	74	74
Interest on loans to finance the project	22	22	22	22
Total cost	496	648	728	616
Profit (loss)	(96)	(48)	72	(16)

Later that day, Davis met the production manager, who explained that if the new machine was installed, it would have sufficient capacity to enable an existing machine to be sold immediately for £20,000 and to create annual cash benefits of £18,000. However, the accountant had told him that, with the machine currently standing in the books at £50,000, the company simply could not afford to write off the asset against this year's slender profits. 'We'd do better to keep it operating for another four years, when its scrap value will produce about £8,000,' he said.

Davis then raised the proposal with the marketing director. It was not long before two new pieces of information emerged:

(a) To stand a realistic chance of hitting the sales forecast for the proposal, marketing would require £40,000 for additional advertising and sales promotion at the start of the project and a further £8,000 a year for the remainder of the project's life. The sales forecast and advertising effort had been devised in consultation with marketing consultants whose bill for £18,000 had just arrived that morning.

(b) The marketing director was very concerned about the impact on other products within the product range. If the investment went ahead, it would lead to a reduction in sales value of a competing product of around £60,000 a year. 'With net profit margins of around 10 per cent and gross margins (after direct costs) of 25 per cent on these sales, this is probably the "kiss of death" for the CNC proposal,' Davis reflected.

The Steel Tube division was a profitable business operating within an attractive market. The investment, which employed new technology, had recently been identified as part of the group's core activities. The chief engineer felt that once they had got to grips with the new technology it should deliver improved product quality, and greater flexibility, enabling shorter production runs and other benefits.

The latest accounts for the division showed a 16 per cent return on assets, but the MD talked about a three-year payback requirement. His phone call to the finance director at head office, to whom this proposal would eventually be sent, was distinctly unhelpful: 'We have, in the past, found that whenever we lay down a hurdle rate for divisional capital projects, it merely encourages unduly optimistic estimates from divisional executives eager to promote their pet proposals. So now we give no guidelines on this matter.'

Davis decided to use 10 per cent as the required rate of return, made up of 6 per cent currently obtainable from risk-free government securities plus a small element to compensate for risk. Davis went home that evening with a very full briefcase and a number of unresolved questions.

1   How much of the information which he had gathered was really relevant to the decision?
2   What was the best approach to assessing the economic worth of the proposal? The company used payback and return on investment, but he felt that discounted cash flow techniques had some merit.
3   Cash was particularly limited this year and acceptance of this project could mean that other projects would have to be deferred. How should this be taken into consideration?
4   How should the strategic factors be assessed?
5   What about tax? Engineering Products plc pays corporation tax at 30 per cent and annual writing-down allowances of 25 per cent on the reducing balance may be claimed. The existing machine has a nil value for tax purposes and tax is payable in the same year as the cash flows to which it relates.

Required
Prepare the case, with recommendations, to be presented by Davis at tomorrow's meeting. The report should address points 1–5 above.

 *Now retake your diagnostic test for Chapter 5 to check your progress and update your study plan.*

6

Investment strategy and process

A Mickey Mouse investment?

The Euro Disney theme park opened with all the razzmatazz of a Disney spectacular. However, from an investment perspective, it was a spectacular flop in its first few years of operation. It planned to make a profit in its opening year. Instead, Euro Disney produced losses in each of its first three years, with the second year producing a staggering loss of FFr 5.3 billion.

The park simply failed to attract sufficient visitors to cover its initial costs. In 2007, it eventually reached the visitor target of 13.3 million set back in 1996. The northern French climate, rising franc, economic recession, Gallic hostility to American culture and high admission cost all contributed to the lack of visitors.

After much huffing and puffing, Euro Disney pulled off in September 2004 its second debt restructuring in a decade, just in time to avoid default by failing to meet the creditor deadline. That financial rescue resolved the immediate crisis at Europe's biggest tourist attraction, but has yet to guarantee its future.

This seemed a far cry from its initial public offer for shares launched on many of the stock exchanges throughout Europe. The offer document indicated an internal rate of return of 13.3 per cent between 1992 and 2017 based on an inflation assumption of 5 per cent per annum. A fairytale start to Euro Disney life, followed by a financially delinquent adolescence – but would Euro Disney ever produce the returns to make investors happy ever after? They are still waiting. The chief executive, Karl Holz, said, 'In 2008, we will continue to execute our growth strategy and remain focused on driving this business toward profitability.'

Learning objectives

This chapter examines strategic issues in investment and the investment process:

■ How strategy shapes investment decisions.

■ Evaluating new technology and environmental projects.

■ The investment decision and control process.

■ Post-audit reviews.

 Complete your diagnostic test for Chapter 6 now to create your personal study plan.

6.1 INTRODUCTION

A company's ability to succeed in highly competitive markets depends to a great extent on its ability to regenerate itself through wealth-creating capital investment decisions compatible with business strategy. In recent years, most of the combined internal and external funds generated by UK firms have been committed to fixed capital investment. Applying such resources to long-term capital projects in anticipation of an adequate return – although a hazardous step – is essential for the vitality and well-being of the organisation.

6.2 STRATEGIC CONSIDERATIONS

Where do positive NPV projects come from? By definition, a positive NPV means that a project offers returns superior to those obtainable in the capital market on investments of comparable risk. In the short run, it is quite feasible to find capital projects that do just this, but in a competitive market it will not be long before other firms make similar investments, thereby ensuring that any superior returns are not perpetuated.

Selecting wealth-creating capital projects is no different from picking undervalued shares on the stock market. Earlier discussion on market efficiency argued that this is possible only if there are capital market imperfections that prevent asset prices reflecting their equilibrium values.

Companies that consistently create projects with high NPVs have developed a sustainable competitive advantage arising from imperfections in the product and factor markets. These imperfections generally take the form of entry barriers that discourage new entrants. Successful investments are therefore investments that help create, preserve or enhance competitive advantage.

Porter (1985) argues that there are really only three coherent strategies for strategic business units:

1 To be the lowest-cost producer.
2 To focus on a niche or segment within the market.
3 To differentiate the product range so that it does not compete directly with lower-cost products.

Investment expenditure that helps achieve the appropriate strategy is likely to generate superior returns. For example, Coca-Cola invests enormous sums into its product differentiation strategy through its brand support.

strategic portfolio analysis
Assessing capital projects within the strategic business context and not simply in financial terms

Capital projects should be viewed not simply in isolation, but within the context of the business, its goals and strategic direction. This approach is often termed **strategic portfolio analysis**.

The attractiveness of investment proposals coming from different sectors of the firm's business portfolio depends not only on the rate of return offered, but also on the strategic importance of the sector. Business strategies are formulated that involve the allocation of resources (capital, labour, plant, marketing support etc.) to these business units. The allocation may be based on analysis of the market's attractiveness and the firm's competitive strengths, such as the **McKinsey–General Electric portfolio matrix** outlined in Figure 6.1.

McKinsey–General Electric portfolio matrix
An approach for assessing projects within the wider strategic context which focuses on the market attractiveness and business strength of the product and business unit relating to the capital proposal

The attractiveness of the market or industry is indicated by such factors as the size and growth of the market, ease of entry, degree of competition and industry profitability for each strategic business unit. Business strength is indicated by a firm's market share and its growth rate, brand loyalty, profitability, and technological and other comparative advantages. Such analysis leads to three basic strategies:

1 Invest in and strengthen businesses operating in relatively attractive markets. This may mean heavy expenditures on capital equipment, working capital, research and development, brand development and training.

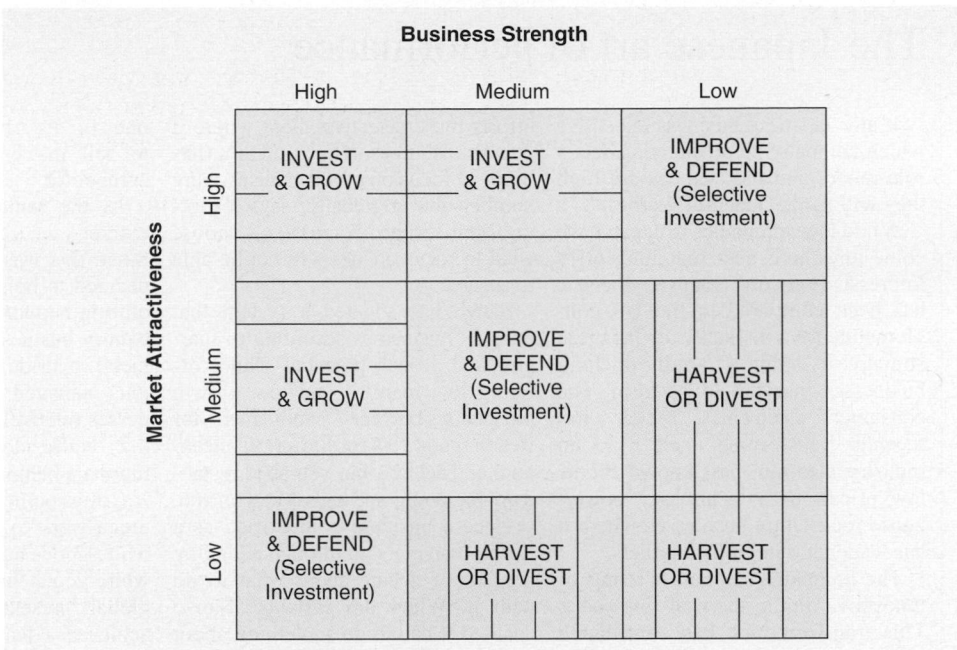

Figure 6.1 McKinsey–GE portfolio matrix

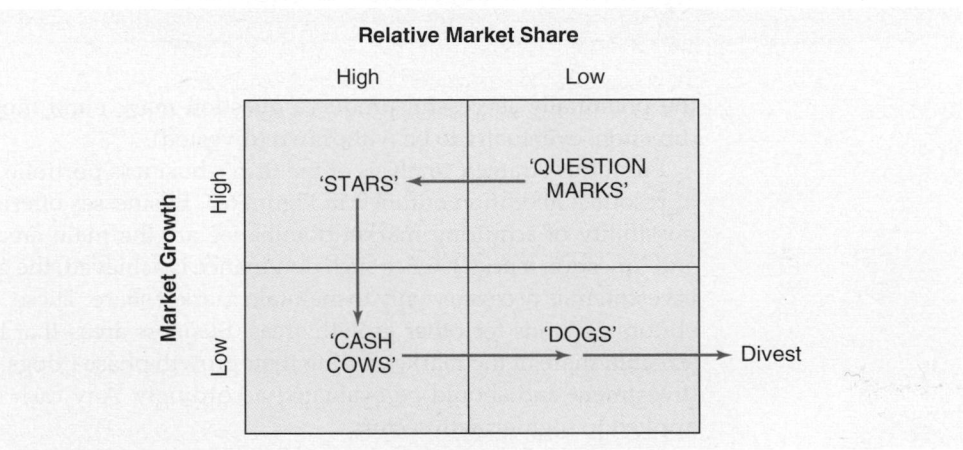

Figure 6.2 Normal progression of product over time

2 Where the market is somewhat less attractive and the business less competitive (the diagonal unshaded boxes), the business strategy is to get the maximum out of existing resources. The financial strategy is therefore to maximise or maintain cash flows, while incurring capital expenditures mainly of a replacement nature. Tight control over costs and management of working capital leads to higher levels of profitability and cash flow.

3 The remaining businesses have little strategic quality and may, in the longer term, be run down or divested unless action can be taken to improve their attractiveness.

Boston Consulting Group approach
An approach for assessing capital proposals based on the market growth and market share of the products relating to the proposal

An alternative is the **Boston Consulting Group approach**, which describes the business portfolio in terms of relative market share and rate of growth (see Figure 6.2). This matrix identifies four product markets within which a firm may operate: (1) 'stars' (high market share, high market growth), (2) 'cash cows' (high market share, low market growth), (3) 'question marks' (low market share, high market growth) and (4) 'dogs' (low market share, low market growth). The normal progression starts with

The Japanese art of performance

Ask any Japanese business executive which company he or she considers a role model and the chances are high they will name General Electric.

While few companies in Japan have come anywhere near matching GE's impressive record, Sanyo Electric has been compared to the US conglomerate for a distinctly un-Japanese strategy: its habit of rapidly ditching businesses that fail to perform. The consumer electronics maker has recently transformed itself from an industry also-ran, best known for its low prices, to a technology powerhouse focused on businesses where it has leadership in global markets.

The company has had a string of innovative hits in the past few years. This transformation has contributed to a 21 per cent rise in operating profits on record sales. Yukinori Kuwano, Sanyo's chief executive, attributes Sanyo's recent success to its recent efforts to be selective about where it puts its resources. 'Our main aim (has been) to focus on products that we are number one in globally,' says a smiling Mr Kuwano. 'Unless you choose what to focus on you will not be able to survive.'

Sanyo has adopted a system that rates its businesses according to margins and growth potential. Those offering low margins and low growth prospects become candidates for weeding out. 'It's easy to say "concentrate and select", but you need a standard for doing so,' says Mr Kuwano.

Once a business is identified as a 'loser' managers go through a lengthy process of debate about what to do with it. When, for instance, Sanyo decided it had to do something about a loss-making vending machine business that was number two in its market, managers considered several options, including an acquisition of one of its rivals, before deciding to sell the business to its leading competitor.

At the same time Sanyo led the market in rechargeable batteries. Since this was a promising sector it decided to bolster its position by acquiring Nippon Batteries' lithium ion battery business and Toshiba's nickel metal hydride business. Sanyo has thus managed to streamline its businesses relatively quickly while ensuring staff clearly understand the rationale behind the decisions.

Critics point out that Sanyo still has much work to do on its loss-making white-goods business. If it can build a white-goods business with a strong global presence, Sanyo will have achieved a feat that so far even GE has failed to pull off.

Source: Michiyo Nakamoto, *Financial Times*, 18 May 2004, p. 12.

the potentially successful product ('question mark') and moves in an anticlockwise direction, eventually to be withdrawn (divested).

From this strategic analysis of the firm's business portfolio, we suggest the pattern of resource allocation outlined in Figure 6.3. Businesses offering high growth and the possibility of acquiring market dominance are the main areas of investment ('stars' and 'question marks'). Once such dominance is achieved, the growth rate declines and investment is necessary only to maintain market share. These 'cash cows' become generators of funds for other growth areas. Business areas that have failed to achieve a sizeable share of the market during their growth phase ('dogs') become candidates for divestment and should be evaluated accordingly. Any cash so generated should be applied to high-growth sectors.

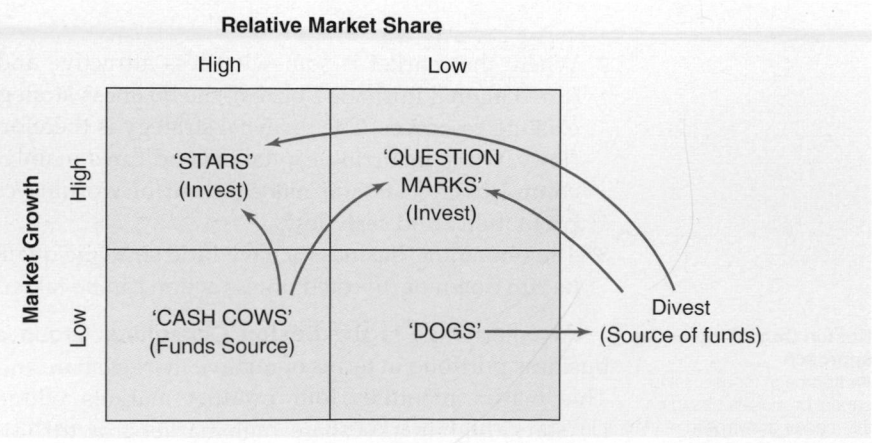

Figure 6.3 Investment strategy

Having developed its investment strategy, management can assess how individual projects fit into the firm's long-term strategic plan. Project appraisal – or, in the case of capital shortage, project ranking – is not only judged according to rates of return. Many companies will reject projects offering high returns because projects fall outside strategic thinking. Ultimately, the capital budget must tie up with corporate strategy so that each project contributes to an element of that strategy.

Shareholder value analysis (SVA) is a valuable planning tool and guide for strategic decision-making. It is basically an extension of the NPV approach where the focus is on business units, strategies and financial goals. A business is viewed as a portfolio of investment projects, but the emphasis is placed on maximising the value of strategic business units, not merely that of the capital projects within them.

Rather than dwell on short-term measures, such as annual earnings per share or return on capital, SVA manages cash flows over time. It is this long-term cash flow that determines the long-term value of the business. The value of adopting a new strategy is assessed in terms of the difference in the value of the business before and after implementation.

Self-assessment activity 6.1

Why is it important to view capital budgeting within a strategic framework?

(Answer in Appendix A at the back of the book)

Corporate mortality at Microsoft

Bill Gates is the world's richest man and head of Microsoft, arguably the world's most successful company. Yet he is brutally aware of his company's vulnerability, even when most people regard the business as an indomitable near monopolist at the heart of the software industry. As he said recently: 'Someday Microsoft will go out of business, it's only a matter of time. Will it happen in my lifetime, while I am still deeply involved in the company? I hope not. I wake up every day working hard to reduce the probability of this. Companies are mortal' (Bennet, 2001).

Gates realises that it is this awareness of corporate mortality that forces the company to continue its punishing innovation cycle.

Over the years, there has been talk that Microsoft will use its vast cash resources and market capitalisation to diversify into other industries such as banking and telecoms. Gates rejects this. So why has it been so successful? Of course, at one level, the main reason is its ability to innovate and keep ahead of the competition. But Gates explains why the company has the highest market capitalisation in terms of its high-volume, low-cost products.

■ Project finance

Large-scale infrastructure projects, such as the construction of tunnels, roads and power stations, are often funded through project finance. Here the operation is financed and controlled separately from the operations of the constructor or user. The obvious benefit to the company is that creditors of the project only have claims on the project's cash flows, not those of the companies involved in the construction process. Following the Private Finance Initiative (PFI), many public sector projects have been funded in this manner.

6.3 ADVANCED MANUFACTURING TECHNOLOGY (AMT) INVESTMENT

An area where strategic decision-making is often required is the evaluation of advanced manufacturing technology projects. Strategic investment appraisal links corporate strategy to the costs and benefits associated with AMT and other strategic decisions. Frequently, it is insufficient to consider only financial issues; many of the benefits are less tangible and hard to quantify.

The past 30 years have seen growth in new technology capital projects, creating different challenges for the decision-maker. AMT projects offer a range of less tangible benefits: for example, greater flexibility with reduced 'downtime' on production changeover. Greater flexibility enables businesses to meet the challenge of increasing competition, shorter product life cycles and satisfying customers' specific requirements. AMT offers a flexible manufacturing system (FMS), in which a sequence of production operations are computer-controlled to respond to ever-changing production and design requirements.

AMT terminology

AMT investment helps companies achieve competitive advantage through a number of computer-controlled automated process technologies:

- **Computer-aided design (CAD)** helps the engineer test and modify a design from any viewpoint.
- **Computer-integrated manufacture (CIM)** brings together the manufacturing process and the computer.
- **Computer-numerically controlled (CNC)** machines can be easily reprogrammed to perform different tasks.
- **Flexible manufacturing systems (FMS)** enable the firm to produce a far greater variety of components quickly.
- **Direct numerical control (DNC)** systems connect a number of numerically controlled machines by computer.

AMT example: Foster Engineering Ltd

Foster Engineering Ltd is considering introducing a flexible manufacturing system (FMS) to modernise production in a department currently using conventional metal-working machinery. The declining market and the awareness that its main competitors have recently introduced new technology have made the need to modernise plant facilities an urgent priority.

An AMT proposal has been put forward, offering an FMS capable of producing the present output. It involves two machining centres with CNC lathes, a conveyor system for transferring components and a computer for scheduling, tooling and overall control. The total investment would cost £2.4 million, half being incurred at the start and the other half after one year, at which point the existing machinery could be sold for £50,000. Any benefit would arise from Year 2 onwards for five years. The two quantified benefits are as follows:

1 A reduction in the number of skilled workers from 50 to 15. The annual cost of a skilled worker is £20,000 (savings of 35 × £20,000 = £700,000 p.a.).
2 Savings in scrap and re-work of £50,000 p.a.

The company requires all projects to offer a positive net present value discounted at 15 per cent.

The accountant produces the following evaluation showing that the FMS proposal has a negative NPV of £159,000 and fails to meet corporate investment criteria:

FMS proposal

		(£000)	
Annual benefit			
PV at Year 1	£700,000 × PVIFA$_{(15\%, 5\text{ yrs})}$		
	£700,000 × 3.352	2,346	
PV at Year 0	£2,346 × PVIF$_{(15\%, 1\text{ yr})}$		
	£2,346 × 0.87		2,041
Initial investment			
Year 0		(£1,200)	
Year 1 (£1,200 − 50) × 0.87		(£1,000)	(2,200)
NPV			(159)

An incensed production engineer in Foster Engineering, on hearing that the proposal is unacceptable, points to the 'intangible' benefits that the FMS will offer:

- Improved quality leading to a significant, but unknown, reduction in sales returns through faulty workmanship.
- Reduced stock and work-in-progress, enabling improved shopfloor layout, greater space and a lower working capital requirement.
- Lower total manufacturing time, enabling the company to respond more quickly to customer orders and to reduce work-in-progress further.
- Significantly improved machine utilisation rates, although the actual degree of improvement is difficult to quantify.
- Increased capacity with the option to operate unmanned night workings.
- Greater flexibility, enabling shorter production runs and faster re-tooling and re-scheduling.

CIM involves the computerisation of functions and their integration into a system that regulates the manufacturing process. It brings together the individual manufacturing techniques referred to earlier under unified computer control.

Many of these benefits could be quantified, at least in part (e.g. the savings in working capital), although the degree of confidence in the underlying assumptions may not be high. But even so, there will still be a large intangible element that cannot be quantified. This has led to the charge that conventional methods of investment appraisal are biased against AMT investments.

Kaplan (1986) raises the question of whether AMT projects must be 'justified by faith alone'. Should managers in Foster Engineering replace the DCF approach with a belief that AMT is the key to the future and that strategic positioning must override economic analysis?

The answer is not to dismiss DCF analysis, but to see it within a wider strategic context. We advocate a three-stage approach to analysing AMT capital projects:

1 Does the project fit well within the company's overall corporate strategy?
2 Does the DCF analysis, based on the quantifiable elements of the decision, justify the investment outlay?
3 Where the net present value calculated in stage 2 is negative, examine the shortfall. Does management believe that the 'value' of the intangible benefits exceeds the shortfall? This last stage is essentially a subjective process whereby managers consider the strategic and operational benefits. No one can put an accurate value on flexibility, for example, but it would be wrong to exclude such a major benefit from consideration in the decision process.

Can firms afford not to invest?

Henry Ford once claimed: 'If you need a new machine and don't buy it, you pay for it without getting it.' The price paid is the loss in competitiveness from not taking advantage of new technology.

In evaluating proposed investments, managers have turned increasingly to sophisticated techniques. Their goal has been greater rationality in making investment decisions, yet their accomplishment has often been quite different – serious under-investment in the capital stock (the productive capacity, technology and worker skills) on which their

companies rest. As a result, they have unintentionally jeopardised their companies' futures.

Ingersoll Milling Machine Company took a strategic view that it needed to invest in the latest technology. Each production department manager annually had to write a justification to keep any machine that was over seven years old. The only generally accepted reason for not replacing equipment was that a new machine did not offer any significant improvements over older models.

6.4 ENVIRONMENTAL ASPECTS OF INVESTMENT

Much like AMT investment, many environmental capital projects have substantial costs or benefits which may not be wholly reflected in conventional net present value analysis. It should be recalled, from Chapter 1, that shareholders are not the only stakeholders in the company and the needs of other stakeholders, including the wider community, should also be incorporated into decisions. Environmental considerations have many dimensions, including economic, political, technological and social.

Pollution issues will be covered by legislation and regulation, but often the directors will want to go beyond the basic statutory requirements. While costs are not difficult to determine, the benefits are harder to quantify. For example, a greater sense of social responsibility may be costly but could have long-term benefits if the enhanced corporate image results in more business and improved shareholder value.

The steps involved in evaluating projects with environmental implications are:

1 Evaluate the projects using conventional capital appraisal methods.
2 Identify and incorporate statutory environmental costs as part of the evaluation.
3 Assess the costs and benefits of other environmental measures. For example, introducing anti-pollution measures should help reduce compensation claims.
4 Specify the internal controls to be introduced to ensure that pollution, etc. is minimised during construction and implementation.
5 Assess the impact of the decision on shareholder wealth, ethical and social responsibility goals.

Does Shell take the longer view?

The Royal Dutch Shell group operates in 140 countries and invests over £8 billion annually in oil exploration, refining and other capital projects. Most of its capital projects have sustainable environmental implications.

Take a look at Shell's annual report (**www.shell.com**) to examine the level of environ-

mental capital investment and provisions for cost of decommissioning and site restoration. What is its policy on environmental investment and sustainable development? To what extent does Shell take a long-term view on investment and consider wider social and environmental aspects?

Table 6.1 presents the results of a survey on the importance of non-financial factors in assessing strategic investment projects. The leading factors are whether the project fits with corporate strategy and customer requirements. Other important factors relate to quality, competitiveness, flexibility and the ability to expand in future.

Table 6.1 The importance of non-financial factors related to strategic investment projects

	Mean score (out of 5)*
Consistency with corporate strategy	4.4
Requirements of customers	4.0
Quality and reliability of outputs	3.7
Keeping up with competition	3.6
Ability to expand in future	3.5
Greater manufacturing flexibility	3.3
Reduced lead times	3.0
Reduced inventory levels	2.9
Experience with new technology	2.7

*5 = maximum importance, 1 = minimal importance
Source: Alkaraan and Northcott (2006).

6.5 THE CAPITAL INVESTMENT PROCESS

So far we have focused on investment *appraisal*. Similar emphasis is found in much of the capital budgeting literature, the assumption being that application of theoretically correct methods leads to optimal investment selection and, hence, maximises shareholders' wealth. The decision-maker is viewed as having a passive role, acting more as a technician than as an entrepreneur. Somehow, investment ideas come to the surface; various assumptions and cash flow estimates are made; and risk is incorporated within the discounting formula to produce the project's net present value. If this is positive, the proposal becomes part of the admissible set of investment possibilities. This set is then further refined by the evaluation of mutually exclusive projects and the appraisal of projects under capital rationing, where appropriate.

Inherent in this approach to capital budgeting are the following assumptions, few of which bear much relevance to the world of business:

1 Investment ideas simply emerge and land on the manager's desk.
2 Projects can be viewed in isolation, i.e. projects are not interdependent.
3 Risk can be fully incorporated within the net present value framework.
4 Non-quantifiable or intangible investment considerations are unimportant.
5 Cash flow estimates are free from bias.

Increasingly, it has become apparent that the emphasis on investment appraisal rather than on the whole capital investment process is misplaced and will not necessarily produce the most desirable investment programme. Investment decision-making could be improved significantly if the emphasis were placed on asking the appropriate strategic question rather than on increasing the sophistication of measurement techniques. Managers need to re-evaluate the investment procedures within their organisations, not to determine whether they are aesthetically and theoretically correct, but to determine whether they allow managers to make better decisions.

Capital budgeting may best be understood as a process with a number of distinct stages. Decision-making is an incremental activity, involving many people throughout the organisational hierarchy, over an extended period of time. While senior management may retain final approval, actual decisions are effectively taken much earlier at a lower level, by a process that is still not entirely clear and that is not the same in all organisations.

Figure 6.4 shows the key stages in the capital budgeting process. The primary aim of such a process is to ensure that available capital resources are distributed to wealth-creating capital projects that make the best contribution to corporate goals. A second

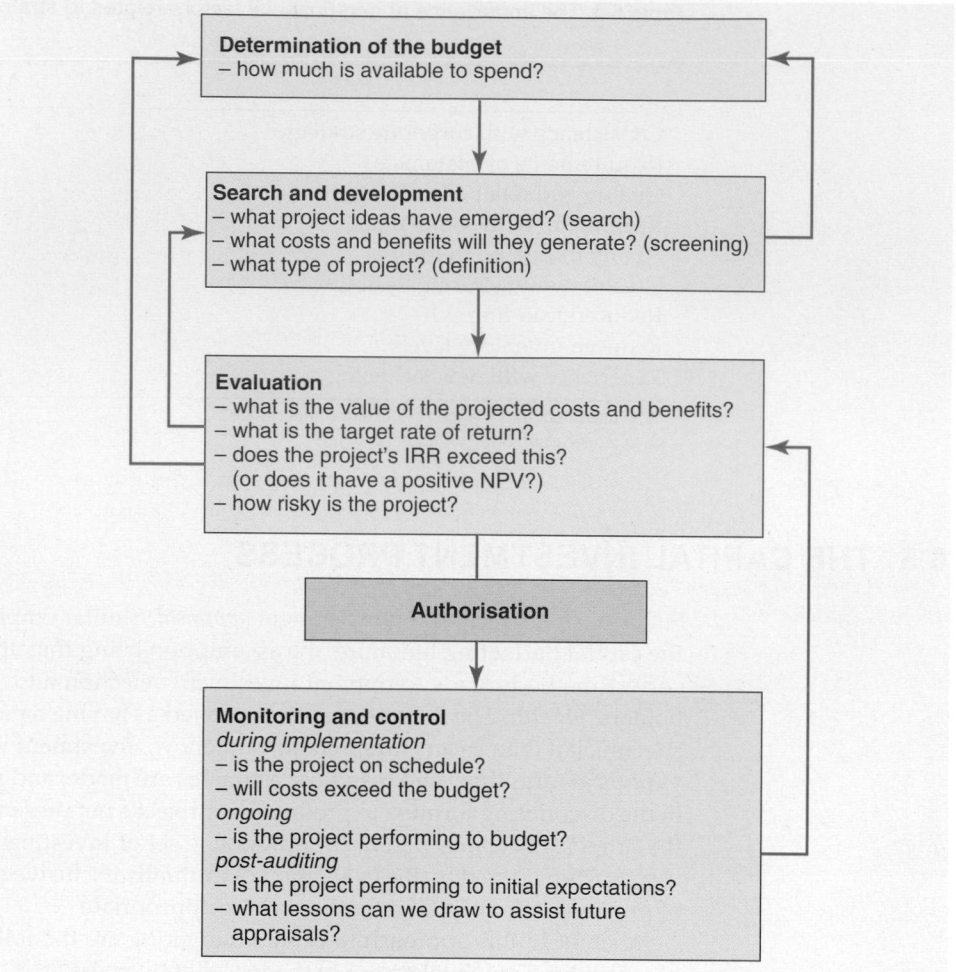

Figure 6.4 A simple capital budgeting system

goal is to see that good investment ideas are not held back and that poor or ill-defined proposals are rejected or further refined. We shall explore the following four stages:

1 Determination of the budget.
2 Search for, and development of, projects.
3 Evaluation and authorisation.
4 Monitoring and control.

■ Determination of the budget

In theory at least, all capital projects could be put to the capital market for funding (individually or collectively as investment programmes), the availability of funds for projects and rate of return required being a function of the market's perception of the prospective returns and associated risks. In practice, multi-divisional organisations operate an internal capital market in which senior management is better informed than the external capital market to assess capital proposals and allocate scarce resources.

If the investment decision-making body is a sub-unit of a larger group, the budget may be more or less rigidly imposed on it from above. However, for quasi-autonomous centres (divisions of larger groups with capital-raising powers) and/or independent units, the amount to be spent on capital projects is largely under their control, subject, of course, to considerations of corporate control and gearing.

■ Search for, and development of, projects

Economic theory views investment as the interaction of the supply of capital and the flow of investment opportunities. It would be wrong, however, to assume that there is a continuous flow of investment ideas. In general, the earlier an investment opportunity is identified, the greater is the scope for reward.

Possibly the most important role which top management can play in the capital investment process is to cultivate a corporate culture that encourages managers to search for, identify and sponsor investment ideas. Questions to be asked at the identification stage include the following:

1 How are project proposals initiated?
2 At what level are projects typically generated?
3 Is there a formal process for submitting ideas?
4 Is there an incentive scheme for identifying good project ideas?

Generating investment ideas involves considerable effort, time and personal risk on the part of the proposer. Any manager who has experienced the frustration of having an investment proposal dismissed, or an accepted proposal fail, is likely to develop an inbuilt resistance to creating further proposals unless the organisation culture and rewards are conducive to such activity. There is some evidence (Larcker, 1983) that firms adopting long-term incentive plans tend to increase their level of capital investment.

For the identification phase of non-routine capital budgeting decisions, especially those of a more strategic nature, to be productive, managers need to conduct environmental scanning, gathering information that is largely externally oriented. We should not expect the formal information system within most organisations, which is set up to help control short-term performance, to be particularly helpful in identifying non-routine investment ideas.

Preliminary screening

At this early stage, a *preliminary screening* of all investment ideas is usually conducted. It is neither feasible nor desirable to conduct a full-scale evaluation of each investment idea. The screening process is an important means of filtering out projects not thought worthy of further investigation. Ideas may not fit with strategic thinking, or may fall outside business units designated for growth or maintenance.

Screening proposals address such questions as the following:

1 Is the investment opportunity compatible with corporate strategy? Does it fall within a section of the business designated for growth, maintenance or divestment?
2 Are the resources required by the project available (expertise, finance, etc.)?
3 Is the idea technically feasible?
4 What evidence is there to suggest that it is likely to provide an acceptable return?
5 Are the risks involved acceptable?

As the quality of data used at the screening stage is generally poor, it makes little sense to apply sophisticated financial analysis. Accordingly, the simple payback method is frequently used at this stage because it offers a crude assessment of project profitability and risk.

Project definition

Any investment proposal is vague and shapeless until it has been properly defined. At the definition stage of the capital investment process, detailed specification of the investment proposal involves the collection of data describing its technical and economic characteristics. For each proposal, a number of alternative options should be generated, defined and, subsequently, appraised in order to create the project offering the most attractive financial characteristics.

Even at this early stage, proposals are gaining commitment. The very act of collecting information necessitates communicating with managers who may either lend support or seek to undermine the proposal. The danger is that, in this process, commitments are accumulated such that investment becomes almost inevitable. The amount of information gathered for evaluation is largely determined by the following:

- The data perceived as desirable to gain a favourable decision.
- The ease and cost of its development.
- The extent to which the proposer will be held responsible for later performance related to the data.

Top management should seek to ensure that the most suitable projects are submitted by managers through establishing mechanisms that induce behaviour congruence. The accounting information system, reward system and capital budgeting procedures should all encourage managers to put forward the proposals that top management is looking for. For many firms, however, the accounting information system and reward mechanism encourage divisional managers to promote their own interests at the expense of those of the organisation, and to emphasise short-term profit performance at the expense of the longer term. Capital budgeting then becomes a 'game', with the accounting and reward systems as its rules. Cash flow estimates are biased to maximise the gains to individuals within such rules.

Self-assessment activity 6.2

Outline the important stages in the capital budgeting process.

(Answer in Appendix A at the back of the book)

Project classification

The information required and method of analysis will vary according to the nature of the project. A suggested investment proposal classification is given below under the headings replacement, cost reduction, expansion or improvement, new products, strategic, and statutory and welfare.

Replacement proposals are justified primarily by the need to replace assets that are nearly exhausted or have excessively high maintenance costs. Little or no improvement may be expected from the replacement, but the expenditure is essential to maintain the existing level of capacity or service (e.g. replacement of vehicles). Engineering analysis plays an important role in these proposals.

Cost reduction proposals (which may also be replacement proposals) are intended to reduce costs through addition of new equipment or modification to existing equipment. Line managers and specialists (such as industrial engineers and work study groups) should conduct a continuous review of production operations for profit improvement opportunities.

Expansion or improvement proposals relate to existing products, and are intended to increase production, service and distribution capacity, to improve product quality, or to maintain and improve the firm's competitive position.

New product proposals refer to all capital expenditures pertaining to the development and implementation of new products.

Strategic proposals are generated at senior management level and involve expenditure in new areas, or where benefits extend beyond the investment itself. A project may appear to offer a negative net present value and yet still create further valuable strategic opportunities. Three examples demonstrate this point:

1 Diversification projects may have the effect of bringing the company into a lower risk category.

2 A patent may be acquired not for use within the firm, but to prevent its use by competitors.
3 Where information is difficult to obtain, such as in overseas markets, it may make sense to set up a small plant at a loss because it places the firm in a good position to build up information and to be ready for minor investment at the appropriate time.

Statutory and welfare proposals do not usually offer an obvious financial return, although they may contribute in other ways, such as enhancing the contentment, and hence productivity, of the labour force. The main consideration is whether standards are met at minimum cost.

Each proposal should be ranked within each category in terms of its effect on profits, its degree of urgency, and whether or not it can be postponed.

■ Evaluation and authorisation

Evaluation

The evaluation phase involves appraisal of the project and decision outcome (accept, reject, request further information, etc.). Project evaluation, in turn, involves the assembly of information (usually in terms of cash flows) and the application of specified investment criteria. Each firm must decide whether to apply rigorous, sophisticated evaluation models, or simpler models that are easier to grasp yet capture many of the important elements in the decision.

The capital appropriation request forms the basis for the final decision to commit financial and other resources to the project. Typical information included in an appropriation request is given below:

1 *Purpose of project* – why it is proposed, and the fit with corporate strategy and goals.
2 *Project classification* – e.g. expansion, replacement, improvement, cost saving, strategic, research and development, safety and health, legal requirements.
3 *Finance requested* – amount and timing, including net working capital, etc.
4 *Operating cash flows* – amount and timing, together with the main assumptions influencing the accuracy of the cash flow estimates.
5 *Attractiveness of the proposal* – expressed by standard appraisal indicators, such as net present value, DCF rate of return and payback period calculated from after-tax cash flows.
6 *Sensitivity of the assumptions* – effect of changes in the main investment inputs. Other approaches to assessing project risk should also be addressed (e.g. best/worst scenarios, estimated range of accuracy of DCF return, discussed in Chapter 7).
7 *Review of alternatives* – why they were rejected and their economic attractiveness.
8 *Implications of not accepting the proposal* – some projects with little economic merit according to the appraisal indicators may be 'essential' to the continuance of a profitable part of the business or to achieving agreed strategy.
9 *Non-financial considerations* – those costs and benefits that cannot be measured.

Following evaluation, larger projects may require consideration at a number of levels in the organisational hierarchy before they are finally approved or rejected. The decision outcome is rarely based wholly on the computed signal derived from financial analysis. Considerable judgement is applied in assessing the reliability of data underlying the appraisal, fit with corporate strategy, and track record of the project sponsor. Careful consideration is required regarding the influence on the investment of such key factors as product markets, the economy, production, finance and people.

Authorisation

Following evaluation, the proposal is transmitted through the various authorisation levels of the organisational hierarchy until it is finally approved or rejected. The driving motive in the decision process is the willingness of the manager to make a commitment to sponsor a proposal. This is based not so much on the grounds of the proposal itself as on whether or not it will enhance the manager's reputation and career prospects. Sometimes those involved in the preliminary investigation and appraisal of major projects are promoted into head office decision-making positions in time to support and assist the approval of the same projects!

In larger organisations, the authorisation of major projects is usually a formal endorsement of commitments already given. Complete rejection of proposals is rare, but proposals are, on occasions, referred back. The approval stage appears to have a twofold purpose:

1 *A quality control function.* As long as the proposals have satisfied the requirements of all previous stages, there is no reason for their rejection other than on political grounds. Only where the rest of the investment planning process is inadequate will the approval stage take on greater significance in determining the destiny of projects.

2 *A motivational function.* An investment project and its proposer are inseparable. The decision-maker, in effect, forms a judgement simultaneously on both the proposal and the person or team submitting it.

Sometimes the costs associated with rejection of capital projects, in terms of managerial motivation, far exceed the costs associated with accepting a marginally unprofitable project. The degree of commitment, enthusiasm and drive of the management team implementing the project is a major factor in determining the success or failure of marginal projects.

Mini case

How SmithKline Beecham makes investments

In recent years, with more projects successfully reaching late-development stage, the demands for funding at SmithKline Beecham were considerable. The pharmaceutical group, which invests more than half a billion dollars annually, had to create more value from its investments to help meet its tough earnings targets.

A new decision-making process was introduced, designed to identify those development projects likely to create the most value and reflect the complexity and risk of its investments. Each project team was asked to develop four alternatives: its current plans, a 'buy-up' option and a 'buy-down' option (where they explored the effects of increasing and reducing the investment outlay) and a minimum plan, where the project was abandoned at minimum cost. These alternatives were generated and valued before a decision was taken.

A variety of decision approaches were taken, focusing on creating net present value. These included decision tree analysis, probability analysis, options analysis and sensitivity analysis (discussed in the following chapters). However, management soon discovered that equally important were softer issues such as information quality, credibility and trust.

It was their initial intention that this new resource allocation process would be useful in cutting the development budget. But they now saw investment decisions in a new light and recognised that the investment portfolio was worth far more than expected. The net result was an increase in capital spending by more than 50 per cent.

Source: Based on Sharpe and Keelin, *Harvard Business Review*, March–April 1998.

Self-assessment activity 6.3

Read the above article and discuss the impact that a new resource allocation system can have on capital investment creation, evaluation and decision-making.

▪ Monitoring and control

The capital budgeting control process can be classified in terms of pre-decision and post-decision controls. Pre-decision controls are mechanisms designed to influence managerial behaviour at an early stage in the investment process. Examples are setting authorisation levels and procedures to be followed, and influencing the proposals submitted by setting goals, hurdle rates and cash limits and identifying strategic areas for growth. Post-decision controls include monitoring and post-audit procedures.

Major investment projects may justify determining the critical path (i.e. a set of linked activities) in the delivery and installation schedule. The critical path is defined as the longest path through a network. Control is established by accounting procedures for recording expenditures. Progress reports usually include actual expenditure; amounts authorised to date; amounts committed against authorisations; amounts authorised but not yet spent; and estimates of further cost to completion.

The case of the disappearing projects

Ameritech, a major US company operating in the electronics industry, invests over $2 billion a year, mostly in thousands of relatively small-scale projects. When the company announced that it proposed to monitor and audit capital projects, that year's budgets had already been submitted. But the company told every division to take back their submissions, think about the fact that everyone who worked on the project was going to be 'tracked', and then resubmit the estimates. Seven hundred projects never came back – they just disappeared. Many others had much lower estimates. We will never know just how many of those 700 projects could have been investment 'winners'.

This illustrates just how influential capital budgeting controls can be on managerial investment behaviour.

Source: Based on Weaver *et al.* (1989).

6.6 POST-AUDITING

post-audit
A re-examination of costs, benefits and forecasts of a project after implementation (usually after 1 year)

The final stage in the capital budgeting decision-making and control sequence is the post-completion audit. A **post-audit** aims to compare the actual performance of a project after, say, a year's operation with the forecast made at the time of approval, and ideally also with the revised assessment made at the date of commissioning. The aims of the exercise are twofold: first, post-audits may attempt to encourage more thorough and realistic appraisals of future investment projects; and second, they may aim to facilitate major overhauls of ongoing projects, perhaps to alter their strategic focus. These two aims differ in an important respect. The first concerns the overall capital budgeting system, seeking to improve its quality and cohesion. The second concerns the control of existing projects, but with a broader perspective than is normally possible during the regular monitoring procedure when project adjustments are usually of a 'fire-fighting' nature.

Self-assessment activity 6.4

What are the main benefits from post-audits?

(Answer in Appendix A at the back of the book)

■ Problems with post-auditing

There are many problems with post-audits:

1 *The disentanglement problem.* It may be difficult to separate out the relevant costs and benefits specific to a new project from other company activities, especially where facilities are shared and the new project requires an increase in shared overheads. Newly-developed techniques of overhead cost allocation (such as Activity-Based Costing) may prove helpful in this respect.

2 *Projects may be unique.* If there is no prospect of repeating a project in the future, there may seem little point in post-auditing, since the lessons learned may not be applicable to any future activity. Nevertheless, useful insights into the capital budgeting system as a whole may be obtained.

3 *Prohibitive cost.* To introduce post-audits may involve interference with present management information systems in order to generate flows of suitable data. Since post-auditing every project may be very resource-intensive, firms tend to be selective in their post-audits.

4 *Biased selection.* By definition, only accepted projects can be post-audited, and often only the underperforming ones are singled out for detailed examination. Because of this biased selection mechanism, the forecasting and evaluation expertise of project analysts may be cast in an unduly bad light – they might have been spot on in evaluating rejected and acceptably performing projects.

5 *Lack of cooperation.* If the post-audit is conducted in too inquisitorial a fashion, project sponsors are likely to offer grudging cooperation to the review team and be reluctant to accept and act upon their findings. The impartiality of the review team is paramount – for example, it would be inviting resentment to draw post-auditors from other parts of the company that may be competitors for scarce capital. Similarly, there are obvious dangers if reviews are undertaken solely by project sponsors. A balanced team of investigators needs to be assembled.

6 *Encourages risk-aversion.* If analysts' predictive and analytical abilities are to be thoroughly scrutinised, they may be inclined to advance only 'safe' projects where little can go awry and where there is less chance of being 'caught out' by events.

7 *Environmental changes.* Some projects can be devastated by largely unpredictable swings in market conditions. This can make the post-audit a complex affair, as the review team is obliged to adjust analysts' forecasts to allow for 'moving of the goalposts'.

■ The conventional wisdom

Studies conducted in North America and the UK have generated a conventional wisdom about corporate post-auditing practices. Its main elements are as follows:

- Few firms post-audit every project, and the selection criterion is usually based on size of outlay.
- The commonest time for a first post-audit is about a year after project commissioning.
- The most effective allocation of post-audit responsibility is to share it between central audit departments and project initiators to avoid conflicts of interest, while using relevant expertise.
- The 'threat' of post-audit is likely to spur the forecaster to greater accuracy, but it can lead to excessive caution, possibly resulting in suppression of potentially worthwhile ventures.

■ When does post-auditing work best?

What guidelines can we offer to managers who wish to introduce post-audit from scratch or to overhaul an existing system? Here are some key points:

1 When introducing and operating post-audit, emphasise the learning objectives and minimise the likelihood of its being viewed as a 'search for the guilty'.

2 Clearly specify the aims of a post-audit. Is it to be primarily a project control exercise, or does it aim to derive insights into the overall project appraisal system?

3 When introducing post-audits, start the process with a small project to reveal as economically as possible the difficulties that need to be overcome in a major post-audit.

4 Include a pre-audit in the project proposal. When the project is submitted for approval, the sponsors should be required to indicate what information would be required to undertake a subsequent post-audit.

SUMMARY

We have examined the strategic framework for investment decisions, paying particular attention to new technology and environmental projects.

The resource allocation process is the main vehicle by which business strategy can be implemented. Investment decisions are not simply the result of applying some evaluation criterion. Investment analysis is essentially a search process: a search for ideas, for information and for decision criteria. The prosperity of a firm depends more on its ability to create profitable investment opportunities than on its ability to appraise them.

Key points

- Investments form part of a wider strategic process and should be assessed both financially and strategically.

- New technology projects are often particularly difficult to evaluate because of the many non-financial values.

- The four main stages in the capital budgeting process are:

  1 Determine the budget.

  2 Search for and develop projects.

  3 Evaluation and authorisation.

  4 Monitoring and control.

  Once a firm commits itself to a particular project, it should regularly and systematically monitor and control the project through its various stages of implementation.

- Post-audit reviews, if properly designed, fulfil a useful role in improving the quality of existing and future investment analysis and provide a means of initiating corrective action for existing projects.

Further reading

Further reading on AMT investment evaluation is found in Alkaraan and Northcott (2006), Pike *et al.* (1989) and Kaplan (1986). Neale and Buckley (1992) consider the practice of post-auditing, while Butler *et al.* (1993) examine strategic investment decisions.

QUESTIONS

 **myfinancelab** | *Questions with an icon are also available for practice in myfinancelab with additional supporting resources.*

Questions with a **coloured number** have solutions in Appendix B on page 733.

1 'Capital budgeting is simply a matter of selecting the right decision rule.' How true is this statement?

2 What are the aims of post-audits?

3 AMT plc is increasing the level of automation of a production line dedicated to a single product. The options available are total automation or partial automation. The company works on a planning horizon of five years and either option will produce the 10,000 units which can be sold annually.

Total automation will involve a total capital cost of £1 million. Material costs will be £12 per unit and labour and variable overheads will be £18 per unit with this method.

Partial automation will result in higher material wastage and an average cost of £14 per unit. Labour and variable overhead are expected to cost £41 per unit. The capital cost of this alternative is £250,000.

The products sell for £75 each, whichever method of production is adopted. The scrap value of the automated production line, in five years' time, will be £100,000, while the line which is partially automated will be worthless. The management uses straight-line depreciation and the required rate of return on capital investment is 16 per cent p.a. Depreciation is considered to be the only incremental fixed cost.

In analysing investment opportunities of this type the company calculates the average total cost per unit, annual net profit, the break-even volume per year and the discounted net present value.

Required

(a) Determine the figures which would be circulated to the management of AMT plc in order to assist their investment analysis.

(b) Comment on the figures produced and make a recommendation with any qualifications you think appropriate.

(Certified Diploma)

4 Bowers Holdings plc has recently acquired a controlling interest in Shaldon Engineering plc, which produces high-quality machine tools for the European market. Following this acquisition, the internal audit department of Bowers Holdings plc examined the financial management systems of the newly acquired company and produced a report that was critical of its investment appraisal procedures.

The report summary stated:

Overall, investment appraisal procedures in Shaldon Engineering plc are very weak. Evaluation of capital projects is not undertaken in a systematic manner and post-decision controls relating to capital projects are virtually non-existent.

Required

Prepare a report for the directors of Shaldon Engineering plc, stating what you consider to be the major characteristics of a system for evaluating, monitoring and controlling capital expenditure projects.

(Certified Diploma)

What procedures should a business adopt for approving and reviewing large capital expenditure projects?

Practical assignment

Read the *Harvard Business Review* article (Sept.–Oct. 1989) 'Must finance and strategy clash?' by Barwise, Marsh and Wensley. Summarise and comment on their views on the question.

 **myfinancelab** | *Now retake your diagnostic test for Chapter 6 to check your progress and update your study plan.*

Part III

VALUE, RISK AND THE REQUIRED RETURN

The preceding analysis of investment decisions has implied that future returns from investment can be forecast with certainty. Clearly, this is unlikely in practice. In Part III we examine the impact of uncertainty on the investment decision, the various approaches available to decision-makers to cope with this problem, and the implications for valuation of assets and companies.

In Chapter 7, we discuss a number of methods that may assist the decision-maker when looking at the risky investment project in isolation. In Chapter 8, we examine the contribution to project appraisal under risk and uncertainty promised by the rapidly developing field of options theory. In Chapter 9, we look at how more desirable combinations of risk and return can be achieved by forming a portfolio of investment activities. In Chapter 10, we examine the contribution to risk analysis of the Capital Asset Pricing Model, which offers a guide to setting the premium required for risk. The earlier study of how capital markets behave is particularly important here. Chapter 10 is highly important because it links the behaviour of individual investors, buying and selling securities, to the behaviour of the capital investment decision-maker. This focus is further developed in Chapter 11, which discusses how to alter the discount rate when faced by projects of degrees of risk that differ from the company's existing activities. Finally, this part of the book culminates in Chapter 12, where many of the concepts of risk and return are brought together in the examination of various approaches to company, or enterprise, value, and equity, or owners', value. The complexities introduced by debt financing are postponed until Part IV.

7

Analysing investment risk

Eurotunnel – a risky investment decision

In 1986, the Anglo-French Treaty authorised the construction, financing and operation of a twin rail tunnel
under the English Channel to be run by Eurotunnel.
The company is effectively a one-project business.

The preliminary prospectus provided forecasts upon
which expected returns and sensitivities could be prepared. Potential investors and lenders were invited to
invest in the highly risky venture that would not pay a
dividend for at least eight years and where the expected
internal rate of return was around 14 per cent. The
Economist commented at the time that the Tunnel
was 'a hole in the ground that will either make or lose
a fortune' for its investors.

Eurotunnel's troubles date back to the project's
launch in the late 1980s. The cost of digging the 30-mile
tunnel between France and Britain was underestimated,
and traffic has consistently fallen short of expectations,
partly because of unforeseen competition from low-cost
airlines. The company also shouldered heavy debt and
interest repayments because at the time it was con-
ceived, the UK government insisted that the project be
entirely funded with private money. On top of this, a
major fire on the supposedly safe freight carriages put it
out of operation and dented public confidence.

By 2000, with things looking a little better, the chairman announced that the first dividend was expected in
2006. However, closer questioning revealed that this
was the 'upper case scenario'. In the 'lower case scenario' dividend payments did not begin until 2010.

Eurotunnel lurched from one debt crisis to another
until it reached a restructuring deal with shareholders
in 2007, allowing it to cut its €9.2 billion debt to
€4.16 billion by repaying banks in the form of shares.
The deal created a new company, Groupe Eurotunnel
SA, and diluted existing shareholders' stake to 13 per
cent. The vast bulk of the debt was held by financial
institutions, but a large number of small shareholders,
mostly in France, bought shares when the company
was floated in 1987 and they have suffered the most,
seeing share price plummeting from 900p to just 20p.

Following the financial restructuring, things are at
last looking a little brighter. The company reported
profits of $1.57 million for 2007, its first annual net
profit, less than a year after the company nearly
drowned in debt. However, the result excludes an
exceptional gain of €3.3 billion from the restructuring
agreement, which halved Eurotunnel's debt and saved
it from bankruptcy.

Learning objectives

The main learning objectives are the following:

- To understand how uncertainty affects investment decisions.
- To explore managers' risk attitudes.
- To appreciate the levels at which risk can be viewed.
- To be able to measure the expected NPV and its variability.
- To appreciate the main risk-handling techniques and apply them to capital budgeting problems.

 *Complete your diagnostic test for Chapter 7 now to create your personal study plan*

7.1 INTRODUCTION

The Channel Tunnel is one of many cases where investment decisions turn out to be far riskier than originally envisaged. The finance director of a major UK manufacturer for the motor industry remarked, 'We know that, on average, one in five large capital projects flops. The problem is: we have no idea beforehand which one!'

Stepping into the unknown – which is what investment decision-making effectively is – means that mistakes will surely occur. Only about 50 per cent of small businesses are still trading three years after start-up. Sir Richard Branson, head of Virgin Atlantic, once said, 'the safest way to become a millionaire is to start as a billionaire and invest in the airline industry.'

This does not mean that managers can do nothing about project failures. In this and subsequent chapters, we examine how project risk is assessed and controlled. The various forms of risk are defined and the main statistical methods for measuring project risk within single-period and multi-period frameworks are described. A variety of risk analysis techniques will then be discussed. These fall conveniently into methods intended to describe risk and methods incorporating project riskiness within the net present value formula. The chapter concludes by examining the extent to which the methods discussed are used in business organisations.

■ Defining terms

At the outset, we need to clarify our terms:

- *Certainty*. Perfect certainty arises when expectations are single-valued: that is, a particular outcome will arise rather than a range of outcomes. Is there such a thing as an investment with certain payoffs? Probably not, but some investments come fairly close. For example, an investment in three-month Treasury Bills will, subject to the Bank of England keeping its promise, provide a precise return on redemption.
- *Risk and uncertainty*. Although used interchangeably in everyday parlance, these terms are not quite the same. Risk refers to the set of unique consequences for a given decision that can be assigned probabilities, while uncertainty implies that it is not fully possible to identify outcomes or to assign probabilities. Perhaps the worst forms of uncertainty are the 'unknown unknowns' – outcomes from events that we did not even consider.

The most obvious example of risk is the 50 per cent chance of obtaining a 'head' from tossing a coin. For most investment decisions, however, empirical experience is hard to find. Managers are forced to estimate probabilities where objective statistical evidence is not available. Nevertheless, a manager with little prior experience of launching a particular product in a new market can still subjectively assess the risks involved based on the information he or she has. Because subjective probabilities may be applied to investment decisions in a manner similar to objective probabilities, the distinction between risk and uncertainty is not critical in practice, and the two terms are often used synonymously.

Investment decisions are only as good as the information upon which they rest. Relevant and useful information is central in projecting the degree of risk surrounding future economic events and in selecting the best investment option.

Self-assessment activity 7.1

Why is risk assessment important in making capital investment decisions?

(Answer in Appendix A at the back of the book)

7.2 EXPECTED NET PRESENT VALUE (ENPV): BETTERWAY PLC

To what extent is the net present value criterion relevant in the selection of risky investments? Consider the case of Betterway plc, contemplating three options with very different degrees of risk. The distribution of possible outcomes for these options is given in Table 7.1. Notice that A's cash flow is totally certain.

Table 7.1 Betterway plc: expected net present values

Investment	NPV outcomes (£)		Probability		Weighted outcomes (£)
A	9,000	×	1	=	9,000
	−10,000	×	0.2	=	−2,000
B	10,000	×	0.5	=	5,000
	20,000	×	0.3	=	6,000
			1.0	ENPV =	9,000
	−50,000	×	0.2	=	−11,000
C	10,000	×	0.5	=	5,000
	50,000	×	0.3	=	15,000
			1.0	ENPV =	9,000

expected net present value
The average of the range of possible NPVs weighted by their probability of occurrence

Clearly, while the NPV criterion is appropriate for investment option A, where the cash flows are certain, it is no longer appropriate for the risky investment options B and C, each with three possible outcomes. The whole range of possible outcomes may be considered by obtaining the **expected net present value** (ENPV), which is the mean of the NPV distribution when weighted by the probabilities of occurrence. The ENPV is given by the equation:

$$\overline{X} = \sum_{i=1}^{N} p_i X_i$$

where $\overline{X}$ is the expected value of event X, X_i is the possible outcome i from event X, p_i is the probability of outcome i occurring and N is the number of possible outcomes.

The NPV rule may then be applied by selecting projects offering the highest expected net present value. In our example, all three options offer the same expected NPV of £9,000. Should the management of Betterway view all three as equally attractive? The answer to this question lies in their attitudes towards risk, for while the *expected* outcomes are the same, the *possible* outcomes vary considerably. Thus, although the expected NPV criterion provides a single measure of profitability, which may be applied to risky investments, it does not, by itself, provide an acceptable decision criterion.

7.3 ATTITUDES TO RISK

Business managers prefer less risk to more risk for a given return. In other words, they are *risk-averse*. In general, a business manager derives less utility, or satisfaction, from gaining an additional £1,000 than he or she forgoes in losing £1,000. This is based on the concept of diminishing marginal utility, which holds that, as wealth increases, marginal utility declines at an increasing rate. Thus the utility function for risk-averse managers is concave, as shown in Figure 7.1. As long as the utility function of the decision-maker can be specified, this approach may be applied in reaching investment decisions.

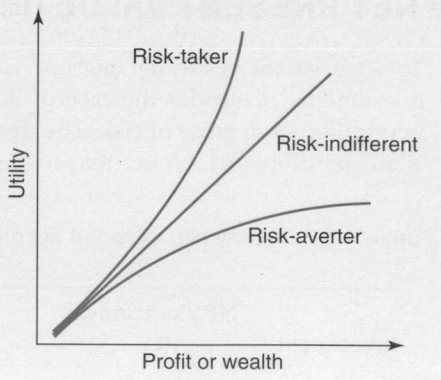

Figure 7.1 Risk profiles

Example: Carefree plc's utility function

Mike Cool, the managing director of Carefree plc, a business with a current market value of £30 million, has an opportunity to relocate its premises. It is estimated that there is a 50 per cent probability of increasing its value by £12 million and a similar probability that its value will fall by £10 million. The owner's utility function is outlined in Figure 7.2. The concave slope shows that the owner is risk-averse. The gain in utility (ΔU_F) as a result of the favourable outcome of £42 million, is less than the fall in utility (ΔU_A) resulting from the adverse outcome of only £20 million.

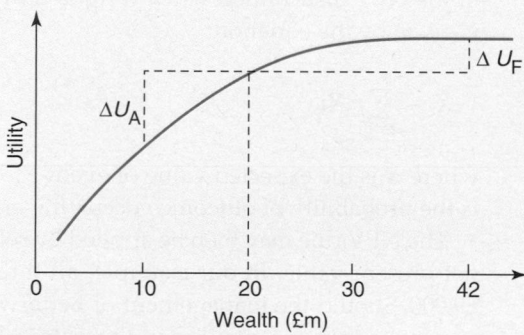

Figure 7.2 Risk-averse investor's utility function

The conclusion is that, although the investment proposal offers £1 million expected additional wealth (i.e. [0.5 × £12 m] + [0.5 × −£10 m]), the project should not be undertaken because total expected utility would fall if the factory were relocated.

While decision-making based upon the expected utility criterion is conceptually sound, it has serious practical drawbacks. Mike Cool may recognise that he is risk-averse, but is unable to define, with any degree of accuracy, the shape of his utility function. This becomes even more complicated in organisations where ownership and management are separated, as is the case for most companies. Here, the agency problem discussed in Chapter 1 arises. Thus, while utility analysis provides a useful insight into the problem of risk, it does not provide us with operational decision rules.

Space tourist insurers eye up the final frontier

Amid the race to send paying passengers into space by the end of the decade there are signs that the final frontier is attracting interest from companies looking to tap the potential for providing personal insurance for those looking to boldly go. Bupa sought to steal a march on rival insurers by announcing it was eyeing the space tourist market and Virgin Galactic, which is planning to carry space tourists by 2009, said space was attracting interest from other insurers. However, Virgin said it was too early for providers to put a price on premiums.

Risks of space travel range from possible loss of life to the negative health implications of exposure to radiation and high G-forces. The potential psychological effects are largely unknown and could necessitate post-trip care. Will Whitehorn, president of Virgin Galactic, has met representatives of Lloyd's of London and said that until Virgin Galactic conducted test flights in 2008, the market would not be able to assess risk conclusively. Some 200 passengers have paid deposits for a place on the company's suborbital flight which will fly at 3,000 miles an hour and provide them with several minutes of weightlessness. The passengers will be required to sign a waiver before they fly after the company has explained all potential health risks.

Source: Elaine Moore, *Financial Times*, 8 March 2007, p. 6.

7.4 THE MANY TYPES OF RISK

Risk may be classified into a number of types. A clear understanding of the different forms of risk is useful in the evaluation and monitoring of capital projects:

1 *Business risk* – the variability in operating cash flows or profits before interest. A firm's business risk depends, in large measure, on the underlying economic environment within which it operates. But variability in operating cash flows can be heavily affected by the cost structure of the business, and hence its **operating gearing**. A company's break-even point is reached when sales revenues match total costs. These costs consist of fixed costs – that is, costs that do not vary much with the level of sales – and variable costs. The decision to become more capital-intensive generally leads to an increase in the proportion of fixed costs in the cost structure. This increase in operating gearing leads to greater variability in operating earnings.

2 *Financial risk* – the risk, over and above business risk, that results from the use of debt capital. **Financial gearing** is increased by issuing more debt, thereby incurring more fixed-interest charges and increasing the variability in net earnings. Financial risk is considered more fully in later chapters.

3 *Portfolio or market risk* – the variability in shareholders' returns. Investors can significantly reduce their variability in earnings by holding carefully selected investment portfolios. This is sometimes called 'relevant' risk, because only this element of risk should be considered by a well-diversified shareholder. Chapters 9 and 10 examine such risk in greater depth.

Project risk can be viewed and defined in three different ways: (1) in isolation, (2) in terms of its impact on the business, and (3) in terms of its impact on shareholders' investment portfolios. One survey (Pike and Ho, 1991) found that 79 per cent of managers in larger UK firms use project-specific risk and 61 per cent consider the impact of business risk, but only 26 per cent consider the impact on shareholder portfolios.

In this chapter, we assess project risk in isolation before moving on to estimate its impact on investors' portfolios (i.e. market risk) in Chapter 10.

Operating gearing example: Hifix and Lofix

Hifix and Lofix are two companies identical in every respect except cost structure. While Lofix pays its workforce on an output-related basis, Hifix operates a flat-rate wage system. The sales, costs and profits for the two companies are given under two economic states, normal and recession, in Table 7.2. While both companies perform equally well under normal trading conditions, Hifix, with its heavier fixed cost element, is more vulnerable to economic downturns. This can be measured by calculating the degree of operating gearing:

$$\text{Operating gearing} = \frac{\text{percentage change in profits}}{\text{percentage change in sales}}$$

$$\text{For Hifix} = \frac{-200\%}{-40\%} = 5$$

$$\text{For Lofix} = \frac{-80\%}{-40\%} = 2$$

The degree of operating gearing is far greater for the firm with high fixed costs than for the firm with low fixed costs. (Chapter 18 further discusses operating gearing.)

Table 7.2 Effects of cost structure on profits (£000)

	Hifix		Lofix	
	Normal	Recession	Normal	Recession
Sales	200	120	200	120
Variable costs	−100	−60	−160	−96
Fixed costs	−80	−80	−20	−20
Profit/loss	20	−20	20	4
Change in sales		−40%		−40%
Change in profits		−200%		−80%

Self-assessment activity 7.2

Which type of risk do the following describe?

1 Risks associated with increasing the firm's level of borrowing.
2 The variability in the firm's operating profits.
3 Variability in the cash flows of a proposed capital investment.
4 Variability in shareholders' returns.

(Answer in Appendix A at the back of the book)

7.5 MEASUREMENT OF RISK

A well-known politician (not named to protect the guilty) once proclaimed, 'Forecasting is very important – particularly when it involves the future!' Estimating the probabilities of uncertain forecast outcomes is difficult. But with the little knowledge the manager may have concerning the future, and by applying past experience backed by historical analysis of a project and its setting, he or she may be able to construct a probability distribution of a project's cash outcomes. This can be used to measure the risks

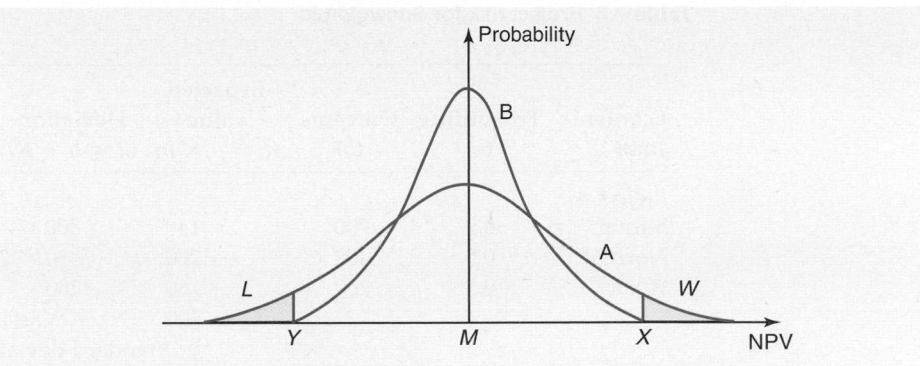

Figure 7.3 Variability of project returns

surrounding project cash flows in a variety of ways. If we assume that the range of possible outcomes from a decision is distributed normally around the expected value, risk-averse investors can assess project risk using expected value and standard deviation. We shall consider three statistical measures: the standard deviation, semi-variance and coefficient of variation for single-period cash flows.

■ Measuring risk for single-period cash flows: Snowglo plc

Standard deviation

We have seen that expected value overlooks important information on the dispersion (risk) of the outcomes. We also know that different people behave differently in risky situations. Figure 7.3 shows the NPV distributions for projects A and B. Both projects have the same expected NPV, indicated by M, but project A has greater dispersion. The risk-averse manager in Snowglo will choose B since he or she wants to minimise risk. The risk-taker will choose A because the NPV of project A has a chance (W) of being higher than X (which project B cannot offer), but also a chance (L) of being lower than Y. Hereafter we make the reasonable assumption that most people are risk-averse.

The standard deviation is a measure of the dispersion of possible outcomes; the wider the dispersion, the higher the standard deviation.

The expected value, denoted by $\overline{X}$, is given by the equation:

$$\overline{X} = \sum_{i=1}^{N} p_i X_i$$

and the standard deviation of the cash flows by:

$$\sigma = \sqrt{\sum_{i=1}^{N} p_i (X_i - \overline{X})^2}$$

Table 7.3 shows the information on two projects for Snowglo plc.

Table 7.3 Snowglo plc project data

State of economy	Probability of outcome	Cash flow (£)	
		A	B
Strong	0.2	700	550
Normal	0.5	400	400
Weak	0.3	200	300

Table 7.4 Project risk for Snowglo plc

Economic state	Probability (a)	Outcome (b)	Expected value (c = a × b)	Deviation (d = b − $\overline{X}$)	Squared deviation (e = d²)	Variance (f = a × e)
Project A						
Strong	0.2	700	140	300	90,000	18,000
Normal	0.5	400	200	0	0	0
Weak	0.3	200	60	−200	40,000	12,000
			$\overline{X}_A = 400$		Variance = σ_A^2	= 30,000
					Standard deviation = σ_A	= 173.2
Project B						
Strong	0.2	550	110	150	22,500	4,500
Normal	0.5	400	200	0	0	0
Weak	0.3	300	90	−100	10,000	3,000
			$\overline{X}_B = 400$		Variance = σ_B^2	= 7,500
					Standard deviation = σ_B	= 86.6

Alternatively:

$$\overline{X}_A = 700(0.2) + 400(0.5) + 200(0.3) = 400$$

$$\sigma_A = \sqrt{[0.2(700 - 400)^2 + 0.5(400 - 400)^2 + 0.3(200 - 400)^2]}$$
$$= 173.2$$

$$\overline{X}_A = 550(0.2) + 400(0.5) + 300(0.3) = 400$$

$$\sigma_B = \sqrt{[0.2(550 - 400)^2 + 0.5(400 - 400)^2 + 0.3(300 - 400)^2]}$$
$$= 86.6$$

Table 7.4 provides the workings for projects A and B.

Applying the formulae, we obtain an expected cash flow of £400 for both project A and project B. If the decision-maker had a neutral risk attitude, he or she would view the two projects equally favourably. But as the decision-maker is likely to be risk-averse, it is appropriate to examine the standard deviations of the two probability distributions. Here we see that project A, with a standard deviation twice that of project B, is more risky and hence less attractive. This could have been deduced simply by observing the distribution of outcomes and noting that the same probabilities apply to both projects. But observation cannot always tell us by how much one project is riskier than another.

Semi-variance

While deviation above the mean may be viewed favourably by managers, it is 'down-side risk' (i.e. deviations below expected outcomes) that is mainly considered in the decision process. Downside risk is best measured by the semi-variance, a special case of the variance, given by the formula:

$$SV = \sum_{j=1}^{K} p_j (X_j - \overline{X})^2$$

where SV is the semi-variance, j is each outcome value less than the expected value, and K is the number of outcomes that are less than the expected value.

Applying the semi-variance to the example in Table 7.4, the downside risk relates exclusively to the 'weak' state of the economy:

$$SV_A = 0.3(200 - 400)^2 = £12,000$$
$$SV_B = 0.3(300 - 400)^2 = £3,000$$

Once again project B is seen to have a much lower degree of risk. In both cases, the semi-variance accounts for 40 per cent of the project variance.

Coefficient of variation (CV)

Where projects differ in scale, a more valid comparison is found by applying a relative risk measure such as the coefficient of variation. The lower the CV, the lower the relative degree of risk. This is calculated by dividing the standard deviation by the expected value of net cash flows, as in the expression:

$$CV = \sigma/\overline{X}$$

The Snowglo example (Table 7.4) gives the following coefficients:

	Standard deviation (1)	Expected value (2)	Coefficient of variation (1 ÷ 2)
Project A	£173.2	£400	0.43
Project B	£86.6	£400	0.22

Both projects have the same expected value, but project B has a significantly lower degree of risk. Next, we consider the situation where the two projects under review are different in scale:

	Standard deviation		Expected value		Coefficient of variation
Project F	£1,000	÷	£10,000	=	0.10
Project G	£2,000	÷	£40,000	=	0.05

Although the absolute measure of dispersion (the standard deviation) is greater for project G, few people in business would regard it as more risky than project F because of the significant difference in the expected values of the two investments. The coefficient of variation reveals that G actually offers a lower amount of risk per £1 of expected value.

Self-assessment activity 7.3

Project X has an expected return of £2,000 and a standard deviation of £400. Project Y has an expected return of £1,000 and a standard deviation of £400. Which project is more risky?

(Answer in Appendix A at the back of the book)

Mean–variance rule

Given the expected return and the measure of dispersion (variance or standard deviation), we can formulate the **mean–variance rule**. This states that one project will be preferred to another if either of the following holds:

1 Its expected return is *higher* and the variance is *equal* to or *less* than that of the other project.
2 Its expected return *exceeds* or is *equal* to the expected return of the other project and the variance is *lower*.

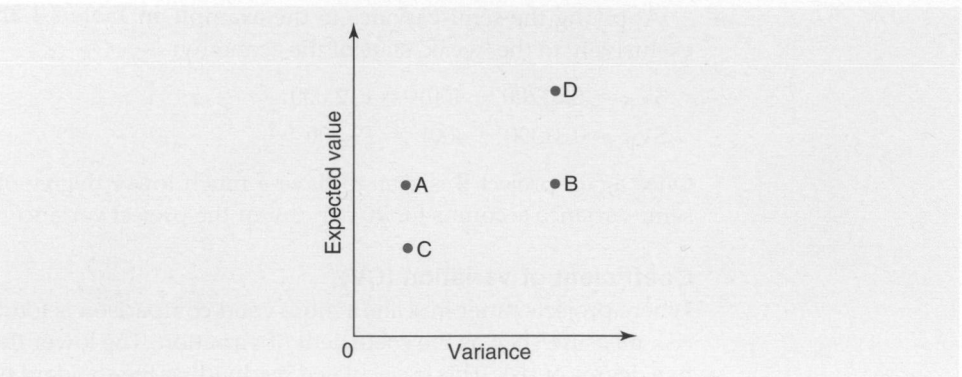

Figure 7.4 Mean–variance analysis

This is illustrated by the mean–variance analysis depicted in Figure 7.4. Projects A and D are preferable to projects C and B respectively because they offer a higher return for the same degree of risk. In addition, A is preferable to B because for the same expected return, it incurs lower risk. These choices are applicable to all risk-averse managers regardless of their particular utility functions. What this rule cannot do, however, is distinguish between projects where both expected returns and risk differ (projects A and D in Figure 7.4). This important issue will be discussed in Chapters 9 and 10.

So far, our analysis of risk has assumed single-period investments. We have conveniently ignored the fact that, typically, investments are multi-period. The analysis of project risk where there are multi-period cash flows is discussed in the appendix to this chapter.

Self-assessment activity 7.4

What do you understand by the following?

(a) risk
(b) uncertainty
(c) risk-aversion
(d) expected value
(e) standard deviation
(f) semi-variance
(g) mean–variance rule

(Answer in Appendix A at the back of the book)

■ Risk-handling methods

There are two broad approaches to handling risk in the investment decision process. The first attempts to *describe* the riskiness of a given project, using various applications of probability analysis or some simple method. The second aims to *incorporate* the investor's perception of project riskiness within the NPV formula.

We turn first to the various techniques available to help describe investment risk.

7.6 RISK DESCRIPTION TECHNIQUES

■ Sensitivity analysis

In principle, sensitivity analysis is a very simple technique, used to isolate and assess the potential impact of risk on a project's value. It aims not to quantify risk, but to identify the impact on NPV of changes to key assumptions. Sensitivity analysis provides the

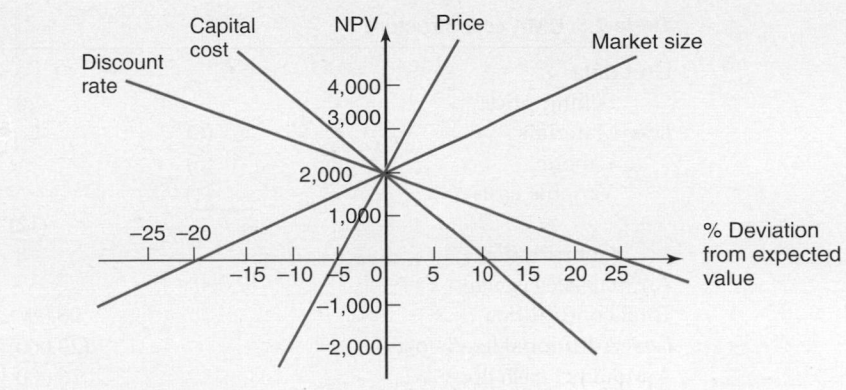

Figure 7.5 Sensitivity graph

decision-maker with answers to a whole range of 'what if' questions. For example, what is the NPV if selling price falls by 10 per cent? What is the IRR if the project's life is only three years, not five years as expected? What is the level of sales revenue required to break even in net present value terms?

Sensitivity graphs permit the plotting of net present values (or IRRs) against the percentage deviation from the expected value of the factor under investigation. The sensitivity graph in Figure 7.5 depicts the potential impact of deviations from the expected values of a project's variables on NPV. When everything is unchanged, the NPV is £2,000. However, NPV becomes zero when market size decreases by 20 per cent or price decreases by 5 per cent. This shows that NPV is very sensitive to price changes. Similarly, a 10 per cent increase in the capital cost will bring the NPV down to zero, while the discount rate must increase to 25 per cent in order to render the project uneconomic. Therefore, the project is more sensitive to capital investment changes than to variations in the discount rate. The sensitivity of NPV to each factor is reflected by the slope of the sensitivity line – the steeper the line, the greater the impact on NPV of changes in the specified variable.

Sensitivity analysis is widely used because of its simplicity and ability to focus on particular estimates. It can identify the critical factors that have greatest impact on a project's profitability. It does not, however, actually *evaluate* risk; the decision-maker must still assess the likelihood of occurrence for these deviations from expected values.

Break-even sensitivity analysis: UMK plc

The accountant of UMK plc has put together the cash flow forecasts for a new product with a four-year life, involving capital investment of £200,000. It produces a net present value, at a 10 per cent discount rate, of £40,920. His basic analysis is given in Table 7.5. Which factors are most critical to the decision?

Investment outlay

This can rise by up to £40,920 (assuming all other estimates remain unchanged) before the decision advice alters. This is a percentage increase of

$$\frac{£40,920}{£200,000} \times 100 = 20.5\%$$

Continued

Table 7.5 UMK cost structure

Unit data	£	£
Selling price		20
Less: Materials	(6)	
Labour	(5)	
Variable costs	(1)	
		(12)
Contribution		8
Annual sales (units)	12,000	
Total contribution		96,000
Less: Additional fixed costs		(20,000)
Annual net cash flow		76,000
Present value (4 years at 10%)		
76,000 × 3.17		240,920
Less: Capital outlay		(200,000)
Net present value		40,920

Annual cash receipts

The break-even position is reached when annual cash receipts multiplied by the annuity factor equal the investment outlay. The break-even cash flow is therefore the investment outlay divided by the annuity factor:

$$\frac{£200,000}{3.17} = £63,091$$

This is a percentage fall of $\dfrac{£76,000 - £63,091}{£76,000} = 17.0\%$

Annual fixed costs could increase by the same absolute amount of £12,909, or

$$\frac{£12,909}{£20,000} \times 100 = 64.5\%$$

Annual sales volume: the break-even annual contribution is £63,091 + £20,000 = £83,091. Sales volume required to break even is £83,091/£8 = 10,386, which is a percentage decline of

$$\frac{12,000 - 10,386}{12,000} \times 100 = 13.5\%$$

Selling price can fall by:

$$\frac{£96,000 - £83,091}{12,000} = £1.07 \text{ per unit}$$

a decline of $\dfrac{£1.07}{£20} \times 100 = 5.4\%$

Variable costs per unit can rise by a similar amount:

$$\frac{£1.07}{£12} \times 100 = 8.9\%$$

Discount rate

The break-even annuity factor is £200,000/£76,000 = 2.63. Reference to the present value annuity tables for four years shows that 2.63 corresponds to an IRR of 19 per cent. The error in cost of capital calculation could be as much as nine percentage points before it affects the decision advice.

Sensitivity analysis, as applied in the above example, discloses that selling price and variable costs are the two most critical variables in the investment decision. The decision-maker must then determine (subjectively or objectively) the probabilities of such changes occurring, and whether he or she is prepared to accept the risks.

■ Scenario analysis

Sensitivity analysis considers the effects of changes in key variables only one at a time. It does not ask the question: 'How bad could the project look?' Enthusiastic managers can sometimes get carried away with the most likely outcomes and forget just what might happen if critical assumptions – such as the state of the economy or competitors' reactions – are unrealistic. Scenario analysis seeks to establish 'worst' and 'best' scenarios, so that the whole range of possible outcomes can be considered. It encourages 'contingent thinking', describing the future by a collection of possible eventualities.

■ Simulation analysis

Monte Carlo simulation
Method for calculating the probability distribution of possible outcomes

An extension of scenario analysis is simulation analysis. **Monte Carlo simulation** is an operations research technique with a variety of business applications. The computer generates hundreds of possible combinations of variables according to a pre-specified probability distribution. Each scenario gives rise to an NPV outcome which, along with other NPVs, produces a probability distribution of outcomes.

One of the first writers to apply the simulation approach to risky investments was Hertz (1964), who described the approach adopted by his consultancy firm in evaluating a major expansion of the processing plant of an industrial chemical producer. This involved constructing a mathematical model that captured the essential characteristics of the investment proposal throughout its life as it encountered random events.

New risks put scenario planning in favour

Who could have predicted the horrific events of September 11, 2001? A 1999 US congressional commission led by former senators Gary Hart and Warren Rudman came close. It warned that the US was 'increasingly vulnerable to attack on our homeland' and that 'rapid advances in information and biotechnologies will create new vulnerabilities'.

But perhaps more important than the commission's prophetic messages was its approach. Instead of forecasting a specific future, it set out a collection of possible attack scenarios. It then evaluated national security by analysing possible policies to prepare for, or respond to them.

This approach – known as scenario planning – has gained renewed popularity among public and private decision-makers.

In January this year, the *New England Journal of Medicine* published a scenario planning analysis on whether US health workers or the whole nation should be vaccinated against smallpox to counter the threat of bio-terrorism. President George W. Bush decided to inoculate 500,000 military personnel and 439,000 health workers.

Scenario planners face three challenges. The first is constructing meaningful scenarios. This requires expert analysis of the factors that affect the outcomes. A second challenge is determining the likelihoods of the scenarios.

Finally, planners must decide on a good criterion for selecting strategies. Most individuals and institutions are risk-averse: they value an uncertain reward at a level significantly below the average level the reward in fact reaches. Strategies with higher average pay-offs often entail greater risks. Hence, scenario planning often involves analysing the reward at different levels of risk – much as is done in financial planning.

What explains the recent interest in scenario planning? For one thing, we live in turbulent times. Terrorism, political instability and threats of war make scenarios of extreme price fluctuations in commodity and energy markets more likely. Severe acute respiratory syndrome, 'mad cow disease' and foot-and-mouth disease have rekindled awareness of the natural biological threats we face. Accounting scandals force us to second-guess what used to be considered accurate information about suppliers and customers. In short, companies face far greater risks than before. Indeed, when Mattel used scenario planning to formulate its 2002 strategy, it considered scenarios with several big customers (such as Kmart, FAO and eToys) going bankrupt and others (Wal-Mart) starting to make their own toys.

Source: Awi Federgruen and Garrett Van Ryzin, *Financial Times*, 19 August 2003, p. 11.

A simulation model might consider the following variables, which are subject to random variation.

Market factors	Investment factors	Cost factors
Market size	Investment outlay	Variable costs
Market growth rate	Project life	Fixed costs
Selling price of product	Residual value	
Market share captured by the firm		

Comparison is then possible between mutually exclusive projects whose NPV probability distributions have been calculated in this manner (Figure 7.6). It will be observed that project A, with a higher expected NPV and lower risk, is preferable to project B.

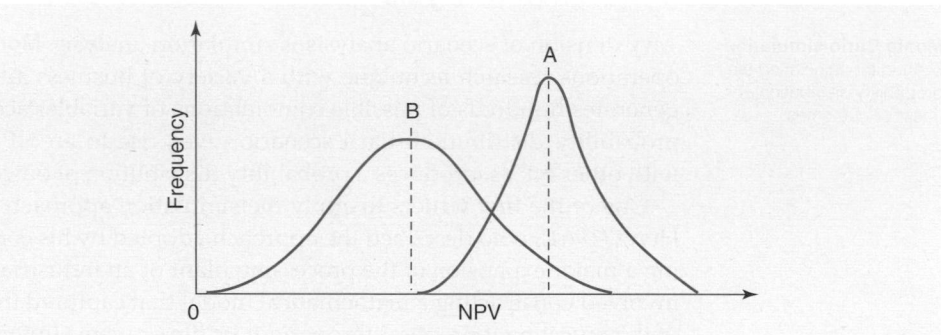

Figure 7.6 Simulated probability distributions

In practice, few companies use this risk analysis approach, for the following reasons:

1 The simple model described above assumes that the economic factors are unrelated. Clearly, many of them (e.g. market share and selling price) are statistically interdependent. To the extent that interdependency exists among variables, it must be specified. Such interrelationships are not always clear and are frequently complex to model.

2 Managers are required to specify probability distributions for the exogenous variables. Few managers are able or willing to accept the demands required by the simulation approach.

Self-assessment activity 7.5

What do you understand by Monte Carlo simulation? When might it be useful in capital budgeting?

(Answer in Appendix A at the back of the book)

7.7 ADJUSTING THE NPV FORMULA FOR RISK

Two approaches are commonly used to incorporate risk within the NPV formula.

■ Certainty equivalent method

This conceptually appealing approach permits adjustment for risk by incorporating the decision-maker's risk attitude into the capital investment decision. The certainty equivalent method adjusts the numerator in the net present value calculation by multiplying the expected annual cash flows by a certainty equivalent coefficient. The revised formula becomes:

$$\overline{NPV} = \sum_{t=1}^{N} \frac{\alpha \overline{X}_t}{(1 + i)^t} - I_o$$

where: $\overline{NPV}$ is the expected net present value; α is the certainty equivalent coefficient, which reflect's management's risk attitude; $\overline{X}_t$ is the expected cash flow in period t; i is the riskless rate of interest; n is the project's life; and I_0 is the initial cash outlay.

The numerator $(\alpha \overline{X}_t)$ represents the figure that management would be willing to receive as a certain sum each year in place of the uncertain annual cash flow offered by the project. The greater is management's aversion to risk, the nearer the certainty equivalent coefficient is to zero. Where projects are of normal risk for the business, and the cost of capital and risk-free rate of interest are known, it is possible to determine the certainty equivalent coefficient.

Example

Calculate the certainty equivalent coefficient for a normal risk project with a one-year life and an expected cash flow of £5,000 receivable at the end of the year. Shareholders require a return of 12 per cent for projects of this degree of risk and the risk-free rate of interest is 6 per cent.

The present value of the project, excluding the initial cost and using the 12 per cent discount rate, is:

$$PV = \frac{£5,000}{1 + 0.12} = £4,464$$

Using the present value and substituting the risk-free interest rate for the cost of capital, we obtain the certainty equivalent coefficient:

$$\frac{\alpha \times £5,000}{1 + 0.06} = £4,464$$

$$\alpha = \frac{(£4,464)(1.06)}{£5,000}$$

$$= 0.9464$$

The management is, therefore, indifferent as to whether it receives an uncertain cash flow one year hence of £5,000 or a certain cash flow of £4,732 (i.e. £5,000 × 0.9464).

■ Risk-adjusted discount rate

Whereas the certainty equivalent approach adjusted the numerator in the NPV formula, the risk-adjusted discount rate adjusts the denominator:

$$\overline{NPV} = \sum_{t=1}^{N} \frac{\overline{X}_t}{(1 + k)^t} - I_o$$

where k is the risk-adjusted rate based on the perceived degree of project risk.

The higher the perceived riskiness of a project, the greater the risk premium to be added to the risk-free interest rate. This results in a higher discount rate and, hence, a lower net present value.

Although this approach has a certain intuitive appeal, its relevance depends very much on how risk is perceived to change over time. The risk-adjusted discount rate involves the impact of the risk premium growing over time at an exponential rate, implying that the riskiness of the project's cash flow also increases over time.

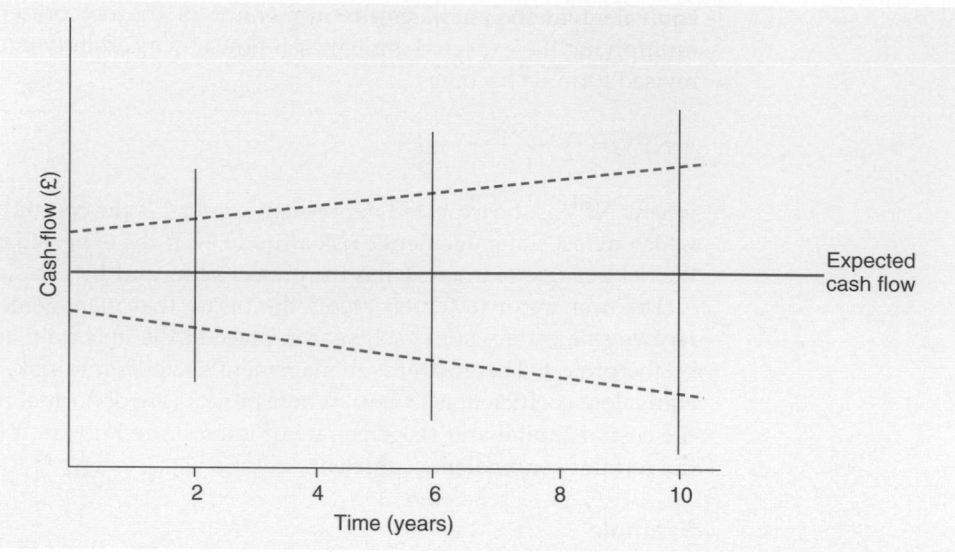

Figure 7.7 How risk is assumed to increase over time

Figure 7.7 demonstrates this point. Although the expected cash flow from a project may be constant over its ten-year life, the riskiness associated with the cash flows increases with time. However, if risk did not increase with time, the risk-adjusted discount rate would be inappropriate.

Adjusting the discount rate: Chox-Box Ltd

Chox-Box Ltd is a manufacturer of confectionery currently appraising a proposal to launch a new product that has had very little pre-launch testing. It is estimated that this proposal will produce annual cash flows in the region of £100,000 for the next five years, after which product profitability declines sharply. As the proposal is seen as a high-risk venture, a 12 per cent risk premium is incorporated in the discount rate. The risk-adjusted cash flow, before discounting at the risk-free discount rate, is therefore £89,286 in Year 1 (£100,000/1.12), falling to £56,742 in Year 5 (£100,000/1.12⁵).

To what extent does this method reflect the actual riskiness of the annual cash flows for Years 1 and 5? Arguably, the greatest uncertainty surrounds the initial launch period. Once the initial market penetration and subsequent repeat orders are known, the subsequent sales are relatively easy to forecast. Thus, for Chox-Box, a single risk-adjusted discount rate is a poor proxy for the impact of risk on value over the project's life, because risk does not increase exponentially with the passage of time, and, in some cases, actually declines over time. The Eurotunnel project provides another illustration of this. By far the greatest risks were in the initial tunnelling and development phases.

A deeper understanding of the relationship between the certainty equivalent and risk-adjusted discount rate approaches may be gained by reading the appendix to this chapter.

7.8 RISK ANALYSIS IN PRACTICE

To what extent do companies employ the techniques discussed in this chapter? Table 7.6 shows changes since 1980.

Use of sensitivity analysis has increased significantly over the period, and is now used in nearly all larger firms. Similar increases here occurred in the use of risk-adjusted discount rates and shortening the payback period. However, the greatest increase has taken place in the use of probability analysis and beta analysis (see Chapters 10 and 11 which cover the Capital Asset Pricing Model).

Table 7.6 Risk analysis in large UK firms

	1980 (%)	1992 (%)	2003 (%)
Sensitivity analysis	42	86	89
Reduced payback period	30	59	75
Risk-adjusted rate	41	64	82
Probability analysis	10	47	77
Beta analysis	–	20	43

Source: Pike (1996), Alkaraan and Northcott (2006)

Mini case

The big gamble: Airbus rolls out its new weapon in its battle with Boeing

The biggest bet placed by Europe's aerospace industry was officially launched in January 2005. The Airbus A380 – a twin-decked behemoth with seats for 555 passengers – was rolled out in Toulouse and gives the company a complete range of models to challenge Boeing, ending the lucrative monopoly Boeing has had in the very large aircraft market for 35 years.

Noel Forgeard, Airbus chief executive, says the company had made a 'successful metamorphosis to a world leader'. He claims the group is almost twice as profitable as Boeing's commercial airplanes division, helped by a 'huge, relentless effort to reduce unit cost and grow our productivity: it is the reason why we can gain market share and grow profitably.'

Boeing has certainly looked increasingly vulnerable. The group's critics say its long years of success led to complacency and it allowed the pace of product innovation to slow as it prioritised short-term earnings over investment. 'Boeing has struggled with the development work needed to take the company into the 21st century,' says Tim Clark, president of Emirates, the Dubai-based airline that is one of the world's most important buyers of long-haul aircraft and will be the biggest operator of the Airbus A380.

Airbus's A380 'will change the game for long-haul airlines and airports,' says Chris Avery, aviation analyst at JP Morgan. 'With operating costs 15 per cent below the B747-400, we believe A380 operations will have an advantage on long-haul services in markets between Europe and Asia, across the Pacific and across the Atlantic.' Boeing, however, thinks the A380 is a white elephant, designed for a world that no longer needs aircraft of such great size. Airbus has 149 orders so far, still short of the 250 that it estimates are needed to give a profit. The $11 billion investment of public and private money is a huge gamble.

Boeing and Airbus agree that air traffic over the next 20 years is expected to increase annually on average by about 5 per cent. But they differ greatly on how airlines will accommodate that. Boeing's vision is based on the 'fragmentation' of aviation markets, reflecting passengers' preference for more point-to-point, non-stop services and more frequent services instead of being routed to destinations via connecting hubs. Airbus accepts that fragmentation, but it also expects consolidation on the main trunk routes.

Airbus seems to be winning the argument. In recent years, it has won over some operators that previously used only Boeing aircraft. But Airbus still has plenty to prove. In Japan, Boeing reigns supreme. There, the government and the aerospace industry are backing Boeing's 7E7.

Another challenge is the weakness of the dollar against the euro. This has the potential to undermine its long-term competitiveness. For most of 2005 and 2006 Airbus is protected, having hedged about $40 billion of revenues at around €1/$1. It has also taken the precaution of pricing most of its purchases in dollars – even in Europe – thereby transferring exchange risk to suppliers. Gerald Blanc, Airbus executive vice-president operations, warned, 'This will probably impair our ability to invest as much in research and development as we have done so far.'

A third challenge is Airbus's ability to show it will not be thrown off course by the change of management at the top. The tussle for supremacy in managing Airbus between the parent company's dominant French and German shareholders means that it is currently without a chief executive at a time when the A380 project is about to enter the crucial phase of flight testing, certification and the build-up of production before the first delivery in 2006.

Source: Based on *Financial Times*, 17 January 2005.

Required

(a) Identify the strategic and financial risks in the Airbus A380 project and suggest how they should be assessed and managed.

(b) Use a search engine to consider the present risks now that the aircraft is in service.

SUMMARY

Risk is an important element in virtually all investment decisions. Because most people in business are risk-averse, the identification, measurement and, where possible, reduction of risk should be a central feature in the decision-making process. The evidence suggests that firms are increasingly conducting risk analysis. This does not mean that the risk dimension is totally ignored by other firms; rather, they choose to handle project risk by less objective methods such as experience, feel or intuition.

We have defined what is meant by risk and examined a variety of ways of measuring it. The probability distribution, giving the probability of occurrence of each possible outcome following an investment decision, is the concept underlying most of the methods discussed. Measures of risk, such as the standard deviation, indicate the extent to which actual outcomes are likely to vary from the expected value.

Key points

■ The expected NPV, although useful, does not show the whole picture. We need to understand managers' attitudes to risk and to estimate the degree of project risk.

■ Three types of risk are relevant in capital budgeting: project risk in isolation, the project's impact on corporate risk and its impact on market risk. The last two are addressed more fully in the following two chapters.

■ The standard deviation, semi-variance and coefficient of variation each measure, in slightly different ways, project risk.

■ Sensitivity analysis and scenario analysis are used to locate and assess the potential impact of risk on project performance. Simulation is a more sophisticated approach, which captures the essential characteristics of the investment that are subject to uncertainty.

■ The NPV formula can be adjusted to consider risk. Adjustment of the cash flows is achieved by the certainty equivalent method. The risk-adjusted discount rate increases the risk premium for higher-risk projects.

Further reading

A fuller treatment of risk is found in Levy and Sarnat (1994). Useful research studies on the use of risk analysis are given in Pike (1988, 1996), Pike and Ho (1991), Mao and Helliwell (1969) and Bierman and Hass (1973).

Appendix
MULTI-PERIOD CASH FLOWS AND RISK

For simplicity, we have so far assumed single-period investments and conveniently ignored the fact that investments are typically multi-period. As risk is to be specifically evaluated, cash flows should be discounted at the risk-free rate of interest, reflecting only the time-value of money. To include a risk premium within the discount rate, when risk is already considered separately, amounts to double-counting and typically understates the true net present value. The expected NPV of an investment project is found by summing the present values of the expected net cash flows and deducting the initial investment outlay. Thus, for a two-year investment proposal:

$$\overline{NPV} = \frac{\overline{X}_1}{1 + i} + \frac{\overline{X}_2}{(1 + i)^2} - I_o$$

where $\overline{NPV}$ is the expected NPV, $\overline{X}_1$ is the expected value of net cash flow in Year 1, $\overline{X}_2$ is the expected value of net cash flow in Year 2, I_o is the cash investment outlay and i is the risk-free rate of interest.

A major problem in calculating the standard deviation of a project's NPVs is that the cash flows in one period are typically dependent, to some degree, on the cash flows of earlier periods. Assuming for the present that cash flows for our two-period project are statistically independent, the total variance of the NPV is equal to the discounted sum of the annual variances.

For example, the Bronson project, with a two-year life, has an initial cost of £500 and the possible payoffs and probabilities outlined in Table 7.7. Applying the standard deviation and expected value formulae already discussed, we obtain an expected NPV of £268 and standard deviation of £206.

Table 7.7 Bronson project payoffs with independent cash flows

Probability	Year 1 cash flow (£)	Year 2 cash flow (£)
0.1	100	200
0.2	200	400
0.4	300	600
0.2	400	800
0.1	500	1,000
Expected value	£300	£600
Standard deviation	£109	£219

Assuming a risk-free discount rate of 10 per cent, the expected NPV is:

$$\overline{NPV} = \frac{300}{(1.10)} + \frac{600}{(1.10)^2} - 500 = £268$$

The standard deviation of the entire proposal is found by discounting the annual variances to their present values, applying the equation:

$$\sigma = \sqrt{\sum_{t=1}^{N} \frac{\sigma_t^2}{(1 + i)^2}}$$

In our simple case, this is:

$$\sigma = \sqrt{\frac{\sigma_1^2}{(1 + i)^2} + \frac{\sigma_2^2}{(1 + i)^4}} = \sqrt{\frac{12,000}{(1.1)^2} + \frac{48,000}{(1.1)^4}} = £206$$

The project therefore offers an expected NPV of £268 and a standard deviation of £206.

■ Perfectly correlated cash flows

At the other extreme from the independence assumption is the assumption that the cash flows in one year are entirely dependent upon the cash flows achieved in previous periods. When this is the case, successive cash flows are said to be perfectly correlated. Any deviation in one year from forecast directly affects the accuracy of subsequent forecasts. The effect is that, over time, the standard deviation of the probability distribution of net present values increases. The standard deviation of a stream of cash flows perfectly correlated over time is:

$$\sigma = \sum_{t=1}^{N} \frac{\sigma_t}{(1+i)^t}$$

Returning to the example in Table 7.7, but assuming perfect correlation of cash flows over time, the standard deviation for the project is:

$$\sigma = \frac{£109}{1.1} + \frac{£219}{(1.1)^2}$$
$$= £280$$

Thus the risk associated with this project is £280, assuming perfect correlation, which is higher than that for independent cash flows. Obviously, this difference would be considerably greater for longer-lived projects.

In reality, few projects are either independent or perfectly correlated over time. The standard deviation lies somewhere between the two. It will be based on the formula for the independence case, but with an additional term for the covariance between annual cash flows.

■ Interpreting results

While decision-makers are interested to know the degree of risk associated with a given project, their fundamental concern is whether the project will produce a positive net present value. Risk analysis can go some way to answering this question. If a project's probability distribution of expected NPVs is approximately normal, we can estimate the probability of failing to achieve at least zero NPV. In the previous example, the expected NPV was £268. This is standardised by dividing it by the standard deviation using the formula:

$$Z = \frac{X - \overline{NPV}}{\sigma}$$

where X in this case is zero and Z is the number of standardised units. Thus, in the case of the independent cash flow assumption, we have:

$$Z = \frac{0 - £268}{£206}$$
$$= -1.30 \text{ standardised units}$$

Reference to normal distribution tables reveals that there is a 0.0968 probability that the NPV will be zero or less. Accordingly there must be a $(1 - 0.0968)$ or 90.32 per cent probability of the project producing an NPV in excess of zero.

It is probably unnecessary to attempt to measure the standard deviation for every project. Even the larger European companies tend to use probability analysis sparingly in capital project analysis. Unless cash flow forecasting is wildly optimistic, or the future economic conditions underlying all investments are far worse than anticipated, the bad news from one project should be compensated by good news from another project.

Sometimes, however, a project is of such great importance that its failure could threaten the very survival of the business. In such a case, management should be fully aware of the scale of its exposure to loss and the probability of this occurring.

■ Probability of failure: Microloft Ltd

Microloft Ltd, a local family-controlled company specialising in attic conversions, is currently considering investing in a major expansion giving wider geographical coverage. The NPV from the project is expected to be £330,000 with a standard deviation of £300,000. Should the project fail (perhaps because of the reaction by major competitors), the company could afford to lose £210,000 before the bank manager 'pulled the plug' and put in the receiver. What is the probability that this new project could put Microloft out of business?

We need to find the value of Z where X is the worst NPV outcome that Microloft could tolerate:

$$Z = \frac{X - \overline{\text{NPV}}}{\sigma}$$

$$= \frac{-£210 - £330}{£330} = -1.8$$

Assuming the outcomes are normally distributed, probability tables will show a 3.6 per cent chance of failure from accepting the project. A family-controlled business, like Microloft, may decide that even this relatively small chance of sending the company on to the rocks is more important than the attractive returns expected from the project.

QUESTIONS

 Questions with an icon are also available for practice in myfinancelab with additional supporting resources.

Questions with a **coloured number** have solutions in Appendix B on page 734.

1 Explain the importance of risk in capital budgeting.

2 Explain the distinction between project risk, business risk, financial risk and portfolio risk.

3 The 'wood pulp' project has an initial cost of £13,000 and the firm's risk-free interest rate is 10 per cent. If certainty equivalents and net cash flows (NCF) for the project are as below, should the project be accepted?

Year	Certainty equivalents	Net cash flows (£)
1	0.90	8,000
2	0.85	7,000
3	0.80	7,000
4	0.75	5,000
5	0.70	5,000
6	0.65	5,000
7	0.60	5,000

4 Mystery Enterprises has a proposal costing £800. Using a 10 per cent cost of capital, compute the expected NPV, standard deviation and coefficient of variation, assuming independent interperiod cash flows.

Probability	Year 1 net cash flow (£)	Year 2 net cash flow (£)
0.2	400	300
0.3	500	400
0.3	600	500
0.2	700	600

5 Mikado plc is considering launching a new product involving capital investment of £180,000. The machine has a four-year life and no residual value. Sales volumes of 6,000 units are forecast for each of the four years. The product has a selling price of £60 and a variable cost of £36 per unit. Additional fixed overheads of £50,000 will be incurred. The cost of capital is 12.5 per cent p.a. Present a report to the directors of Mikado plc giving:

(a) the net present values
(b) the percentage amount each variable can deteriorate before the project becomes unacceptable
(c) a sensitivity graph

6 Devonia (Laboratories) Ltd has recently carried out successful clinical trials on a new type of skin cream, which has been developed to reduce the effects of ageing. Research and development costs in relation to the new product amount to £160,000. In order to gauge the market potential of the new product, an independent firm of market research consultants was hired at a cost of £15,000. The market research report submitted by the consultants indicates that the skin cream is likely to have a product life of four years and could be sold to retail chemists and large department stores at a price of £20 per 100 ml container. For each of the four years of the new product's life sales demand has been estimated as follows:

Number of 100 ml containers sold	Probability of occurrence
11,000	0.3
14,000	0.6
16,000	0.1

If the company decides to launch the new product, production can begin at once. The equipment necessary to make the product is already owned by the company and originally cost £150,000. At the end of the new product's life, it is estimated that the equipment could be sold for £35,000. If the company decides against launching the new product, the equipment will be sold immediately for £85,000 as it will be of no further use to the company.

The new skin cream will require two hours' labour for each 100 ml container produced. The cost of labour for the new product is £4.00 per hour. Additional workers will have to be recruited to produce the new product. At the end of the product's life the workers are unlikely to be offered further work with the company and redundancy costs of £10,000 are expected. The cost of the ingredients for each 100 ml container is £6.00. Additional overheads arising from the product are expected to be £15,000 p.a.

The new skin cream has attracted the interest of the company's competitors. If the company decides not to produce and sell the skin cream, it can sell the patent rights to a major competitor immediately for £125,000.

Devonia (Laboratories) Ltd has a cost of capital of 12 per cent.

Ignore taxation.

Required

(a) Calculate the expected net present value (ENPV) of the new product.

(b) State, with reasons, whether or not Devonia (Laboratories) Ltd should launch the new product.

(c) Discuss the strengths and weaknesses of the expected net present value approach for making investment decisions.

(Certified Diploma)

7 Plato Pharmaceuticals Ltd has invested £300,000 to date in developing a new type of insect repellent. The repellent is now ready for production and sale and the marketing director estimates that the product will sell 150,000 bottles per annum over the next five years. The selling price of the insect repellent will be £5 per bottle and the variable costs are estimated to be £3 per bottle. Fixed costs (excluding depreciation) are expected to be £200,000 per annum. This figure is made up of £160,000 additional fixed costs and £40,000 fixed costs relating to the existing business which will be apportioned to the new product.

In order to produce the repellent, machinery and equipment costing £520,000 will have to be purchased immediately. The estimated residual value of this machinery and equipment in five years time is £100,000. The company calculates depreciation on a straight-line basis.

The company has a cost of capital of 12 per cent. Ignore taxation.

Required

(a) Calculate the net present value of the product.

(b) Undertake sensitivity analysis to show by how much the following factors would have to change before the product ceased to be worthwhile:
   (i)   the discount rate
   (ii) the initial outlay on machinery and equipment
   (iii) the net operating cash flows
   (iv) the residual value of the machinery and equipment

(c) Discuss the strengths and weaknesses of sensitivity analysis in dealing with risk and uncertainty.

(d) State, with reasons, whether or not you feel the project should go ahead.

(Certified Diploma)

8 The managing director of Tigwood Ltd believes that a market exists for 'microbooks'. He has proposed that the company should market 100 best-selling books on microfiche, which can be read using a special microfiche reader that is connected to a television screen. A microfiche containing an entire book can be purchased from a photographic company at 40 per cent of the average production cost of best-selling paperback books.

The average cost of producing paperback books is estimated at £1.50, and the average selling price of paperbacks is £3.95 each. Copyright fees of 20 per cent of the average selling price of the paperback books would be payable to the publishers of the paperbacks plus an initial lump sum that is still being negotiated, but is expected to be £1.5 million. No tax allowances are available on this lump-sum payment. An agreement with the publishers would be signed for a period of six years. Additional variable costs of staffing, handling and marketing are 20p per microfiche, and fixed costs are negligible.

Tigwood Ltd has spent £100,000 on market research, and expects sales to be 1,500,000 units per year at an initial unit price of £2.

The microfiche reader would be produced and marketed by another company.

Tigwood would finance the venture with a bank loan at an interest rate of 16 per cent per year. The company's money (nominal) cost of equity and real cost of equity are estimated to be 23 per cent p.a. and 12.6 per cent p.a., respectively. Tigwood's money weighted average cost of capital and real weighted average cost of capital are 18 per cent p.a. and 8 per cent p.a., respectively. The risk-free rate of interest is 11 per cent p.a. and the market return is 17 per cent p.a.

Corporation tax is at the rate of 35 per cent, payable in the year the profit occurs. All cash flows may be assumed to be at the year end, unless otherwise stated.

Required

(a) Calculate the expected net present value of the microbooks project.

(b) Explain the reasons for your choice of discount rate in the answer to part (a). Discuss whether this rate is likely to be the most appropriate to use in the analysis of the proposed project.

(c) (i) Using sensitivity analysis, estimate by what percentage each of the following would have to change before the project was no longer expected to be viable:

    initial outlay
    annual contribution
    the life of the agreement
    the discount rate

(ii) What are the limitations of this sensitivity analysis?

(d) What further information would be useful to help the company decide whether to undertake the microbook project?

(ACCA)

9 The general manager of the nationalised postal service of a small country, Zedland, wishes to introduce a new service. This service would offer same-day delivery of letters and parcels posted before 10 a.m. within a distance of 150km. The service would require 100 new vans costing $8,000 each and 20 trucks costing $18,000 each. One hundred and eighty new workers would be employed at an average annual wage of $13,000, and five managers on average annual salaries of $20,000 would be moved from their existing duties, where they would not be replaced.

Two postal rates are proposed. In the first year of operation letters will cost $0.525 and parcels $5.25. Market research undertaken at a cost of $50,000 forecasts that demand will average 15,000 letters per working day and 500 parcels per working day during the first year, and 20,000 letters per day and 750 parcels per day thereafter. There is a five-day working week. Annual running and maintenance costs on similar new vans and trucks are estimated to be $2,000 per van and $4,000 per truck, respectively, in the first year of operation. These costs will increase by 20 per cent p.a. (excluding the effects of inflation). Vehicles are depreciated over a five-year period on a straight-line basis. Depreciation is tax-allowable and the vehicles will have negligible scrap value at the end of five years. Advertising in Year 1 will cost $500,000 and in Year 2 $250,000. There will be no advertising after Year 2. Existing premises will be used for the new service, but additional costs of $150,000 per year will be incurred.

All the above cost data are current estimates and exclude any inflation effects. Wage and salary costs and all other costs are expected to rise because of inflation by approximately 5 per cent p.a. during the five-year planning horizon of the postal service. The government of Zedland will not permit annual price increases within nationalised industries to exceed the level of inflation.

Nationalised industries are normally required by the government to earn at least an annual after-tax return of 5 per cent on average investment and to achieve, on average, at least zero net present value on their investments.

The new service would be financed half by internally generated funds and half by borrowing on the capital market at an interest rate of 12 per cent p.a. The opportunity cost of capital for the postal service is estimated to be

14 per cent p.a. Corporate taxes in Zedland, to which the postal service is subject, are at the rate of 30 per cent for annual profits of up to $500,000 and 40 per cent for the balance in excess of $500,000. Tax is payable one year in arrears. All transactions may be assumed to be on a cash basis and to occur at the end of the year, with the exception of the initial investment, which would be required almost immediately.

Required

(a) Acting as an independent consultant, prepare a report advising whether the new postal service should be introduced. Include a discussion of other factors that might need to be taken into account before a final decision was made. State clearly any assumptions that you make.

(b) Monte Carlo simulation has been suggested as a possible method of estimating the net present value of a project. Briefly assess the advantages and disadvantages of using this technique in investment appraisal.

(ACCA)

Practical assignment

Describe the types of risk associated with investment decisions in a firm known to you. (If necessary, read the Annual Report of a major company, like BP plc, to familiarise yourself with a company.) Suggest how these risks should be formally assessed within their investment appraisal process.

 *Now retake your diagnostic test for Chapter 7 to check your progress and update your study plan.*

8

Identifying and valuing options*

The marriage (option) contract

'Wilt thou have this woman to be thy lawful wedded wife?' As the minister posed this question, the groom, a highly trained financial analyst, reflected that marriage bore many characteristics of the investment decisions he faced every day. It was clearly a long-term commitment, involving uncertain costs and benefits. What is more, it was largely irreversible; and the abandonment option of divorce was too costly and painful to contemplate. The follow-on option was interesting. How many kids had she mentioned?

But what about the options he would be sacrificing? The option to remain an independent bachelor, or the option to wait another six months to be better informed as to whether the couple were really right for each other. Just how valuable was the marriage option contract, and was now the right expiry date?

After a moment's hesitation, the groom turned and looked into his bride's adoring eyes. All thoughts of rational economic analysis and option theory evaporated as he found himself saying, 'I will'.

Learning objectives

By the end of this chapter, you should possess a clear understanding of the following:

- The basic types of option and how they are employed.

- The main factors determining option values.

- How options can be used to reduce risk.

- How option values can be estimated.

- The various applications of option theory to investment and corporate finance.

- Why conventional net present value analysis is not sufficient for appraising projects.

 Complete your diagnostic test for Chapter 8 now to create your personal study plan.

---

*The authors are grateful for the contribution to this chapter by Andrew Marshall.

8.1 INTRODUCTION

Business managers like to 'keep their options open'. Options convey the right, but not the obligation, to do something in the future. Like getting married, most business decisions involve closing off certain options while opening up others. Managers should seek to create capital projects or financial instruments with valuable options embedded in them. For example, an investment proposal will be worth more if it contains the flexibility to exit relatively cheaply should things go wrong. This is because the 'downside' risk is minimised. Often it is not possible, or too costly, to build in such options. However, the financial manager can achieve much the same effect by creating options in financial markets.

Options are derivative assets. A 'derivative' is an asset which derives its value from another asset. The primary asset is referred to as the **'underlying' asset**. Option-like features occur in various aspects of finance, and option theory provides a powerful tool for understanding the value of such options.

'underlying' asset
The asset from which option value is derived

This chapter examines the nature and types of options available and how they can reduce risk or add value. It also explains why the conventional net present value approach may not tell the whole story in appraising capital projects. But to cover the basics of option theory we will begin by considering the options as shares.

8.2 SHARE OPTIONS

In finance, *options are contractual arrangements giving the owner the right, but not the obligation, to buy or sell something, at a given price, at some time in the future.* Note the two key elements in options: (1) the right to *choose* whether or not to take up the option, and (2) at an *agreed* price. It is not a true option if I am free to buy in the future at the *prevailing* market price, or if I am *compelled* to buy at an agreed price. Many securities have option features: for example, convertible bonds and share warrants, where options to convert to, or acquire, equity are given to the owner.

■ Share options in Enigma Drugs plc

The simplest form of share option is when a company issues them as a way of rewarding employees. If the current share price of Enigma Drugs plc is £4, it might award share options to some of its employees at, say, the same price. If, over the period in which the shares can be exercised, the shares go up, employees could then purchase shares at a price below market price, either to sell at a profit or gain an equity interest.

Most options relate to assets which already exist. These are termed pure options. To begin with, we will consider pure share options, although much of our analysis could apply equally to interest rates, currency, oil and commodity markets. But first we need to go back to basics. Figure 8.1(a) depicts the payoff line for investing in ordinary shares in Enigma Drugs plc. If the shares are bought today at 400p, the payoff, or gain, from selling at the same price is zero. If, say, three months later, the share price has risen to 450p, the payoff is +50p; but if it has dropped to 350p, the payoff is −50p. The line is drawn at 45 degrees because a 50p increase in share price from its current level of 400p gives a 50p payoff.

We have all seen the warnings accompanying advertisements for financial products, reminding us that share prices can go up or down. But wouldn't it be nice if, whichever way prices moved, you ended up a winner? This can be done if you also acquire share options. With options you can create a 'no lose' option strategy providing protection from a drop in share price, as shown in Figure 8.1(b). The arrowed line represents the payoff from the option to buy shares. If share price increases, so does the option payoff.

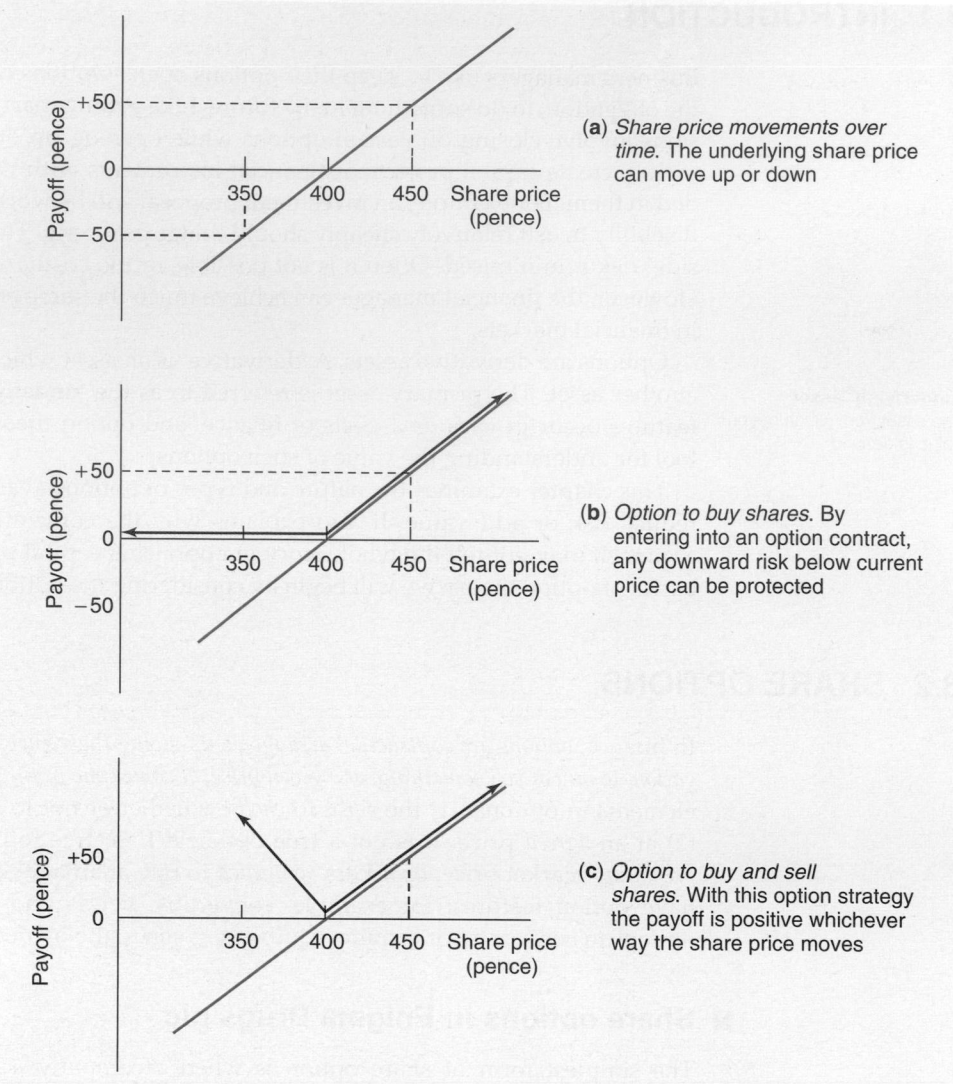

Figure 8.1 Payoff lines for shares and share options in Enigma Drugs plc
(underlying share price is blue, option contract is red)

But if the share price falls below 400p, to say 350p, the option payoff remains at zero. In other words, the holder of the option will not exercise it to buy the shares.

By combining different types of option you can even create a 'win–win' situation. For example, Figure 8.1(c) shows the effects of combining options to buy and sell shares at a fixed price. Here, either a rise or a fall in share price gives you a positive payoff. At this stage, however, you do not need to know how this is achieved, simply that it can be done. Of course, few things in life are free and a share option is not one of them; there is a price for each option. Later we will show how to value such options.

■ Issuing options

Options on shares are not issued by the companies on whose shares they are written but by large financial institutions, such as insurance companies. The companies play no role in the issuing process. For the institutions issuing options the primary motivation is the fee income that their sale generates, but some also use options in their portfolio of management activities to limit their risk exposure.

Considerable interest has developed in recent years in share options, and there is a highly active market on the Stock Exchange for traditional and traded options. **Traditional options** are available on most leading shares and last for three months. A problem with these is that they are not particularly flexible or negotiable: investors must either exercise the option (i.e. buy or sell the underlying share) or allow it to lapse; they cannot trade the option.

To overcome these difficulties, **traded options** markets were established, first in Chicago, then in Amsterdam (the European Options Exchange). In 1978 the London Traded Options Market was established for major companies, now part of the London International Financial Futures and Options Exchange (LIFFE) (see **www.liffe.com**).

An exchange-traded contract is characterised by certain standardised features, particularly the exercise date and the exercise price. This makes it far easier to develop a continuous market in options than was the case for traditional options that were developed and traded on an ad hoc basis.

traditional options
An option available on any security agreed between buyer and seller. It typically lasts for three months

traded options
An option traded on a market

Options terminology

This topic has more than its fair share of esoteric jargon, some of the more essential of which are defined below.

- A **call option** gives its owner the right to *buy* specific shares at a fixed price – the **exercise price** or **strike price**.

- A **put option** gives its owner the right to *sell* (put up for sale) shares at a fixed price.

- A **European option** can be exercised only on a particular day (i.e. the end of its life), while an **American option** may be exercised at any time up to the date of expiry. These terms are a little confusing because most options traded in the UK and the rest of Europe are actually American options!

- The **premium** is the price paid for the option. Option prices are quoted for shares and traded in contracts (or units) each containing 1,000 shares.

- **'In the money'** is where the exercise price for a call option is *below* the current share price. In other words, it makes sense to take up the option.

- **'Out of the money'** is where the exercise price for a call option is *above* the current share price and it is not profitable to take up the option.

- **Long and short positions** – when an investor buys an option the investor is 'long', and when the investor sells an option the investor takes a 'short' position.

option contract
A contract giving one party the right, but not the obligation, to buy or sell a financial instrument or commodity at an agreed price at or before a specified date

There are two parties to an **option contract**, the buyer (or option holder) and seller (or option writer). The buyer has the right, but not the obligation, to exercise the option. One feature of an option is that, if the share price does not move as expected, it can become completely worthless, regardless of the solvency of the company to which it relates. However, if it does move in line with expectations, very considerable gains can be achieved for very little outlay. Such volatility gives share options a reputation as a highly speculative investment. But, as will be seen later, options can also be used to reduce risk.

In return for the option, the purchaser pays a fee or **premium**. The premium is a small fraction of the share price, and offers holders the opportunity to gain significant benefits while limiting their risk to a known amount. The size of the premium depends on the **exercise price** and expected volatility of shares, which, in turn, is a function of the state of the market and the underlying risk of the share. The premium might range from as little as 3 per cent for a well-known share in a 'quiet' market to over 20 per cent for shares of smaller companies in a more volatile market. During past stock market collapses, or where there is substantial volatility, option premiums have shot up dramatically to reflect such uncertainty.

exercise price
Price at which an option can be taken up

call option
The right to buy an asset at a specified price on or before expiry date

put option
The right to sell an asset at a specified price on or before expiry date

A **call option** gives the purchaser the right to buy a share at a given price within a set period, usually three months; a **put option** gives the right to sell. Payment for the option is not immediate, but takes place when the option is exercised (i.e. taken up) or on expiry, if it is not exercised. The seller of an option must meet his or her obligation to buy or sell shares if the right of the purchaser is so exercised. The reward is, of course, the premium received. So a 3-month call option with an exercise price of 220p on a share currently priced at 225p gives you the right to buy the share at 220p at any time before its expiry.

Table 8.1 shows the prices (or premiums) at which options on BP are traded on a particular day on the traded options market of the Stock Exchange. The options are traded over a nine-month period with expiry dates every three months. Two exercise prices are given; the first, at 390p, is below the current share price of 397p ('in the money') and the second, at 420p, is above the current price ('out of the money'). Notice that option prices vary both with the agreed exercise price (the lower the exercise price for a call option, the higher the premium) and the **exercise date** (the longer the period, the higher the premium). To buy a call option on BP shares, at an exercise price of 390p, costs 17p for expiry in April, but $31\frac{1}{2}$p for expiry six months later in October.

exercise date
Final date on which an option is exercised or expires

Table 8.1 Option on BP shares (current price 397p)

Exercise price	Call option prices (p)			Put option prices (p)		
	April	July	Oct	April	July	Oct
390	17	$24\frac{1}{2}$	$31\frac{1}{2}$	$6\frac{1}{2}$	12	17
420	$4\frac{1}{2}$	12	$27\frac{1}{2}$	$24\frac{1}{2}$	28	33

Self-assessment activity 8.1

By now your head may be spinning with all the terms and concepts introduced. It is therefore a good time to take stock of what you should know.

1 Define a call option and a put option.
2 What is the basic difference between European and American call options?
3 Options are available on what types of asset?
4 In relation to the information on traded options in Table 8.1 explain why the following features were observed:
   (a) the lower the exercise price, the higher the value of a call;
   (b) the greater the time to maturity, the higher the price of a call; and
   (c) the price paid for calls exceeds the gross profit that could be made by immediately exercising the call.

(Answer in Appendix A at the back of the book)

A call option has value if the price of the underlying share is *above* the option's exercise price ('in the money'). A put option has value if the price of the underlying share is *below* the option's exercise price. This would allow you to sell shares at a higher price (the exercise price) than they are currently trading at.

Options are only exercised if they have value (i.e. if they are 'in the money'). Conversely, options that are of no value (i.e. 'out of the money') are abandoned.

For a call option:

Profit = share price − exercise price − premium paid

For a put option:

Profit = exercise price − share price − premium paid

Self-assessment activity 8.2

1 Which of the following options has intrinsic value at the start of the contract?
   (a) A 3-month call option with an exercise price of 230p on a share currently priced at 240p.
   (b) A 3-month put option with an exercise price of 230p on a share currently priced at 240p.

2 Which of the following options is 'out of the money'?
   (a) A 3-month call option with an exercise price of 240p on a share currently trading at 230p.
   (b) A 3-month put option with an exercise price of 230p on a share currently priced at 215p.

(Answer in Appendix A at the back of the book)

■ Speculative use of options: Kate Casino

Kate Casino thinks that oil prices will rise in the coming months and that the BP share price will move up sharply from its current level of 397p to a level in April sufficiently above the exercise price of 390p to justify the option price of 17p. Kate instructs her broker to purchase a contract for 1,000 April call options at a cost of (17p × 1,000) = £170. This is termed a *naked* option, held on its own rather than as a hedge against loss. It is a *long* call because Kate is *buying* the option.

The current share price of 397p is above the 390p exercise price, so the option is already 'in the money', since Kate could immediately exercise her option to buy shares at 390p to gain 7p, before transaction costs. Of course, she would not do this because the premium to be paid for the option is 17p.

Let us look at three possible share prices arising in April when the option ends:

Best – if the takeover attempt is made, BP's share price will rise to 460p by April.
Likely – it will do no better than 415p.
Worst – it falls as low as 380p.

Her profit in each case would be:

Kate Casino's profit on the call option (pence)

	Best	Likely	Worst
Share price in April (pence)	460	415	380
Less exercise price (pence)	(390)	(390)	(390)
Profit on exercise (pence)	70	25	Not exercised
Less option premium paid (pence)	(17)	(17)	(17)
Profit (loss) before transaction costs	53	8	(17)
Profit on contract of 1,000 shares	£530	£80	(£170)

Kate would obviously not exercise her option if the price fell to 380p, so the loss in this case would be restricted to the 17p premium paid. The premium is the maximum loss on the contract. If BP's price on expiry is 415p, the contract profit is a modest £80. But if the share price shoots up to 460p, a large gain of £530 is made on the contract.

It is interesting to compare the option returns with those from investing directly in BP shares. Table 8.2 shows that buying a call option has very different effects from buying the underlying share:

1 The capital outlay for the option contract for 1,000 shares is much smaller (£170 compared with £3,970 for the underlying shares).
2 The downside risk on the option contract is far greater in *relative* terms, but not in *absolute* terms. Kate Casino loses all her initial investment on the option contract

Table 8.2 Returns on BP shares and options

Expiry share price	460p	380p
Buy 1,000 shares	£	£
Cost 397p each	(3,970)	(3,970)
Proceeds from sale	4,600	3,800
Profit (loss)	630	(170)
Return over 3 months on original cost	+15.9%	−4.3%
Call option on 1,000 shares	£	£
Cost of option	(170)	(170)
Cost of exercise at 390p	(3,900)	–
	(4,070)	
Proceeds on sale	4,600	–
Profit (loss)	530	(170)
Return over 3 months	+312%	−100%

while the share value declines by only 4 per cent. However, in money terms, the loss is the same for both, £170. Of course, the shareholder does not have to sell at the option expiry date.

3 The return achieved, if the shares reach 460p, is a phenomenal 312 per cent on the options contract compared with 16 per cent on the underlying shares (ignoring dividends).

The payoff chart in Figure 8.2 shows that, if the share price does not rise above the exercise price of 390p, the option is worthless. The option breaks even at 407p (390p + 17p premium) and the potential profit to be made thereafter is unlimited.

In general, the value of a call option at expiry (C_1) with a share price (S_1) and exercise price (E) is:

$$C_1 = S_1 - E, \text{ if } S_1 > E$$

At a share price of 460p, therefore, the option value is

$$C_1 = 460p - 390p$$

$$= 70p, \text{ giving a profit of 53p per share after paying the premium.}$$

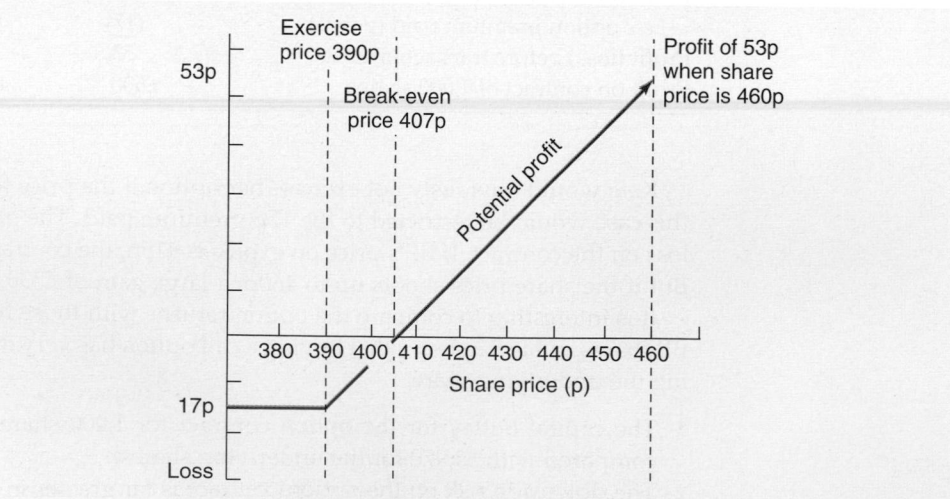

Figure 8.2 BP call option

■ Options as a hedge: Rick Aversion

While options offer an excellent opportunity to speculate, they are equally useful as a means of risk reduction, insurance or **hedging**. Rick Aversion is concerned that the current share price on his BP shareholding will fall over the next two months. Because he wants to keep his shares as a long-term investment, he buys a put option (see Table 8.1), giving the right to sell shares in April at the strike price of 390p. This option costs him $6\frac{1}{2}$p.

By late April, the shares have fallen to 350p, and the option has increased in value to 40p (i.e. 390p − 350p). So Rick sells the option and retains the shares, using the profit on the option to offset the loss on the shares. This is administratively cheaper and more convenient than selling his shares and then buying them back. In this way, investors can capture any profits from a rise in share price and hedge against any price fall.

Figure 8.3 shows that when share prices are above the exercise price, the option value of the put is worthless. It should be exercised when prices fall below the 390p exercise price and breaks even at $383\frac{1}{2}$p (exercise price less premium).

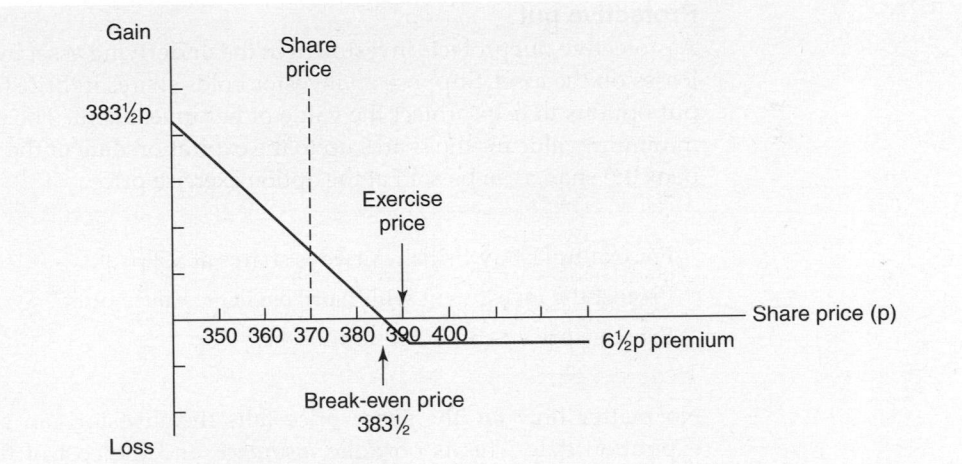

Figure 8.3 BP put option

In general, the value of a put option (P_1) is:

$$P_1 = E - S_1, \text{ if } S_1 < E$$

At a 370p share price, the value of the put option is therefore 20p:

$$P_1 = 390\text{p} - 370\text{p}$$
$$= 20\text{p}$$

This is because the holder could buy the shares in the market at 370p and exercise the option to sell at 390p. The profit on this transaction is $(20\text{p} - 6\frac{1}{2}\text{p}) = 13\frac{1}{2}\text{p}$.

■ Option strategies

Combinations of investments in options and the underlying shares are both of practical and analytical interest. From a practical standpoint a combination can provide a means of reducing exposure to the risks associated with substantial changes in the price of the underlying asset, or, from a speculative perspective, can provide some interesting payoff patterns. The analytical interest stems from the insights provided by such portfolios for the valuation of options.

Straddles or doubles

Combining investments in a put and a call, written on a share at the same exercise price and expiry date, produces what is referred to as a straddle (a long straddle is where you buy both the call and put option and the opposite is true: a short straddle is where you sell both the call and put option). Why should anyone wish to invest in calls and puts simultaneously (a short straddle)? This strategy will be employed by an investor who believes that the price of the underlying share is going to change quite significantly, but is unable to predict the direction of the change. Such an expectation will arise, for example, whenever an investor knows that a company is expected to make an important announcement, but has no knowledge of the content of the announcement.

This is a strategy to adopt whenever there is considerable short-term uncertainty about the price of a share, and it is anticipated that this uncertainty will be resolved before the expiry of the options. A straddle will lose an investor money if there is little change in the share price, but large price changes in either direction will produce gains.

Protective put

A protective put protects investment in the underlying asset by restricting the possible losses on the asset. Suppose an investor holds shares in British Airways. She may buy put options to help protect the value of her investment. The put options guarantee a minimum value for the shares up to the expiration date of the options. Whatever happens the shares can be sold at the option exercise price.

> For example: buy British Airways shares at 579p
>
> Protect the investment with April *out of the money* puts
>
> Exercise price = 550p Premium cost = 33p

No matter how far the share price falls the investor can sell for 550p up to the expiration date. This is portfolio *insurance*, and the cost of the insurance is the put premium of 33p.

Alternatively, the investor might choose to protect the investment with *in the money* puts, e.g. April 600p. Obviously this guarantees a higher minimum selling price up to expiration, but the premium cost is higher.

Covered call

The writing of calls is a risky business. One way of limiting the risk exposure is to buy the share on which the call is being written. When an investor simultaneously writes a call and purchases the underlying asset the resulting combination is known as a covered call. The returns on the written call and the share are negatively correlated as the liability implicit in the written call increases with rises in the value of the share. Covered calls may appeal to risk-averse investors who are mildly pessimistic about the future price performance of a share. The fee income from writing calls is attractive and holding the share implies the risk from an unanticipated rise in the share price is neutralised. (Covered calls are of no interest to really pessimistic investors. They will not wish to purchase the share even if the writing of calls is attractive, and if they already hold the share they will consider its sale or the purchase of puts rather than the writing of calls.)

Covered calls are the combination most frequently employed by financial institutions that regularly write calls. Let us recap on the value of the two main forms of option, the call and put.

Option	Share price at expiry date	Option value at expiry
Call	above exercise price	share price − exercise price
	below exercise price	zero
Put	above exercise price	zero
	below exercise price	exercise price − share price

Consider the implications of buying a call option and selling a put option, where the exercise price is the same for both.

8.3 OPTION PRICING

We can now focus on what determines the value of options. Option prices comprise two elements: intrinsic value and time-value. *Intrinsic value* is what the option would be worth were it about to expire; it reflects the degree to which an option is 'in the money' – in the case of a call option, the extent to which the exercise price is below the current share price. The *time-value* element depends on the length of time the option has to run – the longer the period, the better the chance of making a gain on the contract.

■ Put–call parity

We showed in the previous section how the combination of buying a call option and selling a put option gave the same payoff as the underlying share price. To find a combination that yields a riskless return, we reverse the options.

When a call and a put are written on the same asset with the same exercise price and expiry date a relationship, referred to as the put–call parity, can be expected to hold between their market values. The price of one share(S) plus the price of one put option(P) must equal the value of investing the exercise price(E) until expiry at the risk-free rate of interest (R_f) plus the price of one call option.

> **Put–call parity**
>
> Value of share + Value of put = Present value of exercise price + Value of call
>
> $$S \quad + \quad P \quad = \quad E/(1 + R_f) \quad + \quad C$$
>
> It follows from the above that, given four of the five factors, the fifth can be estimated.

The net cash flow expected from investing in a put and a share is equivalent to that to be expected from investing in a call and placing the present value of the sum necessary to exercise the call in a risk-free investment. In both cases the investor will be left with a sum equivalent to the exercise price if the share price is less than the exercise price, and the value of the share if its price exceeds the exercise price.

? Self-assessment activity 8.3

You take out options contracts to sell a call and buy a put, both at the exercise price of 55p, exercisable one year hence. The cost of the put is 7p and the cost of the call is 1p. The current share price is 44p and the risk-free interest rate is 10 per cent. What is the present value of the exercise price?

(Answer in Appendix A at the back of the book)

Applying the put–call parity model

Melody plc has shares currently trading at 29p. Put and call options for the company are available with an exercise price of 30p expiring in one year. The price of a call option is 6p and the risk-free rate of interest is 6 per cent. What is the price of the put option?

Using the put–call parity model we know that:

$$S + P = E/(1 + R_f) + C \qquad \text{(Eqn 1)}$$

$$\text{or} \qquad P = E/(1 + R_f) + C - S \qquad \text{(Eqn 2)}$$
$$= 30p/(1 + .06) + 6p - 29p$$
$$= 5.3p$$

The price of the put option is 5.3p. At any other price investors would have an arbitrage opportunity to make a profit for zero risk by simultaneously buying and selling identical assets at different prices. Suppose that the actual market price of a put option for Melody plc was 6p. Looking at equation 2 above, we see that an arbitrage opportunity would arise from selling the put option at 6p and investing 30p in a riskless bond, for one year at 6 per cent, and at the same time acquiring a call option at 6p and selling a share priced at 29p.

■ Valuing a call

The following notation is employed with respect to valuing call options:

S_0 = Share price today

S_1 = Share price at expiry date

E = Exercise price on the option

C_0 = Value of call option today

C_1 = Value of call option on expiration date

R_f = Risk-free interest rate

A number of formal statements can be made about call options:

1 *Option prices cannot be negative.* If the share price ends up below the exercise price on the expiration date, the call option is worthless, but no further loss is created beyond that of the initial premium paid. In mathematical terms:

$$C_1 = 0 \quad \text{if} \quad S_1 \leq E \qquad (8.1)$$

This is the case where an option is 'out of the money' on expiry.

2 *An option is worth on expiry the difference between the share price and the exercise price.*

$$C_1 = S_1 - E \quad \text{if} \quad S_1 > E \qquad (8.2)$$

This is the case where an option is 'in the money' on expiry.

Thus far we have found the intrinsic values of the option – what it would be worth were it about to expire. We have previously noted that options with some time still to run will generally be worth more than the difference between current share price and exercise price because the share price may rise further.

3 *The maximum value of an option is the share price itself* – it could never sell for more than the underlying share price value.

$$C_0 \leq S_0 \qquad (8.3)$$

The minimum value of a call today is equal to or greater than the current share price less the exercise price:

$$C_0 \geq S_0 - E \quad \text{if} \quad S_0 > E \qquad (8.4)$$

However the exercise price is payable in the future. It was shown in the previous section that the payoffs from a share are identical to the payoffs from buying a call option, selling a put option and investing the remainder in a risk-free asset that yields the exercise price on the expiry date. In other words, we need to bring the exercise price to its present value by discounting at the risk-free rate of interest. This gives rise to the following revised statement.

4 *The minimum value of an option is the difference between the share price and the present value of the exercise price* (or zero if greater).

$$C_0 \geq S_0 - [E/(1 + R_f)^t] \qquad (8.5)$$

The value of a call option can be observed in Figure 8.4. Bradford plc shares are currently priced at 700p. The diagram shows how the value of an option to buy Bradford shares at 1,100p moves with the share price. The upper limit to the option price is the share price itself, and the lower limit is zero for share prices up to 1,100p, and the share price minus exercise price when share price moves above 1,100p. In fact, the actual option prices lie between these two extremes, on the upward-sloping curve. The curve rises slowly at first, but then accelerates rapidly.

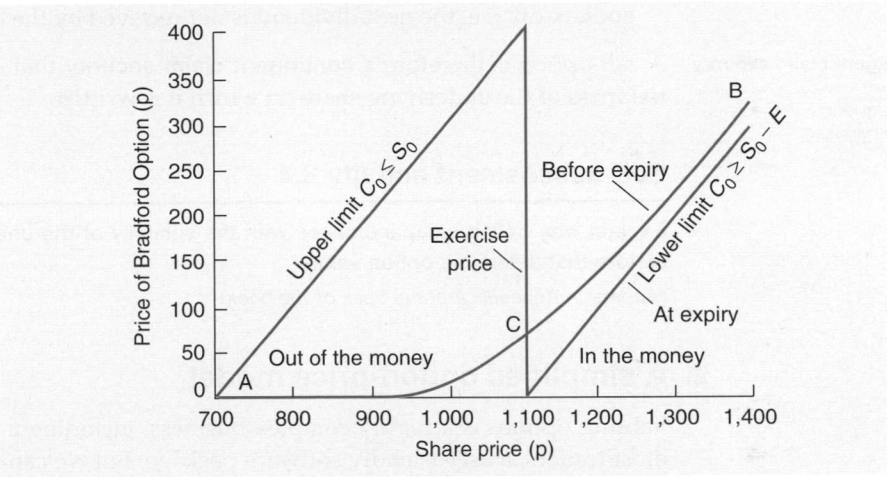

Figure 8.4 Option and share price movements for Bradford plc

At point A on the curve, at the very start, the option is worthless. If the share price for Bradford remained well below the exercise price, the option would remain worthless. At point B, when the share price has rocketed to 1,400p, the option value approximates the share price minus the present value of the exercise price. At point C, the share price exactly equals the exercise price. If exercised today, the option would be worthless. However, there may still be two months for the option to run, in which time the share price could move up or down. In an efficient market, where share prices follow a random walk, there is a 50 per cent chance that it will move higher and an equal probability that it will go lower. If the share price falls, the option will be worthless, but if it rises, the option will have some value. The value placed on the option at point C depends largely on the likelihood of substantial movements in share price. However, we can say that *the higher the share price relative to the exercise price, the safer the option* (i.e. more valuable).

5 *The value of a call option increases over time and as interest rates rise.* Equation 8.4 shows that the value of an option increases as the present value of the exercise price falls. This reduction in present value occurs over time and/or with rises in the interest rate.

6 *The more risky the underlying share, the more valuable the option.* This is because the greater the variance of the underlying share price, the greater is the possibility that prices will exceed the exercise price. But because option values cannot be negative (i.e. the holder would not exercise the option), the 'downside' risk can be ignored.

To summarise, the value of a call option is influenced by the following:

- *The share price.* The higher the price of the share, the greater will be the value of an option written on it.
- *The exercise price of the option.* The lower the exercise price, the greater the value of the call option.
- *The time to expiry of the option.* As long as investors believe that the share price has a chance of yielding a profit on the option, the option will have a positive value. So the longer the time to expiry, the higher the option price.
- *The risk-free interest rate.* As short-term interest rates rise, the value of a call option also increases.
- *The volatility in the underlying share returns.* The greater the volatility in share price, the more likely it is that the exercise price will be exceeded and, hence, the option value will rise.
- *Dividends.* The price of a call option will normally fall with the share price as a share goes ex-div (i.e. the next dividend is not received by the buyer).

contingent claim security
Claim on a security whose value depends on the value of another asset

A call option is therefore a **contingent claim security** that depends on the value and riskiness of the underlying share on which it is written.

Self-assessment activity 8.4

Explain why option value increases with the volatility of the underlying share price. List the factors that determine option value.

(Answer in Appendix A at the back of the book)

■ A simplified option-price model

Valuing options is a highly complex business, including a lot of mathematics or, for most traders, a user-friendly software package. But we can introduce the valuation of options by using a simple (if somewhat unrealistic) example. We argued earlier that it is possible to replicate the payoffs from buying a share by purchasing a call option, selling a put option and placing the balance on deposit to earn a risk-free return over the option period. This provides us with a method for valuing options.

Valuing a call option in Riskitt plc

In April, the share price of Riskitt plc is 100p. A three-month call option on the shares with a July expiry date has an exercise price of 125p. With the current price well below the exercise price it is clear that, for the option to have value, the share price must stand a chance of increasing by at least 25p over the next quarter.

Assume that by the expiry date there is an equal chance that the share price will have either soared to 200p or plummeted to 50p. There are no other possibilities. Assume also that you can borrow at 12 per cent a year, or about 3 per cent a quarter.

What would be the payoff for a call option on one share in Riskitt?

	Best	Worst
Share price	200p	50p
Less exercise price	(125p)	(125p)
Payoff	75p	–

You stand to make 75p if the share price does well, but nothing if it slips below the exercise price. To work out how much you would be willing to pay for such an option, you must replicate an investment in call options by a combination of investing in Riskitt shares and borrowing.

Suppose we buy 200 call options. The payoffs in July will be zero if the share price is only 50p and £150 (i.e. 200 × 75p) if the share price is 200p. This is shown in Table 8.3. Note that the cash flow we are trying to determine is the April premium, represented by the question mark.

To replicate the call option cash flows, you adopt the second strategy in Table 8.3 you need to buy 100 shares and borrow sufficient cash to give identical cash flows in July as the call option strategy. This means borrowing £50. The net cash flows for the two strategies are now the same in July whatever the share price. But the £50 loan repayment in July will include three months' interest at 3 per cent for the quarter. The initial sum borrowed in April would therefore be the present value of £50, i.e. £50/1.03 = £48.54. Deducting this from the share price paid gives a net figure of £51.46, which must also be the April cash payment for 200 call options. The price for one call option is therefore about 26p.

Table 8.3 Valuing a call option in Riskitt plc

| | | Payoff in July if share price is | |
Strategy	Cash flow in April	200p	50p
	£	£	£
1 Buy 200 call options	?	+150	–
2 Buy 100 shares	−100	+200	+50
Borrow	+48.54	−50	−50
	51.46	+150	–
Value of call option = £51.46/200 calls = 25.73p, say 26p			

Risk-neutral method

In the previous Riskitt plc example we assumed that the two possible payoffs each had a 50 per cent probability. The same result is obtained even if we do not know the probabilities for the payoffs by assuming risk neutrality. Although most investors are risk averse, requiring a risk premium for taking on greater risk, both risk-averse and risk-neutral investors place the same value on a risk-free asset and, as was shown earlier in the put–call parity model, both place the same value on a hedged portfolio of options and shares.

Recall from the Riskitt plc example that the share price is 100 pence, a three-month call option has an exercise of 125 pence, and the risk-free rate of interest is 12 per cent a year (or approximately 3 per cent a quarter). The only two possible share price outcomes after three months are 200p and 50 pence, but in this example the probabilities of each are not known.

Assuming no dividends are paid in the period and that investors are risk neutral, the expected share price in three months' time is 3 per cent more than the existing price, or 103 pence. From this we can infer probabilities for the two possible outcomes. Let the probability of the high outcome be p and the probability of the low outcome will therefore be $1 - p$. The expected payoff of 103 pence will be the sum of the two outcomes multiplied by their probabilities:

$$200 \text{ pence} \times p + 50 \text{ pence} \times (1 - p) = 103 \text{ pence}$$
$$150\,p = 53 \text{ pence}$$
$$p = 0.353$$
$$1 - p = 0.647$$

We can now use these probabilities to calculate the value of the call option valued at 125 pence. It is only of value if the high outcome of 200 pence is attained, the net gain being 75 pence. Therefore the expected cash flow is:

$$\text{Expected cash flow} = 75 \text{ pence} \times p = 75 \times 0.353$$
$$= 26.48 \text{ pence}$$

The present value of this is $26.47/(1.03) = 25.7$ pence or approximately 26 pence, as in the previous example.

■ Black–Scholes pricing model

The above example took a highly simplified view of uncertainty, using only two possible share price outcomes. Black and Scholes (1973) combined the main determinants of option values to develop a model of option pricing. Although its mathematics are daunting, the model does have practical application. Every day, dealers in options use it in specially programmed calculators to determine option prices.

For those who like a challenge, the complex mathematics of the Black–Scholes pricing model are given in the appendix to this chapter. However, the key message is that option pricing requires evaluation of five of the variables listed earlier: share price, exercise price, risk-free rate of interest, time and share price volatility.

Acorn plc shares are currently worth 28p with a standard deviation of 30 per cent. The risk-free rate of interest is 6 per cent. What is the value of a call option on Acorn shares expiring in nine months and with an exercise price of 30p?

The fully-worked solution to this problem is given in the appendix to this chapter, but we can identify here the five input variables:

Share price (S) = 28p

Exercise price (E) = 30p

Risk-free rate (k) = 6 per cent p.a.

Time to expiry (t) = nine months

Share price volatility (σ) = 30 per cent

Application of the Black–Scholes formula to the above data (see Appendix) gives a value of the call option of 2.6p.

Black–Scholes option-pricing formula

Value of call option (C) is:

$$C = SN(d_1) - EN(d_2)e^{-tk}$$

where

$$d_1 = \frac{\ln(S/E) + tk}{\sigma t^{1/2}} + \frac{\sigma t^{1/2}}{2}$$
$$d_2 = d_1 - \sigma t^{1/2}$$

$N(d)$ is the value of the cumulative distribution function for a standardised normal random variable and e^{-tk} is the present value of the exercise price continuously discounted.

A simplified Black–Scholes formula can be used as an approximation for options less than one year:

$$\frac{C}{S} \approx \frac{1}{\sqrt{2\pi}} \sigma \sqrt{t}$$

This formula emphasises the impact of volatility and time to expiry on the option price.

Applying the above to the previous example we derive a slightly higher option price:

$$C \approx 0.398 \times 0.3\sqrt{0.75} \times 28 = 2.9\text{p}$$

Although the model is complex, the valuation equation derived from the model is quite straightforward to use, and is widely employed in practice. Four of the five variables on which it is based are observable: the only non-observable variable, the volatility or standard deviation of the return on the underlying asset, is generally estimated from historical data.

The Black–Scholes model is based on the following assumptions:

(a) there are no transactions costs or taxes;

(b) the expected risk-free rate of interest is constant for the period of the option life;

(c) the market operates continuously;

(d) share prices change smoothly over time – there are no jumps or discontinuities in the price series;

(e) the standard deviation of the distribution of returns on the share is known;

(f) the share pays no dividends during the life of the option; and

(g) the option may only be exercised at expiry of the call (i.e. a European-type option).

The assumptions on which the model is based are clearly quite restrictive. However, as these assumptions are consistent with mainstream theorising in finance, the model integrates well into the general body of finance theory. And of more practical importance the model appears to be quite robust: it is feasible to relax many assumptions and incorporate more 'real world' features into the model without changing its overall character.

8.4 APPLICATION OF OPTION THEORY TO CORPORATE FINANCE

Option theory has implications going far beyond the valuation of traded share options. It offers a powerful tool for understanding various other contractual arrangements in finance. Here are some examples:

1 *Share warrants*, giving the holder the option to buy shares directly from the company at a fixed exercise price for a given period of time.

2 *Convertible loan stock*, giving the holder a combination of a straight loan or bond and a call option. On exercising the option, the holder exchanges the loan for a fixed number of shares in the company.

3 *Loan stock* can have a call option attached, giving the company the right to repurchase the stock before maturity.

4 *Executive share option schemes* are share options issued to company executives as incentives to pursue shareholder goals.

5 *Insurance and loan guarantees* are a form of put option. An insurance claim is the exercise of an option. Government loan guarantees are a form of insurance. The government, in effect, provides a put option to the holders of risky bonds so that, if the borrowers default, the bond-holders can exercise their option by seeking reimbursement from the government. Underwriting a share issue is a similar type of option.

6 *Currency and interest rate options* are discussed in later chapters as ways of hedging or speculating on currency or interest rate movements.

7 *Underwriting* a new issue of shares when underwriters must take up any shares not subscribed for by investors.

Two further forms of option are equity options and capital investment options, discussed in subsequent sections.

■ Equity as a call option on a company's assets: Reckless Ltd

Option-like features are found in financially geared companies. *Equity is, in effect, a call option on the company's assets.*

Reckless Ltd has a single £1 million debenture in issue, which is due for repayment in one year. The directors, on behalf of the shareholders, can either pay off the loan at the year end, thereby having no prior claim on the firm's assets, or default on the debenture. If they default, the debenture-holders will take charge of the assets or recover the £1 million owing to them.

In such a situation, the shareholders of Reckless have a call option on the company's assets with an exercise price of £1 million. They can exercise the option by repaying the loan, or they can allow the option to lapse by defaulting on the loan. Their choice depends on the value of the company's assets. If they are worth more than £1 million, the option is 'in the money' and the loan should be repaid. If the option is 'out of the money', because the assets are worth less than £1 million, option theory argues that shareholders would prefer the company to default or enter liquidation. This option-like feature arises because companies have limited liability status, effectively protecting shareholders from having to make good any losses.

Derivatives: a double-edged sword

Three years ago, Jackie Brown, a housewife from Leicestershire who trained as a market researcher, became a full-time day trader in investment derivatives.

Ms Brown is one of the many private investors who have been drawn by the flexibility of derivatives, which allow buyers – usually for a small consideration – to gain exposure to the performance of an underlying share, index or security without physically owning it.

Derivatives are the proverbial double-edged sword. They enable investors to isolate certain risks, such as interest rate risk or credit risk. Investors can then either increase risk or hedge it out of their portfolios altogether.

Unlike buying a share or an asset, these instruments allow investors to go short – sell stock they do not own – in order to profit on falling markets. The danger is that investors can lose more than their original stake.

Not surprisingly, derivatives have been vilified in some quarters and beatified in others. But whatever investors think about them, these tools are becoming impossible to ignore and are fast becoming a part of ordinary investors' everyday life.

There are many hidden risks in the derivatives market, warn experts. Warren Buffett, the investment guru who is famous for his down-to-earth attitude to investing, memorably billed them 'weapons of mass destruction'.

His warning reverberated around the market and was echoed by many others who worry that derivatives markets are opaque and standards of reporting are lax. Investors often do not know who the end-acquirer of the risk is and how much accumulated exposure to one type of risk he might have.

Anyone hoping to delve into spread betting, covered warrants or options, should take heed. As veteran market watchers always say: Do not buy what you do not understand, beware of who you are dealing with, and know that betting with derivatives is seductive but dangerous.

As Mr Buffett says, it is 'like hell – easy to enter and almost impossible to exit'.

Source: Based on Kate Burgess, *Financial Times*, 25 October 2003.

8.5 CAPITAL INVESTMENT OPTIONS

real options
Options to invest in real assets such as capital projects

We can now apply option theory to capital budgeting. Capital investment options (sometimes termed **real options**) are option-like features found in capital budgeting decisions. While discounted cash flow techniques are very useful tools of analysis, they are generally more suited to financial assets, because they assume that assets are held rather than managed. The main difference between evaluating financial assets and real assets is that investors in, say, shares, are generally *passive*. Unless they have a fair degree of control, they can only monitor performance and decide whether to hold or sell their shares.

Chapter 8 Identifying and valuing options 201

Corporate managers, on the other hand, play a far more *active* role in achieving the planned net present value on a capital project. When a project is slipping behind forecast they can take action to try to achieve the original NPV target. In other words, they can create options – actions to mitigate losses or exploit new opportunities presented by capital investments. Managerial flexibility to adapt its future actions creates an asymmetry in the NPV probability distribution that increases the investment project's value by improving the upside potential while limiting downside losses.

We will consider three types of option: the abandonment option, the timing option and strategic investment options.

■ Abandonment option

option to abandon
Choice to allow an option to expire. With a capital investment, abandonment should take place where the value for which an asset can be sold exceeds the present value of its future benefits from continuing its operations

Major investment decisions involve heavy capital commitments and are largely irreversible: once the initial capital expenditure is incurred, management cannot turn the clock back and do it differently. The costs associated with divestment are usually very high. Most capital projects divested early will realise little more than scrap value. In the case of a nuclear power plant, the decommissioning cost could be phenomenal. Because management is committing large sums of money in pursuit of higher, but uncertain, payoffs, the **option to abandon**, without incurring enormous costs if things look grim, can be very valuable. Any project that permits management to extract value when things go bad has an embedded put option. To ignore this is to undervalue the project.

Example: Cardiff Components Ltd

Cardiff Components Ltd is considering building a new plant to produce components for the nuclear defence industry. Proposal A is to build a custom-designed plant using the latest technology, but applicable only to nuclear defence contracts. A less profitable scheme, Proposal B, is to build a plant using standard machine tools, giving greater flexibility in application and having a much higher salvage value than Proposal A.

The outcome of a general election to be held one year hence has a major impact on the decision. If the current government is returned to office, its commitment to nuclear defence is likely to give rise to new orders, making Proposal A the better choice. If the current opposition party is elected, its commitment to run down the nuclear defence industry would make Proposal B the better course of action. Proposal B has, in effect, a put option attached to it, giving the flexibility to abandon the proposed operation in favour of some other activity. We underestimate NPV if we assume that the project must last for the prescribed project life, no matter what happens in the future. A similar type of option is the option to redeploy. A utility company may have the option to switch between various fuels to produce electricity. This dual plant may cost more than one that is only capable of burning a single fuel, but the value of the redeployment option should be considered in reaching a decision.

■ Timing option

timing/delay option
The option to invest now or defer the decision until conditions are more favourable

The Cardiff Components example not only introduces an abandonment option, it also raises the **timing** or **delay option**. Management may have viewed the investment as a 'now or never' opportunity, arguing that in highly competitive markets there is no scope for delay. However, most project decisions have three possible outcomes – accept, reject or defer until economic and other conditions improve. In effect, this amounts to viewing the decision as a call option that is about to expire on the new plant, the capital investment outlay being the exercise price. If a positive NPV is expected, the option will be exercised; otherwise the option lapses and no investment is made.

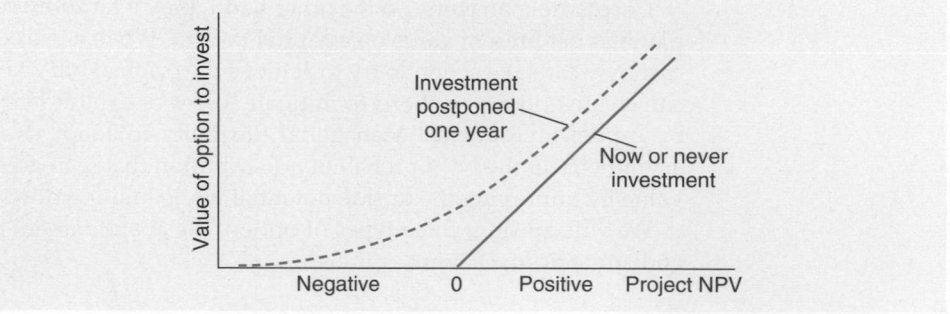

Figure 8.5 The value of the options to delay investments: Cardiff Components Ltd

The option to defer the decision by one year, until the outcome of the general election is known, makes obvious sense. This may look something like the curved line in Figure 8.5.

An immediate investment would yield either a negative NPV – in which case it would not be taken up – or a positive NPV. Delaying the decision by a year to gain valuable new information (the curved broken line) is a more valuable option. Managements sometimes delay taking up apparently wealth-creating opportunities because they believe that the option to wait and gather new information is sufficiently valuable.

Investment as a call option

The five main variables in pricing a share call option can be applied to capital investment (or real) call options.

Share call option	Real call option
Current value of share	Present value of expected cash flows
Exercise price	Investment cost
Time to expiry	Time until investment opportunity disappears
Share price uncertainty	Project value uncertainty
Risk-free interest rate	Risk-free interest rate

■ Strategic investment options

follow-on opportunities
Options that arise following a course of action

Certain investment decisions give rise to **follow-on opportunities** that are wealth-creating. New technology investment, involving large-scale research and development, is particularly difficult to evaluate. Managers refer to the high level of intangible benefits associated with such decisions. What they really mean is that these investments offer further investment opportunities (e.g. greater flexibility), but that, at this stage, the precise form of such opportunities cannot be quantified.

Such valuation calculations applied to strategic investment options raise as many questions as they answer. For example, how much of the risk for the follow-on project is dependent upon the outcome of the initial project? But option pricing does offer insights into the problem of valuing 'intangibles' in capital budgeting, particularly where they create options not otherwise available to the firm.

Example: Strategic options in Harlequin plc

Harlequin plc has developed a new form of mobile phone, using the latest technology. It is considering whether to enter this market by investing in equipment costing £400,000 to assemble and then market the product in the north of England during the first four years. (Most of the product parts will be bought in.) The expected net present value from this initial project, however, is −£25,000. The strategic case for such an investment is that by the end of the project's life sufficient expertise would have been developed to launch an improved product on a larger scale to be distributed throughout Europe. The cost of the second project in four years' time is estimated at £1.32 million. Although there is a reasonable chance of fairly high payoffs, the expected net present value suggests this project will do little more than break even.

'Obviously, with the two projects combining to produce a negative NPV, the whole idea should be scrapped,' remarked the finance director.

Gary Owen, a recent MBA graduate, was less sure that this was the right course of action. He reckoned that the second project was a kind of call option, the initial cost being the exercise price and the present value of its future stream of benefits being equivalent to the option's underlying share price. The risks for the two projects looked to be in line with the variability of the company's share price, which had a standard deviation of 30 per cent a year.

If, by the end of Year 4, the second project did not suggest a positive NPV, the company could walk away from the decision, the option would lapse and the cost to the company would be the £25,000 negative NPV on the first project (the option premium). But it could be a winner, and only 'upside' risk is considered with call options.

Gary knew that Harlequin's discount rate for such projects was 20 per cent and the risk-free interest rate was 10 per cent. Table 8.4 shows his estimation of the main elements to be considered.

Gary Owen then entered these variables into a computer model. He found that the present value of the four-year call option to invest in the follow-on project, with an exercise price of £1.32 million, was worth around £75,000. This is because there is a chance that the project could be really profitable, but the company will not know whether this is likely until the outcome of the first project is known. The high degree of risk in the second project actually increases the value of the call option. It seems, therefore, that the initial project launch, which creates an option value of £75,000 for a 'premium' of £25,000 (negative NPV) may make economic as well as strategic sense.

Table 8.4 Harlequin plc: call option valuation

Initial project	(£000)
Cost of investment	(400)
PV of cash inflows	375
Net present value	(25)
Follow-on-project in Year 4	
Cost of investment	(1320)
PV of cash inflows	1320
Net present value in Year 4	−
Main factors in valuing the call option:	
1 Asset value	PV of cash flows at Year 4 discounted to Year 0
	= £1.32 m/(1.2)^4
	= £0.636 m
2 Exercise price	= cost of follow-on project
	= £1.32 m
3 Risk-free discount rate	= 10%
4 Time period	= 4 years
5 Asset volatility	= standard deviation of 30%

8.6 WHY CONVENTIONAL NPV MAY NOT TELL THE WHOLE STORY

Earlier chapters have rehearsed the theoretical argument that capital projects that offer positive net present values, when discounted at the risk-adjusted discount rate, should be accepted. In Chapter 6 we raised a number of practical shortcomings with discounted cash flow approaches; here we introduce an important theoretical point.

We have noted that orthodox capital projects analysis adopts a 'now or never' mentality. But the timing option reminds us that a 'wait and see' approach can add value. Whenever a company makes an investment decision it also surrenders a call option – the right to invest in the same asset at some later date. Such waiting may be passive, waiting for the right economic and market conditions, or active, where management seeks to gather project-related information to reduce uncertainty (further product trials, competitor reaction, etc.). Hence, the true NPV of a project being undertaken today should include the values of various options associated with the decision:

$$
\text{True NPV} =
\begin{array}{c}\text{NPV of}\\\text{basic}\\\text{project}\end{array}
+
\begin{array}{c}\text{NPV of}\\\text{abandonment}\\\text{option}\end{array}
+
\begin{array}{c}\text{NPV of}\\\text{follow-on}\\\text{projects}\end{array}
+
\begin{array}{c}\text{NPV of}\\\text{option}\\\text{to wait}\end{array}
$$

If the total is positive, the project creates wealth. This is why firms frequently defer apparently wealth-creating projects or accept apparently uneconomic projects. Senior managers recognise that investment ideas often have wider strategic implications, are irreversible and improve with age.

Real options are particularly important in investment decisions when the conventional NPV analysis suggests that the project is 'marginal', uncertainty is high and there is value in retaining flexibility. In such cases, the conventional NPV will almost always understate the true value.

Mini case

Eurotunnel considers all its options

The idea of a road tunnel under the Channel is a legacy of Baroness Thatcher's 11-year reign as the prime minister who got the first tunnel built.

So keen was she on the idea, she insisted Eurotunnel be contractually obliged to submit a feasibility study by 2000, or lose an exclusive option over the second link.

Eurotunnel asked two consultants to investigate seven options for a second link – over and under the water. The study settled on two options: a two-tier road tunnel or a second rail tunnel.

Both would probably run alongside the existing Chunnel; the main difference being that technological advances would make it possible to build a large single-bore tunnel, rather than the existing two main tunnels sandwiching a third service tunnel.

The rail option – to be reserved exclusively for Eurostars and freight trains – sounds safe. For an estimated £3 billion Eurotunnel could simply extend services it and customers already know.

But the report suggests the road tunnel would be more financially viable. Initial studies suggest that a rail option would not make an adequate return unless there was a very significant shift from road to rail.

Whether there will be the passenger demand for a second tunnel of either type is too early to say. Eurotunnel estimates the existing tunnel will reach capacity use in 2025 – but great changes could happen to travel needs and methods over a quarter of a century.

The company has ten years to make up its mind – the deadline is 2010 and it seems in no hurry to be rushed.

Source: Based on Juliette Jowit, *Financial Times*, 6 January 2000.

Self-assessment activity 8.5

What is the type of option available to Eurotunnel and what factors would you consider in assessing its value?

(Answer in Appendix A at the back of the book)

SUMMARY

The options literature has developed highly complex models for valuing options, but insufficient attention has been paid to value creation through options.

Options or option-like features permeate virtually every area of financial management. A better understanding of options and the development of option pricing have made the topic an increasingly important part of financial theory. We have sought to increase your awareness of what options are, where they are to be found, and how managers can begin to value them. The topic is still in its infancy, but its study will yield important insights into financial and investment decisions.

Key points

- Option features are to be found in most areas of finance (e.g. convertibles and warrants, insurance, currency and interest rate management, and capital budgeting).

- Pure options are financial instruments created by exchanges (e.g. stock markets) rather than companies.

- The two main types of option are (1) **call options**, giving the holder the right to buy a share (or other asset) at the exercise price at some future time, and (2) **put options**, giving the holder the right to sell shares at a given price at some future time.

- The minimum value of a call option is the difference between the share price and the present value of the exercise price.

- The value of call options increases as:
  - The underlying share price increases.
  - The exercise price falls.
  - The time to expiry lengthens.
  - The risk-free interest rate rises.
  - The volatility of the underlying share price increases.

- The Black–Scholes Option Pricing Model can be applied to estimate the value of call options.

- Capital investment decisions may have options attached covering the option to (1) abandon, (2) delay or (3) invest in follow-on opportunities.

- Where the value of a company's assets falls below the value of its borrowings, shareholders may not exercise their option to repay the loan, but prefer the company to default on the debt.

Further reading

A more detailed treatment of options is found in Brealey, Myers and Allen (2005) and Bodie and Merton (2000). An introduction to options is given by Redhead (1990). Kester (1984) discusses the topic of real options and Dixit and Pindyck (1995) provide an easy-to-read article on the options approach to capital investment. Brennan and Trigeorgis (2000) offer a number of useful papers on real options. Those who like a mathematical challenge may want to try Black and Scholes' (1973) classic paper or Cox *et al.* (1979). Merton (1998) gives an excellent review of the application of option pricing, particularly to investment decisions.

Useful websites

Futures and Options World: **www.fow.com**
Euronext.liffe: **www.liffe.com**
International Swaps and Derivatives Association: **www.isda.org**

Appendix
BLACK–SCHOLES OPTION PRICING FORMULA

The Black–Scholes formula, for valuing a call option (C), with no adjustment for dividends, is given by:

$$C = SN(d_1) - EN(d_2)e^{-tk}$$

where:

$$d_1 = \frac{\ln(S/E) + tk}{\sigma t^{1/2}} + \frac{\sigma t^{1/2}}{2}$$

$$d_2 = d_1 - \sigma t^{1/2}$$

We already have described S as the underlying share price and E as the exercise price. In addition, σ is the standard deviation of the underlying asset, t is the time, in years, until the option expires, k is the risk-free rate of interest continuously compounded, $N(d)$ is the value of the cumulative distribution function for a standardised normal random variable and e^{-tk} is the present value of the exercise price continuously discounted.

■ Example

Acorn plc shares are currently worth 28p each with a standard deviation of 30 per cent. The risk-free interest rate is 6 per cent, continuously compounded. Compute the value of a call option on Acorn shares expiring in nine months and with an exercise price of 30p.

We can list the values for each parameter: $S = 28$, $\sigma = 0.30$, $E = 30$, $K = 0.06$, $t = 0.75$.

$$\sigma t^{1/2} = (0.3)(0.75)^{1/2} = 0.2598$$

$$d_1 = \frac{\ln(S/E) + tk}{\sigma t^{1/2}} + \frac{\sigma t^{1/2}}{2}$$

$$= \frac{\ln(28/30) + 0.75(0.06)}{0.2598} + \frac{0.2598}{2}$$

$$= -0.2655 + 0.1732 + 0.1299$$

$$= 0.0375, \text{ say } 0.04$$

$$d_2 = d_1 - \sigma r^{1/2} = 0.0375 - 0.2598$$

$$= -002223, \text{ say } -0.22$$

Using cumulative distribution function tables:

$$N(d_1) = N(0.04) = 0.5160$$
$$N(d_2) = N(-0.22) = 0.4129$$

Inserting the above into the original equation:

$$C = SN(d_1) - EN(d_2)e^{-tk}$$

$$= 28(0.5160) - 30(0.4129)e^{-0.045}$$

$$= 2.6p$$

The value of the call is 2.6p.

Strictly speaking, adjustment for dividends on shares should be made by applying the Merton formula, not dealt with in this text.

QUESTIONS

 Questions with an icon are also available for practice in *myfinancelab* with additional supporting resources.

Questions with a coloured number have solutions in Appendix B on page 735.

1 Give two examples where companies can issue call options (or something similar).

2 On 1 March the ordinary shares of Gaymore plc stood at 469p. The traded options market in the shares quotes April 500p puts at 47p. If the share price falls to 450p, how much, if any, profit would an investor make? What will the option be worth if the share price moves up to 510p?

3 What is the difference between traditional and exchange traded options?

4 Explain the factors influencing the price of a traded option and whether volatility of a company's share option price is necessarily a sign of financial weakness.

? 5 Frank purchased a call option on 100 shares in Marmaduke plc six months ago at 10p per share. The share price at the time was 110p and the exercise price was 120p. Just prior to expiry the share price has risen to 135p.

Required
(a) State whether the option should be exercised.
(b) Calculate the profit or loss on the option.
(c) Would Frank have done better by investing the same amount of cash six months ago in a bank offering 10 per cent p.a.?

6 Find the value of the call option given that the present value of the exercise price is 10p, the value of the put option is 15p and the current value of the share on which the option is based is 25p.

7 Find the present value of the exercise price given that the value of the call is 19p, the value of the put is 5p and the current market price of the underlying share is 30p.

8 The current price of a share is 38p and a call option written on this share with six months to run to maturity has an exercise price of 40p. If the risk-free rate of interest is 10 per cent per annum and the volatility of the returns on the share is 20 per cent, use the Black and Scholes model to estimate the value of the call.

9 The current price on British Sky Broadcasting is 420.5p and the price of a call option with a strike price of 420p with six months to maturity is 50.5p. The value of a put option with the same strike price and time to maturity is 38.5p. Determine the annualised rate of interest if put–call parity holds.

10 The following are the closing prices of options on the shares of BAT on Wednesday 10 March 2004.

Exercise Price		Calls			Puts		
		Apr	Jun	Sep	Apr	June	Sep
BAT	800	36.5	53.5	62.5	9.0	20.0	31.5
(*825)	850	11.0	25.5	35.0	33.0	42.0	55.5

*Current price

Refer to the table as required when answering this question.

(a) Explain the fundamental reasons for the large difference between the price of a September 800 call and an April 800 put.

(b) Outline a strategy that combines short calls and short puts. Why would an investor adopt such a strategy? Use data from the table to illustrate some possible payoffs.

11 *Spot the options in Enigma Drugs plc.* The mini-case presented below incorporates five options. Can you identify the type of option, its length and exercise price? Recall that American options offer the holder the right to exercise at any time up to a certain date, while a European option is exercised on one particular date.

> Enigma Drugs plc is an innovative pharmaceutical company. The management team is considering setting up a separate limited company to develop and produce a new drug.
>
> The project is forecast to incur development costs and new plant expenditure totalling £50 million and to break even over the next five years (by which time its competitors are likely to have found a way round the patent rights). Enigma's management is considering deferring the whole decision by two years, when the outcome of a major court case with important implications for the drug's success will be known.
>
> The risks on the venture are high, but should the project prove unsuccessful and have to be abandoned, the 'know-how' developed from the project can be used inside the group or sold to its competitors for a considerable sum. Enigma's management realises that there is little or no money to be made in the initial five years, but it should allow them to gain vital expertise for the development of a 'wonder drug' costing £120 million, which could be launched in four years' time.
>
> The newly-formed company would be largely funded by borrowing £40 million in the first instance, repayable in total after eight years, unless the company prefers to be 'wound up' for defaulting on the loan. Some of the debt raised will be by 9 per cent Convertible Loan Stock, giving holders the right to convert to equity at any time over the next four years at 360p compared with the current price of 297p.

12 A European call option on shares has 3 months to expiry. The current share price is £3 and the exercise price of the option is £2.50. The standard deviation of the share is 20 per cent, and the risk-free rate is 5 per cent. Use the Black–Scholes model to value the call option of a share.

Practical assignment

1 Choose two forms of financial contracting arrangement with option features and show how option pricing theory can help in analysing them.

2 Consider a major capital investment recently undertaken or under review (e.g. the London Olympic games project). Does it offer an option? Could an option feature be introduced? What would the rough value of the option be?

 **myfinancelab** *Now retake your diagnostic test for Chapter 8 to check your progress and update your study plan.*

9

Relationships between investments: portfolio theory

Learning objectives

This chapter is designed to explore the financial equivalent of the maxim 'don't put all your eggs in one basket'. In particular, it aims:

■ To give the reader an understanding of the rationale behind the diversification decisions of both shareholders and companies.

■ To illustrate the mechanics of portfolio construction with a user-friendly approach to the key statistics, using numerical examples.

■ To explain why optimal portfolio selection is a matter of personal choice.

■ To examine the drawbacks of portfolio analysis as an approach to project appraisal.

A good grasp of the principles of portfolio analysis is an essential underpinning to understanding the Capital Asset Pricing Model, to be covered in Chapter 10.

 *Complete your diagnostic test for Chapter 9 now to create your personal study plan.*

9.1 INTRODUCTION

This chapter deals with the theory underlying **diversification** decisions. Diversification is a strategic device for dealing with risk. Whereas the previous chapter examined methods of risk analysis that focused on individual projects, here we study how the financial manager can exploit interrelationships between projects to adjust the risk-return characteristics of the whole enterprise. In the process, we will show why many firms develop a wide spread of activities or **portfolios**. The term 'portfolio' is usually applied to combinations of securities, but we will show that the principles underlying security portfolio formation can be applied to combinations of any type of asset, including investment projects.

Many firms diffuse their efforts across a range of products, market segments and customers in order to spread the risks of declining trade and profitability. If a firm can reduce its reliance on particular products or markets, it can more easily bear the impact of a major reverse in any single market. However, firms do not reduce their exposure to the threat of new products or new competitors for entirely negative reasons. Diversification can generate some major strategic advantages: for example, the wider the spread of activities, the greater the access to star performing sectors of the economy. Imagine an economy divided into five sectors, with one star-performer each year whose identity is always random. A company operating in a single sector is likely to miss out in four years out of five. In such a world, it is prudent to have a stake in every sector by building a portfolio of all five activities.

Diversification is designed to even out the bumps in the time profile of profits and cash flows. The ideal form of diversification is to engage in activities that behave in exactly opposite ways. When sales and earnings are relatively low in one area, the adverse consequences can be offset by participation in a sector where sales and profits are relatively high. With perfect synchronisation, the time profile of overall returns will describe the pattern shown in Figure 9.1. This shows the returns from two activities: A, which moves in parallel with the economy as a whole; and B, which moves in an *exactly* opposite way. The equal and opposite fluctuations in the returns from these two activities would result in a perfectly level profile for a diversified enterprise comprising both activities. In generally adverse economic conditions, the returns from activity A, closely following the economy as a whole, will be depressed, but involvement in activity B has an exactly compensating effect. The reverse applies when the economy is expanding. The returns from B are said to be **contra-cyclical**, and the dampening effect on the variability of returns is called a **portfolio effect**.

For firms planning to diversify, there are two important messages. First, it is not enough simply to spread your activities. Different activities are subject to different types of risk, which are not always closely related. For an internationally-diversified firm, the factors affecting domestic operations may be quite different from those

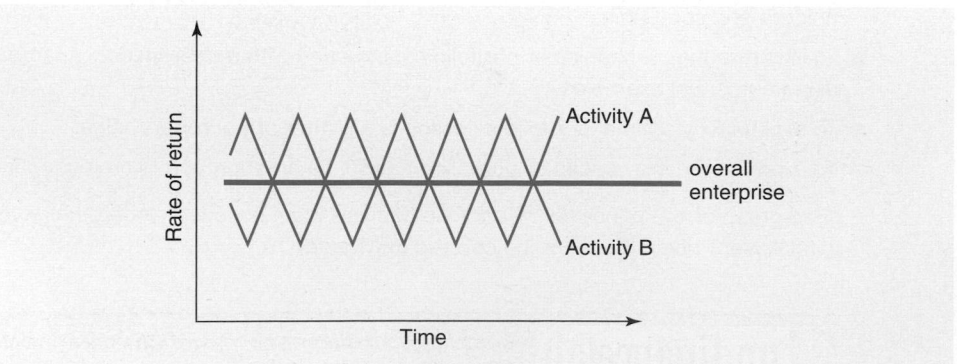

Figure 9.1 Equal and offsetting fluctuations in returns

affecting overseas operations. If changes in these influences are random and relatively uncorrelated, diversification may significantly reduce the variability of company earnings. Second, to generate an appreciable impact on overall returns, diversification must usually be substantial in relation to the whole enterprise. Hence, two key messages of portfolio diversification are: *look for unrelated activities, and engage in significant diversification.*

9.2 PORTFOLIO ANALYSIS: THE BASIC PRINCIPLES

The theory of diversification was developed by Markowitz (1952). It can be reduced to the maxim 'don't put all your eggs in one basket'. This is a simple motto, but one that many investors persistently ignore. How often do we read heart-rending stories of small investors who have lost all their savings in some shady venture or other? Why do more than 50 per cent of private investors persist in holding a single security in their investment portfolios? Perhaps they are unaware of the advantages of spreading their risks, or have not understood the arguments. Perhaps they are not risk-averse or are simply irrational.* Rational, risk-averse investors appreciate that not all investments perform well at the same time, that some may never perform well, and that a few may perform spectacularly well. Since no one can predict which investments will fall into each category in any one period, it is rational to spread one's funds over a wide set of investments.

A simple example will illustrate the remarkable potential benefits of diversification.

■ Achieving a perfect portfolio effect

An investor can undertake one or both of the two investments, Apple and Pear. Apple has a 50 per cent chance of achieving an 8 per cent return and a 50 per cent chance of returning 12 per cent. Pear has a 50 per cent chance of generating a return of 6 per cent and a 50 per cent chance of yielding 14 per cent. The two investments are in sectors of the economy that move in direct opposition to each other. The investor expects the return on Apple to be relatively high when that on Pear is relatively low, and vice versa. What portfolio should the investor hold?

First of all, note that the expected value (EV) of each investment's return is identical:

Investment Apple: EV $= (0.5 \times 8\%) + (0.5 \times 12\%) = (4\% + 6\%) = 10\%$

Investment Pear: EV $= (0.5 \times 6\%) + (0.5 \times 14\%) = (3\% + 7\%) = 10\%$

At first glance, it may appear that the investor would be indifferent between Apple and Pear or, indeed, any combination of them. However, there is a wide variety of possible expected returns according to how the investor 'weights' the portfolio. Moreover, a badly-weighted portfolio can offer wide variations in returns in different time periods.

For example, when Pear is the star performer, a portfolio comprising 20 per cent of Apple and 80 per cent of Pear will offer a return of:

$$\underset{(0.2 \times 8\%)}{\underline{\text{Apple}}} + \underset{(0.8 \times 14\%)}{\underline{\text{Pear}}} = \underset{(1.6\% + 11.2\%)}{\underline{\text{Portfolio}}} = 12.8\%$$

When Apple is the star, the return is only:

$$(0.2 \times 12\%) + (0.8 \times 6\%) = (2.4\% + 4.8\%) = 7.2\%$$

---

*A common reason is probably that they have applied for shares in a privatisation, or been given shares in a building society demutualisation! The biggest demutualisation was that by Halifax Building Society (now HBOS plc, after acquiring Bank of Scotland). 'The Halifax' gave free shares to 7.5 million customers in June 1997. Ten years later, they still accounted for 20% of HBOS's market value.

Although there should be as many good years for Apple as for Pear, resulting over the long-term in an average return of 10 per cent, *in the shorter term*, the investor would be over-exposed to the risk of a series of bad years for Pear. Happily, there is a portfolio which removes this risk entirely.

Consider a portfolio invested two-thirds in Apple and one-third in Pear. When Apple is the star, the return on the portfolio (R_p) is a weighted average of the returns from the two components:

$$R_p = (2/3 \times 12\%) + (1/3 \times 6\%) = (8\% + 2\%) = 10\%$$

Conversely, when Pear is the star, the portfolio offers a return of:

$$R_p = (2/3 \times 8\%) + (1/3 \times 14\%) = (5.33\% \times 4.67\%) = 10\%$$

With this combination, the risk-averse investor cannot go wrong! The portfolio completely removes variability in returns as there are only two possible states of the economy. Any rational risk-averse investor should select this combination of Apple and Pear to eliminate risk for a guaranteed 10 per cent return. Here, the portfolio effect is perfect, like that shown in Figure 9.1. However, not every investor would necessarily opt for this particular portfolio. Super-optimists might load their funds entirely on to Pear, hoping for 14 per cent returns every year. This may work for a year or two, but the chances of achieving a consistent return of 14 per cent year after year are very low. The chance of achieving 14 per cent in the first year is 50 per cent, but the chance of getting 14 per cent in *each* of the first two years is (50 per cent) $\times$ (50 per cent) = 25 per cent and so on. Diversification is usually the safest (and often the most profitable) policy. We will study later in the chapter how different portfolio weightings affect the overall risk and return.

In this example, the opportunity to eliminate all risk arises from the *perfect negative correlation** between the two investments, but this attractive property can only be exploited by weighting the portfolio in a particular way.

Regrettably, cases of perfect negative correlation between the returns from securities are rare. Most investment returns exhibit varying degrees of positive correlation, largely according to how they depend on overall economic trends. This does not rule out risk-reducing diversification benefits, but suggests they may be less pronounced than in our example. As we will see, *the extent to which portfolio combination can achieve a reduction in risk depends on the degree of correlation between returns*. Later in the chapter, we will examine rather more realistic cases, but first we need to explore more fully the nature and measurement of portfolio risk.

Self-assessment activity 9.1

What are the two required conditions for total elimination of portfolio risk?

(Answer in Appendix A at the back of the book)

9.3 HOW TO MEASURE PORTFOLIO RISK

We have just seen the importance of the degree of correlation between the returns from two investments. We saw also how the return from a portfolio could be expressed as a weighted average of the individual asset returns, the weights being the

---

*Readers lacking a grounding in elementary statistics may want to consult an introductory text such as C. Morris, *Quantitative Approaches in Business Studies* (Pearson Education), in order to study the concept of correlation. Correlation is measured on a scale of −1 (perfect negative correlation) through zero to +1 (perfect positive correlation).

proportions of the portfolio accounted for by each of the various components. A similar relationship applies before the event: that is, if we consider the *expected value* of the return from the portfolio. The expected return on a portfolio (ER_p) comprising two assets, A and B, whose individual expected returns are ER_A and ER_B respectively, is given by:

$$ER_p = \alpha ER_A + (1 - \alpha)ER_B \tag{9.1}$$

where α and $(1 - \alpha)$ are the respective weightings of assets A and B, with $\alpha + (1 - \alpha) = 1$.

The riskiness of the portfolio expresses the extent to which the actual return may deviate from the expected return. This may be expressed by the variance of the return, σ_p^2, or by its standard deviation, σ_p.

Portfolio risk

The (rather fearsome!) expression for the standard deviation of a two-asset investment portfolio, σ_p, is:

$$\sigma_p = \sqrt{[\alpha^2\sigma_A^2 + (1 - \alpha)^2\sigma_B^2 + 2\alpha(1 - \alpha)cov_{AB}]} \tag{9.2}$$

where

$\alpha =$ the proportion of the portfolio invested in asset A.

$(1 - \alpha) =$ the proportion of the portfolio invested in asset B.

$\sigma_A^2 =$ the variance of the return on asset A.

$\sigma_B^2 =$ the variance of the return on asset B.

$cov_{AB} =$ the covariance of the returns on A and B.

covariance
A statistical measure of the extent to which the fluctuations exhibited by two (or more) variables are related

We need now to explain the meaning of the covariance. The **covariance**, like the correlation coefficient, is a measure of the interrelationship between random variables, in this case, the returns from the two investments A and B. In other words, it measures the extent to which their returns move together, i.e. their **co-movement** or **co-variability**. When the two returns move together, it has a positive value; when they move away from each other, it has a negative value; and when there is no co-variability at all, its value is zero. However, unlike the correlation coefficient, whose value is restricted to a scale ranging from -1 to $+1$, the covariance can assume any value. It measures co-movement in *absolute* terms, whereas the correlation coefficient is a *relative* measure.

The correlation coefficient between the return on A and the return on B, r_{AB}, is simply the covariance, normalised or standardised, by the product of their standard deviations:

$$\text{Correlation coefficient between A's and B's returns} = r_{AB} = \frac{cov_{AB}}{\sigma_A \times \sigma_B}$$

The covariance, cov_{AB}, between the returns on the two investments, A and B, is given by:

$$cov_{AB} = \sum_{i=1}^{N}[p_i(R_A - ER_A)(R_B - ER_B)] \tag{9.3}$$

where R_A is the realised return from investment A, ER_A is the expected value of the return from A, R_B is the realised return from investment B, ER_B is the expected value of the return from B, and p_i is the probability of any pair of values occurring.

Equation 9.3 tells us first to calculate, for each pair of simultaneously occurring outcomes, their deviations from their respective expected values; next, to multiply these deviations together and then to weight the resulting product by the relevant probability for each pair. Finally, the sum of all weighted products of paired divergences between expected and actual outcomes defines the covariance. This relationship is more easily understood with a numerical example. Table 9.1 shows possible returns from

Table 9.1 Returns under different states of the economy

State of the economy	Probability	Return from A	Return from B
E_1	0.25	−10%	+60%
E_2	0.25	−10%	−20%
E_3	0.25	+50%	−20%
E_4	0.25	+50%	+60%

Table 9.2 Calculating the covariance

R_A	ER_A	R_B	ER_B	$(R_A - ER_A)$	$(R_B - ER_A)$	Product	Probability	Weighted product
−10	20	+60	20	−30	+40	−1200	0.25	−300
−10	20	−20	20	−30	−40	+1200	0.25	+300
+50	20	−20	20	+30	−40	−1200	0.25	−300
+50	20	+60	20	+30	+40	+1200	0.25	+300
							covariance$_{AB}$ =	0

Note: Although the rate of return figures are percentages, they have been treated as integers to clarify exposition.

two assets under four different economic conditions, with associated probabilities. First, do Self-assessment activity 9.2. Then, check through the calculation in Table 9.2.

Self-assessment activity 9.2

With the figures in Table 9.1, check that the expected values for both A and B are 20 per cent, and that their respective standard deviations are 30 per cent and 40 per cent, using the formulae presented in Chapter 7.

(Answer in Appendix A at the back of the book)

In this case, there is no co-variability at all between the returns from the two assets. If the return from A increases, it is just as likely to be associated with a fall in the return from B as a concurrent increase. If the covariance (which measures the degree of co-movement in absolute terms) is zero, we will find the correlation coefficient (the relative measure of co-movement) is also zero. We may now demonstrate this:

$$\text{Correlation coefficient between A's and B's returns} = r_{AB} = \frac{\text{cov}_{AB}}{(\sigma_A \times \sigma_B)} = \frac{0}{(30 \times 40)} = 0$$

The case of zero covariance is a very convenient one, as we can see from looking at the expression for portfolio risk, σ_p (Equation 9.2 from page 213).

$$\sigma_p = \sqrt{[\alpha^2\sigma_A^2 + (1 - \alpha)^2\sigma_B^2 + 2\alpha(1 - \alpha)\text{cov}_{AB}]}$$

When the covariance is zero, the third term is zero, and portfolio risk reduces to:

$$\sigma_p = \sqrt{[\alpha^2\sigma_A^2 + (1 - \alpha)^2\sigma_B^2]}$$

With zero covariance, portfolio risk is thus smaller for any portfolio compared to cases where the covariance is positive. Even better, when the covariance is negative, the third term becomes negative and risk falls even further. In general, the lowest achievable portfolio risk declines as the covariance diminishes: if it is negative, all the better. There is, however, no limit on the covariance value. If we re-express portfolio risk in terms of the correlation coefficient, we can be more specific about the greatest achievable degree of risk reduction. The formula relating covariance and correlation coefficient (Equation 9.3) can be rewritten as:

$$\text{cov}_{AB} = (r_{AB} \times \sigma_A \times \sigma_B)$$

Substituting into the expression for portfolio risk (Equation 9.2), we derive:

$$\sigma_P = \sqrt{[\alpha^2\sigma_A^2 + (1 - \alpha)^2\sigma_B^2 + 2\alpha(1 - \alpha)r_{AB}\sigma_A\sigma_B]}$$

Inspection of this formula shows that when the correlation coefficient is negative, portfolio risk can be lowered by combining assets A and B. From a risk-minimising perspective, the most advantageous value of the coefficient is minus one, since when the portfolio is suitably weighted, the standard deviation of the portfolio return can be reduced to zero. We thus have a formulaic demonstration of the result intuitively obtained in the Apple and Pear example at the start of the chapter. Whether one works in terms of the covariance or the correlation coefficient is generally a matter of preference, but, sometimes, it is dictated by the information available.

■ The optimal portfolio

An obvious question to ask is: which is the best portfolio to hold? In this example, the two investments have the same expected values, so any portfolio we construct by combining them will also offer this expected value. The optimal portfolio is therefore the one that offers the lowest level of risk. Although very few decision-makers are outright risk minimisers, any rational risk-averse manager will adopt the risk-minimising action where every alternative offers an equal expected payoff.

The minimum risk portfolio with two assets

The expression for finding the weightings required to minimise the risk of a portfolio comprising two assets, A and B, where $\alpha_A^* = $ the proportion invested in asset A is:

$$\alpha_A^* = \frac{\sigma_B^2 - \text{cov}_{AB}}{\sigma_A^2 + \sigma_B^2 - 2\,\text{cov}_{AB}} \tag{9.4}$$

Substituting the figures for the AB example into Equation 9.4, we find:

$$\alpha_A^* = \frac{40^2}{30^2 + 40^2} = \frac{1,600}{2,500} = 0.64$$

This formula tells us that, to minimise risk, we should place 64 per cent of our funds in A and 36 per cent in B.

Self-assessment activity 9.3

Verify that the standard deviation of this portfolio is 24%.

(Answer in Appendix A at the back of the book)

In the next section, we analyse the more likely, and more interesting, case where both the risks and expected returns of the two components differ.

9.4 PORTFOLIO ANALYSIS WHERE RISK AND RETURN DIFFER

Suppose we are offered the two investments, Z and Y, whose characteristics are shown in Table 9.3. Which should we undertake? Or should we undertake some combination? To answer these questions, we need to consider the possible available combinations of risk and return. Notice that correlation is negative.

Table 9.3 Differing returns and risks

Asset	Expected return (%)	Standard deviation (%)
Z	15	20
Y	35	40

Correlation coefficient$_{ZY}$ = -0.25; Covariance$_{ZY}$ = $(-0.25) \times (20) \times (40) = -200$

Table 9.4 Portfolio risk–return combinations

Z weighting (%)	Y weighting (%)	Expected return (%)	Standard deviation (%)
100	0	15	20
75	25	20	16
50	50	25	20
25	75	30	29
0	100	35	40

Let us assume that the two assets can be combined in any proportions, i.e. the two assets are perfectly divisible, as with security investments. There is an infinite number of possible combinations of risk and return. However, for simplicity, we confine our attention to the restricted range of portfolios whose risk and return characteristics are shown in Table 9.4.

If we wanted to minimise risk, we would invest solely in asset Z, since this has the lowest standard deviation. However, as we move from the all-Z portfolio to the combination 75 per cent of Z plus 25 per cent of Y the risk of the whole portfolio diminishes and the expected return *increases*. Eventually, though, for portfolios more heavily weighted towards Y, the effect of Y's higher risk outweighs the beneficial effect of negative correlation, resulting in rising overall risk.

opportunity set
The set of investment opportunities (i.e. risk return combinations) available to the investor to select from

Figure 9.2 traces the full range of available opportunities (or **opportunity set**), shaped rather like the nose cone of an aircraft. The profile ranges from point A, representing total investment in Y, through to point C, representing total investment in Z, having described a U-turn at B.

Self-assessment activity 9.4

Verify that the portfolio at B, involving 75 per cent of Z and 25 per cent of Y, is the minimum risk combination.

Not all combinations are of interest to the rational risk-averse manager. Comparing segment AB with the segment BC, we find that combinations lying along the latter are inefficient. For any combination along BC, we can achieve a higher return for the same risk by moving to the combination vertically above it on AB. Point S is clearly superior to T and, applying similar logic to the whole of BC, we are left with the segment AB summarising all efficient portfolios, i.e. those that maximise return for a given risk. AB is thus called the **efficient frontier**. Points along AB are said to dominate corresponding points along BC.

optimal portfolio
The risk–return combination that offers maximum satisfaction to the investor, i.e. his/her most-preferred risk–return combination

However, we cannot specify an **optimal portfolio**, except for the outright risk-minimiser, who would select the portfolio at B, and for the maximiser of expected return, who would settle at point A (all Y). A risk-averse person might select any portfolio along AB, depending on his or her degree of risk aversion: that is, what additional

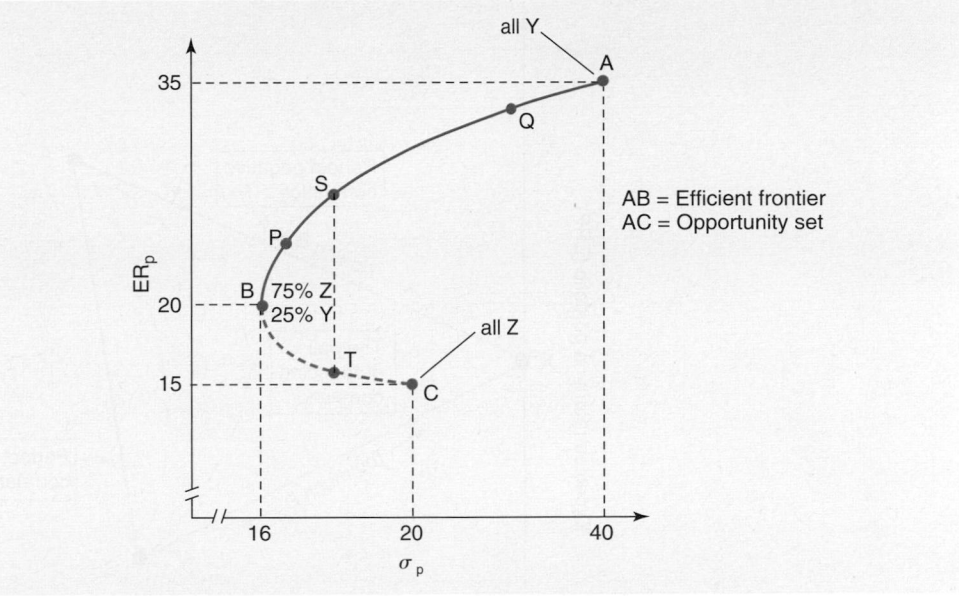

Figure 9.2 Available portfolio risk–return combinations when assets, risks and expected returns are different

return they would require to compensate for a specified increase in risk. For example, a highly risk-averse person might locate at point P, while the less cautious person might locate at point Q.

This is a crucial result. The most desirable combination of risky assets depends on the decision-maker's attitude towards risk. If we knew the extent of their risk-aversion – that is, how large a premium is required for a given increase in risk – we could specify the best portfolio.

Self-assessment activity 9.5

What is meant by an efficient frontier in portfolio analysis?

(Answer in Appendix A at the back of the book)

9.5 DIFFERENT DEGREES OF CORRELATION

Using arbitrary values for the correlation coefficient, we have found that negative correlation offers a handsome portfolio effect, and, to a lesser degree, also zero correlation. It is useful now to consider more carefully the general relationship between risk, correlation and return. To do this, we look at the full range of possible degrees of correlation, extending from perfect negative to perfect positive.

Say we are dealing with two investments, A and B, with Asset A offering the higher expected return but also carrying greater risk. These are shown in Figure 9.3.

Consider the following degrees of correlation:

1 *Perfect positive*. In this case, it is not possible to achieve a portfolio effect at all. Combinations of A and B locate along the straight line AB. To achieve lower risk levels, we would simply invest more in asset B, while the risk-minimising 'portfolio' is simply asset B alone.

2 *Perfect negative*. In this case, combinations along AXB all become possible. With the returns from the two assets moving in perfect opposition to each other, it is possible to eliminate risk by adding B to A, but only by weighting the portfolio correctly is it possible to fully exploit the beneficial effect of correlation. Maximum

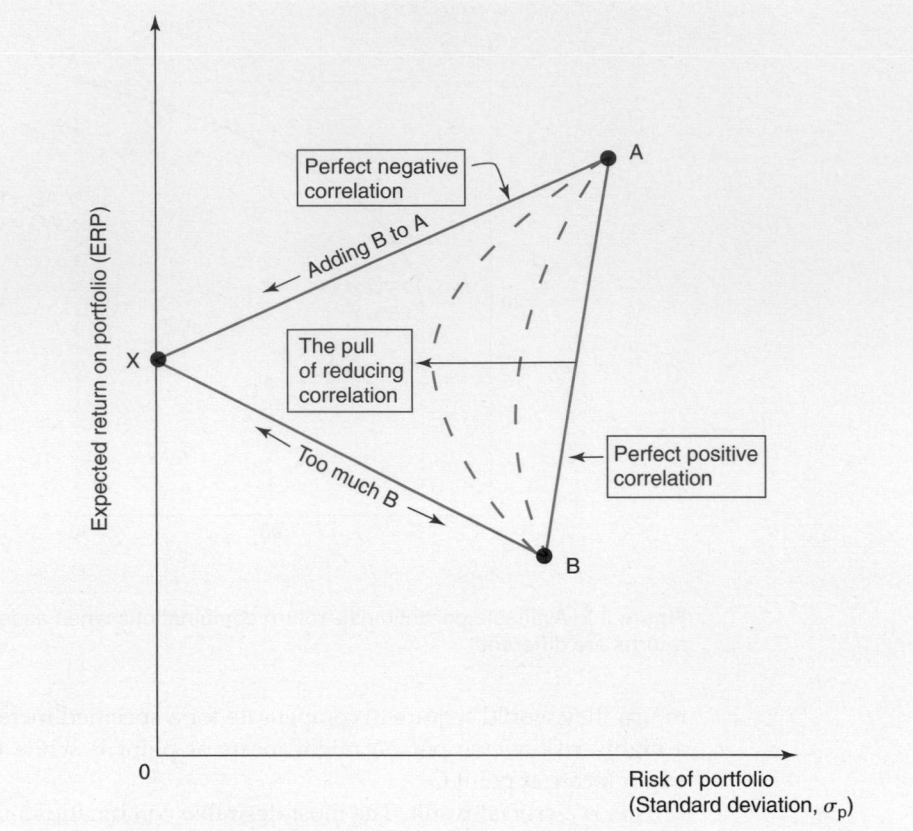

Figure 9.3 The effect on the efficiency frontier of changing correlation

risk-reduction is achieved at point X where the portfolio risk is zero. Combinations along XB are clearly inefficient.

3 *Intermediate values.* For correlation coefficients between +1 and −1, it is still possible to generate a portfolio effect. The lower the correlation, i.e. the further away from +1, the greater the portfolio effect achievable. Two examples are shown in Figure 9.3 as dotted lines between A and B. The characteristic 'bow' shapes result from progressively lower correlation bending the profile from its original position until we start observing the 'nose cones' identified earlier.

9.6 WORKED EXAMPLE: GERRYBILD PLC

Gerrybild plc is a firm of speculative housebuilders that builds in advance of firm orders from customers. It has a given amount of capital to purchase land and raw materials and to pay labour for development purposes. It is considering two design types – a small two-bedroomed terraced town house and a large four-bedroomed 'executive' residence. The project could last a number of years and its success depends largely on general economic conditions, which will influence the demand for new houses. Some information is available on past sales patterns of similar properties in roughly similar locations – the demand is relatively higher for larger properties in buoyant economic conditions, and for smaller properties in relatively depressed states of the economy. Since there appears to be a degree of inverse correlation between demand, and, therefore, net cash flows, from the two products, it seems sensible to consider diversified development. Table 9.5 shows annual net present value estimates for various economic conditions.

Table 9.5 Returns from Gerrybild

State of the economy	Probability	Estimated NPV £ per:	
		Large house	**Small house**
E_1	0.2	20,000	20,000
E_2	0.3	20,000	30,000
E_3	0.4	40,000	20,000
E_4	0.1	40,000	30,000

To analyse this decision problem, we need, first, to calculate the risk–return parameters of the investment, and, second, to assess the degree of correlation. This information may be obtained by performing a number of statistical operations:

1 *Calculation of expected values.* A shortcut is available, since some outcomes may occur under more than one state of the economy. Grouping data where possible:

EV_L = Expected value of a large house = $(0.5 \times £20,000) + (0.5 \times £40,000)$

= £30,000

EV_S = Expected value of a small house = $(0.6 \times £20,000) + (0.4 \times £30,000)$

= £24,000

2 *Calculation of project risks.* We now apply the usual expression for the standard deviation. The calculations for each activity are shown in Table 9.6. Clearly, the relative money-spinner, the large house project, is also the more risky activity.
3 *Calculation of co-variability.* Table 9.7 presents the calculation of the covariance in tabular form, following the steps itemised in Section 9.3.

Table 9.6 Calculation of standard deviations of returns from each investment

Outcome (£)	Probability	EV (£)	Deviation (£)	Squared deviation (£ million)	Weighted squared deviation (£ million)
Large houses					
20,000	0.5	30,000	−10,000	100	50.0
40,000	0.5	30,000	+10,000	100	50.0
				σ_L^2 = Variance =	100.0
				hence σ_L =	$\sqrt{100}$ m
					= 10,000
					i.e. £10,000
Small houses					
20,000	0.6	24,000	−4,000	16	9.6
30,000	0.4	24,000	+4,000	36	14.4
				σ_S^2 = Variance =	24.0
				hence σ_S =	$\sqrt{24}$ m
					= 4,899
					i.e. £4,899

Table 9.7 Calculation of the covariance

Outcomes (£)					Product	Weighted
R_L	R_S	Probability	$(EV_L - R_L)$ (£)	$(EV_S - R_S)$ (£)	(£m)	product (£m)
20,000	20,000	0.2	−10,000	−4,000	40	+8
20,000	30,000	0.3	−10,000	+6,000	60	−18
40,000	20,000	0.4	+10,000	−4,000	40	−16
40,000	30,000	0.1	+10,000	+6,000	60	+6
						$cov_{LS} = -20$
						i.e. −£20 million

The covariance of −£20 million suggests a strong element of inverse association. This is confirmed by the value of the correlation coefficient:

$$r_{LS} = \frac{cov_{LS}}{\sigma_L \times \sigma_S} = \frac{-£20,000,000}{(£10,000)(£4,899)} = -0.41$$

There are clearly significant portfolio benefits to exploit. To offer concrete advice to the builder, we would require information on his risk–return preferences, but we can still specify the available set of portfolio combinations. Rather than compute the full set of opportunities, we will identify the minimum risk portfolio, to enable construction of the overall risk–return profile.

■ The minimum risk portfolio

Using Equation 9.4, and defining α_L^* as the proportion of the portfolio (i.e. proportion of the available capital) devoted to large houses to minimise risk, we have:

$$\alpha_L^* = \frac{(\sigma_S^2 - cov_{LS})}{(\sigma_S^2 + \sigma_L^2 - 2\,cov_{LS})} = \frac{£24\,m + £20\,m}{£24\,m + £100\,m + £40\,m}$$

$$= \frac{£44\,m}{£164\,m} = 0.27$$

If Gerrybild wanted to minimise risk, it would have to invest 27 per cent of its capital in developing large houses and 73 per cent in developing small houses.

Self-assessment activity 9.6

Verify that the lowest achievable portfolio standard deviation is £3,496 and the expected NPV per house built from the minimum risk portfolio is £25,620.

(Answer in Appendix A at the back of the book)

■ The opportunity set

We now have assembled sufficient information to display the full range of opportunities available to Gerrybild. The opportunity set ABC is shown on Figure 9.4 as the familiar nose cone shape. If Gerrybild risk-averts, only segment AB is of interest, but precisely where along this segment it will choose to locate depends on the attitude towards risk of its decision-makers.

Self-assessment activity 9.7

Using Figure 9.4, distinguish between risk minimisation and risk aversion.

(Answer in Appendix A at the back of the book)

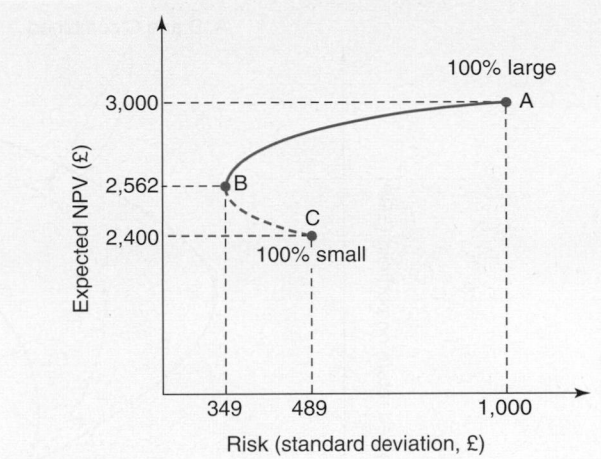

Figure 9.4 Gerrybild's opportunity set

9.7 PORTFOLIOS WITH MORE THAN TWO COMPONENTS*

Having so far looked only at simple two-asset portfolios, it is now useful to extend the analysis to more comprehensive combinations (see Figure 9.5). Imagine three assets are available, A, B and C, for each of which we have estimates of expected return and standard deviation, and also the covariance (and hence correlation) between each pair of assets. Imagine further that, whereas A and B are quite closely correlated, B and C are less so, and that correlation between A and C is even weaker.

Using a technique called Quadratic Programming, developed by Sharpe (1963), we can specify all available portfolios comprising one, two or three assets. Although there are only seven possible combinations of whole investments (A, B and C alone, A plus B, B plus C, A plus C and all three together), there are myriad combinations if we allow for divisibility of assets. The full range of available portfolios, i.e. risk–return combinations, is shown by the opportunity set in the form of an envelope, or 'bat-wing'.

The corners represent individual assets, while two-asset combinations are shown by the solid lines AB and BC and the dotted profile AC. Notice that by combining A and C, the investor can exploit their relative lack of correlation by accessing relatively more attractive portfolios in terms of their respective returns for particular levels of risk. The opportunity set thus moves inwards as assets with lower correlation are included. However, he can now access even more attractive combinations of A and C by combining all three assets. Points inside the envelope, or along the outer boundary, represent all possible combinations of A, B and C.

Notice that the investor now has access to a far wider range of investment combinations. If he is limited to combinations of only two assets, say A and B, as we saw in earlier analyses, he is restricted to risk–return combinations along AB or BC, depending on which two assets are combined. However, if access is opened up to include a third asset, the expanded range of combinations now available allows him to select far superior mixes of risk and return. For example, combinations within the envelope and on its upper bound, AEC, are superior to most of the two-asset portfolios available along AB and BC.

---

*As one might imagine, the mathematics of more complex portfolios becomes more awkward to handle as the number of components increases. The interested reader may wish to consult a more rigorous treatment, such as that given by Copeland, Weston and Shastri (2004). We will rely on an intuitive approach.

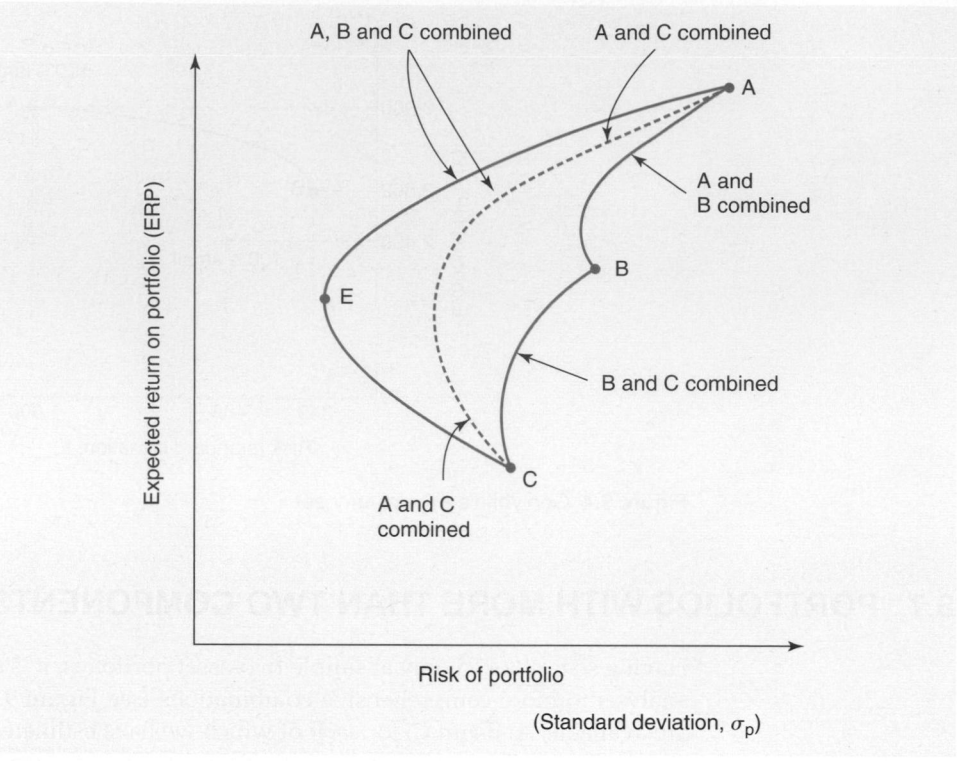

Figure 9.5 Portfolio combinations with three assets

As before, we can differentiate between efficient and inefficient combinations. Clearly, all points lying beneath the upper edge AE and those along the segment EC are inefficient. The efficient set is therefore AE, identical in shape to our earlier profile, except that we are dealing with three-asset combinations (enabling investors to achieve lower levels of risk for specified returns by diversifying away yet more risk). Similar principles would apply if we were dealing with 30 or 300 assets, although the information requirements would become progressively more formidable.

Generally, we can conclude that the more assets that are available, the wider the range of choice open to the investor, and the greater his opportunities to achieve more desirable combinations of risk and return. The more assets under consideration, the nearer to the vertical axis lies the envelope of portfolios. Hence, the higher is the return achievable for a given risk, or conversely, the lower is the risk achievable for a specified expected return.

Notice also that the earlier conclusion about the optimal portfolio remains valid – it still depends on the particular investor's risk–return preferences.

Self-assessment activity 9.8

Draw an envelope of portfolios for the case where four assets are available to invest in, either individually or as portfolios.

(Answer in Appendix A at the back of the book)

9.8 CAN WE USE THIS FOR PROJECT APPRAISAL? SOME RESERVATIONS

The Gerrybild example illustrates some drawbacks with the portfolio approach to handling project risk.

1 Most projects can be undertaken only in a very restricted range of sizes or even on an 'all-or-nothing' basis. This does not entirely undermine the portfolio approach – it simply means that the range of combinations available is much narrower. Besides, enterprises are often undertaken on a joint venture basis (e.g. in large, high-risk activities like Eurotunnel and Airbus and many cross-border automobile operations), where the various parties have some freedom to select the extent of their participation.

2 A more severe problem is the implication of constant returns to scale. Our analyses imply that if a smaller version of a project is undertaken, the percentage returns, or the absolute return per pound invested, will remain unchanged. For example, if the return on a whole project is 20 per cent, the return from doing 30 per cent of the same project is still 20 per cent. This may apply for investment in securities, but is unlikely for investment projects, where there is often a minimum size below which there are zero or negative returns, and, thereafter, increasing returns to scale.

3 We should be wary of any approach that relies on subjective assessments of probabilities, and wary of the probabilities themselves. In the case of repetitive activities, such as replacement of equipment, about which a substantial data bank of costs and benefits has been compiled, the probabilities may have some basis in reality. In other cases, such as major new product developments, probabilities are largely based on inspired guesswork. Different decision analysts may well formulate different 'guesstimates' about the chances of particular events occurring. However, the subjective nature of probabilities used in practice need not be a deterrent if the estimates are well supported by reasoned argument, and therefore instil confidence.

4 Since attitudes to risk determine choice, we need to know the decision-maker's utility function, which summarises his or her preferences for different monetary amounts, if we wanted to pinpoint the optimal (as distinct from the risk-minimising) portfolio. The difficulties of obtaining information about an individual manager's utility function (let alone for a group) are formidable, as Swalm (1966) has shown. Besides, we should really be seeking to apply the risk–return preferences of shareholders rather than those of managers.

5 The portfolio approach to analysing project risk seems unduly management-oriented. Managers formulate the assessments of alternative payoffs, assess the relevant probabilities and determine what combinations of activities the enterprise should undertake. Managers are considerably less mobile and less well diversified than shareholders, who can buy and sell securities more or less at will. Managers can hardly shrug off a poor investment outcome if it jeopardises the future of the enterprise or, more pertinently, their job security. Most managers are more risk-averse than shareholders, resulting in the likelihood of sub-optimal investment decisions. Here, we see another manifestation of the agency problem – how do we get managers to accept the levels of risk that owners are prepared to tolerate?

Capital Asset Pricing Model
The CAPM is a model designed to explain how the stock market values capital assets, including ordinary shares, by assessing their relative risk–return properties

These may appear to be highly damaging criticisms of the portfolio approach, especially as it applies to investment decisions. However, although having limited operational usefulness for many investment projects, it provides the infrastructure of a more sophisticated approach to investment decision-making under risk, the **Capital Asset Pricing Model** (CAPM). This is based on an examination of the risk–return characteristics and resulting portfolio opportunities of securities, rather than physical investment opportunities.

The CAPM explains how individual securities are valued, or priced, in efficient capital markets. Essentially, this involves discounting the future expected returns from holding a security at a rate that adequately reflects the degree of risk incurred in holding that security. A major contribution of the CAPM is the determination of the premium for risk demanded by the market from different securities. This provides a clue as to the appropriate discount rate to apply when evaluating risky projects. The CAPM is analysed in the next chapter.

SUMMARY

This chapter has examined some reasons why firms diversify their activities, and has considered the extent to which the theory of portfolio analysis can provide operational guidelines for diversification decisions.

Key points

- Both firms and individuals diversify investments – firms build portfolios of business activities and individuals build portfolios of securities.

- An important motive for business diversification is to reduce fluctuations in returns.

- Variations in returns can be totally eliminated only if the investments concerned have perfect negative correlation and if the portfolio is weighted so as to minimise risk.

- The expected return from a portfolio is a weighted average of the returns expected from its components, the weights being determined by the proportion of capital invested in each activity or security. For a portfolio comprising the two assets, A and B:

$$ER_p = \alpha ER_A + (1 - \alpha)ER_B$$

- Portfolio risk is given by a square-root formula:

$$\sigma_p = \sqrt{[\alpha^2 \sigma_A^2 + (1 - \alpha)^2 \sigma_B^2 + 2\alpha(1 - \alpha)\mathrm{cov}_{AB}]}$$

- The degree of covariability between the returns expected from the components of the portfolios can be measured by the covariance, cov_{AB}, or by the correlation coefficient, r_{AB}. The lower the degree of covariability, the lower is the risk of the portfolio (for given weightings).

- The available risk–return combinations for mixing investments are shown by the opportunity set.

- Some combinations can be rejected as inefficient. Rational risk-averting investors focus only on the efficient set.

- The optimal portfolio for any investor depends on their attitude to risk, that is, how risk-averse they are.

- In practice, there are serious difficulties in applying the portfolio techniques to physical investment decisions.

Further reading

The classic works on portfolio theory are by Markowitz (1952), Sharpe (1964) and Tobin (1958) (all of whom have won Nobel Prizes for Economics). See also Fama and Miller (1972), Sharpe, Alexander and Bailey (1996), Levy and Sarnat (1994) and Copeland, Weston and Shastri (2004) for more developed analyses, and also proofs and derivations of the formulae used in this chapter. Finally, Markowitz's Nobel address (1991) is well worth reading. Rubinstein (2002) gives a '50 years on' assessment of the impact of the CAPM.

QUESTIONS

 **myfinancelab** *Questions with an icon are also available for practice in myfinancelab with additional supporting resources.*

Questions with a **coloured number** have solutions in Appendix B on page 737.

1 The returns on investment in two projects, X and Y, have standard deviations of 30 per cent and 45 per cent respectively. The correlation coefficient between the returns on the two investments is 0.2. What is the standard deviation of a portfolio containing *equal* proportions of the two investments?

2 Determine the risk-minimising portfolios for the following two asset portfolios.
 (i) $ER_A = 8\%$; $ER_B = 10\%$; $\sigma_A = 3\%$; $\sigma_B = 7\%$; $r_{AB} = +1$
 (ii) $ER_A = 20\%$; $ER_B = 12\%$; $\sigma_A = 12\%$; $\sigma_B = 6\%$; $r_{AB} = +\frac{1}{2}$
 (iii) $ER_A = 11\%$; $ER_B = 5\%$; $\sigma_A = 15\%$; $\sigma_B = 1\%$; $r_{AB} = -\frac{1}{2}$
 (iv) $ER_A = 11\%$; $ER_B = 5\%$; $\sigma_A = 15\%$; $\sigma_B = 1\%$; $r_{AB} = 0$

? 3 Tomb-zapper plc manufactures computer video games. It is considering whether to expand production at its existing site in 'Silicon Glen' in Scotland, or to start production in a 'greenfield site' in China, where labour costs are considerably lower than in Europe. The IRRs for each project depend on average rates of growth in the world economy over the ten-year lifespan of the project. These are expected to be:

World growth	Probability	IRR China	IRR Scotland
Rapid	0.3	50%	10%
Stable	0.4	25%	15%
Slow	0.3	0%	16%

Tomb-zapper wants to exploit the less than perfect correlation between the returns from the two projects, without over-committing itself to the China investment.

Required
(a) What is the expected return and standard deviation of return for each separate project?
(b) Determine the expected return and standard deviation of an expansion programme that involves 25 per cent of available funds in China and 75 per cent in the Scottish location.

4 Nissota, a Japanese-based car manufacturer, is evaluating two overseas locations for a proposed expansion of production facilities at a site in Ireland and another on Humberside. The likely future return from investment in each site depends to a great extent on future economic conditions. Three scenarios are postulated, and the internal rate of return from each investment is computed under each scenario. The returns with their estimated probabilities are shown below:

	Internal rate of return (%)	
Probability	Ireland	Humberside
0.3	20	10
0.3	10	30
0.4	15	20

There is zero correlation between the returns from the two sites.

Required

(a) Calculate the expected value of the IRR and the standard deviation of the return from investment in each location.

(b) What would be the expected return and the standard deviation of the following split investment strategies:
   (i) committing 50 per cent of available funds to the site in Ireland and 50 per cent to Humberside?
   (ii) committing 75 per cent of funds to the site in Ireland and 25 per cent to the Humberside site?

5 The management of Gawain plc is evaluating two projects whose returns depend on the future state of the economy as shown below:

Probability	IRR$_A$ (%)	IRR$_B$ (%)
0.3	27	35
0.4	18	15
0.3	5	20

The project (or projects) accepted would double the size of Gawain.

Required

(a) Explain how a portfolio should be constructed to produce an expected return of 20 per cent.

(b) Calculate the correlation between projects A and B, and assess the degree of risk of the portfolio in (a).

(c) Gawain's existing activities have a standard deviation of 10 per cent. How does the addition of the portfolio analysed in (a) and (b) affect risk?

Practical assignment

Select a company with a reasonably wide portfolio of activities. Such companies do not always give segmental earnings figures, but they usually divulge sales figures for their component activities. By looking at the annual reports for three or four years, you can obtain a series of annual sales figures for each activity.

Assess the degree of past volatility of the sales of each sub-unit and their degree of inter-correlation. Also, see whether you can assess the extent of the correlation between each segment and the overall enterprise. How well diversified does your selected company appear to be? What qualifications should you make in your analysis?

 *Now retake your diagnostic test for Chapter 9 to check your progress and update your study plan.*

10

Setting the risk premium: the Capital Asset Pricing Model

Target practice does not make perfect

Efficient capital markets should generate an upward-sloping risk–return frontier on which all securities locate – the higher the risk, the higher the required return. The Capital Asset Pricing Model (CAPM) explains how great a premium is required for specified risks. Although companies acknowledge the need to discount returns expected from risky activities at higher rates, a survey conducted by the Confederation of British Industry (CBI) revealed some alarming information about how firms approach capital investment decisions.

In 1998, the CBI surveyed 326 firms with turnovers above £20 million and with capital expenditures ranging from rather less than £1 million to well over £25 million. Two points stood out:

1 Firms tended to apply much higher rates than appear warranted by theoretical best practice, with those using IRR setting higher hurdle rates than users of the NPV method.

2 Firms tend not to adjust their hurdle rates when inflation rates change. Only 60 per cent of respondents conducted a regular review of hurdle rates, and there was little evidence that targets had fallen since the previous study in 1994, despite lower inflation.

Setting too high a cut-off rate for investment projects carries two dangers. First, it may curtail the volume of capital expenditure to the detriment of business growth. Second, setting too high a target may lead to over-investment in high risk, speculative projects (albeit potentially lucrative ones) at the expense of more secure 'bread and butter' capital projects.

Source: Target Practice, Confederation of British Industry, 1998.

Learning objectives

This chapter deals with the rate of return required by shareholders of an all-equity financed company, building on the principles of portfolio theory covered in Chapter 9. Its specific aims are:

■ To explain what type of risk is relevant for valuing capital assets.

■ To explain what a 'Beta coefficient' is.

■ To determine the appropriate risk premium to incorporate into a discount rate, whether for investment in securities or in capital projects.

■ To examine the case for *corporate* diversification.

■ To examine some criticisms of the CAPM.

An understanding of the significance of Beta coefficients is particularly important in appreciating how financial managers should view risk.

 *Complete your diagnostic test for Chapter 10 now to create your personal study plan.*

10.1 INTRODUCTION

In Chapters 7 and 9, we examined various methods of handling risk and uncertainty in project appraisal, ranging from sensitivity analysis to diversification to seek to exploit the less than perfect correlation between the returns from risky investments. Most of these approaches aim to identify the sources and extent of project risk and to assess whether the expected returns sufficiently compensate investors for bearing the risk. Utility theory suggests that, as risk increases, rational risk-averse people require higher returns, justifying the common practice of adjusting discount rates for risk. However, none of these approaches offers an explicit guide to measuring the *precise* reward investors should seek for incurring a particular level of risk.

The CAPM is a theory originally devised by Sharpe (1964) to explain how the capital market sets share prices. It now provides the infrastructure of much of modern financial theory and research and offers important insights into measuring risk and setting risk premiums. In particular, it shows how the study of security prices can help in assessing required rates of return on investment projects. However, as we shall see, the CAPM has not gone unchallenged.

10.2 SECURITY VALUATION AND DISCOUNT RATES

Asset value is governed by two factors – the stream of expected benefits from holding the asset and their 'quality', or likely variability. For example, the value of a single-project company is assessed by discounting future project cash flows at a discount rate reflecting their risk. The value, V_o, of a company newly formed by issuing one million shares to exploit a one-year project offering a single net cash flow of £10 million, at a 25 per cent discount rate, is:

$$V_o = \frac{£10 \text{ m}}{(1.25)} = £8 \text{ m}$$

This suggests a market price per share of (£8 m/1 m shares) = £8. This would be the value established by an efficient capital market taking account of all known information about the company's future prospects.

Sometimes, the 'correct' discount rate is unclear to the firm. A major contribution of the CAPM is to explain how discount rates are established and hence how securities are valued. However, from the capital market value of a company, we can 'work backwards' to infer what discount rate underlies the market price. In the example, if we observe a market price of £8, this suggests a required return of 25 per cent.

By implication, if the market sets a value on a security that implies a particular discount rate, it is reasonable to conclude that any further activity *of similar risk* to current operations should offer about the same rate. This argument depends critically on market prices being unbiased indicators of the intrinsic worth of companies, i.e. that the efficient markets hypothesis applies.

Any discount rate is an amalgam of three components:

1 Allowance for the time value of money – the compensation required by investors for having to wait for their payments.
2 Allowance for price level changes – the additional return required to compensate for the impact of inflation on the real value of capital.
3 Allowance for risk – the promised reward that provides the incentive for investors to expose their capital to risk.

Ignoring expected inflation (or assuming that it is 'correctly' built into the structure of interest rates), discount rates have two components – the rate of return required on totally risk-free assets, such as government securities, and a risk premium.

10.3 CONCEPTS OF RISK AND RETURN

In this section, we examine risk and return concepts relevant for security valuation.

■ The returns from holding shares

Investors hold securities because they hope for positive returns. Purchasers of ordinary shares are attracted by two elements: first, the anticipated dividend(s) payable during the holding period; and second, the expected capital gain. Taken together, these elements make up the **Total Shareholder Return (TSR)**.

Total Shareholder Return (TSR)

In general, for any holding period, t, and company, j, the TSR is the percentage return, R_{jt}, from holding its shares:

$$R_{jt} = \frac{D_{jt} + (P_{jt} - P_{jt-1})}{P_{jt-1}} \times 100$$

where D_{jt} is the dividend per share paid by company j in period t, P_{jt} is the share price for company j at the end of period t and P_{jt-1} is the share price for company j at the start of period t.

To illustrate this calculation, consider the following figures for the UK energy firm, Scottish and Southern Energy plc for the calendar year 2007:

Share price at end of December 2006 = £15.54

Share price at end of December 2007 = £16.38

Net dividend paid during 2007 = 55p per share

The percentage return over this year was:

$$\frac{55p + (£16.38 - £15.54)}{£15.54} \times 100 = \frac{(55p + 84p)}{£15.54} \times 100 = 8.9\%$$

However, the TSR data relate to just one year, and may be influenced in either direction by random factors. A more meaningful measure of shareholder return would remove these short-term fluctuations, adverse or favourable. This is done by taking the overall return over a specific period, commonly five years, and converting this into an average annual or annualised equivalent rate of return.

To illustrate this, figures from D.S. Smith, the UK packaging firm, are shown in Figure 10.1. For the period 2002–7, £100 invested in this firm would have grown to about £200, i.e. shareholder wealth grew by 100 per cent, although not in a steady fashion (and indeed at a rate below the growth in the FTSE 250 index of which D.S. Smith is a member). This growth corresponds to an average annual growth rate of about 15 per cent (check: $£100(1.15\%)^5 = £201$).

Firms increasingly use TSR, usually in relation to other firms, as a performance benchmark as part of the executive reward scheme. For example, Barclays Bank plc aims to locate in the top quartile of TSR outcomes in the banking sector.

Self-assessment activity 10.1

Determine the TSR for the year 200X in the following case:

- Share price 1 January: £2.20
- Share price 31 December: £2.37
- Interim dividend paid: £0.035 per share
- Final dividend paid: £0.065 per share

(Answer in Appendix A at the back of the book)

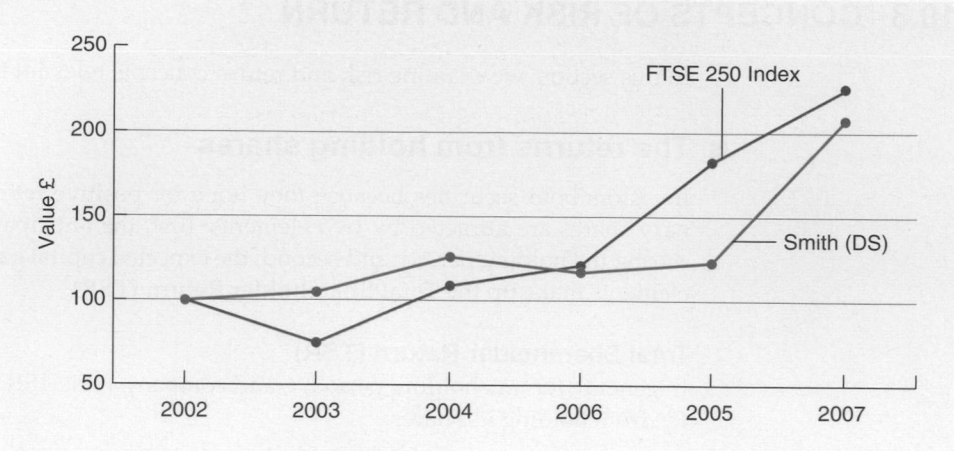

Figure 10.1 TSR of D.S. Smith plc 2002–7

Source: D.S. Smith plc Annual Report 2007.

■ The risks of holding ordinary shares

In Chapter 9, we saw the power of portfolio combination in reducing the risk of a collection of investments. Risk was measured by the variance or standard deviation of the return on the combination. This measure can also be applied to portfolios of securities, with some remarkable results, as shown in Figure 10.2.

As the number of securities held in the portfolio increases, the overall variability of the portfolio's return, measured by its standard deviation, diminishes very sharply for small portfolios, but falls more gradually for larger combinations. This reduction is achieved because exposure to the risk of volatile securities can be offset by the inclusion of low-risk securities *or even ones of higher risk*, so long as their returns are not closely correlated.

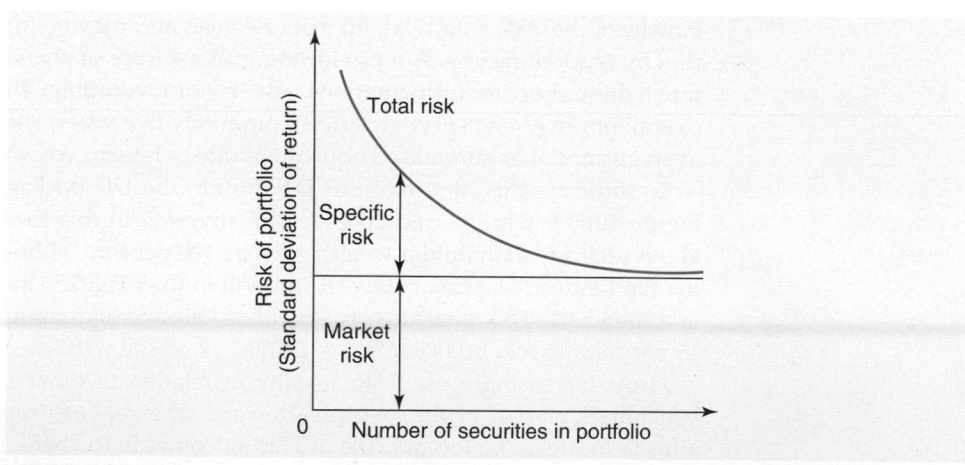

Figure 10.2 Specific vs. market risk of a portfolio

■ Specific and systematic risk

Not all the risk of individual securities is relevant for assessing the risk of a portfolio of risky shares. The total risk of securities (and also of portfolios) has two components:

1 *Specific risk*: the variability in return due to factors unique to the individual firm.
2 *Systematic risk*: the variability in return due to dependence on factors that influence the return on all securities traded in the market.

Table 10.1 How to remove portfolio risk

Number of securities (N)	Reduction in specific risk (%)
1	0
2	46
4	72
8	81
16	93
32	96
64	98
500	99

Source: Fosback (1985).

Specific risk refers to the expected impact on sales and earnings of random events – industrial relations problems, equipment failure, R&D achievements etc. In a portfolio of shares, such factors tend to cancel out as the number of securities included increases.

Systematic risk refers to the impact of movements in the macroeconomy, such as fiscal changes, swings in exchange rates and interest rate movements, all of which cause reactions in security markets. These are captured in the movement of an index reflecting security prices in general, such as the FTSE in the UK or the DAX index in Germany. No firm is entirely insulated from these factors, and even portfolio diversification cannot provide total protection. Because these factors affect all firms in the market, such risk is often called 'market-related' (or just 'market') risk.

Returning to Figure 10.2, we see that the reduction in the total risk of a portfolio is achieved by gradual elimination of the risks unique to individual companies, leaving an irreducible, undiversifiable risk floor. The extent to which specific risk declines for a portfolio comprising N equally-weighted and randomly-selected securities is also shown in Table 10.1.

Substantial reductions in specific risk can be achieved with quite small portfolios, with the bulk of risk reduction being achieved with a portfolio of some 25–30 securities. To eliminate unique risk totally would involve holding a vast portfolio comprising all the securities traded in the market. This construct, called the **'market portfolio'**, has a pivotal role in the CAPM, but for the individual investor, it is neither practicable nor cost-effective, in view of the dealing fees required to construct and manage it. However, since relatively small portfolios can capture the lion's share of diversification benefits, it is only a minor simplification to use a well-diversified portfolio as a proxy for the overall market, such as the FTSE-100, which covers approximately 81 per cent of the market capitalisation of all UK quoted companies (**www.ftse.com**).

Self-assessment activity 10.2

How many shares would an investor have to hold in order to *totally* eliminate specific risk?
(Answer in Appendix A at the back of the book)

■ Implications

Three major implications now follow:

1 *It is clear that risk-averse investors should diversify*. Yet in reality, over half of UK investors hold just one security (usually, shares in a privatised company or a former building society). However, the major players in capital markets, holding well over

two-thirds of all quoted UK ordinary shares (according to the Office of National Statistics, July 2007*) are financial institutions such as pension funds and insurance companies, which do hold highly diversified portfolios.

2 *Investors should not expect rewards for bearing specific risk.* Since risk unique to particular companies can be diversified away, the only relevant consideration in assessing risk premiums is the risk that cannot be dispersed by portfolio formation. If bearing unique risk was rewarded, astute investors prepared to build portfolios would snap up securities with high levels of unique risk to diversify it away, while still hoping to enjoy disproportionate returns. The value of such securities would rise and the returns on them would fall until only systematic risks were rewarded.

3 *Securities have varying degrees of systematic risk.* Few securities exhibit patterns of returns rising or falling exactly in line with the overall market. This is partly because in the short term, unique random factors affect particular companies in different ways. Yet even in the long term, when such factors tend to even out, very few securities track the market. Some appear to outperform the market by offering superior returns and some appear to underperform it. However, performance relative to the market should not be too hastily judged, because the returns on different securities do not always depend on general economic factors in the same way.

For example, in an expanding economy, retail sales tend to increase sharply, but sales in less responsive sectors like water and defence are barely altered. Share prices of retailers usually increase quite sharply in an expanding economy, but the share prices of water companies and armaments suppliers respond far less dramatically. Retail sales are said to be 'more highly geared to the economy'. Systematic or market risk varies between companies, so we find different companies valued by the market at different discount rates. Already, we begin to see that the CAPM, based on the premise that rational investors can and do hold efficiently diversified portfolios, may show us how these discount rates might be assessed. Clearly, we need to measure systematic risk. This is covered in Section 10.5.

Self-assessment activity 10.3

Give three examples of systematic and unique factors that cause the returns on holding ordinary shares to vary over time.

(Answer in Appendix A at the back of the book)

10.4 THE RELATIONSHIP BETWEEN DIFFERENT EQUITY MARKETS

Investors have tended to prefer to invest in their own national stock markets, although this is changing. Reasons for this past parochialism include:

- relative lack of research into overseas markets and firms
- transactions costs, especially connected with foreign exchange
- fear of foreign exchange risk
- legal barriers, e.g. custody regulations
- political risk.

Several studies have shown that international diversification can generate even greater portfolio benefits than investing in purely domestic shares. Recall that the reason that portfolio risk reduces as the number of component shares increases was low correlation between investments, enabling investors to reduce specific risks. However,

---

*UK financial institutions held 46%, foreign investors 41% and private investors 13%. The majority of foreign shareholdings (which have risen from 16% in 1994) is reckoned to represent institutional, rather than private, holdings.

if international stock markets are less than perfectly correlated, it may be possible to lower risk below the level of market risk that defines the floor of the risk profile relating to purely domestic investment.

Indeed, studies pioneered by Solnik (1974) have shown that international markets are not all closely correlated. Kaplanis (1997) showed that between 1990 and 1994, London had the following cross-national correlation coefficients: USA (0.7), Germany (0.4), Italy (0.2), Japan (0.3) and Australia (0.5). However, mainland European markets tended to have higher correlations, e.g. Germany/France (0.7), Netherlands/Germany (0.7), due, presumably, to closer European integration.

Astute investors could exploit these less than perfect correlations by combining investments in two or more markets, thus achieving a bodily shift downwards in the risk profile. The effect is shown in Figure 10.3.

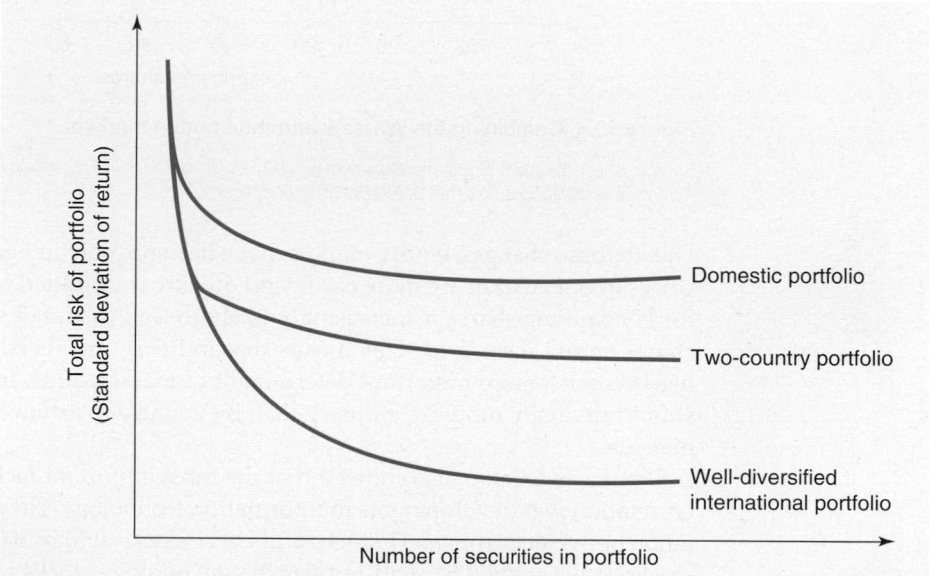

Figure 10.3 The effect of international diversification on portfolio risk

An illustration of this effect is shown in Figure 10.4. The author examined portfolio formation on both the Polish and the London stock exchanges, and found that it would have benefited Polish investors (but not British ones) to combine Warsaw- and London-quoted stocks. Clearly, Warsaw stocks were more risky, possibly due to a lower level of market efficiency, although it is interesting to observe the risk profile flattening out at virtually the same size of portfolio for each stock market.

However, such opportunities may be disappearing. By the mid-1990s, the correlation between changes in US and European share price movements was estimated at around 0.4 – Wall Street movements would 'explain' 40 per cent of movement in the main European indices. But Brooks and Catao (2000) showed that rapid technical and institutional change had raised the correlation to 0.8 by 2000.

They suggested several reasons for this convergence:

- removal of controls on capital movements;
- more efficient trading systems;
- greater cross-border trading volumes;
- more large companies obtaining listings on several stock markets;
- more cross-border mergers and acquisitions with foreign activities accounting for higher proportions of company profits;
- easier access to information on foreign firms via the internet.

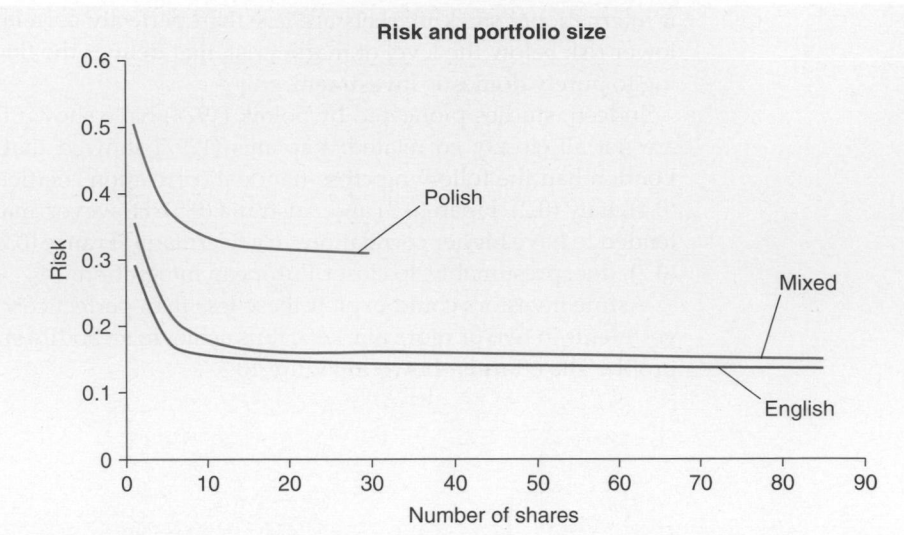

Figure 10.4 Combining the Warsaw and the London markets

Source: Short, T. (2000) 'Should foreign investors buy Polish shares?' in T. Kowalski and S. Letza (eds) *Financial Reform and Institutions* (Poznan University of Economics).

Due to these changes, equity markets have become more integrated, so that changes in prices in one market are more easily and quickly transmitted to others, e.g. good news for US banking shares is increasingly likely to lead to higher share prices for banking shares across the world. This means that industry membership rather than location has become a more important determinant of market value. In other words, investors should diversify more by industry than by country to achieve optimal diversification benefits.

Brooks and Catao also showed that the most important factor explaining increased correlation was developments in information technology. They found an overall correlation between European IT stocks and US IT stocks at May 2000 of 0.85, but for non-IT stocks, it was only 0.54. This implies that technology stocks now constitute a channel whereby shocks in one market are spread throughout the world, e.g. in October 2007, a shock profits warning by Sweden's Ericsson, the world's largest maker of mobile phone networks sent telecoms and technology stocks lower across the world. Caught in the fall-out were Nokia (Finland), Nortel (Canada) and Alcatel-Lucent (France).

10.5 SYSTEMATIC RISK

As specific risk can be diversified away by portfolio formation, rational investors expect to be rewarded only for bearing systematic risk. Since systematic risk indicates the extent to which the expected return on individual shares varies with that expected on the overall market, we have to assess the extent of this co-movement. This is given by the slope of a line relating the expected return on a particular share, ER_j, to the return expected on the market, ER_m. It is important to appreciate that 'returns' in this context include both changes in market price and also dividends as we saw in the Scottish and Southern plc calculation in Section 10.3. For the overall market, dividend returns may be measured by the average dividend yield on the market index.

■ Example: Walkley Wagons

The case of Walkley Wagons is shown in Table 10.2. Investors anticipate four possible future states of the economy. For every percentage point increase in the expected market return (ER_m), the expected return on Walkley shares (ER_j) rises by 1.2 percentage

Table 10.2 Possible returns from Walkley Wagons

State of economy	ER_m(%)	ER_j(%)
E_1	10	12
E_2	20	24
E_3	5	6
E_4	15	18

points. Walkley thus outperforms a rising market. The graphical relationship between ER_j and ER_m, shown in Figure 10.5, is known as the **characteristics line**. Its slope of 1.2 is the **Beta coefficient**. Beta indicates how the return on Walkley is *expected* to vary alongside given variations in the return on the overall stock market.

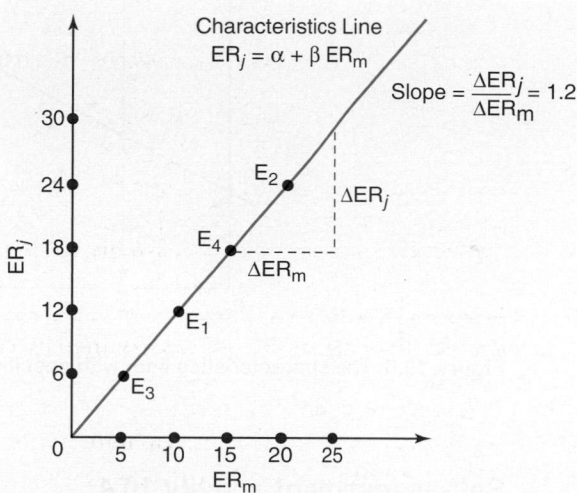

Figure 10.5 The characteristics line: no specific risk

The market model

In practice, because it is not easy to record people's expectations, the measurement of Beta cannot be done by looking forward. We have to measure Beta using *past* observations of the actual values of both the return on the individual company's shares, and also for the overall market, i.e. R_j and R_m respectively. So long as the past is accepted as a reliable indication of likely future events (i.e. people's expectations are moulded by examination of the frequency distribution of past recorded outcomes), observed Betas can be taken to indicate the extent to which R_j may vary for specified variations in R_m. A regression line* is fitted to a set of recorded relationships, as in Figure 10.6. The hypothesized relationship is:

$$R_j = \alpha_j + \beta_j R_m$$

and the fitted line is given by:

$$R_j = \hat{\alpha}_j + \hat{\beta}_j R_m + u$$

---

*Readers unfamiliar with the technique of regression analysis might refer to C. Morris, *Quantitative Approaches in Business Statistics* (Pearson).

where $\hat{\alpha}_j$ and $\hat{\beta}_j$ are estimates of the 'true' values of α and β, and u is a term included to capture random influences, that are assumed to average zero. This regression model is called the **market model**.

The intercept term, α_j, deserves explanation. This is the return on security j when the return on the market is zero, i.e. the return with the impact of market or systematic risk stripped out. Consequently, it indicates what return the security offers for specific risk. We might expect this to average out at zero over time, given the random character of sources of specific risk. However, it is by no means uncommon empirically to record non-zero values for α. Notice that in Figure 10.5 α is zero.

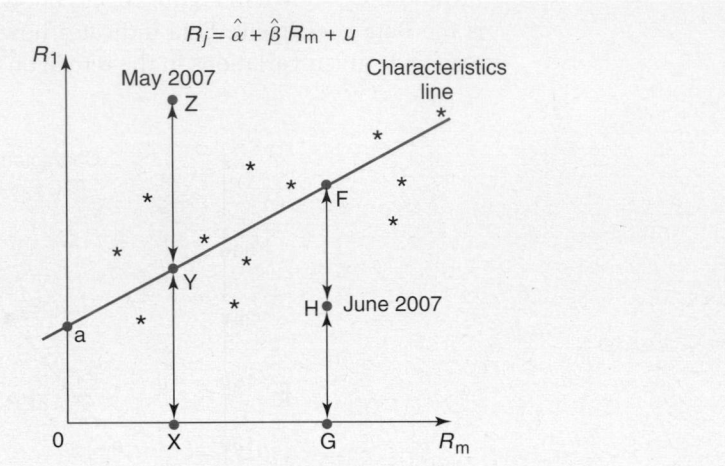

Figure 10.6 The characteristics line: with specific risk

Self-assessment activity 10.4

You read in the financial press that the 'experts' are predicting overall stock market returns of 25 per cent next year. What return would you expect from holding Walkley Wagons ordinary shares?

(Answer in Appendix A at the back of the book)

■ Systematic and unsystematic returns

Figure 10.6 shows an imaginary set of monthly observations relating to a given year, say 2007, to which has been fitted a regression line. Clearly, unlike the *expected* values displayed in Figure 10.5, most values actually lie off the line of best fit. These divergences are due to the sort of random, unsystematic factors suggested in Section 10.3. For example, observation Z relates to the returns in May 2007. The overall return on security j in this month, XZ, can be broken down into the market-related return, XY, due to co-movement with the overall stock market, and the non-market return, or 'excess return', YZ, due to unsystematic factors, which, in this month, have operated favourably. The opposite appears to have applied in June 2007, indicated by point H. The market-related return 'should' have been FG, but the actual return of GH was dampened by unfavourable random factors represented by FH. *This analysis implies that variations in R_j along the characteristics line stem from market-related factors, which systematically affect all securities, and that variations around the line represent the impact of factors specific to company j. The systematic relationship is captured by β.*

Self-assessment activity 10.5

What is the significance of variations around the characteristics line? Relate this to a particular company, say, British Airways.

(Answer in Appendix A at the back of the book)

Beta values: the key relationships

Beta is the slope of a regression line. The slope coefficient relating R_j to R_m equals the covariance of the return on security j with the return on the market (cov_{jm}) divided by the variance of the market return (σ_m^2):

$$\text{Beta}_j = \frac{\text{cov}_{jm}}{\sigma_m^2}$$

Since the covariance is equal to the correlation coefficient times the product of the respective standard deviations ($r_{jm}\sigma_j\sigma_m$) (see Chapter 9), Beta is also equivalent to:

$$\text{Beta}_j = \frac{r_{jm}\sigma_j\sigma_m}{\sigma_m^2} = \frac{r_{jm}\sigma_j}{\sigma_m}$$

Beta is thus the correlation coefficient multiplied by the ratio of individual security risk to market risk. If the security concerned has the same total risk as the market, Beta equals the correlation coefficient. For a given correlation, the greater the security's systematic risk in relation to the market, the greater is Beta. Conversely, the lower the degree of correlation, for a given risk ratio, the lower the Beta. *Therefore, while Beta does not measure risk in absolute terms, it is a risk indicator, reflecting the extent to which the return on the single asset moves with the return on the market*, i.e. it is a measure of relative risk. To obtain a risk measure in absolute terms, we have to examine the total risk of the security in more detail, using a statistical technique called **analysis of variance**. This is explained in the appendix to this chapter.

analysis of variance
A statistical technique for isolating the separate determinants of the fluctuations recorded in a variable over time

■ Systematic risk: Beta measurement in practice

Betas are regularly calculated by several agencies. The Risk Measurement Service (RMS) operated by the London Business School (LBS) is the best known in the UK. The RMS is a quarterly updating service, based on monthly observations extending back over five years, which computes the Betas of all firms listed both on the main market and also on AIM. For each of the preceding 60 months, R_j is calculated for every security and regressed against R_m. An extract from the RMS showing the components of the FT 30 Index of leading industrial shares is given in Table 10.3.

The Beta values of securities fall into three categories: 'defensive', 'neutral' and 'aggressive'. An aggressive security has a Beta greater than 1. Its returns move by a greater proportion than the market as a whole. In the case of GKN, with a Beta of 1.13, for every percentage point change in the market's return, the return on GKN's shares changes by 1.13 points. Such stocks are highly desirable in a rising market, although the excess return is not guaranteed due to the possible impact of company-specific factors. A defensive share is National Grid, with a Beta of 0.52, movements in whose returns tend to understate those of the whole market. The returns on neutral stocks like Vodafone, with its Beta of 1.01, parallel those on the market portfolio.

Notice that the total risk of each security is shown as 'variability', e.g. 28 for GKN. This is a standard deviation. Notice also that this invariably exceeds 'Specific Risk', e.g. 26 for GKN. The difference indicates the market risk that cannot be diversified away. (See the appendix to this chapter for a fuller explanation.)

Self-assessment activity 10.6

Suggest why the Beta values tend to cluster in a range of roughly 0.70 to 1.30.

(Answer in Appendix A at the back of the book)

Table 10.3 Beta values of the constituents of the FT 30 Share Index

Company name	FTSE-ICB classification	Market Capit'n	Beta	Varia-bility	Specific risk	Std Err of Beta	R-Sq'rd
BAE Systems	Defense	17322	1.19	25	22	.21	22
BG Group	Oil + Gas	28471	.96	19	16	.17	27
BP	Oil + Gas	107941	.77	17	15	.16	22
British Airways	Airlines	4418	1.43	33	30	.24	19
British American Tobacco	Tobacco	35501	.59	17	15	.17	13
BT Group	TelecFix	24819	1.04	19	16	.17	31
Cadbury-Schweppes	FoodProd	11926	.97	19	16	.17	28
Compass Group	RestBars	5828	.91	27	25	.22	12
Diageo	Distillr	28098	.52	14	13	.15	14
GKN	AutoPart	2494	1.13	28	26	.22	17
GlaxoSmithKline	Pharmact	72743	.55	15	14	.16	13
Imperial Chemical Industries	ChemSpc	7796	1.33	37	35	.25	13
Invensys	ElectEqp	2471	1.21	54	52	.27	5
ITV	BroadEnt	3990	1.30	31	28	.23	18
Ladbrokes	Gambling	2703	1.02	19	16	.17	30
Land Securities Group	RealEsIT	7823	.89	18	16	.17	25
Lloyds TSB Group	Banks	30620	1.47	22	16	.17	48
LogicaCMG	CompSvs	2234	1.49	34	31	.24	19
Man Group	FinAsMan	10852	.96	26	24	.22	14
Marks & Spencer Group	RetBroad	10474	.82	24	22	.21	12
National Grid	MultUtil	20350	.52	15	14	.16	12
Prudential	InsLife	18558	1.66	26	20	.19	43
Reuters Group	Publishg	8129	1.14	44	43	.26	7
Royal Bank of Scotland	Banks	49647	1.24	19	15	.16	42
Royal & Sun Alliance Ins Grp	InsFull	4963	1.68	36	32	.24	22
Tate & Lyle	FoodProd	1937	.77	31	30	.24	6
Tesco	RetFood	34501	.81	19	17	.18	19
Vodafone Group	TelecMob	93628	1.01	24	21	.20	19
Wolseley	Ind Supp	5465	1.11	22	19	.19	25
WPP Group	MediaAgy	7970	1.34	24	20	.20	31

Source: Risk Measurement Service, London Business School, October–December 2007.

10.6 COMPLETING THE MODEL

The CAPM suggests that only systematic risk is relevant in assessing the required risk premiums for individual securities, and we have established that Beta values reflect the sensitivity of the returns on securities to movements in the market return. However, the size of the risk premium on individual securities (or on efficient portfolios) will depend on the extent to which the return on the investment concerned is correlated with the return on the market. For a security that is perfectly correlated with the market, the market risk premium would be suitable; otherwise, the required return depends on the Beta.

The CAPM concludes that when an efficient capital market is in equilibrium, i.e. all securities are correctly priced, the relationship between risk and return is given by the **security market line (SML)**, as depicted in Figure 10.7.

Self-assessment activity 10.7

Why is the Beta of the overall market equal to 1.0?

(Answer in Appendix A at the back of the book)

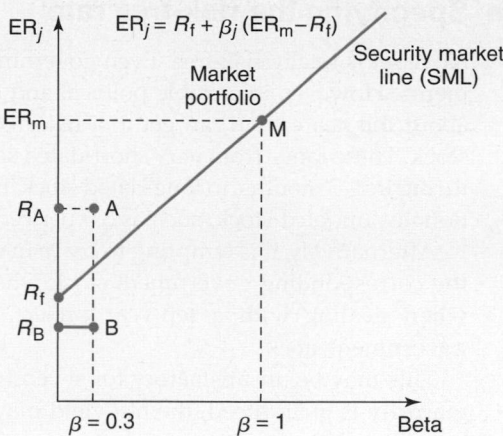

Figure 10.7 The security market line

■ The security market line

The equation of the SML states that the required return on a share is made up of the return on a risk-free asset, plus a premium for risk that is related to the market's own risk premium, but which varies according to the Beta of the share in question:

$$ER_j = R_f + \beta_j(ER_m - R_f)$$

If Beta is 1, the required return is simply the average return for all securities, i.e. the return on the benchmark market portfolio. Otherwise, the higher the Beta, the higher are both the risk premium and the total return required. *A relatively high Beta does not, however, guarantee a relatively high return*. The actual return depends partly on the behaviour of the market, which acts as a proxy for general economic factors. Similarly, expected returns for the individual security hinge on the expected return for the market. In a 'bull', or rising, market, it is worth holding high Beta (aggressive) securities. Conversely, defensive securities offer some protection against a 'bear', or falling, market. *However, holding a single high Beta security is foolhardy, even on a rising market. Undiversified investments, whatever their Beta values, are prey to specific risk factors. Portfolio formation is essential to diversify away the risks unique to individual companies.*

Self-assessment activity 10.8

As explained, the SML is an equilibrium relationship that traces out the set of required returns for securities of different levels of risk which an efficient capital market would demand.

How would you interpret securities such as A and B on Figure 10.7 that lie off the SML, yielding current returns of R_A and R_B respectively?

(Answer in Appendix A at the back of the book)

10.7 USING THE CAPM: ASSESSING THE REQUIRED RETURN

We may now apply the CAPM formula to derive the rate of return required by shareholders in a particular company. To do this, we require information on three components: the risk-free rate, the risk premium on the market portfolio and the Beta coefficient.

■ Specifying the risk-free rate

No asset is totally risk-free. Even governments default on loans and defer interest payments. However, in a stable political and economic environment, government stock is about the nearest we can get to a risk-free asset. Most governments issue an array of stock. These range from very short-dated securities, such as Treasury Bills in the UK, maturing in 1–3 months, to long-dated stock, maturing in 15 years or more and even, exceptionally, undated stock, such as 3.5 per cent War Loan with no stated redemption date.

Alternatively, it is tempting to try to match up the life of the investment project with the corresponding government stock when assessing the risk-free rate. For example, when dealing with a ten-year project, we might look at the yield on ten-year government stock.

This may be unsatisfactory for several reasons. First, although the *nominal* yield to maturity is guaranteed, the *real* yield may well be undermined by inflation at an unknown rate. Second, there is an element of risk in holding even government stock. This is reflected in the 'yield curve', which normally rises over time to reflect the increasing liquidity risk of longer-dated stock. Third, although the yield to maturity is given, a forced seller of the stock might have to take a capital loss during the intervening period, since bond values fluctuate over time with variations in interest rates.

A better way to specify R_f is to take the shortest-dated government stock available, normally three-month Treasury Bills, for which these risks are minimised. The current yield appears in the financial press. This is about the same as LIBOR, the London Interbank Offered Rate, the rate of interest at which banks lend to each other overnight.

■ Finding the risk premium on the market portfolio

The risk premium on the market portfolio, $(ER_m - R_f)$, is an expected premium. Therefore, having assessed R_f, we need to specify ER_m by finding a way of capturing the market's expectations about future returns. An approximation can be obtained by looking at past returns, which, taken over lengthy periods, are quite stable. The usual approach with ordinary shares is to analyse the actual total returns on equities as compared with total returns on fixed-interest government stocks over some previous time period. The results are likely to differ according to the period taken and the type of government stock used as the reference level (e.g. short-term securities such as Treasury Bills or long-term gilts). However, studies seem to come up with quite stable results. For example, Dimson and Brealey (1978), Day *et al.* (1987) and Dimson (1993) for the periods 1918–77, 1919–84 and 1919–92, respectively, showed average annual returns above the risk-free rate of 9.0, 9.1 and 8.7 per cent (before taxes) for the market index in the UK.

Similar estimates have been obtained in the USA. In 1985, Mehra and Prescott found that, after adjusting for inflation, equities delivered average *real* returns of 7 per cent p.a. over a quarter of a century, compared with 1 per cent for Treasury bonds – a real risk premium of 6 per cent. Mehra and Prescott found this premium 'puzzling' on the grounds that it seemed too large a premium for bearing non-diversifiable market risk, especially given international opportunities for diversification. Fama and French (2000) found the equity risk premium averaged 8.3 per cent p.a. over 1950–99, this being well in excess of the 4.1 per cent p.a average for 1872–1949.

Dimson (1993) reported similar premia in Japan (9.8 per cent, 1970–92), Sweden (7.7 per cent, 1919–90) and the Netherlands (8.5 per cent, 1947–89), although the last two estimates were in real terms, i.e. relative to domestic inflation.

A rather lower UK risk premium was recorded by Grubb (1993/4), at 6.2 per cent for 1960–92. Grubb suggests that returns to equities in the 1970s and 1980s were exceptional and that under a 'modern scenario of moderate growth and moderate inflation', a much lower premium on equities of only 2 per cent would be reasonable. This view was supported by Wilkie (1994), who, after exhaustive study of past trends in dividend yields and inflation, argued for a risk premium of 3 per cent for longer-term investment and 2 per cent for the short term. The evidence is inconclusive, but it is unlikely that many finance directors would contemplate recommending projects with such low premia for risk.

However, for shorter periods, say five or ten years (more akin to project lifetimes), returns are highly volatile and sometimes negative. Clearly, people neither require nor expect negative returns for holding risky assets! It therefore seems more sensible to take the long-term average, and to accept that, in the short-term, markets exhibit unpredictable variations.

The investment banking arm of Barclays Bank, Barclays Capital (**www.barcap.com**) publishes an annual analysis of equity and gilt-edged returns for various time periods called the 'Equity–Gilt Study'. Their data show real investment returns on equities and government stock, and also on cash deposits. The long-term (108 years) UK equity risk premium is 4.2 per cent in real terms, and 4.3 per cent above the return on cash deposits.

Like many observers, Barclays Capital suggests that as the world economy moves from the low growth/high inflation phase of the 1970s and 1980s to the high growth/ low inflation experienced more recently, equity returns were untypically high. One reason for expecting lower future returns is technological progress, in general, and the information revolution, in particular, resulting in shorter competitive advantage periods. Firms typically have less time to exploit a 'first mover's advantage' before competitors arrive i.e. entry barriers are lower. Another likely depressant is the increased openness of the world economy due to the activities of the World Trade Organisation. A complicating factor is the 'unusual demographic outlook of a shrinking working population and an expanding dependent population'. This suggests that the prices of financial assets will fall relative to prices of goods and services, so that equities may offer a less effective inflation hedge in the future.

Table 10.4 shows Barclays data for real investment returns for different types of asset in the UK and the US, and also the equity premium. Note the strong similarity between UK and US premia. The data are *real* geometric average annualised returns, i.e. they exclude the effect of inflation.

Whether the real equity premium is entering a period of long-term decline is still a matter of some debate. However, subsequent analysis will build in a risk premium for equities, i.e. the risk premium of the overall market portfolio, of 5%, a 'guesstimate' that is supported by a substantial weight of recent evidence.

In probably the most thorough analysis to date of the equity risk premium, Dimson, Marsh and Staunton (2002) updated and largely corroborated these figures in a study of the equity risk premium for 16 countries, over a full century (1900–2000). They suggested that some earlier studies (including the earlier Dimson Studies!) might have over-estimated the equity premium by excluding the First World War era, when equity returns were poor, and by confining the study to the performance of surviving firms, thus excluding the relatively poor performers that had expired.

They found:

■ The average global real return on equity was 4.6 per cent.
■ Germany had offered the highest risk premium at 6.7 per cent.
■ Denmark offered the lowest risk premium at just 2 per cent.
■ In the US, for every 20-year period examined, equities outperformed bonds.
■ Only four countries – Germany, Netherlands, Sweden and Switzerland – exhibited any 20-year periods over which bonds outperformed equities.
■ It is reasonable to expect a real equity premium of no more than 5 per cent or so in the UK in the future.

The LBS team now offer an annual update of this analysis (www.abn-amro.com).

Table 10.4 Equity–gilts relative returns

(a) Real investment returns (% pa): UK to 2007

	Equities	Gilts	Index-linked	Cash	Equity premium over gilts
1907–17	−3.8	−7.2	–	−3.8	+3.4
1917–27	9.1	6.1	–	5.2	+3.0
1927–37	6.1	7.3	–	2.6	−1.2
1937–47	4.0	1.3	–	−1.8	+2.7
1947–57	2.3	−6.2	–	−2.5	+8.5
1957–67	11.4	0.8	–	2.1	+10.6
1967–77	−0.2	−3.2	–	−2.5	+3.0
1977–87	12.0	4.5	–	3.4	+7.5
1987–97	10.4	6.9	5.0	4.6	+3.5
1997–2007	3.1	3.3	3.7	2.5	−0.2

(b) Real investment returns (% pa): USA to 2007

	Equities	Bonds	Cash	Equity premium over bonds
1927–37	0.4	6.0	3.2	−5.6
1937–47	5.3	−1.5	−4.5	+6.8
1947–57	13.3	−0.1	−0.3	+13.2
1957–67	11.9	−0.4	1.3	+11.5
1967–77	−2.8	−1.0	−0.4	+1.8
1977–87	8.7	3.0	2.7	+5.7
1987–97	13.8	7.3	1.9	+6.5
1997–2007	4.1	4.4	0.8	−0.3

Source: Barclays Capital.

These updates on the real returns on equities and bonds allow us to infer the following risk premia for equities over 1900–2007 for selected countries:

	Real risk premium on equities 1900–2007	
Country	vs. bills*	vs. bonds*
Australia	7.2	6.4
Japan	6.5	5.7
South Africa	6.4	5.7
Sweden	5.8	5.3
USA	5.5	4.5
World average	**4.8**	**4.0**
Netherlands	4.6	4.1
UK	4.4	4.1
Ireland	3.9	3.5
Belgium	2.9	2.7

*Corporate bonds are more risky than short-term government bills, and hence offer a higher return, yielding a smaller equity risk premium.

Source: ABN AMRO/LBS *Global Investment Returns Yearbook* (2008).

Dimson *et al.* have also discussed the 'puzzle' raised by Mehra and Prescott (1985), regarding the size of the equity premium. They suggest that, given the persistent worldwide out-performance by equities, the risk element in equity investment, at least in developed, efficient markets, is overplayed. Prescott and McGrattan (2003) have revisited this puzzle. They found that in the USA, after taking into account certain factors ignored by Mehra and Prescott, e.g. taxes, regulatory constraints, diversification costs, and focusing on long-term rather than short-term saving instruments, the puzzle is solved. Allowing for all these factors, they found that the difference between average equity and debt returns during peacetime is less than 1 per cent p.a., with the average real equity return just under 5 per cent, and the average real return on debt instruments a little under 4 per cent, a far lower premium than other writers have suggested.

■ Finding Beta

Beta values appear to be fairly stable over time, so we can use Beta values based on past recorded data, such as those provided by the RMS, with a fair degree of confidence. This is acceptable so long as the company is not expected to alter its risk characteristics in the future: for example, by a takeover of a company in an unrelated field or a spin-off of unwanted activities.

■ The required return

We now demonstrate the calculation of the required return for the 'aggressive' share British Airways, using the equation for the SML:

$$ER_j = R_f + \beta_j(ER_m - R_f)$$

The Beta recorded by the RMS was 1.43 (Table 10.3). At the same date, the yield on three-month Treasury Bills was about 5.75 per cent. For British Airways, this results in the following required return, assuming a market risk premium of 5 per cent:

$$ER = 5.75\% + 1.43\,(5\%) = 5.75\% + 7.15\% = 12.90\%$$

■ Application to investment projects

As British Airways shareholders appear to require a return of 12.90 per cent, it may seem reasonable to use this rate as a cut-off for new investments. However, two warnings are in order.

First, the discount rate applicable to new projects often depends on the nature of the activity. For example, if a new project takes British Airways away from its present spheres of activity into, say, mobile telephony, its systematic risk will alter, as suggested by the Beta for Vodaphone of 1.01. The relevant premium for risk hinges on the systematic risk of telecommunications rather than of airline operation. This suggests that we 'tailor' risk premiums, and thus discount rates, to particular activities. This aspect is examined in the next chapter.

Self-assessment activity 10.9

What is the implied discount rate for investment by British Airways into retailing?

(Answer in Appendix A at the back of the book)

Second, the appropriate discount rate may depend upon the method of financing used. Until now, we have implicitly been dealing with an all-equity financed company whose premium for risk is a reward purely for the business risk inherent in the company's activity. In reality, most firms are partially debt-financed, exposing shareholders to financial risk. Using debt capital increases the risk to shareholders because of the legally-preferred position of creditors. Defaulting on the conditions of the loan (e.g. failing to pay interest) can result in liquidation if creditors apply to have the company placed into receivership. The more volatile the earnings of the firm, the greater the risk of default.

Financial risk raises the Beta of the equity, as shareholders demand additional returns to compensate. The Beta of the equity becomes greater than the Beta of the underlying activity. In Chapter 19, we shall see that observed Betas have two components, one to reflect business risk and one to allow for financial risk. The Betas recorded by the RMS are actually equity Betas, so the required return computed for British Airways (a highly geared company) is the shareholders' required return, part of which is to compensate for financial risk. However, when a company borrows, only the method of financing changes; nothing happens to alter the riskiness of the basic activity. The cut-off rate reflecting the basic risk of physical investment projects is often lower than the shareholders' own required return.

10.8 WORKED EXAMPLE

An investor holds the following portfolio of four risky assets and a deposit in a risk-free asset. The table shows their respective portfolio weightings and the current returns on the assets, together with their Beta coefficients.

Asset	Weighting (%)	Current return (%)	Beta
A	20	12.0	1.5
B	10	18.0	2.0
C	15	14.0	1.2
D	25	8.0	0.9
Risk-free asset	30	5.0	0

The overall return on the market portfolio of risky assets is 11 per cent, and this is expected to continue for the foreseeable future.

Required

(a) What is the current return on the whole portfolio, and its Beta value?

(b) Which of the four risky assets (if any) appear to be inefficient/efficient/super-efficient?

(c) In view of the answer to part (b), what predictions would you make regarding future asset values and, hence, their rates of return as the market moves to full equilibrium?

(d) What is the equilibrium return on this portfolio? (Assume the weightings remain unchanged.)

Answers

(a) The portfolio return is a weighted average of the individual asset returns, *viz*:

$$R_p = (0.2 \times 12\%) + (0.1 \times 18\%) + (0.15 \times 14\%) + (0.25 \times 8\%) + (0.3 \times 5\%)$$
$$= 2.4\% + 1.8\% + 2.1\% + 2.0\% + 1.5\%$$
$$= 9.8\%$$

The portfolio Beta is a weighted average of the individual asset Betas, *viz*:

$$\text{Beta}_p = (0.2 \times 1.5) + (0.1 \times 2.0) + (0.15 \times 1.2) + (0.25 \times 0.9) + (0.3 \times 0)$$
$$= 0.3 + 0.2 + 0.18 + 0.225 + 0$$
$$= 0.905$$

These results imply that the investor is relatively risk-averse, choosing to combine risky assets and the risk-free asset in such a way as to undershoot the overall market return of 11% and the market Beta of 1.0.

(b) Efficient assets lie on the security market line, thus offering a return consistent with their Beta values. If we compare the actual with the required returns for each asset, we can judge the status of each one. The table shows this evaluation.

Asset	Risk-free rate (%)	Beta	Market premium (%)	Required return (%)	Actual return	Assessment
A	5	1.5	(11% − 5%) = 6%	5 + (1.5 × 6) = 14%	12%	Inefficient
B	5	2.0	6%	5 + (2.0 × 6) = 17%	18%	Super-efficient
C	5	1.2	6%	5 + (1.2 × 6) = 12.2%	14%	Super-efficient
D	5	0.9	6%	5 + (0.9 × 6) = 10.4%	8%	Inefficient

Super-efficient assets offer in excess of what their Beta values warrant. The opposite is true for inefficient assets. Assets A and D are thus inefficient and B and C are super-efficient.

(c) Super-efficient assets are very attractive while they offer abnormal returns, and conversely for inefficient assets. Investors will therefore scramble to buy the former and to sell the latter, triggering windfall gains for those lucky enough to be holding the former and losses for those holding the latter. Prices will adjust until every asset offers a return consistent with its Beta value. Hence, we would predict a rise in price for assets B and C, depressing their returns, i.e. the equilibrium return will be lower than the current return, and price falls for assets A and D until their expected returns increase accordingly.

(d) The equilibrium portfolio return is:

$$R_p = (0.2 \times 14.0\%) + (0.1 \times 17.0\%) + (0.15 \times 12.2\%) + (0.25 \times 10.4\%)$$
$$+ (0.3 \times 5\%)$$
$$= 2.8\% + 1.7\% + 1.83\% + 2.6\% + 1.5\%$$
$$= 10.43\%$$

Thus, the equilibrium portfolio return is a little above its initial level and closer to the market return.

10.9 THE UNDERPINNINGS OF THE CAPM

In the previous sections, we have concentrated on developing the operational aspects of the CAPM, without explaining the underlying theoretical relationships. The underlying theory is explained in Sections 10.9 and 10.10 and brought together in Section 10.11, which you may omit at this stage. Section 10.12 discusses some general issues raised by the CAPM.

All theories rely on assumptions in order to simplify the analysis and expose the important relationships between key variables. In economics and related sciences, it is generally accepted that the validity of a theory depends on the empirical accuracy of its predictions rather than on the realism of its assumptions (Friedman, 1953). However, if we find that the predictions fail to correspond with reality, and we are satisfied that this is not due to measurement errors or random influences, then it is appropriate to reassess the assumptions. The ensuing analysis, based on an amended set of assumptions, may lead to the generation of alternative predictions that accord more closely with reality.

■ The assumptions of the CAPM

The most important assumptions are as follows:

1 All investors aim to maximise the utility they expect to enjoy from wealth-holding.
2 All investors operate on a common single period planning horizon.

3 All investors select from alternative investment opportunities by looking at expected return and risk.
4 All investors are rational and risk-averse.
5 All investors arrive at similar assessments of the probability distributions of returns expected from traded securities.
6 All such distributions of expected returns are normal.
7 All investors can lend or borrow unlimited amounts at a similar common rate of interest.
8 There are no transaction costs entailed in trading securities.
9 Dividends and capital gains are taxed at the same rates.
10 All investors are price-takers: that is, no investor can influence the market price by the scale of his or her own transactions.
11 All securities are highly divisible, i.e. can be traded in small parcels.

Several of these assumptions are patently untrue, but it has been shown that the CAPM stands up well to relaxation of many of them. Incorporation of apparently more realistic assumptions does not materially affect the implications of the analysis. A full discussion of these adjustments is beyond our scope, but van Horne (2000) offers an excellent analysis.

10.10 PORTFOLIOS WITH MANY COMPONENTS: THE CAPITAL MARKET LINE

The theory behind the CAPM revolves around the concept of the 'risk–return trade-off'. This suggests that investors demand progressively higher returns as compensation for successive increases in risk. The derivation of this relationship, known as the **capital market line (CML)**, relies on the portfolio analysis techniques examined in Chapter 9.

The reader may find it useful to re-read Section 9.7, where we explained the derivation of the efficient set available to an investor who can invest in a large number of assets. One conclusion of this analysis was that the only way to differentiate between the many portfolios in the efficient set was to examine the investor's risk–return preferences, i.e. there was no definable optimal portfolio of equal attractiveness to all investors.

■ Introducing a risk-free asset

The above conclusion applies only in the absence of a risk-free asset. A major contribution of the CAPM is to introduce the possibility of investing in such an asset. If we allow for risk-free investment, the range of opportunities widens much further. For example, on Figure 10.8, which is based on Figure 9.5 which showed an efficient frontier of AE, consider the line from R_f, the return available on the risk-free asset, passing through point T on the efficiency frontier. This represents all possible combinations of the risk-free asset and the portfolio of risky securities represented by T. To the left of T, both portfolio return and risk are less than those for T, and conversely for points to the right of T. This implies that between R_f and T the investor is tempering the risk and return on T with investment in the risk-free asset (i.e. lending at the rate R_f), while above T, the investor is seeking higher returns even at the expense of greater risk (i.e. he borrows in order to make further investment in T).

However, the investor can improve portfolio performance by investing along the line R_fV, representing combinations of the risk-free asset and portfolio V. He or she can do better still by investing along R_fWZ, the tangent to the efficient set. This schedule describes the best of all available risk–return combinations. No other portfolio of risky assets when combined with the risk-free assets allows the investor to achieve higher returns for a given risk. The line R_fWZ becomes the new efficient boundary.

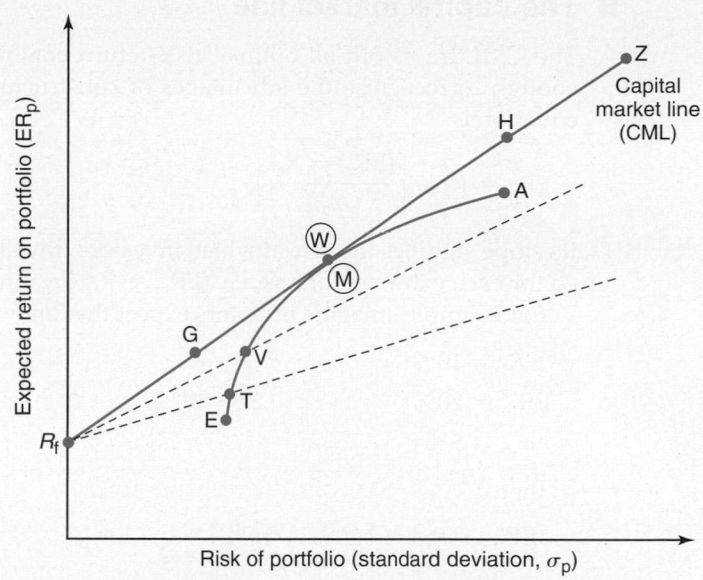

Figure 10.8 The capital market line

Portfolio W is the most desirable portfolio of risky securities as it allows access to the line R_fWZ. If the capital market is not already in equilibrium, investors will compete to buy the components of W and tend to discard other investments. As a result, realignment of security prices will occur, the prices of assets in W will rise and hence their returns will fall; and conversely, for assets not contained in W. The readjustment of security prices will continue until all securities traded in the market appear in a portfolio like W, where the line drawn from R_f touches the efficient set. *This adjusted portfolio is the 'market portfolio' (re-labelled as M), which contains all traded securities, weighted according to their market capitalisations. For rational risk-averting investors, this is now the only portfolio of risky securities worth holding.*

There is now a definable optimal portfolio of risky securities, portfolio M, which all investors should seek, and which does not derive from their risk–return preferences. This proposition is known as the **Separation Theorem** – the most preferred portfolio is separate from individuals' attitudes to risk. The beauty of this result is that we need not know all the expected returns, risks and covariances required to derive the efficient set in Figure 10.8. We need only define the market portfolio in terms of some widely used and comprehensive index.

However, having invested in M, if investors wish to vary their risk–return combination, they need only to move along R_fMZ, lending or borrowing according to their risk–return preferences. For example, a relatively risk-averse investor will locate at point G, combining lending at the risk-free rate with investment in M. A less cautious investor may locate at point H, borrowing at the risk-free rate in order to raise his or her returns by further investment in M, but incurring a higher level of risk. However, we would still need information on attitudes to risk to *predict* how individual investors behave.

The line R_fMZ is highly significant. It describes the way in which rational investors – those who wish to maximise returns for a given risk or minimise risk for a given return – seek compensation for any additional risk they incur. In this sense, R_fMZ describes an optimal risk–return trade-off that all investors and thus the whole market will pursue; hence, it is called the **capital market line** (CML).

If the reader refers back to Figure 1.3 in Chapter 1 s/he will notice that the trade-off schedule R_fMZ is, in fact, a more fully developed version of the upward-sloping relationship in that earlier diagram.

Separation Theorem
A model that shows how individual perceptions of the optimal portfolio of risky securities is independent of (i.e. separate from) individuals' different risk–return preferences

capital market line
A relationship tracing out the efficient combinations of risk and return available to investors prepared to combine the market portfolio with the risk-free asset

■ **The capital market line**

The CML traces out all optimal risk–return combinations for those investors astute enough to recognise the advantages of constructing a well-diversified portfolio. Its equation is:

$$\mathrm{ER}_p = R_f + \left[\frac{(\mathrm{ER}_m - R_f)}{\sigma_m}\right]\sigma_p$$

Its slope signifies the rate at which investors travelling up the line will be compensated for each extra unit of risk, i.e. $(\mathrm{ER}_m - R_f)/\sigma_m$ units of additional return.

For example, imagine investors expect the following:

$$R_f = 10\%$$
$$\mathrm{ER}_m = 20\%$$
$$\sigma_m = 5\%$$

so that

$$\left[\frac{\mathrm{ER}_m - R_f}{\sigma_m}\right] = \left[\frac{20\% - 10\%}{5\%}\right] = 2$$

Every additional unit of risk that investors are prepared to incur, as measured by the portfolio's standard deviation, requires compensation of two units of extra return. With a portfolio standard deviation of 2 per cent, the appropriate return is:

$$\mathrm{ER}_p = 10\% + (2 \times 2\%) = 14\%$$

for $\sigma_p = 3\%$, $\mathrm{ER}_p = 16\%$; for $\sigma_p = 4\%$, $\mathrm{ER}_p = 18\%$; and so on.

Anyone requiring greater compensation for these levels of risk will be sorely disappointed.

To summarise, we can now assess the appropriate risk premiums for combinations of the risk-free asset and the market portfolio, and therefore the discount rate to be applied when valuing such portfolio holdings. The final link in the analysis of risk premiums is an explanation of how the discount rates for individual securities are established and hence how these securities are valued. This was already provided by the discussion of the SML in Section 10.6.

10.11 HOW IT ALL FITS TOGETHER: THE KEY RELATIONSHIPS

The CAPM on first acquaintance may look complex. However, its essential simplicity can be analysed by reducing it to the three panels of Figure 10.9.

Panel I shows the CML, derived using the principles of portfolio combination developed in Chapter 9. The CML is a tangent to the envelope of efficient portfolios of risky assets, the point of tangency occurring at the market portfolio, M. Any combination along the CML (except M itself) is superior to any combination of risky assets alone. In other words, investors can obtain more desirable risk–return combinations by mixing the risk-free asset and the market portfolio to suit their preferences, i.e. according to whether they wish to lend or borrow.

The slope of the CML, given by $[(\mathrm{ER}_m - R_f)/\sigma_m]$ defines the best available terms for exchanging risk and return. It is desirable to hold a well-diversified portfolio of securities in order to eliminate the specific risk inherent in individual securities like C. When holding single securities, investors cannot expect to be rewarded for total risk (e.g. 15 per cent for C) because the market rewards investors only for bearing the undiversifiable or systematic risk. The extent to which risk can be eliminated depends on the covariance of the share's return with the return on the overall market. Hence, the degree of correlation with the return on the market influences the reward from holding a security and thus its price.

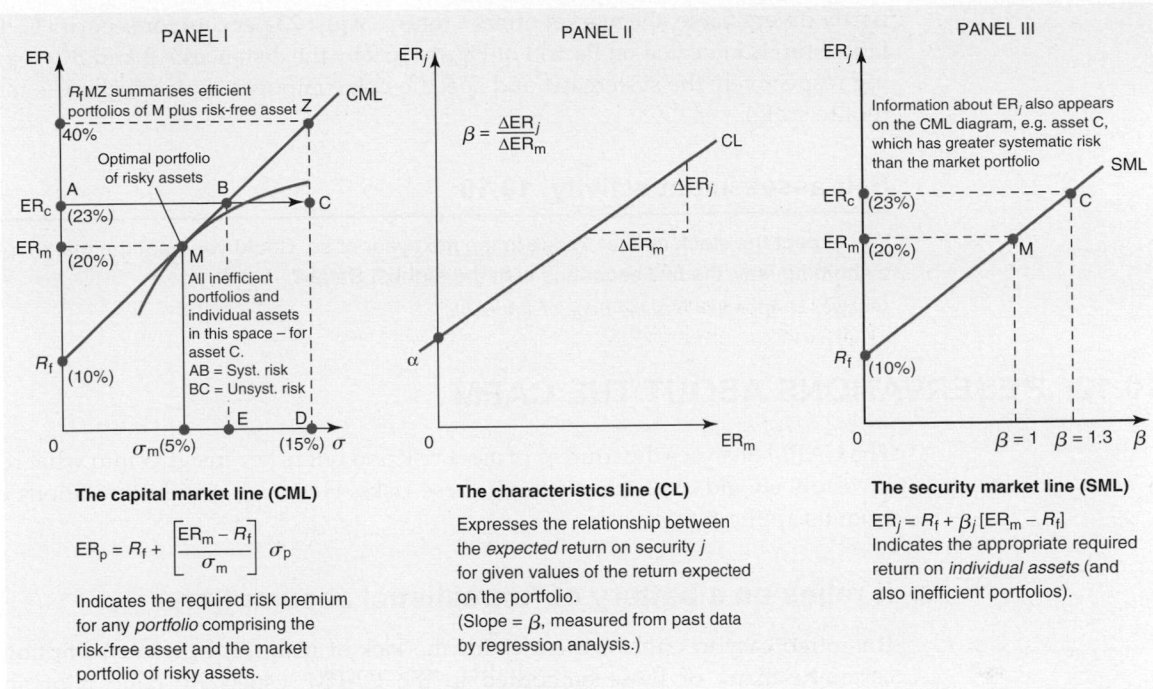

Figure 10.9 The CAPM: the three key relationships

The characteristics line in Panel II shows how the return on an individual share, such as C, is expected to vary with changes in the return on the overall market. Its slope, the Beta, indicates the degree of systematic risk of the security.

The security market line in Panel III shows the market equilibrium relationship between risk and return, which holds when all securities are 'correctly' priced. Clearly, the higher the Beta, the higher the required return. Although Beta is not a direct measure of systematic risk, it is an important indicator of relevant risk.

The decomposition of the overall variability, or variance, of the share's return into systematic and unsystematic components is explained in the appendix to this chapter. It can be demonstrated by focusing on security C in Panel III of Figure 10.9. Security C lies to the north-east of the market portfolio because its Beta of 1.3 exceeds that of the overall market. If the market as a whole is expected to generate a return of 20 per cent, and the risk-free rate is 10 per cent, C's expected return is:

$$ER_C = 10\% + 1.3(20\% - 10\%) = 10\% + 13\% = 23\%$$

This reward compensates only for systematic risk, rather than for the share's total risk. Of the total risk of C, represented by distance OD, only OE is relevant.

The risk–return trade-off, given by the slope of the CML, is $(20\% - 10\%)/5\% = 2$, since the risk of the market itself is 5 per cent. For C, with overall risk of 15 per cent, we would not expect to obtain compensation at this rate (i.e. $2 \times 15\% = 30\%$ giving an overall return of 40 per cent), because much of the total risk can be diversified away.

Observe that a variety of required return figures could have emerged from our calculation – in fact, anything along the perpendicular ZD in Panel I of Figure 10.9, depending on the extent to which security C is correlated with the market portfolio. The nearer C lies to Z, the greater the correlation and the higher the required return, and conversely, should C be nearer to D. This reflects the changing balance between the two risk components along ZD.

If the market rewarded total risk, the return offered on security C would be the risk-free rate of 10 per cent supplemented by the risk–return trade-off ($2 \times$ the total security risk of 15 per cent), yielding a total of 40 per cent. However, because the total risk is partly diversifiable, the market offers a return of just 23 per cent for security C. This relationship is indicated on Panel I of Figure 10.9 by the distances AB and BC, representing respectively the systematic and specific risk components of security C's total risk (not to scale).

Self-assessment activity 10.10

You expect the stock market to rise in the next year or so. Could you beat the market portfolio by holding, say, the five securities with the highest Betas?

(Answer in Appendix A at the back of the book)

10.12 RESERVATIONS ABOUT THE CAPM

The CAPM analyses the sources of asset risk and offers key insights into what rewards investors should expect for bearing these risks. However, certain limitations detract from its applicability.

■ It relies on a battery of 'unrealistic' assumptions

It is often easy to criticise theories for the lack of realism of their assumptions, and certainly, many of those embodied in the CAPM, especially concerning investor behaviour, do not seem to reflect reality. However, if the aim is to provide predictions that can be tested against real world observations, the realism of the underlying assumptions is secondary. Obviously, if the predictions themselves do not accord reasonably closely with reality, then the theory is undoubtedly suspect.

■ Single time period

A key assumption of the CAPM is that investors adopt a one-period time horizon for holding securities. Whatever the length of the period (not necessarily one year), the rates of return incorporated in investor expectations are rates of return over the whole holding period, assumed to be common for all investors. This provides obvious problems when we come to use a required return derived from a CAPM exercise in evaluating an investment project. Quite simply, we may not compare like with like. If an investor requires a return of, say, 25 per cent, over a five-year period, this is rather different from saying that the returns from an investment project should be discounted at 25 per cent p.a. Attempts have been made, notably by Mossin (1966), to produce a multi-period version of the CAPM, but its mathematical complexity takes it out of the reach of most practising managers, especially those inclined to scepticism about the CAPM itself.

10.13 TESTING THE CAPM

Many writers have observed that, in principle, the CAPM is untestable, since it is based on investors' expectations about future returns, and expectations are inherently awkward to measure. Hence, tests of the CAPM have to examine past returns and take these as proxies for future expected returns. This is based on the key premise that if a long enough period is examined, mistaken expectations are likely to be corrected, and people will come to rely on past average achieved returns when formulating expectations. Greatly simplified, the essence of the research methods is as follows.

Research usually proceeds in two stages. First, using time series analysis over a lengthy period applied to a large sample of securities (say 750), researchers estimate both the Beta for each security and its average return. Relying heavily on market efficiency, these estimates are taken to be estimates of the *ex ante* expected return, i.e. it is assumed that rational investors will be strongly influenced by past returns and their variability when formulating future expectations.

Second, the researcher tries to locate the SML to investigate whether it is upward sloping, as envisaged by the CAPM. The 750 pairs of estimates for Beta and the average return for each security are used as the input into a cross-section regression model of the form:

$$R_i = a_1 + a_2\beta_i + u_i$$

where R_i is the expected return from security i, a_1 is the intercept term (i.e. the risk-free rate), a_2 is the slope of the SML and u_i is an error term.

If the CAPM is valid, the measured SML would appear as in the steeper line on Figure 10.10, with an intercept approximating to recorded data for the risk-free rate: for example, the realised return on Treasury Bills.

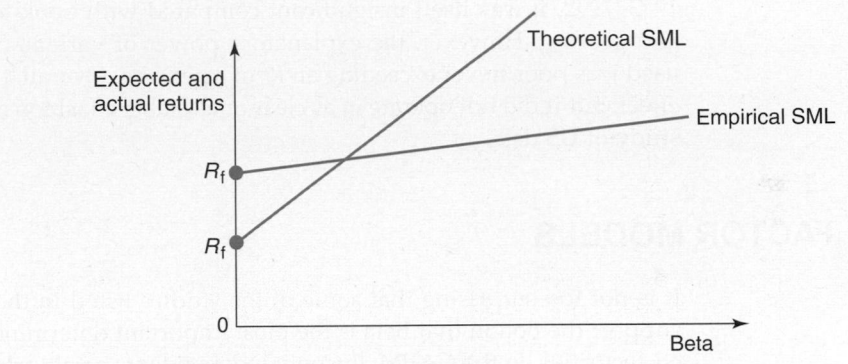

Figure 10.10 Theoretical and empirical SMLs

Several early studies (e.g. Black *et al.*, 1972; Fama and McBeth, 1973) did seem to support the positive association between Beta and average stock returns envisaged by the CAPM for long periods up to the late 1960s. However, evidence began to emerge that the empirical SML was much flatter than implied by the theory and that the intercept was considerably higher than achieved returns on 'risk-free' assets.

Some researchers have continued to test the validity of the CAPM, but others, following Ross (1976), have concluded that some of the 'rogue' results stem from intrinsic difficulties concerning the CAPM that make it inherently untestable. In the process, they have developed an alternative theory, based on the Arbitrage Pricing Model (APM), discussed in Section 10.15.

Other reasons why the CAPM is thought to be nigh impossible to test adequately are as follows:

1 It relies on specification of a risk-free asset – there is some doubt whether such an asset really exists.
2 It relies on analysing security returns against an efficient benchmark portfolio, the market portfolio, usually proxied by a widely-used index. Because no index captures all stocks, the index portfolio itself could be inefficient, as compared with the full market portfolio, thus distorting empirical results.
3 The model is unduly restrictive in that it includes only *securities* as depositories of wealth. A full 'capital asset pricing model' would include all forms of asset, such as real estate, paintings or rare coins – in fact, any asset that offers a future return. Hence, the CAPM is only a *security* pricing model.

Fama and French (1992) made a thorough test of the CAPM, finding no U.S. evidence for the 'correct' relationship between security returns and Beta over the period 1963–90.

The cross-section approach supported neither a linear nor a positive relationship. It appeared that average stock returns were explained better by company size as measured by market capitalisation, large firms generally offering lower returns, and by the ratio of book value of equity to market value, returns being positively associated with this variable. They concluded that rather than being explained by a single variable, Beta, security risk was multi-dimensional.

Neither of the UK studies conducted by Beenstock and Chan (1986) and by Poon and Taylor (1991) found significant positive relationships between security returns and Beta. Acting on Levis' (1985) observation for the period 1958–82 that smaller firms tend to outperform larger firms (although erratically), Strong and Xu (1997) attempted to replicate the Fama and French analysis in a UK context. Specifically, they investigated whether Beta could explain security returns and whether it was outweighed by 'the size effect'.

For the period 1960–92, they found a positive risk premium associated with Beta in isolation, but this became insignificant when Beta was combined with other variables in a multiple regression. For the whole period, market value dominated Beta, but over 1973–1992, it was itself insignificant compared with book-to-market value of equity, and gearing. However, the explanatory power of various combinations of variables used was poor, never exceeding an R^2 of 8 per cent. Overall, there appeared to be a size effect, but it did not operate in as clear or as stable a fashion as in the Fama and French study of US data.

10.14 FACTOR MODELS

It is not too surprising that some of the studies listed in the previous section do not support the notion that Beta is the most important determinant of the return on quoted securities. In the CAPM, the only independent variable driving individual security returns is the return on the market, i.e. there is a single factor at work. In reality, everyone knows there are many factors at work, but the researcher is hoping that their various impacts will all be rolled up into this single market factor.

However, the returns on a share react to general industry or sector changes in addition to general market changes. These aspects are all confused in Beta. This helps explain why the CAPM is such a poor explanatory model. The explanatory power of a regression model like the CAPM is measured by the R-squared, or Coefficient of Determination, which is measured on a scale of zero to +1. These are shown in Table 10.3 in the final column. While expert opinions vary on this, it is commonly accepted that an R-squared of above 50 per cent indicates a strong relationship, i.e. a high degree of explanatory power. The highest figure shown in the table is 48 per cent for Lloyds TSB. The interpretation we have to put on this is that there are other, perhaps many other, factors at work impacting on security returns.

Whereas the CAPM is a single factor model, many researchers like Fama and French (1992) have attempted to develop multi-factor models. A multi-factor model will include two elements:

- a list of factors that have been identified as having a significant influence on security returns
- a measure of the sensitivity of the return on particular securities to changes in these factors.

In the CAPM, there is only the one factor, the return on the market portfolio, and the sensitivity is measured by each security's Beta. As in the CAPM, which distinguishes between specific and market-related risk, there are two types of risk – factor risk, and non-factor risk. Thus, variations in the returns on stocks can be explained by variations in the identified factor(s) (analogous to market risk) and variations due to background 'noise', i.e. changes in factors not included in the model (analogous to specific risk).

■ A two-factor model

In the UK, 60 per cent of the economy is represented by consumer expenditure, which is largely driven by income growth and the 'feel-good factor' from rising house prices. Bear also in mind that the stock market is generally supposed to herald movements in the overall economy one to two years ahead. Therefore, a model devised to explain stock market returns in terms of income growth and house prices would be quite plausible.

This would be a two-factor model of the following form:

$$R_j = a + b_1 F_1 + b_2 F_2 + e_j$$

where R_j is the return on stock j in the usual sense, a is the intercept term, F_1 and F_2 are the two identified factors, income growth and house prices, b_1 and b_2 are the sensitivity coefficients and e_j is an error term.

The values of the parameters a, b_1 and b_2 would be found by multiple regression analysis, while the error term is assumed to average zero. Say the values established by empirical investigation are:

$a = 0.01$

$b_1 = 2.0$

$b_2 = 0.2$

This means that for every 1 per cent point change in income growth, individual security returns change by twice as much, i.e. by two percentage points. Similarly, for every 1 per cent point change in the house price index, security returns change by 0.2 of a percentage point.

It should be stressed that the explanatory factors in the equation would be common to all firms, but the sensitivity coefficients, the 'Betas', would vary according to how closely 'geared' the returns on each firm were to each factor. For example, if one identified factor was the sterling/dollar exchange rate, we would expect to see much higher sensitivity for a firm exporting to, or operating in, the USA, compared to one conducting most of its operations in the domestic arena.

10.15 THE ARBITRAGE PRICING THEORY

The most fully developed multi-factor model is the **Arbitrage Pricing Theory (APT)**, developed by Ross (1976). Unlike the CAPM, APT does not assume that shareholders evaluate decisions within a mean–variance framework. Rather, it assumes the return on a share depends partly on macroeconomic factors and partly on events specific to the company. Instead of specifying a share's returns as a function of one factor (the return on the market portfolio), it specifies the returns as a function of multiple macroeconomic factors upon which the return on the market portfolio depends.

The expected risk premium of a particular share would be:

$$ER_j = R_f + \beta_1(ER_{\text{factor 1}} - R_f) + \beta_2(ER_{\text{factor 2}} - R_f) + \cdots + e_j$$

where ER_j is the expected rate of return on security j, $ER_{\text{factor 1}}$ is the expected return on macroeconomic factor 1, β_1 is the sensitivity of the return on security j to factor 1 and e_j is the random deviation based on unique events impacting on the security's returns. The bracketed terms are thus risk premiums, as found in the CAPM.

Diversification can eliminate the specific risk associated with a security, leaving only the macroeconomic risk as the determinant of required security returns. A rational investor will arbitrage (hence the name) between different securities if the current market prices do not give sufficient compensation for variations in one or more factors in the APT equation.

The APT model does not specify what the explanatory factors are; they could be the stock market index, Gross National Product, oil prices, interest rates and so on. Different companies will be more sensitive to certain factors than others.

In theory, a riskless portfolio could be constructed (i.e. a 'zero Beta' portfolio) which would offer the risk-free rate of interest. If the portfolio gave a higher return, investors could make a profit without incurring any risk by borrowing at the risk-free rate to buy the portfolio. This process of 'arbitrage' (i.e. taking profits for zero risk) would continue until the portfolio's expected risk premium was zero.

The Arbitrage Pricing Theory avoids the CAPM's problem of having to identify the market portfolio. But it replaces this problem with possibly more onerous tasks. First, there is the requirement to identify the macroeconomic variables. American research indicates that the most influential factors in explaining asset returns in the APT framework are changes in industrial production, inflation, personal consumption, money supply and interest rates (McGowan and Francis, 1991).

Tests of the APT, especially for the UK, are still in their relative infancy. However, in an early test, Beenstock and Chan (1986) found that, for the period 1977–83, the first few years of the UK's 'monetarist experiment', share returns were largely explained by a set of monetary factors – interest rates, the sterling M3 measure of money supply and two different measures of inflation, all highly interrelated variables. In 1994, Clare and Thomas reported results from analysing 56 portfolios, each containing 15 shares sorted by Beta and by size of company by value. For the Beta-ordered portfolios, the key factors were oil prices, two measures of corporate default risk, the Retail Price Index (RPI), private sector bank lending, current account bank balances and the yield to redemption on UK corporate loan stock. Using portfolios ordered by size, the key factors reduced to one measure of default risk and the RPI. Again, there was much intercorrelation among variables, but the return on the stock market index, although included in the initial tests, appeared in none of these final lists.

Once the main factors influencing share returns are established, there remain the problems of estimating risk premiums for each factor and measuring the sensitivity of individual share returns to these factors. For this reason, the APT is currently only in the prototype stage, and yet to be accepted by practitioners.

10.16 FAMA AND FRENCH'S THREE-FACTOR MODEL

An approach that marries the APT to the multi-factor approach is the three-factor model developed in a series of papers by Fama and French (1993, 1995, 1996). This has the distinctive merit of an empirical grounding, being based on their paper of 1992. In Section 10.13, we noted that they found that stock returns in the USA were explained better by company size and by the ratio of book value of equity to market value than merely by movements in the return on the whole market, magnified or moderated by Beta, as in the CAPM.

These two additional explanatory variables are utilised in the three-factor model. It states that stock returns above the risk-free rate (i.e. the equity premium) are determined by:

■ The risk premium on the market portfolio.
■ The difference between the return on a portfolio of small company shares and the return on a portfolio of large company shares (small less big, or SLB).
■ The difference between the return on a portfolio of high book-to-market value stocks and the return on a portfolio of low book-to-market value stocks (high less low, or HLL).

The three-factor equation can be written thus:

Expected return on stock$_j$ =
$$ER_j = R_f + \text{Risk premium} = R_f + [Beta_1(ER_m - R_f) + Beta_2(SLB) + Beta_3(HLL)]$$

The logic behind the formulation of the model is that the average small company and its stock is assumed to be more risky than the average large firm and its stock and thus commands a higher risk premium. Larger firms are generally more stable as they are more diversified by products and markets, and have better credit ratings, as their stock of assets is larger. Similarly, a stock with a high book value relative to market value is assumed to be more risky than one with a low book value relative to market value. The former owes its higher valuation rating to a greater growth potential and/or greater endowment of intangible assets such as intellectual capital.

To make the model operational, information is required on the risk premia related to each factor, and for the various Beta factors. For example, imagine that empirical evidence suggests that in past years the risk premium on the market portfolio has averaged 5 per cent, the risk premium for a small company stock compared to a larger firm has averaged 6 per cent, and the risk premium for the stock of a typical firm with a high book-to-market value compared to market price has averaged 4 per cent.

When the risk-free rate is 3 per cent, for a firm of average risk, i.e average sensitivity to each of these three factors, and thus with Beta values of 1.0 across the board, the overall expected return will be:

$$ER_j = 3\% + [(1.0 \times 5\%) + (1.0 \times 6\%) + (1.0 \times 4\%)] = 3\% + 15\% = 18\%$$

In practice, firms exhibit varying sensitivities to these factors depending on their product and market profiles, for example, and thus carry Beta values different from one. Assume that Firm X has a low sensitivity to market movements (Beta = 0.4), a relatively high sensitivity in respect of relative size (Beta = 1.2) and a relatively low sensitivity to the book versus market value factor (Beta = 0.8), then its expected return is:

$$ER_j = 3\% + [(0.4 \times 5\%) + (1.2 \times 6\%) + (0.8 \times 4\%)] = 3\% + 12.4\% = 15.4\%$$

10.17 ISSUES RAISED BY THE CAPM: SOME FOOD FOR MANAGERIAL THOUGHT

The CAPM raises a number of important issues, which have fundamental implications for the applicability of the model itself and the role of diversification in the armoury of corporate strategic weapons.

■ Should we trust the market?

Legally, managers are charged with the duty of acting in the best interests of shareholders, i.e. maximising their wealth (although company law does not express it *quite* like this). This involves investing in all projects offering returns above the shareholders' opportunity cost of capital. The CAPM provides a way of assessing the rate of return required by shareholders from their investments, albeit based partly on past returns. If the Beta is known and a view is taken on the future returns on the market, then the apparently required return follows. This becomes the cut-off rate for new investment projects, at least for those of similar systematic risk to existing activities. This implies that managers' expectations coincide with those of shareholders or, more generally, with those of the market. If, however, the market as a whole expects a higher return from the market portfolio, some projects deemed acceptable to managers may not be worthwhile for shareholders.

The subsequent fall in share price would provide the mechanism whereby the market communicates to managers that the discount rate applied was too low. The CAPM relies on efficiently-set market prices to reveal to managers the 'correct' hurdle rate and any mistakes caused by misreading the market. The implication that one can trust the market to arrive at correct prices and hence required rates of return is problematic for many practising managers, who are prone to believe that the market persistently undervalues the companies that they operate. Managers who doubt the validity of the EMH are unlikely to accept a CAPM-derived discount rate.

■ Should companies diversify?

The CAPM is based on the premise that rational shareholders form efficiently diversified portfolios, realising that the market will reward them only for bearing market-related risk. The benefits of diversification can easily be obtained by portfolio formation, i.e. buying securities at relatively low dealing fees. The implication of this is that *corporate diversification is perhaps pointless as a device to reduce risk because companies are seeking to achieve what shareholders can do themselves, probably more efficiently*. Securities are far more divisible than investment projects and can be traded much quicker when conditions alter. So why do managers diversify company activities?

An obvious explanation is that managers have not understood the message of the EMH/CAPM, or doubt its validity, believing instead that shareholders' best interests are enhanced by reduction of the total variability of the firm's earnings. For some shareholders, this may indeed be the case, as a large proportion of those investing directly on the stock market hold undiversified portfolios.

Many small shareholders were attracted to equity investment by privatisation issues or by Personal Equity Plans and their successor, ISAs (Individual Savings Accounts). Larger shareholders sometimes tie up major portions of their capital in a single company in order to take, or retain, an active part in its management. In such cases, market risk, based on the co-variability of the return on a company's shares with that on the market portfolio, is an inadequate measure of risk. The appropriate measure of risk for capital budgeting decisions probably lies somewhere between total risk, based on the variance, or standard deviation, of a project's returns, and market risk, depending on the degree of diversification of shareholders.

A more subtle explanation of why managers diversify is the divorce of ownership and control. Managers who are relatively free from the threat of shareholder interference in company operations may pursue their personal interests above those of shareholders. If an inadequate contract has been written between the manager-agents and the shareholder-principals, managers may be inclined to promote their own job security. This is understandable, since shareholders are highly mobile between alternative security holdings, but managerial mobility is often low. *To managers, the distinction between systematic risk and specific risk may be relatively insignificant, since they have a vested interest in minimising total risk in order to increase their job security*. If the company flounders, it is of little comfort for them to know that their personal catastrophe has only a minimal effect on well-diversified shareholders.

As we will see in Chapter 20, there are many motives for diversification beyond merely reducing risk. However, it is common to justify diversification to shareholders purely on these grounds, at least under certain types of market imperfection. When a company fails, there are liquidation costs to bear as well as the losses entailed in selling assets at 'knock-down' prices. These costs may result in both creditors and shareholders failing to receive full economic value in the asset disposal. Although this will not devastate a well-diversified shareholder, the resulting hole in his or her portfolio will require filling in order to restore balance. Company diversification may reduce these risks and also the costs of portfolio disruption and readjustment.

St Gobain

Despite contemporary strategic thinking, the conglomerate is not extinct everywhere. In France, famous for its policy of nurturing national champions, the glass-maker, St Gobain, privatised in 1986, has since thrived on a diet of acquisition of often unrelated businesses. The Chairman/CEO, Jean-Louis Beffa, is scornful of the drive for focus as firms try to concentrate operations on 'core' areas of business. M. Beffa has overseen the acquisition of over 900 companies, including many in the distribution of building materials, an activity uncharted by St Gobain until the 1990s.

Beffa says about ideas of focus:

> Look at Siemens. They are better for having a mix of companies from which they can get a strong cash flow.

In support, he points to St Gobain's balancing of distribution operations, covering a broad range of items for the building trade and operated mainly on a regional basis, with the global manufacturing of flat glass (where St Gobain is world number 2 after Asahi of Japan), and containers. Glass production is highly cyclical, changing with the oscillations of the world economy, whereas the distribution of building materials is far more stable because different national markets have their own peculiar patterns of troughs and peaks. St Gobain's diversification strategy gives it the consistent financial fire-power – cash flow of Euros 2.8 billion in 2003 – to finance growth by capital spending and by acquisition.

Beffa also stresses the need to enable executives to build up expertise in certain areas and to transfer skills horizontally across the overall business, for example legal expertise acquired in different fields that can be applied elsewhere, and experience of using specific financial instruments in different parts of the world. It also encourages the flow of ideas between divisions through nine overseas 'delegate offices', which act as collection points for ideas so that executives can transmit them with utmost efficiency.

Of course, one might argue that a growth-oriented policy that makes the firm increasingly important to the national economy also makes it more likely that the state will step in with financial assistance when necessary. St Gobain makes a virtue of this by suggesting that governments should help to fuel national economic growth by state investment, in their case, in developing novel applications for glass structures, for example, for flat-screen TVs.

Source: Based on Peter Marsh, *Financial Times*, 4 January 2005.

Self-assessment activity 10.11

In the light of the St Gobain case, explain why it might be good to be a conglomerate.

(Answer in Appendix A at the back of the book)

SUMMARY

We have examined the nature of the risks affecting the holders of securities and have begun to discuss whether the return required by shareholders, as implied by market valuations, can be used as a cut-off rate for new investment projects.

Key points

- Security risk can be split into two components: risk specific to the company in question, and the variability in return due to general market movements.
- Rational investors form well-diversified portfolios to eliminate specific risk.
- The most efficient portfolio of risky securities is the market portfolio, although investors may mix this with investment in the risk-free asset in order to achieve more preferred risk–return combinations along the capital market line.
- The risk premium built into the required return on securities reflects a reward for systematic risk only.
- The risk premium on a particular share depends on the risk premium on the overall market and the extent to which the return on the security moves with that of the whole market, as indicated by its Beta coefficient.

- This premium for risk is the second term in the equation for the security market line:

$$ER_j = R_f + \beta_j(ER_m - R_f).$$

- Practical problems in using the CAPM centre on measurement of Beta, specification of the risk-free asset and measurement of the market's risk premium.

- In an all-equity financed company, the return required by shareholders can be used as a cut-off rate for new investment if the new project has systematic risk similar to the company's other activities.

- There is some debate about whether managers should diversify company activities merely in order to lower risk.

- Empirical studies seem to throw increasing doubt on the CAPM.

- The main proposed alternative, the Arbitrage Pricing Theory (APT), relies on fewer restrictive assumptions but is still in the prototype stage.

Further reading

As with basic portfolio theory, Copeland *et al.* (2004) offer a rigorous treatment of the derivation of the formulae used in this chapter. Brealey, Myers and Allen (1996) offer an alternative, less mathematical treatment. You should also read the famous critique of the CAPM by Roll (1977). Fama and French's paper (1992), although difficult, is essential reading, as is Strong and Xu (1997), for a UK perspective.

An excellent text on Modern Portfolio Theory is that by Elton *et al.* (2007) which covers the basic theory and includes an up-to-date survey of empirical work. There is a very good resumé of the Fama and French analysis in Ross *et al.* (2005). Meanwhile, they have not gone unchallenged – see, for example, the two articles by Black (1993a, 1993b) and that by Kothari *et al.* (1995).

Fama and French (1995, 2002) have updated the earlier study, reaching essentially similar conclusions, and Fama (2002) has also entered the debate on the equity premium, while Campbell and Vuolteenako (2004) offer a 'modern' critique of the Beta concept in the light of recent empirical studies.

Appendix
ANALYSIS OF VARIANCE

The total risk of a security (σ_T), comprising both unsystematic risk (σ_{USR}), and systematic risk (σ_{SR}), is measured by the variance of returns, which can be separated into the two elements. Imagine an asset with total risk of $\sigma_T^2 = 500$, of which 80 per cent (400) is explained by systematic risk factors, the remainder resulting from factors specific to the firm:

$$\sigma_T^2 = 500 = \sigma_{SR}^2 + \sigma_{USR}^2 = 100 + 400$$

In terms of standard deviations, $\sigma_{SR} = \sqrt{400} = 20$ and $\sigma_{USR} = \sqrt{100} = 10$. Notice that we cannot express the overall standard deviation by summing the two component standard deviations – variances are additive, standard deviations are not – the square root of the total risk is $\sqrt{500} = 22.4$, rather than the sum of σ_{SR} and σ_{USR} (20 + 10 = 30).

In regression models, the extent to which the overall variability in the dependent variable is explained by the variability in the independent variable is given by the

R-squared (R^2) statistic, the square of the correlation coefficient. The R^2 is thus a measure of 'goodness of fit' of the regression line to the recorded observations. If all observations lie on the regression line, R^2 equals 1, and the variations in the market return fully explain the variations in the return on security j. In this case, all risk is market risk. It follows that the lower is R^2, the greater the proportion of specific risk of the security. For investors wishing to diversify away specific risk, such securities are highly attractive. Notice that an R^2 of 1 does not entail a Beta of 1, as Figure 10.11 illustrates. All three securities have R^2 of 1, but they have different degrees of market risk, as indicated by their Betas.

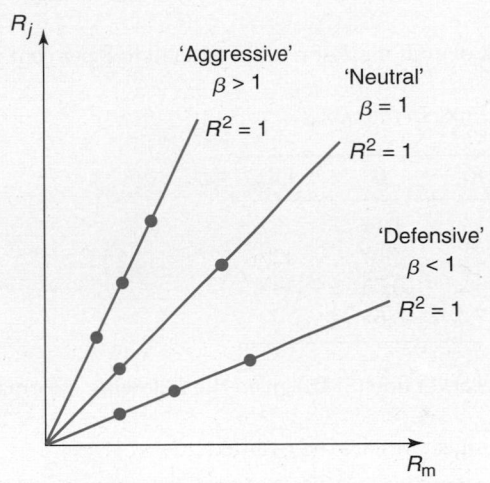

Figure 10.11 Alternative characteristics lines

In the example above, the R^2 of 80 per cent would correspond to a correlation coefficient, r_{jm}, of $\sqrt{0.8} = 0.89$. Looking at the standard deviations, we can infer that 0.89 of the standard deviation is market risk, i.e. $(0.89 \times 22.4) = 19.94$, while the specific risk $= (1 - r_{jm}) \times 22.4 = (0.11 \times 22.4) = 2.46$. Let us re-emphasise these relationships:

Market, or systematic, risk is:

$R^2 \times$ the overall variance, $\sigma_T{}^2$; or ($r_{jm} \times$ the overall standard deviation, σ_T)

$(0.8 \times 500) = 400$; or $(0.89 \times 22.4\%) = 19.94\%$

Specific risk is:

$(1 - R^2) \times$ overall variance, $\sigma_T{}^2$; or $(1 - r_{jm}) \times$ overall standard deviation σ_T

$(0.2 \times 500) = 100$; or $(0.11 \times 22.4\%) = 2.46\%$.

The reader may find it useful to test out these relationships using the data provided in Table 10.3 ('Variability' is total risk expressed as a standard deviation). However, not all cases work out neatly owing to rounding errors.

QUESTIONS

 **myfinancelab** *Questions with an icon are also available for practice in myfinancelab with additional supporting resources.*

Questions with a **coloured number** have a solution in Appendix B on page 739.

1 The ordinary shares of Firm A have a Beta of 1.23. The risk-free rate of interest is 5 per cent, and the risk premium achieved on the market index over the past 20 years has averaged 11.5 per cent p.a. What is the future expected return on A's shares?

  If you believe that overall market returns will fall to 8 per cent in future years, how does your answer change?

2 Supply the missing links in the table:

	ER_j	R_f	β	ER_m
(i)	19%	?	1.10	18%
(ii)	17%	5%	?	12%
(iii)	?	4%	0.75	10%
(iv)	15%	7%	0.65	?

3 Locate the security market line (SML) given the following information: $R_f = 8\%$, $ER_m = 12\%$.

? **4** Which of the following shares are over-valued?

	Beta	Current rate of return
A	0.7	7%
B	1.3	13%
C	0.9	9%

The risk-free rate is 5 per cent, and the return on the market index is 10 per cent.

? **5** The market portfolio has yielded 12 per cent on average over past years. It is expected to offer a risk premium in future years of 7%. The standard deviation of its return is 8 per cent. The risk-free rate is 5 per cent.

(i) What is the expected return from the market portfolio?
(ii) Draw a diagram to show the location of the capital market line.
(iii) What is the expected return on a portfolio comprising 50% invested in the market portfolio and 50% invested in the risk-free asset?
(iv) What is the risk of the portfolio in (iii)?
(v) What is the market trade-off between portfolio risk and return suggested by these figures?

6 The following figures relate to monthly observations of the percentage return on a widely-used stock market index (R_m) and the return on a particular ordinary share (R_j) over a period of six months.

Month	R_m	R_j
1	5	4
2	−10	−8
3	12	9.6
4	3	2.4
5	−4	−3.2
6	7	5.6

(a) Plot these data on a graph and deduce the value of the Beta coefficient.

(b) To what extent are variations in R_m due to specific risk factors?

(c) Calculate the systematic risk of the security. (NB: systematic risk $= \beta^2 \sigma_m^2$)

7 Z plc is a long-established company with interests mainly in retailing and property development. Its current market capitalisation is £750 million. The company trades exclusively in the UK, but it is planning to expand overseas either by acquisition or joint venture within the next two years. The company has built up a portfolio of investments in UK equities and corporate and government debt. The aim of developing this investment portfolio is to provide a source of funds for its overseas expansion programme. Summary information on the portfolio is given below.

Type of security	Value £ million	Average % return over the last 12 months
UK equities	23.2	15.0
US equities	9.4	13.5
UK corporate debt	5.3	8.2
Long-term government debt	11.4	7.4
Three-month Treasury bonds	3.2	6.0

Approximately 25 per cent of the UK equities are in small companies' shares, some of them trading on the Alternative Investment Market. The average return on all UK equities, over the past 12 months, has been 12 per cent. On US equities, it has been 12.5 per cent.

Ignore taxation throughout this question.

Required

Discuss the advantages and disadvantages of holding such a portfolio of investments in the circumstances of Z plc.

(CIMA, November 1997)

 **myfinancelab** *Now retake your diagnostic test for Chapter 10 to check your progress and update your study plan.*

11

The required rate of return on investment

Setting the cost of equity

The following is taken from the analysis of company value by the entity that was attempting to restructure Eurotunnel, GET S.A., in 2007.

We calculated the cost of equity of GET S.A. in accordance with the Capital Asset Pricing Model. This model consists of determining the rate of return required by investors (GET S.A.) by applying the risk-free rate plus a market risk premium that depends on the sensitivity of the rate of return on the asset to the market rate of return. To this effect, we used the following parameters:

■ A risk-free rate of 4.72 per cent corresponding to the average of the rates of yield on OAT (French government bonds) over 30 years (4.21 per cent) and on British gilts over a period of 30 years (4.33 per cent).
■ A prospective market risk premium of 5.0 per cent, determined on the basis of the various studies

carried out on the subject, it being specified that these studies generally point to premiums of between 3 per cent and 7 per cent.
■ A risk coefficient (forecast asset beta) amounting to 0.7. It was selected taking into consideration the two activities of Eurotunnel, and was compared to the beta for infrastructure management companies. To obtain the beta, this coefficient was adjusted each year to take into account the market leverage for the year (debt/equity ratio).

The cost of equity was calculated on an annual basis by adopting a recursive process consisting in valuing equity for the last year of the Financial Forecasts and calculating backwards progressively to 2007. This cost thus varies from 11.73 per cent in 2007 to 7.77 per cent from 2048 onwards.

Source: GET circular to shareholders.

Learning objectives

This chapter applies the models developed in earlier chapters to measuring the required rate of return on investment projects. After reading it, you should:

■ Understand how the Dividend Growth Model is used to set the hurdle rate.
■ Understand how the Capital Asset Pricing Model is also used for this purpose.
■ Be able to apply the required rate of return to firm valuation.
■ Appreciate that different rates of return may be required at different levels of an organisation.
■ Be aware of the practical difficulties in specifying discount rates for particular activities.
■ Appreciate how taxation may influence discount rates.

 **myfinancelab** *Complete your diagnostic test for Chapter 11 now to create your personal study plan.*

11.1 INTRODUCTION

No company can expect prolonged existence without achieving returns that at least compensate investors for their opportunity costs. Shareholders who receive a poor rate of return will vote with their wallets, depressing share price. If its share price underperforms the market (allowing for systematic risk), a company is ripe for reorganisation, takeover or both. A management team, motivated if only by job security, must earn acceptable returns for shareholders. This chapter deals with assessing such rates of return and showing how they can be used in valuing firms. Different returns may be required for different activities, according to their riskiness. Multi-division companies, which operate in a range of often unrelated activities, may require tailor-made 'divisional cut-off rates' to reflect the risk of particular activities.

The return that a company should seek on its investment depends not only on its inherent business risk, but also on its capital structure – its particular mix of debt and equity financing. However, because determining this rate for a geared company is complex, we defer treatment of the impact of gearing until Chapters 18 and 19. *Here, we focus on the return required by the shareholders in an all-equity company.*

Shareholders seek a return to cover the cost of waiting for their returns, plus compensation for inflation, plus a premium to cover the exposure to risk of their capital, depending on the risk of the business activity.

Two widely-adopted approaches are the Dividend Growth Model (DGM), encountered in Chapter 3, and the Capital Asset Pricing Model (CAPM), developed in the last chapter. Under each approach, we determine the return that shareholders demand on their investment holdings. We then consider whether this return should dictate the hurdle rate on new investment projects.

11.2 THE REQUIRED RETURN IN ALL-EQUITY FIRMS: THE DGM

■ The DGM revisited

In Chapter 3, we discussed the value of shares in an all-equity firm which retained a constant fraction, b, of its earnings in order to finance investment. If retentions are expected to achieve a rate of return, R, this results in a growth rate of $g = bR$. The share price is:

$$P_0 = \frac{D_0(1 + g)}{(k_e - g)} = \frac{D_1}{(k_e - g)}$$

where D_0 and D_1 represent this year's and next year's dividends per share respectively, and k_e is the rate of return required by shareholders.

■ The cost of equity

Rearranging the expression, we find the shareholders' required return is:

$$k_e = \frac{D_1}{P_0} + g$$

The shareholders' required return is thus a compound of two elements, the *prospective* dividend yield and the expected rate of growth in dividends.

It is important to appreciate that this formula for k_e is based on the current market value of the shares, and that it incorporates specific expectations about growth, dependent on assumptions about both the retention ratio, b, and the expected rate of return on new investment, R. With b and R constant, the rate of growth, g, is also constant. These are highly restrictive assumptions. Often, the nearest we can get to assessing the likely growth rate is to project the past rate of growth, 'tweaking' it if we believe that a faster or slower rate may occur in future.

For example, assume Arthington plc is valued by the market at £3 per share, having recently paid a dividend of 20p per share, and has recorded dividend growth of 12 per cent p.a. Projecting this past growth rate into the future, we can infer that shareholders require a return of 19.5 per cent, viz:

$$k_e = \frac{20p \, (1.12)}{300p} + 0.12 = (0.075 + 0.12) = 0.195, \text{ i.e. } 19.5\%$$

Self-assessment activity 11.1

Determine the required return by shareholders in the following case:

Share price = £1.80 (ex div)
Past growth = 3%
EPS = £0.36
Dividend cover = 3 times

(Answer in Appendix A at the back of the book)

■ Whitbread plc (www.whitbread.co.uk.)

Let us relate this approach to a real company. Table 11.1 shows the dividend payment record and end-of-financial year share prices for Whitbread, the leisure conglomerate, for the years 2001–7.

The dividend per share (DPS) grew by 70 per cent from 17.80p in 2001/2 to 30.25p by 2006–7. Using discount tables, we find the average annual compound growth rate is about 11.2 per cent.* Applying this result to the share price of 1632p ruling at Whitbread's 2006–7 year end, we find:

$$k_e = \frac{30.25p \, (1.112)}{1632p} + 0.112 = 0.021 + 0.112 = 0.133 \text{ (i.e.) } 13.3\%.$$

Table 11.1 The dividend return on Whitbread plc shares 2001–7

Year	DPS (p)
2001–2	17.80
2002–3	19.87
2003–4	22.30
2004–5	25.45
2005–6	27.30
2006–7	30.25

Source: Whitbread plc, Annual Report and Accounts.

---

*The growth rate, g, is found from the expression:

$$17.80/(1 + g)^5 = 30.25, \text{ or } (1 + g)^5 = 1.6994$$

The growth rate can be found directly from compound interest tables, or by inverting the expression from the present value tables, i.e. $1/(1 + g)^5 = 0.5884$, whence g approximates to 11.2 per cent.

■ Some problems

Apart from the restrictive assumptions of the Dividend Growth Model, some further warnings are in order.

1 The dividend growth depends on the time period used

The choice of time period can have a significant impact on the results. Too short a period and the estimate of growth is distorted by random factors, and too long a period exposes the result to the impact of structural changes in the business, e.g. divestment and acquisitions.

The calculation of g, and hence k_e, should certainly be based on a sufficiently long period to allow random distortions to even out. We may still feel that past growth is an unreliable guide to future performance, especially for a company in a mature industry, growing roughly in line with the economy as a whole. If past growth is considered unrepresentative, we may interpose our own forecast, but this would involve second-guessing the market's growth expectations, which is tantamount to challenging the EMH.

2 The calculated k_e depends on the choice of reference date for measuring share price

Our calculation used the price at the end of the accounting period, but this pre-dates the announcement of results and payment of dividend. Arguably, we should use the ex-dividend price, as this values all future dividends, beginning with those payable in one year's time. This would reduce the distortion to share price caused by the pattern of dividend payment (i.e. the share price drops abruptly when it goes 'ex-dividend', beyond which purchasers of the share will not qualify for the declared dividend). However, the eventual ex-dividend price may well reflect different expectations from those ruling at the company financial year end.

Conversely, in an efficient capital market, share prices gradually increase as the date of dividend payment approaches, so that, especially for companies that pay several dividends each year, some distorting effect is always likely to be present. Our practical advice is to take the ruling share price as the basis of calculation, but to moderate the calculation according to whether a dividend is in the offing. For example, if a 5p dividend is expected in two months' time, a prospective fall in share price of 5p should be allowed for. In our assessment, the error caused by using an out-of-date share price is likely to outweigh that from using a valuation incorporating a forthcoming dividend.

3 The calculation is at the mercy of short-term movements in share price

If, as many observers believe, capital markets are becoming more volatile, possibly undermining their efficiency in valuing companies, the financial manager may feel disinclined to rely on current market prices. Managers are generally reluctant to accept the EMH and commonly assert that the market undervalues 'their companies'. However, there remains a need for a benchmark return to guide managers. One might examine, over a period of years, the actual returns received by shareholders in the form of both dividends and capital gains. One way of conducting such a calculation is to focus on average annual rates of return, based on the analysis adopted in Chapter 10, as applied to Scottish and Southern Energy plc and to D.S. Smith plc. This evens out short-term fluctuations. However, it does not follow that the achieved return matches the required return.

4 Taxation

In Chapter 5, we argued the importance of allowing for taxation in project appraisal when estimating cash flows. Consistency seems to require discounting post-tax cash flows at a tax-adjusted cost of finance.

A project's NPV can be found on a post-tax or a pre-tax basis. If the NPV model is used on a pre-tax basis, both denominator and numerator must be on a pre-tax basis, and vice versa. If, for example, we wish to work in post-tax terms, the standard NPV expression for a one-off end-of-year cash flow, X, is:

$$\text{NPV} = \frac{X(1 - T)}{(1 + k_T)}$$

where T is the rate of corporation tax and k_T is the required return adjusted for tax. If shareholders seek a return of, say, 10 per cent after tax at 30 per cent, the company has to earn a pre-tax return of $10\%/(1 - 30\%) = 14.3$ per cent. In principle, computation on a pre-tax basis should generate the same NPV as that produced by a post-tax calculation, so long as the discount rate is suitably adjusted. However, this relationship is complicated by access to capital allowances. As a result, it is usual to compute NPVs on a post-tax basis.

The rate of tax applicable to corporate earnings might appear to be the rate of corporation tax. However, the picture is clouded by the prevailing type of tax regime (e.g. whether classical or an imputation tax system), and by the forms in which shareholders receive income (i.e. the balance between dividend income and capital gains, and the relevant rates of tax on these two forms of income). In other words, it is important to consider the interaction between the system of corporate taxation and the system of personal taxation.

Under an imputation tax, a shareholder receives a tax credit for the income tax component incorporated into the profits tax. Shareholders subject to tax at the standard rate face no further tax liability, while higher rate taxpayers face a supplementary tax demand. To add to the complexity, some imputation systems allow investors to reclaim all the tax paid on their behalf (full imputation), while others involve a discrepancy between the rate of corporation tax and the relevant rate of income tax (partial imputation). Since partial imputation applies in the UK, we will consider only this form.

When we calculated k_e using the DGM, the computation was based on the net-of-tax dividend payment, so it may appear that we have met the requirement to allow for taxation. However, the UK tax system imposes two possible tax distortions. First, the relative tax treatment of capital gains and dividend income has differed over time, and second, as we have just seen, different shareholders are subject to tax in different ways.

A major problem facing a company is divining the tax status of its shareholders. Inspection of the shareholder register may provide much information, but there is no easy solution to this problem. The share price is set by the market as a result of the interaction of the supply and demand for its shares as expressed by thousands of investors. Although each may well be in a different tax position, the resulting share price is the result of investors assessing whether the shares represent good value or not. In other words, the market automatically takes into account the average tax positions of its participants.

Under this view, it is not the function of the company to gauge the tax requirements of the investor and to adjust the discount rate accordingly. This is impossible in a capital market with large numbers of investors. The market imposes a required return for particular companies, and then it is up to individual investors to make their own arrangements regarding taxation. The market-determined rate of return can be regarded as the return that the company must make on its investments. This becomes the after-tax return that the company should use to discount the after-tax cash flows from capital projects. (The only adjustment that the company should make is to allow for the tax shield on debt, as explained in Chapter 18.)

To summarise: in principle, we could discount pre-tax cash flows, but the identification of the appropriate pre-tax required return is complicated by the existence and timing of capital allowances. Hence, a post-tax computation is preferable. Theoretically, we ought to allow for investors' personal tax positions as well as corporation tax (i.e. discount project cash flows net of both corporation tax and investors' personal tax liabilities). But this requires such detailed knowledge of the relevant tax rates applicable to shareholders as to render it impracticable. As a result, it is usual to discount post-corporation tax cash flows at the market-expressed required return, assuming that shareholders have made their own tax arrangements. This means that shareholders will gravitate to those companies whose dividend policies most suit their tax positions. This personal **clientèle effect** is discussed further in Chapter 17.

Self-assessment activity 11.2

Specify the two situations under which the DGM breaks down completely. (You may have to revisit Chapter 3.)

(Answer in Appendix A at the back of the book)

11.3 THE REQUIRED RETURN IN ALL-EQUITY FIRMS: THE CAPM

In Chapter 10, we saw how the security market line (SML) traces out the systematic risk–return characteristics of all the securities traded in an efficient capital market. The SML equation is:

$$\text{ER}_j = R_f + \beta_j(\text{ER}_m - R_f)$$

ER_j is the return required on the shares of company j, and is therefore the same as k_e, R_f is the risk-free rate of return, and ER_m is the expected return on the market portfolio. We saw in Chapter 10 that, in order to utilise the CAPM, we needed either to measure or to make direct assumptions about these items. (Refer back to the discussion of measurement difficulties and the application to British Airways.)

However, despite these problems, the CAPM has major advantages over the DGM. The DGM usually involves extrapolating past rates of growth and accepting the validity of the market's valuation of the equity at any time. If we suspect that past growth rates are unlikely to be replicated and/or that a company's share price is over- or under-valued, we might doubt the validity of an estimate of k_e derived from the DGM.

The CAPM does not require growth projections; nor does it totally depend on the instantaneous efficiency of the market. Recall that the Beta is derived from a regression model relating the returns from holding the shares of a particular company j to the returns on the market over a lengthy period. Taking, say, monthly observations over five years (60 in all) effectively irons out short-term influences. This requires semi-strong market efficiency for the period and a reasonably consistent relationship between security returns and the returns on the market portfolio.

■ Applying the CAPM to Whitbread plc

The Risk Measurement Service quoted a Beta of 1.14 for Whitbread shares as at Oct–Dec 2007. At that time, the yield on three month Treasury Bills was 5.75 per cent. Using a market risk premium of 5 per cent yields the following required return:

$$\text{ER}_j = R_f + \beta(\text{ER}_m - R_f) = 0.0575 + 1.14(0.05)$$
$$= 0.0575 + 0.0570 = 0.1145, \text{ i.e. } 11.45\%$$

This is somewhat below the DGM result of 13.3 per cent. As the two approaches, in principle, should yield about the same result, some reconciliation is required. At the time of this calculation, market interest rates were historically low, at least in money terms, generating expectations of low interest rates for the future. It is doubtful whether Whitbread can sustain 11 per cent dividend growth in the future, so it might be more prudent to use a rate nearer to that of the industry as a whole.

It appears that estimates of k_e obtained by either method are susceptible to the date of the calculation and prevailing expectations for the future. More fundamentally, whereas the DGM looks at performance over a number of years, the CAPM is essentially a one-period model, although it is commonly used for long-term purposes.

LEX COLUMN

Counting the cost

There are few more essential items in the corporate finance tool-kit than a company's cost of capital – the return its investors expect as compensation for putting their funds in one business rather than another. Estimating this cost of capital, however, involves as much art and guesswork as it does science, and the results can vary widely.

Three years ago, those companies that publish a figure for their cost of capital – usually those which have adopted a form of economic profit or economic value added performance framework – often came out with figures 1–2 percentage points higher than those implied by market values, or estimated by stock market analysts. Today, the gap has in many cases reversed. Lloyds TSB, for example, calculates its economic profit using a cost of equity of 9 per cent. Yet its share price appears to imply, even if you assume it will halve its dividend, a cost of equity in excess of 10 per cent.

Why does this matter? To create value for shareholders, companies need to make returns greater than their cost of capital. If companies are underestimating cost of capital, they will make acquisitions or invest in projects that destroy value. Conversely, if the market is setting the hurdle too high, investors will miss out on value-creating investments.

CAPM

Computing the cost of debt is fairly straightforward, at least for companies whose bonds are traded. The cost of equity is more complicated. The standard formula remains the capital asset pricing model, or CAPM, devised separately by William Sharpe, John Lintner and Jack Treynor. Though many academic studies have raised doubts about its empirical validity, three out of four chief financial officers use CAPM.

CAPM's starting point is the risk-free rate – typically a 10-year government bond yield. To this is added a premium, which equity investors require to compensate them for the extra risk they accept. This equity risk premium is multiplied by a factor, known as beta, to reflect a company's volatility and correlation with the market as a

whole. Beta is designed to capture the risk that an investor cannot diversify away by holding a portfolio of other shares; a company whose share price tends to rise and fall more than the market will have a high beta. There are difficulties with all three of these elements. Government bond yields are currently very low, by historical standards. A company contemplating a long-term investment can lock in these low rates for its debt, but if interest rates then rise so will its cost of equity. It may generate the cash flows it anticipated from its investment, but these will no longer cover its cost of capital. It may be appropriate to use a somewhat higher normalised risk-free rate. Yet it looks as though many equity analysts have taken insufficient account of the fall of risk-free rates in their cost of capital estimates.

The equity risk premium is the element that has generated most controversy. In the early 1990s, most companies used numbers in excess of 6 per cent, drawing on data from Ibbotson Associates and others. Then market analysts started to use equity risk premiums of 3–4 per cent and these numbers began to filter into corporate use. Historical performance data compiled by Elroy Dimson, Paul Marsh and Mike Staunton give a world equity premium over bonds of 3.8 per cent over the last 103 years. Marakon Associates, the strategic consultancy, derives an equity risk premium of 5.3 per cent, rather higher than the recent average, from the implied internal rate of return of 1,190 stocks, but of 3.6 per cent on the basis of dividend yield and growth. Splitting the difference, that gives an estimate of about 4.5 per cent.

Beta

Beta can be even trickier to calculate. Ideally, companies would use a forward looking beta but estimates depend on historical trading data. Yet as McKinsey analysts pointed out in a recent study, the TMT bubble of 1998–2001 has dramatically lowered the apparent betas of unaffected sectors. They calculate an improbably low current beta of 0.02 for the food, beverage and tobacco sector, against an average of 0.85 for 1990–97.

Individual company betas can also deliver counter-intuitive results. An accident-prone company may have a very low beta, because its mishaps mean it shows less correlation with the overall market.

Take Allianz as an example: the German insurer bases its embedded value calculations on an 8.15 per cent risk discount rate for Europe and the US. This is based on a 5 per cent long-term view of risk-free rates, a 3.5 per cent equity risk premium and a beta of 0.9. This beta, in particular, might raise an eyebrow, since the vulnerability of the company's capital base to equity market declines would prompt most investors to call it a high beta stock. Substituting a historical German equity risk premium of 5.7 per cent – according to Dimson, Marsh and Staunton – and a Bloomberg-calculated beta of 1.14 would yield a cost of equity of 11.5 per cent.

The finer points of CAPM mattered less when nominal interest rates were high. Take a company whose cash flows are growing at 3 per cent: using a 12 per cent cost of capital to discount these cash flows, only one third of its value lies more than 10 years out but, at 7 per cent, more than half is accounted for by these more distant years. Small adjustments to the cost of capital will also have a larger impact on the overall valuation at these lower rates. This effect weighs even more on non-financial companies with a significant amount of debt on their balance sheets, as their weighted average cost of capital will be lower than their cost of equity.

In most corporate investment decisions, the odd half point makes little difference, though in pricing acquisitions the precise cost of capital may be more significant. With equity markets still jittery, however, companies are better off setting a higher hurdle rate for investment than a straightforward CAPM calculation would imply. That might not be consistent with academic theory but it will, in practice, make them choose more carefully between their business units in allocating capital and lead to less wasteful investment than in the past.

Source: Financial Times, 24 March 2003.

11.4 USING 'TAILORED' DISCOUNT RATES

Applying the discount rates derived using the CAPM to investment projects assumes that new projects fall into the same risk category as the company's other operations. This might be a reasonable assumption for minor projects in existing areas and perhaps for replacements, but hardly seems justifiable for major new product developments or acquisitions of companies in unrelated areas. If the expected return is positively related to risk, firms that rely on a single discount rate may tend to over-invest in risky projects to the detriment of less risky, though still attractive projects. Many multi-divisional companies are effectively portfolios of diverse activities of different degrees of risk. The Beta of the firm as a whole is thus the weighted average of its component activity Betas. Each division contributes to the firm's overall business risk in a way similar to that in which individual shares contribute to the systematic risk of a portfolio of securities. The dangers of using a uniform discount rate are shown in Figure 11.1.

Figure 11.1 shows the relationship between the rate of return required on a particular project and that expected on the market portfolio, linked by the Beta. The overall portfolio of company activities may have a Beta of, say, 1.2, which is a weighted average of the Betas of component activities. For example, activity A has a greater than average degree of risk, with a Beta of 2.0, and thus a higher than average discount rate would be applicable when appraising new projects in this area, while the reverse applies for activity B, which has a Beta of only 0.8. Clearly, to appraise all new projects using a discount rate based on the overall company Beta of 1.2 would invite serious errors. For example, in area X, application of the uniform discount rate would result in accepting some projects that should be rejected because they offer too low a return for their level of risk, while in area Y, some worthwhile low-risk projects would be rejected. Firms should use 'tailor-made' cut-off rates for activities involving a degree of risk different from that of the overall company.

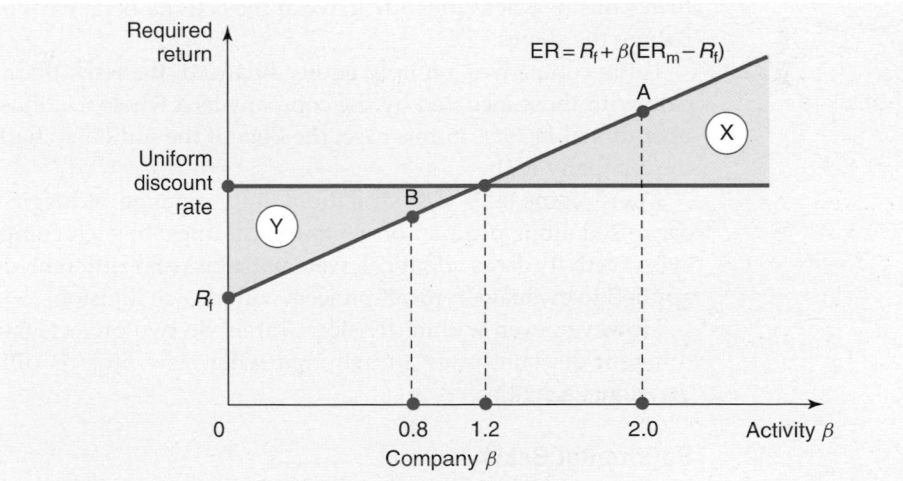

Figure 11.1 Risk premiums for activities of varying risk

Self-assessment activity 11.3

What are the discount rates applicable to the firm as a whole and activities A and B on Figure 11.1, assuming a risk-free rate of 5 per cent, and a market risk premium of 6 per cent?

(Answer in Appendix A at the back of the book)

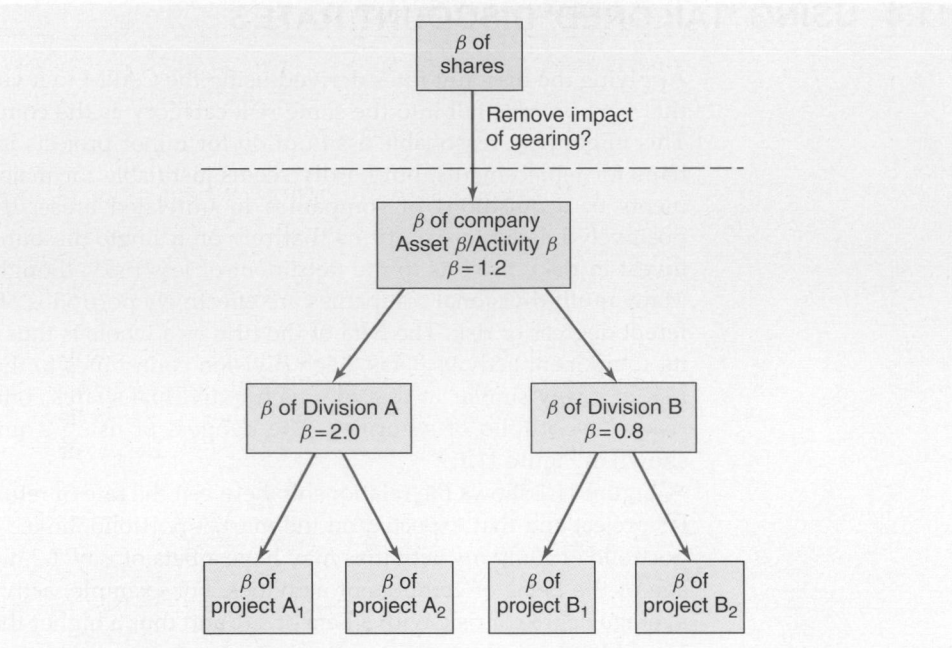

Figure 11.2 The Beta pyramid

Figure 11.2 shows the three levels, or tiers, of risk found in the multi-activity enterprise, each requiring a different rate of return.

In Chapter 19, we will find that there is a fourth tier of risk that uniquely applies to ordinary shareholders. In a geared firm, that faces financial risk, the returns achieved by shareholders are more volatile than the firm's operating cash flows due to the interest payments that must be paid on debt. In response to this higher risk, shareholders demand a higher return. In other words, the Beta of the shares exceeds the Beta of the firm's business activities. To arrive at the activity Beta, we would need to 'ungear' the Beta of the shares.

If the company is entirely equity-financed, the risks that shareholders incur coincide with those incurred by the company as a whole, i.e. those related to trading and operational factors. In this case, the Beta of the ordinary shares coincides with that of the company itself.

Many companies are structured into separate strategic sub-units or divisions, organised along product or geographical lines. In such companies, it is unlikely that every activity faces identical systematic risk. So different discount rates should be applied to evaluate 'typical' projects within each division.

However, even within divisions, rarely do two projects have identical risk. Hence, different discount rates are required when new projects differ in risk from existing divisional activities.

Segmental Betas

The company Beta is a weighted average of component divisional Betas. For a company with two divisions, A and B, the overall Beta is a weighted average given by:

$$\text{Company } \beta = \left(\beta_A \times \frac{V_A}{V_A + V_B} \right) + \left(\beta_B \times \frac{V_B}{V_A + V_B} \right)$$

where the weights represent the proportion of company value accounted for by each segment. A similar expression would apply for each division, where the corresponding weights would represent the contribution to divisional value accounted for by each component activity. Figure 11.2 illustrates these concepts in the form of a 'Beta pyramid.'

Self-assessment activity 11.4

What is the company Beta for the firm shown in Figure 11.1 if activities A and B constitute 65 per cent and 35 per cent of its assets respectively?

(Answer in Appendix A at the back of the book)

Let us use Whitbread plc to illustrate the derivation of the appropriate discount rate at different levels of an organisation. It is organised into four broad product division lines, as shown in Table 11.2, which lists the four operating divisions. These titles suggest quite different activities, although a firm's own description of its division is not always a reliable guide to the nature of those activities.

We performed a CAPM calculation earlier in relation to the equity of Whitbread, obtaining a result of 11.45 per cent. Should we apply this rate to all investments undertaken by Whitbread? The answer is 'no', if we believe there are risk differences between the divisions, in which case, we should calculate tailor-made discount rates.

Table 11.2 Divisional Betas for Whitbread plc

Activity	% share of sales	Surrogate company	Beta	Weighted beta
Premier Travel Inns	33.0%	Millennium and Capthorne	1.02	0.34
Restaurants	37.4%	Regent Inns	1.12	0.42
Costa Coffee	12.6%	Coffee Republic	1.04	0.13
David Lloyd Leisure	17.0%	Powerleague*	1.02	0.17
	100			1.06

*There are no quoted sports clubs on the LSE. Powerleague is an operator of facilities for 5-a-side football leagues.

Source: Whitbread plc Annual Report 2006–7; Risk Measurement Service, Oct–Dec 2007.

■ The divisional cut-off rate

We need now to consider what are suitable Betas for the four Whitbread divisions. However, no Betas are recorded for company divisions, simply because no market trades securities representing title to a firm's divisional assets. Instead, we need to look for four surrogate companies and use their ungeared Betas as the 'stand-in' estimates for the Betas of the four Whitbread divisions. This involves using what Fuller and Kerr (1981) called the **pure play technique**. It relies on the principle that: 'the risk of a division of a conglomerate company is the same as the risk of an undiversified firm in the same line of business (adjusted for financial risk)'.

pure play technique
Adoption of the Beta value of another firm for use in evaluating investment in an unquoted entity such as an unquoted firm, or a division of a larger firm

Consulting the RMS, we look for suitable surrogate companies whose Betas we can use as proxies for those of the Whitbread divisions. The dangers of doing this should not be understated. Ideally, the surrogate should be a close match for the relevant Whitbreads division, i.e. they should conduct the same activity or mix of activities in the same proportions, and should also be ungeared. (If they use debt finance, the gearing effect on their Betas should be stripped out, as explained in Chapter 19.) In principle, the weighted averages of these Beta values will coincide with the overall Beta of Whitbread plc if we have selected good surrogates. The weightings ought to be based on market values, but as these are unknown for company divisions, book values of net operating assets could be used. Not all companies reveal divisional asset values, so a proxy measure such as sales or operating profits may have to be used. For Whitbread, share of sales has been used. This is only a valid proxy for assets if the sales-to-assets ratio is similar from division to division, which is quite unlikely.

Table 11.2 shows that the weighted average Beta for Whitbread is 1.06, a shade below the value given in the RMS. The discrepancy could be due to the following reasons:

- The chosen surrogates are not close enough matches for Whitbread's array of activities.
- Differences in gearing. As we will see in a later chapter, gearing has the effect of raising Beta values as shareholders seek an extra premium to compensate for the financial risk that gearing imposes. The RMS Beta values are all equity Betas – they include the effect of gearing, and, of course, different firms may have different gearing ratios. Hence, if we take a Beta from a low-geared firm and apply it to a high-geared one, our weighted average calculation will understate the true Beta of the focus firm, and vice versa. Ideally, we should compare like with like, either strip out the effect of gearing altogether, and work in terms of pure equity (or activity) Betas, or ungear the Betas of the surrogates, and then re-gear them to reflect the gearing of the focus firm. These issues we defer to Chapter 19.

Self-assessment activity 11.5

During 2007, Whitbread disposed of its David Lloyd Leisure subsidiary to a private equity house.

- By reworking the weighted average Beta calculation, predict the impact on the Whitbread Group Beta.

(Answers in Appendix A at the back of the book)

■ The project cut-off rate

If any division undertakes a new venture that takes it outside its existing risk parameters, clearly we must look for different rates of return – in effect, we need to obtain estimates for individual project Betas. Without access to internal records, our analysis can only be indicative, but the following principles offer broad guidance.

Essentially, we look for sources of risk that make the individual project more or less chancy relative to existing operations. There are two broad reasons why projects have different risks to the divisions where they are based – different revenue sensitivity and different operating gearing.

■ Revenue sensitivity

Imagine Whitbread is looking at developing a new coffee shop brand. The sales generated by the projected facility may vary with changes in economic activity to a greater or lesser degree than existing sales in the relevant division. For example, we may expect that, for a specified rise in the level of GDP, whereas overall retail sales of Whitbread existing outlets increase by 7 per cent, the sales of the new brand rise by 9 per cent.

■ The revenue sensitivity factor

revenue sensitivity factor
The sensitivity to economic fluctuations of a project's sales in relation to that of the division to which it is attached

This magnifying effect is measured by the **revenue sensitivity factor** (RSF). The RSF is calculated as follows:

$$\text{RSF} = \frac{\text{Sensitivity of project sales to economic changes}}{\text{Sensitivity of divisional sales to economic changes}} = \frac{9\%}{7\%} = 1.29$$

This relationship may stem from the nature of the product – if it is pitched at discretionary spenders (e.g. people who frequent more 'upmarket' outlets), it may be more closely geared to the economy as a whole.

■ Operating gearing

This concerns the extent to which the project cost structure comprises fixed charges. The higher the proportion of fixed costs in the cost structure, the greater the impact of a change in economic conditions on the operating cash flow of the project, thus magnifying the revenue sensitivity effect. Again, the project may exhibit a degree of operating gearing different from that of the division as a whole.

To illustrate the impact of operating gearing, consider the figures in Table 11.3, where the firm applies a 50 per cent mark-up on variable cost. An increase in sales revenue of 50 per cent will lead to an increase in net operating cash flow of 67 per cent because of the gearing effect. There is thus a magnifying factor of 1.34. This so-called **project gearing factor** (PGF) may well differ from the gearing factor(s) found elsewhere in the division.

project gearing factor
The proportionate increase in a project's operating cash flow in relation to a proportionate increase in the project's sales

Table 11.3 The effect of operating gearing (£m)

Sales revenue	Variable costs	Fixed cash costs	Operating cash flow
90	60	5	25
60	40	5	15

operating gearing factor
The operating gearing factor of an individual project in relation to that of the division to which it is attached

To measure the relative level of gearing, the **operating gearing factor** (OGF) is used. This is defined as:

$$\text{OGF} = \frac{\text{Project gearing factor}}{\text{Divisional gearing factor}}$$

If the divisional gearing factor is 1.80, for example, the project's OGF = 1.34/1.80 = 0.74.

The second step in assessing the project discount rate brings together these two sources of relative project risk into a project risk factor (PRF).

■ The project risk factor

This is the compound of the revenue sensitivity factor and the operating gearing factor:

Project risk factor = RSF × OGF

In our example, this is equal to (1.29 × 0.74) = 0.95. In this case, the project is less risky than the 'average' project within the division and merits the application of a lower Beta. Based on the Whitbread coffee shop Beta, shown in Table 11.2, this is given by:

Project Beta = (0.95 × 1.04) = 0.99

The final step calculates the required return using the basic CAPM equation, based on a 5.75 per cent risk-free rate and a market risk premium of 5 per cent:

Required return = 0.0575 + 0.99(0.05) = 0.0575 + 0.0495 = 0.107 (i.e. 10.7%)

Self-assessment activity 11.6

Determine the required return on a project whose revenue sensitivity is 50 per cent and operating gearing 80 per cent compared to the division where it is located. The divisional Beta is 1.2, the risk-free rate is 5 per cent and the market risk premium 6 per cent.

(Answer in Appendix A at the back of the book)

■ Project discount rates in practice

Considering the informational requirements for obtaining reliable tailor-made discount rates for particular investment projects, few firms go to these lengths. A far more common practice is to seek an overall divisional rate of return, which becomes the average cut-off rate, but is then adjusted for risk on a largely intuitive basis, according to the perceived degree of risk of the project. For example, many firms group projects into 'risk categories' such as the classification in Table 11.4. For each category, a target or required return is established as the cut-off rate.

Table 11.4 Subjective risk categories

Project type	Required return (%)
Replacement	12
Cost saving/application of advanced manufacturing technology	15
'Scale' projects, i.e. expansion of existing activities	18
New project development:	
Imitative products	20
Conceptually new products, i.e. no existing competitors	25

Imagine the divisional required return is 18 per cent, the rate applicable to projects that replicate the firm's existing activities. Around this benchmark are clustered activities of varying degrees of risk, and as the perceived riskiness increases, the target return rises in tandem.

In all these cases, we are discussing a discount rate derived from the ungeared Beta. In other words, we are separating out the inherent profitability of the project from any financing costs and benefits. Analysis of financing complications is deferred to Chapters 18 and 19.

11.5 WORKED EXAMPLE: TIEKO PLC

Tieko plc is a diversified conglomerate that is currently financed entirely by equity. Its five activities and their respective share of corporate assets are shown in the table below. Also shown are the Beta coefficients of a highly similar surrogate firm operating in the same markets as the Tieko divisions.

Division	% share of book value of Tieko assets	Beta of a close substitute
Electronics	30	1.40
Property	20	0.70
Defence equipment	30	0.20
Durables	20	1.05

The yield on short-term government stock is currently 6 per cent, and people expect the stock market portfolio to deliver an average annual return of 13 per cent in future years.

Required

In each of the following (separate) situations, determine Tieko's company Beta and the return required by shareholders.

(i) As it is currently structured. Also, calculate the hurdle rates for the four divisions.
(ii) If Tieko sells the defence division for book value and returns the cash proceeds to shareholders as a Special Dividend.
(iii) If Tieko sells the defence division for book value and places the cash proceeds on deposit.
(iv) If Tieko acquires a telecommunications firm that has a Beta of 1.60, and total assets equal in value to the defence division.

Solution

In each case, the Beta value is a weighted average of the Betas of the component activities, with each division's share of total assets providing the weights.

(i) At present, the Beta is:

$$(0.3 \times 1.4) + (0.20 \times 0.70) + (0.30 \times 0.20) + (0.20 \times 1.05)$$
$$= (0.42 + 0.14 + 0.06 + 0.21) = 0.83$$

Using the CAPM formula, the required return for the firm as a whole is thus:

$$6\% + 0.83\,[13\% - 6\%] = (6\% + 5.8\%) = 11.8\%$$

The separate divisional hurdle rates are as shown in the following table.

Division	Risk-free rate	Beta	Market premium	Required return
Electronics	6%	1.40	7%	6% + 1.40(7%) = 15.80%
Property	6%	0.70	7%	6% + 0.70(7%) = 10.90%
Defence equipment	6%	0.20	7%	6% + 0.20(7%) = 7.40%
Durables	6%	1.05	7%	6% + 1.05(7%) = 13.35%

(ii) If Tieko 'downsizes' to 70 per cent of its previous size, the weightings for the three remaining divisions become:

Electronics 3/7 = 0.429

Property 2/7   = 0.285

Durables 2/7   = 0.285

The overall Beta becomes:

$$(0.429 \times 1.4) + (0.285 \times 0.70) + (0.285 \times 1.05)$$
$$= (0.60 + 0.20 + 0.30) = 1.10$$

Clearly, Tieko has become more risky and the required return increases to

$$6\% + 1.10\,[13\% - 6\%] = 6\% + 7.7\% = 13.7\%$$

(iii) If it retains the cash, then the total assets remain unchanged but the Beta will alter as the Beta of cash is zero – it is uncorrelated with the risky securities quoted on the stock market.

The overall Beta becomes:

$$(0.3 \times 1.4) + (0.20 \times 0.70) + (0.30 \times 0) + (0.20 \times 1.05)$$
$$= (0.42 + 0.14 + 0 + 0.21) = 0.77$$

Having disposed of its least risky division and replaced it with risk-free cash, Tieko has become much less risky. As a result, its overall required return decreases to:

$$6\% + 0.77\,[13\% - 6\%] = (6\% + 5.39\%) = 11.39\%$$

(iv) If Tieko expands by adding another division of equal size to the electronics arm, i.e. a 30 per cent expansion, the new weightings are:

Electronics 3/13 = 0.230
Property   2/13 = 0.154
Defence   3/13 = 0.230
Durables   2/13 = 0.154
Telecomms 3/13 = 0.230

The overall Beta becomes:

$$(0.230 \times 1.4) + (0.154 \times 0.70) + (0.230 \times 0.20) + (0.154 \times 1.05) + (0.230 \times 1.60)$$
$$= (0.322 + 0.108 + 0.046 + 0.162 + 0.368) = 1.006$$

The required return increases to:

$$6\% + 1.006\,[13\% - 6\%] = (6\% + 7.04\%) = 13.04\%$$

After this expansion, Tieko has a Beta very similar to that of the market portfolio (1.0). It has thus managed to diversify itself into a portfolio of activities of virtually average risk.

11.6 ANOTHER PROBLEM: TAXATION AND THE CAPM

Empirical studies of the risk premium usually reveal gross-of-personal-tax results. To adjust for tax, one might consider the tax status of interest income from the risk-free asset, normally taken as government stock of some form, and the tax status of the return on the market portfolio.

Franks and Broyles (1979) recommended two adjustments. First, adjust the risk-free rate for the shareholders' rate of personal tax (at present, UK basic-rate tax on interest income is 20 per cent), then adjust the risk premium according to the relative proportions of excess return earned in dividends and in capital gain form. Grubb (1993/4) shows that over 1960–92, about half of the return on equities was from dividends and half from capital gain. Two major problems follow. Capital gains tax (CGT) was only introduced in 1965, and has been applied at varying rates, while the basic rate of income tax has also changed many times over this period. Moreover, in reality, very few shareholders are liable to CGT. However, taking 20 per cent as an average rate of CGT and 25 per cent as an average rate of tax on dividends thus yielding a weighted average tax rate (WAT) of $(0.5 \times 0.20) + (0.5 \times 0.25) = 22.5\%$, the calculation of the post-tax required return on Whitbread's ordinary shares (ER_W) would be:

$$ER_W = R_f(1 - \text{present income tax rate}) + \beta[\text{market risk premium}][1 - \text{WAT}]$$
$$= 0.0575(1 - 0.2) + (1.14)(0.05)(1 - 0.225)$$
$$= 0.046 + 0.044$$
$$= 0.09 \text{ i.e. } 9\%$$

There are obvious problems in taking average rates of tax over periods when tax regimes have altered. Wilkie (1994) argues that such a calculation is conceptually flawed, being based on the assumption that individuals would have invested in a tax-inefficient vehicle (government stock) – although many do! Most personal investors would have been subject to higher rates of income tax and thus would have taken steps anyway to shelter their income from tax. Finally, he points out that the securities market historically has been dominated by tax-exempt investors, in terms of both the percentages of share value held and, more crucially, the flow of new funds to the market, which dictates market prices. He concludes that little accuracy is lost by using the gross-of-tax risk premium, and ignoring any tax effect on the risk-free rate, especially as future tax rates on investment income are likely to be lower than past rates, following widespread tax cuts in the 1990s.

This view is very appealing, both in view of the rapid growth in the popularity of tax breaks like the old TESSAs and PEPs, and now ISAs; and also for simplicity. Difficulties over specifying discount rates may go some way to explaining the continuing popularity of the payback method. Significantly, this is usually used in conjunction with the IRR, which does not require the pre-specification of a discount rate. This combination becomes a convenient means of communicating criteria of investment acceptability throughout a company without requiring continuous updating of the discount rate for the tax position. A survey of over 200 firms conducted for the CBI by Junankar (1994) found that 56 per cent of those that used the IRR method applied it on a pre-tax basis.

11.7 PROBLEMS WITH 'TAILORED' DISCOUNT RATES

The pure play technique is an appealing device for estimating discount rates for specific activities, but suffers from a number of practical difficulties.

1 *Selecting the proxy.* To select a proxy, the firm needs to examine the range of apparently similar candidates operating in the relevant sector. However, no two companies have the same business risk due to diversity of markets, management skills and other operating characteristics. How one chooses between a range of 'fairly similar' candidates is essentially an issue of judgement.

2 *Divisional interdependencies.* In practice, it is difficult to make a rigid demarcation of divisional costs and incomes, since most divisionalised companies share facilities, ranging from the highest decision-making level to joint research and development, joint distribution channels and joint marketing activities. Indeed, access to shared facilities often provides the initial motive for forming a diversified conglomerate, enabling the elimination of duplicated services and the exploitation of scale economies. If carefully evaluated and implemented, the merging of activities should create value and reduce business risk. Only when a merger has no operating impact across divisional lines can it be suggested that business risk itself is unaffected. Even so, there may well be synergies at the peak decision-making level.

3 *Differential growth opportunities.* Using a cut-off rate based on another firm suggests that the division in question has the same growth prospects as the surrogate. However, opportunities to grow are determined by dividend policy, the extent of capital rationing and the interaction between divisions, e.g. competition for scarce investment capital. In reality, because the firm's own decision processes help to determine the potential for growth, it is not accurate to assume that growth opportunities are externally derived.

4 *Joint ventures.* The use of differential discount rates may destroy the incentive to cooperate on projects that straddle divisional boundaries. For example, a joint venture whose expected return lies between the cut-off rates of the two divisions will be attractive to one and unacceptable to the other. Here, some form of mediation is

required at peak level, which reassures the 'loser' of the decision that subsequent performance will be assessed after adjusting for having to operate with a project that it did *not* want, or without a project that it *did* wish to undertake.

11.8 A CRITIQUE OF DIVISIONAL HURDLE RATES*

Modern strategic planning has moved away from crude portfolio planning devices such as the Boston Consulting Group's market share/market growth matrix towards capital allocation methods that emphasise the creation of shareholder value. Central to value-based approaches is discounting projected cash flows to determine the value to shareholders of business units and their strategies. *A key feature of the DCF approach is the recognition that different business strategies involve different degrees of risk and should be discounted at tailored risk-adjusted rates.*

However, critics such as Reimann (1990) suggest that differential rates will increase the likelihood of internal dissension, whereby a manager of a 'penalised' division may resent the requirement to earn a rate of return significantly higher than some of his colleague-competitors. This resentment may be worsened by the observation that longer-term developments, especially in advanced manufacturing technology and other risky, but potentially high value-added activities, may be 'unfairly' discriminated against. As a result, managers may be reluctant to propose some potentially attractive projects.

As we saw in Chapter 7, risk-adjusted discount rates have the effect of compounding risk differences, making ostensibly riskier projects appear to increase in risk over time. One school of thought contends that in order to avoid this risk penalty, the attempt to tailor discount rates to divisions should be modified, if not abandoned. For example, instead of using differential discount rates, firms might use a more easily understood and acceptable, company-wide discount rate for projects of 'normal' risk, but appraise high-risk/high-return projects using different approaches.

Underlying these arguments is the familiar assertion that diversification by firms differs crucially from shareholder diversification, so that applying the CAPM to the former could be misleading. If an investor adds a new share to an existing portfolio, the market risk of the portfolio will alter according to the Beta of the new security. If its Beta is higher than that of the existing portfolio, then the portfolio Beta increases, and vice versa. With corporate diversification, however, we are not dealing with a basket of shares of unrelated companies, which may be freely traded on the market. A firm that diversifies rarely adds totally unrelated activities to its core operations. It may add value if the new activity possesses synergy, or detract from value if the market views the combination as merely a bundle of disparate, unwieldy activities that are hard to manage.

Market risk can be altered by strategic diversification decisions at two levels. At the corporate level, decisions concerning business and product mixes, and operating and financial gearing can affect market risk. The effect of both types of gearing can be magnified by the business cycle, so that a firm which engages in contra-cyclical diversification may dampen oscillations in shareholder returns and thus reduce market risk. At the business level, market risk can be reduced by tying up outlets and supplier sources (i.e. by increasing market power), and by developing business activities that enjoy important interrelationships, such as common skills or technologies (i.e. by exploiting economies of scale).

Many managers feel that the emphasis on hurdle rates is probably misplaced insofar as accurate cash flow forecasts are more important to creating business value than the particular discount rate applied to them. This probably helps explain the continuing popularity of the payback method, and the reluctance, at least in the UK, to adopt

---

*This section relies heavily on arguments used by Reimann (1990).

CAPM-based approaches. It may also explain why so many successful firms place great emphasis on post-auditing capital projects in order to sharpen up the cash flow forecasting and project appraisals of subordinate staff. Furthermore, there is evidence (Pruitt and Gitman, 1987; Pohlman *et al.*, 1988) that senior managers manifest their suspicion of subordinates' cash flow predictions by deflating the figures presented to them when projects are submitted for approval.

In view of these arguments, there may be a case for reconsidering the merits of using certainty equivalents – adjusting the cash flow estimates and then discounting at the risk-free rate. However, this has not been widely adopted. Apart from the difficulty of specifying the risk-free asset, there is the problem of determining the certainty equivalent factors, which involves specifying the probabilities of different possible cash flows as a basis for assessing their utility values. While techniques are available for doing this (Swalm, 1966; Chesley, 1975), it has not been practicable in most firms.

Reimann (1990) suggests a 'management by exception' approach. The firm should establish and continuously update a corporate cost of capital, based on CAPM principles. This should be applied as a common hurdle rate for the majority of business activities, which, he argues, typically exhibit very similar degrees of risk. At the business level, major emphasis should be given to careful cash flow estimation, based on evaluation of long-term strategic opportunities and competitive advantage. A key element should be a multiple scenario approach, whereby the implications of 'best', 'worst' and 'most likely' states of the world are examined. For projects that, by their very nature, have a demonstrably greater level of risk, other procedures may be appropriate. Rather than adjust the corporate discount rate, Reimann suggests the risk adjustment be made to the cash flow estimates by the business unit executives themselves, i.e. those with closest knowledge both of the market and of competitors' behaviour patterns. Again, a multiple scenario approach should be adopted. This avoids the effect of compounding risk differences over time and thus penalising longer-term projects, which may have a demotivating effect on staff engaged in pursuing high-risk activities.

This discussion may seem to downgrade the importance of DCF and CAPM approaches in project appraisal. However, it is really intended to remind you that apparently neat mathematical models rarely hold the whole answer. If the rigid application of a numerical routine leads managers to question the basis of the routine itself (one which we believe offers powerful guidance in many situations) and to exhibit dysfunctional behaviour, it is far better to modify the routine itself to reflect real world practicalities.

SUMMARY

We have considered the relative merits of using the DGM and the CAPM to derive the rate of return required by shareholders. The case for and against using tailor-made discount rates for particular business segments and projects was also discussed.

Key points

- The return required on new investment depends primarily on two factors: degree of risk and the method of financing the project.
- The return required by shareholders can be estimated using either the DGM or the CAPM.
- The DGM relies on several critical assumptions: in particular, sustained and constant growth, and the instantaneous reliability of the share price set by the market.
- The CAPM relies on a Beta estimate obtained after smoothing short-term distortions, but the estimated k_e may be affected by random influences on the risk-free rate.

- Application of a uniform company-wide discount rate to all company projects can lead to accepting projects that should be rejected and to rejecting projects that should be accepted.

- To resolve the problem of risk differences between divisions of a company, the Beta of a surrogate firm (adjusted for gearing) can be used to establish divisional cut-off rates.

- If individual projects within the division also differ in risk, the divisional Beta can be adjusted for differences in revenue sensitivity and/or differences in operating gearing.

- Not all academics and business people accept the need to define discount rates so carefully, preferring instead to concentrate on the problems of cash flow estimation.

- Reimann argues that a divisional cut-off rate should be used as a rough benchmark for projects, but alternative methods of risk analysis should be applied to explore more fully the risk characteristics and the acceptability of investment proposals.

Further reading

Analyses of the 'tailored' discount rate can be found in Dimson and Marsh (1982) and Andrews and Firer (1987). Weaver (1989), Gup and Norwood (1982) and Harrington (1983) all provide practical illustrations of how US corporations apply divisional discount rates, while Reimann (1990) gives a critique of the whole approach.

QUESTIONS

 Questions with an icon are also available for practice in *myfinancelab* with additional supporting resources.

Questions with a **coloured number** have solutions in Appendix B on page 741.

1 The ordinary shares of Rasal plc have a market price of £10.50, following a recent dividend payment of £0.80 per share. Dividend growth has averaged 4.5 per cent p.a. over the past five years. What is the rate of return required by shareholders implied by the current share price?

2 Insert the missing values in the following table:

	k_e	P_0	g	D_0
(i)	11%	£8.00	3%	?
(ii)	14%	?	4%	£0.350
(iii)	?	£5.00	6%	£0.155
(iv)	12%	£4.60	?	£0.250

3 Lofthouse plc has paid out dividends per share over the past few years, as follows:

1996	11.0p
1997	12.5p
1998	14.0p
1999	17.0p
2000	20.0p

In March 2000, the market price per share of Lofthouse is £5.00 ex-dividend. What is the rate of return required by investors in Lofthouse's equity implied by the Dividend Growth Model?

4 The all-equity financed Lasar plc has a Beta of 0.8. What rate of return should it seek on new investment:

(i) with similar risk to existing activities?
(ii) with 25 per cent greater risk compared to existing activities?
(iii) with 25 per cent lower risk compared to existing activities?

The risk-free rate of interest is 6 per cent, and the expected return on the market portfolio is 11 per cent.

5 Salas Ltd is an unquoted company that operates four divisions, all focused on single activities as shown in the table below. Salas identifies a proxy quoted company for each activity in order to calculate cut-off rates for new investment.

Division	Proxy Beta	Assets employed (£m)
Construction (C)	0.7	3.00
Engineering (E)	1.1	8.00
Road haulage (R)	0.8	4.00
Packaging (P)	0.6	5.00

The risk-free rate is 7 per cent, and the expected return on the market portfolio is 15 per cent.

Required
(i) Calculate the required return at each division.
(ii) Calculate Salas' overall required rate of return.

6 Megacorp plc, an all-equity financed multinational, is contemplating expansion into an overseas market. It is considering whether to invest directly in the country concerned by building a greenfield site factory. The expected payoff from the project would depend on the future state of the economy of Erewhon, the host country, as shown below:

State of Erewhon economy	Probability	IRR from project (%)
E_1	0.1	10
E_2	0.2	20
E_3	0.5	10
E_4	0.2	20

Megacorp's existing activities are expected to generate an overall return of 30 per cent with a standard deviation of 14 per cent. The correlation coefficient of Megacorp's returns with that of the new project is −0.36, Megacorp's returns have a correlation coefficient of 0.80 with the return on the market portfolio, and the new project has a correlation coefficient of −0.10 with the UK market portfolio.

- The Beta coefficient for Megacorp is 1.20.
- The risk-free rate is 12 per cent.
- The risk premium on the UK market portfolio is 15 per cent.
- Assume Megacorp's shares are correctly priced by the market.

Required

(a) Determine the expected rate of return and standard deviation of the return from the new project.
(b) If the new project requires capital funding equal to 25 per cent of the value of the existing assets of Megacorp, determine the risk–return characteristics of Megacorp after the investment.
(c) What effect will the adoption of the project have on the Beta of Megacorp?

Ignore all taxes.

7 PFK plc is an undiversified and ungeared company operating in the cardboard packaging industry. The Beta coefficient of its ordinary shares is 1.05. It now contemplates diversification into making plastic containers. After evaluation of the proposed investment, it considers that the expected cash flows can be described by the following probability distribution:

State of economy	Probability	Internal rate of return (%)
Recession	0.2	−5
No growth	0.3	8
Steady growth	0.3	12
Rapid growth	0.2	30

The overall risk (standard deviation) of parent company returns is 20 per cent and the risk of the market return is 12 per cent. The risk-free rate is 5 per cent and the FTSE-100 Index is expected to offer an *overall* return of 10 per cent per annum in the foreseeable future.

The new project will increase the value of PFK's assets by 33 per cent.

Required

(a) Calculate the risk–return characteristics of PFK's proposed diversification.
(b) It is believed that the plastic cartons activity has a covariance value of 40 with the company's existing activity.
  (i) Calculate the *total* risk of the company *after* undertaking the diversification.
  (ii) Calculate the new Beta value for PFK, given that the diversification lowers its overall covariance with the market portfolio to 120.
  (iii) Deduce the Beta value for the new activity.
  (iv) What appears to be the required return on this new activity?
(c) Discuss the desirability, from the shareholders' point of view, of the proposed diversification.

You may ignore taxes.

8 Lancelot plc is a diversified company with three operating divisions – North, South and West. The operating characteristics of North are 50 per cent more risky than South, while West is 25 per cent less risky than South. In terms of financial valuation, South is thought to have a market value twice that of North, which has the same market value as West. Lancelot is all-equity-financed with a Beta of 1.06. The overall return on the FT All-Share Index is 25 per cent, with a standard deviation of 16 per cent.

Recently, South has been under-performing and Lancelot's management plan to sell it and use the entire proceeds to purchase East Ltd, an unquoted company. East is all-equity-financed and Lancelot's financial strategists reckon that while East is operating in broadly similar markets and industries to South, East has a revenue sensitivity of 1.4 times that of South, and an operating gearing ratio of 1.6 compared to the current operating gearing in South of 2.0.

Assume: no synergistic benefits from the divestment and acquisition. You may ignore taxation.

Required

(a) Calculate the asset Betas for the North, South and West divisions of Lancelot. Specify any assumptions that you make.

(b) Calculate the asset Beta for East.

(c) Calculate the asset Beta for Lancelot after the divestment and acquisition.

(d) What discount rate should be applied to any new investment projects in East division?

(e) Indicate the problems in obtaining a 'tailor-made' project discount rate such as that calculated in section (d).

Note: More questions on required rates of return can be found in Chapter 19, where the additional complexities of gearing are discussed.

 **myfinancelab** | *Now retake your diagnostic test for Chapter 11 to check your progress and update your study plan.*

12

Enterprise value and equity value

Must do better ...

Provision of training courses and learning materials for students preparing for professional accounting exams is now big business, with a number of listed companies involved. In June 1998, the first quoted operator, Nord Anglia plc, acquired EW Fact, a leading accountancy training firm, for £19 million. After the acquisition, Nord Anglia was dismayed to find that restructuring costs were much higher than expected.

In May 1999, Nord Anglia announced a profits warning and also that it was considering legal action against EW Fact for 'materially overstating' profits for the year before acquisition. Nord Anglia's chairman alleged that pre-tax profits in 1997, posted at £1.4 million, should have been just £80,000. This knowledge would presumably have affected the sum that Nord Anglia would have paid to acquire EW Fact. Shares in Nord Anglia fell from 230p to 187p on the announcement.

The *Daily Telegraph* quipped: 'The sort of mistake any student can make, of course. But not what should be expected from auditors – if properly trained'.

Learning objectives

The ultimate effectiveness of financial management is judged by its contribution to the value of the enterprise. This chapter aims:

■ To provide an understanding of the main ways of valuing companies and shares, and of the limitations of these methods.

■ To stress that valuation is an imprecise art, requiring a blend of theoretical analysis and practical skills.

A sound grasp of the principles of valuation is essential for many other areas of financial management.

 **myfinancelab** *Complete your diagnostic test for Chapter 12 now to create your personal study plan.*

12.1 INTRODUCTION

The concept of value is at the heart of financial management, yet the introductory case demonstrates that valuation of companies is by no means an exact science. Inability to make precisely accurate valuations complicates the task of financial managers.

The financial manager controls capital flows into, within and out of the enterprise attempting to achieve maximum value for shareholders. The test of his/her effectiveness is the extent to which these operations enhance shareholder wealth. He/she needs a thorough understanding of the determinants of value to anticipate the consequences of alternative financial decisions. If there is an active and efficient market in the company's shares, it should provide a reliable indication of value. However, managers may feel that the market is unreliable, and may wish to undertake their own valuation exercises. Indeed, some managers behave as though they doubt the Efficient Markets Hypothesis (EMH), outlined in Chapter 2.

In addition, there are specific situations where financial managers must undertake valuations, for example, when valuing a proposed acquisition, or assessing the value of their own company when faced with a takeover bid. Directors of unquoted companies may also need to apply valuation principles if they intend to invite a takeover approach from a larger firm or if they decide to obtain a market quotation.

Valuation skills thus have an important strategic dimension. In order to advise on the desirability of alternative financial strategies, the financial manager needs to assess the value to the firm of pursuing each option. This chapter examines the major difficulties in valuation and explains the main methods available.

12.2 THE VALUATION PROBLEM

Anyone who has ever attempted to buy or sell a second-hand car or house will appreciate that value, like beauty, is in the eye of the beholder. Value is whatever the highest bidder is prepared to pay. With a well-established market in the asset concerned, and if the asset is fairly homogeneous, valuation is relatively simple. *So long as the market is reasonably efficient, the market price can be trusted as a fair assessment of value.*

Problems arise in valuing unique assets, or assets that have no recognisable market, such as the shares of most unquoted companies. Even with a ready market, valuation may be complicated by a change of use or ownership. For example, the value of an incompetently-managed company may be less than the same enterprise after a shake-up by replacement managers. But by how much would value increase? Valuing the firm under new management would require access to key financial data not readily available to outsiders. Similarly, a conglomerate that has grown haphazardly may be worth more when broken up and sold to the highest bidders. But who are the prospective bidders, and how much might they offer? Undoubtedly, valuation in practice involves considerable informed guesswork. (Inside information often helps as well!)

Regarding the introductory case, we do not know how the valuation was arrived at, but we can see that even the 'experts' can get it wrong. This illustrates an important lesson – the only certain thing about a valuation is that it will be 'wrong'! However, this is no excuse for hand-wringing. A key question is whether the valuations were reasonable in the light of the information then available.

price–earnings multiples
The price–earnings multiple, or ratio (PER), is the ratio of earnings (*i.e. profit after tax*) per share (EPS) to market share price

The three basic valuation methods are **net asset value**, **price–earnings multiples** and **discounted cash flow**. None of these is foolproof, and they often give different answers. Moreover, different approaches may be required when valuing whole companies from those appropriate to valuing part shares of companies. In addition, the value of a whole company (i.e. the value of its entire stock of assets) may differ from the value of the shareholders' stake. This applies when the firm is partly financed by debt capital.

■ Enterprise value vs. equity value: Innogy plc

To persuade the present owners to sell, a bidder has to offer an acceptable price for their equity and expect to take on responsibility for the company's debt. Consider the purchase in 2002, by RWE Ag, the German multi-utility group of the British electricity supplier Innogy, itself a spin-off from the privatised company International Power. RWE's logic was to complement its previous acquisition of Thames Water in 2000 in order to gain access to 10 million customer accounts to which it could offer gas, electricity and water. The overall deal was valued at around £5 billion, comprising some £3 billion of equity and £2 billion of debt.

Innogy's stock of assets was financed partly by equity and partly by debt. To obtain ownership of all the assets, i.e. the whole company, RWE was obliged to offer £3 billion to the shareholders to induce them to sell, *and* either pay off the debt or assume responsibility for it. Although RWE chose the latter route, either course of action made the total cost of the acquisition £5 billion.

Obviously, to make the acquisition worthwhile to RWE, its own (undisclosed) valuation would presumably have exceeded £5 billion. We thus encounter several different concepts of value:

Enterprise value →	Value of whole company to the buyer:	probably more than £5 billion
	Cost to acquire whole company:	£5 billion
Equity value →	Value of equity stake required to clinch sale:	£3 billion
	Value of equity stake perceived by owners:	possibly below £3 billion

net asset value approach
Calculation of the equity value in a firm by netting the liabilities against the assets

The distinction between company or enterprise value and the value of the owners' stake is clarified by considering the first method of valuation, the **net asset value approach**, which is based on scrutiny of company accounts.

Self-assessment activity 12.1

Using the Innogy example, distinguish between the value of a whole company and the value of the equity stake. When would these two measures coincide?

(Answer in Appendix A at the back of the book)

12.3 VALUATION USING PUBLISHED ACCOUNTS

Using the asset value stated in the accounts has obvious appeal for those impressed by the apparent objectivity of published accounting data. The balance sheet shows the recorded value for the total of fixed assets* (sometimes, but not invariably, including intangible assets) and current assets, namely stocks and work-in-progress, debtors, and other holdings of liquid assets such as cash and marketable securities. After deducting the debts of the company, both long- and short-term, from the total asset value (i.e. the value of the whole company), the residual figure is the net asset value (NAV), i.e. the value of net assets or the book value of the owners' stake in the company or, simply, 'owners' equity'.

The balance sheet for D.S. Smith plc, the paper and packaging group, is shown in Table 12.1. The balance sheet in its modern vertical form pinpoints the NAV, the net assets figure, £567.1 million, which, by definition, must coincide with shareholders' funds, i.e. the value of the shareholders' stake net of all liabilities (and, in this case, net

*Fixed assets are nowadays called 'non-current assets' following the adoption of IFRSs. We prefer the old name.

Table 12.1 Balance Sheet of D.S. Smith plc as at 30 April 2007

	£m	£m
Assets:		
Non-current assets		
(including intangibles of £192.9 m)		765.0
Current assets		
Stocks	160.5	
Receivables	350.2	
Cash and cash equivalents	92.4	
Other	1.7	604.8
Total assets		1,369.8
Liabilities:		
Non-current liabilities		
(mainly borrowings of £230.9 m)		(376.6)
Current liabilities		
(mainly trade payables of £384.8)		(426.1)
Total liabilities		(802.7)
Net assets		567.1
Equity		
Issued capital		39.3
Share premium		262.9
Reserves		267.2
Minority interest		(2.3)
Total equity (shareholders' funds)		567.1

Source: D.S. Smith plc Annual Report 2007.

of a small minority item, i.e. residual ownership in an acquired firm). The book value of the whole company, i.e. its total assets, is fixed assets plus current assets = (£765.0 m + £604.8 m) = £1,369.8 m. However, the NAV is a very unreliable indicator of value in most circumstances. Most crucially, it derives from a valuation of the separate assets of the enterprise, although the accountant will assert that the valuation has been made on a 'going concern basis', i.e. as if the bundle of assets will continue to operate in their current use. Such a valuation often, but not invariably, understates the earning power of the assets, particularly for profitable companies.

On 1 August 2007, the market value of D.S. Smith's equity was £958 million (share price of 243.75p times number of 10p shares, i.e. 393 million). Hence, the firm as a going concern with its existing and expected strategies, management and skills, all of which determine its ability to generate profits and cash flows was worth rather more than its net assets. If the profit potential of a company is suspect, however, then asset value assumes greater importance. The value of the assets in their best alternative use (e.g. selling them off) might then exceed the market value of the business, providing a signal to the owners to disband the enterprise and shift the resources into those alternative uses. Sometimes, then, we may be able to adjust the NAV to take into account more up-to-date, or more relevant information, thus obtaining the **adjusted NAV**.

Self-assessment activity 12.2

For D.S. Smith plc, identify:

(i) the value of the whole firm, i.e. enterprise value

(ii) the value of its total liabilities

(iii) the value of the owners' equity.

(Answer in Appendix A at the back of the book)

■ Problems with the NAV

The NAV, even as a measure of break-up value, may be defective for several reasons.

1 Fixed asset values are based on historical cost

Book values of fixed assets, e.g. £765.0 million for D.S. Smith, are expressed net of depreciation, the result of writing down asset values over their assumed useful lives. Depreciating an asset, however, is not an attempt to arrive at a market-oriented assessment of value but an attempt to spread out the historical cost of an asset over its expected lifetime so as to reflect the annual cost of using it. It would be an amazing coincidence if the historical cost less accumulated depreciation were an accurate measure of the value of an asset to the owners, especially at times of generally rising prices. Some companies try to overcome this problem by periodic valuations of assets, especially freehold property. However, few companies do this annually, and even when they do, the resulting estimate is valid only at the stated dates. Whichever way we look at it, fixed asset values are always out of date!

replacement cost
The cost of replacing the existing assets of a firm with assets of similar vintage capable of performing similar functions

A more sophisticated approach (but thus far stoutly resisted by the accounting profession) is to adopt **current cost accounting (CCA)**. Under CCA, assets are valued at their **replacement cost**, i.e. what it would cost the firm now to obtain assets of similar vintage. For example, if a machine cost £1 million five years ago, and asset prices have inflated at 10 per cent p.a., the cost of a new asset would be about £1.6 million, i.e. $£1m \times (1.10)^5$. The historical cost less five years' depreciation on a straight-line basis, and assuming a ten-year life, would be £0.5 million. However, the cost of acquiring an asset of similar vintage would be around £0.8 million.

There are obvious problems in applying CCA. For example, estimating current cost requires knowledge of the rate of inflation of identical assets, and of the impact of changing technology on replacement values. Nevertheless, the replacement cost measure is often far closer to a market value than historical cost less depreciation. Ideally, companies should revalue assets annually, but the time and costs involved are generally considered prohibitive.

Asset values may also fall. Directors are legally required to state in the annual report if the market value of assets is materially different from book value. It is better to 'bite the bullet' and actually reduce the value of poorly-performing assets in the accounts. In July 2007, Metronet, the five-firm consortium formed to upgrade London's underground system with capital of £350 million, collapsed owing over £2 billion, amidst allegations of management incompetence and poor cost control, with the job well short of completion. All five of its constituent shareholders made it clear that no extra funding would be forthcoming, but Bombardier, the Canadian conglomerate, went further by totally writing off its £70 million investment.

The highest write-off to date was the $50 billion write-down in 2003 by Worldcom (later renamed MCI) of assets acquired during an acquisition spree, following which several executives saw the inside of jails after convictions for false accounting. Write-offs are, in effect, an admission that profits have been overstated in the past, i.e. depreciation has been too low. Firms tend to increase write-offs during difficult trading times on the principle of unloading all the bad news in one go.

Listed UK firms adopted International Reporting Standards (IFRSs) in 2005, and no longer have to depreciate goodwill (the difference between the price paid for an acquisition and the book value of the assets acquired), but to carry out an annual 'impairment review'. The results of the switch to IFRSs could be remarkable. In January 2005, Vodafone, which has grown rapidly by acquisition, revealed that its loss of £1.88 billion for the six months ending September 2004 would have been shown as a profit of £4.5 billion under IFRSs.

2 Stock values are often unreliable

Under **Generally Accepted Accounting Practice (GAAP)**, stocks are valued at the lower of cost or net realisable value. Such a conservative figure may hide appreciation in the value of stocks, e.g. when raw material and fuel prices are rising. Conversely, in some activities, fashions and tastes change rapidly, and although the recorded stock value might have been reasonably accurate at the balance sheet date, it may look inflated some time later.

Vanishing stock values (USA: stock = inventory)

In March 2000, shares in New Economy powerhouse Cisco Systems Inc. peaked at $80. Cisco, whose remarkable growth was founded on making gear to power the internet, was now planning to re-focus on selling equipment to new-world telecoms companies planning to supplant 'dinosaurs' like AT&T.

Yet its customers were beginning to complain about long lead times for products. So Cisco entered into long-term supply contracts with suppliers and manufacturers to ensure the availability of customised components. But, already, the US economy was slowing down, reducing demand for Cisco's products. In April 2001, Cisco announced that sales for the current quarter were set to drop by 30 per cent, driving the share price down to a 52-week low of $13.63.

In May 2001, Cisco announced a third quarter loss of $2.7 billion, a loss struck after a write-down of excess stock by $2.2 billion, 70 per cent of this involving telecom gear and parts. The amount and the timing of the write-down surprised many. Cisco's inventory, valued at $4.1 billion for the quarter ending April 2001, was 65 per cent higher than the previous quarter's $2.5 billion, itself up from $1.3 billion a year earlier. Over the whole year, Cisco was clearly adding inventory that it knew it could not sell, given weak demand and rapid technological change. This raised the issue of why it had not disclosed any similar write-downs in previous quarters. The Cisco case clearly illustrates the folly of rapid stock-building of high-tech products based on suspect demand forecasts.

3 The debtors figure may be suspect

Similar comments may apply to the recorded figure for debtors. Not all debtors can be easily converted into cash, since debtors may include an element of dubious or bad debts, although some degree of provision is normally made for these.

The debtor collection period, supplemented by an ageing profile of outstanding debts, should provide clues to the reliability of the debtors position.

4 A further problem: valuation of intangible assets

Even if these problems can be overcome, the resulting asset valuation is often less than the market value of the firm. 'People businesses' typically have few fixed assets and low stock levels. Based on the accounts, several leading quoted advertising agencies and consultancies have tiny or even negative NAVs.

However, they often have substantial market values because the people they employ are 'assets' whose interactions confer earning power – the quality that ultimately determines value. This may be seen most clearly in the case of professional football clubs, few of which place a value for players on their Balance Sheets. Manchester United led the way in this respect when it valued its players prior to flotation on the market in 1991. There are 8 quoted football companies in the *Financial Times* listings, seven English and one Scottish.* Is your club shown in Table 12.2?

Valuation of brands

However, some other companies have attempted to close the gap between economic value and NAV by valuing certain intangible assets under their control, such as brand names.

---

*There used to be as many as 17 listed clubs, but their ranks have been thinned by takeover (e.g. Aston Villa) and worse (e.g. Leeds United).

Table 12.2 Football clubs quoted on the London Stock Exchange (as at 1 October 2007)

■ Birmingham City	■ Sheffield United
■ Celtic	■ Southampton Leisure Holdings
■ Millwall Holdings	■ Tottenham Hotspur
■ Preston North End	■ Watford Leisure

The brand valuation issue came to the fore in 1988 when the Swiss confectionery and food giant Nestlé offered to buy Rowntree, the UK chocolate manufacturer, for more than double its then market value. This generated considerable discussion about whether and why the market had undervalued Rowntree, and perhaps other companies that had invested heavily in brands, either via internal product development or by acquisition. Later that year, Grand Metropolitan Hotels (now Diageo) decided to capitalise acquired brands in their accounts, and were followed by several other owners of 'household name' brands, such as Rank-Hovis-McDougall, which capitalised 'home-grown' brands.

Decisions to enter the value of brands in balance sheets were partly a consequence of the prevailing official accounting guidelines, relating to the treatment of assets acquired at prices above book value, often termed 'goodwill'. These guidelines enabled firms to write off goodwill directly to reserves, thus reducing capital, rather than carrying it as an asset to be depreciated against income in the profit and loss account, as in the USA and most European economies. UK regulations allowed companies to report higher earnings per share, but with reduced shareholder funds, thus raising the reported return on capital, especially for merger-active companies. Such write-offs were stopped by a new accounting standard, FRS10, which also prevented capitalisation of 'home-grown' brands. (FRS10 obliged UK firms to follow US practice by depreciating goodwill. Under IFRSs, adopted by all listed UK firms, acquired goodwill only needs to be depreciated if there is judged to be a 'substantial impairment' in the value of the asset.)

Brand valuation raises the value of the intangible assets in the balance sheet and thus the NAV. Some chairpeople have presented the policy as an effort to make the market more aware of the 'true value' of the company. Under strong-form capital market efficiency, the effect on share price would be negligible, since the market would already be aware of the economic value of brands. However, under weaker forms of market efficiency, if placing a balance sheet value on brands provides genuinely new information, it may become an important vehicle for improving the stock market's ability to set 'fair' prices.

■ Methods of brand valuation

Many methods are available for establishing the value of a brand, all of which purport to assess the value to the firm of being able to exploit the profit potential of the brand.

1 Cost-based methods

At its simplest, the value of a brand is the historical cost incurred in creating the intangible asset. However, there is no obvious correlation between expenditure on the brand and its economic value, which derives from its future economic benefits. For example, do failed brands on which much money has been spent have high values? Replacement cost could be used, but it is difficult to estimate the costs of re-creating an asset without measuring its value initially. Alternatively, one may look at the cost of maintaining the value of the brand, including the cost of advertising and quality control. However, it is difficult to differentiate between expenditure incurred in merely maintaining the value of an asset and investment expenditure which enhances its value.

2 Methods based on market observation

Here, the value of the brand is determined by looking at the prices obtained in transactions involving comparable assets, for example, in mergers and acquisitions. This may be based on a direct price comparison, or by separating the market value of the company from its net tangible assets, or by looking at the P:E multiple at which the deal took place, compared to similar unbranded businesses. Although the logic is more acceptable, the approach suffers from the infrequency of transactions involving similar brands, given that individual brands are supposedly unique.

3 Methods based on economic valuation

In general, the value of any asset is its capitalised net cash flows. If these can be readily identified, this approach is viable, but it requires separation of the cash flows associated with the brand from other company cash inflows. The 'brand contribution method' looks at the earnings contributed by the brand over and above those generated by the underlying or 'basic' business. The identification, separation and quantification of these earnings can be done by looking at the financial ratios (e.g. profit margin, ROI), of comparable non-branded goods and attributing any differential enjoyed by the brand itself as stemming from the value of the brand, i.e. the incremental value over a standard or 'generic' product.

For example, if a brand of chocolates enjoys a price premium of £1 per box over a comparable generic product, and the producer sells ten million boxes per year, the value of the brand is imputed as (£1 $\times$ 10m) = £10m p.a., which can then be discounted accordingly to derive its capital value. Alternatively, looking at comparative ROIs as between the branded manufacturer and the generic, we may find a 5 per cent differential. If capital employed by the former is £100 million, this implies a profit differential of £5 million, which is then capitalised accordingly.

Such approaches beg many questions about the comparability of the manufacturers of branded and non-branded goods, the lifespan assumed, and the appropriate discount rate. Adjustments should also be made for brand maintenance costs, such as advertising, that result in cash outflows.

4 Brand strength methods

Other, more intuitive, methods have been devised which purport to capture the 'strength' of the brand. This involves assessing factors like market leadership, longevity, consumer esteem, recall and recognition, and then applying a subjectively determined multiplier to brand earnings in order to derive a value. Although appealing, the subjectivity of these approaches divorces them from commercial reality.

No broad measure of agreement has yet been reached about the best method to use in brand valuation, or whether the whole exercise is meaningful. Indeed, a report commissioned by the ICAEW (1989), which rejected brand valuation for balance sheet purposes, was said to have been welcomed by its sponsors. The report claimed that brand valuation 'is potentially corrosive to the whole basis of financial reporting', arguing that balance sheets do not purport to be statements of value!

■ The role of the NAV

Generally speaking, the NAV, even when based on reliable accounting data, only really offers a guide to the lower limit of the value of owners' equity, but even so, some form of adjustment is often required. Assets are often revalued as a takeover defence tactic. The motive is to raise the market value of the firm and thus make the bid more expensive and difficult to finance. However, the impact on share price will be minimal unless the revaluation provides new information, which largely depends on the perceived quality and objectivity of the 'expert valuation'.

Capturing the indefinable value of a brand

European companies find themselves having to value their intangibles.

Two-thirds of Coca-Cola's market value is attributable to one asset: the soft drink maker's brand. So said Interbrand, the consulting firm, last year when it ranked the Coca-Cola name as the world's most expensive at $67bn.

Its contribution to the company's worth is far from unusual. Brands, together with other intangibles such as customer relationships and technology, account for an ever-growing proportion of corporate value: 48 per cent, according to PwC research on the American M&A market in 2003.

European Union companies have not had cause to put detailed numbers on what makes them what they are. But that is now changing, because international accounting standards force acquirers to spell out, item by item, the value of the businesses they are buying.

That has created a new market for expertise from the US, where intangibles have been shown separately on balance sheets for several years. Two specialist groups are in expansion mode in Europe – American Appraisal and Standard & Poor's Corporate Value Consulting – while the big four accounting firms are plugging their services more heavily.

But as Sarpel Ustunel, senior manager at American Appraisal in London explains, there is no simple way to put a price on something 'that is difficult to put your arms around'.

Mr Ustunel, one of 200 staff in Europe, says there are several options with brands.

One method is to calculate what proportion of a company's future earnings can be attributed to its property, machinery and other assets. The rest should represent the value of the brand. But this assumes there are already neat values for the other intangibles.

Another way is to estimate how much it would cost to buy the brand if the company did not own it already.

Alternatively, and if possible, valuers look for the equivalent of two tins of soup made to exactly the same specifications, and sold on the same supermarket shelf – but one under a specialist mark and one under the supermarket's own label. 'Whatever the difference in price is attributable to the brand,' says Mr Ustunel.

Valuing intellectual property, too, is vexatious. If a patent for a similar technology has been sold before that price can be a starting point, he says, but such data is difficult to come by.

The solution is to talk to as many people as possible about the technology's importance. 'Engineers,' he cautions, 'can be overenthusiastic in explaining what their technology is about. Once you talk to the acquirer you may find they were unaware it existed.'

Valuing intangibles takes accounting, and the auditors who have to check financial statements, into a murky area. Given the need to make assumptions and estimates, Richard Winter, partner in valuation and strategy at PwC, concedes: 'There is a degree of rattle room.'

People may think there is a definitive answer, but inevitably there is scope for judgment.'

Critics say the whole exercise is misleading because it implies a precision that is not really there.

'The huge danger with going into inordinate detail is that readers of accounts cannot understand how the numbers arise,' says Ian Robertson, president of the Institute of Chartered Accountants of Scotland.

Mr Ustunel accepts there are no black and white answers, but says putting more numbers on the balance sheet is a useful step forward.

'Would you rather I tell you there are three cupboards, a table and a few chairs in this room,' he asks, 'or would you prefer just to know there is some furniture?'

Source: Barney Jopson, *Financial Times*, 9 February 2005.

We conclude that while the NAV may provide a useful reference point, it is unlikely to be a reliable guide to valuation. This is largely because it neglects the capacity of the assets to generate earnings. We now consider the commonest of the earnings-based methods of valuation, the use of price-to-earnings multiples.

12.4 VALUING THE EARNINGS STREAM: P:E RATIOS

price-to-earnings multiple/P:E ratio
Another way of expressing the PER

It is well known that accounting-based measures of earnings are suspect for several reasons, including the arbitrariness of the depreciation provisions (usually based on the historic cost of the assets) and the propensity of firms to designate unusually high items of cost or revenue as 'exceptional' (i.e. unlikely to be repeated in magnitude in future years). Yet we find that one of the commonest methods of valuation in practice is based on accounting profit. This method uses the **price-to-earnings multiple** or **P:E ratio**.

■ The meaning of the P:E ratio

As we saw in Chapter 2, the P:E ratio is simply the market price of a share divided by the last reported earnings per share (EPS). P:E ratios are cited daily in the financial press and vary with market prices. A P:E ratio measures the price that the market attaches to each £1 of company earnings, and thus (superficially at least) is a sort of payback period. For example, for its financial year 2006–7, Severn Trent Water plc reported EPS of 106p. Its share price on 19 July 2007 was 1355p, producing a P:E ratio of 12:8. Allowing for daily variations, the market seemed to indicate that it was prepared to wait about 13 years to recover the share price, on the basis of the latest earnings. So would a higher P:E ratio signify a willingness to wait longer? Not necessarily, because companies that sell at relatively high P:E ratios do so because the market values their perceived ability to grow their earnings from the present level. Contrary to some popular belief, a high P:E ratio does not signify that a company has done well, but that it is *expected* to do better in the future. (Not that they always do – witness the very high P:E ratios among 'dotcom' companies in 1999–2000.)

The P:E ratio varies directly with share price, but it also *derives from* the share price, i.e. from market valuation, so how does this help with valuation? Investment analysts typically have in mind what an 'appropriate' P:E ratio should be for particular share categories and individual companies, and look for disparities between sectors and companies. If, for example, BP is selling at a P:E ratio of say 12:1 with EPS of 50p, and Shell has EPS of 200p with a P:E ratio of 10:1, then their share values may look out of line. Assuming Shell's shares are correctly valued at $(10 \times 200p) = £20$, then BP's shares, priced at $(12 \times 50p) = 600p$, might appear overvalued.

Of course, there is a circularity here – this conclusion relies on the assumption that Shell rather than BP is correctly valued. Moreover, despite the apparent similarity of these two oil majors, there may be very good reasons why they should be valued differently. BP operates further 'upstream' (away from the final consumer) than Shell, and hence sustained upward pressure on oil prices would work to its advantage.

Using P:E ratios to detect under- or over-valuation implies that markets are slow or inefficient processors of information, but there are reliable, rough benchmarks that can be utilised. The industry benchmark is established by one or more transactions, against which other deals in the same industry can be judged, and exceptions identified. In some industries, analysts use benchmarks other than the earnings figure implicit in the P:E ratio. Some examples are multiples of billings in advertising, sale price per room in hotels, price per subscriber in mail order businesses, price per bed in nursing homes, and the more grisly 'stiff ratio' (value per funeral) in the undertaking business. At the height of the 'dotcom boom', some analysts attempted to explain the stratospheric valuations of internet companies in terms of number of 'hits' or visits to the site in question. More analysts are now utilising multiples based on cash flow. This development hints at the major problem with using P:E ratios – it relies on accounting profits rather than the expected cash flows which confer value on any item. We now consider cash-flow-oriented approaches to valuation.

Self-assessment activity 12.3

XYZ plc, which is unquoted, earns profit before tax of £80 million. It has issued 100 million shares. The rate of corporation tax is 30 per cent.

A similar listed firm sells at a P:E ratio of 15:1. What value would you place on XYZ's shares?

(Answer in Appendix A at the back of the book)

12.5 EBITDA – A HALFWAY HOUSE

Cash flows and profits differ due to application of accruals accounting principles, but value depends upon cash generating ability rather than 'profitability'. An intermediate concept currently in vogue is that of **EBITDA**, an unattractive acronym standing for **Earnings Before Interest, Taxes, Depreciation and Amortisation.** EBITDA is equivalent to operating profit with depreciation and amortisation (the writing-down of intangible assets) added back. As such, it is a measure of the basic operating cash flow before deducting tax, but ignoring working capital movements.

Many companies use EBITDA as a measure of performance, especially when related to capital employed. However, being a *performance* measure, it can only be used in valuation if we look at the way in which the market values other companies' EBITDAs. As with P:E ratios, comparison with other companies is needed as a reference point.

When BAT acquired the Turkish state tobacco firm Tekel in 2008, the *Financial Times* noted that, at 11 times the target's EBITDA, this compared well with the 12 times EBITDA that Japan Tobacco paid for Gallaher, and was 'well below Imperial Tobacco's price for Altadis of Spain'.

The 'relative valuation' implicit in the use of EBITDA can also be seen in the extract from the FT below.

Heidelberg/Hanson

How much? It must have taken Hanson's board all of three seconds to decide to recommend Heidelberg-Cement's £11 a share cash offer to shareholders.

This equates to an enterprise value for the UK building materials company of £9.5bn. That is 12.3 times trailing earnings before interest, tax, depreciation and amortisation. Deals in the sector have, on average, been done at 8.4 times over the past decade. Heidelberg is also paying almost twice Hanson's average multiple for the past six years. And Hanson was already expensive before Heidelberg's interest was made public on May 3 – at a share price 30 per cent lower than yesterday's offer.

Source: Financial Times, 16 May 2007.

Like a P:E multiple, an EBITDA multiple used in valuation stems from the value which the market attaches to other companies' EBITDAs, which invites the question of how it values those other companies, i.e. the EBITDA multiple is led by the valuation. Moreover, even when used crudely as a rough-and-ready comparison of value, one should appreciate that it is still based on accounting earnings. Although gross of depreciation and special items, it is still subject to different accounting practices between firms at the operating level, e.g. stock valuation.

Continuing to focus on income-generating methods, we now examine the genuine article, discounted cash flow.

12.6 VALUING CASH FLOWS

The value of any asset depends upon the stream of benefits that the owner expects to enjoy from his or her ownership. Sometimes these benefits are intangible, as in the case of Van Gogh's *Sunflowers*, which simply gives aesthetic pleasure to people looking at it. In the case of financial assets, the benefits are less subjective. Ownership of ordinary shares, for example, entitles the holder to receive a stream of future cash flows in the form of dividends plus a lump sum when the shares are sold on to the next purchaser, or if held until the demise of the company, a liquidating dividend when it is finally wound up. In the case of an all-equity financed company, the earnings over time should be compared on an equivalent basis by discounting them at the minimum rate of return required by shareholders or the **cost of equity capital** (henceforth denoted as k_e).

■ Valuing a newly-created company: Navenby plc

Navenby plc is to be formed by public issue of ten million £1 shares. It proposes to purchase and let out residential property in a prime location. It has been agreed that, after five years, the company will be liquidated and the proceeds returned to shareholders. The fully-subscribed book value of the company is £10 million, the amount of cash offered for the shares. However, this takes no account of the investment returns likely to be generated by Navenby. In the prospectus inviting investors to subscribe, the company announced details of its £10 million investment programme. It has concluded a deal with a builder to purchase a block of properties on very attractive terms, as well as instructing a letting agency to rent out the properties at a guaranteed income of £1.3 million p.a. Based upon past property price movements, Navenby's management estimate 70 per cent capital appreciation over the five-year period. All net income flows (after management fees of £300,000 p.a.) will be paid out as dividends.

In the absence of risk and taxation, Navenby is easy to value. Its value is the sum of discounted future expected cash flows (including the residual asset value) from the project i.e. (£1.3 m − £300,000) p.a., plus the eventual sale proceeds:

	Year				
	1	2	3	4	5
Net rentals p.a. (£m)	+1.0	+1.0	+1.0	+1.0	+1.10
Sale proceeds (£m)					+10.7

If shareholders require, say, a 12 per cent return for an activity of this degree of risk, the present value (PV) of the project is found using the relevant annuity* (PVIFA) and single payment* (PVIF) discount factors, introduced in Chapter 3, as follows:

$$PV = (£1.0\text{ m} \times PVIFA_{(12,5)}) + (£1.7\text{ m} \times PVIF_{(12,5)})$$
$$= (£1.0\text{ m} \times 3.6048) + (£10.7\text{ m} \times 0.5674)$$
$$= (£3.61\text{ m} + £9.64\text{ m}) = £13.25\text{ m}$$

The value of the company is £13.25 million and shareholders are better off by £3.25 million. In effect, the managers of Navenby are offering to convert subscriptions of £1 million into cash flows worth £13.25 million. If there is general consensus that these figures are reasonable estimates, and if the market efficiently processes new information, then Navenby's share price should be (£13.25 m/10 m) = £1.325 when information about the project is released. If so, Navenby will have created wealth of £3.25 million for its shareholders.

Self-assessment activity 12.4

Navenby has a value of £13.25 million, but a major part of this reflects the eventual resale value of the assets. What final asset value would enable investors to just break even?

(Answer in Appendix A at the back of the book)

---

*Remember the notation convention – interest rate first, time period second. Hence, $PVIF_{(12,5)}$ refers to the PV factor at 12% for 5 years.

■ The general valuation model (GVM)

general valuation model
A family of valuation models that rely on discounting future cash flows to establish the value of the equity or the whole enterprise

In analysing Navenby, we applied the **general valuation model**, which states that the value of any asset is the sum of all future discounted net benefits expected to flow from the asset:

$$V_o = \sum_{t=0}^{n} \frac{X_t}{(1 + k_e)^t}$$

where X_t is the net cash inflow or outflow in year t, k_e is the rate of return required by shareholders and n is the time period over which the asset is expected to generate benefits.

It should be noted that for a newly-formed company, such as Navenby, the valuation expression can be written in two ways:

value = cash subscription + NPV of proposed activities

or

value = present value of all future cash inflows less outflows

These are equivalent expressions. The value of Navenby is £13.25 million, and the net present value of the investment is £3.25 million, i.e. it would be rational to pay up to £3.25 million to be allowed to undertake the investment opportunity. Valuation of Navenby is relatively straightforward partly because the company has only one activity, but primarily because most key factors are known with a high degree of precision (although not the residual value). In practice, future company cash flows and dividends are far less certain.

■ The oxygen of publicity

Many corporate managers are somewhat parsimonious in their release of information to the market. Their motives are often understandable, such as reluctance to divulge commercially sensitive information. As a result, many valuations are largely based on inspired guesswork. The value of a company quoted on a semi-strong efficient share market can only be the product of what information has been released, supplemented by intuition.

Yet company chairpeople are often heard to complain that the market persistently undervalues 'their' companies. Some, for example Richard Branson (Virgin) and Andrew Lloyd-Webber (Really Useful Group), in exasperation, even mounted buy-back operations to repurchase publicly held shares. The 'problem', however, is often of their own making. The market can only absorb and process that information which is offered to it. Indeed, information-hoarding may even be interpreted adversely. If information about company performance and future prospects is jealously guarded, we should not be surprised when the valuation appears somewhat enigmatic.

12.7 THE DCF APPROACH

The previous section implies that we should rely on a discounted cash flow approach. After all, it is rational to attach value to future cash proceeds rather than to accounting earnings, which are based on numerous accounting conventions, including the deduction of a non-cash charge for depreciation. Given that depreciation is not a cash item, surely all we need do is to take the reported profit after tax (PAT) figure and add back depreciation to arrive at cash flow and then discount accordingly?

As a first approximation, we could thus value a company by valuing the stream of annual cash flows as measured by:

Cash flow = (operating profit + depreciation)

= (cash revenues − cash operating costs)

The depreciation charge is added back because it is merely an accounting adjustment to reflect the fall in value of assets. If firms did replace capacity as it expired, in principle, this investment should equate to depreciation. In practice, however, only by coincidence does the annual depreciation charge accurately measure the annual capital expenditure required to maintain production, and thus earnings capacity. Moreover, most companies need investment funds for growth purposes as well as for replacement. The value of growing companies depends not simply on the earning power of their existing assets, but also on their growth potential; in other words, the NPV of the cash flows from all future non-replacement investment opportunities.

This suggests a revised concept of cash flow. To obtain an accurate assessment of value, we should assess total ongoing investment needs and set these against anticipated revenue and operating cost flows; otherwise, we might over-value the company.

■ Valuation and free cash flow (FCF)

The inflow remaining net of investment outlays is referred to as **free cash flow** (i.e. 'free' of 'must-do' outlays such as interest, tax and investment). The most common definition of this is:

$$\text{Free cash flow} = [\text{revenues} - \text{operating costs}] - [\text{interest payments}] - [\text{taxes}]$$
$$+ [\text{depreciation}] - [\text{investment expenditure}]$$

Using this measure, the value of the owners' stake in a company is the sum of future discounted free cash flows:

$$V_o = \sum_{t=1}^{n} \frac{FCF}{(1 + k_e)^t}$$

Self-assessment activity 12.5

What is the free cash flow for the following firm?

Operating profit (after depreciation of £2m)	= £25m
Interest paid	= £1m
Tax rate	= 30%
Investment expenditure	= £3m

(Answer in Appendix A at the back of the book)

This approach removes the problem of confining investment financing to retentions, as in the **Dividend Growth Model**. However, we encounter significant forecasting problems in having to assess the growth opportunities and their financing needs in all future years.

Unfortunately, the accounting data for revenues and operating costs upon which this approach is based may fail to reflect cash flows due to movements in the various items of working capital. For example, a sales increase may raise reported profits, but if made on lengthy credit terms, the benefit to cash flow is delayed. Indeed, the net effect may be negative if suppliers of additional raw materials insist on payment before debtors settle.

It is important to mention another distortion. Stock-building, either in advance of an expected sales increase or simply through poor inventory control, can seriously impair cash flow, although the initial impact on profit reflects only the increased stock-holding costs.

For these and similar reasons, accurate estimation of cash flow involves forecasting not merely all future years' sales, relevant costs and profits, but also all movements in working capital. Alternatively, one may assume that these factors will have a net cancelling effect, which may be reasonable for longer-term valuations but much less appropriate for short time-horizon valuations, as in the case of high-risk activities. Figure 12.1 provides a schema to show the calculation of FCF, and how it relates to other cash flow concepts.

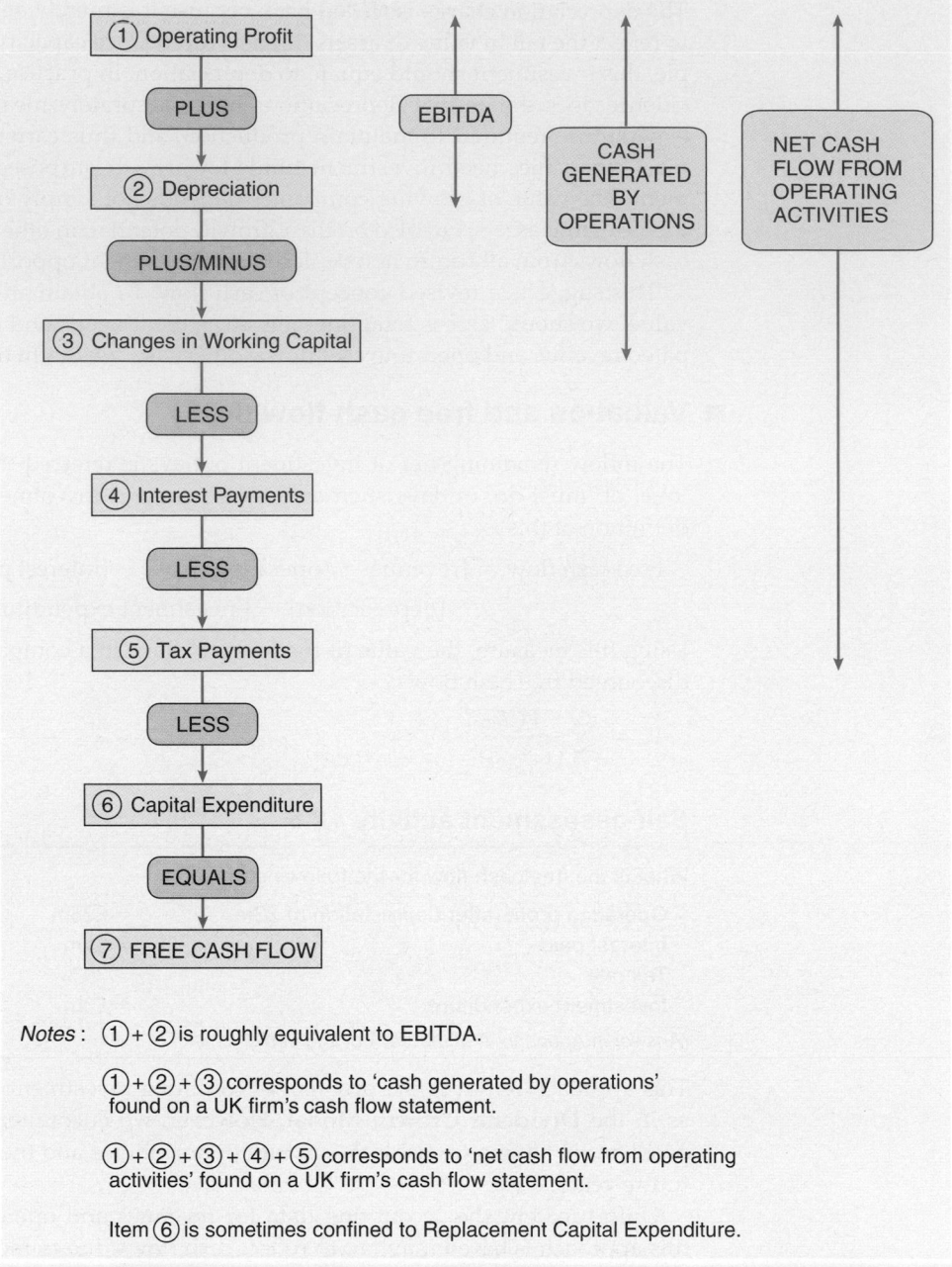

Notes : ① + ② is roughly equivalent to EBITDA.

① + ② + ③ corresponds to 'cash generated by operations' found on a UK firm's cash flow statement.

① + ② + ③ + ④ + ⑤ corresponds to 'net cash flow from operating activities' found on a UK firm's cash flow statement.

Item ⑥ is sometimes confined to Replacement Capital Expenditure.

Figure 12.1 Calculating free cash flow (FCF)

■ A warning!

The term 'free cash flow' is used in a wide variety of ways in practice. Here, we use it to signify cash left in the company after meeting all operating expenditures, all mandatory expenditures such as tax payments, and investment expenditure. It focuses on what remains for the directors to spend either as dividend payments, repayment of debts, acquisition of other companies or simply to build up cash balances. This broad definition is necessary because the cash inflow figure is defined to include revenues from both existing and future operations. Consequently, the investment expenditure required to generate enhancements in revenue must be allowed for. By the same

Table 12.3 D.S. Smith plc: cash flow

	2007 £m	2006 £m
Operating profit before exceptional items	77.7	60.4
Depreciation and amortisation	62.9	67.2
Adjusted EBITDA	140.6	127.6
Working capital movement	8.5	27.4
Exceptional cash costs	(7.2)	(4.6)
Other	(13.9)	(12.2)
Cash generated from operations	128.0	138.2
Capital expenditure payments	(55.8)	(62.7)
Sales of assets	41.0	13.2
Tax paid	(15.1)	(13.5)
Interest paid	(14.1)	(12.0)
Free cash flow	84.0	63.2
Dividends	(32.7)	(32.6)
Net (acquisitions)/disposals	0.2	0.5
Net cash flow	51.5	31.1
Shares issued	3.7	2.6
Net debt acquired	–	(2.6)
Non-cash movements	1.4	(6.1)
Net debt movement	56.6	25.0

Source: D.S. Smith plc, Annual Report and Accounts 2007.

token, a growth factor should be incorporated in the operating profit figures to reflect the returns on this investment.

A narrower definition could be used to confine cash inflows to those relating to existing operations and investment, and expenditures to those required simply to make good wear and tear, i.e. replacement outlays. This has the merit of expressing the cash flow before strategic investment, over which directors have full discretion. It also avoids financing complications, e.g. where a company wishes to invest more than its free cash flows, thus requiring additional external finance, which may distort the actual cash flow figure, as reflected in the cash flow statement.

The data in Table 12.3 relating to D.S. Smith plc are consistent with the first, broader definition, which is probably in widest use in the UK.

However, this yields a very restricted, static vision of the business, neglecting the strategic opportunities and their costs and benefits, which are truly responsible for imparting a major portion of value in practice. Failure to capture these longer-term strategic opportunities could yield a valuation well short of the market's assessment.

The problem of defining free cash flows is compounded by examination of UK company reports. Listed UK companies are obliged to present cash flow statements which report the net change in cash and near cash holdings over the year. This is a backward-looking statement which says more about past liquidity changes than future cash flows. Some firms do report a figure for 'free cash flow', but often without defining it. Jupe and Rutherford (1997) analysed the reports of 222 of the 250 largest listed UK companies. They found that just 21 disclosed a free cash flow figure, although only 14 used the term itself, and few of these supplied either a definition or a breakdown. Analysis of the comments of 13 companies appeared to reveal the use of 13 different definitions. Clearly, this is an area where care is required in definition and usage.

The following article concerning the Swedish manufacturer of mobile phones, Ericsson, highlights some of the accounting problems in measuring cash flow and in clearly distinguishing it from profit.

Ericsson's cash flow

Many investors talk a good game on cash flow but have little enthusiasm for examining it closely. But once a company gets a reputation for failing to convert accounting profits into money in the bank, it can be hard to lose it. Ericsson is grappling with this problem. Sales and profits have soared since the dark days after the telecoms bubble. But using Ericsson's own definition (which, broadly, compares underlying accounting profits with the cash flow available for capital expenditure, dividends and buybacks), cash conversion fell from a mediocre 72 per cent in 2005 to a poor 49 per cent in the first half of this year.

Defenders of Ericsson make three, entirely reasonable, points. First, as a manager of long-term complex projects, when Ericsson grows it is bound to bear upfront working capital outflows and provision movements. Second, these movements have balanced out in the past. Between 1996 and 2000, Ericsson suffered a €4.5bn cash outflow for these items, but this was more than recouped by the €8.1bn inflow between 2001 and 2003. Finally, Ericsson has been above board. It has improved its disclosure this year (although its reporting remains imperfect – it is impossible to reconcile operating profit with cash flow). Executive remuneration schemes treat cash flow as a key metric and are likely to place more emphasis on this.

Unfortunately, the fact that Ericsson's cash flow performance is understandable does not mean that it is irrelevant. The €5bn outflow for provisions and working capital since 2004 is a big number for a company with a market capitalization of €47bn. Most importantly, the company's own target is that cash conversion will rise towards 70 per cent. This implies a persistent gap between earnings and cash flow. There are plenty of reasons to admire Ericsson but for valuation purposes its profit and loss statement should be treated with real caution.

Source: Financial Times, 16 October 2007.

12.8 VALUATION OF UNQUOTED COMPANIES

The inexact science of valuing a company or its shares is made considerably simpler if the firm's shares are traded on a stock market. If trading is regular and frequent, and if the market has a high degree of information efficiency, we may feel able to trust market values. If so, the models of valuation merely provide a check, or enable us to assess the likely impact of altering key parameters such as dividend policy or introducing more efficient management.

With unquoted companies, the various models have a leading rather than a supporting role, but give by no means definitive answers. Attempts to use the models inevitably suffer from information deficiencies, which may be only partially overcome. For example, in using a P:E multiple, a question arises concerning the appropriate P:E ratio to apply. Many experts advocate using the P:E ratio of a 'surrogate' quoted company, one that is similar in all or most respects to the unquoted subject. One possible approach is to take a sample of 'similar' quoted companies, and find a weighted average P:E ratio using market capitalisations as weights.

However, the shares of a quoted company are, by definition, more marketable than those of unquoted firms, and marketability usually attracts a premium, suggesting a lower P:E ratio for the unquoted company. Any adjustment for this factor is bound to be arbitrary, and different valuation experts might well apply quite different adjustment factors.

Furthermore, a major problem in valuing and acquiring unquoted companies is the need to tie in the key managers for a sufficient number of years to ensure the recovery of the investment. The cost of such 'earn-outs', or '**golden handcuffs**', could be a major component of the purchase consideration.

golden handcuffs
An exceptionally good remuneration package paid to executives to prevent them from leaving

In principle, all the valuation approaches explained in this chapter are applicable to valuing unquoted companies, so long as suitable surrogates can be found, or if reliable industry averages are available. If surrogate data cannot be used, valuation becomes even more subjective. In these circumstances, it is not unusual to find valuers convincing themselves that company accounts are objective and reliable indicators of value. While accounts may offer a veneer of objectivity, we need hardly repeat the pitfalls in their interpretation.

12.9 SHAREHOLDER VALUE ANALYSIS

During the 1980s, based on the work of Rappaport (1986), an allegedly new approach to valuation emerged, called **shareholder value analysis (SVA)**. In fact, it is not really novel, but a rather different way of looking at value, based on the NPV approach.

The key assumption of SVA is that a business is worth the net present value of its future cash flows, discounted at the appropriate cost of capital. Many leading US corporations (e.g. Westinghouse, Pepsi and Disney) and a growing number of European companies (e.g. Philips, Siemens) embraced SVA because it provides a framework for linking management decisions and strategies to value creation. The focus is on how a business can plan and manage its activities to increase value for shareholders and, at the same time, benefit other stakeholders.

How is this achieved? Figure 12.2 shows the relationship between decision-making and shareholder value. Key decisions – whether strategic, operational, investment or financial – with important cash flow and risk implications are specified. Managers should focus on decisions influencing the **value drivers**, the factors that have greatest impact on shareholder value. Typically, these include the following:

> **value drivers**
> Factors that have a powerful influence on the value of a business, and the investors' equity stake

1 *Sales growth and margin.* Sales growth and margins are influenced by competitive forces (e.g. threat of new entrants, power of buyers and suppliers, threat of substitutes and competition in the industry). The balance between sales, growth and profits should be based not only on profit impact, but also on value impact.

2 *Working capital and fixed capital investment.* Over-emphasis on profit, particularly at the operating level, may result in neglect of working capital and fixed asset management. In Section 12.7, the free cash flow approach advocated using cash flows after meeting fixed and working capital requirements.

3 *The cost of capital.* A firm should seek to make financial decisions that minimise the cost of capital, given the nature of the business and its strategies. As will be seen later, this does not simply mean taking the source of finance that is nominally the cheapest.

4 *Taxation* is a fact of business life, especially as it affects cash flows and the discount rate. Managers need to be aware of the main tax impact on both investment and financial decisions. (This is not always negative).

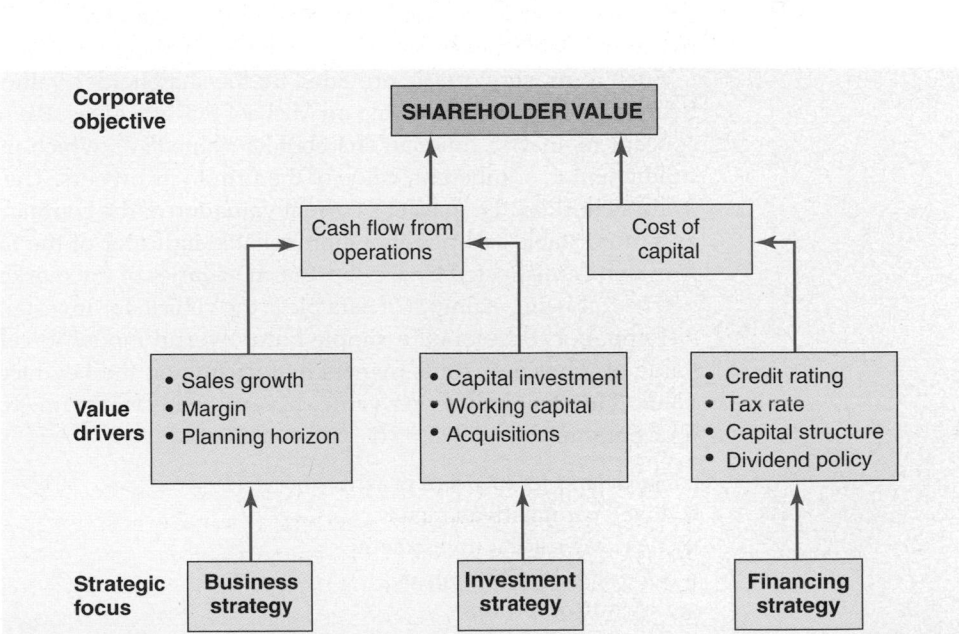

Figure 12.2 Shareholder value analysis framework

SVA requires specification of a planning horizon, say, five or ten years, and forecasting the cash flows and discount rates based on the underlying plans and strategies. Various strategies can then be considered to assess the implications for shareholder value.

A particular problem with SVA is specifying the terminal value at the end of the planning horizon. One approach is to try to predict the value of all cash flows beyond the planning horizon, based on that of the final year. Another is to simply take the value of the net assets predicted at the end of the horizon. None of the methods suggested is wholly satisfactory. It could be argued, however, that SVA does not have to be used to obtain the value of the business – rather, it can estimate the *additional* value created from implementing certain strategies. Assuming these strategies deliver competitive advantage, and therefore returns in excess of the cost of capital over the planning horizon, there is no need to wrestle with the terminal value problem.

The real benefit of SVA is that it helps managers focus on value-creating activities. Acquisition and divestment strategies, capital structure and dividend policies, performance measures, transfer pricing and executive compensation are seen in a new light. Short-term profit-related activities may actually be counter-productive in value-creation terms.

12.10 USING VALUE DRIVERS

As we have observed, in recent years, there has been much greater appreciation of the need for managers to optimise the interests of shareholders. In general terms, this can be achieved by generating a rate of return on investment which, at the very least, matches their required return on investment, i.e. the cost of equity. Remember that shareholders incur an opportunity cost when subscribing capital for firms to use and managers are legally obliged to safeguard those funds with all due diligence.

Sometimes, managers feel that 'their' companies are not 'correctly' valued by the stock market – there was a steady trickle of firms de-listing from the main UK market during 1998–2003, largely for this reason, although the recovery in the stock markets since 2003 arrested this trend. Moreover, share prices can swing quite violently in the short term, which tends to undermine managers' faith in market efficiency. Often nonplussed by such gyrations, both managers and shareholders may require a more objective and reliable measure of value than simply the prevailing market price.

Such a measure can be provided by the shareholder value approach, propounded by Alfred Rappaport, drawing on Michael Porter's ideas. We now use the value driver concept to analyse inherent shareholder value (SV), which may be thought of as the fundamental, or inherent, value of the firm to its owners. The SV figure also provides a cross-check on the market's current valuation of the company. This may be regarded as a more stable and possibly more reliable indicator of the fundamental value of the firm that is unaffected by the short-term vagaries of the market.

The following example of Safa plc is the vehicle for investigating the SVA approach.

Rappaport developed a simple but powerful model to calculate the fundamental value of a business to its owners by focusing on the key factors that determine firm value. He identified seven value drivers, comprising three cash flow variables and four parameters:

- Sales, and its speed of growth
- Fixed capital investment
- Working capital investment
- Operating profit margin
- Tax rate on profits
- The planning horizon
- The required rate of return

In its simplest form, SVA takes the last four drivers as given, and assumes that the first three, the cash flow variables, change at a constant rate. The key to the analysis, as with any budgeting exercise, is the level of sales and the projected rate of increase. From the sales projections, we can programme the operating profits and cash flows over the planning horizon and discount at a suitable rate to find their present value.

In the full model, the value of the firm comprises three elements, the value of the equity, the value of the debt and the value of any non-operating assets, such as marketable securities. However, to keep the analysis simple, we focus on an all-equity-financed company with no holdings of marketable assets. In addition, we need to explain the treatment of investment expenditure. To generate value, firms have to invest, i.e. to generate future cash flows requires preliminary cash outflows. These appear to reduce value in the short term but should generate a more than compensating increase in value via future cash flows.

■ Categories of investment

Investment in working capital, especially inventories, is required to support a planned increase in sales. Often, companies attempt to apply a roughly constant ratio of working capital to sales so that an X per cent sales increase needs an equivalent increase in working capital investment. This is called *incremental working capital investment*.

Replacement investment is undertaken to make good the wear and tear due to using equipment, or 'depreciation'. However, there are phasing issues to consider. In reality, in relation to particular items, the act of replacement is infrequent, occurring in discrete chunks, whereas depreciation in the accounts is an annual provision, so that in all but the year of replacement, depreciation will likely exceed replacement expenditure. However, taken in aggregate, and especially for larger firms, replacement may be closely related to depreciation provisions.

New investment in fixed assets. This has two dimensions. First, if the firm wants to expand sales of existing products, then, unless it has spare capacity, it will need to invest in additional capital equipment to support the planned sales increase. Second, new investment may be undertaken to accompany a major strategic venture such as the development of a new product, which will also generate an increase in sales. Taken together, these may be related to the planned sales increase, although there is likely to be a time lag before strategic investment comes fully 'on stream' and is able to deliver higher sales quantities. Notwithstanding this qualification, we can link the amount of investment in new capacity, for whatever reason, and which adds to the firm's stock of assets, to a planned increase in sales. We call the resulting sum the *incremental fixed capital investment*.

In the following demonstration example of Safa plc, replacement investment is assumed to equal depreciation provisions (which are treated as part of operating expenses in accounting statements), and both working capital investment and incremental fixed capital investment are made a percentage of any planned sales increase.

12.11 WORKED EXAMPLE: SAFA PLC

The board of Safa plc is concerned about its current stock market value of £95 million, especially as board members hold 40 per cent of the existing 100 million ordinary shares (par value £1) already issued. They are vaguely aware of the SVA concept and have assembled the following data:

Current sales	£100 million
*Operating profit margin**	20 per cent

(*After depreciation. On average, depreciation provisions are assumed to match ongoing investment requirements and are fully tax-deductible.)

Estimated rate of sales growth	5 per cent p.a.
Rate of corporation tax	30 per cent (with no delay in payment)
Long-term debt	zero
Net book value of assets	£120 million (net fixed assets plus net current assets)

To support the increase in sales, additional investment is required as follows:

(i) Increased investment in *working capital* will be 8 per cent of any concurrent sales increase.

(ii) Increased investment in *fixed assets* will be 10 per cent of any concurrent sales increase.

The risk-free rate of interest is 7.6 per cent, Safa's Beta coefficient is 0.8 and a consensus view of analyst's expectations regarding the overall return on the market portfolio is 15.6 per cent.

Safa presently pays out 20 per cent of profit after tax as dividend. The board estimate that Safa can continue to enjoy its traditional source of competitive advantage as a low cost provider for a further six years, at the end of which it estimates the net book value of its assets will be £140 million.

What is the inherent underlying value of this company?

Answer and comments

First of all, we need to find the return required by the shareholders of Safa, using the CAPM formula. This is:

$$k_e = R_f + \beta[ER_m - R_f]$$
$$= 7.6\% + 0.8[15.6\% - 7.6\%]$$
$$= (7.6\% + 6.4\%) = 14\%$$

This becomes the appropriate rate at which to value Safa's future cash flows. There is no debt finance so all operating profits (less tax) are attributable to shareholders. There appears to be no long-term strategic investment programme, and wear-and-tear is made good at a rate roughly corresponding to tax-allowable depreciation provisions. This means that free cash flows are equal to operating profits less tax.

The firm enjoys a temporary cost advantage for six years, beyond which cash flows are uncertain. Post-year-six cash flows can be handled in a number of ways:

1 The year six cash flow figure can be assumed to flow indefinitely. This seems quite an optimistic assumption to make both in relation to Safa plc and also more generally.

2 A view can be taken on the firm's efforts to restore competitive advantage and some growth assumption can then be incorporated. Again, this can only be speculative, as there is no information on this issue.

3 Perhaps the most prudent assumption to make is that the expected year six book value of assets will approximate to the value of all future cash flows, i.e. the company has no further supernormal earnings capacity. This implies that any subsequent investment has an NPV of zero.

We adopt the third approach mainly for simplicity.

Table 12.4 shows the cash flows over the 'competitive advantage period', years 1–6 inclusive. The base year (year 0) figures are given to establish a reference line from which future cash flows will grow.

Table 12.4 Cash flow profile[1] for Safa plc (ungeared)

£m	0	1	2	3	4	5	6, etc.
1 Sales (5% growth)	100	105	110.25	115.76	121.55	127.63	134.00
2 Operating profit margin @ 20%[2]	20	21	22.05	23.15	24.31	25.53	26.80
3 Taxation @ 30%	(6)	(6.30)	(6.62)	(6.95)	(7.29)	(7.66)	(8.04)
4 Incremental working capital investment @ 8% of sales increase		(0.40)	(0.42)	(0.44)	(0.46)	(0.49)	(0.51)
5 Incremental fixed capital investment @ 10% of sales increase		(0.50)	(0.53)	(0.55)	(0.58)	(0.61)	(0.64)
6 Free cash flow		13.80	14.48	15.21	15.98	16.77	17.61
7 Present value @ 14%	–	12.11	11.14	10.27	9.46	8.71	8.02

[1]Accuracy of figures influenced by rounding errors.
[2]These can be taken as operating cash flows given the assumption that depreciation = replacement investment.

■ Valuing Safa plc

Taking firstly the value created over the competitive advantage period (years 1–6):

PV of operating cash flows (line 7) = £59.5 m

Second, we add in the estimated residual value, our proxy for all future operating cash flows:

The PV of the residual value = £140 m $\times$ PVIF$_{14,6}$
$$= £140 \text{ m} \times 0.4556 = £63.8$$
Shareholder Value = £59.50 m + £63.8 m
$$= £123.3 \text{ m}$$

■ A note on taxation – two simplifications

You should appreciate how taxation is being handled in this example. All replacement investment is treated as being fully tax-deductible in the year of expenditure. This is a simplification adopted primarily for arithmetic convenience. In reality, the tax relief will be spread out over time as the firm claims the writing down allowance (WDA) each year. In addition, we have ignored the tax saving in relation to the 25 per cent WDA on the incremental fixed capital expenditure.

Correction for the first factor would reduce the valuation simply because delay in taking the tax relief would lower the PV of the stream of tax savings. On the other hand, inclusion of the second set of tax savings would raise the SV figure. If you calculate the 'true' valuation by allowing for these aspects, you will find a net increase in the valuation, although the calculation is a little messy.

We now turn to discuss the actual valuation obtained.

■ Commentary

Looking at the figures as calculated, we find, rather alarmingly, that a large proportion (52 per cent) of the SV is accounted for by the residual value. Moreover, the SV clearly exceeds market value £95 m, itself below the current book value of assets £120 m. This seems to imply that the company might be worth more if it were broken up (although the resale value of the assets may not fetch book value). It is thus possible that the market is valuing Safa for its break-up potential rather than as a going concern.

This raises the obvious question of why the market should place such an apparently low value on Safa. We can consider some possible reasons for the market under-valuation of Safa.

- The market may currently apply a higher discount rate, for example, seeking a higher reward for risk.
- The growth estimate may be regarded as optimistic.
- The flow of information provided to the market may be inadequate – for example, if it does have plans for future investment, are these generally known and under-stood, at least in outline?
- Board control – presumably reflecting domination by members of the founding family – may look excessive. Such enterprises rarely enjoy a good stock market rating, because there is often a suspicion that the interests of family members may be allowed to dominate those of 'outside' shareholders.
- The dividend policy may be thought ungenerous – a 20 per cent payout ratio is low by UK standards, and there appears to be little scope for worthwhile strategic investment. Retentions may simply be going into cash balances.
- There may be doubts about whether Safa can recover some form of competitive advantage.
- The market may be unimpressed with its present cost advantage-based strategy.
- Its gearing – currently, zero – may be thought to be too low. There is no tax shield to exploit (see Chapter 19).

Whatever the reason(s), there is plenty for the board to consider!

12.12 ECONOMIC VALUE ADDED (EVA)

Along with SVA comes another piece of 'alphabet spaghetti', EVA, a concept trade-marked by the US consultancy house Stern Stewart (**www.sternstewart.com**). Whereas SVA is a forward-looking technique devised for assessing the inherent value of the equity invested in a firm, EVA is backward-looking, i.e. a measure of past performance. Like SVA, EVA relies heavily on the concept of the cost of capital. It is used as a device for assessing how much value or wealth a firm actually has created. Its roots lie in the accounting concept of Residual Income (e.g. see Horngren *et al.* 1998), which is simply the accounting profit adjusted for the cost of using the capital tied up in an activity.

However, the Stern Stewart version is rather more sophisticated as it attempts to adjust the recorded profit in various ways. The logic of these adjustments is, broadly, to avoid recording as a cost the items that are value-creating and that should perhaps be treated as capital rather than current expenditure. For example, spending on R & D and on product advertising and promotion contributes to wealth-creation in important ways. In addition, any goodwill that has been written off in relation to previous acquisitions is added back. The general impact of these adjustments – over 150 of these might be required in a full EVA calculation – is to raise the profit measure and also the capital employed.

Relating this, for simplicity, to an all-equity-financed firm, EVA is calculated after making a further adjustment for the opportunity cost incurred by shareholders when entrusting their capital to the firm's directors. The EVA formula can be written as:

$$\text{EVA} = \text{NOPAT} - (k_e \times \text{invested capital})$$

where:

NOPAT = the Net Operating Profit After Tax, and after adjustment for the items mentioned above

k_e = the rate of return required by shareholders

Invested capital = Net assets, or shareholders' funds

Table 12.5 Calculation of EVA

	NOPAT	Equity	k_e	EVA
Firm A	£20 m	£100 m	15%	£20 m − £15 m = £5 m
Firm B	£10 m	£100 m	15%	£10 m − £15 m = (£5 m)

To illustrate the concept, consider the data in Table 12.5.

Both firms have the same equity capital employed of £100 m, and both make positive accounting profits. However, after adjusting for the cost of the equity capital employed, Firm B has effectively made a loss for investors, i.e. the negative EVA indicates that it has destroyed value.

On the face of it, EVA is a simple and powerful tool for assessing performance, explaining why it has been adopted by many firms as an internal performance measurement device, e.g. for determining the performance of different operating units.

However, it is by no means problem-free:

1 Few firms have the resources required to compute EVA, division by division, with the same degree of rigour as the full Stern Stewart model with its myriad required adjustments.
2 It is based on book value, rather than market values (necessarily so for business segments).
3 It relies on a fair and reliable way of allocating shared overheads across business units, the Holy Grail of management accountants.
4 It is difficult to identify the cost of capital for individual operating units.
5 It may be dysfunctional if managers are paid according to EVA, especially short-term EVA. It is quite possible to encounter investment projects that flatter EVA in the short term by virtue of high initial cash flows but to have a negative NPV. Such projects might be favoured by managers who are paid by EVA. Similarly, some long-term projects that take time and money to develop may lower EVA in the early years but have a positive NPV. These, of course, could be rejected under an EVA regime.

The verdict is yet to be delivered on EVA, but like many other management tools, it is probably inadequate when used alone – it is one way of looking at the picture that should be supplemented by other perspectives.

SUMMARY

We have discussed the reasons why financial managers may wish to value their own and other enterprises, the problems likely to be encountered and the main valuation techniques available.

Given the uncertainties involved in valuation, it seems sensible to compare the implications of a number of valuation models and to obtain valuations from a number of sources. A pooled valuation is unlikely to be correct, but armed with a range of valuations, managers should be able to develop a likely consensus valuation. This consensus is, after all, what a market value represents, based upon the views of many times more market participants. There should be no stigma attached to obtaining more than one opinion – doctors do not hesitate to call for second opinions when unsure about medical diagnoses.

Key points

- An understanding of valuation is required to appreciate the likely effect of investment and financial decisions, to value other firms for acquisition, and to organise defences against takeover.

- Valuation is easier if the company's shares are quoted. The market value is 'correct' if the EMH applies, but managers may have withheld important information.

- Using published accounts is fraught with dangers, e.g. under-valuation of fixed assets.

- Some companies attempt to value the brands they control. An efficient capital market will already have valued these, but not necessarily in a fully-informed manner.

- The economic theory of value tells us that the value of any asset is the sum of the discounted benefits expected to accrue from owning it.

- A company's earnings stream can be valued by applying a P:E multiple, based upon a comparable, quoted surrogate company.

- Some observers like to compare the EBITDA (Earnings Before Interest, Tax Depreciation and Amortisation) with share price for different companies as a cross-check on valuation. Market-based EBITDA multiples can be used as valuation tools.

- Valuing a company on a DCF basis requires us to forecast all future investment capital needs, tax payments and working capital movements.

- Valuation of unquoted companies is highly subjective. It requires examination of similar quoted companies and applying discounts for lack of marketability.

- Economic Value Added (EVA) is the residual profit after allowing for the charge for the firm's use of investors' capital.

- The two main lessons of valuation are: use a variety of methods (or consult a variety of experts), and don't expect to get it exactly right.

Further reading

The theory tells us that a company is worth the total amount of cash that it is expected to generate over its lifetime, discounted at the cost of capital to present value. But the theory is the easy part – the ongoing message in valuation is that it is a mix of theory and intuition. Wise birds will remember the words of Warren Buffet, the so-called Sage of Omaha:

> It is far better to buy a wonderful company at a fair price than a fair company at a wonderful price.

Therefore, the good books on valuation are those with a practical bent that look beyond the numbers. The best are reckoned to be the pair from the McKinsey stable, the two books by Koller *et al.* (2005) and by Copeland *et al.* (2000), and that by Damodoran (2002).

Another useful and easy to read book is by Frykman and Tolleryd (2003). Antill and Lee (2005) offer an accounting-based approach that explains how to allow for the effect of International Financial Reporting Standards.

The brand valuation issue is addressed by Murphy (1989) and by Barwise *et al.* (1989).

The text by Young and O'Byrne (2001) is a comprehensive primer on the application of EVA. Young (1997) provides a detailed practical example of the EVA concept related to a particular firm, and Klieman (1999) presents evidence on EVA generation in practice.

QUESTIONS

 *Questions with an icon are also available for practice in myfinancelab with additional supporting resources.*

Questions with a **coloured number** have solutions in Appendix B on page 743.

1 Amos Ltd has operated as a private limited company for 80 years. The company is facing increased competition and it has been decided to sell the business as a going concern.
The financial situation is as shown on the balance sheet:

Balance Sheet as at 30 June 1999

	£	£	£
Fixed assets			
Premises			500,000
Equipment			125,000
Investments			50,000
			675,000
Current assets			
Stock	85,000		
Debtors	120,000		
Bank	25,000		
		230,000	
Creditors: amounts due within one year			
Trade creditors	(65,000)		
Dividends	(85,000)		
		(150,000)	
Net current assets			80,000
Total assets less current liabilities			755,000
Creditors: amounts due after one year			
Secured loan stock			(85,000)
Net assets			670,000
Financed by			
Ordinary shares (50p par value)			500,000
Reserves			55,000
Profit and loss account			115,000
Shareholders' funds			670,000

The current market values of the fixed assets are estimated as:

Premises	780,000
Equipment	50,000
Investments	90,000

Only 90 per cent of the debtors are thought likely to pay.

Required
Prepare valuations per share of Amos Ltd using:
(i) Book value basis
(ii) Adjusted book value

2 The Board of Directors of Rundum plc are contemplating a takeover bid for Carbo Ltd, an unquoted company which operates in both the packaging and building materials industries. If the offer is successful, there are no plans for a radical restructuring or divestment of Carbo's assets.

Carbo's Balance Sheet for the year ending 31 December 2005 shows the following:

	£m	£m
Assets employed		
Freehold property		4.0
Plant and equipment		2.0
Current assets:		
stocks	1.5	
debtors	3.0	
cash	0.1	4.6
Total assets		10.6
Creditors payable within one year		(3.0)
Total assets less current liabilities		7.6
Creditors payable after one year		(1.0)
Net assets		6.6
Financed by		
Ordinary share capital (25p par value)		2.5
Revaluation reserve		0.5
Profit and loss account		3.6
Shareholders' funds		6.6

Further information:

(a) Carbo's pre-tax earnings for the year ended 31 December 2005 were £2.0 million.

(b) Corporation Tax is payable at 33 per cent.

(c) Depreciation provisions were £0.5 million. This was exactly equal to the funding required to replace worn-out equipment.

(d) Carbo has recently tried to grow sales by extending more generous trade credit terms. As a result, about a third of its debtors have only a 50 per cent likelihood of paying.

(e) About half of Carbo's stocks are probably obsolete with a resale value as scrap of only £50,000.

(f) Carbo's assets were last revalued in 1994.

(g) If the bid succeeds, Rundum will pay off the presently highly overpaid Managing Director of Carbo for £200,000 and replace him with one of its own 'high-flyers'. This will generate pre-tax annual savings of £60,000 p.a.

(h) Carbo's two divisions are roughly equal in size. The industry P:E ratio is 8:1 for packaging and 12:1 for building materials.

Required

(a) Value Carbo using a net asset valuation approach.

(b) Value Carbo using a price:earnings ratio approach.

3　Lazenby plc has been set up to exploit an opportunity to import a new product from overseas. It has issued two million ordinary shares of par value 25p, sold at a 25 per cent premium. Its projected accounts show the following annual operating figures:

Sales revenue	£500,000
Operating costs	(£300,000)
(after depreciation of £50,000)	
Operating profit	£200,000
Taxation @ 30%	(£60,000)
Profit after tax	£140,000

Notes:

(i) Shareholders require a return of 10 per cent p.a.

(ii) Replacement investment is financed out of depreciation provisions and is fully tax-allowable.

(iii) 2% of sales should be written off as bad debts.

(iv) Bad debt write-offs are 50 per cent tax-allowable.

Required

Value each share in Lazenby:

(a) assuming perpetual life.

(b) over a ten-year horizon.

4 The most recent balance sheet for Vadeema plc is given below. Vadeema is a stock market-quoted company that specialises in researching and developing new pharmaceutical compounds. It either sells or licenses its discoveries to larger companies, although it operates a small manufacturing capability of its own, accounting for about half of its turnover:

Balance Sheet as at 30 June 2005

Assets employed	£m	£m	£m
Fixed assets			
Tangible	50		
Intangible	120		170
Current assets			
Stock and work in progress	80		
Debtors	20		
Bank	5	105	
Current liabilities			
Trade creditors	(10)		
Bank overdraft	(20)	(30)	
Net current assets			75
10% loan stock			(40)
Net assets			205
Financed by			
Ordinary shares capital (25p par value)			100
Share premium account			50
Revenue reserves			55
Shareholders' funds			205

Further information:

1 In 2004–05, Vadeema made sales of £300 million, with a 25 per cent net operating margin (i.e. after depreciation but before tax and interest).

2 The rate of corporate tax is 33 per cent.

3 Vadeema's sales are quite volatile, having ranged between £150 million and £350 million over the previous five years.

4 The tangible fixed assets have recently been revalued (by the directors) at £65 million.

5 The intangible assets include a major patent (responsible for 20 per cent of its sales) which is due to expire in April 2006. Its book value is £20 million.

6 50 per cent of stocks and work-in-progress represents development work for which no firm contract has been signed (potential customers have paid for options to purchase the technology developed).

7 The average P:E ratio for quoted drug research companies at present is 22:1 and for pharmaceutical manufacturers is 14:1. However, Vadeema's own P:E ratio is 20:1.

8 Vadeema depreciates tangible fixed assets at the rate of £5 million p.a. and intangibles at the rate of £25 million p.a.

9 The interest charge on the overdraft was 12 per cent.

10 Annual fixed investment is £5 million, none of which qualifies for capital allowances:

Required

(a) Determine the value of Vadeema using each of the following methods:

    (i) net asset value

    (ii) price:earnings ratio

    (iii) discounted cash flow (using a discount rate of 20 per cent)

(b) How can you reconcile any discrepancies in your valuations?

(c) To what extent is it possible for the Stock Market to arrive at a 'correct' valuation of a company like Vadeema?

? 5 (a) The directors of Oscar plc are trying to estimate its value under its current strategy using a Shareholder Value Analysis framework. The last reported annual sales of Oscar plc were £30 million.

The key value drivers are estimated as follows:

Sales growth rate	7%
Operating profit margin (before tax)	10%
Corporation tax	30%
Fixed capital investment	15% of sales growth
Working capital investment	9% of sales growth
Planning period	6 years
Weighted average cost of capital	13%

Depreciation is currently charged on a reducing balance basis. The most recent charge was £0.5 m. This is expected to remain constant over the next few years. All depreciation is tax-allowable.

The dividend pay-out ratio is 20%.

Assume marketable securities held are £2.5 m, and debt (in the form of a bank loan) is £6 m (interest payable is at the rate of 8.33% p.a.).

Required

Calculate the overall company (or enterprise) value, *and* the shareholder value.
(Clearly state any assumptions that you make.)

(b) The market value of Oscar's equity is £25 m (lower than the directors' estimate of value), and also below the book value of net assets of £50 m.

Several directors argue that Oscar is undervalued by the stock market, and are wondering how to improve the firm's stock market rating.

Required

Suggest possible reasons for this apparent undervaluation, and evaluate suitable *financial* policies that Oscar's directors might adopt to enhance its value.

? 6 AB is a telecommunications consultancy based in Europe that trades globally. It was established 15 years ago. The four founding shareholders own 25% of the issued share capital each and are also executive directors of the entity. The shareholders are considering a flotation of AB on a European stock exchange and have started discussing the process and a value for the entity with financial advisers. The four founding shareholders, and many of the entity's employees, are technical experts in their field, but have little idea how entities such as theirs are valued.

Assume you are one of AB's financial advisers. You have been asked to estimate a value for the entity and explain your calculations and approach to the directors. You have obtained the following information.

Summary financial data for the past three years and forecast revenue and costs for the next two years is as follows.

Income Statement for the years ended 31 March

	Actual			Forecast	
	2004 **€ million**	**2005** **€ million**	**2006** **€ million**	**2007** **€ million**	**2008** **€ million**
Revenue	125.0	137.5	149.9	172.0	198.0
Less:					
Cash operating costs	37.5	41.3	45.0	52	59
Depreciation	20.0	22.0	48.0	48	48
Pre-tax earnings	67.5	74.2	56.9	72	91
Taxation	20.3	22.3	17.1	22	27

Other information/assumptions:

■ Growth in after tax cash flows for 2009 and beyond (assume indefinitely) is expected to be 3% per annum. Cash operating costs can be assumed to remain at the same percentage of revenue as in previous years. Depreciation will fluctuate but, for purposes of evaluation, assume the 2008 charge will continue indefinitely. Tax has been payable at 30% per annum for the last three years. This rate is expected to continue for the foreseeable future and tax will be payable in the year in which the liability arises.

Balance Sheet at 31 March

	2004 € million	2005 € million	2006 € million
Assets			
Non-current assets			
Property, plant and equipment	150	175	201
Current assets	48	54	62
	198	229	263
Equity and liabilities			
Equity			
Share capital (Shares of €1)	30	30	30
Retained earnings	148	179	203
	178	209	233
Current liabilities	20	20	30
	198	229	263

Note: The book valuations of non-current assets are considered to reflect current realisable values.

- The average P/E ratio for telecommunication entities' shares quoted on European stock exchanges has been 12.5 over the past 12 months. However, there is a wide variation around this average and AB might be able to command a rating up to 30% higher than this;
- An estimated cost of equity capital for the industry is 10% after tax;
- The average pre-tax return on total assets for the industry over the past 3 years has been 15%.

Required

(a) Calculate a range of values for AB, in total and per share, using methods of valuation that you consider appropriate. Where relevant, include an estimate of value for intellectual capital.

(b) Discuss the methods of valuation you have used, explaining the relevance of each method to an entity such as AB. Conclude with a recommendation of an approximate flotation value for AB, in total and per share.

(CIMA – Financial Strategy November 2006)

Practical assignment

Obtain the latest annual report and accounts of a company of your choice.* Consult the Balance Sheet and determine the company's net asset value.

- What is the composition of the assets, i.e. the relative size of fixed and current assets?
- What is the relative size of tangible fixed and intangible fixed assets?
- What proportion of current assets is accounted for by stocks and debtors?
- What is the company's policy towards asset revaluation?
- What is its depreciation policy?

Now consult the financial press to assess the market value of the equity. This is the current share price times the number of ordinary shares issued. (The notes to the accounts will indicate the latter.)

- What discrepancy do you find between the NAV and the market value?
- How can you explain this?
- What is the P:E ratio of your selected company?
- How does this compare with other companies in the same sector?
- How can you explain any discrepancies?
- Do you think your selected company's shares are under- or over-valued?

 *Now retake your diagnostic test for Chapter 12 to check your progress and update your study plan.*

---

*Most large companies post their Annual Reports and Accounts on their websites. The commonest address forms of UK companies are: companyname.co.uk or companyname.com.

Part IV

SHORT-TERM FINANCING AND POLICIES

The acquisition of every asset has to be financed. Companies obtain two forms of finance, short and long term, although, in practice, it is difficult to make a rigid demarcation between them. Part IV is devoted to analysing short-term financing, while the analysis of long-term financing decisions appears in Part V.

Chapter 13 offers an overview of the financing operations of the modern corporation, focusing on balancing the inflows and outflows of funds in the process of treasury management. The chapter examines how the financial manager may use the derivatives market to manage interest rate risk, and explores treasury policy for a variety of issues including working capital management.

Chapter 14 looks at managing short-term assets – cash, stocks, debtors – and the financing implications of different working capital policies. Chapter 15 describes the various forms of short- (and medium-) term sources of finance, especially trade credit and the banking system, and also discusses the analysis of leasing decisions and the finance of foreign trade.

13

Treasury management and working capital policy

Treasury management at D.S. Smith plc

The Group treasury strategy is controlled through a Treasury Committee, which meets regularly and includes the Chairman, the Group Chief Executive and the Group Finance Director. The Group Treasury function operates in accordance with documented policies and procedures approved by the Board and controlled by the Group Treasurer. The function arranges funding for the Group, provides a service to operations and implements strategies for interest rate and foreign exchange exposure management.

The major treasury risks to which the Group is exposed relate to movements in interest rates and currencies. The overall objective of the Treasury function is to control these exposures whilst striking an appropriate balance between minimising risks and costs. Financial instruments and derivatives may be used in implementing hedging strategies, but no speculative use of derivatives or other instruments is permitted.

The Treasury Committee regularly reviews the Group's exposure to interest rates and considers whether to borrow on fixed or floating terms. For the

last few years the Group has generally chosen to borrow on floating rates, which the Committee believes have provided better value. During the year, however, the Group took advantage of the historically low level of medium to long-term sterling interest rates and fixed the interest rate on £40 million of sterling denominated borrowings for a period of five years at an average rate (before margin) of just under 4%.

Group policy is to hedge the net assets of major overseas subsidiaries by means of borrowings in the same currency to a level determined by the Treasury Committee. The borrowings in currency give rise to exchange differences on translation into sterling, which are taken to reserves. A portion of the Group's net borrowings are denominated in euros, which are held to hedge the underlying assets of our eurozone operations. At the year end, these borrowings represented 64% of our eurozone net assets.

Reprinted with permission, D.S. Smith plc, Annual Report and Accounts, 2004.

Learning objectives

Treasury management and working capital policy are central to the whole of corporate finance. After reading this chapter, you should appreciate the following:

- The purpose and structure of the treasury function.
- Treasury funding issues.
- How to manage banking relationships.
- Risk management, hedging and the use of derivatives.
- Working capital policies.
- The cash operating cycle and overtrading problems.

 Complete your diagnostic test for Chapter 13 now to create your personal study plan.

13.1 INTRODUCTION

The introductory case study gives a flavour of the work of the Treasury in a large modern organisation (D.S. Smith is a leading packaging manufacturer). It also identifies some of the areas where things can potentially go wrong.

Treasury management, once viewed as a peripheral activity conducted by back-office boffins, today plays a vital role in corporate management. Most business decisions have implications for cash flow and risk, both of which are of direct relevance to treasury management. Many major firms have experienced problems through poor treasury management in recent years. This area has become a major concern in business, particularly the manner in which companies manage exposure to currency and other risks.

Most companies do not have a corporate treasurer; such a person is usually warranted only in larger companies. However, all firms are involved in treasury management to some degree. Treasury management can be defined in many ways. We will adopt the Association of Corporate Treasurers definition: 'the corporate handling of all financial matters, the generation of external and internal funds for business, the management of currencies and cash flows, and the complex strategies, policies and procedures of corporate finance.'

This chapter seeks to explain the main functions of treasury management and to provide an overview of working capital management. It also acts as an introduction to many of the succeeding chapters in this book.

13.2 THE TREASURY FUNCTION

The size, structure and responsibilities of the treasury function will vary greatly among organisations. Key factors will be corporate size, listing status, degree of international business and attitude to risk. For example, BP plc is a major multinational company with a strong emphasis on value creation, where currency and oil price movements can have a dramatic impact on corporate earnings. It is not surprising that it has a highly developed group treasury function, covering the following:

1 *Global dealing* – foreign exchange, interest rate management, short-term borrowing, short-term deposits.
2 *Treasury services* – cash management systems, transactional banking.
3 *Corporate finance* – capital markets, banking relationships, trade finance, risk management, liability management.
4 *M&A equity management* – mergers and acquisitions, equity markets, investor relations and divestitures (from *The Treasurer*, February 1992).

funding
Cash and liquidity management, short-term financing and cash forecasting

treasury operations
Financial risk management, and portfolio management

In most companies, the treasury department is much simpler, typically with a distinction between **funding** (cash and liquidity management, short-term financing and cash forecasting) and **treasury operations** (financial risk management and portfolio management). The 2007 survey of global treasury management (*The Treasurer*, 2008) found that cash management and cash forecasting are regarded the most important areas of treasury operations, followed by risk management. Treasury departments have come under increasing scrutiny by the financial press. Barely a month passes without some large company announcing hefty losses resulting from some major blunder by its treasury department. In the highly complex, highly volatile world of finance, there are bound to be mistakes; the secret is to set up the treasury function such that mistakes are never catastrophic.

It is the responsibility of the board of directors to set the treasury aims, policies, authorisation levels, risk position and structure. It should establish, for example, the following:

- The degree of treasury centralisation.
- Whether it should be a profit centre or cost centre.

UK TREASURER

International Consumer Products Group

West London

- UK headquartered consumer products group with wide range of household name brands. Turnover exceeds £3 billion from some 40 countries.

- Will be a member of a small team reporting to the Group Treasurer and will have responsibility for all banking, supported by a team of two.

- Principal activities will cover the dealing area; cash management systems and liaison with the Group's bankers; interest risk management, both forex and interest rate; and ad hoc projects including overseas banking reviews

- Graduate, part- or fully-qualified ACT with hands-on dealing room experience. Background is likely to be within a substantial international group. An accountancy qualification would be advantageous.

- Excellent communicator, able to quickly establish credibility and develop sound working relationships across the business. A team-worker with flexibility of approach, committed to technical excellence.

- This is a first-class opportunity within a group which has an excellent reputation for its pro-active approach to treasury management.

A typical job advertisement in the press

- The extent to which the company should be exposed to financial risk.
- The level of liquidity desired.

We pick up the last two points later, but deal with the structural issues in the following sections.

■ Degree of centralisation

Even in the most highly decentralised companies, it is common to find a centralised treasury function. The advantages of centralisation are self-evident:

1 The treasurer sees the total picture for cash, borrowings and currencies and is therefore able to influence and control financial movements on a global basis to achieve maximum after-tax benefit. The gains from centralised cash management can be considerable.

2 Centralisation helps the company develop greater expertise and more rapid knowledge transfer.

3 It permits the treasurer to capture any benefits of scale. Dealing with financial and currency markets on a group basis not only saves unnecessary duplication of effort, but should also reduce the cost of funds.

4 It enables the centre to cover a deficit in one area with surpluses from elsewhere, avoiding the costs of borrowing.

The major benefits from decentralising certain treasury activities are:

- By delegating financial activities to the same degree as other business activities, the business unit becomes responsible for all operations. Divisional managers in centralised treasury organisations are understandably annoyed at being assessed on profit after financing costs, over which they have little direct control.

- It encourages management to take advantage of local financing opportunities of which group treasury may not be aware and be more receptive to the needs of each division.

■ Profit centres and cost centres

In many large multinational firms, there is a substantial flow of cash each year in both domestic and foreign currencies. The volumes involved offer the opportunity to speculate, especially if the more favourable interest rates and exchange rates are available. Moreover, such firms probably employ staff skilled in cash and foreign exchange management techniques and may decide to use these resources pro-actively, i.e. to make a profit.

profit centre treasury (PCT)
A corporate treasury that aims to make a profit from its dealing – managers are judged on profit performance

In a **profit centre treasury (PCT)**, staff are authorised to take speculative positions, usually within clearly specified limits, by trading financial instruments in the same way as a bank. Such 'in-house banks' are judged on their return on capital achieved, although it is difficult to arrive at an accurate measure of capital employed. The main problem with operating a profit centre is that traders may exceed their permitted positions, either through negligence or in pursuit of personal gain.

cost centre treasury (CCT)
A treasury that aims to minimise the cost of its dealings

Conversely, a **cost centre treasury (CCT)** aims at operating as efficiently as possible, and eliminates risks as soon as they arise. D.S. Smith, the firm in the introductory case, clearly operates a CCT, i.e. it hedges rather than speculates, as a matter of policy.

Self-assessment activity 13.1

How would you define treasury management?

(Answer in Appendix A at the back of the book)

Let us now examine the four pillars of treasury management: funding, banking relationships, risk management, and liquidity and working capital.

13.3 FUNDING

Corporate finance managers must address the funding issues of: (1) how much should the firm raise this year, and (2) in what form? We devote two later chapters to these questions, examining long-, medium- and short-term funding. For the present, we simply raise the questions that subsequent chapters will pursue in greater depth.

1 *Why do firms prefer internally generated funds*? Internally generated funds, defined as profits after tax plus depreciation, represent easily the major part of corporate funds. In many ways, this is the most convenient source of finance. One could say it is equivalent to a compulsory share issue, because the alternative is to pay it all back to shareholders and then raise equity capital from them as the need arises. Raising equity capital, via the back door of profit retention, saves issuing and other costs. But, at the same time, it avoids the company having to be judged by the capital market as to whether it is willing to fund its future operations in the form of either equity or loans.

2 *How much should companies borrow*? There is no easy solution to this question. But it is a vital question for corporate treasurers. Borrow too much and the business could go bust; borrow too little and you could be losing out on cheap finance.

The problem is made no easier by the observation that levels of borrowing differ enormously among companies and, indeed, among countries. Levels of borrowing in Italy, Japan, Germany and Sweden are generally higher than in the UK and the USA. One reason is the difference in the strength of relationship between lenders and borrowers. Bankers in Germany and Japan, for example, tend to take a longer-term funding view than UK banks. Japanese banks may even form part of the same group of companies. For example, the Bank of Tokyo, one of Japan's leading banks, is part of the Mitsibushi conglomerate (**www.mitsibushi.com**). We devote Chapters 18 and 19 to the key question of how much a firm should borrow.

3 *What form of debt is appropriate*? If the strategic issue is to decide upon the level of borrowing, the tactical issue is to decide on the appropriate form of debt, or how to

manage the debt portfolio. The two elements comprise the capital structure decisions. The debt mix question considers:

(a) *form* – loans, leasing or other forms?
(b) *maturity* – long-, medium- or short-term?
(c) *interest rate* – fixed or floating?
(d) *currency mix* – what currencies should the loans be in?

The first three issues are discussed in Chapters 15 and 16 and currency issues are dealt with in Chapters 21 and 22.

4 *How do you finance asset growth*? Each firm must assess how much of its planned investment is to be financed by short-term finance and how much by long-term finance. This involves a trade-off between risk and return.
Current assets can be classified into:

(a) *Permanent current assets* – those current assets held to meet the firm's long-term requirements. For example, a minimum level of cash and stock is required at any given time, and a minimum level of debtors will always be outstanding.
(b) *Fluctuating current assets* – those current assets that change with seasonal or cyclical variations. For example, most retail stores build up considerable stock levels prior to the Christmas period and run down to minimum levels following the January sales.

Figure 13.1 illustrates the nature of fixed assets and permanent and fluctuating current assets for a growing business. How should such investment be funded? There are several approaches to the funding mix problem.

First, there is the **matching approach** (Figure 13.1), where the maturity structure of the company's financing exactly matches the type of asset. Long-term finance is used to fund fixed assets and permanent current assets, while fluctuating current assets are funded by short-term borrowings.

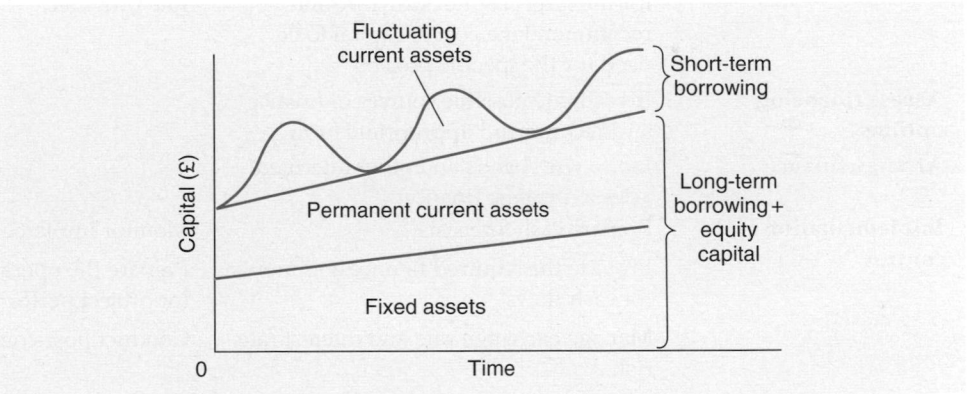

Figure 13.1 Financing working capital: the matching approach

A more aggressive and risky approach to financing working capital is seen in Figure 13.2, using a higher proportion of relatively cheaper short-term finance. Such an approach is more risky because the loan is reviewed by lenders more regularly. For example, a bank overdraft is repayable on demand. Finally, a relaxed approach would be a safer but more expensive strategy. Here, most if not all the seasonal variation in current assets is financed by long-term funding, any surplus cash being invested in short-term marketable securities or placed in a bank deposit.

Self-assessment activity 13.2

What do you understand by the matching approach in financing fixed and current assets?

(Answer in Appendix A at the back of the book)

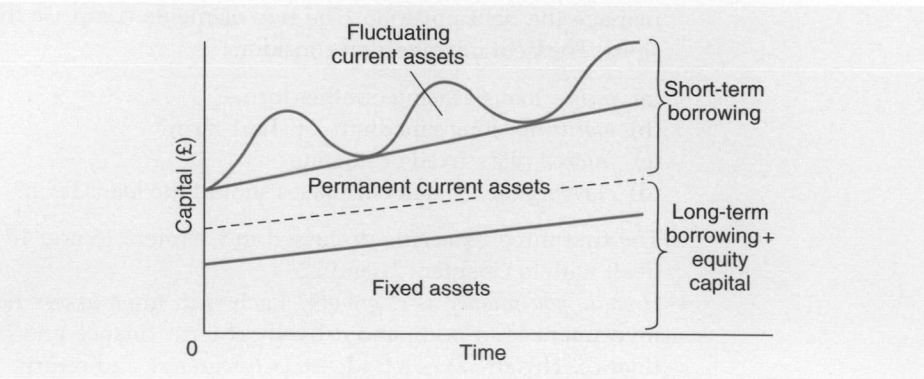

Figure 13.2 Financing working capital needs: an aggressive strategy

Example: Makepeace plc

Makepeace plc is a large UK multinational group. It is considering a major strategic capital project in Australia. What are the respective roles of the group's Treasury and Financial Control departments in appraising and implementing the new investment?

	Treasury	Financial Control
Project evaluation	Identify and quantify risks	Forecast operating costs and revenues
	Consider hedging or managing exchange rate and interest rate risks	Assess risk factors and conduct sensitivity or probability analysis
Discount rate	Advise on the costs of debt and equity for the target capital structure and recommend the cost of capital to be used for the specific project	Evaluate the project using the discount rate provided
Assess financing options	Investigate possible sources of finance and recommend appropriate form	
Arrange finance	Liaise with banks and other intermediaries to arrange finance	
Implementation and control	Prepare cash forecasts	Monitor implementation
	Provide the required finance and monitor cash flows	Prepare the operating budget and monitor project performance against budget
	Manage exchange rate and interest rate risk	Conduct post-completion audit
	Reinvest project cash inflows	

The issue of whether to borrow long term or short term is examined in more detail in the next section.

13.4 HOW FIRMS CAN USE THE YIELD CURVE

In Chapter 3, we examined the term structure of interest rates showing the yields on securities of varying times to maturity. The yield curve offers important information to treasury managers wanting to borrow funds. Although it is based on the structure of yields on government stock, similar principles apply to the market for corporate loans,

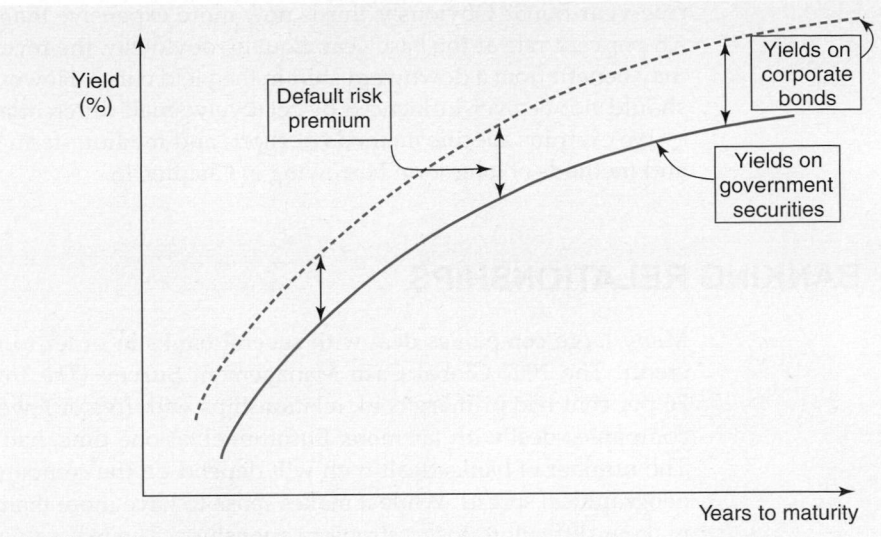

Figure 13.3 Yield curves

or bonds. However, corporates have higher default risk than governments so that markets require higher yields on corporate bonds.

The market for government stock provides a benchmark that dictates the general shape of the yield curve with the curve for corporate bonds located above this. Figure 13.3 reproduces Figure 3.2 with an additional yield curve to describe yields in the market for corporate bonds. The distance between the two lines represents the premium required by the market to cover the risk of default by corporate borrowers. For top-grade corporate borrowers, with a very high credit rating, the premium will be relatively narrow, whereas firms considered to be more risky will be subject to higher risk premia. The corporate versus government yield premium would usually widen with time to maturity as corporate insolvency risk probably increases with time.

Today's yield curve incorporates how people expect interest rates to move in the future. An upward-sloping yield curve reflects investors' expectations of higher future interest rates and vice versa. The action points are clear:

- A rising yield curve may be taken to imply that higher future interest rates are expected. This suggests firms might borrow long-term now, and avoid variable interest rate borrowing.
- A falling yield curve may be taken to imply that lower future interest rates are expected. This suggests firms might borrow short term now, and utilise variable interest rate borrowing.

■ Words of warning

In some circumstances, managers may be deceived by short-term rates. Say, they follow a policy of borrowing at short-term rates while the yield curve is upward-sloping, planning to switch to long-term borrowing when short-term rates exceed long-term rates.

For example, Jordan plc wants to borrow for six years, and the yield curve currently slopes upwards. The yields on five-year and six-year bonds are 5.5 per cent and 5.8 per cent respectively, while the yield on one-year bonds is 5.0 per cent. So, Jordan goes for one-year bonds, planning to issue a five-year bond a year later. But what if, a year later, the whole yield curve shifts upwards due to macroeconomic changes, e.g. a rise in the expected rate of inflation, so that Jordan has to pay say, 7.5 per cent on a five-year bond? Obviously, this is now more expensive than arranging to lock in the 5.8 per cent rate at the base year. Equally obviously, the reverse could apply – Jordan may benefit from a downward shift in the yield curve. However, the point is that firms should not be over-influenced by relatively small differentials along the yield curve.

We examine specific methods of short- and medium-term borrowing in Chapter 15 and methods of long-term borrowing in Chapter 16.

13.5 BANKING RELATIONSHIPS

Many large companies deal with several banks in order to maximise their access to credit. The 2007 Global Cash Management Survey (*The Treasurer*, 2008) found that 76 per cent had primary bank relationships with five or fewer banks. However, same companies deal with far more; Eurotunnel, at one time, had 225 banks to deal with! The number of banks dealt with will depend on the company's size, complexity and geographical spread. While it makes sense to have more than one bank, too many can make it difficult to foster strong relationships. The real value of a good banking relationship is discovered when things get tough and when continued bank support is required.

We often hear the charge, particularly from smaller businesses, that banks are providing an inadequate service or charging too high interest rates. It seems that the banking relationship can be more of a love/hate relationship than a healthy financial partnership.

A flourishing banking relationship requires the company to deal openly, honestly and regularly with the bank, keeping it informed of progress and ensuring there are no nasty surprises.

13.6 RISK MANAGEMENT

The financial manager should recognise the many types of risk to be managed:

- *Liquidity risk* – managing corporate liquidity to ensure that funds are in place to meet future debt obligations. We discuss this later in this chapter.
- *Credit risk* – managing the risks that customers will not pay. We discuss this in the next chapter.
- *Market risk* – managing the risk of loss arising from adverse movements in market prices in interest rates, foreign exchange, equity and commodity prices. It is this form of risk that we now consider.

Every business needs to expose itself to risks in order to seek out profit. But there are some risks that a company is in business to take, and others that it is not. A major company, like Ford, is in business to make profits from making cars. But is it also in business to make money from taking risks on the currency movements associated with its worldwide distribution of cars?

While the risks of business can never be completely eliminated, they can be *managed*. Risk management is the process of identifying and evaluating the trade-off between risk and expected return, and choosing the appropriate course of action.

With the benefit of hindsight, it is all too easy to see that some decisions were 'wrong'. In this sense, errors of commission are more visible than errors of omission; the decision to invest in a risky project which subsequently fails is more obvious than the rejected investment which competitors take up with great success. As with all aspects of decision-making, risk management decisions should be judged in the light of the available information when the decision is made. The treasurer plays a vital role in identifying, assessing and managing corporate risk exposure in such a way as to maximise the value of the firm and ensure its long-term survival.

Self-assesment activity 13.3

Take a look at the latest Annual Report of Cadbury Schweppes (**www.cadburyschweppes .com**). What does the Operating and Financial Review say about its treasury risk management policy?

■ Stages in the risk management process

Identify risk exposure. Taking risks is all part of business life, but businesses need to be quite sure exactly what risks they are taking. For example, while a firm will probably insure against the risk of fire, it may not consider the risk of loss of profits from the resulting disruption of the fire. The Brazilian coffee farmer could see his whole crop wiped out by a late frost. The UK fashion exporter could see her profit margins disappear because of the rising value of sterling against other currencies.

Before any attempt is made to cover risks, the treasurer should undertake a complete review of corporate risk exposure, including business and financial risks. Some of these risks will naturally offset each other. For example, exports and imports in the same currency can be netted off, thereby reducing currency exposure.

Evaluate risks. We saw in Chapter 7, that there are various ways in which the risks of investments can be forecast and evaluated. The decision as to whether the risk exposure should be reduced will depend on the corporate attitude to risk (i.e. its degree of risk aversion) and the costs involved. **Hedgers** take positions to reduce exposure to risk. **Speculators** take positions to increase risk exposure.

Manage risks. The treasurer can manage risk exposure in four ways: risk retention, avoidance, reduction and transfer, each of which is considered below.

hedgers
Hedgers try to minimise or totally eliminate exposure to risk

speculators
Speculators deliberately take positions to increase their exposure to risk, hoping for higher returns

1 *Risk retention.* Many risks, once identified, can be carried – or absorbed – by the firm. The larger and more diversified the firm's activities, the more likely it is to be able to sustain losses in some areas. There is no need to pay premiums to market institutions when the risk can easily be absorbed by the company. Firms may hold precautionary cash balances, or maintain lower than average borrowing levels, in order to be better able to absorb unanticipated losses. It should, of course, be borne in mind that there are costs associated with such action, particularly the lower return to the firm from holding such large cash balances.

2 *Risk avoidance.* Some businesses prefer to keep well away from high-risk investments. They prefer to stick to conventional technology rather than promising new technology manufacture, and to avoid doing business with countries with volatile exchange rates. Such risk-avoiding behaviour may be acceptable in the short term, but, ultimately, it threatens the firm's competitiveness and survival.

3 *Risk reduction.* We all know that by having a good diet and taking the right amount of exercise, we can reduce our risk exposure to a variety of health problems. Similarly, firms can reduce exposure to failure by doing the right things. Risk of fire can be reduced by an effective sprinkler system; risk of project failure can be reduced by careful planning and management of the implementation process and clear plans for abandonment at minimum cost should the need arise.

4 *Risk transfer.* Where a risk cannot be avoided or reduced and is too big to be absorbed by the firm, it can be turned into someone else's problem or opportunity by 'selling', or transferring, it to a willing buyer. Bear in mind that most risks are two-sided. There may be a speculator willing to acquire the very risk that the hedger firm wishes to lose. It is this area of risk transfer which is of particular importance to corporate finance. Whole markets and industries have developed over the years to cater for the transfer of risk between parties.

Risk can be transferred in three main ways:

- *Diversification.* We saw, in Chapter 9, that the risk exposure of the firm or share-holder can be considerably reduced by holding a diversified portfolio of investments. Diversification rarely eliminates all risk because most assets have returns positively correlated with the returns from other assets in the portfolio. It does, however, eliminate sufficient risk for the firm to consider absorbing the remaining risk exposure.
- *Insurance.* This seeks to cover downside risk. A premium is paid to the insurer to transfer losses arising from insured events but to retain any gains. As we saw in Chapter 8, financial options are a form of insurance whereby losses are transferred to others while profits are retained.
- *Hedging.* With hedging, the firm exchanges, for an agreed price, a risky asset for a certain one. It is a means by which the firm's exposure to specific kinds of risk can be reduced or 'covered'. Hence the fashion exporter can now enter into a contract guaranteeing an exchange rate for her exports to be paid in three months' time. Similar hedges can be created for risks in interest rates, commodity prices and many more transactions.

    Hedging has a cost, often in the form of a fee to a financial institution, but this cost may well be worth paying if hedging reduces financial risks. The extent to which an exposure is covered is termed the *hedge efficiency*: eliminating all financial risk is a 'perfect hedge' (i.e. 100 per cent efficiency).

Bako Ltd is a medium-size bakery business. The financial manager has identified that its main risk exposures lie in the following areas:

Risk exposure	Market hedge
Raw material prices – specifically, flour and sugar	Commodity
Currency movements on imports and exports	Currency
Interest rate movements on its variable-rate borrowings	Financial
Loss of profits, e.g. lost production from a possible bakery fire or a bad debt	Insurance

The first three risks can be managed through hedging in the commodity, currency and financial markets, letting the market bear the risks. The last can be covered through various forms of insurance.

■ Derivatives

The financial instruments employed to facilitate hedging are termed **derivatives,** because the instrument derives its value from securities underlying a particular asset, such as a currency, share or commodity. One of the earliest derivatives was money itself, which for centuries derived its value from the gold into which it could be converted. 'Derivative' has today become a generic term that is used to include all types of relatively new financial instruments, such as options and futures.

The esoteric world of derivatives has hardly been out of the news in recent years. Barings Bank, Metallgesellschaft and Kodak are all examples of major businesses whose corporate fingers have been burned through derivative transactions. Société Générale revealed in 2008 that it had lost a staggering €4.9 billion from allegedly rogue derivatives trading. Although sometimes viewed as instruments of the devil, derivatives are really nothing more than an efficient means of transferring risk from those exposed to it, but who would rather not be (hedgers), to those who are not, but would like to be (speculators).

Derivatives are financial instruments, such as options or futures, which enable investors either to reduce risk or speculate. They offer the treasurer a sophisticated

No future in futures for Barings?

When Nick Leeson was posted by the Barings group to work as a clerk at Simex, the Singapore International Monetary Exchange, who would have thought that he would eventually, apparently single-handedly, bring the famous bank to its knees?

He progressed well and by 1993 had risen to general manager of Barings Futures (Singapore), a 25-person operation that ran the bank's Simex activities. The original role of the operation was to allow clients to buy and sell futures contracts on Simex, but the group decided to focus on trading on its own account as part of its group strategy. In the first seven months of 1994, Leeson's department generated profits of US$30.7 million, one-fifth of the whole of Barings' group profits in the previous year.

The bank set up an integrated Group Treasury and Risk function to try to manage its risk exposure better. Leeson adopted a new strategy of buying and selling options (or 'straddles') on the Nikkei 225 index, paying the premium into a secret trading account. In effect, he was betting on the market not having sharp movements up or down. But on 17 January 1995, an earthquake hit Japan, causing immense damage and loss of life. It also led to a collapse of the Nikkei 225 index, exposing Barings to huge losses.

Leeson's response was to invest heavily in buying Nikkei futures contracts in an apparent attempt to support the market price. Some have suggested he was simply applying the traditional 'wisdom' of trying to salvage an otherwise hopeless position by a 'double-or-quits' approach. If so, the high-risk strategy backfired. The result is well known: Barings Futures (Singapore) lost £860 million for the group, leaving the group with no future and resulting in its acquisition by the Dutch bank Internationale Nederlandes Group (ING) for £1.

Nick Leeson left the following fax for his boss in London: 'Sincere apologies for the predicament that I have left you in.'

Was it the use of futures derivatives that brought Barings down? Derivatives were certainly involved, and it is hardly conceivable that such a disaster could have arisen from, say, share dealing. But it was the strategy and lack of controls – not the instrument – that were the real problems. To ban derivatives on the grounds that they are dangerous instruments would be akin to banning cars because they lead to more accidents than bicycles. But we all know that it is usually the person behind the wheel, not the car, that is at fault. Similarly, it is the derivatives trader and his or her trading strategy that are really the problem when spectacular collapses like that of Barings occur.

'tool-box' to manage risk. A risk management programme should reduce a company's exposure to the risks it is *not* in business to take, while reshaping its exposure to those risks it does wish to take. Risk exposure comes mainly in unexpected movements in interest rates, commodity prices and foreign exchange, all of which should be managed.

There are, essentially, four main types of derivative: forwards, futures, swaps and options.

Forward contracts

forward contract
An agreement to sell or buy at a fixed price at some time in the future

A **forward contract** is an agreement to sell or buy a commodity (including foreign currency) at a fixed price at some time in the future. In business, buyers and sellers are often subject to exactly opposite risks. The manufacturer of confectionery is concerned that the price of sugar may rise next year, while the sugar cane producer is concerned that the price may fall. In a world where it is extremely difficult to predict future commodity prices, both parties may want to exchange uncertain prices for sugar delivered next year for a fixed price.

By agreeing a price for sugar delivery next year, the confectionery manufacturer hedges against prices escalating, while the sugar cane producer hedges against prices dropping. They do this by entering into a forward contract, enabling future transactions and their prices to be agreed today, but not to be paid for until delivery at a specified future date.

Forward markets exist for most of the major commodities (e.g. cocoa, metals and sugar), but even more important is the forward market in foreign exchange.

A forward currency contract is when a company agrees to buy or sell a specified amount of foreign currency at an agreed future date and at a rate that is agreed in advance.

For example, if you want to pay US$50,000 in six months' time, you can use a forward contract to hedge against adverse currency movements. You can agree a price today that will pay for the dollars by arranging with your bank to buy dollars forward. At the end of six months, you pay the agreed sum and take delivery of the US dollars (see Chapter 21 for a fuller explanation).

Futures contracts

futures contract
A commitment to buy or sell an asset at an agreed data and price, and traded on an exchange

Like a forward contract, a **futures contract** is a commitment to buy or sell an asset at an agreed date and at an agreed price. The difference is that futures are standardised in terms of period, size and quality and are traded on an exchange. In the UK, this is the London International Financial Futures and Options Exchange (LIFFE).

A chemical company plans to buy crude oil in three months' time. The spot price (i.e. current market price) for Brent crude is $80 a barrel and a three-month futures contract can be agreed at $82 a barrel. To guard against the possibility of an even higher price rise, the company enters a 'long' futures position (i.e. agrees to buy) at $82 a barrel, thereby reducing its exposure to oil price hikes. If, in three months' time, the spot oil price has risen beyond $82, the company will not suffer unforeseen losses.

If, however, just before delivery, the spot price has fallen to $76 a barrel, the company will want to benefit from the lower price. It will buy at the spot price and cancel the long contract by entering into a short contract (i.e. an agreement to sell) at around the $76 spot price. The loss of $6 a barrel on the two contracts is offset by the profit of $6 from buying at the spot rather than the original futures price.

Why might a company prefer a futures contract when a forward contract could be tailor-made to meet its specific requirements? The main reason lies in the obvious benefits from trading through an exchange, not least that the exchange carries the default risk of the other party failing to abide by the contract terms, so-called 'counterparty risk'. For this benefit, both the buyer and seller must pay a deposit to the exchange, termed the 'margin'.

Financial futures have become highly popular among both hedgers and traders, who buy or sell futures in order to profit from a view that the market will go up or down. The main forms of financial futures contracts cover short-term interest futures, bond futures and equity-linked futures using stock market indices.

Swaps

Swaps are arrangements between two firms to exchange a series of future payments. A swap is essentially a long-dated forward contract between two parties through the intermediation of a third party, such as a bank. For example, a company might agree to a currency swap, whereby it makes a series of regular payments in yen in return for receiving a series of payments in US dollars.

An interest rate swap effectively allows a firm to change between fixed and floating rate loans. In the first half of 2007, there was $350 billion of swap contract outstanding (**www.isda.org**).

Options

An **option** gives the right, but not the obligation, to buy or sell an asset at an agreed price at, or up to, an agreed time. It is this right not to exercise the option that distinguishes it from a future. We discussed options in Chapter 8.

A farmer has a ripening crop which he plans to sell in September. He would like to benefit from any price movements but also be 'insured' against any fall in price. A put option (i.e. the right to sell at an agreed price) is rather like insurance. If the price falls, the option to sell at an agreed price is exercised. If the price rises, the option is not exercised, and the spot price at the date of sale is taken.

Example: Interest rate swap

X and Y are companies each looking to borrow £1 million for two years. X wants a floating rate loan while Y is looking for a fixed rate loan. X has a stronger credit rating than Y. The table below shows that X can borrow in the fixed rate market 1.5 per cent more cheaply than Y, but only 0.5 per cent more cheaply in the floating rate market. It therefore is considering a standard interest rate swap whereby X can swap interest payment on its floating rate loan for Y's fixed rate loan interest payments. This produces a total gain of 1 per cent (i.e. 1.5% − 0.5%). The table also shows the rates quoted by a swap dealer, the difference being the dealer's margin. We now must establish how the 1 per cent gain is to be shared between the two companies and the swap dealer.

	Floating	Fixed
X	LIBOR + 0.5%	7%
Y	LIBOR + 1.0%	8.5%
Swap quote	LIBOR% − LIBOR + 0.1%	6.8% − 6.9%

X

Borrows fixed rate at 7% and swaps at 6.8%
This gives a loss of 0.2%
Swaps floating rate at LIBOR + 0.1% giving a gain on floating of 0.4 on the borrowing
    rate (LIBOR + 0.5%)
Net saving = 0.2%

Y

Borrows floating rate at LIBOR + 1% and swaps at LIBOR.
This gives a gain of 1%.
Swaps at 6.9% which is 1.6% less than the cost of a fixed rate loan.
Net saving = 0.6%
The swap dealer makes 0.1% on the two swaps. The total gain of 1% on the swap is
    therefore 0.2% for X + 0.6% for Y + 0.2% for dealer = 1%.

Self-assessment activity 13.4

(a) A futures contract can be customised to fit the particular needs of the customer. True or false?

(b) A currency swap can be used to hedge for a longer period than that offered by forward exchange contracts. True or false?

(c) Max International plc is due to receive €100,000 in three months' time for exports to a German customer and has decided to hedge the currency by taking a forward contract. The following rates have been quoted:

	Spot rate	Three-month forward
Euro per £	1.4925–1.4985	1.4892–1.4898

What is the sterling value that the company will receive from the forward contract?

(d) Consider the following example of a company which plans to buy aluminium. It enters into a call option contract, paying an appropriate premium for the right to buy aluminium at $1,500/tonne in three months' time. If, at the end of the period, the spot price is $1,400/tonne, should the company exercise its option or let it lapse?

(Answers in Appendix A at the back of the book)

■ To hedge or not to hedge

Does hedging enhance shareholder value? Some argue that it helps firms achieve competitive advantage over rivals by cost-effectively reducing risks over which it has little experience and exploiting those risks over which it has strong levels of competence. Pure theorists, on the other hand, argue that corporate hedging is a costly process doing no favours for shareholders. After all, portfolio diversification by investors is one form of hedging. Corporate hedging does nothing that shareholders could not do themselves, employing derivatives in exactly the same way as corporate treasurers to follow their own risk management strategy. So why do many large companies hedge and use derivatives? Consider the following case study.

Yourfired plc is a global recruitment agency with exposure to macroeconomic risks concerning interest and exchange rates. The treasury function is considering who should manage such risks. Should it be the group treasury manager or left to the shareholders investing in the company?

The treasurer rehearsed the arguments for leaving risk management to portfolio managers rather than at the firm level. First, market, or systematic, risk is lower than total risk and therefore it will incur lower transaction costs. Second, companies are less experienced than portfolio managers in using potentially dangerous derivative instruments. There is therefore a good case for firms leaving risk management – particularly of the kind that employs derivatives – to the market, allowing large investors to decide whether and how to manage their portfolio risk.

However, the treasurer then recognised that this ignores some important issues.

1 These risks can bankrupt the company, and the realised value of a failed business is far lower than its value as a profitable going concern. All stakeholders, including investors, have a vested interest in the company's long-run prosperity.
2 Investors prefer steady earnings and cash flows to the more volatile ones likely when currency and interest rates are not managed.
3 The company's management may be reluctant to disclose to investors the full picture on risk exposure, making it difficult for them to make an accurate risk assessment on their portfolios.

The treasurer of Yourfired plc eventually decided that the company's hedging policy would be one of eliminating the worst 'down-side risk', but not attempt to hedge all risks. Whatever the risk management strategy, it is important that the treasurer explains to senior management what has been done and what risk exposure remains.

■ Interest rate management

Every company is exposed to a degree of interest rate risk. This occurs when changes in the interest rate affect a company's profits and/or the value of its assets and liabilities. The nature of the exposure depends on whether the company is a net borrower or net investor. Where a company with a high proportion of floating rate debt is exposed to the risk of rate increases it will adversely affect the volatility of cash flows. Conversely, a company which has a high proportion of fixed interest rate debt would suffer from a loss of competitive advantage relative to floating rate borrowers if the prevailing interest rate falls.

The first form of interest rate risk is *basis risk* – the risk that the level of interest rates will change. A second form of risk relates to changes in the yield curve over time, as discussed earlier, and refers to differences in short- and long-term interest rates. The normal, positive yield curve arises where interest rates increase as the term lengthens. In practice, however, the curve can be flat or even inverted.

Steps to manage interest rate exposure

The treasurer needs to understand the company's interest rate risk exposure, how it is likely to change over time and, where any of these exposures are compensating, how they can be netted off against each other. The three-step process involves:

1 Identify the expected future cash flows that are exposed to interest rate fluctuations.
2 Specify those rates of interest beyond which steps must be taken to reduce exposure.
3 Reduce exposure by:
  – *Natural (or internal) hedging*. This seeks to match liabilities against assets with similar interest rates. For example, an exposure to pay a rate of interest on a loan may be partially offset by an investment linked to the same or a similar rate.
  – *Fixing the interest rate*. Loans can be taken out at a fixed rate rather than a floating rate.
  – *Interest rate swaps*. This is an arrangement whereby two parties agree to exchange interest payments with each other over an agreed period. In other words, Company A agrees to pay the interest on Company B's loan, while Company B reciprocates by paying the interest on Company A's loan. Of course, what they are really swapping is the different characteristics of the two loans. The most common characteristic being exchanged is the fixed or variable interest rate, and this swap is termed a *plain vanilla* or *generic swap*.
     Heavy dependence upon short-term borrowing not only increases the risk of insolvency from funding long-term assets with short-term borrowing, but also exposes the company to short-term interest rate increases.
  – *Hedging contracts*. The corporate treasurer has a variety of techniques available to reduce interest rate risk, many of which have already been discussed. The main methods are forward rate agreements (FRAs), interest rate futures, interest rate options, interest rate swaps and more complex methods, such as options on interest rate swaps ('swaptions'). We are more concerned with the principles of interest rate management than the detailed application. The following example illustrates an approach to managing interest rates.

Table 13.1 summarises the findings of an international study of use of derivatives based on over 7,000 listed firms in 50 countries. The total picture indicates that forward contracts are most popular for foreign exchange hedging (37 per cent of firms), while swaps are most common for interest rate management (29 per cent). Japanese companies have by far the highest proportion of firms using derivatives, while German companies are far more reluctant to employ derivatives.

Table 13.1 International comparison of financial derivative usage

Country	Foreign exchange derivatives			Interest rate derivatives		
	Forward %	Swap %	Option %	Forward %	Swap %	Option %
Australia	48	9	18	4	39	15
Germany	27	11	13	2	18	10
Japan	71	33	18	1	60	14
UK	49	17	8	1	32	11
USA	31	6	7	1	36	7
Others	36	11	11	2	19	5
All firms	37	11	10	1	29	7

Source: Bartram *et al.* (2006).

Managing interest rate risk at MedExpress Ltd

It was Karen Bailey's first day as the financial controller of MedExpress Ltd, a fast-growing business in the medical support industry. A quick look at the balance sheet revealed that the company, although highly profitable, was heavily geared, with large amounts of debt capital repayment due over the coming years. Interest rates had changed little over the past two years, but opinions were divided over whether the Bank of England would have to raise interest rates quite steeply in order to keep inflation within prescribed government limits, or whether rates would hold, or even fall, to stimulate exports currently suffering from the strength of sterling.

To Bailey's surprise, the company had taken no steps to manage its exposure to interest rate movements. Her first step was to identify the exposure to interest rate risk.

1 A £2 million overdraft, with a variable interest rate, would have a significant impact on profits and cash flow if the rate increased in the near future. If the interest rate rise was dramatic, it could seriously affect cash flows and increase the risk of liquidation.
2 The £5 million fixed-rate long-term loan would become much less attractive if interest rates fell. Paying unduly high interest rates adversely affects profitability.
3 £1.8 million of the fixed rate loan would mature shortly and need replacing. The company could choose to repay the loan at any time over the next two years. If rates were expected to rise over that period, early redemption would be preferable.

As Bailey sought to get a grip on the interest rate exposure, she considered the following ways of managing interest rate risk:

(a) *Interest rate mix.* A mix of fixed and variable rate debt to reduce the effects of unanticipated rate movements. She would need to give more thought to whether the existing ratio of £2 million variable/£5 million fixed rate debt was sufficiently well balanced.

(b) *Forward rate agreement (FRA).* Some risk exposure could be eliminated by entering into a forward rate agreement with the bank. This would lock the company into borrowing at a future date at an agreed interest rate. Only the difference between the agreed interest that would be paid at the forward rate and the actual loan interest is transferred.

(c) *Interest rate 'cap'.* It is possible to 'cap' the interest rate to remove the risk of a rate rise. If the cap is set at 11 per cent, an upper limit is placed on the rate the company pays for borrowing a specific sum. Unlike the FRA, if the rate falls, the company does not have to compensate the bank.

(d) *Interest rate futures.* These contracts enable large interest rate exposures to be hedged using relatively small outlays. They are similar in effect to FRAs, except that the terms, the amounts and the periods are standardised.

(e) *Interest rate options.* Also termed interest rate guarantees, these contracts grant the buyer the right, but not the obligation, to deal at a specific interest rate at some future date.

(f) *Interest rate 'swaps'.* These occur where a company (usually very large firms) with predominantly variable rate debt, worried about a rise in rates, 'swaps' or matches its debt with a company with predominantly fixed-rate debt concerned that rates may fall. A bank usually acts as intermediary in the process, but it can be through direct negotiations with another company. Each borrower will still remain responsible for the original loan obligations incurred. Typically, firms continue to pay the interest on their own loan and then, at the end of the agreed period, a cash adjustment will be made between the two parties to the swap agreement. Interest rate swaps can also involve exchanges in different currencies.

Example: Forward rate agreement (FRA)

A forward rate agreement (FRA) is a cash-settled forward contract on a short-term loan. No loan is actually given and contracts settle with a single cash payment on the first day of the underlying loan, termed the settlement date.

Frodo plc needs to raise a £1 million 6-month fixed rate loan from 1 November. On 31 July, it enters into an FRA with a bank that fixes the rate of interest for borrowing for 6 months from 1 November, the relevant rate at that date being 6 per cent. This is termed a '3–9' FRA: it starts in 3 months' time and lasts for 6 months. What is the result of the FRA if the rate has moved to:

(a) 5 per cent?
(b) 8 per cent?

Answer

If the effective loan rate moves to 5 per cent, the company will pay the bank:

	£
FRA payment £1 m × (6% − 5%) × 6/12	(5,000)
Payment on underlying loan 5% × £1 m × 6/12	(25,000)
Net payment on loan	30,000
Effective interest rate	6%

If the effective loan rate moves to 8%, the bank will pay the company:	£
FRA receipt £1 m × (8% − 6%) × 6/12	10,000
Payment on underlying loan at market rate 8% × £1 m × 6/12	(40,000)
	(30,000)
Effective interest rate	6%

Interest rate futures can be employed to hedge against changes in short-term interest rates for up to approximately two years. They are similar to FRAs, except that the terms, amounts and periods are standardised. This can make them less flexible as a hedging instrument because the corporate treasurer cannot always match them with specific interest rate exposures.

Lenders will want to hedge against the possibility of falling interest rates which would reduce the interest they receive. They can do this by purchasing futures now and selling futures on the date the lending commences. Conversely, borrowers can hedge against interest rate rises by selling futures now and buying futures on the date that the interest rate is fixed.

Example: Interest rate futures hedging strategy for X-factor plc

On 30 July, the three-month sterling interest rate futures are quoted as follows:

September	92.50
December	92.70
March	92.80

Each futures contract has a notional value of £500,000. X-factor plc is looking to raise a £1 million floating rate loan at the end of December for three months.

> Suggest a safe hedging strategy to reduce interest rate risk if the company is concerned that the interest rate may rise.

X-factor plc could sell the December futures contract for 92.70. If each contract has a notional value of £500,000 it would need to sell:

£1,000,000/£500,000 = 2 contracts

Not everyone likes derivatives

Warren Buffet, the so-called 'Sage of Omaha', has an excellent track record in managing his investment vehicle, Berkshire Hathaway, having outperformed the S&P 500 index in 34 of the past 39 years (up to 2003). His success is based largely on sticking to firms that produce simple basic products for which there is always likely to be a demand. 'If you don't understand it, don't invest in it' is one of his mottos – he is famed for *not* investing in technology stocks during the internet boom.

He is also very scathing about the relative freedom of companies and dealers to value positions in swaps, options and other complex products whose prices are not listed on exchanges, thus giving a potentially misleading picture of a firm's true future liabilities. According to Buffet, derivatives are 'Weapons of Mass Financial Destruction', time bombs waiting to explode in the faces of the parties that deal in them, and for the whole economic system. Designed as risk management devices, he says they actually pose risks that central banks and governments have so far found no effective way to control, or even monitor.

Source: Based on Warren Buffet's annual letter to shareholders, as reported in an article in the *Economist*, 15 March 2003.

Self-assessment activity 13.5

Define in your own words the main forms of derivatives – forwards, futures, swaps and options.
(Answer in Appendix A at the back of the book)

13.7 WORKING CAPITAL MANAGEMENT

The last main area of treasury management is the management of working capital, including liquidity management. We devote the remainder of this chapter to working capital policy and the following chapter to short-term asset management. Let us first clarify the basic terms and ratios employed in working capital management.

net working capital
Current assets less current liabilities

Net working capital (or simply working capital) refers to current assets less current liabilities – hence its alternative name of net current assets. Current assets include cash, marketable securities, debtors and stock. Current liabilities are obligations that are expected to be repaid within the year.

Working capital management refers to the financing, investment and control of net current assets within policy guidelines. The treasurer acts as a steward of corporate resources and needs to devise and operate clear and effective working capital policies.

liquidity management
Planning the acquisition and utilisation of cash, i.e. cash flow management

Liquidity management is the planned acquisition and utilisation of cash – or near cash – resources to ensure that the company is in a position to meet its cash obligations as they fall due. It requires close attention to cash forecasting and planning. If the wheels of business are oiled by cash flow, the cash forecast, or cash budget, gauges how much 'oil' is left in the can at any time. Any predicted cash shortfall may require the raising of additional finance, disposal of fixed assets or tighter control over working capital requirements in order to avoid a liquidity crisis.

Various ratios are useful in assessing corporate liquidity, the following being the most commonly employed:

current ratio
Current assets divided by current liabilities

1   The **current ratio** is the ratio of current assets to current liabilities. A high ratio (relative to the industry) would suggest that the firm is in a relatively liquid position. However, if much of the current assets are in the form of raw materials and finished stocks, this may not be the case.

quick/'acid test' ratio
Current assets minus stocks, divided by current liabilities

2 The **quick** or **'acid test' ratio** recognises that stocks may take many weeks to realise in cash terms. Accordingly, it is computed by dividing current liabilities into current assets excluding stock.

Example: The General Eclectic Company (GEC)

The working capital of GEC is as follows:

	£m
Current assets	
Stocks and contracts in progress	1,195
Debtors	1,572
Investments	400
Cash at bank and in hand	1,009
	4,176
Less creditors due within one year	(2,037)
Net current assets	2,139

Notice that current assets are ranked in descending order of liquidity. The liquidity ratios for GEC and the industry are:

		GEC	Industry average
Current ratio	(4,176/2,037)	2.05	1.6
Acid test	(4,176 − 1,195)/2,037	1.46	1.2

GEC's current and acid test ratios are both higher than the industry averages, reflecting the company's healthy liquidity position. But what would the position look like if the £1 billion of cash were already committed, say, for major capital expenditure? If you recalculate the current and acid test ratios, you will find that the liquidity position then falls below the industry average.

days cash-on-hand ratio
Cash and marketable securities divided by daily cash operating expenses

3 **Days cash-on-hand ratio** is found by dividing the cash and marketable securities by projected daily cash operating expenses. As its name implies, it indicates the number of days for which the firm could meet its cash obligations, assuming that no further cash is received during the period. Daily cash operating expenses should be based on the projected cash flows from the cash budget, but a somewhat cruder approach is to divide the annual cost of sales, plus selling, administrative and financing costs, by 365.

Self-assessment activity 13.6

Which areas of treasury management would you say are most neglected by smaller firms?

(Answer in Appendix A at the back of the book)

13.8 PREDICTING CORPORATE FAILURE

Excessive levels of gearing are often responsible for corporate failure. However, very highly geared companies do survive and, conversely, some low-geared companies fail. This suggests that there are many other clues to the viability of a company, and it is not enough simply to examine a single Balance Sheet ratio when attempting to predict financial failure.

The **Z-score** method, developed by Altman (1968), attempts to balance out the relative importance of different financial indicators. This was based on examining the financial characteristics of two samples of failed and surviving US companies to detect which ratios were most important in discriminating between the two groups. For example, were past failures characterised by low liquidity ratios? What other ratios were important discriminators, and what was their relative importance?

Using a technique called discriminant analysis, the relative significance of each critical ratio can be expressed in an equation that generates a 'Z-score', a critical value below which failed firms typically fall, and above which survivors are located. In general terms, the equation is:

$$Z = a + bR_1 + cR_2$$

In this equation, a, b and c are constants derived from past observations and R_1 and R_2 are two identified key discriminatory ratios.

A Z-score model using data for UK firms was developed by Marais (1982), an extension of which is currently used by the Datastream database. For Datastream, Marais examined over 40 ratios before settling on four critical ones in his final model:

1 Profitability: $$\frac{\text{Pre-tax profit} + \text{depreciation}}{\text{Current liabilities}}$$

2 Liquidity: $$\frac{\text{Current assets less stocks}}{\text{Current liabilities}}$$

3 Gearing: $$\frac{\text{All borrowing}}{\text{Total capital employed less intangibles}}$$

4 Stock Turnover: $$\frac{\text{Stock}}{\text{Sales}}$$

Other analysts, using different samples of firms, employ different ratios and weightings in the equation for Z. In Marais' model, the critical Z-value is zero. This does not prove that an existing company displaying a Z-score of around zero is on the brink of insolvency, merely that the firm is displaying characteristics similar to previous failures. Given that there are accounting policy differences between companies, it may be more useful to look at changes in the Z-score over time. A declining Z-score suggests a worsening financial condition, while an improving Z-score indicates strong corporate financial management.

Corporate failure models, such as Z-scores, have their weaknesses (e.g. see Grice and Ingram, 2001):

(a) 'Failure' is difficult to define. Usually its definition is wider than liquidation, but all sorts of restructuring and rescue operations arise for a variety of reasons.

(b) All models are based on the past, when macroeconomic conditions were different from the present.

(c) Companies employ different accounting policies, making comparison difficult.

Z-scoring is used primarily for credit risk assessment by banks and other financial institutions, industrial companies and credit insurers. While it does not tell the whole story behind the company's prospects, it is widely regarded as an important indicator of a company's financial health and hence its credit status.

13.9 CASH OPERATING CYCLE

For a typical manufacturing firm, there are three primary activities affecting working capital: purchasing materials, manufacturing the product and selling the product. Because these activities are subject to uncertainty (delivery of materials may come late, manufacturing problems may arise, sales may become sluggish, etc.), the cash flows associated with them are also uncertain. If a firm is to maintain liquidity, it needs to invest funds in working capital, and to ensure that the operating cycle is properly controlled.

The **cash operating cycle** is the length of time between the firm's cash payment for purchases of material and labour, and cash receipts from the sale of goods. In other words, it is the length of time the firm has funds tied up in working capital. This is calculated as follows:

Cash operating cycle = stock period + customer credit period
                         − supplier credit period.

■ The cash operating cycle: Briggs plc

Briggs plc, a manufacturer of novelty toys, has the following working capital items in its balance sheet at the start and end of its financial year:

	1 January	31 December
Stock	£5,500	£6,500
Debtors	£3,200	£4,800
Creditors	£3,000	£4,500

Turnover for the year, all on credit, is £50,000 and cost of sales is £30,000. For how many days is working capital tied up in each item? What is the cash operating cycle period?

Our first task is to calculate the turnover ratios for each:

$$\text{Stock turnover} = \frac{\text{Cost of sales}}{\text{Average stock}} = \frac{£30,000}{£6,000} = 5 \text{ times p.a.}$$

$$\text{Debtors' turnover} = \frac{\text{Sales}}{\text{Average debtors}} = \frac{£50,000}{£4,000} = 12.5 \text{ times p.a.}$$

$$\text{Creditors' turnover} = \frac{\text{Cost of sales}}{\text{Average creditors}} = \frac{£30,000}{£3,750} = 8 \text{ times p.a.}$$

To find the number of days each item is held in working capital, we divide the turnover calculations into 365 days:

$$\text{Stock period} = 365/5 = 73 \text{ days}$$
$$\text{Debtors (customer credit) period} = 365/12.5 = 29.2 \text{ days}$$
$$\text{Creditors' (supplier credit) period} = 365/8 = 45.6 \text{ days}$$

The cash operating cycle is therefore:

$(73 + 29.2 - 45.6) = 56.6$ days

This is illustrated in Figure 13.4.

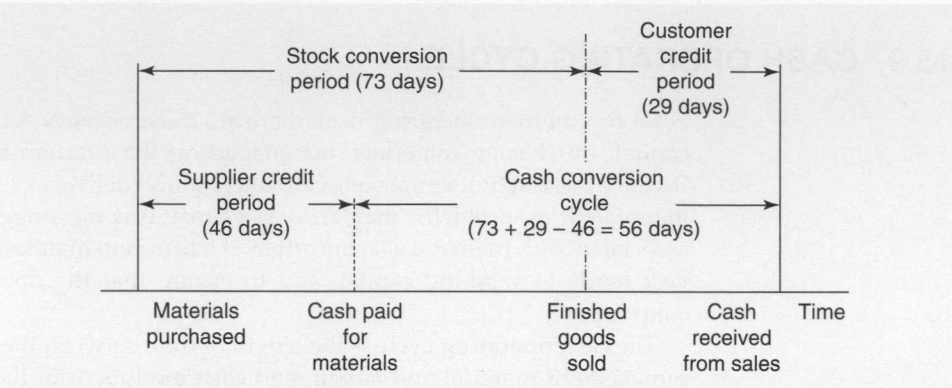

Figure 13.4 Cash conversion cycle

Self-assessment activity 13.7

Explain why two firms in the same industry could have very different cash operating cycles. What are the financial implications?

(Answer in Appendix A at the back of the book)

13.10 WORKING CAPITAL POLICY

The treasury manager should ensure that the firm operates sound working capital policies. These policies cover such areas as the levels of cash and stock held, and the credit terms granted to customers and agreed with suppliers. Successful implementation of these policies influences the company's expected future returns and associated risk, which, in turn, influence shareholder value.

Failure to adopt sound working capital policies may jeopardise long-term growth and even corporate survival. For example:

1 Failure to invest in working capital to expand production and sales may result in lost orders and profits.
2 Failure to maintain current assets that can quickly be turned into cash can affect corporate liquidity, damage the firm's credit rating and increase borrowing costs.
3 Poor control over working capital is a major reason for **overtrading** problems, discussed later in this chapter.

Typical questions arising in the working capital management field include the following:

■ What should be the firm's total level of investment in current assets?
■ What should be the level of investment for each type of current asset?
■ How should working capital be financed?

We now consider how firms establish and finance the levels of working capital appropriate for their businesses, and how they impact on profitability and risk. The level and nature of working capital within any organisation depend on a variety of factors, such as the following:

- The industry within which the firm operates.
- The type of products sold.
- Whether products are manufactured or bought in.
- Level of sales.
- Stock and credit policies.
- The efficiency with which working capital is managed.

We saw in Chapter 1 that the relationship between risk and the required financial return is central to financial management. Investment in working capital is no exception. In establishing the planned level of working capital investment, management should assess the level of liquidity risk it is prepared to accept, risk in the sense of the possibility that the firm will not be able to meet its financial obligations as they fall due. This is a further dimension of financial risk.

■ Working capital strategies: Helsinki plc

Helsinki plc, a dairy produce distributor, is considering which working capital policy it should adopt.

Figure 13.5 shows the two working capital strategies under consideration. Notice that both schedules are curvilinear, suggesting that economies of scale permit working capital to grow more slowly than sales. The firm operates with lower levels of stock, debtors and cash under a more *aggressive* approach than under a more relaxed strategy.

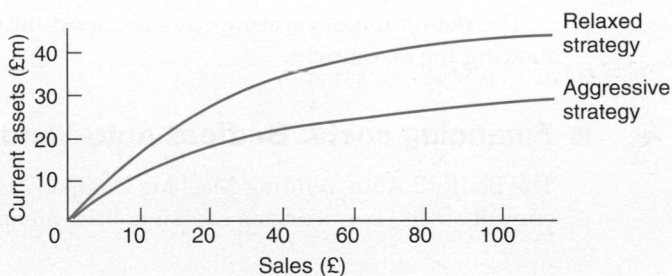

Figure 13.5 Helsinki plc working capital strategies

A *relaxed*, lower risk and more flexible policy for working capital means maintaining a larger cash balance and investment in marketable securities that can quickly be turned into cash, granting more generous customer credit terms and investing more heavily in stock. This may attract more custom, but will usually lead to a reduction in profitability for the business, given the high cost of tying up capital in relatively low profit-generating assets. Conversely, an aggressive policy should increase profitability, while increasing the risk of failing to meet the firm's financial obligations.

In Table 13.2, the relaxed working capital strategy involves a further £20 million investment in current assets. The additional stocks and more generous credit facilities enable Helsinki's management to attain an additional £5 million sales over the aggressive policy. This gives a 19.5 per cent return on capital employed and a secure current ratio of 2.7.

Table 13.2 Helsinki plc: profitability and risk of working capital strategies

	Relaxed (£m)	Aggressive (£m)
Current assets (CA)	40	20
Fixed assets	25	25
Total assets	65	45
Current liabilities (CL)	(15)	(15)
Capital employed (net assets)	50	30
Planned sales	65	60
Planned profit (15% of sales)	9.75	9.0
Return on capital employed	19.5%	30.0%
Net working capital (CA − CL)	£25 m	£5 m
Current ratio (CA/CL)	2.7	1.3

A more aggressive working capital strategy is likely to improve the return on capital. In Helsinki's case, the rate increases to 30 per cent. But this is achieved by increasing liquidity risk. Net working capital falls to only £5 million and the current ratio to 1.3.

■ Working capital costs

carrying costs
Stock costs that increase with the size of stock investment

shortage costs
Stock costs that reduce with size of stock investment

Managing working capital involves a trade-off not only between risk and required return, but also between costs that increase and costs that fall with the level of investment. Costs that increase with additional investment are termed **carrying costs**, while costs that fall with increases in investment are termed **shortage costs**. These two types of cost may be found in most forms of current assets, but particularly in stocks and cash.

The main form of carrying costs is opportunity costs associated with the cost of financing the investment.

■ Financing costs: Bedford Auto-Vending Machine Company

The Bedford Auto-Vending Machine Company is considering how much to invest in current assets. Two working capital policies are under investigation.

	Relaxed policy (£m)	Aggressive policy (£m)
Stock	32	25
Debtors	28	22
Cash and marketable securities	12	–
	72	47

It will be seen that the relaxed policy requires a further £25 million investment in working capital over and above that required for the aggressive policy. What is the cost of carrying this £25 million additional working capital? The main carrying cost is the return that could be earned by investing the additional £25 million in financial assets outside the business. If these could generate 10 per cent p.a., the additional earnings would be £2.5 million (less any interest earned on short-term cash and securities). Other carrying costs include the additional storage and handling costs for stock.

Aggressive or restrictive working capital policies are more susceptible to incurring shortage costs. These costs are usually of two types:

1 **Ordering costs** – costs incurred in placing orders for stock, cash, etc. (in the case of stocks this may also include the production setup costs). Operating a restrictive policy means ordering stock more regularly and in smaller amounts than for more relaxed policies.

2 **Costs of running out of stock or cash** – the most obvious costs here are the loss of business and even the possible liquidation of the firm. Less tangible costs are the loss of customer goodwill, the disruption to the production schedule, and the time and cost of negotiating alternative sources of finance.

The trade-off between carrying costs and shortage costs is shown in Figures 13.6 and 13.7. In Figure 13.6, carrying costs are seen to increase steadily as current assets grow. Conversely, shortage costs fall with the level of investment in current assets. The cost of holding current assets is the combined cost of the two, *the minimum point being the optimal amount of current assets held*. For simplicity, we have shown current assets in total. Later, we consider each element, such as cash or stock, separately.

Different businesses will be more sensitive to certain types of cost. An aggressive policy is more appropriate when carrying costs are high relative to shortage costs, as in Figure 13.7. For example, a major car manufacturer like Ford will not want to hold excessive quantities of raw material stocks, but will buy in materials and parts just before they are to be used in car production, reflecting the Just-in-Time philosophy. Often there will be penalty clauses for non-delivery of such materials to the manufacturer by agreed dates. A flexible policy tends to be more suited to low carrying costs relative to shortage costs.

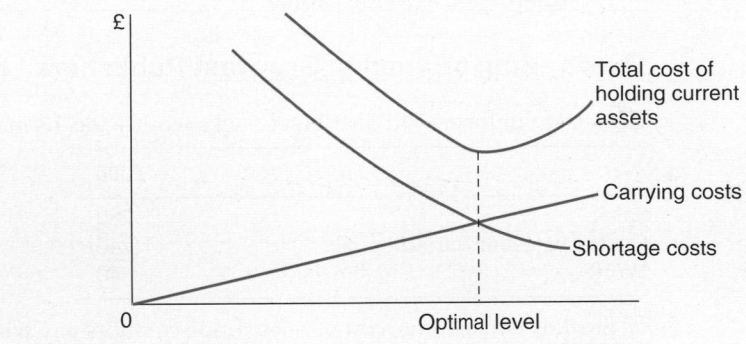

Figure 13.6 Optimal level of working capital for a 'relaxed' strategy

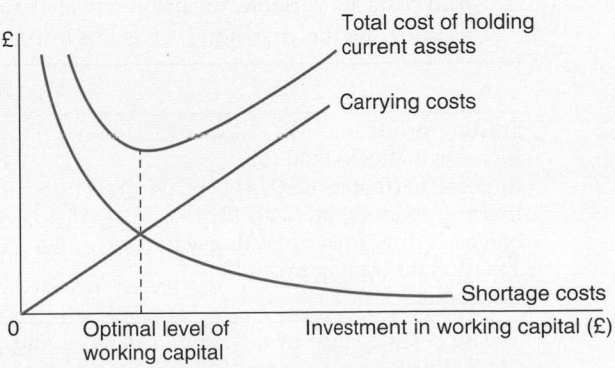

Figure 13.7 Optimal level of working capital for an 'aggressive' strategy

13.11 OVERTRADING PROBLEMS

overtrading
Operating a business with an inappropriate capital base, usually trying to grow too fast with insufficient long-term capital

Here, we address the problems arising from operating a business with an inappropriate capital structure, a phenomenon known as **overtrading**.

Overtrading arises from at least three serious managerial mistakes:

1 *Initial under-capitalisation.* Many businesses experience overtrading problems from the very start because they never invested sufficient equity at the time of formation to finance the anticipated level of trading. Experience suggests that the early years of trading are often difficult years, and shareholders will probably want some incentive in the form of dividends.

2 *Over-expansion.* When a business expands to such a degree that its capital base is insufficient to support the new level of activity, the business is overtrading or, to put it another way, under-capitalised. In many cases, the business looks healthy, in that the level of activity is growing and the business is profitable. But unless sufficient cash is generated to finance the anticipated increase in working capital and fixed investment, the business may encounter serious overtrading problems.

3 *Poor utilisation of working capital resources.* Even when a business has been adequately capitalised and is not over-expanding its activity, overtrading can still occur in several ways:

(a) Failure to achieve planned profit and cash flow levels may mean that debt capacity, originally intended for working capital needs, is used to replace lost earnings.

(b) Cost overruns on fixed capital projects and other unanticipated capital investment can swallow up finance intended for working capital needs.

(c) Similarly, strategic decisions, such as a major acquisition, can have adverse effects on working capital finance unless the capital basis is adequately enlarged.

(d) Higher dividends mean reduced profit retentions, often the major source of finance for working capital.

Overtrading problems: Growfast Publishers Ltd

Growfast Publishers Ltd distributes books worldwide. Its most recent accounts reveal:

	£000
Sales	280
Cost of sales (all variable)	(240)
Profit	40

Stock is two months' cost of sales. Trade creditors pay within one month, while debtors take three months to pay.

Growfast Publishers has recently gained exclusive rights to publish *Corporate Finance and Investment* in Mongolia and is confident that this will lead to a doubling in total sales. Because all costs are variable, the profit will also double, giving a healthy £80,000 profit.

Cash flow for the year, however, is less impressive:

	£000
Trading profit	80
Increase in stocks (240/6)	(40)
Increase in debtors (280/4)	(70)
Increase in creditors (240/12)	20
Net cash flow from operating activities	(10)
Taxation on trading profit (30%)	(24)
Consequence	(34)

The consequence of a doubling in sales and profits is actually a reduction in cash by £34,000. If the increased working capital is thought to be permanent, it should be funded by longer-term finance.

■ Consequences of, and remedies for, overtrading

The consequences of overtrading can be extremely serious and possibly fatal. As the pace of activity increases, working capital needs will also increase. Without the necessary capital structure and cash flow, serious liquidity problems will arise. Business life then becomes a matter of crisis management: finding the cash to meet the wage bill, the creditors' claims and the tax charges. Such myopic behaviour takes attention away from the business of creating wealth and will, ultimately, lead to a decline in competitiveness and profitability.

What can management do to remedy the cash flow problems caused by overtrading?

1 The most drastic step is to *reduce the level of business activity*. Profitable orders may be rejected due to insufficient capital to finance additional working capital needs. If the alternative is to accept the order and, in so doing, jeopardise the business by exceeding the overdraft limit, a slower rate of growth is the preferred course of action.

2 The most obvious remedy is to *increase the capital base*. Figure 13.2 showed that an aggressive strategy for financing working capital operates on a lower long-term capital base, thus making overtrading more likely. Movement towards a matching approach is perhaps called for, where permanent increases in current assets are matched by the injection of permanent capital, preferably in the form of equity or long-term loans, perhaps with a moratorium on repayments in the early years.

3 Finally, steps should be taken to maintain tight control over working capital. Constant review of the working capital policy and its cash flow implications can allow the firm to minimise the extra capital resources required to fund expansion.

Self-assessment activity 13.8

Define overtrading. How does it arise and what are its consequences?

(Answer in Appendix A at the back of the book)

SUMMARY

Treasury management is central to corporate finance in practice. Even in smaller businesses, where no formalised treasury function exists, the main treasury activities of managing corporate funding, risk, banking relationships, liquidity and working capital will still be conducted. This chapter has introduced the reader to those treasury activities, but most of them receive extensive treatment in subsequent chapters.

Key points

■ Treasury management is the efficient management of liquidity and financial risk in the business.

■ Each company should establish whether it requires a separate treasury function and whether it should be a cost or profit centre.

■ Clear treasury policies are required for funding, banking relationships, risk management and working capital management.

■ In general, long-term finance should be used to fund both fixed assets and permanent current assets, fluctuating current assets being funded by short-term borrowing.

■ Hedging can take various forms, but derivative instruments, such as futures, forward contracts, options and swaps, are the most common.

■ Working capital policy trades off expected profitability and risk. An 'aggressive' working capital policy, which seeks to employ the minimum level of net current assets (including cash and marketable securities), will probably achieve a higher return on investment, but may jeopardise the financial health of the business.

■ The cash operating cycle (the length of time between cash payment and cash receipt for goods) should be regularly reviewed and controlled.

■ The consequences of overtrading (or under-capitalisation) can be extremely serious, if not fatal, for the firm.

Further reading and website

Treasury management is a highly practical topic and *The Treasurer* is a useful guide. Collier *et al.* (1988) cover the subject of treasury management, while Smith (1988) is a helpful book of readings and Gentry (1988) a good article on short-term financial management. A pioneer paper on predicting corporate failure is Altman (1968). Risk management and the benefits of hedging are discussed in *Mastering Finance* (1997). A useful website is the Association of Corporate Treasurers: **www.corporate-treasurers.co.uk**.

QUESTIONS

 **Questions with an icon are also available for practice in *myfinancelab* with additional supporting resources.**

Questions with **coloured numbers** have solutions in Appendix B on page 743.

1 Atlas Ltd is a newly-formed digital media company with a number of locations in the UK, France and Germany. The board of directors is currently discussing whether the finance function should be centralised or decentralised. What advice would you offer?

2 What are the risks that a manufacturing company might encounter as a result of interest rate movements? Describe two financial instruments the company could use to reduce such risks.

3 ABC plc is a UK-based service company with a number of wholly-owned subsidiaries and interests in associated companies throughout the world. In response to the rapid growth in the company, the Managing Director has ordered a review of the company's organisation structure, particularly the finance function. The Managing Director holds the opinion that a separate treasury department should be established. At present, treasury functions are the responsibility of the chief accountant.

Required
(a) Describe the main responsibilities of a treasury department in a company such as ABC plc and explain the benefits that might accrue from the establishment of a separate treasury function.
(b) Describe the advantages and disadvantages which might arise if the company established a separate treasury department as a profit centre rather than a cost centre.

(CIMA, November 1995)

4 (a) (i) Discuss the theories, or arguments, which suggest that financial analysis can be used to forecast the probability of a given firm's failure; and
(ii) explain why such an analysis, even if properly applied, may not always predict failure.
(b) Discuss the following statement: 'It is always a sound rule to liquidate a company if its liquidation value is above its value as a going concern.'

5 (a) Explain, with the use of a numerical example, the meaning of the term 'cash operating cycle' and its significance in relation to working capital management.
(b) Delcars plc own a total of ten franchises, in a variety of United Kingdom locations, for the sale and servicing of new and used cars. Six of the franchises sell only second hand vehicles, with the remaining four operating a car service centre in addition to retailing both new and used vehicles. Delcars operate different systems for banking of sales receipts, depending on the type of sale. All monies from new car sales must be banked by the garage on the day of the sale; receipts from second hand car sales are banked once a week on Mondays, and receipts from car servicing work are banked twice a week on Wednesdays and Fridays. No banking facilities are available at the weekend, i.e. Saturdays and Sundays. The sales mix of the three elements (as a percentage of Delcars' total revenue) is as follows: 60 per cent new vehicles; 25 per cent second hand vehicles; 15 per cent servicing. Total sales for all three business areas amounted to £25 million in 1999. Delcars pays interest at a rate of 8.5 per cent per annum on an average overdraft of £65,000, and the company's finance director has suggested that the company could significantly reduce the interest charge if all sales receipts were banked on the day of sale. All the garages are open every day except Sunday. Assume that the daily sales value (for all three areas of business) is spread evenly across the week.
  Calculate the value of the annual interest which could be saved if all ten franchises adopted the finance director's suggestion of daily banking.
(c) Using the example of a car dealership such as Delcars, as given in **(b)** above, outline the advantages and disadvantages of centralisation of the treasury function.

(ACCA)

6 Hercules Wholesalers Ltd has been particularly concerned with its liquidity position in recent months. The most recent profit and loss account and balance sheet of the company are as follows:

Profit and Loss Account for the year ended 31 May 199X

	£	£
Sales		452,000
Less: Cost of sales		
Opening stock	125,000	
Add purchases	341,000	
	466,000	
Less: Closing stock	(143,000)	323,000
Gross profit		129,000
Expenses		(132,000)
Net loss for the period		(3,000)

Balance Sheet as at 31 May 199X

	£	£	£
Fixed assets			
Freehold premises at valuation			280,000
Fixtures and fittings at cost less depreciation			25,000
Motor vehicles at cost less depreciation			52,000
			357,000
Current assets			
Stock		143,000	
Debtors		163,000	
		306,000	
Less creditors due within one year			
Trade creditors	(145,000)		
Bank overdraft	(140,000)	(285,000)	
			21,000
			378,000
Less creditors due after more than one year			
Loans			(120,000)
			258,000
Capital and reserves			
Ordinary share capital			100,000
Retained profit			158,000
			258,000

The debtors and creditors were maintained at a constant level throughout the year.

Required

(a) Explain why Hercules Wholesalers Ltd is concerned with its liquidity position.

(b) Explain the term 'operating cash cycle' and state why this concept is important in the financial management of a business.

(c) Calculate the operating cash cycle for Hercules Wholesalers Ltd based on the information above. (Assume a 360-day year.)

(d) State what steps may be taken to improve the operating cash cycle of the company.

(Certified Diploma)

7 Micrex Computers Ltd was established in 1989 to sell a range of computer software to small businesses. Since its incorporation, the business has grown rapidly and demand for its products continues to rise. The most recent financial accounts for the company are set out below:

Balance Sheet as at 31 May 199X

	£	£	£
Fixed assets			
Freehold land and buildings at cost		55,000	
Less: Accumulated depreciation		(4,000)	51,000
Equipment and fittings at cost		20,000	
Less: Accumulated depreciation		(5,000)	15,000
Motor vehicles at cost		24,000	
Less: Accumulated depreciation		(6,000)	18,000
			84,000
Current assets			
Stocks		26,000	
Trade debtors		59,000	
		85,000	
Less creditors: amounts falling due within one year			
Trade creditors	(88,000)		
Proposed dividend	(1,000)		
Taxation	(6,000)		
Bank overdraft	(10,000)	(105,000)	(20,000)
			64,000
Less creditors: amounts falling due beyond one year			
14% bank loan (secured on freehold property)			(20,000)
			44,000
Capital and reserves			
Ordinary £1 shares			25,000
Retained profit			19,000
			44,000

Profit and Loss Account for the year ended 31 May 199X

	£	£
Sales		660,000
Less: Cost of sales		
Opening stock	22,000	
Purchases	426,000	
	448,000	
Less: Closing stock	(26,000)	422,000
Gross profit		238,000
Less: Selling and distribution expenses	(176,000)	
Administration expenses	(38,000)	
Finance expenses	(7,000)	(221,000)
Net profit before taxation		17,000
Corporation tax		(6,000)
Net profit after taxation		11,000
Proposed dividend		(1,000)
Retained profit for the year		10,000

The company is family-owned and controlled and, since incorporation, has operated without qualified finance staff. However, the managing director recently became concerned with the financial position of the company and therefore decided to appoint a qualified finance director to help manage the financial affairs of the business. Soon after joining the company, the finance director called a meeting of his fellow directors and at this meeting, stated that, in his opinion, the company was overtrading.

Required

(a) What do you understand by the term 'overtrading' and what are the possible consequences of this type of activity?

(b) What are the main causes of overtrading and how might the management of a business overcome the problem of overtrading?

(c) Use financial ratios for Micrex Computers Ltd that you believe would be useful in detecting whether the company was overtrading. Explain the significance of each ratio you calculate.

(Certified Diploma)

Practical assignment

Look at the annual reports of two companies in the same industry. What do they say about the treasury function and treasury or risk management policies? How do the two companies differ and what might be the implications of such differences in treasury management?

 Now retake your diagnostic test for Chapter 13 to check your progress and update your study plan.

14

Short-term asset management

Chart of shame for slow payers

The first league table charting the time every single UK public company takes to pay its bills was launched in March 2008 in an effort to shame those companies that delay so long in paying suppliers they place some at risk of going bust.

Philip King, director of ICM, said: 'Payment times are likely to lengthen in difficult times, as companies want to hold on to their profits or are struggling to pay. This will give suppliers an idea of who is prompt and who is not.'

Companies that take the longest to pay are in the construction, manufacturing, pharmaceuticals and retail sectors. These sectors typically consist of a large number of small suppliers that rely on a handful of large companies for their business. The average payment time is 44 days for all plcs, while for the largest 350 companies it is 34 days.

Source: Based on article by David Oakley, *Financial Times*, 4 March 2008.

Learning objectives

Having read this chapter, you should have a good appreciation of the importance of short-term asset management in corporate finance and of the basic control methods involved. Specific attention will be paid to the following:

- Managing trade credit.
- Inventory management.
- Cash management.

 *Complete your diagnostic test for Chapter 14 now to create your personal study plan.*

14.1 INTRODUCTION

It is a common mistake to assume that financial management concerns only long-term financial decisions, such as capital investment, capital structure and dividend policy decisions. In reality, much of financial management addresses issues of shorter duration, such as short-term financing and working capital management. In this chapter, we examine how short-term assets, such as debtors, inventory and cash, can be managed to maximise shareholder wealth.

14.2 MANAGING TRADE CREDIT

Trade credit can be both a source and a use of finance because it can be received (via trade creditors or payables) and offered (via trade debtors or receivables). We will concentrate on the extension of trade credit and its management, although many of the issues raised apply also to the receipt of trade credit (discussed further in Chapter 15 because it is a form of finance).

Debtors represent the currently unpaid element of credit sales. While the extension of credit is accepted practice in most industries, credit is essentially an unproductive asset (unless it generates additional business) which both ties up scarce financial resources and is exposed to the risk of default, particularly when the credit period taken by customers is lengthy. Effective management of debtors is therefore an essential element of sound financial management practice.

■ Why offer trade credit?

Approximately one-third of the assets of UK businesses is in the form of trade debt – money to be paid at some future time for goods or services already received. The benefits to the customer are obvious, but why should the seller incur financing and other costs in extending credit to selected customers?

1 *Investment and marketing.* Trade credit should be viewed as an investment forming part of the sales package, the payoff being profitable repeat business. Most companies would lose a significant proportion of their customer base to their competitors were they to demand cash on delivery. As with all investment, there are risks involved. Credit risk exists when the company offering credit is exposed to the possibility that the debt will not be paid on time or at all.

The decision to grant credit involves a trade-off between the credit risk and the reward from the profit margin. A common mistake is to assume that a credit sale is a 'one-off', ignoring potential repeat business. If a firm loses business from refusing a customer £1,000 credit, what is the effect? It is more than simply the lost profit margin of, say, 40 per cent, or £400, on the sale. The business from many new customers will grow in time and offer significant repeat business. Assuming they would have entered into a very long-term relationship and ordered £10,000 p.a., growing at 3 per cent a year, the present value of the lost business (given an 8 per cent interest rate) could exceed £80,000:

$$\text{PV} = \frac{£10,000 \times (1.03)}{0.08 - 0.03} \times 0.4 = £82,400$$

2 *Industry and competitive pressures.* It is difficult for firms to offer credit terms that are less generous than their competitors' offerings.
3 *Finance.* Certain types of firm have better access to capital markets and can raise finance more cheaply than others. This competitive advantage can be reflected in offering generous credit to customers who experience greater difficulties in raising finance.

4 *Efficiency.* Information asymmetry exists between buyer and seller. The buyer does not know whether the product delivered is of the quality ordered until it has been thoroughly inspected. The credit period therefore provides a valuable inspection and verification period. Many companies deliver to customers on a daily basis. Trade credit is therefore a convenient means for separating the delivery of goods from the payment of deliveries.

Customer credit mission and goals for Makebelieve Ltd

Mission: To maintain and protect a portfolio of high-quality accounts receivable and to develop sound credit policies and administer credit operations in a manner that increases sales, contributes to profits, aids customer loyalty and improves shareholder value.

Goals:

1 To restrict monthly debtors to 45 days.

2 To achieve agreed monthly cash collection targets.

3 To limit overdue debts to 30 per cent of sales.

4 To limit bad debts to 1 per cent of sales.

5 To resolve credit-related customer queries within 3 days.

6 To improve the relationship between the credit function and major customers through regular contact and visits.

7 To convert 20 per cent of existing customers to direct debit in the year.

The aims of trade credit management are the following:

■ To safeguard the firm's investment in debtors.
■ To maximise operational cash flows by assessing customer credit risks, agreeing appropriate terms and collecting payments in accordance with these terms.

The level of debtors in a company will depend on its terms of sale, credit-screening, cash discounts offered and cash collection procedures.

Effective debtor control policy requires careful consideration of the following:

■ Credit period.
■ Credit standards.
■ Cost of cash discounts.
■ Collection policy.

Each of these are discussed in the following section.

While the main responsibility for setting credit policy lies within financial management, other functions should be involved, particularly marketing. However, all too often, this collaboration is lacking. The credit management process is shown in Figure 14.1.

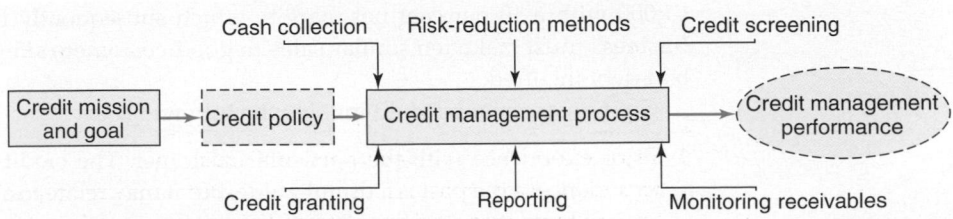

Figure 14.1 The credit management process

■ Credit period

The main factors influencing the period of credit granted to customers are:

1 *The normal terms of trade for the industry.* It is difficult to operate a trade credit policy where the period offered is considerably below the normal expectation for the industry unless the company has another clear competitive advantage, such as a recognised better quality product.

2 *The importance of trade credit as a marketing tool.* Determining the optimum credit period requires the finance manager to identify the point where the costs of increased credit are matched by the profits made on the increased sales generated by the additional credit. The more vital the perception of credit as a marketing tool, the longer the likely period of credit offered.

3 *The individual credit ratings of customers.* Most firms operate regular credit terms for good-quality customers and specific credit terms for higher-risk customers. The credit quality of customers is based on the credit standards addressed in the 'Pickles Ltd' worked example.

credit limits
The maximum amount of credit that a firm is willing to extend to a customer

Credit limits should be set for each customer based on their credit-worthiness. The firm should consider:

1 *Customer payment record*: is the customer a prompt payer?

2 *Financial signals*: is there evidence of the customer running up losses or having liquidity problems?

Very high-risk customers may be reviewed monthly and have to pay, in full or part, with order. Other customers may be granted credit on the basis of percentage of annual purchases.

Commonly quoted trade credit terms

- **Cash before delivery (CBD)**
- **Cash on delivery (COD)**
- **Invoice terms** (e.g. 2/10, net 30). Payment term offering a 2 per cent cash discount for payment within 10 days, otherwise the net amount is due after 30 days.
- **Consignment sales** – pay for goods when used or sold.
- **Periodic statement** – payment by a specific date for all invoices up to a cut-off date.
- **Seasonal dating** – payments due at specific dates to match the buyer's seasonal income.

■ Credit standards

We have noted that granting trade credit is partly a marketing exercise designed to increase sales. However, at the individual customer level, it is essentially a credit assessment and control exercise. In this sense, extending trade credit is no different from a bank granting a loan to a customer. The risk of granting trade credit can be seen when we consider the effect on profit of customer default. If a company sells a product for £1,000 with a 10 per cent net margin, which subsequently becomes a bad debt, the business must make ten similar sales to good customers simply to recover the £1,000 bad debt incurred.

Credit assessment should involve the following:

1 Prior experience with the particular customer. The credit extended and payment experience in the past is a useful guide, but it may relate to a time when the customer was not experiencing financial difficulty. Even so, it is wise to have more rigorous procedures for assessing new accounts.

2   Analysis of the customer's accounts and credit reports. Profit and Loss Accounts and Balance Sheets are available from the company's registered office, but can more easily be taken from computer databases. Credit reports include:
   (a) Bank references
   (b) Trade references expressing the views of other businesses trading with the customer
   (c) Credit bureau reports. Credit-reporting agencies (such as Dun & Bradstreet) provide data and credit ratings that can be used in credit analysis. It is common practice for firms to offer credit agencies full disclosure of financial and trading information in order to gain a good rating. From an assessment of the customer's creditworthiness, it is possible to establish appropriate credit rules covering the terms of sale:
      (i) the maximum period of credit granted;
      (ii) the maximum amount of credit;
      (iii) the payment terms, including any discounts for early payment and interest charges on overdue accounts.

The businesses most vulnerable to late payment are often those that do least to vet their customers. According to the Confederation of British Industry, many small firms fail to chase their late payers with any degree of urgency, partly because their credit management systems are not good enough to support such activity.

In evaluating customer creditworthiness, it is useful to remember the five Cs of credit: capacity, character, capital, collateral and conditions.

1   *Capacity* – does the customer have the capacity to repay the debt within the required period? This may require examination of the past payment record of the customer.
2   *Character* – will the customer make a serious effort to repay the debt in accordance with the terms agreed? Bank and trade references will be useful here.
3   *Capital* – what is the financial health of the customer? Is the firm profitable and liquid? Is it borrowing beyond its means? Financial accounts and credit agency reports will help here.
4   *Collateral* – should some form of security be required in return for extending credit facilities? Alternatively, should part payment in advance, or retention of title be specified?
5   *Conditions* – what are the normal terms for the industry? Are our main competitors offering more generous terms?

■ Cash discounts

The longer a customer's account remains unpaid, the greater the risk that it will never be paid. But the cost of financing late payments is often greater than the cost of bad debts. Surveys suggest that customers, on average, take 30 days' extra credit beyond the payment terms.

Cash discounts are financial inducements for customers to pay accounts promptly. Such discounts can be very costly.

Self-assessment activity 14.1

A survey of large UK companies (Pike *et al*., 1998) found that the normal credit period granted was 30 days, but the average credit period taken by customers was 46 days. Only 20 per cent of firms offered prompt payment cash discounts, with the most common terms being $2\frac{1}{2}$ per cent/net 30 days. For a company offering those terms, what would be the effective interest rate for granting cash discounts assuming that firms would otherwise pay within 46 days?

(Answer in Appendix A at the back of the book)

Example: Yorko plc

Yorko plc offers terms of trade which are '2/10 net 30'. This means that a 2 per cent discount is offered for all accounts settled within ten days, otherwise payment in full is to be made in 30 days. A 2 per cent discount may not seem much until one realises that it is given for a payment in advance of just 20 days (i.e. 30 − 10). The annualised cost is actually over 37 per cent, calculated by the formula below:

The cost of a discount

$$\text{Cost of cash discount} = \frac{\text{Discount \%}}{(100 - \text{discount \%})} \times \frac{365}{(\text{Final date} - \text{discount period})}$$

In the Yorko example, the annualised cost of forgoing the cash discount is:

$$\text{Cost} = \frac{2}{(100 - 2)} \times \frac{365}{(30 - 10)}$$
$$= 0.0204 \times 18.25$$
$$= 37.23\%$$

A more precise calculation is to find the **effective annual rate of return**. We have already calculated for Yorko the two elements:

Cost of discount $= 2/(100 - 2) = 0.0204\%$ per 20-day period

Number of 20-day periods a year $= 365/20 = 18.25$ periods

Effective annual interest rate $= (1.0204)^{18.25} - 1 = 44\%$

This expression of the cost is greater than in the first calculation because it assumes compound rather than simple interest which is more accurate.

Where such generous terms are available, it probably makes sense for customers to opt for the discount even if it means borrowing, as long as the cost of finance is clearly below the annualised cost of discount. So why should firms offer such inducements? First, early payment can significantly improve cash flow and reduce bad debt risk. Second, cash discounts can encourage new customers who are attracted by the discounts. However, the financial manager should be aware of the true cost of such discounts and be able to justify why terms should be offered costing more than the cost of capital.

■ Credit collection policy

A good credit collection policy is one in which procedures are clearly defined and customers know the rules. Debtors who are experiencing financial difficulties will always try to delay payment to companies with poor or relaxed collection procedures. The supplier who insists on payment in accordance with agreed terms, and who is prepared to cut off supplies or take action to recover overdue debts, is most likely to be paid in full and on time.

Figure 14.2 shows the debt collection cycle, starting with the customer order and ending with the cash received. Any speeding up of the order will reduce the required working capital. Late payment by major customers often has a knock-on effect throughout the supply chain. For example, if a customer of company A pays its debts 60 days late, this may force Company A to pay its bills late to Company B, which might create sufficient cash flow pressures for B to go out of business.

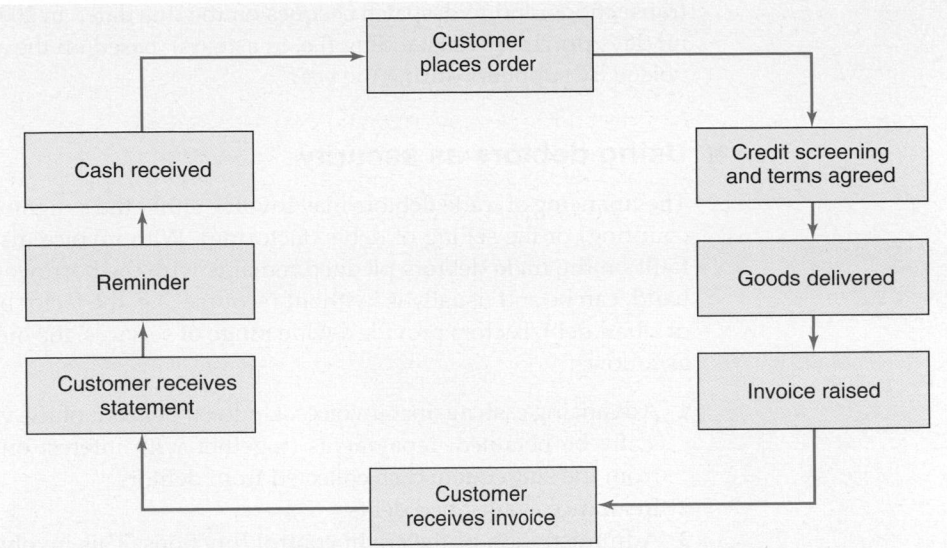

Figure 14.2 Ordering and debt collection cycle

It is a sad fact that firms usually only run out of cash once. Second chances are rare when it comes to cash failure. So getting on top of the credit-screening and control process is vital. Smaller businesses often complain that some larger companies take an unduly long time to settle their accounts. There is a real problem in British industry that far too much time and energy has to be devoted to chasing debts, for no apparent net gain to the business community. The CBI has introduced a Code of Practice, *Prompt Payers – In Good Company*, where firms agree to pay within the agreed payment terms. Businesses have a statutory right to charge larger customers interest on overdue accounts. The interest rate is set high (currently, Bank of England base rate +8%) because most firms must finance late payment from bank overdrafts.

CBI prompt payment code

This states that a responsible company should:

- Have a clear, consistent policy of paying bills in accordance with contract.
- Ensure that the finance and purchasing departments are both aware of this policy and adhere to it.
- Agree payment terms at the outset of a deal and stick to them.
- Not extend or alter payment terms without prior agreement.
- Provide suppliers with clear guidance on payment procedures.
- Ensure that there is a system for dealing quickly with complaints and disputes, and advise suppliers without delay when invoices are contested.

The CBI has joined forces with other interested parties (e.g. the DTI, the British Chambers of Commerce, the British Bankers Association, the Institute of Credit Management) to form the Better Payment Practice Group, which provides a set of best practice guidelines for both buyers and sellers. Its website (**www.payontime.co.uk**) gives a listing of the average payment times of public companies to enable small suppliers, in particular, to monitor and compare the payment practices of these firms. Most listed firms state their payment policy in their annual reports.

For example, Corus plc, the steel-making firm, now part of Tata Steel, declared its policy as to 'establish payment terms with suppliers when agreeing the terms of business transactions, and to despatch cheques on the due date.' In 2006, Corus claimed to have nil days purchases outstanding (i.e. in arrears) 'based on the average daily amount invoiced by suppliers during the year'.

■ Using debtors as security

The financing of trade debtors may involve either the assignment of debts (invoice discounting) or the selling of debts (factoring). With invoice discounting, the risk of default on the trade debtors pledged remains with the borrower. Factoring, on the other hand, can be and usually is 'without recourse', i.e. the factor bears the loss in the event of a bad debt. Factors provide a wide range of services, the most common of which are as follows:

1 Advancing cash against invoices. Up to 80 per cent of the value of invoices can typically be obtained; repayments (together with interest on the advances) are paid from the subsequent cash collected from debtors.
2 Insurance against bad debts.
3 Administration of the credit control functions. This involves sending out invoices, maintaining the sales ledger and collecting payments.

We return to this topic in Chapter 15.

The subtle art of getting paid: late payment

Some small businesses develop creative ways to pursue customers who are paying their bills late.

An antique fireplace shop in north London until recently kept on call a 6ft 3in ex-con who had two fingers missing on his left hand and halitosis. His job was simple: to persuade defaulting customers to pay up by going to their workplace and sitting quietly, but unpleasantly, in the lobby. He seldom had to stay long before the promised cheque appeared.

Another small businessman, this time in advertising, was owed money by a smart furniture shop. He took the afternoon off to stand in the customer's doorway telling people coming in that they would be ripped off. He had his cheque within an hour.

Neither approach would feature in a business school textbook on credit management, but both were effective. One spent money on paying someone to chase the debt, the other judged it an effective use of his time to do it himself. Both related to a simple business problem: staying afloat when customers delay paying invoices as long as possible.

Each year, 10,000 UK businesses fail because their invoices are paid late, according to Dun & Bradstreet, the credit management consultancy. Out of £17 billion owed to UK small businesses last year, £6.8 billion was paid after the due date. Yet few small businesses make use of legislation that penalises late payers, and most believe the law can be of little help when withholding payment appears to be becoming the norm. As an economic downturn approaches the situation is bound to deteriorate.

To address this, the Late Payment of Commercial Debts (Interest) Act 1998 allows creditors to add interest to unpaid invoices without having to go to court. A European Community directive, for which the UK consultation period ends on Friday, would allow companies to claim compensation as well as interest from late paying customers.

Trade credit is a loan to your customer, yet customer/supplier contracts can be surprisingly vague on the terms of payment. There are three steps to managing trade credit:

■ Sell the payment terms at the same time as you sell the product, agree those terms and get to know the person who actually signs the cheque.
■ Eliminate 'own goals' such as delivering the product late or sending an invoice that does not match the delivery note.
■ Be prepared to ask for the money you are owed. Big companies, which are organised, will introduce interest on overdue accounts automatically. Small companies will not have the resources to chase up interest payments.

Source: Based on article in *Financial Times*, 26 April 2001.

Self-assessment activity 14.2

What are the main elements in a firm's credit policy?

(Answer in Appendix A at the back of the book)

■ Worked example: Pickles Ltd

Pickles Ltd produces a single product sold throughout the UK. Its profit analysis is given below:

	Per unit	
	£	£
Selling price		40
Variable costs	(36)	
Fixed cost apportionment	(3)	(39)
Net profit per unit		1

Pickles has an annual turnover of £4.8 million and an average collection period for debtors of one month. It has conducted a study on entering new European markets and believes that this would produce an additional 25 per cent of sales, but the new business would require three months' credit. Stocks and creditors would rise by £400,000 and £200,000 respectively. The cost of financing any increase in working capital is 10 per cent.

Operating profit before finance costs increases as a result of the new business by £120,000:

Sales increase (25% × £4.8 m) = £1.2 m

Contribution/sales ratio (40 − 36)/40 = 10%

Increase in profit = £120,000

The question of whether profits increase as a result of the expansion into European markets very much rests on whether the existing UK customers also demand more favourable terms.

1 *Assuming only new customers take three months' credit*

	£000
New business (£4.8 m × 25%)	1,200
New debtors (£1.2 m × 3/12)	300
Additional stocks	400
Additional creditors	(200)
Required increase in working capital	500
Increase in operating profit (£1.2 m × 10%)	120
Less financing cost (£500,000 × 10%)	(50)
Net profit increase	70

Thus net profits increase by £70,000. But what happens if *existing* customers demand the same credit terms?

2 *Assuming existing customers take three months' credit*

	£000
Sales (£4.8 m + £1.2 m)	6,000
New debtors level (3/12 × £6 m)	1,500
Less existing debtors (1/12 × £4.8 m)	(400)
	1,100
Additional stocks	400
Additional creditors	(200)
Additional working capital	1,300
Operating profit increase (as above)	120
Less financing cost (10% × £1.3 m)	(130)
Net profit reduction after financing costs	(10)

After charging the cost of finance on additional debtors, stocks and creditors, the extra business does not increase profits.

14.3 INVENTORY MANAGEMENT

Inventory, or stock, may be classified into the following:

1 *Pre-production inventory* – stocks of raw materials and bought-in parts.
2 *In-process inventory* – work-in-progress at various stages of the production process.
3 *Finished goods inventory* – manufactured goods ready for sale.

In most cases, finished goods will convert most rapidly into cash; but where customer tastes change rapidly, such as in the fashion trade, this stock can also be the most risky.

Inventory is the least liquid of current assets. It is therefore vital to manage it in such a way that it can be converted from raw material to work-in-progress and finished goods as quickly as possible.

Stock is carried for two reasons:

1 *Business is uncertain.* Consumer demand and production requirements are difficult to forecast, and suppliers may not always be reliable in meeting delivery requirements. The cost of being out-of-stock, in terms of lost sales, profits and goodwill, is generally very high.
2 *Economies in ordering.* Every business needs to determine its economic order quantity for its main stock items.

Inventory control is an important topic for both production management and financial management, which should work closely to establish an inventory policy that meets customers' requirements while operating at optimum stock levels. It should avoid the twin evils of overstocking and understocking.

Overstocking results in the following:

- An unduly high level of working capital investment.
- Additional storage space requirements and greater handling and insurance costs.
- Possible deterioration and increased obsolescence risk.

Understocking reduces the working capital required, but can lead to out-of-stock situations ('stockouts') with orders unfulfilled, idle machines and underemployed workers.

SOS from ASOS: from hero to zero

For retailers, the most important current asset is stock. Failure to have on hand the right amount of stock at the right time results in lost opportunities to make profits. For a clothing retailer, this is especially important if the product quickly goes out of fashion, as these opportunities may never reappear.

ASOS (formerly As Seen On Screen), the online fashion retailer that specialises in selling celebrity-style clothes to 20-something shoppers, was the top performer on the London Stock Exchange during 2004, when its shares rose from 5p to 78p. However, in March 2005, it was forced to issue a profits warning. As a result of problems with distribution of merchandise, winter stock that should have been sold over Christmas had become backed up, necessitating sharp price cuts to shift excess produce. ASOS's Chief Executive said, 'This discounting had led to a significant increase in sales, well beyond budgeted levels. As a consequence, we are bearing the costs associated with very high sales volumes, but without the gross margin to support them.'

In fact, average gross margin fell from 50% to about 30%. He added that ASOS would have done even better than its 70% sales leap over Christmas, had it not been working out of four dispersed warehouses, when it needed a centralised strategic site. The difficulties in coordinating distribution resulted in delays in items appearing on the ASOS website, causing the backlog of stock. Happily, he was able to report the appointment of a new general manager to oversee distribution, and that ASOS had found a 70,000 sq. ft. warehouse expected to come into use in three months. This mixed message probably helped to moderate the market's reaction to the profit warning, limiting the share price fall to 11%.

Source: Based on article by Lisa Urquhart, *Financial Times*, 4 March 2005.

In the past, carrying higher than necessary stocks has been a way of compensating for inefficient production and distribution or poor forecasting. But in today's highly competitive global markets, with Japanese and other overseas businesses operating efficient production schedules and minimal stock levels, European companies have been forced to examine their inventory management processes more closely.

■ Approaches to inventory management

There is now a whole variety of methods for improving stock control, some simple, others more sophisticated, using computer software. We will limit our discussion to three forms of stock control:

1 'Broad-brush' approaches.
2 Economic order quantity models.
3 Computer-based material requirements and just-in-time methods.

Broad-brush approaches

A simple, but useful, starting point is to consider the total stock position using the number of days' stock ratio:

$$\text{Number of days' stock} = \frac{\text{Average stock}}{\text{Cost of sales}} \times 365$$

Consider the stock levels of two companies, based on latest accounts (Table 14.1). U-Save, a discount supermarket chain, carries only finished stocks. Its generic strategy – to be the lowest-cost grocery retailer – requires tight control over its ordering, deliveries and stocks. Its stockholding period is 22 days, which means that stock will probably be turned into cash before the invoice for the goods is paid. By contrast, a major diversified producer like Unicom has very significant raw material stocks and a stockholding period which is double that of U-Save.

Table 14.1 Total inventory levels and stockholding periods

	U-Save (£m)	Unicom (£m)
Raw materials	—	1,380
Work-in-progress	—	175
Finished goods	148	1,894
Total stock	148	3,449
Cost of sales	2,403	25,926
Days' stock	22.5	48.5

While such cross-industry comparisons are interesting, U-Save will want to compare its stockholding period against Tesco and other competitors to see whether it is more efficient in its inventory control processes.

Major companies may well have thousands of items in stock. How should they determine the appropriate level of inventory control for each item? A simple stock classification, often called the **ABC system**, can help identify how closely stock items should be controlled. It divides a company's inventory into three groups according to importance to sales value, with high-value stocks requiring the highest stock control attention.

ABC system
A system of stock management that prioritises items accounting for greatest stock value

ABC stock classification in Boris plc

An analysis of stock items in Boris plc revealed the following:

Short-term financing and policies

Category	Stock items (%)	Stock value (%)
A	12	72
B	38	18
C	50	10
	100	100

Category A stock items have only 12 per cent of the total number of items in stock, but account for 72 per cent of stock value. It was decided that these items required a considerable degree of stock control attention, regular forecasting and monitoring, carefully assessed economic order quantities and an appropriate level of buffer stocks.

Category B stock items cover 38 per cent of total items, but only 18 per cent of stock value. The inventory policy for these items would be less sophisticated; forecasting would be simpler and less frequent.

Category C, which covers half the stock items but only 10 per cent of stock value, requires much simpler treatment, with few stock records and less regular monitoring. For example, stocks of nuts and bolts might simply require that an ample supply is always on hand.

Economic order quantity models

The costs of holding high levels of stocks include the interest lost in tying up capital in such assets, the costs of storing, insuring, managing and protecting stock from pilferage, deterioration, etc., and obsolescence costs. Against this, there are costs involved in holding low levels of stock or running out of stock:

1 Loss of goodwill from failure to deliver by the date specified by customers.
2 Lost production and disruption due to essential items being unavailable.
3 More frequent re-order costs (buyer's and storekeeper's time, telephone, postage, invoice-processing costs, etc.).

A variety of stock management models have been developed to help managers determine the optimal level of stock that balances holding costs against shortage costs. One way of addressing the issue is to determine the **economic order quantity (EOQ)** for the stock required.

Every firm should operate a clear stock control policy, which specifies for its main items the timing of stock replenishments, re-order quantities, safety stock levels and the implications of being out of stock. Figure 14.3 depicts the inventory cycle for a simple stock control model. It assumes a single product, immediate stock replenishment, constant usage and certainty.

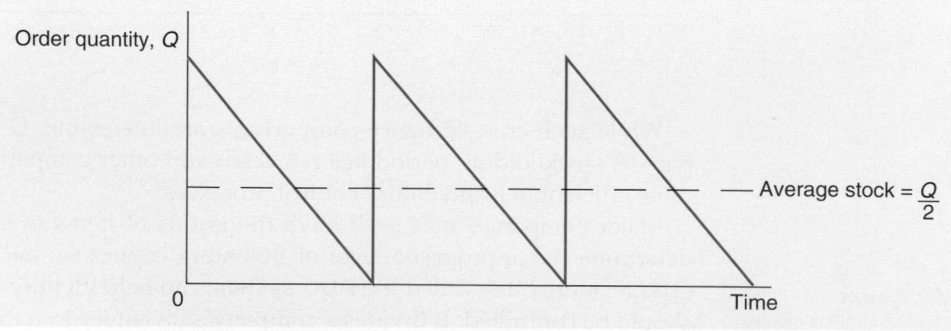

Figure 14.3 The inventory cycle

The total cost is:

Total cost = Ordering costs + Holding costs

Algebraically,

$$\text{Total cost} = \left(\frac{C}{Q} \times A \right) + \left(\frac{Q}{2} \times H \right)$$

where Q is the quantity ordered, C is the cost of placing an order, H is the cost of holding a unit of stock for one year and A is the annual usage of stock.

The economic order quantity is that quantity which minimises the total cost. We examined a similar cost function when we discussed total working capital investment in the previous chapter.

At its simplest, the economic order quantity (EOQ) can be calculated as follows:*

$$\text{EOQ} = \sqrt{\frac{2AC}{H}}$$

Self-assessment activity 14.3

What do you understand by carrying costs and ordering costs? How do they fit into the economic order quantity formula?

(Answer in Appendix A at the back of the book)

EOQ example: Ivan plc

Ivan plc uses 2,000 units of stock item KPR each year. The cost of holding a single item for a year is £2 and the cost of placing each order is £45. The current order quantity is 200 units, but the company is considering changing to batches of 400. Is this the optimum re-order quantity?

Using the cost equation above:

$$\text{Total cost for 200} = \left(\frac{£45}{200} \times 2,000 \right) + \left(\frac{200}{2} \times £2 \right) = £650$$

$$\text{Total cost for 400} = \left(\frac{£45}{400} \times 2,000 \right) + \left(\frac{400}{2} \times £2 \right) = £625$$

$$\text{Total cost for 300} = \left(\frac{£45}{300} \times 2,000 \right) + \left(\frac{300}{2} \times £2 \right) = £600$$

The minimum total cost is achieved by ordering 300 units.

This is confirmed as the most economic order quantity by using the EOQ model:

$$EOQ = \sqrt{\frac{2 \times 2,000 \times £45}{£2}} = \sqrt{90,000}$$

$$= 300 \text{ units}$$

Each order will be placed for 300 units, which implies that orders will be placed every 55 days (i.e. 300/2,000 × 365).

---

*Mathematically, the EOQ is the value of Q that minimises the sum of ordering and holding costs. This is found by the technique of differentiation.

This simple model has two important limitations:

1 Demand, and therefore stock usage, may be seasonal. Hence the constant usage rate for stock assumed here may be unrealistic. Alternatively, demand may be difficult to predict, which necessitates holding **safety** or **buffer stocks**, and calls for a modification of the model.
2 Only the more easily quantifiable costs are included. Many of the other costs referred to earlier (lost goodwill, lost production, etc.) should also be considered.

We have so far assumed that stocks are used up at a constant rate and are replenished when the old level falls to zero. A **stockout** occurs when a firm is unable to deliver a product due to the lack of a specific inventory item. It is therefore tempting for firms to hold large levels of safety stocks to reduce this risk. In effect, this is a form of just-in-case management, as opposed to just-in-time management (see below).

Holding costs will rise through carrying safety stocks, but costs associated with stockouts will fall. The level of safety stock will be affected by management's ability to forecast stock usage and lead time replenishment. Lead time is the delay between ordering and arrival of stock. Each stock item requires a re-order point to be set to cover safety stocks and lead time. The re-order point will be:

$$R = LW + S$$

In words, the re-order point (R) equals the lead time (L) times the weekly stock demand (W) plus the average safety stock (S).

Returning to the example of Ivan plc, if the stock item under consideration has a three week re-order lead time and an average level of safety stocks is set at 40 units, we can determine the re-order point, assuming a 50-week working year.

$$R = LW + S$$
$$= 3 \times \frac{2,000}{50} + 40$$
$$= 160 \text{ units}$$

In practice, rarely is demand uniform, and usage and lead times are uncertain. Determining the optimal stock levels and order quantities under conditions of uncertainty requires probabilistic inventory control models, which are beyond the scope of this book. However, the fundamental point remains that determining the optimal stock level involves balancing the expected costs of ordering and stockouts against the cost of holding additional stocks.

Materials requirement planning (MRP)

MRP is a computer-based planning system for scheduling stock replenishment, ensuring that adequate materials are always available for production purposes. Raw materials are determined from production schedules and lead times for replenishment. MRP can greatly reduce stockholding costs where the finished product requires a multi-stage production process with a large number of components and sub-assemblies, such as in motor car manufacture.

MRPII is a more comprehensive manufacturing resource planning system, which integrates all the resource requirements of the company. In addition to stocks, it also encompasses labour and machine requirements.

Just-in-time

In recent times, managers in some manufacturing firms have been aiming for 'stockless production' and just-in-time (JIT) deliveries. JIT aims for an 'ideal' level of zero stocks, but with no hold-ups due to stock shortages. Materials and parts are delivered from suppliers just before they are needed, and products are manufactured just before they are needed for sale to customers. Where such an operation is successful, the consequent reduction in inventory and the cash operating cycle can be very considerable. Indeed, trade creditors can virtually match the current asset investment, thus enabling the business to operate with the minimum of working capital.

While it focuses on minimising stock levels, JIT forms part of a total quality production programme and rarely works well in isolation. A number of conditions are necessary for JIT to operate successfully:

- Strong links and shared information with suppliers and customers.
- Satisfied customers.
- A quality production process in a 'right-first-time' culture.
- Computerised ordering and inventory tracking systems.
- Smooth movement of materials from process to process.

Suppliers are typically located close to the manufacturer, making regular (often daily) deliveries in small quantities. They are tightly managed and deliver quality assured components to meet agreed production schedules.

JIT was first introduced by Toyota in 1981, using the famous Kanban (card) system. Cards are attached to component containers to monitor the flow of production through the factory, then are returned to signal the need for more supplies. It is particularly suited to high-volume products where assembly line schedules operate continuously.

The main benefits of JIT, experienced by a growing number of companies, are:

1 Drastically reduced stock levels with commensurate savings in storage space, staff and financing costs.
2 A 'right-first-time' culture.
3 Reduced stock defects.
4 Increased productivity.

14.4 CASH MANAGEMENT

Throughout this book, we have emphasised the importance of cash – rather than profit – in financial management. We now consider why cash has such a vital role to play, and how cash flow forecasts are prepared and used to help manage businesses operating in uncertain environments.

■ Why hold cash?

The word *cash* is something of a misnomer. While some 'cash' will be in the form of notes and coins, or bank accounts giving immediate access, much will be invested in short-term bank deposits.

Why should a company hold sums of money in cash or short-term deposits when the return is often quite low? There are a number of reasons why companies hold cash balances:

1 *Transactions motive*. Day-to-day cash inflows and outflows do not match perfectly; cash serves as a buffer to ensure that transactions occur at the appropriate time. Cash balances are particularly important where the patterns of cash inflows and outflows differ greatly, e.g. where business is highly seasonal.
2 *Precautionary motive*. Cash flows are often difficult to predict. Cash balances are required to cater for unanticipated cash disbursements.
3 *Speculative motive*. Cash allows the business to be highly flexible and to exploit wealth-creating opportunities more easily. Large cash balances are common among acquisitive companies where a cash alternative to a takeover bid is required.
4 *Compensation balances motive*. Banks provide a range of financial services, many of which are 'free' as long as the company keeps a positive bank balance.

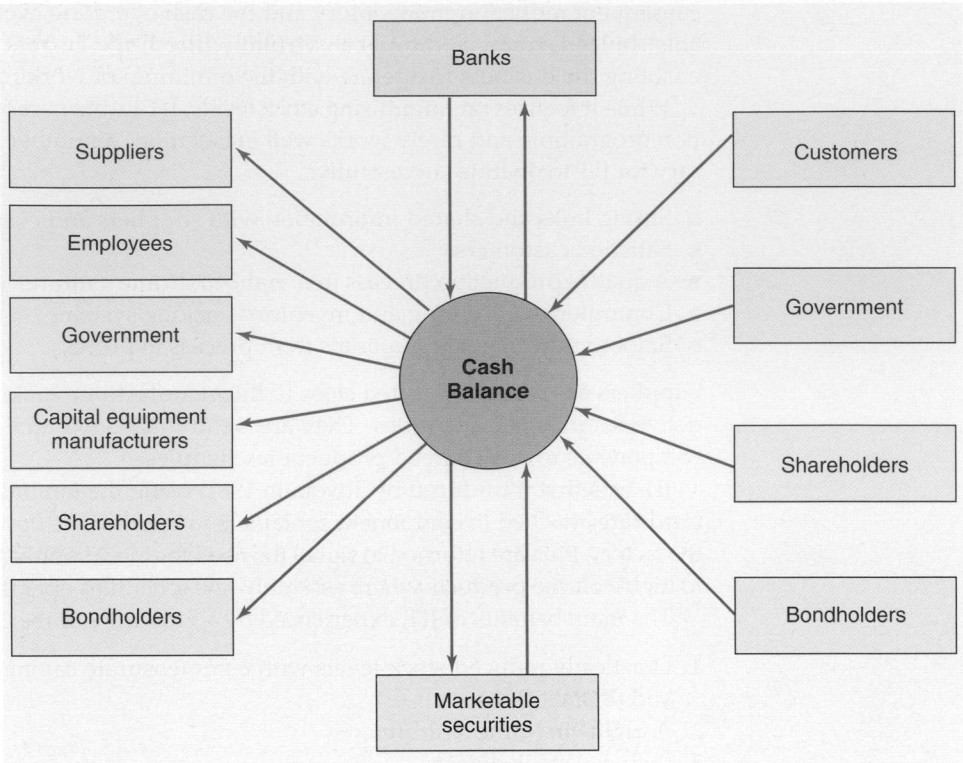

Figure 14.4 Cash flow activity for main stakeholders

Surplus cash is not always reinvested immediately in the business. The cash – or near cash – balance in some companies can be far greater than that required for normal trading purposes. The financial press publishes the main types of short-term financial investment opportunities available to companies, showing the relationship between maturity and interest rates.

Figure 14.4 illustrates the pivotal role played by cash in a typical firm. The cash balance is the result of the interactions of various activities with stakeholders.

■ *Operating activities* – cash from customers less payments to employees and suppliers.
■ *Servicing finance* – dividends and interest on loans.
■ *Taxation* – corporation tax and VAT.
■ *Investing activities* – purchase and sale of fixed assets.
■ *Financing activities* – new finance from shareholders and bondholders, and loan repayments.

The cash balance is restored to its appropriate level by short-term bank borrowing or repayment and the sale or purchase of marketable securities. The financial manager should therefore project the firm's ability to finance its operations and to manage corporate cash flow.

Self-assessment activity 14.4

What are the main motives for holding cash?

(Answer in Appendix A at the back of the book)

Even in well-diversified firms, it makes sense to centralise cash management:

1 It allows the treasurer to operate on a larger scale, which should lead to more competitive interest rates and lower staffing costs.

2 Specialist staff can be employed to work in cash management.

3 Negative cash flows from one operating unit may be offset by positive cash flows from others, thus avoiding additional financing and loan-raising costs. This may well mean that the overall level of cash required to cover unanticipated cash shortfalls is reduced.

4 Banking operations become faster and more efficient, giving rise to advantageous banking arrangements.

■ Cash flow statements

In 1991, the Accounting Standards Board in the UK issued a standard (FRS1) requiring companies to present a cash flow statement within their published accounts. The intention was to move away from over-reliance on profits.

A summarised cash flow statement for the chocolate manufacturer and retailer, Thorntons plc, is given in Table 14.2. The starting point in the statement is the company's ability to generate cash from its operations. A shareholder reading the cash flow statement can identify the reasons for the change in cash position over the year. For 2004, Thorntons achieved £18.132 million positive cash flow from operating activities and £6.765 million before financing. However, the requirement to repay loans resulted in a decrease in cash in the year of £2.453 million.

■ Cash flow forecasting

The cash flow forecast, or cash budget, is the primary tool in short-term financial planning. It helps identify short-term financial requirements and surpluses based on the firm's budgeted activities. Cash budgeting is a continuous activity with budgets being rolled forward, usually in weeks or months, over time.

Preparation of the cash budget involves four distinct stages:

1 *Forecast the anticipated cash inflows.* The main source of cash is usually sales, and the sales forecast will therefore be the primary data source. Sales can be divided into cash sales and credit sales, the timing of the cash flow arising from the latter depending on the agreed credit terms. Thus, for example, the sales forecast for January would appear as a cash receipt in March if all sales were on credit terms of 60 days.

Table 14.2 Thorntons plc consolidated cash flow statement

Year ended 26 June:	2004	2003
	£000	*£000*
Cash flow from operating activities	18,132	24,860
Returns on investment and servicing of finance	(2,825)	(3,225)
Taxation	(1,627)	(2,727)
Capital expenditure and financial investment	(2,492)	(2,337)
Equity dividends paid	(4,423)	(4,422)
Cash inflow (outflow) before use of liquid resources and financing	6,765	12,149
Management of liquid resources	(500)	(1,792)
Financing		
Issue of shares	90	—
(Decrease)/increase in debt	(9,808)	(11,132)
Decrease/increase in cash in the period	(2,453)	(2,809)

Source: Thorntons plc, Annual Report, 2004 (**www.Thorntons.com**).

Other cash inflows might be income from investments, cash from disposal of fixed assets, etc.

2 *Forecast the anticipated cash outflows.* The main payment is generally the payment of trade purchases. Once again, the credit period taken must be allowed for. Other cash outflows include wages and salaries, administrative costs, taxation, capital expenditure and dividends.

3 Compare the anticipated cash inflows and outflows to determine the *net cash flow* for each period.

4 Calculate the *cumulative cash flow* for each period by adding the opening cash balance to the net cash flow for the period.

■ Float

The 'cash at bank' position shown in a company's books will not usually be the same as that shown on the bank statement. Float is the money arising from the time lag between posting a cheque and it being cleared by the bank.

Float management in Marcus Ltd

On 1 July, the bank statement and cash account in the ledger of Marcus Ltd both show £40,000. The company pays suppliers £20,000 by cheque and receives cheques from suppliers for £15,000. The net cash balance in the accounts is therefore (£40,000 − £20,000 + £15,000) = £35,000. But the bank position is still £40,000. The cheques from customers will take three days to clear and the cheques paid will take more like 6–8 days to clear, including postal delay and time taken to pay in the cheque.

The cash controller must, therefore, regularly reconcile the two positions, but also manage the float by recognising that the actual cleared bank balance, upon which interest is calculated, is likely to be somewhat higher than the balance in the company's accounts. If Marcus can get its customers to pay by **direct debit**, this will speed up the banking process and further improve the bank position.

direct debit
An automatic payment from a customer's bank account pre-arranged with the bank by both trading partners

Another method of speeding up collections is **concentration banking**, where customers in a geographical area pay a local branch office rather than head office. The cheque is then deposited in the local bank branch. Where both the customer and the company have local banks, this can reduce both postal and clearing time.

concentration banking
Where customers in a geographical area pay bills to a local branch office rather than to the head office

Electronic funds transfer has certain benefits over cheque payment by post. Cost savings arise from the reduction in administration effort, time and postage. The transfer is instantaneous, which means that cash can stay in the company's bank account longer. The main disadvantage is that 'the cheque is in the post' excuse can no longer be employed. If payment is made on the same day as the cheque used to be posted, it impacts on the bank account quicker and results in more interest charges. BACS (Bankers' Automated Clearing Services) enables computerised funds transfers between banks. Corporate customers can use BACS, particularly for payment of salaries, by providing details of payments. Payment is made in two days. For large payments, same day clearance can be made through CHAPS (Clearing House Automated Payment System).

electronic funds transfer
Instantaneous transfer of money from a debtor's bank account to a creditor

14.5 WORKED EXAMPLE: MANGLE LTD

Mangle Ltd produces a single product – a manually operated spindryer. It plans to increase production and sales during the first half of next year; the plans for the next eight months are shown in Table 14.3.

The selling price is £100, with an anticipated price increase to £110 in June. Raw materials cost £20 per unit; wages and other variable costs are £30 per unit. Other fixed costs are £1,800 a month, rising to £2,200 from May onwards. Forty per cent of sales are

Table 14.3 Mangle Ltd: production and sales

Month	Production	Sales
November	70	70
December	80	80
January	100	80
February	120	100
March	120	120
April	140	130
May	150	140
June	150	160

for cash, the remainder being paid in full 60 days following delivery. Material purchases are paid one month after delivery and are held in stock one month before entering production. Wages and variable and fixed costs are paid in the month of production.

A new machine costing £10,000 is to be purchased in February to cope with the planned expansion of demand. An advertising campaign is also to be launched, involving payments of £2,000 in January and March. The directors plan to pay a dividend of £1,000 in May. On 1 January the firm expects to have £2,000 in the bank. How will the cash position appear over the following six months?

Table 14.4 reveals that the first step is to determine the sales revenue each month. Forty per cent of sales are for cash and are therefore received in the selling month, while the remaining 60 per cent are received two months after the month of sale.

Purchases are made in the month prior to entering production, but because a month's credit is taken, the payment to creditors is in the same month as production.

After including all cash flows, the net cash flow for each of the six months shows that, in the first three months, Mangle Ltd has a negative cash flow, and a negative cash balance for the February to May period. The company may decide that this is a

Table 14.4 Mangle Ltd: cash budget for six months to June (£)

		Jan.	Feb.	Mar.	Apr.	May	June
Inflows							
Receipts from cash sales		3,200	4,000	4,800	5,200	5,600	7,040
Receipts from debtors		4,200	4,800	4,800	6,000	7,200	7,800
	(A)	7,400	8,800	9,600	11,200	12,800	14,840
Outflows							
Payments to creditors		2,000	2,400	2,400	2,800	3,000	3,000
Variable costs		3,000	3,600	3,600	4,200	4,500	4,500
Fixed costs		1,800	1,800	1,800	1,800	2,200	2,200
Advertising		2,000		2,000			
Capital expenditure			10,000				
Dividend						1,000	
	(B)	8,800	17,800	9,800	8,800	10,700	9,700
Net cash surplus (deficit) in month (A − B)		(1,400)	(9,000)	(200)	2,400	2,100	5,140
Opening cash balance		2,000	600	(8,400)	(8,600)	(6,200)	(4,100)
Closing cash balance		600	(8,400)	(8,600)	(6,200)	(4,100)	1,040

seasonal business and that it is acceptable to operate with such monthly cash flow figures. In this case, the firm should seek to negotiate a loan to cover the shortfall. However, it would do well to consider ways of minimising the monthly deficits. For example, could cash receipts from debtors be collected more quickly? Could payment of creditors be deferred for a few weeks? Could the price increase be brought forward? Could payment terms on the capital equipment be extended over a longer period, possibly by leasing the equipment? The cash budget allows finance managers to consider such questions well ahead of events. Frequently, forward planning avoids the need to raise external finance through astute management of working capital.

14.6 CASH MANAGEMENT MODELS

We have already examined inventory control models. Cash is, in many ways, simply a form of inventory, the stock of cash required to enable a business to operate effectively.

William Baumol (1952) recognised the similarity between cash and inventory for control purposes, particularly when the bank balance is simply a draw-down account. This is when a firm has a cash balance upon which it draws steadily, and which it replenishes to the original balance at some point before running out (see the sawtooth diagram for stock in Figure 14.3). Any surplus cash is invested in interest-bearing short-term securities.

In this case, the economic order quantity (EOQ) model can be employed, where EOQ represents the short-term securities to be sold to replenish the balance.

Recall that the equation is:

$$EOQ = \sqrt{\frac{2AC}{H}}$$

where C now represents the transaction cost for selling securities, H is the holding cost of cash (i.e. the interest rate) and A is the annual cash disbursements.

Example: Cash management model for Bizarre plc

The treasurer of Bizarre plc, a company specialising in unusual gifts for eccentric business managers (a rapidly growing market), has a sizeable sum invested in short-term investments, earning 6 per cent interest. Every time she sells investments to top up the bank balance the transaction cost is £25. Monthly cash payments are around £200,000. How often and by how much should she transfer money to the bank account?

The EOQ model gives an indication of the most economic amount of cash to be drawn each time:

$$EOQ = \sqrt{\left(\frac{2 \times \text{annual cash payments} \times \text{cost of selling securities}}{\text{annual interest rate}}\right)}$$

$$= \sqrt{\left(\frac{2 \times £2,400,000 \times £25}{0.06}\right)} = £44,721, \text{ say } £45,000$$

The frequency with which the treasurer will transfer cash is (£2.4 million/£45,000) = 53 times a year, or approximately weekly.

Self-assessment activity 14.5

As interest rates increase, it obviously becomes more attractive to transfer smaller quantities. Try reworking the Bizarre problem assuming a doubling of the interest rate to 12 per cent p.a.

(Answer in Appendix A at the back of the book)

The above cash management model works quite well when daily cash drawings on the bank account are uniform. This is rarely the pattern of actual cash flows in business. Cash flows go up and down in what often appears a fairly random manner. Miller and Orr (1966) suggested that, instead of assuming a constant rate of cash payment, perhaps we should assume that daily balances cannot be predicted – they meander in a random fashion. This is further discussed in the Appendix to this chapter.

■ Short-term investment

When the cash budget indicates a cash surplus, the financial manager needs to consider opportunities for short-term investment. Any cash surplus beyond the immediate needs should be put to work, even if just invested overnight. The following considerations should be made in assessing how to invest short-term cash surpluses:

1 The length of time for which the funds are available.
2 The amount of funds available.
3 The return offered on the investment in relation to the investment involved.
4 The risks associated with calling in the investment early (e.g. the need to give three months' notice to avoid losing the interest).
5 The ease of realisation.

Examples of short-term investment opportunities are as follows:

■ *Treasury Bills* – issued by the Bank of England and guaranteed by the UK government. No interest as such is paid, but they are issued at a discount and redeemed at par after 91 days. At any time, the bills can be sold on the money market.
■ *Bank deposits* – a wide range of financial instruments is available from banks, but the more established investment opportunities are:
   (a) term deposits, where for a fixed period (usually from one month to six years) a fixed rate is given. For shorter periods (typically up to three months), the interest may be at a variable rate based on money market rates
   (b) Certificates of Deposit, issued by the banks at a fixed interest rate for a fixed term (usually between three months and five years), but which can be sold on the money market at any time.
■ *Money market accounts* – most major financial institutions offer schemes for investment in the money market at variable rates of interest (e.g. treasury accounts).

SUMMARY

The management of working capital is a key element in financial management, not least because, for most firms, current assets represent a major proportion of their total investment. Working capital policy is concerned with determining the total amount and the composition of a firm's current assets and current liabilities.

Key points

■ An effective debtor control policy should cover the credit period, credit standard, cost of cash discounts and collection policy.
■ Credit terms should reflect the customer's credit rating, normal terms of the industry and the extent to which the firm wishes to use credit as a marketing tool.
■ In evaluating customer credit worthiness, remember the five Cs: capacity, character, capital, collateral and conditions.

- Trade debtors can be used to raise finance through invoice discounting and factoring.
- Inventory management involves determining the level of stock to be held, when to place orders and how many units to order at a time.
- Inventory costs can be classified into carrying or holding costs, which increase as the level of stock rises, and shortage costs (or stockout costs and ordering costs), which fall as stock levels increase.
- A variety of economic order quantity models is available for determining the order quantity that will minimise total inventory cost. The basic model is:

$$EOQ = \sqrt{\frac{2AC}{H}}$$

where C is the cost of placing an order, A is the annual usage of stock and H is the cost of holding a unit of stock for one year.

- The cash flow forecast, or cash budget, is a vital tool in short-term financial planning.
- Cash management models (e.g. Baumol and Miller–Orr models) are useful in setting limits that trigger cash adjustment.

Further reading

Pike *et al.* (1998) report findings on trade credit management in large companies.

Brigham and Gapenski (1996) and Brealey, Myers and Allen (2005) provide fuller discussions on the issues addressed in this chapter. Sartoris and Hill (1981) provide a present value approach to credit evaluation. The pioneering work on cash management models is found in Baumol (1952) and Miller and Orr (1966).

Useful websites

Better Payments Practice Group: **www.payontime.co.uk**
Credit checks: **www.checkit.co.uk**
Companies House: **www.companieshouse.co.uk**
Dun & Bradstreet: **www.dnb.com**

Appendix
MILLER–ORR CASH MANAGEMENT MODEL

When daily cash flows are very difficult to predict, it may be sensible to assume that the pattern of cash flows is random. This gives rise to the cash position shown in Figure 14.5. Rather than decide how often to transfer cash into the account, the treasurer sets upper and lower limits that trigger cash adjustments, sending the balance back to the return point by selling short-term investments. The diagram prompts two questions:

1 How are the limits set?
2 Why isn't the return point midway between the two?

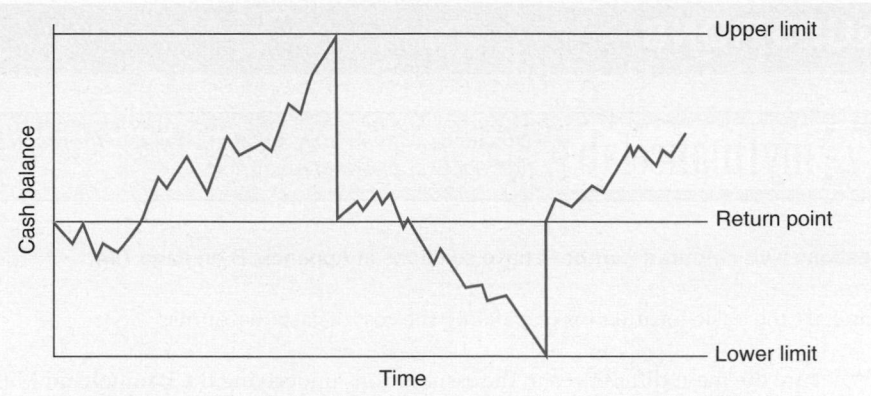

Figure 14.5 Miller–Orr cash management model

In general, the limits will be wider apart when daily cash flows are highly variable, transaction costs are high and interest on short-term investments is low. The Miller and Orr formula for setting the limits is:

$$\text{Range between upper and lower limits} = 3\left(\frac{3}{4} \times \frac{\text{transaction cost} \times \text{cash flow variance}}{\text{interest rate}}\right)^{1/3}$$

The cash balance does not return to a point mid-way between the upper and lower limits. The return point is:

$$\text{Return point} = \text{lower limit} + \frac{\text{range}}{3}$$

By having a return point below the mid-point between the two limits, the average cash balance on which interest is charged is reduced.

The Baumol and Miller–Orr models are really two extremes: the former assumes cash flows are constant, while the latter assumes they are unpredictable. In practice, an experienced cash controller, using a detailed cash budget, should perform better than a cash management model and should enable the company to operate on a lower cash balance. The primary value of simple cash management models may be to evaluate how much better the 'hands-on' expert can perform than the 'hands-off' control model.

Example

The financial manager at Millor Ltd believes that cash flows are almost impossible to predict on a daily basis. She knows that a minimum cash balance of £20,000 is required and transferring money to or from the bank costs £50 per transaction. Inspection of daily cash flows over the past year suggests that the standard deviation is £3,000 a day (i.e. £9 million variance). The interest rate is 0.03 per cent per day.

The Miller–Orr model specifies the range for the cash balance as below:

$$\text{Range} = 3\left(\frac{3}{4} \times \frac{50 \times 9,000,000}{0.0003}\right)^{0.3333} = \text{£31,200 approx.}$$

The upper limit = lower limit + range = (£20,000 + £31,200) = £51,200

$$\text{The return point} = \text{lower limit} + \frac{\text{range}}{3} = \text{£30,400}$$

The decision rule is: if the cash balance reaches £51,200, buy (£51,200 − £30,400) = £20,800 of marketable securities. If the cash balance falls to £20,000, sell £10,400 of marketable securities for cash to return it to £30,400.

QUESTIONS

 **myfinancelab** | *Questions with an icon are also available for practice in myfinancelab with additional supporting resources.*

Questions with **coloured numbers** have solutions in Appendix B on page 744.

1 Specify the basic formula for calculating the cost of cash discounts.

2 What are the main differences in the assumptions underlying the Baumol and Miller–Orr cash models?

3 Hunslett Express Company specifies payment from its customers at the end of the month following delivery. On average, customers take 70 days to pay. Sales total £8 million per year and bad debts total £40,000 per year.

The company plans to offer cash discounts for payment within 30 days. It is estimated that 50 per cent of customers will take up the discount, but that the remaining customers will take 80 days to pay. The company has an overdraft facility costing 13 per cent p.a. If the proposed scheme is introduced, bad debts will fall to £20,000 and savings in credit administration of £12,000 p.a. are expected.

Should the company offer the new credit terms?

4 Salford Engineers Limited, a medium-sized manufacturing company, has discovered that it is holding 180 days' stock while its main competitors are holding only 90 days' stock.

Required

(a) Discuss what you consider to be the most important factors determining the optimum level of stockholding for the company.

(b) What action would you take if you were asked to investigate the reasons for Salford's high level of stock?

(Certified Diploma)

5 Torrance Ltd was formed in 1988 to produce a new type of golf putter. The company sells the putter to wholesalers and retailers and has an annual turnover of £600,000. The following data relate to each putter produced:

	£	£
Selling price		36
Variable costs	(18)	
Fixed cost apportionment	(6)	(24)
Net profit		12

The cost of capital (before tax) of Torrance Ltd is estimated at 15 per cent.

Torrance Ltd believes it can expand sales of this new putter by offering customers a longer period in which to pay. The average collection period of the company is currently 30 days. The company is considering three options in order to increase sales. These are as follows:

	Option		
	1	2	3
Increase in average collection period (days)	10	20	30
Increase in sales (£s)	30,000	45,000	50,000

Torrance Ltd is also reconsidering its policy towards trade creditors. In recent months, the company has suffered from liquidity problems, which it believes can be alleviated by delaying payment to trade creditors. Suppliers offer a 2.5 per cent discount if they are paid within 10 days of the invoice date. If they are not paid within 10 days, suppliers expect the amount to be paid in full within 30 days. Torrance Ltd currently pays suppliers at the end of the 10-day period in order to take advantage of the discounts. However, it is considering delaying payment until either 30 or 45 days after the invoice date.

Required
(a) Prepare calculations to show which credit policy the company should offer its customers.
(b) Discuss the advantages and disadvantages of using trade credit as a source of finance.
(c) Prepare calculations to show the implicit annual interest cost associated with each proposal to delay payment to creditors. Discuss your findings.

(Certified Diploma)

6 (a) Discuss:
(i) The significance of trade creditors in a firm's working capital cycle, and
(ii) the dangers of over-reliance on trade credit as a source of finance.
(b) Keswick plc traditionally follows a highly aggressive working capital policy, with no long-term borrowing. Key details from its recently compiled accounts appear below:

	£m
Sales (all on credit)	10.00
Earnings before interest and tax (EBIT)	2.00
Interest payments for the year	0.50
Shareholders' funds (comprising £1 million issued share capital, par value 25p, and £1 million revenue reserves)	2.00
Debtors	0.40
Stocks	0.70
Trade creditors	1.50
Bank overdraft	3.00

A major supplier which accounts for 50 per cent of Keswick's cost of sales is highly concerned about Keswick's policy of taking extended trade credit. The supplier offers Keswick the opportunity to pay for supplies within 15 days in return for a discount of 5 per cent on the invoiced value.

Keswick holds no cash balances but is able to borrow on overdraft from its bank at 12 per cent. Tax on corporate profit is paid at 33 per cent.

Required
Determine the costs and benefits to Keswick of making this arrangement with its supplier, and recommend whether Keswick should accept the offer. Your answer should include the effects on:

- The working capital cycle
- Interest cover
- Profits after tax
- Earnings per share
- Return on equity
- Capital gearing

7 International Golf Ltd operates a large warehouse, selling golf equipment direct to the public by mail order and to small retail outlets. The cash position of the company has caused some concern in recent months. At the beginning

of December 1993, there was an overdraft at the bank of £56,000. The following data concerning income and expenses has been collected in respect of the forthcoming six months:

	December £000	January £000	February £000	March £000	April £000	May £000
Expected sales	120	150	170	220	250	280
Purchases	156	180	195	160	150	160
Advertising	15	18	20	25	30	30
Rent	40			40		
Rates		30				
Wages	16	16	18	18	20	20
Sundry expenses	20	24	24	26	26	26

The company also intends to purchase and pay for new motor vans in February at a cost of £24,000 and to pay taxation due on 1 March of £30,000.

Sales to the public are on a cash basis and sales to retailers are on two months' credit. Approximately 40 per cent of sales are made to the public. Debtors at the beginning of December are £110,000, 70 per cent of which are in respect of November sales.

Purchases are on one month's credit and, at the beginning of December, the trade creditors were £140,000. The purchases made in December, January and February are considered necessary to stock up for the sales demand from March onwards.

All other expenses are paid in the month in which they are incurred. Sundry expenses include £8,000 per month for depreciation.

Required

(a) Explain the benefits to a business of preparing a cash flow forecast.

(b) Identify and discuss the costs to a business associated with:
  (i) holding too much cash;
  (ii) holding too little cash.

(c) Prepare a cash flow forecast for International Golf Ltd for the six months to 31 May 1994, which shows the cash balance at the end of each month.

(d) State what problems International Golf Ltd is likely to face during the next six months and how these might be dealt with.

(Certified Diploma)

8 (a) The Treasurer of Ripley plc is contemplating a change in financial policy. At present, Ripley's balance sheet shows that fixed assets are of equal magnitude to the amount of long-term debt and equity financing. It is proposed to take advantage of a recent fall in interest rates by replacing the long-term debt capital with an overdraft. In addition, the Treasurer wants to speed up debtor collection by offering early payment discounts to customers and to slow down the rate of payment to creditors.

As his assistant, you are required to write a brief memorandum to other Board members explaining the rationales of the old and new policies and pinpointing the factors to be considered in making such a switch of policy.

(b) Bramham plc, which currently has negligible cash holdings, expects to have to make a series of cash payments (P) of £1.5 million over the forthcoming year. These will become due at a steady rate. It has two alternative ways of meeting this liability.

Firstly, it can make periodic sales from existing holdings of short-term securities. According to Bramham's financial advisers, the most likely average percentage rate of return (i) on these securities is 12 per cent over the forthcoming year, although this estimate is highly uncertain. Whenever Bramham sells securities, it incurs a transaction fee (T) of £25, and places the proceeds on short-term deposit at 5 per cent per annum interest until needed. The following formula specifies the optimal amount of cash raised (Q) for each sale of securities:

$$Q = \sqrt{\frac{2 \times P \times T}{i}}$$

The second policy involves taking a secured loan for the full £1.5 million over one year at an interest rate of 14 per cent based on the initial balance of the loan. The lender also imposes a flat arrangement fee of £5,000, which could be met out of existing balances. The sum borrowed would be placed in a notice deposit at 9 per cent and drawn down at no cost as and when required.

Bramham's Treasurer believes that cash balances will be run down at an even rate throughout the year.

Required

Advise Bramham as to the most beneficial cash management policy.

Note: ignore tax and the time-value of money in your answer.

(c) Discuss the limitations of the model of cash management used in part (b). (ACCA)

Practical assignment

1 Sound credit management can play an important role in the financial success of a business.

Required

(a) Explain the role of the credit manager within a business.

(b) Discuss the major factors a credit manager would consider when assessing the creditworthiness of a particular customer.

(c) Identify and discuss the major sources of information that may be used to evaluate the creditworthiness of a commercial business.

(d) State the basis upon which any proposed changes in credit policy should be evaluated.

(Certified Diploma)

2 If you are based in a firm where credit management is important, apply the above question to your organisation.

 myfinancelab

Now retake your diagnostic test for Chapter 14 to check your progress and update your study plan.

15

Short- and medium-term finance

Naming and shaming

In recent years, several large firms have earned adverse publicity by attempting to impose new terms of trade on suppliers. In 2005, Sainsburys attracted substantial opprobrium in attempting to lengthen the credit period taken from its suppliers. The Forum of Private Business (FPB) has opened a Hall of Shame of alleged offenders to which the name of Inbev, the giant Belgian brewing firm that manufactures Stella Artois and Beck's, was added in 2007. Inbev, which claims to be the biggest brewer in the industry, was reported by an FPB member who had her terms for payment hiked from 30 days from the end of the month of invoice to 60 days.

The FPB's chairman, Mr Len Collinson, said Inbev's behaviour was 'reprehensible', adding: 'Not only was this decision taken without the consent of Inbev's suppliers but also they were given less than a month's notice of the change. Many suppliers will find it difficult to adapt, and will have problems with their cash flow.'

In a letter to suppliers, Inbev stated: 'We rely on strong working relationships with our suppliers, and we thank you for your efforts so far in helping us towards this vision to move from "Biggest to Best".' It maintains that the action is part of a move to harmonise payment terms across Western Europe to 'allow the company to . . . place maximum efforts in connecting and investing in our consumers and therefore develop further our ability to offer the potential of long-term sustainable business growth with our suppliers.'

Mr Collinson questions whether suppliers will share that vision, asserting:

This is an abuse of buying power. Suppliers are unlikely to stand up against such unilateral action on payment terms for fear of losing Inbev's custom completely. They will have no choice but to accept these changes and the consequences for their firms. This is nothing short of making suppliers pay for savings at Inbev. To try to pass it off as in the best interests of suppliers in the long-term is particularly galling.

Source: FPB website: www.fpb.org.uk.

Learning objectives

This chapter aims to evaluate the advantages and disadvantages of the following means of short- and medium-term finance:

- Trade credit.
- Bank finance.
- Factoring and invoice discounting.
- Bills of exchange and acceptance credits.
- Hire purchase.
- Leasing.

Particular attention is given to leasing in view of its importance as a method of financing the acquisition of a wide range of assets.

In addition, the commonest ways of financing foreign trade are also described.

 **myfinancelab** *Complete your diagnostic test for Chapter 15 now to create your personal study plan.*

15.1 INTRODUCTION

While banks have taken a number of steps to address the criticisms raised by their detractors, many firms still find it difficult to arrange short-term finance, and many remain critical of the banking system. (For news of developments in the field of finance for small firms, see the website of the British Bankers Association: **www.bba.org.uk**.) Despite such problems, the banks remain the most important source of external finance for small and medium-sized firms, providing more than 50 per cent of their external finance. However, it is not only smaller firms that tend to rely on shorter-term finance; firms of all sizes use these sources to varying degrees. For example, many larger companies arrange access to overdraft finance to tide them over temporary liquidity shortages.

In this chapter, we examine the nature and characteristics of alternative sources of finance, ranging from very short-term facilities (e.g. trade credit) to rather longer-term ones (e.g. bank loans and finance leasing). Generally, we classify these under the heading of short and medium term, although the distinction is somewhat arbitrary. For example, lease finance can be used for varying time periods, ranging from a few weeks (operating leases) to as long as 15 years or more in the case of some finance leases. In addition, bank overdrafts are essentially short term in nature, but are often used continuously for lengthy periods.

15.2 TRADE CREDIT

Trade credit is finance obtained from suppliers of goods and services over the period between delivery of goods (or provision of a service) and the subsequent settlement of the account by the recipient. During this time, the company can enjoy the goods or benefit from the service provided without having to pay up. Granting customers a credit period is part of normal trading relationships throughout most of UK industry. For this reason, it is sometimes called 'spontaneous finance'. Additional features of trade credit packages include the amount of credit that a company is allowed to obtain, whether interest is paid on overdue accounts, and whether discounts are offered for early payment.

A common way of expressing credit terms is as follows:

'2/10: net 30'

This means that the supplier will offer a 2 per cent discount for early settlement (in this case, within 10 days); otherwise, it expects payment of the invoice in full within 30 days. As shown below, *not* to take a discount is often an expensive option. Moreover, a very effective way of antagonising suppliers is to delay payment *and* attempt to claim the discount.

The length of the trade credit period offered depends partly on the following factors:

- *Industry custom and practice.* Terms of trade credit typically reflect traditional norms built up over years of trading. Although these terms often vary between industries, they are quite uniform within industries. Any supplier wanting to depart from the industry norm has to compensate with some other product offering, say, speed of delivery, to avoid losing sales.
- *Relative bargaining power.* If the supplier has a large range of customers, none of whom are crucial to it, and if the product is essential to buyers, the supplier has great power to impose its own terms. This power is enhanced by lack of strong competitors.
- *Type of product.* Products that turn over rapidly are often sold on short credit terms because they command small profit margins. Delay in settlement would severely erode the margin.

A firm's trade credit position is volatile, as it depends on which of its suppliers are awaiting payment and for how much, and these factors change continuously with the flow of business transactions. It is useful to express the average trade credit period in days calculated as follows:

$$\text{Creditor days} = \frac{\text{Trade creditors}}{\text{Credit purchases}} \times 365$$

However, for the outside observer, this figure is only observed at the balance sheet date, and even so, could have been 'window-dressed' by accelerated settlements immediately prior to drawing up the accounts. It is sometimes expressed in terms of total purchases and sometimes, when data for purchases is unavailable, in terms of overall cost of sales.

Self-assessment activity 15.1

Trade credit is often called 'spontaneous finance' or 'automatic finance'. Can you see why?

(Answer in Appendix A at the back of the book)

Because trade credit represents temporary borrowing from suppliers until invoices are paid, it becomes an important method of financing investment in current assets. Firms may be tempted to view trade creditors as a cheap source of finance, although statutory rights to claim interest on late payments now exist. Having a debtors' collection period shorter than the trade collection period may be taken as a sign of efficient working capital management. However, trade credit is by no means free – it carries both hidden and overt costs.

Excessive delay in settling invoices can undermine the stability of a business in a number of ways. Existing suppliers may be unwilling to extend more credit until existing accounts are settled. They may start to assign lower priority to future orders placed by the culprit, they may raise prices in the future or they may simply not supply at all. In addition, if the firm acquires a reputation as a bad payer among the business community, its relationships with other suppliers may be soured.

Finally, by delaying payment of accounts due, the company may be passing up valuable discounts, thus effectively increasing the cost of goods sold. This can be shown with a simple example.

Martock plc is offered a discount of 2.5 per cent on an invoice of £100,000 by a major supplier if it settles the account within 10 days, rather than taking the normal credit period of 30 days. Martock, which has at present a zero cash balance, can borrow from its bank at an interest rate of 15 per cent p.a. Should it borrow and exploit the discount, or take the full credit period of 30 days?

If it takes the discount and pays on (but not before!) day 10, it will have to borrow (97.5 per cent × £100,000) = £97,500 for an additional period of 20 days, since it would have to settle anyway after 30 days. If it waits until day 30, the cost of settling the bill is effectively the lost discount of £2,500. By advancing payment, Martock is borrowing £97,500 for 20 days in order to save £2,500, an interest rate over 20 days of 2.56 per cent. Expressed as an annual interest rate, this approximates to:

$$\frac{£2,500}{£97,500} \times \frac{365}{20} \times 100 = 46.8\%$$

A more accurate solution can be obtained by compounding over the number of 20-day periods in a year (18.25). The true rate is:

$$[(1.0256)^{18.25} - 1] = 58.6\%$$

As this compares very favourably with the 15 per cent cost of borrowing, Martock should borrow in order to advance this payment.

During the recession of the early 1990s, there was widespread concern over the tendency for more firms to delay payment of accounts. Larger firms allegedly exploited their industrial muscle by simultaneously spinning out their trade credit periods while insisting on prompter payment from small firm customers. This problem led the UK Government in 1998 to introduce legislation for a statutory right to demand interest on overdue accounts, initially for smaller companies only.

Until the passage of the Late Payments of Commercial Debts (Interest) Act in 1998, interest could only be claimed on late debts if it was included in the contract, or if awarded by a court. The Act enabled small businesses (50 or fewer employees) to claim interest from large businesses and the public sector. From November 2000, they were entitled to claim interest from other small businesses. From November 2002, any business obtained the statutory right to claim interest from any other firm or from the public sector.

Firms that suffer from late payment face a difficult choice – delay settlement to their own suppliers or fall back on their banks for supplementary finance, via either overdraft or loan facilities. In the next section, we consider bank lending.

Self-assessment activity 15.2

What is the true annual interest rate paid when a firm delays payment under the following credit terms: '$1\frac{1}{2}$ /15: net 40'?

(Answer in Appendix A at the back of the book)

15.3 BANK CREDIT FACILITIES

Major commercial banks extend a variety of credit facilities, ranging from short-term overdrafts to long-term loans of varying terms. The interest rate generally increases with the term of the advance, the actual rate being linked to the bank's base rate, which in turn depends on the base rate set by the national monetary authority (in the UK, the Monetary Policy Committee of the Bank of England).

Self-assessment activity 15.3

What do you think are the main considerations a banker makes in assessing an application for a loan or overdraft?

(Answer in Appendix A at the back of the book)

■ Overdrafts

The best-known form of bank finance is the overdraft, a facility available for specified short-term periods such as six months or a year. This facility specifies a maximum amount that the firm can draw upon either via direct cash withdrawals or in payments by cheque to third parties. Interest is paid on the negative balance outstanding at any time rather than the maximum advance agreed. Compared with many other forms of finance, it is relatively inexpensive, with the interest cost set at some two to five percentage points above base rate, although most banks also levy an arrangement fee (perhaps 1 per cent) of the maximum facility.

In principle, overdrafts are repayable at very short notice, even on demand, although unless the company abuses the terms of the facility by exceeding the agreed overdraft limit, the overdraft is unlikely to be called in. Besides, it is rarely in the best interests of a bank to do this suddenly, as it could exert such severe financial pressure on the client as to force it into liquidation.

Nevertheless, the bank retains the right to appoint a receiver if the client defaults on the debt. In practice, well-behaved clients can roll forward overdrafts from period to period. As a result, the overdraft effectively becomes a form of medium-term finance. Even in these cases, it is wise policy not to use an overdraft to invest in long-term assets that would be difficult to liquidate at short notice if the bank suddenly decided to call in the debt.

To protect against risk of loss, the bank will usually demand that the overdraft be secured against company assets, i.e. in the event of default, the receiver will reimburse the bank out of the proceeds of selling these assets. Security can be in two forms: a **fixed charge**, where the overdraft is secured against a specific asset, or a **floating charge**, which offers security over all of the company's assets, i.e. those with a ready and stable second-hand market. A floating charge therefore ranks behind a fixed charge in the queue for payment. For trading companies, overdrafts are often secured against the inventory that the company purchases with the funds borrowed, or even against debtors. In this respect, the overdraft is '*self-liquidating*' – it can be reduced as the company sells goods and banks the proceeds.

Alternatively, and more to the liking of most bankers, overdrafts are secured against property. However, this created problems in the recession of the early 1990s, when the unprecedented collapse of property market prices often reduced the value of assets upon which overdrafts were secured to below the balance outstanding. Many banks made major provisions against the increasing likelihood of bad debts. In addition, they incurred much ill-will by allegedly recalling overdrafts prematurely, thus exacerbating the liquidity difficulties of their clients, already seriously affected by falling sales. Many critics accused the banks of forcing many essentially sound companies out of business.

■ Term loans

term loans
Loans made by a bank for a specific period or term, usually longer than a year

Term loans are loans for a year or longer. UK banks have traditionally been reluctant to lend on a long-term basis, mainly because the bulk of their deposit liabilities are short-term. In the event of unexpectedly high demand by the public to withdraw cash, this could leave them vulnerable if they were unable to recall advances quickly from borrowers. This low exposure to default risk is generally regarded as the reason why banking collapses are relatively uncommon in the UK.

However, because of criticism by a series of official reports on the financial system and the advent of intensive competition from London branches of overseas banks, the main UK banks are now far more willing to lend long-term. Term loans can be arranged at variable or fixed rates of interest, although the interest cost is usually higher in the latter case. For variable rate loans, the rate set may be two to five percentage points above the bank's base rate, depending on the credit rating of the client and the quality of the assets offered as security. In addition, an arrangement fee is usually charged.

Self-assessment activity 15.4

Which are normally more expensive for firms – overdrafts or term loans? Why?

(Answer in Appendix A at the back of the book)

balloon loan
Where increasing amounts of capital are repaid towards the end of the loan period

bullet loan
Where no capital is repaid until the very end of the loan period

Tailor-made facilities are available to some firms, with repayment terms designed to suit their expected cash flow profiles. Sometimes, the bank may grant a 'grace period' at the outset of the loan when no capital is repayable and interest may be charged at a relatively low, but increasing, rate. This is particularly suitable for a small, developing company trying to establish itself. Similarly, a **balloon loan** is where increasing amounts of capital are repaid towards the end of the loan period, whereas a **bullet loan** is where no capital is repaid until the very end of the loan period.

The proportion of borrowing on overdraft by small businesses is in long-term decline. In its *Finance for Small Firms* report in 1995, the Bank of England estimated that fixed-term lending to small firms had overtaken overdraft finance, the proportion of total lending in the form of term loans then being 60 per cent, compared with 40 per cent only two years previously. By 2006, the British Bankers Association (**www.bba.org**) was reporting that overdraft lending accounted for only 19 per cent of its members' support for small businesses, with over 75 per cent of term lending having a maturity of more than three years, with interest charged on a variable rate basis.

■ Revolving credit facilities (revolvers)

A term loan generally specifies an agreed payment profile and the amounts repaid cannot normally be re-borrowed. A revolver allows the borrower to borrow, repay and re-borrow over the life of the loan facility, rather like a continuous overdraft. Like an overdraft, it is frequently secured on the borrower's working capital, e.g. using debtors and stocks as collateral, although very large firms may not be asked for any security. The advantage of revolvers is the enhanced flexibility provided, i.e. funds can be re-used in a continuous credit line. The commitment by the bank thus 'revolves' – the borrower can continue to ask for loans, subject to giving suitable notice, so long as the committed total is not exceeded. The fees charged include:

- A front-end or facility fee for setting up the loan.
- A commitment fee to compensate the bank for having to commit some of its loan capacity by setting aside reserve assets to meet capital adequacy rules.
- The interest cost, usually expressed as so many basis points (one bp = 0.01%) above LIBOR, the rate at which London-based banks lend to each other.

In April 2007, Kingfisher plc, the stores group, disclosed that its 'committed banking facilities' included a £500 million revolving credit facility provided by a number of banks, originally due to mature in August 2010, had been extended to mature in August 2011. In addition, it had bilateral revolving credit facilities provided by a number of banks due to mature in March 2010. All these facilities paid interest 'based on LIBOR', fixed for periods between 1 and 6 months. The facilities were available to be drawn down for 'general corporate purposes including working capital requirements'.

■ The Small Firms Loan Guarantee Scheme (SFLGS)

First known as the Loan Guarantee Scheme, this was introduced in 1981 following the Report of the Committee to Review the Functioning of Financial Institutions, set up to investigate the provision of finance to business. The Committee pointed to the difficulty faced by smaller firms with little or no track record in obtaining suitable longer-term finance.

Such firms are reluctant to release control by issuing equity, while investors are reluctant to purchase equity owing to the risks involved, especially the difficulty of liquidating their investment on acceptable terms. Firms may also have difficulty persuading banks of the inherent viability of their businesses, and are frequently unable to offer sufficient and suitable security. Where banks do offer loan finance, it is often on less advantageous terms than those extended to larger firms and less likely to be augmented in times of difficult trading.

The SFLGS is financed by the Department of Business, Enterprise and Regulatory Reform. In its initial form, the scheme was supposed to be self-financing. The government originally guaranteed that, in the event of failure of a business, 80 per cent of the loan would be repaid to the financial institution making the loan. The cost of meeting guarantees would be met by borrowers paying to the government a premium of three percentage points above the normal commercial rate applied by the bank. Loans were available for periods of two to seven years. Losses on the scheme turned out much higher than anticipated, largely, it was alleged, because banks contrived to shift existing shaky clients on to it. After many modifications over the years, it now (2008–9) embodies the following features:

- Small and medium-size firms of any age may apply for loans directly to banks and other lenders.
- Applicants must demonstrate that they have applied for a conventional loan and have been rejected for lack of security.
- Only firms with annual turnovers no more than £5.6 million are eligible.
- 75 per cent of the loan is guaranteed. Total DBERR funding is £360 million.
- Applicable for sums between £5,000 and £250,000.
- Borrowers pay a 'normal commercial' interest rate to the bank, plus a premium of 2.0 per cent p.a. on the outstanding amount of the loan, payable to the Department of Business, Enterprise and Regulatory Reform.
- Term of loan available from 2 to 10 years.

For updates on the LGS, see **www.bba.org.uk** and **www.berr.gov.uk**.

15.4 INVOICE FINANCE (OR 'ASSET-BASED FINANCE')

factors
Organisations that offer to purchase a firm's debtors for cash

Some companies, which need to offer trade credit to customers for competitive reasons, find that they need payment earlier than agreed in order to assist their own cash flow. Institutions called **factors**, mainly subsidiaries of the major banks and members of the Asset-Based Finance Association (**www.abfa.org.uk**) (formerly known as the Factors and Discounters Association), exist to help such companies. Factors do not always provide new finance, but can accelerate the cash conversion cycle for client companies, allowing them to gain access to debtors more quickly than if they waited for the normal trade credit period to unwind. The essence of both factoring and invoice discounting is to use debtors (i.e. invoices) to provide security for financing, hence the term 'invoice finance'. Invoice finance thus includes both factoring proper and invoice discounting.

Between 1993 and 2007, total domestic invoice financing, measured by ABFA clients' turnover, grew from £19 billion to £192 billion, 89 per cent of which was invoice discounting. In 2007, invoice discounting itself grew by 10 per cent from £145 billion to £159 billion.

■ Factoring

Factoring involves raising immediate cash based on the security of the company's debtors, thus accelerating payment from customers. A factor provides three main services – sales administration, credit protection and provision of finance, commonly 80–85 per cent of the value of approved invoices.

Sales administration
A factor assumes the various functions of sales ledger administration, ranging from recording sales details to sending out invoices and reminders and collecting payment. The benefits for the client are the cost savings from reducing in-house administration and access to a more efficient, specialist debtor management team. This is particularly valuable to a young fast-growing company, which may outgrow its administration system and otherwise be exposed to the liquidity risks of overtrading. The fee for such an administration service would lie typically in the range of 0.75–2.5 per cent of the value of turnover handled, depending on whether credit protection against bad debt losses is included.

Credit protection

The factor may provide a credit evaluation service for clients, analysing customer characteristics before deciding on their creditworthiness. When all risks are borne by the factor, the service is termed *'without recourse'*, i.e. the factor has no 'comeback' on the client if customers default. Where this applies, the factor requires total control of credit approval, monitors customers' payments and attempts to collect payment. This suggests a possible problem with factoring – the intervention of the factor between the factor's client and the debtor company could endanger trading relationships and damage goodwill. For this reason, some clients prefer to retain responsibility for collection of problem debtors. This is known as 'undisclosed factoring'. In the case of *'with recourse'* factoring, the factor will call upon its client to reimburse the funds advanced on an invoice relating to a delinquent account.

Provision of finance

A factor will also advance funds to a client, based on a proportion, say 80 per cent, of approved (i.e. reliable) invoices. For example, a company with sales on 30-day credit terms from reliable customers of £500,000 per month would receive an advance of 80 per cent, i.e. £400,000 each month. The interest rate would be related to bank base rate, probably slightly above the cost of an overdraft. The client would receive the balance of the payment less interest and an administration charge, perhaps equal to 0.5 per cent of turnover.

Although factors provide valuable services, companies are sometimes wary of using them for reasons other than cost. Besides possible difficulties over collection of payment, widespread knowledge that a company is using a factor may arouse fears that the company is beset by cash flow problems. If so, its suppliers may impose more stringent payment terms, thus negating the benefits provided by the factor. Companies concerned by these risks may seek the more widely used, but less comprehensive service, of **invoice discounting**.

invoice discounting
Where a factor purchases selected invoices from a client firm, without providing debt collection or account administration services

Self-assessment activity 15.5

Summarise the benefits of factoring for a small firm.

(Answer in Appendix A at the back of the book)

■ Invoice discounting

Although factors provide a range of services, a company seeking merely to improve its cash flow is likely to use an invoice discounting facility. This involves the purchase of selected invoices, sometimes just one, by the discounter. The discounter will advance immediate cash up to 85 per cent of face value. It assumes no responsibility for the administration of the accounts receivable, or the collection of the debts. The service is totally confidential, the client's debtors being unaware of the existence of the discounter. It is therefore equivalent to the financing service provided by a factor, although restricted to a narrower range of invoices. Invoice discounting business in the UK amounted to over £159 billion in 2007.

Administration charges for this service are around 0.5 per cent of a client's turnover. It is more risky than factoring, since the client retains control of its credit policy. Consequently, such facilities are usually confined to established companies with turnovers above £1 million. Interest costs are usually 3–6 per cent above base rate, although larger companies and those that arrange credit insurance may receive keener terms.

How invoice discounting works

At the beginning of February, Menston, plc sells goods for a total value of £300,000 to regular customers, but decides that it requires payment earlier than the agreed 30-day credit period for these invoices. A discounter agrees to finance 80 per cent of their face value, i.e. £240,000. Interest is set at 13 per cent p.a. The invoices were due for payment in early March, but were subsequently settled in mid-March, exactly 45 days after the initial transactions. The service charge is set at 1 per cent. As usually applies, a special account is set up with a bank, into which all payments are made. The sequence of cash flows is as follows:

February:
  Menston receives cash advance of £240,000
Mid-March:
  Customers pay up £300,000
  Invoice discounter receives the full £300,000
  Menston receives the balance less charges, i.e.

  Service fee = 1% × £300,000   = £3,000

  Interest = (13% × £240,000 × 45/365) = £3,847

  Total charges   = £6,847

  Net receipts = (20% × £300,000) − £6,847

             = (£60,000 − £6,847) = £53,153

  Total receipts by Menston = (£240,000 + £53,153) = £293,153

In effect, Menston has settled for a discount of £6,847/£300,000 = 2.3 per cent over 45 days for early receipt of 80 per cent of the accounts payable. As there are about eight 45-day periods in a year, this corresponds to an annual interest rate of:

$$(1.023^8 - 1) = 0.1995, \text{ i.e. } 19.95\%$$

To view the offerings of a typical factor, visit the website of RBS Commercial Services Ltd, a subsidiary of the Royal Bank of Scotland (**www.rbcs.co.org**).

15.5 USING THE MONEY MARKET: BILL FINANCE

A bill is often likened to a post-dated cheque or an IOU, as it represents a commitment to pay out a specific sum of money after a specified period of time. Bills are traded on the money market, which specialises in providing funds repayable over periods of less than a year. The major players in the money market are the commercial banks, which lend to each other, often overnight, to cover temporary cash shortages, at the London Inter-Bank Offered Rate (LIBOR), and to the discount houses. The latter exist primarily to deal in short-term bills issued by the government (Treasury Bills), local governments and companies. They borrow on a very short-term basis from the commercial banks, usually at a very keen interest rate, and use the proceeds to purchase bills, which they may hold to maturity or sell at a profit in the money market.

A company can use two main types of bill to raise finance: a trade bill (Bill of Exchange) or a bank bill (acceptance credit).

■ Bills of Exchange

Bills of Exchange are generally trade-related, connected to specific trading transactions. The trader purchasing goods draws up a bill stating a promise to repay at some future date, and then conveys the bill to the supplier of the goods.

The supplier may then hold the bill until maturity or sell it in the market to a discount house, if cash is required earlier. The terms of the bill usually include an implicit interest charge, although no interest as such is paid, as the following example indicates.

A trader wishes to acquire goods to the value of £1 million. He draws up a Bill of Exchange, promising to pay £1 million in three months' time. The bill is immediately sold by the seller of the goods to a discount house, which pays out £975,000, i.e. a discount of £25,000. If the discount house holds the bill to its maturity, it earns a profit of £25,000. This is equivalent to an interest rate of (£25,000/£975,000) = 2.56% over three months, or in annual terms:

$$(1.0256)^4 - 1 = (1.1064 - 1) = 0.1064, \text{ i.e. } 10.64\%$$

If the discount house can borrow at less than this rate, it stands to make a profit. If interest rates fall, the discount house's profit margin widens. Alternatively, the discount house may sell the bill on the money market, its value rising as it nears maturity. The ultimate holder will present the bill to the trader, who assumes responsibility for payment and, by this time, hopes to have sold the goods at a profit.

Bills of Exchange are drawn up for periods of 60–180 days, for values of at least £75,000, reflecting interest rates based on LIBOR, but dependent on the riskiness of the companies concerned and their respective credit ratings.

■ Acceptance credits

These are often called 'bank bills' – whereas trade Bills of Exchange are drawn on the purchaser of goods by a supplier, an acceptance credit is drawn by a company on a bank. Acceptance credits were originally developed by merchant banks, but all large banks now offer this facility. The bank grants a credit facility whereby a client company can draw bills (up to an agreed limit) that the bank will accept, i.e. agree to honour, when presented for payment at a future specified date. The client company may not use the facility immediately, but may treat it as a standby to be used when required, usually in minimum tranches of £250,000, over a period of up to five years. Accepted bills are sold on the discount market by the bank on behalf of the client company at a relatively 'fine', or low, discount. The company thus effectively obtains finance from the purchaser of the discounted bills, using the name and reputation of the accepting bank as security. It is thus a somewhat roundabout way of one company lending to another, using the intermediary services of the bank. Bank bills have a period to maturity of 30, 60, 90 or 180 days. At the maturity date, the bank can either 'accept' a new bill (i.e. effectively roll over the old one) or receive the full value of the bill from the client's account, using the funds to pay the bearer of the bill. The cost to the client is the amount of discount on the bill plus a fee payable to the bank of around 0.6 per cent. It is not unusual for banks to require security for such facilities – the nature and type depending on the size and credit standing of the client.

Acceptance credits have become increasingly popular in recent years. Among their advantages are the following:

1 The 'interest' cost is relatively low, often below that of an overdraft, because the bills are backed by a bank.
2 The cost involved with the bill is known when it is discounted, and is not affected by subsequent interest rate changes, allowing greater accuracy in cash budgeting.
3 Acceptance credits can be negotiated for longer periods than overdrafts, thus offering more security to the borrower.
4 Unlike traditional Bills of Exchange, they are not tied to specific transactions.

However, they are available only to large companies with sound credit ratings, and they lack the flexibility of overdrafts, which a firm can reduce if it wishes to lower interest costs.

■ Commercial paper (CP)

Another financial innovation imported from the USA that UK firms have been allowed to use since 1986, CP is a means whereby the treasurer can circumvent the banking system by issuing promissory notes – effectively IOUs – directly to large financial institutions such as pension funds and insurance companies, or to corporates with temporary excess liquidity.

disintermediation
Business-to-business lending that eliminates the banking intermediary

CP is an example of **disintermediation** as it cuts out the banking intermediary as middleman and thus avoids paying the spread on the difference between the bank's lending rate and the rate it pays depositors, i.e. it is cheaper. In addition, CP avoids having to submit to the restrictive covenants that banks often impose on borrowers. However, because it is unsecured, the ability to issue CP is confined to the largest firms with the highest credit ratings, i.e. 'blue chips'.

Specifically, to be allowed to issue CP, a UK firm must:

- be listed on the London Stock Exchange
- have net assets in excess of £50 million
- issue CP that matures between 7 and 365 days (2–270 in the USA)
- issue CP with a minimum denomination of £500,000 (usually, $100,000 in the US).

Interest is not usually payable on CP – it has a maturity value greater than the amount lent, the difference being the cost to the issuer. There is virtually no secondary market in CP.

15.6 HIRE PURCHASE (HP)

With HP (also known as asset purchase), the user of the asset will eventually own it. Small and medium-sized firms (SMEs) are particularly reliant on HP, which, although often an expensive option, is readily available because the loan is secured on the asset acquired. HP of equipment by firms is very similar to HP by consumers of durables such as washing machines. The equipment is purchased initially by specialist institutions called finance houses, most of which are subsidiaries of commercial banks and members of the Finance and Leasing Association (FLA, **www.fla.org.uk**). The hirer usually makes a down-payment and then signs a commitment to a series of (usually) monthly hire charges over a specified period, at the end of which the legal title to the article passes to the user. The hire charge contains two elements: an interest charge to reflect the borrowing of the capital involved; and a capital repayment element.

Two major advantages of HP are the avoidance of a major cash outlay at the outset of the project and the immediate availability of the asset for use, although the user

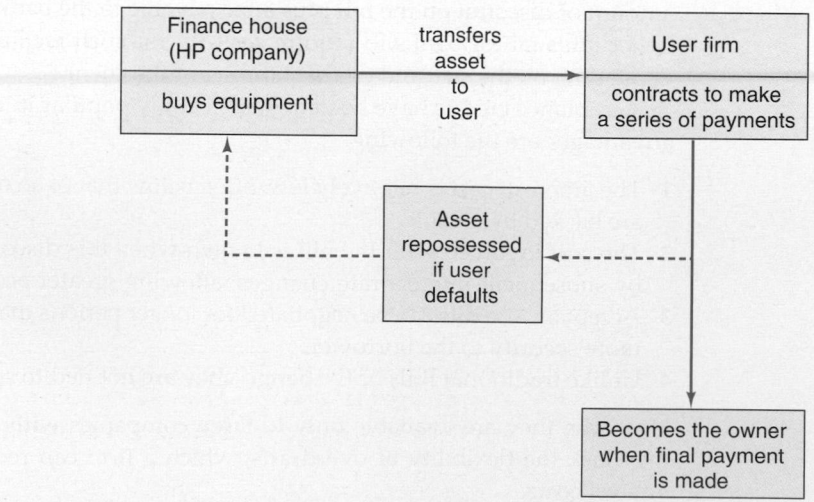

Figure 15.1 How hire purchase works

assumes all the responsibilities of maintenance and insurance. However, if the user fails to fulfil the payment schedule, the owner can repossess the asset. The user then loses all title to the asset and obtains no credit for payments already made. HP tends to be used by smaller and riskier firms. Because of the default risk, the contract period is invariably less than the asset lifespan, so that in the event of repossession, the owner knows that it will hold a saleable asset with some working life remaining. Conversely, over longer-term HP contracts, the owner is exposed to the risk of technological progress outdating the asset and reducing its marketability. As a result of these considerations, HP is expensive.

How HP works

Boston Builders, wishing to obtain an earthmover using HP, contacts a manufacturer, which arranges an HP contract with a finance house. The finance house buys the asset, which has an expected useful life of four years, for £120,000 and arranges the following contract:

- The asset will be purchased over three years but operated for four years.
- A down-payment of £20,000 is required.
- Interest is charged at 15 per cent on the initial loan of (£120,000 − £20,000) = £100,000.
- Capital will be repaid in three equal instalments.

Note that, even though the balance outstanding declines over the three-year period, interest is applied to the initial loan of £100,000. The annual payments are thus:

Interest = (15% × £100,000) = £15,000 p.a.

Capital = £100,000/3 = £33,333 p.a.

Total   = £48,333 p.a. (or £4,028 if paid monthly)

Because the interest is paid in this way, the true interest rate is roughly double the quoted rate. The annual interest charge of £15,000, applied to the average capital outstanding of £50,000, yields an effective interest rate of (£15,000/£50,000) = 30 per cent, although the precise rate depends on the actual timing of payments.

An advantage of HP is that the user qualifies for tax relief on the interest element in the repayment profile and also a writing-down allowance (WDA) on the capital expenditure component (20 per cent reducing balance), based on the total cash price of the asset. Table 15.1 shows the profile of tax reliefs for the Boston Builder's example,

Table 15.1 Tax relief on 3-year HP contract with 4-year asset lifetime (£)

Year	Interest @ 15%	+	WDA @ 20%*	=	Total tax reliefs	Tax saving @ 28%
1	15,000		20,000		35,000	9,800
2	15,000		16,000		31,000	8,680
3	15,000		12,800		27,800	7,784
4	–		51,200**		51,200	14,336
Totals	45,000		100,000		145,000	40,600

* The outlay is assumed to occur on Day 1 of Year 1. Expenditure on the last day of year 0 would bring forward the tax reliefs by a full year and raise the PV of the tax savings.

***Note:* Year 4 allowable depreciation = [Outlay less depreciation claimed for years 1–3 less sale value]

= £120,000 − [£20,000 + £16,000 + £12,800] − £20,000
= £51,200

assuming tax is paid at 28 per cent, with no delay, and that the asset has a sale value of £20,000 after four years, i.e. tax allowable depreciation totals £100,000 over the four-year lifespan of the asset.

Tax relief is of value only to profitable companies with taxable capacity. Many of the smaller and higher-risk companies (often start-ups) that use HP are unable to exploit the tax breaks, often making HP a very expensive form of finance.

Self-assessment activity 15.6

What is the monthly payment on the following HP contract?

- Total cost of equipment = £40,000
- Down payment = 30%
- Interest rate = 10%
- Hire period = 4 years

(Answer in Appendix A at the back of the book)

15.7 LEASING

Leasing resembles both HP and conventional borrowing, but deserves separate and extensive analysis for two reasons. First, it is highly significant as a means of financing fixed capital investment, having grown substantially since the 1970s. In 2006, members of the FLA provided £27 billion of new finance to the UK business and public sectors, representing 30 per cent of all fixed capital investment (excluding real estate).

Second, it provides an example of the interaction between investment and financing decisions, and an opportunity to show how DCF principles can be applied to financing as well as investment decisions.

■ What is leasing?

According to the International Accounting Standards Board:

> A leasing transaction is a commercial arrangement whereby an equipment owner conveys the right to use the equipment in return for payment by the equipment user of a specified rental over a pre-agreed period of time.

Thus, leasing is a way for companies to obtain the use of equipment when, for varying reasons, they may wish to avoid acquiring it outright using other financing methods. Leasing is a distinctive method of finance because it involves important interactions between the investment and financing decisions.

■ How a lease works

the lessee
A firm that leases an asset from a lessor

the lessor
A firm that acquires equipment and other assets for leasing out to firms wishing to use such items in their operations

Most leasing activity is undertaken by banking and similar institutions, such as HSBC Equipment Finance (UK) Ltd. In addition, some manufacturers, such as IBM and John Deere, operate leasing companies to market their own products. A company wishing to obtain the use of an asset (**the lessee**), such as an oil company wishing to lease a tanker, or a Development Corporation wishing to lease property, will approach the leasing specialist (**the lessor**) with its requirements. The deal will involve the lessor purchasing the tanker, or the site, and renting it to the lessee in return for a specified series of rental payments over an agreed time period.

■ Types of lease

Where the agreed term of the lease approximates to the expected lifetime of the asset, the lessee is clearly using the lease arrangement as an alternative form of finance to outright acquisition. As a result, it avoids having to incur the perhaps substantial cash outlay required at the outset of the project. Hence this type of lease is called a **finance lease**, (or a **capital lease**, or a **full payout lease**).

In the UK, it is important that the asset remains the legal property of the lessor, otherwise certain tax advantages may be lost. At the expiration of the lease contract, the two parties may negotiate a **secondary lease**, or the owner may otherwise dispose of the asset. For assets with long lives, the lessor may ignore the potential resale value of the asset when setting the rentals, since it is too distant in time to predict accurately. Instead, the lessor may agree to reimburse the lessee with a proportion, often over 90 per cent (but never 100 per cent) of the resale value, an agreement known as a **rebate clause**. Rebates are taxable in the hands of the lessee. Secondary leases are often undertaken at nominal or 'peppercorn' rentals to reflect their bonus nature – the owner will already have received back its outlay plus target profit once the contract has reached its full term.

However, not all forms of lease operate over long time periods. The user may not wish to incur the long-term contractual liability to pay rentals, especially if it wishes to obtain the use of the asset only to perform a specific job, e.g. drilling equipment to bore out a specific oil well. Lessors are willing to rent equipment to such firms on the basis of an **operating lease**. This is usually job-specific and can be cancelled easily, whereas cancellation of a finance lease usually involves financial penalties so severe that termination is rarely worthwhile. The lessor will hope to arrange a series of such contracts in order to recover its capital outlay and achieve a profit. For this reason, the operating lease is called a **part-payout lease**. Unlike the finance lease, where the user bears the full risks both of 'downtime' (inability to use the asset) and obsolescence, in an operating lease, the owner incurs the brunt of these risks. To compensate for the risk of having a yard full of idle and rusting equipment, the lessor will apply a rental that is higher per unit of time than that for a financial lease, thus incorporating a risk premium.

Operating leases have another advantage for lessees. They are usually negotiated on a 'maintenance and insurance' basis, whereby the owner undertakes to insure and service the asset. This is normally the responsibility of the user in the case of the finance lease, although the owner may actually perform the servicing functions for a fee. The suitability of the operating lease for short-term projects explains its popularity in the construction industry under the guise of **plant hire**, and for assets with a rapid rate of technological advance such as photocopiers and computers, often called **contract hire**. To compensate for the risks of leasing out high-technology assets, lessors are sometimes able to protect themselves by using specialist computer leasing insurers, although premium rates are generally high.

■ The characteristics of a finance lease

Until 1984, UK companies were able to disguise their true indebtedness by undertaking financial leases. Neither the asset acquired nor the contractual liability incurred had to appear on the balance sheet, although the rental payment obligation had to be stated in the notes to the accounts. To ensure that balance sheets would give a truer picture of a company's asset/liability position, the Accounting Standards Committee issued SSAP 21, 'Accounting for Leases and Hire Purchase'. This clarified the definition of a finance lease and issued instructions on how to account for leases.

Since the period of a finance lease usually matches the expected lifetime of an asset, the old SSAP 21 defined a finance lease as one 'that transfers substantially all the risks and rewards of ownership to the lessee'. It assumes that such a transfer takes place if, at the start of the lease, the present value of the lease amounts to 90 per cent or more of

capital lease/full payout lease
A lease that transfers most of the benefits and risks of ownership to the lessee

secondary lease
A second lease arranged to follow the termination of the initial lease period

rebate clause
An arrangement whereby the lessor pays a proportion of the resale value of an asset to the lessee

operating lease
A job-specific lease contract, usually arranged for a short period, during which the lessor retains most of the benefits and risks of ownership

part-payout lease
A lease contract which recovers a return lower than the capital outlay made by the lessor

plant hire
Hiring of construction equipment

the fair value of the asset (normally its cash price). Leases are still treated as financial leases if they meet one or more of the following criteria:

1 At the end of the lease period, the lessee is given the option to buy the asset at a price below its anticipated market value.
2 The lease period is no less than 75 per cent of the estimated working life of the asset.
3 Ownership of the asset can be transferred to the lessee at the end of the lease.

If, at the termination of a finance lease, the ownership of the asset does pass from lessor to lessee, all previously exploited tax benefits will be clawed back by the Inland Revenue.

The rules laid down for the accounting treatment of a finance lease are as follows:

1 The fair value is capitalised as a fixed asset in the Balance Sheet and is depreciated over its useful life (or lease term, if shorter).
2 Rental payments are treated as comprising two elements – a finance charge and a capital repayment. The finance charge is written off over the lease period at a constant periodic rate.
3 The obligation to pay the capital element of the future rentals is recorded as a long-term creditor in the balance sheet. At the outset of the lease period, the capital element should equal the cash value of the asset.
4 Every year, the appropriate finance charge is recorded in the profit and loss account.

The International Accounting Standards Board is currently working on a project to investigate the feasibility of aligning the accounting treatment of finance and operating leases, i.e. to require that all leases are shown on the balance sheet (see **www.fla.org.uk** and **www.asb.org.uk** for developments).

Self-assessment activity 15.7

What are the main differences between an operating lease (OL) and a finance lease (FL)?

(Answer in Appendix A at the back of the book)

15.8 LEASE EVALUATION: A SIMPLE CASE

We now formally evaluate the decision of whether to lease or purchase an asset. We begin by establishing the basic principles.

A lease requires a series of fixed rentals. This is a major appeal of a lease – the lessee can predict with certainty its future lease payments, and budget accordingly. Because leasing effectively offers fixed-rate finance, we examine the merits of a lease against the yardstick of the cheapest alternative form of borrowing that would otherwise be used to acquire the asset, normally a bank loan. In other words, *leasing is an alternative to borrowing at the risk-free rate (or, more realistically, the bank's lending rate) in order to buy the asset outright.* Because leasing involves incurring a fixed liability, astute lenders regard leases and debts as substitutes, so that an increase in the former should lead to an exactly compensating decrease in the other. While this 'one-for-one debt displacement hypothesis' is not universally accepted, we will assume that lenders recognise leasing for what it is. The leasing decision then amounts to evaluating the question: 'Is it preferable to lease or borrow-to-buy?'

It follows that the *appropriate rate of discount to use in lease evaluation is the lessee's cost of borrowing.* This is shown in the following example.

■ Lease evaluation: Hardup plc

Hardup plc wishes to lease an executive jet aircraft from Flush Ltd, the leasing subsidiary of Moneybags Bank plc. The aircraft would otherwise cost £13.75 million to purchase via a bank loan at a 12 per cent interest rate. Flush quotes an annual rental of £5 million over three years, with the first instalment payable immediately, and the rest annually thereafter. Should Hardup lease or borrow-to-buy? With a lease, Hardup avoids the immediate cash outlay of £13.75 million, but loses out on any resale value (unless there is a rebate clause).

We may assess the value of the lease on an incremental basis by finding the NPV of the decision to lease rather than borrow-to-buy, sometimes called the **net advantage of the lease (NAL)**. Table 15.2 sets out the relevant cash flows. For the purposes of this simple example, tax has been ignored (we will see later that the tax regulations have had an important bearing on the growth of leasing). We also ignore any resale value.

Table 15.2 Hardup plc's leasing analysis

	Year		
Item of cash flow (£m)	0	1	2
Lease			
Rentals	−5.00	−5.00	−5.00
Buy			
Outlay	−13.75		
Incremental cash flows	+8.75	−5.00	−5.00
Present value at 12%	+8.75	−4.46	−3.99
Net present value = £0.30 m			

The incremental cash flow profile shows that, by leasing, Hardup effectively obtains net financing of £8.75 million in exchange for debt service costs of £5 million in each of the following two years. (Notice that the timing of rental payments conflicts with our usual assumption of year-end payments. The lease rentals are actually paid at the start of Years 1, 2 and 3, respectively, which has negligible impact on the present value computations, but could have important tax implications.) After allowing for the 12 per cent cost of borrowing, the NPV of +£0.30 million indicates that the optimal form of financing arrangement is to obtain a lease from Flush. The same result could have been obtained by separately discounting the two cash flow streams and choosing the one with the lowest present value. This is simply a comparison between an outlay of £13.75 million and three payments of £5 million p.a. beginning now, with a present value of £13.45 million, discounted at 12 per cent. For many purposes, the incremental layout is easier and clearer: for example, it focuses attention directly on the relative merits of the two options.

■ An alternative method of evaluation: the equivalent loan

Another approach to lease evaluation is to compare the purchase price of the asset concerned with the **equivalent loan**, defined as the loan that would involve the same schedule of interest and repayments as the profile of rentals required by the lessor. The lease is adjudged worthwhile if the lease rental schedule provides more finance than the loan that would be required to purchase the asset outright. The appeal of this approach is that it emphasises the financing function of a lease. In the Hardup example, the equivalent loan is the maximum loan at 12 per cent that could be supported by a payment of £5 million now and two further payments of £5 million, after one and two years respectively. To find the equivalent loan, we simply calculate the present value of a two-year annuity of £5 million and add the undiscounted first payment of £5 million, yielding a total of:

$$PV = £5\,m + £5\,m\,(PVIFA_{(12,2)}) = (£5\,m + £8.45\,m) = £13.45\,m$$

Table 15.3 The behaviour of the equivalent loan (£m)

Year	0	1	2	3
1 Balance of loan at start of year	13.45	8.45	4.46	0
2 Payments, of which:	5.00	5.00	5.00	–
3 Interest at 12%	–	1.01	0.54	–
4 Capital	5.00	3.99	4.46	–
5 Balance of loan at end of year (1–4)	8.45	4.46	0	–

Table 15.3 shows the behaviour of a loan of this amount, serviced by the same profile of payments as required by the lease itself.

All we have done here is to express the calculation in a different way. We have found that the equivalent loan is £13.45 million, while the lease itself provides the ability to acquire an asset whose price is £13.75 million, i.e. for the same profile of payments, the lease allows the firm to 'borrow' an extra £0.30 million, which is precisely the NAL. We can now see that the lease evaluation effectively involves computing a loan equivalent. A lease is worthwhile if the required payment of cash flows could service a loan higher than the outlay required to undertake the project.

In other words, a lease is worthwhile if the effective financing obtained is higher than the equivalent loan. Many writers and analysts prefer to evaluate leases in terms of equivalent loans, but, in more realistic cases, the phasing of rental payments, intermingled with tax complications, can make the computation of an equivalent loan highly complex. Generally, we favour the incremental cash flow approach.

■ Leasing as a financing decision: the three-stage approach

In this section, we offer a note of warning. In the Hardup example, we found leasing was preferable to borrowing-to-buy. However, in many companies, the analysis might not have got this far. For example, had the underlying project (the acquisition of the jet) been revealed as unattractive, the financial manager might not have bothered to undertake a financing analysis. In other words, firms may evaluate decisions in two stages: first, to assess the basic desirability of the activity; and second, if the project is deemed acceptable, to assess the optimal form of financing. The danger in this sequence is that some projects that might be rendered worthwhile by especially favourable financing packages could be rejected. It may make more sense to *evaluate projects as linked investment and financing packages*. To explore further this interaction between investment and financing, we need to explain the *three-stage approach*.

In analysing Hardup's jet acquisition, it was assumed that the investment was inherently worthwhile, and that the remaining issue was how best to finance it. Let us now examine the underlying investment decision. Imagine that Hardup is all-equity financed and that its shareholders require a return of 15 per cent. Assume further that acquisition of the jet would result in annual benefits (such as savings in executive time and in travel expenses, and income from hiring) of £6 million p.a. for three years, treated as year-end lump-sum payments, and which, while uncertain, are no more risky than existing company operations. The NPV of the jet purchase is:

$$\text{NPV} = -\text{£}13.75 \text{ m} + \text{£}6 \text{ m} (\text{PVIFA}_{(15,3)}) = -\text{£}13.75 \text{ m} + \text{£}13.70 \text{ m} = -\text{£}0.05 \text{ m}$$

The project would be (just) rejected on the NPV criterion and the issue of how best to finance it might never arise (unless there were non-financial motives, such as corporate prestige, to justify it).

However, it could be acceptable using other methods of finance, such as borrowing or leasing. Of these, lease financing is the most attractive because we know the NAL is

£0.30 million. In this example, the acquisition should be undertaken, since the NAL exceeds the negative NPV anticipated from all-equity financing. This suggests how and why a three-stage analysis should be applied. The three stages are as follows:

1 Determine whether the project is inherently attractive.
2 Even if the NPV is negative, evaluate the NAL to assess whether to lease or borrow-to-buy.
3 Assess the value of the project with the chosen financing method.

(N.B. Stage 2 could precede stage 1, of course, but all three should be undertaken.)

To examine stage 3, we compare the benefits anticipated from the project, discounted at the 'risky' rate of 15 per cent, with the costs associated with the cheapest financing method, in this case, the stream of rental payments.

As found earlier, the present value of project benefits is £13.70 million. Now applying the 12 per cent cost of debt finance to the rental stream, we have:

PV of rentals = £5 m + £5 m (PVIFA$_{(12,2)}$) = £13.75 m

Since the project benefits exceed the costs associated with the cheapest financing method, the overall investment-cum-financing package is worthwhile. This is an important result. When judged on its intrinsic merits, and evaluated using 'normal' criteria, the project reduces shareholder wealth by £0.05 million. Yet when evaluated using alternative financing methods, it becomes worthwhile.

Thus we find:

PV of project cash flows = £13.70 m

PV of lease rentals   = (£13.45 m)

NPV of project if leased = £0.25 m

This equals the NPV of the basic project (−£0.05 m) plus the NAL (+£0.30 m).

Not all marginal investment decisions can be turned around by clever financing decisions, but a three-stage approach can help to avoid rejecting some projects that might be worth undertaking if financed in particular ways. Sometimes, the lease may yield net benefits due to the borrowing advantage of the lessor (e.g. a bank), which is passed on to the lessee. Alternatively, it is possible to exploit tax advantages for the mutual benefit of the two parties. Finally, in some cases, the lessor may have bulk-buying advantages not possessed by the lessee.

We will continue to assume that projects are worthwhile in their own right when their benefits are discounted at the appropriate risk-adjusted rate, although it is implicit that a full three-stage analysis is undertaken.

15.9 MOTIVES FOR LEASING

Some writers suggest that leasing is undertaken primarily to exploit tax advantages. Large numbers of firms in the 1970s and early 1980s found that their desired capital expenditures exceeded their taxable earnings, and as a result they could not take advantage of the 100 per cent First Year Allowances then available, at least until their profitability had recovered. Under these circumstances, leasing was often a more cost-effective form of asset finance than borrowing-to-buy.

Many firms possessing taxable capacity set up leasing subsidiaries in order to shelter their own profits from tax by purchasing capital equipment on behalf of tax-exhausted firms. As a result, the list of active lessors included such odd-looking bedfellows as Tesco, Mothercare, Ladbrokes and Marks & Spencer, all highly profitable companies during this period. Such companies were able to obtain the tax benefits from equipment purchase considerably earlier than their tax-exhausted clients could expect to. In effect, lessors bought equipment on behalf of clients, took the tax benefits and passed these on to clients in the form of reduced rentals.

The extent to which tax benefits are actually passed on depends on the state of competition in the market for leasing and how near to the end of the lessor's tax year the negotiations take place. Sometimes, very attractive lease terms can be obtained from a lessor anxious to qualify for tax reliefs as soon as possible. In these cases, the lessor can profit from the contract, and the lessee may find leasing more attractive than outright purchase. Therefore, both parties can gain from the arrangement at the expense of the taxpayer.

While tax breaks have been important in explaining the rise of leasing, they do not account for the continuing popularity of leasing after the phased abolition of First Year Allowances between 1984 and 1986. Beyond the tax system, a variety of reasons have been proposed to explain the continuing popularity of leasing.

■ 'Leasing offers an attractive alternative source of funds'

For firms subject to capital rationing, leasing may offer an attractive means to access capital markets. This applies especially to small, growing businesses that lack a sufficiently impressive track record to satisfy lenders, or possess inadequate assets upon which to secure a loan. With a lease, no security is required, since if the lessee defaults, the owner simply repossesses the asset and looks for another client. For this reason, it is unusual to find restrictive clauses in lease contracts, in contrast to debt covenants where the lender may stipulate, for example, that the borrower should not exceed a specified gearing ratio. In addition, few lenders will offer 100 per cent debt financing. They prefer instead to see the client inject a significant amount of equity. This is not the case with a lease contract, which may thus be seen as a 'back-door' method of obtaining total debt financing for the equipment needs of a project.

Some organisations, such as local authorities and government departments, persistently suffer from constraints on capital expenditure and may find leasing an appealing device. In such organisations, there is often a rigid distinction between 'revenue' budgets and 'capital' budgets, which can be exploited by managers aware of the leasing alternative. Equipment may be acquired not by using the tightly controlled capital budget, but by undertaking a lease contract where the rentals are paid out of the revenue budget. Indeed, in the short term, leasing may even be presented as a way of 'saving money'. In 1979, the ability of lessors to obtain tax relief on equipment purchase, regardless of the tax status of the lessee, was removed. Since then, tax relief has been available only in cases where the lessee is normally liable for corporation tax, even if, perhaps temporarily, tax-exhausted. However, the public sector remains an important source of leasing business.

■ 'Leasing has cash flow planning advantages'

Leasing removes the need for a substantial cash outlay at the outset of a project in return for a series of contractually agreed, predictable cash flows over the term of the lease contract. A lease thus has the effect of smoothing out cash flows, which facilitates budgetary planning. However, this is a rather spurious argument as the same effect could be achieved with a bank loan. So this argument only applies if a bank loan is not available. This would of course render the 'lease vs. borrow-to-buy' mode of evaluation redundant.

■ 'Leasing provides off-balance sheet financing'

Until SSAP 21 made capitalisation of leases mandatory, companies were not required to show lease obligations in published accounts. This had the effect of disguising their true indebtedness by lowering the recorded gearing ratio, and also raising the return on capital employed. However, lease obligations had to be mentioned in the notes to the accounts, but it may have been fanciful to imagine that lenders (perhaps with their own leasing subsidiaries) were unaware of this form of window-dressing when assessing corporate performance and borrowing levels. However, this argument does apply for operating leases.

■ 'Leasing is cheaper than other forms of finance'

As well as its pragmatic attractions, leasing is often a more cost-effective way of acquiring an asset. We have seen how leasing can be a profitable alternative to bank borrowing under certain conditions. This is equivalent to saying that the effective rate of interest on a lease contract is lower than that on a bank loan. Indeed, many firms evaluate leases by comparing the bank's effective lending rate with the rate of interest implicit in the lease contract, i.e. the internal rate of return on the profile of lease payments, including the 'up-front' financing. Of particular interest is the case of an 'end-year lease', written on or just before the end of the lessor's tax year. At this juncture, the lessor is anxious to get the contract drawn up so as to claim the tax relief in the year just ending, rather than having to wait a further year before reaping the tax advantage. Many lessors have borrowing advantages owing to their size, which they may pass on to lessees in the form of lower rental charges.

A survey of quoted UK companies by Drury and Braund (1990) found that the cost of leasing, corporation tax considerations and, to a lesser extent, conservation of working capital were the most important factors in the lease or borrow-to-buy decision.

15.10 ALLOWING FOR CORPORATION TAX IN LEASE EVALUATION

The Hardup example omitted the impact of taxation. Lease rentals qualify for tax relief, as do interest payments on loans, while expenditures to purchase capital equipment generally attract capital allowances against corporation tax. Including tax is particularly important, since the UK tax system was widely believed to be largely responsible for the original upsurge in leasing activity in the UK.

Retaining the figures from the Hardup example, we now assume corporation tax is paid at 28 per cent with no delay, and the availability of a 20 per cent writing-down allowance (WDA). A firm with sufficiently high taxable profits can set off 20 per cent of its outlay on capital equipment against profits in the year of expenditure, and in each subsequent year, based on a reducing balance. Consequently, by careful timing of expenditures, a company with sufficient taxable capacity can enjoy significant tax savings.

Assuming that Hardup is not tax-exhausted, it can shelter its profits from tax by acquiring the aircraft. In effect, the taxpayer subsidises the required outlay, making equipment purchase a more attractive proposition for the tax-paying enterprise. Conversely, tax relief on rental payments lowers their effective cost. (For exposition, we assume that all of the rental is wholly tax-allowable as a normal business expense. However, the regulations in this area were changed in 1991, as explained below.) The final tax adjustment is to the discount rate.

The after-tax cost of borrowing

When interest payments qualify for tax relief, the effective rate of interest, r^*, is deflated by the rate of Corporation Tax (T), applicable to company profits:

$$r^* = r(1 - T)$$

where r is the nominal or quoted pre-tax interest rate applied by the lender.

Table 15.4 Hardup's leasing decision with tax

Item of cash flow (£m)	Year 0	Year 1	Year 2	Year 3
Lease:				
Rental	−5.00	−5.00	−5.00	–
Tax saving at 28%[1]		+1.40	+1.40	+1.40
Net lease cash flows (L)	−5.00	−3.60	−3.60	+1.40
Borrow-to-buy:				
Outlay	−13.75			
Tax saving[2]	+0.77	+0.62	+0.49	+1.97
Net purchase cash flows (B)	−12.98	+0.62	+0.49	+1.97
Net incremental cash flows (L − B)	+7.98	−4.22	−4.09	+0.57
Present value at 8.64%	+7.98	−3.88	−3.47	+0.44
Net present value (NAL) =	+1.07			

[1]Tax savings on the rentals are actually delayed by a year owing to the timing of the rental payments, i.e. on the first day of each of Years 1, 2 and 3.
[2]The profile of tax savings is based on setting the allowable expenditure against profits in Year 0 and in three subsequent years. The undepreciated balance is set against profits in the last year. No salvage value is assumed. Insurance and maintenance costs are also ignored.

For Hardup, facing a pre-tax borrowing rate of 12 per cent, this produces a tax-adjusted effective interest rate of about 8.64 per cent, i.e.

$$r^* = 0.12\,(1 - 0.28) = 0.0864 \quad \text{i.e. } 8.64\%$$

Retaining our previous incremental format, and allowing for the various tax complications, Table 15.4 shows the relevant cash flows.

The figures indicate that the lease is still worthwhile for the lessee. We could have undertaken a parallel analysis from the lessor's standpoint to examine whether it was worthwhile for it to purchase the asset for leasing to Hardup. In fact, the computation has effectively been done for us, because all the cash flows have the same numerical values, except that their signs are reversed. As a result, the NPV of the project to Moneybags is −£1.07 million (assuming that both firms could obtain capital at 12 per cent pre-tax). On this basis, a leasing contract between tax-paying companies is a zero-sum game. In reality, however, the lessor is likely to have special advantages, like access to cheaper finance, reducing its required return to below that of the lessee, and preferential buying terms. Economies of bulk purchase are common for lessors of motor cars, with 40 per cent discounts on list price not unknown.

The relative desirability of leasing and borrowing-to-buy depends largely on the tax regime. If the company is in a non-tax-paying situation (and expects to be so indefinitely), the lease evaluation should be conducted on a pre-tax basis.

Self-assessment activity 15.8

What is the tax-adjusted interest rate when the nominal rate is 9 per cent for:

(i) a small firm paying tax at 10 per cent?
(ii) a tax-exhausted firm?

(Answer in Appendix A at the back of the book)

■ The current tax treatment of lease rentals

In the Hardup example, we treated the rental payment as wholly tax-allowable. In 1991, the UK tax authorities made the rules more complex, to bring the tax treatment of leases into line with SSAP 21. This dictates that finance leases be stated on the balance sheet and depreciated accordingly. For leases taken out after April 1991, the Inland Revenue requires a separate treatment for the implicit finance charge and the implicit capital charge. The former depends on the interest rate implicit in the lease contract, and is tax-allowable in full in the relevant accounting period. The capital repayment element is derived by spreading the total capital charge – normally the asset's cash price – over its useful life. In other words, the depreciation charge in the accounts becomes the tax-allowable figure, dependent on the method of depreciation used and its acceptability to the Inland Revenue. An example should clarify this.

Example

Consider the case of a four-year lease contract to finance acquisition of an asset costing £10 million. The rental is £3.15 million p.a., paid at each year-end. The effective interest rate is the solution rate in the IRR expression formed by setting the finance raised equal to the sum of discounted payments:

$$\pounds10 \text{ m} = \frac{\pounds3.15 \text{ m}}{(1 + R)} + \frac{\pounds3.15 \text{ m}}{(1 + R)^2} + \frac{\pounds3.15 \text{ m}}{(1 + R)^3} + \frac{\pounds3.15 \text{ m}}{(1 + R)^4}$$

R is exactly 10 per cent. The annual interest charge can now be found by analysing the implicit loan of £10 million as shown in Table 15.5.

Table 15.5 Interest charges on a lease contract (figures in £m)

Year	Opening balance	Interest @ 10%	End of year debt	Repayment	Closing balance
	10.00	1.00	11.00	3.15	7.85
	7.85	0.78	8.63	3.15	5.48
	5.48	0.55	6.03	3.15	2.88
	2.88	0.29	3.17	3.15	0.02*

*Rounding errors prevent this reducing exactly to zero.

Applying straight-line depreciation to the asset cost, the annual charge is £10 m/4 = £2.5 m. This is fully allowable against tax in each of the four years. Table 15.6 compares the profiles of tax-allowable expenditures, and thus the impact of the 1991 changes. The present system accelerates the tax relief, and reduces the effective cost of the lease (for the tax-paying company). The interest charge is included under interest payments on the firm's profit and loss account, while the depreciation element is applied to lower the value of the asset in the usual way.

Table 15.6 Changes in tax-allowable lease costs (figures in £m)

Year	*Then* Rental	*Now* Interest	+	capital charge	=	Total
1	3.15	1.00	+	2.50	=	3.50
2	3.15	0.78	+	2.50	=	3.28
3	3.15	0.55	+	2.50	=	3.05
4	3.15	0.29	+	2.50	=	2.79
Totals	12.60	2.62	+	10.00	=	12.62*

*Rounding error.

15.11 WORKED EXAMPLE: LEE/LOR

This question appeared in the CIMA Strategic Financial Management exam paper in May 2007.

LEE is a manufacturing entity located in Newland, a country with the dollar ($) as its currency. LOR is a leasing entity that is also located in Newland.

LEE plans to replace a key piece of machinery and is initially considering the following two approaches:

- Alternative 1 – purchase the machinery, financed by borrowing for a five-year term.
- Alternative 2 – lease the machinery from LOR on a five-year operating lease.

The machinery and maintenance costs

The machinery has a useful life of approximately 10 years, but LEE is aware that the industry is facing a period of intense competition and the machinery may not be needed in five years' time. It would cost LEE $5,000 to buy the machinery, but LOR has greater purchasing power and could acquire the machinery for $4,000.

Maintenance costs are estimated to be $60 in each of years 1 to 3 and $100 in each of years 4 and 5, arising at the *end* of the year.

Alternative 1 – purchase financed by borrowing for a five-year term

$ interbank borrowing rates in Newland are currently 5.5% per annum. LEE can borrow at interbank rates plus a margin of 1.7% and expects $ interbank rates to remain constant over the five-year period. It has estimated that the machinery could be sold for $2,000 at the end of five years.

Alternative 2 – five-year operating lease

Under the operating lease, LOR would be responsible for maintenance costs and would charge LEE lease rentals of $850 annually *in advance* for five years.

LOR knows that LEE is keen to lease rather than buy the machine and wants to take advantage of this position by increasing the rentals on the operating lease. However, it does not want to lose LEE's custom and requires advice on how high a lease rental LEE would be likely to accept.

Tax regulations

Newland's tax rules for operating leases give the lessor tax depreciation allowances on the asset and give the lessee full tax relief on the lease payments. Tax depreciation allowances are available to the purchaser of a business asset at 25% per annum on a reducing balance basis. The business tax rate is 30% and tax should be assumed to arise at the end of each year and be paid one year later.

Alternative 3 – late proposal by production manager

During the evaluation process for Alternatives 1 and 2, the production manager suggested that another lease structure should also be considered, to be referred to as 'Alternative 3'. No figures are available at present to enable a numerical evaluation to be carried out for Alternative 3. The basic structure would be a five-year lease with the option to renew at the end of the five-year term for an additional five-year term at negligible rental. LEE would be responsible for maintenance costs.

Required:

(i) Use discounted cash flow analysis to evaluate and compare the cost to LEE of each of Alternatives 1 and 2.

(ii) Advise LOR on the highest lease rentals that LEE would be likely to accept under Alternative 2.

Answer to LEE/LOR leasing question

(i) As this is a borrow-to-buy versus a lease decision, we must first calculate the after-tax discount rate. This is:

Pre-tax cost $(1 - \text{tax rate}) = (5.5\% + 1.7\%)(1 - 30\%) = 7.2\% \times 0.7 = 5.04\%$, rounded to 5%

Second, depreciation allowances and tax savings are calculated:

$	0	1	2	3	4	5
			Year			
Balance b/f	5,000	5,000				
Depreciation @ 25%		(1250.00)	(937.50)	(703.13)	(527.34)	
Balance c/f		3,750.00	2,812.50	2,109.37	1,582.03	
Residual value					(2,000.00)	
					417.97	
Tax saving @ 30%		375.00	281.25	210.94	158.20	
Balancing charge						(125.39)

(It is assumed that LEE has sufficient taxable capacity to absorb the depreciation allowance.)

Evaluation of the purchase option now follows:

$	0	1	2	3	4	5	6
				Year			
Outlay/Residual value	(5,000)					2,000	
Maintenance costs		(60)	(60)	(60)	(100)	(100)	
Tax relief			18	18	18	30	30
Tax savings on outlay			375	281	211	158	(125)
Net cash flows	(5,000)	(60)	333	239	129	2,088	(95)
Discount factor @ 5%	1.000	0.952	0.907	0.864	0.823	0.784	0.746
Present value	(5,000)	57	302	206	106	1,637	(71)

Present value of purchase option = ($2,763)

The lease is much simpler. There is an annuity starting in year zero (i.e. 'one year in advance') of the $850 rentals, followed by another annuity of delayed tax savings of $850(1 − 30%) = $255 over years 2–6.

The relevant annuity factors are:
Years 0–4: (1.000 + 3.546) = 4.546
Years 2–6: 4.329/(1.05) = 4.123

The calculation is thus:
PV = [(850) × 4.546] + [255 × 4.123] = (3,864) + 1,051 = ($ 2,813)

These figures suggest that the purchase option is cheaper, so that the break-even rental would be lower than $850.

(ii) Setting R = highest acceptable lease rental, we need to solve for R in the expression
PV of lease rental flows = PV of purchase cash flows
i.e.
4.456R − (4.123 × 0.3R) = 2,763

$$3.309R = 2,763$$
$$\text{Whence } R = (2,763/3.309) = \$835$$

To break even, the pre-tax lease rental must be *lowered* to $835.

15.12 POLICY IMPLICATIONS: WHEN SHOULD FIRMS LEASE?

There are two lessons to be drawn from this analysis of leasing decisions:

1 Lease evaluation involves analysing a financing decision, as borne out by the three-stage analysis. Sometimes, especially favourable financing arrangements, available via a lease contract, may tip the balance between project rejection and acceptance. But these cases are rare, and the margin of acceptance offered by attractive financing will probably be fairly narrow. If the project is worthwhile only because of the financing package, this should be explicitly recognised and the concessionary finance regarded as a 'one-off', which may not be repeated in other cases. It seems unduly purist to argue that, if the financing deal is all that makes the project acceptable, then it should be rejected. Few financial managers would pass up an attractive financing opportunity. *However, policy is more profitably directed at finding genuinely worthwhile projects, rather than diverting resources to obtaining marginal financing advantages.*

2 The analysis enables us to pinpoint the factors that suggest the relative attractiveness of leasing as compared to other financing arrangements. We now list the factors that impact on the leasing decision, but note that these are mainly 'other things being equal' prognoses. 'Look before you lease' is a sensible motto.

Taxable capacity

Leasing is a means whereby lessors can exploit their own taxable capacity and pass on any tax savings to firms in less favourable tax positions. Hence, the greater the taxable capacity of would-be lessors and the lower that of users of capital equipment, the greater the attractions of leasing.

Competition among lessors

A major factor in the development of the UK leasing market was the entry of new players, eager to exploit their taxable capacity and thus shelter their profits from corporation tax. This had the effect of increasing competition for available leasing business and reducing the general level of lease rentals. Therefore, the greater the competition among lessors, the greater the attractiveness of leasing.

Investment incentives

Another major factor in the growth of the leasing industry was the availability of especially generous inducements to encourage firms to acquire plant, machinery and industrial buildings. Although UK incentives are now less attractive than those existing prior to 1984, there was no fall-off in leasing after the removal of First Year Allowances, probably owing to the other attractions of leasing. However, it seems reasonable to argue that, the greater the generosity of the tax authorities, the greater the attractiveness of leasing because tax-breaks lower the effective cost of equipment purchase for lessors, which may then pass on the benefits to lessees, whether or not they are tax-exhausted.

Corporation tax

Investment incentives are most valuable when high rates of corporation tax raise the value of the tax savings. Until 1984, with 100 per cent First Year Allowances, the total (undiscounted) tax saving with corporation tax payable at 52 per cent for a £10 million investment would have been £5.2 million. This falls to £2.8 million with the present tax rate of 28 per cent and, when discounted, even lower with 20 per cent annual WDAs. The higher the rate of corporate profits tax, the more attractive leasing becomes, because lessors can shelter their profits to a greater extent.

Inflation

Rising price levels reduce the real value of future payments, such as a series of fixed rental payments. Therefore, the higher the expected rate of inflation, the greater the likely appeal of leasing. However, the inflation effect will benefit the lessee only if it correctly anticipates a higher rate of inflation than the lessor. As in all contractual arrangements, unanticipated inflation is what does the damage, and it is by no means certain that the lessee will be more successful than the lessor in forecasting inflation. If lessors feel confident in their ability to predict inflation, if the rentals they set incorporate their expectations, and if they are more or less correct, the benefits expected by lessees will evaporate. In addition, lease contracts may incorporate some form of inflation adjustment.

Interest rates

The relevant rate of discount in lease evaluation is the rate applicable to the best alternative bank loan. The higher the rate of interest, the greater the discounting effect on future contractual lease obligations. Generally, therefore, the higher the interest rate, the more attractive a lease appears, since the present value of a given set of rentals will become lower. However, this effect may be diluted by the impact of lessors applying higher rentals in order to cover their own increased borrowing costs. If there is any tendency for the spread of interest rates to widen at higher levels of interest rates, this effect may still operate, since lessors usually have access to borrowed capital at more advantageous rates than lessees.

In the next section, we describe a range of methods for financing foreign trade.

15.13 FINANCING INTERNATIONAL TRADE

Obviously, it takes two parties to trade across country borders, the importer and the exporter, but here, we adopt the perspective of the exporter. Exporting carries particular risks, e.g. the risk of slow (or even non-) payment by customers and the risk of adverse currency movements. Hence, export finance is rather more complex than the finance of domestic trade. The banking system, both domestic and foreign, usually has a pivotal part to play both in arranging contract terms and making arrangements for the exporter to be paid.

open account
Where trading partners agree settlement terms with no formal contract

The simplest basis of trading is **open account** where the parties simply make a 'gentleman's agreement' about the settlement date, prior to which the goods are shipped, received by the importer and used for production or sale. The exporter thus loses control over the goods and has recourse to the legal system if the importer fails to pay. Open account is thus confined to deals involving established, reliable and highly creditworthy customers located in countries that have rapid, fair and reliable payment and legal procedures. Much of the trade involving large UK firms with large EU-based firms is conducted on open account. The advent of the euro as the common currency of the EU is likely to increase the proportion of intra-EU trade conducted on open account.

With less reliable, or new, customers based in other parts of the world, more formal procedures are required. As a first requisite, to overcome the fear of non-payment by a customer in a foreign country, the exporter enlists the help of a well-respected bank to act as intermediary. Using the reputation and good offices of the bank, the trade-cum-financing package typically works as follows:

■ the importer secures the promise of the bank to pay on its behalf;
■ the bank promises the exporter to pay on behalf of the importer;

- the exporter ships the goods, trusting to the bank's promise to pay;
- the bank pays the exporter at a pre-agreed juncture;
- the bank passes title to goods to the importer;
- the importer pays the bank, often using the sale proceeds of the goods shipped.

■ Trade documents

Most export deals involve three key documents:

- the bank's promise to pay is called a **Letter of Credit**.
- when the exporter ships the goods to the importer's location, title to the goods is conveyed to the bank by a **Bill of Lading**.
- When the exporter seeks payment from the bank, it presents a '**sight draft**'. When the bank has paid the exporter, title to ownership of the goods passes to the importer who duly pays the bank.

sight draft
A document presented to a bank by an exporter seeking payment for an export deal

There is obvious potential for delay in these procedures, e.g. in trans-shipment of goods, in inspecting the goods, in acceptance of the documents, etc. Hence the bank plays a key role in 'holding the ring' between importer and exporter and providing a source of short-term finance. The three documents help protect the two parties from the risk of non-completion of the contract.

We now pay particular attention to the Letter of Credit, as this represents the bank's promise to pay.

■ Letters of credit (LOC)

An LOC, or documentary letter of credit (DLOC), is a document drawn up by an importer giving its bank detailed instructions as to the circumstances in which the credit can be honoured by the importer's bank (the 'opening bank') in favour of the exporter. It details the nature of the goods, their quality and price and the dates between which the LOC is valid. A DLOC enables the exporter to receive payment for goods in its country of location once shipment has taken place. The burden of financing is thus on the importer who gains the comfort of knowing a definite date by which the exporter must ship the goods. The exporter's risk is reduced insofar as the creditworthiness of the involved banks is substituted for that of the buyer.

irrevocable DLOC
A written authority for a bank to make specified payments to an exporter, whose terms cannot be varied

There are two types of DLOC. An **irrevocable DLOC** is a written authority from the opening bank to its correspondent bank in the exporter's country (the 'advising bank') to make specified payments provided that the documents specified are presented between the specified dates. If requested, the advising bank will add its own undertaking to that of the opening bank by confirming the DLOC, which becomes a 'confirmed DLOC.' This type of DLOC is legally binding and thus cannot be modified or cancelled except with the consent of all parties involved.

revocable DLOC
A letter of credit whose terms can be varied without consulting the exporter

Conversely, a **revocable DLOC** can be altered or cancelled at any time without having to give prior notice to the beneficiary, i.e. the exporter. It cannot be confirmed and is not legally binding.

Drawbacks with letters of credit

DLOCs are not problem-free:

- They can be expensive. The importer is likely to demand a lower price if it uses a DLOC, as insurance against potential hitches in the transaction, removing the need for the exporter to build these into the price.
- The advising bank has to check that the documents presented exactly conform to the specifications stipulated. If they do not, payment will be refused.

- Banks do not accept responsibility for the goods shipped. A DLOC arrangement is simply about transferring documents at arm's length from the trading activity. If the documents are in order, the credit will be honoured.
- The opening bank bears a credit risk while the credit is open. It may thus demand a cash deposit or lower the importer's other borrowing facilities.

Other methods of financing international trade are: Bills of Exchange, documentary collections, forfaiting and export factoring.

■ Bills of Exchange in export trade

The use of Bills of Exchange in export trade is similar in essence to their domestic use. An exporter can send a Bill of Exchange for the value of goods shipped through the banking system for payment by a foreign buyer when presented. The exporter usually prepares the bill, drawn on the foreign-based importer, for the amount specified in the export contract.

Using the bill along with the other shipping documents via the banking system, the exporter retains greater control over the goods because, until the bill is paid or accepted by the foreign buyer, they cannot be released. The importer acknowledges his agreement to pay on the due date by writing an acceptance on the bill, but does not have to pay, or agree to pay, until delivery of the goods by the exporter.

The exporter may pass the bill to a domestic bank, which forwards it to a foreign branch of the same bank or to a correspondent bank in the importer's country. The so-called 'collecting bank' presents the bill to whoever it is drawn upon, either for immediate payment, if it is a 'sight draft', or for acceptance, if it is a 'term draft', payable after a specified credit term. Where no shipping documents are required, this is called a 'clean bill collection'.

Where the importer's financial standing is doubtful, the bill may be used in a 'documentary credit collection'. The exporter sends the bill to its bank which in turn sends it to its foreign-based correspondent, along with the shipping documents, including the title to the goods, represented by the Bill of Lading. The bank releases the documents to the importer only on payment or acceptance of the bill by the foreign buyer. In this way, greater control of the goods is achieved.

Bills of Exchange have several advantages:

- They are cheaper than letters of credit.
- Because the title documents can only be released on payment, the exporter has a stronger position than in open account trades.
- The bills can be discounted via the banking system to release finance for other uses. Effectively, the bank buys the customer's account payable, represented by the proceeds of the export deal at the time that the collection is remitted abroad.
- Banks may also give an advance based on an outward collection of Bills of Exchange, i.e. the anticipated payment.
- Bills of Exchange reduce the risk of non-payment, given that banks do not want to be associated with clients of poor credit standing.

■ Forfaiting

Forfaiting is a form of medium-term export financing that involves the purchase by a bank (the 'forfaiter') of a series of promissory notes, usually due at six-month intervals over perhaps three to four years, signed by an importer in favour of an exporter. The notes are usually guaranteed ('avalised') by the importer's bank and then sold by the exporter to the forfaiting bank at a discount. The bank pays the exporter, allowing it to finance the production of the goods destined for export, and enabling the importer to settle later. The promissory notes are held by the forfaiter for collection as they mature, on a without-recourse basis, thus the exporter is not liable in case of default by the importer.

The rate of discount applied in forfaiting depends on the terms of the notes, the currencies of denomination, the credit ratings of the importer and of the bank that guarantees the notes, as well as the country risk of the importer's base. Because the forfaiting bank will quote a discount rate on demand, the exporter is able to quote a selling price to foreign customers that allows for financing costs.

■ Export factoring and invoice discounting

Export factoring is similar in essence to domestic factoring, with the added bonus that the factor usually assumes the foreign exchange risk. If an export receivable is to be settled on open account, rather than by DLOC or Bill of Exchange, the exporter can offer the receivable to a factor in exchange for domestic currency, thus offering protection against foreign exchange rate movements. The factor provides finance and also absorbs credit risk. Export factoring may therefore be expensive.

Factors may have their own offices abroad where they can investigate directly the credit status of firms' clients locally, or may be a member of a network of independent factoring organisations. The largest of these is Factors Chain International (FCI) (**www .factors-chain.org**), founded in 1968. FCI now has around 220 member firms, operating in about 60 of the world's main commercial centres.

As with domestic factoring, recourse and non-recourse versions are available, the latter being very convenient for export business, although more expensive. As well as a charge for providing finance, the factor applies a service fee of between 0.5 per cent and 2 per cent for running the sales ledger and collecting debts. In 2006, UK export invoice discounting grew 31 per cent to £5.7 billion, while export factoring grew a massive 116 per cent to £2.3 billion.

SUMMARY

We have described a variety of short- and medium-term methods of financing company operations: trade credit, bank lending, factoring and invoice discounting, money market finance, HP and leasing. The key features of each source were examined, and particular emphasis was given to leasing and HP, where the tax implications were analysed. Several ways of financing foreign trade were also outlined.

Key points

- Companies have access to significant amounts of trade credit through normal trading relationships.

- Abuse of trade credit facilities can lead to severe liquidity problems.

- Companies are using term loans more extensively, even though these are generally more expensive than overdraft facilities.

- The money market can offer significant amounts of credit through bill finance.

- Hire purchase is often used by smaller, less creditworthy, companies to purchase equipment. The asset becomes the property of the user upon completion of the contract as scheduled. Users can exploit capital allowances in relation to assets acquired via HP.

- Leasing is a way of obtaining the use of an asset without incurring the initial 'lump' of capital outlay required for outright purchase.

- Leases may be 'job-specific', contracts applicable for periods less than the lifespan of assets (operating leases), or for periods coinciding with the asset's expected lifetime (financial leases).

- A lease contract normally involves a commitment to pay a series of fixed rental charges, which qualify the lessee for tax relief. A finance lease is an alternative to borrowing in order to purchase an asset, and the firm should expect its ability to borrow to fall by an equivalent amount.

- In the UK, if the ownership of a leased asset passes to the lessee, the tax breaks are clawed back by the Inland Revenue.

- SSAP 21 stipulates that leased assets acquired via finance leases must be included among fixed assets on the firm's balance sheet and that future rental obligations be recorded as liabilities.

- Evaluation of a lease, i.e. whether to lease or to borrow-to-buy, may be undertaken in three equivalent ways:

  1 by comparing the present values of the respective cash flow streams;

  2 by assessing what equivalent loan could be raised with the same stream of payments entailed by the lease;

  3 by comparing the effective rate of interest payable on the lease with the costs of raising an equivalent loan.

- Lease evaluation usually assumes that the asset is worth obtaining in its own right, regardless of financing method, but an unattractive investment could be rendered worthwhile by leasing. To investigate this, a three-stage analysis should be undertaken.

- Even though financial criteria may point to a definite preference, consideration of non-financial factors may reverse the lease or borrow-to-buy decision.

- Banks perform a pivotal role in financing international trade.

Further reading

The classic works on leasing are by Clark (1978) and Tomkins *et al.* (1979). Rutterford (1992) gives a more up-to-date treatment, while successive annual reports of the Finance and Leasing Association (**www.fla.org.uk**) will keep you abreast of developments. Drury and Braund (1990) survey UK practice. See Bowman (1980) and Narayanaswamy (1994) for empirical work that supports the leasing–debt equivalence view in the USA and the UK respectively, and Ang and Peterson (1984) for an exception. Myers *et al.* (1976) develop the standard lease evaluation model on which the analysis in the chapter is based. Jarvis *et al.* (2000) have conducted a historical appraisal of UK government policy in this area. See Koh (2006) for an overview of trends in the European leasing industry.

QUESTIONS

 **myfinancelab** | *Questions with an icon are also available for practice in myfinancelab with additional supporting resources.*

Questions with a coloured number have solutions in Appendix B on page 745.

1 A supplier offers you the following trade credit terms: '3/15: net 45'.
  If you delay payment until day 45, what effective annual interest rate are you paying for additional trade credit?

2 A bank offers a client a choice between two financing options over a one-year period:

Option 1: a bullet loan for the full year of £500,000 to be repaid at end year with interest fixed at 12 per cent p.a.

Option 2: An overdraft, with a quoted rate of 14 per cent p.a, with interest charged quarterly on the average balance.
  The firm expects to need finance of £400,000 in the first quarter, £500,000 in quarter 2, £500,000 in quarter 3 and only £200,000 in the final quarter due to the seasonal nature of its business. (These are all quarterly averages).
  Unused funds can be invested at 2 per cent per quarter. The bank will not charge interest on accumulated quarterly interest charges.

  (a) What advice would you give?
  (b) What is the break-even rate on the overdraft, assuming the interest rate on the loan is fixed at 12 per cent?

? 3 A trader receives from a customer a Bill of Exchange set to mature in six months, and decides, after two months, to sell it to a bank. The face value is £200,000 and the bank discounts the Bill for £195,500.
  What effective annual interest rate is the trader paying to accelerate receipt of his money?

? 4 What is the monthly interest payment on the following HP contract?

  ■ Total purchase cost of equipment = £100,000
  ■ Down payment = 15 per cent
  ■ Interest rate = 7.5 per cent
  ■ Equal monthly payments over four years.

5 Haverah plc is a manufacturer and distributor of denim garments. It employs a highly aggressive working capital policy, and uses no long-term borrowing. Highlights from its most recent accounts appear below:

	£m
Sales	45.00
Purchases	20.00
Earnings before interest and tax	5.00
Interest payments	2.00
Shareholder funds (comprising £1 m issued shares, par value 50 p, and £3 m reserves)	4.00
Debtors	2.50
Stocks	1.00
Trade creditors	6.00
Bank overdraft	5.00

  Killinghall is a supplier of cloth to Haverah and accounts for 40 per cent of its purchases. It is most anxious about Haverah's policy of taking extended trade credit, so it offers Haverah the opportunity to pay for supplies within 25 days in return for a discount on the invoiced value of 4 per cent.
  Haverah is able to borrow on overdraft from its bank at 10 per cent. Tax on corporate profit is paid at 30 per cent.

Required

If Haverah made this arrangement with its supplier, what would be the effect on its working capital cycle and key accounting measures such as interest cover, profit after tax, earnings per share, return on equity and gearing? Should it accept the offer?

6 Raphael Ltd is a small engineering business which has annual credit sales of £2.4 million. In recent years, the company has experienced credit control problems. The average collection period for sales has risen to 50 days even though the stated policy of the business is for payment to be made within 30 days. In addition, 1.5 per cent of sales are written off as bad debts each year.

The company has recently been in talks with a factor who is prepared to make an advance to the company equivalent to 80 per cent of debtors, based on the assumption that customers will, in future, adhere to a 30 day payment period. The interest rate for the advance will be 11 per cent per annum. The trade debtors are currently financed through a bank overdraft which has an interest rate of 12 per cent per annum. The factor will take over credit control procedures of the business and this will result in a saving to the business of £18,000 per annum. However, the factor will make a charge of 2 per cent of sales for this service. The use of the factoring service is expected to eliminate the bad debts incurred by the business.

Raphael Ltd is also considering a change in policy towards payment of its suppliers. The company is given credit terms which allow a 2.5 per cent discount providing the amount due is paid within 15 days. However, Raphael Ltd has not taken advantage of the discount opportunity to date and has, instead, taken a 50 day payment period even though suppliers require payment within 40 days. The company is now considering the payment of suppliers on the fifteenth day of the credit period in order to take advantage of the discount opportunity.

Required

(a) Calculate the net cost of the factor agreement to the company and state whether or not the company should take advantage of the opportunity to factor its trade debts.

(b) Explain the ways in which factoring differs from invoice discounting.

(c) Calculate the approximate annual percentage cost of forgoing trade discounts to suppliers and state what additional financial information the company would need in order to decide whether or not it should change its policy in favour of taking the discounts offered.

(d) Discuss any other factors which may be important when deciding whether or not it should change its policy in favour of taking the discounts offered.

(Certified Diploma, June 1996)

7 Amalgamated Effluents plc, a chemical company currently in legal difficulties over its pollution record, is considering a proposal to acquire new equipment to improve waste generation. The equipment would cost £500,000 and have a working life of four years. At the end of Year 4, the disposal value of the equipment is expected to be £50,000. The machine is expected to generate incremental cash flows of £200,000 for each of the four years.

Amalgamated can acquire the equipment in two ways:

(a) Outright purchase via a four-year bank loan at a pre-tax interest cost of 7 per cent.

(b) A financial lease with rentals of £70,000 at the end of each of the four years.

Amalgamated is presently ungeared. Its shareholders seek a return of 10 per cent after allowing for all taxes. Corporation Tax is paid at 30 per cent with no tax delay. If the equipment is purchased, a 25 per cent writing-down allowance (reducing balance) is available.

Required

Should Amalgamated acquire the equipment and, if so, how should it be financed?

 **myfinancelab** | *Now retake your diagnostic test for Chapter 15 to check your progress and update your study plan.*

Part V

STRATEGIC FINANCIAL DECISIONS

Financial managers face two key decisions: 'Which assets to invest in?' and, 'How to finance them?' Earlier in the book, we discussed strategic investment decisions, and we examined short-term financing in Part IV.

Also in Part IV, there was strong emphasis on the 'Golden Rule' of financing – that the term of financing should be matched to the term of the asset acquired finance (although we did encounter alternative stances). It is now appropriate to look at methods of financing long-term strategic investments, i.e. long-term sources of finance. This is done in Chapter 16.

Choices between alternative forms of finance essentially reduce to choices between different mixes of borrowing and equity capital in the firm's capital structure. We therefore examine, in Chapter 17, the factors that determine a firm's dividend policy – a decision to distribute higher dividends has implications for the debt/equity mix, but there are important constraints on dividend policy.

In Chapters 18 and 19, we examine how the use of borrowed funds can affect the value of the company and the return required on new investments. Chapter 18 presents the 'traditional' view on these issues, while Chapter 19 presents the 'modern' theory of capital structure, which emphasises the essential underlying relationships. We will also discover how borrowing affects the Beta coefficient, and how to tackle cases of investment evaluation where a firm alters its gearing and its (systematic) risk. We extend the analysis to cover interactions between investment and financing decisions and develop a decision rule – the adjusted present value – to handle complex interactions.

In Chapter 20, we examine mergers and takeovers – why they occur, how they are financed, and the effects they have on organisational structure and shareholder wealth. Other forms of restructuring are also considered.

Long-term finance

Search engine trouble

Successful start-ups that grow and prosper usually aspire to a stock exchange floatation, partly to enable the founders to realise some of their investment, and partly to raise new capital for further expansion.

The Initial Public Offering (IPO) by Google, founded by Sergey Brin (in his bedroom) and Larry Page in August 2004, was one of the most interesting and controversial floatations in recent years for several reasons:

- Google is one of the few dotcom firms with a sustainable and profitable business model.
- This was one of the largest technology issues ever, in a sector where firms are notoriously difficult to value.
- The founders attempted to minimise the financial pickings of the Wall Street establishment. They tried to moderate likely first-day excess demand in a complex Dutch auction process that favoured small investors, and was designed to slash investment banks' fees by over 50 per cent (2.8 per cent instead of a more usual 7 per cent).
- Google made a number of highly publicised errors in the run-up to the floatation. Notable among these gaffes were failure to register shares awarded to employees (called 'googlers'), and giving an interview to *Playboy* magazine that appeared to breach the rules of the Securities and Exchange Commission, the US stock market regulator, regarding promotion of a company's shares in the lead-up to an IPO.

- In addition, it was reticent in its prospectus about the intended use ('general corporate purposes') of the cash to be raised but, bizarrely, promised 'not to do evil'.
- On the day it announced the target price range, its operating system all but crashed, overloaded by the MyDoom internet virus.

Concerned that sentiment was turning against it, Google, having set its original price range between $108 and $135, was forced to cut this back to $85 a few days before the float, as well as reducing the number of shares on offer from 25.7 million to 19.6 million. On the launch day, Google shares closed at an 18 per cent premium, suggesting a successful issue but also implying that Google's maladroitness had depressed the issue price, thus reducing the proceeds of the issue from a hoped-for $3.5 billion to less than $2 billion.

Nevertheless, founders Brin and Page each banked around $40 million as the payoff for their entrepreneurship, and their remaining stakes were now valued at $3 billion each.

Following the IPO, the shares of Google took off, and in October 2007, broke through the $600 barrier as the US market surged to a record high.

Source: Several articles in the *Financial Times* during July and August 2004.

Learning objectives

After reading this chapter, you should understand the following:

- The key characteristics of the main forms of long-term finance.
- The benefits and drawbacks of each capital form.
- The factors that influence the choice between the various forms.

 **myfinancelab** *Complete your diagnostic test for Chapter 16 now to create your personal study plan.*

16.1 INTRODUCTION

The next four chapters analyse the factors that determine a firm's long-term financing decisions. Before we can do this, it is important to present the main sources of finance and their essential characteristics. With an understanding of the pros and cons of different financial instruments, the reader will more fully appreciate the choices open to companies and the pressures that drive their choices.

The basic distinction in long-term financing is that between debt and equity, or, more accurately, between the various forms of debt and the various forms of equity. In recent years, the demarcation line between debt and equity has become increasingly blurred by the development of '**hybrid**' forms of finance, such as warrants and convertibles, as companies and their advisers have sought to exploit the advantages of each form of capital without incurring all the disadvantages.

hybrid
A security that combines features of both equity and debt

In 1994, the UK accounting authorities issued a standard, FRS4 Capital Instruments, designed to clarify these differences and to offer guidance to companies in constructing and presenting their balance sheets. FRS4 divides capital instruments into debt and shareholder funds. Debt is any instrument that creates a liability, including the right to demand cash or shares at some future date. Shareholder funds are instruments that are not debt. These are split into equity and non-equity interests: broadly, ordinary shares and preference shares respectively.

This distinction continues to apply under International Financial Reporting Standards.

16.2 GUIDING LIGHTS: CORPORATE AIMS AND CORPORATE FINANCE

Strategists often divide companies into two categories: niche and global. A **niche company** sells high-quality products at a price that offers healthy profits. Such companies exist by exploiting a product differentiation advantage. A **global company** is able to compete in world markets with leading international firms by exploiting its own products at a competitive cost of production. Global companies thus combine product and cost advantages.

Not all companies aspire to global status; some are run by the proprietor to provide a livelihood for his or her family. Such **proprietorial companies** rarely offer sufficient potential to interest outside investors. Conversely, the **entrepreneurial company** has high growth potential and is driven by the desire of the owners to generate substantial wealth – in crude terms, to be 'seriously rich'.

Yet the predominant motive of many entrepreneurs is not money but the creative urge, the desire to build a thriving company, often for dynastic purposes. However, for the owners of such companies to realise their full potential, most of them eventually have to offer a share of the action to external participants, thus relinquishing a degree of control. Selling shares of equity involves releasing voting rights (it is possible to issue non-voting shares, but the Stock Exchange bans these for companies seeking a listing), and exposing the company's operations to outside scrutiny. Issuing debt usually involves some form of restrictive covenant over company financial policy, such as limits on subsequent borrowing and on dividend policy.

We now examine a short case to give a flavour of the strategic dimension of long-term financing.

■ Mitre Ltd: strategic financing issues

Mitre is a young, rapidly growing company, operating in the highly competitive computer software market. Its directors are the five founder members, all former employees of a giant US-owned computer manufacturer. All five found that 'working for a master' inhibited their creative energies. The company was set up on a shoestring, using the personal financial resources of the founders, supplemented by small inputs of capital from trusting close friends and relatives. After two years of struggle in rented premises, without paying its owners a salary, it managed to show an operating profit, on the basis of which it persuaded the local bank to extend overdraft facilities.

In subsequent years, Mitre has financed its operations through ploughing back the bulk of its profits, by further borrowing from the bank and by taking as much trade credit as possible from suppliers. It expects further rapid growth in the next few years, but is most concerned to avoid the problems of overtrading. It is, therefore, seeking to obtain long-term funding to support this growth.

Choosing the financing mix of short- and long-term debt and equity that best meets the investment requirements of a business is a key element of strategic financial management. Four issues need to be addressed.

Risk

How uncertain is the environment in which the business operates? How sensitive is it to turbulence in the economy? Mitre Ltd would probably be viewed by potential investors as having relatively high risk, particularly if the existing level of borrowing was high. Such a company will probably have a relatively inflexible cost structure, at least in the short term. As most of its costs will be fixed (unless it subcontracts work), it exhibits a high level of **operating gearing**.

operating gearing
The proportion of fixed costs in the firm's operating cost structure

Ownership

A major injection of equity capital by financiers would dilute the control currently exercised by the founder members/directors. The desire to retain control of the company's activities may well make them prefer to borrow.

Duration

The finance should match the use to which it is put. If, for example, Mitre Ltd required finance for an investment in which no returns were anticipated in the early years, it might be desirable to raise capital that has little, if any, further drain on cash flow in these years. Conversely, it would be unwise for Mitre Ltd to raise long-term finance if the projects to be funded have a relatively short life. This could result in the business being over-capitalised, and unable to generate returns sufficient to service and repay the finance.

Debt capacity

If Mitre Ltd has a low level of borrowing at present, it has a greater capacity to raise debt than a similar firm with a higher borrowing level. However, debt capacity is not just a function of current borrowing levels, but also depends on factors such as the type of industry, and the security that the company can offer. An important benefit of borrowing is that the interest paid attracts tax relief and, hence, lowers effective financing charges, although this advantage can only be exploited when the company is profitable.

16.3 HOW COMPANIES RAISE FINANCE IN PRACTICE

The financial manager can raise long-term funds internally, from the company's cash flow, or externally, via the capital market – the market for funds of more than a year to maturity. This exists to channel finance from persons and organisations with temporary cash surpluses to those with, or expecting to have, cash deficits. A critical intermediary function is provided by the major institutions such as pension funds, insurance companies and various types of bank. These collect relatively small savings and channel them to companies and other organisations seeking capital. As a result, the institutions are now the major holders of securities, both debt and equity, issued by companies.

The record of business financing patterns over the past 50 years or so shows several clear features:

- The majority of funds comes from internal sources – retained earnings and depreciation provisions (more strictly, cash flow), depending on company profitability and dividend policies.
- Bank borrowing plays a major role but is highly volatile, even negative in some years as firms repay debt. Firms tend to use bank finance as a buffer – borrowing heavily when interest rates are low and repaying when rates rise.
- Sales of shares on the New Issue Market are relatively unimportant, although many firms try to exploit high and rising stock markets by making rights issues.
- Long-term debt issues are also small contributors, as are preference share issues.

We now consider the main forms of raising long-term finance to cast some light on these trends. In Chapters 18 and 19, we will examine several theories that attempt to explain firms' long-term financing decisions.

Self-assessment activity 16.1

What impact might the events of 11 September 2001 have had on corporate financing decisions?

(Answer in Appendix A at the back of the book)

16.4 SHAREHOLDERS' FUNDS

■ Ordinary shares

Share ownership lies at the heart of modern capitalism. By purchasing a portion or 'share' of the ownership of a firm, an investor becomes a shareholder with some degree of control over a company. A share is therefore a 'piece of the action' in a company. When a company is formed, its **Articles of Association** will specify how many ordinary shares it is authorised to issue. This maximum can be varied only with the agreement of ordinary shareholders. Meanwhile, the issued share capital is the cornerstone of a company's capital structure. Ordinary shareholders carry full rights to participate in the business through voting in general meetings. They are entitled to payment of a dividend out of profits and ultimately, repayment of capital in the event of liquidation, but only after all other claims have been met. As owners of the company, the ordinary shareholders bear the greatest risk, but also enjoy the fruits of corporate success in the form of higher dividends and/or capital gains.

Authorised and issued share capital

The accounts for Scottish and Southern Energy (SSE) plc (**www.scottish-southern.co.uk**) as at 31 March 2007 showed:

Authorised	£m
1,200 million ordinary shares of 50p each	600
Allotted, called up and fully paid	
861.9 million ordinary shares of 50p each	431

SSE thus has scope to issue about a further 338 million shares without seeking amendment to its Articles. Notice that share capital is recorded in the accounts at par value.

When ordinary shares are first issued, they are given nominal or par values. For example, a company could issue £5 million ordinary share capital in a variety of configurations, such as 10 million shares at a nominal value of £0.50 each, or 500,000 shares of £10 each. The main consideration is marketability. Investors tend to view shares with lower nominal values (e.g. 50p) as more easily marketable, particularly when, after a few years, shares often trade well above their nominal value. The most common unit nowadays is 25p. Issued share capital appears on the balance sheet at its par value, which explains why it often seems a minor item. For example, SSE's issued share capital (i.e. issued and paid for), denominated in 50p units, accounts for only £431 million (16 per cent) of shareholders' equity (£2,596 million). However, the market value of these shares in August 2007 was many times higher than the par value, at 1400p each.

Once a company is established, its shares usually trade above the basic par value. A company wanting to sell new shares is therefore likely to issue them not at their nominal value, but at a **premium**. Share premium forms an integral part of the share capital, £99 million in the case of SSE. Most companies are limited liability companies incorporated under the Companies Acts. This implies that the owners (or shareholders) have obligations limited to the amount they have invested. If the company were ever wound up or liquidated, leaving outstanding debts, the shareholders would not be liable to meet such claims. Limited liability companies in the UK with a capitalisation over £50,000 and shares held by the public are termed public liability companies and denoted by the letters 'plc'. Public status does not necessarily mean that their shares are quoted on a stock exchange, simply that the shares can be traded among members of the public.

premium
The difference between the issue price of an ordinary share and its par (or nominal) value

■ Preference shares

Preference shares also constitute part of shareholders' funds – designated 'non-equity shareholders' funds', as they are hybrids falling between pure equity and pure debt. Holders receive an annual dividend, usually a fixed percentage of the par value. Holders also have preferential rights over ordinary shareholders. Preferred dividends are paid before ordinary share dividends, and preference capital precedes ordinary share capital when assets are sold in a liquidation and the sale proceeds are distributed.

In the UK, they normally carry no voting rights except in the case of a proposed liquidation or a takeover, or when the company passes the dividend. Preference share dividends do not qualify for tax relief. This lack of tax relief explains why preference shares are relatively unattractive to companies compared to other forms of fixed rate security. They may have some appeal to risk-averse investors looking for a relatively reliable income stream and a limited degree of participation in company affairs.

Types of preference share
Firms can issue many varieties of preference share. They can be:

- convertible
- cumulative
- participating
- redeemable,

or some combination.

These terms have the following meanings:

- **Convertible** means that the preference shares can be converted into ordinary shares at some future date.
- **Cumulative** means that if a dividend is passed (i.e. not paid), it will be carried forward for payment at some future date.
- Some preference shares are **redeemable**, meaning that holders will eventually be repaid their capital, usually at par.

redeemable
Repayable (usually at nominal value)

- Preference shares may also be '**participating**'. In an exceptionally good year, the directors may decide to declare an extra dividend for these preference shareholders above the regular fixed return, i.e. holders participate in the profits otherwise attributable to ordinary shareholders.

Maybe one can now appreciate why these are called hybrids – they can exhibit characteristics of both equity and debt, but the question often arises as to which they *most* resemble, for example, when calculating gearing ratios. The answer really depends on what type of preference share we are dealing with, and the strength of the commitment made to investors:

- If they are cumulative, so that a passed dividend will eventually be paid, they resemble debt more than equity.
- If convertible, and the conversion date is near, they look more like equity than debt.
- If redeemable, they look more like debt.
- If participating, they look more like equity.

At one extreme, non-cumulative, convertible, irredeemable, participating preference shares look very much like equity, while the cumulative, convertible, redeemable, non-participating variety look very much like debt. However, there is plenty of room along the spectrum for preference shares with less clear-cut combinations of features.

■ Reserves

Reserves are also part of shareholders' funds. It is a common mistake to assume that reserves represent cash balances. There is a fundamental difference between cash, which is an asset, and reserves, which represent part of the firm's financing. Some reserves may once have had a cash counterpart, but this will almost certainly already have been re-invested in wealth-creating assets such as plant and machinery, stocks and debtors. In addition to share premium, reserves are created to account for profits retained from the recently ended, and preceding, years. These may be called revenue reserves, retained earnings or, simply, Profit and Loss Account, to indicate their origin. This reserve sets the limit on the amount of dividend that a company can pay (assuming it has the cash available!). A **revaluation reserve** is created when a revaluation of assets reveals a surplus – without such a reserve to represent enhanced value of the shareholders' stake, the balance sheet would not balance.

Internal finance, including retained profits, provides the main source of new capital for companies in general. This is partly because it is less costly than selling new shares, as it avoids expensive issuing costs. It may seem that retained earnings are a free source of finance and far more attractive to management than, say, raising interest-bearing loans. However, retention of earnings imposes an opportunity cost – if returned to shareholders as a dividend, the cash could be invested to yield a return. The cost of retained earnings is therefore the return that could be achieved by shareholders on investments of comparable risk to the company: in other words, the usual return required by shareholders, or cost of equity. However, the cost of using retained earnings can be adjusted as follows to reflect the issue costs saved:

$$\text{Cost of retained earnings} = k_{RE} = \frac{\text{normal cost of equity}}{(1 + f)}$$

where f is the percentage costs of issue, or flotation, costs that are avoided.

Self-assessment activity 16.2

Criticise the following statement made by your marketing director:

'Well, we can always finance the new product development from our reserves.'

(Answer in Appendix A at the back of the book)

16.5 METHODS OF RAISING EQUITY FINANCE BY UNQUOTED COMPANIES

It is useful to make a distinction between companies not quoted on the Stock Exchange and those which are listed. A listing opens up far greater access to capital of all kinds, and to a wider pool of shareholders. This is one reason why many private and other unquoted companies aspire to an eventual listing.

The supply of equity to an unquoted company depends partly on its size and partly on its stage of development. Most new companies find it virtually impossible to raise equity finance except of a 'personal' nature, i.e. from supportive friends and relatives. Some venture capitalists (see below) will provide start-up finance for highly promising activities, but most invariably look for a track record on the part of the entrepreneur(s). Until this is established, such companies need to rely on additional supplies of personal finance and retained earnings for further equity.

Having established some sort of record of operating success and potential, further avenues open up. Among these are the following.

Business angels

A business angel is a private equity investor with spare funds to invest who wishes to gamble on the future prospects of young companies, often start-ups, managed by entrepreneurs lacking a track record but possessing a good idea plus enthusiasm. Angels are motivated partly by the prospect of riches and partly out of a desire to nurture the spirit of entrepreneurship. Bank of England data suggests that 20,000 angels invest some £1 billion in 6,000 UK firms each year. Unlike venture capitalists, which typically seek an exit in 3–5 years, angels tend to invest for much longer periods. However, despite attractive tax relief (see below) angel investment is not for the faint-hearted. Mason and Harrison (2002) estimate that angels lose money in 40 per cent of their investments and make returns exceeding 50 per cent in less than a quarter of them.

Several networks or 'marriage bureaux' are designed to bring together prospective investors who seek high returns but are prepared to tolerate a high risk of total loss, and entrepreneurs needing capital and prepared to release an equity stake. Many are localised, and some are highly informal, operated by accountants and solicitors; others are more organised. The umbrella organisation for introducing capital-hungry entrepreneurs to wealthy individuals is the British Business Angels Association (BBAA) (**www.bbaa.org.uk**). This evolved from the National Business Angels Network (NBAN) formed in 1999 by the major high street banks and leading firms of solicitors and accountants. Backed by the Department of Trade and Industry (now the Department of Business, Enterprise and Regulatory Reform), it grew out of the former Local Investment Networking Company (LINC), which had operated for ten years, and absorbed the NatWest Bank's own scheme, Network Angels, set up in 1996.

Increasingly, it is recognised that many businesses fail to raise equity, not merely because of 'finance gaps' on the supply side, but also because they are not 'investment ready' (Mason and Harrison, 2004). Accordingly, enterprise support agencies are now giving increasing attention to provision of information on sources of finance and presentational skills, e.g. what to include in a business plan, and how to present it to potential investors in a convincing way.

Government-backed schemes

Business Start-up Scheme (BES)
A now-defunct scheme set up to provide 'seed-corn' finance for new businesses

In 1981, the **Business Start-up Scheme** (later named the **Business Expansion Scheme** or **BES**) was introduced to assist small, newly-formed and hence high-risk companies to raise equity finance, and was extended to include existing companies in 1983. Investors could invest up to £40,000 p.a. in such companies and qualify for relief from income tax and also capital gains tax on disposal of shares if held for five years.

The BES had only modest success in its early years, but grew rapidly in the late 1980s and early 1990s after it was extended to include investment in companies investing in residential property on short leasehold terms. This shift was ostensibly to increase the supply of rented accommodation to encourage greater labour mobility. Whereas orthodox companies were restricted to raising just £0.5 million p.a. under the BES, residential property companies were allowed to raise up to £5 million p.a. This change spawned an upsurge in speculative property schemes, and because it became clear that company sponsors were using the BES primarily as a tax avoidance device, it was abolished as from December 1993.

It was superseded in January 1994 by the **Enterprise Investment Scheme** (EIS) with similar objectives to the original version of the BES, but offering less attractive tax reliefs. Investors holding no more than a 30 per cent stake in the business may invest up to £500,000 in any tax year in qualifying companies, receive relief from income tax at 20 per cent and freedom from capital gains tax if the investment is held for three years. Unlike the old BES, losses realised under the scheme are allowable against income tax. Companies cannot raise more than £1 million under the scheme in any one tax year, and investment in shipbuilding and coal and steel production was excluded from 2008.

Raising equity under the EIS

In July 2006, Carnaby Feature Films plc advertised an issue of 5 million 'B' shares at a price of £1 each to finance, produce and exploit a full-length British feature film entitled *The Bridge of Lies*.

The 'B' shares would be subordinate to 100,000 'A' shares, held by the firm's promoters who would retain 51 per cent of voting rights in the firm, and after payment of a £1 dividend per share, would receive 50 per cent of all subsequent dividends.

Other firms were also invited to subscribe for shares under the Corporate Venturing Scheme.

The directors of Carnaby were also the promoters of a series of film companies set up on a limited liability basis by Carnaby Film Companies plc (**www.carnabyfilms.com**).

Venture Capital Trusts
Stock-market listed financial vehicles set up to invest in a spread of highly risky new or young firms

Venture Capital Trusts (VCTs) are a form of investment trust set up specifically to provide equity capital to small and growing companies. To attract investors concerned at the high risks of EIS companies, VCTs offer investors a wider spread of investments than is possible under the EIS, thus operating in a similar way to orthodox investment trusts. At present (2008–9), investors can obtain income tax relief at 30 per cent in the year of investment plus freedom from capital gains tax and tax on dividends if the funds remain invested for at least five years. The maximum investment in any tax year is £200,000. Each company assisted can have no more than 50 employees, have gross assets of no more than £7 million and cannot have raised more than £2 million from VCT schemes in the previous 12 months. The VCT itself must be quoted on the Stock Exchange.

More information on VCTs can be found on the website of the British Venture Capital Association (**www.bvca.co.uk**), and **www.allenbridge.co.uk**, a leading organiser of VCT schemes. Both websites emphasise the risk factors involved in VCT investment, especially the illiquidity of the shares.

Corporate venturing

The Confederation of British Industry describes corporate venturing as: 'a formal, direct relationship usually between a larger, and an independent smaller company, in which both contribute financial management or technical resources, sharing risks and rewards equally for mutual growth'. The larger partner aims to foster the development of the small firm for a variety of reasons, partly altruistic. The arrangement can take a number of forms. The two parties could simply agree to cooperate in joint distribution or production, they could form a joint venture and/or the major company could take a minority equity stake in the small firm and inject cash into it. Aside from idealistic notions of nurturing the capitalist ethic, there are sound business reasons behind developing such links. For example:

- the large firm can tap into innovative R&D at low cost
- it may be able to use spare managerial capacity for training purposes
- it may develop a new source of supply
- it may gain access to a new market.

Corporate Venturing Scheme
A device to enable larger firms to take an equity stake in much smaller businesses with attractive prospects

The UK government provides fiscal incentives to encourage corporate venturing under the **Corporate Venturing Scheme**, for example, Corporation Tax relief at 20 per cent on corporate venturing investments in new ordinary shares held for at least three years, and tax relief for capital losses on disposals of shares. In addition, the 'investing company' must not hold more than 30 per cent of the 'issuing company's' ordinary share capital over the three-year qualifying period.

Venture capital (see also www.bvca.co.uk)

Venture capital (VC) is funding mainly for the development of existing companies with sound management and high growth potential, but sometimes for especially attractive start-ups. Funds are usually provided in large packages (typically over £250,000) of debt and equity, split this way partly to offer a degree of security, but mainly to allow the VC company to participate in any major success. VC companies will be hoping for an eventual flotation of the companies they assist, are usually prepared to see their investments 'locked in' for five to ten years, but will anticipate an annualised return of 30 per cent or more. VC companies often provide managerial assistance, and some (e.g. the biggest VC company, 3i, itself a listed company following its own Stock Exchange flotation in 1994) reserve the right to place their own appointee on the Board of Directors, although this is infrequently done.

independent companies
Investment companies set up as 'standalone' entities by venture capitalists to invest money in risky but potentially attractive businesses

captive firms
Subsidiaries of banks and other institutions established to invest in risky but attractive businesses

VC companies fall into two groups: **independent companies** and **captive firms**. Independent funds are set up by private venture capitalists, raising funds from a variety of sources to invest in projects for a specified time and then to liquidate the investments. Captive funds are subsidiaries of major financial institutions, such as banks and insurance companies, set up to channel a portion of their capital into risky enterprises. Some VC firms operate with a strong social perspective. An example is Bridges Community Ventures which is partly government-funded. In 2002, it invested £345,000 in Simply-Switch, the internet price comparison site set up by Karen Darby in a deprived area of Croydon. Its faith was rewarded in 2006 when Darby sold out to Daily Mail and General Trust for £22 million of which £8 million went to the VC firm.

VC organisations devote much of their energies and funding to large **management buy-out (MBO)** investments, rather than concentrating on smaller and riskier start-up and development capital. In an MBO, the existing management of a company undertakes to purchase the firm from the present owners. Most MBOs involve managers buying out an unwanted, often underperforming, subsidiary from a larger parent, which feels that the business unit no longer fits its strategy. Exceptionally, the buy-out may involve taking an existing quoted company off the Stock Exchange, usually financed by large amounts of debt, but with equity provided by the managers, who themselves borrow heavily to provide these funds. There is an element of '**moral hazard**' here – managers wishing to buy out a company or division may be tempted to depress its performance so as to lower the price they need to pay to acquire it. Moreover, there is a problem of **information asymmetry** – managers are far better able to judge the value of a company or operating unit than an outside analyst or the head office of a larger company.

Nevertheless, given the structure of the funding arrangements, the managers of MBOs are under considerable pressure to perform in order to service debts out of cash flow and to enhance the values of their own equity stake (and that of the venture capitalist). Many VC firms find the risk–return prospects of these deals more appealing than those of orthodox businesses. This is largely because the time taken before the venture capitalist can make an exit is usually much shorter.

Piercing the jargon

Venture capitalists, suspended between the worlds of heady financiers and down-to-earth businessmen, have developed their own incomprehensible jargon. Here's an introduction to the terms that make their hearts beat faster:

Business plan: needed to apply for financing. A good plan should outline the business opportunity, the management's track record, the company's past and present performance, market demand, competition and trends, the structure of the company and the shareholders, long-term plans and goals, and the proposed role of venture capital (how much and what for). See *Due diligence*.

Control: one of the most hotly debated topics. Even where venture capitalists have a minority share, they want to be able to influence and/or have veto rights over decisions on strategy, like taking out credits, setting up subsidiaries or signing licensing agreements. This issue often creates conflict with entrepreneurs, but venture capitalists disagree among themselves as to whether control requires a majority share or just a good shareholders' agreement, and where the balance lies between supporting and stifling the entrepreneurial spirit. See *Hands on*.

Deal flow: the number and quality of applications for investment that come to the venture capitalist. Deal flow can come from referrals (banks, accountants, advisory services), walk-ins, or from hard work pounding the streets.

Development capital: also known as expansion capital. Venture capital financing used for expansion of an already established company.

Due diligence: several months of scrutiny designed to dig up all the necessary information for evaluating the business plan and establishing whether the entrepreneur is a solid investment.

Exit: the all-important sale of the shareholding in a company which enables the venture capitalist to (at least) recoup his investment and pacify his investors. Most exits in Central Europe will be by trade sale, but some might be by management buy-out or flotation on the stock market – or by write-off, if the investment is a disaster. The usual time-frame for exit is around five years, meaning there has been very little so far in the region.

Hands on: what most venture capitalists in Central Europe claim to be. It means that they're not interested in a passive shareholding but want to support the management with expertise, usually in marketing, finding export partners, or in financial planning.

Living dead: a company that is not a failure but which is making the sort of insignificant profits that make it almost impossible to sell, and even then would yield low returns.

Management buy-out: the purchase of a business by the existing management. The venture capitalist pays proportionately much more for his stake to recognise the value the management brings to the enterprise. Also used to describe a possible exit route, by which the management buys out the venture capitalist's share. The question is: at what price?

Second-round financing: most companies need more than the initial injection of capital, whether to enable them to expand into new markets, develop more production capacity, or to overcome temporary problems. There can be several rounds of financing.

Seed capital: capital used to turn a good idea into a commercially viable product or service; a very risky form of investment, although it generally involves small sums.

Start-up capital: capital used to establish a company from scratch or within the first few months of its existence. Also risky, but with huge returns for the few successes.

Trade sale: the sale of a company (or part of it) to a larger corporation. This is the main source of exits envisaged for Central Europe, but it means that the entrepreneur may have to become an employee or sell out his share.

Venture capital: equity finance in an unquoted, and usually quite young, company to enable it to start up, expand or restructure its operations entirely. It's cheaper than bank finance initially because paying dividends can be deferred; it also provides a strategic partner – but it implies handing over some control, a share of earnings and decisions over future sales.

Source: Business Central Europe, February 1995. Reprinted with permission.

In a **management buy-in (MBI),** an outside group of managers buys a stake in an existing company and assumes managerial responsibility for its operation. MBIs are largely financed in the same ways as MBOs, and occur for similar motives.

Private placing (or 'placement')

A **private placing** is a means whereby well-established companies can widen their shareholder base without going for a Stock Exchange listing. It is arranged through a stockbroker or issuing house, which buys the shares and then 'places' them with (i.e. sells them to) selected clients.

16.6 WORKED EXAMPLE: YZ AND VCI

This question appeared as Question 3 on CIMA Paper 9 Financial Strategy, November 2006.

VCI is a venture capital investor that specialises in providing finance to small but established businesses. At present, its expected average pre-tax return on equity investment is a nominal 30% per annum over a five-year investment period.

YZ is a typical client of VCI. It is a 100% family-owned transport and distribution business whose shares are unlisted. The company sustained a series of losses a few years ago, but the recruitment of some professional managers and an aggressive marketing policy returned the company to profitability. Its most recent accounts show revenue of $105 million and profit before interest and tax of $28.83 million. Other relevant information is as follows:

- For the last three years dividends have been paid at 40% of earnings and the directors have no plans to change this payout ratio.
- Taxation has averaged 28% per annum over the past few years and this rate is likely to continue.
- The directors are forecasting growth in earnings and dividends for the foreseeable future of 6% per annum.
- YZ's accountants estimated the entity's cost of equity capital at 10% some years ago. The data they worked with was incomplete and now out of date. The current cost could be as high as 15%.

Extracts from its most recent balance sheet *at 31 March 2006* are shown below.

	$ million
ASSETS	
Non-current assets	
Property, plant and equipment	35.50
Current assets	4.50
	40.00
EQUITY AND LIABILITIES	
Equity	
Share capital (Nominal value of 10 cents)	2.25
Retained earnings	18.00
	20.25
Non-current liabilities	
7% Secured bond repayable 2016	15.00
Current liabilities	4.75
	19.75
	40.00

Note: The entity's vehicles are mainly financed by operating leases.

YZ has now reached a stage in its development that requires additional capital of $25 million. The directors, and major shareholders, are considering a number of alternative forms of finance. One of the alternatives they are considering is venture capital funding and they have approached VCI. In preliminary discussions, VCI has suggested it might be able to finance the necessary $25 million by purchasing a percentage of YZ's equity. This will, of course, involve YZ issuing new equity.

Required:

(a) Assume you work for VCI and have been asked to evaluate the potential investment.

   (i) Using YZ's forecast of growth and its estimates of cost of capital, calculate the number of new shares that YZ will have to issue to VCI in return for its investment and the percentage of the entity VCI will then own. Comment briefly on your result.

   (ii) Evaluate exit strategies that might be available to VCI in five years' time and their likely acceptability to YZ.

(b) Discuss the advantages and disadvantages to an established business such as YZ of using a venture capital entity to provide finance for expansion as compared with long-term debt. Advise YZ about which type of finance it should choose, based on the information available so far.

Answer

(a) (i) First, it is necessary to calculate dividends and retained profit:

	$m
Revenue	105.00
Profit before interest and tax (PBIT)	28.83
Interest ($15.0m × 7%)	(1.05)
Profit before tax (PBT)	27.78
Tax @ 28%	(7.78)
Profit after tax (PAT)	20.00
Dividends @ 40% pay-out	(8.00)
Retained profit	12.00

Using the dividend growth model, and assuming constant growth of 6%, the equity is valued as follows:

Cost of equity	10%	15%
Value $= \dfrac{D_1}{(k_e - g)}$	$\dfrac{(\$8.00m \times 1.06)}{(0.10 - 0.06)}$ = $212 million	$\dfrac{(\$8.00m \times 1.06)}{(0.15 - 0.06)}$ = $94 million
Implied P:E ratio	$212m/$20m = 10.6 times	$94m/$20m = 4.7 times
Value per share $ = Value/PAT Given 22,500,000 shares in issue	$212m/22.5m = $9.42	$94m/22.5m = $4.19
Shares to issue to VCI	2,653,928 ($25m/$9.42)	5,966,587 ($25m/$4.19)
Total shares in issue after new issue	25,153,928	28,466,587
Proportion owned by VCI	10.6% (2.654/25.154 × 100)	21.0% (5.967/28.467 × 100)

These figures suggest that YZ would need to issue between approximately 2.7 million and 6 million new shares depending on the valuation adopted, resulting in VCI owning between around 11% and 21% respectively of YZ. Even 21% is not a particularly high percentage, so long as VCI does not end up with the highest single shareholding. If this were the case, YZ managers would become vulnerable to pressure from VCI.

(ii) Possible exit strategies for VCI:

- *Sell back to YZ shareholders, perhaps via an MBO.* As founding shareholders, they might value the business more highly than a third-party investor. This would be an advantage for VCI giving it a good base to negotiate a higher price. This might be a disadvantage to YZ shareholders, who might not be in a position to raise the necessary finance.
- *Encourage YZ to apply for a stock market listing.* YZ is too small for a main market listing, so it would have to be on the secondary market (AIM in the UK). This could be an administratively lengthy and expensive process. Also, YZ may not wish to make available the percentage of shares necessary (10% on AIM) to allow a market in its shares. However, there would be many advantages of listing at this stage in YZ's development.
- *Sell to a ready buyer, for example a trade sale.* If VCI is seeking a quick sale, it may be easier to do this, although it might require a lower price for speed and ease of disposal. YZ may be unhappy with the new shareholder unless it has some right of veto built into the initial deal with VCI.

VCI financing versus long-term debt:

Advantages

- The money appears to be readily available.
- VCI may bring much-needed management expertise and, possibly, take a seat on the board (this could also be seen as a disadvantage/interference by YZ).
- It lowers, rather than raises, gearing.

Disadvantages

- VCI may want more control than management wish to cede, and may push for higher risk strategies than YZ is comfortable with in order to pursue its required rate of return.
- VCI may demand a seat on the board (but this could be an advantage, as noted above).
- VCI may eventually sell shares to an unwanted (to YZ) buyer, or push for an early flotation.
- There are no tax advantages with dividend payments, as compared to debt.
- It is difficult to value the shares. In these circumstances, we are valuing only part of the entity and estimates of value might need to be adjusted for a part-sale. Any adjustment will inevitably be subjective, but in some way, it is no different from flotation where founding shareholders issue less than 50% of the share capital in order to retain control.
- There may be higher set-up fees.

YZ has high gearing (based on book values) and appears dangerously illiquid, with a current ratio of less than 1. Borrowing from a bank could be difficult and expensive in these circumstances. However, although YZ is 'well established', thus meeting one of VCI's investment criteria, it has lower growth than would normally be expected by venture capitalists. New finance from either route might therefore be problematic.

16.7 GOING PUBLIC

Seeking a quotation: from unquoted to quoted status

Eventually, the unquoted company may require such a large amount of new capital that it may decide to 'go public' by issuing shares through a Stock Exchange. When a firm obtains a listing on a Stock Exchange by selling shares, this is referred to as an **Initial Public Offering (IPO)**. Such issues are managed by sponsors, such as a merchant bank or a member of the Stock Exchange, which advises on aspects such as the timing and the price of the shares to be issued.

To enhance the prospects of a successful IPO, the company ought to show a record of consistent and increasing profitability, and that it is managed by respected, experienced directors. There are numerous other criteria to qualify for a 'full listing' on the main London market that must also be satisfied, principally:

- It must provide fully audited accounts covering at least three years.
- It must be an independent business activity that has earned revenue for at least three years.
- The senior management and key directors should not have changed significantly throughout the three-year period, and should possess appropriate expertise and experience.
- If the company has a controlling shareholder, this must not prevent it from operating and making decisions independently.
- At least 25 per cent of the company's ordinary shares must be in public hands after listing.

The Stock Exchange cites a seven-point list of benefits and possible drawbacks involved in obtaining a quotation:

- *Prestige.* Being seen to comply with the rules of the Stock Exchange may enhance the company's standing in the business community. This may lead to better relationships with suppliers, creditors and customers.
- *Growth.* The initial flotation can be used to raise cash, which can be used to lower gearing. Flotation may also lower the cost of using bank facilities. Access to the wider markets can be exploited if the company wishes to raise more finance in the future. Greater marketability of the shares increases the ability to conduct takeovers by offering equity in exchange for the target company's shares.
- *Access.* As well as giving companies greater access to fresh supplies of capital, flotation gives shareholders access to a wider market, enabling existing owners to convert shares into cash, and new ones to buy and sell more readily, i.e. it makes the company's shares more liquid.
- *Visibility.* The initial flotation gives the company publicity and more regular coverage by the media as it announces subsequent results. Greater awareness among the securities industry may enhance the ability to raise further capital on favourable terms.
- *Accountability.* Quotation imposes new responsibilities on directors and increases the need to consult shareholders before taking major decisions, such as a major acquisition.
- *Responsibility.* Directors must ensure that the company meets the listing rules of the Exchange, and that price-sensitive information is released in a timely and orderly way, so that every shareholder has equal access to it.
- *Regulation.* The Stock Exchange, as the only competent authority for listing, rigorously screens applications of companies seeking a listing, monitors companies' ongoing compliance with its rules and deals with breaches of its rules.

Source: London Stock Exchange

■ Methods of obtaining a listing

There are four main methods of obtaining a quotation on the Stock Exchange:

offer for sale by prospectus
An issue of ordinary shares through an issuing house that promotes the shares in a detailed prospectus aimed at the public in general

1 **Offer for sale by prospectus**. Shares are sold to an issuing house, generally a merchant bank, and then are offered at a fixed price to the public at large, including both institutions and private individuals. Application forms and a prospectus, setting out all relevant details of the company's past performance and future prospects, as stipulated by Stock Exchange regulations laid down in 'The Yellow Book', must be published in the national press. An offer for sale is obligatory for issues involving £30 million or more.

Usually such an issue is underwritten, i.e. the issuing house guarantees to buy up any shares not taken up by the public so as to ensure that the company receives the monies required. To spread the risk of being called upon to buy up possibly substantial blocks of shares, the lead underwriter usually makes sub-underwriting arrangements with other financial institutions.

Offers for sale are usually made at a fixed price, which is determined before the offer period based on the expertise and knowledge of the company's financial advisers. There is a tendency to price prospectus issues conservatively to ensure their success – companies whose shares have to be bought up by the underwriters often find it difficult to raise further capital through the Stock Exchange on attractive terms.

issue by tender
A share issue where prospective investors are invited to bid or 'tender' for shares at a price of their own choosing

2 A variant on this method is the **issue by tender**, where no prior issue price is announced, but prospective investors are invited to bid for shares at a price of their choosing. The eventual 'striking price' at which shares are sold is determined by the weight of applications at various prices. Essentially, the final price is set by supply and demand. A tender is often used when there is no comparable company already listed to use as a reference point in valuing the company. However, by underlining the uncertainty in valuation, it may deter investors. In a tender, the shares are underwritten at a certain minimum price. This is the most expensive form of issue, although it may be argued that there is less risk of underpricing.

placing
An issue of ordinary shares directly to selected institutional investors, who may re-sell them on the stock market when dealings officially commence

3 In a **placing**, shares are 'placed', or sold to, institutional investors, such as pension funds and insurance companies, selected by the merchant bank advising the company and the company's stockbroker. In this case, the general public has to wait until official dealing in the shares begins before it, too, can buy the shares. Placings are geared to smaller companies and involve relatively little publicity and, hence, limited expense. Conversely, although placings should aim at securing a wide distribution of shareholdings to promote liquidity on the secondary market, the resulting spread of holdings is inevitably far narrower than with offers for sale.

intermediaries offer
A placing made to stock brokers other than the one advising the company making the issue

An **intermediaries offer** is a placing with financial intermediaries that allows brokers other than the one advising the issuing company to apply for shares. These brokers are allocated shares that they can subsequently distribute to their clients.

Stock Exchange introduction
Where an established firm obtains a Stock Market listing without selling any shares

4 **Stock Exchange introduction**. This is applicable when the shares are already widely held, the proportion in public hands already exceeds 25 per cent, and existing shareholders do not intend to dispose of shares at the time of flotation. No money is actually raised from the public – the purpose of the exercise is merely to create a wider market in the company's shares. Because no underwriting is required, and advertising costs are minimal, this is the cheapest form of issue.

■ Continuing obligations

Once a company is listed, its directors have to obey a strict set of rules in order to safeguard its continued listing. In particular, it must observe the regulations relating to disclosure and directors' dealings.

disclosure
Release of financial and
operating information about a
firm into the public domain

1 **Disclosure**. A quotation places considerable demands on directors, especially in the release of price-sensitive information about the company's activities. The main occasions on which announcements are required through the Stock Exchange include:

- major developments in a company's activities, e.g. new products, contracts or customers
- decisions to pay (or not pay) a dividend
- preliminary announcements of profits for the year or half-year
- an acquisition or disposal of major assets
- a change in directors, or directors' responsibilities
- decisions to make major capital issues.

directors' dealings
Share sales and purchases
by senior officers of the firm

2 **Directors' dealings**. Share dealings by the directors of a listed company are subject to the Criminal Justice Act 1993 and the Exchange's 'Model Code for Directors' Dealings', aimed to prevent insider dealing. Directors are precluded from dealing for a minimum period (normally two months) prior to an announcement of recurrent information such as trading results, or dealing in advance of the announcement of extraordinary events involving the publication of price-sensitive information. Companies whose directors infringe these rules are likely to jeopardise the continued listing of the company and also to sour relationships with investors, especially the financial institutions.

Hargreaves Lansdown soars on debut

Shares in stockbroker Hargreaves Lansdown soared by more than 30 per cent on their London Stock Exchange debut, as investors piled in hoping to profit from increasing savings rates in the UK.

Shares opened at 160p and rose to close at 209½p, adding £232.4m in value to the 26-year-old Bristol-based financial adviser and stockbroker which finished the day with a market capitalisation of £991.4m.

Founders Peter Hargreaves and Stephen Lansdown gained an extra £139m more than anticipated, as the 60 per cent holding they retained increased in value from £455.4m to £594.8m.

The price-earnings ratio implied by 160p was about 14 times 2009

earnings, roughly in the middle of the range between the highest and lowest rated UK stockbrokers and asset managers. But at the end of trading, the company was trading at more than 19 times estimated 2009 earnings.

Investors said that Hargreaves Lansdown merited a premium rating because it was serving a market that is not well targeted by other companies: investors with between £50,000 and £200,000 of assets to invest. Private client stockbrokers tend to target wealthier clients.

"There are 6.9m mass-affluents (in the UK), many fed up with years of poor service from life (insurance)

companies," analysts at company broker Citigroup said in a note.

They added that Hargreaves Lansdown has the only quoted platform in the UK that markets directly to investors. It has 350,000 clients. The main products that Hargreaves Lansdown sells to its clients are unit trusts.

Clients often pay 5 per cent of the investment upfront when they invest in unit trusts through brokers. But about ten years ago Hargreaves Lansdown innovated and started to offer unit trusts with large discounts to this 5 per cent rate and now charges no upfront fees on unit trusts.

Source: Sarah Spikes, *Financial Times*, 16 May 2007.

■ The Alternative Investment Market (AIM)

The AIM was set up by the Stock Exchange in 1995 to replace the declining Unlisted Securities Market. Its purpose is to provide a market for the shares of companies that are too young or too small to qualify for, or benefit from, a full listing. The Exchange organises a service called SEATS PLUS, which displays information about orders to buy and sell shares, to enable matching of buyers and sellers. To qualify for the AIM, companies have to satisfy certain criteria – in the main, less demanding than those of the full market – and to observe certain 'ongoing obligations' regarding the release of key information at appropriate times and the conduct of directors. In addition, the company must appoint and retain a nominated adviser (or 'Nomad') to assist with continuing compliance with the AIM's rules and a nominated broker to organise share transactions. By December 2007, there were 1,694 companies including 347 foreign ones with an AIM listing with total market capitalisation of £97 billion. The spectacular rise of the AIM to a position of international pre-eminence has been chronicled by Owen *et al.* (2007).

Aim flotation for London's Capital Pub Company

David Bruce, the founder of the real ale-themed Firkin pub chain, and Clive Watson, a former Regent Inns finance director, return to the quoted arena today with the flotation of their London-based pub company on Aim.

Capital Pub Company was set up in 2001 as an Enterprise Investment Scheme, a vehicle to help start-up businesses raise funds by giving lucrative tax breaks to investors. It raised £15.4m, which was used to buy 13 pubs.

"When we set up the company we decided that the pubs market had become very branded and there were too many pubs on the high street. A lot of branded pubs were very similar so we decided to set up a pub company that kept the individuality of the

original pub," said Mr Bruce, who's past ventures have included the Firkin, Slug & Lettuce and Hogshead & Hedgehog chains.

The decision to float Capital Pub Company, which owns 23 unbranded and mainly freehold pubs in central London, is intended to crystallise the value of the stakes of hundreds of private investors who backed the group through the EIS.

The Aim listing is to be done via an introduction to the market but, despite expectations that the company would raise between £10m and £20m on admission, it has decided not to seek new funds.

The company, which will have a market capitalisation of about £40m, said it was not raising money via the

float because it recently completed a £10m sale-and-leaseback of its flagship Hog in the Pound public house, opposite London's Bond Street Underground station.

The money raised from that property deal plus existing funds and an existing banking facility means that Capital Pub Company has enough money to continue with its existing expansion plans.

The company has appointed Grant Thornton to be its nominated adviser and Fairfax, the boutique advisory finance house, to be its broker.

Source: Lucy Warwick-Ching, *Financial Times*, 17 May 2007.

16.8 EQUITY ISSUES BY QUOTED COMPANIES

pre-emption rights
The right for existing shareholders to be offered newly issued shares before making them available to outside investors

Once a company has achieved a quotation, it will find it easier to raise further equity, assuming a successful trading and profit record. The commonest method of raising new equity is by a **rights issue** (see below). The Companies Act of 1985 gives existing shareholders the right to subscribe to new share issues in proportion to their existing holdings. This generally rules out a public issue although these **pre-emption rights** can be waived with the agreement of shareholders at a properly convened meeting. With such agreement, a placing may be arranged whereby shares are sold to participating institutions provided that the price involves no more than a 10 per cent discount to the market price.

vendor placing/placing with clawback
A placing of new shares with financial institutions where existing investors have the right to purchase the shares from the institutions concerned to protect their rights

Shares can also be issued as full or partial consideration when acquiring another company. In some cases, this may be done via a **vendor placing**, or **placing with clawback**. In a vendor placing, the acquiring company places the new shares with a group of institutions, thus diluting the ownership and earnings of existing shareholders. For sufficiently large issues, existing shareholders have the right to reclaim the shares they would have been entitled to, had there been a rights issue. If they do not, they receive no compensation for the loss in value of their holdings as there are no detachable rights to sell (see below).

In view of their importance, we now give detailed consideration to rights issues.

Investcorp puts Welcome Break up for sale with £500m price tag

Investcorp, the Bahrain-based private equity firm, has put Welcome Break, the UK's second-largest motorway service area operator behind Moto, up for sale with a price tag of about £500m.

Investcorp bought Welcome Break from Granada in 1997 for £473m, and three years ago sold nine of the chain's buildings for £270m to Robert Tchenguiz, the property investor.

Philip Yea, now chief executive of 3i, joined Investcorp in 1999 with the goal of resuscitating Welcome Break before Rod McKie, a former Pret a Manager executive, joined to help run the business in 2001.

In February, Investcorp recapitalised Welcome Break with £300m, valuing the business at about £500m. At the time of the recapitalisation, Mr McKie said sales were rising at 8 per cent a year.

It is understood a buyer would not need to renegotiate the recapitalisation finance because Royal Bank of Scotland and Calyon designed it to be transferable to a new owner.

Earnings in the year to September were £40m, as Welcome Break's 23 sites – which have average leases of 33 years – continued to generate stable cash flow from the 80m people who pass through the stores each year.

The long-term stability of Welcome Break's cash flow is part of the reason infrastructure investors, such as Reef, 3i, and Goldman Sachs could show interest in the service station operator.

Welcome Break is also expected to attract trade buyers. Autogrill, the world's biggest provider of food and drink services at motorway service stations, airports and railway stations, recently raised capital to make investments.

In 1995, Granada, which already operated a large share of the UK service area market, acquired Forte, owners of the Welcome Break Group, UK Travelodge chain and a number of roadside restaurants.

Granada then had to sell the Welcome Break chain to comply with competition rules. But in 2000 Granada merged with Compass, which also had a service business, thereby creating the UK's largest network of motorway service areas.

The combined company changed its name to Moto, and was last year purchased by Australia's Macquarie Bank. Macquarie's market share makes it an unlikely contender for Welcome Break.

Rothschild is advising Investcorp on the sale.

Source: Sarah Spikes, Nicola Cappin and Martin Arnold, *Financial Times*, 20 October 2007.

■ Rights issues

It is much easier for a quoted firm to make a rights issue than for an unlisted firm. In a rights issue, shareholders are granted the right to subscribe for shares (or less commonly, for other types of security) in proportion to their existing holdings, thus enabling them to retain their existing share of voting rights. Apart from the control factor, rights issues have certain other attractions:

prospectus
A document setting out the existing financial situation of a firm and its future prospects that is published to accompany a share issue

1 They are far cheaper than a public share issue. Provided the issue is for less than 10 per cent of the class of capital, there is no need for a **prospectus**, although a brochure must still be made available.

2 They may be made at the discretion of the directors without consent of the shareholders or the Stock Exchange. At one time, a queuing system for all new issues was operated by the government broker, acting for the Bank of England, in order to ensure a measured flow of new securities on to the market.

3 When stock market prices are generally high, companies have been known to raise cash through rights issues and to place it on deposit while seeking suitable candidates for acquisition. This gives a high degree of flexibility in timing a bid, i.e. the cash is already to hand.

4 The finance is guaranteed, either from existing shareholders or from the underwriters. Existing shareholders are given an incentive either to take up their rights or to sell them. It is not a sensible option to do nothing: this effectively reduces their wealth, as shares are typically offered at a discount of about 20 per cent below the current market price. If, as is usual, they are underwritten, the company is guaranteed to receive the cash, although it is embarrassing to have to call upon the underwriters to fulfil their obligations.

However, underwriting is costly. It has been estimated that companies typically 'lose' around 2 per cent of the funds raised in an equity issue shared out as follows: lead underwriter, usually the lead investment bank that organises the issue, taking 0.5 per cent, the firms' stockbrokers taking 0.25 per cent, and the remaining 1.25 per cent split among the various institutional shareholders, who are in many cases already shareholders. Conversely, putting out the underwriting business to tender can halve the underwriting fees. Nevertheless, most issues are still not put out to tender, despite a report by the Monopolies and Mergers Commission in 1999, which stated that a 'complex monopoly' existed in the awarding of sub-underwriting contracts. Firms now have to justify the mode of underwriting to shareholders under revised Stock Exchange listing rules. It is likely that the persistence of traditional ways of handling the underwriting process is due to firms wishing to build a new, or protect an existing, relationship with key institutional shareholders whose support might be critical in a possible future crisis.

Why even discounted rights issues need underwriting

The ill wind blowing a flock of rights issues into UK markets brings with it the old debate about how much – if anything – underwriters and sub-underwriters should earn for insuring cash calls against failure.

Shareholders in Royal Bank of Scotland have just won themselves improved terms for insuring part of the bank's £12bn whopper share issue. They'll get 1 per cent of the sum insured for sub-underwriting it, while the principal underwriters receive 1.5 per cent plus an additional discretionary 0.25 per cent.

There's little high ground in the debate about how these fees are divvied up. Institutional investors complain that investment banks make a fat profit from underwriting, even after paying the institutions to take on much of the risk. But if the same institutions have already decided to subscribe for their rights – and are big enough to influence others to do the same – they have virtually guaranteed its success. At which point sub-underwriting is itself almost free of risk.

The bigger question is whether deeply discounted rights issues need underwriting at all. Asked to look into the system nine years ago, the old Monopolies and Mergers Commission expressed the wish for more frequent non-underwritten deeply discounted issues.

The problem is that the deep discount sends an emergency signal and

is usually applied at times when equities are already in turmoil. This is an underwriters' market: capacity is low, demand high. Confidence is all. It would be reckless of RBS not to pay for its cash call to be underwritten. But could you not argue that the chance of RBS falling from 340p to below the 200p price of the new shares is negligible? Of course you could. But for anyone to have confidence in your argument, you would have to be the only person who successfully predicted both the run on Northern Rock and the bail-out of Bear Stearns.

Source: Andrew Hill, *Financial Times*, 25 April 2008.

■ Shareholders' choices in a rights issue: Grow-up plc

Grow-up plc decides to make a rights issue of one new share for every three held. The share price prior to the issue is 200p and the new shares are to be offered at 160p. In practice, rights issues are made at a discount, partly to make them *look* attractive and thus encourage shareholders to subscribe, and partly to safeguard against the risk of a fall in the market price during the offer period. The **theoretical ex-rights price (TERP)** is the price at which shares are expected to trade after the rights issue has been completed. It is calculated below at 190p:

theoretical ex-rights price (TERP)
The share price that should in theory be established, other things being equal, after a rights issue is completed

Effect of a 1-for-3 rights issue		
Before	3 old shares prior to rights issue at 200p each:	600p
	1 new share at 160p:	160p
After	4 shares worth:	760p
	1 share is therefore worth (760p ÷ 4) = TERP =	190p

nil paid price of rights
The market value of the right to subscribe for new shares offered in a rights issue

The value of the rights is the difference between the pre-rights share price and the TERP. In the case of Grow-up plc, this is (200p − 190p) = 10p for every existing share held. This 30p is termed the '**nil paid price of rights**'. The first option for shareholders is to sell their rights, obtaining 10p per share, less any dealing costs. A shareholder with 3,000 shares in the company would have a holding with market value prior to the rights issue of (3,000 × £2) = £6,000. After the issue, the value will fall to

(£3,000 × £1.90) = £5,700, a decline of £300, which is the amount he or she would receive for the rights sold.

The formula for the TERP is thus:

$$\text{TERP} = \frac{(N \times \text{cum rights price}) + \text{issue price}}{N + 1}$$

where N = the base number of shares held (i.e. number of rights required to buy one share).

In this example, TERP is thus:

$$\frac{(3 \times 200p) + 160p}{(3 + 1)} = \frac{760p}{4} = 190p$$

Similarly, the value of a right = (TERP − issue price).

In the example, this is:

$$(190p - 160p) = 30p$$

The nil paid price can be expressed per existing share, (10p) or more usually, per block of shares required to acquire one new share (3 × 10p) = 30p, i.e. the difference between the TERP and the issue price.

The second option is to subscribe for the new shares by *taking up the rights*. This should happen only if the shareholder has the resources to acquire the additional shares and believes this is the best way to invest such money. Additional reasons for taking up the rights are the fact that no stamp duty or broker's commission is payable, and the desire to maintain one's existing share of voting power.

A third option is to *sell sufficient rights to provide the cash to take up the balance*. This option, known as 'tail-swallowing', makes sense for shareholders who want to maintain their existing investment in the company in value terms.

This was done by Daedulus Projects Ltd, a major shareholder in Costain, the construction firm that made a rights issue in autumn 2007 (see below).

The formula for calculating the number of shares for a tail-swallower to buy is:

$$\frac{\text{Nil paid price}}{\text{Ex-rights price}} \times \text{Number of shares allotted}$$

As noted, the *nil paid price* is the difference between the TERP and the subscription price, i.e. (190p − 160p) = 30p. The number of new shares to which our investor with 3,000 existing shares retains acquisition rights is:

$$\frac{30p}{190p} \times 1,000 = 157 \text{ shares}$$

To buy 157 shares at 160p will cost £251.20, funded from (843/1,000) rights sold at 30p = £252.90. The total investment is now worth (3,157 × 190p) = £5,998, which (when rounded) is equivalent to the original investment of £6,000.

The final option is to *let the rights lapse* by doing nothing. In this case, the company may sell the new shares in the market and reimburse the shareholder net of dealing fees. Alternatively, the issuer may conduct an auction of rights not taken up to avoid the need to appoint underwriters.

The real message from rights issues is that shareholders cannot expect to receive something for nothing. The apparent gain from the invitation to purchase new shares at a discount on the existing price is more illusory than real.

To some, a rights issue may look damaging because the share price (in theory) has to fall due to the sale of shares at a discount, but again this apparent damage is illusory. Of course, the EPS, based on the last reported profits, will fall, as there are more shares in issue. But if people are bullish about the firm's prospects then the post-issue price may exceed the TERP (and vice versa). In this case, the market would be pricing in the expected returns from new investment, i.e. adding in the NPV of the new project (and v.v.). In effect, investors are saying that the cash raised is worth more than its nominal value as it brings with it the promise of positive investment returns (and v.v.); similarly, if the post-issue price is equal to the TERP, investors are assessing the NPV of the investment project at zero.

■ Open offers

An open offer, or 'entitlement offer', may also be made by a quoted company to its existing shareholders. Like a rights issue, it invites shareholders to buy new shares at a specified price, normally lower than the going market price. The investor's entitlement to buy is also based on his/her existing holdings. However, there is one important difference – an open offer cannot be traded on the market – if the offer is not taken up it lapses. An additional difference is that the firm may invite investors to apply for more than their strict entitlement – a so-called 'excess application', although there is no guarantee that this excess will be satisfied, as demand for shares may exceed the amount the firm wishes to issue.

Open offers are not that common. One such was announced in March 2007 by Mecom plc, the AIM-listed regional newspaper publishing and media group with operations in Norway, the Netherlands, Poland and Denmark. Mecom proposed to raise £570 million by way of a placing and 7-for-10 open offer priced at 78p to fund acquisitions that would drive its ambition to become a 'pan-European publishing group'. The offer was made at a small discount (1.5p) to the market price, but was so well received that the market price rose to 84p on the announcement day.

In the event, 84 per cent of the shares offered were taken up by the institutional investors who underwrote the offer.

Self-assessment activity 16.3

What is the TERP in the following case?

■ pre-announcement share price = £5.
■ rights issue of 1-for-6 at £3.50 issue price.

(Answer in Appendix A at the back of the book)

Anatomy of a rights issue

Costain Group plc, the construction firm, formally announced a rights issue on Friday 14 September 2007, avowedly to 'strengthen its balance sheet to take advantage of opportunities in its key sectors of operation', and also to provide some financial elbow-room for restoring the dividend.

Costain had been something of a basket case in recent years, after breaching banking covenants twice in 2006, making contract write-downs of over £47 million and closing its international business at a cost of £27 million. Its pre-tax loss was £62 million in 2006. Rumours of the rights issue had circulated on the preceding Tuesday when its shares fell over 4 per cent to 38p on market talk of a deeply-discounted rights issue, usually a sign of trouble. Costain confirmed that a rights issue would indeed be made on the following day, but waited until the Friday before making a formal announcement of the terms. The amount to be raised via a 3-for-4 rights offer, underwritten at an issue cost of £4 million, would be £64 million at an issue price of 24p, representing a 36 per cent discount to the closing price.

The pill was sweetened by the announcement of first-half pre-tax profits of £8 million compared to a loss of £21 million a year earlier.

Based on the 'undisturbed' share price ruling before the rumours hit the market, the theoretical ex-rights price (TERP) was:

4 shares @ 39.75p	= 159p
Cash: 3 shares @ 24p =	72p
Total	231p
TERP	= (231p/7 shares) = 33p

With the Friday closing market price at 37.5p, some 4.5p above the TERP, this suggested a nil-paid rights price of about 4p, and that people viewed the issue favourably. Indeed, some brokers raised their ratings on Costain to a 'buy'. Moreover, on the announcement day, the market as a whole had fallen sharply in the wake of the problems besetting Northern Rock.

Source: Based on articles in the *Financial Times*, especially by Toby Shelley.

■ Scrip issues and bonus issues

Whereas a rights issue raises new finance, a **scrip issue** simply gives shareholders more shares in proportion to their existing holdings. As a result, the value of their total holdings is unchanged, but the share price will fall due to earnings dilution. Scrip issues are often used by companies whose unit share price is 'high' – a high or '**heavyweight**' share price (in the UK, £10 or above) is regarded as a deterrent to trading. This was the reason given by the German biotechnology company, Geneart AG, the global market leader in gene synthesis technology in 2007, when it made a 'one-for-one' scrip issue. For every share held, owners were given a free share. According to the CFO of Geneart, the firm aimed 'to significantly increase the number of tradable shares . . . to support the liquidity and tradability of our shares.'

Companies like Geneart have built up substantial reserves by retention of earnings, making their issued share capital look relatively small. In the case of Geneart, the issued share capital was €2.243 million and the capital reserve was €16.900 million. The effect of the scrip issue was to double the issued share capital and to reduce the reserve by €2.243 million to €14.657 million. In other words, Geneart converted, or 'capitalised', its reserves into issued share capital, hence the common use of the synonym '**capitalisation issue**'.

Scrip issues do not always involve such a drastic reorganisation of shareholder funds. They are often given as '**bonus issues**' in addition to cash dividends, and are often taken by the market as a signal of higher future dividends. If a company makes, say, a one-for-ten scrip and maintains the dividend per share, this is tantamount to a future increase in dividends of 10 per cent (the new shares do not normally qualify for the dividend immediately). This signifies the company's expectation of greater capacity to pay dividends in the future, i.e. higher future earnings. In such cases, the share price may not fall quite so far as the simple arithmetic may suggest, i.e. by 1/11th, but may even increase as the market responds to the 'signals' emitted by the company.

■ Share splits ('stock splits' in the USA)

An alternative way of addressing the heavyweight status of a share is to split the ordinary shares into a larger number with lower par value. For example, one additional share may be given for every existing share in a '2-for-1' split (i.e. one share becomes two). In theory, this has no effect on the accounting numbers, i.e. the book value of the share capital. Nor should it affect the share price since no additional funds are raised and each shareholder's interest in future profits is unchanged.

In 2007, Iberdrola SA, the Spanish power utility that acquired Scottish Power, announced a 4-for-1 split whereby the new shares would have their nominal value cut from 3 euros to 0.75 euros. Noting that total share capital would not change from its current 3.75 billion euros, Iberdrola stated the aim of the exercise was to 'offer more liquidity and trading opportunities to the stock'.

Microsoft Inc. has made nine stock splits since its IPO in March 1986, the most recent being in February 2003, as shown in Table 16.1, when 5.4 billion shares were multiplied into a total of 10.8 billion in a 2-for-1 split.

Self-assessment activity 16.4

In a share split, e.g. '2-for-1', what is the effect on:

■ the number of shares issued?
■ the shareholders' capital in the balance sheet?
■ the firm's assets?
■ its market value – per share? in total?

(Answer in Appendix A at the back of the book)

Table 16.1 History of Microsoft common stock splits

Split	Payable date	Type of split	Closing price before/after
First	Sept. 18, 1987	2 for 1	Sept. 18–$114.50/ Sept. 21–$53.50
Second	April 12, 1990	2 for 1	April 12–$120.75/ April 16–$60.75
Third	June 26, 1991	3 for 2	June 26–$100.75/ June 27–$68.00
Fourth	June 12, 1992	3 for 2	June 12–$112.50/ June 15–$75.75
Fifth	May 20, 1994	2 for 1	May 20–$97.75/ May 23–$50.63
Sixth	Dec. 6, 1996	2 for 1	Dec. 6–$152.875/ Dec. 9–$81.75
Seventh	Feb. 20, 1998	2 for 1	Feb. 20–$155.13/ Feb. 23–$81.63
Eighth	March 26, 1999	2 for 1	March 26–$178.13/ March 29–$92.38
Ninth	Feb. 14, 2003	2 for 1	Feb. 14–$48.30 Feb. 18–$24.96

Source: www.microsoft.com.

Equity capital: checklist of key features

For:

- No fixed charges (e.g. interest payments). Dividends are paid if the company generates sufficient cash, the level being decided by the directors.

- No repayment is required. It is truly permanent capital.

- In the case of retained profits and rights issues, directors have greater control over the amount and timing.

- It carries a higher return than loan finance and acts as a better hedge against inflation for investors.

- Shares in most listed companies can be easily disposed of at a fair value.

Against:

- Issuing equity finance can be cost-effective (as in the case of retained profits or a rights issue), but it is expensive in the case of a public issue (often 5 per cent or more of the finance raised).

- Issuing ordinary shares to new shareholders dilutes the degree of control of existing members.

- Dividends are not tax-deductible, making equity relatively more expensive than borrowing.

- A higher proportion of equity can increase the overall cost of capital for the company (see Chapter 18).

- Shares in unlisted companies are difficult both to value and to dispose of.

■ Reversing the flow: going private again

The last decade has seen an upsurge in the number of firms being taken off the stock market by so-called private equity firms, generally specialist funds that are subsidiaries of banks or syndicates set up by a number of banks. Traditionally, they have specialised in funding management buy-outs or spin-offs of unwanted divisions of larger firms, but more recently, they have been active in taking quoted firms off the stock market.

Sometimes, they are set up specifically to acquire one particular firm. Some observers estimated that by the end of 2007, firms controlled by private equity firms accounted for around 20 per cent of private sector employment in the UK.

Not being quoted themselves, they do not face the same public scrutiny or continuous pressure to perform. Their aim is to restructure the acquired firm and sell it on, either in a trade sale or by a refloatation. Some spectacular successes have been achieved with substantial increases in the value of firms taken private and then refloated a few years later being recorded.

Among the reasons for the rise in the private equity sector are:

■ *The weight of regulation and disclosure that listed firms have to bear,* for example, the move to International Reporting Standards in 2005, the ongoing requirements of the Combined Code, and the introduction of the Sarbanes-Oxley Act affecting firms with a US listing. Regarding the Combined Code, many firms do not see the need to separate the roles of Chairman and Chief Executive, arguing that it leads to lack of flexibility, which hampers swift and effective decision-making.
■ *Greater liquidity among financial institutions.* Many institutions have curtailed their investment of new money into the stock market, and others have cut back their exposure, creating vacuums that they have filled by investing in private equity funds.
■ *The ability to tolerate higher gearing.* With no public scrutiny, the amount of debt that they can carry is greater than for an equivalent listed firm. Private equity firms tend to concentrate on asset-rich firms with solid cash flows, most notably firms in the retail store sector, which often need a re-vamp.

16.9 DEBT INSTRUMENTS: DEBENTURES, BONDS AND NOTES

The array of instruments for raising debt finance is even greater than for equity finance. Firms can raise long-term debt via the banking system, e.g. by a term-loan, or via the money and bond markets, by issuing a security that can be traded rather than held to maturity. The money markets supply short-term borrowing while the bond markets supply medium-to-long term finance.

The word 'bond' is a general term used to describe a variety of longer-term loans to companies. In some markets, they are described as '**loan stock**', or, especially where the interest payable is variable, as '**notes**'. A bond is simply a receipt or promise to repay money on a loan, usually with interest i.e. it binds the borrower to a commitment that can range between one and 30 years.

notes
Loan securities in general, but often referred to securities that carry a floating rate of interest

Characteristics of a bond are:

■ the nominal or par value in the currency of denomination.
■ the redemption value – usually the par value, but other possibilities include a stated premium or index-linking.
■ the rate of interest payable – known as the coupon – expressed as a percentage of the nominal value.
■ the redemption date.

For example, Compass Group plc, the food services contractor, is committed to the following bond:

Sterling Eurobond 7% 2014 £250 million

The £250 million will be repaid in full in 2014; and £7.00 interest is payable per £100 of stock each year in two stages.

■ Debentures

In strict legal terms, a debenture is a document acknowledging that the firm has borrowed money, whether or not any security has been given to back the loan. However, in normal business usage, this term is used to describe a loan which is secured on the assets of the company by mortgage deeds – a secured debenture is often called a **mortgage debenture**. If the issuer goes into liquidation, or defaults on interest or capital payments, the holders can apply for a court ruling to order the sale of either specified assets (called a **fixed charge**) or any of the firm's assets (a **floating charge**). The firm cannot dispose of assets subject to a fixed charge without the permission of the creditors.

As regards priority for payment, debts rank in order of issue – holders of the earliest issued debentures must be paid interest before the later comers. Where a firm has issued a series of bonds, a *pari passu* clause is inserted into the document acknowledging the debt. Debentures that rank lower down the priority list are called 'junior' or 'subordinated' stock.

The Compass plc stock referred to above is unsecured, and thus ranks behind its secured borrowing, but ahead of a number of subsequently-issued unsecured bonds. Unsecured stock is riskier than secured stock and investors thus require a higher coupon rate.

The pejorative term '**junk bond**' is applied to the unsecured loan stock of a borrower that merits sub-investment grade by a bond-rating agency. The credit-rating agency Standard & Poor's investment grade is BBB or above – any security rated below this 'is regarded as having predominantly speculative characteristics with respect to capacity to pay interest and repay principal'. Obviously, junk carries much higher than average yields, hence the euphemism 'high yield bonds'.

Loan agreements usually specify **restrictive covenants**. Such conditions might include the following:

1 *Dividend restrictions* – limitations on the level of dividends a company is permitted to pay. This is designed to prevent excessive dividend payments, which may seriously weaken the company's future cash flows and thereby place the lender at greater risk.
2 *Financial ratios* – specified levels below which certain ratios may not fall, e.g. current ratio.
3 *Financial reports* – regular accounts and financial reports to be provided to the lender to monitor progress.
4 *Issue of further debt* – the amount and type of debt that can be issued may be restricted. Subordinated loan stock (i.e. stock ranking below the existing unsecured loan stock) can usually still be issued.
5 *Asset backing* – a specified minimum level of tangible fixed assets.

mortgage debenture
A loan instrument under which the ownership of selected assets is mortgaged to the lender – in a default, the title passes to the lender

restrictive covenants
Limitations on managerial freedom of action, stipulated as conditions of making a loan

Debentures and unsecured loan stock: checklist of key features

For:
■ Most corporate loan stocks give ten or more years before repayment is due. A 'bullet' loan is where there is just one final repayment, and a 'balloon' loan is where increasing amounts of capital are repaid towards the end of the period of the loan. Bullet and balloon loans give attractive cash flow benefits in the early years, where little or no interest is payable.
■ A successful company may eventually be able to redeem the loan stock through a new issue, without drawing upon operating cash flows (although the company is exposed to the risk of higher interest rates).
■ Interest is tax-deductible.

Against:
■ Restrictions are placed on the company in terms of either the charge over assets or the restrictive covenants imposed.
■ Unsecured loan stock may impose demanding performance requirements.
■ Greater monitoring and control takes place over a public issue such as a debenture than with, say, a term loan with a bank.

■ Deep-discount bonds

Some debt instruments are sold at a price well below the par value, with a so-called deep-discount. An extreme case of this is zero-coupon bonds, e.g. a bond issued at £70 with a five-year life to maturity when it will be repaid at par of £100. Such bonds carry no entitlement to interest as such, thus appealing to investors who would normally pay income or corporation tax on interest income, but who may not be liable to capital gains tax, or who wish to defer it. In this example, the annualised rate of return from the capital gain if held to maturity for the full five years is represented by the rate r in the following compound interest expression:

$$£70(1 + r)^5 = £100$$

This can be written as:

$$\frac{1}{(1 + r)^5} = \frac{£70}{£100} = 0.700$$

The discount tables (PVIF) can be used to find r (about 10.75 per cent p.a.).

Self-assessment activity 16.5

What is the yield to maturity on a zero-coupon bond issued at £50, repayable at par of £100, in ten years' time?

(Answer in Appendix A at the back of the book)

■ Asset-Backed Securities (ABSs)

In recent years, some companies – and even certain individuals – have issued a new breed of securities, backed not by physical assets but by a reliable long-term stream of future earnings. A category of assets commonly utilised has been intellectual property represented by patents and copyrights. Like most security issues, ABSs are sold essentially to raise cash for investing in other activities.

Organisations effectively *capitalise* their future income into a single lump sum and sell it on the financial markets to generate immediate cash. The firm's financial advisers set up a **Special Purpose Vehicle (SPV)**, as shown in Figure 16.1. This is effectively a 'bank' which handles the bond issue and into which the designated income stream is paid and from which is paid the stream of interest payments needed to service the borrowing.

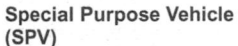

Special Purpose Vehicle (SPV)
A financial vehicle set up to manage the issue of Asset-Based Securities and arrange for payment of interest and eventual redemption

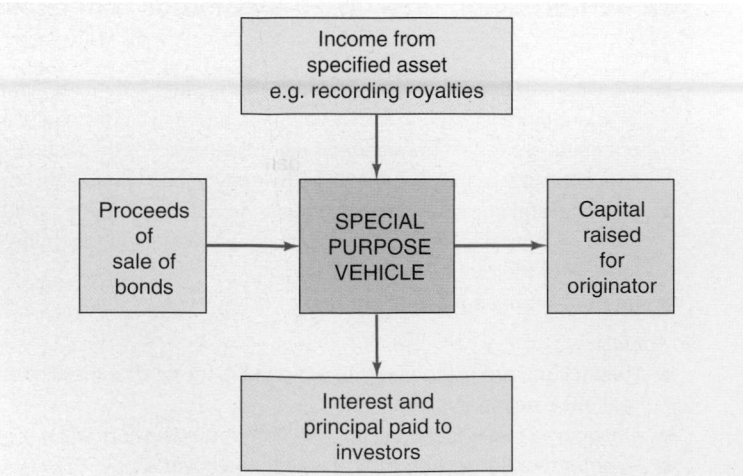

Figure 16.1 How an SPV works

This process of converting non-tradable claims into tradable ones is called **securitisation**. Like most financial innovations, securitisation originated in the USA. Banks parcelled up mortgage commitments made by house purchasers into bundles of mortgages to sell as interest-bearing securities, originally known as collateralised mortgage obligations (CMOs) now known as collateralised debt obligations (CDOs). Having both liquidity and a bank's guarantee, these could be offered at a lower interest rate than that charged on the underlying mortgages, the difference representing profit for the bank. This practice also became widespread in Europe, where it was increasingly seen as a cheaper alternative to unsecured bond issues.

The following examples of the ABS principle (not all of which involved SPVs) demonstrate its flexibility and versatility:

- In 1992, the Disney Corporation issued $400 million in seven-year notes with a variable rate of interest to be paid from royalties receivable from its portfolio of film copyrights, a path followed also by News Corporation in 1996.
- In 1997, David Bowie raised $55 million by selling bonds backed by his music copyright portfolio, with an average bond life of ten years. This tactic was also adopted by Rod Stewart and Michael Jackson, using similar security.

CDOs and the credit crisis

In recent years, CDOs have received a bad reputation, especially those based on private mortgages. During the early 2000s, there was a surge in sales of such CDOs, especially in the USA, an estimated $600 million being sold in 2007. Mortgage CDOs were sold as packages of mortgages of varying quality (i.e. risk). They were said to be 'structured' according to riskiness, hence the term 'structured finance'. They were assigned risk ratings by the credit rating agencies depending on the risk of default of the underlying assets.

Banks were keen on CDOs because they offered a way to exchange relatively illiquid loan assets for cash, thus strengthening their balance sheets, thus enabling them to take on more lending activities. The major banks also acquired CDOs from specialist mortgage banks and sold on CDO tranches to a wide range of investors across the financial system.

However, the quality of these CDOs was often overstated, with a larger proportion than investors had realised based on 'sub-prime mortgages' – loans made to high-risk borrowers with a high probability of default, and often, it was alleged, not adequately credit-screened by the original lender. As defaults occurred, the value of many CDOs plummeted, making it very difficult for holders to sell them on. As some CDOs use the same mortgage pool as collateral, any defaults in the pool causes a ripple effect through several investments. It was estimated that defaults of, say, $100 million could trigger losses throughout the financial system of $500 million. Investors who were attracted by relatively high rates of return for supposedly moderate risk investments were stunned to find that the market value of their CDOs had nose-dived, often to as low as 60 per cent of face value.

Many financial institutions that had acquired CDOs had to make substantial write-offs, the largest being $38 billion by Swiss bank UBS, $21 billion by Citigroup and $25 billion by Merrill Lynch. However, there was widespread suspicion that too few banks had 'come clean' about the scale of their losses with the result that the inter-bank lending market that provides temporary liquidity to banks with over-stretched balance sheets effectively dried up, which had a domino impact on other areas of the financial markets. In the UK, the most famous casualty was former building society Northern Rock, which was unable to refinance market borrowing that was due for repayment. Facing insolvency, it was offered for sale via the Bank of England, but with no acceptable offers forthcoming, the UK government was forced to nationalise it to prevent it collapsing.

At the time of writing (April 2008), the CDO market had effectively seized up, no issues having been made since August 2007. Only time will tell whether the market will recover.

- Holland's De Nationale Investeringsbank NV (DNIB) is a major player. Its ABS issue in March 1999, worth 290 million euros, was its fourth inside two years.
- Calvin Klein, GE Capital and Nestlé have all issued ABSs secured on trademarks.
- In 2004, British football club Leeds United issued bonds secured on future revenue from 'gate money', i.e. ticket sales, as part of a rescue package.
- Also in 2004, in the USA, Florida's Seminole Indian tribe sold $410 million of bonds to fund development of its gaming resorts in Tampa and Hollywood, Fa. The bonds were secured on future gambling takings, and income from a joint venture with Hard Rock Café.

Convertibles

A convertible begins life as a form of debt, but carries the right, at the holder's option, to convert into ordinary shares at some specified date in the future and on specified terms, e.g. how many new ordinary shares can be obtained on conversion per unit of convertible stock.

Firms that issue convertibles increase their gearing ratios and may be viewed as being more risky. Yet the greater risk is not always reflected in a higher coupon rate. As there is a prospect of making a capital gain should the share price market perform strongly, convertibles can usually be issued at a lower rate of interest than straight or 'plain vanilla' debt. Until the date of conversion, the holder receives a fixed rate of interest and is a long-term creditor of the company.

Convertibles are particularly suitable for companies facing relatively high business risks but strong potential growth because they offer investors the possibility of participating in future prosperity. This explains the ease with which many 'dotcom' companies were able to issue so much convertible debt. The downside for companies is that interest, although tax-deductible, must be paid every year, good or bad, and the principal requires repayment if holders do not convert.

The downside for existing shareholders is the prospect of dilution of their equity, and hence a fall in EPS, as and when conversion occurs. Dilution is especially damaging if the conversion terms are misjudged, e.g. if growth is a lot stronger than expected, the conversion terms may be over-generous to convertible holders. It often makes sense for existing shareholders to hedge against this risk by acquiring the convertibles themselves. Indeed, convertibles may be issued initially to owners in a rights issue.

The language of convertibles

Convertible conversion terms can be complex.

- the **conversion date** (or range of dates) tells you when it can be converted.
- the **conversion rate** tells you the terms on which conversion can be made. This is stated either as a **conversion price** – the nominal value of loan stock that can be converted into one ordinary share – or as a **conversion ratio** – the number of ordinary shares that will be obtained from one unit of loan stock.
- the **conversion value** is the market value of ordinary shares into which a unit of convertible loan stock can be converted. This is equal to the conversion ratio times the current market price per ordinary share.
- the **conversion premium** is the difference between the market price of the convertible and its conversion value.
- the **rights premium** is the difference between the market value of the convertible and its value as straight debt. Each of the last two terms can be expressed as an absolute value or per share.

An example will help clarify this terminology.

Example: Cannon plc

Cannon plc's balance sheet shows 10 per cent convertible loan stock, par value of £100, redeemable at par in seven years. Each unit of stock can be converted at any time in the next three years into 20 ordinary shares. The debenture currently trades at £117, interest has just been paid and the current ordinary share price is £3.60. The ex-interest market price of the debentures of a company of similar risk is £109.

Current conversion value   = (20 × £3.60) = £72
Current conversion premium = (£117 − £72) = £45 (or £2.25 per share)
Current rights premium   = (£117 − £109) = £8 (or £0.40 per share)

At the initial issue date, the conversion value will be less than the issue price. Investors hope that as the conversion date nears, and as the market price of the underlying shares increases, the conversion value will rise accordingly i.e. conversion becomes more attractive to investors. The conversion premium is proportional to the time remaining before conversion occurs. As conversion approaches, the market value and the conversion value converge until the conversion premium disappears. With no conversion premium, the value of the convertible is simply its value as straight debt with a similar coupon and maturity.

The market value of the convertible thus depends on:

- the current conversion value
- the time remaining to conversion
- the market's expectations regarding the expected returns
- the degree of risk of the underlying ordinary shares.

The 2006–7 accounts of Scottish and Southern Energy plc disclosed a convertible issue of 3.75 per cent convertible bonds with par value £1. Holders of the stock were entitled until October 2009 to convert these into ordinary shares on the basis of one ordinary share for each bond held. The applicable conversion price payable was £9.00 per ordinary share at the date of the issue. Conversion is at the option of the bond holder. The share price was around 1400p in August 2007, reflecting a prospective capital gain of about £5 per unit.

Convertible loan stock: key features

- Convertible loan stock can be issued more cheaply than a 'straight' loan because it offers an equity incentive.
- Companies perceived as relatively high risk can attract loan finance by offering the possibility of participating in future growth.
- Interest on the loan (while it is a loan) is tax-deductible.
- The bonds can also be traded on the Stock Exchange.
- Where it is believed that the true worth of the company is not adequately reflected in the share price, convertibles provide a means of raising capital that may eventually become equity without diluting the value of existing equity.
- Convertibles offer the benefits of both equity and loan stock, thereby attracting additional investors.
- If all goes as planned, the conversion to equity will occur, reducing the gearing ratio (but also lowering Earnings Per Share).
- If the conversion price is misjudged, the company is left with unwanted debt. If the equity growth is faster than expected, conversions will take place on over-generous terms at the expense of existing shareholders.

■ Bond yields

interest yield
The annual interest on a bond or similar security divided by its market price

Investors who buy debentures and loan stock will want to know the rate of return, or yield, on their investments. Two ratios are of interest to investors. The **flat yield** or **interest yield** is the gross interest receivable, expressed as a percentage of the current market value of the stock. Thus a 7 per cent stock with a market value of £85 and a nominal value of £100 has a flat yield of:

$$\frac{£7}{£85} \times 100 = 8.2\%$$

This represents the gross yield. The net-of-tax yield to the investor depends on his or her tax position:

$$\text{Net interest yield} = \text{Gross yield} \times (1 - t_p)$$

where t_p is the personal rate of tax incurred by the bond-holder.

redemption yield
The interest yield adjusted for any capital gain or loss if the security is held to maturity

The **redemption yield** combines the income accruing from interest payments with the capital gain or loss on maturity. It will be greater than the flat yield where the current value is below the redemption value because the investor will also receive a capital gain if the bond is held until maturity.

If the above stock has five years life to maturity, someone who buys it for £85 will receive a capital gain on redemption of £15. Averaged over five years, this represents an additional gain of £3 p.a. Based on the purchase price of £85, this raises the yield to redemption thus:

$$\text{Flat yield} + £3/£85 = 8.2\% + 3.5\% = 11.7\%$$

However, this is only an approximation to the true redemption yield, which depends on the precise timing of the investor's returns. For example, assume he or she buys the stock now, having just missed out on the most recent interest payment, thus anticipating five future annual interest payments plus the redemption payment of £100 at the end of the fifth year. The yield to maturity is the solution R in the following internal rate of return expression:

$$£85 = \frac{£7}{(1 + R)} + \frac{£7}{(1 + R)^2} + \cdots + \frac{(£7 + £100)}{(1 + R)^5}$$

gross redemption yield
The redemption yield before allowing for income tax payable by investors

The precise solution is 10.9 per cent. As above, this would be offset by the investor's liability to tax. As this figure is gross-of-tax, it is also called the **gross redemption yield**. Similar principles apply in calculating yields on government securities.

Self-assessment activity 16.6

A bond with nominal value of £100 and coupon rate of 8.3 per cent has market value of £110. What is:

(a) its flat yield?
(b) its yield to maturity in three years?

(Answer in Appendix A at the back of the book)

■ Warrants

A **warrant** is an option to buy ordinary shares. We include warrants in the section on debt finance because they are more frequently linked to debt issues than to equity issues (although many companies distribute warrants to shareholders as a 'sweetener'). The warrant holder is entitled to buy a stated number of shares at a specific price up to a certain date. Each warrant will state the number of shares the owner may purchase and the

perpetual warrant
A warrant with no time limit for exercising it

time limit (unless it is a **perpetual warrant**) within which the option to purchase can be exercised.

Companies issue warrants for a number of reasons. They can be attached to loan stock, thus providing loan stock holders with an opportunity to participate in the future growth and prosperity of the company, or, alternatively, used to attract investors by new and expanding companies. They may be also part of the purchase consideration in a takeover. In both cases, they act as a 'sweetener' to the investor. Frequently, such an inducement enables the company to obtain a lower rate of interest or less restrictive conditions in the debenture agreement. Whether or not warrants eventually give rise to additional finance by holders taking up their option to purchase depends, of course, on the future trading success of the company and the exercise price.

How warrants work

In 2008, when its ordinary shares carried a market price of £2.00, XYZ plc issued debenture stock with one warrant attached to each unit of stock giving the right to buy one ordinary share at a price of £2.50 in 2012. If, at the exercise date, the market share price is, say, £4.00, we would expect investors to exercise their rights, thus making a capital gain of £1.50 per share purchased. As a result of their purchases, the earnings of existing shareholders will be diluted – the company is, in effect, giving away £1.50 per share, which is not a problem if the same shareholders also hold the debentures, to which the warrants are attached, although this is rather unusual. The accounting mechanism would involve a reduction in the company's reserves. Warrants are often called 'time bombs' for this reason. This also explains why the issue of warrants may be used as a takeover defence tactic – warrants implant a 'poison pill' for the predator to swallow.

Warrants can be traded separately from the securities to which they were originally attached. In the XYZ example, if the ordinary shares are trading at £3.00, the warrants will be worth 50p each, because they embody the right to buy ordinary shares at a 50p discount. They also possess the attractive property of gearing. If the ordinary shares rise in value by 5 per cent from £3.00 to £3.15, the warrant would also rise by 15p in value, but by the considerably greater proportion of (15p/50p) = 30 per cent. For this reason, they are referred to as 'geared plays'.

Self-assessment activity 16.7

What is the market value of a warrant that gives the right to buy one new share for every four shares held, given the market price per share is £8, and the exercise price is £5?

(Answer in Appendix A at the back of the book)

■ Mezzanine finance

Mezzanine finance is frequently used in the financing of management buy-outs (MBOs). It is often described as a bridge between the secured debt that a business can raise and pay interest thereon with reasonable comfort plus the equity that the management team can raise, and the purchase price. Although it lies nearer the former, it ranks behind more formal borrowing contracts, which is why it is called 'subordinated' or 'intermediate' debt. Because of its low priority for payment, it attracts a relatively high interest rate, usually 3–5 per cent above LIBOR, and usually carries warrants and/or the right to convert into ordinary shares. The appeal of mezzanine finance to investors is that it offers investors exposure to the upside potential of the venture – if the MBO performs well, the warrants offer the prospect of capital gain. However, these are chancy investments – both the company and the instrument carry major risks.

■ Foreign bonds

These are domestic issues by non-residents, e.g. an issue of stock by a US company in London (a 'bulldog'), or in Tokyo (a 'samurai') or in Australia (a 'kangaroo'). Such bonds are domestic bonds in the local currency – only the issuer is foreign. If British Airways makes a bond issue in New York, this is a 'yankee' bond.

■ Eurobonds ('International bonds')

The term **Eurobond** is a misnomer – they need not be issued in Europe nor be denominated in a European currency! They are most easily understood as international loans denominated in a currency other than that of the issuer, but technically they are any long-term bonds issued outside the country of the currency in which the bonds are denominated. An example is a Eurodollar issue denominated (and thus repayable) in dollars by a US corporation, issued via the Frankfurt capital market. They are issued only by large, credit-worthy companies, development banks and state-owned corporations, and are generally unsecured. The vast majority of Eurobond issues are denominated in euros and in US dollars, e.g. the issue by Ghana Telecommunications, in 2007, for $200 million (see cameo below). However, issues in other currencies are not uncommon, e.g. in 2007, Nordic Investment Bank of Iceland raised 5 billion Icelandic krone.

Eurobond issues are mainly organised and underwritten by international syndicates of investment banks acting on behalf of borrowers. The market originally developed in response to the desire by US companies to avoid the now defunct withholding tax, which they were obliged to pay to the tax authorities on behalf of investors who purchased bonds sold locally.

Eurobonds are bearer bonds (i.e. they are transferable), with interest paid annually, and gross of tax, which may appeal to investors eager to delay (or evade) tax. Empirical evidence suggests that this feature may allow companies to borrow at rates lower than on their own domestic markets, although any such yield differential would be eliminated by arbitrageurs if the international bond market were efficient and unsegmented.

Lower overseas interest rates are not necessarily good news. Many corporate treasurers who try to take advantage of relatively low overseas interest rates often overlook the reasons why interest rates are lower overseas. Domestic interest rates are linked to future expected inflation rates and to expected exchange rate movements. If the inflation rate in Switzerland is lower than in the UK, the Swiss franc will appreciate against sterling and current interest rates in London will exceed those in Zurich. This is to compensate both domestic investors for inflation and also overseas investors from, say, Switzerland for the prospective reduction in the Swiss franc value of investments in London. If a British corporate treasurer borrows 'cheap' in Zurich, he or she should not be surprised to find that, when it comes to repay the loan in Swiss francs, the sterling cost has increased, following depreciation of sterling. What is won from the interest rate savings will probably be lost from the capital value change. However, for treasurers who believe that exchange rate movements can be predicted in advance, Eurobonds may offer speculative opportunities. For those who simply want to create an overseas liability to offset an exposure in relation to an overseas income flow, Eurobonds may present an attractive way of hedging. (These aspects are examined in more detail in Chapter 21.)

In addition to interest cost advantages, Eurobonds usually involve fewer, if any, restrictive covenants, and usually require less disclosure of information than is required for similar issues on domestic markets. The unregulated nature of the market has probably contributed to its innovativeness in terms of the features attaching to many issues, whereby bonds can be tailored to specific corporate requirements.

GT achieves a first with $200m issue

Ghana Telecommunications has raised $200m with the sale of the west African country's first dollar-denominated corporate bond.

The bond issue from GT comes less than three months after the government made its debut in the international bond markets with its sale of a bond to raise $750m – and is the latest sign of robust investor appetite for African assets.

"This is quite an impressive placement, considering the issue size and tight spread over the government's recent eurobond, as well as the widespread view that the company faces many operational and structural challenges," said one emerging markets analyst.

The notes issued by GT, a leading provider of fixed telephony and the third-largest mobile operator in Ghana, mature in 2012, carry a coupon of 8.5 per cent and yield 9.5 per cent – or 100 basis points above the original issue yield of the sovereign eurobond.

The notes, which are unrated, will amortise annually in five equal instalments of $40m and the issuer may call some or all of the notes at any time following the second anniversary of the issue date at a price above par.

More than 12 investors took part in the deal, including UK and European fund managers and hedge funds, said Peter Bartlett, managing director of

Exotix, the London broker that managed the deal with Iroko Securities, which is based in Mauritius.

The deal would help refinance the company's existing portfolio of maturing short-term commercial obligations and generally extend the maturity profile, according to Joe Owusu-Ansah, chief financial officer of GT.

He also said it restored the company's financial flexibility and access to the vendor financing necessary to support a much-needed upgrade and extension of the company's fixed and mobile networks.

Source: Joanna Chung, *Financial Times*, 18 December 2007.

■ Floating rate notes (FRNs)

These are Eurobonds that pay a variable rather than a fixed interest rate. FRNs are especially favoured by financial institutions that conduct a great part of their business at a floating rate and therefore value the protection given by the ability to borrow at a floating rate. As rates of interest rise, banks' income rises, as do their costs – FRNs are thus a way of achieving a neat match between assets and liabilities.

HSBC bank is an active player in the Eurobond market, borrowing in several currencies, e.g. US dollars and euros, at both fixed and floating rates. The reason why it borrows in many currencies and at fixed rates is bound up with the swaps market. HSBC may not want the currency for itself nor even want a variable rate liability. It is borrowing in the currency in which, at prevailing market conditions, it is easiest and cheapest to borrow, say, the Eurodollar market. It can then use the swap market to exchange the US$ for the currency it really requires.

A swap works in two stages. Take the case of HSBC's issue of US$750 million FRNs repayable in 2015. This could have been swapped into sterling via another bank (the 'swap bank') at the time of issue, perhaps at a fixed rate. The parties would agree to swap back at maturity so that HSBC could repay the bond. The swap bank then passes to HSBC a stream of sterling until 2015 with which to pay the interest, in return for which it would receive from HSBC a fluctuating stream of USD with which to pay its own interest liability up to 2015.

In such a swap, there is an exchange not only of currencies but also a floating liability for a fixed one. This is called a Combined Interest Rate and Currency Swap (CIRCUS). The point of this is to enable the two parties to raise money where the terms are most attractive before swapping it into the currency they really want. At other times, HSBC itself might borrow at floating rates both in sterling and in other currencies.

A new financial market takes wing

In July 2001, Emirates, the Dubai-based airline (and Chelsea FC sponsor) launched a Floating Rate Note (FRN) issue, amounting to dirhams* 1.5 billion, repayable at par in 2006, and denominated in 100,000 dirham units. The minimum subscription was 5 million dirhams, targeting the issue at corporate and institutional investors plus (very) high net worth individuals. The issue was over-subscribed two-and-a-half times and taken up by pension funds (7%), corporates (6%), professional investors (14%) and banks (73%).

The finance was needed for Emirates' aggressive expansion plans, including investment of 14.5 billion dirhams in fleet and equipment modernisation. Its growth ambitions involve trebling its fleet size over 2001–10, including the acquisition of 43 Airbus A380 Superjumbos.

This issue was the first bond issue made on the nascent Dubai Financial Market, the largest dirham-denominated bond issue to date and the first made by a United Arab Emirates corporate borrower. Interest on the notes is paid semi-annually at 70 basis points (0.7%) above the six-month Emirates Interbank Offered Rate (EBOR), a formula resulting in a rate of 4.675% for the first interest period.

*3.678 United Arab Emirate dirhams = one US dollar

16.10 LEASING AND SALE-AND-LEASEBACK (SAL)

Lease contracts were discussed in Chapter 15 where long-term leases for certain assets, such as ships and aircraft, or for property were discussed.

SAL involves selling assets (usually property) to a financial institution seeking good quality investments with potential for long-term growth in capital value. The seller, while giving up the ownership rights to the property, will then arrange to lease the premises from the new owner. SAL, therefore, is a transfer of ownership with retention of rights to use for a specified period. Its attraction to the vendor is the raising of capital, although its obvious disadvantage is usually the loss of any entitlement to capital appreciation. However, for a company eager to grow, the returns in the early years of strategic growth projects could outweigh the lost capital gains. This was the reasoning behind the SAL by Tesco, the supermarket chain, of many of its store sites in the 1980s when it was eager to claw back the market share advantage held by Sainsbury (it subsequently grew to number one in the supermarket league by 1996). The finance generated by a series of SALs was used to purchase and develop new sites for stores.

Having cemented its No. 1 position in the UK, Tesco is now pursuing a policy of overseas expansion. To fund this, it announced a SAL in March 2007. It plans to sell over £500 million of property to British Land in a complex joint venture operation. The 21 stores involved accounted for 3 per cent of Tesco's UK floor space. In 2007, Tesco planned to open 400 stores overseas bringing its foreign portfolio to 1,200 outlets.

■ Sale and manage-back

A variant of the SAL is the sale and manage-back (SAMB) tactic adopted by several hotel chains in 2004–5. Rather than lease back the sold property, the vendors undertake the management of it. One of the first exponents was Whitbread plc in October 2004, when it announced a wide-ranging disposal of properties, including its historic site in the City of London, for around £800 million. Half of the cash generated was to be returned to shareholders, and the remainder would be used to reduce its pension fund deficit and to pay down debt. Among the disposals were 12 Marriott-branded hotels (later increased to 46) that it would continue to manage.

In early 2005, the Hilton Group, Intercontinental and the French Accor group all announced SAMB programmes. The two British firms intended to return the sale proceeds to shareholders, while Accor planned to accelerate its expansion into budget hotels. In all these cases, while the planned use of cash differed, the common thread was that the vendors felt they could earn higher returns on hotel management rather than owning upmarket hotels.

Sources: Whitbread plc press release 28 October 2004, articles in the *Financial Times* 29 October 2004, and in *The Economist*, 19 March 2005.

Self-assessment activity 16.8

What is the effect of a sale and lease back on a firm's balance sheet?

(Answer in Appendix A at the back of the book)

16.11 ISLAMIC FINANCE*

A relatively new phenomenon in the international financial markets has been the rise of Islamic banks offering *shariah*-compliant products that conform to the teaching and interpretation of the Koran by Islamic scholars. The key principles in Islamic finance are the strict, explicit prohibition of *riba*, or the earning of interest, which is, of course, at the core of the traditional banking and finance, and the exclusion of transactions involving gambling, tobacco, alcohol, porcine products and pornography.

In recent years, Islamic finance is reckoned to have grown at 10–15 per cent p.a. and to be now worth £250 billion globally, with some 300 financial institutions offering Islamic products. Its growth has been attributed to various factors, principally, the heightened awareness and observance of Islamic principles among Muslims, the increase in liquidity in Middle Eastern nations following the increase in oil prices in the 1970s and more recently, in the 2000s, the presence of certain Western banks, such as HSBC, in the Middle East anxious to develop new products utilising their resources and expertise, including close knowledge and experience of local markets. London, where many international banks have located their Islamic banking and advisory services is pitching to become the leading centre for development of Islamic products, helped along by tax changes introduced in the 2007 Finance Act that clarify the treatment of costs of servicing loans. The UK was the first European country to authorise an exclusively Islamic bank (see cameo below).

Under *shariah*, interest is reckoned to 'unfairly' reward the provider of capital in a loan for little or no effort or risk undertaken. The Islamic economic model is based on a risk and profit/loss-sharing contract. Islamic financial products allow interest income to be replaced with sharing of the cash flows earned from profit-making activities, in effect, converting lending and borrowing into equity-based transactions. Because of this, most conventional equity products are judged to be acceptable under *shariah* (convertibles and warrants would be exceptions), although the distribution of profit is based more on reward for effort rather than for mere ownership of capital. The cameo records the listing on the AIM of Britain's first fully *shariah*-compliant bank in 2004. Clearly, issue of, and dealings in, equity-based paper are not problematic.

In terms of the provision of corporate finance, the main contribution of Islamic finance is in the bond market via the issue of *shariah*-compliant bonds, called *sukuks*.

---

*This section uses material from the paper published by the Financial Services Authority in November 2007, authored by Ainley *et al*.

Salaam Alaykum to the AIM

In August 2004, the first fully Islamic British bank, the Islamic Bank of Britain, but with origins in the tiny desert state of Qatar, was given permission by the FSA to offer a range of consumer banking products compliant with Sharia, the code of laws that govern Islam. None of its products would involve the taking or paying of interest, or investing in *haram* (prohibited) activities such as alcohol, tobacco or pornography. Depositors in such banks are offered a share in profit from the bank's operations (rather like the Co-operative dividend). Formed with £14 million of seed capital, raised largely from the Qatari royal family and other wealthy Arabian Gulf investors, the biggest shareholders were the Emir of Qatar, Sheikh Hamad Khalifa bin Hamad al Thani,

and the Qatar International Islamic Bank, both with around 17 per cent of the equity.

Later that month, it announced details of a floatation to raise £40 million by the issuance of 160 million shares at 25 pence through a combination of a public issue on the AIM market and a private placing, with existing investors invited to participate to avoid diluting their holdings. This issue price valued the business at £105 million. The proceeds were to be used to open new branches in London and other cities with large Muslim populations, such as Leicester and Bradford, and to develop new products such as mortgages by the end of 2004, and an internet banking service in 2005.

Source: Based on *Financial Times*, 9 August 2004, and www.ft.com 27 August 2004.

The volume of outstanding *sukuks* is estimated at US$70 billion, most of which is listed in Bahrain and Dubai. London's first *sukuk* was listed in July 2006. In December 2006, London established the world's first secondary market in *sukuks*. Over the period 2000–2006, the leading industrial sector in *sukuk* issuance was infrastructure activities (39 per cent), followed by financial services (18 per cent), energy (16 per cent), real estate (11 per cent), manufacturing (9 per cent), utilities (5 per cent) and transport and shipping (2 per cent).

There remains much controversy over Islamic financial products as the cameo suggests. What is or what is not *shariah*-compliant continues to be open to the varying interpretations among Islamic scholars of Islamic law, or *Fiqh*, as set out in the Koran. Similarly, there are many variants of the *sukuk* model. However, they all share the common feature that the returns to the investor are linked to the performance of a real asset, i.e. what is bought with the money raised. The word *sukuk* itself means certificates, namely, the documents issued when such a deal is set up. The design of the security resembles the more conventional securitisation process that sets up an SPV to acquire assets, issue financial claims on the asset and arrange for payment of returns. The certificates that encapsulate these financial claims represent a proportionate beneficial ownership for a stated period. The risk and the returns connected with the cash flows from the underlying asset are passed to the investors in the *sukuk*. The equivalent agency to the SPV is called a Special Purpose *Mudaraba* (SPM).

A *mudaraba sukuk* essentially records agreement between two parties whereby one party provides the capital required to finance the venture for the other party (the *Mudarib*) to work with, on the condition that the profit will be shared in accordance with a pre-agreed ratio, and that the capital will be returned when the *sukuk* are surrendered. They represent units of equal value in the equity of the business venture, and all investors receive returns in relation to their proportional ownership.

Modaraba sukuks are used to enhance public participation in large investment projects, such as oil-field development. *Sukuks* that provide medium-term asset-term finance require a further device based on leasing principles (*Ijara'*). The cash flows paid over to the SPM that manages the lease contract are passed over to investors as payments that include elements of rental and principal. While this may be thought to come close to a conventional lease (or perhaps an HP) contract, it should be remembered that the *sukuk* gives the investor beneficial ownership of the underlying asset(s), and thus participation in the *'usufructs'* (earning or fruits) of the assets, *and also in any losses*. The anatomy of an *ijara'*-based *sukuk* is explained more fully in Iqbal (1999).

Islamic bonds hit by growing religious concerns

The fast-growing Islamic bond industry has been seized by a fit of religious doubt.

The Islamic credentials of the bonds, called sukuks, have faced growing questioning in recent months, forcing financial engineers back to the drawing board in search of structures more compliant with Islam.

Criticism of sukuks as "un-Islamic" was first voiced late last year by Sheikh Muhammad Taqi Usmani, a prominent religious scholar who heads the Bahrain-based Accounting and Auditing Organisation for Islamic Financial Institutions.

Bankers say the views reflected a growing unease among religious scholars, whose support for sukuks has fuelled phenomenal growth in the industry.

"It [the criticism] has had a huge impact," said Hussein Hassan, head of Islamic finance at Deutsche Bank, on the sidelines of a Euromoney conference in London. "A lot of structuring of sukuks was put on hold until the issue was clarified."

After consultations with bankers, the 18-member Sharia, or Islamic law, board of the AAOIFI will give its say on the sukuk debate next week, according to Mohamad Nedal al-Chaar, secretary-general of the organisation.

Total sukuk issuance surged 73 per cent last year to more than $47bn (£24bn), according to the Islamic Finance Information Service, a data provider. Sukuk issuance has reached $1.3bn this year, according to IFIS. But some bankers say sukuks will now require more stringent structures to appeal to buyers, regardless of the AAOIFI ruling.

The debate over the purity of sukuks underlines the wider problem of a lack of standardisation in Islamic finance. Each financial institution relies on its own Sharia board to sign off on products. Different scholars can disagree on what is "Islamic", even within one country.

Sukuks come in different structures but the most popular form during the past two years involves a repurchase undertaking where the issuer promises to pay back the face value of the bond when it matures or in the event of a default. This structure, however, looks to some like a guaranteed return, which goes against the spirit of Islamic finance where interest is banned and buyers should share risk and profit.

Until now, most Sharia scholars have approved the controversial structure as they sought to expand the market.

"They're saying we've given leeway to develop the market and now we need to be stricter in our approval," Mr Hassan said.

Source: Roula Khalaf, *Financial Times*, 7 February 2008.

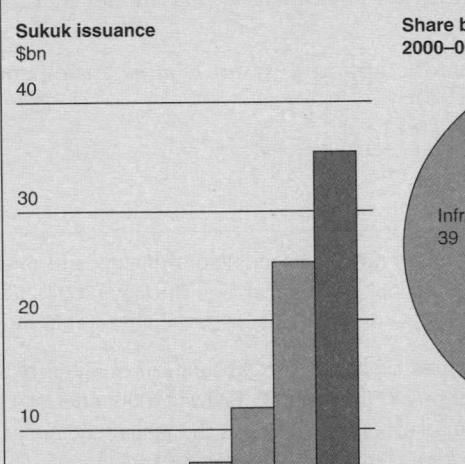

Sukuk issuance
$bn

* First 9 months
Source: IFSL

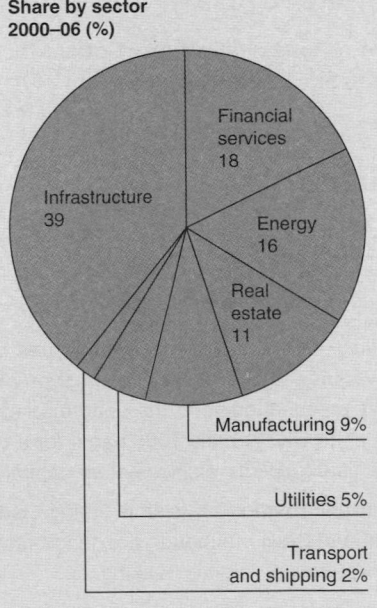

Share by sector 2000–06 (%)

Financial services 18
Infrastructure 39
Energy 16
Real estate 11
Manufacturing 9%
Utilities 5%
Transport and shipping 2%

SUMMARY

Chapter 16 has discussed the features of the principal forms of long-term finance available to companies, and their benefits and drawbacks.

Key points

- The main factors in considering the appropriate source of finance are risk, ownership, duration and debt capacity.

- Over half of the new finance raised for UK companies is usually through retained profits. This is not a free source of finance as it involves an opportunity cost in terms of the return that shareholders would have obtained had they received the profit as dividends and re-invested it elsewhere.

- New shares are issued through a private placing, stock exchange placing, public issue (prospectus issue, offer for sale or issue by tender) or by a rights issue.

- The enhanced ability to make a rights issue is a major attraction of a Stock Exchange listing.

- Equity capital is an attractive form of finance to companies because there are no interest charges or capital repayments; to investors, it offers a hedge against inflation, and a higher yield than loan stock. On the other hand, equity capital can be expensive to raise, and new issues to the public dilute the control of existing members.

- Debt finance (debentures, loan stock, etc.) is flexible, offering a wide range of financial products to the corporate treasurer. Interest payments are tax-deductible, but restrictions (e.g. charges over assets and monitoring of activities) are common practice.

- Convertible loan stock is a debt instrument that can, at the option of the holder, be converted into equity. It offers investors and firms the benefit of loan stock in the early years and, if all goes to plan, will enable the benefits of equity to be captured when the business is better established.

- The Eurobond market offers the opportunity for large firms to raise money in foreign markets.

- The rapidly-growing *sukuk* market offers the opportunity for Islamic borrowers to raise money in Sharia-compliant forms.

Further reading

Most UK finance textbooks carry a 'straight down the middle' description of methods of raising long-term finance in the UK. For a broader international perspective, see Buckley (2004). For a US perspective, see Emery and Finnerty (1997), Ross *et al.* (2005), or Block and Hirt (1994).

An excellent survey of the market for new issues of equity (IPOs), and an analysis of the tendency for new issues to be under-priced is given by Ritter (2006). Brown (2006) offers an up-to-date and highly practical guide to the bond markets. Colin Mason is the leading authority on business angels in the UK. See his 2006 paper for a comprehensive review of the angels' literature. A highly practical guide to raising venture capital is given by Pearce and Barnes (2006).

A useful source of information on Islamic finance is the *Handbook of Islamic Banking* (2005), edited by Hassan and Lewis (although it extends far beyond mere banking issues), while Arar (1998) gives an overview of Islamic banking.

Both *The Economist* and the *Financial Times* provide a regular supply of examples of financing sources and methods covered in this chapter.

QUESTIONS

 *Questions with an icon are also available for practice in myfinancelab with additional supporting resources.*

Questions with a **coloured number** have solutions in Appendix B on page 747.

1 The ordinary shares of Anglia Paper Company are currently trading at £3.20. Existing shareholders are offered one new share at £2 for every three held.

(i) What is the theoretical ex-rights price?
(ii) What is the nil-paid rights price?

2 Cambridge Castings Ltd plans a major expansion to modernise its manufacturing plant, thereby improving productivity and reducing unit costs. The existing capital base is fairly evenly divided between equity and debt, and it is clear that the capital investment programme can only partly be funded through profit retention.

It is suggested that the additional finance could be raised through a preference share issue. You are required to evaluate this source of finance for the company, compared with equity or debt:

(a) from the company's point of view;
(b) from the viewpoint of investors.

3 Shaw Holdings plc has 20 million ordinary shares of 50p in issue. These shares are currently valued on the Stock Exchange at £1.60 per share. The directors of Shaw Holdings believe the company requires additional long-term capital and have decided to make a one-for-four rights issue at £1.30 per share.

An investor with 2,000 shares in Shaw Holdings has contacted you for investment advice. She is undecided whether to take up the rights issue, sell the rights, or allow the rights offer to lapse.

Required
(a) Calculate the theoretical ex-rights price of an ordinary share.
(b) Calculate the value at which the rights are likely to be traded.
(c) Evaluate each of the options being considered by the owner of 2,000 shares.
(d) Explain why rights issues are usually made at a discount.
(e) From the company's viewpoint, how critical is the pricing of a rights issue likely to be?

(ACCA Certified Diploma)

4 Burnsall plc is a listed company which manufactures and distributes leisurewear under the brand name Paraffin. It made sales of 10 million units worldwide at an average wholesale price of £10 per unit during its last financial year ending at 30 June 1995. In 1995–96, it is planning to introduce a new brand, Meths, which will be sold at a lower unit price to more price-sensitive market segments. Allowing for negative effects on existing sales of Paraffin, the introduction of the new brand is expected to raise total sales value by 20 per cent.

To support greater sales activity, it is expected that additional financing, both capital and working, will be required. Burnsall expects to make capital expenditures of £20 million in 1995–96, partly to replace worn-out equipment but largely to support sales expansion. You may assume that, except for taxation, all current assets and current liabilities will vary directly in line with sales.

Burnsall's summarised Balance Sheet for the financial year ending 30 June 1995 shows the following:

Assets employed	£m	£m	£m
Fixed (net)			120
Current:			
stocks	16		
debtors	23		
cash	6		
	—	45	
Current liabilities:			
Corporation tax payable	(5)		
Trade creditors	(18)		
		(23)	
Net current assets			22
Long-term debt at 12%			(20)
Net assets			122
Financed by			
Ordinary shares (50p par value)			60
Reserves			62
Shareholders' funds			122

Burnsall's profit before interest and tax in 1994–95 was 16 per cent of sales, after deducting depreciation of £5 million. The depreciation charge for 1995–96 is expected to rise to £9 million. Corporation tax is levied at 33 per cent, paid with a one-year delay. Burnsall has an established distribution policy of raising dividends by 10 per cent p.a. In 1994–95, it paid dividends of £5 million net.

You have been approached to advise on the extra financing required to support the sales expansion. Company policy is to avoid cash balances falling below 6 per cent of sales.

Required

(a) By projecting its financial statements, calculate how much additional *external* finance Burnsall must raise.

Note: You may assume that all depreciation provisions qualify for tax relief.

(b) Offer advice as to the appropriate method of financing for Burnsall's sales expansion.

(ACCA)

5 The managing director of Lavipilon plc wishes to provide an extra return to the company's shareholders and has suggested making either:

(i) a two-for-five bonus issue (capitalisation issue) in addition to the normal dividend.
(ii) a one-for-five scrip dividend instead of the normal cash dividend.
(iii) a one-for-one share (stock) split in addition to the normal dividend.

Summarised Balance Sheet of Lavipilon plc (end of last year)

Fixed assets		65
Current assets	130	
Less: current liabilities	(55)	
Net current assets		75
Total assets less current liabilities		140
Less: Long-term liability		
11% debenture		(25)
Net assets		115
Capital and reserves:		
Ordinary shares (50 pence par value)	25	
Share premium account	50	
Revenue reserves	40	
Shareholders' funds		115

The company's shares are trading at 300 pence before the dividend is paid, and the company has £50 million of the (post-tax) profit from this year's activities available to ordinary shareholders, of which £30 million will be paid as a dividend if options (i) or (iii) are chosen. None of the £40 million revenue reserves would be distributed. This year's financial accounts have not yet been finalised.

(a) For each of the three proposals, show the likely effect on the company's Balance Sheet at the end of this year, and the likely effect on the company's share price.

(b) Comment on how well these suggestions fulfil the managing director's objective of providing an extra return to the company's shareholders.

(c) Discuss reasons why a company might wish to undertake:
  (i) a scrip dividend,
  (ii) a share (stock) split.

(ACCA)

6 Netherby plc manufactures a range of camping and leisure equipment, including tents. It is currently experiencing severe quality control problems at its existing fully depreciated factory in the south of England. These difficulties threaten to undermine its reputation for producing high-quality products. It has recently been approached by the European Bank for Reconstruction and Development, on behalf of a tent manufacturer in Hungary, which is seeking a UK-based trading partner which will import and distribute its tents. Such a switch would involve shutting down the existing tent manufacturing operation in the United Kingdom and converting it into a distribution depot. The estimated restructuring costs of £5 million would be tax-allowable, but would exert serious strains on cash flow.

Importing, rather than manufacturing, tents appears inherently profitable, as the buying-in price, when converted into sterling, is less than the present production cost. In addition, Netherby considers that the Hungarian product would result in increased sales, as the existing retail distributors seem impressed with the quality of the samples which they have been shown. It is estimated that for a five-year contract, the annual cash flow benefit would be around £2 million p.a. before tax.

However, the financing of the closure and restructuring costs would involve careful consideration of the financing options. Some directors argue that dividends could be reduced, as several competing companies have already done a similar thing, while other directors argue for a rights issue. Alternatively, the project could be financed by an issue of long-term loan stock at a fixed rate of 12 per cent.

The most recent Balance Sheet shows £5 million of issued share capital (par value 50p), while the market price per share is currently £3. A leading security analyst has recently described Netherby's gearing ratio as 'adventurous'. Profit after tax in the year just ended was £15 million and dividends of £10 million were paid.

The rate of Corporation Tax is 33 per cent, payable with a one-year delay. Netherby's reporting year coincides with the calendar year and the factory will be closed at the year end. Closure costs would be incurred shortly before deliveries of the imported product began, and sufficient stocks will be on hand to overcome any initial supply problems. Netherby considers that it should earn a return on new investment of 15 per cent p.a. net of all taxes.

Required
(a) Is the closure of the existing factory financially worthwhile for Netherby?

(b) Explain what is meant when the capital market is said to be information-efficient in a semi-strong form.
  If the stock market is semi-strong efficient and without considering the method of finance, calculate the likely impact of acceptance and announcement of the details of this project to the market on Netherby's share price.

(c) Advise the Netherby board as to the relative merits of a rights issue rather than a cut in dividends to finance this project.

(d) Explain why a rights issue generally results in a fall in the market price of shares. If a rights issue is undertaken, calculate the resulting theoretical ex-rights share price of issue prices of £1 per share and £2 per share, respectively. (You may ignore issue costs.)

(e) Assuming the restructuring proposal meets expectations, assess the impact of the project on earnings per share if it is financed by a rights issue at an offer price of £2 per share, and loan stock, respectively. (Again, you may ignore issue costs.)

(f) Briefly consider the main operating risks connected with the investment project, and how Netherby might attempt to allow for these.

(ACCA)

Practical assignment

Consider the long-term financing of a company with which you are familiar. Evaluate each of the main sources of finance and suggest, with reasons, two methods of finance that are not currently used, but which may prove attractive to the company.

 *Now retake your diagnostic test for Chapter 16 to check your progress and update your study plan.*

17

Returning value to shareholders: the dividend decision

The first cut is the deepest

When firms list on the Stock Exchange, it is advisable to declare a dividend policy and stick to it. China Shoto, a maker of batteries for the booming electric bicycle market, listed on the AIM in December 2005, and duly paid a maiden interim dividend the following summer. However, in September 2007, it decided to pass the next interim, citing the mantra that 'investing for continued growth rather than distributing profits is in the best interests of the company and its shareholders at the present time.'

Despite the favourable outlook for sales, the dividend cut was made against the background of a falling share price as the price of its main input, lead, rose sharply, along with other commodities around this time.

The higher cost of acquiring and holding stocks was reflected in an increase in net debt from £1.3 million to £14 million in the six months to the end of June 2007.

Therefore, for several reasons, the dividend cut looked like good business sense, but the question remained whether China Shoto should have entered the dividend lists in the first place, only to resort to cutting it so soon. The negative signals emitted by the dividend cut sparked a share price fall of about 10 per cent. Investor confidence is easy to lose and difficult to win back, although China Shoto continued to command a 'Strong Buy' rating from analysts.

Source: Based on an article in the *Financial Times*, 21 September 2007.

Learning objectives

After reading this chapter, you should:

■ Understand the competing views about the role of dividend policy.

■ Understand what factors a financial manager should consider when deciding to recommend a change in dividend payouts.

■ Understand what is meant by the 'information content' of dividends.

■ Know what alternatives to cash dividends may be used to deliver value to owners.

■ Appreciate the impact of taxation on dividend decisions.

■ Understand why changes in dividend payments usually lag behind changes in company earnings.

 *Complete your diagnostic test for Chapter 17 now to create your personal study plan.*

17.1   INTRODUCTION

This chapter will help you to appreciate the factors that drive dividend decisions.

Most quoted companies pay two dividends to ordinary shareholders each year: an interim, or 'taster', based on half-year results, followed by the main, or final, dividend, based on the full-year reported profits. The amount of dividend is determined by the board of directors, advised by financial managers, and presented to the Annual General Meeting of shareholders for approval. The board and their advisers thus face a twice-yearly decision about what percentage of post-tax profits to distribute to shareholders (the 'payout ratio') and hence what percentage to retain (the 'retention ratio').

Until a specified **Record Day**, the shares are traded **cum-dividend**: that is, purchasers will be entitled to receive the dividend. The approved dividends are paid to all shareholders appearing on the share register on the Record Day, after which the shares are quoted **ex-dividend**, i.e. without entitlement to the dividend. In practice, there is a time lag between the shares going ex-dividend and the Record Day to allow the company's Registrar to update the shareholders' register to reflect recent dealings. The Financial Calendar of Scottish and Southern Energy plc, the electricity supply utility, is shown in Table 17.1. People purchasing Scottish and Southern Energy plc shares whose names did not appear on the register by 24 August 2007 would not have received the final dividend payable per share (39.9p).

ex-dividend
A share trades ex-dividend (xd) when people who buy are no longer entitled to receive the upcoming dividend payment

Table 17.1 Scottish and Southern Energy plc Financial Calendar 2007

Annual general meeting	26 July 2007
Ex-dividend date	22 August 2007
Record date	24 August 2007
Final dividend payment date	21 September 2007

Source: Scottish and Southern Energy plc, Annual Report and Accounts, 2007 (www.scottish-southern.co.uk).

When the shares first trade ex-dividend, the share price will fall by the amount of the dividend. As dividends are usually a small proportion of the share price anyway, this is not a substantial fall. For example, in the case of SSE, the dividend of 39.9p compared to a share price of well over £14. However, sometimes, firms with a surplus of liquidity decide to pay a substantial special dividend (SD). This is good news at the time of announcement, and the shares usually rise, but it stores up the problem of a substantial fall in share price when the dividend is paid, which may alarm some investors. For example, WH Smith plc announced an SD of 33p in February 2008, around 10 per cent of the then share price. To counter the ex-dividend effect, Smith planned to consolidate the share structure by giving 67 new shares for every 74 that shareholders currently held with the aim of leaving the share price, EPS and (normal) dividend per share unaffected by the SD payment.

How should top management approach the dividend decision? Should it be generous and follow a high payout policy, or retain the bulk of earnings? The pure theory of dividend policy shows that, under certain conditions, it makes no difference what they do! One authority argues: 'to the management of a company acting in the best interests of its shareholders, dividend policy is a mere detail' (Miller and Modigliani, 1961). However, the conditions required to support this conclusion are highly restrictive and unlikely to apply in real-world capital markets. Indeed, many financial managers and investment analysts take the opposite view, appearing to believe that the dividend payout decision is critical to company valuation and hence a central element of corporate financial strategy. These are the extreme views – an evaluation of the case for and against dividend generosity leads to more pragmatic 'middle-of-the-road' conclusions.

In this chapter, we consider the strategic, theoretical and practical issues surrounding dividend policy, and discuss some of the alternatives to dividend payment. *The basic message for management is: define the dividend policy, make a smooth transition towards it, and think very carefully before changing it, and avoid cutting dividends!*

Few people doubt that dividend levels influence share prices in some way or other. Indeed, a common method of valuation, the Dividend Valuation Model (introduced in Chapter 3) relies on discounting the future dividend stream. However, debate centres on what is the most attractive *pattern* of dividend payments, and the effects of *changes* in dividend policy.

Shareholders can receive returns in the form of dividends now and/or capital appreciation, which is the market's valuation of future expected dividend growth. But are shareholders more impressed by the higher near-in-time dividends than by the capital gain generated by a policy of low current payments, with retentions used to finance worthwhile investment? Graham *et al.* (1962) claimed that $1 of dividend was valued four times more highly by shareholders than $1 of retained earnings. Yet it is not uncommon to observe quite parsimonious (or even zero, e.g. Dell, Ryanair) payout ratios. Conventional wisdom warns companies subject to volatility of earnings to operate relatively high dividend covers to safeguard dividend payments in the event of depressed profitability. Similarly, smaller companies with less easy access to fresh supplies of capital are also advised to conserve cash and liquidity by operating a conservative dividend policy.

Over the past two decades, attitudes in the UK to dividend payouts have appeared to change in favour of higher payouts. Why is this?

1 Companies may be more aware of the threat of takeover, and use high payouts as a pre-emptive defence to buy shareholder loyalty.
2 The tax changes in the 1988 Finance Act largely removed the tax discrimination against dividend payments, although the pendulum has swung back in favour of capital gains, following the introduction of an across-the-board rate of capital gains tax of 18 per cent in 2008.
3 Reflecting a short-termist perspective, it may be that companies feel under greater pressure from institutional shareholders to pay out higher dividends.

Whatever the reasons, it has become more difficult to generalise about optimal dividend policy, except that dividend cuts are generally perceived as conveying bad news.

In the following sections, we consider strategic and legal issues before examining the theory of dividend policy and, in particular, the 'irrelevance' hypothesis. We then consider the major qualifications to the theory, suggesting reasons why some shareholders may, in practice, prefer dividend income. You will see that dividend policy is an enigma with no obvious optimal strategy capable of general application. However, it will be possible to make some broad recommendations to guide the financial manager.

17.2 THE STRATEGIC DIMENSION

Formulating corporate strategy requires specification of clear objectives and delineation of the strategic options contributing to the achievement of these objectives. Although maximisation of shareholder wealth may be the paramount aim, there may be various routes to it. For example, diversifying into new industrial sectors involves a choice between internal growth and growth by acquisition. Whichever alternative is selected, the enterprise will need to consider both the level of required financing and the possible sources. The main alternatives for the listed company are short- and long-term debt capital, new share issues (normally rights issues) and internal financing (via retention of profits and depreciation provisions).

This is where we encounter the role of dividend policy. The amount of finance required to support the selected strategic option may exceed the borrowing capacity of the company, necessitating the use of additional equity funding. A capital-hungry firm therefore faces the choice between retention of earnings, i.e. restricting the dividend payout, or paying out high dividends but then clawing back capital via a subsequent rights issue. Neither policy is risk-free, as both may have undesirable repercussions on the ability to exploit strategic options. Retention may offend investors reliant on dividend income, resulting in share sales and lower share price, conflicting with the aim of wealth-maximisation, as well as exposing the company to the threat of takeover. Alternatively, the dividend payment-plus-rights issue policy incurs administrative expenses and the risk of having to sell shares on a flat or falling market. To this extent, the dividend decision is a strategic one, since an ill-judged financial decision could subvert the overall strategic aim. Consequently, financial managers must carefully consider the likely reaction of shareholders, and of the market as a whole, to dividend proposals.

17.3 THE LEGAL DIMENSION

Legal factors impose further constraints on managers' freedom of action in deciding dividends. Although shareholders are the main risk-bearers in a company, other stakeholders, such as creditors and employees, carry a measure of risk. Accordingly, shareholders can be paid dividends only if the company has accumulated sufficient profits. They cannot be paid out of capital, except in a liquidation, as to do so would mean they would be paid ahead of prior claims on the business.

The Companies Act 1985 states that, in general, companies are restricted to accumulated realised profits. However, public companies are further restricted to realised profits less unrealised losses. Furthermore, bondholders may insist on restrictive covenants being written into loan agreements to prevent large dividends being declared.

Of course, paying the maximum legally permitted dividend is rarely likely to make strategic sense. Moreover, dividends are paid *out of profits* but *with* cash (or borrowings). It is quite common for companies to report low or negative profits, yet to maintain the previous year's dividend. This is both feasible and acceptable if the firm in question has built up sufficient reserves for previous years' retentions and has a reasonably healthy cash position. Indeed, the share prices of some struggling companies actually rise when they pay a maintained dividend out of reserves. This is because a maintained dividend is believed to be a signal to the market of expectations of better times ahead. Remember that managers have more information to hand than investors. So, in this case, many investors may believe that directors are conveying favourable information (the concept of information asymmetry again).

17.4 THE THEORY: DIVIDEND POLICY AND FIRM VALUE

The critical issue is how, if at all, does dividend policy affect the value of the firm? This section shows that the answer depends on whether or not a firm has access to external financing. In the absence of external financing, dividend payment may damage company value if the company has better investment opportunities than its shareholders. Payment of dividends may prevent access to worthwhile investment. With external financing, however, at least in a perfect market, dividends become totally irrelevant, since payment no longer precludes worthwhile investment, simply because the firm can recoup the required finance by selling shares.

In Chapter 3 we used the Dividend Valuation Model to show that the value of a company* ultimately depends on its dividend-paying capacity. For a company with constant and perpetual free cash flows[†] of E_t in any year t, paid wholly as dividend, D_t, the market will discount this stream at the rate of return required by shareholders, k_e, so that:

$$\text{Value of equity} = V_o = \sum_{t=1}^{N} \frac{E_t}{(1 + k_e)^t} = \frac{E_t}{k_e} \left(\text{or } V_o = \frac{D_t}{k_e} \text{ since } E_t = D_t \right)$$

If a proportion, b, of earnings is retained each year, beginning in period 1, for example, thus reducing the next dividend payable to $E_1 (1 - b)$, company value is given by the forthcoming dividend divided by the cost of equity less the growth rate. The growth rate, g, is given by the retention ratio, b, times the return on reinvested funds, R (as explained in Chapter 3):

$$\text{Value of equity} = V_o = \frac{E_1 (1 - b)}{(k_e - g)} = \frac{D_1}{(k_e - g)} = \frac{D_1}{(k_e - bR)}$$

Self-assessment activity 17.1

Using the following figures, remind yourself why the growth rate of earnings = $(b \times R)$.

Earnings in latest year = £1,000; b = 60%; R = 15%.

(Answer in Appendix A at the back of the book)

These expressions can cause confusion. In the first case, the market appears to value the stream of earnings, while in the second, the valuation is based upon the stream of dividends. Indeed, in the early stages of the debate about the importance and role of dividend policy, much attention was paid to the issue of whether the market values earnings or dividends. This apparent dichotomy can be easily resolved if we focus on the reasons for earnings retention. Retention may occur for two main reasons.

First, the company may wish to *bolster its holdings of liquid resources*. For example, dividend distribution may run down current assets or perhaps increase borrowings. Many profitable companies borrow to avoid having to reduce, or omit (or 'pass') a dividend, for reasons explained later.

However, the primary reason for retention is to *finance investment in fixed and other assets* to generate higher future earnings and, hence, enhance future dividend-paying capacity. By introducing retention and growth, the Dividend Valuation Model becomes the **Dividend Growth Model** (DGM).

It is important to remember that the DGM assumes a constant retention ratio and a constant return on new investment projects. If these assumptions hold, both earnings and dividends grow at the same rate. Whether this rate is acceptable to shareholders depends on the return they require from their investments, k_e. The relationship between k_e and R, the return on reinvested earnings, provides the key to resolving the valuation dispute.

---

*In this chapter, we assume no company borrowing, hence the value of the company is synonymous with the value of the equity. When referring to the present value of the whole equity, we use the symbol V_o, and when referring to the value of an individual share, P_o.

[†]In this and subsequent chapters, the letter E is used to denote a free cash flow concept of earnings, as explained in Chapter 12. Here, free cash flow is calculated *after* replacement investment but *before* strategic investment. Free cash flow is likely to deviate from accounting earnings, so when the latter accounting concept is intended, the phrases 'accounting profit' or 'reported earnings' are used.

■ Dividend irrelevance in perfect capital markets

In 1961 Miller and Modigliani (MM) pointed out that earnings retention was simply one way of financing investment. If the company has access to better investment opportunities than its shareholders, under perfect capital market conditions, investors may benefit from retention.

The original MM analysis proves dividend irrelevance in terms of a single dividend cut to finance worthwhile investment. MM envisaged an all-equity-financed company that has previously paid out its entire annual net cash flow as dividend. To illustrate their analysis, let us examine the case of Divicut plc.

■ One-off dividend cuts: Divicut plc

Divicut currently generates a perpetual free cash flow of £1,000 p.a. Shareholders require a return of 10 per cent. The market will value Divicut at (£1,000/10%) = £10,000 on unchanged policies. If the management now decides to retain the whole of next year's earnings in order to invest in a project offering a single cash flow of £1,200 in the following year, the new market value of Divicut equals the present value of the revised dividend flow, assuming it reverts to 100 per cent payouts:

Year	1	2	3	4	etc.
Dividend (£)	0	2,200	1,000	1,000	⟶

This revised dividend flow is acceptable only if the resulting market value is at least equal to the pre-decision value, i.e. if the NPV of the project is at least zero. This is clearly the case, as the PV of the extra £1,200 in two years is greater than the PV of £1,000 less in one year, i.e.:

$$\frac{£1,200}{(1.1)^2} > \frac{£1,000}{(1.1)}$$

or, £992 > £909

Divicut's shareholders are thus better off by £83.

Shareholder wealth is enhanced, since Divicut has used the funds released by the dividend cut to finance a worthwhile project. If these funds had been invested at only 10 per cent, the net effect would have been zero, while if investment had occurred at a rate of return of less than 10 per cent, shareholders would have been worse off. This simple example shows that the impact on company value is attributable to the investment, rather than the dividend, decision.

Self-assessment activity 17.2

Rework the Divicut example for the case where the returns on the new investment are £1,080 rather than £1,200 in year 2.

(Answer in Appendix A at the back of the book)

MM's dividend irrelevance conclusion was obtained in a world of certainty, and then extended to the case of risk/uncertainty, where the same conclusions emerge so long as investor behaviour and attitudes conform to conditions of 'symmetric market rationality'. This requires the following:

1 All investors are maximisers of expected wealth.
2 All investors have similar expectations.
3 All investors behave rationally.
4 All investors believe that other market participants will behave rationally and that other investors expect rational behaviour from them.

These assumptions were spelled out by Brennan (1971) and form part of the battery of conditions required for a perfect capital market. The additional assumptions required to support the 'irrelevance' hypothesis are as follows:

1 No transaction costs or brokerage fees.
2 All investors have equal and costless access to information.
3 All investors can lend or borrow at the same rate of interest.
4 No buyer or seller of securities can influence prices.
5 No personal or corporate income or capital gains taxes.
6 Dividend decisions are not used to convey information.

The full significance of some of these assumptions will be highlighted when we examine some of the reasons why shareholders may have a definite preference for either dividends or retentions.

The kindest cut

Go on, say it: British Telecommunications is going to cut its dividend. The company more or less admitted as much to buyers of yesterday's £6.4 billion bond issue. With BT paying 40 to 50 basis points more than the market price of Deutsche Telekom's debt and facing stiff interest penalties if it does lose its single-A credit rating, it would be senseless not to.

Scrapping the dividend would send a stark message to the market and infuriate legions of retail investors. But a business should not be imprisoned by its old dividend policy when its circumstances have changed. It is far better for BT to stop paying out cash than to sell more assets than it needs when prices are at rock bottom – a certain recipe for value destruction.

Source: Based on Lex Column, *Financial Times*, 19 January 2001.

■ Permanent dividend cuts: more Divicut

This argument can also be applied to the case of permanent retentions using the Dividend Growth Model, although the analysis is a little more complex. Retention may lower the dividend payment only temporarily, resulting in a rate of growth yielding higher future dividends. This is shown in Figure 17.1. At point-in-time t_1, the

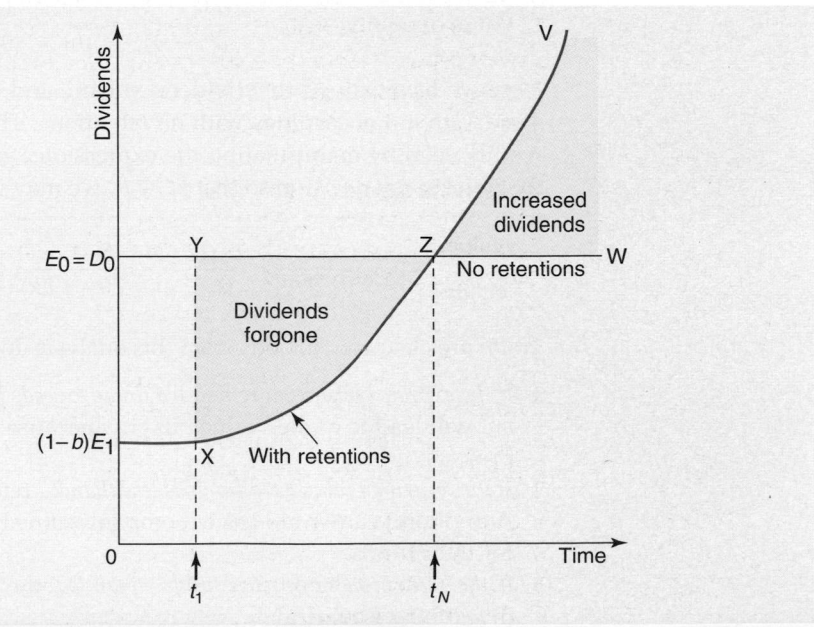

Figure 17.1 The impact of a permanent dividend cut

company, which currently pays out all its earnings, E_0, plans to cut dividends from D_0 to $(1 - b)E_1$, where b is the retention ratio.

Dividends do not regain their former level until time t_N, involving a cumulative loss in dividend payments equal to the area XYZ. Beyond time t_N, dividends exceed their original level as a result of continued reinvestment. Clearly, shareholders will be better off if the area of higher future dividends, VZW, exceeds XYZ (allowing for the discounting process). This holds if R, the return on retained funds, exceeds k_e. This contention can be illustrated using the example of Divicut again.

Recall that, prior to alteration in dividend policy, the value of Divicut was £10,000, derived by discounting the perpetual earnings stream. Imagine the company announces its intention to retain 50 per cent of earnings in all future years (i.e. from and including Year 1) to finance a series of projects offering a perpetual yield of 15 per cent. Market value according to the dividend growth model is:

$$\text{Value of equity} = V_o = \frac{E_1(1-b)}{(k_e - g)} = \frac{E_1(1-b)}{(k_e - bR)} = \frac{£1,000(1 - 0.5)}{0.1 - (0.5 \times 0.15)}$$

$$= \frac{£500}{0.025} = £20,000$$

This may seem a remarkable result. The decision to retain and reinvest doubles company value! Does this mean that dividend payments make shareholders worse off? The answer is simple.

Payment of dividends may make shareholders worse off than they otherwise might be if distribution results in failure to exploit worthwhile investment. In other words, the beneficial impact on Divicut's value is achieved because of the inherent attractions of the projected investments. Conversely, if the funds had been invested to yield only 5 per cent, market value would have fallen to £6,666.

The in-between case, where the return on reinvested funds, $R = k_e = 10$ per cent, leaves company value unchanged. This suggests that, if we strip out the effects of the investment decision, the dividend decision itself has a *neutral* effect. This can be done by assuming retentions are used to finance projects that yield an aggregate NPV of zero, i.e. yielding a return of k_e. With this assumption, dividend decisions are irrelevant to shareholder wealth. With a return on reinvested funds of 10 per cent, the company's value is:

$$\text{Value of equity} = V_o = \frac{E_1(1-b)}{(k_e - g)} = \frac{£1,000(1 - 0.5)}{0.1 - (0.5 \times 0.1)} = \frac{£500}{0.05} = £10,000$$

Here we have valued the dividend stream, and the result is equivalent to valuing the steady stream of earnings with no retentions. This equivalence may perhaps be more readily seen by manipulating the expressions for value. If we neutralise the effect of the investment decision so that $k_e = R$, we may write:

$$\text{Value of equity} = V_o = \frac{D_1}{(k_e - g)} = \frac{E_1(1-b)}{(k_e - g)} = \frac{E_1(1-b)}{(k_e - bR)} = \frac{E_1(1-b)}{(k_e - bk_e)} = \frac{E_1(1-b)}{k_e(1-b)} = \frac{E_1}{k_e}$$

There are clear conclusions from this analysis. In the absence of external financing:

1 *If the expected return on reinvested funds exceeds k_e*, it is beneficial to retain. A dividend cut will lead to higher value but only because funds are used to finance worthwhile projects.
2 *If the return on reinvested funds is less than k_e*, retention damages shareholder interests. A dividend cut would lower company value because shareholders have better uses for their funds.
3 *If the return on reinvested funds equals k_e*, the impact of a dividend cut to finance investment is neutral.

Self-assessment activity 17.3

How can dividend policy damage shareholder interests when no external financing is available?

(Answer in Appendix A at the back of the book)

■ Dividends as a residual

The dividend decision is simply the obverse of the investment decision. As observed earlier, we are examining the impact of one of the various ways in which proposed investment may be financed. Divicut is a case where the company is forced to retain funds through lack of alternative financing options. With the explicit assumption that the firm is capital-rationed with access only to internal sources of finance, we can illustrate the **residual theory of dividends**.

This argues that dividends should be paid only when there are no further worthwhile investment opportunities. Having decided on the optimal set of investment projects, and determined the required amount of financing, the firm should distribute to shareholders only those funds not required for investment financing. This idea is shown graphically in Figure 17.2, using the **marginal efficiency of investment (MEI)** model, a construct borrowed from economic theory. The MEI traces out the rate of return on the last £1 invested, and thus shows investment opportunities ranked in declining order of attractiveness.

With free cash flow of OE_0, there is scope for dividend payments. The limit of worthwhile new investment is at X, where the return on the last unit of investment is equal to the minimum required return k_e. The company can now make residual dividend payments of XE_0. However, if the free cash flow is only OE_1, distribution would impose an opportunity cost on shareholders. The whole cash flow should be reinvested, but it falls short of the finance required to support the optimal programme. The shortfall is E_1X, which could only be plugged by external financing if available.

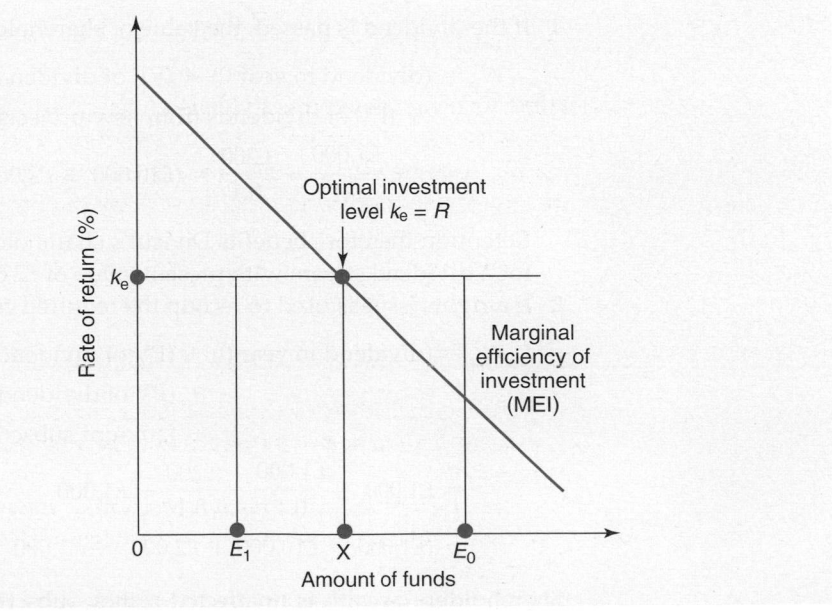

Figure 17.2 Dividends as a residual

Self-assessment activity 17.4

What dividend, if any, should the firm pay in the following situation? Earnings are £1 million, $k_e = 12$ per cent.

Project	Required outlay	IRR
A	£300,000	18%
B	£400,000	16%
C	£700,000	14%
D	£200,000	10%

(Answer in Appendix A at the back of the book)

■ External equity financing: yet more Divicut

We now consider the case where the firm has access to external financing. Rather than use earnings retention to finance investment, the firm may make a rights issue of shares, i.e. offer existing shareholders the right to buy new shares in proportion to their present holdings. If they all exercise their rights, the existing balance of control is unchanged, i.e. everyone ends up holding the same proportionate share of the firm. To sustain the irrelevance conclusion, we need to show that shareholder wealth is unaffected by choice of financing method, retention or new issues.

Consider the case where Divicut contemplates a one-off dividend cut to finance a project that requires a cash outlay of £1,000, and offers perpetual earnings of £200 p.a. The shares are not yet quoted ex-dividend. The options are to:

1 Retain the earnings.
2 Pay the dividend out of earnings and then recoup the cash via a rights issue.

The project itself is worthwhile, since its NPV is:

$$\text{NPV} = -£1,000 + \frac{£200}{0.1} = -£1,000 + £2,000 = +£1,000$$

The implications for shareholder wealth of each financing alternative are respectively:

1 If the dividend is passed, the value of shareholder wealth (W_o) is:

$$W_o = \text{(dividend in year 0)} + \text{(PV of dividends from existing projects)}$$
$$+ \text{(PV of dividends from new projects)}$$
$$= 0 + \frac{£1,000}{0.1} + \frac{£200}{0.1} = (£10,000 + £2,000) = £12,000$$

Retention therefore benefits Divicut's shareholders, since £1,000 now is exchanged for a dividend stream with present value of £2,000.

2 If a rights issue is used to recoup the required capital, the wealth of shareholders is:

$$W_o = \text{(dividend in year 0)} + \text{(PV of dividends from existing projects)}$$
$$+ \text{(PV of dividends from new projects)}$$
$$- \text{(amount subscribed for new shares)}$$
$$= £1,000 + \frac{£1,000}{0.1} + \frac{£200}{0.1} - £1,000$$
$$= (£1,000 + £10,000 + £2,000) - £1,000 = £12,000$$

Shareholders' wealth is unaffected if they subscribe for the new shares, anticipating the higher future stream of dividends implied by the new venture. The attraction of the rights issue to some people is that it offers a choice between further investment in the company and an alternative use of capital (although this second option may be irrational if k_e accurately measures the opportunity cost of capital). Either way, in practice, companies often ensure they obtain the required funds by employing specialist financial institutions to underwrite such issues, as explained in Chapter 16.

■ Summary

This section has demonstrated that dividends are relevant to company valuation in the absence of external financing, but only in the sense that, if the company has better investment opportunities than its shareholders, payment of dividends prevents worthwhile investment. With external financing, however, in a perfect capital market, dividends become totally irrelevant since dividend payment no longer precludes worthwhile investment. The firm can simply recoup the required finance via a rights issue.

In the next section, we begin to unpick some of the assumptions underpinning the irrelevance theory and consider the implications of doing so.

17.5 OBJECTIONS TO DIVIDEND IRRELEVANCE

This section examines some of the arguments advanced against the irrelevance theory. Financial managers should weigh up several important considerations before deciding upon the appropriate dividend payout policy:

- To what extent do shareholders rely on dividend income?
- Are nearer-in-time dividends less risky than future dividends?
- Do market imperfections lead companies to adopt policies that attract a particular clientèle of investors?
- Would investors accept a rights issue?
- How does taxation affect dividend policy?

Let us deal with these questions in turn.

■ Do shareholders rely on dividend income to support expenditure?

Shareholders who require a steady and reliable stream of income from dividends may be concerned by a sudden change in dividend policy, especially a dividend cut, albeit to finance worthwhile projects. Some groups of shareholders may well have a definite preference for current income: for example, the elderly, and institutions such as pension funds, that depend on a stable flow of income to meet their largely predictable liabilities. However, in an efficient capital market, such shareholders should be no worse off after a dividend cut, since the value of their holdings will rise on the news of the new investment. They can realise some or all of their gains, thus converting capital into income. The capital released is called a '**home-made dividend**'. (This is similar to the 'equity release schemes' whereby UK home-owners extract some of the value locked up in their homes.) A worked example of this procedure is given in the appendix to this chapter.

There are several criticisms of the validity of the home-made dividends mechanism. Even in the absence of market imperfections, the investor is forced to incur the inconvenience of making the required portfolio adjustments. Allowing for brokerage and other transactions costs, the net benefits of the project for the income-seeking investor are reduced. Also, if capital gains are taxed, the enforced share sale may trigger a tax liability. In the case of only marginally attractive projects, these effects may be sufficient to more than offset the benefits of the project, at least for some investors. Conversely, it can be argued that payment of a dividend to investors who then incur brokerage fees in reinvesting their income is equally disadvantageous.

■ Are future dividends seen as more risky by shareholders?

The practical limitations on the unfettered ability to home-make dividends were spelt out by Myron Gordon (1963) in a ringing attack on MM's irrelevance conclusion. However, Gordon extended his critique of MM to argue that $1 of dividend now is necessarily valued more highly than $1 of retained earnings because investors regard the (albeit higher) future stream of dividends stemming from a new project as carrying a higher level of risk. In other words, investors prefer what Gordon called an 'early resolution of uncertainty'. Gordon was, in effect, arguing that shareholders evaluate future expected dividends using a set of rising discount rates. If present dividends are reduced to allow greater investment, thus shifting the dividend pattern into the future, company value will fall.

Keane (1974) refined Gordon's position by suggesting that it is secondary whether, in fact, future dividends are more risky than near ones. If investors *perceive* them to be riskier, a policy of higher retentions, while not actually increasing risk, may unfavourably alter investor attitudes. In capital markets where full information is not released about investment projects, investors' subjective risk assessments may result in low payout companies being valued at a discount compared to high payout companies: that is, investors' imperfect perceptions of risk may lead them to undervalue the future dividend stream generated by retentions.

The 'bird-in-the-hand fallacy'

If the firm's dividend policy does alter the perceived riskiness of the expected dividend flow, there may be an optimal dividend policy that trades off the beneficial effects of an enhanced growth rate against the adverse impact of increased perceived risk, so as to maximise the market value. However, advocates of dividend irrelevance argue that Gordon's analysis is inherently fallacious. More distant dividends are more risky only if they stem from inherently riskier investment projects. Risk should already have been catered for by discounting cash flows at a suitably risk-adjusted rate. To deflate future dividends for risk further would involve double-counting. There is no reason why risk necessarily increases with time – a model based on this supposition incorporates the **'bird-in-the-hand fallacy'**. According to MM (1961), dividend policy remains a 'mere detail' once a firm's investment policy, and its inherent business risks, is made known. This argument is examined more rigorously by Bhattacharya (1979).

Self-assessment activity 17.5

Why is Gordon's 'early resolution of uncertainty' argument in favour of paying early dividends logically flawed?

(Answer in Appendix A at the back of the book)

■ Market imperfections and the clientèle effect

The extent to which investors are willing and able to home-make dividends, and thus adjust the company's actual dividend pattern to suit their own personal desired consumption plans, depends on the degree of imperfection in the capital market. In practice, numerous impediments, especially when aggregated, may significantly offset the benefits of exploiting a profitable project. Some of these have already been mentioned, but the main ones are as follows:

- Brokerage costs incurred when shares are sold.
- Other transaction costs incurred, e.g. the costs of searching out the cheapest brokerage facilities.
- The loss in interest incurred in waiting for settlement.
- The problem of indivisibilities, whereby investors may be unable to sell the precise number of shares required, forcing them to deal in sub-optimal batch sizes.
- The sheer inconvenience of being forced to alter one's portfolio.
- Share sales may trigger a capital gains tax liability.
- If the company is relatively small, its shares may lack marketability, requiring a significant dealing spread and hence a leakage of shareholder capital.
- If the company is unquoted, it may be difficult or impossible to find a buyer for the shares.

Under such imperfections, maximising firm value may not be the unique desire of all shareholders; the pattern of receipt of wealth may become equally or more important. Some shareholders may prefer companies that offer dividend flows which correspond to their desired consumption, perhaps being prepared to pay a premium to hold these shares. In this way, they avoid having to make their own adjustments. The vehicle for aiding such investors is to provide a stable and known dividend stream. Shareholders can then perceive the nature of the likely future dividend pattern and decide whether or not the company's policy meets their requirements. In other words, the company attracts, and attempts to cater for, a **clientèle** of shareholders.

However, such a policy has costs, including the benefits forgone from projects that have to be passed over, the costs of borrowing if debt finance is used, and/or the issue expenses of a rights issue if external financing is employed. The implications for control if rights are not fully exercised by the existing shareholders may be another problem.

The key difficulty facing the financial manager is lack of knowledge of shareholder preferences, without which it is difficult to balance the two sets of costs.

Self-assessment activity 17.6

What is meant by a shareholder clientèle? Which shareholders are most likely to prefer near-in-time dividends?

(Answer in Appendix A at the back of the book)

■ Problems with rights issues: Rawdon plc

Among the effects of a rights issue are the costs incurred. These can be substantial, and will affect the required return on new investment. Among the costs are the administrative expenses, the costs of printing brochures and circulation of these to shareholders, and also the underwriters' fees. The impact of these costs is examined using the case of Rawdon plc, whose details are shown in Table 17.2.

Although Rawdon's shareholders require a return of 20 per cent, the new project has to offer a return of *over* 20 per cent. This is because some of the finance raised by the share issue is required to meet the costs of the issue, but the holders of those shares will nevertheless demand a return. Total share capital is now £6 million, and earnings of £1.2 million are now required to generate a 20 per cent return overall. However, the required increase in earnings of £0.2 million must be generated by an investment of

Table 17.2 Rawdon plc

- $k_e = 20\%$
- Rawdon's value = £5 m
- 1 m shares have been issued in the past
- Market price per share is £5
- Current company earnings = £1 m (EPS = £1)
- Proposed investment outlay = £950,000
- The project has a perpetual life.
- Terms of rights issue:
  One share for every five held, i.e. 200,000 new shares.
  Purchase price £5: gross proceeds = £1 m
  Issue costs: 5% of gross receipts = (0.05 × £1 m) = £50,000.
  Net proceeds = £950,000

£950,000, necessitating a return of (£0.2 m/£0.95 m) = 21.1 per cent. Rawdon's required return on this project is:

$$\frac{\text{Normally required return}}{(1 - \% \text{ issue costs})} = \frac{k_e}{(1 - c)} = \frac{0.2}{(1 - 0.05)} = 0.211, \text{ i.e. } 21.1\%$$

However, the problem of transactions costs may operate in another direction. Some shareholders, when paid a dividend, may incur some reinvestment costs. The higher the proportion of investors who wish to reinvest, either in the same or in another company, the greater the total saving of brokerage and other fees enjoyed by shareholders when the company retains earnings. Hence, the greater is the attraction of retained earnings to finance investment.

Bearing in mind the impact on share price, the announcement of a rights issue is not always greeted with delight. The shareholder is forced either to take up or sell the rights in order to avoid losing money (even sale of rights will result in brokerage fees), and thus incur the inconvenience involved. If the market takes a dim view of the proposed use of the capital raised (for example, if the funds are required to finance a takeover whose benefits look distinctly speculative), the share price may fall. *But it is important to realise that this would be a consequence of a faulty investment decision rather than of the financing decision itself.*

■ The impact of taxation

In many economies, the tax treatment of dividend income and realised capital gains differs, either via differential tax rates or via different levels of exemption allowed, or both. In such regimes, the theoretical equivalence between dividends and retention becomes further distorted. Typically, tax on capital gains is lower than tax on income from dividends. In some countries there is *no* tax on capital gains. Some shareholders might prefer 'home-made dividends' (i.e. selling shares) now or in the future when the rate of capital gains tax (CGT) is lower than the marginal rate of income tax applied to dividends. Others may prefer dividend payments because their income tax liability (plus any reinvestment costs) is lower than CGT payments.

In addition to considerations of personal taxation, where the financial manager is often unaware of the particular tax positions of shareholders (a major exception to this is the case of institutional investors), complications may also be imposed by the corporate tax system. In an imputation tax system, shareholders receive dividend income net of basic rate tax, while the rate of tax applied to corporate profits is designed to include an element of personal tax, thus making the distribution of dividends tax-neutral.

There are so many different tax regimes operated throughout the world, that it is difficult to generalise about the effect of taxation on dividend policy. However, the following simple example captures the flavour of some of the issues.

Barlow plc

Barlow plc is financed entirely by equity and its future cash flows have present value of £500 million at the start of 2008. During 2008, it earns £100 million – for simplicity, we assume that all transactions are in cash, so it now holds cash of £100 m. Profits are taxed at 30 per cent so Barlow must set aside £30 million for tax payments, leaving (70% × £100 m) = £70 m available for distribution. According to the EMH, the firm will be valued at (£500 m + £70 m) = £570 m.

Should Barlow distribute or retain? The answer depends on three factors:

- shareholders' marginal rates of tax.
- the relative rate of tax on dividends vs. tax on capital gains.
- The type of tax regime.

classical tax system
A system where dividends are effectively taxed twice – the firm pays profits tax and then investors pay income tax on any dividend payment

Under a **classical tax system**, profits are taxed twice if distributed, firstly, as profits tax, and secondly, as income tax paid by investors. Imagine the firm makes a full distribution. Consider two rates of investor income tax:

(i) if investors pay tax at 10%, income tax is (10% × £70 m) = £7 m, and the total tax charge is (£30 m + £7 m) = £37 m (or 37% of pre-tax earnings).

(ii) if investors pay tax at 50%, income tax is (50% × £70 m) = £35 m, and the total tax charge is (£30 m + £35 m) = £65 m (or 65% of pre-tax earnings).

It looks better to retain in the second case, but the decision also depends on the rate of capital gains tax (CGT).

Assume the CGT rate is 20 per cent. If we also assume that the firm invests in zero-NPV projects (unlikely, but a necessary assumption to strip out the effect of the investment decision), the value of the firm will have risen from £500 million at start-year to £570 million at end-year. The CGT payable is thus (20% × £70 m) = £14 m. Along with the profits tax, the total tax payable is (£30 m + £14 m) = £44 m.

Obviously, shareholders paying income tax at 50 per cent would prefer retention and vice versa.

Under an **imputation tax system**, the relative attractiveness of distribution and retention depends not only on the relative tax rates, but also on whether there is full or partial imputation.

With full imputation, investors get full credit for corporate profits tax already paid.

In case (i) above, the investor would face no further tax on income and may even get a tax rebate, depending on the tax regime, because the rate of corporate tax exceeds the rate of personal tax.

In case (ii), the investor obtains credit for the corporate tax already paid, and thus faces an additional income tax charge of (50% − 30%) × £70 m = £14 m. With these particular figures, investors would be indifferent between distribution and retention.

With partial imputation, it is less clear-cut – the relative desirability of distribution and retention depends on the degree of imputation, as well as the respective tax rates.

Current UK rates

Under the present UK system of partial imputation, income from dividends and capital gains are taxed at different rates, and different tax thresholds also apply. Following reforms implemented in 2008, there is now a clear discrimination in favour of capital gains.

Income tax on dividends

Dividends are paid net of tax, assumed payable at 10 per cent, thus generating a tax credit. Current (2008–9) UK income tax rates applicable to gross dividend income (ignoring other sources of income) are:

> 10% for gross (i.e. including, the tax credit at 10%)
> dividend income of £5,435–£41,435,
> and at 40% where total income exceeds £41,435.

As the tax is applied to gross incomes, investors in the starting rate band of income pay no further tax, and those in the higher rate band pay a further 32.5 per cent (to make their overall charge 40 per cent). Income tax becomes payable at incomes above £5,435 but dividends are treated as the top slice of income, thus bearing the highest rate of tax.

Capital gains tax

Gains above the threshold £9,600 are taxed as if additional income at 18%. The indexation and taper reliefs that mitigated tax were abolished in 2008, although husbands and wives continue to enjoy a double allowance for jointly-owned assets.

■ The tax irrelevance thesis

Some argue that, under perfect capital markets, the question of relative tax rates is irrelevant because of arbitrage. Zero/low-rate taxpayers prefer to hold shares in firms with high payouts, and those with high marginal rates of tax prefer low payout firms. The process of tax-driven arbitrage would lead each group to bid up the share prices of the firms whose distribution policy suits their own particular tax positions, until in equilibrium, post-tax rates of return are equalised. This means there is no share price advantage to be obtained from following a particular dividend policy.

Elton and Gruber (1970) noted the importance of marginal tax brackets in determining the return required by shareholders. They defined the cost of using retained earnings as 'that rate which makes a firm's marginal shareholders indifferent between earnings being retained or paid out in the form of dividends'. Under differential tax treatments of dividend income and capital gains, the cost of using retained earnings is a function of the shareholder's marginal tax bracket. We might also expect companies whose shareholders incur high rates of income tax to exhibit low payout rates. Such a relationship between corporate dividend policy and shareholder tax brackets would support the notion of tax 'clientèles', whereby companies seek to tailor their payout policies to the tax situation of particular shareholders. Elton and Gruber's empirical work seemed to indicate that firms attract rational clientèles – *shareholders gravitate to companies whose distribution policy is compatible with their personal tax situations*.

17.6 THE INFORMATION CONTENT OF DIVIDENDS: DIVIDEND SMOOTHING

Possibly the most important consideration for corporate financial decision-makers when framing dividend policy is the information-processing capacity of the market. Any dividend declaration can be interpreted by the market in a variety of ways. In an uncertain world, information regarding a company's prospects is neither generally available nor costless to acquire. Managers possess more information about the company's trading position than is available to investors as a whole. This so-called **information asymmetry** may mean that the announcement of a new or changed company policy may be interpreted by the market as a *signal* conveying particular information about a company's prospects, as in the Charter plc case below.

For example, the decision to pay an unexpectedly high dividend may be seized upon as evidence of greater expected dividend-paying capacity in the future. It may be taken as guaranteeing an ability to sustain at least the higher declared dividend and probably more. As a result, the financial manager should consider carefully how the

Charter set for dividend pay-out

Charter, the engineering group, beat market expectations with a strong half-year performance and said it expected to pay its first dividend for seven years at the end of this years.

The resumption of dividend payments will be a milestone for Charter, which has not paid one since 2000, when the share price collapsed after a takeover of Lincoln Electric, a US rival, fell through.

What was then a debt-laden, sprawling and diversified group now has two businesses: Esab, a welding, cutting and automation division; and Howden, an air and gas handling operation.

Both fared well in the first half despite dollar-related currency losses,

which trimmed sales by 5 per cent and operating profits by 4 per cent. Group revenues rose 10.7 per cent to £691.5m and pre-tax profits were up more than 30 per cent at £92.1m, ahead of market expectations.

Charter's share price, which fell below 60p in 2003, rose 35p to £11.75.

Michael Foster, chief executive, said the board had intended to pay a dividend when shareholder funds reached "something in the order of half a billion pounds".

He expects shareholder funds to equal around £450m by the end of the year.

"We've got to remember that just two years ago this company had no shareholder funds at all," he said.

Esab, which accounts for two-thirds of revenue, saw strong performances in Europe, South America and the Middle East but also benefited from a growing business in the US, in spite of the weakness of the dollar.

Howden's order book rose 17.7 per cent in the first half and continued rising in July and August.

Charter has appointed Lars Emilson, chief executive of Rexam, the beverage can business, from 2004 to 2007, as non-executive director. In November he will take over as non-executive chairman from David Gawler, widely credited with turning the company round.

Source: Toby Shelley and William MacNamara, *Financial Times*, 13 September 2007.

market is likely to decode the signals contained in the dividend decision, and the likely consequences for share price.

Company finance directors (FDs) are well aware of the signalling power of dividends and their capacity to influence share price. This awareness is implied by the recent earnings and dividend record of the two firms, shown in Table 17.3.

Table 17.3 shows the EPS and DPS record of two firms in quite different operating environments. Kelda plc (since taken over by private equity firm Saltaire Holdings Ltd) supplies water to customers in Yorkshire and, apart from regulatory constraints, faced a very stable marketplace, with demand highly predictable and operating costs easy to manage. D.S. Smith plc, in contrast, faced a far more turbulent environment. Its main markets are subject to far more volatile influences as it supplies packaging products for final consumers, while its cost structure can rapidly be disrupted by rises in

Table 17.3 Kelda and D.S. Smith: dividend smoothing

Year	EPS (p)	% change	DPS (p)	% change
Kelda				
2003	42.4	–	27.05	–
2004	46.2	9.0	27.83	2.9
2005	52.1	12.8	29.00	4.2
2006	54.0	3.6	30.35	4.7
2007	61.4	13.7	32.25	6.3
Overall		44.8		19.2
D.S. Smith				
2003	16.8	–	8.2	–
2004	13.6	(19.4)	8.2	–
2005	14.4	5.9	8.4	2.4
2006	10.0	(29.6)	8.4	–
2007	13.1	31.0	8.8	4.8
Overall		(20.2)	7.3	

Source: Respective company Annual Reports.

input costs, especially that of energy. Not surprisingly, we find that Kelda was able to pursue a far more progressive (but not particularly generous) dividend policy than D.S. Smith, but it is clear that both firms endeavoured to smooth their DPS payment, i.e. both had far more stable DPS profiles than their corresponding EPS records. This 'smoothing' of dividends can be found among many other firms and has been identified empirically. Lintner (1956) showed that companies appear to raise dividends only on the basis of reported earnings increases that they expect to be sustainable in the long term. In a survey of 179 Finance Directors of large quoted companies conducted by 3i (1993), 43 per cent said that the single most important factor influencing dividend policy was prospective long-term profit growth, while 93 per cent agreed with the statement that 'dividend policy should follow the long-term trend in earnings'.

When dividend increases lag behind increases in EPS, the payout ratio falls and vice versa. However, average payout ratios have risen in recent years, owing to a number of factors:

1 Increasing corporate profits.
2 Hence, increased confidence in companies' ability to sustain dividends.
3 The increasing pressures exerted by the ever-more powerful institutional investors, especially the tax-exempt funds.
4 The use of pre-emptive dividend increases to ward off takeover raiders.
5 The tax changes of 1988, which largely removed the tax discrimination for higher-rate taxpayers in favour of retentions.

So do companies ever cut dividends? Obviously, in adversity, dividend cuts are forced on firms in order to preserve cash. In the 3i survey, 55 per cent of FDs agreed with the statement that 'any cut in dividend payout sends adverse signals to the market and should be avoided'. Dividend-cutting is far more common in the USA than in the UK. Empirical work by Ghosh and Woolridge (1989) suggests that even when this was motivated by the need to conserve funds for investment purposes, shareholders suffered significant capital losses, despite the merits claimed for the proposed investments. In the UK, it is almost unheard of (almost – but remember the China Shoto case at the start of the chapter) for companies to cut dividends to finance investment – the dividend payment-plus-rights issue is invariably the preferred alternative. In the 3i survey, one FD said that 'a dividend cut is a sign of fundamental management failure – when management fails and the Board should change'.

17.7 WORKED EXAMPLE

The data given below relates to three firms' dividend payouts over the last six years. None of the companies has issued, or cancelled, any shares over the period. All three firms are listed on the London Stock Exchange.

		2003	2004	2005	2006	2007	2008
Company A							
Shares issued	1,200 m						
Profit after tax (£m)		600	630	580	600	640	660
Dividends declared (£m)		240	252	232	240	256	264
Company B							
Shares issued	2,000 m						
Profit after tax (£m)		1,200	1,300	1,580	1,800	1,240	1,460
Dividends declared (£m)		120	132	145	160	176	194
Company C							
Shares issued	3,500 m						
Profit after tax (£m)		2,200	1,400	2,100	1,950	2,200	2,560
Dividends declared (£m)		200	0	100	0	200	560

Required

Showing appropriate calculations, describe the dividend policy which each of the companies shown above appears to be following.

How would you justify each of these policies to the shareholders of each of the companies concerned?

Answer

General comments

Company A

This firm pays out a constant percentage (40 per cent) of profit after tax as dividend. It is thus prepared to tolerate fluctuations in the actual amount paid and, hence, DPS, as corporate profits, and thus EPS, oscillates. DPS fluctuates from a high of 20p in 2003 to a low of 19.3p in 2005, ending up at 22p in 2008. The range of fluctuation is thus quite narrow.

>*In brief:* constant payout (and cover)
>   variable DPS

Company B

This firm increases the amount of dividend and thus the DPS by a constant 10 per cent p.a., come what may, in terms of PAT and EPS. The dividend cover thus takes the strain, although it is never less than 7 times (2007).

>*In brief:* rising DPS
>   variable payout (and cover)

Company C

This firm pays out any excess of PAT above a base level of £2 billion, accepting a zero payout if PAT undershoots this figure.

>*In brief:* variable DPS
>   variable payout (and cover)

Interpretation

The nature of the industry in which the firms operate could well be an important influence on dividend policy. Profits are relatively stable for Company A, but oscillate wildly in the case of Company C. Hence, A is able to pay out a relatively high proportion in dividends reasonably safe in the knowledge that profits will soon recover after a bad year as they did in 2006, exceeding the previous peak of 2004 shortly afterwards in 2007. The maintenance of a high payout (if not the actual DPS) might be taken as an expression of confidence in the future and thus perform an important signalling function. Indeed, the fall in DPS in 2005 might be taken as a sign of financial prudence – conserving cash at a time of faltering performance.

It is likely that Company A has a clientèle that relies on a high level of dividend income, although one that is prepared to accept some belt-tightening in lean years. A fluctuating DPS does impact on liquidity, e.g. in the leaner years, despite the cut in DPS, the amount of retained earnings and, by implication, free cash flow is reduced, perhaps impacting on investment expenditure. However, it is unlikely that firms would want to push ahead with substantial investments in the lean years (although there is a strong argument in favour of this, i.e. to be primed for the fat years when they arrive). Also, if the industry is stable, and the firm has a solid record of increasing profits over time, then it may have good access to capital markets. A high payout also has agency advantages – managers have less access to discretionary spending, which may not necessarily be in the best interests of the owners.

Looking at Company C, its base profits level of £2 billion before a dividend payment is made suggests that it has an ongoing investment programme to fund, and is thus concerned about liquidity, or that it is highly geared and faces debt covenants that stipulate this profits threshold before a dividend can be paid. The fluctuations in profits and the dividend policy implies that the shareholder clientèle is relatively unconcerned by dividend income, but is more interested in profits and value growth. The industry is probably a high-risk one, perhaps a New Economy one, with attractive long-term prospects. Clearly, this is a residual dividend policy where other demands on cash flow take precedence over distribution of profits. Eventually, however, like Microsoft, the firm will exhaust its (organic) growth potential and will have to frame a coherent dividend policy.

Finally, Company B has a highly predictable dividend policy that allows shareholders to anticipate the next year's income with great accuracy and confidence. This will suit a variety of clientèles that rely on a predictable stream of income such as institutional investors and personal investors advanced in years. However, the payout ratio is not high, so the firm is taking no risks in terms of the sustainability of its policy, i.e. there is plenty of slack to enable an increase in DPS even in bad years such as 2007. This suggests that it is using retained earnings to a significant extent to finance growth, possibly because it faces difficult access to the capital markets – perhaps it is a relatively young firm or has high gearing. Either way, it is unlikely to have to endure the ignominy of cutting the dividend in the future – profits would have to fall over 80 per cent before the dividend was uncovered.

17.8 ALTERNATIVES TO CASH DIVIDENDS

The motive for cutting a dividend is to save cash outflows. This suggests that liquidity considerations are another influence on the dividend decision. This section discusses how firms that wish to preserve liquidity can still offer 'dividends', and also how firms can cope with excessive liquidity. Alternatives to cash dividends are thus divided into those that preserve liquidity and those that reduce liquidity.

■ Liquidity-saving alternatives

(i) Scrip dividends

Most major companies offer shareholders scrip dividends. This is an opportunity to receive new shares instead of a cash dividend payment. It has certain advantages for both the shareholder and the company.

From the firm's point of view, the scrip alternative preserves liquidity, which may be important at a time of cash shortage and/or high borrowing costs, although it could face a higher level of cash outflows if shareholders revert to preference for cash in the future. With more shares issued, the company's reported financial gearing may fall, possibly enhancing borrowing capacity. In this respect, the scrip dividend resembles a mini-rights issue.

For shareholders wishing to expand their holdings, the scrip is a cheap way into the company as it avoids dealing fees and Stamp Duty (currently, 0.5 per cent in the UK). A scrip dividend has no tax advantages for shareholders as it is treated as income for tax purposes.

In an efficient capital market, there is no dilution effect on the share price, because it would have fallen anyway with a cash dividend because of the 'ex-dividend effect'. However, if the additional capital retained is expected to be used wisely, the share price may be maintained or may even rise, giving shareholders who elect for the scrip dividend access to higher future dividends and capital gains. As a result, there may be a longer-term tax advantage for shareholders. However, most of these effects are marginal. Perhaps the main benefit of a scrip dividend is that, by giving shareholders a choice, it makes the shares more attractive to a wider clientèle

(ii) Enhanced scrip dividends

To overcome the low take-up rate of scrip dividends, in 1993, several companies with chronic liquidity problems began to offer 'enhanced scrip dividends' (ESDs). The first was BAT Industries (**www.bat.com**), which announced a scrip alternative 50 per cent above the equivalent cash dividend, designed to be so generous that shareholders could not refuse. BAT declared a 22.6p final dividend, and then, later that month, offered an ESD alternative of 33.9p. Had all of BAT's shareholders taken up the ESD, the cash saving would have been £423 million.

The ESD mechanism is a rights issue in disguise. To the extent that shareholders take the scrip alternative, the company is left with as much cash as if it had paid a cash dividend and then clawed it back from the same shareholders via a rights issue. As with a rights issue, the number of shares issued will rise. Finally, because the scrip is voluntary, the control of shareholders taking cash is diluted.

Self-assessment activity 17.7

How many shares would an investor receive in lieu of cash dividends of £1,000 if offered:

(i) a pro rata scrip alternative?
(ii) an ESD of 20%?

Share price is £20.

(Answer in Appendix A at the back of the book)

Due to tax changes enacted in 1997, some firms experienced a sharp rise in the take-up rate of scrips – e.g. some 20 per cent of Boots plc's dividend outflow was in scrip form. This was unpalatable to firms that were concerned about the effect on the total number of shares issued by the company, e.g. it reduced the scope for any future rights issue.

In April 1998, GKN plc (**www.gkn.plc**) responded to this problem by withdrawing its scrip dividend scheme because 'recent changes in tax legislation affecting institutional shareholders make it more likely that a scrip dividend could lead to the issue of large numbers of new shares which, we believe, is not in shareholders' best interests'. Instead, GKN introduced its Dividend Re-investment Plan (DRIP) as from May 1998, whereby shareholders can purchase existing shares of equivalent value to the cash dividend at a preferential dealing cost (0.5 per cent of the dividend displaced). Many firms now offer similar facilities.

■ Liquidity-reducing alternatives

The dividend decision is contorted by the case of companies with too much cash.

As an alternative to paying cash dividends, the facility to repurchase shares, subject to shareholders' approval, was introduced in the UK by the 1981 Companies Act. It resulted from pressure from the small firms lobby, concerned that low marketability of unquoted firms' equity was hampering their development. Certain conditions were stipulated, designed principally to safeguard the position of creditors, and further restrictions were imposed by the Stock Exchange and the Takeover Panel, aimed at reducing the risk of market-rigging during takeover battles. The first major company to mount a buy-back was GEC in 1984, ostensibly to raise the EPS of the remaining shares. A component of Marks and Spencer's defence package against a takeover bid in 2004 was share repurchase, designed to raise the cost of the bid.

The Bank of England (1988) cited five possible reasons for repurchases:

- To return surplus cash to shareholders.
- To increase underlying share value.

- To support share price during periods of temporary weakness.
- To achieve or maintain a target capital structure.
- To prevent or inhibit unwelcome takeover bids.

The hoped-for increase in share price works through the effect on EPS. A company may attempt a buy-back when it has a cash surplus beyond what it needs to finance normal business operations. This situation is assumed in Table 17.4, which shows the before-and-after situation of a company sitting on a cash pile of £100 million earning interest at 5 per cent. It buys in 20 million shares at 180p, above the market price of 172p, the discrepancy being due to the required premium required to attract enough sellers. The total buy-back cost is therefore £36 million, on which the company will lose interest income.

Table 17.4 Analysis of a share repurchase

	Before	After
Trading profit (£m)	200.00	200.00
Interest income at 5% (£m)	5.00	3.20
Pre-tax profit (£m)	205.00	203.20
Tax at 30% (£m)	(61.50)	(60.96)
Profits available for shareholders (£m)	143.50	142.24
Number of shares in issue	500 m	480 m
EPS (p)	28.70	29.63
P:E Ratio	6:1	6:1
Share price (p)	172	178
Cash balances (£m)	100.00	64.00

In the UK (unlike the USA), when a company buys back shares it has to cancel them. This suggests a problem with buy-backs – directors may buy in shares and might subsequently have to make a rights issue, the costs of which are far higher than those involved in repurchasing shares. US evidence suggests a strong signalling effect, whereby executives can express confidence in their firms by investing in equity. Yet there remains a strong suspicion in the UK that such messages are likely to be decoded adversely. To many observers, buy-backs are tantamount to an admission of managerial failure, signalling lack of confidence in their ability to identify and exploit wealth-creating projects.

However, investment expenditure is rarely a continuous process and small buy-backs may be a useful way of investing temporary cash surpluses, with the beneficial effects on EPS and share price providing a platform for a subsequent rights issue if development funds are needed. In addition, buy-backs may be a useful alternative to paying higher but unsustainable dividends, while preserving choice for investors.

Buy-backs: a word of warning

One of the unfortunate consequences of buy-backs is that (admittedly with the benefit of hindsight) they can make directors look rather foolish. If, after the buy-back, the share price falls, it means that more money has been expended on the buy-back than would had been required at the lower price (see the FT article by Chris Hughes).

An example of directors looking red-faced in retrospect was the case of Next plc, the fashion retail chain. In 2007, it announced a buy-back programme at a cash cost of £513 million, some 11 per cent of the shares then in issue, for an average of £19.74 per share. In accounting terms, this pressed the right buttons, as, assisted by a modest 4 per cent profits increase, EPS rose by 15 per cent in 2008. However, amid descending

GSK doubles buy-back pledge

GlaxoSmithKline, the UK-based pharmaceuticals group, yesterday more than doubled its planned share buy-back programme to £12bn to appease investors after sales setbacks on Avandia, its diabetes drug.

Unveiling second-quarter sales of Avandia down 22 per cent to £349m, the company said it would buy back £12bn in shares over the next two years, compared with a previous £6bn pledge over three years.

Jean-Pierre Garnier, GSK's chief executive, said: "We have a very pristine balance sheet and a policy of actively returning cash to shareholders ... to help left our share price. We were doing well until Avandia and outpacing the FTSE, then everything came to a halt."

The move for greater buy-backs – coupled with increased dividends – reflects a growing trend for drug companies to attempt to woo disgruntled investors after consistent underperformance because of rising generic competition and insufficient experimental medicines to fill the pipeline of those with patents coming up for expiry.

GSK's shares in London rose more than 2 per cent on the news after weeks of poor performance since concerns over Avandia were raised in May, boosting the FTSE 100 index.

Standard & Poor's, the rating agency, said that in spite of the buy-back increase, GSK's financial position was strong enough to maintain its AA/A-1+ long-term and short-term corporate credit ratings.

Mr Garnier would not comment on the outcome of a definitive ruling expected next Monday on the fate of Avandia, under intense scrutiny since a "meta analysis" in May suggested a risk of increased cardiac problems.

But he said he was confident in the company's own recently submitted analysis from 400,000 patients, which should support the drug's continued use in the US – as it had been approved elsewhere – as long as decision-makers looked at the science.

His comments come as critics argue that the FDA, under political pressure, has placed too much emphasis on safety, although the US regulator has issued only a slightly increased safety warning and has rejected calls for the drug's withdrawal.

Mr Garnier stressed that Avandia represented "less than 5 per cent of the GSK story", and should be measured against a pipeline which has included six product launches so far this year.

GSK's earnings per share rose 11 per cent to 24p for the quarter, on sales up 3 per cent to £5.7bn. It proposed a dividend of 12p.

Source: Andrew Jack, *Financial Times,* 26 July 2007.

Chevron to spend $15bn on buying back shares

Chevron, the second-largest US oil company, is to spend $15bn on its own shares within the next three years, extending a $5bn-a-year programme that began in 2005.

The move reflects big oil companies' difficulties in finding uses for their large cash flows, boosted by high oil prices.

Some of Chevron's rivals among the "big five" international oil companies have been returning even larger sums to shareholders.

The rise of resource nationalism among oil and gas-rich countries, and growing competition from ambitious national oil companies from emerging economies, has made it harder for the international oil companies to find investment projects with attractive commercial terms.

Dave O'Reilly, Chevron's chairman and chief executive, said: "Our continuing strong cash flows have enabled us to fund a significant capital programme budgeted at almost $20bn in 2007, increase dividends to our stockholders, repurchase our shares in the market and reduce the company's debt."

Chevron's buy-backs are dwarfed by ExxonMobil's and BP's. Exxon's gross share purchases were worth $16bn in the first half of 2007, reducing the shares outstanding by 3.2 per cent. It spent $29.6bn in 2006. BP, which has a market capitalisation roughly in line with Chevron at about $220bn, is the next heaviest spender. It bought back $15.5bn-worth of shares last year. European rival Royal Dutch Shell was more modest, buying back $8.2bn worth of shares.

There have been signs that both Shell and BP have been reducing their repurchasing this year.

BP's buy-backs for the year to date have been worth about $5.9bn, and Shell's were just $1.4bn in the first half.

Shell has questioned the value of buy-backs and some analysts and investors agree: there appears to be little correlation between the scale of the buy-back and share price performance.

During 2005–06, BP's shares underperformed the sector by the greatest amount since the company's financial crisis of 1992–93.

Yet, over those two years, BP bought more than $27bn-worth of its own shares.

In the year to date, Chevron's shares have outperformed Exxon's, rising more than 25 per cent compared with Exxon's gain of 20 per cent, in spite of the much larger scale – both in absolute terms and relative to the company's market capitalisation – of Exxon's buy-back programme.

Shell has significantly outperformed BP this year.

Source: Ed Crooks, *Financial Times,* 27 September 2007.

Ill-judged buy-backs are 'collectively destroying billions in shareholder value'

The extra investment performance created by share buy-backs has dwindled to "negligible" levels, with many companies destroying shareholder value by ill-judged share purchases, according to new research.

Share buy-backs have become increasingly popular in recent years as the expanding economy has strengthened corporate cashflow and boards have bowed to perceived investor pressure to hand cash back. The relatively low level of debt carried by listed companies has also provided scope to take on more leverage and return cash through share buy-backs.

When a company pursues a share buy-back, it nominates a broker to buy its shares in the stock market. Normally these shares are then cancelled, so the company's earnings are then distributed less widely, and this boosts earnings per share (EPS).

It is estimated that in 2007 almost 15 per cent of Europe's large and mid-cap companies have undertaken buy-backs exceeding 2 per cent of their market capitalisation.

However, Collins Stewart, the broker, says many companies are ignoring the "essential valuation criteria" that determine whether or not a buy-back creates value for a company's investors.

It says companies focus too much on boosting EPS, when the fundamental question should be the price at which shares are repurchased. It also blames investment banks for advising companies to do ill-judged buy-backs just for the fees.

"We find numerous examples where the rationale for a buy-back – usually involving EPS enhancement – ignores the basic principles that will best ensure positive returns to shareholders," says Collins Stewart. "A substantial number of buy-backs are, collectively, destroying billions in shareholder value through ill-judged decisions."

The research found that in aggregate, there was less than a 50 per cent chance that a share buy-back would now lead to a company's share price outperforming.

The general problem, Collins Stewart argues, is that shares are less cheap

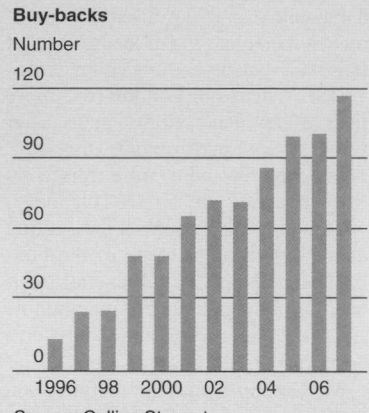

Buy-backs
Number

Source: Collins Stewart

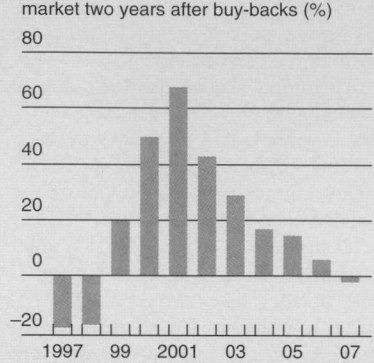

Performance of stocks relative to the market two years after buy-backs (%)

than they used to be. However, the broker singles out several UK companies for particular criticism, including Next, Tate & Lyle and LogicaCMG.

Tale & Lyle, Europe's largest sugar refiner, conducted a buy-back in August after issuing two profit warnings and ignoring "the near-inevitability of a third", says the study. When a third profit warning happened, it knocked the share price and generated a paper loss of £10m on the shares Tate had bought back.

LogicaCMG, the information technology services group, similarly bought back shares after its May profit warning, but in November issued downbeat guidance following the appointment of a new chief executive, further depressing the shares and creating a paper loss of £26m.

Meanwhile, Next, the retailer dubbed by Collins Stewart as the "buy-back junkies par excellence", destroyed a notional £70m of value on its buy-back this year – in spite of a good record in previous years.

Tate & Lyle says its buy-back was launched with the support of its shareholders.

"They welcomed it because it made efficient use of our balance sheet after our sale of European starch assets. Less than 20 per cent of the buy-back had been completed when the share price fell in the autumn. Since our results announcement in October, we have resumed the buy-back with the full support of our shareholders," he says.

Next is also unapologetic. The retailer says: "Next has a stated policy,

the board firmly believes that it is by improving EPS over a period of time that drives the share price and shareholder value. Next does have quite a lot of rules about the circumstances for buying back shares."

LogicaCMG declined to comment.

Collins Stewart says an emphasis on EPS-based performance hurdles in executive bonuses could create an incentive for boards to sanction share buy-backs even when their share price is too high.

It blames guidance from the Association of British Insurers, dating back to 1990, that share buy-backs are to be exercised only "if to do so would result in an increase in earnings per share and is in the best interests of shareholders generally".

The study says the ABI's guidance is "like a red rag to a bull for managements who may find it, perhaps coincidentally, to be closely aligned with their own interests".

However, Peter Montagnon, head of investment affairs at the ABI, rejects the accusation and says Collins Stewart is jumping to conclusions that are not justified, especially given that the ABI guidance stresses that buy-backs should always be in shareholders' interest.

"It is simply wrong to say we are encouraging buy-backs and falling into the trap of encouraging executives to use them to meet their performance hurdles," he says.

Source: Chris Hughes, *Financial Times*, 16 December 2007.

gloom for the retail sector, the share price had fallen to £11.08. The amount spent acquiring 26 million shares at the old price would have bought 46 million at the new, lower price. Obviously, those shareholders who did sell benefited significantly, but the remaining 89 per cent would have been far better off had the firm distributed its surplus cash as a dividend. In 2008, Next did raise its dividend by 15 per cent.

Further consolation for shareholders was that directors missed out on performance-related bonuses amounting to around £13 million.

Self-assessment activity 17.8

Suggest some arguments *against* share repurchases.

(Answer in Appendix A at the back of the book)

17.9 THE DIVIDEND PUZZLE

At a practical level, we encounter a dilemma known as the 'dividend puzzle'. Dividend cuts are generally viewed as undesirable, yet dividend payments may not always be in the best interests of shareholders. For example, under many tax regimes, dividends are immediately taxed, while the tax on capital gains can be deferred indefinitely. Dividend distribution may force a firm subsequently to issue equity or debt instruments which incur issue costs and interfere with gearing ratios (see Chapter 18). However, markets do react favourably to news of unexpectedly high dividend increases despite the acknowledged costs of dividends. The message conveyed by firms when raising dividends seems to be:

> 'Despite the cash flow costs of dividend payments, we are prepared to pay higher dividends because we are confident of our ability to withstand these costs and at least to maintain these higher disbursements.'

■ The case for a stable dividend policy

Astute financial managers appreciate that different shareholders have different needs. A financial institution reliant on a stream of income to match its stream of liabilities will prefer stable dividends, while a 'gross fund' (i.e. one exempt from tax on its income) will prefer shares that offer a high level of dividends. The private individual in the 40 per cent tax bracket will prefer capital gains, at least up to the £9,600 exemption limit (2008–9), while the old-age pensioner with a relatively short time horizon is likely to seek income rather than capital appreciation.

Given the wide diffusion of shareholdings, it is almost impossible for most financial managers to begin to assess the needs of all their shareholders. This is a powerful argument for a stable dividend policy, e.g. the application of a fairly constant rate of dividend increase, implying that dividends rise at the same rate as corporate post-tax earnings, subject perhaps to the proviso that dividends should not be allowed to fall, unless earnings suffer a serious reverse. This will enable shareholders to gravitate towards companies whose payout policies suit their particular income needs and tax positions. In this way, companies can expect to build up a clientèle of shareholders attracted by a particular dividend pattern. Hardly surprising, then, that 75 per cent of FDs in the 3i survey said that 'Companies should aim for a consistent payout ratio.'

Self-assessment activity 17.9

Rehearse the arguments in favour of a stable dividend policy.

(Answer in Appendix A at the back of the book)

Buy-back to basics?

FT

BP's decision to raise its dividends at the expense of share buy-backs marks a shift in the way the oil group rewards investors.

It also comes amid mounting scepticism about whether buy-backs are an effective way of boosting investment performance.

In a buy-back, a company purchases its own shares for cancellation, via a broker, reducing the shares in issue and so boosting earnings per share.

BP has been one of the UK's leading proponents, buying back almost $50bn (£25.5bn) of its own shares since 2000, equal to 16 per cent of its shares in issue.

However, it said yesterday it would "shift the balance" between buy-backs and dividends as it raised its quarterly dividend by 31 per cent amid growing confidence that higher oil prices are here to stay.

Although BP has pledged to continue buying back, some analysts argue the method is ineffective.

Keith Macquarie, an analyst at Collins Stewart, recently accused companies of "destroying billions in shareholder value through ill-judged buy-back decisions" after concluding

that the excess returns associated with buy-backs were "negligible".

Figures from Morgan Stanley have shown the companies that pursued buy-backs in 2006 collectively underperformed the wider market. "Just doing a buy-back is no guarantee of outperformance. In fact, on average, stocks that have conducted buy-backs have actually underperformed," said Graham Secker, strategist at Morgan Stanley.

The popularity of buy-backs has surged in recent years as improved cash-flows have allowed companies to return billions of pounds to shareholders. Mr Secker estimates UK companies bought back about £32bn of their own shares last year, while others put the figures as high as £45bn.

However, the flexibility that makes buy-backs popular with company boards is also its weakness. Mr Secker said: "Dividends are a much better indicator of corporate confidence. Companies only cut their dividends as a last resort, whereas buy-backs are more transitory."

BP is a case in point: the company is acutely aware of the embarrassment

Buy-backs

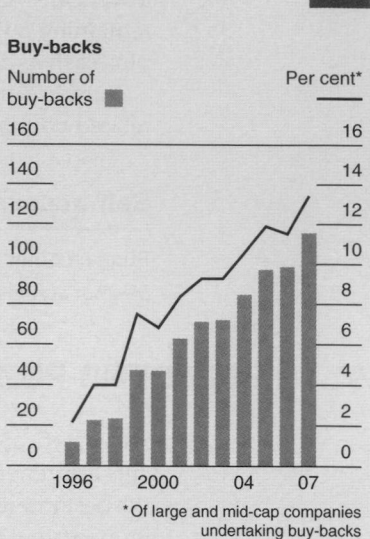

* Of large and mid-cap companies undertaking buy-backs

Source: Collins Stewart

caused by its humiliating decision in 1992 to cut its dividend in half.

Mr Macquarie applauded the BP move. He said the dividend rise "sends a positive message, improves certainty of returns and equality of treatment for all shareholders".

Source: Robert Orr, *Financial Times*, 6 February 2008.

17.10 CONCLUSIONS

What advice can be offered to practising financial managers? We suggest the following guidelines:

1 Remember the capacity of ill-advised dividend decisions to inflict damage. In view of the market's often savage reaction to dividend cuts, managers should operate a safe dividend cover, allowing sufficient payout flexibility should earnings decline. This suggests a commitment to a clientèle of shareholders who have come to expect a particular dividend policy from their company. This, in turn, suggests a long-term payout ratio sufficient to satisfy shareholder needs, but also generating sufficient internal funds to finance 'normal' investment requirements. Any 'abnormal' financing needs can be met by selling securities, rather than by dividend cuts. In the long term, it seems prudent to minimise reliance on external finance, but not to the extent of building up a 'cash mountain'. This may merely signal to the financial world that the company has run out of acceptable investment projects. If there is no worthwhile alternative use of capital, then dividends should be paid.

2 If an alteration in dividend policy is proposed, the firm should minimise the shock effect of an unanticipated dividend cut, e.g. by explaining in advance the firm's investment programme and its financing needs.

3 In an efficient capital market, the dividend announcement, 'good' or 'bad', will be immediately impounded into share price in an unbiased fashion. News of higher dividends will lead to a higher share price because it conveys the information that management believes there is a strong likelihood of higher future earnings and dividends. In view of this, many observers argue that the primary role of dividend policy should be to communicate information to a security market otherwise starved of hard financial data about future company prospects and intrinsic value, while pursuing a year-by-year distribution rate that does not conflict with the interests of the existing clientèle of shareholders.

Dividends are likely to be higher:

- The greater the shareholders' reliance on current income.
- The more difficult it is to generate home-made dividends.
- The greater the impact of imperfections, e.g. brokerage fees.
- The lower are income taxes compared to capital gains taxes.
- The lower the costs of a rights issue.
- The greater the ease of reinvestment by shareholders.
- The more often that past dividend increases have heralded subsequent earnings and dividend increases.

4 While it is impossible to be definitive in this area, the box lists the circumstances in which shareholders are likely to prefer dividends to retention and hence, when a dividend increase may raise share price.

5 Finally, the dividend decision is 'only' a financing decision, insofar as paying a dividend may necessitate alternative arrangements for financing investment projects. There can be few companies (at least, quoted ones) for which dividends have constrained really worthwhile investment. For most companies, it is often sensible to adopt a modest target payout ratio. It is not wise to pay large dividends and then have to incur the costs of raising equity from the very same shareholders.

SUMMARY

We have reviewed the competing arguments regarding the relative desirability of paying out dividends to shareholders and retaining funds to finance investment, resulting in capital gains.

Key points

- The market value of a company ultimately depends on its dividend-paying capacity.
- The irrelevance or residual theory of dividends argues that, under perfect capital market conditions, an alteration in the pattern of dividend payouts has no impact on company value, once the effect of the investment decision is removed.
- If retentions are used to finance worthwhile investment, company value increases, both for one-off retentions and for sustained retention to finance an ongoing investment programme.
- Failure to retain profits can damage shareholder interests if the company has better investment opportunities than shareholders, unless outside capital sources are available and utilised.

- Dividend irrelevance implies that shareholders are indifferent between dividends and capital gains, because the latter can be converted into income.

- In practice, various market imperfections, especially transactions costs and taxes, interfere with this conclusion, although it is difficult to be categoric about the net directional impact.

- Companies are generally unable to detect shareholder preferences, so they should follow a stable dividend policy, designed to suit a particular category, or 'clientèle', of shareholders.

- In practice, companies are reluctant to cut dividends for fear of adverse interpretation by the market of the information conveyed in the announcement.

- Similarly, companies are reluctant to increase dividends too sharply for fear of encouraging over-optimistic expectations about future performance.

- This 'information content' in dividend decisions provides a further argument for dividend stability.

- The advice regarding share re-purchases has to be 'use with care'.

Further reading

Firstly, a couple of health warnings. Fashions change in distribution policy, so important references that report the 'state-of-the-art' can quickly become out of date. Similarly, because of different tax regimes, the relevance of, for example, US studies in the UK context may be limited. However, some references are timeless. The best starting point is Gordon's first article (1959), followed by the original MM analysis, and Gordon's rejoinder (1963). Lintner (1956) was the first to reveal the smoothing strategy, and Brennan (1971) and Black (1976) present interesting discussions of the 'state of play' as the debate on dividend policy developed.

Bhattacharya (1979) gives a good analysis of the 'bird-in-the-hand fallacy'. Among the important studies that analyse signalling and the information content of corporate distribution announcements are Pettit (1972), Watts (1973), Vermaelen (1981), Healy and Palepu (1984), Asquith and Mullins (1986) and Bernatzi et al. (1997). Among more recent contributions that examine current trends are Fama and French (2001), an article and a case study respectively by Dobbs and Rehm (2006), and Wood (2006), published in Rutterford et al. (2006) give an up-to-date perspective on alterative distribution methods in the UK.

Benito and Young (2001) analyse reasons for dividend cuts and omissions. Pettit (2001) gives a highly readable assessment of the advantages and dangers of share buy-backs. Copeland et al. (2004) give a thorough analysis of the competing theories. Emery and Finnerty (1997) also contains two good chapters.

Appendix

HOME-MADE DIVIDENDS

Kirkstall plc is financed solely by equity. Its cash flow from existing operations is expected to be £24 million p.a. in perpetuity, all of which has hitherto been paid as dividend. Shareholders require a return of 12 per cent. Kirkstall has previously issued 50 million shares. The value of the dividend stream is:

$$V_o = \frac{£24 \text{ m}}{0.12} = £200 \text{ m, yielding share price of } \frac{£200 \text{ m}}{50 \text{ m}} = £4 \text{ (ex-div)}$$

Imagine you hold 1 per cent of Kirkstall's equity, worth £2 million. This yields an annual dividend income of £240,000, all of which you require to support your lavish lifestyle. Kirkstall proposes to pass the dividend payable in one year's time in order to invest in a project offering a single net cash flow after a further year of £40 million, when the previous 100 per cent payout policy will be resumed. Your new expected dividend flow is:

Year 1	Year 2	Year 3
0	£240,000 + (1% × £40 m) = £640,000	£240,000

The market value of Kirkstall rises to:

$$V_o = 0 = \frac{£64 \text{ m}}{(1.12)^2} + \left[\frac{£24 \text{ m}}{(0.12)} \times \frac{1}{(1.12)^2} \right]$$

$$= (£51.02 \text{ m} + £159.44 \text{ m}) = £210.46 \text{ m}$$

The increase in value reflects the NPV of the project. Share price rises from £4 to £4.21, thus raising the value of your holding from £2 million to £2.104 million. To support your living standards, you could either borrow on the strength of the higher expected dividend in Year 2 or sell part of your share stake in Year 1, when no dividend is proposed.

In Year 1, the value of your holding will be £2.356 million, and each share will be priced at £4.71. To provide sufficient capital to finance expenditure of £240,000, you need to sell £240,000/£4.71 = 50,955 shares, reducing your holding to (500,000 − 50,955) = 449,045 shares. In Year 2, Kirkstall will distribute a dividend of £64 million (£1.28 per share), of which your share will be £574,778. Out of this, you will require £240,000 for immediate consumption, leaving you better off by £334,778. This may be used to restore your previous shareholding, and thus your previous flow of dividends. The ex-dividend share price in Year 2 will settle back to £4, at which price, the repurchase of 50,955 shares will require an outlay of £203,820, leaving you a surplus of (£334,778 − £203,820) = £130,958. The net effect of your transactions is to yield an income flow of:

Year	1	2	3, etc.
Dividend paid by company	0	£574,778	£240,000
Market transaction	£240,000	(£203,820)	
Net income	£240,000	£370,958	£240,000

Overall, your shareholding remains the same, and you earn extra income in Year 2 of £130,958. This has a present value of £104,000, the very amount of your wealth increase when Kirkstall first announced details of the project, i.e.

1% × change in market value = 1% × (£210.41 m − £200 m) = £104,000

(Note that rounding errors account for minor deviations.)

QUESTIONS

 myfinancelab

Questions with an icon are also available for practice in myfinancelab with additional supporting resources.

Questions with a coloured number have solutions in Appendix B on page 749.

1 Tom plc follows a residual dividend policy. It has just announced earnings of £10 m and is to pay a dividend of 20p per share. Its nominal issued share capital is £5 m with par value of 50p. What value of capital expenditure is it undertaking? (You may ignore taxes.)

2 Dick plc faces the following marginal efficiency of investment profile. All projects are indivisible.

Project	IRR (%)	Required outlay (£m)
A	15	3
B	24	5
C	40	2
D	12	4
E	21	7

Dick's shareholders require a return of 20 per cent. Dick has just reported earnings of £9 m.
What amount of dividend would you recommend if Dick is unable to raise external finance?

3 Harry plc prides itself on its consistent dividend policy. Past data suggests that its target dividend payout ratio is 50 per cent of earnings. However, when earnings increase, Harry invariably raises its dividend only halfway towards the level that the target dividend payout ratio would indicate. The last dividend was £1 per share and Harry has just announced earnings for the recently ended financial year of £3 per share.
   What dividend per share would you expect the Board to recommend?

4 Tamas plc, which is ungeared, earned pre-tax accounting profits of £30 million in the financial year just ended. Replacement investment will match last year's depreciation of £2 million. Both are fully tax-allowable.
   Corporation tax is payable at 30 per cent. Tamas operates a 50 per cent dividend payout policy, and has previously issued 100 million shares, with par value of 25 pence each. Its shareholders require a return of 15 per cent p.a.
   Tamas holds £15 million cash balances.

Required
Determine the market price per ordinary share of Tamas, both cum-dividend and ex-dividend.
(N.B. Use the perpetuity formula to value Tamas' shares.)

5 Galahad plc, a quoted manufacturer of textiles, has followed a policy in recent years of paying out a steadily increasing dividend per share as shown below:

Year	EPS	Dividend (net)	Cover
1986	11.8p	5.0p	2.4
1987	12.5p	5.5p	2.3
1988	14.6p	6.0p	2.4
1989	13.5p	6.5p	2.1
1990	16.0p	7.3p	2.2

Galahad has only just made the 1990 dividend payment, so the shares are quoted ex-dividend. The main board, which is responsible for strategic planning decisions, is considering a major change in strategy whereby greater financing will be provided by internal funds, involving a cut in the 1991 dividend to 5p (net) per share. The investment projects thus funded will increase the growth rate of Galahad's earnings and dividends to 14 per cent. Some operating managers, however, feel that the new growth rate is unlikely to exceed 12 per cent. Galahad's shareholders seek an overall return of 16 per cent.

Required

(a) Calculate the market price per share for Galahad, prior to the change in policy, using the Dividend Growth Model.
(b) Assess the likely impact on Galahad's share price of the proposed policy change.
(c) Determine the break-even growth rate.
(d) Discuss the possible reaction of Galahad's shareholders and of the capital market in general to this proposed dividend cut in the light of Galahad's past dividend policy.

6 Laceby manufactures agricultural equipment and is currently all-equity financed. In previous years, it has paid out a steady 50 per cent of available earnings as dividend and used retentions to finance investment in new projects, which have returned 16 per cent on average.

Its Beta is 0.83, and the return on the market portfolio is expected to be 17 per cent in the future, offering a risk premium of 6 per cent.

Laceby has just made earnings of £8 million before tax and the dividend will be paid in a few weeks' time. Some managers argue in favour of retaining an extra £2 million this year in order to finance the development of a new Common Agricultural Policy (CAP) surplus crop disposal machine. This may offer the following returns under the listed possible scenarios:

	Probability	(£m p.a.) Cash flow (pre-tax)
WTO talks succeed	0.2	0.5
No reform of CAP	0.6	1.5
CAP extended to Eastern Europe	0.2	4.0

The project may be assumed to have an infinite life, and to attract an EU agricultural efficiency grant of £1 million. Corporation Tax is paid at 33 per cent (assume no tax delay).

Required

(a) What is the NPV of the proposed project?
(b) Value the equity of Laceby:
  (i) before undertaking the project,
  (ii) after announcing the acceptance of the project.
(c) Assuming you have found an increase in value in (b)(ii), explain what conditions would be required to support such a conclusion.

? 7 Pavlon plc has recently obtained a listing on the Stock Exchange. Ninety per cent of the company's shares were previously owned by members of one family, but, since the listing, approximately 60 per cent of the issued shares have been owned by other investors.

Pavlon's earnings and dividends for the five years prior to the listing are detailed below:

Years prior to listing	Profit after tax (£)	Dividend per share (pence)
5	1,800,000	3.60
4	2,400,000	4.80
3	3,850,000	6.16
2	4,100,000	6.56
1	4,450,000	7.12
Current year	5,500,000	(estimate)

The number of issued ordinary shares was increased by 25 per cent three years prior to the listing and by 50 per cent at the time of the listing. The company's authorised capital is currently £25,000,000 in 25p ordinary shares, of which 40,000,000 shares have been issued. The market value of the company's equity is £78,000,000.

The board of directors is discussing future dividend policy. An interim dividend of 3.16p per share was paid immediately prior to the listing and the finance director has suggested a final dividend of 2.34p per share.

The company's declared objective is to maximise shareholder wealth.

The company's profit after tax is generally expected to increase by 15 per cent p.a. for three years, and 8 per cent per year after that. Pavlon's cost of equity capital is estimated to be 12 per cent per year. Dividends may be assumed to grow at the same rate as profits.

Required

(a) Comment upon the nature of the company's dividend policy prior to the listing and discuss whether such a policy is likely to be suitable for a company listed on the Stock Exchange.

(b) Discuss whether the proposed final dividend of 2.34 pence is likely to be appropriate:

    (1) if the majority of shares are owned by wealthy private individuals;

    (2) if the majority of shares are owned by institutional investors.

(c) Using the Dividend Valuation Model give calculations to indicate whether Pavlon's shares are currently undervalued or overvalued.

(d) Briefly outline the weaknesses of the Dividend Valuation Model.

(ACCA)

8 Mondrian plc is a newly-formed company which aims to maximise the wealth of its shareholders. The board of directors of the company is currently trying to decide upon the most appropriate dividend policy to adopt for the company's shareholders. However, there is strong disagreement between three of the directors concerning the benefits of declaring cash dividends:

Director A argues that cash dividends would be welcome by investors, and that as high a dividend payout ratio as possible would reflect positively on the market value of the company's shares.

Director B argues that whether a cash dividend is paid or not is irrelevant in the context of shareholder wealth maximisation.

Director C takes an opposite view to Director A and argues that dividend payments should be avoided as they would lead to a decrease in shareholder wealth.

Required

(a) Discuss the arguments for and against the position taken by each of the three directors.

(b) Assuming the board of directors decides to pay a dividend to shareholders, what factors should be taken into account when determining the level of dividend payment?

(ACCA Certified Diploma June 1996)

Practical assignment

Obtain the Annual Report and Accounts of any firm of your choice. Look at the five-year record provided, and work out the dividend cover each year and the annual rate of dividend per share increase as compared to the annual increase in EPS. Is there any obvious pattern to the dividend decisions over this period? Does the firm have a clear dividend policy?

 myfinancelab

Now retake your diagnostic test for Chapter 17 to check your progress and update your study plan.

18

Capital structure and the required return

The dangers of debt

Debt-bloated Premier Foods will have to count every calorie as it fights its way back to fitness. The UK food manufacturer has avoided a painful rights issue and forced asset sales by halving the dividend and renegotiating terms with its creditor banks, as announced at its full-year results on Tuesday. Debt covenants have been reset to provide some breathing space while the group integrates the RHM and Campbell's businesses it swallowed in the past two years.

So is it in the clear? Investors' relief prompted a 7 per cent bounce in the share price. But even with that, the group's £830m market capitalisation is still dwarfed by net debt of £1.62bn (not including the £123m pension deficit). And the headroom granted by the bankers is not generous. Net debt at the end of 2008 cannot exceed 4.5 times earnings before interest, tax, depreciation and amortisation. So Premier needs to generate at least £360m of EBITDA this year,

compared with the £380m it would have made in 2007 with a full 12-month contribution from RHM.

That is pretty close for comfort, given that the consensus forecast for 2008 EBITDA has already slipped from almost £500m just nine months ago, to about £420m now. There are £45m of annual cost savings from integration targeted this year, but cutting costs in a tough trading environment is difficult. Furthermore, raw material prices are rising and Premier has the burden of rolling out a new IT platform. A quarter of the company's revenues comes from supplying own-label products, and Marks & Spencer has already stated that it intends to squeeze suppliers.

If it can scrape through this year, Premier is cheap – its shares are trading on just six times forward earnings with a 6 per cent dividend yield, even adjusting for the cut. But it requires a big leap of faith to believe that the Premier diet will achieve results.

Source: Financial Times, 4 March 2008, and www.ft.com.

Learning objectives

This chapter has the following aims:

- To explain some of the ways of measuring gearing.
- To enable you to understand more fully the advantages of debt capital.
- To explain the meaning of, and how to calculate, the WACC.
- To enable you to understand the likely limits on the use of debt, and the nature of 'financial distress' costs.
- To help you understand the issues involved in financing foreign operations.
- To enable you to understand the factors that a finance manager should consider when framing capital structure policy.

 Complete your diagnostic test for Chapter 18 now to create your personal study plan.

18.1 INTRODUCTION

Most financing decisions in practice reduce to a choice between debt and equity. The finance manager wishing to fund a new project, but reluctant to cut dividends or to make a rights issue, has to consider the borrowing option. In this chapter, we further examine the arguments for and against using debt to finance company activities and, in particular, consider the impact of gearing on the overall rate of return that the company must achieve.

The main advantages of debt capital centre on its relative cost. Debt capital is usually cheaper than equity because:

1 The pre-tax rate of interest is invariably lower than the return required by shareholders. This is due to the legal position of lenders who, have a prior claim on the distribution of the company's income and who, in a liquidation, precede ordinary shareholders in the queue for the settlement of claims. Debt is usually secured on the firm's assets, which can be sold to pay off lenders in the event of **default**, i.e. failure to pay interest and capital according to the pre-agreed schedule.

2 Debt interest can be set against profit for tax purposes, reducing the effective cost.

3 The administrative and issuing costs are normally lower, e.g. underwriters are not always required, although legal fees are usually involved.

The downside of debt, as illustrated by the case of Premier Foods, is that excessively high borrowing levels can lead to the risk of inability to meet debt interest payments in years of poor trading conditions. Shareholders are thus exposed to a second tier of risk above the inherent business risk of the trading activity. As a result, rational shareholders seek additional compensation for this extra exposure. In brief, debt is desirable because it is relatively cheap, but there may be limits to the prudent use of debt financing because, although posing relatively low risk to the lender, it can be highly risky for the borrower, i.e. the risk is two-sided.

In general, larger, well-established companies are likely to have a greater ability to borrow because they generate more reliable streams of income, enhancing their ability to service (make interest payments on) debt capital. Ironically, in practice, we often find that small developing companies that should not over-rely on debt capital are forced to do so through sheer inability to raise equity, while larger enterprises often operate with what appear to be very conservative gearing ratios compared to their borrowing capacities. Against this, we often encounter cases of over-geared enterprises that thought their borrowing levels were safe until they were caught out by adverse trading conditions.

So is there a 'correct' level of debt? Quite how much companies should borrow is another puzzle in the theory of business finance. There are cogent arguments for and against the extensive use of debt capital and academics have developed sophisticated models, which attempt to expose and analyse the key theoretical relationships.

For many years, it was thought advantageous to borrow so long as the company's capacity to service the debt was unquestioned. The result would be higher earnings per share and higher share value, provided the finance raised was invested sensibly. The dangers of excessive levels of borrowing would be forcibly articulated by the stock market by a downrating of the shares of a highly geared company. The so-called 'traditional' view of gearing centres on the concept of an **optimal capital structure** which maximised company value. However, while the critical gearing ratio is thought to depend on factors such as the steadiness of the company's cash flow and the saleability of its assets, it has proved to be like the Holy Grail, highly desirable, but illusory, and difficult to grasp. Some academics felt that a firmer theoretical underpinning was needed to facilitate the analysis of capital structure decisions and to offer more helpful guidelines to practising managers.

■ The Modigliani and Miller contribution

When Nobel laureates Modigliani and Miller (MM) published their seminal paper in 1958, finance academics began to examine in depth the relationship between borrowing and company value. MM's work on the pure theory of capital structure initially suggested that company value was unaffected by gearing. This conclusion prompted a furore of critical opposition, leading eventually to a coherent theory of capital structure, the current version of which looks remarkably like the traditional view.

Because this is a complex topic, we have organised the treatment of the impact of gearing into two chapters. This chapter is mainly devoted to the 'traditional' theory of capital structure and the issue of how much a company should borrow. In Chapter 19, we deepen the analysis by discussing the 'modern' theory. However, the present chapter gives a strong flavour of the main issues involved.

The two chapters together are designed to examine the following issues:

- How is gearing measured?
- Why do companies use debt capital?
- How is the cost of debt capital measured?
- What are the dangers of debt capital? How do shareholders react to 'high' levels of gearing?
- What do the competing theories of capital structure tell us about optimal financing decisions?
- How does taxation affect the analysis?
- What overall return should be achieved by a company using debt?
- What practical guidelines can we offer to financial managers?

18.2 MEASURES OF GEARING

There are two basic ways to express the indebtedness of a company. **Capital gearing** indicates the proportion of debt capital in the firm's overall capital structure. **Income gearing** indicates the extent to which the company's income is pre-empted by prior interest charges. Both are indicators of **financial gearing**.

■ Capital gearing: alternative measures

A widely-used measure of capital gearing is the ratio of all long-term liabilities (LTL), i.e. 'amounts falling due after more than one year', to shareholders' funds, as shown in the balance sheet. This purports to indicate how easily the firm can repay debts from selling assets, since shareholder funds measure net assets:

$$\text{Capital gearing} = \frac{\text{LTL}}{\text{Shareholders' funds}}$$

There are several drawbacks to this approach.

First, the market value of equity may be considerably higher than the book value, reflecting higher asset values, so this measure may seem unduly conservative. However, the notion of market value needs to be clarified. When a company is forced to sell assets hurriedly in order to repay debts, it is by no means certain that buyers can be found to pay 'acceptable' prices. The **break-up values** of assets are often lower than those expressed in the accounts, which assume that the enterprise is a going concern. However, using book values does at least have an element of prudence. In addition, the oscillating nature of market values may emphasise the case for conservatism, even for companies with 'safe' gearing ratios.

A second problem is the lack of an upper limit to the ratio, which hinders inter-company comparisons. This is easily remedied by expressing long-term liabilities as a fraction of all forms of long-term finance, thus setting the upper limit at 100 per cent. The gearing measure would become:

$$\frac{LTL}{LTL + Shareholders'\ funds}$$

A third problem is the treatment of provisions made out of previous years' income. Technically, provisions represent expected future liabilities. Companies provide for contingencies, such as claims under product guarantees, as a matter of prudence. Provisions thus result from a charge against profits and result in lower stated equity. However, some provisions turn out to be unduly pessimistic, and may be written back into profits, and hence equity, in later years. A good example is the provision made for deferred taxation. This is a highly prudent device to provide for possible tax liability if the firm were to sell its fixed assets.

Provisions could thus be treated as either equity or debt according to the degree of certainty of the anticipated contingency. If the liability is 'highly certain', it is reasonable to treat it as debt, but if the provision is the result of ultra-prudence, it may be treated as equity. For example, deferred taxation is a provision against the possibility of incurring a corporation tax charge if assets are sold above their written-down value for tax purposes. For most firms, this risk will diminish over time and the provision could safely be treated as equity. In practice, company accounts carry a mixture of provisions of varying degrees of certainty, and it is tempting to delete provisions from liabilities but not to include them in equity when expressing the gearing ratio. However, the nature of provisions should be questioned when the item appears substantial. Adjusted to exclude provisions, the capital gearing ratio becomes:

$$\frac{Long\text{-}term\ borrowings\ (LTB)}{Shareholders'\ funds} \quad or \quad \frac{LTB}{LTB + Shareholders'\ funds}$$

Arguably, any borrowing figure should take into account both long-term and short-term borrowing. Many companies depend heavily on short-term borrowing, especially bank overdrafts, and having to repay these debts quickly would place a significant burden on both the cash flow and liquidity of such companies. For this reason, some firms present their gearing ratios inclusive of such liabilities. For example, BP's measure of gearing focuses on 'finance debt', i.e. borrowing via the financial markets from financial institutions:

$$\frac{Finance\ debt}{Finance\ debt + Shareholders'\ funds}$$

There are two objections to this approach. First, since short-term borrowing can be volatile, the year-end figure in the balance sheet is not always a reliable guide to short-term debts. However, many companies effectively use their bank overdraft as a long-term form of finance. In other words, the actual bank overdraft figure may include a hard core element of long-term debt and a fluctuating component, although it is not easy to separate these two items from external examination of the accounts.

Second, it may be argued that any holdings of cash and highly liquid, marketable securities ('near cash' assets) should be offset against short-term debt, to yield a measure of 'net debt'. For example, BP expresses it thus:

$$Net\ debt = \frac{Finance\ debt - (cash\ and\ liquid\ resources)}{Equity}$$

In fact, financial commentators increasingly use this measure of gearing, which is shown in the annual reports of many companies, expressed either in absolute terms or in relation to equity.

Self-assessment activity 18.1

Determine the following gearing ratios using the information supplied.

(i) debt-to-equity ratio
(ii) debt-to-debt plus equity ratio
(iii) net debt

Equity = £100 m
Long-term debt = £50 m
Short-term debt = £20 m
Cash = £10 m

(Answer in Appendix A at the back of the book)

All the above measures are commonly used in the UK. In some other countries, a more direct measure of gearing (or leverage) is used. The UK ratios tend to focus on the relationship between debt and equity capital, which is reasonable since equity represents net assets. However, when necessary, debts are repaid by liquidating the assets to which the capital relates. Rather than this roundabout focus on capital, a direct focus on assets available to repay debts may give a clearer picture of ability to repay. For this purpose, many US commentators use an 'American gearing' ratio, such as:

$$\frac{\text{Total liabilities (including short-term liabilities)}}{\text{Total assets}}$$

■ Interest cover and income gearing

All the above measures purport to express the ability of the company to repay loans out of capital. However, they are only really helpful if book values and market values of the assets that would have to be sold to repay creditors approximate to each other. Yet, as we have noted, the market value of assets is volatile and difficult to assess. Moreover, capital gearing only indicates the security of creditors' funds in a crisis and may be an unduly cautious way of viewing debt exposure.

The trigger for a debt crisis is usually inability to make interest payments, and the 'front line' is therefore the size and reliability of the company's income in relation to its interest commitments. Although, in reality, cash flow is the more important consideration, the ability of a company to meet its interest obligations is usually measured by the ratio of *profit* before tax and interest, to interest charges, known as **interest cover**, or **'times interest earned'**:

times interest earned
The ratio of profit before interest and tax (operating profit) to annual interest charges, i.e. how many times over the firm could meet its interest bill

$$\text{Interest cover} = \frac{\text{Profit before interest and tax}}{\text{Interest charges}}$$

Strictly, the numerator should include any interest received and the denominator should become interest outgoings. This adjustment is rarely made in practice; net interest charges are commonly used as the denominator (as we do for D.S. Smith plc below).

income gearing
The proportion of profit before interest and tax (PBIT) absorbed by interest charges

The inverse of interest cover is called **income gearing**, indicating the proportion of pre-tax earnings committed to prior interest charges. If a company earns profit before interest and tax of £20 million, and incurs interest charges of £2 million, then its interest cover is (£20 m/£2 m) = 10 times, and 10 per cent of profit before interest and tax is pre-empted by interest charges.

Arguably, cash flow-to-interest is a better guide to financial security, given that profits are expressed on the accruals basis, i.e., profit is recognised even though cash may not have been received yet for sales. Hence, the formula below is sometimes used:

$$\text{Cash flow cover} = \frac{\text{Operating cash flow}}{\text{Net interest payable}}$$

Self-assessment activity 18.2

Distinguish between capital gearing and income gearing. How is each measured?

(Answer in Appendix A at the back of the book)

■ Example: D.S. Smith plc's borrowings

The figures in Table 18.1 are taken from the Annual Report for D.S. Smith plc (**www .dssmith.uk.com**) for the year ended 30 April 2007. With these data, we can calculate various gearing indicators:

$$\frac{\text{LT borrowing}}{\text{Shareholder's funds}} = \frac{£231\,m}{£567\,m} = 41\%$$

$$\frac{\text{LT borrowing}}{\text{All LT funds}} = \frac{£231\,m}{£231\,m + £567\,m} = \frac{£231\,m}{£798\,m} = 29\%$$

$$\frac{\text{All borrowing}}{\text{Shareholder's funds}} = \frac{£231\,m + £12\,m}{£567\,m} = \frac{£243\,m}{£567\,m} = 43\%$$

$$\text{Net debt} = (£231\,m + £12\,m - £92\,m) = £151\,m$$
$$(27\% \text{ of equity})$$

$$\text{Interest cover} = \frac{£78\,m}{£15\,m} = 5.2 \text{ times}$$

$$\text{Income gearing} = \frac{1}{5.2} = 19\%$$

$$\frac{\text{Cash generated from operations}}{\text{Net interest payable}} = \frac{£128\,m}{£15\,m} = 8.5 \text{ times}$$

$$\frac{\text{Total liabilities}}{\text{Total assets}} = \frac{£803\,m}{£1,370\,m} = 59\%$$

Table 18.1 Financial data for D.S. Smith plc

	£m*
Shareholders' funds	567
Cash and short-term investments	92
Long-term borrowing (LTD)*	231
Short-term borrowing	12
Group profit before interest and tax	78
Net financing costs	15
Total assets	1,370
Total liabilities*	803
Cash generated from operations	128

*Provisions of £9 m are treated as part of liabilities, but not as part of long-term borrowing.

Self-assessment activity 18.3

Obtain the accounts of a firm of your choice and conduct a similar exercise to the D.S. Smith plc calculations.

18.3 OPERATING AND FINANCIAL GEARING

A major reason for using debt is to enhance or 'gear up' shareholder earnings. When a company is financially geared, variations in the level of earnings due to changes in trading conditions generate a more than proportional variation in earnings attributable to shareholders if the interest charges are fixed. This effect is very similar to that exerted by **operating gearing**. We will now examine these two gearing phenomena and illustrate them numerically.

Most businesses operate with a combination of variable and fixed factors of production, giving rise to variable and fixed costs respectively. The particular combination is largely dictated by the nature of the activity and the technology involved. Operating gearing refers to the relative importance of fixed costs in the firm's cost structure, costs that have to be met regardless of output and revenue levels. In general, the higher the proportion of fixed to variable costs, the higher the firm's break-even volume of output. As sales rise above the break-even point, there will be a more than proportional upward effect on profits before interest and tax, and on shareholder earnings.

Firms with high operating gearing, mainly capital-intensive ones, are especially prone to fluctuations in the business cycle. In the downswing, as their sales volumes decrease, their earnings before interest and tax decline by a more than proportional amount; and conversely in the upswing. Hence, such companies are regarded as relatively risky. If such companies borrow, they add a second tier of fixed charges in the form of interest payments, thus increasing overall risk – the higher the interest charges, the greater the risk of inability to pay. Consequently, the risk premium required by investors in such companies is relatively high. It follows that companies that exhibit high operating gearing should use debt finance sparingly.

■ Operating and financial gearing: Burley plc

Burley plc produces and sells briefcases. It has issued 4 million 50p shares and has £2 million loan finance. Its gearing ratio, measured by the ratio of debt to equity at book values, is one-to-one (i.e. (4 m × 50p)/£2 m). Last year, Burley sold 60,000 units to large retailers at £30 per unit.

Its profit and loss account is:

		£000	£000
Sales			1,800
Less:	Variable costs (VC)	(720)	
	Fixed operating costs (FC)	(480)	(1,200)
	Profit before interest and tax (PBIT)		600
Less:	Interest payable @ 10%		(200)
	Profit before taxation (PBT)		400
Less:	Corporation tax @ 30%		(120)
	Profit after tax (PAT)		280

The earnings per share (EPS) are:

$$\frac{\text{Profit after tax}}{\text{No of ordinary shares}} = \frac{\pounds280,000}{4 \text{ m}} = 7.0\text{p}$$

contribution
Revenues (turnover) less variable costs, i.e contribution to meeting fixed costs

Notice that the P & L hides the distinction between gross and net profit. The gross profit is the sales less variable costs, i.e. (£1,800,000 − £720,000) = £1,080,000, also called **contribution** because the fixed costs are paid from this amount. Any surplus over FC is the net or operating profit (PBIT). Operating profit is thus £600,000.

Now let us consider the break-even volume, initially ignoring the debt interest obligation. Recall that breaking even means just covering fixed operating costs and variable costs. In Burley's case, this requires an output sufficient to generate a gross margin high enough to cover the fixed operating costs of £480,000. The unit variable cost is:

$$(£720,000/60,000) = £12$$

Were it financed entirely by equity, Burley's break-even output would be found by dividing fixed cost by the gross profit margin of (£30 − £12) = £18:

$$(£480,000/£18) = 26,667 \text{ units}$$

Allowing for the interest commitments of £200,000, Burley has to cover total fixed charges of (£480,000 + £200,000) = £680,000. This requires the higher output of (£680,000/£18) = 37,778 units to break even. Hence, using debt finance raises the break-even volume of production because fixed obligations are higher.

We can use this example to distinguish between operating and financial gearing. **Operating gearing** can be expressed in a variety of ways. Most simply, it is the proportion of total production cost accounted for by fixed costs: (£480,000/£1,200,000) = 40 per cent. Allowing for interest payments, Burley needs to generate a gross margin or contribution of (£480,000 + £200,000) = £680,000 to cover total fixed charges. At present, it is doing this fairly comfortably, since in percentage terms, fixed charges account for (£480,000 + £200,000) ÷ £1,080,000 = 63 per cent of the contribution. Looking at the importance of financial gearing, out of its profit before interest and tax of £600,000, a third (£200,000) is required to cover interest payments, i.e. the interest cover is 3 times.

A more sophisticated way of viewing the impact of fixed charges is to calculate leverage ratios. **Operating leverage** is the number of times the contribution covers the profit before interest and tax (PBIT), i.e. a multiple of:

$$\frac{\text{Contribution}}{\text{PBIT}} = \frac{(\text{Sales} - \text{VC})}{\text{PBIT}} = \frac{£1,080,000}{£1,600,000} = 1.8 \text{ times}$$

This indicates the leeway between contribution and the PBIT, and hence, the extent to which the fixed costs can increase without forcing the company into an operating loss. More significantly, the multiplier of 1.8 signifies the relationship between a given increase in sales and the resulting effect on PBIT. As we show below, a 10 per cent increase in sales will result in an increase in PBIT of 18 per cent.

Similarly, **financial leverage** is the number of times the PBIT covers the profit before tax (PBT), i.e. a multiple of:

$$\frac{\text{PBIT}}{\text{PBT}} = \frac{£600,000}{£400,000} = 1.5 \text{ times}$$

The difference between PBIT and PBT is the interest charge, so this multiple indicates the extent to which interest charges can rise without forcing the company into pre-tax loss. More significantly, the multiplier of 1.5 magnifies the effect of operating leverage – the effect of a sales increase on PBT is greater in a financially geared firm than in one with no borrowing. Taking the two multipliers together, we obtain a combined leverage effect. In this case, a sales increase of 10 per cent will result in an increase in PBT of (1.8 × 1.5) = 2.7 times as great, i.e. 27 per cent. For a given tax rate, here 30 per cent, the profit after tax and, hence, the EPS, will also rise by the same proportion.

operating gearing
The relationship between fixed and variable cost in a firm's cost structure

operating leverage
The ratio of contribution to profit before interest and tax

financial leverage
The ratio of profit before interest and tax (PBIT) to profit before tax (PBT)

To clarify these relationships, it is helpful to demonstrate the impact of Burley experiencing a sales increase of 10 per cent. Assuming no change in unit variable costs, the profit and loss account becomes:

		£000	£000	
Sales			1,980	(10% increase)
Less:	Variable costs (VC)	(792)		
	Fixed costs (FC)	(480)	(1,272)	
	Profit before interest and tax (PBIT)		708	(18% increase)
Less:	Interest payable @ 10%		(200)	
	Profit before taxation (PBT)		508	(27% increase)
Less:	Corporation tax @ 30%		(120)	
	Profit after tax (PAT)		356	(27% increase)

The new EPS is:

PAT/No. of shares = (£356,000/4 m) = 8.9p (27% increase)

The increase in EPS of 27 per cent (rounded) is a far greater proportion than the sales increase, illustrating the operation of the combined gearing multiplier. It follows that the higher the proportion of fixed costs in overall costs, and the greater the commitment to interest charges, the greater will be the combined gearing effect. This may suggest that using fixed factors of production and using debt capital are both desirable things. However, as the following example demonstrates, financial gearing is double-edged. It is beneficial in favourable economic conditions, but because the gearing effect also works in reverse, it can spell trouble in adverse trading conditions.

Self-assessment activity 18.4

Show the effect on the combined gearing multiplier in the Burley example if fixed costs are £530,000 and interest charges are £240,000.

(Answer in Appendix A at the back of the book)

18.4 FINANCIAL GEARING AND RISK: LINDLEY PLC

Lindley plc retains no profit and its shareholders require a 20 per cent return. Issued share capital is £100 million, with par value of £1. Lindley's operating profit can vary as shown in Table 18.2, according to trading conditions characterised as bad, indifferent and good. These are denoted by scenarios A, B and C, which have probabilities of 0.25, 0.50 and 0.25, respectively.

After all costs, but before deducting debt interest, earnings are £5 million, £20 million and £35 million under scenarios A, B and C, respectively. This measure of earnings is termed net operating income (NOI). (For simplicity, taxation is ignored.) Let us examine shareholder returns with gearing ratios of zero, 25 per cent and 50 per cent, measured by long-term debt (interest rate 10 per cent) to total long-term finance that is held constant at £100 million.

Notice that for a given increase in income, shareholder earnings rise by a greater proportion: for example, with gearing of 25 per cent, if NOI rises by 300 per cent from £5 million to £20 million, shareholder earnings increase by 600 per cent from £2.5 million to £17.5 million. It is easy to see why adding debt to the capital structure is called gearing – the change in earnings is magnified by a factor of 2.0 in shareholders' favour. Unfortunately, this effect also applies in a downward direction – a given proportionate fall in earnings generates a more pronounced *decrease* in shareholder earnings. Indeed, with 50 per cent gearing, under scenario A, shareholder earnings are entirely wiped

Table 18.2 How gearing affects shareholder returns in Lindley plc

	Trading conditions		
	Scenario A ($p = 0.25$)	Scenario B ($p = 0.50$)	Scenario C ($p = 0.25$)
Profit before interest (PBIT)* (Net operating income)	£5 m	£20 m	£35 m
Zero gearing (100 m equity, £0 m debt)			
Debt interest at 10%	–	–	–
Shareholder earnings	£5 m	£20 m	£35 m
Return on equity (ROE)	5%	20%	35%
25% gearing $\left(\text{Debt/Equity} = \frac{1}{3}\right)$ (£75 m equity, £25 m debt, interest 10%)			
Debt interest at 10%	£2.5 m	£2.5 m	£2.5 m
Shareholder earnings	£2.5 m	£17.5 m	£32.5 m
Return on equity (ROE)	3.3%	23.3%	43.3%
50% gearing $\left(\text{Debt/Equity} = \frac{1}{1}\right)$ (£50 m equity, £50 m debt, interest 10%)			
Debt interest at 10%	£5 m	£5 m	£5 m
Shareholder earnings	0	£15 m	£30 m
Return on equity (ROE)	0	30%	60%

*Taxes are ignored.

out by prior interest charges. The return on equity would be negative at any higher gearing level under this scenario.

Negative returns are not necessarily fatal – companies often survive losses in especially poor trading years – but the likelihood of survival when continued trading losses combine with high fixed interest charges is lowered if the company cannot pay interest charges. In these cases, the enterprise is technically insolvent, although creditors may agree to restructure the company's capital, e.g. by converting debt into equity or preference shares. There is, however, an effective upper limit of gearing for Lindley. Beyond 50 per cent gearing, it may be unable to meet interest charges out of earnings. For practical purposes, the lower limit of earnings will dictate maximum borrowing capacity, although, in reality, this lower earnings limit is highly uncertain. This is why it is usually argued that the more reliable the company's expected cash flow stream, the greater its borrowing capacity.

Our Lindley example demonstrates that, under debt financing, although shareholders may achieve enhanced returns in good years, they stand to receive much lower returns in bad years. In other words, the residual stream of shareholder earnings exhibits greater variability. This can be examined by computing the expected value and the range, or dispersion, of the return on equity (ROE) with each of the three gearing levels.

Self-assessment activity 18.5

Calculate the expected value of Lindley's ROE under each scenario.

(Answer in Appendix A at the back of the book)

Table 18.3 shows that, although the expected value of the return on equity is greater at higher levels of gearing, the dispersion, or range, of possible returns is also wider,

Table 18.3 How gearing affects the risk of ordinary shares

Gearing (%)	Expected ROE (%)	Total dispersion (%)	Due to: Business risk (%)	Financial risk (%)
0	20	30	30	0
25	23.3	40	30	10
50	30	60	30	30

which might concern risk-averting shareholders. Notice also that we can decompose the overall risk incurred by shareholders into its underlying business and financial elements. Business risk refers to the likely variability in returns for an equivalent all-equity financed company, i.e. the dispersion of returns is due to underlying business-related factors. Financial risk is the additional dispersion in net returns to shareholders due to the need to meet interest charges whatever the trading conditions. At every gearing ratio, the range of returns due to business risk is unchanged – nothing has happened to its product range, its customer base or any other aspect of its trading activities. Lindley would simply share out the proceeds of its operations in different ways at different gearing ratios.

It is also helpful to show the effect on ROE graphically. Figure 18.1 shows the data for Lindley's ROE for the three different capital structures. Clearly, the higher the debt-to-equity ratio, the greater the ROE for any level of profit before interest. Figure 18.1 also shows how the break-even value of profit before interest increases as gearing rises. As gearing increases from zero to 1:3 and to 1:1, the break-even earnings increase from 0 to 0X and to 0Y, corresponding to the three interest payment levels of zero, £2.5 million and £5 million respectively (Scenario A in Table 18.2 and Figure 18.1). Notice finally that earnings of £10 million would generate the same ROE under all three capital structures.

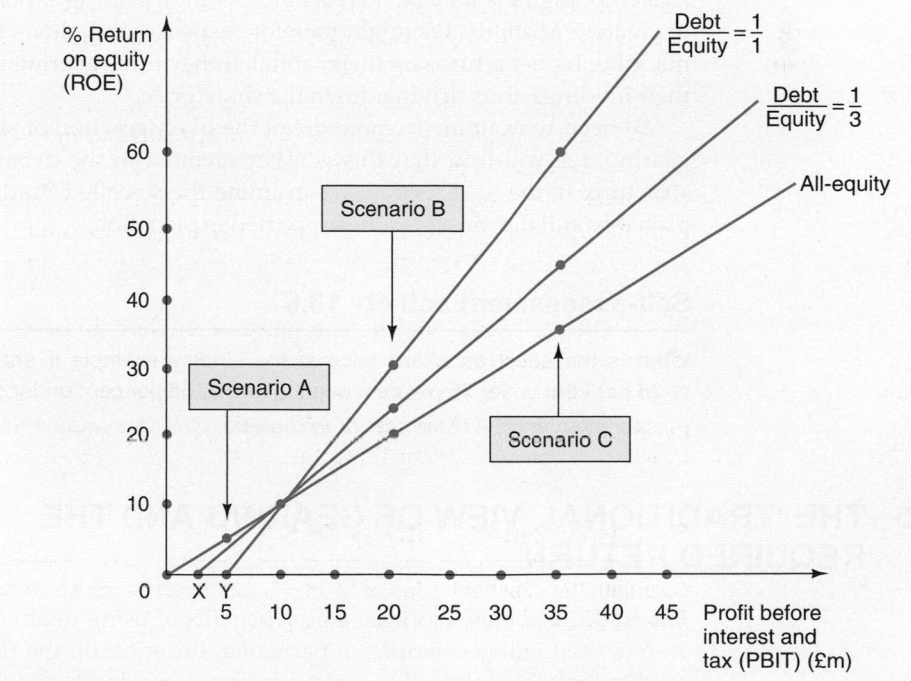

Figure 18.1 How gearing affects the ROE

This discussion of the impact of gearing is incomplete in one important respect. The analysis has been based on book values, despite earlier remarks that gearing ratios may often be better measured in terms of market values. We have yet to consider the effect of gearing on the value of the firm – does gearing actually make shareholders better off?

To examine the effect on share price, we need to focus on the expected earnings figure and recall that the value of a share can be found by discounting its stream of earnings, in the simplest case, as a perpetuity. (No distinction is needed between earnings and dividends, as Lindley makes no retentions.) The expected values of shareholder earnings for each of the three gearing ratios are shown in Table 18.4. Recalling that Lindley's shares have a nominal value of £1, we can specify the number of shares corresponding to each gearing ratio and hence the expected value of the EPS. Share prices are found by applying the valuation formula, discounting the perpetual EPS at the 20 per cent return required by Lindley's shareholders.

Table 18.4 How gearing can affect share price

Gearing %	Number of shares	Expected value of shareholder earnings	EPS	Share price*
0	100 m	£20 m	20.0p	20.0p/0.2 = £1.00
25	75 m	£17.5 m	23.3p	23.3p/0.2 = £1.17
50	50 m	£15 m	30.0p	30.0p/0.2 = £1.50

*Share price is found by discounting the perpetual and constant EPS at 20 per cent

It appears that, by using debt capital, financial managers can achieve significant increases in shareholder wealth. However, we ought to be suspicious of this effect. Why should shareholders' wealth increase when there have been no changes in trading activity or in expected aggregate income?

The analysis assumes that shareholders are prepared to accept a return of 20 per cent at all permissible gearing levels – they seem to be unconcerned by financial risk. Even though there may be no risk of insolvency, gearing exposes shareholder earnings to greater variability. We might therefore expect shareholders to react to gearing by demanding higher returns on their capital. If they think gearing is too risky, they may sell their holdings, thus driving down the share price.

We need to examine in more detail the likely reaction of shareholders to increased gearing; we will find that this is a key element in the debate about optimal capital structure. In the next section, we examine the so-called 'traditional' view of gearing, probably still the most widely-supported explanation.

Self-assessment activity 18.6

What is the effect on share price in the Lindley example if shareholders require returns of 25 per cent under 25 per cent gearing and of 35 per cent under 50 per cent gearing?

(Answer in Appendix A at the back of the book)

18.5 THE 'TRADITIONAL' VIEW OF GEARING AND THE REQUIRED RETURN

The traditional view emphasises the benefits of using relatively cheap debt capital as seen in the Lindley example; in particular, the effect on the rate of return required on investment. To analyse this approach in greater depth, we first need to make some definitions.

■ Value of an ungeared company

For an ungeared company, market value is found by discounting (or capitalising) its stream of annual earnings, E, at the rate of return required by shareholders, k_e.

The value of an ungeared company, V_u, is simply the value of the ordinary shares, V_S. For a constant and perpetual stream of annual earnings, E:

$$V_u = V_S = \frac{E}{k_e} \qquad \text{so that } k_e = \frac{E}{V_S}$$

Much of the argument about capital structure centres on what happens to the discount rate (or capitalisation rate) as gearing increases. If the analysis is conducted in terms of *substituting* debt for equity, i.e. keeping size of firm constant as we did in the Lindley example, the effect of gearing can be examined while holding E constant. In this case, gearing simply rearranges the share-out of E among the company's stakeholders. We denote the book value of borrowings as B and the interest rate as i, thus involving a prior interest charge of $(i \times B)$. Gearing splits the earnings stream of the company into two components, the prior interest charge and the portion attributable to shareholders, the net income (NI) of $(E - iB)$. The overall value of the company is the value of the shares (V_S) plus the value of the debt, each capitalised at its respective rate of discount. For debt, where there is no discrepancy (as in this example) between book value (B) and market value (V_B), the capitalisation rate is simply the nominal interest rate. The overall value of the geared company, V_g, is the combined value of its shares and its debt:

$$V_g = V_S + V_B = \frac{(E - iB)}{k_e} + \frac{iB}{i} = \frac{(E - iB)}{k_e} + V_B$$

The *overall* capitalisation rate (denoted by k_0) for a company using a mixed capital structure is a weighted average, whose weights reflect the relative importance of each type of finance in the capital structure, i.e. V_S/V_g and V_B/V_g for equity and debt, respectively:

$$k_0 = \left(k_e \times \frac{V_S}{V_g} \right) + \left(i \times \frac{V_B}{V_g} \right)$$

Bearing in mind that $(k_e V_S)$ and iB (or iV_B, when $V_B = B$) represent the returns to shareholders and lenders respectively, i.e. their respective shares of corporate earnings, E, the weighted average expression simplifies to:

$$k_0 = \frac{k_e V_S + iB}{V_g} = \frac{E}{V_g}$$

For both ungeared and geared firms alike, k_0 is found by dividing the total required earnings by the value of the whole firm. k_0 is also known as the **weighted average cost of capital (WACC)**, since it expresses the overall return required to satisfy the demands of both groups of stakeholders. The WACC may be interpreted as an average discount rate applied by the market to the company's future operating cash flows to derive the capitalised value of this stream, i.e. the value of the whole company.

It looks as if a company could lower the WACC by adding 'cheap' debt to an equity base. For instance, in the Lindley example, while the required return for the all-equity case is 20 per cent, i.e. the cost of equity, with gearing at 25 per cent, the WACC, using book value weights, becomes:

$$k_0 = (0.75 \times 20\%) + (0.25 \times 10\%) = 17.5\%$$

Apparently, gearing can lower the overall cost of capital if both k_e and i remain constant. The effect of this is highly significant. In the traditional view of gearing, shareholders are deemed unlikely to respond adversely (if at all) to minor increases in gearing so long as the prospect of default looks remote. If the cost of equity remains static, substitution of debt for equity will lower the overall cost of capital applied by the market in valuing the company's stream of earnings. This is shown in Figure 18.2 by the decline in the k_0 schedule between A and B. Corresponding to this fall in k_0 is an increase in the value of the whole geared company, V_g, in relation to that of an equivalent ungeared company, V_u.

This benign impact of gearing has already been shown in the Lindley example. Looking back to Tables 18.3 and 18.4, consider the switch from 0 to 25 per cent gearing. Assuming shareholders continue to seek a return of 20 per cent, the EPS discounted to infinity yields a share price of £1.17. The market value of equity becomes (£1.17 × 75 m) = £88 m, and the overall company value is:

$$V_g = V_S + V_B = (£88 \text{ m} + £25 \text{ m}) = £113 \text{ m}$$

Gearing up to 25 per cent raises market value by £13 million above book value, thus demonstrating the benefits of gearing to shareholders. The market value of the whole company rises because the value per unit of the residual equity increases due to the increase in EPS. Without gearing, each share would sell at £1.

However, sooner or later, shareholders will become concerned by the greater financial risk to which their earnings are exposed and begin to seek higher returns. In addition, providers of additional debt are likely to raise their requirements as they perceive the probability of default increasing. The k_e schedule will probably turn upwards before any upturn in i, given the legally-preferred position of debt-holders, although the phasing of these movements is not clear in this model. Whatever the sequence of the upward revisions in required returns, the WACC profile will eventually be forced to rise, and the value of the company will fall. The model involves a clear optimal debt/equity mix, where company value is maximised and the WACC is minimised. This is gearing ratio 0X in Figure 18.2.

To financial managers, a major disappointment of this approach is its failure to pinpoint a specific optimal gearing ratio for all firms in all circumstances. The optimal ratio is likely to depend on the nature of the industry (e.g. whether the activity generates

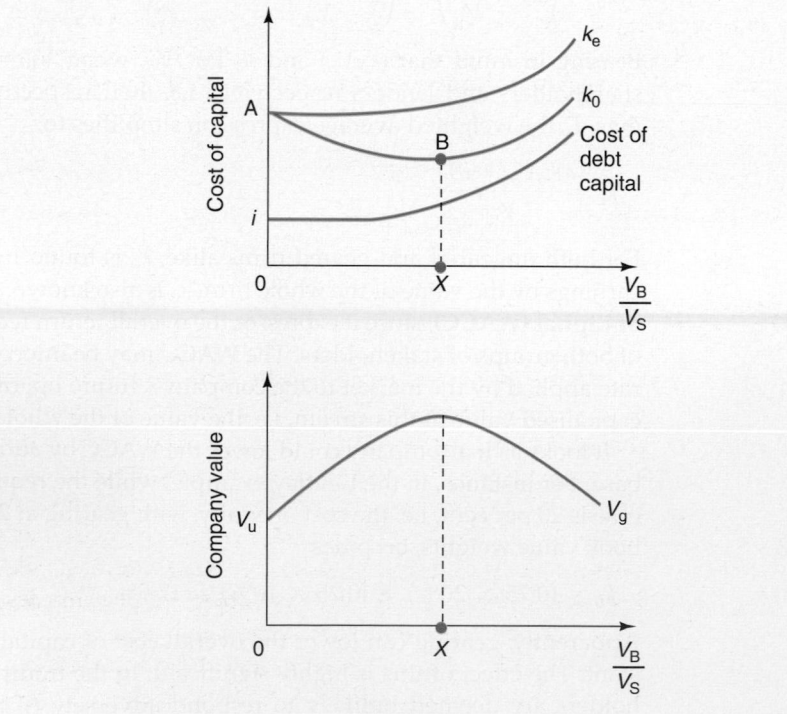

Figure 18.2 The 'traditional' view of capital structure

strong cash flows), the general marketability of the company's assets, expectations about the prospects for the industry, and the general level of interest rates. Clearly, many of these factors vary over time as well as between industries. However, a few pointers are possible.

For example, a supermarket chain, characterised by strong cash flow, can sustain a higher level of gearing than a heavy engineering enterprise, where the working capital cycle is lengthy. Similarly, an airline, for whose assets there is a ready and active second-hand market, might withstand a higher gearing level (especially as flights are often paid for well in advance) than a steel company with both a high level of operating gearing and assets that are highly specific and difficult to sell.

Self-assessment activity 18.7

What is this firm's WACC? (ignore tax)

■ debt-to-equity ratio = 40%
■ cost of equity = 18%
■ cost of debt = 8%

(Answer in Appendix A at the back of the book)

Calculating the WACC at Tomkins plc

Tomkins plc, the UK-based engineering and motor components group, calculates its WACC as follows:

'The weighted average cost of capital is the weighted average by value of the after-tax costs of each of the elements of the Group's capital structure.

The cost of equity is calculated using the capital asset pricing model, with the risk-free rate based on a 10-year sterling government bond rate, an equity beta of 1.0 and an equity market risk premium of 4%. The value of equity is the current market value of the ordinary shares of Tomkins.

The cost of debt comprises both the cost of the perpetual preference shares and the after-tax cost of the group's net debt, taking into account a normalised tax rate of 30%. The value of net debt used is the actual current sterling value of the Group's net debt. The value of the preference shares is the current sterling equivalent nominal value of the preference shares.'

Notes

1 The cost of preference shares is simply the dividend divided by the market price if the stock is traded, or the nominal value if not traded. Note that there is no tax relief on preference shares. This calculation is equivalent to calculating the pre-tax cost of a perpetual government bond.

2 Tomkins' preference shares are denominated in US dollars.)

Source: Tomkins plc Annual Report 2006.

18.6 THE COST OF DEBT

Our analysis has so far assumed that market and book values of debt coincide. This is by no means always the case. Corporate bond values behave in a similar way to the market prices of government stock (gilt-edged securities). When general market interest rates increase, the returns on previously-issued bonds may look unattractive compared with the returns available on newly-issued ones. As a result, bond dealers mark down the value of existing stocks until they offer the same yield as investors can obtain by purchasing new issues. In other words, equilibrium in the bond market is achieved when all stocks that are subject to the same degree of risk and that have the same period to redemption offer the same yield.

The simplest case is perpetual (irredeemable) stock such as the UK government's 3.5 per cent War Loan. These were issued, never to be repaid, to support the British war effort between 1939 and 1945. They offer the holder a return of 3.5 per cent (the nominal rate of interest, or '**coupon rate**') on the par value of the stock, i.e. £3.50 per £100 of stock. With higher market rates, say, 7 per cent, War Loan would look unattractive, and its value would fall, e.g. a £100 unit would have to sell at £50 to generate a yield of 7 per cent. An inverse relationship applies between fixed-interest bond prices and interest rates:

$$\text{The market value of an irredeemable bond} = \left(\text{nominal value} \times \frac{\text{coupon rate}}{\text{market rate}} \right)$$

Self-assessment activity 18.8

What is the market value of 3.5 per cent War Loan when market rates are:

(i) 11%?
(ii) 3%?

(Answer in Appendix A at the back of the book)

Perpetual corporate bonds are very rare indeed. However, the longer the term of the bond, the more it resembles a perpetuity. For example, in February 2007 Tesco plc, the supermarket giant, issued a 50-year bond, maturing in 2057, to finance its overseas expansion plans. This is effectively a perpetuity (for proof, examine the size of the discount factors in the 0–50-year range for a plausible discount rate, say 15 per cent – they are very small indeed). The Tesco issue raised £500 million and was four times oversubscribed. It offered a yield of 5.23 per cent, compared to 4.06 per cent on an equivalent government bond, a quite modest risk premium of 0.83 per cent.

In practice, the calculation is more complex when we consider the far more common case of bonds with limited lifetimes until maturity. In assessing the value of such bonds, the market value will also include the eventual capital repayment.

For example, if the market rate is 10 per cent, a ten-year bond with a coupon rate of 10 per cent, denominated in £100 units, would have the following (present) value:

PV =	discounted interest payments over 10 years	+	discounted capital repayment in year 10
=	$(10\% \times £100) \cdot (\text{PVIFA}_{(10,10)})$	+	$(£100) \cdot (\text{PVIF}_{(10,10)})$
=	$(£10 \times 6.1446)$	+	$(£100 \times 0.3855)$
=	£61.45	+	£38.55 = £100

The market value coincides with the par value because the coupon rate equals the going market rate. If, however, the market rate were to rise to 12 per cent, all future payments to the bond-holder, both capital and interest, would be more heavily discounted, i.e. at 12 per cent, reducing the market value to £88.70.

Values of corporate bonds behave in essentially the same way, although, since companies are more risky than governments, they have to offer investors a rather higher rate of interest. This allows us to identify the cost of corporate debt capital. A company can infer the appropriate rate of interest at which it could raise debt by looking at the market value of its own existing debt or that of a similar company. For example, if the market value of each £100 unit of debenture stock is £95 and has to be repaid in full in two years' time, the cost of debt can be found by solving a simple IRR expression. Someone who decided to purchase the stock in the market would anticipate two interest payments of £10 and a capital repayment of £100. The return expected is denoted by k_d in the following IRR expression:

$$£95 = \frac{£10}{(1 + k_d)} + \frac{(£10 + £100)}{(1 + k_d)^2}$$

The solution for k_d is 13 per cent. The market signals this rate as the cost of raising further debt.

There is another adjustment to make for tax relief on debt interest payments. To allow for tax, we look at the cost of debt from the company's perspective, since it is the company that enjoys the tax break. With a corporation tax rate of 30 per cent, each £10 interest payment will generate a tax saving of £3 for the company, reducing the effective interest cost to £10(1 − 30%) = £7.0. The IRR equation becomes:

$$£95 = \frac{£7.0}{(1 + k_d)} + \frac{(£7.0 + £100)}{(1 + k_d)^2}$$

Self-assessment activity 18.9

Verify that the solution rate in the above IRR equation is 9.9 per cent.

The tax benefits from using debt can be substantial. Take the case of a 10-year bond, issued and redeemable at a price of £100, with a coupon rate of 10 per cent. The value of the tax savings on interest payments, or the '**tax shield**', is:

$$\text{Tax shield} = \text{interest charge} \times (\text{tax rate}) \times \text{PVIFA}_{(10,10)}$$
$$= (10\% \times £100) \times (30\%) \times 6.1446$$
$$= £18.43$$

In practice, this value is reduced by any delay in tax payments, which in turn delays the receipt of tax benefits. It is also assumed that the company always has sufficient taxable profits to benefit from the tax relief on interest payments.

Self-assessment activity 18.10

What is the value of the tax shield for £10 million debt, coupon rate 6 per cent, tax rate 30 per cent:

(a) if the debt is perpetual?
(b) if it is to be repaid in 20 years?

(Answer in Appendix A at the back of the book)

Cost of capital in Europe: the importance of the tax shield

Lee et al (2006) estimated the cost of equity capital and weighted average cost of capital (WACC) for a large sample of companies from 17 European countries between 1995 and 2005. They found that the equity premium (i.e. cost of equity minus risk-free rate of interest) was systematically lower in the UK (4.64 per cent) than in the rest of Europe (5.84 per cent). This is seen as the result of higher disclosure requirements and greater use of equity which leads to a lower cost of equity. However, firms in other European countries typically have much higher borrowings, offering a greater tax shield. This resulted in the WACC premium (WACC minus risk-free rate of return) for UK companies being no different to other European countries. The study also indicated that costs of capital in the IT, steel and materials, and mining sectors were consistently high while utilities and beverage sectors had lower equity and weighted average costs of capital.

Source: E. Lee, M. Walker, and H. Christensen, 2006.

18.7 THE OVERALL COST OF CAPITAL

Section 18.5 discussed the weighted average cost of capital (WACC) concept, illustrated in Figure 18.2. This was interpreted as the overall rate of return required in order to satisfy all stakeholders in the company. It described a U-shaped profile as the firm's level of gearing increased. It fell initially, as cheap debt was added to the capital structure, reached a minimum at the optimal gearing ratio, then rose as gearing came to be regarded as 'excessive'. The behaviour of this schedule provides a clue to the appropriate rate of return required on the company's activities, and, by implication, on new investment projects. We will examine this issue using the Lindley example covered on pages 493–496.

■ Lindley plc and the cut-off rate for new investment

Lindley's shareholders require a 20 per cent return and its pre-tax cost of debt is 10 per cent. Let us make the simplifying assumptions that Lindley's debt is perpetual and sells at par. Adjusting for tax at 30 per cent, as explained above, this corresponds to an after-tax cost of 7 per cent. What return on investment should Lindley achieve when issuing debt to finance a new project?

It is tempting to argue that the cut-off rate on this new project should be the cost of servicing the finance raised specifically to undertake the project. However, this is probably erroneous because using debt has an opportunity cost. The use of 'cheap'debt now may erode the company's ability to undertake worthwhile projects in the future by the depletion of credit lines. For example, assume that in 2007 Lindley used debt costing 7.0 per cent after tax to finance a project offering a post-tax return of 12 per cent, but this exhausted its credit-raising capacity. As a result, it was unable to exploit a project available in 2008 that offered 14 per cent. This suggests that the 'true' cost of the finance used in 2007 exceeds 7.0 per cent. Hence, to assess the 'correct' cost of capital really requires forecasting all future investment opportunities and capital supplies.

In addition, our previous analysis leads us to expect, at some level of gearing, an adverse reaction by shareholders, who may demand higher returns to compensate for higher financial risk. Consider two possible cases, denoted by points A and B, respectively, on the WACC profile in Figure 18.2. Note that A corresponds to zero gearing and B to the critical ratio.

Case A

Lindley has no debt at present and shareholder capital is £100 million. A new project with perpetual life is to be financed by the issue of £10 million debt at an after-tax cost of 7 per cent. No impact on the cost of equity is expected. In this case, the company will have to generate additional post-tax annual returns of (7% × £10 m) = £0.70 million in order to meet the extra financing costs associated with the new project, so that the hurdle rate for the new project is 7 per cent. Here, with the explicit assumption that shareholders will not react adversely, it may be reasonable to use the cost of debt as the cut-off rate. In this case, the required return would be simply the interest cost divided by the debt financing provided, i.e. the interest rate:

Required return = $iB/B = i$

However, this position is unlikely to be tenable, except for very small projects, and hence small borrowings, since significant changes in gearing (in either direction) are likely to provoke a market reaction.

Case B

We will assume that the optimal gearing ratio involves a capital structure with £50 million of each type of capital. Any further debt financing, even at a constant debt cost, will cause the cost of equity to increase. Assume that the extra £10 million debt financing will provoke shareholders to demand a return of 24 per cent. This would be expressed by downward pressure on share price until the return on holding Lindley's shares became 24 per cent. Now, the project has to meet not only the debt financing costs, but also the additional returns required by shareholders. The total additional required income is:

$$\begin{array}{ll} \text{Required} & = \text{debt financing} + \text{extra return required} \\ \text{extra income} & \quad\text{costs} \qquad\qquad \text{on equity} \end{array}$$

$$= (7\% \times £10\text{ m}) + (4\% \times £50\text{ m})$$

$$= £0.70\text{ m} + £2\text{ m}$$

$$= £2.70\text{ m}$$

Instead of an apparent cost of just 7 per cent, the true cost of using debt to finance this project is actually $(£2.70\text{ m}/£10\text{ m}) = 27$ per cent. This figure of 27 per cent is the **marginal cost of capital (MCC)**.

marginal cost of capital (MCC)
The extra returns required to satisfy all investors as a proportion of new capital raised

In the next section, we pinpoint the conditions under which it is acceptable to use the WACC as the cut-off rate for evaluating new investment.

■ Required conditions for using the WACC

Some major requirements have to be satisfied before use of the WACC can be justified:

1 The project is a marginal, scalar addition to the company's existing activities, with no overspill or synergistic impact likely to disturb the current valuation relationships.
2 Project financing involves no deviation from the current capital structure (otherwise the MCC should be used).
3 Any new project has the same systematic risk as the company's existing operations. This may be a reasonable assumption for minor projects in existing areas and perhaps for replacements, but hardly for major new product developments.
4 All cash flow streams are level perpetuities (as in the theoretical models).

In the short term, at least, firms are almost certain to deviate from the target structure, especially as market values fluctuate and financial managers perceive and exploit ephemeral financing bargains, e.g. an arbitrage opportunity in an overseas capital market. It is thus unrealistic to expect the hurdle rate for new investment to be adjusted for every minor deviation from the target gearing ratio. To all intents and purposes, the capital structure is given – only for major divergences from the target gearing ratio should the discount rate be altered. Similarly, even where a project is wholly financed by debt or equity, so long as the project is a minor one with no appreciable impact on the overall gearing ratio, then it is appropriate to use the WACC as the cut-off rate.

The preceding discussions suggest that the marginal cost of capital (MCC), rather than the cost of debt or the WACC, should be used as the cut-off rate for new investment. However, the MCC does have operational limitations. In particular, we are required to anticipate *how* the capital market is likely to react to the issue of additional debt. Given that we seem unable to define the WACC profile or pinpoint the optimal gearing ratio at any one time, this presents a problem. We could assume that the present gearing ratio is optimal, but this prompts the question of why different firms in the same industry have different gearing ratios.

■ The target capital structure: a solution?

target capital structure
What the firm regards as its optimal long-term ratio of debt to equity (or debt to total capital)

A solution commonly adopted in practice is to specify a **target capital structure**. For example, BHP Billiton (**www.bhpbilliton.com**), the world's largest mining company, states in its 2007 'Credit Summary' that it has a target gearing range of 35–40 per cent, (defined as net debt/net debt + book value of equity), among other key financial targets (including a minimum EBITDA to interest cover of eight times). The firm defines what it regards as the optimal long-term gearing range or ratio, and then attempts to adhere to this ratio in financing future operations. If the optimal ratio is deemed to involve, say, 50 per cent debt and 50 per cent equity (i.e. a debt-to-equity ratio of 100 per cent), any future activities should be financed in these proportions. For example, a £10 million project would be financed by £5 million debt and £5 million equity, via retained earnings or a rights issue. The corollary is to use the WACC as the cut-off rate for new investment. When shareholders require 20 per cent and debt costs 7 per cent post-tax, the WACC is:

(cost of equity × equity weighting) + (post-tax cost of debt × debt weighting)
= (20% × 50%) + (7% × 50%) = (10% + 3.5%) = 13.5%

The WACC is recommended because it is difficult to anticipate with any precision how shareholders are likely to react to a change in gearing. The somewhat pragmatic solution proposed assumes that the new project will have no appreciable impact on gearing: in other words, that the company already operates at the optimal gearing ratio and does not deviate from it. Obviously, the WACC and the MCC will coincide in this case.

Cost of capital

The cost of capital employed is determined by calculating the weighted-average cost of equity and debt. This average represents the market-rate returns expected by stockholders and creditors. The cost of equity is the return expected by an investor in E.ON stock. The cost of debt equals the long-term financing terms (after taxes) that apply in the E.ON Group. The premises of the cost of capital determination are reviewed on an annual basis. The cost of capital is adjusted if there are significant changes.

The table at right illustrates the derivation of the cost of capital before and after taxes. For 2006, the E.ON Group's average cost of capital was unchanged at 5.9 percent after taxes and 9 percent before taxes. The individual market units' minimum ROCE requirement varies between 8 percent and 9.2 percent before taxes.

Source: E.ON, Annual Report 2006, p. 38.

Cost of capital

	2006
Risk-free interest rate	5.1%
Market premium[1]	5.0%
Beta factor[2]	0.7
Cost of equity after taxes	**8.6%**
Cost of debt before taxes	5.6%
Tax shield (tax rate: 35%)[3]	−2.0%
Cost of debt after taxes	**3.6%**
Share of equity	45%
Share of debt	55%
Cost of capital after taxes	**5.9%**
Tax rate	35%
Cost of capital before taxes	**9.0%**

[1]The market premium reflects the higher long-term returns of the stock market compared with German treasury notes.
[2]The beta factor is used as an indicator of a stock's relative risk. A beta of more than one signals a higher risk than the risk level of the overall market, a beta factor of less than one signals a lower risk.
[3]The tax shield takes into consideration that the interest on corporate debt reduces a company's tax burden.

18.8 WORKED EXAMPLE: DAMSTAR PLC

Damstar plc produces and sells computer modems. The company obtained a stock market quotation four years ago, since when it has achieved a steady annual return for its shareholders of 14 per cent after tax. It has an issued share capital of 2 million 50p ordinary shares. The ordinary shares sell at a P:E ratio of 11:1. In the year ended 31 March 2008, the company sold 20,000 units.

The profit and loss account for the year to 31 March 2008 is as follows:

	£000	£000
Sales		1,600
Less: Variable expenses	(880)	
Fixed expenses	(350)	(1,230)
Profit before interest and taxation		370
Less: interest payable (10% loan stock)		(150)
Profit before taxation		220
Less: corporation tax (at 30%)		(66)
Profit after taxation		154
Dividend		(120)
Retained profit		34

In recent months, the company has been experiencing labour problems. As a result, it has decided to introduce a new highly-automated production process in order to improve efficiency. The production process is expected to increase fixed costs by £140,000 (including depreciation), but will reduce variable costs by £19 per unit.

The new production process will be financed by additional debt in the form of a secured £1,000,000 debenture issue at an interest rate of 12 per cent. If the new production process is introduced immediately, the directors believe that sales for the forthcoming year will be unchanged.

Required

(a) Determine how the proposal affects Damstar's break-even volume.
(b) Assuming no change in P:E ratio, calculate the change in EPS and share price if Damstar introduces its new production process immediately.
(c) What is the effect of the new process on Damstar's weighted average cost of capital using market value weights?

Answer to Damstar example

(a) First, we establish the revenue and cost parameters. Currently, the price is (£1.6 m/20,000) = £80 per unit. The average variable cost (AVC) = (£0.88 m/20,000) = £44. Hence, the gross profit margin (GPM) = (£80 − £44) = £36.

The break-even volume (BEV) can be expressed in operating terms, i.e. before fixed financing charges, and also after allowing for interest. Ignoring interest, the BEV is:

(Fixed costs of £0.35 m/£36) = 9,722 units

Allowing for interest of £0.15 million, the BEV is:

(£0.35 m + £0.15 m/£36) = 13,889 units.

With the new process, the AVC becomes (£44 − £19) = £25, and the GPM becomes (£80 − £25) = £55.

The new level of fixed operating costs is £0.49 million.

The BEV ignoring interest = (£0.49 m/£55) = 8,909 units. Allowing for the interest, increased by (12% × £0.12 m) to (£0.15 m + £0.12 m) = £0.27 m, the BEV is:

(£0.49 m + £0.27 m)/£55 = 13,818 units.

Notice that despite the increase in interest charges and the increase in fixed operating costs, the BEV has actually fallen due to the substantial fall in AVC. This warns that higher fixed costs does not always raise the BEV – it depends on what other changes are occurring at the same time.

(b) Currently, the EPS = (£154,000/2 m) = 7.7p. With the 11:1 P:E ratio, the share price is (7.7p × 11) = 85p.

Predicted profit and loss account for year ending 31 March 2009

	£	£
Sales		1,600
Variable Costs (20,000 × £25)	(500)	
Fixed Costs (£0.35 m + £0.14 m)	(490)	(990)
PBIT		610
Interest (£0.15 m + £0.12 m)		(270)
PBT		340
Taxation @ 30%		(102)
PAT		238

EPS now becomes (£238,000/2 m) = 11.9p, and with the same P:E ratio, the new share price would be (11.9p × 11) = £1.31.

(c) Based on market value weights, the WACC is:

	£	Weight
Value of equity = (£0.85 × 2 m) =	1.7 m	53
Value of debt = (£150,000/0.1) =	1.5 m	47
(unknown market value, so book value used)		
Total	3.2 m	100

WACC = (cost of equity × % of equity) + (post-tax debt cost × % of debt)

= (14% × 53%) + (10%[1 − 30%] × 47%)

= (7.4% + 3.3%) = 10.7%

After the issue of £1 million debt at 12 per cent interest, there are now two categories of debt, senior and junior.

	£	Weight
Value of equity = (£1.31 × 2 m) =	2.62 m	54
Vaue of senior debt = (10/12 × £1.5 m) =	1.25 m	26
Value of junior debt =	1.00 m	20
Total	4.87 m	100

WACC = (14% × 54%) + (10%[1 − 30%] × 26%) + (12%[1 − 30%] × 20%)

= (7.6% + 1.8% + 1.7%) = 11.1%

Comment: In this example, the increase in borrowing hardly affects the WACC, as the equity value increases to compensate. This result, of course, depends on the P:E ratio remaining at 11:1. Note also that the market value of the existing debt falls if interest rates rise. If the market rate of interest is now 12 per cent, existing debt with a coupon of 10 per cent must sell at a discount.

A practical application of the WACC

The UK's regulated industries live in a wacky world. While other companies try to make profits, they try to beat the cost of capital set by their regulator. This hurdle varies idiosyncratically across industries, from a real pre-tax rate of 6 per cent proposed for the gas distributors, to up to 9.4 per cent for BT Group, the telecoms company. Estimates for weighted average cost of capital across industries are not as dispersed as they were just after privatisation, but the remaining differences are still hard to justify.

Take gearing. In 2004 the airports' regulator assumed BAA, which operates London's Heathrow and Gatwick, was 75 per cent financed by equity and 25 per cent by debt. Its proposals for 2008–12 assume 60 per cent gearing. While BAA, under new management, has hugely increased its debt, a regulator should ideally form a view of an industry's optimal capital structure and stick to it. Working this out is tricky, even if estimates are market based. Smithers & Co, the consultancy, has found that the belief that electricity distribution companies might be bailed out by their regulator if they went bankrupt has distorted the market's perception of risk.

Other regulatory assumptions are even more controversial. The equity risk premium – the return equities are expected to generate over government bonds – is an integral part of the WACC calculation. The Competition Commission's proposed ERP for BAA is 3.5 per cent, while Ofgem suggests 5 per cent for the gas distributors. Economists disagree on the right number but there is no reason that some industries should benefit from a more generous assumption than others. Correcting anomalies is a challenge. Regulatory settlements are reflected in stock prices, and companies' investment plans are long term, so regulators are nervous about making dramatic changes. Still, the difficulties should not stop them aiming for greater consistency.

Source: Financial Times, 19 October 2007.

18.9 MORE ON ECONOMIC VALUE ADDED (EVA)

In Chapter 12, we explained the concept of EVA, a popular tool for assessing the amount of value created by a firm. In that chapter, we confined the analysis to all-equity-financed firms. In that form, the resulting EVA measured the wealth created for the owners of the business. EVA can also be calculated in a broader sense. Instead of focusing on wealth creation for the owners, one can focus on total economic value added for distribution to all the investors in the business. As such, for a geared firm, the EVA is arguably a better way of assessing managers' performance, i.e. the efficiency with which they utilise financial resources, because it is not distorted by the particular method of financing the business.

The adjustment is simple. Instead of applying the cost of equity to the net assets, we apply the WACC to the total capital employed to measure the capital cost, and deduct this from the NOPAT. For consistency, since the cost of debt appears in the WACC, the returns to debt should appear in the NOPAT, i.e. in this case the NOPAT is before charging interest, and thus includes returns to all types of investor.

The EVA formula thus becomes:

$$\text{EVA} = \text{NOPAT} - (\text{WACC} \times \text{Capital employed}).$$

It is usual to apply the WACC to long-term capital, given the relative volatility of short-term assets and liabilities.

Example

EVA plc's assets have a total book value of £900 million. It is financed by 60 per cent equity and 40 per cent debt. The costs of equity and debt are 12 per cent and 7 per cent respectively. Operating profits are £100 million, and the tax rate is 30 per cent.

$$\text{NOPAT} = (\text{Operating profit} - \text{tax}) = £100\,\text{m}\,(1 - 30\%) = £70\,\text{million}$$
$$\text{WACC} = (60\% \times 12\%) + (40\% \times 7\%[1 - 30\%])$$
$$= 7.2\% + 1.96\% = 9.16\%$$
$$\text{EVA} = £70\,\text{million} - (9.16\% \times £900\,\text{million}) = (£70\,\text{m} - £82.44\,\text{m}) = -£12.44\,\text{m}$$

As this result is negative, EVA plc appears to have destroyed value.

Readers may notice that under the traditional view of gearing, as a firm raises the debt/equity ratio, the EVA is likely to increase, so long as the WACC declines.

18.10 FINANCIAL DISTRESS

This is an appropriate stage to clarify our terminology. Up to this point, we have implied that the reason for the upturn in the WACC profile is the threat of 'bankruptcy' resulting in 'liquidation', i.e. a company that fails to meet its debts will be forced by its creditors to liquidate. The term 'bankruptcy' in the UK strictly applies to personal insolvency. Individuals go bankrupt, and firms become insolvent. But what happens to a firm in severe financial distress? Broadly speaking, there are two main forms of treatment. The first is called 'receivership'.

Creditors – and it only takes one of many – may apply to a court to appoint a 'receiver' to recover the debt if the firm defaults on interest or capital repayments.

Burning before earning

Although accountancy firms often do well from insolvencies, they are not immune from failure themselves. Ascot Drummond is a case in point. It pioneered the concept of online accountancy services including company formation, accounts, payroll, book-keeping and taxation accessible on clients' computers. It obtained financing of over £6 million from venture capitalists Mercury

Private Equity in October 2001, yet having burnt its way through the first *tranche* of £3 million, it was in receivership less than a year later. After pitching for a client base of over 10,000 by end-2001, it failed to break the 1,000 mark, and was unable to turn this into significant revenues, let alone profit.

Source: Based on an article by L. Meall, 'Dot.com dot.gone', *Accountancy*, August 2001.

The receiver may sell the business, or parts of it, as a going concern, which continues to trade in a different guise, often involving a reduced scale of operation. However, the receiver's primary duty is to the appointing bank, and once sufficient funds have been realised to repay the loan, the receiver is under no obligation to maximise the proceeds of the sale of the remaining assets, or even to keep them operating. The receiver may choose to liquidate them *in toto* and disburse the net proceeds to remaining creditors and then any residue to shareholders.

administration
An attempt to reorganise an insolvent firm under an administrator, rather than liquidate it

The Insolvency Act 1986 introduced a new procedure, **administration**, as an attempt to rescue ailing companies and to protect employment. The company is allowed to continue trading under the overall control of an administrator, who will attempt to reorganise the company's finances and its operating structure. The administrator is appointed by a court at the request of the directors and has an equal duty to all creditors. In effect, administration, rather like filing for Chapter 11 bankruptcy under the US Bankruptcy Code, is an attempt to protect the company from its creditors, thus giving the administrator a breathing space during which it can attempt to secure the company's survival as a reorganised going concern. The main difference compared with the US equivalent is that Chapter 11 bankruptcy allows the incumbent managers and owners to retain control. In addition, Chapter 11 enables the firm to impose a moratorium on interest payments on existing debts for a specified period and also to borrow more funds as money lent after the filing has a prior claim on assets.

Delphi struggling to finance its way out of Chapter 11

Last spring, before turmoil hit the debt market, the bonds of Delphi, a car parts company operating under bankruptcy protection, were changing hands at well over 100 cents in the dollar.

Today, the market values that same debt at about 40 cents in the dollar.

Delphi may not be able to emerge from Chapter 11 because it is struggling to raise exit finance from its bankers at Citigroup and JPMorgan.

"Companies in bankruptcy assume that there will be a liquid market to fund them when they exit but that hasn't happened," says one banker familiar with Delphi.

Delphi and its advisers at Rothschild have spent the past two years trying to fix the group.

The company has shut high-cost plants in the US and sold non-core operations, such as its steering division.

It has reached understandings with the union and dealt with its pension obligations.

The court has already approved Delphi's plans. But those plans, crafted months ago, reflect yesterday's conditions in financial markets, not today's.

They value Delphi at about $10bn (£5.1bn) and say that lenders and bondholders will get 100 cents in the dollar, not a fraction of that. The plans also assume something will be left for the shareholders, who are at the bottom of the heap. This now looks unrealistic.

Last month, Delphi sought $4.5bn of exit finance to repay money it borrowed when it filed for bankruptcy protection and for working capital. However, it met little enthusiasm from the financial community and, without the money, its plans could unravel, leaving creditors fighting over what is left.

The situation is especially fraught for an investment group led by Appaloosa, which includes Harbinger, Pardus and Goldman Sachs.

That investment group has agreed to back a $2.55bn rights issue if other investors don't.

They can walk away if the exit finance does not materialise but could face substantial losses on other positions they hold in the company's securities.

Today, with the credit markets totally frozen, the best hope is that General Motors, Delphi's former owner and best customer, will put up a big share of the exit finance.

Source: Henry Sender, *Financial Times*, 5 March 2008.

Not everyone likes Chapter 11 (nor 22, and 33, and 44)

The Chapter 11 arrangements are not universally admired. Of particular concern is that in industries such as airlines and telecoms, which seem, in the US at least, particularly prone to insolvency, bankrupt firms will reappear with more manageable debts and thus be better able to compete with the industry survivors. This means they could drive healthier rivals into failure in turn.

Europeans are often aghast at a procedure that puts the debtor in control of the insolvency procedure. Few managers in such a situation are likely to opt for full liquidation, resulting in many 'walking dead' firms living on to haunt their competitors. In the US, there are also strong suspicions that judges in the bankruptcy courts tend to identify more closely with incumbent executives than with creditors, especially if they view the latter as being over-assertive of their rights.

In principle, it is desirable to keep an essentially sound business in operation if it is still capable of creating wealth, positive operating profits often being taken as an indicator of this. However, few if any 'zombie' firms have emerged from Chapter 11 sufficiently reinvigorated to challenge the leading players in their industries. Indeed, many firms have gone bust twice (the so-called Chapter 22s), or even three times, for example, the now defunct airline TWA. Harvard Industries, an electronics firm that started life as a brewery, holds the record with four bankruptcies, before going into liquidation in 2002, apparently to avoid pension and healthcare obligations.

Source: Some of the material quoted here appeared in an article in *The Economist*, 'The Night of the Killer Zombies', 14 December 2002.

Any visitor to an auction of bankrupt stock will have no difficulty in appreciating the importance of postponing the break-up decision. Similarly, when repossessed assets, such as consumer durables and houses, are sold by creditors, they rarely fetch 'market values'. This is partly because the vendor often only needs, and expects, to recover an amount less than the market value, having deliberately set the loan itself at less than the market value of the asset upon which it is secured. The vendor is interested in a quick sale to minimise depreciation, interest and other carrying costs. Moreover, when it is generally known that the assets are offered under distressed conditions, asset values usually head South! In January 2002, Britain's BG Group plc announced it would pay distressed US energy firm Enron $350 million for its stake in oil and gas fields off the coast of India. The 30 per cent stake in the Tapti and Panna-Mukta fields plus 63 per cent of a further untapped field, was valued by Dutch bank ABN AMRO at $450 million, 30 per cent above the agreed price.

The costs incurred at and during liquidation are called the 'direct' costs of financial distress. Empirical studies (e.g. van Horne, 1975; Sharpe, 1981) have suggested that liquidation costs, including legal and administrative charges, may lower the resale value of distressed companies by 50 per cent or more.

However, a more recent US study suggests that distress costs of this magnitude may be an overestimate. In a study of 31 Highly Leveraged Transactions (HLTs) occurring between 1980 and 1989, Andrade and Kaplan (1998) tried to differentiate between the costs of dealing with economic distress, e.g. reacting to loss of contracts, and direct costs of financial distress. Comparing enterprise values at the date of the HLT and at the date of resolution of the distress, they estimated an average loss in firm value of 38 per cent, of which 26 per cent was due to economic distress and 12 per cent due to financial distress.

A more insidious form of financial distress is the impact of increasing gearing on managerial decision-making and the performance of the firm – the so-called 'indirect' costs. As a firm's indebtedness begins to look excessive, it may develop an overriding concern for short-term liquidity. This may be manifested in reduced investment in training and R&D, thus damaging long-term growth capability, and reducing credit periods and stock levels, which may hamper marketing efforts. Supplier power triggered the collapse of US discount retailer K-Mart in January 2002. Its sole supplier of grocery products suspended payments after K-Mart failed to make a regular weekly payment.

More obviously, a distressed firm may sell established operations at bargain prices, sell or abandon promising new product developments, and, to the extent that it does continue to invest, may express a preference for short-payback projects, cash-generating projects, rather than strategic activities. Troubled companies often cut their dividends to preserve liquidity, but this often signals to the market the extent of their difficulties. Finally, there may be a pervasive 'corporate gloom effect', which saps morale internally and damages public image externally.

Such costs are likely to be encountered well before the trigger point of cash flow crisis, and, of course, many firms have successfully surmounted them, but not without an often prolonged dip in the value of the company. In other words, both actual and anticipated liquidation costs detract from company value, lowering the effective limit to debt capacity.

The practical importance of the facility to appoint an Administrator before creditors can appoint receivers may now be seen. Administration enhances the probability of survival of a company unable to meet its immediate liabilities and may thus lead to lower costs of financial distress. However, there may be an element of **'moral hazard'** to the extent that financial managers might undertake more dangerous levels of debt, knowing that there is a more relaxed legal procedure in the event of insolvency.

Self-assessment activity 18.11

Identify examples of distressed behaviour by highly geared firms from your reading of the financial press.

Kwik Save poised to enter administration

Kwik Save is expected to enter administration today with the loss of 1,100 jobs and the closure of 90 stores.

The grocery chain, sold by Somerfield in February last year, had teetered for months with shelves bare and staff complaining they were not being paid.

In April, Paul Niklas, chief executive announced a £50m rescue package via a consortium, saying: "Kwik Save has a great future."

An administration order was granted yesterday at Chancery Court in Manchester; the grocer will enter administration at midday today with KPMG, which has been advising the company on a restructuring, to be appointed administrator.

In May the group announced the closure of 81 of its 273 stores. It is understood that Somerfield's owners are exposed to lease liabilities of up to £10m related to Kwik Save.

Usdaw, the retail union, said it had "kept the pressure on Kwik Save throughout their recent difficulties and this announcement is no way to treat loyal and hard-working staff who have gone way beyond what could be reasonably expected to keep the company afloat".

In the restructuring, Kwik Save will try to save 56 stores, which are expected to be rebranded as Fresh Express.

In a grocery market dominated by Tesco and including discounters such as Aldi, some analysts doubt the new entity can prosper.

Bryan Roberts of Planet Retail, a research firm, said shopping at Kwik Save had become a "thoroughly depressing experience".

"The out-of-stocks were horrendous, particularly on basics such as bread and milk, and staff morale was visibly at rock-bottom.

"Obviously the Kwik Save brand has been damaged beyond repair".

Kwik Save was not available for comment.

Source: Tom Braithwaite, *Financial Times*, 6 July 2007.

18.11 TWO MORE ISSUES: SIGNALLING AND AGENCY COSTS

An unexpected reduction in indebtedness is usually greeted with pleasure by the market, whereas a debt increase can be regarded in a favourable or unfavourable light depending on the accompanying arguments. According to Ross (1977), managers naturally have a vested interest in not making the company insolvent, so an increase in

Rights issue to cut SMG debt by £91m

SMG, the embattled media company, sought the sanctuary of its investors yesterday, announcing a two-for-one rights issue to reduce debt from £131m to £40m.

The issue, fully underwritten, was also aimed at giving Rob Woodward, the chief executive, a period of calm in which to sell SMG's Virgin Radio at a better price than offered so far during a protracted process.

It is understood that after market volatility sank the option of a flotation, a sale of Virgin stalled because no bidder would offer more than £57m for the stations. SMG wrote down the value of Virgin to £85m in its financial results.

Mr Woodward said he had the support of his main shareholders for the rights issue, which will see investors offered two new shares at 15p for each one they hold and aims to raise

£95m. After expenses, £91m will go towards reducing the debt, which will then be renegotiated with bankers.

Mr Woodward said this would save £20m yearly and free him to concentrate on the core business, SMG's two Scottish ITV franchises, while also avoiding a fire-sale of Virgin Radio.

"Rather than sell Virgin in this atmosphere, we have focused on ensuring that we can deliver the right level of value for shareholders," he said.

"As the costs of our debt became obvious, it was overshadowing the financial health or otherwise of the company."

All but £15m of the price obtained for Virgin, after expenses, would be returned to investors after the eventual sale, he added.

Mr Woodward said ITV, which holds a 17 per cent share, was "uncer-

tain" as to whether it would take up the rights issue.

The shares, which initially fell sharply, recovered to close up 1½p at 30p yesterday.

FT Comment

It is certainly welcome to get some clarity over SMG's debt, which was threatening to drown the group. But a rights issue reflects the distress in which the company found itself. The waters remain choppy and Mr Woodward will still find himself with a lot of explaining to do if the eventual sale of Virgin falls far short of £85m. Neverthless, it is hard to find anyone in the radio industry who thinks it is worth much more than £60m right now.

Source: Ben Fenton, *Financial Times*, 7 November 2007.

gearing might be construed by the market as signalling a greater degree of managerial confidence in the ability of the company to service a higher level of debt. This argument relies on asymmetric information between managers and shareholders, and reflects the pervasive principal/agent problem.

Financial managers, as appointees of the shareholders, are expected to maximise the value of the enterprise, but it is difficult for the owners to devise an effective, but not excessively costly, service contract to constrain managerial behaviour to this goal. In the context of capital structure theory, the financial manager acts as an agent for both shareholders and debt-holders. Although the latter do not offer remuneration, they do attempt to limit managers' freedom of action by including restrictive covenants in the debt contract, such as restrictions on dividend payouts, to protect the asset base of the company.

Such restraints on managerial decision-making may adversely affect the development of the firm and, together with the monitoring costs incurred by the shareholders themselves, may detract from company value. Conversely, it is possible that the close monitoring by a small group of creditors, aiming to protect their capital, may induce managers to pursue more responsible policies likely to enhance the wealth of a widely-diffused group of shareholders.

18.12 CONCLUSIONS

What conclusions does this body of analysis lead to, and how does it help financial managers?

1 Gearing can lower the overall or weighted average cost of capital that the company is required to achieve on its operations, and can raise the market value of the enterprise. However, this benign effect can be relied upon only at relatively safe gearing levels. Companies can expect the market to react adversely to 'excessive' gearing ratios. The implications for project appraisal are reasonably clear. Strictly, the appropriate cut-off rate for new investment is the marginal cost of capital, but if no change in gearing is caused by the new activity, the WACC can be used.

2 Considerable care should be taken when prescribing the appropriate use of debt that will enhance shareholder wealth without ever threatening corporate collapse. Levels of gearing that look quite innocuous in calm trading conditions may suddenly appear ominous when conditions worsen. Corporate difficulties do not usually occur singly, and highly geared companies are relatively less well placed to surmount them.

3 The capital structure decision, like the dividend decision, is a secondary decision – secondary, that is, to the company's primary concern of finding and developing wealth-creating projects. Many people argue that the beneficial impact of debt is largely an illusion. Clever financing cannot create wealth (although it may enable exploitation of projects that would not otherwise have proceeded). It may, however, transfer wealth if some stakeholders are prepared, perhaps due to information asymmetry, to accept too low a return for the risks they incur, or if the government offers a tax subsidy on debt interest.

4 The decision to borrow should not be over-influenced by tax considerations. There are other ways of obtaining tax subsidies, such as investing in fixed assets, which qualify for tax allowances. A highly geared company could find itself unable to exploit the other tax-breaks offered by governments when a favourable opportunity is uncovered.

5 Remember that interest rates fluctuate over time. If interest rates move from what seems a 'high' level, financial managers should take advantage of the reduction. For example, if 10 per cent seems like the 'normal' long-term level of interest rates, when rates next fall below 10 per cent, and bankers are offering variable rate loans

at, say, 9 per cent, one should not be afraid to take a fixed rate loan at, say, 9.5 per cent. Readiness to work with a slightly higher than minimum rate in the short term could have significant payoffs in the longer term. Anyone who thinks that rates will continue to fall should reserve some borrowing capacity to retain flexibility.

6 Firms should avoid relying on too many bankers, as with syndicated loans, despite the benefits of access to a variety of banking facilities. If the company hits trading and liquidity problems, it is hard enough to convince one banker that the company should be saved. But if it has to persuade 10 or 20, and their decision has to be unanimous, it is virtually impossible to reach a satisfactory conclusion about capital restructuring. The ill-feted Eurotunnel had to deal with 225 at one stage of its troubled existence, later reduced to a mere 200. It was reported that Nissan had 170 banking relationships before implementing its recovery plan after the merger with Renault.

7 The finance manager should question whether debt is the most suitable form of funding in the circumstances. For example, there should be a clear rationale to support the case for debt rather than retentions (i.e. lower dividends) or a rights issue. He or she should recognise the value of retaining reserve borrowing capacity to draw upon under adverse circumstances or when favourable opportunities, like falling interest rates, arise.

8 These considerations are reflected in two popular theories that attempt to explain how firms address long-term financing decisions. These are:

- The Trade-off Theory. This recognises that firms seek to exploit the lower cost benefits of borrowing, especially the tax shield, but at the same time, they are reluctant to increase the financial risk entailed in entering contractual commitments to make ongoing interest and capital repayments. In other words, they trade-off the returns (the cost benefits) against the risks. We might thus expect to find that firms enjoying higher and more stable profit levels, which offer greater scope to shelter profits from tax, should operate at higher borrowing levels.

- The Pecking Order Theory. This suggests that firms have an order of priorities in selecting among alternative forms of finance:

  – First, they prefer to use the internal finance generated by operating cash flow.
  – Second, they prefer to borrow when internal sources are drained.
  – Third, they regard selling new shares almost as a last resort.

information asymmetry
The imbalance between managers and owners of information possessed about a firm's financial state and its prospects

The reason for this order of preference lies yet again in **information asymmetry** – managers know far more about the firm's performance and prospects than outsiders. They are unlikely to issue shares when they believe shares are 'undervalued', but more inclined to issue shares when they believe they are 'overvalued'. Naturally, shareholders are aware of this likely managerial behaviour and thus regard equity issues with suspicion. For example, they may interpret a share issue as a signal that management thinks the shares are overvalued and mark them down accordingly – a very common occurrence – thereby increasing the cost of equity. Investors would expect managers to finance investment programmes, first, using internal resources, second, via borrowing up to an appropriate debt/equity combination, and finally through equity issues. Yet again, signalling considerations are crucial.

SUMMARY

We have explained the meaning of gearing, its likely benefits to shareholders, its dangers and its possible impact on the required return on investment projects.

Key points

- Borrowing often looks more attractive than equity due to its lower cost of servicing, tax-deductibility of interest and low issue costs.
- A company's indebtedness is revealed by its capital gearing and by its income gearing.
- The sum of discounted tax savings conferred by the tax-deductibility of debt interest is called the tax shield.
- In a geared company, variations in earnings before interest and tax generate a magnified impact on shareholder earnings.
- The downside of gearing is the creation of a prior charge against profits, which results in the risk of possible default as well as greater variability of shareholder earnings.
- Default risk is likely to impose further costs on the geared company's shareholders, referred to as the 'costs of financial distress'.
- An insolvent company, i.e. one unable to meet its immediate commitments, is unlikely to achieve full market value in a sale of assets.
- For companies using a mixture of debt and equity, there may be an optimal capital structure at which the overall cost of capital (WACC) is minimised.
- The WACC is found by weighting the cost of each type of finance by its proportionate contribution to overall financing, and may fall as gearing increases.
- The increased risks imposed by gearing are likely to cause lenders and shareholders eventually to demand a higher rate of return, raising the WACC.
- The WACC is the appropriate cut-off rate for new investment so long as the company adheres to the optimal capital proportions.
- When companies deviate from the optimal capital structure, the marginal cost of capital becomes the correct cut-off rate.
- Because the optimal gearing ratio is difficult to identify in practice, many firms aim for a target gearing ratio, which they regard as 'acceptable'.
- In view of the risks of gearing, an increase in borrowing may be a way of signalling to the market greater confidence in the future.

Further reading

Because Chapters 18 and 19 are 'paired', the reading guides have been amalgamated. A composite guide is given at the end of Chapter 19.

Appendix
CREDIT RATINGS

Standard & Poor's credit rating system is shown below, together with that of the other main credit rating agency, Moody's.

Credit risk	S & P	Moody's
Prime	AAA	Aaa
Excellent	AA	Aa
Upper Medium	A	A
Lower Medium	BBB	Baa
Speculative	BB	Ba
Very Speculative	B, CCC, CC	B, Caa
Default	C, D	Ca, C

Source: www.moodys.com, www.standardpoor.com, www.ratings.com

Both agencies make further differentiation on the quality of bonds within each category. Moody's uses a numerical system (1,2,3) and S & P uses a plus or minus. For example, a rating of Aa1 from Moody's is a superior rating than Aa3, and an A+ from S & P is a better rating than A−.

Both rating systems are based on 'default risk and an assessment of the likelihood of recovery'. In each case, the cut-off between investment grade bonds and speculative ones is critical – anything of a speculative nature ('significant speculative characteristics' according to Moodys, and having 'speculative elements whose future cannot be considered as well-assured', according to Standard and Poor) is critical. Anything speculative attracts the unattractive label 'junk bonds', which, of course, incur higher rates of interest.

QUESTIONS

 Questions with an icon are also available for practice in myfinancelab with additional supporting resources.

Questions with a **coloured number** have solutions in Appendix B on page 751.

1 Using the accounting information provided below, calculate the following measures of gearing:

- long-term debt (LTD) to equity
- LTD to LTD plus equity
- total debt to equity
- net debt in absolute terms
- net debt to equity
- total debt to total assets
- interest cover
- income gearing
- total liabilities to total assets

Shareholders' funds	£500 m
Cash	£20 m
Short-term deposits	£40 m
Short-term bank borrowing	£50 m
Debentures and other long-term debts	£200 m
Total assets	£800 m
Total liabilities	£300 m
Profit before interest and tax	£120 m
Net interest payable	£25 m

2 Calculate the cost of debt facing a firm that issued £50 million in debentures in £100 units two years ago at a nominal interest rate of 8 per cent p.a., in each of the following cases:
(i) market value of debt is £45 million; perpetual life.
(ii) as **(i)**, but allowing for Corporation Tax at 30 per cent.
(iii) as for **(i)**, but lifespan of debt is 8 years.
(iv) as for **(iii)**, but with tax payable at 30 per cent.

3 Darnol plc is currently ungeared and is considering a buy-back of ordinary shares via an open market purchase, borrowing in order to do so. You have been commissioned to report on the likely impact of two alternative policies, depending on the level of sales and operating profit for its products. You are given the following information:

Level of sales	PBIT	Probability
Weak	£5 m	0.3
Average	£50 m	0.5
Strong	£150 m	0.2

Equity is currently £200 million at book value. Tax is paid at 30 per cent. Two alternative share buy-back programmes are under consideration:
(i) Borrowing £40 million at 8 per cent.
(ii) Borrowing £80 million at 8.5 per cent.

Required

(a) Calculate the current, and potential expected annual return on equity (ROE) under each programme.

(b) Calculate the standard deviation of the ROE in each case.

(c) Using the figures you have obtained, explain and illustrate the distinction between business and financial risk.

4 (a) Calculate the value of the tax shield in each of the following cases, all based on borrowing of £100 million at 10 per cent interest p.a., pre-tax.

  (i) Perpetual life debt, tax rate is 30 per cent.

  (ii) Debt repayable in full after five years.

  (iii) Debt repayable in equal tranches over five years, interest paid on the declining balance.

(b) Specify the factors that determine the value of the tax shield that a firm can exploit.

5 Calculate the weighted average cost of capital for the following company, using both book value and market value weightings.

	Balance sheet values	Cost of finance
Ordinary shares: (par value 50p)	£10 m	20%
Reserves:	£20 m	20%
Long-term debt:	£15 m	10%

The debt is permanent and its market value is equal to book value.
The rate of Corporate Tax is 30%.
The ordinary shares are currently trading at £4.50.

6 The directors of Zeus plc are considering opening a new manufacturing facility. The finance director has provided the following information:
Initial capital investment: £2,500,000.
Dividends for the last five years have been:

Year	1993	1994	1995	1996	1997
Net dividend per share (pence)	10.0	10.8	11.4	13.2	13.7

The following is an extract from the Balance Sheet of Zeus plc for the year ended December 31 1997:

Creditors due in more than one year

8% Debenture	£700,000
Long-term loan (variable rate)	£800,000
Capital and reserves	
2,000,000 shares of 25p each	£500,000

The authorised share capital is 4 million shares, the current market price per share at December 31 1997 is 136p ex-dividend.

The current market price of debentures is £60 (ex-interest) and interest is payable each year on 31 December.
The interest rate on the long-term loan is 1 per cent above LIBOR, which at present stands at 16 per cent.
The debentures are irredeemable.
Ignore taxation.

Required
Calculate the weighted average cost of capital (WACC) for Zeus plc at 31 December 1997.

7 RH plc manufactures machine tools. It has issued two million ordinary shares, quoted at 168 pence each, and £1 million 10 per cent secured debentures quoted at par. To finance expansion, the directors of the company want to raise £1 million for additional working capital.

Cash flow from trading before interest and tax is currently £1 million per annum. It is expected to rise to £1.3 million per annum if the expansion programme goes ahead. To simplify placing a valuation on the company's equity, you should assume that:

- The forecast level of cash flow, and a tax rate of 33 per cent, will continue indefinitely.
- The required rate of return on the market vale of equity, 18 per cent post-tax, will be unaffected by the new financing.
- There is no difference between taxable profits and cash flow.

The company's directors are considering two forms of finance – equity via a rights issue at 15 per cent discount to current share price, or 12 per cent unsecured loan stock at par.

Required

(a) Calculate for both financing options, the expected
  (i) increase in the market value of equity
  (ii) debt/(debt + equity) ratio
  (iii) weighted average cost of capital.
(b) Assume you are the financial manager for RH plc. Write a brief report to the board advising which of the two types of financing is to be preferred. Include in your report brief comments on non-financial factors which should be considered by the directors before deciding how to raise the £1 million finance.

(CIMA)

8 Celtor plc is a property development company operating in the London area. The company has the following capital structure as at 30 November 1993:

	£000
£1 ordinary shares	10,000
Retained profit	20,000
9% debentures	12,000
	42,000

The equity shares have a current market value of £3.90 per share and the current level of dividend is 20 pence per share. The dividend has been growing at a compound rate of 4 per cent per annum in recent years. The debentures of the company are irredeemable and have a current market value of £80 per £100 nominal. Interest due on the debentures at the year end has recently been paid.

The company has obtained planning permission to build a new office block in a redevelopment area. The company wishes to raise the whole of the finance necessary for the project by the issue of more irredeemable 9 per cent debentures at £80 per £100 nominal. This is in line with a target capital structure set by the company where the amount of debt capital will increase to 70 per cent of equity within the next two years.

The rate of corporation tax is 25 per cent.

Required

(a) Explain what is meant by the term 'cost of capital'. Why is it important for a company to calculate its cost of capital correctly?
(b) What are the main factors which determine the cost of capital of a company?
(c) Calculate the weighted average cost of capital of Celtor plc which should be used for future investment decisions.

9 Redley plc, which manufactures building products, experienced a sharp increase in operating profit (i.e. profits before interest and tax) from £27 million in 1995–6 to £42 million in 1996–7 as the company emerged from recession and demand for new houses increased. The increase in profits has been entirely due to volume expansion, with margins remaining static. It still has substantial excess capacity and therefore no pressing need to invest, apart from routine replacements.

In the past, Redley has followed a conservative financial policy, with restricted dividend payouts and relatively low borrowing levels. It now faces the issue of how to utilise an unexpectedly sizeable cash surplus. Directors have made two main suggestions. One is to redeem the £10 million of the secured loan stock issued to finance investment several years previously, the other is to increase the dividend payment by the same amount.

Redley's present capital structure is shown below:

	£m
Issued share capital (par value 50p)	90
Reserves	110
Creditors due after more than one year:	
9% secured loan stock 2004	30

Further information
(i) Redley has no overdraft.
(ii) Redley pays corporate tax at a rate of 33%.
(iii) The last dividend paid by Redley was 1.45 pence per share.
(iv) Sector averages currently stand as follows:

dividend cover	2.5 times
gearing (long-term debt/equity)	48%
interest cover	5.9 times

(v) Redley's P:E ratio is 17:1.

Required
(a) Calculate **(i)** the dividend cover and **(ii)** the dividend yield for both 1995–6 and for the reporting year 1996–7, if the dividend is raised as proposed.
(b) You have been hired to work as a financial strategist for Redley, reporting to the Finance Director. Using the information provided, write a report to your superior, which identifies and discusses the issues to be addressed when assessing the relative merits of the two proposals for reducing the cash surplus.

(ACCA)

Practical assignment

For a company of your choice, undertake an analysis of gearing similar to that conducted in the text for D.S. Smith plc. Pay particular attention to the treatment of provisions and other components of long-term liabilities. If you decide to include short-term indebtedness in your capital gearing measure, would you include trade and other creditors as well?

Try to form a view as to whether your company is operating with high or low gearing.

 *Now retake your diagnostic test for Chapter 18 to check your progress and update your study plan.*

19

Does capital structure really matter?

Travelling in hope

MyTravel, the package holiday firm, formerly known as Airtours, was hit heavily by the downturn in travel following the atrocities of 9/11. This all occurred just as it had significantly increased capacity via a series of takeovers, mainly financed by borrowing, resulting in high costs and many unsold holidays. Despite a major programme of cost-cutting, and several sell-offs of underperforming businesses, MyTravel headed rapidly into the red. Matters were not helped by discovery of a financial black hole in the accounts, necessitating substantial write-offs, and propelling the shares on a downward trajectory.

In October 2004, it announced an £800 million debt-for-equity swap that would virtually wipe out the equity of the existing shareholders, leaving them with just a 4 per cent stake in the restructured business. Creditors agreed to accept 88 per cent of the firm's shares and bond-holders 8 per cent. Following the swap, MyTravel was left with debts of £140 million, mainly in aircraft leases taken out to finance the MyTravelLite airline. However, it claimed that it was on course to break even at the operating level in the current financial year, and its banks arranged a new five-year overdraft facility of £167 million.

The ordinary shares which were trading at 140p less than a year previously now fell further to 4.75p, leaving several pension funds nursing significant losses.

After all but sinking under the weight of its own debt, MyTravel embarked on a recovery programme involving cost-cutting, closure of many Going Places high-street shops and cutting back on the number of holidays offered. By 2007, it was able to report profit before tax of £44 million, its first since 2001. Almost simultaneously, it was taken over by the German firm, Thomas Cook, then co-owned by Lufthansa and stores chain Karstadt Quelle, based in Essen. Cook's, now quoted on the London Stock Exchange, was attracted by a reported £75 million p.a. in cost savings, and estimated tax losses of €1.2 billion euros. Soon after the takeover, Cooks announced 2,800 job losses in the UK, mainly in the Manchester area, and closure of 150 Going Places outlets.

Source: Based on an article in the *Financial Times* by Matthew Garrahan, 14 October 2004, and Thomas Cook Group plc (**www.thomascookgroup.com**).

Learning objectives

This chapter offers a more rigorous analysis of capital structure decisions. After reading it, you should:

■ Understand the theoretical underpinnings of 'modern' capital structure theory.

■ Appreciate the differences between the 'traditional' view of gearing and the Modigliani–Miller versions.

■ Appreciate how the CAPM is integrated into capital structure analysis.

■ Be able to identify the extent to which a Beta coefficient incorporates financial risk.

 Complete your diagnostic test for Chapter 19 now to create your personal study plan.

19.1 INTRODUCTION

This chapter begins with a question. In the last chapter, we warned that debt could be lethal to company survival, yet here we find an effectively insolvent firm 'rising from the dead'. This survival instinct may thus pose a puzzle – how can insolvent firms be worthless one moment but still survive?

The answer lies in recognising that insolvency does not necessarily mean total loss of *enterprise value*. Insolvency is the formal acceptance that the business entity cannot meet its financial obligations, whether payment to creditors for supplies, or payments to lenders. Yet it is possible for insolvent firms to retain value as operating entities even though their owners' equity may have been wiped out. A few figures may help.

XYZ owes £10 million to lenders and its assets are only £8 million, so it is technically insolvent, unable to cover its debts with its assets, i.e. its net assets are negative – minus £2 million. However, if its operating activities generate more cash inflows than outflows, then as a debt-free entity, it would be viable and have value. Hence, it might look attractive as a restructured going concern to other investors prepared to take responsibility for the debts. Creditors might be prepared to exchange debt for equity or preference shares, or take a discount on their principal, accepting, say, 30p in the pound, just to salvage something from the mess. The equity value has disappeared but the enterprise still has value as a going concern if investors can be found to refinance it and reorganise it into a viable operation.

Even the heavily indebted (€4.2 billion in 2008) Eurotunnel is able to deliver a positive EBITDA (estimated at €439 million for 2007). On this basis, it has an enterprise value even though the equity has been all but wiped out (share price down by over 90 per cent since floatation in 1987).

One might then conclude that indebtedness does not really matter – a firm that cannot pay its way can be restructured and the jobs of the workers, if not the management, can be preserved.

A few years ago, former England football manager Terry Venables bought Portsmouth FC for £1 – a remarkable bargain, you might think.* However, he was buying the club's assets encumbered with debts, which he and his backers were hoping to refinance or to pay off. The club as a bundle of assets was worth more than £1, and Venables was really paying far more than this. Had he bought the club free of debt, he would have had to pay out more.

So long as assets can be sold at their full economic value, i.e. reflecting operating cash flows, then debt is of most consequence to the hapless owners whose equity is usually all but obliterated. Meanwhile, the business can proceed, often in a slimmed-down form and usually under different management, with a new set of backers hoping to do better next time round. This suggests that the sting of insolvency can be drawn.

The traditional theory of gearing says that debt should be handled with great caution but there is a body of analysis that proves that, under certain conditions, debt is truly irrelevant in determining company value and the cost of capital. This is the famous theory developed by Franco Modigliani and Merton Miller (MM), both Nobel prizewinners for Economics. Our next task is to explain this theory.

19.2 THE MODIGLIANI–MILLER MESSAGE

To Modigliani and Miller (MM), the traditional perception of the impact and desirability of gearing seemed unsupported by a theoretical framework. In particular, there seemed little reason, apart from some form of market imperfection such as information

---

*This is by no means a unique event in football. Ken Bates also paid £1 for Chelsea FC, before selling his stake in the subsequently-floated Chelsea Village plc to Roman Abramovich for £17 million (and then be bought into Leeds United FC!).

deficiency, why merely altering the capital structure of a firm should be expected to alter its value. *After all, neither its earnings stream nor its inherent business risk would alter –* it would remain essentially the same enterprise, operating under the same managers and in the same industry.

MM contended that, in a perfect capital market, the value of a company depended simply on its income stream and the degree of business risk attaching to this, regardless of the way in which its income was split between owners and lenders, i.e. its capital structure. Therefore, any imbalance between the value of a geared company and an otherwise identical ungeared company could only be a temporary aberration and would be quickly unwound by market forces. The mechanism for equalising the values of companies, identical except for their respective gearing, was the process of 'arbitrage', a feature of all developed financial markets which ensures that assets with the same risk–return characteristics sell at the same prices.

To support these contentions, some algebraic analysis is required, although readers will find that it is much less complex than may appear at first sight.

■ The analytical framework

No distinction is made between short- and long-term debt and we assume that all borrowing is perpetual. The company is expected to deliver constant and perpetual estimated annual earnings, described by a normally-distributed range of possible outcomes. Investors are assumed to have homogeneous expectations, i.e. they all formulate similar estimates of company earnings, E, the net operating income (NOI) before interest and tax, or, more simply, revenues less variable costs less fixed operating costs. It is important to note that we are using **free cash flow** concept of earnings as explained in Chapter 12, i.e. income net of any investment required to rectify wear-and-tear on capital equipment and hence maintain annual earnings at E.

The discount rate applied to the stream of expected earnings depends on the degree of business risk incurred by the enterprise. MM used the concept of 'equivalent risk classes', each one containing firms whose earnings depend on the same risk factors and from which the market expects the same return. In terms of the Capital Asset Pricing Model, this means that the earnings streams from firms in the same risk category are perfectly correlated and that member companies have identical activity Betas.

For consistency, we use the same definitions and notation as in Chapter 18. The key definitions are reproduced in Table 19.1.

The overall rate of return that the company must achieve to satisfy all its investors is the weighted average cost of capital (WACC), denoted by k_o. This can be expressed as:

$$k_o = \left(k_e \times \frac{V_S}{V_o} \right) + \left(k_d \times \frac{V_B}{V_o} \right) = \frac{E}{V_o}$$

The WACC equals E/V_o, since net operating income is composed of earnings attributable to shareholders, $k_e V_S$, plus payments to lenders, iB.

Using this set of definitions, we now examine the impact of variations in capital structure on V_o and k_o. The *original* MM analysis did not apply directly to the UK context, being expressed in terms of substituting debt for equity, i.e. using debt to repurchase ordinary shares. This was not generally possible in the UK until the Companies Act 1981, which enabled firms, subject to shareholder approval, to undertake such repurchases. Now this is permitted, the original MM analysis is more readily applicable to the capital structure decisions of UK firms. Using debt-for-equity substitution rather than adding debt to equity has the major advantage of enabling us to hold constant both the book values of assets and capital employed and also the PBIT, a device used in the Lindley example in Chapter 18. Any gearing change alters only the company's capital structure, with no effect on company size or the level and riskiness of operating earnings. As a result, we can focus directly on the relationship between V_o and k_o.

Table 19.1 Key definitions in capital structure analysis

V_o = the overall market value of the whole company
V_S = the value of the shareholders' stake in the company
V_B = the market value of the company's outstanding borrowings
B = the book value of borrowings (generally assumed equal to its market value)
k_e = the rate of return required by shareholders
k_d = the rate of return required by providers of debt capital
k_o = the overall (weighted average) cost of capital
i = the coupon rate on debt
iB = annual interest charges (i.e. payments to lenders, based on book value)
$k_e V_S$ = payments to shareholders = $(E - iB)$, so that
E = annual net operating income (NOI) = $(iB + k_e V_S)$

It should be stressed that we are assuming no retention of earnings, i.e. D(dividends) = E, and hence no growth, and, for the moment, no taxes on corporate profits.
The value of equity in an all-equity firm is:

$$V_S = \frac{D}{k_e} = \frac{E}{k_e}$$

The value of a geared firm making interest payments of iB is:

$$V_S = \frac{(E - iB)}{k_e}$$

The value of the whole firm in either case is:

$$V_o = V_S + V_B$$

■ MM's assumptions

The MM thesis did not go unchallenged. Much criticism of MM's analysis stemmed from failure to understand positive scientific methodology. Their analysis attempted to isolate the critical variables affecting firm value under the restrictive conditions of a perfect capital market. This provided a systematic basis for examining how imperfections in real world markets could influence the links between value and risk. The key assumptions are:

- All investors are price-takers, i.e. no individual can influence market prices by the scale of his or her transactions.
- All market participants, firms and investors, can lend or borrow at the same risk-free rate.
- There are neither personal nor corporate income taxes.
- There are no brokerage or other transactions charges.
- Investors are all rational wealth-seekers.
- Firms can be grouped into 'homogeneous risk classes', such that the market seeks the same return from all member firms in each group.
- Investors formulate similar expectations about future company earnings. These are described by a normal probability distribution.
- The assets of an insolvent firm can be sold at full market values.

19.3 MM'S PROPOSITIONS

MM's analysis was presented as three propositions, the first being the crucial one.

■ Proposition I

The central proposition is that *a firm's WACC is independent of its debt/equity ratio, and equal to the cost of capital that the firm would have with no gearing in its capital structure*. In other words, the appropriate capitalisation rate for a firm is the rate applied by the market to an ungeared company in the relevant risk category, i.e. that company's cost of equity. The arbitrage mechanism will operate to equalise the values of any two companies whose values are temporarily out of line with each other. The example of Nogear plc and Higear plc will illustrate this.

Nogear plc and Higear plc

Nogear plc is ungeared, financed by 5 million £1 shares, while Higear plc's balance sheet shows £1 million debt, interest payable at 10 per cent, and 4 million £1 shares. Higear's debt/equity ratio is thus (£1 m/£4 m) = 25 per cent, at book values. The two firms are identical in every other respect, including their business risks and levels of annual expected earnings (E) of £1 million. The market requires a return of 20 per cent for ungeared streams of equity income of this risk.

Imagine that, temporarily, the market value of Nogear is £4 million and that of Higear is £6 million. Higear's equity is thus valued by the market at (£6 m − £1 m) = £5 m. (Its debt/equity ratio expressed in terms of *market* values is thus £1 m/£5 m = 20 per cent.) These market values correspond to respective share prices of (£4 m/5 m shares) = 80p for Nogear and (£5 m/4 m shares) = £1.25 for Higear.

The different share values conform to the traditional relationship at relatively low gearing ratios. Higear has a greater value presumably due to its gearing. Also, it appears that Nogear is undervalued by the market since, at a required return of 20 per cent, its value should be (£1 m/0.2) = £5 m.

MM argue that such imbalances can only be temporary and the benefit obtained by Higear for its shareholders is largely illusory. It will pay investors to sell their holdings in the overvalued company and buy stakes in the undervalued one. Specifically, shareholders can achieve a higher return by selling holdings in Higear, and simultaneously, replicate its gearing (MM call this **'home-made gearing'**) and achieve a higher overall return. This process of arbitrage will force up the value of Higear and lower Nogear's value, until their values are equalised. There is thus little point in a firm borrowing to gear-up its capital structure when investors can achieve the same benefits by acting independently.

'home-made gearing'
Where an investor borrows to arbitrage between two identical but differently-valued assets

Home-made gearing

Consider the case of an investor with a 1 per cent equity stake in Higear. At present, this stake is worth (1 per cent of £5 m) = £50,000, attracting an income of 1 per cent of (£1 million less interest payments of 10% × £1 m), i.e. (1% × £900,000) = £9,000. This investor could realise his or her holdings for £50,000 and duplicate Higear's debt/equity ratio of 20 per cent by borrowing £10,000 at 10 per cent and investing the total stake of £60,000 in Nogear shares. This would buy (£60,000/£4 m) = 1.5% of Nogear's equity, to yield a dividend of (1.5% × £1 m) = £15,000. Personal interest commitments amount to (10% × borrowings of £10,000) = £1,000, for a net return of (£15,000 − £1,000) = £14,000. Clearly, it would pay all investors to undertake this arbitrage exercise, thus pushing down the value of Higear and pushing up the value of Nogear until there was no further scope to exploit such gains. This point would be reached when the market values of the two companies were equal and when each offered the appropriate 20 per cent return required by the market:

$$\text{Value of Nogear} = \text{Value of Higear} = \frac{E}{k_e} = \frac{£1 \text{ m}}{0.2} = £5 \text{ m}$$

At this equilibrium relationship, the price of each company's shares is £1. For Nogear, the calculation is (£5 m/5 m shares) = £1, while for Higear, the relevant figures are

(£5 m − £1 m debt) divided by 4 m shares = £1. *In an MM world, there are no prolonged benefits from gearing, and any short-term discrepancies between geared and otherwise identical ungeared companies quickly evaporate. As a result, MM concluded that both company value and the overall required return, k$_o$, are independent of capital structure.*

In reality, not all of the conditions required to support the arbitrage process may apply, suggesting that any *observed* benefits may derive from imperfections in the capital market. Moreover, if gearing does result in higher company value, there must have been a wealth transfer, since nothing has occurred to alter the fundamental wealth-creating properties of the company.

■ Proposition II: the behaviour of the cost of equity

Underpinning Proposition I is a statement about the behaviour of the relevant cost of capital concepts – in particular, the rate of return required by shareholders. This is expressed in MM's second proposition which states *'the expected yield of a share of equity is equal to the appropriate capitalisation rate, k$_e$, for a pure equity stream in the class, plus a premium related to the financial risk equal to the debt/equity ratio times the spread between k$_e$ and k$_d$'*. This proposition can be expressed as:

$$k_{eg} = k_{eu} + (k_{eu} - k_d)\frac{V_B}{V_S}$$

where k_{eg} and k_{eu} denote the returns required by the shareholders of a geared company and an equivalent ungeared company, respectively. The expression is easily obtained from Proposition I. (See Appendix I to this chapter.) It simply tells us that *the rate of return required by shareholders increases linearly as the debt/equity ratio is increased,* i.e. the cost of equity rises exactly in line with any increase in gearing to offset precisely any benefits conferred by the use of apparently cheap debt. The relevant relationships are shown in Figure 19.1.

If you check back to Chapter 18, which covered the traditional view of gearing, and to Figure 18.2 in particular, you will find that the behaviour of k$_e$ is the critical difference between the MM version and the traditional theory. In the latter, there is little or no reaction by shareholders to an increase in debt-to-equity ratio over 'modest' levels of gearing. They presumably are not alarmed by the 'judicious' use of debt. By contrast, shareholders, in the MM view, respond immediately when any gearing is undertaken, i.e. to them, *any* use of debt introduces an element of risk.

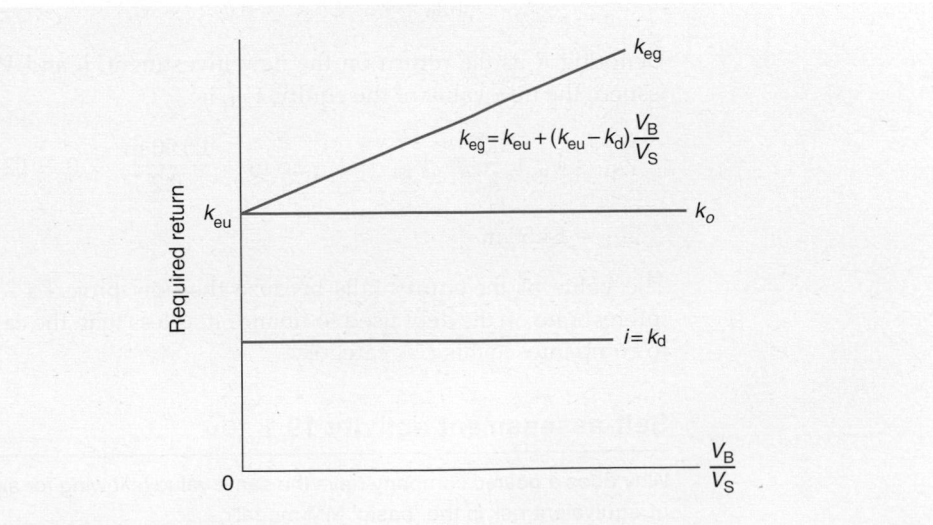

Figure 19.1 MM's Propositions I and II

It should now be appreciated that in the Nogear/Higear example, Higear shareholders were seeking too low a rate of return, i.e. Higear was overvalued, and the market was temporarily offering Nogear's shareholders too high a return, i.e. Nogear was undervalued. Via the process of arbitrage, their values were brought back into line, and appropriate rates of return on equity were established, reflecting their respective levels of gearing. The correct rate of return for Higear's equity, for its particular debt/equity ratio of 25 per cent (at equilibrium market values) is:

$$k_{eg} = k_{eu} + (k_{eu} - k_d)\frac{V_B}{V_S} = 20\% + (20\% - 10\%)\frac{£1 \text{ m}}{£4 \text{ m}} = 22.5\%$$

■ Proposition III: the cut-off rate for new investment

MM's third proposition asserts that '*the cut-off rate for new investment will in all cases be k_o and will be unaffected by the type of security used to finance the investment*'.

A proof of this proposition is given in Appendix II to this chapter, but it is quite easy to justify intuitively. Proposition I states that the WACC, k_o, is constant and equal to the cost of equity in an equivalent ungeared company. Since k_o is invariant to capital structure, it follows that however a project is financed, it must yield a return of at least k_o, the overall minimum return required to satisfy stakeholders as a whole.

It is worth illustrating this contention for the case where a company invests to yield a return *above* the cost of the debt used to finance the project, but *below* the cost of equity in an ungeared company.

Nogear: right and wrong investment cut-off rates

Nogear decides to raise £2 million via a debt issue at 10 per cent to finance a new project expected to yield an annual return of 15 per cent for many years into the future. Is this an acceptable project? Proposition I tells us that the initial value of the company, V_o, and hence the equity, V_{So}, prior to the issue is:

$$V_o = V_{So} = \frac{E}{k_e} = \frac{£1 \text{ m}}{0.2} = £5 \text{ m}$$

Incorporating the new project's earnings, the post-issue value of the whole company, V_1, is:

$$V_1 = \frac{£1 \text{ m} + (15\% \times £2 \text{ m})}{20\%} = \frac{£1.30 \text{ m}}{0.2} = £6.50 \text{ m}$$

Denoting R as the return on the new investment, I, and V_B as the value of the debt issued, the new value of the equity, V_{S1}, is:

$$V_{S1} = V_o + \frac{RI}{k_o} - V_{Bo} - I = £5 \text{ m} + \frac{£0.30 \text{ m}}{0.2} - 0 - £2 \text{ m} = (£6.50 \text{ m} - £2 \text{ m})$$

$$= £4.50 \text{ m}$$

The value of the equity falls because the new project's return, although above the interest rate on the debt used to finance it, is less than the capitalisation rate applicable to companies in this risk category.

Self-assessment activity 19.1

Why does a geared company have the same value (allowing for size) as an ungeared company of equivalent risk in the 'basic' MM model?

(Answer in Appendix A at the back of the book)

19.4 DOES IT WORK? IMPEDIMENTS TO ARBITRAGE

The operation of the arbitrage process requires that corporate and personal gearing are perfect substitutes in a perfect capital market. The Nogear/Higear example showed how individual investors could replicate corporate gearing to unwind any transitory premium in the share price of a geared company. Much criticism of MM centres on the perfect capital market assumptions and hence the extent to which the arbitrage process can be expected to operate in practice.

In reality, brokerage fees discriminate against small investors, and other transaction costs limit the gains from arbitrage. Moreover, if companies can borrow at lower rates than individuals, investors may prefer the equity of geared companies as vehicles for obtaining benefits otherwise denied to them. It is well known that, for reasons of size, security and convenience, large firms can borrow at lower rates than small firms and individuals. In addition, some major UK investors (e.g. pension funds) face restrictions on their borrowing powers, limiting their scope for home-made gearing. Finally, whereas the shareholders in a geared firm have the protection of limited liability, personal borrowers enjoy no such protection in the event of bankruptcy.

Some authors suggest that such imperfections may foster investor demand for the equity of geared companies. However, to sustain this argument, we would need to produce evidence that relatively (but safely) geared companies are more attractively rated by the market. There is little evidence that such firms sell at relatively high P:E ratios. Indeed, UK investment trust companies, which invest in equities, often using substantial borrowed capital, typically sell at significant *discounts* to their net asset values – discounts far higher than can be plausibly explained by the transactions costs that would be incurred in liquidating their portfolios.

Self-assessment activity 19.2

What factors restrict the ability of investors to arbitrage in the way envisaged by MM?

(Answer in Appendix A at the back of the book)

19.5 MM WITH CORPORATE INCOME TAX

The analysis of MM's three propositions in Section 19.3 is a theoretical exercise, designed to isolate the key variables relating company value and gearing. This only becomes operational when 'real-world' complications are introduced. Perhaps the most important of these is corporate taxation. In most economies, corporate interest charges are tax-allowable, providing an incentive for companies to gear their capital structures. In a taxed world, the MM conclusions change significantly.

Because corporation tax is applied to earnings after deducting interest charges, the value of a geared company's shares is the capitalised value of the after-tax earnings stream (net income), i.e. $(E - iB)(1 - T)$:

$$V_S = \frac{(E - iB)(1 - T)}{k_{eg}}$$

where k_{eg} is the return required by shareholders, allowing for financial risk, and T is the rate of tax on corporate profits.

Assuming that the book and market values of debt capital coincide ($B = V_B$), so that the cost of debt, k_d, equates to the coupon rate, i, the value of debt is the discounted interest stream, i.e. $V_B = iB/i$. The value of the whole company is thus:

$$V_o = V_S + V_B = \frac{(E - iB)(1 - T)}{k_{eg}} + \frac{iB}{i}$$

It can be shown that geared companies will sell at a premium over equivalent ungeared companies because of the benefits of tax-allowable debt interest. The post-tax annual expected earnings stream, E_T, comprises the earnings attributable to shareholders plus the debt interest:

$$E_T = (E - iB)(1 - T) + iB$$

This simplifies to:

$$E_T = E(1 - T) + TiB$$

This second expression is very useful: the first element is the net income that the shareholders in an equivalent ungeared company would receive, while the second element is the annual tax benefit afforded by debt interest relief. The total value of the geared company, V_g, is found by capitalising the first element at the cost of equity capital applicable to an ungeared company (k_{eu}), while the second is capitalised at the cost of debt, which we have assumed equals the nominal rate of interest, i:

$$V_g = \frac{E(1 - T)}{k_{eu}} + \frac{TiB}{i} = \frac{E(1 - T)}{k_{eu}} + TB = V_u + TB$$

Self-assessment activity 19.3

What are the respective values of geared and ungeared firms if:

- Earnings = £100 m before tax
- Tax rate = 30%
- k_{eu} = 15%
- The geared firm borrows £200 m?

(Answer in Appendix A at the back of the book)

This is a highly significant result. The expression for the value of the geared company comprises the value of an equivalent ungeared company, V_u, plus a premium derived by discounting to perpetuity the stream of tax savings that can be claimed so long as the company has sufficient taxable capacity, i.e. if $E > iB$. The introduction of this second term, TB, the discounted value of future tax savings, or the **tax shield**, is a major modification of MM's Proposition I, as shown in Figure 19.2.

tax shield
The tax savings achieved by setting tax-allowable expenses such as interest payments against profits

The company value profile now rises continuously with gearing. Proposition II also needs modification. With no corporate tax, this stated that the shareholders in a geared company require a return, k_{eg}, of:

$$k_{eg} = k_{eu} + (k_{eu} - k_d)\frac{V_B}{V_S}$$

However, in a taxed world, the return required by shareholders becomes:

$$k_{eg} = k_{eu} + (k_{eu} - k_d)(1 - T)\frac{V_B}{V_S}$$

The return required by the geared company's shareholders is now the cost of equity in an identical ungeared company plus a financial risk premium related to the corporate tax rate and the debt/equity ratio.

The premium for financial risk required by shareholders is lower in this version owing to the tax deductibility of debt interest, making the debt interest burden less onerous. This relationship is also shown by Figure 19.2. It follows that if, at every level of gearing, the cost of equity is lower and also the cost of debt itself is reduced by interest deductibility, the WACC (k_o) is lower at all gearing ratios, and declines as gearing increases. Figure 19.2 shows the effect on the WACC.

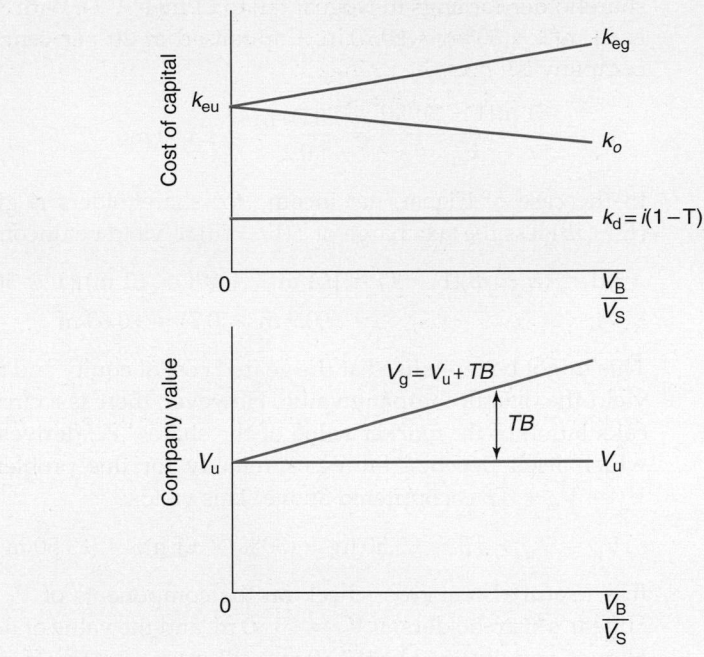

Figure 19.2 The MM thesis with corporate income tax

The tax advantage of debt financing is incorporated in the revised equation for the WACC:

$$k_o = \left[k_{eg} \times \frac{V_S}{V_S + V_B} \right] + \left[i(1 - T) \times \frac{V_S}{V_S + V_B} \right]$$

This can also be written as:

$$k_o = k_{eu} \left[1 - \frac{T \times V_B}{V_S + V_B} \right]$$

Clearly, there are significant advantages from gearing, with the implication that companies should gear up until debt provides almost 100 per cent of its financing. However, this does not seem plausible. Surely there are practical, 'sensible' limits to company gearing, given the risks involved? More of this later!

Self-assessment activity 19.4

Compare the overall required return in geared and ungeared firms if:

■ k_{eu} = 15%
■ Tax rate = 30%
■ The geared firm has borrowed £200 m at 7% interest, and has issued equity of £400 m.

(Answer in Appendix A at the back of the book)

■ Example of the impact of corporate taxation

It is now helpful to demonstrate 'with-tax' relationships using the examples of Nogear and Higear. Recall that both companies had E of £1 million and their equilibrium market values were £5 million under the 'no-tax' version of the MM thesis. After taxation, shareholder earnings in Nogear fall to £1 m$(1 - T)$. With 30 per cent corporate tax, this is £1 m$(1 - 30\%)$ = £0.70 m. Capitalised at 20 per cent, the value of the ungeared company is:

$$V_u = \frac{£1\,m(1 - 30\%)}{k_{eu}} = \frac{£0.70\,m}{0.2} = £3.50\,m$$

In the case of Higear, net income for shareholders is given by taxable earnings of $(E - iB)$ less the tax charge of $T(E - iB)$ to yield net income of:

$$NI = (E - iB)(1 - T) = [£1\,m - (10\% \times £1\,m)](1 - 30\%)$$
$$= (£0.9\,m \times 0.7) = £0.63\,m$$

This might be capitalised at the geared cost of equity and added to the value of debt to yield the overall company value. However, there is a circular problem here, since the calculation of the market value of the shares, V_S, derives from the calculation of k_{eg}, which itself depends on V_S. A remedy for this problem is to use the expression $V_g = V_u + TB$ encountered above. This yields:

$$V_g = V_u + TB = £3.50\,m + (30\% \times £1\,m) = (£3.50\,m + £0.30\,m) = £3.80\,m$$

It is useful also to cross-check on the components of V_g and the return required by Higear's shareholders. If V_g = £3.80 m, and the value of debt is £1 million, the value of Higear's equity must be (£3.80 m − £1 m) = £2.80 m. Using the revised expression for the return required by the shareholders of a geared company, we find:

$$k_{eg} = k_{eu} + (k_{eu} - i)(1 - T)\frac{V_B}{V_S}$$
$$= 20\% + (20\% - 10\%)(1 - 30\%)\frac{£1\,m}{£2.8\,m}$$
$$= (20\% + 2.5\%) = 22.5\%$$

The geared company clearly has a greater market value – it is worth more due to the value of the tax shield. The size of this tax shield depends on the gearing ratio, the rate of taxation and the taxable capacity of the enterprise. Since gearing has raised company value, the earlier conclusion, that the benefits of gearing are illusory, must be modified. The reason is that the stakeholders of Higear benefit at the expense of the taxpayer due to the tax deductibility of debt interest. (Whether this is desirable or not in a wider context depends on the value of the forgone tax revenues in their alternative use, which is an issue for welfare economists.)

In its tax-adjusted form, the MM thesis looks rather more like the traditional version, in so far as the WACC declines over some range of gearing. However, the benefits from gearing clearly derive from the tax system, rather than from the apparent failure of the shareholders to respond fully to financial risk by seeking higher returns. We will discover that the similarity becomes even closer when we allow for financial distress. Before doing this, we will show how the MM approach can be integrated with the CAPM.

19.6 CAPITAL STRUCTURE THEORY AND THE CAPM

A feature of MM's initial model was the classification of firms into 'homogeneous risk classes' as a way of controlling for inherent operating or business risk. The modern distinction between systematic and specific risk makes this device unnecessary, as relevant business risk is expressed by the Beta. The key point is that gearing introduces additional risk so that shareholders require additional compensation. Whereas in an ungeared firm the cost of equity is:

$$k_{eu} = R_f + \beta_u(ER_m - R_f)$$

in a geared firm this becomes

$$k_{eg} = R_f + \beta_g (ER_m - R_f)$$

with

$$\beta_g > \beta_u$$

and

$$k_{eg} > k_{eu}$$

Clearly, gearing increases the equity Beta. It is a relatively simple task to integrate the MM analysis with the CAPM. This was first performed by Hamada (1969), who demonstrated that the required return on the equity of a geared firm in a CAPM framework is:

$$k_{eg} = R_f + (ER_m - R_f) \times \beta_u \times \left[1 + \frac{V_B (1 - T)}{V_S}\right]$$

where β_u is the Beta applicable to the earnings of an ungeared company, or the pure equity Beta. Multiplying out, we derive:

$$k_{eg} = R_f + \beta_u (ER_n - R_f) + (ER_m - R_f) \times \beta_u \times \left[\frac{V_B (1 - T)}{V_S}\right]$$

This looks unwieldy, but is a useful vehicle for making the distinction between business and financial risk. The Betas recorded by the London Business School, are geared equity Betas, incorporating elements of both types of risk. Given that, and using Hamada's revised CAPM expression, the geared Beta, β_g, is:

$$k_{eg} = R_f + \beta_g (ER_m - R_f)$$

$$\beta_g = \beta_u \left[1 + \frac{V_B (1 - T)}{V_S}\right]$$

The ungeared equity Beta is therefore:

$$\beta_u = \frac{\beta_g}{\left[1 + \dfrac{V_B (1 - T)}{V_S}\right]}$$

This can also be written as:

$$\beta_u = \beta_g \times \left[\frac{V_S}{V_S + V_B (1 - T)}\right]$$

The shareholders of a geared company seek compensation for two separate types of risk – the underlying or basic risk of the business activity, and also for financial risk. The rewards for bearing these two forms of risk are the respective premiums for business risk and for gearing.

■ Higear and Nogear: separating the risk premiums

To explore this distinction, consider again the example of Nogear and Higear. Assume that the ungeared Beta applicable to this risk class is 1.11, the risk-free return is 10 per cent, the return expected on the market portfolio is 19 per cent and the corporate tax rate is 30 per cent. Recall that when we last encountered these companies (see Section 19.5) their respective values were:

Nogear: $V_u = V_S = £3.50$ m

Higear: $V_g = V_u + TB = (£3.50 \text{ m} + £0.30 \text{ m}) = £3.80$ m

$\qquad V_B = £1$ m

$\qquad V_S = (£3.80 \text{ m} - £1 \text{ m}) = £2.80$ m

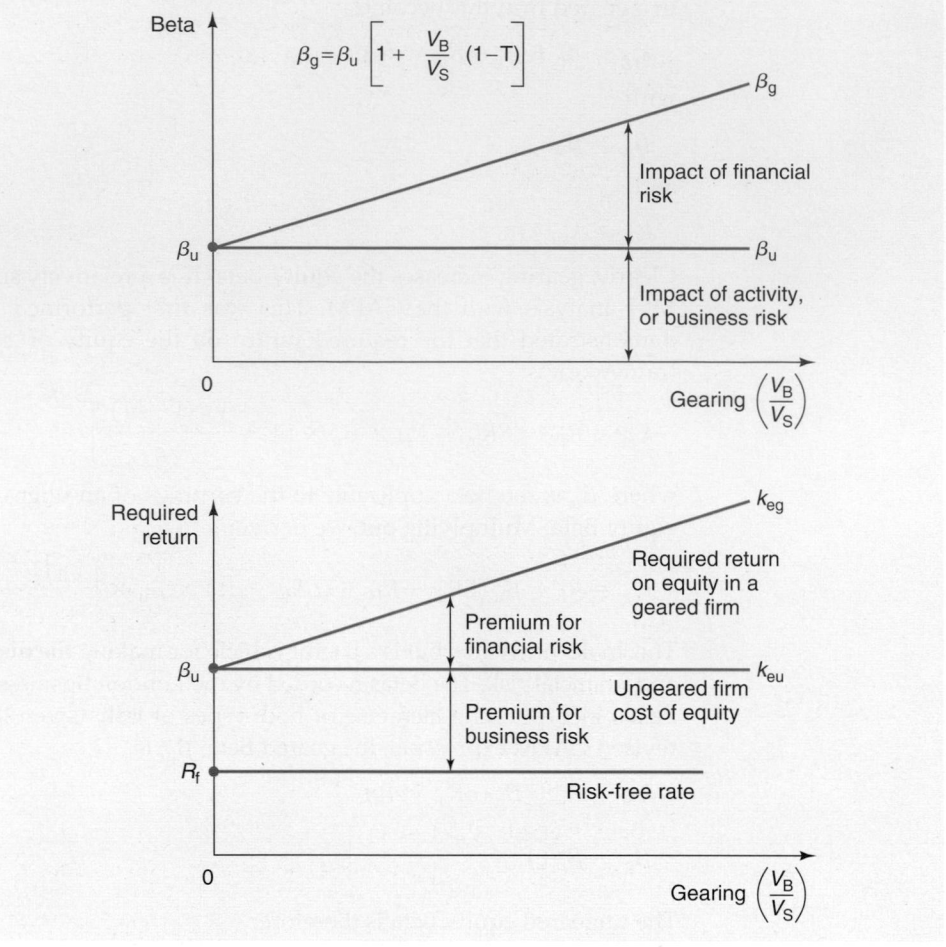

Figure 19.3 Business and financial risk premia and the required return

First, we can verify the return required by Nogear's shareholders. This is:

$$k_{eu} = R_f + \beta_u [ER_m - R_f]$$
$$= 10\% + 1.11[19\% - 10\%] = (10\% + 10\%) = 20\%$$

Second, we can analyse the composition of the return required by Higear's shareholders. To find the overall return they seek, we need to know the geared Beta. This is given by:

$$\beta_g = \beta_u \left[1 + \frac{V_B (1 - T)}{V_S} \right] = 1.11 \times \left[1 + \frac{£1\,m(1 - 30\%)}{£2.80\,m} \right] = 1.3875$$

For $\beta_g = 1.3875$, the return required by Higear's shareholders is:

$$k_{eg} = R_f + \beta_g [ER_m - R_f] = 10\% + 1.3875 [19\% - 10\%]$$
$$= (10\% + 12.5\%) = 22.5\%$$

Analysing the cost of equity for Higear into its components, we find:

$$k_{eg} = \text{Risk-free rate} + \text{Business risk premium} + \text{Financial risk premium}$$
$$= R_f + \beta_u[ER_m - R_f] + [ER_m - R_f]\beta_u \times \frac{V_B (1 - T)}{V_S}$$
$$= 10\% + 1.11 [19\% - 10\%] + [19\% - 10\%]\,1.11 \times \frac{£1\,m(1 - 30\%)}{£2.80\,m}$$
$$= (10\% + 10\% + 2.5\%) = 22.5\%$$

This corresponds to the result obtained more directly with the CAPM formula. The two separate components of the geared Beta are shown in Figure 19.3. The increase in the geared Beta, as the debt/equity ratio increases, drives up the additional required premium *pro rata*.

19.7 LINKING THE BETAS

There is a useful expression available to show how the various Betas are linked together. It is important to recall the MM message that underlying business or activity risk is unaffected by the method of financing. If a firm chooses to borrow, thus introducing financial risk, the shareholders will respond by looking for a higher return as they perceive greater financial risk affecting their future income, but the risk attaching to the firm's actual operating activities is untouched – it is the same firm operating in the same business environment and operated by the same managers. All that has happened is a repackaging of the firm's flow of operating income resulting in lenders now having a prior claim. The size of the operating income itself is unaffected, only its distribution changes.

Given that the activity risk is unaffected by gearing, we can use the accounting equation to show the linkages. The accounting equation tells us that the assets are equal to the methods of financing. Translating this into CAPM terms, the asset Beta (i.e. the activity Beta) equals the Beta of the methods of finance used to acquire those assets. In other words, the asset Beta equates to a weighted average of the Betas of the various methods of financing, according to the importance of each source of finance in the capital structure.

Algebraically, this is given by:

Beta of assets

= (Equity Beta × proportion of equity) + (Debt Beta × proportion of debt)

$$\text{Beta}_A = \left(\text{Beta}_S \times \frac{V_S}{V_S + V_B(1-T)} \right) + \left(\text{Beta}_B \times \frac{V_B(1-T)}{V_S + V_B(1-T)} \right)$$

Notice that the tax shield is reflected in applying the term $(1-T)$ to the debt component. Notice also that, as the debt proportion increases, the equity Beta must increase to preserve the constant asset Beta. It is usual to assume that the debt Beta is zero, although there is some evidence that corporate debt has a very low Beta, around 0.1 to 0.2.

However, if we do assume a debt Beta of zero, this becomes a very versatile expression, e.g. when moving into a new activity we can take a firm's equity Beta and ungear it to reveal the underlying activity Beta. This is particularly useful when diversifying into a new activity – we might borrow a Beta from another firm, whose gearing may differ from our own. In this case, we might ungear the borrowed Beta to strip out that firm's financial risk, and then re-gear to incorporate our own firm's gearing ratio.

To illustrate this, assume we have the following data:

Equity Beta of firm operating in new activity = 1.35
Gearing ratio (debt/equity) of this firm   = 40%
(i.e. debt proportion = 40:100)
Tax rate   = 30%
Own gearing ratio   = 10% (debt/equity)

Ungearing the other firm's equity Beta, assuming the debt Beta is zero, we have:

$$\text{Beta}_A = \text{Beta}_S \times \left(\frac{V_S}{V_S + V_B(1-T)} \right) = 1.35 \times \left(\frac{60}{60 + 40(1-T)} \right)$$

$$= 1.35 \times 60/88 = 0.92$$

Re-gearing to incorporate our own gearing, the equity Beta is given by:

$$0.92 = \text{Beta}_S \times \frac{100}{100 + 10(1 - T)} = \text{Beta}_S \times 100/107$$

Whence, equity Beta $= 0.92 \times 107/100 = 0.98$

Self-assessment activity 19.5

Ungear a Beta of 1.45 if:

- Tax rate $= 30\%$
- The debt–equity ratio $= 1:2$

(Answer in Appendix A at the back of the book)

19.8 MM WITH FINANCIAL DISTRESS

In Section 19.5, we saw how including corporate taxation in the MM model implied that companies should rely on debt for nearly 100 per cent of their financing. This implication is clearly at odds with observed practice – few companies gear up to extreme levels, through both their own and lenders' fear of insolvency, and its associated costs. MM's omission of liquidation costs from their analysis was a logical consequence of their perfect capital market assumptions. In such a market, where investors are numerous and rational, and have homogeneous expectations and plentiful access to information, the resale value of assets, even those being sold in a liquidation, will reflect their true economic values. Investors will recognise the worth of such assets as measured by the present values of their future income flows, and be prepared to bid up to this value, so that the price realised by a liquidator should not involve any discount.

costs of financial distress
The costs incurred as a firm approaches, and ultimately reaches, the point of insolvency

In effect, liquidation costs and the other **costs of financial distress** introduce a new imperfection into the analysis of capital structure decisions: namely the actual or expected inability to realise 'full value' for assets in a distress sale and the costs of actions taken to forestall this contingency.

Incorporating financial distress

Denoting the 'costs of financial distress' by *FD*, the value of a geared company becomes:

$$V_g = V_u + [TB - FD]$$

From this, we may conclude that the *financial manager should attempt to maximise the gap between tax benefits and financial distress costs, i.e. (TB − FD), and that there exists an optimal capital structure where company value is maximised.* This occurs where the marginal benefit of further tax savings equals the marginal cost of anticipated financial distress. This occurs with debt of X^* in Figure 19.4.

The costs of financial distress rise with gearing once the market starts to perceive a substantially increased risk of financial failure. The likelihood of *FD* being non-zero depends on the probability distribution of the firm's earnings profile. For example, in the Lindley example in Chapter 18, for gearing ratios up to 50 per cent the probability of inability to meet interest payments is zero, but it would be 0.25 for any higher gearing ratio. For most companies, the probability, p, of financial distress will increase with the book values of debt, B, so that the *FD* function increases with gearing. If d denotes the expected percentage discount on the pre-liquidation value in the event of a forced sale, the expected costs of financial distress are:

$$FD = (p \times d \times V_g)$$

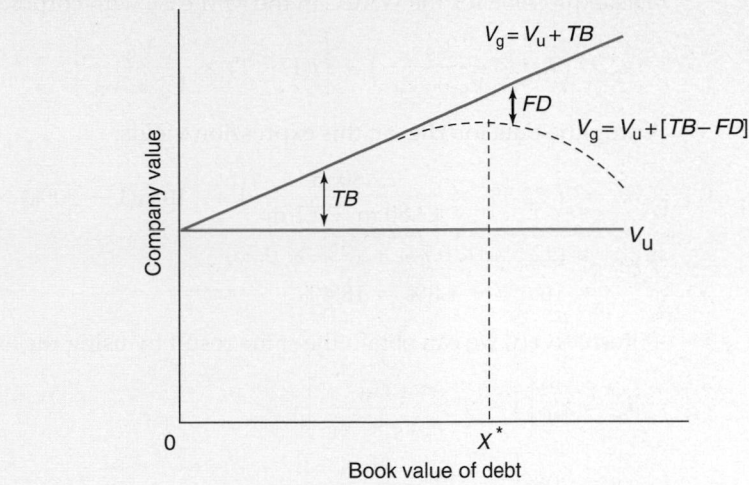

Figure 19.4 Optimal gearing with liquidation costs

and the value of the geared firm is:

$$V_g = V_u + (TB - p \times d \times V_g)$$

This suggests that market imperfections can be exploited to raise company value so long as TB exceeds ($p \times d \times V_g$). Notice that the inverted U-shaped value profile now appears remarkably similar to the traditional version and, of course, is associated with a mirror-image WACC schedule.

You may recall our earlier comment that, after introducing market imperfections such as tax, the MM model begins to look more like the traditional version. With the inclusion of financial distress costs, this resemblance is closer still. However, the discussion of the impact of personal taxation in Appendix III shows that the debate is not yet dead.

19.9 CALCULATING THE WACC

Before progressing, you may find it useful to reread Chapter 11, where we discussed the hierarchy of discount rates and required rates of return, but deferred consideration of the problems posed by mixed capital structures until Chapter 18.

The WACC is the overall required return needed to satisfy all stakeholders. It is also the required return on the assumption that new projects are financed in exactly the same way as existing ones. If the company is all-equity financed, then the WACC is simply the return required by shareholders.

Gearing does not affect the underlying risk of the company's business activities. If a company uses debt capital, it is merely repackaging its operating income into different proportions of debt interest and equity income, but not influencing the size or the riskiness of this income before appropriation. What does change is the riskiness of the stream of residual equity income, which is why the equity Beta rises, pulling up with it the return required by shareholders.

We can explore this proposition with the case of Higear. The relevant figures for Higear were:

Value of debt	$= V_B$	$= £1$ m
Value of equity	$= V_S$	$= £2.80$ m
Shareholders' required return	$= k_{eg}$	$= 22.5\%$
Interest cost of debt	$= i$	$= 10\%$
Rate of corporate tax	$= T$	$= 30\%$

The expression for the WACC in the MM case with corporate tax is:

$$k_o = \left(k_{eg} \times \frac{V_S}{V_S + V_B} \right) + \left[i(1 - T) \times \frac{V_B}{V_S + V_B} \right]$$

Using the data for Higear, this expression yields:

$$k_o = \left(22.5\% \times \frac{\pounds2.80\text{ m}}{\pounds2.80\text{ m} + \pounds1\text{ m}} \right) + \left[10\%(1 - 30\%) \times \frac{\pounds1\text{ m}}{\pounds2.80\text{ m} + \pounds1\text{ m}} \right]$$

$$= (22.5\% \times 0.74) + (7\% \times 0.26)$$

$$= 16.6\% + 1.8\% = 18.4\%$$

Alternatively, we can obtain the same result by using the expression:

$$k_o = k_{eu} \left[1 - T \times \frac{V_B}{V_S + V_B} \right]$$

$$= 20\% \left[1 - 30\% \frac{\pounds1\text{ m}}{\pounds3.80\text{ m}} \right]$$

$$= 20\% \times 0.92$$

$$= 18.4\%$$

■ Relaxing critical assumptions

Two important questions now arise. First, what happens to the discount rate if a company diversifies into an activity with a risk profile different from existing operations? Second, what happens if the gearing ratio is altered? The first issue is easier to handle.

Allowing for different risks

Imagine Higear proposes to diversify into a higher risk business. Because the discount rate applicable to evaluating this project should reflect the systematic risk involved, the required return previously calculated is no longer appropriate. To cope with this problem, the following procedure is suggested:

1 Select a company already operating in the target activity, ideally, one with operating characteristics very similar to those exhibited by the project, and identify its Beta coefficient, e.g. by using the RMS.
2 If the surrogate company's gearing differs from that of Higear, the Beta must be adjusted by removing the effect of the surrogate's own gearing, and then superimposing Higear's gearing on the resulting ungeared Beta.
3 Calculate the WACC incorporating the surrogate activity Beta, adjusted for Higear's own gearing.

Assume Higear plans to enter an activity already served by Supergear, whose equity Beta is 1.8, and which has a debt/equity ratio of 1:2. Supergear's Beta is ungeared as follows:

$$\beta_u = \frac{\beta_g}{1 + \frac{V_B}{V_S}(1 - T)} = \frac{1.8}{1 + \frac{1}{2}(1 - 30\%)} = \frac{1.8}{1.35} = 1.33$$

The geared Beta applicable to Higear's capital structure (i.e. £1 million debt and £2.80 million equity) is:

$$\beta_g = \beta_u \left[1 + \frac{V_B}{V_S} \times (1 - T) \right]$$

$$= 1.33 \left[1 + \frac{\pounds1\text{ m}}{\pounds2.80\text{ m}} \times (1 - 30\%) \right]$$

$$= 1.33[1.25] = 1.663$$

For this risk, and with a 9 per cent market risk premium, Higear's shareholders require a return of:

$$ER_j = R_f + \beta_g [ER_m - R_f] = 10\% + 1.663[9\%] = 25\%$$

Finally, the WACC applicable to this activity risk and Higear's own gearing is:

$$(25.0\% \times 0.74) + (10\% [1 - 30\%] \times 0.26) = 18.5\% + 1.8\% = 20.3\%$$

The second issue, the effect of a change in gearing, poses more of a conundrum.

Allowing for a change of gearing

In the Higear example, no change in gearing was envisaged when financing new projects. However, as we have repeatedly warned, a significant change in gearing affects the market values of both debt and equity capital: for example, shareholders may respond adversely to higher gearing and the higher financial risk. Also, the value of debt may be marked down in the market. To compute the WACC, we would have to assess the new return required by shareholders, k_{eg}, given by:

$$k_{eg} = k_{eu} + (k_{eu} - i)(1 - T)\frac{V_B}{V_S} = R_f + \beta_g (ER_m - R_f)$$

where

$$k_{eu} = R_f + \beta_u (ER_m - R_f)$$

β_u = the ungeared Beta coefficient

To value the equity, i.e. to derive a measure for V_S, we would need to apply the (perpetuity) expression for valuing a stream of post-tax geared equity income:

$$V_S = \frac{(E - iB)(1 - T)}{k_{eg}}$$

We now encounter a circular problem, since the market value depends on k_{eg}, and to find k_{eg}, we need to know the market value!

A possible solution is to work in terms of a 'tailor-made' WACC based on the project's characteristics (i.e. its systematic risk, allowing for any divergence from existing operations) and on the project's own financing. For example, imagine the project in the previous example were to be financed 20 per cent by debt and 80 per cent by equity. You should verify that with $\beta_u = 1.33$, $\beta_g = 1.56$, that shareholders would seek a return of 24 per cent, and that the WACC is:

$$(24\% \times 4/5) + (10\% [1 - T] \times 1/5) = (19.2\% + 1.4\%) = 20.6\%$$

As it happens, use of the WACC in this situation may be inappropriate anyway, since unless the firm is at, and adheres to, the target ratio, the WACC and the marginal cost of capital (MCC) will diverge. If the firm is below the optimal capital structure, the MCC is less than WACC, and the MCC exceeds the WACC when it overshoots the optimal gearing ratio. We found, in Chapter 18, that when the firm departs from the optimal gearing ratio, the appropriate required return is the MCC:

$$MCC = \frac{\text{Change in total returns required by shareholders and lenders}}{\text{Amount available to invest}}$$

However, to calculate the MCC we again need to know the market values of both equity and debt at the higher level of gearing, i.e. we encounter the circular problem described earlier. It is clear that the WACC is suitable only for small-scale projects that do not materially disturb the gearing ratio, and that the theoretically more correct MCC is also problematic.

An 'off-the-cuff' solution is to work in terms of book values. This pragmatic approach has the merit of simplicity, as book values do not vary with gearing, and it might be appropriate for unlisted firms, which by definition have no market values. Nevertheless, it is desirable to work, whenever possible, in terms of market values, given that most investors are more concerned with the current values of their investments, and the returns thereon, than with historic balance sheet values.

Fortunately, as we shall see in the next section, help is at hand.

19.10 THE ADJUSTED PRESENT VALUE METHOD (APV)

adjusted present value
The inherent value of a project adjusted for any financial benefits and costs stemming from the particular method(s) of financing

The **adjusted present value** (APV) of a project is simply the 'essential' worth of the project, adjusted for any financing benefits (or costs) attributable to the particular method of financing it. The rationale for the APV method was provided by Myers (1974), using MM's gearing model with corporate tax, but is valid only so long as the WACC profile is declining due to the value of the tax shield. In Section 19.5, we saw that the value of a geared firm, V_g, is the value of an equivalent all-equity-financed company, V_g, plus a tax shield, TB, which is the discounted tax savings resulting from the tax-deductibility of debt interest:

$$V_g = V_u + TB$$

This can be translated from the value of a firm to the value of an individual project. However, different projects can probably support different levels of debt. For example, they may involve different inputs of easily resaleable fixed assets and may also have different levels of operational gearing. As a result, it may be more appropriate to evaluate the effects of the financing of each project separately.

The APV is calculated in three steps:

Step 1 Evaluate the 'base case' NPV, discounting at the rate of return that shareholders would require if the project were financed wholly by equity. This rate is derived by ungearing the company's equity Beta.

Step 2 Evaluate separately the cash flows attributable to the financing decision, discounting at the appropriate risk-adjusted rate.

Step 3 Add the present values derived from the two previous stages to obtain the APV. The project is acceptable if the APV is greater than zero.

A simple example will illustrate the use of the APV.

19.11 WORKED EXAMPLE: RIGTON PLC

Rigton plc has a debt/equity ratio of 20 per cent. The equity Beta is 1.30. The risk-free rate is 10 per cent and a return of 16 per cent is expected from the market portfolio. The rate of corporate tax is 30 per cent. Rigton proposes to undertake a project requiring an outlay of £10 million, financed partly by equity and partly by debt. The project, a perpetuity, is thought to be able to support borrowings of £3 million at an interest rate of 12 per cent, thus imposing interest charges of £0.36 million. It is expected to generate pre-tax cash flows of £2.3 million p.a.

Required
Using the APV method, determine whether this project is worthwhile.

Answer
Using the formula developed earlier for the ungeared Beta:

$$\beta_u = \frac{\beta_g}{\left[1 + \dfrac{V_B}{V_S} \times (1 - T)\right]} = \frac{1.30}{1 + 0.20(1 - 0.30)} = \frac{1.30}{1.14} = 1.14$$

This yields a required return on ungeared equity of:

$$ER_j = R_f + \beta_u (ER_m - R_f) = 0.10 + 1.14 (0.16 - 0.10) = (0.10 + 0.068)$$
$$= 0.168, \quad \text{i.e. } 16.8\%$$

The base case NPV is:

$$NPV = -£10 \text{ m} + \frac{£2.3 \text{ m}(1 - 0.30)}{0.168} = -£10 \text{ m} + \frac{£1.61 \text{ m}}{0.168}$$
$$= -£10 \text{ m} + £9.58 \text{ m}$$
$$= -£0.42 \text{ m}$$

The present value of the tax savings, i.e. the tax shield, TB, is given by:

$$\frac{TiB}{i} = \frac{(0.30)(0.12)(£3 \text{ m})}{0.12} = \frac{(0.30)(£0.36 \text{ m})}{0.12} = \frac{£0.108 \text{ m}}{0.12} = £0.9 \text{ m}$$

The adjusted present value is thus:

$$APV = -£0.42 \text{ m} + £0.90 \text{ m} = +£0.48 \text{ m}$$

and the project appears worthwhile. The significance of this result is that, although the base case NPV is negative, the project is rescued by the tax shield of £0.90 million. An essentially unattractive project is rendered worthwhile by the taxation system.

In the Rigton example, the project creates wealth only for Rigton's shareholders. From the perspective of the overall economy, it is wealth-reducing and, unless there are compelling 'social' reasons to justify it, should not be undertaken. This sort of reasoning led the UK government in 1984 to reduce the rate of corporation tax in order to lower the tax advantage of debt financing, and hence reduce the extent to which investment decisions were likely to be distorted by the system of tax breaks.

Self-assessment activity 19.6

What is the APV and how is it calculated?

(Answer in Appendix A at the back of the book)

19.12 FURTHER ISSUES WITH THE APV

Before leaving the APV, several related issues are worth examining.

1 The APV in practice is affected by the terms and conditions of a pre-arranged schedule for debt interest and capital repayment. Sometimes, the calculations can be exceptionally tedious. Rather than using the convenient assumption of perpetual debt financing, let us assume that the debt plus interest must be repaid over two years, with interest and two equal capital payments occurring at end-year. Table 19.2 shows the repayment schedule and the resulting tax savings.
With no tax delay assumed, the present value of the tax savings is:

$$\frac{£0.108 \text{ m}}{(1.12)} + \frac{£0.054 \text{ m}}{(1.12)^2} = (£0.096 \text{ m} + £0.043 \text{ m}) = £0.139 \text{ m}$$

Table 19.2 The tax shield with finite-life debt

Balance of loan at start of year	Interest at 12%	Tax saving ($T = 30\%$)	Repayment	Balance of loan at end of year
£3.0 m	£0.36 m	(30% × £0.36 m) = £0.108 m	£1.5 m	£1.5 m
£1.5 m	£0.18 m	(30% × £0.18 m) = £0.054 m	£1.5 m	0

Obviously, the value of the tax shield is much lower with the shorter payment profile.

2 Although our example focused on the side-effects of debt financing, the APV routine can be easily applied to any other financing costs and benefits, many of which are awkard to handle with the simple WACC. For example, if equity capital is externally raised, normally there are various issuing and underwriting costs to bear. Including these would alter the APV formula as follows:

$$\text{APV} = \text{Base case NPV} + \text{Tax shield} - \text{PV of issue costs}$$

A similar treatment would be applied to subsidised borrowing costs, investment grants and tax savings from exploiting investment allowances.

3 Tax savings are not certain because they depend on the inherent profitability of the company. As this is a random variable, the company's ability to set off interest payments (and other tax reliefs) against income is also random. Our examples assume continuous profitability, but if there are periods during which the company is expected to be tax-exhausted, this should be allowed for in the computation of the APV. If the future pattern of liability to tax is uncertain, then it is not appropriate to use a risk-free rate to discount the tax savings.

4 Finally, we have glossed over the issues that impact on the debt-supporting capacity of particular projects. In principle, the debt capacity of a project is given by the present value of future expected earnings from the firm as a whole, taking into account any existing borrowings. It might seem obvious that more profitable companies are able to borrow relatively more than unprofitable companies. However, this assumes that there are no costs of financial distress. Enhanced borrowing ability for more profitable companies is not universal, since a would-be lender would still look at the break-up value of the enterprise. In the final analysis, the crucial factor which governs debt capacity is how much can be raised by a distress sale of assets.

Self-assessment activity 19.7

How would you identify the point beyond which a firm would be unable to borrow?

(Answer in Appendix A at the back of the book)

19.13 WHICH DISCOUNT RATE SHOULD WE USE?

Specifying the correct discount rate to use when a new project involves financing and other differences from parent company activities is something of a puzzle. Now that we have examined the main variations on the discount rate theme, this check-list should help.

If the new project has a:

Case 1 *Similar business risk and capital structure as the parent company.*
Use the parent's WACC.

Case 2 *Higher/lower business risk than the parent but similar financing mix.*
Adjust the Beta, using a surrogate firm's Beta as a basis but adjust for relative gearing, i.e. ungear the surrogate Beta and gear up the residual equity Beta. Then use the parent's capital structure weights to calculate the WACC.

Case 3 *Similar business risk, but capital structure different from that of the parent.*
Use the parent's equity Beta, gear it for the project financing mix and then use the project's financing mix to find the project WACC.

Case 4 *Higher/lower business risk, and a different capital structure.*
Use the project Beta, and, as in Case 2, gear it for the project financing, and calculate the WACC using the project financing mix.

Case 5 *Complex mixture of risk, financial structure, and side-effects.* Use the APV method.

SUMMARY

Chapters 18 and 19 have covered extensive ground, attempting to isolate the critical variables relating company value to capital structure. In this process, we have moved from the somewhat crude 'traditional' version to the pure and less pure MM analyses, before arriving at the model displayed in Figure 19.4. This closely resembles the traditional theory itself, with its U-shaped cost of capital schedule and optimal capital structure. We have established that *the benefits of debt stem mainly from market imperfections, especially the tax relief on debt interest, but that a different type of imperfection, distress costs, can offset these tax breaks at higher levels of gearing.* In addition, even the tax benefits of gearing may be overstated as they depend on the particular mix of personal and corporate tax rates faced by the company and its stakeholders (see Appendix III).

So in response to the question posed at the start of the chapter, 'Does capital structure matter?' the answer seems to be 'yes', but in a number of complex ways. Debt, or rather, excessive debt, certainly matters to the owners but it may not destroy value. Distressed, but operationally viable, companies can still survive. For non-distressed companies, debt can offer significant tax advantages.

Key points

- MM argue that, as the method of financing a company does not affect its fundamental wealth-creating capacity, the use of debt capital, under perfect market conditions, has no effect on company value.

- Shareholders respond to an increase in the likely variability of earnings, i.e. financial risk, by seeking higher returns to offset exactly the apparent benefits of 'cheap' debt.

- The appropriate cut-off rate for new investment is the rate of return required by shareholders in an equivalent ungeared company.

- When corporate taxation is introduced, the tax-deductibility of debt interest creates value for shareholders via the tax shield, but this is a wealth transfer from taxpayers.

- The value of a geared company equals the value of an equivalent ungeared company plus the tax shield:

$$V_g = V_u + TB$$

- With corporate taxation, the rate of return required by the geared company's shareholders is less than that in the all-equity company, reflecting the tax benefits.

- A further effect of corporation taxation is to lower the overall cost of capital, which appears to fall continuously as gearing increases.

- However, this result relies on the absence of default risk and the consequent costs of financial distress incurred as a company reaches or approaches the point of insolvency.

- For geared companies, the required return can be derived by combining k_e with the after-tax debt cost to obtain the WACC.

- However, the WACC is acceptable only under restrictive conditions: in particular, when project financing replicates existing gearing, and when project risk is identical to that of existing activities.

- To resolve the problems of the WACC, the adjusted present value can be used. This is the 'basic' worth of the project, i.e. the NPV assuming all-equity financing, adjusted for any financing benefits such as tax savings on debt interest, or costs such as issue expenses.

- Eventually, the costs of financial distress may begin to outweigh the benefits of the tax shield. A major cost of financial distress is the inability to achieve 'full market value' in a 'distress sale'.

- There is, in theory, an optimal capital structure where the marginal benefit of tax savings equals the marginal cost of financial distress.

- In reality, while companies should balance the benefits of the tax shield against the likelihood of financial stress costs, most finance directors will restrain gearing levels, especially as tax savings are uncertain, depending on fluctuations in corporate earnings.

Further reading

Similar health warnings apply here as with dividend policy i.e. the fashion aspect, and the different taxation and institutional regimes that apply in different countries. But, as with dividend policy, the relevance of key papers is timeless and universal.

Look at the original articles by Modigliani and Miller (1958, 1963). Other important articles are those by Myers (1974, 1984), which analyse the interactions between financing and investment decisions, and Miller's attempt to resurrect the capital structure irrelevance thesis (1977) and his subsequent Nobel lecture (1991). As ever, Copeland *et al.* (2004) offer a more rigorous, mathematical development. Resumés of current thinking on capital structure theory can be found in Barclay *et al.* (1995) and Barclay and Smith (2006). Luehrman (1997a,b) offers two articles on the present state of valuation theory and analysis, with strong emphasis on APV, and also on strategic options.

As well as Copeland *et al.* (2004), good textbook treatments can be found in Emery and Finnerty (1997), Ross *et al.* (2005) and Block and Hirt (1994).

Important articles include DeAngelo and Masulis (1980) on taxation and capital structure, Bradley *et al.* (1984) and Rajan and Zingales (1995) for empirical evidence, Ross, S. (1977) on signalling, Warner, J. (1977) on bankruptcy costs, Marsh (1982) on target debt ratios, Harris and Raviv (1990) on debt signalling, and (1991) for an overview of the debate, and Myers and Majluf (1984) on information asymmetry.

Appendix I
DERIVATION OF MM'S PROPOSITION II

Given that:

$$\frac{E}{V_{\text{S}} + V_{\text{B}}} = \frac{E}{V_o} = k_o$$

and

$$k_{\text{e}} = \frac{(E - iB)}{V_{\text{S}}}$$

we may write

$$E = k_o V_o = k_o(V_{\text{S}} + V_{\text{B}})$$

Substituting for E,

$$k_e = \frac{k_o(V_S + V_B) - iB}{V_S} = \frac{k_o V_S + k_o V_B - iB}{V_S} = k_o + (k_o - i) \times \frac{V_B}{V_S}$$

Since Proposition I argues that k_o equals the return required by shareholders in an equivalent ungeared company, k_{eu}, and so long as the book and market values of debt capital coincide, thus ensuring that $i = k_d$, then this expression may be written as:

$$k_{eg} = k_{eu} + (k_{eu} - k_d)\frac{V_B}{V_S}$$

as in the text. In other words, the return required by shareholders is a linear function of the company's debt/equity ratio.

Appendix II
MM'S PROPOSITION III: THE CUT-OFF RATE FOR NEW INVESTMENT

MM's third proposition asserts that 'the cut-off rate for investment will in all cases be k_o and will be unaffected by the type of security used to finance the investment'.

To show this, consider a firm whose initial value, V_o, is:

$$V_o = V_{So} + V_{Bo} = \frac{E_o}{k_o} \qquad (A)$$

It contemplates an investment project, with outlay £I, involving a perpetual return of R per £ invested. After the investment is accepted, the new value of the firm, V_1, is:

$$V_1 = \frac{E_1}{k_o} = \frac{E_o + RI}{k_o} = V_o + \frac{RI}{k_o}$$

Assuming the project is debt financed, the post-project acceptance value of the shares is:

$$V_{S1} = (V_1 - V_{B1}) = V_1 - (V_{Bo} + I) \qquad (B)$$

Substituting Equation A into Equation B yields:

$$V_{S1} = V_o + \frac{RI}{k_o} - V_{Bo} - I$$

and since

$$V_{So} = (V_o - V_{Bo})$$

the change in V_S equals

$$(V_{S1} - V_{So}) = \frac{RI}{k_o} - I$$

This exceeds zero only if $R > k_o$. Hence, *a firm acting in the best interests of its shareholders should only undertake investments whose returns at least equal k_o, the weighted average cost of capital, which itself is invariant to gearing according to Proposition I.*

Appendix III
ALLOWING FOR PERSONAL TAXATION: MILLER'S REVISION

The MM analysis including corporate earnings taxation still leaves something of a 'puzzle'. The expression for the value of a geared company indicates that the tax shield is equal to the corporate tax rate (T) times the book value of corporate debt (B), i.e. TB. With the present UK rate of corporation tax of 30 per cent, for every £1 of corporate debt the value of the company would be increased by £0.30. If such tax benefits can stem from corporate gearing, why do we find widely dispersed gearing ratios even in the same industry? And why are some of these so much lower than the MM theory (even allowing for the costs of financial distress) might suggest? According to Miller (1977), the answers to such questions lie in the interaction of the corporate taxation system with the personal taxation system, an issue omitted from the MM analysis.

Miller's agenda was to re-establish the irrelevance of gearing for company value, thus explaining why US firms did not appear to exploit apparently highly valuable tax shields. Miller argued that if individuals and corporations can borrow at the same rate, and if individuals invest in corporate debt as well as equity, there are no advantages to corporate borrowing because corporations that borrow are simply doing what personal investors can do for themselves. Any temporary premium in the market valuation of a geared company will be quickly unwound by the usual arbitrage process. However, this presupposes that individuals also can obtain tax relief on their personal borrowing (as applies in the USA, but not generally in the UK). Intuitively, we may expect to find some benefit to corporate borrowing in the UK because tax breaks on personal borrowing are not available.

Greatly simplifying, the Miller position can be expressed by the simple expression:

$$\text{Post-tax cost of debt} = \text{pre-tax cost } [1 - (T_c - T_p)]$$

where T_c is the tax rate at which corporations enjoy relief on debt interest and T_p is the tax rate at which individuals enjoy relief on debt interest.

If $T_c = T_p$, then there is no tax advantage of corporate debt and hence no tax shield to exploit.

Only if T_c and T_p differ is there a tax shield. Note that for $T_c > T_p$ the tax shield is positive, and for $T_c < T_p$, the tax shield appears to be negative, as might apply for shareholders subject to very high rates of tax.

Miller introduced a further mechanism to support the irrelevance of gearing for company value. He argued that if there is a (temporary) tax advantage relating to debt financing, this will lead firms to increase their demand for debt (i.e. increase the supply of debt instruments), thus exerting upward pressure on interest rates until the advantage of issuing further debt disappears. If the effective tax rate on equity income were zero, and personal investors paid tax on debt interest income, companies would have to compensate investors for switching from untaxed equity to taxed debt investments by a higher interest rate. This would stop when the net-of-tax cost of debt to companies equalled the cost of equity. Miller concludes that movement to capital market equilibrium would eliminate any tax advantage of debt, so that $V_g = V_u$.

Ashton and Acker (2003) have undertaken an assessment of the average tax advantage of debt in a UK context, and conclude that it is 'likely to be no more than 13% of the value of debt'.

QUESTIONS

 **Questions with an icon are also available for practice in myfinancelab with additional supporting resources.**

Questions with a **coloured number** have solutions in Appendix B on page 753.

1 With the following information about Rushden plc, determine its cost of equity according to the MM no-tax model.

$$k_{eu} = 20\%; \quad k_d = 8\%; \quad \frac{V_B}{V_B + V_S} = 20\%$$

2 Diamonds plc estimates its costs of debt and equity for different capital structures as follows:

% Debt	% Equity	k_d	k_e	WACC
–	100	–	20%	?
25	75	8%	24%	?
50	50	8%	32%	?
75	25	8%	56%	?

Required
(i) What theory of capital structure is portrayed? (Complete the WACC column.)
(ii) Restate the table allowing for taxation of corporate profits (hence, tax relief on debt) at 30 per cent. Assume Diamonds plc always has sufficient taxable capacity to exploit the tax shield.
Identify the relevant theory of capital structure.

3 Demonstrate how the process of home-made gearing-cum-arbitrage would operate in an MM world so as to equalise the values of the following two firms. The companies are identical in every respect except their capital structures.

	Geared	Ungeared
Expected earnings	£100	£100
Debt finance (nominal)	£200	–
Interest rate	5%	–
Market value of equity	£900	£950
Market value of company	£1,100	£950

Assume that the market value of geared debt is equal to the nominal value, and the investor holds 10 per cent of Geared's equity.

4 Kipling plc is a food manufacturer which has the following long-term capital structure:

	£
£1 ordinary shares (fully paid)	2,500,000
Share premium account	1,000,000
Retained profit	1,400,000
8% preference shares	1,200,000
10% debentures (secured)	2,600,000
	8,700,000

The directors of the company wish to raise further long-term finance by the issue of either preference shares or debentures. One director, who supports the issue of debentures, believes that, although a debenture issue will increase the company's gearing, it will reduce the overall cost of capital.

Required

(a) Discuss the arguments for and against the view that the company's overall cost of capital can be reduced in this way. The views of Modigliani and Miller should be discussed in answering this part of the question.

(b) Discuss the major factors which the directors should consider when deciding between preference shares and debentures as a means of raising further long-term finance.

(c) Identify and discuss the major factors which will influence the amount of additional debenture finance that Kipling plc will be able to raise.

(ACC Certified Diploma)

5 (a) Berlan plc has annual earnings before interest and tax of £15 million. These earnings are expected to remain constant. The market price of the company's ordinary shares is 86 pence per share cum div and of debentures £105.50 per debenture ex-interest. An interim dividend of six pence per share has been declared. Corporate tax is at the rate of 35 per cent and all available earnings are distributed as dividends. Berlan's long-term capital structure is shown below:

	£000
Ordinary shares (25 pence par value)	12,500
Reserves	24,300
	36,800
16% debenture 31 December 1994 (£100 par value)	23,697
	60,497

Required

Calculate the cost of capital of Berlan plc according to the traditional theory of capital structure. Assume that it is now 31 December 1991.

(b) Canalot plc is an all-equity company with an equilibrium market value of £32.5 million and a cost of capital of 18 per cent per year. The company proposes to repurchase £5 million of equity and to replace it with 13 per cent irredeemable loan stock.

Canalot's earnings before interest and tax are expected to be constant for the foreseeable future. Corporate tax is at the rate of 35 per cent. All profits are paid out as dividends.

Required

Using the assumptions of Modigliani and Miller, explain and demonstrate how this change in capital structure will affect Canalot's:

(i) market value

(ii) cost of equity

(iii) cost of capital

(c) Explain any weaknesses of both the traditional and Modigliani and Miller theories and discuss how useful they might be in the determination of the appropriate capital structure for a company. (ACCA)

6 The ordinary shares of Stanley plc are quoted on the London Stock Exchange. The directors, who are also major shareholders, have been evaluating some new investment opportunities. If they go ahead with these, new capital of £38 million will be required. The directors expect the new projects to earn 15 per cent per annum before tax. Financial information about the company for 1996 is as follows:

EBIT (existing operations)	£79.50 million
Number of shares in issue (par value £1)	50 million

The company is at present all-equity financed. It has the choice of raising the £38 million new capital by an issue of equity or debt. Equity would be issued by a new issue at a 15 per cent discount to current market price. Debt will be raised by an issue at par of 12 per cent unsecured loan stock.

If the finance is raised via equity, the company's P:E ratio is likely to rise from its current level of 9 to 9.5. However, if debt is introduced into the capital structure, the company's financial advisors have warned the two directors that the market is likely to lower the P:E ratio of the company to 8.5.

The company's marginal tax rate is 33 per cent.

Issue costs should be ignored.

(a) Determine the expected share price, total value of equity and value of the firm under the two financing options and comment briefly on which financing option appears the most advantageous.

(b) Assume the company's average cost of equity as an ungeared firm is 14 per cent and it expects to continue to pay tax at 33 per cent. The estimated cost of bankruptcy or financial distress is estimated at £5 million. According to Modigliani and Miller, what would be the value of equity and the firm if the company finances the expansion by (i) equity or (ii) debt?

(c) Explain the basic assumptions underlying MM's theories of capital structure and why, in an efficient market with no taxes, capital structure can have no effect on the value of the firm.

7 You are given the following information about Electronics plc. It has a payout ratio of 0.6, a return on equity of 20 per cent, an equity beta of 1.33 and is expected to pay a dividend next year of £2.00. There are 1 million shares outstanding and it is fairly valued. It also has nominal debt of £20 million issued at 10 per cent and maturing in 5 years. Yields on similar debt have since dropped to 8 per cent. The risk free rate is 6 per cent and the expected market return is 13.5 per cent.

(a) Find Electronics' cost of capital and cost of equity.

(b) The company decides to retire half its debt at current prices. Find the company's cost of capital and equity and explain your results.

(c) The company decides to diversify into a completely different business area and decides to look at Betas of firms currently trading in the new business area. The information is given below.

Company	Beta	Debt/Equity	Market capitalisation
A	1.5	1:2	£20 million
B	1.8	1:1	£30 million
C	1.2	No debt	£50 million

What discount rate should the company use for the new business?

8 Claxby is an undiversified company operating in light engineering. It is all-equity financed with a Beta of 0.6. Total risk is 40 (standard deviation of annual return). Management want to diversify by acquiring Sloothby Ltd, which operates in an industrial sector where the average equity Beta is 1.2 and the average gearing (debt to total capital) ratio is 1:3. The standard deviation of the return on equity (on a book value basis) for Sloothby is 25%. The acquisition would increase Claxby's asset base by 40 per cent. The overall return on the market portfolio is expected to be 18 per cent and the current return on risk-free assets is 11 per cent. The standard deviation of the return on the market portfolio is 10%. The rate of corporation tax is 33 per cent.

(a) What is the asset Beta for Sloothby?

(b) Analyse both Sloothby's and Claxby's total risk into their respective specific and market risk components.

(c) What would be the Beta for the expanded company?

(d) Using the new Beta, calculate the required return on the expanded firm's equity.
    Under what conditions could this be taken as the cut-off rate for new investment projects?

(e) In the light of the figures in this example, discuss whether the acquisition of Sloothby may be expected to operate in the best interests of Claxby's shareholders.

9 The managing director of Wemere, a medium-sized private company, wishes to improve the company's investment decision-making process by using discounted cash flow techniques. He is disappointed to learn that estimates of a company's cost of equity usually require information on share prices which, for a private company, are not available. His deputy suggests that the cost of equity can be estimated by using data for Folten plc, a similar sized company in the same industry whose shares are listed on the AIM, and he has produced two suggested discount rates for use in Wemere's future investment appraisal. Both of these estimates are in excess of

17 per cent p.a. which the managing director believes to be very high, especially as the company has just agreed a fixed rate bank loan at 13 per cent p.a. to finance a small expansion of existing operations. He has checked the calculations, which are numerically correct, but wonders if there are any errors of principle.

Estimate 1: Capital Asset Pricing Model
Data have been purchased from a leading business school
Equity Beta of Folten: 1.4
Market return: 18%
Treasury Bill yield: 12%

The cost of capital is $18\% + (18\% - 12\%)1.4 = 26.4\%$. This rate must be adjusted to include inflation at the current level of 6 per cent. The recommended discount rate is 32.4 per cent.

Estimate 2: Dividend Growth Model

Folten plc

Year	Average share price (pence)	Dividend per share (pence)
1985	193	9.23
1986	109	10.06
1987	96	10.97
1988	116	11.95
1989	130	13.03

The cost of capital is: $D_1/(P - g)$, where D_1 is the expected dividend, P is the market price and g is the growth rate of dividends $(= 14.20\text{p}/(138\text{p} - 9) = 11.01\%)$.
When inflation is included, the discount rate is 17.01 per cent.
Other financial information on the two companies is presented below:

	Wemere £000	Folten £000
Fixed assets	7,200	7,600
Current assets	7,600	7,800
Less: Current liabilities	(3,900)	(3,700)
	10,900	11,700
Financed by:		
Ordinary shares (25 pence)	2,000	1,800
Reserves	6,500	5,500
Term loans	2,400	4,400
	10,900	11,700

Notes
1 The current ex div share price of Folten plc is 138 pence.
2 Wemere's board of directors has recently rejected a takeover bid of £10.6 million.
3 Corporate tax is paid at the rate of 35 per cent.

Required
(a) Explain any errors of principle that have been made in the two estimates of the cost of capital and produce revised estimates using both of the methods.
State clearly any assumptions that you make.
(b) Discuss which of your revised estimates Wemere should use as the discount rate for capital investment appraisal.

(ACCA)

Practical assignment

Reread the exposition in Chapter 11 of how we obtained tailored discount rates for the Whitbread plc divisions. How close do you think our surrogates were?

For another divisionalised company of your choice (try to find a two- or three-division company):

1 Consult the Risk Measurement Service for an up-to-date estimate of the equity Beta, and use the CAPM to assess the shareholders' required rate of return.
2 Estimate discount rates for each division. You will need to select surrogate companies, record their Betas, and obtain an indication of their own asset Betas by ungearing their equity Betas.
3 Determine whether the weighted average Beta for the company corresponds to its ungeared Beta. You will probably have to use weights based on earnings or sales as very few companies report book values (let alone market values!) of their segments.

 *Now retake your diagnostic test for Chapter 19 to check your progress and update your study plan.*

20

Acquisitions and restructuring

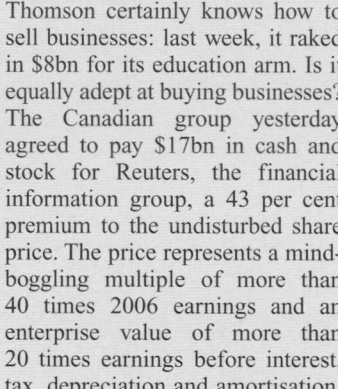

Screen saver

Thomson certainly knows how to sell businesses: last week, it raked in $8bn for its education arm. Is it equally adept at buying businesses? The Canadian group yesterday agreed to pay $17bn in cash and stock for Reuters, the financial information group, a 43 per cent premium to the undisturbed share price. The price represents a mind-boggling multiple of more than 40 times 2006 earnings and an enterprise value of more than 20 times earnings before interest, tax, depreciation and amortisation.

That sounds like a better deal for the seller. Although Tom Glocer, the Reuters chief executive who will take the helm of the new group, has received due credit for rescuing it from the dark days of the early part of the decade, his

vision of a dynamic new Reuters never quite rang true. Blending Reuters into the much broader Thomson business information group, which serves the legal and healthcare as well as the financial services industries, solves a couple of problems that had never really gone away. Reuters has tended to be highly cyclical. In the last financial downturn, it suffered double-digit annual sales declines. It has also been vulnerable to consolidation among its customers.

Thomson, meanwhile, enjoys both higher growth and higher margins. If the new entity can deliver the synergies it believes are feasible – annual cost synergies alone are estimated at $500m after three years – the combined

group should be able to replicate these characteristics.

It is easy to see the appeal of Reuters, with its superior brand, broader client base and global reach, for Thomson. The cost-cutting objectives look achievable and Thomson leaps from a weakish third place to being a market leader. But boosting growth through new products in areas such as science and healthcare is ambitious and untested. With formidable regulatory hurdles to clear, just joining the two businesses together is a daunting challenge for a team that appears competent but not brilliant. Compared with this, Thomson's previous experience of buying and selling businesses was child's play.

Source: Financial Times, 16 May 2007.

Learning objectives

A major aim of this chapter is to emphasise the interaction between the financial and strategic dimensions of takeovers. Having read it, you should understand the following:

- Why firms select acquisitions rather than other strategic options.
- How acquisitions can be financed.
- How acquisitions should be integrated.
- How the degree of success of a takeover can be evaluated.
- How corporate restructuring can enhance shareholder value.

 Complete your diagnostic test for Chapter 20 now to create your personal study plan.

20.1 INTRODUCTION

The introductory cameo gives a good idea of the motivations that underpin takeovers. But it is essential to remember that the underlying motivation of acquisition is to generate higher cash flows for shareholders.

The jury is still out on the acquisition of Reuters by Thomson, but if it fails, then this will not be an unusual outcome.

Acquisitions of other companies are investment decisions and should be evaluated as (if not more) thoroughly and on essentially the same criteria as, say, the purchase of new items of machinery. However, there are two important differences between takeovers and many 'standard' investments.

First, because takeovers are frequently resisted by the target's managers, bidders often have little or no access to intelligence about their targets beyond published financial and market data, and any inside information they may glean. (As and when takeover is accepted as inevitable, the defending board is obliged to provide key information to enable the bidder to conduct 'due diligence' examinations. This is essentially a search for 'skeletons in the cupboard'. See Sudarsanam (2004) for due diligence procedures.)

In August 2007, British chemicals group Imperial Chemical Industries (ICI) finally agreed to the takeover by Dutch rival Akzo Nobel for an offer price of 670 pence per ICI share, having previously rejected bids of 600p and 650p. Once it accepted that the bid was now in the best interests of shareholders, ICI opened its books to Akzo to allow the Dutch firm to perform due diligence.

In 1997, Vereinsbank of Germany, in an effort to avoid being taken over by the much larger Deutsche Bank, reportedly cobbled together over a single weekend a merger agreement with Hypobank, a smaller rival. The penalty for undue haste in addressing the financial, legal and human issues involved was the discovery a year later, by the auditors of the new entity (HypoVereinsBank), of huge losses in Hypobank's property portfolio, necessitating loan loss provisions of DM 3.5 billion (information from the *Economist*, 2000).

Second, many takeovers are undertaken for longer-term strategic motives, and the benefits are often difficult to quantify. It is common to hear the chairmen of acquiring companies talk about an acquisition opening up a 'strategic window'; what they often do not add is that the window is usually not only shut, but has thick curtains drawn across it! To a large extent, a takeover is a shot in the dark, partly explaining why so many firms that launch giant takeovers come to grief.

But there are other reasons. Targets are often too large in relation to bidders, so that excessive borrowings or unexpected integration problems throttle the parent. The demise of Marconi (formerly General Electric), whose share price fell by 97 per cent during 2000–1, was largely the result of headlong expansion into telecoms equipment manufacturing just before the sector entered recession. Its debt financing proved an albatross which, despite a debt-for-equity swap, was a major factor in its eventual demise. It was acquired by Swedish telecoms giant Ericsson in 2005 for a fraction of the value it once commanded as the UK's leading electricals and electronics company.

There are important lessons to be learned from risk analysis and portfolio theory. When acquisitions have highly uncertain outcomes, the larger they are, the more catastrophic the impact of any adverse outcomes. As a result, it may be rational and less risky to confine takeover activity to small, uncontested bids. Alternatively, a spread of large acquisitions might confer significant portfolio diversification benefits, so long as the components have low cash flow correlation. However, the greater the scale of takeover activity, the greater the resulting financing burden placed on the parent, and the greater the impact of diverting managerial capacity into solving integration problems.

The acquisition decision is thus a complex one. It involves significant uncertainties (except in purely **asset-stripping** takeovers), it often requires substantial funding and it may pose awkward problems of integration. Yet, as some takeover 'kings' have shown, spectacular payoffs can be achieved. These are some of the themes of this chapter – how to evaluate a takeover, how to finance it and how to integrate it. But first, we examine the phenomenon of takeover surges.

20.2 TAKEOVER WAVES

Although the terms 'takeover' and 'merger' are used as synonyms, there is a technical difference. A **takeover** is the acquisition by one company of the share capital of another in exchange for cash, ordinary shares, loan stock or some mixture of these. This results in the identity of the acquired company being absorbed into that of the acquirer (although, of course, the expanded company may continue to use the acquired company's brand names and trademarks). A **merger** is a pooling of the interests of two companies into a new enterprise, requiring the agreement of both sets of shareholders. The ill-fated combination of Daimler–Benz and Chrysler in 1998 was presented (initially, at least) as a merger of friendly partners. In 2001, an unusual three-way merger of leading steel firms Usinor of France, Arbed of Luxembourg and Aceralia of Spain was forged under the name Arcelor. By definition, mergers involve the friendly (initially, at least) restructuring of assets into a new organisation, whereas many takeovers are hotly resisted. In practice, the vast majority of business amalgamations are takeovers rather than mergers.

Table 20.1 shows the path of takeover activity of both kinds, looking at acquisitions by UK firms of other UK firms during 1970 to 2007. There are clear examples of waves in motion here, for example, in terms of number of firms, that of the early 1970s, the late 1980s and the mid-2000s, with the all-time peak occurring in 1972. However, in terms of amount expended, although the same waves appear in the 1970s, late 1980s and late 1990s, the year 2000 stands out as the all-time peak, before collapsing in 2001 (note that the figures are not inflation-adjusted). The late 1990s saw the crescendo of the 'dotcom boom' in the world stock markets with share prices reaching highly-inflated price: earnings ratios. When share prices are high and rising, it becomes easier to conduct a takeover bid by exchange of shares. Accordingly, the proportion of acquisitions completed in this way rose to historically high levels at this time. So, here we have one reason for takeover waves – takeover booms tend to reflect general stock market activity, so when the stock market booms, listed firms tend to become more acquisitive.

Yet there may be a 'chicken-and-egg' argument here. It has been argued that takeover activity often provides the trigger for a stock market recovery. When share prices fall, and the market value of firms looks low in relation to the replacement cost of their assets (i.e. the cost of setting up an equivalent operating facility), acquisition may seem the cheaper way for a firm wishing to expand compared to internal (or 'organic') growth. This explanation appears to lie behind the recovery of takeover activity in 2004. It does look as if the mid-2000s may be associated with another takeover boom (although the 'credit crunch' of 2007-8 may stifle this upsurge).

Table 20.2 shows how the acquisitions by UK firms ('acquirors') of other UK companies ('acquirees') were split into purchases of other independent firms and acquisitions of subsidiaries of other firms, or 'trade sales'. It can be seen that the bigger mergers tend to occur among the latter category, looking at average deal size. This is despite a number of factors tending to push up the price of independent firms. Bidding is more public, resistance by incumbent directors is often encountered, competition is more likely and a premium above the market price must be offered to encourage present owners to sell.

According to the late Peter Doyle, the eminent marketing academic (1994), the motives for the mega-mergers of more recent years differ from those of the 1980s. In the earlier wave, companies like Hanson and BTR were looking to exploit financial economies by restructuring badly-run companies and giving managers incentives to

Table 20.1 The scale and financing of takeover activity in the UK by UK firms

Year	Number acquired	Outlay (£m)	Cash (%)	Ordinary shares (%)	Fixed interest (%)
1970	793	1,122	22	53	25
1971	884	911	31	48	21
1972	1,210	2,532	19	58	23
1973	1,205	1,304	53	36	11
1974	504	508	68	22	9
1975	315	291	59	32	9
1976	353	448	72	27	2
1977	481	824	62	37	1
1978	567	1,140	57	41	2
1979	534	1,656	56	31	13
1980	469	1,475	52	45	3
1981	452	1,144	68	30	3
1982	463	2,206	58	32	10
1983	477	2,343	44	54	2
1984	568	5,474	54	34	13
1985	474	7,090	40	52	7
1986	842	15,370	26	57	17
1987	1,528	16,539	35	60	5
1988	1,499	22,839	70	22	8
1989	1,337	27,250	82	13	5
1990	779	8,329	77	18	5
1991	506	10,434	70	29	1
1992	432	5,941	63	36	1
1993	526	7,063	81	16	3
1994	674	8,269	64	34	2
1995	505	32,600	78	20	1
1996	584	30,457	63	36	1
1997	506	26,829	41	58	1
1998	635	29,525	41	58	1
1999	493	26,163	62	37	1
2000	587	106,916	37	62	1
2001	492	28,994	n/a	n/a	n/a
2002	430	25,236	70	27	3
2003	558	18,679	86	9	5
2004	741	31,408	63	33	4
2005	769	25,134	87	11	2
2006	779	28,511	n/a	n/a	2
2007	825	26,300	75	19	6

Source: National Statistics, March 2008 (First Release, www.statistics.gov.uk).

deliver strong cash flows to create value. By contrast, recent mergers are more likely to be driven by strategic factors. Prominent among these are the increased globalisation of markets, with greater exposure to more aggressive international competition.

According to Doyle, this process was fuelled by deregulation and privatisation in many countries, which have freed companies in the telecommunications and airline industries, in particular, to seek out global strategic alliances. In addition, technological change raised the investment expenditures required to research and market new products, so that size of firm conferred a major advantage in industries like pharmaceuticals. Moreover, distance was no longer a barrier, given the improvements in transportation and information technology; hence the wave of banking mergers in North America and

Table 20.2 Acquisition according to status of acquiree

Year	Total acquisitions		Independent firms		Inter-company sales of subsidiaries	
	Number	Value (£m)	Number	Value (£m)	Number	Value (£m)
1992	432	3,941	232	4,108	200	1,833
1993	526	7,063	337	2,986	189	4,078
1994	674	8,269	465	5,743	209	2,526
1995	505	32,600	299	25,647	206	6,953
1996	584	30,742	336	23,348	248	7,394
1997	506	26,829	384	22,453	122	4,376
1998	635	29,525	485	24,086	150	5,439
1999	493	26,163	400	22,211	93	3,952
2000	587	106,916	466	100,513	121	6,403
2001	492	28,994	319	21,029	173	7,965
2002	430	25,236	323	16,998	107	8,238
2003	558	18,679	392	10,954	166	7,725
2004	741	31,408	577	22,882	164	8,256
2005	769	25,134	604	16,276	165	8,858
2006	779	28,511	628	20,180	151	8,331
2007	825	26,300	662	19,388	163	6,912

Source: National Statistics, March 2008 (First Release, www.statistics.gov.uk).

Europe in the late 1990s, and the flurry of mergers in the US telecommunications industry in 2005.

The importance of cross-border acquisitions involving UK firms can be seen in Table 20.3, which shows data on acquisitions of UK firms by foreign entities, and acquisitions by UK firms of overseas enterprises.

Table 20.3 Cross-border acquisitions involving UK companies

Year	UK firms acquired by foreign firms		UK firms' acquisitions of foreign firms	
	Number	Value	Number	Value
1992	210	4,139	679	7,264
1993	267	5,187	521	9,213
1994	202	5,213	422	15,164
1995	131	12,817	365	11,967
1996	133	9,513	442	13,377
1997	193	15,717	464	19,176
1998	252	32,413	569	54,917
1999	252	60,860	590	111,193
2000	227	64,618	557	181,285
2001	162	24,382	371	41,473
2002	117	16,798	262	26,626
2003	129	9,309	243	20,756
2004	178	29,928	305	18,709
2005	242	50,280	365	32,732
2006	259	77,750	405	37,412
2007	259	81,399	441	58,128

Source: National Statistics, March 2008 (First Release, www.statistics.gov.uk).

The international data clearly show the fall-back in activity following the collapse of the 'dotcom' boom. UK firms were net acquirors in value terms until 2004, since when foreign entities have conducted a high level of net acquisitions of UK firms, which in value terms easily outweighs the acquisition of UK firms by other UK firms recorded in Table 20.1. In this sense, internal merger activity by UK firms has become a relative sideshow, although UK firms continue to spend large amounts on foreign firms as they increasingly globalise their activities.

The surge in foreign acquisitions has involved several high-profile, very large deals such as Tata Steel (India)/Corus, Telefonica (Spain)/O$_2$, Ferrovial (Spain)/British Airports Authority, and Dubai Ports World (United Arab Emirates)/P&O. In some of these cases, the acquiror was a 'sovereign wealth fund (SWF)' set up by the foreign government to invest income from oil or other sources. The increased ownership and involvement of such investors (accounting for $61 billion investment globally in 2007) has raised issues of potential foreign influence in the economic affairs of the UK (and other countries), and appears to contradict the UK government's desire to reduce state ownership of industry. Major acquisitions of US firms included the $12 billion acquisition of GE Plastics by Saudi Arabian Basic Industries Corporation (Sabic), and of Dow Chemical by Kuwait Petroleum Corporation (KPC) for $9.5 billion.

On the global front, 2007 was a record year for M&A activity. The level of announcements (i.e. including deals yet to be completed), according to the consultancy Dealogic (**www.dealogic.com**), was $4.83 trillion, surpassing the previous peak of $3.91 trillion recorded in 2006. Cross-border M&A also reached a new peak at $1.99 trillion, 41 per cent of total global volume, and an increase of 78 per cent from the $1.12 trillion recorded in 2006. The USA is the most targeted location by foreign acquirers with $363 billion in announcements, while the UK is the foremost cross-border acquirer with $307 billion, an increase from $83 billion in 2006. The UK figure for 2007 was boosted by five $10 billion plus deals, including the $96 billion bid for the Dutch bank ABN Amro by a consortium led by the Royal Bank of Scotland (which saw off a competing bid from Barclays Bank). However, the largest announced deal of all was that by Australian mining group BHP Billiton for the UK-owned Rio Tinto Zinc at $152 billion, although this had yet to be completed by mid-2008.

Average deal size was $222 million, compared to $198 million in 2006. There were 36 deals valued at over $10 billion, and 949 hostile bids against 374 in 2006.

■ The regulation of takeovers

UK takeovers are regulated in three ways.

The first mode of regulation is under the competition policy of the European Union, set out in Articles 81 (formerly Article 85) and 82 (formerly 86) of the Treaty of Rome. Article 82 prohibits the abuse of a dominant firm position insofar as it may affect trade between member states. The EC Merger Regulation (ECMR) provides that a merger that creates a dominant position, as a result of which competition would be significantly impeded, shall be declared incompatible with the common market. The Regulation applies to all mergers with a 'Community Dimension', defined in terms of turnover levels. The ECMR was designed to provide 'one-stop' merger control to avoid the risk of mergers being investigated under two or more jurisdictions. National authorities may not normally apply their own competition laws to mergers falling within the ECMR, which are investigated by the Competition Commission.

In 2002, the EU was forced to overhaul its procedures after losing three court cases in which the plaintiffs had challenged the prohibition of their respective mergers. The Court of First Instance (CFI) criticised both the EU's procedures and also the quality of its economic analysis, especially its reliance on the theory of 'collective dominance'. The EU's interpretation of this was that a reduction in the number of competitors in an industry would necessarily lead to anti-competitive behaviour by the survivors, which was *not* necessarily so, according to the CFI. In response, the EU has introduced a tighter Merger Regulation that came into force in May 2004, which incorporated clearer guidelines for firms wishing to merge, including access to official files and to the investigating officials themselves. The Competition Commission also created a new post of Chief Economist to enhance the economic expertise at its command. (For fuller details on EU merger policy, see **www.europa.eu.int**.)

Mergers falling outside the ambit of the ECMR are the responsibility of the Department for Business Enterprise and Regulatory Reform (DBERR), previously known as the Department of Trade & Industry (DTI). Mergers qualify for investigation if UK turnover of the target enterprise exceeds £70 million, or if the merger creates or increases a 25 per cent share in a market for goods or services in the UK, or in a substantial part of it (i.e. local monopolies can qualify).

Qualifying mergers are investigated by the Competition Commission, the replacement for the old Monopolies and Mergers Commission in 1999 (operating under the Fair Trading Act of 1973), following the Competition Act 1998. This legislation was superseded by the Enterprise Act of 2002 that provided for the continued enforcement of undertakings and orders made under the 1973 Act, and transferred responsibility for enforcement of these to the competition authorities.

The UK competition policy is effectively run at two levels:

1 **The Office of Fair Trading (OFT)** exists 'to make markets work better' by addressing anti-competitive practices and consumer empowerment, using a combination of enforcement and communication. The OFT investigates all mergers in the first instance, and decides whether they warrant further investigation by the CC. There are three ways of treating a proposed merger:

   ■ It may be referred to the CC for further investigation
   ■ It may be cleared
   ■ Undertakings may be sought instead of a reference to the CC

   It is possible (as with the ECMR) for firms to seek informal guidance and advice from the OFT on whether or not a potential merger is likely to be referred.

2 **The Competition Commission.** This body, with around 150 staff, including lawyers, economists, accountants and support staff, conducts in-depth inquiries into mergers, markets and the operation of the major regulated industries such as electricity and gas.

When responding to a merger reference, the CC is required 'to determine whether the merger has resulted or may be expected to result in a substantial lessening of competition, and to take the action it considers reasonable and practicable to address any adverse effects of the merger that it has identified.' The CC's legal role is squarely focused on competition issues, replacing the wider public interest test of the previous regime. The Enterprise Act gives the CC remedial powers to direct companies to take certain actions to improve competition. Under the previous structure, its role was merely to make recommendations to government. As well as prevention of a merger from proceeding, remedial action open to the CC includes requiring a firm to sell off part of its business, and requiring firms to behave in ways that safeguard competition (so-called 'undertakings').

The third control on takeovers is operated by the **Takeover Panel** (**www .thetakeoverpanel.org.uk**), formed in 1968 to counter the perceived inadequacy of the statutory mechanisms for regulating the conduct of both parties in the takeover process. The Panel consists of representatives from City and other leading business institutions, such as the CBI, the Stock Exchange and the ICAEW accounting body, thus representing the main associations whose members are involved in takeovers, whether as advisers, shareholders or regulators. The Panel promulgates and administers the City Code on Takeovers and Mergers, known as the **City Code**, a set of rules originally with no force of law, reflecting what those most closely involved with takeovers regard as best practice. It did, however, have some sanctions to enforce its authority, such as public reprimands, which damage the reputation of violators of the Code, perhaps leading to the collapse of the bid and, for financial advisers, to long-term loss of business. The Panel's ultimate sanction was to request its members to withdraw the facilities of the City from offenders, although this is extremely rare.

In 2006, the EU Takeover Directive came into force. This is very largely based on the UK City Code, but with one important difference in that it is has statutory backing. The Panel can now order compensation to be paid in certain cases, and can pursue miscreants in the UK courts. It can also ask the Financial Services Authority to take enforcement action in cases of market abuse for which penalties include unlimited fines.

■ The chronology of a hostile bid

The following schedule details the necessary timing of bids and provision of information as required by the City Code.

Day 1: Bid announced. Bidder has 28 days in which to make a formal offer to target's shareholders.

Day 14 after formal offer is made: Deadline for target company to publish its 'defence document'.

Day 21 after formal offer: First date at which the contest can be ended. Bidder must disclose how many of target's shares have been voted in its favour. If over 50 per cent, bidder has won; if less, it may choose to walk away.

Day 39: First day on which offer can close.

Day 57: Last day for defender to produce new arguments ('material new information') to encourage shareholder loyalty.

Day 60: Accepting target shareholders may withdraw their acceptances if offer not unconditional as to acceptances.

Day 74: Last day for offeror to revise its offer.

Day 88: Last day for offer to be declared unconditional as to acceptances.

Day 109: Last day for offer to be declared wholly unconditional.

Day 123: Last day for paying the offer consideration to target shareholders who accepted by day 109.

Normally, the maximum time span allowed for the whole process is thus 109 days, although the Takeover Panel may 'stop the clock' pending clarification of key points. In the event of a reference to the CC, the process is halted *sine die* to await its report. This can take upwards of six months, during which the initial 'urge to merge' has been known to evaporate.

The key requirements of the City Code, now into its 8th edition (2006), are:

- All shareholders must be offered equally good terms, as defined by the code.
- All shareholders must be given equal access to information.
- A timetable is adhered to that sets time limits for each phase of the bid.
- Bidders and members of a **concert party** (a group acting together) must disclose their dealings.
- The bidder must set an acceptance level (of over 50 per cent) at which the bid becomes unconditional.
- There are limits on the conditions attached to a bid.
- A **mandatory offer** must be made if a shareholder's or concert party's holdings exceed 30 per cent.
- The board of the target company may not use **poison pills** (see below) and other actions to frustrate a *bona fide* bid, unless they have shareholder approval.

In addition to these the Companies Act imposes its own requirements: all shareholdings of above 3 per cent must be disclosed, and any changes of more than 1 per cent in such shareholding must also be disclosed, whether or not they are related to a bid.

20.3 MOTIVES FOR TAKEOVER

Managers seeking to maximise the wealth of shareholders should continually seek to exploit value-creating opportunities. There are two situations when managers feel able to enrich shareholders via takeovers:

1 *When managers believe that the target company can be acquired at less than its 'true value'.* This implies disbelief in the ability of the capital market consistently to value companies correctly. If a company is thought to be undervalued on the market, there may well be opportunities for 'asset-stripping', i.e. selling off the components of the taken-over company for a combined sum greater than the purchase price.

2 *When managers believe that two enterprises will be worth more if merged than if operated as two separate entities.* Thus for two companies, A and B:

$$V_{A+B} > V_A + V_B$$

value additivity
The notion that other things being equal, the combined present value of two entities is their separate present values added together

The principle of **value additivity** would refute this unless the amalgamation resulted in some form of synergy or more effective utilisation of the assets of the combined companies.

In practice, it is very difficult to differentiate between these two explanations for merger, especially as many mergers result in only partial disposals, when activities that appear to fit more neatly into existing operations are retained. Companies are valued by the market on the basis of information that their managements release regarding market prospects, value of assets, R&D activity, and so on. Market participants may suspect that an under-performing company could be operated more efficiently by an alternative management team, but until a credible bidder emerges, poor results may simply be reflected in a poor stock market rating.

■ How different types of acquisition create value

Acquisitions can be split into three types:

1 **Horizontal integration** – where a company takes over another from the same industry and at the same stage of the production process: for example, a brewery acquiring a competitor e.g. Greene King's acquisition of Belhaven Breweries in 2005. The motivation is usually enhancement of market power and/or to obtain production economies.

2 **Vertical integration** – where the target is in the same industry as the acquirer, but operating at a different stage of the production chain, either nearer the source of materials (backward integration) or nearer to the final consumer (forward integration), e.g. Ford's takeover of Kwikfit, the car spares firm.

3 **Conglomerate or unrelated diversification** – where the target is in an activity apparently dissimilar to the acquirer although some activities such as marketing may overlap (known as concentric diversification in this case). These takeovers are often said to lack 'industrial logic', but can lead to economies in the provision of company-wide services such as Head Office administration and access to capital markets on improved terms, i.e. financial economies.

In reality, most mergers are difficult to classify into such neat categories, as they are motivated by a complex interplay of factors, which it is hoped will enhance the value of the bidder's equity. The more specific reasons cited for launching takeover bids usually reflect the anticipated benefits that a merger is expected to generate:

1 *To exploit scale economies.* Larger size is usually expected to yield production economies if manufacturing operations can be amalgamated, marketing economies if similar distribution channels can be utilised, and financial economies if size confers access to capital markets on more favourable terms. The Shed Productions takeover of Twenty Twenty Productions (see cameo) was scale-driven. Akzo Nobel specified a cost-savings target of €2.5 billion in its campaign to acquire ICI, generated by plant closures and redundancies.

TV producer Shed buys Twenty Twenty

The door opened at Shed Productions yesterday and the company ushered in another independent television programme-maker, buying Twenty Twenty Productions for up to £19m.

Aim-listed Shed, maker of *Footballers' Wives* and *Bad Girls*, said the acquisition would add significantly to its top and bottom line figures.

Twenty Twenty, which has produced programmes such as *The Choir, Bad Lads Army* and *Grandad's Back in Business*, currently showing on BBC2, had turnover of £9.5m last year, with an adjusted profit of £2.1m.

Shed's revenues to February 2007 were £19.1m with a pre-tax profit of £2.3m.

Independent producers have been involved in a consolidation process, with Shed having bought Ricochet, which makes programmes such as *Supernanny*, in November 2005. They are looking to establish scale in order to deal with the big broadcasters, ITV and Channel 4, on as strong a footing as possible.

Analysts said that would become increasingly important after Michael Grade, ITV's chairman, announced last week that he aspired to cut the proportion of programmes made by independents from 46 per cent to 25 per cent.

Eileen Gallagher, chief executive of Shed, said: "This purchase will broaden our revenue base."

Shed, which is understood to be on the verge of announcing new commissions, is 60 per cent-owned by directors. Ms Gallagher said that ruled out any likelihood of it being swallowed up unwillingly by another independent company.

"There is a growing realisation in this business that there are not that many really good, quality, creative production companies out there, and this is one of them," she added.

Shares in Shed, which floated in March 2005 at a price that gave it a market capitalisation of £44m, closed unchanged at 91½p, valuing the equity at £59.5m.

Source: Ben Fenton, *Financial Times,* 20 September 2007.

2 *To obtain synergy.* This term is often used to include any gains from merger, but, strictly, it refers to benefits unrelated to scale. Gains may emerge from a particular way of combining resources. One company's managers may be especially suited to operating another company's distribution systems, or the sales staff of one company may be able to sell another company's, perhaps closely related, product as part of a package. Akzo Nobel predicted annual revenue synergies of €375 million in addition to cost savings.

3 *To enter new markets.* For firms that lack the expertise to develop different products, or do not possess the outlets required to access different market segments, takeover may be a simpler, and certainly a quicker, way of expanding, as with JCB.

JCB digs up its second takeover

JCB, the maker of bright yellow construction machines, is buying Germany's Vibro-max in a deal that is only the second in JCB's 60-year history and its first since 1968.

The move also suggested that Sir Anthony Bamford, chairman and owner of the UK group, has overcome his antipathy to investing in Germany and France.

JCB, one of the world's top five makers of construction machines, had sales last year of £1.15bn, two-thirds outside Britain.

In recent years the company has blazed an often unfashionable trail for maintaining most of its manufacturing in the UK – even though it has opened plants in the US, India and Brazil and is opening one in China. Of its 5,000 global employees, 4,000 are based in Britain.

Sir Anthony – son of Joe Bamford, JCB's founder – has never hidden his distaste for investing in most of continental Europe, which he has regarded as being hide-bound by labour regulations and high wage rates.

The JCB chairman is very keen on selling to continental Europe – a large market for JCB machines – but is one of the UK business world's strongest advocates of keeping Britain out of the euro. However, Sir Anthony has been won over by the opportunity of acquiring Leipzig-based Vibromax, the price of which has not been disclosed.

John Patterson, JCB's chief executive, said the company had made the move because Vibromax's products fitted into the group's existing range of machines and used "very good" engineering.

With annual sales of €40m (£27.5m), half in the US, the privately owned

German manufacturer is the world's fourth-biggest maker of specialised rolling machines used to prepare earth prior to road building.

He indicated that he hoped to double Vibromax's sales in the next few years by stepping up production in Leipzig and using JCB's global distribution network to sell more products.

He said a factor in the acquisition was that wage rates in the Leipzig area – formerly part of East Germany – were lower than those in western Germany. Also, Vibromax's 170-strong workforce – who are mainly employed in Germany – are not members of a union.

Mr Patterson said the worldwide market for construction machines remained "fairly buoyant" this year.

Source: Peter Marsh, *Financial Times,* 20 July 2005.

The Daimler–Chrysler merger in 1998 was driven by the desire by each firm to 'fill in' its product line – Daimler was strong in highly-engineered luxury vehicles while Chrysler's expertise lay in volume production of automobiles and the fast-growing market for sports utility vehicles (SUVs).

4 *To provide 'critical mass'.* As many product markets have become more global and the lifespan of products has tended to diminish, greater emphasis has to be placed on R&D activities. In some industries, such as aerospace, telecommunications and pharmaceuticals, small enterprises are simply unable to generate the cash flows required to finance R&D and brand investment. This factor was largely responsible for the sale by Fisons and Boots of their drug-development activities in 1994 to much larger German companies. There is also a credibility effect. For example, companies may be unwilling to use small firms as a source of components when their future survival, and hence ability to supply, is suspect.

5 *To impart or restore growth impetus.* Maturing firms whose growth rate is weakening may look to younger, more dynamic companies both to obtain a quick, short-term growth 'fix', and also for entrepreneurial ideas to achieve higher rates of growth in the longer term. For some years, British American Tobacco has been using its substantial cash flows to push into markets such as Serbia where the health lobby is weaker than in Western Europe. The cameo portrays its latest foray.

BAT Tekel deal to draw market share from Turkey's Marlboro men

British American Tobacco has strengthened its presence in Turkey, the eighth-largest tobacco market in the world, by outgunning rivals to win the auction for state-owned Tekel Cigarette.

The maker of Lucky Strike emerged as the winner with a $1.72bn (£873m) bid after Citigroup Ventures, Cinven and Dogan, a Turkish media and consumer goods conglomerate, bowed out of the auction process.

The price is towards the high end of market expectations that had put the value of Tekel at $1.5bn to $1.8bn.

However, analysts said the deal, struck on a multiple of 11.4 times Tekel's earnings before interest, tax, depreciation and amortisation in 2007, was in line with recent sector takeovers.

Jonathan Fell, analyst at Deutsche Bank, said: "This looks like a pretty sensible deal to me. The multiples are not extreme and it's a chance to get an equalish market share with Philip Morris in a country where we are still seeing growth."

Turkey's cigarette market is dominated by Philip Morris, the US maker of Marlboro, which has 40 per cent of the market.

If successful, the transaction will lift BAT's share from 7 per cent to 36 per cent.

Like its competitors, BAT has sought to build sales in emerging markets as tobacco consumption in western European markets such as the UK decline after governments banned smoking in public places and launched anti-smoking health campaigns.

About half of BAT's revenues and pre-tax profits come from emerging markets.

Tekel employs about 15,000 people, but its operations are lossmaking. Last year it incurred a loss of nearly TL340m (£143m) on sales of TL916m.

Nonetheless, Paul Adams, BAT chief executive, said he expected the deal to enhance earnings from 2009, as improvements in the supply chain and savings in administrative costs fed into the company's overall results.

This would come in spite of an expected 5 per cent decline in the Turkish market as a public smoking ban is implemented in the next 18 months, Mr Adams said.

"We are buying into a declining asset. But the fall in volume will be compensated by the growing population, better pricing and consumers trading up."

Source: Pan Kwan and Vincent Boland, *Financial Times*, 23 February 2008.

6 *To acquire market power.* Obtaining higher earnings is easier if there are fewer competitors. Competition-reducing takeovers are likely to be investigated by the regulatory authorities, but are often justified by the need to enhance ability to compete internationally on the basis of a more secure home market, as in the case of the three-way merger of European steel firms mentioned earlier. In addition, backward vertical integration, mergers undertaken to capture sources of raw

materials (e.g. US oil firm Chevron's acquisition of Unocal in 2005 to increase its exploration and production capability), and forward vertical integration to secure new outlets for the company's products have the effect of increasing the firm's grasp over the whole value chain, and are thus competition-reducing in a wider sense. Many past brewery takeovers were mounted not to obtain production capacity, but to secure access to the target's estate of tied public houses, and to acquire brands, as in the case of Scottish and Newcastle's purchase of Theakstons.

7 *To reduce dependence on existing, perhaps volatile activities.* In Chapter 10, we concluded that risk reduction *per se* as a motive for diversification may be misguided. There is no reason why two enterprises owned by one company should have greater value unless the amalgamation produces scale economies or some other synergies. If shareholder portfolio formation is a substitute for corporate diversification, there is no point in acquiring other companies to reduce risk – rational shareholders will already have diversified away specific risk, and market risk is undiversifiable. There are two major qualifications to this argument. First, diversification into overseas securities may lower market risk, given that different economies, and hence stock markets, are not perfectly correlated (Madura, 1995). Second, it is possible that achieving greater size via conglomerate diversification may lower the costs of financial distress.

8 *To obtain a stock market listing.* This is achieved via a 'reverse takeover' in which an unlisted firm acquires a smaller listed firm. This 'back-door' method of achieving a listing is conducted by the listed firm issuing new shares in order to acquire the unlisted firm. Because of the difference in size, the bidder has to issue so many shares that the shareholders in the unlisted company emerge with a majority stake in the expanded firm.

Eddie Stobart drives on to LSE

Eddie Stobart, the haulage group renowned for its distinctive green lorries, is planning to list on the London Stock Exchange through a reverse takeover to create a £250m transport and logistics business.

The group has agreed to be acquired by Westbury Property Fund, the listed commercial property, port and rail operator for £137.7m in cash and shares, continuing a trend towards consolidation and scale in the logistics industry.

The merged entity, called Stobart Group, will combine Eddie Stobart's 900-vehicle haulage fleet with Westbury's port and rail assets, to create a transport and logistics business with net assets of more than £250m. It also plans to acquire O'Connor, a rail freight handling business.

Speculation about Eddie Stobart's future began in February when the group brought in advisers to consider its options. Norbert Dentressangle, a French rival, expressed interest in acquiring the business.

William Stobart, a son of the founder, who owns 27 per cent of the group and Andrew Tinkler, who owns the other 73 per cent, together expect to hold 28.5 per cent of the enlarged group in the same proportions and would respectively become chief executive and chief operating officer.

They also intend to acquire Westbury's commercial property portfolio through WADI Properties, a separate wholly owned company, for £142m in cash and assumed debt, completing what is in effect an asset swap.

Analysts expect the merged group to have revenues of £250m in its first year of trading and earnings before interest, tax, depreciation and amortisation of £25m. Eddie Stobart last year had revenues of £168m and ebitda of £14.1m.

In addition to its vehicle fleet, Eddie Stobart operates a rail freight service between Daventry and Grangemouth and owns 2.6m square feet of storage facilities at 27 sites round the country. Big customers include Tesco, Johnson & Johnson, Coca-Cola and Nestlé.

The company also has a prolific marketing arm, which sells branded Eddie Stobart toys and paraphernalia in three stores and online.

The 25,000-strong Eddie Stobart fan club provides a loyal customer base and fans who compete to "spot" its trucks. Each truck bears a different woman's name.

"Already we've had quite a few [fan club members] contact us this morning looking to buy some shares," Mr Tinkler said. "We're really excited about that."

Westbury owns a port in Runcorn, Cheshire, and a rail terminal operator and warehousing business in neighboring Widnes.

Source: Chris Bryant, *Financial Times*, 16 August 2007.

■ The 'market for management control'

Several of the above motives for merger suggest that some companies can be more efficiently operated by alternative managers. A more general motive for merger is thus to weed out inefficient personnel. There are three ways in which the market mechanism can penalise managerial inefficiency:

1 Insolvency, which usually involves significant costs.
2 Shareholder revolt, which is difficult to organise given the diffusion of ownership and the general reluctance of institutional investors to interfere in operational management.
3 The takeover process, which may be regarded as a 'market for managerial control'. The threat of takeover provides a spur to inefficient managers, while removing inefficient managers lowers costs and removes barriers to more effective utilisation of assets. Theory suggests that incompetently managed firms will be acquired at prices that ensure the owners of the acquirer suffer no loss in value. If a bid premium over the market price is payable, this should be recoverable from the higher cash flows generated from more efficient asset utilisation. To this extent, takeover activity is seen by authors such as Jensen (1984) as a perfectly healthy expression of the workings of the market system, potentially benefiting all parties.

■ Managerial motives for takeover

The motive of diversification to reduce risk suggests a second possible explanation for takeover activity. With the divorce of ownership and control, and the consequent high level of managerial autonomy, managers are relatively free to follow activities and policies, including acquisition of other firms, which enhance their own objectives, both in monetary and non-pecuniary forms.

Managerial salaries and perquisites are usually higher in large and growing firms, and since growth by acquisition is usually easier and swifter than organic growth, managers may view acquisition with some eagerness. If acquisitions are 'managerial' in this sense, then acquirers may be prepared to expend 'excessive' amounts to gain ownership of target companies simply to secure deals that promote managerial well-being, but at the expense of shareholder value. If this explanation is correct, acquisitions may result in a transfer of wealth from shareholders of acquiring firms to shareholders of acquired companies, even when presented as promoting the best interests of the former.

Takeovers may also be related to the way managers are remunerated. In the 1980s, UK managers increasingly came to be paid by results, with the commonest criterion of performance being growth in EPS. This is a notoriously unreliable measure of performance, as it is not only dependent on accounting conventions, but relatively easy to manipulate and also easy to increase by takeover. For example, shutting down a loss-making activity can raise reported EPS.

Self-assessment activity 20.1

Suggest some 'managerial' motives for growth by takeover.

(Answer in Appendix A at the back of the book)

■ How to increase EPS by takeover: Hawk takes over Vole

A common means of increasing EPS has been to acquire other companies with lower P:E ratios than one's own, these being companies out of favour with the market, either through poor performance or because too little was known about them. The acquisition of such companies, in certain conditions, can raise both EPS and share price. Consider the example in Table 20.4. Hawk, with a P:E ratio of 20, reflecting strong growth

Table 20.4 Hawk and Vole

	Pre-bid		Post-bid
	Hawk	**Vole**	**Hawk + Vole**
Number of shares	100 m	20 m	100 m + 5 m = 105 m
Earnings after tax	£20 m	£2 m	£20 m + £2 m = £22 m
EPS	20p	10p	£22 m ÷ 105 m = 21p
P:E ratio	20:1	10:1	20:1
Share price	£4	£1	20 × 21p = £4.20
Capitalisation (market value)	£400 m	£20 m	105 m × £4.20 = £441 m

expectations, contemplates the takeover of Vole, whose P:E ratio is only 10. Hawk proposes to make an all-share offer. If it were able to obtain Vole at the current market price, it would have to issue 5 million shares to Vole's shareholders in exchange for their 20 million shares, i.e. (5 million × £4) = (20 million × £1) = £20 million.

Table 20.4 shows the impact of the exchange if the P:E ratio of the expanded company were to remain at 20. The new EPS is (£22 m/105 m) = 21p, resulting in a post-bid share price of £4.20, and an overall market value of £441 million. This apparently magical effect seems to have generated wealth of £21 million. If it works out this way, the beneficiaries are the two sets of shareholders: Hawk's existing shareholders find their 100 million shares valued at a price higher by 20p, i.e. £20 million in total, and Vole's former shareholders find they now hold shares valued at £21 million, rather than the value of £20 million placed on Vole prior to the bid, i.e.:

Gains to Hawk's shareholders = £20 m

Gains to Vole's shareholders = £1 m

Total gain = £21 m

This so-called **'boot-strapping'** effect may simply be 'financial illusion' because it is unlikely to occur quite like this in reality. First, it assumes the absence of a bid premium. In practice, Hawk would have to offer above the market price to tempt Vole's shareholders into selling, thus altering the balance of gain. Second, it assumes that the market applies the same P:E ratio to the expanded group as the pre-bid ratio for Hawk. If no synergies were expected, then the likely post-bid P:E ratio is the total pre-bid value of the two firms relative to their total pre-bid earnings, i.e.:

$$\frac{(£400 \text{ m} + £20 \text{ m})}{(£20 \text{ m} + £2 \text{ m})} = \frac{£420 \text{ m}}{£22 \text{ m}} = 19.09$$

However, if Hawk is expected to reorganise Vole and impart the same growth impetus expected from Hawk itself, the P:E ratio post-bid could exceed this figure, and approach Hawk's pre-bid P:E value of 20. If this occurs, then both groups of shareholders can enjoy the value created by the expectation of more efficient operation of Vole's assets and higher cash flows thereafter. Conversely, expectations of integration difficulties might offset such gains.

It does not follow that a higher EPS will lead to a higher share price. If the acquisition moved Hawk into riskier areas of operation, its activity Beta should rise accordingly and the higher expected cash flows will be discounted at a higher required return. Similarly, if instead of financing the bid by a share exchange, Hawk had borrowed the required £20 million, then the share price might not rise if the greater gearing and accompanying financial risk resulted in a higher equity Beta. The suspicion remains that many acquisitions ostensibly undertaken to raise the acquirer's share price, are really undertaken for 'managerial' reasons (see Gregory, 1997).

Certainly, the subsequent difficulties experienced in post-merger integration and operation do not support the view that mergers are always in the best interests of the bidders' shareholders (see below).

Self-assessment activity 20.2

Suggest how managerial pay schemes might encourage takeovers against the interests of shareholders.

(Answer in Appendix A at the back of the book)

20.4 ALTERNATIVE BID TERMS

Table 20.1 showed data on the three main ways of financing takeovers: cash, issue of ordinary shares and fixed-interest securities (loan stock, convertibles and preference shares). Clearly, the first two methods predominate, although their relative importance varies over time. As a rule of thumb, share exchange is favoured when the stock market is high and rising, while cash offers are used more when interest rates are relatively low or falling, given that many cash offers are themselves financed by the acquirer's borrowing. This pattern is illustrated clearly by the figures for the early 2000s, when the stock market was depressed and interest rates low and falling. Increasingly, however, bidders offer their targets a choice of cash or shares, or even a three-way choice between straight cash, cash with shares, or shares alone.

For example, when bidding for a smaller property company, City North plc, a former BES company, in 2005, the considerably larger Grainger Trust plc offered two alternatives. Shareholders of City North could either opt for a full cash consideration of 270 pence per share, or accept 180 pence in cash plus 0.2423 of a Grainger share, worth 95 pence on the last dealing day prior to the bid announcement. The cash plus paper offer was thus worth 275 pence, a little higher than the straight cash offer, which is typically the case, as the target shareholder bears the twin risks of exposure to share price falls and the danger of projected synergies not materialising.

Such complex offers are designed to appeal to the widest possible body of shareholders. The chosen package depends on the balance of relative advantages and disadvantages of the different methods, from both the bidder's and the target shareholders' viewpoints.

■ Cash

The previously-mentioned acquisition of ICI by Akzo Nobel was an all-cash offer of £8 billion. Everyone understands a cash offer. The amount is certain, there being no exposure to the risk of adverse movement in share price during the course of the bid. The targeted shareholders are more easily able to adjust their portfolios than if they received shares, which involve dealing costs when sold. Because no new shares are issued, there is no dilution of earnings or change in the balance of control of the bidder (unless, in the case of borrowed capital, creditors insist on restrictive covenants). Moreover, if the return expected on the assets of the target exceeds the cost of borrowing, the EPS of the bidder may increase, although perceptions of increased financial risk may mitigate this apparent benefit. A disadvantage from the recipient's viewpoint is possible liability to capital gains tax (CGT). However, in the case of cross-border takeovers there are major advantages in receiving cash rather than the shares of the bidder, traded only on a foreign stock exchange, one reason why Akzo Nobel offered cash to ICI shareholders. Nevertheless, the cash has to be found from somewhere.

■ Share exchanges

Any liability to CGT is delayed with a share offer, and the cash flow cost to the bidder is zero, apart from the administration costs involved. However, equity is more costly to service than debt, especially for a company with taxable capacity, and an issue of new shares may interfere with the firm's gearing ratio. There could be an adverse impact on the balance of control if a major slice of the equity of the bidder came to be held by institutions looking for an opportunity to sell their holdings. The overhanging threat of a substantial share sale may depress the share price of the bidder.

■ Other methods

The use of other financing instruments is comparatively rare. When fixed-interest securities are used, they are usually offered as alternatives to cash and/or ordinary shares. Convertibles have some appeal because any diluting effect is delayed and the interest cost on the security, which qualifies for tax relief, can usually be pitched below the going market rate on loan stock, due to the expectation of capital gain on conversion. Preference share financing in general is comparatively rare, owing to the lack of tax-deductibility of preferred dividends and to limited voting rights.

Fortis sets out rights issue for ABN bid

Fortis, part of the Royal Bank of Scotland-led consortium vying with Barclays to buy ABN Amro, yesterday unveiled the €13.4bn (£9.4bn) deeply discounted rights issue needed to buy the Dutch bank.

The proceeds will partly finance Fortis's €24.7bn share of the joint bid for ABN Amro by a consortium also consisting of Santander of Spain.

The rights issue, one of the biggest ever in Europe, consists of a two-for-three share rights issue of 896,181,684 new shares at €15 each.

The price represents a near 50 per cent discount to Thursday's closing price of €26.63. Fortis shares rose slightly yesterday to €26.65 – up

from €25.30 at the beginning of the week.

Fortis has long been viewed as the weak link in the consortium and there had been questions about whether there was sufficient investor appetite for such a large rights issue.

However, Fortis shareholders voted overwhelmingly in favour of the transaction last month. People familiar with the situation expect 95 per cent of the shares to be taken up in the rights issue.

The rights issue has been fully underwritten by investment banks led by Merrill Lynch. Shares not taken up by existing shareholders will be sold on a nil paid basis.

RBS and Santander this week initiated funding drives to help pay for their share of the ABN Amro bid.

The European Commission gave its approval for the Scottish and Spanish banks' involvement in the consortium.

RBS said it planned to issue an unquantified amount of dollar preference shares priced at $25 each that will make up some portion of the €5bn–€6bn of preference shares it will issue in total across the US, European and UK markets.

Santander unveiled plans for up to €7bn of convertible bonds to be sold to help fund its part of the deal.

Source: Jane Croft, *Financial Times,* 22 September 2007.

The Fortis cameo details how one takeover bidder decided to finance an acquisition. The cameo also indicates how its fellow consortium members – Royal Bank of Scotland and Santander – were planning to arrange their financial inputs.

Self-assessment activity 20.3

Why might the shareholders of a target company prefer to be paid in cash rather than shares?

(Answer in Appendix A at the back of the book)

20.5 EVALUATING A BID: THE EXPECTED GAINS FROM TAKEOVERS

Evaluating an acquisition is little different from other investments, assuming the motive of the bid is economic rather than managerial, i.e. designed to maximise the post-bid value of the expanded enterprise. It would be worthwhile Company A taking over Company B so long as the present value of the cash flows of the enlarged company exceeds the present value of the two companies as separate entities:

$$V_{A+B} > V_A + V_B$$

Thus, $[V_{A+B} - (V_A + V_B)]$ measures the increase in value. The net cost to the bidder is the value of the amount expended less the value of the target as it stands:

$$\text{Net cost} = [\text{Outlay} - V_B]$$

so that the net present value of the takeover decision is the gain less the cost, i.e.:

$$\begin{aligned} \text{NPV} &= V_{A+B} - (V_A + V_B) - [\text{Outlay} - V_B] \\ &= V_{A+B} - V_A - \text{Outlay} \end{aligned}$$

The NPV will depend on the method of financing and, of course, the terms of the transaction. Essentially, the bidder is hoping to extract the maximum value of any expected cost savings and synergies from the takeover for its own shareholders. Conversely, the offer must be made attractive to the owners of the target to induce them to sell.

■ Fewston plc and Dacre plc

An example will illustrate the way in which the division of the spoils can depend on the method of financing. Fewston plc is launching a cash bid for Dacre plc, both are quoted companies and both are ungeared. The market value of Fewston is £200 million (100 million 50p shares, market price £2) and that of Dacre is £40 million (10 million 50p shares, market price £4). Fewston hopes to exploit synergies, etc., worth £20 million after the takeover. It offers the shareholders of Dacre £50 million in cash. The NPV of the bid to Fewston is thus:

$$\begin{aligned} \text{NPV} &= V_{A+B} - V_A - \text{Outlay} \\ &= (\text{£200 m} + \text{£40 m} + \text{£20 m}) - \text{£200 m} - \text{£50 m} = \text{£10 m} \end{aligned}$$

The overall gain from the takeover (i.e. the synergies of £20 million) is split equally between the two sets of shareholders. The need to make a higher bid or the appearance of another bidder would probably tilt the balance of gain towards Dacre's shareholders.

If the bid is made in the form of a share-for-share offer to the same value, the arithmetic alters. In this case, Fewston is giving up part of the expanded firm and hence a further share of the gains to Dacre's shareholders. Assuming a bid of the same value, Fewston must offer them (£50 m/£2) = 25 m shares. This would result in a total share issue of 125 million shares, i.e. Fewston is handing over 20 per cent of the expanded company to Dacre's shareholders. In this case, the gain enjoyed by Fewston's shareholders will be lower. The NPV of the takeover is still the gain less the cost; but the cost is greater, i.e. the proportion of the expanded company handed over less the value of Dacre as it stands:

$$\begin{aligned} \text{Cost} &= \left(\frac{25\text{ m}}{100\text{ m} + 25\text{ m}} \times \text{£260 m} \right) - \text{£40 m} \\ &= (\text{£52 m} - \text{£40 m}) = \text{£12 m} \end{aligned}$$

Hence, the NPV of the takeover from Fewston's perspective is:

$$NPV = (gain\ in\ value - cost) = (£20\ m - £12\ m) = £8\ m$$

Fewston's shareholders are thus left with only £8 million of the net gains from the takeover, 20 per cent lower than in the cash offer case, which is the same proportion as the share of the expanded company handed to Dacre's shareholders.

A share exchange of equivalent value to a cash bid generally leaves the bidder's shareholders worse off compared to a cash deal because their share of both the company and the gains from the takeover are diluted among the larger number of shares. The post-bid share prices in these two cases are as follows:

Cash bid: $(£260\ m/100\ m) = £2.60$

Share exchange: $(£260\ m/125\ m) = £2.08$

Against this, given that takeovers carry risks, for example, the risk of inability to capture the anticipated synergies, a share-based deal has the merit of transferring a portion of these risks to the targets' former owners.

However, if Fewston has to borrow in order to make the cash bid, the increase in gearing may result in shareholders seeking a higher return, thus lowering the market price. In addition, the analysis hinges on the existence of an efficient capital market whose assessment of the gains from takeover corresponds with that of the two parties.

Self-assessment activity 20.4

Predator is valued on the market at £1,000 million, and Prey at £200 million. Predator values the expected post-merger synergies at £50 million. If it bids £230 million for Prey what is the NPV of the bid? What is the share of the gains for each firm?

(Answer in Appendix A at the back of the book)

20.6 WORKED EXAMPLE: ML PLC AND CO PLC

The following question appeared on the CIMA Strategic Financial Management examination paper, May 1999. (It carried 15 of the total 20 marks available.)

■ Question

ML plc is an expanding clothing retailer. It is all-equity financed by ordinary share capital of £10 million in shares of 50p nominal. The company's results to the end of March 1999 have just been announced. Pre-tax profits were £4.6 million. The Chairman's statement included a forecast that earnings might be expected to rise by 5 per cent per annum in the coming year and for the foreseeable future.

CO plc, a children's clothing group, has an issued share capital of £33 million in £1 shares. Pre-tax profits for the year to March 31 were £5.2 million. Because of a recent programme of reorganisation and rationalisation, no growth is forecast for the current year but, subsequently, constant growth in earnings of approximately 6 per cent per annum is predicted. CO plc has had an erratic growth and earnings record in the past and has not always achieved its often ambitious forecasts.

ML plc has approached the shareholders of CO plc with a bid of two new shares in ML plc for every three CO plc shares. There is a cash alternative of 135 pence per share.

Following the announcement of the bid, the market price of ML plc shares fell while the price of shares in CO plc rose. Statistics for ML plc and two other listed companies in the same industry immediately prior to the bid announcement are shown below. All share prices are in pence.

1998				
High	Low	Company	Dividend % Yield	PER
225	185	ML plc	3.4	15
145	115	CO plc	3.6	13
187	122	HR plc	6.0	12
230	159	SZ plc	2.4	17

Both ML plc and CO plc pay tax at 33 per cent.

ML plc's cost of capital is 12 per cent per annum and CO plc's is 11 per cent per annum.

Required

Assume you are a financial analyst with a major fund manager. You have funds invested in both ML plc and CO plc.

- Assess whether the proposed share-for-share offer is likely to be beneficial to the shareholders in ML plc and CO plc, and recommend an investment strategy based on your calculations.
- Comment on other information that would be useful in your assessment of the bid. Assume that the estimates of growth given above are achieved and that the new company plans no further issues of equity.

State any assumptions that you make.

Answer

First of all, some introductory calculations are needed, before we can analyse the impact of the bid.

Basic information

	ML	CO	Combined
Profit after tax (PAT) for each firm is			
ML: (0.67 × £4.6 m) CO: (0.67 × £5.2 m)	£3.082 m	£3.484 m	£6.566 m
Given respective P:Es, market values are:			
ML: (15 × £3.082) CO: (13 × (£3.484)	£46.230 m	£45.290 m	£91.520 m
Given the number of shares, share price is:			
ML: (£46.230 m/20,000) CO: (£45.290 m/33,000)	£2.31	£1.37	
EPS:			
ML: (£3.082 m/20,000) CO: (£3.484 m/33,000)	15.41p	10.56p	

Analysis

No. of shares post-bid: 20,000 + (2/3 × 33,000) = 42,000

Expected market prices post-bid = Total market value/No of shares

$$= (£91.520 \text{ m}/42,000) = £2.18$$

Value of bid at post-issue price   = (2 shares × £2.18) = £4.36

Cash value of bid per 3 shares offered: (£1.35 × 3)   = £4.05

Assessment

Assuming no changes in the level of market prices, and no re-rating of the sector, ML's share price would fall post-acquisition to £2.18. At this price, the value of the 2-for-3 share offer should attract CO shareholders. They would get shares worth (2 × £2.18) = £4.36 in exchange for shares currently worth (3 × £1.37) = £4.11.

The share-for-share offer is also worth more than the cash alternative: £4.36 vs. £4.05.

This is a 'reverse takeover', where the shareholders of the target end up holding a majority stake in the expanded company – but who gains from this?

Former CO shareholders would hold (22,000/42,000) × £91.520 m = £47.939 m of the value of the expanded firm, a gain of (£47.939 m − £45.290 m pre-bid value of CO) = £2.649 m.

ML shareholders would lose £2.649 m, making the share-financed deal distinctly unattractive to them.

Conversely, the cash offer would create wealth for ML shareholders, i.e. they give £4.05 for something worth £4.11 post-bid.

The advice to the fund manager is: 'accept the bid in respect of CO shares and sell ML shares in the market if you can achieve a price above £2.18'.

Commentary on other information required

The advice given above hinges on the behaviour of ML's share price – it has already fallen on the announcement, but by how much? It may already be too late if the market is efficient, as it would already have digested the information contained in the announcement.

Also:

- What benefits are expected from the merger, i.e. cost savings and synergies? To make sense of the bid, ML must be setting the PV of these benefits above £2.649 m to yield a positive NPV for the acquisition.
- How quickly are these benefits likely to show through? Any delay in exploiting these lowers the NPV.
- It is feasible that the market might apply a higher PER to the expanded company – maybe not as high as ML's but possibly at the market average, currently 14.25, compared to the weighted average PER for ML/CO of 14.
- Is ML likely to sell part of CO's operations? And to whom? If ML has already lined up a buyer, it must expect to turn a profit on the deal.
- Is the bid likely to be defended by the target's managers, fearful for their jobs? If so, a higher bid might be expected.
- Is a White Knight likely to appear with a higher bid on more favourable terms?
- Are there competition implications likely to attract the interest of the authorities?

20.7 THE IMPORTANCE OF STRATEGY

Considerable evidence has emerged that acquisitions have less than an even chance of success. Although definitions of 'success' may vary, any activity that fails to enhance shareholder interests is unlikely to be regarded favourably by the stock market. While it is often difficult to assess what would have happened had a company not embarked on the takeover trail, it is difficult to argue that the acquisition has not been a failure if post-acquisition performance is inferior to pre-acquisition performance, or if the acquisition actually leads to a fall in shareholder wealth.

The McKinsey firm of management consultants studied the 'value-creation performance' of the acquisition programmes of 116 large US and UK companies, using financial measures of performance. The criterion of success used was whether the company earned at least its cost of capital on funds invested in the acquisition process. On this basis, a remarkable 60 per cent of all acquisitions failed, with large unrelated takeovers achieving a failure rate of 86 per cent.

Acquisitions fail for numerous reasons:

1 Acquirers often pay too much for their targets, either as a result of a flawed evaluation process that overestimates the likely benefits or as a result of getting caught up in a competitive bidding situation, where to yield is regarded as a sign of corporate weakness.

2 'Skeletons' appear in cupboards with alarming frequency. The infamous and disastrous takeover by Ferranti of International Signal Corporation (see Appendix to this chapter) was a good example of a badly-researched acquisition that ultimately destroyed the acquirer.

3 Acquirers often fail to plan and execute properly the integration of their targets, frequently neglecting the organisational and internal cultural factors. Inadequate knowledge about the target's business should be corrected in the process of due diligence. Lees (1992) explains how all too often this aspect is overlooked.

Yet many companies have sound acquisition records. Reckitt Benckizer, the Anglo-Dutch producer of household products is often held up as a prime example. Firms like Reckitt have several things in common. Their targets are carefully selected, they rarely get involved in competitive auctions, they often have the sense to walk away from deals when they realise the gravity of the likely integration problems, and they seem able quickly and successfully to integrate acquisitions once deals are completed. What these companies have in common is a coherent strategic approach to acquisitions.

20.8 THE STRATEGIC APPROACH

Most successful acquirers see their acquisitions as part of a long-term strategic process, designed to contribute towards overall corporate development. This requires acquirers to approach acquisitions only after a careful analysis of their own underlying strengths, and to identify candidates that satisfy chosen criteria and, most importantly, provide 'strategic fit' with the company's existing activities.

Figure 20.1 displays a simple strategic framework within which a thorough-going acquisition programme might be conducted. It begins with a full strategic review of

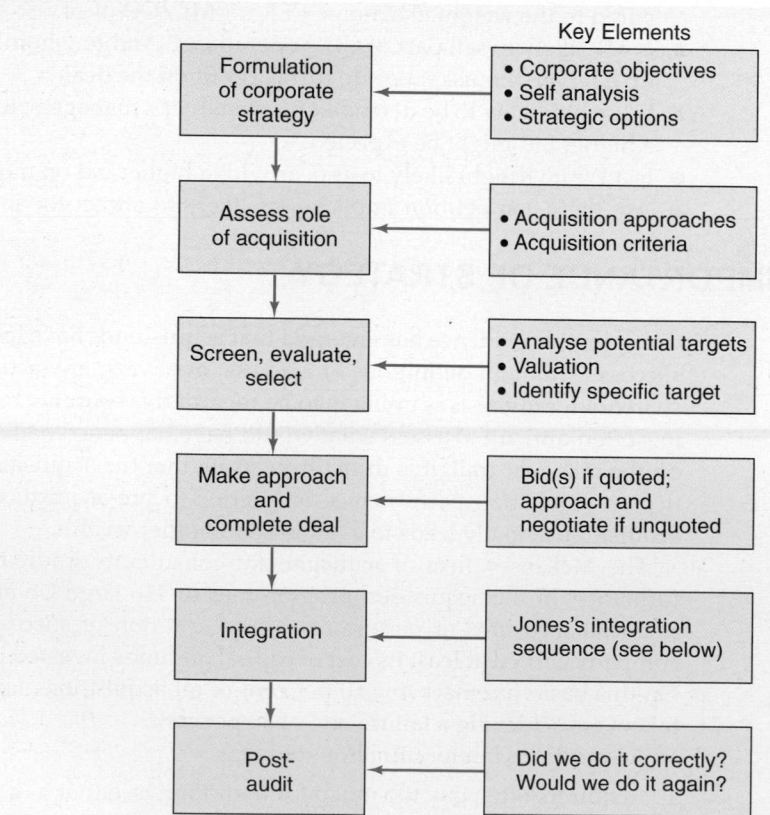

Figure 20.1 A strategic framework (based on Payne, 1985)

the company as it stands, and its strategic options, followed by a detailed consideration of the role of acquisitions (i.e. the reasons why an acquisition target may be selected), leading to the process of selecting and bidding for the chosen prey, and culminating in the often neglected activities of post-merger integration and post-audit.

■ Objectives

Formulating strategy should begin with an expression of corporate objectives, concentrating on maximising shareholder wealth. Many firms now publish mission statements, but these are usually somewhat vague expressions of the image that the company would like to portray, often largely for internal consumption in order to motivate staff (Klemm *et al.*, 1991). If, in building the desired image, the company's managers fail to earn at least the cost of equity, they will themselves invite the risk of takeover. Strategy concerns the examination of alternative routes to achieving the ultimate aim, and then the optimal way of executing the chosen path. Achieving

---

Pfizer's strategy a sign that 'bigger is rarely better'

Pfizer and Merck, the biggest US drugmakers, have maintained dramatically opposing views on large acquisitions since the former embarked on the first of two big takeovers about eight years ago. But Merck's view appears to be winning out.

Over that time, despite the third anniversary yesterday of its withdrawal of painkiller Vioxx and associated litigation crisis, shares in Merck have performed better than Pfizer's.

A challenging new environment for drugmakers including pricing and generic pressures have put both in the red but, dating back to February 2000 when Pfizer agreed to buy Warner-Lambert for $90bn, Merck's shares have fallen 20 per cent versus Pfizer's decline of about 30 per cent.

There is a rarely discussed but tangible cultural rivalry – science versus business – between Merck and Pfizer.

But what really has made them different over the past decade is that Pfizer has led the drumbeat of consolidation in Big Pharma with its acquisition of Warner-Lambert and then Pharmacia for $60bn in 2003.

Merck has consistently said it would avoid large mergers because it did not help in making drugs.

Merck and its shares have had to overcome a plunge after Vioxx's withdrawal due to heart risks. It still faces almost 30,000 lawsuits.

Pfizer, under Hank McKinnell, former chief executive, based its strategy on bigger is better. Its two deals made it the world's largest pharmaceuticals company by sales.

But Pfizer's size has not only failed to make it a fast-growing drug company but may have left it in need of a dramatic change under Jeff Kindler, its new chief executive since July last year when the board moved to replace Mr McKinnell and change the group's direction.

Jon LeCroy, analyst at Natixis Bleichroeder, says Pfizer's strategy is evidence that "bigger is rarely better". Growth is stagnant but the biggest challenge will be a record loss of revenues in 2011 when the patent expires on Lipitor, the cholesterol drug, with more than $12bn in sales this year, or about one-quarter of its total sales.

"Kindler is in a bit of a bind because the company is so big there's not much they could acquire to make a difference," Mr LeCroy says.

Mr Kindler, speaking recently, said his mission was to "change the focus, structure and culture", to "reset Pfizer – strategically and rationally – for the future" and make it the "right size".

Planning for Lipitor's patent expiration, Mr Kindler said Pfizer would have four product buckets: current products, new drugs from its labs, new drugs from external deals, and extending the shelf life of profitable off-patent products.

Shortly after taking over, Mr Kindler presented his plan to change Pfizer's reputation into a scientific collaborator and leader. That is a step towards Merck's philosophy and recent drive to build external drug collaborations rather than embark on large acquisitions.

Despite questions about its pipeline after two high-profile failures in 2003, Merck is developing new medicines including Isentress, an HIV therapy, and Cordaptive to raise good cholesterol.

It also faces significant patent expirations as the decade ends. But Mr LeCroy rates its ability to make up for lost sales with new products as second in the industry only to Schering-Plough.

Chris Schott, analyst at Bank of America, said: "Merck's focus on new mechanisms of action [for novel drugs] positions the company well for the current difficult FDA environment."

Consolidation is still considered likely in the drug industry and one drug company chief executive stresses that it might not be the big merger at fault but what management does with it. But a look at Merck and Pfizer offers a cautionary tale.

Source: Christopher Bowe, *Financial Times*, 1 October 2007.

long-term goals usually involves expansion of the enterprise, a route often preferred by managers for personal motives.

■ Internal or external growth?

There are two main ways of achieving growth: (1) by self-development of new products, markets and processes (internal growth) and (2) by acquisition (external growth). Although both of these routes are usually expensive in executive time and resources, external growth has the advantage of securing quick access to new markets or productive capacity. However, firms should not overlook intermediate strategies, such as **licensing**, whereby a royalty is paid to the developer of new technology in exchange for rights to exploit it, or joint ventures, where an existing company could be partially acquired, or a totally new one set up in partnership with another firm.

The decision to grow internally or externally will depend partly on an analysis of the strengths, weaknesses, opportunities and threats (SWOTs) of the firm. This self-analysis should make the potential acquirer aware of any competitive advantages it enjoys over rival companies. Competitive advantage stems from two sources: cost advantage, where products are virtually similar, and product differentiation. Exploitation of each of these creates value for shareholders. When areas of competitive advantage have been identified, the company can decide whether to build upon existing strengths or to attempt to develop distinctive competence in areas of perceived weakness. This evaluation may also result in a decision to divest certain activities where no obvious advantage is possessed, or where too many resources would be required to sustain an advantage.

Porter (1987) examined the acquisition record of 33 large diversified US companies. The criterion for judging 'success' was the subsequent divestment rate of earlier acquisitions. The main finding was that successful acquirers almost invariably diversify into related fields, and vice versa. In other words, diversifications into activities unrelated to the core business of the acquirer carry much greater risks of failure. Even companies with successful 'related diversification' records achieved poor results when they wandered into unrelated fields. Porter concluded that the corporate portfolio strategy of many diversifying companies had failed because most diversifiers fail 'to think in terms of how they really add value'.

■ Acquisition criteria

The bidder should next assess what specific role it hopes the acquisition will perform. Table 20.5, drawn from a publication (*Making an Acquisition*) by the investment bank 3i (Investors in Industry), which specialises in offering acquisition advice, lists possible strategic reasons for acquisition with suggested routes to achieving the stated aims.

At this stage, the company should reassess the alternatives to merger, in view of the many difficulties involved. Taking over another company is rather like moving to a larger, more expensive house. Mergers involve considerable disruptions during the planning and bidding phase; costs, such as legal advice and the printing and publishing of documents; possible exposure to increased financial risk; and the upheavals of integration. Just as some marriages do not survive the strains of house-moving, some companies often fail to recover after the stress of merger. Having identified the specific role of the acquisition, the company can now consider whether it can be achieved in other, perhaps more cost-effective ways (see Pfizer case above).

Harrison (1987) suggests that, for every merger motive, there are several alternative ways of achieving the same end. For example, if the aim is sales growth, this can be

Table 20.5 Strategic opportunities

Where you are	How to get to where you want to be
■ Growing steadily but in a mature market with limited growth prospects	■ Acquire a company in a younger market with a higher growth rate
■ Marketing an incomplete product range, or having the potential to sell other products or services to your existing customers	■ Acquire a company with a complementary product range
■ Operating at maximum productive capacity	■ Acquire a company making similar products operating substantially below capacity
■ Under-utilising management resources	■ Acquire a company into which your talents can extend
■ Needing more control of suppliers or customers	■ Acquire a company which is, or gives access to, a significant customer or supplier
■ Lacking key clients in a targeted sector	■ Acquire a company with the right customer profile
■ Preparing for flotation but needing to improve your Balance Sheet	■ Acquire a suitable company which will enhance earnings per share
■ Needing to increase market share	■ Acquire an important competitor
■ Needing to widen your capability	■ Acquire a company with the key talents and/or technology

Source: 3i (Investors in Industry).

achieved by internal expansion or by a joint venture. If the aim is to improve earnings per share, a loss-making subsidiary can be shut down or efficiency-enhancing measures can be implemented. If it is wished to use spare cash, this can be invested in marketable securities and trade investments, or even returned to shareholders as dividends, or in the form of share repurchases. If an improvement in management skills is sought, appropriately skilled personnel can be bought in to replace existing managers, outside consultants can be used for advice, or incentive and bonus schemes can be introduced. In short, if the decision to grow by acquisition is made, the potential acquirer must be very sure that the stipulated aims are unattainable by alternative measures.

Most firms with corporate planning departments exercise a continuous review of the key members of the industry in which they operate and also of related and, often, unrelated areas. Some firms are known to 'track' several dozen potential takeover candidates, assessing their various strengths and weaknesses, and estimating the likely net value obtainable if they were acquired. Such target companies are continually cross-checked against a set of possible acquisition criteria.

When the decision to expand by acquisition is taken, the corporate planning staff should be able rapidly to provide a short-list of candidates, expressing the SWOTs of each, especially its vulnerability to takeover at that time. It is common for defending managements to dismiss takeover bids as 'opportunistic' in a pejorative way. For an acquisitive company that adopts the strategic approach, this means 'well-timed', as such companies are continually seeking opportune moments to launch a bid, especially when the stock market rating of the target appears low. The joint takeover by the former GEC and Siemens of Plessey in 1989 was opportunistic, in the sense that the target's return on capital was relatively low due to a recent substantial investment programme. Whether the market had correctly valued Plessey is arguable, but the bidders undoubtedly spotted a favourable opportunity to acquire Plessey at a time when its performance looked weak in relation to the market, thus eliminating a major competitor for lucrative British Telecom contracts. Although the word 'opportunistic' was not explicitly used, the directors of Dutch company Corporate Express clearly took a dim view of the bid by Staples, the office supplies company, in 2008 when the former's share price was languishing.

Staples offers €2.5bn for Dutch rival

Staples, the US office supplies company, yesterday offered to buy Corporate Express, a Dutch rival with a significant US customer base, for €7.25 per share, valuing the Dutch group at about €2.5bn ($3.7bn).

Staples said its all-cash offer was not subject to financing and wrote to Peter Ventress, chief executive of Corporate Express, to urge him to enter takeover talks.

The Dutch company rejected the offer, saying it was unsolicited and undervalued the company.

"Corporate Express is of the opinion that this proposal significantly undervalues the company and fails to reflect Corporate Express's prospects," it said in a statement.

"We do not believe the proposal is in the best interests of our shareholders and other stakeholders."

"We therefore reject this proposal and reiterate our commitment to pursuing our declared strategy."

Corporate Express shares jumped nearly 40 per cent to €7.59 following news of the offer.

Its depressed stock market price has long made the company the subject of takeover speculation.

Ronald Sargent, chief executive of Staples, said in the letter to Mr Ventress that Staples had made "repeated" attempts over several months to discuss a takeover.

"We have been disappointed that you have not been willing to do so," he wrote.

Corporate Express, formerly known as Buhrmann, is one of four companies – along with Staples, Office Depot and Office Max – that control 25 per cent of the global office supplies market.

The Staples offer represents a 67 per cent premium to the share price before a report in early February, denied by Corporate Express, that Staples was in early talks with the company.

Source: Michael Steen and Jonathan Birchall, *Financial Times*, 20 February 2008.

■ Bidding (and defending)

Bidding is an exercise in applied psychology. Readiness to bid implies an assessment that the target is either undervalued as it stands or would be worth more under alternative management. In such cases, the bid itself provides new information about prospective value, and the bidder should expect to have to pay above the market price to secure control. However, it is often unclear before the event how much of a bid premium, if any, is already built into the market price as the market attempts to assess the probability of a bidder emerging and succeeding with its offer. The trick in mounting profitable takeover bids is to promise to use assets more effectively in order to entice existing shareholders to sell, without making such extravagant claims that the target's market price moves up too sharply before the acquisition is completed. Conversely, to accentuate the difficulties of reorganising the target could be regarded as disingenuous or even call into question the wisdom of the bid itself, leading to a fall in the bidder's own share price. The following box summarises some of the defence tactics which a bidder may encounter.

Choose your weapons! Takeover defence tactics

Takeover strategy is not a one-sided affair. Very few takeovers are recommended at once by the directors of target companies. Even if they expect to lose the fight, the incumbents usually reject the initial bid in the hope of attracting better terms. The first line of defence, therefore, is rejection because the bid is too low, or because the proposed union 'lacks industrial logic'.

Once defenders have had time to marshal their resources and get their public relations act together, more effective defences can be adopted. Some are more credible than others, and some are illegal in the UK but common in the USA. Typical defence ploys by UK firms are:

- Revalue assets – this is often a waste of time, as the market should already have assessed the market value.
- Denigrate the profit and share price record of the bidder, and hence the quality of its management. This invites retaliation in kind.
- Promise a dividend increase – this calls past dividend policy into question, and bidders usually offer this anyway.
- Publish improved profit forecasts – a dangerous ploy, since the forecast has to be plausible yet attractive, and once made, the company has to deliver. Companies that repel raiders but fail to meet profit forecasts are susceptible to further bids.
- Seek a White Knight – an alternative suitor that will acquire the target on more favourable terms (mainly for the management?).
- Lobby the competition authorities.

The following defences mostly originated in the USA and some are difficult to reconcile with the City Code:

- The Crown Jewels defence – selling-off the company's most attractive assets.
- Issuing new shares into friendly hands – this, of course, requires shareholder agreement.
- The Pacman defence – launching a counter-bid for the raider.
- Golden Parachutes – writing such attractive severance terms for managers that the bidder will recoil at the prospective expense.
- Tin Parachutes – offering excessively attractive severance terms for blue collar workers.
- Launching a bid for another company – if successful, this will increase the size of the firm, making it less digestible for the bidder.
- Leveraged buy-out – the purchase of the company by its own management using large amounts of borrowed funds.
- Poison pills – undertaking methods of finance that the bidder will find unattractive to unwind, e.g. large issues of convertibles that the bidder will have to honour (see cameo below).
- Repurchasing of shares to drive up the share price and increase the cost of the takeover (common now in the UK), although not allowed once an informal approach has been made (although it can be promised if the bid fails).

Examples of poison pills are mechanisms to trigger deeply-discounted rights issues if predator holdings rise above a certain level. Both Newscorp and Yahoo have implemented such arrangements. In the case of Yahoo's 'stockholder rights plan', a rights issue can be made if another firm builds a stake in Yahoo above 15 per cent. In Germany, the so-called 'VW law' of 1960 limits any investor's voting rights in the carmaker to 20 per cent, and confers a blocking vote on the minority stake held by the state of Lower Saxony (this has been declared illegal by the European Court of Justice).

Japan hit by new 'poison pill' concerns

Japanese companies are turning to a financing scheme that includes what many investors fear is a hidden "poison pill" aimed at thwarting takeovers and reviving the *zaibatsu* industrial conglomerate system.

The new scheme combines loans with warrants, which if exercised would substantially dilute the stakes of shareholders and defend against unwanted bidders, bankers say.

Japanese companies have been building up their defences against unsolicited bids and the prospect has raised eyebrows in the investment community. "The conditions of the warrants give the company a lot of flexibility to dilute the shareholders in the case of a takeover bid," one western banker said. "It's certainly enough to stop any potential buyer."

An official at a large Japanese investment bank said: "It is a very opaque takeover defence."

Two Sumitomo Group companies recently raised funds by taking out loans with warrants. Sumitomo Realty, a real estate group, this month raised Y120bn ($1bn) through a subordinated loan from Sumitomo Mitsui Banking Corp, with warrants that have a moving strike price and could result in dilution of 11 per cent to 23 per cent.

Sumitomo Metal Mining has taken out a Y100bn loan from SMBC with warrants.

Real estate companies and those with undervalued assets are considered attractive targets.

More than 400 listed companies have adopted poison pills and many are increasing cross-shareholdings with business partners and friendly banks.

The fund-raising structure raises concerns that Japan's business groups are rebuilding pre-war industrial conglomerates.

Japanese banks are prohibited from owning more than a 5 per cent stake in a company, but warrants could be parcelled out to other group companies.

Under the terms of the warrants, if there were a takeover bid SMBC would exercise its right to the underlying shares, "giving any potential bidder a bad impression", said Sadakazu Osaki, head of research of Nomura.

"In that respect it can be called a takeover defence measure."

SMBC denied the scheme would result in a revival of the *zaibatsu* system. "We have no intention of holding the shares we receive after exercising our rights," SMBC said.

SMM and Sumitomo Realty denied that the financing was designed to protect against takeover bids.

"It is true that this raises misunderstandings [but] we have no intention to have the warrants converted into equity," SMM said.

Source: Michiyo Nakamoto, *Financial Times*, 25 February 2008.

How and when to concede

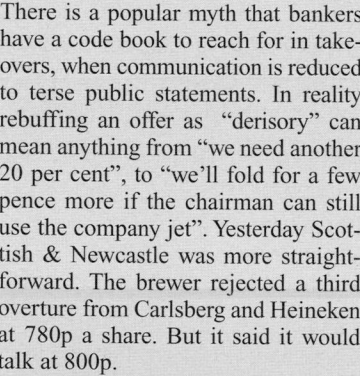

There is a popular myth that bankers have a code book to reach for in takeovers, when communication is reduced to terse public statements. In reality rebuffing an offer as "derisory" can mean anything from "we need another 20 per cent", to "we'll fold for a few pence more if the chairman can still use the company jet". Yesterday Scottish & Newcastle was more straightforward. The brewer rejected a third overture from Carlsberg and Heineken at 780p a share. But it said it would talk at 800p.

Being so explicit is unusual. Spain's Altadis nailed its colours to the mast by setting a minimum bid level of €50 a share last year. After a long negotiation, Imperial Tobacco did eventually pay up, but there was competing interest from private equity outfit CVC to generate price tension. Setting an absolute price level can backfire.

The Sainsbury family indicated that they would only accept an offer over 600p. Unfortunately the Qatar-backed bidder withdrew rather than pay this price. It often pays to take the opposite stance and refuse to define "fair value" other than somewhere above the latest offer. The London Stock Exchange fought off multiple bids on this basis and has seen its shares rise from £3 to £20.

So has S&N made a mistake? Probably not. The "here is my price" defence can work under two conditions. First, the bidder must be keen and not just an opportunist. Imperial had been stalking Altadis for years while Carlsberg has long been a reputed buyer of S&N, and, judging from their rhetoric, both companies are emotionally involved. Second, the level of increase asked for must be small. Altadis held its ground for a final boost of 6 per cent from Imperial. S&N's insistence on an extra 2.6 per cent is embarrassingly modest by comparison.

Source: Financial Times, 11 January 2008.

20.9 POST-MERGER ACTIVITIES

Probably the most difficult part of takeover strategy and execution is the integration of the newly-acquired company into the parent. In the case of contested bids, the acquirer will normally have only a limited amount of information to guide its integration plans and should not be too shocked to encounter unforeseen problems regarding the quality of the target's assets and personnel. The difficulty of integration depends on the extent to which the acquirer wants to control the operations of the target. If only limited control is required, as in the case of unrelated acquisitions, integration will probably be restricted to meshing the financial reporting systems of the component companies. Conversely, if full integration of common manufacturing activities is required, integration assumes a different order of complexity.

Jones (1982, 1986) points out that the degree of complexity of integration depends on the type of acquisition: for example, whether it is a horizontal takeover of a very similar company, requiring a detailed plan for integrating supply, production and distribution; or, at the other extreme, a purely conglomerate acquisition where there is little or no overlap of functions. The relationship between type of acquisition, overlap of activity (split into financial, manufacturing and marketing) and the resulting degree of integrative complexity is shown in Figure 20.2. Because we believe that integration is perhaps the most important part of the acquisition process, we devote most of the following section to further analysis of this issue.

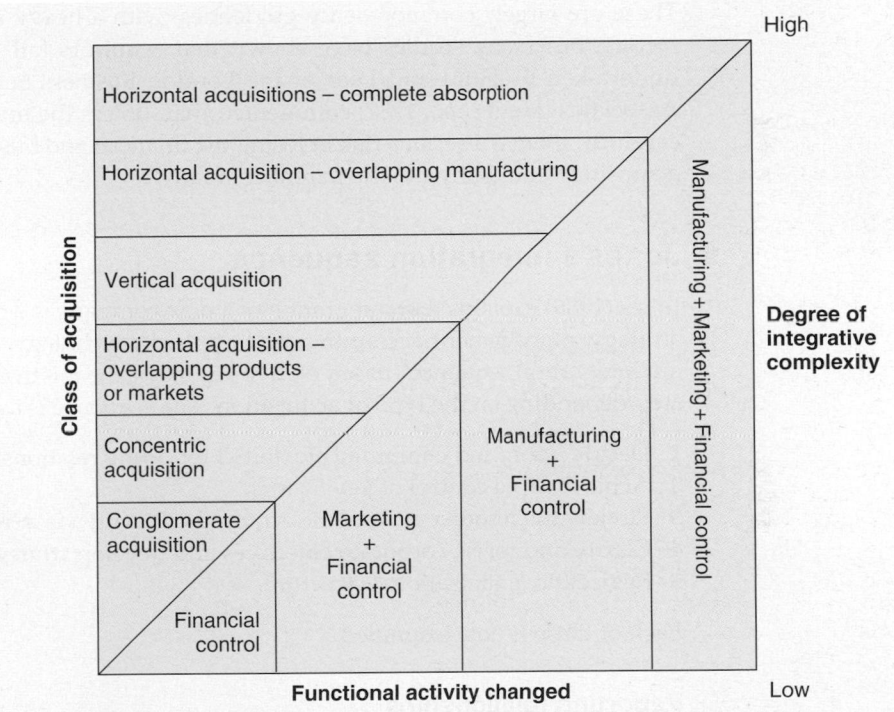

Figure 20.2 Type of acquisition and integrative complexity (Jones, 1986)

Finally, the acquisition should be post-audited. The post-audit team should review the evaluation phase to assess whether, and to what extent, the appraisal was under- or over-optimistic, and whether appropriate plans were formulated and executed. The review should centre on what lessons can be learned to guide any subsequent acquisition exercise.

Poor planning and poorly-executed integration are two of the commonest reasons for takeover failure. All too often, acquisitive companies focus senior management attention on the next adventure rather than devoting adequate resources to absorbing the newly-acquired firm carefully. It is rash to lay down optimal integration procedures in advance, because the appropriate integration procedures are largely situation-specific. The 'right' way to approach integration depends on the nature of the company acquired, its internal culture and its strengths and weaknesses (Lees, 1992).

However, Drucker (1981) contends that there are Five Golden Rules to follow in the integration process:

1 Ensure that acquired companies have a 'common core of unity' with the parent. In his view, mere financial ties between companies are insufficient to obtain a bond. The companies should have significant overlapping characteristics like shared technology or markets in order to exploit synergies.
2 The acquirer should think through what potential skill contribution it can make to the acquired company. In other words, the takeover should be approached not solely with the attitude of 'what's in it for the parent?' but also with the view 'what can we offer them?'
3 The acquirer must respect the products, markets and customers of the acquired company. Disparaging the record and performance of less senior management is likely to sap morale.
4 Within a year, the acquirer should provide appropriately skilled top management for the acquired company.
5 Again, within a year, the acquirer should make several cross-company promotions.

These are largely common-sense guidelines, with a heavy emphasis on behavioural factors, but many studies have shown that acquirors fail to follow them. A study undertaken by Hunt and Lees of the London Business School with Egon Zehnder Associates (Hunt *et al.*, 1987) commented that 'unless the human element is managed carefully, there is a serious risk of losing the financial and business advantages that the acquisition could bring to the parent company'.

■ Jones's integration sequence

Jones (1986) explains that integration of a new company is a complex mix of corporate strategy, management accounting and applied psychology. Acquirors should follow an 'integration sequence', based on five key steps, the relative weight attaching to each step depending on the type of acquisition. The sequence is as follows:

1 Decide upon, and communicate, initial reporting relationships.
2 Achieve rapid control of key factors.
3 Review the resource base of the acquired company via a 'resource audit'.
4 Clarify and revise corporate objectives and develop strategic plans.
5 Revise the organisational structure.

Each of these is now examined.

Reporting relationships

Clear reporting relationships have to be established in order to avoid uncertainty. An important issue is whether to impose reporting lines at the outset or whether to await the new organisational structure. In resolving this issue, it is desirable to avoid managers establishing their own informal relationships, and to stress that some changes may only be temporary.

Control of key factors

Control requires access to plentiful and accurate information. To control key factors, acquirers should rapidly gain control of the information channels that export control messages and import key data about resource deployment. It may sometimes be desirable not to introduce controls identical to those of the parent, first, because group controls may not be appropriate for the acquired company, and second, because those group controls may no longer be appropriate for the revised organisation. If the acquired company's existing control systems are thought to be adequate, it may be worth retaining them. Two important financial controls are the setting of clear borrowing limits and an early review of capital expenditure limits and appraisal procedures.

Jones notes that poor financial controls are often found within newly-acquired companies, and indeed, have often contributed to their acquisition. Examples are over-reliance on financial rather than management accounting systems (MAS), a MAS that provides inappropriate information in an inappropriate format, poor use of the MAS, and distortions in the overhead allocation mechanism, making it difficult to pinpoint unprofitable products and customers. The net result is often poor budgetary control, inadequate costing systems and inability to monitor and control cash movements.

Resource audit

The resource audit should examine both physical and human assets to obtain a clear picture of the quality of management at all levels. The extent of the audit required will depend on how much information is made available prior to the acquisition, but auditors should not be surprised if 'skeletons' are found in cupboards, requiring a reappraisal of the value of the acquired firm and possibly a different way of integrating it into the parent's future strategy and plans. For example, a business that was meant to be absorbed into the parent's operations may be divested if its capital equipment is unexpectedly dilapidated.

Corporate objectives and plans

These should be harmonised with those of the parent, but should also reflect any differences due to industrial sector, such as different 'normal' rates of return or profit margins in different industrial sectors. Managers of acquired firms should have some freedom to formulate their plans to meet the stated aims, but the degree of freedom should depend on the complexity of the merger. For example, in a conglomerate acquisition, where the primary aim is to secure financial control, it may be appropriate to allow executives to develop a system of management control suitable for their own operating patterns, so long as these are consistent with the aims of the takeover. In cases where cash generation is the main spur, all that may be needed is centralised cash management plus control over capital allocations.

Revising the organisational structure

A discussion of organisational design is beyond our scope, but obviously a demoralised labour force is unlikely to offer optimal performance.

Two important factors enhancing the success of a takeover are the thoroughness of the resource audit and the degree of senior management contact in the very early stages of the takeover. Employees of acquired companies seek a rapid resolution of uncertainty, especially regarding how they and their company fit into the future structure and strategy of the acquirer, and how soon the new management team will assume control. Particularly important for morale are the lifting of any previous embargo on capital expenditure and the provision of improved performance incentives, pension schemes and career prospects.

Some of these aspects are illustrated in the following short cases.

■ Case 1: Post-acquisition integration at GE Capital

According to Ashkenas, DeMonaco and Francis (1998) the process of melding one company into another is frequently badly done because few firms go through the process often enough to develop a pattern. They suggest that many firms see it as a one-off event rather than a replicable process – 'something to get finished, so everyone can get back to business'. Because 'the acquisition event is painful and anxiety-generating, involving change, disruption and job loss, most managers think how to get it over with, not how to do it better next time around'.

They reported on acquisition integration at GE Capital, a subsidiary of General Electric Corporation, founded in 1933 to provide consumers with credit to purchase GEC appliances. GE Capital has grown by acquisition, including over 100 in the five years prior to their report. From interviews with managers and key staff involved in acquisitions, both as former targets and subsequently as acquirors at GE Capital, they devised a framework for acquisition management based on four 'Lessons':

Lesson 1

Acquisition management is not a discrete phase of a deal and does not begin when the documents are signed. Rather, it is a process that begins with due diligence and runs through the ongoing management of the new enterprise.

Lesson 2

Integration management is a full-time job and needs to be recognised as a distinct business function, just like operations, marketing, or finance.

Lesson 3

Decisions about management structure, key roles, reporting relationships, lay-offs, restructuring, and other career-affecting aspects of the integration should be made, announced and implemented as soon as possible after the deal is signed – within days, if possible. Creeping changes, uncertainty, and anxiety that last for months are debilitating and immediately start to drain value from an acquisition.

Lesson 4

A successful integration melds not only the various technical aspects of the businesses but also the different cultures. The best way to do it is to get people working together quickly to solve business problems and accomplish results that could not have been achieved before.

Source: Ashkenas *et al.* (1998).

■ Case 2: UK bank mergers bear fruit

In early 2002, two British banking groups reported strongly improved performance resulting from the respective mergers that had created them. The source of the benefits was broadly similar – large cost savings combined with buoyant demand for retail financial services.

Royal Bank of Scotland, the UK's second largest by market value after its acquisition of NatWest Bank, was able to shrug off increased bad debt provisions and exposure to Enron, the insolvent US energy trader, after exploiting more synergies than originally forecast from the acquisition. Royal Bank shares had outperformed the UK banking sector by 18 per cent as news of rapidly increased revenue growth from NatWest synergies leaked on to the market.

HBOS plc, number five in the banking league table, was formed by Halifax's takeover of Bank of Scotland. It also achieved strong gains from the merger in the area of savings and mortgage loans. It had achieved a 25 per cent market share in the mortgage market via a policy of interest rate cutting, after years of decline following the demutualisation of Halifax Building Society in 1997. It was also threatening the market dominance of the 'Big Four' banks in the small business sector by offering higher interest rates on credit balances (4 per cent against 0.1 per cent!). HBOS had outperformed the banking sector by 27 per cent over the previous year.

Smooth integration of GB helps EasyJet soar

Shares in EasyJet rose sharply after the leading UK low-cost carrier said that it was making strong progress in integrating GB Airways, the former British Airways franchise carrier that it acquired last month for £103.5m.

EasyJet is establishing a new domestic stronghold at Gatwick supported by the takeover of GB Airways. It has 35 aircraft operating at the airport and has built a presence to rival the retrenching operation of British Airways.

It said its forward bookings for the former GB Airways operation were "encouraging" with more than 30 per cent of its summer seats already booked.

Through the takeover EasyJet has gained access to more take-off and landing slots at the congested airport and is adding 19 new destinations for a total of 62 routes.

The enlarged EasyJet has 24 per cent of the slots at Gatwick – up from 17 per cent before the takeover – compared with BA's 25 per cent. It also has 29 per cent of short-haul passengers at the airport – at about 8m – compared with BA's 23 per cent.

EasyJet is briefly continuing the BA franchise until March 29. The operation will then be changed overnight at the start of the summer season to become part of the low-cost carrier working under the EasyJet air operator's certificate.

The full rebranding to EasyJet and the reconfiguration of the aircraft to remove the business-class seating will be completed by May.

The GB Airways head office at Gatwick will be closed in July with the loss of about 220 jobs and a further 50 transferring to EasyJet.

EasyJet said there would be savings of £10m by eliminating the BA franchise fee; £10m from closing the GB Airways head office; £5m–£10 m in ground-handling; a 20 per cent reduction in insurance fees; and in the longer term savings of £10m–£20 m a year in cheaper aircraft ownership costs.

It said it was in negotiation with BAA, the operator of Gatwick, with the long-term aim of being able to combine its operations at the airport at one terminal.

EasyJet shares rose 5.6 per cent to close 23¼p higher at 435¾p.

Source: Kevin Done, *Financial Times*, 20 February 2008.

Self-assessment activity 20.5

What are the key elements of an acquisition strategy?

(Answer in Appendix A at the back of the book)

20.10 ASSESSING THE IMPACT OF MERGERS

The impact of mergers can be assessed at various levels. At the macroeconomic level, if takeover activity is performing its function of weeding out inefficient managements, we might expect to find takeovers resulting in superior economic performance. However, over the long haul, the two economies where takeover activity is most prevalent – the USA and the UK – have underperformed economies where growth by takeover is less common (e.g. until the mould-breaking takeover of Mannesmann by Britain's Vodafone in 2000, there had been only four completed hostile takeovers in Germany since 1945 and not one completed by a foreign bidder). Measures of economic performance such as growth in gross domestic product (GDP) per head and capital formation as a percentage of GDP were, certainly until the 1990s, considerably lower in the UK than in Japan and Germany, as were figures for 'growth drivers' like R&D expenditure as a percentage of GDP.

Although such associations do not prove causation, the presumed link is via the impact of horizontal mergers reducing competition and through the pressures for short-term performance. It is often argued that managers in the 'Anglo-Saxon' economies are forced to pay out higher dividends to bolster share price in order to deter prospective raiders. Ever higher payouts represent cash that could have been used to finance investment and growth. However, it seems implausible to argue that the institutions would withhold funds from companies that showed a willingness to perform according to short-termist rules. Nevertheless, after reviewing the evidence, Peacock and Bannock (1991) concluded that mergers and takeovers did not create wealth, but merely transferred ownership of assets. The full explanation of why merger-active economies underperform is probably less simple, as Porter (1992) explains, involving a complex interplay of economic, social and political factors.

Another level of analysis is that of the performance of individual companies. If takeovers are beneficial, we should expect to see merger-active firms improving their performance post-merger.

Investigating the effects of merger activity is one of the busiest areas of contemporary applied finance research. There are two main ways of attempting to assess the impact of mergers.

The first, **the financial characteristics approach**, is based on examining the key financial characteristics of both acquiring and acquired firms before the takeover, to study whether they are more or less profitable (taking profitability as an indicator of efficiency) than firms not involved in acquisitions, and whether their profitability improves after the acquisition. The second, 'the capital markets approach', is based on examining the impact of the takeover on the share prices of both acquired and acquiring firms, to assess the extent to which expected benefits from merger are impounded in share prices and how these are shared between the two sets of shareholders.

The first method, the financial characteristics approach, suffers from severe limitations:

1 Different accounting conventions used by different firms (e.g. treatment of R&D) often make comparisons misleading.

2 Measures of profitability may have been distorted due to the application of acquisition accounting procedures. Prior to the issue of FRS 10 in 1997, firms in the UK were allowed to write off goodwill (the excess of purchase price over 'fair value') against reserves. This lowered their equity bases and raised their recorded return on investment. Goodwill now has to be shown on the balance sheet as an asset and amortised over its useful life up to 20 years. This abrupt change in the rules of accounting for goodwill means that pre- and post-takeover measurement of performance of firms active in takeovers before and after 1997 could be seriously distorted.

3 To assess properly the impact of the takeover requires an extended analysis ranging over, say, five to ten years. Many acquisitions are undertaken for 'strategic' purposes, the benefits of which may emerge only after several accounting periods, perhaps following lengthy and costly reorganisation. Very frequently, when 'efficient' companies take over 'inefficient' companies, the group's return on net assets and fixed asset turnover ratios automatically fall.

4 Accounting studies are not capable of assessing what the performance of the expanded group would have been in the absence of the merger, and are thus unable to assess what improvement in performance (if any) was due to factors beyond the merger. This problem increases with the time period used for the post-merger investigation.

5 Any improvement in profitability may simply be due to a restriction of competition, rather than more efficient use of resources.

6 The approach does not allow for risk. If the aim of many mergers is to lower total risk (possibly for managerial reasons), or to shift the company into a lower Beta activity, a lower return post-merger is not especially surprising, since according to the EMH/CAPM, relatively low-risk investments offer relatively low returns.

The second method, **the capital markets approach**, caters for many of these difficulties, and is thus the most frequently used mode of analysis. By adopting a CAPM framework, it enables the returns on shares of acquiring firms to be examined prior to and following the merger. As noted in Chapter 10, the market model indicates that the expected return on any security, j, in any time period t, ER_{jt}, is a linear function of the expected return on the market portfolio:

$$ER_{jt} = \alpha_{jt} + \beta_j ER_{mt}$$

However, the actual return in any time period, R_{jt}, results from a compound of market-related and company-specific factors:

$$R_{jt} = \alpha_{jt} + \beta_j R_{mt} + u_{jt}$$

u_{jt} is an error term with zero expected value, indicating that random company-specific factors are expected to cancel out over several time periods.

If, over a suitable period, and allowing for overall movements in the market, we examine the differences between the expected returns and the actual returns, the 'residuals' or 'abnormal returns' should sum to zero, i.e. the expected value of the cumulated differences between ER_{jt} and R_{jt} is zero.

A takeover bid is a company-specific event likely to raise the share price, and hence, the return on holding the shares. When a bid occurs, the increase in returns can be attributed to the market's assessment of the impact of the bid, i.e. the evaluation of its likelihood of success, and if successful, its appraisal of the benefits likely to ensue. Therefore, both in the period leading up to the bid, as the potential bidder builds a stake, and also on the day of announcement of the bid, we might expect the residuals to be non-negative. For example, if the market thinks the takeover is a mistake, we may find negative residuals for the bidder but positive residuals for the target, if, as usually happens, the share price of the target rises sharply. Hence, the cumulated residual returns may be taken as the stock market's assessment of the value of the takeover to the shareholders of the acquirer and acquired companies, respectively.

To illustrate this use of the market model, consider the data in Table 20.6. This relates to the successful takeover bid for the Kenning Motor Group by the conglomerate Tozer Kemsley Milbourn (TKM) in 1986. There was a degree of industrial logic in this bid, as both TKM and Kennings retailed motor cars and Kennings operated a substantial car hire business. Kennings had attempted in previous years to diversify, but owing to the haphazard selection of targets and poorly-executed integration, its core business was suffering, reflected in declining profitability and weakening cash flow.

Table 20.6 Pre- and post-bid returns

Kenning Motor Group					
Time period (months)	Share price (p)	Actual return (%)	Expected return (%)	Residual	Cumulated residuals
−3	140	−3.45	−0.29	−3.16	−3.16
−2	160	14.29	1.67	12.61	9.45
−1	158	1.25	−6.64	−7.89	1.56
0	212	34.18	7.63	26.54	28.10
1	268	26.42	−0.19	26.60	54.70
2	310	15.67	−3.04	18.71	73.41
3	310	0	3.08	−3.08	70.33
Tozer Kemsley Milbourn					
−3	66	−2.94	−0.32	−2.62	−2.62
−2	81	22.73	1.87	20.86	18.24
−1	108	33.33	7.42	25.91	44.15
0	136	25.93	8.53	17.39	61.54
1	177	30.15	−0.21	30.36	91.90
2	172	−2.82	−3.40	0.58	92.48
3	180	4.65	3.44	1.21	93.69

Source: Data collected by Krista Bromley.

The data show the residual returns for both companies in the three months prior to, and also following, the bid; month 0 is the day of the bid, eventually completed at a price of £3.10 per share.

The data clearly show positive residuals for Kennings, before the bid, on the day of the bid itself, and also in the following periods as TKM raised its bid. In this case, both sets of shareholders enjoyed substantial returns. After allowing for the movement in the market as a whole, the returns on the shares of both companies were substantially positive, indicating market expectations that this merger would be wealth-creating. This proved to be the case, following very swift and effective reorganisation of Kennings by TKM.

Until quite recently, the bulk of the empirical evidence (e.g. see surveys by Jensen and Ruback (1983); Department of Trade and Industry (1988); Mathur (1989)) suggested that positive gains from takeovers accrue almost entirely to the shareholders of target firms. While the average abnormal returns (the cumulated residuals for all firms divided by the number of firms examined in the study) recorded in these studies are invariably positive and statistically significant, returns to the shareholders of bidding firms are negative for mergers and not significantly different from zero for takeovers. In other words, on average, takeovers and mergers are not wealth-creating, but the acquisition process transfers wealth from the shareholders of acquirers to those of the acquired. These are very important results as they seem to question the judgement or the motives, or both, of the instigators of takeover bids.

However, a very significant study by Franks and Harris (1989), based on both UK and US data, contradicted much earlier work in an important respect. The study is especially important for two reasons. First, the authors took mergers and takeovers over a much longer period (1955–85) than most other studies; and second, they examined a considerably larger sample than any previous study. For the UK, 1,900 acquisitions involving 1,058 bidders were studied and the US sample was 1,555 acquisitions involving 850 bidders. The targets were all publicly traded, facilitating a capital markets approach. Table 20.7 shows their results in terms of excess stock market returns as compared to the market portfolio, and allowing for systematic risk.

Table 20.7 The gains from mergers

	UK (months)		US (months)	
	0	−4 to +1	0	−4 to −1
Targets	+24%	+31%	+16%	+24%
Bidders	+1%	+8%	+1%	+4%

Source: Franks and Harris (1989).

As with earlier studies, these data record a substantial increase in wealth for the shareholders of target firms, but unlike earlier ones, they reveal a relatively smaller, but statistically significant, increase in wealth for the shareholders of acquirers over the whole period leading up to and just after the bid.

A subsequent study by Limmack (1991), using only UK data for the period 1977–86, suggested that 'the gains made by target company shareholders are at the expense of shareholders of bidder companies'. He also suggested that the average wealth decreases suffered by the shareholders of bidding companies were mainly confined to the period 1977–80, and that bids made in the years 1981–86 produced no significant wealth decrease for shareholders of bidding firms.

Limmack's study seemed to imply that bidding firms and the capital market in general might have learned from earlier mistakes, and that some of the gains from merger might be retained by the bidder.

However, Sudarsanam *et al.* (1996), in a study of returns around bid announcement dates over the period 1980–90, found significant negative CARs of around 4 per cent for the announcement period minus 20 days to plus 40 days. In a subsequent study of 398 takeovers during 1984–92 involving bid values of over £10 million, Gregory (1997) found 'the post-takeover performance of UK companies undertaking large domestic acquisitions is unambiguously negative on average, in the longer term'. The results were clearly 'not compatible with shareholder wealth maximisation on the part of acquiring firms' management', Gregory found that acquiring firms underperformed the stock market by 18 per cent on average for two years following a bid. The method of financing the takeover was also discovered to be significant, with share exchanges being especially poor for shareholders. Agreed bids were less successful for acquiring companies than hostile takeovers, and companies making their first takeovers were likely to be poorer performers than more experienced bidders. Finally, companies bidding to diversify their interests were less likely to succeed than takeovers in related business areas (which echoes many previous analyses).

US studies, by both academics and firms of consultants, have found parallel results. Research in 1995 by the Mercer consultancy and *Business Week* of 150 deals valued at $500 million or above during 1990–95 showed that about half destroyed shareholder wealth and only 17 per cent created results that were more than 'marginal'. These and similar studies appear to confirm the view that bidders overpay for their targets (Gregory found an average bid premium of around 30 per cent) and that managers underestimate the work required to make a takeover succeed. Moreover, many managers seem determined to ignore the warning signs.

Some researchers in both the USA and the UK have studied the post-merger behaviour of corporate cash flows. This has the merit of avoiding the problems of accounting measurement and policy when profitability data are used, and also avoids relying on the efficiency of the capital market if share price data is used. For example, Manson *et al.* (1994) studied 38 UK mergers over 1985–87, and found that, on average, takeovers produced significant improvements in operating performance, reflected in higher cash flow. This supports the view that the market for corporate control provides a discipline for managers, and is consistent with US studies using similar methods.

Franks and Mayer (1996a, b) undertook an explicit test of the disciplinary role of hostile takeovers. They hypothesised that hostile takeovers would have a disciplinary impact, shown by association with, first, a high rate of managerial turnover, second, with substantial internal restructuring (asset sales exceeding 10 per cent of the value of post-acquisition fixed assets) and third, with weak firm performance post-acquisition. Their study of hostile UK takeovers completed in 1985–6 found:

- 50 per cent of directors resign soon after a friendly bid as compared to 90 per cent after a hostile bid.
- Asset sales exceeded 10 per cent of the total fixed asset value in 26 per cent of friendly bids compared to 53 per cent of hostile bids.
- Bid premia averaged 30 per cent of firm value for hostile bids compared to 18 per cent for friendly bids.
- Bid premia on hostile bids were not correlated with rate of management turnover.
- Using four measures of performance – abnormal share price performance, dividend payments, pre-bid cash flow rates of return and Tobin's Q (the ratio of market value to replacement cost of assets) – they found little significant evidence that recipients of hostile bids were poor performers prior to the bid.

Overall, the evidence was not consistent with hostile takeovers exerting a form of natural selection in which underperforming managements are supplanted by alternative teams offering improved results.

Finally, what about international mergers? In 2000, the accounting firm KPMG reported a study of 700 of the largest cross-border takeovers occurring between 1996 and 1998 in which it attempted to assess the value created. KPMG measured companies' share price performance before and after the deal, and compared the post-deal performance with trends in each of the firm's industries.

The results were:

- 17 per cent of takeovers added value to the combined company.
- 30 per cent produced no significant difference.
- 53 per cent actually destroyed value.

Commenting on why 83 per cent of mergers apparently fail to generate net benefits for shareholders, it was suggested that many firms concentrate too heavily on the business and financial mechanics and overlook the personnel-related issues, echoing the earlier study by Lees (1992).

KPMG identified six factors that merit close attention:

- evaluation of synergies
- integrated project planning
- due diligence
- selection of the right management team
- dealing with culture clashes
- communication with staff.

Not much new here, readers may think, but it is remarkable how often over-ambitious managers have to be reminded of the ingredients of a successful merger.

In a further survey released in 2002, KPMG claimed that over a third of giant international takeovers completed at the peak of the bull market were being unwound, that 32 per cent of chief executives or Finance Directors responsible for planning the original deals had moved on, and that two-thirds of firms acquired during 1996–98 still needed to be properly integrated. The lesson appears to be that, like Chris Gent of Vodafone, following the acquisition of Mannesmann, top managers should negotiate their bonuses after completion of the mega-deal, but before the problems begin to appear.

The KPMG study suggests the importance of internal cultural issues. One might be forgiven for thinking that acquisitions by British firms of US enterprises would have a greater chance of success given the similarity in language and corporate cultures. Yet research by Gregory and McCorriston (2004), based on a study of 197 major British takeovers in the USA (excluding banking – itself a disaster area), suggests exactly the opposite. See the box below for more details.

Failure to address cultural issues was also highlighted by a report by the Hay Group (Dion *et al.*, 2007). Its study of European mergers concluded that only 9 per cent of mergers achieve their pre-stated aims. The researchers found that only 29 per cent of acquirors had conducted a cultural audit of the target, suggesting that firms had over-prioritised the 'hard keys' like financial and IT due diligence.

However, although echoing many earlier findings, a slightly less gloomy picture was obtained in the most recent KPMG survey. Using a telephone interview approach, this covered a sample of global companies that had conducted deals worth over US$100 million between 2002 and 2003. The KPMG team came to six broad conclusions.

1 More deals enhanced value (31 per cent), in terms of shareholder value relative to the relevant industry average, than in its earlier study, and more deals enhanced value than reduced value (26 per cent). Yet, still over two-thirds of deals failed to increase value. By contrast, executives from 93 per cent of firms claimed that their deals did enhance value. The 'perception gap' suggested that firms were not yet prepared to make an honest assessment of their deals.
2 There was strong correlation between firms that enhanced value and those that met or exceeded their synergy and performance improvement targets. This suggests that firms that set out to outperform their targets achieved the targets. KPMG surmised that synergies arrive from unexpected areas, and that narrow pursuit of cost savings may fail to capture the full benefit of revenue synergies.

3 Almost two-thirds of acquirors failed to meet their synergy targets, 43 per cent of which was included in the purchase price. This suggests that firms typically overpay for synergies.

4 Firms found they did not start post-deal planning early enough. Although 59 per cent of firms had started planning for the integration process, the most commonly cited actions that firms said they would adopt on their next deal was earlier planning. Stated advantages of early planning were limiting the risk of losing customers, bringing forward synergy delivery, and avoiding a communication vacuum, in which rumour and speculation are rife.

5 Despite citing differences in organisational culture as the second greatest post-deal challenge, 80 per cent of firms said they had not been well prepared to handle this. The top three post-deal challenges were stated as: complex integration of two businesses, dealing with different organisational cultures and difficulty in integrating IT and reporting systems. About half said they had been well-prepared for challenges one and three, but over two-thirds had placed low emphasis on addressing people and cultural issues. Only 20 per cent said these issues were less important than expected. The top three actions that executives said they would adopt in their next deal were: to plan earlier, perform additional cultural due diligence, and to set up a dedicated team to handle the post-deal tasks.

6 It required on average nine months for companies to feel they had obtained control of the key issues facing the business post-deal. A third said it had taken longer than expected, and over 10 per cent took more than two years before being able to feel in control. More than two-thirds of firms stated that the two critical activities were gaining understanding of the target's finances and reporting systems, and understanding and overcoming the cultural difficulties between the two companies.

Plus ça change!

Acquisitions in US 'disastrous' for British companies

The US has been a graveyard for acquisitive UK companies, losing shareholders vast sums of money on projects picked for 'the wrong reasons', says new UK academic research.

Alan Gregory, professor of corporate finance at Exeter University, said: 'The research shows on average that UK companies make disastrous acquisitions in the US.'

Five-year returns from UK companies acquiring US companies between 1985 and 1994 underperformed stay-at-home companies by 27 per cent. British acquisitions of EU companies, by comparison, have yielded slightly negative returns short-term but paid off over longer periods.

Prof Gregory, who is also on the panel of the UK's Competition Commission, said: 'The research findings tell us that UK companies are attracted to buy businesses abroad for the wrong reasons: short-term events such as exchange rate fluctuations, growth in share prices and stock markets, and policy decisions by governments, often drive a wave of cross-border acquisitions.'

The research, covering UK acquisitions in the US of more than £10 m ($17.9 m), was prompted by the growth in cross-border acquisitions by UK groups. Prof Gregory said UK companies accounted for more than a third of the developed world's cross-border transactions, which have risen markedly in recent years. Most research on the success of acquisitions has focused on short-term effects, for example, share price movements immediately following takeovers.

The exceptions to Exeter University's findings were acquisitions in industries rich in research and development, for example technology or pharmaceuticals, which have specific advantages such as unique products or patents.

One example, said Prof Gregory, was SmithKline Beecham's purchase of Sterling Winthrop in 1993 which added 'substantial value'.

This compared, he said, with less successful deals, including Hanson's £2.1 bn acquisition of Quantum Chemical, English China Clay's deal to buy Calgon, and Rolls Royce's £335 m acquisition of Allison Engine. 'This underperformance is despite cultural similarities between the US and UK, which might make the job easier,' he said.

The Exeter team plans further research to explain why UK companies have done so badly in the US but Prof Gregory attributed some of it to companies overpaying for US opportunities and inadequate local knowledge.

Source: Kate Burgess, *Financial Times*, 11 October 2004.

In the next section, we return to a recurrent theme in this book: the meaning and reliability of the values placed on companies by the stock market.

Self-assessment activity 20.6

What are the main causes of 'failure' of takeovers?

(Answer in Appendix A at the back of the book)

20.11 VALUE GAPS*

Evidence from some studies indicates that there may be net gains from merger, while most surveys indicate that the shareholders of target companies experience a beneficial wealth effect. The near certainty that shareholders of targets will benefit suggests that market values typically fall short of the value that potential or actual bidders would place on them. These disparities in value are called value gaps, and there are four main explanations of how they arise.

■ Poor corporate parenting

Value gaps may arise because some business segments do not make their maximum possible cash or profit contributions to the parent. Ultimately, this is a reflection of poor central management, which is thus failing to add value to the group or actually reducing value. The following are some examples of management deficiencies:

1 Some assets, such as land and premises, may not be fully utilised by either the parent or its subsidiaries.
2 The parent pursues too many ventures of dubious value, perhaps intending to gain entry into other areas of business, but in which it does not possess appropriate expertise.
3 HQ may fail to take sufficiently decisive action to prevent or correct poor profitability in business segments.
4 HQ may indulge in costly central activities or services that are a net burden rather than of benefit to business units.
5 Poor group structure may leave business units at a disadvantage compared with competitors. For example, a business unit may be too small to compete effectively in its main markets, or it may be denied sufficient capital resources to develop its activities. As a result, it may have a greater value under alternative, more perceptive, management.

■ Poor financial management

The HQ corporate finance department might have followed a gearing policy that fails to fully exploit its ability to borrow and gear up returns to equity; or alternatively, it may be severely over-borrowed. Similarly, its past dividend policy may have been over- or under-generous.

■ Over-enthusiastic bidding

It has been said that takeover bidders' greatest victims are themselves. Many bids are undoubtedly successful, such as the acquisition of Arcelor by Mittal. Lakshmi Mittal, CEO of the new entity, was able to report cost savings of US$973 million after just

---

*This section relies on ideas presented in Young and Sutcliffe (1990).

Marrying in haste

Merger and acquisitions continue apace in spite of an alarming failure rate and evidence that they often fail to benefit shareholders. Last week's collapse of the planned Deutsche-Dresdner Bank merger tarnished the reputation of both parties. Deutsche Bank's management was exposed as divided and confused. Dresdner Bank lost its chairman, who resigned. Senior members of Dresdner Kleinwort Benson, its investment banking unit, walked out.

Deutsche-Dresdner was a fiasco that damaged both parties. But even if the takeover had gone ahead, it would probably still have claimed its victims. A long list of studies have all reached the same conclusion: the majority of takeovers damage the interests of the shareholders of the acquiring company. They do, however, often reward the shareholders of the acquired company, who receive more for their shares than they were worth before the takeover was announced. Mark Sirower, visiting professor at New York University, claims that 65 per cent of mergers fail to benefit acquiring companies, whose shares subsequently underperform their sector. Yet the evidence of failure has done nothing to dim senior managers' enthusiasm for takeovers.

Why do so many mergers and acquisitions fail to benefit shareholders? Colin Price, a partner at McKinsey, the management consultants, who specialises in mergers and acquisitions, says the majority of failed mergers suffer from poor implementation. And in about half of those, senior management failed to take account of the different cultures of the companies involved.

Melding corporate cultures takes time, which senior management does not have after a merger. 'Most mergers are based on the idea of "let's increase revenues", but you have to have a functioning management team to manage that process. The nature of the problem is not so much that there's open warfare between the two sides. It's that the cultures don't meld quickly enough to take advantage of the opportunities. In the meantime, the marketplace has moved on.'

Many consultants refer to how little time companies spend before a merger thinking about whether their organisations are compatible. The benefits of mergers are usually couched in financial or commercial terms: cost-savings can be made or the two sides have complementary businesses that will allow them to increase revenues.

Mergers are about compatibility, which means agreeing whose values will prevail and who will be the dominant partner. So it is no accident that managers as well as journalists reach for marriage metaphors in describing them. Merging companies are said to 'tie the knot'. When mergers are called off, as with Deutsche Bank and Dresdner Bank, the two companies fail to 'make it up the aisle' or their relationship remains 'unconsummated'. Yet the metaphors fail to convey the scale of risk companies run when they launch acquisitions or mergers. Even in countries with high divorce rates, marriages have a better success rate than mergers. And in an age of frequent pre-marital cohabitation, the bridal couple usually know one another better than merging companies do.

Mr Sirower rejects the view that the principal problem is 'post-merger implementation'. 'Many large acquisitions are dead on arrival, no matter how well they are managed after the deal is done,' he says. He asks why managers should pay a premium to make an acquisition when their shareholders could invest in the target company themselves. How sure are managers that they can extract cost savings or revenue improvements from their acquisition that match the size of the takeover premium?

Perhaps it would help if senior managers abandoned the marriage metaphor in favour of the story of the princess and the toad. Warren Buffett, the US investor, said nearly 20 years ago that many acquisitive managers appeared to see themselves as princesses whose kisses could turn toads into handsome princes. Investors, he observed, could always buy toads at the going price for toads. 'If investors instead bankroll princesses who wish to pay double for the right to kiss the toad, those kisses better pack some real dynamite. We've observed many kisses, but very few miracles. Many managerial princesses remain serenely confident about the future potency of their kisses, even after their corporate backyards are knee-deep in unresponsive toads.'

Source: Based on Michael Skapinker, *Financial Times*, 12 April 2000.

a year of operation, around two-thirds of the synergies of $1.6 billion promised at the time of the takeover. Reported earnings rose 42 per cent in 2006–7. However, many others are outright failures; and some, such as Morrisons' purchase of Safeway, are totally disastrous. Perhaps, at the time of the bid, the assessment of the bidding management was correct, but they were caught out by changed circumstances. A more likely explanation is that they were buoyed-up with excessive enthusiasm about the bid. Although some bidders have the sense to walk away from a bid (e.g. Comcast's aborted bid for Disney in 2004), in too many cases, bidders delude themselves that the proposed takeover is vital to the development of the group.

Referring to such cases, cynics use the term 'winners' curse', for obvious reasons.

Ebay writes down Skype value by \$1.4bn

Ebay yesterday conceded that its controversial acquisition of the internet telephone service Skype had fallen well short of its hopes, writing down the value of its investment in the company by \$1.43bn (£699.5m), or nearly 50 per cent.

At the same time, Niklas Zennstrom, one of the founders of Skype, quit as the unit's chief executive after missing out on a pay-day that could have earned him and a handful of other shareholders an extra \$1.2bn.

The writedown and management upheaval come two years after Ebay unveiled its controversial plan to make Skype the third leg of an expanding internet conglomerate, alongside its e-commerce and online payments businesses.

However, while the internet telephone service has continued to add users at a fast rate, it has so far failed to produce the sort of e-commerce and advertising revenues that Ebay had hoped.

"At the time, we thought strategically it didn't make a lot of sense and the price seemed excessive," said Scott Kessler, an analyst at Standard & Poor's, echoing a view of the deal that was widely held on Wall Street.

Yesterday, however, Ebay continued to insist that its long-term hopes for Skype were unchanged, and said it had no plans to sell the business.

The news represents a setback for two of Europe's most prominent technology entrepreneurs.

Skype has been the most successful of a string of high-profile start-ups by Mr Zennstrom and partner Janus Friis, whose other ventures have included peer-to-peer music service Kazaa and the online video venture Joost.

Ebay said the management change was amicable, and that it was looking outside the company for a new Skype chief executive.

Ebay agreed two years ago to pay \$2.6bn for Skype and spend up to €1.2bn (£836m) more if the unit hit performance targets over the following three years.

In the event, the company said that it had made an early payment of only €375m to end all its obligations, taking its total purchase price, based on current exchange rates, to about \$3.13bn.

Skype revenues reached \$168m in the first half of this year, but the company admitted yesterday that this was shy of its short-term goals.

Source: Richard Waters, *Financial Times*, 2 October 2007.

■ Stock market inefficiency

Does the stock market fail to assess the full value of a business, perhaps because it belongs to a sector that is 'out of favour', or because the market adopts too short-term a view of the prospects of the company?

Assessing the relative weights of these arguments is a major challenge for finance and business researchers.

20.12 CORPORATE RESTRUCTURING

Corporate *restructuring* is an important vehicle by which managements can enhance shareholder value by changing the ownership structure of the organisation. It involves three key elements:

1 Concentration of equity ownership in the hands of managers or 'inside' investors well-placed to monitor managers' efforts.
2 Substitution of debt for equity.
3 Redefinition of organisational boundaries through mergers, divestment, management buy-outs, etc.

In the dynamic environment within which companies operate, financial managers should be ever-alert to new and better ways of structuring and financing their businesses. The value-creation process will involve the following:

■ Review the corporate financial structure from the shareholders' viewpoint. Consider whether changes in capital structure, business mix or ownership would enhance value.
■ Increase efficiency and reduce the after-tax cost of capital through the judicious use of borrowing.

- Improve operating cash flows through focusing on wealth-creating investment opportunities (i.e. those having positive net present values), profit improvement and overhead reduction programmes and divestiture.
- Pursue financially-driven value creation using various new financing instruments and arrangements (i.e. financial engineering).

■ Types of restructuring

Recent years have seen the development of new and elaborate methods of corporate restructuring. Restructuring can occur at three different levels:

1 *Corporate restructuring* refers to changing the ownership structure of the parent company to enhance shareholder value. Such changes can arise through diversification, forming strategic alliances, leveraged buy-outs and even liquidation.

2 *Business restructuring* considers changing the ownership structure at the strategic business unit level. Examples include acquisitions, joint ventures, divestments and management buy-outs.

3 *Asset restructuring* refers to changing the ownership of assets. This can be achieved through sale and leaseback arrangements, offering assets as security, factoring debts and asset disposals.

The levels of restructuring are summarised below:

Corporate	Business unit	Asset
Diversification/demerger	Acquisitions	Pledging assets as security
Share issues	Joint ventures	Factoring debts
Share repurchase	Management buy-outs	Leasing and HP
Strategic alliances	Sell-offs	Sale and leaseback
Leveraged buy-outs	Franchising	Divestment
Liquidation	Spin-offs	

Most of these devices have been covered elsewhere in this book. We devote the next sections to a brief discussion of divestments and management buy-outs (MBOs), which have so far received only passing mention.

Divestment

A divestment is the opposite of an investment or acquisition; it is the sale of part of a company (e.g. assets, product lines, divisions, brands) to a third party. The heavy use of divestment as a means of restructuring reflects the continuing efforts of corporate management to adjust to changing economic and political environments.

One of the motives for acquisition, identified earlier, is the managerial belief that the two businesses are worth more when combined than separate. A form of reverse synergy is a major reason for divestment: the two elements of an existing business are worth more separated than combined. Whereas the arithmetic of synergy for acquisitions argues that (2 + 2) = 5, the arithmetic behind divestment, or reverse synergy, argues that (5 − 1) can be worth more than 4. In other words, part of the business can be sold off at a greater value than its current worth to the company. The management team may not relish the prospect of divesting itself of certain business activities, but it is often necessary as the strategic focus changes.

Self-assessment activity 20.7

Suggest reasons why a firm may choose to divest part of its business.

(Answer in Appendix A at the back of the book)

Two particular forms of divestment are sell-offs and spin-offs. Sell-offs involve selling part of a business to a third party, usually for cash. The most common reason for sell-offs is to divest less profitable, non-core business units to ease cash flow problems. In spin-offs, there is no change in ownership. A new company is created with assets transferred to it, as in BT's spin-off of MMO$_2$ in 2002, but the shareholders now have holdings in two companies rather than one.

In theory, the value of the two companies should be no different from that of the single company prior to spin-off. But numerous US studies suggest that spin-offs usually result in strong positive abnormal returns to shareholders. Why should this be so?

1 It enables investors to value the two parts of the business more easily. Poor-performing business units are more exposed to the stock market, necessitating appropriate managerial action. For example, some high-street retailers have spun-off their properties into separate companies, so that investors can judge performance on retailing and property activities more easily.

2 The creation of a clearer management structure and strategic vision of the two companies should result in greater efficiency and effectiveness.

3 Spin-offs reduce the likelihood of a predatory takeover bid where the bidder recognises underperforming assets in a single company. They also make it easier for the group to sell a clearly-defined part of its business at a price better reflecting its true worth.

Divestments enable companies to move their resources to higher-value investment opportunities. They should be evaluated along exactly the same lines as investment decisions, based on the net present value resulting from the divestment.

The motives for a demerger are broadly the same as for divestment discussed earlier. However, with the current vogue to focus on core businesses, it is easy to forget the drawbacks of demerger or divestment:

■ Loss of economies of scale where the demerged business shared certain activities, including central overheads.

■ A smaller firm may find it harder to raise finance or may incur a higher cost of capital.

■ Greater vulnerability to takeover.

Divestment by Smiths Group plc

Not all bands have a Yoko moment but there often comes a point when it doesn't make sense to stick together any more. With earnings expectations cut three times in recent months and the shares down almost a fifth from their June high, is it time to say the same for Smiths Group?

Since its aerospace division was sold to General Electric this year, the logic for maintaining the status quo has looked increasingly tenuous.

Aside from the decent price tag, the reason for the sale was Smiths' lack of scale. The same argument might equally apply to its medical equipment arm. This should be an attractive area as western populations age, but organic profit growth has been lacklustre. As corporate America finally starts to tackle rising healthcare liabilities, prompting downward pressure on pricing, the danger is that the division gets squeezed in its largest market.

The John Crane unit, which sells industrial seals, needs attention too. Performance has been respectable but lacks the pep of peers also benefiting from the capital spending boom in the oil and gas industry.

One of the historical advantages of Smiths has been its ability to generate cash flow, but what both businesses now need is more investment – for research and development in the medical business and to improve prof-itability at John Crane. However, the programme to cut central costs and integrate the joint venture in detection with GE will put further demands on bosses' time.

Assuming trade buyers can be found for the individual parts, a break-up valuation in the region of £12 a share would be reasonable. By selling the trophy asset, aerospace, Smiths' management has already shown an open mind. With Keith Butler-Wheelhouse, the long-standing chief executive, due to step down next year, a break-up could be a parting gift to shareholders.

Source: Financial Times, 14 September 2007.

Management buy-outs

Management buy-outs (MBOs) occur when the management of a company 'buys out' a distinct part of the business that the company is seeking to divest. MBOs usually arise because a parent company decides to divest a subsidiary for strategic reasons. For example, it may decide to exit a certain activity; to sell off an unwanted subsidiary acquired by a takeover of the parent company; to improve the strategic fit of its various business units; or simply to concentrate on its core activities. MBOs can also be purchaser-driven, where the local management recognises that the business has greater potential than the parent company management realises, or where the alternative is closure, with high redundancy and closure costs.

Once a company decides to divest itself of part of its activities, it will usually seek to sell it as an ongoing business rather than selling assets separately. With an MBO, the firm sells the operation to its managers, who put in some of their own capital and obtain the remainder from venture capitalists.

The growth of MBO activity has been fuelled by venture capital firms enabling managers to raise large sums of capital through borrowings (leveraged buy-outs), particularly mezzanine finance, a cross between debt and equity, offering lenders a high coupon rate and, frequently, the right to convert to equity should the company achieve a quotation. The Finance Act 1981 gave considerable help by allowing finance raised to be secured on the acquired assets.

An MBO typically has three parties: the directors of the group looking to divest, the management team looking to make the buy-out and the financial backers for the buy-out team. A private company may agree to an MBO because the directors wish to retire or because it needs cash for the remaining operations.

It is quite common for the new management team to obtain a better return on the business than the old company. Reasons for this include:

- the greater personal motivation of management
- flexibility in decision-making
- lower overheads
- negotiating a favourable buy-out price.

Branson lets go of record store chain

Sir Richard Branson, the entrepreneur, has cut ties with one of his first businesses by selling the UK and Irish Virgin Megastore retail chains to a management buy-out team.

The deal, for an undisclosed sum, was concluded late on Friday evening. Virgin stores will be rebranded "zavvi", after the business was bought by Zavvi Entertainment Group.

Zavvi is a vehicle for a management buy-out team led by Simon Douglas, managing director, and Steve Peckham, finance director.

Sir Richard has with this deal parted with an asset that was seminal in the development of the Virgin brand. In 1970 he founded Virgin as a mail-order record retailer and a year later he opened a record shop on Oxford Street.

"We now choose to franchise our global entertainment retail operations, rather than own them, and this was the last significant Virgin wholly owned retail business in the world," Sir Richard said.

"The Virgin brand will continue to be represented in-store through the Virgin Mobile and Virgin Media in-store concessions."

Some 125 stores will transfer and be renamed by November. The websites and Irish stores will be rebranded in January.

Virgin had also held talks with HMV Group, although a potential deal with the high street rival foundered on valuation and structure.

Music retail's traditional model is under threat from online alternatives, both legal and illegal, and by supermarkets, which now devote more space to cheap CDs and DVDs. This supplies the motivation for market consolidation.

The sale of Virgin Megastores comes as Sir Richard is making a number of strategic departures for the group. He has several high-profile launches coming up, such as Virgin America, a US airline, and a business that aims to fly commercial passengers into space by the end of 2008.

Virgin Galactic, the space venture, in particular is at the heart of Virgin's ambition to become the world's most respected brand.

Source: Philip Stafford, *Financial Times*, 17 September 2007.

Checklist for a successful buy-out

The venture capitalist will ask a number of penetrating questions in evaluating whether the MBO is worth backing.

1 Has the management team got the right blend of skill, experience and commitment? The financial backers may require changes in personnel, frequently the introduction of a finance director.
2 What are the motives for the group selling and the management team buying? If the business is currently a loss-maker, how will the new company turn it round? A convincing business plan with detailed profit and cash flow projections will be required.
3 Is it the assets or the shares of the company that are to be purchased?
4 Will assets require replacement? What are the investment and financing needs?
5 What is an appropriate price?
6 Is there an exit strategy?

Buy-out failure is often the result of a wrongly priced bid, lack of expertise in key areas, loss of key staff or lack of finance.

Once the financial backers are satisfied that the MBO is worth backing, a financial package will be agreed. The management team will typically have a minority share-holding, with the financial backers (often more than one) taking the majority stake. While the venture capitalist company views the investment as long-term, it will look for a potential exit route, frequently through a stock market flotation.

The management team will usually be expected to demonstrate commitment by investing personal borrowings in the business. Redeemable convertible preference shares often form part of the financial arrangements. These shares give voting rights should the preference dividend fall into arrears and enable the holder to redeem shares should the investment fail, or to convert to equity if the business performs well.

Venture capitalists make handsome returns from MBOs and similar deals – much higher than the returns on the stock market investments. The question must be asked as to whether large companies are acting in their shareholders' best interests in permitting buy-outs on such favourable terms. It is all very well to say that a particular division is a non-core activity; but firms are constantly changing their definitions as to what exactly is core. Restructuring, whether through MBOs or other forms, is wasteful and non-value-adding if the group selling the business could have achieved the same efficiency and other gains as the new company.

Management buy-ins are the opposite of buy-outs. A group of business managers with the necessary expertise and skills to run a particular type of business search for a business to acquire. The ideal candidate is a business, or part of a business, with strong potential, but which has been underperforming or is in financial difficulty, perhaps because of poor management. The new team rarely has the necessary capital to buy in and often requires the backing of a venture capitalist.

The two cameos below show examples of management successfully exiting a buy-out in different ways.

Joint ventures and strategic alliances

Unlike mergers or acquisitions, **joint ventures** and other strategic alliances enable both sides to retain their separate identities. They have been employed to good effect to achieve a variety of objectives, but have become a particularly popular way of developing new products and entering new markets, especially overseas. One of the financial benefits is that the strength of two organisations coming together for some specific strategic purpose can often lower capital costs associated with the new investment.

Getting out of a Buy-out I: Safestore to unlock value with a £449m flotation

Safestore, the UK's biggest self-storage company, will be floated on London's main market with a capitalisation of about £449m under the pricing terms of its initial public offering.

The company said yesterday the IPO would be priced at £2.40 per share, giving an enterprise value, which includes debt, of £661m.

Pricing was midway in an estimated range of £2.10 to £2.70, according to Steve Williams, chief executive. "Given the market over the last week or two, it is quite good," he said.

Conditional dealings are expected to begin with admission to the official list on March 14. Citigroup and Merrill Lynch advised.

Safestore has 99 self-storage facilities in the UK and Paris, with about 4m sq ft of space.

The 19 of the total that are in France are branded Une Piece en Plus.

The flotation comes amid enthusiasm for the sector. Shares in Big Yellow – a domestic rival – doubled in price last year.

Safestore made earnings before interest, tax, depreciation and amortisation of £33.5m (£27m) on turnover of £64.3m (£52.9m) in the year to October 31 and has a property portfolio worth £475.2m.

In the pipeline are 13 new stores, nine of which have planning permission.

The management, which owned 15 per cent of the company, will see its stake crystallised at about £67m.

About £27m of this stake is held by Mr Williams, who led a management buy-out backed by Bridgepoint nearly four years ago. He will sell a quarter

of this, worth about £6.7m.

Safestore originally listed on Aim in 1998 after demerging from Safeland, its parent company, which also incubated Bizspace.

The company was taken private in September 2003 for just £39.8m.

Since then, however, it has expanded rapidly. In June 2004 it bought Mentmore, a larger listed rival, for £209m. A year later it expanded to mainland Europe with the purchase of a portfolio from rival Access.

Source: Jim Pickard, *Financial Times*, 10 March 2007.

Getting out of a Buy-out II: Holidaybreak acquires PGL

Holidaybreak, the specialist holiday company, is extending its presence in the educational travel market with a £100m deal to acquire PGL, which runs residential trips for school groups.

The acquisition, Holidaybreak's biggest, means educational travel will make up about 14 per cent of the group's pro-forma sales. Carl Michel, Holidaybreak's chief executive, said PGL was a good fit given its high margins, market leading position and strong growth prospects.

PGL, which is 50 years old and owned by its management, runs 26 activity centres in Britain, France and Spain that are attended by 250,000 children each year. Last year, its earnings before interest and tax were £6.3m out of sales of £50.6m.

Holidaybreak will pay £50m for PGL and refinance £50m of PGL's

debt, with the PGL management set to remain with the company.

Holidaybreak said the deal was partly underpinned by PGL's property portfolio, valued at £93m, and by a high level of advance bookings, with 90 per cent already taken for the year ending next February.

Mr Michel also expects government education policy to generate more interest in structured activity trips.

Synergies are expected, with PGL's peak season in late spring complementing the main summer season for Holidaybreak's camping division, which owns brands such as Eurocamp and Keycamp. The deal will require shareholder approval.

Holidaybreak already owns two German specialist educational holiday businesses. However, its main focus has been on short hotel breaks,

adventure travel – where Mr Michel said there could be other acquisitions – and camping, which is a more mature market and was the subject of private equity interest last year.

Holidaybreak added that recent trading was in line with expectations as it announced a 13 per cent rise in first-half revenues, from £88.7m to £100.6m. The pre-tax loss for the six months to the end of March widened from £6.3m to £7.9m. The group usually makes an interim loss because of the seasonality of its business. The interim dividend was raised from 8p to 8.8p. The shares rose 53p to 851½p.

Source: James Wilson, *Financial Times*, 19 May 2007.

There are two main types of joint venture. An *industrial cooperation* joint venture is for a fixed period of time, where the responsibilities of each of the parties are clearly defined. These are particularly popular in the emerging mixed economies of Central and Eastern Europe, and China. A *joint-equity* venture is where two companies make significant investments in a long-term joint activity. These are more common as a means of investing in countries where foreign ownership is discouraged, such as in Japan and parts of the Middle East.

Like any other investment, the potential partners need to assess the costs and benefits of the joint venture and identify and manage the activities critical to success. One problem can be the inability of the joint venture management team to make decisions without the approval of parent companies. This can be overcome by structuring the alliance with its own board of directors and financial reporting system.

How does restructuring enhance shareholder value? We suggest four ways in which value can be created.

The Road to Morocco

Renault and Nissan will produce up to 400,000 vehicles at a low-cost complex in Morocco in the biggest joint manufacturing investment by the car-making alliance.

The operation, to be built in the new free-trade port near Tangiers, will produce a new generation of light commercial vehicles for Nissan and variants of Renault's low-cost Logan car, mostly for export.

Carlos Ghosn, Renault and Nissan's chief executive, said that the facility would be more competitive than Renault's plants in Romania and Turkey, and at least as cheap as Nissan's in China. "We're talking about a whole set of cars that are competitive and will be exported from this platform," Mr Ghosn told the Financial Times.

At the weekend, the companies signed a memorandum of understanding with Morocco's government for the project, the country's biggest car facility yet and one of the Mediterranean region's largest. They expect to sign a final agreement by the end of the year.

Renault and Nissan will invest €600m ($818m) in the project, plus another €200m to €400m depending on the vehicles produced. The complex will have capacity to build 200,000 vehicles from 2010, with planned capacity later rising to 400,000.

Mr Ghosn faces pressure to squeeze more cost savings out of Renault and Nissan's eight-year-old alliance as the companies struggle to meet earnings targets in western Europe and the US.

He said that Renault remained on track to achieve its goal of adding 800,000 units per year by 2009. "We're working on it, and there is no change," he said.

Renault and Nissan purchase jointly and share engines and other parts for some vehicles, but the alliance is seen as underexploited in the area of common manufacturing. The carmakers are building a plant in Chennai with India's Mahindra & Mahindra, and make vehicles together on a small scale in Mexico.

Odile Desforges, Renault-Nissan's head of purchasing, said the Moroccan project would push the two groups' engineering departments to co-operate more closely.

Renault and its Romanian Dacia brand control about one-third of the Moroccan market, and it makes Kangoo minivans and Logans at a plant in Casablanca.

To date Renault has sold about 575,000 Logans, which it produces in sedan, estate and two-box versions. Future variants might include a 4x4 or pick-up versions, Mr Ghosn said.

Source: John Reed, *Financial Times*, 3 September 2007.

1 Business fit and focus

As we saw with divestments, a business unit may 'fit' one company better than another. Management should review their strategic business units and ask whether they operate best under the present ownership or whether they would create more value under some other ownership through an external acquisition or management buy-out. When unrelated activities have been divested, management has a much better focus on its core businesses and can concentrate on pursuing wealth-creating investment opportunities and improving efficiencies.

2 Eliminate sub-standard investment

Managers commonly enter into investments that do not enhance shareholder value:

1 A decision to reduce reliance on a single business may lead to diversification. Quite apart from the additional overheads that may be created from diversification and the lack of managerial expertise, such diversification may have no real benefit for shareholders. As we saw in Chapters 9 and 10, shareholders can often achieve the same, or better, risk-reduction effects by creating diversified portfolios.
2 Pursuit of growth in sales and earnings brings power and, possibly, protection from takeover, but does little for shareholders. Rather than pay out larger dividends, management may be tempted to reinvest in projects or acquisitions that do not add value.
3 While a strategic business unit may be profitable, it is often an amalgamation of profitable and unprofitable projects, the former subsidising the latter. Restructuring the business creates a leaner operation with no room for cross-subsidisation.

3 Judicious use of debt

Cautious managers argue that borrowing should be minimised, as it increases financial risk and leaves little room for errors. Aggressive managers take a very different view.

Debt provides a powerful incentive to improve performance and minimise errors. The consequences of management's successes and mistakes are magnified through gearing, leaving little room for error. Managerial mediocrity is no longer acceptable. Cash flow – not profit – becomes the all-important yardstick, for it is cash flow that must be generated to service the debt and meet repayment schedules. In this respect, incurring debt obligations may provide an important signal to the market concerning the resolve of the management team.

Furthermore, debt is a cheaper source of finance because interest is tax-deductible, while dividends on equity are not. Restructuring the balance sheet by substituting debt for equity, within acceptable gearing limits, creates a tax shield and increases the company's market value.

4 Incentives

Raising debt to realise equity can be a powerful incentive to both shareholders and managers. Equity is concentrated in the hands of fewer shareholders, providing a greater incentive to monitor managerial actions. This often leads to the creation of managerial incentives to enhance shareholder value, through executive share options or profit-sharing schemes. Remuneration packages may increase profit-related pay at the expense of salaries and wages. This will also benefit loan stock holders, who have priority ahead of profit-sharing, but after employees' wages and salaries.

20.13 PRIVATE EQUITY

One of the most important mechanisms for industrial restructuring is the investment activity of private equity houses. Private equity investment is conducted by private equity fund management firms, usually in the form of a Limited Liability Partnership. The partners consist of syndicates of financial institutions, such as pension funds. They specialise in buying majority stakes in high growth businesses or firms where greater efficiencies can be achieved by restructuring activities. In principle, private equity covers all investment activities in unquoted firms including development and venture capital, management buy-outs and buy-ins. However, in recent years, the term has become associated, often in a pejorative context, with the practice of private equity firms buying out quoted companies, thus taking them off the Stock Exchange into private hands. These deals are usually associated with high levels of borrowing to gear up the capital structure while restructuring the firm, and paying the interest out of both existing cash flow and the proceeds of more efficient operations. In most cases, the ultimate aim is to refloat the firm on the Stock Exchange and make a capital gain. Until the enterprise is refloated, the private equity firm is the only shareholder in the business, and has total control over strategy and operations.

Despite achieving some notable successes, private equity has received considerable bad publicity for various reasons (see below). A flavour of the opprobrium can be gleaned from two cases.

In 2003, Debenhams, the stores group, was taken private, only to reappear on the Stock Exchange barely two years later with a market value greater than it was sold for, but loaded up with debt. The greater part of its property assets had been sold off, with the proceeds paid as dividends to the investors in the private equity firm, and in bonuses to the private equity fund managers. Since 2005, Debenhams has struggled to assert itself in the marketplace and its shares have generally traded below the (re)floatation price of 190p.

In July 2006, Boots, the chain of chemists, and Alliance, a firm of drug distributors, completed a £7 billion 'nil-premium' merger. Less than a year later, in March 2007, a private equity firm, Kohlberg Kravis Roberts (KKR), and Stefano Pessina, the former CEO of Alliance, made an indicative offer of £10 per share for Alliance Boots, as it was known. This was rejected, the offer was increased to £10.40, and accepted. Meanwhile, Terra Firma, another private equity firm, and the Wellcome Trust, a privately-owned charitable organisation, made a counter-bid of £10.85 per share. KKR/Pessina responded with a bid of £10.90 that was accepted. Terra Firma then responded with a higher offer worth £11.15 per share, only to be trumped by KKR/Pessina with a winning bid of £11.39 per share, valuing the equity at £11 billion, making this Europe's biggest private equity deal, and the first time a FTSE 100 firm had been taken out by a private equity firm. The bid raised several ethical issues.

No details were given until after the battle was over as to how the bid would be financed, or what the winning consortium planned to do with its prize, and it was suggested that Pessina could well double his money in just five years if the venture paid off. As one observer put it: 'There remains a strong feeling among fund managers [other than private equity firms] that they may have been mugged by an inside job.' The books had been opened to KKR/Pessina at £10.40 per share, almost £1.00 below the final bid, suggesting the Board of Alliance Boots (many of whom were former colleagues of Pessina) had been far too ready to capitulate to Pessina and his partners. The box on page 599 gives a 'back-of-the-envelope' evaluation of the possible arithmetic underlying the bid.

■ Criticisms of private equity

Against this background, it may be easier to appreciate some of the criticisms of private equity firms, or at least in the buy-out arena:

- They are only interested in 'quick flips', stripping out company assets, and cutting jobs before selling firms on, or closing them down.
- They increase unemployment.

- They may curtail the pension benefits of existing members.
- Returns paid out to investors and managers are 'excessive'.
- They pay little tax. Until 2008, UK private equity firms and their partners could pay capital gains tax on disposal at a concessionary rate applicable to business sales of 10 per cent, that often fell to as low as 5 per cent after allowances. CGT was raised to 18 per cent in 2008, but was still below the new threshold rate of income tax of 20 per cent.
- They lower government tax revenues – £130 million in corporation tax being at stake in the case of Alliance Boots. As equity is replaced by debt, so pre-tax profits reduce, thus reducing the firm's tax bill, given the tax shield.
- As they do not have to report to anyone other than their small number of shareholders, their activities are not transparent, allegedly unacceptable for a sector that accounts for such a high proportion of the economy.

Filling your Boots – did the numbers stack up?

- At the final bid price of 11.39 per share, making the equity worth £11 billion, and with net debt of £1.2 billion, the enterprise value was £12.2 billion.

- The projected restructuring thus appeared to involve £7.6 billion of debt plus £4.6 billion of direct equity investment (KKR £3.6 billion plus Pessina's £1.0 billion).

- At an interest rate of, say, 7.5 per cent (2 per cent above LIBOR), the interest bill would be about £500 million p.a.

- It could sell off property assets (100 freehold stores plus the Nottingham HO) to pay down debt. The property portfolio value was estimated at over £1 billion, but valued in the books at £400 million. This might lower debt to, say, £6.5 billion.

- After property disposals, the interest bill could fall by about (7 per cent × £1 billion =) £70 million to £430 million, compared to pre-tax profit estimated for 2007 of £600 million, a comfortable enough cushion, ignoring the rent that now became payable.

■ Rejoinders

Such criticisms have not gone unchallenged. For example, using US data, Lerner (2008) has shown that the private equity investment period is much longer than popularly imagined – in the US, this averages over 5 years and is increasing. There is a lower insolvency rate among private equity-owned firms than for orthodox firms, suggesting superior management. Against this, private equity-owned firms did cut more jobs in the two years after takeover – possibly because many taken-over firms were already distressed. However, private equity-owned firms tend to grow faster, and may create more employment as they recover in the longer term, as well as saving jobs in firms that might otherwise have failed.

In the UK, the trade association for the industry, the British Private Equity and Venture Capital Association (**www.bvca.co.org**) produces an annual report to chart the contribution of the industry. Its latest evidence (BVCA, 2008) suggests that firms with private equity involvement now account for about a fifth of the whole private sector force, and grew considerably faster than their listed peers and than the overall economy in terms of sales, employment and exports over the previous five years. However, it does acknowledge that the industry has an 'image problem', and, partly at the urging of the Financial Services Authority, has commissioned a report on the industry, planned for publication in autumn 2008. The committee, chaired by City 'grandee' Sir David Walker, is briefed to make recommendations about higher standards of transparency, and disclosure of operations and results, by private equity-backed firms.

This is doubtless an ongoing debate, and time will tell whether private equity-inspired restructuring is, on balance, good or bad. For the moment, it may be useful to peruse Table 20.8 that shows the claimed advantages of private equity-backed firms over orthodox public companies.

Table 20.8 Public vs. Private equity

Public companies	Private equity-backed firms
■ Large number of shareholders, many of them very small.	■ Small number of large shareholders.
■ Most shareholders have no say in strategy or operations, except at the AGM.	■ All private equity shareholders intimately involved in strategy and operations.
■ Shareholders may have conflicting personal investment aims and views on company strategy.	■ Shareholders usually have common interests.
■ Incentives of managers could lead to conflict with owners' interests.	■ Management highly incentivised; incentives closely aligned with investors' aims.
■ Public companies often over-focused on short-term earnings; may take decisions that damage long-term prospects to safeguard short-term earnings.	■ Freedom from stock-market pressures may encourage a more long-term focus.
■ Need shareholder approval for major decisions, e.g. mergers and divestments; may slow decision-making and execution.	■ Decision-making and implementation can be quicker; more likely to capture market opportunities.
■ Tend to be relatively low geared; WACC may be sub-optimal.	■ Able to employ much higher levels of gearing, so long as cash flow 'comfortably' exceeds interest commitments.
■ Difficult for shareholders to remove under-performing management.	■ Management changes can be quickly and easily made.
■ Possible that talented managers may seek higher rewards attainable in private equity.	■ Prospect of higher rewards can lure talented personnel.
■ Increasing burden of regulation and disclosure requirements, e.g. to meet corporate governance standards.	■ Less regulation and lower disclosure requirements.
■ Pressure to increase, or at least to maintain, dividends.	■ No external pressure to pay dividends thus reserving cash flow for other uses.

Source: Based on JPMorgan Cazenove Ltd (2008).

Private equity and the secret recipe of success

Two years ago, after 30 years as the chief executive of a public company, latterly with United Business Media, I crossed over to the world of private equity and became a partner of Kohlberg Kravis Roberts & Co (KKR).

I was already well-acquainted with private equity. As an investor in several funds and a member of two advisory boards, I had witnessed consistently good investment returns, seen how investment decisions were made and how companies were managed. There was evidently much more to their success than simply piling on debt and reducing corporate overheads.

The failure of the private equity industry to explain itself and, in the case of many privately owned companies, to provide information beyond the statutory minimum, has unsurprisingly led to a caricatured version of private corporate life where calculating investors sanction unsupportable levels of debt, a squeezed cash flow and a low level of investment.

But private equity specialists do not operate this way. They know it would be a sure-fire recipe for disaster – destroying value and fatally compromising the private equity business model. And research in the US demonstrates this: initial public offerings of private equity-backed firms perform better than IPOs in general.

So what is the private equity specialist's recipe for success? It is this: first find a good company, add talented management and then implement a bold plan to invest in growth and optimise operating and financial performance.

As with public companies, privately owned companies have to contend with the competitive dynamics of the marketplace and the impact globalisation. But there are some significant benefits to being – or going – private. Free from the short-term perspective of quarterly reporting and the aversion of some institutional shareholders to sensible levels of debt and risk investment in growth, privately owned companies can adopt a long-term approach and embrace an entrepreneurial growth culture. Moreover, the governance structure is based upon ownership: all directors have a significant stake in the success of the business and board meetings can concentrate on products, customers, people and performance.

Of course, private equity ownership is not going to be to every executive's liking. Expectations are high – in line with rewards – and underperformance is not indulged. The chief executive used to quasi-imperial powers and supported by a pliant bureaucracy would be in for a rude awakening. Those self-important chief executives who seek newspaper profiles, the adulation of the gabfest crowd and the unquestioning support of their colleagues should stay put. Executives should leave their ego at the door and prepare to answer tough questions from private equity executives often less than half their age but with formidable intellects.

But if the private equity culture is bracing and challenging, it is also highly supportive.

We are constantly meeting chief executives and chief financial officers who either want to discuss the merits of private ownership for all or part of their business or are looking to move and lead a privately owned company. If you are such a chief executive or chief financial officer, then the first thing to do is to take a leaf out of the private equity investment handbook and conduct extensive due diligence and rigorous benchmarking.

Not all firms offer the same things. A few firms, such as KKR, invest in big companies and operate on a global basis. At the other end of the scale, there are many firms that are highly skilled at start-ups and growing small enterprises. Then there are those that invest in just one country or region or in one or two sectors of the economy such as technology or infrastructure.

Seek out those firms with a deep knowledge and experience of your sector. Many firms are organised on a sector basis and have recruited senior business leaders to help bring the right blend of investment and commercial skills to each company in their portfolio.

While making your choice, ensure that you spend time with the sector teams and probe the depth of their knowledge and experience. Remember, you will be partners with them for the next five or six years, so you need to be satisfied that you can work effectively with them on both a professional and personal level. It is certainly worth seeking the advice of a couple of the former chief executives who have worked with the firm for a number of years.

Once you have chosen the private equity house, you need to develop the business plan. You should propose challenging but realistic targets in order to optimise performance, to grow the company and to focus on the core businesses. Make sure that the proposed capital structure ensures that you have the financial resources to deliver the plan.

The final ingredient is the management equity plan. This should include all key executives – each of whom will be required to make a meaningful investment within the context of their personal financial circumstances.

When I chose a private equity house, I made a judgement on the quality of the people, their integrity, their record of innovation and the depth of their industry knowledge. I have not been disappointed.

1 Conduct extensive research: Always speak to a number of firms before making your choice. Look for companies with a proficient understanding of your sector.

2 Spend time with the team: You are looking for people who you can work with for five or six years – professionally and personally. Ask for the advice of past chief executives who have spent time with the firm.

3 Develop the business plan: Set challenging but realistic targets. Ensure you have the financial resources to sustain an economic cycle and deliver the plan.

4 The management equity plan: This must require all key executives to make a meaningful investment within the context of their own financial circumstances.

Source: Financial Times, 28 March 2007.

A 3-eyed venture capitalist (www.3i.com)

3i is way out in front as the UK's leading venture capital investment company, accounting for €2.7 billion of venture capital investment in 2007. Its roots go back to 1929, when a report under the chairmanship of Lord Macmillan identified the 'Macmillan gap' – a shortage in the financing of small and medium-sized companies. Little was done, however, until 1945, when the Industrial and Commercial Finance Corporation, backed by the Bank of England, was created to bridge this gap with just £10 million of capital to invest. In 1973, it united with the Finance Corporation for Industry to form Investors in Industry, which later became simply 3i. It was floated on the stock market in 1994 in a £1.6 billion flotation, making it a FTSE Top 100 company. At early March 2008, its market value was €5.6 billion.

During its long history, spanning some 65 years, it has played a major role in helping to bridge the finance gap experienced by small companies and bringing 150 firms to their IPOs. Among its major successes were the sale in 2005 of foreign exchange specialist Travelex for a tenfold return on investment, and the sale of Go Fly, the low-cost airlines in 2002 to easyJet for £37 million. But venture capital is a risky business, and 3i has had its share of failures, the largest of which was its investment in Isosceles, the company set up to mount a contested £2.4 billion management buy-in of the Gateway supermarket chain. The deal backfired, leaving 3i with £83 million debt and equity to write off. Many well-known businesses, like Waterstones, Geest and Laura Ashley, have benefited from its financial backing.

3i now has a major international presence, operating in Germany, France, the USA, Scandinavia and the Far East.

SUMMARY

We have explored various motives for merger and takeover activity, and have argued the importance of a coherently-structured strategic approach to acquisitions, including planned integration that emphasises human and organisational factors. Finally, we briefly discussed other forms of corporate restructuring.

Key points

- The decision to acquire another company is an investment decision and requires evaluation on similar criteria to the purchase of other assets.

- Added complications are the resistance of incumbent managers to hostile bids and the presence of long-term strategic factors.

- The takeovers most likely to succeed are those approached with a strategic focus, incorporating detailed analysis of the objectives of the takeover, the possible alternatives and how the acquired company can be integrated into the new parent.

- If the takeover mechanism works well, it is an effective and valuable way of clearing out managerial dead wood.

- Many takeovers appear to be launched for 'managerial' motives, such as personal and financial aggrandisement.

- The main reasons for failure of takeovers are poor motivation and evaluation, excessive outlays (often with borrowed capital) and poorly planned and executed integration.

- The complexity of takeover integration is related to the motive for the takeover itself, ranging from cash generation, requiring only a loose control over operations, to economies of scale, requiring highly detailed integration.

- The impact of mergers can be studied by comparing the financial characteristics of merger-active and merger-inactive firms to assess any performance differentials, but this approach suffers from many problems.

- The main alternative is a capital market-based approach to assess how the market judges a merger in terms of share price movements.

- The available evidence suggests that the bulk of the gains from mergers accrue to shareholders of acquired companies, although some evidence suggests that shareholders of acquirers can also share in the benefits, presumably if the takeover is well-considered.

- Corporate restructuring enhances shareholder value through (i) improving the business fit and focus, (ii) judicious use of debt and (iii) providing incentives for management.

- Private equity houses are increasingly important agents for restructuring, but their activities have prompted mixed reactions.

Further reading

There has been an explosion in books on M&A in recent years, reflecting the ongoing high level of M&A activity, its increasingly cross-border nature and the amount of academic research resource devoted to understanding the whole process.

The most comprehensive UK text is Sudarsanam (2003), while a more international flavour can be gleaned from de Pamphilis (2001). Neither book is especially strong on valuation for takeover, but Arzac (2004), Damadoran (2002), and the two 'McKinsey' books, Koller *et al.* (2005) and Copeland *et al.* (2000) fill this gap. A deal-based study of valuation of merger targets is given by Arzac (2004), while Feldman (2005) focuses on valuation of private firms. Rankine *et al.* (2003) examine the due diligence process.

As ever, Grant (2005) gives an excellent analysis of the strategic factors underpinning M&A. An important article that explains the market for corporate control is Jensen (1984).

As evidence accumulates about the disappointing results of M&A, more attention is being given to post-merger integration. Books that concentrate on this aspect are those by Gaughan (2005), Very (2004), Bruner and Perella (2003), Habeck *et al.* (2000), Angwin (2000), Haspeslagh *et al.* (1991) and Pritchett *et al.* (1997).

Devine (2002) analyses the stages of the merger process with particular emphasis on 'people issues'. *The Economist* published a set of case studies in 2000 that sheds great light on the critical success factors in making mergers work. An up-to-date collection of leading papers in the M&A field is Gregoriu and Neuhauser (2007). Two useful papers that focus on post-merger integration are by Angwin (2004) and Quah and Young (2005).

Appendix
PHANTOM DEBTORS: A CASE TO REMEMBER

Numerous 'phantom' debtors were found by the electronics and defence contractor Ferranti, when it consolidated the accounts of its 1987 acquisition, US armaments manufacturer International Signal Corporation, ISC. ISC had carried some highly questionable debtors in its accounts for several years. As a result of this revelation, and allowing also for significant over-valuation of work-in-progress, Ferranti was forced, in 1990, to write off £215m of shareholder funds, when it finally realised the magnitude of the problem within its new subsidiary. ISC's auditors subsequently paid £40m to Ferranti to settle litigation over this affair.

The former chairman of ISC was jailed for fifteen years in 1992 for fraud, accused of misrepresenting ISC's true financial position. Ferranti never recovered from this fiasco, being forced to sell off its radar division to GEC for £310m, and then experiencing a severe reduction in orders due to the ending of the Cold War. Mounting losses, worsening liquidity and the failure to secure a defence contract in Bahrain which would have yielded a much-needed advance payment, forced it to call an emergency general meeting (EGM) of shareholders in October 1993, when net worth fell below half of its share capital. Shortly afterwards, it received a bid for a nominal 1p per share from GEC. This valued the equity at a mere £10m compared with a mighty £845m just before the ISC scandal was revealed. GEC were attracted by the sonar division, a joint venture with the French company Thomson, a substantial pension fund surplus and unutilised tax losses. The bid, pitched well below the current market price of 9p, was probably designed to fail, but to draw more attention to Ferranti's parlous position. The shares were quickly re-rated by the market to around 3p.

The bid was recommended by the Ferranti board, reckoning that the alternative was liquidation, but in the face of mounting shareholder opposition, it became clear to GEC that they could not achieve the 90 per cent acceptances which they demanded. GEC withdrew the bid, and Ferranti was duly placed into receivership. GEC subsequently purchased a range of Ferranti's defence interests for less than they would have had to pay for the whole business, allowing for the Ferranti debt which they would have had to assume (about £100m, dating back to the ISC acquisition).

QUESTIONS

Questions with a **coloured number** have solutions in Appendix B on page 755.

1 As treasurer of Holiday Ltd you are investigating the possible acquisition of Leisure Ltd. You have the following basic data:

	Holiday	Leisure
Earnings per share (expected next year)	£5	£1.50
Dividends per share (expected next year)	£3	£0.80
Number of shares	1 million	0.6 million
Share price	£90	£20

You estimate that investors currently expect a steady growth of about 6 per cent in Leisure's earnings and dividends. Under new management, this growth rate would be increased to 8 per cent per year, without any additional capital investment required.

Required

(a) What is the gain from the acquisition?
(b) What is the cost of the acquisition if Holiday pays £25 in cash for each Leisure share? Should it go ahead?
(c) What is the cost of the acquisition if Holiday offers one of its own shares for every three shares of Leisure? Should it go ahead?
(d) How would the cost of the cash offer and the share offer alter if the expected growth rate of Leisure were not changed by the takeover? Does it affect the decision?

2 The directors of Gross plc have made a 850p per share cash bid for Klinsmann plc, a company that is in a similar line of business. The summarised accounts of these two companies are as follows:

	Gross £m		Klinsmann £m	
Sales (all credit)		216		110
Operating costs		(111)		(69)
Operating profit		105		41
Interest		(8)		(10)
Earnings before tax		97		31
Tax		(25)		(10)
Earnings for shareholders		72		21
Fixed assets		76		50
Current assets				
Stock	20		25	
Debtors	40		24	
Cash	8		1	
	68		50	
Current liabilities				
Creditors	(28)		(12)	
Bank overdraft	—		(8)	
	(28)		(20)	
Net current assets		40		30
Total assets less current liabilities		116		80
Long-term liabilities		(60)		(50)
Net assets		56		30

Included in the operating costs for each company are the purchases made during this year – £100 million for Gross and £70 million for Klinsmann.

The number and market value of each company's shares are:

	Gross	**Klinsmann**
No. of shares issued	100 m	20 m
Share price	600p	700p

Required

Analyse this bid to include:

(a) Possible ways in which Gross may hope to recoup the bid premium when operating Klinsmann.

(b) The final and strategic effects on Gross if the bid is accepted by Klinsmann's shareholders.

3 Dangara plc is contemplating a takeover bid for another quoted company, Tefor plc. Both companies are in the leisure sector, operating a string of hotels, restaurants and motorway service stations. Tefor's most recent balance sheet shows the following:

	£m	£m
Fixed assets (net)	800	
Current assets *less*		
Current liabilities	50	
Long-term debt		
(12% debenture 2002)	(200)	
		650
Issued share capital (25p units)	80	
Revenue reserves	420	
Revaluation reserve	150	
		650

Tefor has just reported full-year profits of £200 m after tax.

You are provided with the following further information:

(a) Dangara's shareholders require a return of 14 per cent.

(b) Dangara would have to divest certain of Tefor's assets, mainly motorway service stations, to satisfy the competition authorities. These assets have a book value of £100 million, but Dangara thinks they could be sold on to Lucky Break plc for £200 million.

(c) Tefor's assets were last revalued in 1992, at the bottom of the property market slump.

(d) Dangara's P:E ratio is 14:1, Tefor's is 10:1.

(e) Tefor's earnings have risen by only 2 per cent p.a. on average over the previous five years, while Dangara's have risen by 7 per cent p.a. on average.

(f) Takeover premiums (i.e. amount paid in excess of pre-bid market values) have recently averaged 20 per cent across all market sectors.

(g) Many 'experts' believe that a stock market 'correction' is imminent, due to the likelihood of a new government, led by Bony Clair, being elected. The new government would possibly adopt a more stringent policy on competition issues.

(h) If a bid is made, there is a possibility that the Chairman of Tefor will make a counter-offer to its shareholders to attempt to take the company off the Stock Exchange.

(i) If the bid succeeds, Tefor's ex-chairman is expected to offer to repurchase a major part of the hotel portfolio.

(j) Much of Tefor's hotel asset portfolio is rather shabby and requires refurbishments, estimated to cost some £50 million p.a. for the next five years.

Required

As strategic planning analyst, you are instructed to prepare a briefing report for the main board, which:

(i) assesses the appropriate value to place on Tefor, using suitable valuation techniques. (State clearly any assumptions you make.)

(ii) examines the issues to be addressed in deciding whether to bid for Tefor at this juncture.

4 Larkin Conglomerates plc owns a subsidiary company, Hughes Ltd, which sells office equipment. Recently, Larkin Conglomerates plc has been reconsidering its future strategy and has decided that Hughes Ltd should be sold off. The proposed divestment of Hughes Ltd has attracted considerable interest from other companies wishing to acquire this type of business.

The most recent accounts of Hughes Ltd are as follows:

Balance Sheet as at 31 May 1995

	£000	£000	£000
Fixed assets			
Freehold premises at cost		240	
Less Accumulated depreciation		(40)	200
Motor vans at cost		32	
Less Accumulated depreciation		(21)	11
Fixtures and fittings at cost		10	
Less Accumulated depreciation		(2)	8
			219
Current assets		34	
Stock at cost		22	
Debtors		20	
Cash at bank		76	
Creditors: amounts falling due within one year			
Trade creditors	(52)		
Accrued expenses	(14)	(66)	10
			229
Creditors: amounts falling due beyond one year			
12% Loan – Cirencester Bank			(100)
			129
Capital and reserves			
£1 ordinary shares			60
General reserve			14
Retained profit			55
			129

Profit and Loss Account for the year ended 31 May 1995

	£000
Sales turnover	352.0
Profit before interest and taxation	34.8
Interest charges	(12.0)
Profit before taxation	22.8
Corporation tax	(6.4)
Profit after taxation	16.4
Dividend proposed and paid	(4.0)
	12.4
Transfer to general reserve	(3.0)
Retained profit for the year	9.4

The subsidiary has shown a stable level of sales and profits over the past three years. An independent valuer has estimated the current realisable values of the assets of the company as follows:

	£000
Freehold premises	235
Motor vans	8
Fixtures and fittings	5
Stock	36

For the remaining assets, the balance sheet values were considered to reflect their current realisable values.

Another company in the same line of business, which is listed on the Stock Exchange, has a gross dividend yield of 5 per cent and a price:earnings ratio of 12.

Assume a standard rate of income tax of 25 per cent.

Required

(a) Calculate the value of an ordinary share in Hughes Ltd using the following methods:
  (i) net assets (liquidation) basis
  (ii) dividend yield
  (iii) price:earnings ratio
(b) Briefly evaluate each of the share valuation methods used above.
(c) Identify and discuss four reasons why a company may undertake divestment of part of its business.
(d) Briefly state what other information, besides that provided above, would be useful to prospective buyers in deciding on a suitable value to place on the shares of Hughes Ltd.

5 The directors of Fama Industries plc are currently considering the acquisition of Beaver plc as part of its expansion programme. Fama Industries plc has interests in machine tools and light engineering while Beaver plc is involved in magazine publishing. The following financial data concerning each company is available:

Profit and Loss Accounts for the year ended 30 November 1995

	Fama Industries £m	Beaver £m
Sales turnover	465	289
Profit before interest and taxation	114	43
Interest payable	(5)	(9)
Profit before taxation	109	34
Taxation	(26)	9
Net profit after taxation	83	25
Dividends	(8)	(12)
Retained profit for the year	75	13

Balance Sheets as at 30 November 1995

	Fama Industries £m	Beaver £m
Fixed assets	105	84
Net current assets	86	38
	191	122

	Fama Industries £m	Beaver £m
Less creditors due beyond one year	(38)	(58)
	153	64
Capital and reserves		
Ordinary shares	50	30
Retained profit	103	34
	153	64
Price:earnings ratio prior to bid	16	12

The ordinary share capital of Fama Industries plc consists of 50p shares and the share capital of Beaver plc consists of £1 shares. The directors of Fama Industries plc have made an offer of four shares for every five shares held in Beaver plc.

The directors of Fama Industries plc believe that combining the two businesses will lead to after-tax savings in overheads of £4 million per year.

Required
(a) Calculate:
  (i) the total value of the proposed bid
  (ii) the earnings per share for Fama Industries plc following the successful takeover of Beaver plc
  (iii) the share price of Fama Industries plc following the takeover, assuming that the price:earnings ratio is maintained and the savings are achieved.
(b) Comment on the value of the bid from the viewpoint of shareholders of both Fama Industries plc and Beaver plc.
(c) Identify, and briefly discuss, two reasons why the managers of a company may wish to take over another company. The reasons identified should not be related to the objective of maximising shareholder wealth.

6 Europium plc is a large conglomerate which is seeking to acquire other companies. The Business Development division of Europium plc has recently identified an engineering company – Promithium plc – as a possible acquisition target.
  Financial information relating to each company is given below:

Profit and Loss Account for the year ended 30 November 1997

	Europium plc	Promithium plc
Turnover	820	260
Profit on ordinary activities before tax	87	33
Taxation on profit on ordinary activities	(27)	(9)
Profit on ordinary activities after tax	60	24
Dividends	(15)	(5)
Retained profit for the year	45	19
Price:earnings ratio	16	10
Capital and reserves		
£1 ordinary shares	80	30
Retained profits	195	124
	275	154

The Business Development division of Europium plc believes that shares of Promithium plc can be acquired by offering its shareholders a premium of 25 per cent above the existing share price. The purchase consideration will be in the form of shares in Europium plc.

Required
(a) Calculate the rate of exchange for the shares and the number of shares of Europium plc which must be issued at the anticipated price in order to acquire all the shares of Promithium plc.
(b) Suggest reasons why Europium plc may be prepared to pay a premium above the current market value to acquire the shares of Promithium plc.
(c) Calculate the market value per share of Europium plc following the successful takeover and assuming the P:E ratio of Europium plc stays at the pre-takeover level. Would you expect the P:E ratio of Europium plc to stay the same?
(d) State what investigations Europium plc should undertake before considering a takeover of Promithium plc.

7 As a defence against a possible takeover bid the managing director proposes that Woppit make a bid for Grapper plc, in order to increase Woppit's size and, hence, make a bid for Woppit more difficult. The companies are in the same industry.
  Woppit's equity Beta is 1.2 and Grapper's is 1.05. The risk-free rate and market return are estimated to be 10 and 16 per cent p.a. respectively. The growth rate of after-tax earnings of Woppit in recent years has been 15 per cent p.a. and of Grapper 12 per cent p.a. Both companies maintain an approximately constant dividend payout ratio.

Woppit's directors require information about how much premium above the current market price to offer for Grapper's shares. Two suggestions are:

(i) The price should be based upon the balance sheet net worth of the company, adjusted for the current value of land and buildings, plus estimated after tax profits for the next five years.

(ii) The price should be based upon a valuation using the Dividend Valuation Model, using existing growth rate estimates.

Summarised financial data for the two companies are shown below:

Most recent Balance Sheets (£m)

		Woppit		Grapper
Land and buildings (net)[a]		560		150
Plant and machinery (net)		720		280
Stock	340		240	
Debtors	300		210	
Bank	20	660	40	490
Less: Trade creditors	(200)		(110)	
Overdraft	(30)		(10)	
Tax payable	(120)		(40)	
Dividends payable	(50)	(400)	(40)	(200)
Total assets less current liabilities		1,540		720
Financed by:				
Ordinary shares[b]		200		100
Share premium		420		220
Other reserves		400		300
		1,020		620
Loans due after one year		520		100
		1,540		720

[a]Woppit's land and buildings have been recently revalued. Grapper's have not been revalued for four years, during which time the average value of industrial land and buildings has increased by 25 per cent p.a.

[b] Woppit 10p par value, Grapper 25p par value

Most recent Profit and Loss Accounts (£m)

	Woppit	Grapper
Turnover	3,500	1,540
Operating profit	700	255
Net interest	(120)	(22)
Taxable profit	580	233
Taxation	(203)	(82)
Profit attributable to shareholders	377	151
Dividends	(113)	(76)
Retained profit	264	75

The current share price of Woppit is 310 pence and of Grapper 470 pence.

Required

(a) Calculate the premium per share above Grapper's current share price that would result from the two suggested valuation methods. Discuss which, if either, of these values should be the bid price. State clearly any assumptions that you make.

(b) Assess the managing director's strategy of seeking growth by acquisition in order to make a bid for Woppit more difficult.

(c) Illustrate how Woppit might achieve benefits through improvements in operational efficiency if it acquires Grapper.

(ACCA)

Practical assignment

Select one of the merger/takeover situations that has been given prominence recently in the media. Analyse your selected case under the following headings (indicative guidelines are provided).

1 *Strategy* – How does the 'victim' appear to fit into the acquirer's long-term strategy?
2 *Valuation and bid tactics* – Has the acquirer bid or paid 'over the odds'? What were the pros and cons of the financing package?
3 *Defence tactics* – Were the tactics employed sensible ones? Were the managers of the target company genuinely resisting or simply seeking to squeeze out a higher offer?
4 *Impact* – Will the acquired company be difficult to integrate? Are any sell-offs likely?

 *Now retake your diagnostic test for Chapter 20 to check your progress and update your study plan.*

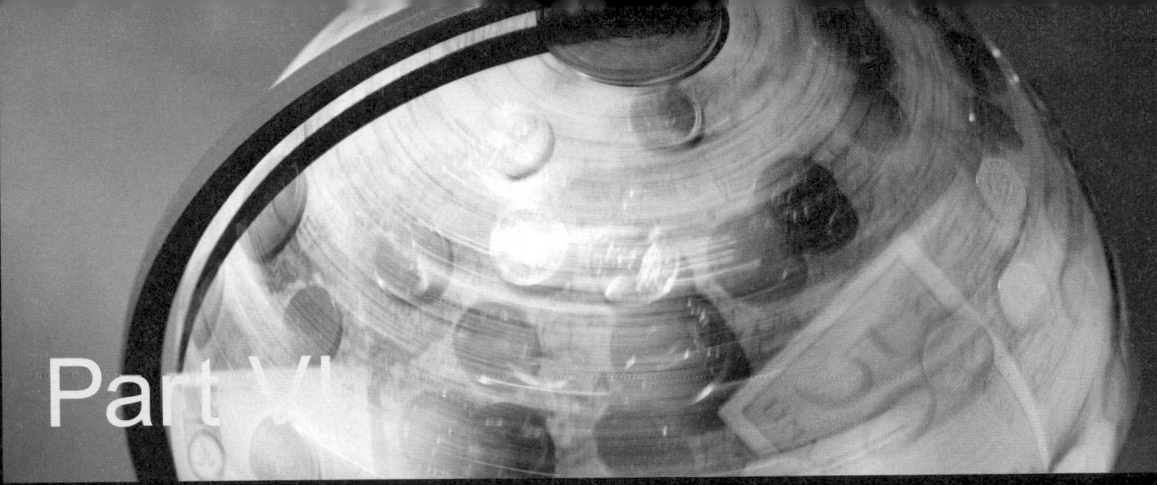

INTERNATIONAL FINANCIAL MANAGEMENT

This section contains two chapters that deal with issues of international financial management (IFM). IFM adopts essentially the same perspective as domestic financial management, i.e. it looks at how decisions in the areas of investment appraisal, financing and dividend policy can be used to create wealth for the owners of the firm. However, there is a major difference – cash flows expected from foreign trading and investment activities are subject to exchange risk, the risk that the domestic currency value may be undermined by adverse exchange rate changes.

Chapter 21 explores the various types of exposure and explains how exporters and importers can, if they so choose, take precautionary measures against such exposures.

Chapter 22 examines how to evaluate foreign investment decisions (FIDs), using principles developed in earlier chapters, but focusing on the strategic motives for undertaking FIDs, and the particular problems, such as exchange exposure and multiple tax regimes, facing firms when evaluating FIDs. In particular, the issue of managing long-term operating exposure is examined in a strategic context.

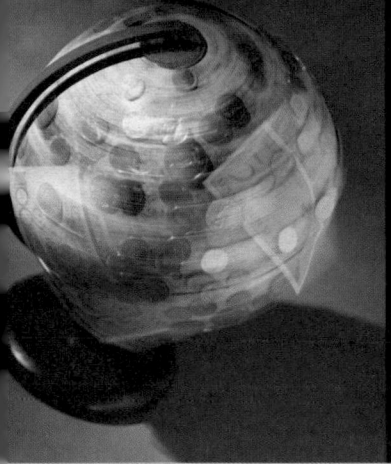

21

Managing currency risk

Losing an ARM and a leg

The pound's strength against the dollar is a boon for British consumers planning luxury mini-breaks to New York. But some UK companies are cursing the toll the exchange rate is taking on their profits and competitiveness.

The issue came to prominence last month when the pound breached $2 for the first time since the UK withdrew from the exchange rate mechanism in 1992. In fact, the currency has been strengthening steadily over the past five years, and especially the past nine months.

About 22 per cent of the sales of the FTSE-350 are directly exposed to the US, while a further 11 per cent come from regions closely tied to the dollar, according to Citigroup.

So how is UK plc holding up? The victims will fall into one or more of three categories. First, ex-porters serving dollar markets are at a competitive disadvantage if their cost base is located in the UK and denominated in sterling. The weak dollar crimps their revenues, but wages still have to be paid in pounds. That creates a painful squeeze.

Second, companies that serve US markets through local operations will see dollar costs and dollar revenues move in tandem, but their dollar profits still take a hit when converted into sterling.

Finally, the exchange rate reduces the sterling dividends of companies that report their financial results in dollars. That applies to many large UK companies, such as GlaxoSmithKline, BP and HSBC.

It is companies in the first camp that are being hit the hardest. The strong pound is not just reducing the profits and dividends paid to investors. It is affecting their long-term competitiveness and may even lead to loss of market share to US or Asian rivals.

One such victim is ARM Holdings, the Cambridge-based designer of semiconductors, an industry whose global currency is dollars. Warren East, chief executive, estimates that the dollar has wiped £1bn from the company's stock market value. "We are at the acute end of the scale. The last three years have been pretty horrid," he says. "The only way you can respond is by matching your cost base [to the dollar]. That means jobs going outside the UK. We have gone as far as we can and about 50 per cent of our costs are now overseas."

ARM has also taken advantage of relatively low-cost skilled labour in India. Headcount there has risen from two to 230 since 2004.

Source: Chris Hughes, *Financial Times*, 13 May 2007.

Learning objectives

This chapter explains the nature of the special risks incurred by companies that engage in international operations:

- It explains the economic theory underlying the operation of international financial markets.
- It examines the three forms of currency risk: translation risk, transaction risk and economic risk.
- It explains how firms can manage these risks by adopting hedging techniques internal to the firm's operations.
- It explains how firms can use the financial markets to hedge these risks externally.

 *Complete your diagnostic test for Chapter 21 now to create your personal study plan.*

21.1 INTRODUCTION

With the huge growth in world trade over the last few decades, companies increasingly deal, as buyer, seller or investor, in foreign currency, making it a key factor in financial management. For competitive reasons, exporters are commonly obliged to invoice in the customer's currency – the greater the strength of competition from exporters based in other countries, the greater the likelihood of a UK exporter having to accept the foreign exchange risk.

Foreign currency can change in value relative to the home currency to significant degrees over a short time.

Such changes can seriously undermine the often wafer-thin profit margin of a trader, say, a Japanese car exporter awaiting payment in foreign currency. If the yen appreciates, the yen value of the deal can evaporate before its eyes, while the likelihood of repeat business diminishes unless it lowers price, i.e. takes a smaller profit margin in yen terms. It is easy to understand the concern of a major exporter like Toyota, a great proportion of whose export trade is priced in dollars.

Yen rise to have 'big impact' on Toyota

Toyota faces tough conditions caused by an excessive appreciation of the yen and an unexpectedly steep downturn in the US, the chief executive of Japan's largest carmaker said yesterday.

"The yen has strengthened too much and will have a big impact on us," said Katsuaki Watanabe, as the currency jumped from about Y105 to the dollar just days ago to a more than three-year high of almost Y102 yesterday.

Many analysts expect the yen to strengthen above Y100 to the dollar, a level which, if breached, could lead to a further rapid appreciation.

A rapid rise of the yen to more than Y100 to the dollar could tip the Japanese economy into recession, some analysts say. "If it went to Y95 [to the US dollar] and stayed there, that would probably shave another two-tenths to four-tenths off real growth," said John Richards, manag-

ing director at RBS Securities in Tokyo.

Mr Watanabe said that it would be difficult to offset the effect of the yen's rapid rise in spite of Toyota's legendary ability to cut costs and buoyant demand from emerging markets.

Every rise of Y1 to the US dollar cuts Toyota's annual operating profits by Y35bn ($341bn), said the company.

Mr Watanabe's grim assessment of the challenges facing the carmaker raises concerns that Toyota could revise down its forecasts if the US economy deteriorates further.

The Toyota chief indicated that North American vehicle demand this year could be weaker than the group had expected last year, although he suggested the second half of the year should see an improvement.

Mr Watanabe's downbeat assessment of the outlook for the year contrasts with Toyota's confidence just

weeks ago that it could weather the current global downturn.

Toyota, which is competing with General Motors for the title of world's largest carmaker, had until recently indicated strongly that it did not expect a significant drop in North American sales.

Toyota also indicated it would be able to offset the US slump with increased sales in emerging markets.

Just last month, the group posted record third-quarter profits as strong sales in emerging markets offset a US slowdown.

Unit sales in China increased 62 per cent in the third quarter.

Even in North America, where sales in the third quarter were down, Toyota achieved record sales in the calendar year for the 12th consecutive year.

Source: Michiyo Nakamoto, *Financial Times*, 8 March 2008.

In this chapter, we explain both the theory of foreign exchange markets and also how they work in practice, and how exporters and importers can protect themselves against the risks of foreign exchange rate variations. There are two key issues for the treasurer of a company with significant foreign trading links to address:

1 *Whether* to seek protection against these variations, i.e. to '*hedge*', or to ride the risks, on the basis that in the long-term they will even out. Most companies do seek hedges, being risk-averters. Yet some actively seek out foreign exchange risk, and use dealing opportunities as a source of profit by deliberately taking 'positions' in particular currencies. BP plc (**www.bp.com**) for example, exploits its position as a multinational with a substantial two-way flow in several currencies to operate its currency dealing activity as a separate profit centre. Such companies are called '*speculators*'.

hedge
A hedge is an arrangement effected by a person or firm attempting to eliminate or reduce exposure to risk – hence to hedge and hedger

2 The second issue concerns the *extent* to which the firm wants to **hedge** – whether to totally avoid exposure to exchange rate risk or to control the degree of exposure.

Some firms attempt to eliminate their exposures by matching their operating inflows in a foreign currency with operating outflows in that currency to achieve a perfect hedge. However, it is difficult to keep these two flows perfectly in unison due to timing differences in receipt and disbursement of cash.

A variation on this policy is to match operating cash flows with borrowings in the same currency. Such an approach is adopted by Compass Group plc. Its 2006 Annual Report stated that:

> The Group's policy is to match its principal cash flows by currency to . . . borrowings in the same currency. As currency earnings are generated, they are used to service and repay debt in the same currency. For the period of the currency loans, therefore, the objective is to achieve an effective foreign currency hedge in real economic terms.

Some firms actively court foreign exchange risks. In October 2003, Nintendo, the Japanese videogame producer, reported its first-ever loss of £16 million, largely as a result of the strength of the yen. Nintendo kept much of its foreign earnings in local currencies to take advantage of better interest rates outside Japan. This policy resulted in losses of some £215 million on foreign currency transactions as the yen rose strongly against the US dollar. In the following year, BMW announced that it would stop hedging USD income, gambling on a fall in the euro from its level of $1.30 per euro. By 1 April 2008, the euro stood at $1.56 (see the cameo below). We will see later why such policies are misguided.

The task of this chapter is to explain the various types of exchange risk and the various ways how they can be managed.

BMW bets on rebound for falling US dollar

BMW, the German luxury carmaker, has stopped all long-term hedging of the dollar, seeing an end to the US currency's two-year decline.

The company is one of Europe's heaviest users of currency hedging to protect its revenues from volatile foreign exchange markets. But it now believes the US currency is "significantly" undervalued and must bounce back.

The dollar has fallen by 29 per cent against the euro in the past two years, pricing many European exporters out of US markets. As the US currency approached the $1.30 mark against the euro earlier this year, European politicians clamoured for a cut in interest rates to make the eurozone more competitive.

BMW said it believed the "correct" value for the dollar was $1.10 to the euro compared with $1.22 – the level it reached in late trading yesterday.

But the carmaker could be premature in its belief in a dollar rebound as

few strategists are confident of a dollar bounce in the near-term, and currency traders remain concerned about the twin US deficits.

Bob Sinche, head of currency strategy at Citigroup, said the "panic mentality" that set in as the dollar fell last year was diminishing, but few companies seemed ready to go completely unhedged. "We have not seen a lot of discussion [from companies] about whether the process of dollar weakening has come to an end," he said. "The general notion remains one of concern about the dollar on a medium-term basis, and corporates are using periods of dollar strength to put on some hedging."

BMW said it was limiting its use of derivatives to protect against the weak dollar to short-term "buying on the dips".

"We think that the euro will go down again," said Stefan Krause, finance director. "In such a period of significant under-valuation of the

US dollar it is important to remain consistent and to have the courage not to hedge at unattractive currency rates."

Hedging the dollar has become important to BMW because the US last year passed Germany as the company's largest market. But the strength of the euro against the dollar is also a wider issue for the German economy.

Mr Krause said BMW remained "widely" hedged this year, with between two-thirds and all of the US turnover covered. He also said the company had other hedging options, such as cutting the allocation of vehicles to sell.

The company still has short-term hedges in place for next year, but surprised analysts by saying it had not increased these beyond the one-third of turnover already covered.

Source: James Mackintosh and Steve Johnson, *Financial Times*, 18 March 2004.

Self-assessment activity 21.1

What is the distinction between a foreign exchange 'speculator' and a 'hedger'? How would you describe Nintendo and BMW?

(Answer in Appendix A at the back of the book)

21.2 THE STRUCTURE OF EXCHANGE RATES: SPOT AND FORWARD RATES*

Most currency transactions are conducted between firms and individuals on one hand, and banks which make a market (i.e. quote an exchange rate in a variety of currencies) on the other. As in any other market, the two parties set a price – in this case, the exchange rate is the price of one currency in terms of another. There are two ways of quoting the resulting price, which is often a source of confusion:

- **The direct quote** gives the exchange rate in terms of the number of units of the home currency required to purchase one unit of the foreign currency.
- **The indirect quote** gives the price in terms of how many units of the foreign currency can be bought with one unit of the home currency.

In London, dealers usually use the indirect quote, (although this is changing). When we hear that the sterling/US dollar exchange rate (the so-called 'cable rate') is $2.00, this means that each pound can buy two units of the 'greenback', the US dollar. The corresponding direct quote would be £0.50 which indicates how many units of sterling that one US dollar can purchase. The direct quotation is simply the reciprocal of the indirect quotation.

In continental Europe, the direct quotation is used. In the USA, dealers generally use the indirect quotation when dealing with European banks, except for ones in London.

It is also misleading to talk of '*the* exchange rate' between currencies because there always exists a spectrum of rates according to when delivery of the currency traded is required.

The simplest rate to understand is the **spot market rate** that the bank quotes for 'immediate' (in practice, within two days) delivery. For example, on 28 August 2007, the closing quotation for the spot rate for Swiss Francs (CHF) against sterling (GBP) was

$$2.4100 - 2.4110$$

The first figure is the rate at which the currency can be purchased from the bank and the higher one is the rate at which the bank sells CHF. The difference (0.10 centimes), or **spread**, provides the bank's profit margin on transactions. At times of great volatility in currency markets, the spread usually widens to reflect the greater risk in currency trading.

It is also possible to buy and sell currency for delivery and settlement at specified future dates. This can be done via the **forward market**, which sets the rate applicable for advance transactions. On the above day, the following terms were quoted for CHF delivery in one month:

$$77 - 63 \text{ pr}$$

The numbers are referred to as 'points' with each point representing 1 per cent of a centime, or 0.0001 of a CHF. The 'pr' means that the CHF is selling at a *forward premium*, i.e. it is 'predicted' to appreciate versus sterling.

---

*Throughout the following sections, we use standard international abbreviations for currencies (based on SWIFT money transmission codes), e.g. pound sterling = GBP, US dollar = USD, etc. We also frequently use the abbreviation FX to denote foreign exchange (rates).

The quotation given is not an exchange rate as such, but a 'prediction' of how the CHF spot exchange rate will change over the relevant period: in this case, appreciate against sterling. The rate itself (called an *'outright'*) is found by deducting the expected premium from the spot rate (or adding a discount to it). In this case, subtraction is required because the market expects that one unit of sterling will purchase fewer CHF in the future, i.e.

Spot	2.4100 − 2.4110
F/w premium	(0.0077 − 0.0063)
F/w outright	2.4023 − 2.4047

Notice that the spread widens from 0.10 centimes (or 10 points) to 0.24 centimes (24 points). This is a reflection of the greater risk associated with more distant transactions. An important point to note is that, when a forward transaction is entered into, there exists a contractual obligation to deliver the currency that is legally binding on both parties. The rate of exchange incorporated in the deal is thus fixed. Hence, a forward contract is a way of locking in a specific exchange rate, and is appealing when there is great uncertainty about the future course of exchange rates.

From spot to forward

Spot and forward rates for other currencies against GBP are thus connected as follows:

$$\text{Forward rate} = \text{spot rate} \begin{cases} \textit{plus} \text{ forward discount} \\ \qquad \text{OR} \\ \textit{minus} \text{ forward premium} \end{cases}$$

Forward rates, therefore, appear to be an assessment of how the currency market expects two currencies to move in relation to each other over a specified time period, and are sometimes regarded as a prediction of the future spot rate at the end of that period. As we shall see, this is not entirely a correct interpretation.

The reader may wish to visit the website (**www.bis.org**) of the Bank of International Settlements (BIS) for statistics on the volume of trading on these markets. The BIS conducts a tri-ennial survey of foreign currency trading activity on a specified trading day. In April 2007, 54 central banks and monetary authorities participated. The average *daily* turnover in April 2007 was US$1,005 billion in spot transactions and US$362 billion in outright forward transactions (BIS, 2007).

By currency, 43 per cent of trades involved the US$, 18.5 per cent the euro, 8 per cent the Japanese yen and 7.5 per cent sterling. The most frequent trading pair was US$/euro, while by location 34 per cent of activity was conducted via London, 17 per cent New York, and 6 per cent for each of Tokyo and Singapore.

Self-assessment activity 21.2

The closing spot rates and forward quotations on 28 August 2007 for GBP versus two other currencies were as shown below. Calculate the forward outrights.

	Closing rates	Forward quotation (1 month)
Eurozone	1.4721 − 1.4729	26 − 25 pr
Canada	2.1315 − 2.1323	24 − 15 pr

Source: The Times, 29 August 2007.

(*Answer in Appendix A at the back of the book*)

21.3 FOREIGN EXCHANGE EXPOSURE

Foreign exchange exposures occur in three forms:

1 Transaction exposure
2 Translation exposure
3 Economic exposure

■ Transaction exposure

Transaction exposure is concerned with the exchange risk involved in sending money over a currency frontier. It occurs when cash, denominated in a foreign currency, is contracted to be paid or received at some future date.

For example, a UK company might contract to buy US$45 million worth of computer chips from a US company over a three-year period. When the contract is set up, the rate of exchange between the dollar and the pound is US$2.00 to £1, but what will happen in a year or two's time? What if the rate of exchange alters to US$1.75 to £1 in a year's time?

The US$45 million was equivalent to $45 m/2.00 = £22.50 m at the beginning of Year 1, but after the fall in the value of the pound against USD, the cost of the contract in GBP rises to $45 m/1.75 = £25.71 m. Such a substantial rise in costs could easily eliminate the UK company's profit margin.

Similar risks apply to expected cash inflows. If the UK company was due to receive 50 million Canadian dollars (CAD) and the CAD actually rose from C$2.2 to £1 to C$2.0 to £1, the UK company would gain £2.28 million on the contract (i.e. the difference between the expected income of £22.72 m (C$50/2.2) and the actual income of £25 m (C$50/2.0)).

Thus, unexpected changes in exchange rates can inflict substantial losses (and provide unexpected gains) unless action is taken to control the risk.

■ Translation exposure

translation exposure
Exposure to the risk of adverse currency movements affecting the domestic currency value of the firm's consolidated financial statements

Translation exposure is the exposure of a multinational's consolidated financial accounts to exchange rate fluctuations. If the assets and liabilities of, say, the Australian subsidiary of a UK parent firm are translated into sterling at year-end at a rate different from the start-year rate, exchange losses or gains will be reflected in the new balance sheet, and will also affect the profit and loss account. Similarly, the earnings of the subsidiary when translated into sterling are also affected by exchange rate changes.

Whereas transaction exposure is concerned with the effect on *cash flows* into the parent company's currency, translation exposure affects *balance sheet values*, and to a lesser extent (because assets typically exceed profits or cash flow in magnitude) the profit and loss account.

Examples of items that a treasurer might consider to be subject to translation exposure if denominated in foreign currency are debts, loans, inventory, shares in foreign companies, land and buildings, plant and equipment, as well as the subsidiary's retained profits.

Not everyone accepts that this risk is important. If the CAD falls in value by 3 per cent between the date an export contract is signed and the date the dollars are received in the UK, this represents a real loss to the UK company if no action is taken to hedge the exchange risk. But is a real loss sustained by a UK company with a Canadian subsidiary if C$30 million of its capital stock or C$10 million of its inventory are being held in Toronto at the time of a devaluation of the CAD against GBP? This question has been much debated during the last 30 years or so.

It is often argued that translation risk is a purely accounting issue, i.e. it relates to past transactions, so it has no impact on the economic value of the firm and thus there is no need to hedge, i.e. people already know about it in an efficient market. However, it may become a problem if there are plans to realise assets held overseas and/or if earnings cannot be profitably reinvested in the location where they arise, and the parent wishes to repatriate them. (Arguably, these upcoming cash movements essentially reflect a transaction exposure rather than a translation exposure.) Moreover, a policy of 'benign neglect' tends to overlook possible effects on key performance measures and ratios, especially EPS, in relation to reporting overseas earnings, and gearing, via reported asset and liability values.

A multinational company may have significant borrowings in several currencies. If foreign currencies have been used to acquire assets located overseas, then, should the GBP decline in value, any adverse effect on the GBP value of borrowing will be offset by a beneficial effect on the sterling value of overseas assets. In this respect, the overseas borrowing is 'naturally' hedged, and no further action is required.

However, the UK company may face limits on its total borrowing which could be violated by adverse foreign exchange rate movements. For example, a weaker domestic currency, relative to currencies in which debt is denominated, could adversely affect borrowing capacity and the cost of capital.

Say a company has debt expressed in both GBP and USD, as in the following capital structure:

	£m
Equity	350
Loan stock: sterling	50
Loan stock: (US$80 m)	40
Total	440

The valuation of the USD loan is translated at the exchange rate of $2.00:£1, the rate ruling at the end of the financial year. At this juncture, the gearing ratio (debt-to-equity) is (£90 m/£350 m) = 25.7%. Imagine there is a covenant attaching to the sterling loan which limits the gearing ratio to 30 per cent. If GBP falls to, say, $1.50:£1, the company has a problem. Its USD-denominated debt now represents a liability of $80 m/1.50 = £53.33 m, and the debt-to-equity ratio rises to:

$$(\text{£50 m} + \text{£53.33 m}) \div \text{£350 m} = \frac{\text{£103.33 m}}{\text{£350 m}} = 29.5\%$$

The firm is now on the verge of violating the covenant. To avoid this situation occurring, the company could borrow in a range of currencies that might move in different directions relative to GBP, with adverse movements offset by favourable ones. For example, Compass Group plc borrows in yen, euros and US dollars as well as in sterling (see later), thus mixing a so-called 'currency cocktail'.

■ Economic exposure

Economic exposure is also known as long-term cash flow or operating exposure. Imagine a UK company which buys goods and services from abroad and sells its goods or services into foreign markets. If the exchange rate between sterling and foreign currencies shifts over time, then the value of the stream of foreign cash flows in sterling will alter through time, thus affecting the sterling value of the whole operation.

In general, a UK company should try to buy goods in currencies falling in value against GBP and sell in currencies rising in value against GBP.

Of course, the transactions exposure could be eliminated by denominating all its contracts in GBP, which shifts the risk to the trading partner. However, this tactic cannot remove economic exposure. The foreign company will convert the GBP cost of purchases and sales into its own currency for comparison with purchases or sales from companies in other countries using other currencies. Management of economic exposure involves looking at long-term movements in exchange rates and attempting to hedge long-term exchange risk by shifting out of currencies that are moving to the detriment of the long-term profitability of the company. It is worth noting that many economic exposures are driven by political factors, e.g. changes in overseas governments resulting in different economic policies such as taxation.

Self-assessment activity 21.3

Distinguish between translation, transaction and economic exposure.

(Answer in Appendix A at the back of the book)

21.4 SHOULD FIRMS WORRY ABOUT EXCHANGE RATE CHANGES?

According to the theory of **Purchasing Power Parity** (PPP), the answer to this question is 'no'.

PPP says that the purchasing power of any currency should be equivalent in any location. It is based on the **Law of One Price**, which asserts that identical goods must sell at the same price in different markets, after adjusting for the exchange rate. For example, if the market rate of exchange between USD and GBP is $2.00:£1, a microcomputer could not sell for very long at simultaneous prices of, say, £1,500 in London and $2,000 (i.e. £1,000) in New York. People would buy in the 'cheap' market (New York) and ship the goods to London, thus tending to equalise the two prices at, say, a London price of £1,200 and a New York price of $2,400 (£1,200). (In reality, transport and other transaction costs may prevent the precise operation of PPP.)

The Law of One Price states that, for tradeable goods and services, the

$$(£ \text{ price of a good} \times \$/£ \text{ exchange rate}) = \text{USD price of a good}$$

However, part of the adjustment will occur via the effect on the exchange rate itself.

■ Absolute and relative PPP

In fact, the Law of One Price only applies under a rarified form of PPP called absolute PPP (APPP). For the law to apply (i.e. for prices of similar products to be equal after adjusting for currency values), resources would have to be perfectly mobile so that the levels of supply and demand should equalise at just the appropriate level. In reality, the existence of transport costs, different local taxes such as VAT, barriers to trade such as tariffs, quotas and other government restrictions, not to mention sheer inertia, prevent APPP from existing, and price discrepancies will persist. One only has to think about land and property to appreciate that some resources are quite difficult to move (although titles to land can be traded).

In practice, market theorists put their faith in a more limited theory, the relative PPP theory (RPPP), which allows for all the market imperfections listed above. RPPP theorists accept that the Law of One Price will probably not hold in its purest form, but it suggests that rates of change of prices will be similar when expressed in common currency terms, for a given set of trade restrictions. Take the previous example, with the computer selling at two different prices in New York and London ($2,000 and £1,500 respectively, or in terms of GBP, at the ruling exchange rate of $2.00 = £1, £1,000 and £1,500). It may be that this pair of prices reflects the market equilibria after all scope for price equalisation is exhausted. It should be noted that the prices infer a 'correct' exchange rate of:

US price/UK price = $2,000/£1,500 = US$1.33 per £1

This compares to the market rate of US$2.00 = £1. This implies the USD is under-valued or, to say the same thing, that the GBP is over-valued.

Obviously, the Law of One Price does not apply here as the purchasing power of sterling is lower in London (where £1,000 buys only 2/3 of a computer) than in New York (where £1,000 buys a whole computer).

RPP focuses on the differential in inflation through time – if prices rise faster in one location, say London, the local currency would decline in relation to others such as the USD. If, say, prices rose by 10 per cent in London and only 5 per cent in the USA, the local prices for the computer would be £1,500(1.1) = £1,650 and $2,000(1.05) = $2,100 respectively. To preserve the purchasing power of both currencies, the exchange rate would have to alter to:

2.00 × US inflation/UK inflation = 2.00 × (1.05)/(1.10) = 1.91

This reflects a sterling depreciation of about 5 per cent, the inflation differential. Thus RPPP involves the assertion that a currency will fall (or is expected to fall) in relation to another according to the difference in actual (or expected) inflation rates between the two countries.

■ Exchange rate changes

If foreign exchange markets operate freely without government intervention, goods that can be easily traded on international markets, such as oil, are highly likely to obey the Law of One Price, although transport costs between markets may explain a continued price discrepancy. However, not all goods can be easily transported. Most notably, with land and property, which are physically impossible to shift, a sustained price discrepancy may apply between markets. In the longer term, however, even these differentials may close as investors and property speculators perceive that one market is cheap relative to the other.

PPP may be expected to operate broadly in the longer term for most goods and services, although it can be distorted by government intervention in the foreign exchange markets and the formation of currency blocs. The authorities in these cases are attempting to smooth out the effects and hence minimise the dislocation to business activity that sudden swings in currency values might cause. However, while exchange controls and official intervention can delay any adjustment necessary to reflect differential rates of inflation, the required change will eventually take place.

Accepting PPP and the Law of One Price, we arrive at a remarkable conclusion regarding the need to hedge FX risks – there is no need to worry! The Stonewall plc example explains the mechanics.

Example: Stonewall plc

A British-based firm, Stonewall plc has a factory in Baltimore, USA. It plans to produce and sell goods to generate a net cash inflow of $180 million at today's prices over the coming trading year. For simplicity, we assume all transactions are completed at year-end, and that any price adjustments resulting from inflation also occur at year-end.

At the current exchange rate of US$2.00 vs. £1, the sterling value of its planned sales = ($180 m/2.00) = £90 m. Stonewall is worried about the USD falling due to the annual rate of inflation in the US of 6 per cent compared to 3 per cent in the UK.

Concern about exposure to foreign exchange risk seems justified – with these inflation rates, PPP predicts the USD will decline to:

$2.00 × (1.06/1.03) = $2.058 after one year.

At this exchange rate, the sterling value of the USD cash flow is ($180 m/2.058) = £87.46 m, a fall of about 3 per cent on the start-year valuation. But should sleep be lost over this?

The answer is 'yes' if selling prices within the USA remained static. However, prices within the USA are not static – the reason why the FX rate will change is due to inflation at a higher rate in the USA relative to the UK.

With US prices rising at 6 per cent, the US$ cash flow *ought* to rise to $180 m (1.06) = $190.8 m. Converted to sterling at the year-end rate, this is worth ($190.8 m/2.058) = £92.70 m. This is precisely equal to its sterling-denominated value at the end of one year with UK price inflation at 3 per cent (£90 m × 1.03 = £92.70 m).

So what has been lost from inflation affecting the relative value of sterling and $US? The answer is nothing if PPP operates! Should the firm take precautions against FX exposure? The answer is 'no' – why should it bother when it is automatically protected by market adjustments? Should the firm try to forecast future rates of exchange, e.g. by comparing the respective inflation rates? It could, but again, it is a waste of time, at least in theory, as the rate of $2.058 should already be quoted in the market for one-year forward deals.

However, it is not always this simple. In reality, prices rise in a continuous process rather than in a series of end-year adjustments. The policy of benign neglect only works if prices of the traded goods are adjusted *pari passu* as prices in general alter and the exchange rate 'crawls' in the appropriate direction, by the appropriate amount, and if the movement is synchronised.

In reality, FX rates adjust in response to relative inflation rates at the national level, as measured by a basket of goods. The basket may well inflate at a different rate from the goods traded. Indeed, competitive conditions (i.e. strategic considerations) may be so powerful that firms may be unable to raise prices even to compensate for inflation. For these reasons, most firms seek protection against FX movements.

21.5 ECONOMIC THEORY AND EXPOSURE MANAGEMENT

The first step in currency management is to identify the transaction, translation and economic exposure to which the company is subject. The second step is to decide how the exposure should be managed. Should the risk be totally hedged, or should some degree of risk be accepted by the company?

The international treasurer must devise a hedging strategy to control exposure to exchange rate changes. The precise strategy adopted is likely to be influenced by several economic theories that have evolved over the last century, and the extent to which they are considered valid. These theories are as follows:

1 The Purchasing Power Parity Theory (PPP).
2 The Expectations Theory.

3 The Interest Rate Parity Theory (IRP).
4 The Open, or International, Fisher Theory.
5 The international version of the efficient markets hypothesis (EMH).

We will provide brief sketches of these important contributions to the literature of international economics.

■ Purchasing Power Parity (PPP)

In the last section, we encountered the Law of One Price and Purchasing Power Parity. PPP and the Law of One Price have important implications for the relationship between spot and forward rates of exchange. If people possessed perfect predictive ability, and the rates of inflation were certain, the market could specify with total precision the appropriate exchange rate between USD and GBP for delivery in the future (i.e. the *forward rate* of exchange).

More specifically, PPP states that foreign exchange rates will adjust in response to international differences in inflation rates and so maintain the Law of One Price. Thus the forward rate should be:

$$Forward\ rate\ =\ Spot\ rate\ \times\ \frac{(1\ +\ US\ inflation\ rate)}{(1\ +\ UK\ inflation\ rate)}$$

If the spot rate between sterling and the US$ is $2.00 vs. £1, and people expect UK inflation at 10 per cent and only 3 per cent in the USA, this implies a one-year *forward rate* of:

$2.00:£1$ spot rate $\times (1.03)/(1.10) = $1.87:£1$

■ Expectations Theory

In the above example, the forward rate is predicting the spot rate that *should* apply in the future. If buyers and sellers of foreign exchange can rely on the currency markets to operate in this way, the risks presented by differential inflation rates could be removed by using the forward market. Forecasting future spot rates would then be a trivial exercise.

Unfortunately, the forward rate has been shown to be a poor predictor of the future spot rate. Yet it has also been shown to be an **unbiased predictor** in that, although the forward rate often underestimates and often overestimates the future spot rate, it does not consistently do either. In the long run, the differences between the forward rate's 'prediction' for a given date in the future and the actual spot rate on that date in the future should sum to zero. If the forward market operates in this way, firms can regard today's forward rate as a reasonable expectation of the future spot rate. This is the **Expectations Theory**.

Levich (1989) found that in the early 1980s the forward rate of GBP vs. USD tended to underestimate the strength of the USD, but during 1985–87, the forward rate overestimated the strength of the dollar. However, taking the 1980s as a whole, the data suggested that the forward rate on average was very close to the future spot rate.

Self-assessment activity 21.4

Use the Law of One Price and PPP to predict the relative local prices of a cup of coffee and the future sterling/dollar spot rate under the following conditions:

■ Price now in New York = $2.00
■ Price now in London = £1.00
■ Exchange rate for USD vs. GBP = $2.00:£1
■ UK inflation is 4 per cent; US inflation is 2 per cent

(Answer in Appendix A at the back of the book)

■ Interest Rate Parity (IRP)

Interest Rate Parity is concerned with the difference between the spot exchange rate (the rate applicable for transactions involving immediate delivery) and the forward exchange rate (the rate applicable for transactions involving delivery at some future specified time) between two currencies. Suppose the spot rate for USD to GBP is $2.00:£1, and the one-year forward rate is $1.85:£1. Here, the USD is selling at a 15 cent premium – it is more expensive in terms of GBP for forward deals. The currency market thus expects the USD to rise in value against GBP during the year by about 7.5 per cent.

IRP converts this expected rise in the value of the USD against GBP into a difference in the rate of interest in the two countries. The rate of interest on one-year bonds denominated in USD will be lower than bonds otherwise identical in risk, but denominated in GBP. The difference will be determined by the premium on the forward exchange rate. If depreciation of GBP against USD is expected, this should be reflected in a comparable interest rate disparity as borrowers in London seek to compensate lenders for exposure to the risk of currency losses. In other words, interest rates offered in different locations tend to become equal, to compensate for expected exchange rate movements.

The equilibrium relationship that operates under IRP is given by:

$$Forward\ rate = \text{Spot rate} \times \frac{(1 + \text{US interest rate})}{(1 + \text{UK interest rate})}$$

For example, if the interest rate available in London is 12 per cent p.a., the figures in our example will indicate a US interest rate as follows:

$$(1 + \text{US interest rate}) = 1.12 \times \frac{1.85}{2.00} = 1.036$$

So the US interest rate is $(1.036 - 1) = .036$ (i.e.) 3.6% p.a.

This is an interesting result. A New Yorker attracted by high UK interest, who is tempted to place money on deposit for a year in London, will find that what is gained on the interest rate differential will be lost on the adverse movement of GBP against USD over the year. To appreciate this 'swings and roundabouts' argument, consider the following figures, which relate to the two investment options faced by a US investor wanting to deposit $1,000:

1 *Invest in GBP:*

January	Convert $1,000 into GBP @ 2.00 = £500.
	Invest for one year in London at 12%:£500(1.12) = £560.
December	Convert back to USD @ 1.85 = $1,036.

vs.

2 *Invest in USD:*

January   Invest $1,000 in New York @ 3.6% = $1,036 in December.

Clearly, the rational investor should be indifferent between these two alternatives, unless interest rates are expected to fall in New York relative to those in London, or the forward rate is not a good predictor of the spot rate in one year's time.

One reason why this predictive ability is weakened in practice is intervention in foreign exchange markets by governments. In the absence of such intervention, exchange rates seem to operate so as to smooth out interest rate disparities, but with the creation of artificial market inefficiencies, there often exist opportunities to arbitrage: for example, borrowing money at low interest rates in one market, hoping to repay it before IRP fully exerts itself. However, in the past, many UK corporate treasurers were wrong-footed by borrowing apparently cheap money overseas, but having to repay at exchange rates quite different from those envisaged when raising the loan, because market forces have eventually asserted themselves to remove the interest rate discrepancy.

arbitrageurs
Arbitrageurs attempt to exploit differences in the values of financial variables in different markets e.g. borrowing in a low-cost location and investing where interest rates are relatively high (interest arbitrage)

This equalising process is effected by financial operators called **arbitrageurs**, who act upon any short-term disparities. For example, if in the previous example the interest rate disparity were 3 per cent, it would pay to borrow in GBP and purchase US bonds in London.

Checking the agios: the scope for arbitrage

When currency and money markets are in equilibrium, any difference in interest rates available through investment in two separate locations should correspond to the differential between the spot and forward rates of exchange. The interest rate differential is called the **interest agio**, and the spot/forward differential is called the **exchange agio**. If these are not equal, arbitrageurs have scope to earn profits.

Consider this example. An investor has £1 million to invest for a year. The interest rate is 5 per cent in London and 8 per cent in New York. The current spot rate of exchange (ignoring the spread) is $2.00:£1 and the dollar sells at a one year forward discount of 5 cents, i.e. the forward outright is $2.05:£1. What is the best home for the investor's money?

He could invest the £1 million on deposit in London to earn £50,000 interest over one year, thus increasing his cash holding to £1.05 million. Alternatively, he could engage in **covered interest arbitrage**. This works as follows:

covered interest arbitrage
Using the forward market to lock in the future domestic currency value of a transaction undertaken to exploit an interest arbitrage opportunity

1 Convert £1 m at spot into USD, i.e.

£1 m × 2.00 = $2.00 m

2 Invest $2.00 m at 8 per cent for one year in the USA, i.e.

$2.00 m × 1.08 = $2.16 m

3 Meanwhile, sell this forward over one year, i.e. for delivery in one year:

$2.16 m/2.05 = £1,053,658

The guaranteed proceeds from arbitrage are greater by £3,658. However, this so-called 'carry trade' cannot last for very long. As other investors spot the scope for risk-free profits and rush into the market, their actions will quickly eliminate the opportunity. This is why spot/forward relationships almost always reflect prevailing interest rate differentials.

For this reason, the forward rate is the product of a technical relationship linking the spot rate to relative interest rates, rather than a prediction in the true sense.

Equilibrium requires equality between the exchange agio and the interest agio, i.e. the spot/forward differential should equal the interest rate differential:

$$\frac{F_0 - S_0}{S_0} = \frac{i_\$ - i_\pounds}{1 + i_\pounds}$$

where F_0 is today's forward quotation, S_0 is today's spot quotation, $i_\$$ is the interest rate available by investment in USD, and $i_\pounds$ is the interest rate available by investment in GBP.

Note that the interest agio is found by discounting the interest differential over one year at the UK interest rate. If the period concerned were less than a year – say, three months – the equivalent three-monthly interest rate would be used.

In the above example, the two agios are:

$$\frac{2.05 - 2.00}{2.00} \quad \text{vs.} \quad \frac{0.05 - 0.08}{1.08}$$

i.e.:

−2.5% vs. −2.78%

uncovered arbitrage
Interest arbitrage without the use of the forward market to lock in future values of proceeds

This inequality signifies the scope for risk-free profit via covered interest arbitrage. **Uncovered arbitrage** is where the arbitrageur does not sell forward, but takes a gamble on how the spot rate changes over the year. In the example, he or she would earn bigger profits if the spot rate in one year turned out to be lower than \$2.05 : £1 (e.g. \$2.02). This distinction highlights the difference between hedging and speculation. However, although differences in agios can persist for a while, transactions costs may preclude profitable arbitrage (but see the cameo below!).

Self-assessment activity 21.5

If interest rates are higher in London than New York by 2.5 per cent p.a. and today's spot rate is \$2.0250 vs. £1, what would you expect the three-month forward quotation to be if IRP applied?

(Answer in Appendix A at the back of the book)

This shouldn't happen!

According to IRP, if the interest rate in one location is higher than in another one, this is due to inflation differentials between the two countries concerned, which will soon result in a decline in the relative value of the first currency. But if for a given rate of inflation, interest rates are higher in, say, Country A than country B, then it pays to borrow in B and deposit in A and exploit the differential. However, according to IRP, any short-term gains would be wiped out as the currencies alter in relative value – i.e. the currency in country A will eventually fall as the greater influx of money leads to inflationary pressures.

Yet in 2006–7, what should happen in theory just has not happened in practice. Many investors have made money by the so-called 'carry-trade' – they borrow at low rates of interest in one currency and invest in higher rates in another. The most common carry trade in recent years has been in yen. With interest rates in Japan barely above zero, specula-

tors have been borrowing there to deposit in countries like the UK and the US, where rates were above 5 per cent for long periods. Another currency to have 'benefited' from this trade has been the New Zealand dollar (NZ\$) which reached its highest level since it floated in 1985, causing the Reserve Bank of New Zealand to intervene by selling its own currency.

The risk incurred by carry-traders is that the exchange rate will move against them, as indeed eventually did happen during the financial turmoil of summer 2007, when investors unloaded risky positions and fled into yen, causing the yen to surge 10 per cent in as many days.

(If the reader consults the Big Mac Index on Page 635, s/he will see that yen was undervalued by 33 per cent against the US\$ – at least some of this would have been caused by speculative carry-trading.)

Galati *et al.* (2007) explain popular carry-trading strategies.

■ The Open Fisher Theory

The 'Open Fisher' Theory, sometimes called the 'International Fisher' Theory, claims that the difference between the interest rates offered on identical bonds in different currencies represents the market's estimate of the future changes in the exchange rates over the period of the bond. The theory is particularly important in the case of fixed-rate bonds having a long life to maturity, say, five to fifteen years' duration.

Suppose that a firm wishes to raise £50 million for a one-year period. It approaches a bond broker and is offered the following loan alternatives:

1 A loan in GBP at 12 per cent p.a.
2 A USD loan at 5 per cent p.a.

The Open Fisher Theory asserts that the interest rate difference represents the market's 'best estimate' of the likely future change in the exchange rates between the currencies over the next year. In other words, the market expects GBP to depreciate by around 7 per cent against USD over the next year.

But it does! 'Carry on Speculating'

No comment on the financial markets these days is complete without mention of the "carry trade", the borrowing or selling of currencies with low interest rates and the purchase of currencies with high rates. The trade is often blamed for the weakness of the Japanese yen and the unexpected enthusiasm of investors for the New Zealand and Australian dollars.

But why does the carry trade work? In theory, it shouldn't – or not for as long as it has. Foreign-exchange markets operate under a state of "covered interest parity". In other words, the difference between two countries' interest rates is exactly reflected in the gap between the spot, or current, exchange rate and the forward rate. High-interest-rate currencies are at a discount in the forward market; low-rate currencies at a premium.

If that were not so, it would be possible for a Japanese investor to sell yen, buy dollars, invest those dollars at high American interest rates for 12 months and simultaneously sell the dollars forward for yen to lock in a profit in a year's time. The potential for arbitrage means such profits cannot be earned.

However, economic theory also suggests that "uncovered interest parity" should operate. Countries that offer high interest rates should be compensating investors for the risk that their currency will depreciate. In other words, the forward rate should be a good guess of the likely future spot rate.

In the real world, uncovered interest parity has not applied over the past 25 years or so. A recent academic study has shown that high-rate currencies have tended to appreciate and low-rate currencies to depreciate, the reverse of theory. Carry-trade strategies would have brought substantial profits, not far short of stockmarket returns, although dealing costs would have limited the size of the bets traders could make.

Academics have struggled for some time to explain this discrepancy. One possibility is that investors demand a risk premium, separate from the better interest rate, to compensate them for investing in a foreign currency. As this risk premium varies, it might overwhelm the effects of interest-rate changes. For example, American investors might worry about the credibility of the Bank of Japan, but Japanese investors may regard the dollar as a "safe haven". This would drive the dollar up and the yen down.

However, according to Andrew Scott, of the London Business School, it has been a struggle to find risk premiums that are large enough to explain exchange-rate volatility. So academics have been looking at the structure of foreign-exchange markets, to see if behavioural factors might be at work.

One obvious possibility is that the actions of carry traders are self-fulfilling; when they borrow the yen and buy the dollar, they drive the former down and the latter up. If other investors follow "momentum" strategies – jumping on the bandwagon of existing trends – this would tend to push up currencies with high interest rates.

Financial jaywalking

Such a strategy has its dangers. It has been likened to "picking up nickels in front of steamrollers": you have a long run of small gains but eventually get squashed. In the currency markets, this would mean a steady series of profits from the interest-rate premium that are all wiped out by a large, sudden shift in exchange rates: think of the pound's exit from the European exchange-rate mechanism in 1992. The foreign-exchange markets have been remarkably calm since the Asian crisis of 1998 (when the yen rose sharply, hitting many carry traders). So a whole generation of investors may have grown up in a state of blissful innocence, unaware that their carry strategy has severe dangers.

Inflation may provide an alternative explanation. The theory of purchasing-power parity (PPP) implies that high-inflation currencies should depreciate, relative to harder monies. In other words, while nominal exchange rates might vary, real rates should be pretty constant. And over the very long term, this seems to happen. A study by the London Business School, with ABN Amro, a Dutch bank, found that real exchange rates in 17 countries moved by less than an average of 0.2% a year over the period 1900–2006.

Other things being equal (such as roughly similar real interest rates across countries) nominal interest rates should be higher in countries with higher inflation rates. So this should give support to uncovered-interest parity and deter the carry trade. Clearly, though, PPP has not been a useful guide over the past ten years, as the deflation-prone yen has declined against the dollar.

Perhaps the success of the carry trade reflects biases built up in an earlier era, during the inflationary 1970s and 1980s. Currencies prone to inflation back then, such as sterling and the dollar, have had to pay higher interest rates to compensate investors for their reputation. In fact, because inflation has declined, investors in Britain and America have been overcompensated for the risks – a windfall gain that has been exploited by followers of the carry trade. However, it is hard to believe that this effect could have lasted for as long as it has. So the reasons for the success of the carry trade remain a bit of a mystery.

What does seem plain, however, is that the carry trade tends to break down when markets become more turbulent. In such conditions, those who borrowed yen to buy other assets (such as emerging-market shares) might face a double blow as the yen rose while asset prices fell. If the turbulence were sufficiently large, many years' worth of profits from the carry trade might be wiped out. A steamroller could yet restore the reputation of economic theory.

Source: The Economist, 24 February 2007.

To understand this, recall the relationship between 'real' and 'money' interest rates encountered in Chapter 5. The Fisher Effect concerns the relationship between expectations regarding future rates of inflation and domestic interest rates – investors' expectations about future price level changes will be translated directly into nominal market interest rates. In other words, rational lenders will expect compensation not only for waiting for their money, but also for the likely erosion in real purchasing power. For example, if in the UK, the real rate of interest that balances the demand and supply for capital is 5 per cent, and people expect inflation of 10 per cent p.a., then the nominal rate of interest will be about 15 per cent (actually 15.5 per cent). Recall that real and nominal interest rates are connected by the Fisher formula:

$$(1 + P)(1 + I) = (1 + M)$$

where P is the real interest rate, I is the expected general inflation rate and M is the market interest rate.

The Open Fisher Theory asserts that all countries will have the same real interest rate, i.e. in real terms, all securities of a given risk will offer the same yield, although nominal or market interest rates may differ due to differences in expected inflation rates. It can be more precisely expressed by amalgamating the PPP and IRP theories:

$$\frac{(1 + \text{US interest rate})}{(1 + \text{UK interest rate})} \times \text{Spot rate} = \textit{Forward rate}$$

$$= \frac{(1 + \text{US inflation rate})}{(1 + \text{UK inflation rate})} \times \text{Spot rate}$$

For example, suppose the London and New York interest rates are 12 per cent and 5 per cent, respectively, as quoted by our bond brokers, and the respective expected rates of inflation are 10 per cent and 3 per cent. If the spot rate is $2.00:£1, then the Open Fisher Theory predicts a depreciation in the pound as expressed by the forward rate thus:

$$\frac{1.05}{1.12} \times 2.00 = \mathbf{1.87} = \frac{1.03}{1.10} \times 2.00$$

In other words, when the spot rate is $2.00:£1, this combination of inflation rates and interest rates is consistent with a forward rate of $1.87:£1, as calculated earlier.

These economic theories are interlocking or mutually reinforcing, as shown by the 'equilibrium grid' in Figure 21.1. Several other factors, such as the timing of the change, tax and exchange controls can also affect the relative movement of currencies, but the major factor influencing the movements in exchange rates is claimed to be the expected future movement in inflation rates, which is signalled by current differentials in interest rates.

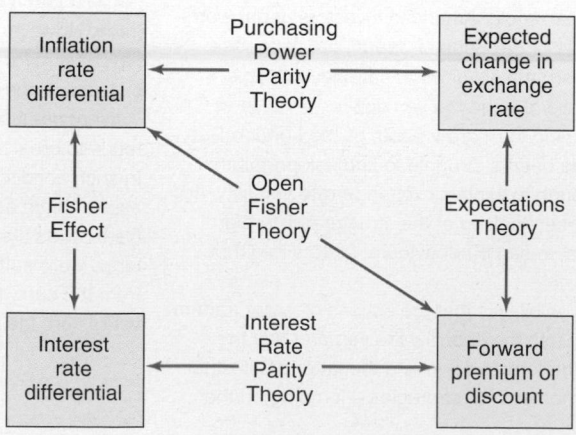

Figure 21.1 Interlocking theories in international economics

■ The international efficient markets hypothesis (EMH)

The EMH claims that, in an efficient market, all publicly-available information is very quickly incorporated into the value of any financial instrument. In other words, past information is of no use in valuation. Any change in value is due to future events, which are, by definition, unknowable at the present time. Past trends in exchange rates cannot provide any useful information to assist in predicting future rates.

This theory applies only to information-efficient markets. Currencies operating within a system of fixed average rates (or maximum permitted bands of fluctuation), such as the former European exchange rate mechanism (ERM), are operating within a controlled market, so the EMH will not apply fully. *Where markets are information-efficient, the EMH casts doubt on the ability of treasurers to make profits out of using exchange rate forecasts.*

This section of the chapter has provided a brief sketch of some economic theories relevant to devising a foreign exchange management strategy. We will shortly try to design such a strategy by applying these theories to the various types of foreign exchange exposure outlined earlier. But because these theories may not always apply (and some people think they rarely, if ever, apply), it is helpful to examine approaches to forecasting FX rates.

21.6 EXCHANGE RATE FORECASTING

First of all, consider why firms may want to forecast future exchange rates. There are both short-term and long-term reasons for this:

- To help decide whether to protect outstanding current assets and liabilities from potential foreign exchange losses.
- To assist in quoting prices in foreign currency when constructing an international price list.
- To aid working capital management, e.g. accurate exchange rate forecasts may assist the decision regarding the most efficient timing of transmitting currency in situations where the firm is able to lead and lag payments.
- To evaluate foreign investment projects requiring exchange rate forecasts over several years.

Because FX forecasts are required for both short- and long-term purposes, they may require continuous revision. In addition, long-term forecasts for investment appraisal purposes often require more intensive analysis of a range of different scenarios. In general, the firm's FX forecasting needs hinge on:

- The pattern of its trading and investment activities, i.e. its degree of globalisation.
- The required frequency of forecast revision.
- The internal resources and expertise available for forecasting analysis.

■ Approaches to FX forecasting

There are two broad approaches to FX prediction: **(a) fundamental analysis**, which bases forecasts on the financial and economic theories outlined earlier and **(b) technical analysis**, which is based on analysis and projection of time series trends.

technical analysis
The intensive scrutiny of charts of foreign exchange rate movements attempting to identify persistent patterns

(a) Fundamental analysis
This approach is sub-divided into two analytical perspectives:

(i) The balance of payments (BOP) perspective
This regards a country's BOP (more accurately, its balance of payments on current account) as an indicator of likely pressure on its exchange rate. When a country, say

the UK, spends more on foreign-produced goods and services than its export earnings, the resulting deficit on current account increases the probability of depreciation of its currency. Overseas residents accumulate monetary claims on sterling – when they convert into their own currencies, this will exert downward pressure on the GBP (and vice versa for a surplus of exports over imports).

Analysts who focus on the BOP try to evaluate not only the country's ongoing BOP performance but also the determinants of international competitiveness, such as prospects for inflation, e.g. a government budget deficit and how it is financed, and underlying productivity movements.

(ii) The asset market approach

This examines the willingness of foreign residents to hold claims on the domestic currency in monetary form. Their willingness depends on relative real interest rates and on a country's prospects for economic growth and the profitability of its industry and commerce. The asset market perspective could explain the continuing strength of the USD during the 'Greenspan Boom' of the 1990s, during which the USA received a massive inflow of overseas funds seeking a home in the stock markets, helping to offset the continuing gaping US current account deficit.

Any factor expected to increase real returns on investment, e.g. technological progress such as the rise of e-commerce, promising higher corporate profitability is thus likely to lead to relative exchange rate appreciation (and vice versa).

In practice, it is difficult to disentangle the various fundamental pressures on exchange rates to *identify* the true reasons for their movements. Some argue that short-term movements are largely determined by the relative attractiveness of international asset markets, interest rates and the expectations of market players plus a dose of speculation, while in the long-term, equilibrium exchange rates depend on PPP.

(b) Technical analysis

Technical analysts conduct intensive scrutiny of charts to identify trends in foreign exchange rate movements. These **chartists** focus on both price and volume data to ascertain whether past trends are likely to persist into the future. The underlying premise behind chartism is that future FX rates are based on past rates. Chartists assert that FX movements can be split into three temporal categories:

(i) day-to-day movements, mainly random 'noise'
(ii) short-term movements, which extend from a few days to periods lasting several months
(iii) long-term movements, characterised by persistent upward and/or downward trends.

The longer the forecasting time horizon, the less accurate the prediction is likely to be. However, for most firms, a major part of their forecasting needs are short-to-medium term, so 'expert' forecasting may have some role to play. Forecasting for the long-term, however, depends on the economic fundamentals of exchange rate determination, although some people believe in the existence of long-term waves in currency movements (at least, when they float!). A major flaw of technical analysis is that it is purely mechanical with no attempt to provide supporting theory regarding explanation of causation.

Research by Chang and Osler (1999) suggests that technical analysis is largely a waste of time and money for trades in most currency pairs. They studied the performance of a particular dealing rule, the so-called 'head and shoulders' pattern. This pattern is formed when a market price forms three peaks, a high one (the head) flanked by two lower ones (the shoulders). When the price rises through the second neckline, many technical analysts treat this as a buy signal. During 1973–94, this rule would not have worked except for JPY–USD and DEM–USD trades, where profits at annualised rates of 19 per cent and 13 per cent would have been made. Possibly, this is because with the widespread use of these techniques, such patterns often become self-fulfilling prophecies. However, even for these trading pairs, much simpler trading rules, such as buying when a price was above its recent trading levels, would have generated superior returns.

■ Forecasting in practice

Most leading banks offer FX forecasting services and many MNCs employ in-house forecasting staff. The value of these activities is open to question, but this really depends on the motivation for forecasting. A long-term forecast may be needed to underpin an investment decision in a foreign country. A forecast based on long-term fundamentals may not need to be perfectly accurate but may help in analysing more fully the risks surrounding the decision and its implementation.

Conversely, short-term forecasts may be needed to hedge debtors or creditors for settlement in a month or so. In such cases, long-term fundamentals may be less important than market-related technical factors, e.g. closing of positions, political factors or 'sentiment' in the market. The required degree of accuracy increases as the prospect of loss is more immediate and less remedial action is possible. In general, long-term forecasts are based on economic models reflecting fundamentals, while short-term forecasts tend to rely on technical analysis. Chartists often attempt to correlate exchange rate changes with various other factors regardless of the economic rationale for the co-movement.

■ FX forecasting and market efficiency

The likelihood of forecasts being consistently useful or profitable depends on whether the FX markets are efficient. The more efficient the market, the more likely that FX rates are random walks, with past behaviour having no bearing on future movements. The less efficient the market, the more likely that forecasters will 'get lucky' and stumble on a key relationship that happens to hold for a while. Yet, if such a relationship really exists, others will soon discover and exploit it, and the market will regain its efficiency regarding that item of information.

■ The role of central banks

A key requirement of market efficiency is that all market players are rational wealth seekers. This is often not the case with a major market participant, the central bank that tries to raise or lower its currency by buying or selling in the open market, often in defiance of market trends and sentiment. Evidence exists that at times when central banks intervene, markets become less efficient and it is possible to make money by betting against them. This happened most notably on 'Black Wednesday' in 1992 when sterling was evicted from the ERM despite the Bank of England spending billions of GBP worth of foreign exchange and raising bank base rate from 10 per cent to 15 per cent. However, these opportunities are likely to be only very short-term phenomena.

A forecasting fiasco

During mid-to-late 1999, after an early flurry, the euro fell sharply from its opening value of €1.17 against the USD, reaching €1.05. It was widely felt that the euro had been oversold and would recover rapidly during 2000. Table 21.1 shows a selection of forecasts made by 17 leading banks in November 1999 of the euro's value twelve months ahead, i.e. for November 2000.

The actual spot rate on 1 November 2000 was €0.86 per USD. Similar shortfalls were recorded against other major currencies. Subsequently, the euro did recover to 0.96 in

Table 21.1 Twelve-month forecasts to 1 November 2000

Bank	euro vs. US$
American Express	1.15
Bank of Montreal	1.09
Barclays Bank	1.10
CCF	1.10
Citibank	1.15
Commerzbank	1.18
Dresdner Kleinwort Benson	1.18
Goldman Sachs	1.22
Hanseatic Bank	1.20
HSBC Midland	1.20
Lloyds TSB	1.18
NatWest Group	1.16
Royal Bank of Scotland	1.15
Societé Générale	1.08
Warburg Dillon Read	1.14
Mean	1.15
Range	1.22–1.08
Spot rate November 1 1999	1.05
Forward rate (one year)	1.08

Source: Corporate Finance, December 1999.

early 2000, but slipped back to 0.88 in April that year. The forward market was predicting in November 1999 a 3 cents appreciation in the euro. These figures remind us that the 'experts' like the FX markets sometimes get things wrong, but warn us that, at times, all the experts get things badly wrong – and so do the markets.

The jury is still out on FX forecasting – it should not be possible to outguess the market, but sometimes it works: the question is 'when?' Meanwhile, some businesspeople derive comfort from having 'expert' forecasts available, possibly as a focus of blame! The implications for hedging are hazy – if one believes in PPP in the short term, then hedging is pointless, but PPP seems only to operate in the longer term and with unclear time lags. So, for peace of mind, most firms try to devise a hedging strategy.

The *Economist* magazine publishes an annual survey on PPP. This is based on the global price of a Big Mac (The Big Mac Index), and purports to identify – in a tongue-in-cheek fashion – cases where PPP does not apply between currencies. This has recently been extended to cover the worldwide price of Starbucks coffee, to widen the range of goods examined. Results are broadly similar although, of course, a more rigorous approach (as the *Economist* concedes) would scrutinise the prices of a standard 'basket' of goods, as in national price indices (see below).

21.7 DEVISING A FOREIGN EXCHANGE MANAGEMENT (FEM) STRATEGY

■ Hedging translation exposure: balance sheet items

Total exchange exposure is made up of cash flowing across a national frontier plus the assets and liabilities of the company that are denominated in a foreign currency.

An international treasurer who does not believe the theories outlined above, might decide to hedge all foreign currency transactions plus the total net worth of all foreign subsidiaries. This strategy is over-elaborate and very expensive, but is adopted by many companies, particularly those dealing in currencies that fluctuate widely in value over short periods.

The Big Mac Index

Food for thought about exchange-rate controversies

American politicians bash China for its policy of keeping the yuan weak. France blames a strong euro for its sluggish economy. The Swiss are worried about a falling franc. New Zealanders fret that their currency has risen too far.

All these anxieties rest on a belief that exchange rates are out of whack. Is this justified? *The Economist*'s Big Mac

Cash and carry: The hamburger standard

	Big Mac prices in dollars*	Implied PPP† of the dollar	Under (−)/ over (+) valuation against the dollar, %
United States‡	3.41	–	–
Argentina	2.67	2.42	−22
Australia	2.95	1.01	−14
Brazil	3.61	2.02	+6
Britain	4.01	1.71§	+18
Canada	3.68	1.14	+8
Chile	2.97	459	−13
China	1.45	3.23	−58
Czech Republic	2.51	15.5	−27
Denmark	5.08	8.14	+49
Egypt	1.68	2.80	−51
Euro area**	4.17	1.12††	+22
Hong Kong	1.54	3.52	−55
Hungary	3.33	176	−2
Indonesia	1.76	4,663	−48
Japan	2.29	82.1	−33
Malaysia	1.60	1.61	−53
Mexico	2.69	8.50	−21
New Zealand	5.89	1.35	+73
Peru	3.00	2.79	−12
Philippines	1.85	24.9	−46
Poland	2.51	2.02	−26
Russia	2.03	15.2	−41
Singapore	2.59	1.16	−24
South Africa	2.22	4.55	−35
South Korea	3.14	850	−8
Sweden	4.86	9.68	+42
Switzerland	5.20	1.85	+53
Taiwan	2.29	22.0	−33
Thailand	1.80	18.2	−47
Turkey	3.66	1.39	+7
Venezuela	3.45	2,170	+1

*At current exchange rates
†Purchasing-power parity; local price divided by price in United States
‡Average of New York, Chicago, Atlanta and San Francisco
§Dollars per pound
**Weighted average of prices in euro area
††Dollars per euro

Sources: McDonald's; *The Economist*

Index, a light-hearted guide to how far currencies are from fair value, provides some answers. It is based on the theory of purchasing-power parity (PPP), which says that exchange rates should equalise the price of a basket of goods in any two countries. Our basket contains just a single representative purchase, but one that is available in 120 countries: a Big Mac hamburger. The implied PPP, our hamburger standard, is the exchange rate that makes the dollar price of a burger the same in each country.

Most currencies are trading a long way from that yardstick. China's currency is the cheapest. A Big Mac in China costs 11 yuan, equivalent to just $1.45 at today's exchange rate, which means China's currency is undervalued by 58%. But before China's critics start warming up for a fight, they should bear in mind that PPP points to where currencies ought to go in the long run. The price of a burger depends heavily on local inputs such as rent and wages, which are not easily arbitraged across borders and tend to be lower in poorer countries. For this reason PPP is a better guide to currency misalignments between countries at a similar stage of development.

The most overvalued currencies are found on the rich fringes of the European Union: in Iceland, Norway and Switzerland. Indeed, nearly all rich-world currencies are expensive compared with the dollar. The exception is the yen, undervalued by 33%. This anomaly seems to justify fears that speculative carry trades, where funds from low-interest countries such as Japan are used to buy high-yield currencies, have pushed the yen too low. But broader measures of PPP suggest the yen is close to fair value. A New Yorker visiting Tokyo would find that although Big Macs were cheap, other goods and services seemed pricey. A trip to Europe would certainly pinch the pocket of an American tourist: the euro is 22% above its fair value.

The Swiss franc, like the yen a source of low-yielding funds for foreign-exchange punters, is 53% overvalued. The franc's recent fall is a rare example of carry traders moving a currency towards its burger standard. That is because it is borrowed and sold to buy high-yielding investments in rich countries such as New Zealand and Britain, whose currencies look dear against their burger benchmarks. Brazil and Turkey, two emerging economies favoured by speculators, have also been pushed around. Burgernomics hints that their currencies are a little overcooked.

Source: The Economist, 7 July 2007.

Figure 21.2 illustrates a more systematic approach. The basic strategy is to remove from consideration all items that are self-hedging so far as exchange rate risk is concerned, and to concentrate attention on those cash flows, assets and liabilities that are subject to exchange rate risk in the short term.

We start with a position where all cash flows, assets and liabilities denominated in foreign currency values are assumed to be subject to exposure. Let us now try to

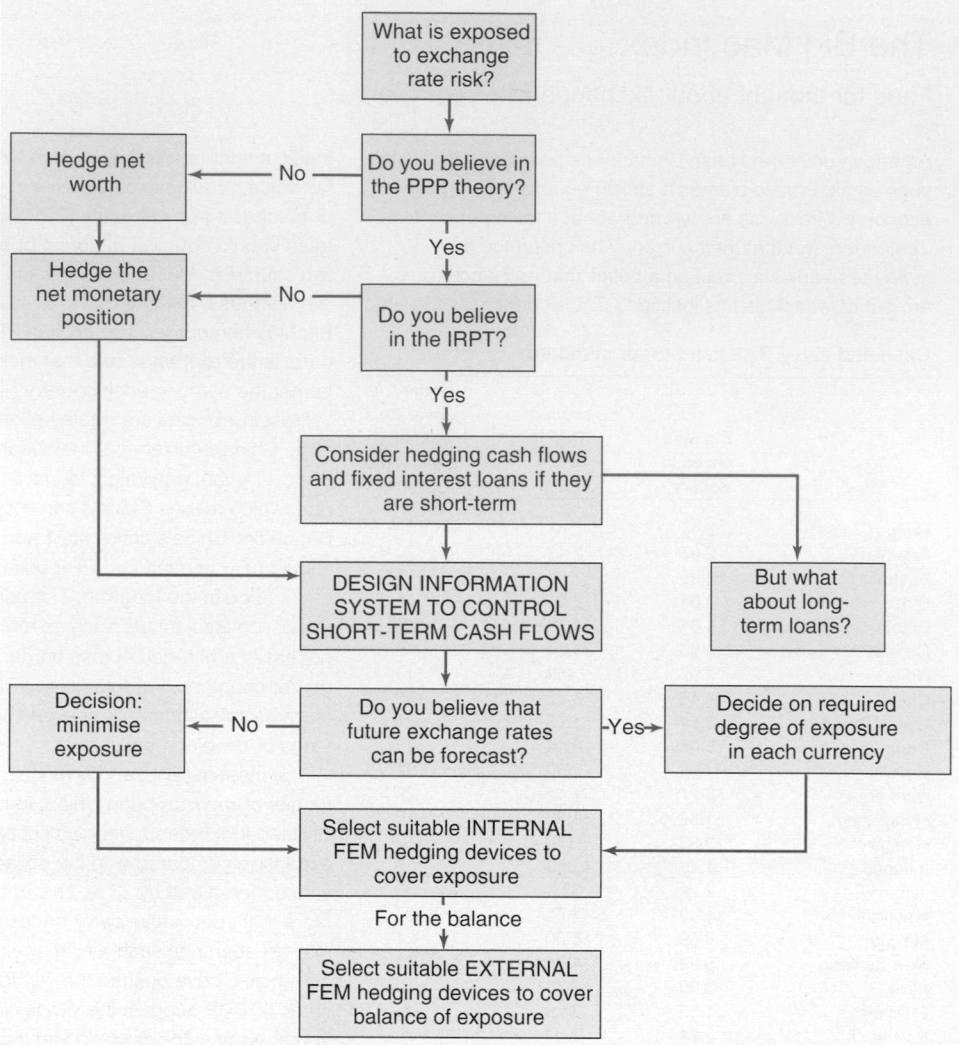

Figure 21.2 Flow chart demonstrating a logical approach towards devising a foreign exchange management strategy (based on McRae, 1996)

eliminate some of these items from the exposure equation. First, we eliminate all non-monetary assets such as land, buildings and inventory. These should float in value with internal inflation. The rate of adjustment in value will vary, internationally-traded goods will jump in local value faster than the value of land, but eventually the prices of all of these non-monetary assets should rise to compensate for the fall in value of the local currency. PPP relates inflation differences to changes in exchange rates. In time, the asset or liability denominated in the foreign currency will rise in value sufficiently to compensate for the fall in the foreign currency value. In other words, the owner of the asset could sell it for more foreign currency units, each commanding a lesser external value than before. The total in terms of home currency will remain unchanged.

Self-assessment activity 21.6

Langer plc, a UK firm, is worried about a fall in the Australian dollar by 5 per cent, compared to the present A$2.75 per £1, that might inflict translation losses regarding its A$100 million assets located in Adelaide. Why should it not worry?

(UK inflation is 2 per cent. Australian inflation is 7 per cent.)

(Answer in Appendix A at the back of the book)

Non-monetary assets are thus self-hedging at least in the long term. If the asset has to be sold in the short term and the foreign cash exchanged into local currency, the amount then becomes a part of transaction exposure because a real loss might be involved.

Short-term loans can, for the most part, also be considered self-hedged. The higher or lower interest rate on the foreign currency loan is a kind of insurance policy against the future fall or rise of the 'away' currency in terms of the 'home' currency. A forward contract could be taken out to cover the risk, but this would be a needless expense (given that spreads are wider on forward transactions), since the forward rate is an unbiased predictor of the future spot rate. On average, the forward contracts would make neither a profit nor a loss.

Long-term loans are more problematic. A fervent believer in the Open Fisher Theory would claim that the long-term loan, like the short, is also self-hedged. The interest rate difference is the market's best guess as to the future changes in the value of the currency. A lower-rate loan suggests a higher capital sum to repay in the home currency. A higher-rate loan suggests a smaller capital sum.

If in doubt about monetary assets or liabilities being self-hedging, one solution is to calculate the 'net monetary asset position' in each currency and make sure it is either in balance or in the 'right' direction. *In other words, if it is predicted that a currency will fall in value against GBP, the firm should owe money in that currency. If it is predicted that a currency will rise in value against GBP, then it should be owed money in that currency.* This might require some juggling with the financing mix of the firm via 'currency swaps', which we discuss later in the chapter.

The key problem in currency risk management is thus to identify the various types of exposure facing the company and then to hedge any unwanted exposure risks. Non-monetary assets and short-term loans in foreign currency are for the most part self-hedged. The exchange risk involved in financing with foreign loans and bonds is less clear. With regard to transaction exposure, a currency information system needs to be designed and installed to identify estimated short-term cash flow exposure in each currency.

■ Transaction exposure: hedging the cash flows

currency information system
An information system set up to identify values that are exposed to currency risk, e.g. cash in- and outflows and asset and liability values

The first step in identifying and hedging cash flow exposure in foreign currency is thus to set up a **currency information system**. The control of currency risk is much simplified if this information system is centralised, but this is not a necessary condition of efficient currency management.

Once this system is in place, the company must decide whether it (1) believes that future exchange rates can be forecast, and (2) will permit speculation in currency. If the answer to either question is 'no', then the company must seek to minimise the exposure position in all currencies. If a profit-maximising strategy is adopted, the company will use currency forecasts to decide on an optimal position in each foreign currency. If the company believes that currency forecasting is impossible, or not profitable, then it has to adopt a **risk-minimising policy**. The aim will be to reduce exposure in all currencies to a minimum unless the cost of this policy is prohibitive.

risk-minimising policy
A foreign exchange policy designed to eliminate, as far as possible, the firm's exposure to currency risk

Once the estimated cash flows in each currency have been identified, the next step is to consolidate the data. The individual flows are netted to arrive at the estimated net balance in each currency for each future period. Monthly estimates for six months ahead are the most common requirement, but large companies holding, or trading in, many currencies may require weekly or even daily reports (especially if speculative positions are opened).

If the company believes that currency forecasting is both possible and profitable, it must decide, in the light of current currency forecasts, the degree of imbalance desirable in each currency in which it trades. Even if forecasting is thought to be possible and profitable, the company might decide to prohibit currency speculation as a matter of principle. Many UK multinationals take this position. In the past, US multinationals have been more willing than similar UK companies to speculate in currency, but research by Belk and Glaum (1990) suggests that attitudes among UK treasurers have changed.

The next step is to convert the 'natural' exposure position arising from normal trading into the 'desired' exposure position. This is done by using various currency hedging devices, some of which are internal to the firm and others external. Prindl (1978), who introduced the distinction between internal and external hedging, also pointed out that internal hedging is almost invariably cheaper than external hedging. The international treasurer should first adjust the 'natural' exposure position using internal techniques and use the more expensive external techniques only after the internal hedging possibilities have been exhausted.

BMW steers a tricky course

The German car industry has lost billions of euros in the last few years because of the strength of the European currency. BMW is a perfect example of why and how it happened.

The Munich-based luxury carmaker has lost €1.6bn ($2.3bn) over the past four years due to currency effects, and admits it could have been twice as much if it were not for hedging.

Several hundred million euros more of currency losses are expected this year. "You have to weather this period," says Erich Ebner von Eschenbach, head of group treasury.

The biggest problems have come in the US, pushing the issue of the dollar to the top of the list of worries for German carmakers.

All the manufacturers are examining their US strategies and in particular the issue of local production. BMW is perhaps the furthest through this process and the most explicit about what it will mean.

"We have been successful in tapping markets like the US and UK but it has been difficult to adapt our structures so quickly," says Mr Ebner von Eschenbach.

BMW sold 340,000 cars last year in North America, but produced only about 140,000 at its sole US plant of Spartanburg, South Carolina.

It is planning to raise production there to 250,000 by 2012, but its US sales are also expected to increase, with executives underlining that the

market share of premium brands in the US is just 11 per cent, compared with 30 per cent in Germany.

In fact, production is relatively unimportant in quantifying the dollar risk at BMW. Production accounts for about 10–15 per cent of its exposure to the dollar and is far outweighed by purchasing, which is responsible for about 60–80 per cent.

BMW is only keen on producing models in the US that will have strong sales there, largely because of the high duties it pays if it exports them. Some 10 per cent is due on cars imported from the US into Europe, against just 2.5 per cent in the other direction.

BMW has just publicly recognised the relative importance of purchasing by creating a management board to look for opportunities for more global sourcing of components.

An example of the problems is the X5, BMW's large sports utility vehicle that is built in Spartanburg.

A total 60 per cent of the content is locally produced, while about 40 per cent – predominantly the engine and gearbox – is shipped from Europe.

Overall North America accounts for only 9 per cent of purchasing, but about a quarter of sales. Meanwhile 84 per cent of purchasing but only about a half of sales were in Europe.

BMW is hoping to boost its purchasing in the US just as it is in Japan and other markets. That should help

it boost its local competitiveness against US and Japanese carmakers, which are undoubtedly benefiting from their weak currencies.

BMW estimates that Toyota enjoys a 40 per cent cost advantage in the US with its motorcycles and Lexus luxury brand, solely through currency issues, compared with the models the German group has to export from Europe. Similarly many analysts see the weak dollar as being at least partly behind the rise in General Motor's sales in the US.

BMW is renowned in the industry for its reluctant – and some would say rather dogmatic – use of hedging, especially when the euro is relatively strong. BMW bases all its decisions on the purchase-price parity currency rate, which is currently just below $1.20 per euro, against yesterday's intraday low of $1.4438.

Each cent the euro rises is estimated to cost BMW and its German rivals Mercedes-Benz and Audi about €50m a year, making the euro's recent rise expensive.

In spite of this, BMW still hints it is profitable in the US. "The US is still an attractive market for us," says Mr Ebner von Eschenbach. He also underlines the essentially long-term view of the family-controlled company: "We can handle a three-year period of currency volatility from a strategic point of view."

Source: Richard Milne, *Financial Times*, 30 October 2007.

21.8 INTERNAL HEDGING TECHNIQUES

Internal hedging techniques exploit characteristics of the company's trading relationships without recourse to the external currency or money markets. Most are simple in concept and operation.

Netting applies where the head office and its foreign subsidiaries net off intra-organisational currency flows at the end of each period, leaving only the balance exposed to risk and hence in need of hedging. Netting is illustrated in the following simple example.

A UK-based multinational has a German operating subsidiary. In a particular month, it transfers components worth €20 million to Germany. In the same month, the subsidiary transfers finished goods worth €40 million to the UK. With netting, the company need only make a net currency transfer of €20 million, rather than making two separate transactions totalling €60 million. As well as reducing exposure, netting saves transfer and commission costs, but it requires a two-way flow in the same currency.

Bilateral netting applies where pairs of companies in the same group net off their own positions regarding payables and receivables, often without the involvement of the central treasury. If the previous company also had a Swiss subsidiary, a bilateral netting arrangement could operate between the German and the Swiss subsidiaries. **Multilateral netting** is performed by the central treasury where several subsidiaries interact with the head office. Subsidiaries are required to notify the treasury of the intra-organisational flows of receivables and payments. Again, a common currency is required. To illustrate this, Oilex is a UK-based oil company with an exploration division based on Norway, major interests in the USA and chemical plants in the UK. The group treasury 'holds the ring' at the centre of this nexus, as shown in Figure 21.3. All intra-group transactions are conducted in USD, which is the operating currency of the oil industry.

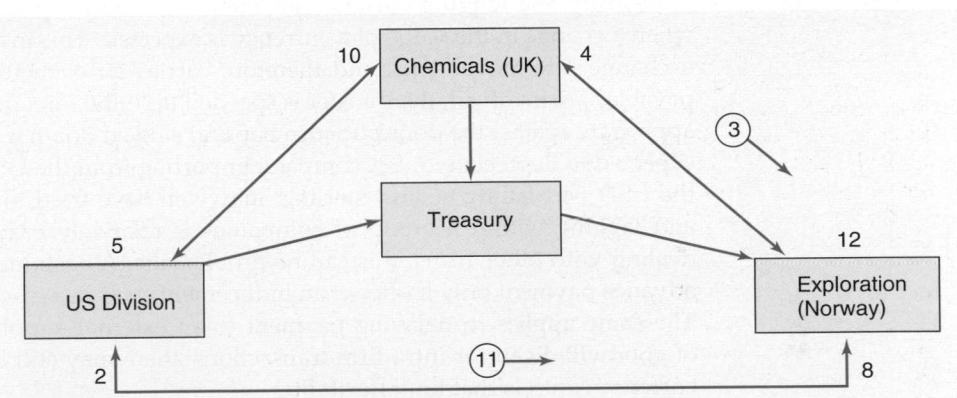

Figure 21.3 Illustration of multilateral netting

Table 21.2 shows transactions expected for one particular month. In total, currency flows of $41 million would be required with no treasury intervention. By multilateral netting, the treasury can reduce the exposed flows by $27 million. Such a system

Table 21.2 Oilex's internal currency flows

		Paying subsidiary ($m)				
		UK	US	Norway	Total	Net
Receiving	UK	–	10	4	+14	−3
subsidiary	US	5	–	2	+7	−11
($m)	Norway	12	8	–	+20	+14
	Total	−17	−18	−6	41	
	Net	−3	−11	+14		

produces greatest benefits when the inter-subsidiary positions are most similar, and where payments are made directly to the relevant subsidiaries, thus avoiding cash transfers into and then out of the treasury. In this case, chemicals would transfer $3 million direct to exploration and the US operation would transfer $11 million likewise, resulting in total flows of only $14 million.

Some experts dispute whether netting is a true hedging technique, rather than a cost-saving device, especially where the netted currency differs from the parent's reporting currency. However, if it does result in lower values of currency being shipped across the exchanges, then it is undeniable that it is capable of saving considerable banking and money transmission costs.

Matching is similar in concept to netting, but involves third parties as well as intra-group affiliates. A company tries to match its currency inflows by amount and timing with its expected outflows. For example, a company exporting to the USA and thus anticipating USD receipts could match this payable by arranging a USD outflow, perhaps by contracting to import from the same country. Clearly, as with netting, a two-way flow of currency is desirable – '**natural matching**'. 'Parallel matching' can be achieved by matching in terms of currencies that tend to move closely together over time, e.g. matching USD outflows to Canadian dollar inflows. Matching can also be achieved by offsetting balance sheet items against profit and loss account items. For example, a company with a long-term cash inflow stream in USD may also borrow in USD, to create an offsetting outflow of interest and capital payments. Earlier in the chapter we observed Compass Group plc doing exactly this.

Leading and lagging currency payments is done to speed up or delay payments when a change in the value of a currency is expected. This involves forecasting future exchange rate movements, and therefore carries an element of speculation. Where payables are involved, the transfer is speeded up if the foreign currency is expected to appreciate against the domestic currency and slowed down if the overseas currency is expected to depreciate. A UK company importing from the USA during 2006–7, when the USD was falling against sterling, may well have tried to lag payments. Leading and lagging within a group of companies is relatively easy to arrange, but when dealing with other firms, this can be problematic. A customer buying on credit will advance payment only if offered an inducement such as a discount for early payment. The same applies to delaying payment to an external supplier – the danger is loss of goodwill. Even for intra-firm transactions, there may still be local regulations and currency controls that limit flexibility.

Currency transfers by companies into and out of less-developed countries, whose currencies tend to be weak, are closely scrutinised by the governments of those countries because of the destabilising effect they may have on their currency. In some cases, they are illegal, both for their ability to exacerbate currency weakness and also because of the effect on local minority shareholders of an overseas subsidiary. Leading a payment from the overseas subsidiary to the UK parent will raise the GBP profits of the parent, but lower the overseas currency profits of the subsidiary thus damaging local shareholders' interests, which risks alienating local opinion and antagonising the host government. This is one reason why repatriation of profits from overseas subsidiaries is often closely controlled by foreign governments.

> **natural matching**
> A natural match is achieved where the firm has a two-way cash flow in the same currency due to the structure of its operations, e.g. selling in a currency in which it sources supplies

Self-assessment activity 21.7

Delete as appropriate.

Leading is advancing outflows in a *strong/weak* currency and advancing inflows in a *strong/weak* currency. Lagging is delaying inflows in a *strong/weak* currency and delaying outflows in a *strong/weak* currency.

(Answer in Appendix A at the back of the book)

price variation
Adjustment of a firm's pricing policy to take into account expected foreign exchange rate movements

The UK exporter might also consider a pre-emptive **price variation**. If it expects GBP to strengthen against the currency of an overseas customer, it may raise the contract price. However, this may have adverse consequences for sales, especially if competitors are prepared to shoulder currency risk by accepting payment in the overseas currency. Conversely, the acceptability of this ploy may be greater if the exporter quotes a price based on the forward rate rather than the spot rate when setting the value of the contract. Generally, however, such price variations require a strong competitive position in overseas markets. For this reason, another such device, switching the currency in which the contract is denominated to a third currency, say USD, also has to be used with caution. However, traders in basic commodities (most notably, oil) have no such flexibility, since most of these are priced in USD.

risk-sharing
An arrangement where the two parties to an import/export deal agree to share the risk, and thus the impact of unexpected exchange rate movements

Risk-sharing is a contractual arrangement whereby the buyer and seller agree in advance to share between them the impact of currency movements. This is recommended when the two parties want to build a long-term relationship. However, if exchange rate variations exceed tolerable limits, the arrangement may have to be renegotiated.

It might work like this. Firm X supplies Firm B in another country. They may agree that all transactions will be made at the ruling spot rate between the two parties' respective currencies. If, however, the rate at settlement varies by up to, say, 5 per cent either side of the original spot rate, X may accept the transaction exposure. If the rate varies by, say, 5–10 per cent of the original spot, they may share the difference equally, but for variations in excess of 10 per cent, the agreement may become void. Harley Davidson is known to operate this policy with foreign importers.

re-invoicing centre
A corporate subsidiary set up usually in an off-shore location to manage transaction exposure arising from trade between separate divisions of the parent firm

Re-invoicing centres. A re-invoicing centre (RIC) is a separate corporate subsidiary that manages from one location, often off-shore, all the transaction exposure arising from intra-company trading.

For example, a manufacturing unit may sell goods to distribution subsidiaries of its parent firm indirectly by selling first to the re-invoicing centre, which then re-sells the goods to the distribution subsidiary. Title to the goods passes to the RIC but the goods are shipped directly from the manufacturing subsidiary to the distributor. The RIC thus manages the transactions on paper but keeps no physical stocks. All transactions exposure resides with the RIC.

A problem may arise due to allegations of profit-shifting via transfer pricing. To avoid such allegations, the RIC may sell at cost plus a commission for its services. The resale price is commonly the manufacturer's price times the forward exchange rate for the date when settlement by the distributor is expected.

RICs offer the major benefit of concentrating the management of all FX transactions in one location. As a result, the multinational corporation (MNC) can develop specialist expertise in judging which hedging technique is optimal at any one time. However, it should avoid conducting business with other firms in its country of location in order to establish non-resident status.

21.9 SIMPLE EXTERNAL HEDGING TECHNIQUES

credit risk
The risk that a foreign customer might not pay up as agreed on time or at all

counterparty risk
The risk that the bank which is party to a hedging transaction such as a forward contract may not deliver the agreed amount of currency at the agreed time

The most widely used external hedging technique is the **forward contract**. It involves pre-selling/buying a specific amount of currency at a rate specified now for delivery at a specified time in the future. It is a way of totally removing risk of currency variation by locking in the rate quoted today by the forward market. However, there remain the risks of the trading partner (**credit risk**) defaulting and that of failure of the bank that arranges the deal (**counterparty risk**).

Consider the case of a UK exporter entering an export contract in February for $10 million with a company in Denver. The companies agree on payment in three months time, i.e. in May. The current spot rate is $2.00:£1, valuing the contract at

$10 m/2.00 = £5 m. If the exporter is concerned by the possibility of a decline in the USD versus GBP, it will look carefully at the rate quoted for 3-month delivery of USD. Assume the forward market quotes '2c discount'. The forward outright is thus:

Spot $2.00 plus 2.0c = $2.02 : £1

If the exporter believes in the predictive accuracy of the forward market, it may decide to sell forward the anticipated $10 million receipt for $10 m/2.02 = £4.95 m. This involves taking a discount on the current spot value of the deal. Hedging costs the exporter £50,000, 1 per cent of the original value of the deal (although a higher proportion of his profits), but this may look trivial beside the losses that could materialise if GBP strengthens further than this. Conversely, the exporter is excluded from any gains if the USD appreciates in value.

If the exporter is unsure about the precise payment date by its customer, it may enter a **forward option**. In this case, the bank leaves the currency settlement date open, but books the deal at the worst forward rate ruling over the period concerned. Say the two companies had agreed on payment 'sometime over the next three months', but the exporter knows that the customer may delay payment for six months. The relevant forward quotations are:

1 month:	0.5c dis	Outright: 2.005
2 months:	1.0c dis	Outright: 2.010
3 months:	2.0c dis	Outright: 2.020
6 months:	3.0c dis	Outright: 2.030

The worst rate for the exporter is the six-month rate, so the deal will be booked for $10 m/2.03 = £4.93 m, again a minor increase in cost. If the customer pays up at any other time, the bank is committed to paying the exporter the amount agreed in the forward contract when the $10 million is handed over.

Another way of covering uncertainty over settlement dates is to undertake a **foreign currency swap**. The Bank of International Settlements (BIS, **www.bis.org**) defines a swap as follows:

> Foreign exchange swaps commit two counterparties to the exchange of two cash flows and involve the sale of one currency for another in the spot market with the simultaneous repurchase of the first currency in the forward market.

foreign currency swap
A way of using the forward markets to adjust the maturity date of an initially agreed contract with a bank

An exporter can take forward cover to a specified date, but if a later settlement date than this is agreed, it can extend the contract to the newly-agreed date. For example, a **forward–forward swap** is needed if our exporter covers ahead from February until May, but if in March, a firm settlement date is agreed for June. Contractually, it has to meet the first contract maturing in May, and then take cover for a further month. This is done in March by buying $10 million two months forward, i.e. for delivery in May to meet the existing contract, and by selling $10 million three months forward for delivery in June. In this case, the exporter swaps the maturity date and ends up holding three separate contracts. Instead, it could adopt the riskier alternative of a **spot–forward swap**, fulfilling the May contract by buying the $10 million on the spot market, and also arranging to sell $10 million one month forward, i.e. in June. The BIS estimated average daily foreign exchange swap transactions at US$1,714 billion in its 2007 survey, 53 per cent of total non-derivatives trading (US$3,210 billion).

forward–forward swap
Where the original forward contract is supplemented by new contracts that have the effect of extending the maturity date of the original one

spot–forward swap
A less comprehensive forward swap that involves speculation on the future spot market

Money market cover involves the exporter creating a liability in the form of a short-term loan in the same currency that it expects to receive. The amount to borrow will be sufficient to make the amount receivable coincide with the principal of the loan plus interest. Assume the Eurodollar rate of interest, the annual rate payable on loans denominated in USD, is 8 per cent, i.e. 2.00 per cent over three months. The UK exporter would borrow ($10 m/1.02 = $9.80 m). This would be converted into GBP at the spot rate – in our example, $2.00 : £1 – to realise ($9.80 m/2.00) = £4.90 m. This looks like a considerable discount on the spot value of the export deal (£5.00 m), but the GBP proceeds of this operation can be invested for three months to defray the cost. Obviously, if the exporter could invest at a rate in excess of 8 per cent p.a., it would profit from this, but IRP should make this impossible, i.e. if USD sells at a forward discount, interest rates in New York should exceed those in London. If USD should unexpectedly fall in value against GBP, lower than expected receipts from the US contract are offset by the lower GBP payment required to repay the Eurodollar loan.

An alternative to a one-off loan to cover a specific contract is for the exporter to operate an overdraft denominated in one or a set of overseas currencies. The trader will aim to maintain the balance of the overdraft as sales are made, and use the sales proceeds as and when received to reduce the overdraft. This is a convenient technique where a company makes a series of small overseas sales, many with uncertain payment dates. A converse arrangement, i.e. a currency bank deposit account, may be arranged by a company with receivables in excess of payables.

International invoice finance is a fast-expanding business among UK traders, amounting to a total of £10.8 billion (measured by client turnover figures) in 2007. This comprised £9.2 billion of export invoice discounting (39 per cent growth on 2006) and £1.6 billion in export factoring (13 per cent growth on 2006). The international factor can provide many services to the small company, including absorbing the exchange rate risk. Once a foreign contract is signed, the factor pays, say, 80 per cent of the foreign value to the UK exporter in GBP. If the exchange rate moves against the UK company before receipt of the foreign currency, the factor absorbs the loss. In compensation, the factor also takes any gain arising from a change in rates. Factors make use of overseas 'correspondent' factors, enabling clients to benefit from expert local knowledge of overseas buyers' credit-worthiness. Overseas factoring is usually expensive but offers the benefits of lower administration and credit collection costs.

Export receivables that involve settlement via **Bills of Exchange** can also be discounted with a bank in the customer's country and the foreign currency proceeds repatriated at the relevant spot rate. Alternatively, the bill can be discounted in the exporter's home country, enabling the exporter to receive settlement directly in home currency.

The most sophisticated external hedging facilities involve derivatives such as options, futures and swaps. These are treated in more detail below.

■ Currency options

A currency option confers the right, but (unlike the forward contract) not the obligation, to buy or sell a fixed amount of a particular currency at or between two specified future dates at an agreed exchange rate (the **strike price**).

call option
A financial derivative that gives the buyer the right but not the obligation to buy a particular commodity or currency at a specific future date

A **call option** gives the purchaser of the option the right to buy, while a put option gives the right to sell. In each case, the buyer of the option pays the 'writer' of the option a premium. Options traded through exchanges are written in specified contract sizes: for example, on the Philadelphia Stock Exchange (PHLX), GBP is dealt in lot or contract sizes of 31,250 units, i.e. £31,250. Most exchanges offer a limited number of alternative exercise prices and maturity dates. PHLX also trades USD options against the Australian dollar (contract size A$100,000), Canadian dollar (C$100,000), Japanese Yen (12,500,000) and the Swiss Franc (125,000), as well as euro contracts (62,500). Delivery dates are for the quarter months of March, June, September and December plus the two immediately upcoming months. For non-standard options, the would-be purchaser may have to shop around for a customised quotation on the **over-the-counter**

over-the-counter
An over-the-counter transaction, e.g. the purchase of an option, where the terms are tailor-made to suit the requirements of the purchaser

(OTC) market. This may be necessary for unusual or 'exotic' currencies.

A '**European option**' can be exercised only at the specified maturity date, while an '**American option**' can be exercised at any time up to the specified expiry date. If an option is not exercised, it lapses and the premium is lost. However, the appeal of an option is that the maximum loss is limited to the cost of the premium, while the purchaser retains the upside potential. The size of the premium depends on the difference between the current exchange rate and the strike price (for an unlikely strike price, premiums will be very low), the volatility of the two currencies, the period to maturity and, for OTC contracts, the size of the contract.

Here is an example of how a trader could use a currency option for hedging purposes. A UK exporter sells goods for $12.312 million to a customer in Baltimore in April 2008 for settlement in June. The contract is worth £6.25 million when valued at spot of $1.97 : £1, but the exporter is concerned that sterling might appreciate before settlement, thus eroding the profit margin. In this case, it might purchase a call option on GBP, i.e. an option to sell USD in exchange for GBP. The premiums in cents per option unit that were available at the Philadelphia Stock Exchange on 14 April 2008 are shown in Table 21.3.

Table 21.3 Sterling/US$ options

Strike price 14 April 2008						
Opening Quotation	**Calls**			**Puts**		
$: £1	**May**	**June**	**September**	**May**	**June**	**September**
1.96	–	4.65	6.75	2.22	2.30	4.55
1.97	–	2.52	4.66	2.44	4.15	3.15
1.98	1.72	3.35	5.50	3.40	3.60	6.35
1.99	1.80	3.50	3.85	4.30	4.22	6.15
2.00	1.63	2.59	4.15	3.05	5.70	7.39

Source: Philadelphia Stock Exchange (www.phlx.com).

The exporter can lock in the spot rate, or purchase protection against sterling rising above the current spot by various amounts. This requires it to take a view on the most likely adverse movements. Choosing the $1.99 strike price gives the exporter insurance against the value of sterling going higher than $1.99. The cost of purchasing June options is:

$$\text{Number of lots required} = \frac{\$12.312\,\text{m}}{\$1.97} \div £31,250 = 200^*$$

Cost of option = £31,250 × 3.50c × 200 = $218,750

i.e. £111,000 at spot.

If the spot rate in June is less than $1.99, the option is not worth exercising and is said to be 'out of money'; if spot is $1.99, it is 'at the money'; while if spot is over $1.99, say $2.05, the option is 'in the money'. In the last case, export earnings of $12.312 million can be sold at $1.99 to realise £6.187 million, compared with a spot value of $\frac{\$12.312}{\$2.05}$ = £6.006 million. Although the option has cost £111,000, it has prevented the exporter from losing (£6.187 m − £6.006 m) = £181,000. In this case, there is a net gain, allowing for the premium, of (£181,000 − £111,000) = £70,000.

It is important to appreciate that options are 'zero-sum games' – what the holder wins, the writer loses, and vice versa. It is relatively unusual to use currency options to hedge ongoing trading exposures – options are complex, they are often expensive compared to using the forward market and it is time-consuming to monitor an

---

*The reader may now see why the 'awkward' sales value of US$12.312 m was used, i.e. to achieve a round number of option lots. In practice, such perfectly hedged positions are rare.

American option to judge whether it is worth closing out the position prematurely. Options tend to be used to cover major isolated expenditures, e.g. the cost of completing the acquisition of an overseas company or the phased payments in a major overseas construction project.

The PHLX website (**www.phlx.com**) gives a history of option trading and also a trading guide.

21.10 WORKED EXAMPLE: HOGAN PLC

Hogan plc exports computer components valued at 28 million Australian dollars (AUD) to Dundee Proprietary in Australia on three months' credit. The current spot exchange rate is A$2.80 vs. £1. Because of recent volatility in the foreign exchange markets, Hogan's directors are worried that a fall in the AUD could wipe out their profits on the deal. Three alternative hedging strategies have been suggested:

(i)   using a forward market hedge
(ii)   using a money market hedge
(iii) using an option hedge.

Hogan's treasurer discovers the following information:

- The three-month forward rate is A$2.805 vs. £1.
- Hogan could borrow in AUD at 9 per cent interest (annual rate), and could deposit in London at 4 per cent p.a.
- A three-month American put option to sell A$28 million at an exercise rate of A$2.81 vs. £1 could be purchased at a premium of £200,000 on the London OTC option market.

Required

Show how each hedge would work out, assuming the following spot rates apply in three months' time:

(a) A$2.78 vs. £1
(b) A$2.82 vs. £1.

In the course of your answer, consider whether interest rate parity applies as between these two currencies.

Answer

(i) The forward hedge

The bank contracts to buy A$28 m for GBP in three months' time. The sterling value = (A$28 m/2.805) = £9.982 m.

(a) At future spot of A$2.78, Hogan could have received (A$28 m/2.78) = £10.072 m, involving an opportunity cost of:

(£10.072 m − £9.982 m) = £0.090 m (0.9% of contract value)

(b) with future spot at A$2.82, Hogan would receive (A$28 m/2.82) = £9.929 m. The hedge offers a gain of:

(£9.982 m − £9.929 m) = £0.053 m

(ii) The money market hedge

The three-month interest rate is (9%/4) = 2.25%. Hogan will borrow in AUD sufficient to accumulate at 2.25% to the AUD value of its receipts, i.e.:

(A$28 m/1.0225) = A$27.384 m

Exchanged for GBP at today's spot rate, this is worth:

(A$27.384 m/2.80) = £9.78 m

This can be invested at 4% p.a. i.e. (4%/4) = 0.01% over three months, accumulating to £9.78 m(1.01) = £9.878 m.

(a) At future spot of A$2.78 vs. £1, the unhedged income = £10.072 m. The loss using the money market hedge is:

(£10.072 m − £9.878 m) = £0.194 m. (1.9% of the original contract value)

(b) At future spot of A$2.82 vs. £1, the unhedged income = £9.929 m. Hence, the hedge still loses *viz.*:

(£9.929 m − £9.878 m) = £0.051 m

Apparently, IRP does not apply! With a difference in interest rates of 5% p.a. the forward rate should show the AUD trading at a greater discount than just half a cent, i.e. the AUD is over-valued on the forward market. If these interest rates prevail and IRP does reassert itself, we might conclude that the forward market is overstating the future spot rate (from an AUD perspective).

(iii) The OTC option hedge

(a) At future spot of A$2.78 vs. £1, the option is 'out of the money', i.e. it is better to sell AUD at spot rather than exercise the option. Hogan's unhedged income is £10.072 m. Net of the option premium, the proceeds are:

(£10.072 m − £0.200 m) = £9.872 m

Hogan's loss through the option hedge is simply the option premium.

(b) At future spot of A$2.82 vs. £1, the option is 'in the money', i.e. it is better to exercise it than sell GBP at spot to yield (A$28 m/2.81) = £9.964 m. This nets:

(£9.964 m − £0.200 m) = £9.764 m

Self-assessment activity 21.8

What is the net cost/benefit of an option that is 'at the money'?

(Answer in Appendix A at the back of the book)

21.11 MORE COMPLEX TECHNIQUES: FUTURES AND SWAPS

■ Currency futures

In principle, a futures contract can be arranged for any product or commodity, including financial instruments and currencies. A currency futures contract is a commitment to deliver a specific amount of a specified currency at a specified future date for an agreed price incorporated in the contract. It performs a similar function to a forward contract, but has some major differences.

Currency futures contracts have the following characteristics:

1 They are marketable instruments traded on organised futures markets.
2 They can be completed (liquidated) before the contracted date, whereas a forward contract has to run to maturity.
3 They are relatively inflexible, being available for a limited range of currencies and for standard maturity dates. The world's largest market for currency futures is the Chicago Mercantile Exchange (CME). It trades futures in eleven different currencies for delivery four times each year: March, June, September and December.
4 They are dealt in standard lot sizes, or contracts.
5 The CME requires a down-payment called a 'performance bond' or 'margin' of about 1.5 per cent of the contract value, whereas forward contracts involve a single payment at maturity.

6 They also involve 'variation payments', essentially the ongoing losses on the contract to be paid to the exchange on which the contract is dealt.
7 They are usually cheaper than forward contracts, requiring a small commission payment rather than a buy/sell spread.

It is difficult to 'tailor make' a currency future to the precise needs of the parties involved, which explains why some exchanges have now stopped currency futures trading.

How a currency future works

This is best shown with an example. In June, a UK importer agrees to buy goods worth $10 million from a firm in Detroit. The sterling/dollar spot rate is $1.50 : £1, valuing the deal at $10 m/1.50 = £6,666,666. Settlement is agreed for 15 August, but the importer is concerned that appreciation of USD will undermine its profitability (it will have to find more GBP to meet the import cost). On the CME, the market price (i.e. the exchange rate) for September GBP futures is $1.48, suggesting that the market expects the USD to appreciate.

The importer needs eventually to acquire GBP to pay for its imports, so it should sell (i.e. go short of) GBP by selling GBP futures contracts at $1.48. With the standard contract size of £62,500, the number of whole contracts required is:

[$10 m/(£62,500 × 1.48)] = 108 approx.

Note the indivisibility problem − 108 contracts covers exposure of only (108 × £62,500 × 1.48) = $9,990,000. This makes hedging by futures unattractive to small exporters. There is also a timing problem, as the importer has to supply USD before the expiry of the contract in August. When payment is due, the importer will close out the contract by arranging a reverse trade, i.e. one with exactly opposite features, which means buying 108 September GBP futures at the ruling market price. If USD has strengthened against GBP between June and August, the importer will make a profit on the futures contract.

Imagine this does happen and the spot rate on 15 August is $1.49 and the September futures price is $1.475. As payment for the goods is required, the importer converts GBP for USD on the spot market at a cost of ($10 m/1.49) = $6,711,409. Compared with the cost of the deal at the June spot rate, it has made a loss of £44,743, owing to the feared USD appreciation. However, the importer holds 108 futures contracts, enabling it to sell GBP at $1.48. To close its position, the importer can buy the same number of contracts at an exchange rate of $1.475. It will thus make a profit on the futures market of:

Sells	108 × £62,500 × 1.480 =	$9,990,000
Buys	108 × £62,500 × 1.475 =	$9,956,250
Profit		$33,750

Valued at spot of $1.49, this is worth ($33,750/1.49) = £22,651, leaving a net loss of (£44,743 − £22,651) = £22,088. This demonstrates the difficulty of achieving a perfectly hedged position with currency futures. Moreover, the futures market may not always move to the same degree as the spot market, owing to expectations about future exchange rate movements.

The CME website (**www.cme.com**) provides a beginner's guide to using the futures markets.

■ Currency swaps

The BIS defines currency swaps as follows:

> A currency swap (or cross-currency swap) commits two counterparties to several cash flows, which in most cases involve an initial exchange of principal and a final re-exchange of principal upon maturity of the contract, and in all cases, several streams of interest payments.

currency swaps
Where two or more parties swap the capital value and associated interest streams of their borrowing in different currencies

Currency swaps originated from controls applied by the Bank of England over foreign exchange movements prior to 1979. Firms wishing to obtain foreign currency to invest overseas, say in the USA, found they could avoid these controls by entering an agreement with a US company that operated a subsidiary in the UK. In return for receiving a loan from the US company to finance its own activity in the USA, the UK company would lend to the UK-based subsidiary of the US company. The two firms would agree to repay the loans in the local currency after an agreed period, thereby locking in a particular exchange rate. The interest rate would be based on prevailing local rates. Such arrangements were called '**back-to-back loans**' – from the UK company's perspective, they involved agreeing to make a series of future USD payments in exchange for receiving a flow of GBP income.

back-to-back loans
A simple form of a swap where firms lend directly to each other to satisfy their mutual currency requirements

After exchange controls were removed in 1979, such loans were replaced by currency swaps. These need not involve two companies directly. In general terms, a currency swap is a contract between two parties (e.g. between a bank and an overseas investor) to exchange payments denominated in one currency for payments denominated in another. A simple example will illustrate this.

How currency swaps work

Currency swaps are complex. In particular, they require matching up two companies' mutual requirements in terms of type and amount of currency required and term of financing. The final agreement will reflect the bargaining power of the parties involved and is most viable when each party has a differential borrowing advantage in one currency which it can transfer to the other. This point is important – a currency swap almost invariably involves an interest rate swap.

To illustrate the process, consider the example of two companies, ABC and XYZ. ABC, which can borrow in Swiss Francs (CHF) at 5.5 per cent, is seeking USD financing of $40 million for three years. The Dutch subsidiary of a US bank is prepared to act as intermediary, which involves finding a suitable matching company which has a borrowing advantage in USD and is seeking CHF finance. Until the match is found, the bank will be exposed to currency and interest rate risk, which it may cover by entering the spot market, or possibly using the options market. Company XYZ emerges as a suitable swap candidate. (More complex swaps might involve several participant companies if a directly corresponding currency requirement cannot be identified.)

XYZ has a borrowing advantage in USD, being able to borrow at 7 per cent, compared to ABC's borrowing rate of 7.75 per cent. Conversely, XYZ would have to borrow in CHF at 6 per cent. XYZ is seeking CHF finance of 52 million. At the ruling exchange rate of 1.3 CHF per USD, this is an exact match. With the bank's intermediation, the two companies now agree to swap currencies and assume each other's interest rate obligations over a three-year term, with transactions conducted via the bank. Figure 21.4 shows the structure of the swap and the sequence of transactions. In reality, the two companies would have to pay rather higher interest rates than those shown in order to yield a profit margin for the bank, sufficient to compensate it for assuming the risks of either company defaulting on interest payments or re-exchange of principal.

There are three legs to such deals:

1 Exchange of principal at spot (either notional or a physical transfer) in order to provide a basis for computing interest.
2 Exchange of interest streams.
3 Re-exchange of principal on terms agreed at the outset.

The principal is fully hedged, unlike the interest rate payments, which may require hedging perhaps via the forward market.

In the example, the company benefits from lower interest rates and effectively uses the superior credit rating of the swap specialist to access cheaper finance. Most

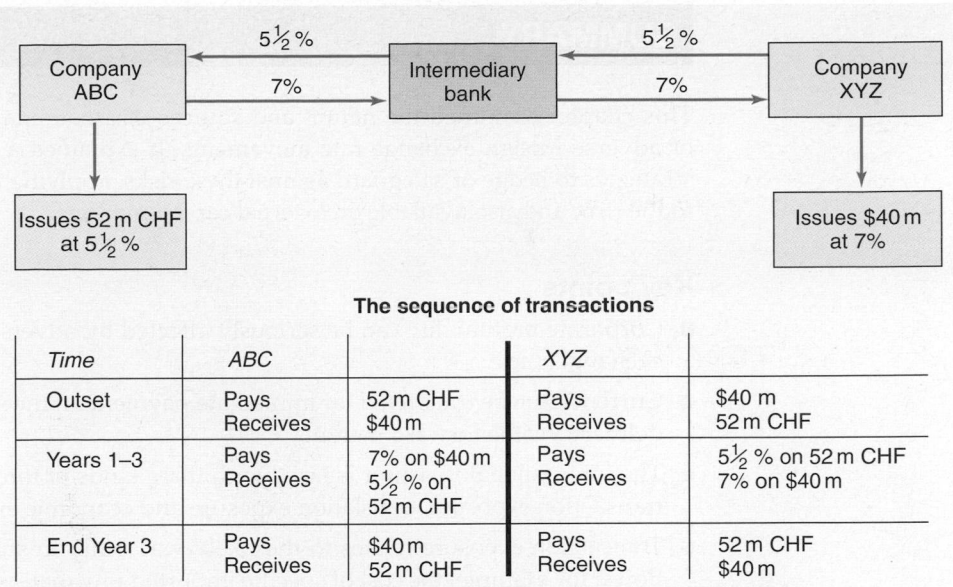

The sequence of transactions

Time	ABC		XYZ	
Outset	Pays Receives	52 m CHF $40 m	Pays Receives	$40 m 52 m CHF
Years 1–3	Pays Receives	7% on $40 m 5½ % on 52 m CHF	Pays Receives	5½ % on 52 m CHF 7% on $40 m
End Year 3	Pays Receives	$40 m 52 m CHF	Pays Receives	52 m CHF $40 m

Figure 21.4 Achieving the swap

fixed/fixed swap
A swap agreement where the parties agree to swap fixed interest rate commitments

cross-currency interest swap
A swap agreement where the parties agree to swap a fixed interest rate commitment for a floating interest rate

currency swaps are undertaken to exploit such interest rate disparities, whereby one party can pass on to another the benefit of superior credit-worthiness. There are two main forms of currency swap. In a **fixed/fixed swap**, one party swaps a stream of fixed interest payments for a corresponding stream of fixed interest payments in another currency (as in the above example). In a **fixed/floating swap**, or **cross-currency interest swap**, one or both payment streams are on a variable basis, e.g. linked to LIBOR.

21.12 CONCLUSIONS

The globalisation of world trade has forced financial managers to take a keener interest in managing foreign exchange exposure. There are three types of exposure: transaction exposure, affecting the flow of cash across a currency frontier; translation exposure, affecting the value of assets and liabilities denominated in a foreign currency; and economic exposure, which is the impact on long-term cash flows of possible changes in exchange rates.

Not all transactions, assets and liabilities denominated in foreign currencies are necessarily exposed to exchange rate risk. The essential skills in currency management are to identify the assets and cash flows which are at risk and to devise suitable means of hedging the risks. It is important to differentiate between hedging techniques internal and external to the firm. Several financial markets have been developed that allow the international treasurer to hedge foreign exchange risk, and financial instruments such as swaps, options, futures and forwards can be used for this purpose.

The international treasurer must decide whether or not exchange rates can be forecast with any degree of reliability. With exchange rates floating freely, research suggests that forecasting is not profitable. However, when governments begin to interfere with the free market, forecasting has proved to be a profitable activity. The dogged, but ultimately doomed, commitment by international monetary authorities to support artificially high or low exchange rates may make forecasting worthwhile.

SUMMARY

This chapter examined the nature and sources of a company's exposure to the risk of adverse foreign exchange rate movements. It explained a number of widely-used strategies to hedge or safeguard against these risks, applying techniques both internal to the firm and also available on external capital markets.

Key points

- Corporate profitability can be seriously affected by adverse movements in foreign exchange rates.

- Currency can be transacted for immediate payment on the spot market or for future delivery via the forward market.

- The international treasurer is faced with three kinds of foreign exchange exposure: transaction exposure, translation exposure and economic exposure.

- Transaction exposure relates to the likely variability in short-term operating cash flows: for example, the cost of specific imported raw materials and the income from specific exported goods.

- Translation exposure relates to the risk of exchange rate movements altering the sterling value of assets located overseas or the sterling value of liabilities due to be settled overseas.

- Economic exposure refers to the ongoing risks incurred by the company in its choice of long-term contractual arrangements, such as licensing deals or decisions to invest overseas. These risks are the long-term equivalent of transaction exposure.

- Companies that trade internationally should devise a foreign exchange strategy.

- The strategy might depend on the treasurer's belief in the validity of various international trade theories: Purchasing Power Parity (PPP), Interest Rate Parity (IRP), the Expectations Theory and the Open Fisher Theory.

- PPP states that, allowing for the prevailing exchange rate, identical goods must sell for a common price in different locations. If inflation rates differ between locations, exchange rates will adjust to preserve the Law of One Price.

- IRP asserts that any differences in international interest rates are a reflection of expected exchange rate movements, so that the interest rate offered in a location whose currency is expected to depreciate will exceed that in an appreciating currency location by the amount of the expected exchange rate movement.

- The forward premium or discount should equal the expected rate of appreciation or depreciation of a currency.

- The Open Fisher Theory asserts that investment in different countries will offer the same expected real interest rate, so that differences in nominal rates of interest can be explained by expected differences in rates of inflation.

- Once the exposure position of the company is identified and measured, the treasurer must devise a hedging strategy to control the foreign exchange risk faced by the firm.

- Many apparent exposures are often self-hedging: for example, holdings of plant and machinery that can be traded internationally.

- Generally, internal hedging techniques are cheaper to apply than using the external markets, which offer various financial instruments for hedging currency risks.

Further reading

There has been an upsurge in texts on international financial management since the 1970s, but the leading book in the field is generally reckoned to be that by Shapiro (2006), which also contains excellent bibliographies. The texts by Madura (2006) and Eiteman *et al.* (2003) are also highly regarded. The leading British text is Buckley (2003), while Pilbeam (2006) and MacRae (1996) are also recommended.

Many of these texts cover aspects of the international financial markets, such as the determination of exchange rates and the institutional structures that are beyond our scope here. However, this is a fast-moving field and older texts may quickly become outdated, although those by Prindl (1978) and Kindleberger (1978) give good treatments of the basic theory. Some more general books on international business may give quite accessible treatments of the broader issues, although they tend to be briefer on currency hedging. Giddy (1994) contains some excellent chapters on the use of derivatives in the international field.

To keep abreast of trends and emerging issues in the field, regular reading of *Euromoney* and *The Economist* magazines is recommended.

QUESTIONS

 myfinancelab | *Questions with an icon are also available for practice in myfinancelab with additional supporting resources.*

Questions with a coloured number have solutions in Appendix B on page 757.

1 The Local Bank plc quotes the following rates for the euro versus sterling:

1.6296 − 1.6320

(a) How many euros would a firm receive when selling £10 million?
(b) How much sterling would it receive when selling €12 million?

2 On 30 November, a UK exporter sells goods worth £10 million to a French importer on three months' credit. The customer is billed in euros, for which the spot rate versus sterling is €1.6 vs. £1. The three-month forward rate is €1.62 per £1.

(a) What is the amount invoiced?
(b) If the spot rate is €1.7 vs. £1 at the settlement date, what is the exporter's gain or loss, assuming it does not hedge?
(c) If the spot rate is €1.5 vs. £1 at the settlement date, what is the exporter's gain or loss, assuming it does not hedge?
(d) If the exporter takes forward cover, what is the cost of the hedge
  (i) compared to the current spot rate?
  (ii) compared to case (b)?
  (iii) compared to case (c)?

? **3** Work out the forward outrights from the exchange rates versus sterling given in the table below.

Country/Currency	Closing market rates	One-month forward rates
Denmark/krona	10.960 − 10.967	25–11 pr
Japan/yen	230.11 − 230.16	110–97 pr
Norway/kronor	11.717 − 11.723	21–4 pr

Source: The Times, 29 August 2007.

? **4** A selected bundle of goods costs £100 in the UK, and a similar bundle costs 1200 krona (DKR) in Denmark. People generally expect the rate of inflation to be 5 per cent in the UK and 3 per cent in Denmark.

(a) Assuming that PPP applies, what is the current exchange between GBP and DKR?
(b) Assuming that PPP will continue to hold, what spot exchange rate would you predict for twelve months hence?
(c) Again assuming PPP, what exchange rate should be quoted for three-month forward transactions?

? **5** The respective interest rates in the USA and the UK are 9 per cent and 10 per cent respectively in annual terms. If the spot exchange rate is US$1.6000 to £1, what is the forward rate if IRP applies?

6 Assume you are the treasurer of a multinational company based in Switzerland. Your company trades extensively with the USA. You have just received US$1 million from a customer in the USA. As the company has no immediate need of capital you decide to invest the money in either US$ or Swiss Francs for 12 months. The following information is relevant:

■ The spot rate of exchange is CHF1.3125 to US$1.
■ The 12-month forward rate is CHF1.275 to US$1.
■ The interest rate on a 1-year Swiss Franc bond is 4⁹⁄₁₆ per cent.
■ The interest rate on a 1-year US$ bond is 7⅝ per cent.

Assume investment in either currency is risk-free and ignore transaction costs.

Required

Calculate the returns under both options (investing in US$ or Swiss Francs) and explain why there is so little difference between the two figures.

Your answer may be expressed either in $US or in Swiss Francs.

(CIMA May 1998)

7 The following situation is observed in the money markets.

Sterling: US dollar exchange rates:

	Spot	$1.6550–1.6600
	Forward (1 Year)	$1.6300–1.6450
Interest rates (fixed):	New York	5½–5¼%
	London	5¾%–5⅝%

(a) Calculate *both* the interest and the exchange agios.

(b) Using the figures provided, investigate whether an arbitrageur operating in London could profit from covered interest arbitrage.

Notes:

(a) Assume he borrows £100,000.

(b) You may ignore commission and transaction costs other than the spreads.

8 Europa plc is a UK-based import–export company. It is now 15 May and the following transaction has been agreed:

The sale of pine furniture worth $1,250,000 to the USA receivable on 15 August.

$/£ FX rates
Spot   1.6480–1.6490
2 months forward   0.88–0.76c premium
3 months forward   1.30–1.24c premium
Three-month money market interest rates

£10%–8%
$8%–6%

Required

(a) Hedge Europa's risk exposure on the forward market.

(b) Hedge Europa's risk exposure on the money market.

(c) Which is the more favourable hedge from Europa's point of view?

9 Slade plc is a medium-sized UK company with export and import trade links with US companies. The following transactions are due within the next six months. Transactions will be in the currency specified.

Purchase of components, cash payment due in 3 months:	£116,000
Sale of finished goods, cash receipt due in 3 months:	$197,000
Purchase of finished goods for resale, cash payment due in 6 months:	$447,000
Sale of finished goods, cash receipt due in 6 months:	$154,000
Exchange rates quoted on London market	$/£
Spot	1.4106–1.4140
3 months forward	0.82–0.77
6 months forward	1.39–1.34

Interest rates (annual)		
3 months or 6 months	Borrowing	Lending
Sterling	7.5%	4.5%
Dollars	6.0%	3.0%

Required

Calculate the *net* sterling receipts which Slade might expect for both its three and six month transactions if it hedges foreign exchange risk on:

(a) the forward foreign exchange market;

(b) the money market.

10 Exchange-traded foreign currency option prices in Philadelphia for dollar/sterling contracts are shown below:

	Sterling (£12,500) contracts			
	Calls		Puts	
Exercise price ($)	December	September	December	September
1.90	5.55	7.95	0.42	1.95
1.95	2.75	3.85	4.15	3.80
2.00	0.25	1.00	9.40	–
2.05	–	0.20	–	–

Option prices are in cents per £. The current spot exchange rate is $1.9405–$1.9425/£.

Required

Assume that you work for a US company that has exported goods to the United Kingdom and is due to receive a payment of £1,625,000 in three months' time. It is now the end of June.

Calculate and explain whether your company should hedge its sterling exposure on the foreign currency option market if the company's treasurer believes the spot rate in three months' time will be:

(a) $1.8950–1.8970/£

(b) $2.0240–2.0260/£

(ACCA)

11 Ashton plc, a UK-based firm, imports computer components from the Far East. The trading currency is Singapore dollars and the value of the deal is 28 m Singapore dollars (S$). Three months' credit is given. The current spot exchange rate is S$2.80 vs. £1. Because of recent volatility in the foreign exchange markets, Ashton's directors are worried that a rise in the S$ could wipe out the profits on the deal. Three alternative hedging methods have been suggested:

■ a forward market hedge
■ a money market hedge
■ an option hedge.

Ashton's treasurer discovers the following information:

■ The three-month forward rate is S$2.79 vs. £1
■ Ashton can borrow in S$ at 2% interest (annual rate), and in London at 5% p.a.
■ Deposit rates are 1% p.a. in Singapore and 3% p.a. in London
■ A three-month American call option to buy S$28 m at an exercise rate of S$2.785 vs. £1 could be purchased at a premium of £200,000 on the London OTC option market

Required

Show how *each* hedge would operate for *each* of the following spot rates in three months' time:

(a) S$2.78 vs. £1

(b) S$2.82 vs. £1

12 TLC Inc manufactures pharmaceutical products. The organisation exports through a worldwide network of affiliated organisations. Its headquarters are in the USA, but there is a large volume of inter-organisation sales, dividend flows and fee and royalty payments. These inter-organisational payments are made in US$. An example of payment flows among four affiliates for the past 6 months is shown on the diagram below:

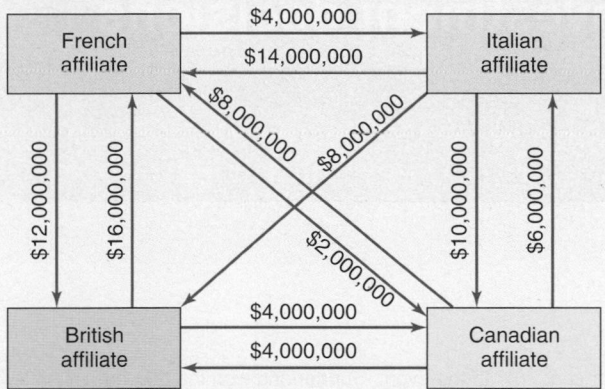

At present, each affiliate has control over its own cash management and foreign currency hedging decisions. The Corporate Treasurer is considering centralising cash and foreign currency management.

Required

(a) Explain, briefly, the characteristics of a bi- or multi-lateral netting system and, using the information in the diagram, advise the Corporate Treasurer of the pattern of cash flows that would have been evident if such a system had been in operation in TLC Inc.

(b) Discuss the advantages TLC Inc might obtain from centralising international cash management and foreign exchange management and advise on the potential disadvantages of such a change in policy.

(CIMA 2004)

Practical assignment

Many companies experience accounting problems in recording overseas transactions and in translating foreign currency values into their accounts. Select a company involved in international trade. Examine its report and accounts to determine its policy with regard to foreign exchange, and the extent of foreign currency losses or gains for the year concerned. Do you think these were 'real money' losses or gains, or merely accounting entries?

 *Now retake your diagnostic test for Chapter 21 to check your progress and update your study plan.*

22

Foreign investment decisions

Cut it or Shut it

In early 2001, Japanese motor firm Nissan, 37 per cent owned by its French strategic partner, Renault, was mulling over whether to locate its new Micra assembly plant in Sunderland, UK, or at Renault's Flins plant, near Paris. In the 1990s, Nissan had spent several hundred million pounds building the Sunderland site into a world-class manufacturing plant, reputedly the most efficient in Europe. A European base to get behind the EU's tariff wall, the flexible working attitudes, low labour costs and less onerous social welfare overheads made the UK an attractive location.

By 2001, things had changed. Despite Sunderland's productivity advantage, Flins had become a serious contender for the new plant due to the 27 per cent appreciation of sterling against European currencies. A Nissan executive claimed: 'A currency movement of 20 per cent or more cancels out all the gross margin on a B-class (small) car'. The sterling appreciation had already put paid to car assembly at Dagenham (Ford) and at Luton (Vauxhall), as well as scuppering Siemens' semi-conductor plant near Sunderland.

In the event, Sunderland won the day, but only after management pledged a 30 per cent cut in costs over three years. Meanwhile Nissan announced that it would henceforth increase its components sourcing in euros from 25 per cent to 65 per cent.

Four years later, in February 2005, Nissan announced that it would invest £200 million at Sunderland to build its new four-wheel drive compact crossover car that was designed and developed in the UK. Currency fears appeared to have receded – sterling had fallen 9 per cent against the euro over this period. But apart from this, Sunderland production had become 'exchange rate neutral'. Nissan had achieved a balance between costs and revenue that neutralised the impact of changes in the sterling/euro rate. In large measure, this was achieved by requiring UK suppliers to invoice in euros, thus shouldering the foreign exchange risk (actually to their advantage over this period).

By 2007, Sunderland had become the biggest car assembly plant in Europe.

Learning objectives

This chapter focuses on foreign investment decisions by multinational corporations (MNCs). It will:

- Study the advantages of MNCs.
- Discuss different ways of entering foreign markets.
- Consider the complexities of foreign direct investment (FDI).
- Analyse the appraisal of foreign FDI.
- Consider the impact of foreign exchange variations on foreign projects.
- Analyse ways of insulating projects against foreign exchange risk.
- Study political and country risk, and how to cope with it.

 *Complete your diagnostic test for Chapter 22 now to create your personal study plan.*

22.1 INTRODUCTION

portfolio investment
Investment in financial securities such as bonds and equities, with no stake in management

direct investment
Investment in tangible and intangible assets for business operating purposes

Foreign investment may be divided into **portfolio** and **direct investment**. Portfolio investment involves the purchase of shares or loan stock in an overseas organisation, usually without control over the running of the business. Foreign direct investment (FDI) is a lasting interest in an enterprise in another economy where the investor's purpose is to have an effective voice in the management of the enterprise. Such direct investment may arise from the acquisition of a controlling interest in an overseas business or from setting up an overseas branch or subsidiary. Firms such as Nissan that invest in foreign countries are examples of multinational corporations (MNCs).

■ The multinational corporation

A working definition of a multinational corporation is: a firm that owns production, sales and other revenue-generating assets in a number of countries (although a wider definition would include firms that simply sell home-produced goods overseas). Foreign direct investment by MNCs includes establishment and acquisition of overseas raw material and component operations, production plants and sales subsidiaries. This occurs because of potentially greater cost-effectiveness and profitability in sourcing inputs and servicing markets through a direct presence in a number of locations, rather than relying solely on a single 'home' base and on imports and exports to support operations. A global firm is one that trades *and* invests abroad.

It is tempting to think of MNCs as Western firms that invest in developing countries. Not so: only about two-thirds of cross-border investment occurs between developed countries. In 2006, the United Nations Conference on Trade and Development (UNCTAD) (UNCTAD, 2007) estimated that direct investment flows to developing and transition countries amounted to $449 billion out of a global total of $1,306 billion with the USA

First tea, then steel, now Scotch: what's next?

Vijay Mallya, chairman of United Spirits, is "seriously considering" listing his international spirits business on the London Stock Exchange following yesterday's acquisition of Whyte & Mackay for £595m.

Mr Mallya said he wanted to make more purchases in the Scotch whisky industry and that a listing would make it easier to raise capital to fund international deals.

United Spirits, whose brands include Bagpiper Whisky and McDowell's, plans to use W&M – which owns single malts as well as blended Scotch brands – to supply Scotch to its Indian blends.

It also wants to import W&M's brands into India and sell them in other international markets such as China. "There is strong demand worldwide and strong growing demand within India," Mr Mallya said.

United Spirits yesterday claimed the acquisition would help position it as a "global player" that derived a quarter of its sales from international markets.

W&M is the world's fourth-largest producer of Scotch behind Diageo, Pernod Ricard and William Grant.

The sale of W&M comes as the number of Indian takeovers of foreign companies last year exceeded the number of foreign acquisitions of Indian companies for the first time. The bid for W&M adds to the tally of 72 foreign takeovers by Indian companies, worth $24.4bn (£12bn), in the first four months of this year, according to Grant Thornton, the advisory firm.

In the same period, there were 38 foreign deals for Indian companies, worth $17bn.

Unlike Tata Steel's £6.7bn acquisition of Anglo-Dutch rival Corus earlier this year, United Spirits was as

interested in buying the Whyte & Mackay brand to help it expand its domestic Indian business as it was in adding overseas production capacity.

"This is a deal done completely for the domestic market," said Sandeep Gill, managing director of Deloitte Corporate Finance.

Mr Gill added that the last deal of this type was Tata Tea's acquisition of Tetley Tea, one of the world's premium tea brands. "If you want to enter the premium market for Scotch, you have to acquire a distillery in Scotland," he said.

United Breweries, parent company of United Spirits, last July bought sparkling winemaker Bouvet-Ladubay for nearly €15m (£10.3m) after losing the bid for Bouvet-Ladubay's former parent, the Taittinger champagne house.

Source: Andrew Bolger, Amy Yee and Joe Leahy, *Financial Times*, 17 May 2007.

and the UK being the largest recipients. Moreover, acquisitions of firms in developed countries by those in developing countries are becoming increasingly common. The acquisition of the UK's Tetley tea operation by the Tata Corporation of India was an early example in 2001. More spectacular were the later developments in the steel industry – Mittal Steel's acquisition of Arcelor in 2005, and Tata Steel's acquisition of Corus in 2006. The cameo documents another Indian incursion into a traditional British beverage. (Answer to question: Land Rover and Jaguar)

Self-assessment activity 22.1

What is meant by a multinational corporation?

(Answer in Appendix A at the back of the book)

22.2 ADVANTAGES OF MNCS OVER NATIONAL FIRMS

The MNC may be in a position to enhance its competitive position and profitability in four main ways:

1 It can take advantage of differences in country-specific circumstances. In a world where countries are at different stages of economic evolution (some industrially advanced, others mainly primary producers), certain advantages in a country may have knock-on effects that the MNC can exploit on a global basis. For example, the MNC may locate its R&D establishments in a technologically advanced country in order to draw on local scientific and technological infrastructure and skills. Similarly, it may locate its production plants in a less-developed country in order to take advantage of lower input costs, especially cheap labour. JCB, the UK firm famous for its yellow diggers operates plants in Brazil, India and China. Alternatively, the MNC may continue to produce its outputs in its 'home' country, but seek to remain competitive by sourcing key components from subsidiary plants based abroad.

2 MNCs can choose the appropriate mode of serving a particular market. For example, exporting may provide an entry route into a low-price, commodity-type market, with the MNC taking advantage of marginal pricing and the absence of set-up costs; licensing may be an appropriate mode if market size is limited or market niches are being targeted; direct investment in production and sales subsidiaries may be a more effective way of capturing a large market share where proximity to customers is important, or where market access via exporting is limited by tariffs. These various routes enable an MNC to pursue a complex global market servicing strategy. For example, Ford makes petrol engines in its plant at Bridgend, Wales, and manual gearboxes in its German factories, which are shipped (along with other parts) to Valencia in Spain for final assembly. The cars are then exported to other European markets.

3 'Internalisation' of the MNC's operations by foreign direct investment provides a unique opportunity for the firm to maximise its global profits by using transfer pricing policies. The transfer price is 'the price at which one affiliate in a group of companies sells goods and services to another affiliated unit' (Buckley, 2004). While a national, vertically-integrated firm needs to establish transfer prices for components and finished products that are transferred between component and assembly plants, and between assembly plants and sales subsidiaries, the greater scale of cross-frontier transactions by MNCs makes these transfer prices more significant because of the impact on relative profitability and tax charges in different locations.

4 An international network of production plants and sales subsidiaries enables an MNC to protect component supplies. The Swedish/Swiss engineering conglomerate Asea Brown Boveri (ABB) has built up a network of component-manufacturing

subsidiaries in the Baltic region, including the former Soviet bloc, in order to diversify sources of supply. Networks also help in the simultaneous introduction of new products in several markets. This is important (where products have a relatively short life cycle and/or patent protection) in order to maximise sales potential. Equally importantly, it spreads the risk of consumer rejection across a diversified portfolio of overseas markets, so that failure in one market may be offset (or perhaps more than compensated by) rapid acceptance in another. Additionally, it enables the MNC to develop a 'global brand' identity (e.g. Coca-Cola, Levis and Subway) or to 'customise' a product more effectively to suit local demand preferences.

22.3 FOREIGN MARKET ENTRY STRATEGIES

Firms enter foreign markets in pursuit of incremental profits and cash flows by exploiting advantages over local producers and other MNCs. The two basic vehicles for foreign market entry are, first, via transactions, and second, via Foreign Direct Investment (FDI). Each mode can be pursued in a number of ways. Figure 22.1 shows the spectrum of entry modes arranged by degree of commitment.

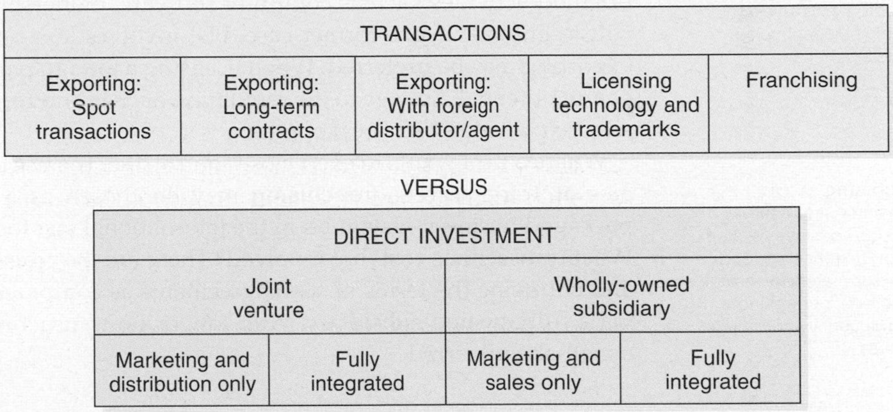

TRANSACTIONS				
Exporting: Spot transactions	Exporting: Long-term contracts	Exporting: With foreign distributor/agent	Licensing technology and trademarks	Franchising

VERSUS

DIRECT INVESTMENT			
Joint venture		Wholly-owned subsidiary	
Marketing and distribution only	Fully integrated	Marketing and sales only	Fully integrated

Figure 22.1 Alternative modes of market entry
Source: Grant (2005), p. 424.

At one extreme, exporting (one-off or 'spot' transactions) involves least commitment, as it is relatively inexpensive and withdrawal is easy. At the other extreme, establishing a fully-owned foreign operating subsidiary involves managing a range of complex functions including production, marketing and distribution. This is relatively expensive and requires substantial long-term commitment.

exporting
Sale of goods and services to a foreign customer

Exporting includes both indirect export of products from the home country via independent agents or distributors, and direct export of products through the firm's own export division to foreign markets. With exporting, most value-adding activity takes place in the home country, while FDI transfers many of these activities to the foreign location. Both exporting and FDI involve **internalisation**, i.e. retaining value-adding activities within the firm (although to different degrees).

Licensing is often a halfway house that results in **externalising** most of these activities. It involves transferring to a licensee the right to use corporate assets, such as a brand name, and often the sale of intermediate goods for the licensee to use in production. Licensing enables a firm to gain rapid overseas market penetration when it lacks the resources to set up overseas operations.

■ Choosing between entry modes: the determinants

The choice of entry mode hinges on several factors:

1 Whether the firm's source of competitive advantage is based on location-specific factors, e.g. low labour costs. If so, the firm is more likely to export. If managerial skills are transferable to other locations, FDI is more likely.

2 Whether the product is tradable, and whether barriers to trade exist. If import restrictions such as tariff barriers and government-imposed regulations make trade infeasible, licensing or FDI becomes more likely.

3 Whether the firm possesses the required skills and resources to build and exploit competitive advantage abroad. Marketing and distribution capabilities are essential for foreign operation, which argues in favour of appointing a distributor or agent when these skills are lacking. If a wider range of manufacturing and/or marketing skills is needed, the firm may license its product and/or technology to a local operator.

In marketing-intensive industries, MNCs may offer their brands to local firms by trademark **licensing**. Licensing involves assigning production and selling rights to producers located in foreign locations in return for royalty payments. If tighter control over operations is required, a joint venture with a local firm may be set up.

patents
A legal device giving the holder the exclusive right to exploit the technology described therein

4 Whether the resources are 'appropriable' – can the technology be stolen? In some industries, technology can be closely guarded via **patents**, but in more service-oriented activities such as computer software, ownership rights are far more difficult to enforce. In the former case, FDI involves less risk, while in the latter case, exporting may be preferred. When licensing a foreign partner, the firm must consider any potential damage to the reputation of its brand resulting from poor quality or service provided by a local operator.

Where a firm wants to exert close control over the use of its trademarks, technologies or trade secrets, **franchising** may be chosen as a way of licensing a fully-packaged business system, as in the international fast food business.

franchising
Licensing out of a fully-packaged business system, including technology and supply of materials, to an entrepreneur operating a separate legally-constituted business

5 What transactions costs are involved? These are the costs of negotiating, monitoring and enforcing the terms of such agreements as compared with internationalisation via a fully-owned subsidiary. With low or no transactions costs, exporting would usually be preferred.

■ Factors favouring foreign direct investment (FDI)

International expansion through FDI is an alternative to growth focused on the firm's domestic market. A firm may choose to expand horizontally on a global basis by replicating its existing business operations in a number of countries, or via international vertical integration, backwards by establishing raw material/components sources, or forwards into final production and distribution. Firms may also choose product diversification to develop their international business interests. Firms may expand internationally by greenfield (new 'start-up') investments in component and manufacturing plants, etc.; takeover of, and merger with, established suppliers; or forming joint ventures with overseas partners. In the case of foreign market servicing, the MNC may also choose to complement direct investment with some exporting and licensing.

FDI also provides opportunities for exploiting competitive advantages over rival suppliers. A firm may possess advantages in the form of patented process technology, know-how and skills, or a unique branded product that it can better exploit and protect by establishing overseas production or sales subsidiaries. A production facility in an overseas market may enable a firm to reduce its distribution costs and keep it in closer touch with local market conditions – customer tastes, competitors' actions, etc. Moreover, direct investment enables a firm to avoid government restrictions on market access, such as tariffs and quotas, and the problems of currency variation. For example, the growth of protectionism by the European Union and the rising value of the yen were important factors behind increased Japanese investment in the EU, especially in the UK motor and electronics sectors.

Firms may benefit from grants and other subsidies by 'host' governments to encourage inward investment. Much Japanese investment (e.g. Nissan's plant near Sunderland) was attracted into the UK by regional selective assistance. In the case of sourcing, direct investment allows the MNC to take advantage of some countries' lower labour costs or provides them with access to superior technological know-how, thereby enhancing their international competitiveness.

Moreover, direct investment, by internalising input sourcing and market servicing within one organisation, enables the MNC to avoid various transaction costs: the costs of finding suppliers and distributors and negotiating contracts with them; and the costs associated with imperfect market situations, such as monopoly surcharges imposed by input suppliers, unreliable sources of supply and restrictions on access to distribution channels. It also allows the MNC to take advantage of the internal transfer of resources at prices that enable it to minimise its tax bill ('transfer pricing') or practise price discrimination between markets.

Finally, for some products (e.g. flat glass, metal cans, cement), decentralised local production rather than exporting is the only viable way an MNC can supply an overseas market because of the prohibitive costs of transporting a bulky product or one which, for competitive reasons, has to be marketed at a low price.

22.4 THE INCREMENTAL HYPOTHESIS

Johanson and Wiedershcim-Paul (1975) observed that firms tend to progress through several stages of foreign market servicing, typically beginning with exporting to obtain a 'bridgehead' and eventually graduating to FDI. A major factor in the progression is the relative cost of each servicing mode, which is a function of market size, or volume of sales.

This thesis can be illustrated with a simple break-even diagram, using Buckley and Casson's (1981) model. Two possible entry modes, exporting and FDI, are displayed in Figure 22.2.

These activities have quite different cost structures:

■ Exporting involves relatively low fixed costs (OA) and relatively high unit variable costs, witness its steeper total cost profile.
■ FDI involves relatively high investment fixed costs (OB) and relatively low variable costs due to lower production and distribution costs, reflected in a flatter TC profile.

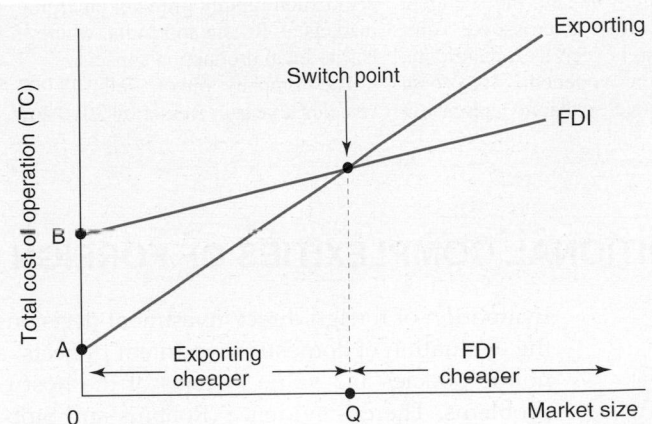

Figure 22.2 Exporting vs. FDI
Source: Buckley and Casson (1981).

For relatively low volumes, exporting may be more appropriate but the firm will eventually encounter a 'crossover' point where it becomes cheaper to switch to FDI. The switchover occurs at market size OQ. Below this, exporting involves lower total operating costs, but beyond OQ, a switch to FDI becomes preferable.

In practice, the switch decision is less clear-cut. Switchover costs would 'step' the FDI function, and most firms would wait a while before switching to ascertain whether the market expansion was permanent. Moreover, some firms 'jump' the intermediate stages if the foreign market size is already deemed sufficiently large.

Self-assessment activity 22.2

Consider how licensing would fit into Figure 22.2 – i.e. what is its likely cost structure relative to exporting and FDI?

(Answer in Appendix A at the back of the book)

Peugeot to build plant in Russia　　FT

PSA Peugeot Citroën is to invest €300m ($442m) in a greenfield assembly plant in Russia in a big push into one of the world's fastest-growing car markets.

Europe's second-largest automaker said it would start construction by the summer on the factory, which will have capacity to produce 150,000 cars a year and be located on a 200-hectare site in Kaluga, 180km south-west of Moscow.

Peugeot said it would use the plant to produce unspecified mid-size vehicles, which account for about 60 per cent of sales on the Russian market. The carmaker's mid-range includes the Peugeot 308 and Citroën C4 small family cars.

A company executive said the carmaker was not ruling out taking a partner at the plant. Mitsubishi Motors, which makes the Peugeot 4007 and Citroën C crossover vehicles in Japan, is seen as a candidate.

Volkswagen opened its own 150,000-vehicle capacity plant in

Kaluga last November and Peugeot said it chose the location over a rival site in Nizhny Novgorod mainly because it had space for an adjacent suppliers' park.

Swedish truckmaker Volvo last October began construction on a plant in the city that, when completed in 2009, will build heavy trucks under its own brand and that of Renault, its shareholder.

Christian Streiff, Peugeot's chief executive, has identified Eastern Europe, along with South America and China, as a priority market for development as it seeks to capitalise on growth away from its core volume-vehicle business in France and Western Europe.

Peugeot, which made 32 per cent of its sales outside Western Europe last year, is trailing other global carmakers in such rapidly growing emerging markets as Russia and India, where it has no local production capacity.

The company aims to sell 100,000 vehicles a year in Russia by 2010 and

300,000 over the medium term. Peugeot's vehicle registrations there rose by a third to 36,000 units last year, behind the market's growth rate.

Renault, Peugeot's closest rival, sold more than 100,000 vehicles in Russia last year, including low-cost Logan sedans. It made 35 per cent of its overall group sales outside Western Europe.

Renault last month announced an agreement with Avtovaz, Russia's largest domestic carmaker, on a partnership that would boost both companies' local production capacity and make Russia a larger market for Renault than France.

General Motors, which already produces cars in Russia, and its US competitor Chrysler have held separate talks with Gaz, another Russian automaker, about possible joint production.

Source: John Reed, *Financial Times*, 30 January 2008.

22.5 ADDITIONAL COMPLEXITIES OF FOREIGN INVESTMENT

Evaluation of foreign direct investment decisions is not fundamentally different from the evaluation of domestic investment projects, although the political structures, economic policies and value systems of the host country may cause certain analytical problems. There is evidence (Robbins and Stobaugh, 1973; Wilson, 1990; Neale and Buckley, 1992) that the majority of MNCs use essentially similar methods for evaluation and control of capital investment projects for overseas subsidiaries and for domestic operations, although they may well apply different discount rates.

Appraising foreign investment involves financial complexities not encountered in evaluating domestic projects. The main ones are:

1 Fluctuations in exchange rates over lengthy time periods are largely unpredictable. On the one hand, these may enhance the domestic currency value of project cash flows, but depreciation of the currency of the host country will reduce the domestic currency proceeds.

2 A foreign investment project may involve levels of risk quite different from those of the equivalent project undertaken in the domestic economy. This poses the problem of how to estimate a suitable required rate of return for discounting purposes.

3 Once up-and-running, the foreign investment is exposed to variations in economic policy by the host government (e.g. tax changes), which may reduce net cash flows.

4 Investment incentives provided by the host country government. Several countries in the 'New Europe' have been able to attract new investment from MNC firms, especially in the motor industry. Slovakia, with a tax rate of 19 per cent can offer definite advantages over established EU members such as Germany (30–33 per cent) and France (33⅓ per cent), among the alleged losers in this process. Other EU countries with low tax rates (as at 2008) include Cyprus (10 per cent), Ireland (12.5 per cent), Hungary (16 per cent), and Bulgaria (10 per cent), while Estonia levies no tax on retained profits (21 per cent on distributions). 'Tax competition' and 'tax harmonisation' issues are likely to remain on the EU agenda for some time (**www.deloitte.com**).

5 Overspill effects on the firm's existing operations; e.g. goods produced overseas may displace some existing sales (referred to as "cannibalisation".)

6 The host government may block the repatriation of profits to the home country. A project that is inherently profitable may not be worth undertaking if the earnings cannot be remitted. This raises the issue of whether the evaluation should be conducted from the standpoint of the subsidiary (i.e. the project itself), or from that of the parent company. But for a company pursuing shareholder value, the relevant evaluation is from the parent's standpoint. Some companies have adopted ingenious ways of repatriating profits. Pepsico, which invested in a bottling plant in Hungary, found it difficult to repatriate profits from this operation. To overcome this problem, it financed the local shooting of a motion picture (*The Ninth Configuration*), which was then exported to the West. (It was not a box-office hit.)

Usually, only remittable cash flows (whether or not they are actually repatriated) should be considered, and the project accepted only if the NPV of the cash flows available for investors exceeds zero. Thus items like management fees, royalties, interest on parent company loans, dividend remittances, and loan and interest payments to the parent should all be included. In effect, a two-stage analysis is applied:

(i) specify the project's own cash flows.
(ii) isolate the cash flows remittable to the parent.

Due allowance should be made for taxation in the foreign location.

■ Overcoming exchange controls

Possible ways of minimising the impact of controls over cash repatriation include:

(i) paying interest on loans or dividends on equity. Maximising dividend flows may involve 'creative accounting' to inflate the local profits, although this may be counter-productive if local tax rates exceed those in the home country. It may also antagonise local interests.

(ii) paying royalty fees where the foreign project utilises any process over which the parent claims proprietary rights, e.g. control of a patent or trade mark, such as Levis.

(iii) transfer pricing policies that involve charging the overseas subsidiary high prices for components and other supplies.

(iv) applying a management charge if the senior managers are seconded from the parent. Similar charges can be used to make the foreign subsidiary pay for other services provided by the parent, e.g. IT and treasury costs. However, this may provoke close scrutiny by the host authorities.

Self-assessment activity 22.3

How does FDI differ from domestic investment?

(Answer in Appendix A at the back of the book)

22.6 THE DISCOUNT RATE FOR FOREIGN DIRECT INVESTMENT (FDI)

Opinions differ as to the appropriate discount rate to apply when discounting net cash flows from foreign investment. 'Gut feelings' may suggest that FDI should be evaluated at a higher discount rate than domestic investment 'simply because it involves more risk'. But is this valid?

Assuming all-equity finance, two commonly-suggested possibilities are:

■ Use a required return based on the risk of similar activities in the home country. This would be the equity cost for projects in activities similar to existing ones, or a 'tailored' project-specific rate for ventures into new spheres.

■ Use a required return comparable to that of local firms.

Before deciding which approach is preferable, there are several issues to consider:

1 Foreign projects generally involve higher levels of risk than domestic ones. However, much of this risk can be dealt with in better ways than simply 'hiking' the discount rate. For example, currency risk can be handled by the sort of hedging techniques mentioned in Chapter 21 (if thought necessary), and political risk can be handled in the ways discussed later.

2 Although the total risk of FDI is often very high, the relevant risk is generally much lower, and sometimes lower than for comparable investment at home. FDI involves diversifying into overseas markets in the same way as an investor might diversify shareholdings across international markets. Solnik (1974) showed that the correlation between national stock markets is generally much less than one (although it may have increased in recent years). If investors can lower relevant risk by cross-border portfolio diversification, why should firms not do so? This is especially relevant for firms whose investors cannot diversify internationally, due to exchange controls or transactions costs, or into countries where no organised stock exchange operates.

The relevant Beta value for, say, a UK firm operating abroad may not be the Beta calculated by reference to the UK market portfolio, but the Beta in relation to the local market (if there is one). Consider the following example.

Example: Malaku mining

Malaku Mining is the newly-formed Indonesian subsidiary of Mowmack plc, an all equity-financed UK firm, whose shares have a Beta value of 0.9 relative to the UK market portfolio. The total risk (standard deviation) of the Malaku project is 30 per cent and the risk of the Jakarta Stock Exchange is 20 per cent. The UK stock market has a 50 per cent correlation with the Jakarta market.

Here, the parent Beta of 0.9 is inappropriate. Instead of using this value, it is more appropriate to consider the risk of the proposed activity in relation to the local market and allow for the low correlation between the London and Jakarta exchanges. We can do this by using a variation of the formula found in Section 10.5:

Project Beta = correlation coefficient × (risk of the activity/local market risk)

This yields a project-specific Beta of:

$(0.5 \times 30\%/20\%) = (0.5 \times 1.5) = 0.75$

The cost of equity would be calculated using the UK risk-free rate of, say, 5 per cent and the UK-equity market risk premium of, say, 5 per cent, as the firm has a UK investor base. Hence, the required rate of return for all-equity funding would be found from the usual CAPM formula:

$k_e = R_f + \text{Beta}[ER_m - R_f] = 5\% + 0.75[5\%] = 8.75\%$

Self-assessment activity 22.4

An ungeared UK firm, with a Beta of 1.4, plans to invest in an emerging country that has no stock exchange. Its economy has a weak correlation (0.4) with the UK. Due to operating gearing, the foreign project is 25 per cent more risky than the UK parent. The risk-free rate is 5 per cent and the expected overall return on the UK stock market is 11 per cent p.a. What return should the UK firm seek on this project?

(Answer in Appendix A at the back of the book)

22.7 EVALUATING FDI

Under certain conditions, analysing foreign project cash flows will yield the same result as analysing the cash flows to the parent. The key conditions are:

(i) Exchange rates adjust to reflect inflation differences between the parent country and the foreign location, i.e. PPP applies.

(ii) Project cash inflows and outflows move in line with prices in general in both locations.

(iii) No tax differentials exist between the two countries.

(iv) No exchange controls.

Because one or more of these conditions probably will not apply in practice, the following steps are generally recommended:

1 Predict local cash flows in money terms, i.e. including local inflation.

2 Allow for any 'overspill' effects like the 'cannibalisation' of existing exports. The opportunity cost is neither the sales revenue nor indeed the profit lost, but the gross margin or contribution.

3 Calculate the project's NPV using a discount rate reflecting the cost of finance in host country terms.

4 Allow for any management charges and royalties.

5 Estimate parent company cash flows by applying the expected future exchange rate to host country cash flows if there are no blocks on remittances, or to net remittable cash flows if exchange controls operate.

6 Allow for both local and parent country taxation.

7 Calculate the project's NPV.

■ Differences between host and parent country taxation

Quite apart from different tax rates, taxation issues can complicate FDI in several ways, most notably, if there are different systems of investment incentives in the two countries, and whether or not a Double Taxation Agreement (DTA) operates. Under a DTA, tax paid in the host country is credited in calculating tax in the parent country. The generally-recommended procedure is to:

1 Allow for host country investment incentives before applying the local tax rate to local cash flows.

2 Apply the relevant UK rate of tax to remitted cash flows only.

3 Adjust stage 2 for any double tax rules. For example, with a DTA and host country tax payable at 15 per cent and where the rate of tax applicable in the UK is 30 per cent, the relevant rate of tax to apply to remittances is (30% − 15%) = 15%.

Self-assessment activity 22.5

A UK MNC earns cash flow (all taxable) of $100 million in the USA. What is its overall tax bill if the rate of tax on profits in both the USA and the UK is 30 per cent and a DTA applies?

(Answer in Appendix A at the back of the book)

A full examination of the complexities of FDI is beyond our scope. However, several of these features are brought out in the example below.

22.8 WORKED EXAMPLE: SPARKES PLC AND ZOLTAN KFT

A UK company, Sparkes plc, is planning to invest £5 million in the Zoltan consumer electronics factory in Hungary. The project will generate a stream of cash flows in the local currency, Forints, which have to be converted into sterling as in Table 22.1. Should Sparkes invest?

First, we must consider the time dimension. The project is capable of operating for ten years, but the host government has expressed its desire to buy into the project after four years. This may signal to Sparkes the possibility of more overt intervention, possibly extending to outright nationalisation, perhaps by a successor government. It seems prudent to confine the analysis to a four-year period and to include a terminal value for the project based on net book values. If we assume a ten-year life, straight-line depreciation and ignore investment in working capital, the NBV after four years will be 60 per cent of Ft1,000 m = Ft600 m. Half of this can be treated as a cash inflow

Table 22.1 Sparkes and Zoltan: project details

■ Expected net cash flows from Zoltan in millions of Hungarian Forints (HUF) (at current prices):

Year	0	1	2	3	4
	−1,000	+400	+400	+400	+400

■ The project may operate for a further six years, but the local government has expressed its desire to purchase a 50 per cent stake at the end of Year 4. The purchase price will be based on the net book value of assets.

■ The spot exchange rate between sterling and Forints is 200 per £1. The present rates of inflation are 25 per cent in Hungary and 5 per cent for the UK. These rates are expected to persist for the next few years.

■ For this level of risk, Sparkes requires a return of 10 per cent in real terms.

paid by the host government and half as a (perhaps conservative) assessment of the value of Sparkes' continuing stake in the enterprise.

In practice, we often encounter complications in assessing terminal values. For example, the assets may include land, which may appreciate in value at a rate faster than general price inflation. If so, there may be holding gains to consider, gains which may well be taxable by the host government. However, it is unwise to rely overmuch on terminal values – if project acceptance hinges on the terminal value, it is probably unwise to proceed with this sort of project.

Second, how should we specify the cash flows? Here, we have two problems: first, divergence between UK and Hungarian rates of inflation; and second, the need to convert locally-denominated cash flows into sterling. To be consistent, we should discount nominal cash flows at the nominal cost of capital or real cash flows at the real cost. Each will give the same answer, but we conduct the analysis in nominal cash flows, thus incorporating the effect of inflation. Hence all cash flows are inflated at the anticipated Hungarian rate of inflation of 25 per cent.

As it is assumed that we are evaluating this project from the standpoint of Sparkes' owners, we need to obtain a sterling NPV figure. There are two ways of doing this.

The inflated cash flows in HUF are shown in Table 22.2. These are converted into sterling using forecast future spot rates. According to PPP, sterling will appreciate by the ratio of the respective inflation rates, i.e. $(1.25)/(1.05) = 19\%$ p.a. The predicted future spot rates are also shown in Table 22.2. The resulting sterling cash flows are then discounted at 10% to obtain a positive NPV.

Table 22.2 Evaluation of the Zoltan project

Year	Un-inflated cash flow in HUFm	Inflated at 25% (HUFm)	Forecast future spot rates: HUF vs. £1	Cash flows in sterling (£m)	PV in £m @ 10%
0	(1,000)	(1,000)	200	(5.00)	(5.00)
1	400	500	238	2.10	1.91
2	400	625	283	2.21	1.83
3	400	781	337	2.32	1.74
4	400	977	401	2.44	1.67
4	600*	600	401	1.50	1.02
				NPV =	+2.17
				i.e.	+£2.17 m

*not inflated

Alternatively, we could proceed by discounting the inflated cash flows at a discount rate applicable to a comparable firm in Hungary, thus arriving at an NPV figure in local currency, and then convert to sterling. The local discount rate using the Fisher formula (I_H) = Hungarian inflation) is:

$$(1 + P)(1 + I_H) - 1 = (1.10)(1.25) - 1 = (1.375 - 1) = 0.375, \text{i.e. } 37.5\%.$$

To obtain a Sterling NPV, we adjust the NPV in HUF terms at *today's* spot rate of 200HUF vs. £1. Table 22.3 shows the result of this operation. Allowing for rounding, the NPVs are identical, i.e. the project is worth £2.18 m to Sparkes' shareholders.

■ The two equivalent approaches: which is best?

Consider the following example. Say a UK firm is investing in Canada. The two approaches are:

A The predicted C$ cash flows are converted to sterling cash flows, using expected future spot rates. These sterling cash flows are discounted at the sterling discount rate to yield a sterling NPV, numerically the same as in Approach B.

Table 22.3 Alternative evaluation of the Zoltan project

Year	Inflated cash flows (HUFm)	PV in HUFm @ 37.5%
0	(1,000)	(1,000)
1	500	364
2	625	331
3	781	300
4	977	273
4	600	168
		NPV = 436
	In sterling, @ spot of 200HUF vs. £1	= 436/200 = £2.18 m

B   The cash flows denominated in C$ are discounted at the local Canadian discount rate to generate a C$ NPV. This is then converted at the current spot rate for C$ against sterling to yield a sterling NPV.

Each approach has the same departure point, i.e. the C$ cash flows, and ends up at the same place, i.e. the sterling NPV. Which approach is 'better'? This depends on what information is available. Both approaches require forecasting cash flows in C$, so forecasts of Canadian inflation are required. Beyond this, it depends whether the financial manager is happier in forecasting future FX rates than in forecasting the required return in local currency terms. Approach B requires merely a one-year forecast of FX rates to derive the required return in C$ terms, whereas Approach A requires forecasts of FX rates over all future years of the project.

Remember that the equivalence of each approach depends on several factors, in particular, the operation of PPP and the existence of project inflation rates similar to those experienced at the national economy level. PPP ensures that if, say, the Canadian inflation rate exceeds the UK rate, the exchange rate of C$ vs. sterling will deteriorate, i.e. sterling will strengthen to ensure the parity of purchasing power of each currency in each country.

Either approach is acceptable, although the first is preferable as it has the advantage of allowing cash flows to be adjusted at inflation rates specific to the project where these may differ from the national rate, although this does require more detailed forecasting.

22.9 EXPOSURE TO FOREIGN EXCHANGE RISK

You may appreciate now that FX variations are not always disastrous. The extent of FX exposure, and thus the urgency of dealing with it, depends on the structure of the firm's net cash flows in terms of its FX denomination. Firms with naturally-hedged cash flows may be relatively unconcerned by FX variations. However, firms differ in the extent to which they are naturally hedged.

■ The four-way classification

Figure 22.3 shows a schema for classifying the extent of a firm's exposure to FX variations. Essentially, this depends on the sensitivity of their domestic currency cash flows to FX movements. Net cash inflows are broken down into their revenue and operating cost components, in order to focus on firms' net exposure. Classified by corresponding sensitivity, the four types of firm are:

■ *Domestics* generate little or no income from abroad and source mainly from local suppliers. Their net exposure is indirect and usually low, stemming from the exposure suffered by their competitors on the UK markets, and by their local suppliers.

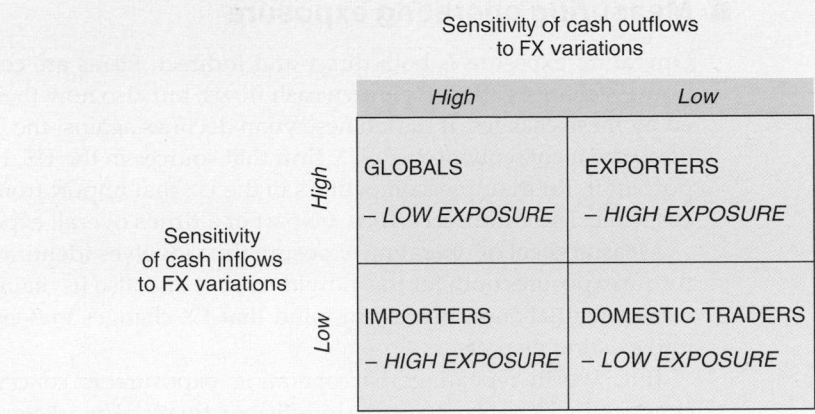

Figure 22.3 Classification of firms by extent of operating exposure

- *Exporters* source mainly from their own country, have high direct net exposures as their cash inflows and outflows are not naturally hedged, being in different currencies. They might consider adjusting their operating and/or financial strategies to achieve more insulated positions.
- *Importers* are in a similar position to exporters but in reverse – their net exposure is high because they source from abroad and sell on domestic markets.
- *Globals* have the lowest, and often minimal, net exposure. They have structured their operations so as to match as far as possible the currencies in which their inflows are denominated with those in which they incur costs. The match may not be perfect, given the indivisibility of some types of operation, e.g. production facilities, but regarding the overall profile of activities, their portfolios of cash inflow currencies should correlate highly with their portfolios of outflow currencies. At the group level, global firms should have little concern about FX exposures.

This is a powerful set of distinctions, but is counter-intuitive to many people. Firms heavily engaged in foreign operations may actually have low net exposures while many domestics, blissfully thinking they are insulated from overseas-generated exposures may, in reality, be more highly exposed. A high indirect exposure could conceivably outweigh a low direct exposure.

■ Operating/economic/strategic exposure

If PPP always worked, forecasting FX rates would be very simple – in practice, prolonged uncertainty over future exchange rates and thus the effect on the firm's future cash flows in domestic currency terms greatly concerns many financial managers. The longer the time horizon that the firm works to, the greater is its concern. Continuing exposure over a period of years is called **economic exposure**. This refers to the effect of changing FX rates on the value of a firm's operations, generally the result of changing economic and political factors, hence the alternative label **operating exposure**. Because these variations will also affect the firm's competitive position, and because protecting or enhancing that position often provokes a change in strategy, it is also called **strategic exposure**.

These three terms are often used as synonyms. Whatever we call it, the impact is felt on the present value of the firm's operating cash flows over time, and thus the value of the whole enterprise. To prevent or mitigate damage, the firm can adopt various strategies to protect its inherent value. Measurement of exposure is the first step.

■ Measuring operating exposure

Operating exposure is both direct and indirect. Firms are concerned not only about how FX changes affect their own cash flows, but also how their competitors are affected by these changes. If the Chinese yuan declines against the US dollar, this may seem of no great consequence to a UK firm that sources in the US. However, it becomes important if, for example, competitors in the US that import from China see their import costs fall. These indirect effects are part of a firm's overall exposure.

Measurement of operating exposure thus involves identification and analysis of all future exposures both for the individual firm and also its main competitors, actual and even potential ones, bearing in mind that FX changes may even entice new entrants into existing markets.

It is worth repeating that operating exposure is concerned less with *expected* changes in FX rates, because, in efficient financial markets, both managers and investors will have already incorporated these into their anticipation of parent company currency cash flows. If the markets expect sterling to decline vs. the US dollar, the likely higher future sterling cash inflows of UK firms that export to the USA will have been factored into company valuations. In this situation, it is generally advisable to incorporate the forward rate of exchange into projections for future planning purposes. The damage is done when expectations are not fulfilled and/or when changes result from totally unexpected factors.

Example: Pitt plc

Pitt plc produces half its output in the USA, valued at today's exchange rate (US$1.50 vs. £1) at £100 million. The other half is sold in the UK. About 25 per cent of Pitt's supplies are sourced from the USA, valued at £15 million. Labour costs are £10 million per annum, and cash overheads are £5 million per annum. Shareholders require a return of 12 per cent per annum.

Required
(a) Determine the present value of Pitt's operating cash flows in sterling terms over a 10-year time horizon.
(b) Identify Pitt's direct and indirect operating exposures.
(c) What is the effect on Pitt's PV if the sterling/dollar exchange rate changes to US$1.40 vs. £1?

Solution
(a) At the present exchange rate, Pitt's cash flows are:
- Cash inflows: $2 \times £100$ m p.a.   = £200 m
- Cash outflows:
  Supplies ($4 \times £15$ m p.a.)   = (£60 m)
  Labour   = (£10 m)
  Overheads   = (£5 m)
                             Net Cash Flow   = £125 m

Present value = (£125 m × 10-year annuity factor at 12%)
              = (£125 m × 5.650) = £706 m

(b) Direct exposures:
- 50 per cent of cash inflows are exposed.
- 25 per cent of payments to suppliers are exposed.
- Any US$ content of labour input and cash overheads would also be exposed.

Indirect exposures stem from the extent to which:

- US competitors are exposed to currency fluctuations.
- US suppliers are exposed.
- UK competitors are exposed.
- UK suppliers are exposed.

As discussed, it is important to realise that overall exposures transcend variations in the US$/£ rate. If, for example, suppliers in the UK import from India, they face exposure from the exchange rate of the rupee vs. sterling. Adverse variations are likely to spur them to recoup cost increases from their customers. Obviously, the extent to which they can achieve this depends on their own market power, e.g. the number and relative size of their own competitors and the importance of the components to customers like Pitt plc.

(c) Sterling depreciation to US$1.40 vs. £1 will increase the sterling value of the net cash inflows, because, at present, annual USD cash inflows exceed USD cash outflows. At the current exchange rate of US$1.50 vs. £1, the annual difference is (£100 m − £15 m) = £85 m p.a.

Revised valuation:
- Inflows: [£100 m + (£100 m × 1.50/1.40)] = £207 m p.a.
- Outflows:
  Supplies: (3 × £15 m) + (£15 m × 1.50/1.40) = (£61 m)
  Labour = (£10 m)
  Overheads = (£5 m)
  Net Cash Inflow = £131 m
- PV @ 12% = (£131 m × 5.650) = £740 m

In this example, depreciation of sterling by (1−1.40/1.50) = 7% has resulted in an increase in firm value of about 5 per cent. The sensitivity of a firm's value will depend on the structure of the firm's cash inflows and outflows – the greater the net foreign currency component, the greater the sensitivity of firm value.

Comment

The result is somewhat oversimplified for several reasons:

It assumes no change in the volumes of US-generated business in response to sterling depreciation. In reality, Pitt may lower the US$ price to stimulate sales as it can now afford a price cut of up to 7 per cent and still achieve the same sterling cash inflow, after converting US$ into sterling at the more favourable rate.

The effect will depend on:

(i) the extent of the price cut, i.e. whether Pitt matches the 7 per cent fall in sterling or takes some or all of this as windfall profit.

(ii) the elasticity of US demand for the product, i.e. the extent to which demand is stimulated by a price cut.

(iii) whether Pitt can produce enough to satisfy the demand increase.

(iv) the extent to which both US competitors and also other foreign firms supplying the US market follow a price cut.

Consequently, a full answer would depend on more rigorous strategic evaluation of the various consequences of the sterling depreciation. Similar comments apply to the increased sterling costs of supplies. Will Pitt try to absorb these costs? Will it try to pass them on wholly or partly? Will it seek alternative UK suppliers? In addition, there may be wider effects – will Pitt win sales from US competitors in the UK and elsewhere? How will fellow UK rivals respond to the sterling depreciation? How will non-exporting UK firms that import from the USA respond?

Having to grapple with these issues provides a powerful stimulus for seeking ways of trying to negate the effects of FX exposure.

Self-assessment activity 22.6

A UK-owned MNC produces in the USA and also exports to South American countries direct-ly from the UK where it is paid in US$. Its US revenue is $50 million and its operating costs are $30 million. Export sales to South America are $100 million. What is its exposure to the US$/sterling exchange rate?

(Answer in Appendix A at the back of the book)

22.10 HOW MNCs MANAGE OPERATING EXPOSURE

Managing operating exposure involves taking steps to insulate the firm's operating cash flows as far as possible from the effect of unexpected FX changes, so as to min-imise the effect on the value of the whole firm.

In Chapter 21, we explained a number of theories of the operation of FX markets. The upshot of these is that, if they are valid, firms need not worry about FX variations. However, because foreign exchange markets cannot always be relied upon to move quickly to new equilibria, it is often considered prudent to safeguard against the impact of future contingencies. Moreover, given the strongly competitive nature of many international markets, it is sensible to anticipate how competitors are likely to behave when operating conditions change.

There are two broad ways in which operating exposure can be minimised. The first involves structuring the firm's operations to insulate it from damage, and the second, structuring its financial policy to this end.

The general aim is to construct a **natural hedge**. This occurs when there is little or no exposure because the adverse impact on cash inflows is exactly offset by the benefi-cial impact on cash outflows. A British firm producing and selling in the USA has a high degree of protection against sterling/dollar variations. If sterling appreciates, thus reducing the sterling value of dollar inflows, the adverse effect is largely counter-balanced by the corresponding fall in the sterling value of its US$-denominated inputs. Only the profit element is unhedged.

In November 2007, Rolls-Royce announced plans to invest in aero-engine facilities in the US and Singapore to reduce its exposure to the US$, the currency in which planes and engines are priced. In 2008, Airbus will open an assembly line for the A320 in China, and is considering an investment in Alabama in the USA, also as a response to the weak US$. Both firms are attempting to 'self-hedge'.

Firms that service overseas locations can achieve this 'self-hedging' effect in various ways:

- Source components and other inputs in the countries where sales are made.
- Open an operating subsidiary there to manufacture or assemble the product.
- Borrow in the same currency as that of cash inflows. The stream of outflows (inter-est and capital repayments) will help to match the series of inflows.
- The firm may pay suppliers in other countries with the currency received from sales. If it sells in US$, and sources from Poland, it can pay the Polish suppliers in US$ rather than zlotys. This is called **currency switching**.

currency switching
Where an exporter pays for imported supplies in the currency of the export deal

The hedge constructed may not be a match in the same currency. As some currencies tend to move together due to the interdependence of the associated economies, e.g. Canada and the USA, the MNC could offset, say, a US$ revenue stream with a C$ cost stream. This is termed **parallel matching**.

Where several locations are serviced, the key is diversification of operations. A diversified firm is better placed both to recognise dis-equilibrium situations in foreign exchange markets and also to adapt accordingly. For example:

- If relative costs in sterling terms alter as between different locations, a diversified MNC can arrange to reschedule production between locations.

- If product prices change in different markets, the MNC may strengthen its marketing efforts to exploit greater profit opportunities where prices are higher.
- If raw material prices alter as between different locations, the MNC can alter its sourcing policy.

Admittedly, such adjustments are likely to trigger certain conversion or switch-over costs, but the MNC may still benefit from favourable 'portfolio effects'. The variability of cash flows in domestic currency terms is likely to fall if the firm receives income in a variety of currencies – foreign exchange rate variations may increase competitiveness in some markets to offset lower competitiveness in others. This, of course, underpins the diversification motive for foreign investment, mentioned earlier in the chapter. Portfolio diversification was considered more fully in Chapter 9. The cameo below illustrates some of the economic risks facing Japanese car-makers, most of which are very well diversified by market and by location of production.

Self-assessment activity 22.7

Revisit SAA 22.6.

Suggest three ways for the firm to lower its exposure.

(Answer in Appendix A at the back of the book)

Yen's rise puts brakes on Japan carmakers

Until recently, Japanese carmakers were cruising comfortably to higher profits, courtesy of a favourable exchange rate and firm demand in the US, and booming sales in emerging markets.

But as yesterday's 3.87 per cent fall in Mazda's share price highlights, the smooth road has given way to a more hazardous path that threatens to slow their progress.

In an abrupt change from a year ago, Japanese carmakers face the threat of a rising yen and a slowing US market at a time when their home market shows no signs of shaking off its slump.

A slowdown in US demand was evident in last month's sales performances. Toyota said its unit sales in the US last month fell 2 per cent while Nissan's dipped 7.3 per cent. Honda's US sales fell 2.3 per cent, and Suzuki's plunged 13 per cent.

Honda, which relies on the US for about 60 per cent of sales, blamed the January decline on higher gasoline prices leading to weaker demand for its light trucks, such as the Pilot and Ridgeline, which comprise 40 per cent of its US sales.

At the same time, the yen's rise threatens to dent whatever profits Japanese automakers generate in the US and Europe.

Toyota, for example, suffers a fall in operating profits of Y35bn ($328m) for every one yen rise against the US dollar, while a one yen increase against the euro hits profits by Y5bn.

Koji Endo, auto analyst at Credit Suisse in Tokyo, says that for the industry as a whole, each one yen rise against the dollar results in an operating profit decline of 1.2 per cent to 1.5 per cent, depending on factors such as hedging strategies.

This means profits will fall 12–15 per cent if the yen rises Y10 to average Y105 against the dollar this year.

An additional concern is the higher price of materials such as steel and plastics. This comes as the Japanese market is expected to remain sluggish. Last year, sales of registered vehicles fell for a third successive year, dropping 6.3 per cent to 5.32m units, according to the Japan Automobile Manufacturers Association. JAMA expects a further slump this year.

Against this background, a slowdown in overseas markets will have a big impact on domestic production, which was resilient last year thanks to booming exports.

For the time being, the automakers are putting on a brave face. While none has announced forecasts for the fiscal year starting in April, Toyota and Nissan have not changed their view that they will sell slightly more cars in the US in the year to March than last year.

There is optimism as well that strength in emerging markets will offset the slowdown in the US and Japan.

With the US market expected to slump to between 15.5m and 16m units this year – its lowest level in a decade – Brazil, Russia, India and China, which are expected to see combined sales of 16.5m, are expected to comprise a larger market for automakers than the US.

Mr Endo says that while Japanese carmakers have made in-roads in some emerging markets, none are strong in all. The automakers' subdued optimism about the US economy also depends on a recovery in the second half, which may or may not materialise.

"Higher profits next year are extremely difficult. It is likely that Japanese carmakers will see lower profits for the first time in years," says Mr Endo.

Source: Michiyo Nakamoto, *Financial Times*, 8 March 2008.

22.11 HEDGING THE RISK OF FOREIGN PROJECTS

This issue has already been addressed. Operating a foreign investment involves both translation exposure and economic exposure. The translation risk stems from exposure to unexpected exchange rate movements: for example, a fall in the value of the Australian dollar (AUD) against GBP will reduce the GBP value of assets appearing in the Australian subsidiary's balance sheet. When consolidated into the parent's accounts, this will require a write-down of the value of assets in GBP terms. This problem can be avoided if the exposed assets are matched by a corresponding liability. For example, if the initial investment is financed by a loan denominated in AUD, the diminution in the sterling value of assets will be matched by the diminution in the GBP value of the loan.

Perfect matching of assets and liabilities is not always possible. Many overseas capital markets are not equipped to supply the required capital. Besides, it is probably politic to provide an input of parent company equity to signal commitment to the project, the government and the country. However, perfect matching is probably unnecessary, since some assets such as machinery can be traded internationally. The cause of the exchange rate depreciation (i.e. higher internal prices) will lead to a rise in prices, and if the Law of One Price holds, this appreciation in the value of locally-held assets will compensate for the reduced GBP value of the currency. As a result, the need to match probably applies only to property assets and items of working capital such as debtors, which cannot readily be traded on international markets. (Note the obvious attraction of operating with a sizeable volume of short-term creditors if sterling is strengthening, especially at the financial year-end.)

Economic exposure is the long-term counterpart of transaction exposure – it applies to a stream of cash inflows and outflows. In theory, the problem of variations in the prices of inputs and outputs should also be solved by the operation of the Law of One Price. For example, local price inflation at a rate above that prevailing in the parent company's country will be exactly offset by depreciation in the local currency, thus maintaining intact the GBP value of locally-produced goods.

However, in practice, problems arise when PPP does not apply in the short term and when *project* prices alter at different rates from *prices in general*. The movement in the local price index is only an average price change, hiding a wide spread of higher and lower price variations. In principle, the firm could use the forward market to remove this element of unpredictability in the value of cash flows, but in practice, the forward market has a very limited time horizon, or is non-existent for many currencies.

Nevertheless, the parent company with widely-spread overseas operations can adopt a number of devices. It can mix the project's expected cash flows and outflows with those of other transactions to take advantage of netting and matching opportunities. It can lead and lag payments when it expects adverse currency movements, although host governments usually object to this. It can also use third-party currencies. For example, if it invests in oil extraction, its output will be priced in USD and the otherwise exposed cost of inputs may be sourced or invoiced in USD or in a currency expected to move in line with USD. A more aggressive policy might involve invoicing sales in currencies expected to be strong and sourcing in currencies expected to be weak, perhaps including the local one.

Another tactic is to use the foreign project's net cash flows to purchase goods produced in the host country that are exportable, or can be used as inputs for the parent's own production requirements. This converts the foreign currency exposure of the project's cash flow into a world price exposure of the goods traded. This may be desirable if the degree of uncertainty surrounding the relevant exchange rate is greater than that attaching to the relevant product price.

Much world trade is conducted on the stipulation that the exporter accepts payment in goods supplied by the trading partner, or otherwise undertakes to purchase goods and services in the country concerned. This linking of export contracts with reciprocal agreements to import is known as **counter-trade**. It is usually found where the importer suffers from a severe shortage of foreign currency or limited access to bank credits.

One form of counter-trade is **buy-back**, which is a way of financing and operating foreign investment projects. In a buy-back, suppliers of plant, equipment or technical know-how agree to accept payment in the form of the future output of the investment concerned. This long-term supply contract with the overseas partner raises some interesting principal/agent issues, concerning in particular the quality of the output and the management of the operation. Ideally, the output should be a product for which a ready market is available or that the exporter can use as an input to its own production process. In recent years, Iran has signed buy-back deals with European energy majors Shell, TotalFinaElf and ENI to finance the development of oil and gas projects.

The advantage of buy-backs for a Western company is that they secure long-term supplies and obviate any need to worry about exchange rate movements. The effective cost is the cost incurred in financing the original construction, and perhaps an opportunity cost if world prices of the goods received fall. Buy-backs thus offer a way of locking into the present world price for the goods transferred, which has some appeal in markets where prices fluctuate widely, for example, oil.

22.12 POLITICAL AND COUNTRY RISK

According to *Euromoney*, which conducts a regular analysis of political risk: 'Political Risk is the risk of non-payment or non-servicing of payment for goods and services, or trade-related finance and dividends, and the non-repatriation of capital.'

The definition and the above discussion imply a distinction between economic and political risk, the latter resulting from governmental interference and the former from general economic turbulence. But because these are usually inter-twined, the two risks are often grouped under the heading of **country risk**.

It is not surprising that FDI by MNCs involves strong elements of political risk. Their very size and strength in relation to host countries creates the possibility of politically-inspired action, whether favourable, e.g. granting generous incentives, or adverse, e.g. expropriation of oil company assets as in Venezuela in 2006. Where the goals of the host government and the MNC conflict, the political risk escalates. Political risk is heightened where political and social instability prevails, and host government objectives are unclear.

The task of the MNC's planners is to define, identify and predict these sources of instability. Instability results from internal pressures or civil strife that may be caused by factors such as inequities, actual or perceived, between internal factions (whether racial, religious, tribal, etc.), extremist political programmes, forthcoming independence or impending elections.

Any MNC that is considering FDI may observe the signals of political instability, but to measure its extent is a complex task. A major cause of political and social instability is due to economic influences. Factors such as oscillating oil prices, banking crises, foreign exchange crises and rampant inflation all promote instability.

As in Argentina, Turkey and Egypt in the early 2000s, economic instability often necessitates heavy overseas borrowing to finance reconstruction. Because reforms take time to implement, risk of default is ever-present. Default risk can be gauged by factors such as a country's debt service ratio (debt service payments relative to exports) and the debt age profile. The political risks of such pressures can provoke actions such as:

- exchange controls
- restrictions on registration of foreign companies
- restrictions on local borrowing
- expropriation or nationalisation

- tax discrimination
- import controls
- limitation on access to strategic sectors of the economy.

Expropriation – confiscation of corporate assets with or without compensation – asset freezing – loss of control over asset management – and outright nationalisation represent the greatest political threat to foreign investors. The main risk is less that of expropriation *per se*, but the risk that compensation will be inadequate or delayed.

'Creeping' expropriation may also occur where mounting restrictions on prices, issue of work permits, transfer of shares, imports and dividends become likely when a nation feels threatened by the size and influence of MNCs. Hence, prior to deciding whether to operate in a new country, a pertinent question to consider is whether the MNC, either individually, or collectively with other MNCs, will dominate the industry, as occurs when oligopolistic rivals follow the strategic entry of a leader. If this is a strong probability, the political risk is greater than when penetration is low.

■ Assessing political and country risk (PCR)

There are several ways of assessing PCR:

1 Scoring systems

The magazines *Institutional Investor* and *Euromoney* produce country risk ratings on a regular basis. *Euromoney*'s ratings are based on a weighted average method, using the following scoring system, based on percentages:

- Political risk – 25 per cent
- Economic performance – 25 per cent
- Debt indicators – 10 per cent
- Debt in default or rescheduled – 10 per cent
- Credit ratings – 10 per cent
- Access to bank finance – 5 per cent
- Access to short-term finance – 5 per cent
- Access to capital markets – 5 per cent
- Discount on forfaiting (see Chapter 15) – 5 per cent

Scores and rankings for 'safe' and 'unsafe' countries as at September 2007 are shown in Table 22.4 (Hexter, 2007).

Table 22.4 Country risk scores

Top Ten*			Bottom Ten		
Rank	Country	Score	Rank	Country	Score
1	Luxembourg	99.59	176	Micronesia	21.16
2	Norway	99.51	177	Zimbabwe	19.80
3	Switzerland	99.04	178	Liberia	17.77
4	Denmark	95.78	179	Zaire (D. R. Congo)	17.59
5	Sweden	94.92	180	Marshall Islands	13.93
6	Ireland	94.65	181	Cuba	12.34
7	USA	94.54	182	Somalia	10.72
8	Netherlands	94.39	183	Iraq	5.97
9	Finland	94.11	184	Afghanistan	5.78
10	Austria	93.33	185	North Korea	4.92

* The UK placed 11th.

Source: Euromoney, September 2007.

Although there is a high element of arbitrariness and subjectivity in such systems, there appears to be a strong degree of correlation between competing league tables for fairly obvious reasons. Given the subjectivity, consideration of movements of countries up and down the tables over time is worth study.

2 Delphi technique

Also known as 'consulting the oracle', this involves canvassing a panel of experts for their opinions, via questionnaires or direct personal or telephone contact, and then aggregating the replies. The *Euromoney* analysis is essentially a combination of checklisting and the Delphi technique.

3 Inspection visits

Key staff from the MNC's Head Office make a 'Grand Tour' to the prospective host country, often accompanied by local embassy officials.

4 Using local intelligence

Consulting local experts, e.g. official insiders and credit analysts from banks, to advise on prevailing local trends.

■ Corruption

A major issue in dealing with foreign officials is often corruption. Several agencies attempt to assess the extent to which local officials misuse their positions for personal gain when dealing with MNCs. For example, Transparency International publishes a regular league table, based on a scale of 0–10 (10 = minimum corruption). In its table published in 2007, Denmark and Finland were reckoned to be the least corrupt in the world. Other European countries ranked as follows: Britain (12), Germany (16), France (19), Italy (41), while Somalia and Myanmar were reckoned to be the most corrupt.

Corruption conundrum

Forget about not being on the same page. These folks are reading different books. US enforcers last year brought cases against a record number of multinational companies for paying bribes in countries including Kazakhstan, Indonesia and Nigeria. Last year's 38 civil and criminal cases more than doubled 2006's total, according to Gibson, Dunn & Crutcher, the law firm. By contrast, Nigeria's top anti-corruption policeman has been seconded on a year-long "policy and strategic studies" course after daring to arrest top officials.

Since 1977 the US has barred companies that do business there from engaging in bribery elsewhere but it has dramatically stepped up enforcement recently against non-US companies

with an American presence. Between 1977 and 2006, US companies accounted for 88 per cent of the overseas corruption cases. Today nearly a third of all open investigations involve non-US companies, including BAE Systems, according to the law firm Shearman & Sterling. Many European companies seem to have been caught by surprise, even though countries there started to ban overseas bribery in the 1990s.

The crackdown comes at a difficult time for western companies in commodity sectors where corruption is common. Under scrutiny from regulators, they are encountering new competition from Chinese and Russian companies which may not be similarly constrained. Transparency Institute,

the watchdog, ranked 30 big countries and found that Russian and Chinese corporate executives were among the most likely to pay bribes.

Some multinationals have decided the risks are too high. Swiss logistics company Panalpina, whose dealings are under US investigation, has suspended operations in Nigeria until at least the end of 2008, citing regulatory uncertainty. Willbros, a Panama-based oil services company also seeking to settle with US regulators, sold its Nigerian subsidiary. If the crackdown ends up driving out the cleaner companies, the whole effort may prove self-defeating.

Source: Lex column, *Financial Times*, 10 January 2008.

22.13 MANAGING POLITICAL AND COUNTRY RISK (PCR)

Protection against the adverse consequences of changes in the political or economic complexion of a host country can be achieved in four ways:

- Pre-investment negotiation
- Laying down operating strategies
- Preparation of a contingency crisis plan
- Insurance

These aspects are now considered in turn.

■ Pre-investment negotiation

The best approach to PCR management is to anticipate problems and negotiate an understanding beforehand. Different countries apply different codes of ethics regarding the honouring of prior contracts, especially if concluded under a previous regime. Nevertheless, pre-negotiation does provide a better basis for subsequent wrangling.

An investment agreement sets out respective rights and obligations on both the MNC and the host government. It could cover aspects such as:

- The basis whereby financial flows, such as dividends, management fees, royalties, patent fees and loan repayments may be remitted back to the home country.
- The basis for setting transfer prices used for costing inputs delivered from the home country.
- Rights relating to third-country markets, i.e. who can serve them.
- Obligations to build or fund social infrastructure like schools and hospitals.
- Methods of taxation – rates and calculation procedures.
- Requirements for **offset**, i.e. local sourcing.
- Access to host country financial markets.
- Employment practices, especially regarding openings for nationals. This is very common in the Middle East, e.g. Oman has an 'Omanisation' policy, and the United Arab Emirates operates an 'Emiritisation' policy.

offset
The requirement for an MNC to undertake a proportion of local sourcing as condition of the award of an export deal (very common in the armaments industry)

■ Operating strategies

Flexible, risk-averting strategies that enhance the MNC's bargaining position can be devised in several areas.

Production and logistics

- Local sourcing increases local employment and may head off trouble if local interests would be damaged thereby.
- Siting production facilities to minimise risk, e.g. siting oil refineries in low-risk locations.
- Retaining control of transportation facilities like oil tankers and pipelines.
- Control of technology embodied in patents and the appointment of home country staff to manage complex technological processes. Coca-Cola has never divulged its magic formula, reputedly locked in a bank vault in Atlanta, Georgia.

Marketing

- Controlling markets by eliminating competition – locals are often happy to sell out.
- Controlling brand names and trademarks. McDonald's franchises only operations to local entrepreneurs.

Financial

- Issuing equity on the local stock market to extend ownership to locals. A Joint Venture is also an effective way to promote local participation.
- Restricting parental equity input – local debt financing is preferable to equity funding as it creates a hedge to offset local inflation and exchange rate depreciation, and also exerts leverage on local politicians if local banks stand to suffer from political intervention.
- Multi-source borrowing – raising loans from banks in several countries, and perhaps international development agencies, will build a wide nexus of vested interests in keeping the MNC healthy.

■ Preparation of a contingency crisis plan

Contingency planning helps in two ways, first, by providing an action plan to implement if things do go wrong, and second, it forces managers to think about the contingencies to which their foreign operation is most vulnerable.

Ikea shelves Thai store plans after new curbs on foreign ownership

IKEA, the Swedish retailer known around the world for its flatpack furniture, has put plans to expand into Thailand firmly on the shelf amid an increasingly uncertain environment for multinational investors.

The Daily Telegraph can reveal that the lingering after-effects of last year's military coup and a series of tougher laws on foreign ownership have prompted Ikea to postpone a move into one of South-East Asia's most important economies.

The delay reflects growing nervousness among overseas investors following draconian revisions to Thailand's Foreign Business Act and the publication of draft legislation covering the retail sector, which may have a serious impact on Tesco and Carrefour, the French supermarkets group.

Since the ousting of Thaksin Shinawatra – the former prime minister who now owns Manchester City Football – a year ago, the Thai government has proposed changes to the definition of a "foreign" company to mean one which is not controlled or majority-owned by Thais.

It has also vowed to stamp out the use of nominee shareholders for the subsidiaries of multinationals operating there, a move that has alarmed many overseas firms that have largely relied on the use of such structures.

"The foreign business ownership laws have always been complicated, but there has always been a measure of understanding," said Alastair Henderson, managing partner of law firm Herbert Smith, in Bangkok.

"The latest proposals have meant great uncertainty for companies about the regulatory climate they are going to face and whether they will be able to retain control of their investments."

According to a spokesman for Ikea, the Thai market "is still under evaluation" by Inter Ikea Systems, the organisation that owns the home improvement retailer's concept and trademark.

An unnamed franchisee has been selected to partner Ikea, she added.

"This partner will further investigate the market for final approval by the franchisor.

"At this point in time, it is premature to confirm if and when an Ikea store will open in Thailand."

A continued toughening of foreign ownership laws could affect a string of the UK's biggest companies, including British American Tobacco, Alliance Boots and HSBC.

The country's interim government, which has said it is likely to hold democratic elections in December, has held numerous rounds of talks with officials from the European Commission and overseas chambers of commerce stationed in Bangkok.

One senior official said the Thai government's actions were evidence of "a clear protectionist backlash" and warned that its stance could lead to Thailand being "cut out of the global economy".

Among the foreign investors with most at stake is Tesco, which operates 400 stores and employs more than 28,000 people in Thailand.

"We continue to invest in the country, opening new stores," said a spokesman for Britain's biggest retailer.

"However, we are concerned that the current uncertainty in Thailand may deter new foreign investors from entering the country."

Source: Mark Kleinman, *Daily Telegraph*, 29 September 2007.

■ Investment insurance

MNCs may be able to shift risk to a home country agency that specialises in accepting international risks. In the UK, the Export Credit Guarantee Department (ECGD) offers confiscation cover for new overseas investments and Lloyds offers insurance facilities for existing and new investments in a comprehensive and non-selective form. In the USA, the government-owned Overseas Private Investment Corporation (OPIC) will cover risks relating to inability to convert overseas earnings into dollars, expropriation, war and revolution, and loss of business income arising from events of political violence that damage MNC assets.

22.14 FINANCING FDI

A multinational company (MNC) may have more opportunities to lower the overall cost of capital than a 'domestic'. This is often due to its larger size, and partly due to greater access to international financial markets, allowing it to exploit any temporary disequilibria, as well as receiving host government concessions.

The international financing decision has three elements:

1 *Whether* to borrow?
2 *How much* to borrow?
3 *Where* to borrow?

■ Key issues

The issues that determine these decisions are:

- **Gearing ratio.** The debt–equity mix selected for overseas activities is influenced by the gearing of the parent, or by the debt–equity ratios of comparable, competing firms in the location of the FDI. If the parent guarantees the subsidiary's borrowing, so long as the existing gearing of the parent is 'reasonable', the subsidiary's borrowing may be a separate issue, especially if a joint venture is formed and the borrowing of the overseas affiliate can be kept off-balance sheet. This enables fuller exploitation of subsidised loans and lower tax rates.
- **Taxation.** MNCs need to examine differences between the treatment of withholding taxes, losses, interest and dividend payments. Tax-deductibility of interest provides an incentive to borrow locally.
- **Currency risk.** Local borrowing can also reduce foreign exchange exposure by enabling a match of interest and capital repayments against locally-generated cash inflows denominated in local currency.
- **Political risk.** Expropriation of assets or other interference is less likely if the MNC borrows from local banks or from the international markets. The host government is unlikely to want to offend the international financial community as it would damage its own credit standing. If the MNC borrows via the World Bank, it can include a **cross-default clause**, so that a default to any creditor automatically triggers default on a World Bank loan.

■ The case for borrowing

- Less risky than using equity which puts owners' capital at risk.
- Debt service is based on a strict schedule, helping cash flow planning and currency hedging. Dividends are more erratic.
- Debt service payments are less likely to antagonise host governments. Erratic dividend flows may interfere with a host government's attempts to manage the external value of its currency, especially if the MNC is a major contributor to local GDP.

- Tax relief on interest.
- Opportunity to hedge foreign exchange risk by matching.
- Protection against political risk.
- Access to concessionary local finance.
- Overseas equity markets (if they exist) can be highly inefficient.

■ The case against borrowing

- The parent's debt–equity ratio may already be high.
- There may be only limited local borrowing facilities.
- Local borrowing may entail more intensive credit risk investigation fees, and higher interest rates if the MNC has no local credit rating.
- The host government may place an upper limit on local borrowing to restrict tax avoidance.

Tesco launches first dollar bond

Tesco has launched its first dollar-denominated bond as the UK supermarket group seeks to raise its profile in advance of the opening of its first stores in the US next week.

Britain's biggest retailer, which raised $2bn in bonds, plans to expand aggressively in the US by opening 50 neighbourhood stores.

Tesco, which has about £5bn of bonds outstanding, said it chose to issue in the US dollar market because it wants to introduce US bond investors to Tesco, to establish itself in a new market and diversify its investor base.

"[We wanted] greater awareness of our product in the US," Tesco said. "The bond markets also look pretty healthy at the moment."

The retailer, which is to roll out its Fresh & Easy chain in Los Angeles, San Diego, Phoenix Arizona and Las Vegas, issued $850m in 10-year notes and $1.15bn in 30-year paper.

The bonds were heavily oversubscribed as investors were impressed by the company's performance and its plans to develop small local stores rather than trying to compete directly with the so-called big box retailers in the US such as Safeway, Wal-Mart, Costco and Target, which operate huge, largely out-of-town supermarkets, bankers said.

Tesco, which dominates the UK retail scene, reported a rise of 18 per cent in pre-tax profits to £1.29bn in the six months to August 25 on the back of a 9.2 per cent increase in sales to £24.7bn.

The bonds, which will pay coupons of 5.5 per cent on the 10-year and 6.15 per cent on the 30-year, will include a so-called "poison put" that would allow investors to sell the

bonds back at 101 cents on the dollar if there was a change of control at the company.

Moody's Investors Service gave the bonds a rating of A1, its fifth-highest, and Standard & Poor's assigned them an equivalent A+. Citigroup and JPMorgan managed the sale.

Tesco has priced a euro-denominated bond and a sterling bond this year.

In March it priced a €600m 40-year issue – the longest dated euro issue from a retailer.

The company also priced a £500m 50-year bond in February. The bond ranks alongside only a handful of other 50-year sterling corporate bonds, including a 2002 deal by French rail network operator Reseau Ferre de France and a 1994 bond issue by British Gas.

Source: David Oakley, *Financial Times*, 31 October 2007.

■ Assessment

The balance of argument usually points to borrowing to finance foreign subsidiaries; indeed braver corporate treasurers may try to exploit perceived disequilibria in global financial markets to access 'cheap' finance. However, the International Fisher Effect (explained in Chapter 21) should caution against this. The benefit of borrowing at 'low' interest rates should always (eventually at least) be offset by the appreciation in currency in which debts are to be repaid.

Special Purpose Vehicles
An entity set up specifically for managing a firm's financial requirements and obligations

Larger MNCs may set up their own financial subsidiaries as **Special Purpose Vehicles**, established largely for the purpose of obtaining the funds required to finance the entire firm's ongoing growth needs. This avoids problems over costs and access to capital in the host country. The SPV simply borrows on world markets using the credit rating of the parent. Firms that use SPVs include General Electric Corporation (via GE Capital), BMW and Ford.

■ The currency cocktail

There is a strong argument for borrowing in a range or 'cocktail' of currencies to spread the risk over a diversified portfolio of borrowed currencies. By diversifying sources of finance, the MNC may take advantage of what it perceives as unusually low rates in certain financial centres. This requires the MNC to be well-known in international financial markets and to have established, sound banking relationships. Thus a firm needing to refinance a medium term loan maturing in London where interest rates are 7 per cent may decide to borrow in Japan where interest rates for corporate borrowing are, say, 2–3 per cent.

To illustrate a currency cocktail, consider Compass Group plc. Table 22.5 shows the composition of group borrowing. This includes both short-term and long-term borrowing. It also includes finance leases and hire purchase arrangements, but excludes accruals and provisions.

Table 22.5 Compass Group plc borrowings as at 30 September 2006

Currency	Amount (£m)	%
Sterling	852	44
US$	506	26
Euro	547	28
Yen	23	1
Other	26	1
Total	1,954	100

Source: Compass Group Annual Report 2006 (www.compass-group.com)

Compass policy is to match projected cash flows by currency to borrowing in the same currency. It also uses earnings generated in foreign currency to service and repay debt in the same currency.

22.15 THE WACC FOR FDI

In Section 22.6, we explained how a project-specific discount rate could be obtained for a foreign investment. At that point, we were assuming all-equity funding. But what if the project is partly debt financed? This issue is often highly relevant where foreign governments offer concessionary interest rates, often prompting a level of local debt financing well above the parent's own gearing. The most appropriate solution is to use a tailor-made WACC calculated to reflect the particular financing mix of the project.

To illustrate this, let us revisit the case of Malaku Mining from page 665. Recall that it was decided that a local Beta value of 0.75 was appropriate to use, resulting in a project-specific equity cost of 8.75 per cent.

Now assume that Malaku can borrow in Indonesia at the concessionary rate of 3 per cent and that local regulations allow it to offset interest charges against local taxation, paid at 40 per cent. The WACC for the Indonesian mining project is found in the usual way by weighting the cost of each form of finance by its contribution to financial structure, in this case, project financing. Now assume that the project is financed 25 per cent by parental equity and 75 per cent by local borrowing. The WACC is thus:

(8.75% × equity weight) + (3% [1 − 40%] × debt weight)

= (8.75% × 25%) + (1.8% × 75%) = (2.2% + 1.4%) = 3.6%

This is an interesting result indeed. The combination of low interest rate and high debt proportion results in a remarkably low required return. These sorts of calculation, although based on sound CAPM logic, can often generate much lower project cut-off rates than for comparable UK investment. To many executives, it is inconceivable that high risk, foreign projects should be evaluated at lower discount rates than UK projects.

Financial managers often have some difficulty in explaining and justifying lower WACCs to sceptical executives whose natural inclination is to use *higher* rates for foreign projects. The response is that although foreign projects often have higher total risks, additional risks are usually project- and country-specific, and are thus very different from those affecting UK projects. These risks are best handled in more effective ways, e.g. by hedging foreign exchange risk, than by simply hiking the discount rate. Raising the hurdle rate, and possibly excluding an attractive, albeit risky, project is inferior to a considered policy of risk management.

22.16 APPLYING THE APV TO FDI

Discounting at the WACC implies that all the complex interactions involved in the investment can be factored into a single discount rate. In no case is this more difficult than for foreign investment projects that differ from domestic activities in aspects such as taxation, foreign exchange rate variability, concessionary financing and numerous additional dimensions of risk. This makes FDI a prime candidate for evaluation by the APV method.

For FDI, the evaluation procedure is:

1 Evaluate the core project assuming it is financed entirely by owners' equity to find the base case, as if the project were undertaken in the home country.
2 Separately evaluate the 'extras' such as the tax breaks and subsidies offered by the host government, and any spill-over effects on other activities, e.g. lost export trade.
3 Calculate APV = [base case NPV − PV of extras]
4 Accept the project if the APV is positive.

A simple APV model is shown in Figure 22.4. The APV is defined as the inherent value of the foreign-located project to the firm's owners, adjusted for all positive and negative side-effects. For foreign projects, the APV is particularly useful when project financing differs from the parent firm's capital structure, but suffers the limitation that calculation of these side-effects and their associated degrees of risk is difficult in practice. Inevitably, a high degree of judgement is required.

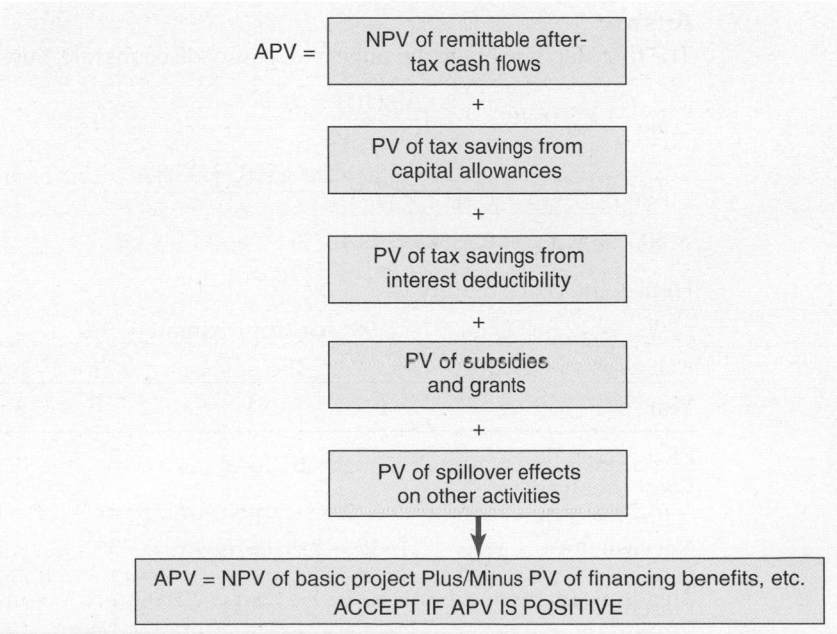

Figure 22.4 A simple APV model

22.17 WORKED EXAMPLE: APPLYING THE APV

(This question was part of a CIMA question in the May 2006 P9 Financial Strategy exam paper, in which it carried 15 of the available 25 marks.)

Question

GHI is a mobile phone manufacturer based in France with a wide customer base in France and Germany, with all costs and revenues based in the euro (€). GHI is considering expanding into the Benelux countries and has begun investigating how to break into this market.

After careful investigation, the following project cash flows have been identified:

Year	€million
0	(20)
1	5
2	5
3	4
4	3
5	3

The project is to be funded by a loan of €12 million at an annual interest rate of 5% and repayable at the end of five years. Loan issue costs amount to 2% and are tax-deductible.

GHI has a debt: equity ratio of 40:60 (at market values), a pre-tax cost of debt of 5.0% and a cost of equity of 10.7%.

Under the terms of a Double Tax Treaty, tax is payable at 25% wherever the earnings of a project are made, and not taxed again when cash is repatriated. The initial investment of €12 million will qualify for full tax relief.

Required

Advise GHI on whether or not to accept the project, based on a calculation of its Adjusted Present Value (APV), and comment on the limitations of an APV approach in this context.

Answer

The first step is to obtain an ungeared, equity discount rate. Substituting into the formula:

$$k_{eg} = k_{eu} + [k_{eu} - k_d]\frac{V_D(100 - t)\%}{V_S}$$

$$10.7\% = k_{eu} + [k_{eu} - 5\%] \times \frac{4 \times (100 - 35)\%}{6}$$

$$10.7\% = k_{eu} + 0.433k_{eu} - 2.167\%$$

Hence, the cost of equity is:

$$k_{eu} = 12.867\%/1.433 = 8.98\% \text{ or, approximately, } 9\%$$

Year	0	1	2	3	4	5
Project cash flows	−12	5	5	4	3	3
Local tax on cash flows @ 25%		−1.25	−1.25	−1	−0.75	−0.75
Net cash flows	−12	3.75	3.75	3	2.25	2.25
Discount factor @ 9%	1	0.917	0.842	0.772	0.708	0.650
Present Value	−12	3.43	3.15	2.32	1.59	1.46

NPV of basic project at 9% = −€12 million + €8.95 million = −€3.05 million

The next step is to find the PV of the 'extras'. Beginning with the tax shield on investment expenditure (assumed offsettable against earnings from other activities)

= 25% × €12 million = €3 million

The tax shield on debt is:

Tax relief on debt interest = €150,000 each year for five years (= €12 million × 5% × 25%)
PV of tax relief on debt interest = €0.67 million (= €150,000 × 4.452)

Finally, the issue costs:

Issue costs = €240,000 (= €12 million × 2%)
PV of tax relief on issue costs = €0.06 (= €240,000 × 0.25 × 0.952)

The full APV calculation is thus:

Adjusted present value	€000
Base case NPV	(3.05)
Tax relief on initial outlay	3.00
PV of tax relief on debt interest	0.67
Issue costs	(0.24)
PV of tax relief on issue costs	0.06
Adjusted present value	0.44

Hence, although the basic project is not worthwhile, adding in the present value of the financing costs and benefits and the investment tax break, the APV is positive. The project should thus be accepted.

Limitations of APV

- Determining the costs and benefits involved in the financing method to be used can be difficult (especially where they are based on an estimate of the enhanced debt capacity provided by the project as in this case).
- Finding a suitable cost of equity for the base NPV calculation is subjective and may not truly reflect the risk associated with the new project, especially if the project is based abroad (as here – although there are no currency complications).
- Whether the tax rate will remain at this level, and the tax treaty will stay in force.
- Whether to discount the tax savings from the debt at the post-tax rate of interest or at the cost of equity.

Self-assessment activity 22.8

List eight ways in which MNCs can lower political risk.

(Answer in Appendix A at the back of the book)

SUMMARY

This chapter has examined the strategic motivations that drive firms to enter foreign markets, and methods of effecting entry, in particular, direct investment. Evaluating FDI is a complex process that differs in several important respects from evaluation of home country investment, not least the exposure to FX rate variations. As a result, FDI evaluation is more art than science, especially as it involves so many unquantifiable aspects such as the prevailing political mood of the host country, assessing political and other country risks and devising appropriate safeguards. Clearly, this topic transcends purely *financial* strategy.

Key points

- Foreign direct investment (FDI) may be undertaken for a variety of strategic reasons: for example, globalisation of component sources or meeting the threat of a competitor already based overseas.

- FDI is generally undertaken when exporting (with relatively high variable costs, but low fixed costs) becomes more expensive than overseas production (with relatively high fixed costs but low variable costs).

- In principle, the valuation of FDI is similar to appraisal of domestic investment, but there are important additional complications that may cloud the analysis.

- Among such complications are exchange rate complications, political risk, such as the risk of expropriation, exceptional inducements to invest in a particular location, and cannibalisation of existing export sales.

- Most of these complexities can be alleviated by (i) evaluating the project from the perspective of the parent company's shareholders, and (ii) as a first approximation at least, relying on the four-way parity relationships.

- If the parity relationships hold, then there are two equivalent ways of evaluating FDI: (i) predicting all future cash flows in the foreign currency, and discounting these at the local (i.e. overseas) cost of capital, and then converting to domestic currency at spot; and (ii) converting all the predicted foreign currency cash flows to the domestic currency using estimates of the future spot exchange rates, and discounting these at the domestic cost of capital.

- A tailor-made discount rate can be found by using the cost of capital of a comparable overseas firm, but allowing for the likely less than perfect correlation between the foreign economy and the domestic one.

- Foreign investment risk is mainly economic risk. This can be alleviated by strategic decisions, e.g. location of production facilities in a range of countries, and hence, operating in a variety of currencies.

- Similarly, diversification of markets can produce the same sort of currency cocktail.

- The extent to which a firm needs to hedge investment project risk depends on its international profile of activities, and thus the extent to which operating risks net off against each other in natural hedges, e.g. a firm producing in Italy and selling in France has no direct operating exposure as all transactions are conducted in euros.

- Counter-trade, i.e. making reciprocal dealing arrangements with firms and governments in other countries, can be used to mitigate foreign exchange risk. Available methods extend from crude barter to buy-back facilities.

- Political risk can be moderated in various ways, e.g. borrowing from local banks, but the guiding principle is to make the MNC indispensable to the local economy and society.

- Corruption is endemic in some foreign countries, so a sad fact of life is that failure to play the local game can lead to loss of business.

- If the complexities of evaluating FDI make the conventional DCF model too cumbersome, a short-cut approach is offered by the adjusted present value (APV) approach which separates out the cash flows from basic operations from those attributable to financing and other complexities.

Further reading

Books on international/multinational financial management (see Chapter 21) invariably carry chapters on evaluation of FDI – Shapiro (2006) has two excellent chapters on this topic.

More general texts on international business such as those by Daniels and Radebaugh (2004) and Rugman *et al.* (1985) include extensive treatments of international strategy and operations, and usually cover appraisal of foreign projects also, but to a lesser depth. Grant (2005) has an excellent chapter on strategic aspects of FDI. Buckley (1995) gives probably the most comprehensive coverage of overseas capital budgeting. Excellent analyses of operating exposure can be found in Lessard and Lightstone (2006) and in Grant and Soenen (2004). Both Buckley (1995) and Lessard (1985) give a thorough treatment of the APV method.

Dunning and Lundan (2008) is perhaps the most authoritative text on the role of the MNC in the global economy.

The World Investment Report, published each autumn (usually September) by UNCTAD (**www.unctad.org**), is a mine of information on trends in FDI.

QUESTIONS

 **myfinancelab** *Questions with an icon are also available for practice in myfinancelab with additional supporting resources.*

Questions with a **coloured number** have solutions in Appendix B on page 758.

1 The USD vs. GBP exchange rate is $1.50 vs. £1. A UK MNC operating in the US plans to sell goods worth $100 million at today's prices to US customers. Show that its GBP revenue *in real terms* will not be affected if PPP applies under each of the following conditions:

(i) UK and US inflation at 5% p.a.
(ii) UK inflation 5%, US inflation 2%.
(iii) UK inflation 2%, US inflation 5%.

2 OJ Limited is a supplier of leather goods to retailers in the UK and other Western European countries. The company is considering entering into a joint venture with a manufacturer in South America. The two companies will each own 50 per cent of the limited liability company JV (SA) and will share profits equally. £450,000 of the initial capital is being provided by OJ Limited, and the equivalent in South American dollars (SA$) is being provided by the foreign partner. The managers of the joint venture expect the following net operating cash flows which are in nominal terms:

	SA$ 000	Predicted future rates of exchange to £ sterling
Year 1	4,250	10
Year 2	6,500	15
Year 3	8,350	21

For tax reasons, JV (SA), the company to be formed specifically for the joint venture, will be registered in South America.
  Ignore taxation in your calculations.
  Assume you are a financial adviser retained by OJ Limited to advise on the proposed joint venture.

Required
(a) Calculate the NPV of the project under the two assumptions explained below. Use a discount rate of 16 per cent for both assumptions, and express your answer in sterling.

  Assumption 1: The South American country has exchange controls which prohibit the payment of dividends above 50 per cent of the annual cash flows for the first three years of the project. The accumulated balance can be repatriated at the end of the third year.
  Assumption 2: The government of the South American country is considering removing exchange controls and restrictions on repatriation of profits. If this happens, all cash flows will be distributed as dividends to the partner companies at the end of each year.

(b) Comment briefly on whether or not the joint venture should proceed based solely on these calculations.
(CIMA)

3 PG plc is considering investing in a new project in Canada that will have a life of four years. Initial investment is C$150,000, including working capital. The net cash flows that the project will generate are C$60,000 per annum for years 1, 2 and 3 and C$45,000 in year 4. The terminal value of the project is estimated at C$50,000, net of tax.
  The current spot rate for C$ against the pound sterling is 1.70. Economic forecasters expect the pound to strengthen against the Canadian dollar by 5 per cent per annum over the next four years. The company evaluates UK projects of similar risk at 14 per cent per annum.

Required

Calculate the NPV of the Canadian project using the following two methods:

(i) Convert the foreign currency cash flows into sterling and discount at a sterling discount rate.

(ii) Discount the cash flows in C$ using an adjusted discount rate that incorporates the 12-month forecast spot rate.

(CIMA)

4 Kay plc, a UK-based chemical firm but with plants in Germany and the Netherlands, manufactures man-made fibres. It would like to expand its exports to Latin America and the country of Copacabana, in particular. However, Copacabana is unable to pay in Western currency and its own currency, the poncho, is subject to rapid depreciation, due to high local inflation. One solution to this problem is an arrangement whereby Kay manages and pays for the construction of a fibres plant and accepts payment in the form of the finished product of fibres (a so-called buy-back).

Construction will take two years and expenditures can be treated as four equal half-yearly payments of 10 million ponchos at today's prices, beginning in six months' time. The plant will have a 15-year life, but will attract no local investment incentives. The inflation rate in Copacabana is expected to average 20 per cent p.a. over the construction period. The current exchange rate of the poncho vs. sterling is 1:4 and inflation in the UK has recently averaged 5 per cent.

The fibres produced and taken as payment can be traded on world markets, probably in Europe, where the present price is €500 per tonne. Kay is not prepared to accept payment in this way for more than five years. The expected production rate of the plant is 20,000 tonnes per annum, and Kay would take 40 per cent of this in payment.

The current euro vs. sterling rate is €1.60 per £1, and sterling is expected to depreciate by 5 per cent per annum prior to joining the euro bloc.

Further information

■ The project will be financed by equity only.

■ Kay is at present debt-free. Its shareholders seek a return of 20 per cent p.a. for projects of this degree of risk.

■ Profits from the operation will be taxed at 30 per cent when repatriated to the UK. Assume no delay in tax payment. All development costs will qualify for UK tax relief.

■ Any losses will be carried forward to qualify for tax relief.

■ There will be no tax liability in Copacabana.

Required

Determine whether Kay should undertake this project.

5 Brighteyes plc manufactures medical and optical equipment for both domestic and export sale. It is investigating the construction of a manufacturing plant in Lastonia, a country in the former Soviet bloc. Initial discussions with the Ministry of Economic Development in Lastonia have met with favourable response, providing the project can generate a 10 per cent pre-tax return. Shareholders look for a return of 15 per cent in real terms.

The investment will be partly import-substituting and partly export-based, selling to neighbouring countries. The project has been offered a local tax holiday, exempting it from all taxes for the first ten years, except for cash remittances, for which a 20 per cent withholding tax will apply. Modern factory premises on an industrial estate with convenient road and rail links have been offered at a reasonable rent.

The initial investment will be £10 million in plant, machinery and set-up costs, all payable in sterling by the parent company. Additional funds will come from a bank loan of 20 million latts, the local currency (4 latts = £1), negotiated with a local bank, at a concessionary rate of interest of 10 per cent p.a. This will be used to finance working capital. Operating cash flows, the basis for calculating tax, are estimated at L10 million in year 1 and L22 million thereafter until year 5.

The whole of the parent's earnings after payment of local interest and taxation will be repatriated to the UK. The Lastonia withholding tax is to be allowed as a deduction before calculating the UK corporation tax, currently at the rate of 30 per cent. All transfers can be treated as occurring on the final day of each accounting period, when all taxes become due.

The new venture is expected to 'cannibalise' exports that Brighteyes would otherwise have made to neighbouring countries, resulting in post-tax cash flow losses of £0.5 million in each of years 2 to 5. For planning purposes, year 5 is the cut-off year, when the realisable value of the plant and equipment is estimated at L24 million. The working capital will be realised, subject to losses of L2 million on stocks and L2 million on debtors. Funds realised will be used to repay the local borrowing, and the balance transferred to the UK without further tax penalty or restriction.

The exchange rate is forecast to remain at L4 vs. £1 until year 2, when the Latt is expected to fall to L5 vs. £1.

Required

(i) Is the project acceptable from the Lastonian Ministry's point of view?

(ii) Is it worthwhile from the viewpoint of the foreign subsidiary?

(iii) Does it create wealth for Brighteyes' shareholders?

6 Palmerston plc operates in both the UK and Germany. In attempting to assess its economic exposure, it compiles the following data:

■ UK sales are influenced by the euro's value as it faces competition from German suppliers. It forecasts annual UK sales based on three possible scenarios:

Euro: sterling exchange rate	Revenue from UK business
1.65:1	£200 m
1.60:1	£215 m
1.55:1	£220 m

■ Revenues from sales made in Germany are expected to be £120 m p.a.

■ Expected cost of goods sold is £120 m p.a. from UK materials purchases, and €200 m from purchases in Germany.

■ Estimated cash fixed operating expenses are £50 m p.a.

■ Variable operating expenses are estimated at 20 per cent of total sales value including German sales translated into sterling).

■ Palmerston is financed entirely by equity and shareholders require a return of 15 per cent p.a.

Required

(i) Construct a forecast cash flow statement for Palmerston under each scenario.

(ii) Value Palmerston's equity under each scenario, assuming a ten-year operating time horizon. Ignore terminal values.

(iii) Suggest how Palmerston might restructure its operations to lower its sensitivity to exchange rate movements.

Ignore taxation.

7 A professional accountancy institute in the UK is evaluating an investment project overseas in Eastasia, a politically stable country. The project involves the establishment of a training school to offer courses on international accounting and management topics. It will cost an initial 2.5 million Eastasian dollars (EA$) and it is expected to earn post-tax cash flows as follows:

Year	1	2	3	4
Cashflow (EA$000)	750	950	1,250	1,350

The following information is available:

■ The expected inflation rate in Eastasia is 3 per cent a year.

■ Real interest rates in the two countries are the same. They are expected to remain the same for the period of the project.

■ The current spot rate is EA$2 per £1 sterling.

■ The risk-free rate of interest in Eastasia is 7 per cent and in the UK 9 per cent.

■ The company requires a sterling return from this project of 16 per cent.

(CIMA)

Required

Calculate the sterling net present value of the project using *both* the following methods:

(i) by discounting annual cash flows in sterling,

(ii) by discounting annual cash flows in Eastasian $.

Practical assignment

Inspect the Report and Accounts for a company of your choice, to examine how its international profile of activities has changed over the years. You may find difficulty in obtaining a full set of accounts reaching very far back in time, but examination of a sample should give you a flavour of the company's policy regarding internationalisation.

Look also at the chairman's statements to glean an indication of the importance attached to overseas operations in the company's strategy.

 *Now retake your diagnostic test for Chapter 22 to check your progress and update your study plan.*

23

Key issues in modern finance: a review

Market inefficiencies prove we're only human

Investors have an insatiable appetite for information. Company announcements, macro economic variables and the latest political news are just some of what they digest on a daily basis. It is almost impossible for investors to assimilate and process the information tidal wave of data that faces them every day. But this does not have to be a problem for market efficiency. Classical economics tells us that market efficiency is not driven by the activities of one participant but rather the overall effect of many self-serving individuals – Adam Smith's invisible hand.

More information should increase transparency and promote efficiency. However, evidence suggests this is not happening in the stock market. Factors such as cheap valuations or earnings upgrades turn out to have predictive power for future share price movements. This runs counter to the concept of efficient

markets where all information should already be reflected in the price.

An increasingly persuasive explanation for market inefficiency comes from the field of behavioural finance. Not only do individuals make mistakes when analysing masses of data but, more fundamentally, they all seem to make the same mistakes. Errors are therefore magnified rather than negated by the combined efforts of many market participants. Faced with complex financial decisions, investors often employ heuristics, or rules-of-thumb, when making decisions. Heuristics can be useful in everyday life, but these inbuilt tendencies are too blunt a tool for the complex environment of financial decision-making.

Source: Based on article by James Hand and Greg Davies, *Financial Times*, 25 October 2004, p. 6.

23.1 INTRODUCTION

This book has presented the theory and practice of modern financial management. This final chapter summarises the main principles of finance underpinning the book and develops several key areas, including those relating to market efficiency and behavioural finance.

■ Financial theory and practice

A good finance theory is one that offers useful explanations of existing behaviour and provides a guide to future behaviour. We frequently find that rather restrictive assumptions are made in developing financial models. For example:

1 All markets – not just capital markets – are perfectly competitive.
2 Information is perfect and costless.
3 Transaction costs are zero.
4 No taxes exist.

These assumptions lead naturally to certain propositions that can be questioned. First, only shareholders really matter. A perfect labour market implies that managers and workers have sufficient mobility and can always find other equally attractive alternative employment. Second, shareholders are only interested in maximising the market value of their shareholdings. Given perfect, costless information, managers are tightly controlled by the shareholders to implement and pursue value-maximising strategies. Third, the pursuit of shareholder wealth is achieved by instructing managers to invest only in those projects that are worth more than they cost. Financing strategies, whether concerning dividends, capital structure or leasing, are largely irrelevant as they do little to increase shareholders' wealth.

The assumptions underlying the theory of finance appear to be at odds with reality. Information is imperfect; transaction costs and information costs may be sizeable. Markets are frequently highly imperfect; management will usually have a good deal of interest in the firm – an interest that may well conflict with that of shareholders. Managers have far from complete knowledge of the set of feasible financing strategies available, their cash flow patterns and impact on market values. Shareholders are even less well informed. Taxation policy, bankruptcy costs and other factors can have a major influence on financial strategies.

Whether we talk about markets, firms or managers, we are essentially looking at behaviour – the behaviour of individual managers, investors or groups.

Most readers will be familiar with the popular board game Monopoly. There is more than a passing resemblance between this game and corporate finance. Both are about maximising investors' wealth in risky environments, making investment decisions with uncertain payoffs, raising finance and managing cash flow. Players and managers must stick to the rules of governance and seek to devise appropriate investment, financing and trading strategies to gain competitive advantage. While rational analysis and sound judgement are essential, there remains room for sentiment, psychology, fun, and, to make the game interesting, a generous portion of luck!

Some aspects of finance, particularly routine finance decisions, can be operationalised through clear rules and procedures. But good finance managers look for something more than rules and procedures. They seek to understand the behaviour of markets and companies. Theories of finance seek to provide explanations for such behaviour – the better the theory, the better we understand how to make financing and investment decisions and set appropriate policies. Throughout this text we have sought to combine the 'why' with the 'how' in the theory and practice of corporate finance. This final chapter reminds the reader of the main pillars on which much of finance theory rests, many of which have been recognised as significant economic developments through the award of the Nobel Prize in finance, and reflects on their practical relevance.

23.2 UNDERSTANDING INDIVIDUAL BEHAVIOUR

To understand how organisations function we must first understand individual behaviour. There are many models of human behaviour (e.g. sociological, psychological and political); we shall restrict our examination to two models of most relevance to finance.

The first is the traditional economic model of human behaviour. Here the manager is seen as a short-run wealth maximiser. The model is a useful starting point in studying finance because it offers a simple approach to model building, using only the pursuit of wealth as a goal. Much of the argument underlying the theories discussed in this book is based on this model. But we all know that this is a poor explanation of many aspects of human behaviour. For many people, and in many situations, money may not come before morality, honesty, love, altruism or having fun. As the song said 'Money can't buy me love'.

This leads us to develop a more realistic model of human behaviour which Jensen and Meckling (1994) term the resourceful, evaluative, maximising model (or REMM). This model assumes that people are resourceful, self-interested maximisers, but rejects the notion that they are only interested in making money. They also care about respect, power, quality of life, love and the welfare of others. Individuals respond creatively to opportunities presented, seeking out opportunities, evaluating their likely outcomes, and working to loosen constraints on their actions.

Neither of the above models place much emphasis on psychological factors in human information-processing and decision-making. This growing area of finance, termed behavioural finance, is discussed in a later section.

To sum up the forgoing discussion, money is not the only, or even the most important, thing in life. But when all else is equal, we act in a rational economic manner, choosing the course of action that **most** benefits us financially. Two fundamental concepts naturally follow.

Managers should only consider present and future costs and benefits in making decisions

This is the principle of incrementalism – only the additional costs or benefits resulting from a choice of action should be considered. For example, expenditures already incurred are not relevant to the decision in hand; they are **sunk costs**.

Choices often involve trade-offs, denying the possibility of other alternatives. The **opportunity cost** of making one decision is the difference between that choice and the next best alternative.

Managers are risk averse

Most managers are risk averse; given two investments offering the same return, they would choose the one with least risk. Unlike the risk-seeker or gambler, most managers try to avoid unnecessary risks. Risk aversion is a measure of a manager's willingness to pay to reduce exposure to risk. This could be in the form of insurance or other 'hedging' devices. Alternatively, it could be by preferring a lower-return investment because it also has a lower risk.

In business, risk and the expected return are usually related. Rational managers do not look for more risk unless the likely benefits are commensurately greater. This is the principle of **risk aversion**. One way in which investors can reduce risk is by spreading their capital across a range (portfolio) of investments. This is the principle of

diversification, or not putting all your eggs in one basket. However, the key to risk management should not simply be to reduce it, but to take decisions in a risky business environment that create value.

23.3 UNDERSTANDING CORPORATE BEHAVIOUR

■ Managers are agents for shareholders

A firm may be viewed as a collection of individuals and resources. More precisely, it is a set of contracts that bind individuals together, each with their own interests and goals. Agency theory explores the relationship between the shareholders (the principals) and the agents (e.g. board of directors) responsible for taking actions on their behalf.

Shareholders want managers to maximise firm value. It follows that to understand how firms behave, and whether managers pursue this goal, we must first understand the nature of the contracts, monitoring procedures and reward mechanisms employed.

Information is not available to all parties in equal measure. For example, the board of directors will know more about the future prospects of the business than the shareholders, who have to rely heavily on published information. This information asymmetry means that investors not only listen to the board's rhetoric and confident projections, but also examine the **information content** in its corporate actions. This **signalling effect** is most commonly seen in the reaction to dividend declarations and share dealings by the board. An increase in dividends signals that the company is expected to be able to sustain the level of cash distribution in the future, because it is usually regarded as the height of financial incompetence to be forced to cut a dividend.

■ Apply the NPV rule

A company will have perhaps thousands of shareholders, all with different levels of wealth and different risk attitudes. How can a firm make decisions that satisfy all of them? **The NPV rule states that decisions should be taken which maximise wealth and this can be achieved by accepting all positive net present value projects.** A firm's value is largely determined by two things: the cash generated over the life of the company and the risk of those cash flows. Shareholder value analysis advocates that firms should seek to maximise cash flows and manage risks. They should invest in those areas in which their firms have some competitive advantage, giving rise to superior return or positive net present value when discounted at the rate commensurate with the perceived level of risk. This rate reflects the return on risk-free investments plus a risk premium to compensate for the additional risk in the company's cash flow stream.

When properly applied, the NPV approach is the best method for evaluating certain forms of investment, where cash flows are fairly predictable and there are few investment options. However, two points are worth mentioning:

1 In long-run equilibrium all projects have a zero NPV. In other words, it may be possible for a firm to achieve positive NPVs because they are the first to spot a market opportunity, or have built entry barriers, but, ultimately, these will be overcome and competitors will continue to enter the market until the benefits no longer exist.

2 The DCF approach is most useful when evaluating bonds and financial leases, where cash flows are highly predictable. They become less relevant as risk and growth opportunities increase. Table 23.1 illustrates this.

■ Options have value

The DCF framework has come under increasing criticism in recent years for failing to consider the options embedded in investment opportunities. **An option allows an investor to buy or sell an asset at a fixed price during a given period.** A firm should

Table 23.1 Usefulness of DCF methods

Usefulness of DCF methods	Financial investment	Capital investment projects
Very	Bonds and fixed-income securities	Financial leases
Moderately	Shares paying regular dividends	1 Replacement decisions 2 Businesses where there are no strategic options ('Cash Cows')
Limited	Shares where there are significant growth opportunities	Business with significant growth opportunities
Very little	Derivatives, e.g. share options	Pure research and development expenditure

only exercise its option if it adds value to the business. Unlike the conventional NPV approach, however, the more volatile the option, the more valuable it is, because the 'good news' is taken up. With capital projects, selecting projects offering the highest NPV at a specific point in time, conventional investment appraisal ignores the possibility that projects may have valuable options (called real options). Such options include greater flexibility for management in terms of growth, delay or abandonment options. The traditional approach of selecting the project with the highest NPV at a particular point in time disadvantages those projects offering greater flexibility to management and the benefits of possible add-on investments in the future. The price paid for an acquired company may look too high based on conventional NPV calculations. But this ignores any valuable strategic options embedded in the decision which may well justify the price paid. As a minimum, management should look to identify and evaluate subjectively any options linked to projects, particularly when projects with known options cannot be financially justified on conventional NPV grounds.

The increasing importance of options means that investors need to understand how they are valued. Black and Scholes first developed a formula for option pricing. In summary, it argues that option value is based on the current market value of the underlying share, the striking price, the time to contract maturity, and the risk-free rate of interest.

Options and other derivative instruments offer considerable scope for corporate treasurers in managing risk, such as foreign exchange and interest rate risk, by developing appropriate hedging strategies.

■ Look for an optimal capital structure and dividend policy

Managers need to know the cost of capital, or cut-off rate, for capital projects. Traditionally it was argued that the cost of capital depended, in part at least, on the mix of equity and debt in the company. The Modigliani–Miller propositions offer an elegant theory demonstrating that how the firm finances its investment schedule is irrelevant. The cost of capital for determining the cut-off rate for capital projects depended only on the risk class of the projects. In other words, financing decisions do not increase a firm's overall value. Value is therefore independent of capital structure.

For any company operating below the capital structure deemed 'appropriate', the private equity industry, with huge amounts of capital at their disposal, is ready to take over the business, introduce far higher gearing levels and return cash to shareholders. This is well illustrated by the car manufacturer Volvo. The company is known as a well-managed company maintaining prudent cash balances and relatively low gearing levels. In 2006, a venture capitalist, who had built up a 5 per cent stake in Volvo, began pressing the company to return to shareholders part of its sizeable cash balance of over £1 billion. The Volvo board responded by setting higher targets for profits, cash flow, and gearing, the latter being an increase in the debt-to-equity ratio from 30 per cent to 40 per cent, and it stated that it would look favourably on opportunities to return cash to investors while also looking for suitable acquisitions.

Although the underlying assumptions are far removed from the world of everyday finance, even in practice, it is generally agreed that, in a no-tax world, issuing low-cost debt does not automatically reduce the overall cost of capital because the cost of equity will rise to compensate for the increased financial risk. However, in practice, we see that higher borrowing levels is a powerful incentive for managers to work harder and smarter, and offers tax advantages for firms through the tax relief on interest payments. On the other hand, excessive levels of debt threaten the firm's very existence and could lead to costly financial distress.

Related to the above is the need to identify the **optimal working capital** to be employed in the firm. This involves striking the right balance between minimising the cash conversion cycle and maintaining good supplier/customer relations. Many of the improvements in working capital management come through supply chain efficiencies and sound management of cash and accounts receivables/payables.

Much the same argument as for capital structure was made by Modigliani and Miller for the **irrelevance of dividends**. Under given assumptions, dividend policy does nothing to improve the value of the firm.

In a perfect capital market, rational investors are indifferent between a cash dividend and retention, once the impact of the investment decision is stripped out (i.e. assuming investment in a zero-NPV decision. If there is an increase in value when a firm retains, it is because investors expect the project to be wealth-creating. If a dividend increase raises value in such a market, it must be because investors have an extremely powerful desire for dividends rather than capital gains (i.e. an overwhelming preference for current dividends).

In imperfect markets, different conclusions apply. For example, dividends are said to be tax-disadvantaged compared to retention-plus-capital gains, as taxes on gains are delayed, and are often levied at a lower rate (as in the UK) than the rate most investors face on their dividends income. Moreover, **dividends have a powerful signalling effect**. In a world where managers possess more information than investors, a dividend increase conveys the information that directors are confident about the future, while a dividend cut signals pessimism about the future. However, one should not exaggerate the signalling effect. When dividend changes are made, they are usually well-telegraphed – the market is usually aware that the firm is struggling and that a dividend cut is both sensible and likely. When the crunch comes, it is usually confirmation of what people generally expect. Indeed, failure to cut a dividend is often taken more seriously than a dividend cut in such circumstances. Converse arguments apply to the case of a dividend increase. This explains the ecstatic reaction given to a firm that rejoins the list of dividend payers, as the attached cameo reporting on Costain's resurrection illustrates. It is unexpected changes that have the most impact on price.

So, as dividend cuts are to be avoided, and most shareholders like dividends, for most firms, the sensible policy is to follow a 'progressive' dividend policy, steadily increasing payment, but only if supported by earnings increases that are regarded as sustainable. 'Declare a policy and stick to it' seems to be the message in order to attract and retain a clientèle of shareholders who find the policy in question suits their needs. However, as ever, there are exceptions. Firms with turbulent operating environments will be forgiven if dividend payouts fluctuate, while firms with highly stable environments, e.g. water and tobacco, had better tell a good story if they retain a high

Costain announces its first dividend in 17 years

Shares in Costain rose 10.1 per cent yesterday as it committed to paying a dividend for the first time since 1991. A highlight of the year, Costain said, was the completion of St Pancras International, the new London terminus for Eurostar trains. '2007 was a year of performance,' said Andrew Wyllie, chief executive, 'and as a measure of confidence, we are recommending our first dividend after quite a gap.' The group is one of the grand old names of the UK construction industry but has suffered years of poor performance.

Xavier Gunner at Arbuthnot said: 'The company delivered its promises. Their balance sheet now gives added opportunity to win new business, hence their order book is looking good for 2008. An added bonus is that their customers will not slow spending if the economy turns.'

Source: Maggie Urry and Stanley Pignal, *Financial Times,* 13 March 2008.

proportion of their earnings. Effective and clear communication, as in most walks of life, is essential.

■ Hedging adds value

Derivative instruments – like options, forward and futures contracts and swaps – can be used for speculation or for hedging. The main difference is that speculation is effectively a bet on a price move over time, whereas **hedging is used in combination with the underlying security; any gain or loss in the cash position is offset by an equivalent loss or gain in the derivatives position**.

At first sight it may seem questionable whether hedging is an appropriate tool for increasing the value of the firm. After all, if investors can eliminate all specific risk by holding a well-diversified investment portfolio, does incurring costs to hedge financial risks do anything for the shareholder? It seems that it does in the following ways:

1 Hedging reduces the probability of corporate failure. The greater security thereby extended to managers, suppliers and employees enables them to take a longer-term view in their decision-making. This is consistent with a net present value approach.
2 Hedging reduces the probability of financial distress which, in turn, reduces the firm's cost of capital.
3 Investors prefer a steady stream of corporate cash flows above the more volatile pattern likely without managing risk exposure.
4 Management may be reluctant to disclose to investors the full picture on risk exposure, making it difficult for investors to make an accurate assessment of the risk on their portfolios.

23.4 UNDERSTANDING HOW MARKETS BEHAVE

To make sound financial decisions consistently, managers need to have some understanding of how financial markets operate and behave. This involves understanding the time-value of money, how risk affects value, and the efficiency of markets.

■ Money has a time-value

To make sound financial decisions consistently, we need to understand how financial markets deal with the transfer of wealth between individuals or firms. This may take the form of investors lending money to borrowers in exchange for future interest and capital repayment, or purchasing a share of a company and participating in its future profits. This inevitably involves the transfer of wealth from one time period to another, based on the **time-value of money**. Put simply, we cannot add current and future money together in a meaningful way without first converting it into a common currency. This currency can be expressed in future value terms, but far more relevant to decisions being considered today is to express all future cash flows in **present value** terms. The value of money changes with time because it has an alternative use (opportunity cost): it can be invested in financial markets or elsewhere to earn a rate of return.

The present value of a cash flow is the amount of money today that is equivalent to the given future amount after considering the rate of return that can be earned. The higher the discount rate and the further the cash flow is from today, the lower its present value. In well-functioning financial markets, all assets of equivalent risk offer the same expected return. An assets positive net present value implies that, after considering the time-value of money, its benefits outweigh the costs. This important concept allows investors and corporate managers alike – regardless of their attitudes to risk – to advocate a simple decision rule: accept investment opportunities that maximise net present value.

■ Non-diversifiable risk matters

In 1952, Harry Markowitz published a paper that provided a more accurate definition of risk and return for shares. The expected return on an investment is the weighted average of the returns of its possible outcomes, while the risk is the variance of those outcomes around the mean. These days we simply talk of the share's expected return and its variance (squared deviations around the expected mean). The significance of this is that risk is no longer seen as purely 'downside' risk, but also includes the likelihood of exceeding the expected value.

From this, Markowitz was able to calculate the risk and return for a portfolio which, in turn, could be used to select efficient portfolios which offered the best combinations of risk and return for investors. **Portfolio risk** is the weighted sum of the variance plus twice the weighted sum of the covariance. The implication of such a model is that investors should focus on portfolio risk rather than the risk of the individual shares within that portfolio. This diversification effect is not simply a matter of how many different shares are included in the portfolio, but the covariance between such shares.

While the mean–variance model offered insights into portfolio risk and return, it was less successful in practice in portfolio selection. However, it formed the basis for William Sharpe to develop his **Capital Asset Pricing Model (CAPM)**. Risk can be categorised into that which can diversify away and that which remains (non-diversifiable or market risk). The theory argues that investors should only really be concerned about the latter type of risk and the required return on an asset is commensurate to the amount of non-diversifiable risk. Sharpe's argument went as follows. Suppose every investor holds portfolios that are mean–variance efficient, and that they all have the same risk–return expectation. In such a world, all investors will hold the same portfolio of risky assets, termed the **market portfolio**. However, these investors may have very different risk attitudes – some may be more risk averse than others.

Different risk profiles can be accommodated by holding different combinations of riskless investments (such as Treasury Bonds) and the market portfolio. It is even possible to further leverage risk and return by borrowing riskless assets and investing in the market portfolio. The investment strategy of investing in the market portfolio is followed by many institutions who invest in index (or tracker) funds employing passive investment strategies.

The attraction of the CAPM lies in its simplicity; the relationship between market (non-diversifiable) risk and the expected return on risky assets is reflected in the **security market line**, which is linear. The only risk that really matters is that which cannot be diversified away. The risk inherent in the market (the sensitivity of each share's returns to the market portfolio) is the only thing that changes with each investment. The expected return, ER_i, is a function of the risk-free rate of interest, R_f, and the risk premium on the market portfolio, $[ER_m - R_f]$, adjusted by the investment's Beta value, β_i.

$$ER_i = R_f + [ER_m - R_f] \times \beta_i$$

Everyone recognises that many of the assumptions behind the CAPM are far removed from the real world. But the issue is whether it is useful as a predictive model. Forty years of empirical observation suggests that while the model is fairly robust, it does not tell the whole story, as the *Financial Times* article below illustrates. Academics and practitioners may dispute just how deficient the model is, but most agree that it fails to capture the full effects of differences in:

- *Size* – why do smaller firms seem to offer higher returns than large firms after adjusting for risk?
- *Market-to-book ratio* – why do firms with high market-to-book ratios tend to exhibit higher returns than others?

Various models have been suggested that move away from a single factor pricing model (for example, Arbitrage Pricing Theory) but these too have their own problems.

From looking at financial markets over a lengthy period of time we observe that there is a reward for bearing risk and that the greater the potential reward, the greater is the risk. For example, Ibbotson and Rex (2006) show that over an 80-year period to 2005 on the NYSE, the risk premium earned from investing in a typical large company stock was 8.5 per cent.

■ Capital markets are efficient

Many of the theories in finance assume that capital markets are reasonably efficient in reflecting all available information. An efficient market is one where there are large numbers of rational profit-maximisers actively competing, with each trying to predict future market values of individual securities, and where important current information is almost freely available to all participants. The efficient markets hypothesis (EMH) argues that in such markets, new information on the intrinsic value of shares will be reflected instantaneously in actual prices.

The exact form of market efficiency in financial markets, in developed and developing countries, has been the subject of much debate and research. For major European stock markets, however, the consensus is that they exhibit efficiency in both the weak form (i.e. share prices contain all past data and superior returns cannot consistently be achieved from trading rules based on past stock market data) and the semi-strong form (i.e. share prices contain all publicly available information, and superior returns cannot consistently be achieved from trading rules based on such information).

Is it true that stock prices are essentially random walks? While the study of past stock prices may well produce interesting patterns, if they arise randomly they cannot by definition have predictive power when it comes to share prices or returns. This implies that, based on available information, there is no simple rule to generate above-average returns.

The time has come for the CAPM to RIP

Few theories are more influential or important in driving financial markets as the inelegantly-named capital asset pricing model. Too bad it does not appear to work very well.

The CAPM, as it is widely known, is a cornerstone of modern financial market analysis, studied like a rosary by analysts and executives at business school. Most financial directors use it to assess everything from the viability of a new project to their cost of capital. Most stock market analysts consider it an essential tool. But it has faced increasing criticism in recent years as unworkable in the real world, even from luminary market academics such as Harry Markowitz who laid the groundwork for the CAPM with research in 1950s on efficient portfolios.

CAPM is basically a model for valuing stocks or securities by relating risk and expected return. Developed separately by William Sharpe, John Lintner and Jack Treynor, it is based on the idea that investors demand additional expected return to take on additional risk. It then assumes markets are efficiently priced to reflect greater returns for greater risk. The risk is assessed on a stock or security's so-called beta, a measure of a company's volatility and correlation with the market as a whole. A company with a share price that tends to rise and fall more than the market will have a high beta and vice versa.

It is a seductively simple, catch-all theory to quantify risk and forecast returns. It has spurred the development of quantitative investing. But there is a problem. James Montier, analyst at Dresdner Kleinwort, says CAPM has become the financial theory equivalent of Monty Python's famous dead parrot sketch. He says the model is empirically bogus – it does not work in any way, shape or form. But like the

shopkeeper who insists to a customer with a dead parrot in the sketch that the bird is merely resting, financial markets are in denial. 'The CAPM is, in actual fact, Completely Redundant Asset Pricing (CRAP),' he says.

Some of the most damning evidence came from an exhaustive 2004 study by Eugene Fama and Kenneth French, the academics who helped develop the efficient markets theory in the early 1970s, that argued stocks are always correctly priced as everything that is publicly known about the stock is reflected in its market price. The study looked at all stocks on the New York Stock Exchange, the American Stock Exchange and Nasdaq from 1923 to 2003. As Montier states, the study shows CAPM woefully underpredicts the returns to low beta stocks and massively overstates the returns to high beta stocks. 'Over the long run there has been essentially no relationship between beta and return,' he says.

Fama and French themselves concluded that while CAPM was a theoretical tour de force, its empirical track record was so poor that its use in 'applications' was probably invalid. In others words, CAPM is a fine theory but useless in the real world. A similar study of the 600 largest US stocks by Jeremy Grantham, the value investor, last year yielded similar results. It showed from 1969 to the end of 2005, the lowest decile of beta stocks – notionally the lowest risk – outperformed by an average 1.5 per cent a year. The highest beta stocks, or the riskiest, actually underperformed by 2.7 per cent a year.

The problems in the CAPM lie in its assumptions, particularly those used to derive the efficient portfolio that is used as a benchmark for the model in theory. The most commonly-cited criticism is an implicit assump-

tion that all investors can borrow or lend funds on equal terms. Other assumptions that have been criticised include: that there are no transaction costs, that all investors have a 'homogeneity' of expectations and risk appetites and that investors can take any market exposure without affecting prices. It also assumes no taxes so investors are indifferent between dividends and capital gains.

Markowitz himself noted that the CAPM is like studying 'the motions of objects on Earth under the assumption that the Earth has no air'. 'The calculations and results are much simpler if this assumption is made. But at some point, the obvious fact that on Earth, cannonballs and feathers do not fall at the same rate should be noted,' he says.

Grantham says the flaws in the CAPM are probably inconvenient enough for the academic financial establishment to want to ignore it. But there ought to be more debate, particularly in using beta as a risk benchmark. The concept of pursuing absolute returns rather than relative performance is now widely debated. There needs to be a similar evolution in market thinking on how risk is defined, measured and dealt with. Montier cites a quote from legendary investor Ben Graham: 'What bothers me is that authorities now equate the beta with the concept of risk. Price variability, yes; risk, no. Real investment risk is measured not by the per cent a stock may decline in price in relation to the general market in a given period but by the danger of a loss of quality and earning power through economic changes or deterioration in management.'

Source: Tony Tassell, *Financial Times,* 10 February 2007.

However, empirical studies suggest that share price movements are not truly random and the EMH moved its position to one of that no gain can be made after allowance for transaction costs in buying/selling and changes in risk over time. The undeniable economic logic of the EMH is that if anyone finds a trading rule that consistently 'beats the market' by giving above-normal returns after all costs, it will quickly be imitated by others until the benefits from such a rule evaporate.

This does not mean that the market share price is perfectly 'correct' at any point in time. It does, however, imply that the market share price is an unbiased estimate of the true value of the share. While actual market prices may fluctuate in a random fashion around the 'true' value, investors cannot consistently outperform the market.

■ Criticisms of the EMH

Michael Jensen, a leading financial economist, argued in 1978 that 'the efficient markets hypothesis is the best-established fact in all of social science'. Why then is the EMH debate still hotly disputed? The main issue is whether investors react correctly to new information or whether they make systematic errors by over- or under-reacting. The **overreaction hypothesis** argues that share prices tend to overshoot the true value due to excessive optimism or pessimism by investors in their initial reactions to new information. There is some evidence for this in UK financial markets (Dissanaike, 1997).

Much criticism of the EMH is misplaced because it is based on a misconception of what the hypothesis actually says. For example, it does not mean that financial expertise is of no value in stock markets and that a share portfolio might as well be selected by sticking a pin in the financial pages. This is clearly not the case. It does suggest, however, that in an efficient market, after adjusting for portfolio risk, fund managers will not, on average, achieve returns higher than that of a randomly selected portfolio. Roll (see Ross *et al.*, 2002) makes the point that all publicly-available information need not be reflected in share prices. Instead, the link 'between unreflected information and prices is too subtle and tenuous to be easily or costlessly detected'.

Market efficiency also suggests that share prices are 'fair' in the sense that they reflect the value of that stock given the available information. So shareholders need not be unduly concerned with whether they are paying too much for a particular share.

The fact that many investors have done very well through investing on the stock market should not surprise us. For much of the last century, the market generated positive returns. Most investment advice, if followed over a long period of time, is likely to have done well; the point is that, in efficient markets, investors cannot consistently achieve above-average returns except by chance.

The strange existence of market anomalies

Stock markets are efficient machines, populated by rational investors seeking to make the best returns that they can. As evidence, look at the difficulty the average fund manager faces in trying to beat the market. But if that is so, how do you explain the dotcom bubble, when companies with no profits and barely any sales had billion-dollar valuations? And what lies beyond the continued existence of market anomalies, such as the tendency for smaller companies to outperform?

Academics have been discovering these effects for decades. There are seasonal patterns (stocks tend to do well in January and poorly during the summer). There are also valuation discrepancies (growth stocks tend to underperform). Some of these effects may be random. Analyse enough data and a few oddities will show up; plenty of people think some lottery numbers are 'lucky' because they occur more often – though it would be odder still if they all turned up the same number of times. Other effects are real, but may be costly to exploit. For small companies, higher returns may be negated by higher costs, reduced liquidity and higher risk (smaller firms are more likely to go bust).

The underperformance of growth stocks is linked to an over-enthusiasm for extrapolation. A company increases its profits at 20 per cent a year for five years and investors are tempted to believe it can do it for 15; historically, however, such paragons are about as rare as vegetarian cats. In contrast, the prices of 'value' stocks (which have a poorer record but lower ratings) perform better than expected.

Academics have just about abandoned the idea that all investors are rational; there are too many examples of psychological quirks (such as an aversion to recognising losses) for that to be the case. In short, it is very hard to quantify the precise irrationality of investors. That is why investors get lured into buying dotcom stocks in the hope that a 'greater fool' will purchase them at a higher price. And that is why there will always be anomalies for academics to discover.

Source: Financial Times, 23 February 2008 (abridged).

■ A few apparent anomalies in the EMH

There appear to be three main anomalies in the EMH; the effects of size and timing, and the periodic emergence of 'bubbles'.

Size effects

Market efficiency seems to be less in evidence among smaller firms. Shares of smaller companies tend to yield higher average returns than those of larger companies of comparable risk. Dimson and Marsh (1986) found that in the UK, on average, smaller firms outperformed larger firms by around 6 per cent per annum. Some of the difference can be accounted for by the higher risk and trading costs involved in dealing with smaller companies. Another explanation is institutional neglect. Financial institutions dominating the stock market often neglect small firms offering what appear to be high returns because the maximum investment is relatively small (if they are not to exceed their normal 5 per cent maximum stake). The costs of monitoring and trading may not warrant the sums involved.

Timing effects

In the longer term, disparities in share returns seem to correct themselves. A share performing poorly in one year is likely to do well the following year. Seasonal effects have also been observed. At the other extreme, it has been observed that share performance is related to the day of the week or time of the day. Prices tend to rise during the last fifteen minutes of the day's trading, but the first hour of Monday trading is generally characterised by heavy selling. Investors may evaluate their portfolios over the weekend and decide what to sell first thing on Monday, but are more cautious in their buying decisions, preferring to take their broker's advice.

Stock market surges and bubbles

An investor holding a wide portfolio of shares (e.g. the FTA All-Share Index) for, say, 25 years, would have been rewarded handsomely. But the capital growth was not a steady monthly appreciation; the bulk of it came in just a fraction of the investment period through stock market surges. In an efficient market, few – if any – are clever enough to be able to predict short-term stock market surges.

The famous South Sea Bubble of 1722 was one of the early speculative stock market 'bubbles' where investors adopt the 'herd' instinct and drive up prices well above any

Black Monday

In October 1987, on 'Black Monday', share prices fell by 30 per cent or more on most of the world's stock markets. Had this collapse been triggered by some cataclysmic event, shareholders' reactions could be easily explained as the efficient market reacting to new information. However, Black Monday was not a reaction to external events, but rather a recognition that the prolonged bull market had ended and that the speculative share price bubble had burst. This brings into question the validity of the simple EMH, which implies that share prices cannot rise to the artificially high levels observed prior to the 1987 crash. The newly-introduced computer trading methods, which automatically sell shares when they fall below a predetermined level, were unable to cope with such an adverse situation. It is now generally accepted that the EMH failed to explain why the Dow Jones Industrials index plummeted 23 per cent in just a few hours.

This enigma has led to a re-evaluation of the simple EMH and the assumption that there is a single 'true' value for shares; there may be a very wide range of plausible values. The EMH, if it operates at all, does so in the weakest of forms and is most efficient when conditions are stable.

Black Monday's crash adds credence to the Speculative Bubble theory. Stock market behaviour is based on inflating and bursting speculative bubbles, rather than fundamental analysis based on new information. Investors buy shares because they believe that others will pay yet more for them later, thus creating a bull market. Eventually, the bubble bursts and the market corrects itself or crashes, depending on the size of the bubble.

rational valuation based on economic fundamentals. The economist J.M. Keynes described this in terms of a 'beauty contest' where investors are not following their own judgments but trying to guess how other investors are going to behave. The Internet Bubble of 1999 shows that speculative bubbles are still with us and the cost of following the trend can be considerable.

Self-assessment activity 23.1

If the stock market is efficient, can no one beat the market average return?

(Answer in Appendix A at the back of the book)

■ A modern perspective – Chaos Theory

The EMH is based on the assertion that rational investors rapidly absorb new information about a company's prospects, which is then impounded into the share price. Any other price variations are attributable to random 'noise'. This implies that the market has no memory – it simply reacts to the advent of each new information snippet, registers it accordingly and settles back into equilibrium; in other words, all price-sensitive events occur randomly and independently of each other.

The crash of 1987, possibly attributed to the market's realisation that shares were overvalued and triggered by the collapse of a relatively minor management buy-out deal, has provoked more detailed scrutiny of the pattern of past share prices. This has uncovered evidence that share price movements do not always conform to a 'random walk'. For example, significant downturns happen more frequently than significant upturns.

A new branch of mathematics, chaos theory, has been harnessed to help explain such features. Observations of natural systems such as weather patterns and river systems often give a chaotic appearance – they seem to lurch wildly from one extreme to another. However, chaos theorists suggest that apparently random, unpredictable patterns are governed by sets of complex sub-systems that react interdependently. These systems can be modelled, and their behaviour forecast, but predictions of the behaviour of chaotic systems are very sensitive to the precise conditions specified at the start of the estimation period. An apparently small error in the specification of the model can lead to major errors in the forecast.

Edgar Peters (1991) has suggested that stock markets are chaotic in this sense. Markets have memories, are prone to major price swings and do not behave entirely randomly. For example, in the UK, he found that today's price movement is affected by price changes that occurred several years previously. The most recent changes, however, have the biggest impact. In addition, he found that price moves were persistent, i.e. if previous moves in price were upwards, the subsequent price move was more likely to be up than down. Yet chaos theory also suggests that persistent uptrends are also more likely ultimately to result in major reversals!

Peters' work suggests that world stock markets exhibit patterns that are overlaid with substantial random noise. The more noise, the less efficient the market. In this respect, the US markets appear to be more efficient than those in the UK and Japan. Other observers suggest that markets are essentially rational and efficient, but succumb to chaos on occasions, with bursts of chaotic frenzy being attributed to speculative activity. This suggests some scope for informed insiders to outperform the market during such periods.

Which view is right? Are stock markets efficient, chaotic or somewhere in between? Pending the results of further research, it seems that corporate financial managers cannot necessarily regard today's market price as a fair assessment of company value, but that the market may well correctly value a company over a period of years. Examination of long-term trends gives more insight than consideration of short-term oscillations. For example, if a company's share price persistently underperforms the market, then perhaps its profitability really is low, or its management poor, or it has failed to release the right amount of information.

To conclude, it seems that the efficient markets hypothesis does not hold, except perhaps in its very weakest form, in today's capital markets. Evolving from both the EMH and chaos theory is a promising successor termed the coherent market hypothesis (CMH) based on a combination of fundamental factors and market sentiment or technical factors (see Vaga, 1991). The CMH argues that capital markets are, at any point in time, in one of the following states, depending on a combination of economic fundamentals and 'crowd behaviour' in the market:

- Random walks – market efficiency with neutral fundamentals
- Unstable transition – market inefficiency with neutral fundamentals
- Coherence – crowd behaviour with bullish fundamentals
- Chaos – crowd behaviour with bearish fundamentals.

We will have to wait to see how well it helps explain stock market behaviour.

23.5 BEHAVIOURAL FINANCE

Behavioural finance is the study of how psychological and sociological factors influence financial decision-making and financial markets. Financial economics traditionally assumes that people behave rationally. We have focused in this book on market efficiency and predictive models based on rational economic choices. This assumes that people have the same preferences, perfect knowledge of all alternatives, and understand the consequences of their decisions. The reality is frequently somewhat different. Behavioural finance relaxes the tight assumptions of financial economics to incorporate models based on observable, systematic and human departures from rationality. Its adherents claim that it helps understand stock market anomalies, including stock market overreaction, underreaction, bubbles and irrational pessimism. Some of the most successful investors have long held the view that to understand the stock market you must first understand the psychology of investors.

Investors show various traits of behaviour

Is it possible to use the principles of behavioural finance to make money on the stock markets? Some investors are certainly putting it to the test. Behavioural finance is the study of certain psychological traits that investors display, which prevent them from acting in a purely rational manner.

Investors display a whole range of traits. These include:

- Loss aversion: an unwillingness to accept losses that causes them to hang on to losers and sell winners.
- Overconfidence: this causes them to trade too often.
- Confirmation bias: this causes them to listen only to such evidence as confirms their original view.

Fuller & Thaler Asset Management is a US fund management group that is attempting to exploit these investor weaknesses. One of the group's directors is Daniel Kahneman, who received the Nobel Prize for economics for his behavioural work.

The company attempts to exploit market anomalies associated with anchoring. This occurs when investors develop a fixed view about a company's prospects and are thus slow to react to new and contradictory information. This can work in two ways. The group has a growth fund that capitalises on the market's under-reaction to new positive information. When companies beat profit forecasts by a wide margin, analysts can be slow to upgrade their forecasts; they hate to admit they were wrong. The next set of results then confirms the faster trend and forecasts get upgraded. This shows up in a 'momentum effect' on the stock market, when shares that have recently performed well continue to do so.

The value fund looks for companies and shares that have been beaten down by the market, but are showing signs of improvement. Investors perceive the company as a basket case and when earnings prove better than expected, they dismiss the result as a fluke. Eventually, as the improvement process becomes permanent, analysts and investors catch on, upgrading their earnings forecasts and applying a higher rating to those earnings.

These approaches have been fairly successful so far. The growth fund returned an annualised 17.3 per cent, after fees, between the start of 1992 and the end of last year; the value fund returned 18.8 per cent per year between the start of 1996 and the end of last year.

Source: Philip Coggan, *Financial Times,* 27 March 2004, p. 10.

Behavioural finance draws on the work of psychologists such as Kahneman and Tversky (1979, 1982) on how human decision-making varies from rational decision-making. Examples of the main differences are:

■ *Information processing.* One example where humans typically have a bias in information processing relates to *loss aversion.* This arises where investors or decision-makers view gains and losses differently. This was observed in Chapter 7 where utility functions were examined. The very word 'loss' is associated with psychological feelings of responsibility, blame and shame. This is called **regret** – the feeling of bereavement when a wrong alternative is chosen, as measured by the difference between the payoff received and what could have been achieved. Typically, we find that an expected loss has more than double the impact on us as a gain of the same magnitude. Shareholders holding diversified portfolios would not wish corporate managers to exhibit such strong loss aversion in their decision-making.

■ *Self-deception.* Most drivers are convinced that they are better than the average driver! Similarly, managers and investors can easily deceive themselves regarding their capabilities. This can be seen in overconfidence or overoptimism, leading to systematic overestimation of what they can achieve, known as hubris. This is frequently encountered in the field of merger and acquisition activity, as seen in Chapter 20. Market traders may deceive themselves that they can consistently beat the market, and ascribe above-average returns to their own skills but below-average returns to bad luck.

Similarly, corporate managers may consistently set unrealistically high targets. Related to this is the mistake that the more the information gathered and time spent analysing a decision, the more control we have over the outcome and, therefore, the more confident that the outcome will be successful. This is termed the **illusion of control bias**. It seems that overconfidence and the illusion of control lead stock market investors to be overactive in their trading and incur high transaction costs, resulting in poorer returns than had they traded less actively. Good decision-making means knowing the limits to one's knowledge, and the limitations imposed by one's endowment of resources and capabilities (Grant, 2004).

■ *Representativeness.* Managers tend to make decisions based on stereotypes formed from experience. They look for patterns and use charts to compare recent stock performance with earlier patterns. Such an approach may lead managers to place excessive trust in patterns repeating themselves rather than focusing on the fundamentals. The poor stock market performance of 'glamorous' shares may be because investors overreact to successful companies thus inflating their share price and reducing the investment yield.

■ *Social effects.* Perhaps the best-known social effect is herd-like behaviour or 'following the crowd', where a choice is made because everyone else seems to be doing it.

■ *Anchoring and adjustment.* In reaching decisions, managers often place undue weight on the first information received. The assessment made from the initial information then acts as a data anchor, subsequent information being used to make minor adjustments. Traditional budgeting is a good example of this approach, where the current year's budget, or actual performance, form the basis upon which incremental adjustments are made to produce next year's budget.

Behavioural finance raises some important issues.

1 Is it possible to exploit irrational behaviour when it arises? If any investor can identify that the market is acting irrationally, giving rise to a gap between market share price and underlying value, he or she has the potential to exploit this profitable opportunity.

2 How can investors avoid making irrational decisions, and so achieve returns superior to other investors?

We consider these as they relate to market efficiency and corporate finance.

■ Behavioural finance and market efficiency

Behavioural finance has examined how investors react to new information. Stock prices appear to underreact to financial news such as earnings announcements, but overreact to a series of good or bad news. Adherents of the efficient markets hypothesis argue that:

■ Investors, in the main, value securities rationally.

■ Even if some investors do not act rationally, their irrational behaviour is random and therefore cancels out.

■ But even if most investors act irrationally, the market will be rectified by rational arbitrageurs who profit from the irrationality of others. Fama (1998) examined the impact of stock market anomalies on market efficiency. He concluded that 'market efficiency survives the challenge from the literature on long-run return anomalies. Consistent with the market efficiency hypothesis that the anomalies are chance results, apparent overreaction to information is about as common as under-reaction, and post-event continuation of pre-event returns is about as frequent as post-event reversal.'

The psychology literature shows that people are irrational in a systematic manner.

Collective behaviour relates to the irrational behaviour of groups. Typically, this gives rise to excessive market swings. Two examples of such are 'herding' and 'price bubbles'. These arise when a large group of investors make the same choice based on the actions of others, which cannot be explained by fundamentals. The 'dotcom' price bubble of the 1990s is a clear example of this type of behaviour. Most individuals find comfort in being part of a crowd, rather than acting independently. Of course, more often than not, the crowd can be right for a while, at least, until an overvalued market lurches downwards into its long-overdue correction phase.

Most financial practitioners are subject to bias. Overconfidence, and emotion cloud their judgement and misguide their actions. The question is whether they recognise this behaviour and take steps to minimise this bias. Practical steps that can be taken to reduce such bias include:

1 Recognise the circumstances leading to overconfidence.
2 Have a written plan for each position, especially exit strategies.
3 Review actions.

Watch the herd, but don't join it

'Honey, it's only a paper loss,' is the classic financial self delusion. Yet many investors cling desperately to bad investments rather than realising the losses and moving on.

This process of suppression is well known to psychologists and sociologists. Apart from the losses themselves, people are dominated by a fear of making mistakes and of losing prestige. These factors all play their part in distorting people's sense of perception.

The two disciplines of economics and psychology, separate for so many decades, are now united under the banner of behavioural finance. This combination is potentially extremely useful to investors. The objective of behavioural finance is to give investors insight into their own mistakes and those of others. The subject of behavioural finance has its origins in the research in the late 1970s of Americans Daniel Kanemann and Amos Tversky. Their work made waves at the time and the surf is still up.

Joachim Goldberg, a German behavioural finance guru, says many people are psychologically unable to follow rule Number One: 'Cut losses and let profits run.' This fundamental principle is broken time and time again because people are governed by the wrong psychology. Goldberg urges investors to forget what they paid because it is no longer relevant. It is a false reference point.

Likewise, the same misconceived psychology often encourages people to invest still more in a failed investment, thinking they can recoup their losses by buying at a bargain-basement price in a bear market. However, a bad investment does not get better when you pour more money into it.

Economists have always tended to underrate the importance of psychology and focus on so-called fundamentals. Yet, there are many unpredictable economic variables that affect stock markets. These range from political events, 'real' infections such as Sars and Chicken Flu to interest rates, the gold price, the impact of the weather on commodity prices and a host of others. Psychology is simply another variable.

Source: Brian Bloch, *Financial Times*, 3 July 2004, p. 25.

■ Behavioural finance and corporate managers

While much of the behavioural finance literature considers the psychological aspects of stock market trading, it also applies to investment and financing decisions within firms, which is the main focus of this book. If finance managers are able to recognise the biases in their judgements they will be better prepared to avoid or manage such bias in future.

Behavioural finance helps us understand:

■ Why most boards believe the market undervalues 'their' firm's shares.
■ Why, more often than not, acquisitions fail to deliver the hoped-for financial benefits.
■ Why financial projections in capital investment proposals are usually overoptimistic.
■ Why boards find it difficult to terminate (or even decide to escalate) unprofitable projects or strategies.

Share valuation

Good financial public relations and communication with shareholders is vital for any listed company, not least because the market should reflect all relevant information in the company's share price. Typically, we find that senior executives believe that the market tends to underprice 'their' shares and fails to recognise the true worth of the company. But in efficient markets the market price is an unbiased estimate of the true value of the investment. Corporate managers may be prone to biases of overconfidence, overoptimism and the illusion of control resulting in the self-deception that they are better judges of share price than the market, and that they can control outcomes.

In recent years, there has been a flurry of firms exiting the stock market, whereby managers use finance provided by private equity capital specialists. Although the motivation for this is diverse, it is common to hear directors complaining that the market does not correctly value their firm, and/or analysts are insufficiently interested to generate much research about their prospects (Evans, 2005).

Acquisitions

The bias of overoptimism and overcompetence may also lead boards to believe that they can acquire a firm and produce greater returns than the previous management. As discussed in Chapter 20, there may be sound strategic and economic motives for mounting a takeover bid, but the evidence suggests that, on average, acquisitions are bad news for the shareholders of the acquiring firm.

When takeover bids are contested, the bidding company may end up paying far more for the target firm than was originally intended. In the language of behavioural finance, this may be seen as the loss aversion bias – the strong desire to have something because it looks like it is being taken away from you – and the associated desire to avoid regret from takeover failure.

Unprofitable projects and strategies

Many companies continue to operate projects or strategies long after they cease to become profitable or value-creating. Similarly, firms may continue to invest heavily in a corporate turnaround even though there is little likelihood of it ever recovering such investment. Overconfidence in management's ability to improve performance is one explanation. Another is termed **entrapment**; managers become entrapped in a strategy or project to which they have committed not just corporate capital, but also personal capital. Against all economic logic, they postpone decisions to terminate such projects or strategies in the hope that they will eventually come good. Sometimes, the only way to change direction is to remove the managers entrapped in this mindset.

There is considerable evidence (e.g. Staw, 1976, 1981) that managers are often reluctant to 'pull the plug' on failing projects, or may even decide, in the teeth of adversity, to even escalate their commitment, hoping for an eventual turnaround, when the 'rational' thing to do is to abandon (Staw and Ross, 1987). However, for managers, blessed with information advantages, the rational course of action is to prolong their employment. Indeed, because they often have a reputation to protect or at least not to sully, prolongation is the lesser evil (Kanodia *et al.*, 1989).

Firms need to have in place clear guidelines preventing the above arising through such mechanisms as regular post-audit reviews (Neale and Holmes, 1990). These mechanisms should specifically look for overoptimistic forecasts, irrational escalation of commitment, entrapment and other systematic biases in behavioural finance.

We are only human

Faced with complex financial decisions, investors often employ heuristics, or rules-of-thumb, when making decisions. Heuristics can be useful in everyday life, but these inbuilt tendencies are too blunt a tool for the complex environment of financial decision-making.

One such heuristic is mental accounting, whereby people separate their finances into distinct accounts in order to make decision-making easier. For example, would you treat a gambling win in the same way as a salary increase? Most people would not, as they see it as a windfall gain despite the fact that in both instances they are equally well off. The problem is that this may not lead to an efficient allocation of resources (people often keep betting if they think they are on a winning streak).

This has important ramifications. Investors only psychologically realise a loss on an investment completely when they sell and thereby 'close' the mental account. When coupled with loss aversion – a tendency for people to react far more strongly to losses than to gains – this results in a reluctance to sell badly performing investments. Although the economic loss already exists, the pain associated with it is only psychologically 'booked' at the actual sale. As a result, investors are prone to selling their winners and riding their losers.

A related error is self-deception. This occurs as investors selectively reinterpret the success of their own previous decisions, giving them an illusion of control in an uncertain world. With hindsight bias, for example, we reconstruct our own previous beliefs, thereby deceiving ourselves into believing we are right more often than we actually are.

Overconfidence can also lead investors to underreact to new information. An investor who has built up a good investment story, and shared it with colleagues, instinctively reacts negatively to contradictory news. Because many analysts make the same mistake the share price may respond gradually, and inefficiently, rather than immediately.

These failings are just some of those uncovered by behavioural finance. An understanding of this relatively new field, at the very least, provides investors with a more realistic view of how the market works. Where these errors are exploitable it may even provide investors with investment opportunities. But it does not seem that investors can change their instinctive reactions to become rational, unemotional financial machines. Inefficient markets are here to stay.

Source: Based on article by James Hand and Greg Davies, *Financial Times*, 25 October 2004, p. 6.

Psychology will contribute to the next big breakthrough

Bill Sharpe, the brains behind the Capital Asset Pricing Model and Nobel laureate, is at Stanford University's Graduate School of Business, where he grapples with such gnomic questions as 'is beta dead?'

One of the conclusions of the CAPM is that asset returns are a function not of total risk, but rather the risk of doing badly in bad times. Prof Sharpe suggests that a simple measure is an asset's 'beta' – the expected change in returns given a 1 per cent change in the return from the overall market.

The challenge to beta in the 1990s came from Eugene Fama and Ken French, who did empirical work that seemed to explain stock returns by the over-return on equities, the size of the company and whether it was a value or growth stock. Consequently, there was no room for beta in this model. Prof Sharpe's response is that empirical research on financial markets is notoriously difficult. 'If you don't like an empirical result, wait till someone does it in another time period using a different method or another country,' he says. 'I have done empirical work – lots of it. But I think empiricists need to be modest.'

Prof Sharpe's analysis of markets and finance has yet to come to rest and he has several answers as to where

financial economics is heading. First, he argues that finance has become too obsessed with mathematics. 'We have got so intent on having elegant solutions to closed-form equations that we have tolerated some really stupid assumptions about people's preferences,' he says.

Linked to this, he wants financial economists to strive for a better understanding of how people really act. Does that make Prof Sharpe a closet fan of behavioural finance, which tries to explain financial markets by looking at human psychology?

'I'm a fan of good behavioural finance. It is not a question of trying to show that people are irrational or throwing out all the models that involve rationality. The interesting thing is to find out what kinds of decisions people make under conditions of uncertainty if they know what they are doing.'

It is from this marriage of psychology and economics that Prof Sharpe expects the next breakthrough in finance. Fund managers, watch this space.

Source: Based on article by Simon London, *Financial Times*, 29 July 2002, p. 4.

SUMMARY

In this final chapter of the book we have reviewed the main theories behind the corporate finance principles and practices outlined in previous chapters. The key material introduced was on the contribution that the emerging area of behavioural finance makes to our understanding of the subject.

Key points

- Financial management is more than applying rules and procedures. It explores the behaviour of markets, firms and individuals.

- Various theories explaining such behaviour include those of agency, risk aversion, present value, portfolio, information asymmetry, options, capital structure and market efficiency.

- Behavioural finance is the study of psychological traits that investors and managers display that prevent them acting in a purely rational manner.

- Examples of behavioural finance bias include regret, overconfidence, over-optimism, the illusion of control, herd behaviour, and anchoring and adjustment.

- Behavioural finance is particularly in evidence in stock market anomalies, share valuation, acquisitions, and loss-making projects.

Further reading

Behavioural finance is a fast-developing topic. A good review of this may be found in Shefrin (1999) and Smith (2001). Key articles on this subject are Barber and Odean (2000) and Fama (1998).

A key article on the development of finance theory is Miller (1999).

Solutions to self-assessment activities

CHAPTER 1

1.1 Tangible real assets: machinery and equipment, vehicles, stock
Intangible assets: patents
Financial assets: debtors, cash and building society deposits
Financial claims: trade creditors, loans, shareholders' equity

1.2 The financial manager has two broad responsibilities:

(1) Providing financial information and advice for internal and external users. The financial accountant prepares the statutory financial accounts and deals with the auditors; the management accountant provides information for decision-making, planning and control. Other departments, such as taxation, may also report to the controller (or chief accountant).

(2) Managing cash and raising finance at the best possible rate. This is the responsibility of the treasurer or financial manager. Typical functions within the treasury area are raising finance, the management of cash, credit and inventory, foreign currency management and financial risk management.

1.3 The past ten years have seen a much greater emphasis on investor-related goals, such as earnings per share and shareholder wealth. This is mainly attributed to the growth in shareholdings by institutional investors (e.g. pension funds). The recognition that profitability is not necessarily the same as shareholder wealth has led to greater emphasis on shareholder goals.

1.4 (1) The goal of maximising owners' wealth is the normally accepted economic objective for resource allocation decisions. Rather than concentrate on the organisation, it evaluates investments from the viewpoint of the organisation's owners – usually shareholders. Any investment that increases their stock of wealth (the present value of future cash flows) is economically acceptable.

In practice, many of the assumptions underlying this goal do not always hold (e.g. shareholders are only interested in maximising the market value of their shareholdings). In addition, owners are often far removed from managerial decision-making, where capital investment takes place. Accordingly, it is common to find that more easily measurable criteria are used, such as profitability and growth goals. There are also non-economic considerations, such as employee welfare and managerial satisfaction, which can be important for some decisions.

(2) Separation of ownership from management can give rise to managerialism – self-serving behaviour by managers at the shareholders' expense, e.g. pursuing managerial 'perks' (company cars, etc.), adopting low-risk survival strategies, settling for less than the best.

CHAPTER 2

2.1 Financial intermediaries are the various financial institutions, such as pension funds, insurance companies, banks, building societies, unit trusts and specialist investment institutions. Their role is to accept deposits from personal and corporate savers to lend to customers (e.g. companies) via the capital and money markets.

Financial intermediaries perform a vital economic service:

(a) Re-packaging finance: collecting small amounts of finance and re-packaging into larger bundles for specific lending requirements (e.g. banks).

(b) Risk reduction: investing sums, on behalf of individuals and companies, into large, well-diversified investment portfolios (e.g. pension funds and unit trusts).

(c) Liquidity transformation: bringing together short-term lenders and long-term borrowers (e.g. building societies).

(d) Cost reduction and advice: minimising transaction costs and providing low-cost services to lenders and borrowers.

2.2 (a) Companies using the main market are generally those with a lengthy track record (i.e. a history of stable/growing sales and profits), seeking large amounts of capital, where the board is prepared to release a sizeable proportion of its controlling interest.

(b) AIM is designed to appeal to smaller and growing companies which seek access to risk capital or where the board wishes to sell some of its holding, but cannot meet all the requirements for a full listing and prefers less regulation.

(c) The OTC market is for the remaining, smaller companies where shareholders occasionally wish to dispose of shares. This is performed on a 'matched bargain' basis, using facilities offered by authorised securities dealers.

2.3 A major function of an active capital market is to provide a mechanism whereby investors can realise their holdings by selling securities, and, obviously, for every seller, there has to be a buyer. Investors will be reluctant to commit their funds to the capital market by subscribing to new share issues if they doubt their ability to find a willing buyer as and when they decide to sell their holdings. The more liquid the market, the greater its ability to entice firms to make new share issues and investors to subscribe to them. Where market liquidity is poor, companies will have to offer much higher returns, making share issues uneconomic.

2.4 If these rules ever applied, investors would have soon realised the potential, and would have bought in November rather in advance of the expected price increase in December, thus creating a 'November effect', and so on.

Try the same argument on the old stock market advice 'Sell in May, and go away, and come back on St Leger Day' (the date of a horse race in the UK).

Many statistical tests have shown that all such dealing rules are usually inferior, and never superior, to a simple 'buy and hold' strategy.

2.5 (a) On the face of it, this represents insider dealing, by people in the know.

(b) It could represent speculation and rumour. If the company is known to be weak, then it is a candidate for takeover and people begin to speculate about how much the company may be worth in the hands of an alternative owner.

(c) Most bidders build up a 'strategic stake' in a takeover target prior to formal announcement. The present UK rules (*The City Code*) allow a holding of up to 3% without declaration of beneficial ownership. The upward movement could simply be due to this buying pressure. In reality, 'abnormal' buying also promotes speculation.

2.6 The reader is referred to Section 2.7.

CHAPTER 3

3.1 Most likely, the banker would wish to reflect on the rate of interest required:

(a) the rate of interest available from a risk-free investment,
(b) the expected changes in purchasing power over the five years, and
(c) the risk that you may not be able to repay.

We consider other factors a banker will consider in a later chapter.

3.2 $PV = \dfrac{£623}{(1.07)^8} + \dfrac{£1,092}{(1.07)^{16}} = £732$

3.3 Discounting at 15 per cent, a pound halves in value every five years. The present value of the purchase cost is therefore £750,000 (£500,000 today and £250,000 in year 5).

3.4 $PV = \dfrac{£1,000}{(1.12)^{12}} = £257$

3.5 Using the tables:

$$PV = (£250 \times 8.0751) + (£1,200 \times 0.10067) = £2,140$$

3.6 Price per share $= \dfrac{D_1}{k_e - g} = \dfrac{D_o(1 + g)}{k_e - g}$

$D_o = (25\% \times 16p) = 4p$

growth $= (0.75 \times 0.18) = 0.135$,   i.e.   13.5% and $D_1 = 4p\,(1.135) = 4.54p$

Share price $= \dfrac{4.54}{(0.15 - 0.135)} = £3.03$

With shareholders seeking a 20% return, share price reduces to: $\dfrac{4.54p}{(0.20 - 0.1350)} = £0.70$

Because the return on reinvestment now is less than the cost of equity the firm should stop reinvesting, at least in the short term.

CHAPTER 4

4.1 Your answer could include the costs of developing a new product; a marketing campaign designed to increase long-term brand awareness; investment in training and management development; acquisition of other businesses; reorganisation and rationalisation costs (frequently in the form of redundancy payments), or research costs incurred in developing a strategic advantage.

4.2 The main elements in the capital budgeting decision are found in the formula

$$NPV = \frac{X_1}{1 + k} + \frac{X_2}{(1 + k)^2} + \cdots + \frac{X_n}{(1 + k)^n} - I$$

where X is the net cash flow arising at the end of year, k is the minimum required rate of return (or discount rate), I is the initial cost of the investment and n is the project's life.

4.3 The net present value rule states that a project is acceptable if the present value of anticipated cash flows exceeds the present value of anticipated cash flows. Wealth is maximised by accepting all projects offering positive net present values when discounted at the required rate of return.

4.4

	Accept when
Net Present Value	NPV > 0
Internal Rate of Return	IRR $> k$
Profitability Index	PI > 1 (or PI > 0 if based on NPV)

4.5

For	Against
Payback period	
Simple to calculate	Ignores time-value of money
Easily understood	Ignores cash flows beyond payback period
Crude measure of profitability	Problems in setting payback requirement
Emphasises liquidity	
Net present value	
Theoretically sound	Not well understood by non-financial managers
Based on cash flows	
Incorporates the time-value of money	Estimating the appropriate discount rate is difficult in practice
Internal rate of return	
Based on cash flows	Can incorrectly rank projects
Incorporates the time-value of money	Difficult to calculate without a computer
Accounting rate of return	
Can be related to accounts	Ignores time-value of money
Based on profits not cash flows	Problems in setting the required return

4.6 Evaluation of mutually exclusive projects using IRR and NPV approaches can produce different recommendations. This is particularly the case where projects are very different in scale or where the cash flow profiles of the various alternatives differ significantly. In such circumstances, the NPV approach is the better method.

4.7 Project Y offers the highest NPV for all discount rates up to 17 per cent. At the 12 per cent cost of capital, it offers a better cash return than Project X.

4.8 Soft capital rationing is internally imposed by the firm. This may be because management is unwilling to borrow or wishes to pursue a policy of stable growth. Hard rationing is externally imposed by the capital market. No additional external finance is available to the firm. Capital for a single period may be resolved by applying the profitability index to investment cash flows. Multi-period rationing requires a form of mathematical programming.

4.9 The modified IRR is that rate of return which, when the initial outlay is compared with the terminal value of the project's cash flows reinvested at the cost of capital, gives an NPV of zero. Whereas the IRR method implicitly assumes that cash flows generated by the project are reinvested at the project's internal rate of return, the modified IRR assumes reinvestment at the cost of capital. This means that it gives the same investment advice as the NPV approach.

CHAPTER 5

5.1 Incremental cash flows, applied to capital budgeting, are the additional cash flows created as a direct result of making an investment decision. Frequently, identifying these is not as straightforward as one might think, particularly where replacement decisions are involved or where decisions in one part of the business have ramifications for other parts.

5.2 The original cost of developing the drug is a sunk cost and should not form part of the analysis. Any adverse effects on other parts of the business resulting from the decision to manufacture the product are an associated cash flow and should be included in the analysis. The external sale value of the patent is an opportunity cost of proceeding with production. The £10 million cost should be deducted from the project's cash flows.

5.3 To: Sid Torrance

From: Rick Faldo

Clubs for Beginners Proposal

I have re-examined the points raised in your e-mail and discussed them with our accountant.

In analysing capital projects, only future investment cash flows incremental to the business are relevant to the decision.

1 Depreciation is not a cash flow – it is a charge against profits. By comparing operating cash flows against initial outlay, the need for depreciation becomes unnecessary.
2 Only additional fixed costs resulting from the introduction of the new project should be charged. I have checked that no extra overheads are incurred.
3 The market research is a past cost. Its existence is not dependent upon the outcome of the decision, so it should not be included.
4 I agree that finance costs are important, but the cost of finance has already been accounted for within the discount rate.

My 'Clubs for Beginners' proposal is, I suggest, an attractive one, from which the business should benefit considerably.

5.4 Many companies have failed to take account of the relatively low inflation and interest rates when calculating the required return on investment projects. For example, a nominal rate of 20 per cent, which many firms use, may not be unreasonable during times of relatively high inflation, but is probably excessive during periods of low inflation. This makes it more difficult for economically attractive projects to be accepted.

5.5 Tax is a cash flow and shareholders want firms to maximise their after-tax cash flows. It affects companies in different ways, depending on their tax situations and the capital allowances available to them. Some types of business attract more generous allowances than others. However, for the majority of capital projects, the tax effect will have a relatively minor impact on the project outcome. Good projects are usually viable both before and after tax.

5.7 Weighted average cost of capital: Major plc

Capital	Market value £m	Weight	Cost %	Weighted cost
Ordinary shares	48	0.6	11	6.6
Loan	32	0.4	6	2.4
				9.0

CHAPTER 6

6.1 The attractiveness of any project depends, in large measure, on how well it helps the business implement agreed strategies to achieve agreed goals. Investment is one of the main vehicles for implementing strategy.

It is quite possible for capital projects to be rejected because they are not compatible with long-term corporate intentions. An important element of the investment decision process is to ensure that there is a good 'fit' between projects proposed and corporate strategy.

6.2 Stages in the investment process:

Search for opportunities
Screening (is it worth evaluating further?)
Defining the project
Evaluation
Transmission through the organisation
Authorisation
Monitoring
Post-audit

6.4 Post-auditing may confer substantial benefits on the firm. Among these are the following:

1 The enhanced quality of decision-making and planning, which may stem from more carefully and rigorously researched project proposals.
2 Tightening of internal control systems.
3 The ability to modify or even abandon projects on the basis of fuller information.
4 The identification of key variables on whose outcome the viability of the current and similar future projects may depend.

CHAPTER 7

7.1 While a capital project may have a high expected return, the risks involved may mean that there is the possibility that it will be unsuccessful – even to the extent of putting the whole business in jeopardy.

7.2 (1) Financial risk; (2) Business risk; (3) Project risk; (4) Portfolio or market risk.

7.3 X has the lower degree of risk relative to the expected returns.

Coefficient of variation: X = 400/2,000 = 0.2
Y = 400/1,000 = 0.4

7.4 **(a)** Risk – a set of outcomes for which probabilities can be assigned.
(b) Uncertainty – a set of outcomes for which accurate probabilities cannot be assigned.
(c) Risk-aversion – a preference for less risk rather than more.
(d) Expected value – the sum of the possible outcomes from a project each multiplied by their respective probability.
(e) Standard deviation – a statistical measure of the dispersion of possible outcomes around the expected value.
(f) Semi-variance – a special case of the variance which considers only outcomes less than the expected value.
(g) Mean-variance rule – Project A will be preferred to Project B if either the expected return of A exceeds that of B and the variance is equal to or less than that of B, or the expected return of A exceeds or is equal to the expected return of B and the variance is less than that of B.

7.5 Monte Carlo simulation involves constructing a mathematical model which captures the essential characteristics of the proposal throughout its life as it encounters random events. It is useful for major projects where probabilities can be assigned to key factors (e.g. selling price or project life) which are essentially independent.

CHAPTER 8

8.1 **(1)** A call option gives the owner the right to buy shares (or whatever) at a fixed price within a set period. A put option is the right to sell at a fixed price.
(2) A European option can only be exercised on the expiry date, while an American option can be exercised any time over the life of the option. Most traded options in Europe are actually the American variety.
(3) A wide range including shares, bonds, currency, interest rates, gold, silver, wool, soya beans, etc. Options are also available on interest rates.
(4) **(a)** The lower the exercise price, the more likely it is that it will be profitable to exercise the call, and therefore the more investors are prepared to pay for the call.
(b) The longer the time that a call has to run to maturity, the greater the scope for the price to drift above the exercise price. Of course there is also more time for prices to fall below the exercise price. However, the potential gains and losses are not symmetrically distributed. There are limits to the losses but not to the gains.
(c) The price exceeds the profit that can be made immediately by exercising the call as over its remaining life there will be the opportunity to capitalise on further price movements. These may be upwards or downwards but we have already noted a bias in the consequence of price changes – this offers a higher expected value of potential gains than losses.

8.2 **(1)** **(a)** It has intrinsic value because you could buy the share (exercise the option) for less than the share price.
(2) **(a)**

8.3 Applying the equation below, we find that the put–call parity relationship holds:

Share price + value of put − value of call = E/(1 + R_f)

44   7   1   = 55/1.10
PV exercise price = 50p

8.4 Option value increases with the volatility of the underlying share price because the greater the variability in share price, the greater the probability that the share price will exceed the exercise price. Because option values cannot be negative, only the probability of exceeding the exercise price is considered.

8.5 Option values are determined by five main factors:

- share price
- exercise price of the option
- time to expiry of the option
- the risk-free rate of interest
- volatility of the underlying share price

In addition, the payment of dividends and underlying stock market trends have some influence on option values.

CHAPTER 9

9.1 To eliminate the risk of, say, a two-asset portfolio, there would have to be perfect negative correlation between the returns from the two assets, and also the portfolio would have to be 'correctly' weighted.

9.2 Expected values:

A: EV $= (0.5 \times -10\%) + (0.5 \times 50\%) = 20\%$

B: EV $= (0.5 \times -20\%) + (0.5 \times 60\%) = 20\%$

Standard deviations

A: $\sigma_A = \sqrt{[(0.5)(-10\% - 20\%)^2 + (0.5)(50\% - 20\%)^2]}$

$\quad = \sqrt{[(0.5 \times 900) + (0.5 \times 900)]}$

$\quad = \sqrt{900} = 30$, i.e. 30%

B: $\sigma_B = \sqrt{[(0.5)(60\% - 20\%)^2 + (0.5)(-20\% - 20\%)^2]}$

$\quad = \sqrt{[(0.5 \times 1600) + (0.5 \times 1600)]}$

$\quad = \sqrt{1600} = 40$, i.e. 40%

9.3 The expected value $= 20\%$ as explained. The standard deviation is:

$\sqrt{[(0.64)^2(30)^2 + (0.36)^2(40)^2 + 0]}$

$= \sqrt{(368.64) + (207.36)}$

$= \sqrt{576}$

$= 24$

9.4 To minimise portfolio risk, let α^* = the proportion invested in asset Z. Using the risk-minimising formula 9.4:

$$^*\alpha = \frac{\sigma_y^2 - \text{cov}_{zy}}{\sigma_z^2 + \sigma_y^2 - 2\,\text{cov}_{zy}}$$

$$= \frac{40^2 - (-200)}{200^2 + 40^2 - (2 \times -200)}$$

$$= \frac{1600 + 200}{400 + 1600 + 400}$$

$$= 1800/2400 = 0.75, \text{ i.e. } 75\%$$

9.5 An efficient frontier is a schedule tracing out all the available portfolio combinations that either minimise risk for a given expected return or maximise expected return for a given risk.

9.6   Expected NPV = $(0.27 \times £30{,}000) + (0.73 \times £24{,}000)$

$$= £25{,}620$$

Standard deviation = $\sqrt{[(0.27)^2(10{,}000)^2 + (0.73)^2(4{,}899)^2 + 2(0.27)(0.73)(-20{,}000{,}000)]}$

$$= \sqrt{7.29\text{ m} + 12.79\text{ m} - 7.88\text{ m}}$$

$$= \sqrt{12.22\text{ m}}$$

$$= £3{,}496$$

9.7   The outright risk-minimiser locates at point B, the portfolio with minimum possible risk. However, risk-averters are willing to accept higher levels of risk if offered sufficient additional rewards, i.e. higher returns. Hence, any portfolio along AB is consistent with risk aversion. The lower the investor's concern with risk, the nearer to point A he/she will locate.

9.8   With four separate assets to choose from, the investor faces a wider array of available portfolios. The envelope that summarises his/her opportunities has the same basic shape as in the text, but with an extra 'corner' representing the fourth asset, denoted by point D in Figure A.1. Notice that all sorts of configurations of the envelope are possible, depending on the location of the four assets.

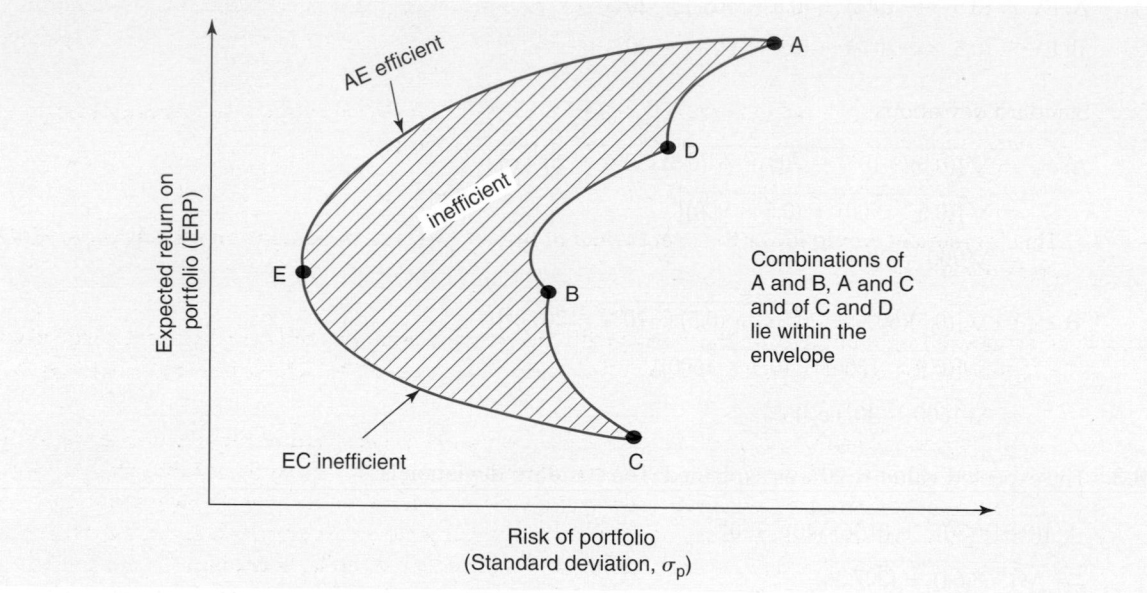

Figure A.1 Portfolio combinations with four assets

CHAPTER 10

10.1   TSR = [Dividend + Capital Gain]/Opening share price, expressed as a percentage.

$$= [£0.01 + £0.17]/£2.20 = £0.27/£2.20 = 0.123 \text{ i.e. } 12.3\%$$

10.2   To eliminate the specific risk, the investor would have to hold every share quoted on the market i.e. the market portfolio.

10.3   Systematic risk: political turmoil
exchange rate fluctuation
interest rate changes.

Unique risk: labour relations problems
announcement of a major new contract
discovery of a defect in a key product.

10.4 According to the CAPM, the Beta of Walkley Wagons is 1.2. Hence, the predicted return on its shares would be 1.2 × the predicted market return of 25%, i.e. 30%. (If you believe the experts!)

10.5 Variations around the characteristics line reflect the impact of factors unique to the firm. For BA, this could be due to a pilots' strike, pressure to relinquish landing slots at Heathrow, sale of its stake in Qantas, competition authorities blocking a proposed strategic alliance, etc.

10.6 The Beta values cluster in a relatively narrow range because these are large firms that are, themselves, mostly well diversified, and also because they constitute a major part of the overall market portfolio, which has a Beta of 1.0.

10.7 The market portfolio has a Beta of 1.0 simply because it varies in perfect unison with itself!

10.8 The SML traces out all combinations of risk and return that are efficient – anything currently located above or below the line represents an aberration from equilibrium that will be eliminated as market players realise the degree of mis-pricing. As they buy or sell, they help to move the market to equilibrium.

In the case of security A, the actual return is 'super-efficient' i.e. in excess of the return warranted by its Beta of 0.3. Conversely, security B is inefficient as it offers too low a return (actually below the risk-free rate) given its Beta. In fact, both securities should offer the same return as they share a Beta. To achieve equilibrium, the price of A must rise thus lowering its return, and the price of B must fall to raise its return.

10.9 Here, we need a surrogate Beta. Taking the Marks & Spencer value of 0.82, using a risk-free rate of 5 per cent and the market risk premium of 5 per cent, the required return is:

$$5\% + 0.82(5\%) = 5\% + 4.1\% = 9.1\%$$

This investment would lower the overall Beta of British Airways, depending on the relative size of the two areas of activity.

10.10 An investor might outperform the market with this policy, assuming it did actually rise. However, with such a narrowly-diversified portfolio, he/she would be unduly exposed to risk factors unique to these five firms.

10.11 Long-term performance. Some of a firm's businesses may be cyclical and during the downswings, the other businesses with more stable characteristics can protect overall performance. St Gobain's glass manufacturing activity is cyclical whereas its building materials business is more stable over the long term.

Cash flow for expansion. A diversified company can generate strong cash flow that can enable it to expand further (only half of St Gobain's cash flow is allocated to capital spending).

Shared expertise. St Gobain operates the nine 'delegate offices' that act as collection points for ideas across the company.

Candidate for state support. The strongest companies – the 'national champions' – often receive support from central governments. St Gobain hopes for support to allow it to develop new glass structures for flat-screen TVs, in particular.

CHAPTER 11

11.1 With EPS = 36p, and the dividend covered three times, the dividend per share must be 12p (36p/12p = 3). Hence, the cost of equity is:

$$k_e = \frac{12p\,(1 + 3\%)}{£1.80} + 3\% = (6.9\% + 3\%) = 9.9\%$$

11.2 The DVM breaks down totally:

   (i) when the firm pays no dividend

   (ii) when the growth rate of dividends exceeds the cost of equity.

11.3 Overall, the required return = 5% + 1.2[6%] = 12.2%

For activity A, it is = 5% + 2.0[6%] = 17%

For activity B, it is = 5% + 0.8[6%] = 9.8%

11.4 Company Beta = (0.65 × Beta of A) + (0.35 × Beta of B)

= (0.65 × 2.0) + (0.35 × 0.8) = 1.58

11.5 The sale will take out 17% of the sales of the group, so that the new weights are fractions of the remaining 83%. The calculations as follows

Activity	Weighting	Beta	Weighted Beta
Hotels	33/83	1.01	0.402
Restaurants	37.4/83	1.02	0.460
Coffee Shops	12.6/83	1.15	0.175
Total			1.037

The change in the structure of the Whitbread group is unlikely to reflect quickly in its recorded Beta value as the LBS exercise is based on 60 monthly observations, only a couple of which cover the 'new look' Whitbread.

11.6 The project Beta = RSF × OGF × divisional Beta

= (0.5 × 0.8 × 1.2) = 0.48

Hence, the required return = 5% + 0.48[6%] = 5% + 2.9% = 7.9%

CHAPTER 12

12.1 The value of a whole company, or enterprise value, is the value of all its assets, whether measured at book value or market value (£5 billion in the case of Innogy plc). The value of the equity is the value of the owners' investment in the firm, or shareholders' funds (£3 billion in the case of Innogy). These are equal only when the firm is financed entirely by equity.

12.2 **(i)** D.S. Smith's enterprise value is the total value of its assets, i.e. £1,369.8 million.

(ii) Total liabilities are long-term and short-term creditors, valued at £376.6 million and £426.1 million respectively, total £802.7 million. Minority interests represent remaining shares in firms previously taken over by Smith's. These are neither Smith's liabilities nor Smith's owners' equity. They represent 'outsiders' share of the total assets. Strictly, the figure given for total assets in (i) should be adjusted for this small item.

(iii) The value of owners' equity = shareholders' funds = net assets = NAV = £567.1 million. These four terms are synonymous.

12.3 Profit after tax = [1 − 30%] × £80 m = £56 m

EPS = (£56 m/100 m) = 56p

Implied share value = EPS × surrogate P:E ratio = (56p × 15) = £8.40.

12.4 Break-even value is £1 million, of which £0.361 represents the PV of the rental income. To break even, the re-sale value must have a PV of (£1 m − £0.361 m) = £0.639 m. Reversing the discounting process, this is a value of £0.639 × $(1.12)^5$ = £1.126 m. Hence, the property must rise in value by about 13% to prevent investors from losing out.

12.5 Taxation = 30% × (£25 m − £1 m) = £7.2 m

Free cash flow = Operating Profit + depreciation − interest − tax − investment expenditure

(£25 m + £2 m − £1 m − £7.2 m − £3 m) = £15.8 m.

CHAPTER 13

13.1 Treasury management involves the efficient management of liquidity and risk in the business. The treasurer usually has primary responsibilities for funding, risk management, working capital management and liquidity, and managing banking relationships.

13.2 The matching approach to funding is where the maturity structure of the company's financing matches the cash flows generated by the assets employed. In simple terms, this means that long-term finance is used to fund both fixed assets and permanent current assets, while fluctuating current assets are funded by short-term borrowings.

13.3 Over to you!

13.4 **(a)** True
      (b) False
      (c) The company is selling euros so the higher three-month forward rate applies, i.e. 100,000/1.4898 = £67,123.
      (d) It should let it lapse as the spot price is cheaper than the exercise price.

13.5 *Forward* – commits the user to buying or selling an asset at a specific price on a specific future date.
      Future – a forward contract traded on an exchange.
      Swap – a contract by which two parties exchange cash flows linked to an asset or liability.
      Option – the right to buy or sell at an agreed price.

13.6 Smaller firms will rarely have a separate treasury function, the accountant or managing director having to perform much of this role. There is a real danger that key areas will be neglected. For example, liquidity management or banking relationships may be largely neglected. Funding may be through short-term overdraft facilities when a larger, more secure form of funding may prove beneficial. Other dangers are that small businesses involved in exporting or importing may neglect the need to manage or hedge currency risks, or corporate borrowers may neglect exposure to interest rate fluctuations.

13.7 A company with a simple production process, that makes to order, enjoys generous supplier credit terms and offers cash discounts for early payment could have a much shorter operating cycle than the industry average. A firm with a longer cycle demands more capital and is more exposed to bad debt and stock losses.

13.8 Overtrading arises when businesses operate with inadequate long-term capital. It occurs when firms:
      (a) are set up with insufficient capital;
      (b) expand too rapidly without a commensurate increase in long-term finance;
      (c) utilise net current assets in an inefficient manner. The consequences can be extremely serious and possibly fatal unless the problem is addressed.

CHAPTER 14

14.1 *Rough approximation*

$$\text{Cost of discount} = \frac{\text{Discount }\%}{100 - \text{discount }\%} \times \frac{365}{\text{Final date} - \text{discount date}}$$

$$= \frac{2.5}{100 - 2.5} \times \frac{365}{46 - 30} = 58.5\%$$

APR method

Interest = 2.5/(100 − 2.5) = 2.56% per period

APR = $1.0256^{365/16} - 1 = 78\%$

In both cases, the cost of granting the cash discount terms looks prohibitive, unless it is thought likely that the customer would take much longer to pay than most other customers.

14.2 (a) Credit terms, including credit period.
   (b) Credit standards for offering credit to existing and new customers, including credit risk screening.
   (c) Credit collection policy, including use of cash discounts and collection agents.
   (d) Credit reporting.

14.3 Carrying costs include the cost of storing, insuring and maintaining stocks as well as the lost interest tied up in holding such assets.

   Ordering costs include not only the obvious administrative cost of making regular orders, but also the costs associated with running out of stock (lost orders, goodwill, etc.). These two types of cost are traded off to find the optimum order quantity, given by the formula:

$$EOQ = \sqrt{\frac{2AC}{H}}$$

   where C is the cost of placing an order, A is the annual stock usage and H is the cost of holding a unit of stock.

14.4 The main motives for holding cash are:

   (a) to act as a buffer to ensure that transactions can be paid for (transactions motive);
   (b) to cater for unanticipated cash outflows (precautionary motive);
   (c) to permit companies to take advantage of profitable opportunities (speculative motive);
   (d) to take advantage of 'free' banking services for firms with positive cash balances (compensation balances motive).

14.5 With interest at 12% p.a., the EOQ formula becomes:

$$EOQ = \sqrt{\frac{2 \times £2,400,000 \times £25}{0.12}} = £10,000$$

CHAPTER 15

15.1 Trade credit results from the time lag between receiving goods and having to pay for them. For as long as it takes the recipient of the goods to settle the account, the supplier is effectively financing the client. Hence, each order and delivery triggers a 'spontaneous' supply of finance.

15.2 Effective annual interest:

   Discount lost/Extra finance = $(1.5\%/98.5\%) = 1.523\%$

   Number of 25 day periods in a year = $(365/25) = 14.6$

   Effective interest rate = $(1 + 0.01523)^{14.6} - 1 = 0.247$, i.e. 24.7%.

15.3 The mnemonic PARTS is sometimes used to help remember this:
   Purpose: Is the purpose of the loan acceptable to the bank?
   Amount: Has the financial requirement been correctly specified (e.g. will additional working capital be required later)?
   Repayment: How will the loan be repaid? Will the funds generate sufficient income to enable repayment?
   Term: What is the duration of the loan?
   Security: What, if any, is the proposed security?

15.4 Overdrafts are normally cheaper than term loans because:

   ■ They can be recalled at short, or no, notice; term loans are for agreed durations.
   ■ Arrangement fees are higher for a term loan.
   ■ With an overdraft, interest is only paid on the credit used, not the credit available.

15.5 It receives cash to pay suppliers promptly and take advantage of any early payment discounts available.

   ■ Growth can be financed through revenue from sales rather than through additional capital.
   ■ Management need not devote so much time to chasing debtors and running the sales ledger.

■ Finance is directly linked to sales. (Overdrafts are linked to the balance sheet and the security offered by assets.)

■ If factoring is 'without recourse', any bad debts are no longer the firm's problem.

15.6 Loan = (70% × £40,000) = £28,000

Total interest = (4 × 10% × £28,000) = £11,200

Total to be repaid = (£28,000 + £11,200) = £39,200

Monthly payment = £39,200/(4 × 12) = £816.67

15.7 ■ *Purpose* – an OL is undertaken in order to perform a specific job over a short period of time, whereas an FL is usually contracted over a term which matches the expected working life of an asset.

■ *Termination* – an OL can be easily terminated whereas cancellation penalties on an FL are prohibitive.

■ *Obsolescence risk* – with an OL, this is borne by the lessor, while the lessee bears the risk with an FL.

■ *Cost* – OLs are generally more expensive than FLs for the same period.

15.8 (i) 9% [1 − 10%] = 8.1%

(ii) The full 9%, as the firm cannot use the tax break.

CHAPTER 16

16.1 The events of 9/11 have made people more fearful for the future, and less inclined to invest. Equity would look safer for firms as it carries no interest obligations, although as the authorities used interest rate reductions to support their economies, this might encourage more borrowing, e.g. as firms re-finance existing loans at more favourable rates. So, on balance, probably a shift towards fixed interest financing to lock in historically low interest rates, certainly for as long as it takes fragile equity markets to recover.

16.2 The statement is nonsense. Reserves reflect financing of past investment, i.e. they represent funds already invested. The speaker is probably confused between reserves and 'cash reserves', common parlance for 'cash balances'.

16.3 Pre-issue, a shareholder holds (6 × £5) = £30 in shares plus £3.50 in cash, totalling £33.50. Post-issue, this value is spread over 7 shares (6 + 1) so the TERP = £33.50/7 = £4.79.

16.4 In a '2-for-1' split:

■ the number of shares doubles, i.e. two new shares for each old one.

■ both capital and assets are unchanged – no further cash is raised.

■ the market value of the whole equity is unchanged but the value per share is halved.

16.5 £50 grows to £100 over 10 years, i.e. £50$(1 + g)^{10}$ = £100.

Inverting, $1/(1 + g)^{10}$ = (£50/£100) = 0.5000. From the present value tables, g = 7.2%.

16.6 (a) Flat yield = interest/market price = (£8.30/£110) = 7.5%

(b) Yield to maturity is the interest rate that satisfies:

£110 = (£8.3 × 3-year annuity factor) + (£100 × 3-year discount factor).

The solution is 5.35%.

16.7 The warrant gives the right to buy something worth £8 for £5 – its value is thus £3. The one-for-four terms are irrelevant to the value per warrant. Someone holding 400 shares would have 100 warrants valued at (100 × £3) = £300, etc.

16.8 Logic might suggest the following:

A SAL generates cash which raises assets, but as the lease is long-term it must be capitalised. It will thus appear (remain as) as a fixed asset, but is offset by the corresponding PV of rental commitments, shown as long-term debt. Overall, it leaves net assets unchanged but, initially, it increases the calculated gearing ratio unless and until the cash generated is used to repay existing debt.

However, the reality is rather different, especially for property assets, which are usually classified as operating leases as the lessor retains the risks and rewards of ownership. Hence, most rented properties do not appear on the tenant's balance sheets. The balance sheet effect is largely cosmetic ('smoke and mirrors'), except that it does increase liquidity.

CHAPTER 17

17.1 Earnings in year zero = £1,000, of which 60%, i.e. £600, are retained. Returns on new investment = (15% × £600) = £90.

Next year's earnings = (£1,000 + £90) = £1,090.

$$\text{Growth rate} = \frac{(£1,090)}{(£1,000)} - 1 = 9\% = (60\% \times 15\%)$$

17.2 Divicut invests £1,000 in year 1 to return £1,080 in year 2.

The NPV of this is £1,080/$(1.1)^2$ vs. £1,000/(1.1) or £893 vs. £909, thus a negative NPV of £16, because the return on reinvested earnings (8%) falls short of the required 10%.

17.3 Paying out dividends rather than investing in worthwhile projects otherwise inaccessible to shareholders reduces the PV of their future income below what it could have been, i.e. there is an opportunity cost imposed on them.

17.4 Projects A, B and C are all attractive, but C appears unavailable as it would lead to exceeding the budget. However, if C is divisible, the firm should undertake A, B and 3/7 of C; otherwise, it should pay a dividend of £300,000.

17.5 Gordon confuses the risk of the dividends with the risk of the underlying cash flows resulting from new investment. If risk (and any increase in it) is suitably allowed for in the discount rate used to deflate future cash flows, then to discount the more distant future dividend flow would entail double-counting for risk. (If the firm moves into a low risk area, the future dividends could even have lower risk than near-in-time dividends.)

17.6 A clientèle is a set of investors whose interests the firm tries to serve via its particular dividend policy. Investors with short time horizons are likely to include the relatively elderly or infirm, and those with pressing needs for rapid income payments, e.g. to repay debts or to fund a daughter's wedding.

17.7 **(i)** £1,000/£20 = 50 shares
(ii) 50 plus 20% = 60 shares.

17.8 Share repurchases may signal that the firm has exhausted investment opportunities or has become too risk-averse to devoting funds to R&D. They may also trigger CGT liabilities for some investors. It may also cause embarrassment to directors if share prices subsequently fall.

17.9 A stable dividend policy gives reliability and security. Investors can plan their future income and expenditures more easily. Sharp cuts in dividends, as well as causing alarm, may force investors to sell shares on a weak market. Similarly, a sharp increase in dividends may force investors to reconfigure their portfolios, as well as triggering a CGT liability.

CHAPTER 18

18.1 **(i)** debt/equity = £50 m/£100 m = 50%.
(ii) debt-to-debt plus equity = £50 m/(£50 m + £100 m) = 33.3%.
(iii) net debt = debt less cash = (£70 m − £10 m) = £60 m, or as a percentage of equity = (£60 m/£100 m) = 60%.

18.2 Both capital and income gearing are forms of financial gearing, i.e. they stem from the firm's financial structure. Capital gearing is obtained from the Balance Sheet and is measured by the amount of borrowing in relation to owners' equity. Income gearing can be gleaned from the P and L, and is measured by interest cover, or its inverse, the proportion of PBIT accounted for by interest charges.

18.3 Over to you!

18.4 Contribution is, of course, unaffected, but operating profit or PBIT would fall by £50,000 to £550,000, and PBT would fall a further £40,000 to £310,000. The two multipliers become:

Contribution/PBIT = (£1,080,000/£550,000) = 1.96

PBIT/PBT = (£550,000/£310,000) = 1.77

The combined multiplier = (1.96 × 1.77) = 3.47

18.5 **Scenario A:** EV = (0.25 × 5%) + (0.50 × 20%) + (0.25 × 35%) = 20%
Scenario B: EV = (0.25 × 3.3%) + (0.50 × 23.3%) + (0.25 × 43.3%) = 23.3%
Scenario C: EV = (0.25 × 0%) + (0.50 × 30%) + (0.25 × 60%) = 30%

18.6 With 25% gearing, share price = (23.3p/0.25) = £0.93
With 50% gearing, share price = (30p/0.35) = £0.86
It seems that shareholders' demand for a higher return outweighs the increase in EPS as gearing increases.

18.7 Debt-to-equity ratio of 40% means that debt is 2/7 of the total long-term funds and equity is 5/7. The WACC is thus:
(5/7 × 18%) + (2/7 × 8%) = (12.9% + 2.3%) = 15.2%

18.8 **(i)** With 11% interest, market value = (£100 × 3.5%/11%) = £31.82
(ii) With 3% interest, market value = (£100 × 3.5%/3%) = £116.67.

18.9 Over to you!

18.10 **(a)** Perpetual debt: (30% × £10 m debt) = £3 m
(b) 20 years to maturity:
(20 year annuity of 6% × £10 m × T) = (11.4699 × £600,000 × 30%)
= £2.06 m

18.11 Over to you again!

CHAPTER 19

19.1 Under MM assumptions, firms identical in all respects apart from capital structure (allowing for size), should have the same value. Value is determined by the stream of operating cash flows and the degree of business risk attaching to these, regardless of how the cash flows are 'packaged' or shared out between different classes of investor.

19.2 Arbitrage in pursuit of 'home-made gearing' is restricted by:

- Taxes on capital gains.
- Differences between the borrowing and lending rates available to all parties.
- Different borrowing rates available to firms and private investors.
- Restrictions on some institutions' ability to borrow.
- Transactions costs.

19.3 Value of ungeared firm $= \dfrac{£100\ m(1-T)}{0.15} = \dfrac{£70\ m}{0.15} = £466.7\ m$

Value of geared firm $= V_u + TB = £466.7\ m + (30\% \times £200\ m)$
$= (£466.7\ m + £60\ m) = £526.7\ m$

19.4 In the ungeared firm, the overall cost of capital is simply the cost of equity, i.e. 15%. In the geared firm, the overall cost of finance is:

$$k_{ou} = 15\%\left(1 - \frac{T \times V_B}{V_S + V_B}\right) = 15\% \, [1 - (30\% \times £200 \, m/£600 \, m)]$$

$$= 15\%(1 - 1.5\%) = 13.5\%$$

19.5 Beta ungeared $= \dfrac{1.45}{\left[1 + (1 - 30\%) \times \frac{1}{2}\right]} = \dfrac{1.45}{1.35} = 1.074$

19.6 The APV is the value of an activity, based on its inherent worth as given by the PV of the stream of operating cash flows, adjusted for any 'special factors' such as tax concessions, financing costs, etc. It is thus the 'basic' NPV plus the PV of non-operating factors.

19.7 In theory, the limit to a firm's taxable capacity is where its interest charge equals the lowest possible level of future cash flows. In reality, it is much less than this. Based on asset-backing, the theoretical limit is where the book value of assets equals the book value of borrowings.

CHAPTER 20

20.1 Managerial motives reduce to the three Ps – power, pay and prestige, all of which are enhanced by size of firm.

20.2 If managers are paid according to growth in EPS, a takeover that exploits synergies may enhance their bonuses. However, if financed by debt, the value of the firm could actually fall. Takeovers that aim to reduce risk rarely benefit shareholders for whom systematic risk is normally more relevant than total risk.

20.3 Cash is more certain, whereas the value of shares is volatile – by the time payment is made, the share value might have fallen. Investors wishing to liquidate the shares received will incur dealing fees.

20.4 The NPV of the bid for Predator's shareholders = (£1,000 m + £200 m + £50 m) − (£1,000 + £230 m) = £20 m.

 Prey's shareholders benefit by £30 m (60 per cent of gains), and those of Predator by £20 m (40 per cent), if the bid is completed at this price.

20.5 The strategic approach to takeover analysis and execution involves the following steps:
- Formulate corporate strategy.
- Assess the role of acquisition candidates in achieving that strategy.
- Screen, value and select from among possible candidates.
- Plan for future integration and/or disposals.
- Make informal approach.
- Announce hostile bid if necessary.
- Complete the deal.
- Integrate the acquisition following a detailed resource audit.
- Post-audit the acquisition and integration processes.

20.6 Takeovers fail to achieve the anticipated benefits due to:
- Managerial motivation rather than shareholder orientation.
- Inadequate evaluation of the target.
- Over-payment for the target.
- Failure to plan the integration process.
- Poor integration, e.g. neglect of the human factor.

20.7 There are many motives for divestment including:

1 Dismantling conglomerates originally created by merger activity through defensive diversification strategies.
2 A change in strategic focus. This may involve a move away from the core business to new strategic opportunities. Alternatively, a business may decide that it is engaged in too wide a range of activities and seek to concentrate its efforts and resources on a narrower range of core activities. Non-core activities will then be divested.
3 Harvesting past successes, making cash flow available for new opportunities.
4 Selling off unwanted businesses following an acquisition. This is called 'asset-stripping'. Such sell-offs are often planned at the time of the bid.
5 Reversing (or learning from) mistakes.

CHAPTER 21

21.1 A speculator deliberately risks losing money by taking a long or short 'position' in a particular currency. A hedger tries to minimise exposure to risk. Obviously, Nintendo and BMW are (or were) speculators.

21.2

	Eurozone	Canada	
Spot	1.4721 − 1.4729	2.1315 − 2.1323	
Forward points	26 − 25	(24−15)	[Deduct as both currencies sell at a premium]
F/ward outright	1.4695 − 1.4704	2.1291 − 2.1308	

21.3
- Transaction exposure is the risk of loss associated with short-term contractual obligations.
- Translation exposure is the risk of loss when constructing end-of-year financial statements.
- Economic, or operating, or strategic, exposure is the risk of the whole value of the firm falling due to adverse foreign exchange rate changes affecting the PV of future cash flows.

21.4 New York price in one year = $2.00(1.02) = $2.04

London price in one year = £1.00(1.04) = £1.04

Exchange rate under PPP = ($2.04/£1.04) = $1.96 per £1

21.5 Annual interest rate differential of 2.5% = 2.5%/4 over 3 months = 0.625%. This is the interest *agio*. Forward rate should be spot less forward *agio* = (£2.0250 × 99.375%) = $2.0123. The US$ trades at a premium of 127 points.

21.6 Langer's assets are now worth (A$100 m/2.75) = £36.36 m. If all prices inflate at 7% in Australia, Langer's assets will appreciate to A$100 m (1.07) = A$107 m in one year. The AUD vs. GBP exchange rate should become (2.75 × 1.07/1.02) = A$2.884 per £1. At this rate, the sterling value of these assets = (A4107 m/2.884) = £37.1. This is the level to which they would have appreciated had they been located in Britain and experienced 2% UK inflation i.e. £36.36(1.02) = £37.1 m.

21.7 Leading is advancing outflows in a strong currency and advancing inflows in a weak one. Lagging is delaying inflows in a strong currency and delaying outflows in a weak one.

21.8 There is no point exercising an option 'at the money'. The purchaser simply loses the premium.

CHAPTER 22

22.1 Broadly, an MNC is a firm whose activities span national borders, but the term usually applies to firms that invest in foreign locations. A global firm both trades and invests abroad.

22.2 Licensing tends to have very low variable costs. Its fixed costs depend on the degree of supervision applied to the overseas operator (the agency costs). Fixed costs are probably higher than for exporting but lower than for FDI.

22.3 FDI differs from domestic investment for many reasons:

- Exposure to FX risk.
- Likelihood of inflation abroad occurring at rates different from home inflation.
- Risk of political intervention, leading to blocked funds, etc.
- Access to concessionary finance and grants.
- Overspill effects on existing operations.

22.4 Beta of project = $(1.4 \times 1.25) = 1.75$
Adjusted for correlation = $(1.75 \times 0.4) = 0.7$
Required return = $5\% + 0.7 [11\% - 5\%] = 5\% + 4.2\% = 9.2\%$

22.5 The firm will pay tax only in the US, i.e. (US$100 m $\times$ 30%) = US$30 m, which is allowable in full against UK tax liability.

22.6

Exposure of US operations = US$(50 m − 30 m)	=	US$20 m
Exposure in South America	=	US$100 m
Total exposure		US$120 m

22.7 The net exposure could be reduced by:

- Producing more output in the US and shipping direct to South America.
- Sourcing more US$-denominated inputs to support UK production.
- Borrowing in US$.

22.8
- Source materials and services from local suppliers.
- Employ locals in key management posts.
- Invest in training programmes for locals.
- Invest in sports and health facilities, open to the wider population.
- Sponsor local cultural events.
- Undertake joint ventures with local firms.
- Reinvestment rather than repatriation of profits.
- Use local sources of finance.

CHAPTER 23

23.1 **(1)** Some people are simply lucky.
(2) Some people stay lucky for several time periods, although shrewd ones remember to get out while ahead!
(3) In any time period, 50 per cent of investors ought to beat the market average.
(4) Some people deliberately assume high levels of risk for which a relatively high return is appropriate. We should therefore talk about 'risk-adjusted returns' rather than simple returns.
(5) Some people have access to inside information. This, of course, contravenes strong-form efficiency but it happens, although such information is unlikely to arrive in a steady stream over time.

Appendix B

Solutions to selected questions

CHAPTER 1

1 The goal of maximising owners' wealth is the normally accepted economic objective for resource allocation decisions. Rather than concentrate on the organisation, it evaluates investments from the viewpoint of the organisation's owners – usually shareholders. Any investment that increases their stock of wealth (the present value of future cash flows) is economically acceptable.

In practice, many of the assumptions underlying this goal do not always hold (e.g. shareholders are only interested in maximising the market value of their shareholdings). In addition, owners are often far removed from managerial decision-making, where capital investment takes place. Accordingly, it is common to find that more easily measurable criteria are used, such as profitability and growth goals. There are also non-economic considerations, such as employee welfare and managerial satisfaction, which can be important for some decisions.

CHAPTER 2

2 (c) (i) *For the cash offer*
On Day 1, the total value of each firm is:

> A: £2 × 2 m shares = £4 m
>
> B: £3 × 6 m shares = £18 m

Company B is making an offer of £6 m for Company A which is apparently worth only £4 million – this will reduce the market value of B by £2 million to £16 million or £16 million/6 = £2.67 per share. Against this, the anticipated savings would raise B's value to (£16 m + £3.2 m) = £19.2 m or £3.20 per share (assuming the market accepts this assessment).

1 *Semi-strong form efficiency*
Under semi-strong form efficiency, the market prices will only react when the information about the bid enters the public domain. The advent of new information will produce the following share prices:

		Share price	
		A	**B**
Day 2	No new information	£2.00	£3.00
Day 4	Takeover bid announced. B appears to be paying £6 m for assets worth £4 m	£3.00	£2.67
Day 10	Information available which revises the market value of B	£3.00	£3.20

2 *Strong form efficiency*
If the market is strong – form efficient, all information is reflected in the share price even if it is not publicly available information. This will mean that on Day 2, when the management of B decides to offer £3.00 for A, the

share prices will then react to reflect the full impact of the bid on both shares, perhaps via leakage of information by an informed insider. This will produce the following share prices:

		Share price	
		A	**B**
Day 2	Full impact of decision to bid and make savings reflected in share price	£3.00	£3.20
Day 4	Public announcement of bid, i.e. information of which the market is aware and therefore has no new information content	£3.00	£3.20
Day 10	Public announcement of savings to be derived from bid, i.e. further information of which the market is aware and therefore has no new information content	£3.00	£3.20

(ii) *For the share exchange*

Prior to the bid, the combined value of the two companies is (£4 m + £18 m) = £22 m. If a share-for-share exchange were made on a one-for-one basis, the value per share of the expanded company would become £22 m/8 m = £2.75 m, until further information about prospective savings emerged. Under semi-strong form efficiency, this will not happen until Day 10, but will leak out on to the market immediately under strong form efficiency. Spread over the whole 8 m shares, the savings are worth £3.2 m/8 m = £0.4 per share.

The sequence of share price movements is thus:

	Level of efficiency			
	Semi-strong		**Strong**	
	A	**B**	**A**	**B**
Day 2	£2	£3	£3.15	£3.15
Day 4	£2.75	£2.75	£3.15	£3.15
Day 10	£3.15	£3.15	£3.15	£3.15

Notice that the ultimate share price under the share exchange is lower than for the cash offer. This is because the benefits of the merger are spread out over a larger number of shares post-bid. In effect, the shareholders of B will have released part of the benefit they expect to receive from the bid to the former shareholders of A.

CHAPTER 3

1 Accounting profit is the excess of income over expenditure. Income and expenses relate to a specific period (e.g. the sales and costs for the month of January) based on accounting conventions such as depreciation. Cash flow is the cash receipts and payments from all operations including capital investment.

2 Using the table in Appendix D, the annuity factor for ten years and $i = 20\%$ is 4.1925:

$$PV = £100 \times 4.1925 = £419.25$$

3 Savings: £500,000 × 3.7908   £18,954
  Residual value: £1,000 × 0.62092   £621
                                                £19,575
  Less: initial cost: (£20,000)
  NPV   £(425)

The NPV is negative. Recommend the project is rejected.

7 Free cash flow per share $= (£5\,m/10\,m) = £0.50$ or 50p

(i) $P_o = (£0.5/0.12) = £4.17$

(ii) $P_o = \dfrac{50\% \times 50p\,(1 + [15\% \times 50\%])}{12\% - [15\% \times 50\%]} = \dfrac{26.88p}{0.045} = £5.97$

(iii) $P_o = \dfrac{50\% \times 50p\,(1 + [10\% \times 50\%])}{12\% - [10\% \times 50\%]} = \dfrac{26.25p}{0.07} = £3.75$

(iv) Present value of dividends over years 1–3:

Year 1 Dividend $= 26.88p$, as per part (ii) PV @ 12% $= £0.239$
Year 2 Dividend $= 26.88p\,(1.075) = 28.89p$ PV @ 12% $= £0.230$
Year 3 Dividend $= 28.89p\,(1.075) = 31.05p$ PV @ 12% $= £0.221$
PV of dividends beyond year 3:

$$= \dfrac{31.05p\,(1.05)}{(12\% - 5\%)} \times \text{(3-year PV factor)}$$

$= (£4.657 \times 0.7118) = \underline{£3.315}$
Share price $=$ Total PV $= £4.005$

8 (i) £0.37955

(ii) $(0.6 \times 16\%) = 9.6\%$

(iii) £0.65

(iv) £0.20

(v) $b = g/R = (2\%/10\%) = 20\%$

(vi) $k_e = ((D_1/P_o) + g) = (£0.054/£0.60) + 8\% = 17\%$
As $g = 10.5\%$, and
$\quad b = 0.7,$
$\quad R = \dfrac{g}{b} = \dfrac{0.105}{0.7} = 15\%$

(vii) 15%

CHAPTER 4

2 Microtic Ltd

		Project A (£)		Project B (£)
1	*Payback period*	1,616/500		556/200
	Outlay/annual flow	= 3.2 Years		= 2.8 Years
2	*Net present value (15%)*			
	Annual cash flow five years £500 × 3.352	1,676,000	200 × 3.352	670,400
	Scrap value £301 × 0.497	149,600	56 × 0.497	27,800
	Outlay	(1,616,000)		(556,000)
	NPV	209,600		142,200
3	*Internal rate of return*			
	(use trial and error to obtain NPV of zero)	try 20%		try 25%
	Annual cash flow 500 × 2,991	1,495,500	200 × 2,689	537,800
	Scrap value 301 × 0.402	121,000	56 × 0.328	18,400
	Outlay	(1,616,000)		(556,000)
	NPV	500		200
4	*Accounting rate of return*			
	Average profit before depreciation	500,000		200,000
	Depreciation (1616 − 301)/5	(263,000)	(556 − 56)/5	(100,000)
		237,000		100,000
	Average capital employed (1616 + 301)/2	958,500	(556 + 56)/2	306,000
	Rate	24.7%		32.7%

Investment advice
All appraisal methods apart from the NPV approach recommend acceptance of Project B. This is because it generates a higher return for every £1 invested. The question is, however, which of the two projects creates most wealth for the owners. Clearly, the much larger Project A has the higher NPV. Unless the firm is experiencing severe capital rationing problems, Project A should be accepted.

3 Mace Ltd (£000)

Project	NPV per £ outlay	Ranking	Fraction accepted	Required capital (£)	NPV (£)
1	1.6/60 = 0.027	4	0	–	–
2	1.3/30 = 0.043	3	1/3	10	0.43
3	8.3/40 = 0.207	1	1	40	8.3
4				0	0.9
5	7.9/50 = 0.158	2	1	50	7.9
				100	17.53

CHAPTER 5

1 The preference for IRR in practice is because:

(a) It is easier to understand (this is debatable).
(b) It is useful in ranking projects (although not always accurate).
(c) Lower-level managers do not need to know the discount rate. Where a risk-adjusted hurdle rate is used, there may be considerable negotiation over the appropriate rate.
(d) Psychological. Managers prefer a percentage.

2 (a) Payback = four years. The reciprocal is 25 per cent, which is the IRR for a project of infinite life.
(b) For a 20-year life, IRR = 24%. For an eight-year life, IRR = 19%.
(c) The annual cash flows are approximately the same for a long-lived project, the payback reciprocal is a reasonable proxy for the IRR. The actual IRR is always something less than that given by the payback reciprocal.

4 The two basic approaches in handling inflation are:

(a) forecast cash flows in money terms (i.e. including inflation) and discount at the money cost of capital; or
(b) forecast cash flows in current terms and discount at the real cost of capital.
The relationship between the two is given by:
$$(1 + P) = \frac{(1 + M)}{1 + I}$$
where M is the money cost of capital, I is the inflation rate and P is the real cost of capital.

5 Bramhope Manufacturing
(i) (a) Additional investment: £123,500 − £15,000 = £108,500. Additional annual inflow = £24,300 (see below).
Therefore payback period = £108,500/£24,300 = 4.5 years.

(b) Time	Cash flow (£)	DF at 15%	PV (£)	DF at 17%	PV (£)
0	(108,500)	1	(108,500)	1	(108,500)
1–8	24,300	4.48732	109,042	4.20716	102,234
8	20,500	0.32690	6,701	0.28478	5,838
			7,243		(428)

(c) NPV AT 15% = £7,243; IRR = approx. 17%.
Workings: Existing project's annual cash inflow = £200,000 × (0.95 − 0.12 − 0.48) = £70,000.
Net project's annual cash inflow = £230,000 × (0.95 − 0.08 − 0.46) = £94,300.
Incremental cash flow = £94,300 − £70,000 = £24,300.

(ii) The project appears to offer a positive net present value and should be accepted. However, an NPV of £7,000 on a project costing £123,500 is relatively small, and questions should be asked as to how sensitive the key assumptions are to uncertainty. For example, is it realistic that the additional capacity can be sold at the current price? Will there really be no increase in fixed overheads? If an advanced machine has been developed after just two years, is an eight-year economic life optimistic?

CHAPTER 6

1 Capital budgeting involves the whole investment process from the original idea to the final post-audit.

The resource allocation process is the main vehicle by which business strategy can be implemented. Investment decisions are not simply the result of applying some evaluation criterion. The investment decision process is essentially one of search: search for ideas, search for information, search for alternatives and search for decision criteria. The prosperity of a firm depends more on its ability to create profitable investment opportunities than on its ability to appraise them.

Once a firm commits itself to a particular project, it must regularly and systematically monitor and control the project through its various stages of implementation. Post-audit reviews – if properly designed – fulfil a useful role in improving the quality of existing and future investment analysis and provide a means of initiating corrective action for existing projects.

2 Aims of post-audits:

(a) improve the quality of existing decisions and underlying assumptions
(b) improve the quality of future decisions
(c) basis for corrective action

3 (a)

	Total automation £ per unit	Partial automation £ per unit
Sales price	75	75
Variable cost	(30)	(55)
Contribution	45	20
Fixed cost (see note 1)	(18)	(5)
Net profit	27	15
Total cost per unit	£48	£60
Annual net profit	£270,000	£150,000
Break-even units (p/a)	4,000	2,500
Net present value:		

	Cash flow £	Cash flow £
Outlay	−1,000,000	−250,000
Inflows years 1–5	+450,000	+200,000
Scrap value year 5	+100,000	−

	Present value £	Present value £
Outlay	−1,000,000	−250,000
Inflows at 3.27	+1,471,500	+654,000
Scrap value at 0.48	+48,000	−
Net present value	+519,500	+404,000

Note 1

$$\frac{(1,000,000 - 100,000)}{5 \times 10,000} \qquad \frac{(250,000)}{5 \times 10,000}$$

(b) The total automation option produces a lower product cost. As the sales price and volume are identical for both alternatives a higher annual profit is also reported by this alternative. Partial automation involves lower fixed costs (depreciation), but variable costs are substantially greater. With the cost pattern above, the total automation alternative results in a higher break-even point.

Remember that there is a higher initial capital cost with total automation. The accrual accounting assumptions applied above acknowledge this with an annual charge for depreciation. The net present value technique takes account of the capital cost in a different manner, charging the full cost immediately against discounted future cash flows. In this case, total automation generates a higher NPV but a higher initial outlay.

When the amount of the initial capital investment is considered, the option to undertake partial automation looks attractive. It achieves a net present value of 80 per cent of total automation with only one quarter of the capital expenditure. Partial automation shows a profitability index of 2.62, compared with 1.52 for the total automation, which emphasises the attractiveness of the lower investment. The recommendation of the NPV technique must, however, be interpreted with care and considered in the light of the overall company situation. Investment in total automation is the course of action which will maximise the wealth of the owners (the extra benefit exceeds the extra capital cost at 16 per cent). If the company is not in a capital-constrained situation then total automation should be pursued. If the company has a number of projects, all competing for limited capital, then we would need to know the return from the other investment opportunities and compare them with the incremental investment being directed to total automation.

The data omit certain other factors which may be pertinent to the analysis, for example, quality differences, maximum machine capacities, maximum sales demand, the flexibility and interchangeability of parts and equipment which may be beneficial.

For a lower risk investment with an attractive return, the company should opt for partial automation. If they have extra capital available, electing for total automation of the product line will earn above their extra cost of capital but nothing is known about other projects competing for this extra capital.

CHAPTER 7

1 While a capital project may have a high expected return, the risks involved may mean that there is the possibility that the project will be unsuccessful – even to the extent of putting the whole business in jeopardy.

2 Project risk – variability in the projects' cash flows; business risk – variability in operating earnings of the firm; financial risk – risk resulting from the firm's financing decisions, e.g. the level of borrowing.

3 Woodpulp project

Year	CE	NCF(£)	ENCF(£)	10% DF	PV(£)
1	0.90	8,000	7,200	0.90	6,480
2	0.85	7,000	5,959	0.83	4,938
3	0.80	7,000	5,600	0.75	4,200
4	0.75	5,000	3,750	0.68	2,550
5	0.70	5,000	3,500	0.62	2,170
6	0.65	5,000	3,250	0.56	1,820
7	0.60	5,000	3,000	0.51	1,530
				PV	23,688

NPV = £23,688 − £13,000 = £10,688. Accept the project. ENCF = Expected net cash flow.

4 Mystery Enterprises

Expected value Year 1 (£) $= 0.2(400) + 0.3(500) + 0.3(600) + 0.2(700) = £550$

Variance (£) $= 0.2(400 - 550)^2 + 0.3(500 - 550)^2$
$\quad\quad + 0.3(600 - 550)^2 + 0.2(700 - 550)^2 = £10,500$

Standard deviation $= £102$

Expected value Year 2 (£) $= 0.2(300) + 0.3(400) + 0.3(500) + 0.2(600) = £450$

Variance (£) $= 0.2(300 - 450)^2 + 0.3(400 - 450)^2$
$\quad\quad + 0.3(500 - 450)^2 + 0.2(600 - 450)^2 = £10,500$

Standard deviation $= £102$

Assuming a discount rate of 10 per cent and independent cash flows:

$$\text{NPV} = \frac{\pounds550}{1.1} + \frac{\pounds450}{(1.1)^2} - \pounds800 = \pounds71$$

$$\text{SD} = \sqrt{\frac{(\pounds102)^2}{(1.1)^2} + \frac{(\pounds102)^2}{(1.1)^4}} = \pounds125$$

$$\text{Coefficient of variation} = \frac{\pounds125}{\pounds71} = 1.76$$

CHAPTER 8

1 Companies issue a type of call option when issuing share warrants (giving the holder the right to buy shares at a fixed price) and convertible loan stock (giving the holder the right to exchange the loan for a fixed number of shares).

2 Gaymore plc
Traded options give the holder the right, but not the obligation, to buy (a call option) or sell (a put option) a quantity of shares at a fixed price on an exercise date in the future. They are usually in contracts of 1,000 shares and for three, six or nine months.

Holders of a put option in Gaymore plc have the right to sell shares in April at 500p. For this right they currently have to pay a premium of 47p, or £470 on a contract of 1,000 shares.

If the share price falls below 453p (i.e. 500p–47p), the shares become profitable and the holder is 'in the money'. So if they fall to 450p, the investor can buy shares at this price and exercise his or her put option to sell shares for 500p, a profit of 50p per share which, after the initial cost of the option, gives a net profit of 3p per share or £30 on the contract.

If the share price moves up to 510p by April, the option becomes worthless, and the investor loses his or her 47p premium.

Options such as this one can be used either to speculate or to hedge on share price changes for a relatively low premium.

3 The terms for an exchange traded option are standardised whereas the terms for traditional options can vary from contract to contract. It is the development of standardised options that has facilitated the trading of these instruments.

6 The put–call parity relationship is:

$S + P - C = E/(1 + R_f)$

From this:

$C = S - (E/(1 + R_f) + P$

where $S = 25, P = 15, E/(1 + R_f) = 10$.

$C = 25 - 10 + 15 = 30$

The value of the call option is 30p.

7 The put–call parity relationship is:

$S + P - C = E/(1 + R_f)$
$E/(1 + R_f) = 30 + 5 - 19 = 16p$

8 Inputs:

$S_T = 38p \quad X = 40p$
$T = 0.5 \quad r_f = 0.10$
$SD(R) = 0.20$

Valuation equations:

$$C_t = S_t N(d_1) - Xe^{-rT} N(d_2)$$

$$d_1 = \frac{\ln(S_1/X) + (r + VAR(R)/2)t}{SD(r)\sqrt{t}}$$

$$d_2 = \frac{\ln(S_1/X) + (r + VAR(R)/2)t}{SD(r)\sqrt{t}} = d_1 - SD(r)\sqrt{t}$$

$$d_1 = [\ln 38/40 + (0.10 + 0.20^2/2)0.5]/0.2/\sqrt{0.5}$$

$$= [-0.05129 + 0.06]/[0.2 \times 0.7071]$$

$$= 0.06156$$

$$d_2 = [\ln 38/40 + (0.10 - 0.20^2/2)0.5]/0.2\sqrt{0.5}$$

$$= [-0.05129 + 0.04]/[0.2 \times 0.7071]$$

$$= -0.07986$$

$$N(d_1) = N(0.0615) = 0.5245$$

$$N(d_2) = N(-0.0798) = 0.4682$$

$$C_1 = 38.00 \times 0.5245 - 40.00e^{-0.10 \times 0.5} \times 0.4682$$

$$= 19.93 - 17.81$$

$$= 2.12$$

Calculations:
The estimated value of the call is 2.12p. This implies that if the call is bought at this price the share price would have to increase from 38.00p to 40.12p by the end of the six-month period for the investor to break even.

9 Put–call parity

$$S + P = C + Ke^{-rt}$$
$$\Rightarrow 420.5 + 38.5 = 50.5 + 420e^{-0.5r}$$
$$\Rightarrow 0.9726 = e^{-0.5r}$$
$$\Rightarrow -0.5r = -0.02776$$
$$\Rightarrow r = 0.0555 = 5.55\%$$

10 (a) Intrinsic value: The intrinsic value of an option is the difference between the value of the underlying asset and the exercise price of the option. September 800 call is in the money while the April put is out of the money.

Time value: Options are more valuable the greater is the likelihood of a significant change in the value of the underlying asset. If an option expires tomorrow we can certainly say that there is a smaller likelihood of a significant change than if the option expires in six months. Quite simply, the longer the time to expiration, the more opportunity there is for the price to move and therefore at March the September option is more valuable than the April option.

(b) Short straddle (written straddle)
E.g. June 800 call + put: written premiums = 53.5 + 20 = 73.5p

The option writer receives 73.5p. If neither option is exercised, the writer gains all of this. However, this is the outcome only if $S_T = 800$.

If $S_T > 800$ the call is exercised, if $800 < S_T$ the put is exercised.

The writer will gain from the strategy if the intrinsic value of the exercised option is less than the 73.5 premiums received. Hence the writer will only profit if S_T is within the range 726.5 to 873.5.

If $S_T = 726.5$ the put is exercised, the writer must take delivery at 800, therefore losing 73.5, which exactly cancels the written premiums of 73.5, so net profit = 0.

If $S_T = 873.5$ the call is exercised, the writer must deliver the share at price 800, therefore losing 73.5, which exactly cancels out the written premiums, so net profit $= 0$.

Thus at any share price outside of that range the writer of the short straddle has an overall loss. If share price at expiration is within the range the writer makes a net profit.

In general, the profit to short straddle $=$ written premiums $-$ intrinsic value of the option that is exercised.

Losses *could* be very high, e.g. if $S_T = 650$ loss $= 76.5$ (i.e. $73.5 - 150$), if $S_T = 1000$ loss $= 126.5$ ($73.5 - 200$). It is extremely risky.

12 Using the Black–Scholes formula

$$
\begin{aligned}
d_1 &= \frac{\ln(P_s/X) + 0.5\sigma}{\sigma\sqrt{T}}\sqrt{T} \\
&= \frac{\ln(3/2.50) + (0.5 \times 0.25)}{0.2 \times \sqrt{0.25}} + (0.5 \times 0.2 \times \sqrt{0.25}) \\
&= \frac{0.1823 + 0.0125}{0.1} + 0.5 \\
&= 1.9982
\end{aligned}
$$

$$
\begin{aligned}
d_2 &= d_1 - \sigma\sqrt{T} \\
&= 1.9982 - 0.2\sqrt{0.25} \\
&= 1.9982 - 0.1 \\
&= 1.8982
\end{aligned}
$$

$$
N(d_1) = 0.5 + 0.4767 + \frac{82}{100}(0.4772 - 0.4767)
$$

$$
= 0.9771
$$

$$
N(d_2) = 0.5 + 0.4706 + \frac{82}{100}(0.4713 - 0.4706)
$$

$$
= 0.9712
$$

$$
\begin{aligned}
P_c &= P_sN(d_1) - Xe^{-rT}N(d_2) \\
&= (3 \times 0.9771) - (2.50e^{-0.05 \times 0.25} \times 0.9712) \\
&= 2.9313 - (2.50 \times 0.9876 \times 0.9712) \\
&= 53.35\text{p}
\end{aligned}
$$

CHAPTER 9

1 Portfolio standard deviation $= \sqrt{[(0.5)^2\sigma_x^2 + (0.5)^2\sigma_y^2 + 2(0.5)(0.5)(r_{xy}\,\sigma_x\,\sigma_y)]}$

$$
\begin{aligned}
&= \sqrt{[(0.5)^2\,30^2 + (0.5)^2\,45^2 + 2(0.5)(0.5)(0.2)(30)(45)]} \\
&= \sqrt{[225 + 506.25 + 135]} \\
&= \sqrt{866.25} = 29.4
\end{aligned}
$$

This portfolio has a risk only slightly below that of investment X, even though the correlation coefficient is quite low. This suggests that the portfolio weighting involves too little of investment X if the intention is to lower portfolio risk.

2 The appropriate formula is:

Proportion invested in asset $A = \dfrac{\sigma_B^2 - \text{cov}_{AB}}{\sigma_B^2 + \sigma_A^2 - 2\text{cov}_{AB}}$

Also, remember that $\text{cov}_{AB} = r_{AB}\,\sigma_A\,\sigma_B$

(i) The covariance = −36; % in A = 29%; % in B = 71%

(ii) The covariance is −15; % in A = 6%; % in B = 94%

(iii) The covariance is 0; % in A = 0.5%; % in B = 99.5%

(iv) The covariance = −21; % The weights are misleading as the answer is indeterminate under perfect positive correlation.

3 (a) China:

Expected value = $(0.3 \times 50\%) + (0.4 \times 25\%) + (0.3 \times 0) = 25\%$

Growth	Probability	IRR%	Deviation	Squared deviation	Times probability
Rapid	0.3	50	25	625	187.50
Stable	0.4	25	–	–	–
Slow	0.3	0	−25	625	187.50
					Sum = 375.00

Standard deviation = $\sqrt{375.0}$ = 19.4, i.e. 19.4%

Scotland:

Expected value = $(0.3 \times 10\%) + (0.4 \times 15\%) + (0.3 \times 16\%) = 13.8\%$

Growth	Probability	IRR%	Deviation	Squared deviation	Times probability
Rapid	0.3	10	−3.8	14.44	4.33
Stable	0.4	15	1.2	1.44	0.58
Slow	0.3	16	2.2	4.84	1.45
					Sum = 6.36

Standard deviation = $\sqrt{6.36}$ = 2.5%

(b) Covariance calculation:

Covariance = −45

Expected portfolio return

$= (0.75 \times 13.8\%) + (0.25 \times 25\%) = 16.6\%$

Standard deviation

$= \sqrt{[(0.25^2 \times 19.4^2) + (0.75^2 \times 2.5^2) + (2 \times 0.25 \times 0.75 \times -45)]}$

$= \sqrt{[(23.5) + (3.5) + (-16.9)]}$

$= \sqrt{10.1}$ = 3.2, i.e. 3.2%

4 (a) *Eire*: EV of IRR = $(0.3 \times 20\%) + (0.3 \times 10\%) + (0.4 \times 15\%)$

$= 6\% + 3\% + 6\% = 15\%$

Outcome (%)	Deviation	Sq'd Dev.	p	Sq'd Dev. × p
20	+5	25	0.3	7.5
10	−5	25	0.3	7.5
15	0	0	0.4	0
			Variance = Total	15 σ = 3.87

Humberside: EV of IRR $= (0.3 \times 10\%) + (0.3 \times 30\%) + (0.4 \times 20\%)$
$$= 3\% + 9\% + 8\% = 20\%$$

Outcome (%)	Deviation	Sq'd Dev.	p	Sq'd Dev. $\times p$
10	−10	100	0.3	30
30	+10	100	0.3	30
20	0	0	0.4	0
			Variance = Total	60 $\sigma = 7.75$

(b) (i) For a 50/50 split investment:

EV of IRR $= (0.5 \times 15\%) + (0.5 \times 20\%) = 17.5\%$
$$\sigma = \sqrt{(0.5)^2(15) + (0.5)^2(60) + 2(0.5)(0.5)(0)(3.87)(7.75)}$$
$$= \sqrt{[3.75 + 15]} = \sqrt{18.75} = 4.33$$

(ii) 75/25 split:

EV $= (0.75 \times 15\%) + (0.25 \times 20\%) = 11.25\% + 5\% = 16.25\%$
$$\sigma = \sqrt{(0.75)^2(15) + (0.25)^2(60)}$$
$$= \sqrt{[8.44 + 3.75]} = \sqrt{12.19} = 3.49, \text{ i.e. lower risk than either project}$$

CHAPTER 10

1 Expected return $= 5\% + 1.23[11.5\% - 5\%] = 13\%$.
With more pessimistic expectations about market returns, this drops to:

ER $= 5\% + 1.23[8\% - 5\%] = 8.7\%$

2 (i) R_f $= 9\%$
  (ii) Beta $= 1.71$
  (iii) ER_j $= 8.5\%$
  (iv) $ER_m = 19.3\%$

3 The intercept of the Security Market Line is the risk-free rate. It passes through the market portfolio which has a Beta of 1.0. In diagrammatic terms:

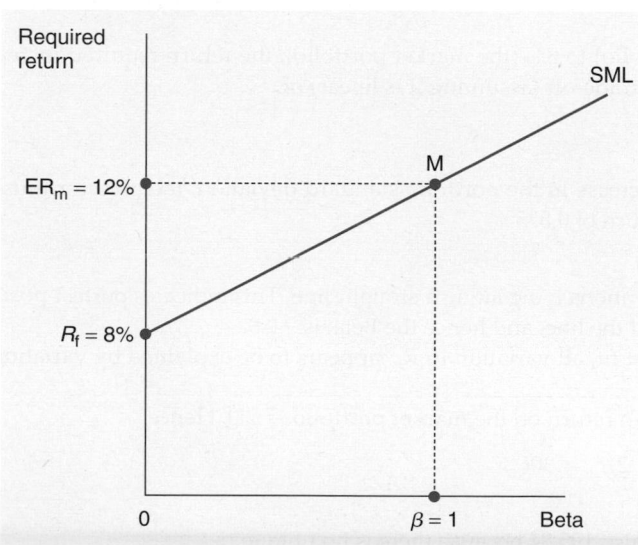

4 $ER_A = 5\% + 0.7[10\% - 5\%] = 8.5\%$

$ER_B = 5\% + 1.3[5\%] \qquad = 11.5\%$

$ER_C = 5\% + 0.9[5\%] \qquad = 9.5\%$

Comparing the expected returns with those currently achieved, A and C generate returns lower than expected, which suggests they are overvalued. (B looks undervalued.)

5 (i) Projecting past returns into the future, the expected return on the whole market is $(5\% + 7\%) = 12\%$.

(ii)

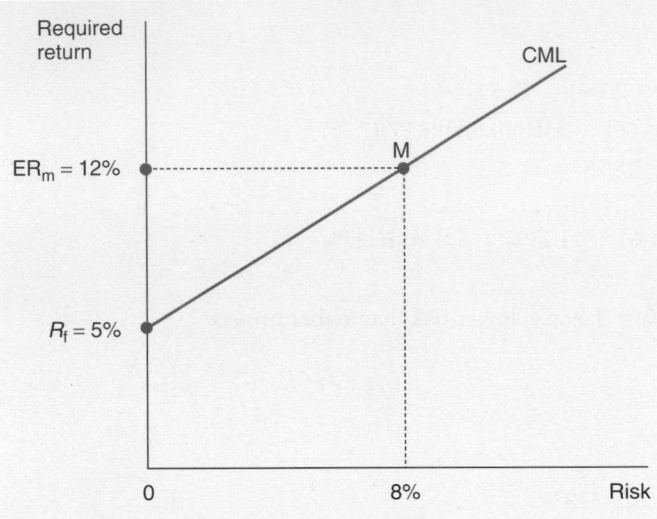

(iii) For a 50/50 portfolio:

$ER_p = (0.5 \times 12\%) + (0.5 \times 5\%) = 8.5\%$

(iv) Bearing in mind there is no correlation between the risk-free rate and the market portfolio, the expression for portfolio risk is;

$O_p = \sqrt{(0.5)^2 (8)^2}$

$\quad = \sqrt{16} = 4$, i.e. 4%

Alternatively, we could use the weighted average expression:

$(0.5 \times 8\%) + (0.5 \times 0) = 4\%$

(v) As risk increases from 4% (the 50/50 portfolio) to 8% (the market portfolio), the return required increases from 8.5% to 12%, suggesting a risk–return trade-off (assuming it is linear) of:

$(3.5\%/4\%) = 0.875$

Thus, for every one percentage point increase in the portfolio standard deviation, the investor must be compensated by an increase in portfolio return of 0.875%.

6 (a) Your graph should show every pair of observations lying along a straight line. This indicates perfect positive correlation between R_j and R_m. The slope of the line, and hence the Beta, is +0.8.

(b) Because all observations lie on the line of best fit, all variation in R_j appears to be explained by variation in R_m, i.e. there is no specific risk.

(c) σ_m^2 = variance of the market return. The mean return on the market portfolio = 2.0. Hence

$\sigma_m^2 = (5 - 2)^2 + (-10 - 2)^2 + \cdots + (6 - 2)^2 = 306$

Thus, systematic risk = $(0.8)^2 (306) = 195.84$.

You should expect to find the total risk is also 195.84 because there is no unique risk.

CHAPTER 11

1 The shares are ex-dividend, so the future required return is:

$$k_e = \frac{£0.80(1.045)}{£10.50} + 0.045 = (0.08 + 0.045)$$

$$= 0.125, \text{ i.e. } 12.5\%$$

2 (i) D_o = £0.62
(ii) P_o = £3.64
(iii) k_e = 9.3%
(iv) g = 6.2%

3 Dividend growth is found by solving:

$$11p (1 + g)^4 = 20.0p$$

$$(1 + g)^4 = \frac{20.0p}{11.0p} = 1.8182$$

From tables, g = 15.8% approx.

$$\text{Implied } k_e = \frac{20p(1.158)}{£5.50} + 0.158$$

$$= 0.042 + 0.158 = 0.20, \text{ i.e. } 20\%$$

4 (i) k_e = 6% + 0.8[11% − 6%] = (6% + 4%) = 10%

(ii) Beta = (0.8 × 1.25) = 1.0
k_e = 6% + 1.0[11% − 6%] = (6% + 5%) = 11%

(iii) Beta = (0.8 × 0.75) = 0.6
k_e = 6% + 0.6[11% − 6%] = (6% + 3%) = 9%

5

Division	R_f	Beta	Market Premium	k_e	% of assets
C	7%	0.7	8.00	12.6%	15
E	7%	1.1	8.00	15.8%	40
R	7%	0.8	8.00	13.4%	20
P	7%	0.6	8.00	11.8%	5
Total	/	0.855	/	13.84%	100

Overall Beta = (0.15 × 0.7) + (0.04 × 1.1) + (0.20 × 0.8) + (0.25 × 0.6)

= (0.105 + 0.44 + 0.16 + 0.15) = 0.855

Overall required return = 7% + 0.855[8%] = (7% + 6.84%) = 13.84%

6 (a) EV of return = (0.6 × 10%) + (0.4 × 20%) = 14%.

Outcome	Deviation	Squared deviation	Prob.	Sq'd dev. × p
10%	−4%	16	0.6	9.6
20%	+6%	36	0.4	14.4
			Variance = Total	24.0 σ = 4.9

(b) Megacorp ER = 30%   s.d. = 14%   proportion = 80%

Erewhon ER = 14%   s.d. = 4.9%   proportion = 20%

$ER = (0.8 \times 30\%) + (0.2 \times 14\%) = 24\% + 2.8\% = 26.8\%$

$\text{S.d.} = \sqrt{[(0.8)^2 (14)^2 + (0.2)^2 (4.9)^2 + 2(0.8)(0.2)(-0.36)(14)(4.9)]}$

(c) The present Beta = 1.20.

$$\text{Beta of project} = \frac{\text{cov}_{jm}}{\sigma_m^2} = \frac{r_{jm}\sigma_j\sigma_m}{\sigma_m^2} = \frac{r_{jm}\sigma_j}{\sigma_m}$$

What is the risk of the market (σ_m)?
Rearranging:

$$\sigma_m = \frac{\sigma_j r_{jm}}{\text{Beta}_j}$$

For Megacorp:

$$\sigma_j = \frac{(14)(0.8)}{1.2} = 9.33$$

$$\text{Project Beta} = \frac{(-0.1)(4.9)}{9.33} = -\frac{0.49}{9.33} = -0.05$$

New Beta for Megacorp = $(0.8 \times 1.2) + (0.2 \times -0.05) = 0.95$. Therefore, the new project lowers Megacorp's Beta.

8 (a) Since the asset Beta is a weighted average of the component segment Betas:

$$\beta_A = \left(\frac{1}{4} \times \beta_N\right) + \left(\frac{1}{4} \times \beta_W\right) + \left(\frac{1}{2} \times \beta_S\right) = 1.06$$

where $\beta_N = \beta$ of North, $\beta_W = \beta$ of West, $\beta_S = \beta$ of South.
Since North is 50 per cent more risky than South, and West is 25 per cent less risky than South, it follows that:

$$\frac{(1.5\beta_S)}{4} + \frac{(0.75\beta_S)}{4} + \frac{\beta_S}{2} = 1.06$$

whence $\beta_S = 1.00$, $\beta_N = 1.50$, $\beta_W = 0.75$.

(b) The asset Beta for East (β_E) is:

$\beta_E = \beta_S \times \text{Relative risk factor}$

$\quad = \beta_S \times \text{Revenue sensitivity factor} \times \text{Operational gearing factor}$

$\quad = 1.0 \times 1.4 \times \dfrac{1.6}{2.0} = 1 \times 1.12 = 1.12$

(c) The asset Beta for Lancelot after the divestment and acquisition is again a weighted average of the component asset Betas:

$$\beta_A = \left(\frac{1}{2} \times \beta_E\right) + \left(\frac{1}{4} \times \beta_W\right) + \left(\frac{1}{4} \times \beta_N\right)$$

$$= \left(\frac{1}{2} \times 1.12\right) + \left(\frac{1}{4} \times 0.75\right) + \left(\frac{1}{4} \times 1.5\right)$$

$$= 0.56 + 0.1875 + 0.375 = 1.12$$

(d) If we evaluate projects in East on the assumption of all-equity financing, the cut-off rate is

$R_f + \beta_E (ER_m - R_f) = 10\% + 1.12(15\%) = 26.8\%$

(e) There are numerous problems involved in obtaining tailor-made project discount rates. First, given that we may have to take the Beta from a surrogate company, ungearing the Beta of its shares requires an accurate assessment of the values of the equity and debt. If the company is quoted, the market valuation of equity can be taken, but not all corporate debt is quoted. Thus the debt:equity ratio used will often be a mixture of market and book values, even for quoted companies. Second, decomposing the ungeared Beta into segmental asset Betas strictly requires weighted average calculations based on market values. Since corporate segments are not generally quoted on stock markets, book values invariably have to be used. Third, to measure project Betas requires consideration of whether the project is of a different order of risk from 'typical' projects in the division. If so, the revenue sensitivity factor requires estimation, mainly based on guesswork for unique projects. In addition, the project gearing factor must also be estimated. Fourth, the general problems relating to specification of the risk-free return and the risk premium on the market portfolio still have to be addressed.

CHAPTER 12

3 Free cash flow (£) = Revenue less bad debts less operating costs + depreciation less replacement investment less tax (allowing for relief on bad debts)

$$= (0.98 \times 500,000) - (300,000) + (50,000)$$
$$- (50,000) - (60,000 + [0.3 \times 50\% \times 2\% \times 500,000])$$
$$= (490,000 - 300,000 + 50,000 - 50,000 - 60,000 + 1,500)$$
$$= £131,500$$

(a) valued @ 10% as a perpetuity:

$V = (£131,500/0.1) = £1.315$ million

i.e. (£1.315 m/2 m) = £0.6575 per share (65.75p)

(b) valued over 10 years:

$V = (£131,500 \times$ ten year annuity factor 10%)

$= (£131,500 \times 6.1446) = £808,015$ m, or £0.405 per share (40.5p).

CHAPTER 13

6 **Hercules Wholesalers Ltd**

(a) The liquidity position may be examined by calculating the following ratios:

$$\text{Current ratio} = \frac{\text{Current assets}}{\text{Creditors due for repayment within one year}}$$

$$= \frac{£306,000}{£285,000}$$

$$= 1.1{:}1$$

$$\text{Acid-test ratio} = \frac{\text{Current assets (less stocks)}}{\text{Creditors due for repayment within one year}}$$

$$= \frac{£163,000}{£285,000}$$

$$= 0.6{:}1$$

The current ratio reveals that the current assets exceed the short-term liabilities of the company. However, the ratio of 1.1:1 seems rather low. If current assets were to be liquidated, they would only have to be sold off at a small discount on the cost to be insufficient to meet the short-term liabilities. The acid-test ratio excludes stocks from the calculation and represents a more stringent test of liquidity. The ratio of 0.6:1 is also low and suggests the company has insufficient liquid assets to meet its maturing obligations. When interpreting these ratios, it should be borne in mind that they are based on Balance Sheet figures and are therefore representative of only one particular moment in time. It would be useful to monitor the trends in these ratios over time. It would also be useful to prepare a cash flow forecast in order to gain a better understanding of the likely liquidity position of the business in the future. The bank overdraft is the major form of short-term finance and the continuing support of the bank is likely to be of critical importance to the company.

(b) The operating cash cycle of a business represents the time period between the outlay of cash on the purchase of stocks and the receipt of cash from trade debtors. In the case of a wholesale business it can be calculated as follows:

Average holding period for stocks + Average settlement period for debtors − Average settlement period for creditors

The operating cash cycle is important because the longer this period is, the greater the financing requirements of the business and the greater the risks involved.

(c) The operating cash cycle is:

No. of days

$$\text{Average period stocks are held} = \frac{\text{Average value of stocks held}}{\text{Average daily cost of sales}}$$

$$= \frac{(£125,000 + £143,000)/2}{(£323,000/360)} \qquad 149$$

$$\text{Average settlement period for debtors} = \frac{\text{Average level of debtors}}{\text{Average daily sales}}$$

$$= \frac{£163,000}{(£452,000/360)} \qquad \frac{130}{279}$$

Less:

$$\text{Average settlement period for creditors} = \frac{\text{Average level of creditors}}{\text{Average daily purchases}}$$

$$= \frac{£145,000}{(£341,000/£360)} \qquad \frac{(153)}{126}$$

(d) The operating cash cycle of the company seems to be quite long. It may be reduced by a reduction in the stocks, improving the collection period from debtors, extending further the average settlement period for creditors or some combination of these measures. The stockholding period and average settlement period for debtors seems high and might be reduced without difficulty. As the average settlement period of creditors appears to be high, it may be difficult to extend this further without incurring problems for the company.

CHAPTER 14

1 Formula for the cost of cash discounts:

$$\frac{\text{Discount\%}}{(100 - \text{Discount\%})} \times \frac{365}{(\text{Final date} - \text{Discount period})}$$

2 The Baumol cash management model assumes cash is drawn down at an even rate and, at some point, is replenished to its original balance. The Miller–Orr model assumes that short-term cash movements cannot be predicted since they meander in a random fashion.

3 Hunslett Express Company

	(£000s)
Average level of debtors – current policy	
70/365 × £8 m	1,534
Average levels of debtors – proposed policy	
50% × 30/365 × £8 m	329
50% × 80/365 × £8 m	876
	1,205
Reduction in debtors under new policy	329
Financing cost savings (13% × £329,000)	43
Bad debt savings	20
Administration cost savings	12
	75
Cost of cash discounts	
50% × £8 m × 2%	(80)
Estimated cost of scheme	(5)

The net cost of the proposed scheme is £5,000.

4 Salford Engineers Ltd

(a) The optimum stockholding level is a trade-off between the cost of holding stocks and the cost of not holding stock.

Stockholding costs include:

(i) Storage costs. Where stock is valuable these costs can be large.
(ii) Financing costs. Excessive stock requires unnecessary and expensive working capital.
(iii) Insurance costs against theft or damage.
(iv) Obsolescence costs. Stock held for long periods of time may become obsolete through new products coming to the market.

Costs of holding too little stock include:

(i) Loss of customer goodwill and business through not being able to supply goods on time.
(ii) Production stoppages. A 'stockout' can mean costly and harmful disruptions to the production process.
(iii) Lost flexibility. Shortage of stock makes it difficult for a firm to respond to unexpected demand or to extended production runs.

(b) In investigating the reasons for large stock levels for Salford Engineering, the following action should be taken:

(i) Examine the stock re-order levels for each stock line.
(ii) Examine how the optimum stock level is determined. Is any technique for assessing the optimum level employed?
(iii) Are stock requirements carefully budgeted?
(iv) Are ratios used (e.g. stock turnover) to monitor stock levels in total and by stock lines?
(v) Are stock records reliable and adequate?

CHAPTER 15

1 Cost of extra trade credit = lost discount of 3 per cent , i.e. £30 on an invoice of face value of £1,000. Over the extra 30 days during which you delay payment, you are paying a cost of roughly:

$$(£30/\text{extra finance}) \times 365/30 = (£30/£970) \times 12.167$$
$$= 3.09 \times 12.167 = 37.6\%$$

More accurately, the annual rate is

$$(1 + £30/£970)^{12.167} - 1 = (1.0309)^{12.167} - 1 = 1.448 - 1 = 0.448, \quad \text{i.e. } 44.8\%$$

2 (a) Option 1: The net cost of a full-year loan.

Interest £500,000 × 12%   = £60,000

Less interest received on surplus funds

Q1: (£500,000 − £400,000) × 2% = £2,000

     Compounded for Qs 2–4: = £2,000$(1.02)^3$ = (£2,122)

Q4: (£500,000 − £200,000) × 2%   = (£6,000)

                                         £51,878

Option 2: The cost of a fluctuating overdraft.

Quarterly interest rate = 14%/4 = 3.5%

Q1:£400,000 × 3.5% = £14,000

Q2:£500,000 × 3.5% = £17,500

Q3:£500,000 × 3.5% = £17,500

Q4:£200,000 × 3.5% = £7,000

        Total = £56,000

    Even without any interest charges on accumulated quarterly overdraft interest, the loan is the cheaper option. However, the overdraft offers possible access to lower interest rates. (Equally, they could rise!).
(Let I = average monthly overdraft rate

£51,878 = I × [£400,000 + £500,000 + £500,000 + £200,000]

     I = [£51,878/£1,600,000]

     I = 0.0324, rounded to 0.0325,   i.e. 3.25%, or 13%p.a.)

(b) An average annual interest rate of 13% on the overdraft (3.25% per month) is the break-even rate.

3 Cost of early discounting = (£200,000 − £195,500) = £4,500. This corresponds to an effective annual interest cost of roughly:

(£4,500/£195,500) × 12/4 = 6.9%

More accurately, the annual rate is:

$(1.023)^3$ − 1 = 1.0706 − 1 = 0.0706,   i.e. 7.06%

4 Net amount borrowed = (£100,000 − £15,000)   = £85,000

Total interest cost = (7.5% × £85,000 × 4)   = £25,500

Total required payments   = £110,500

Required monthly payment = £110,500/(4 × 12) =   £2,302.

7 Amalgamated Effluents plc

The evaluation should be conducted in three stages.

(i) Is the project inherently worthwhile? The basic project must first be evaluated assuming no change in financing, i.e. using the equity cost of capital. Assuming the outlay can be set against year zero profit, cash flows are:

	Cash flows (marked by *) (£m)				
	0	1	2	3	4
Operating cash inflow*		0.200	0.200	0.200	0.200
Equipment cost*	(0.500)				0.050
WDA	0.125	0.094	0.070	0.053	0.108
WDV	0.375	0.281	0.211	0.158	0.050
Taxable income	(0.125)	0.106	0.130	0.147	0.092
Tax impact @ 30%	0.038	(0.032)	(0.039)	(0.044)	(0.028)
Net cash flows*	(0.462)	0.168	0.161	0.156	0.172
PV @ 10%	(0.462)	0.153	0.133	0.117	0.117

The NPV = +0.58, i.e. + £58,000, hence the project is worthwhile.

(ii) Should the equipment be leased or purchased? The relevant discount rate is the post-tax cost of borrowing, i.e.

$7\%(1 - 30\%) = 5\%$

The respective cash flow implications of leasing and outright purchase are:

The PV of the rental stream $= 0.070(\text{PVIFA}_{5/4}) = 0.070(3.546) = 0.248$

The PV of the tax savings $= 0.248(0.30) = 0.074$, i.e. £74,000

Hence PV of leasing costs $= 0.248 - 0.074 = 0.174$, i.e. (£0.174).

Purchase via bank loan:

The relevant cash flows are those connected with the purchase itself rather than the operation of the project, i.e. the outlay less tax savings and the residual value.

	Cash flows (marked by *) (£m)				
	0	**1**	**2**	**3**	**4**
Outlay*	(0.500)				0.050
WDA	0.125	0.094	0.070	0.053	0.108
Tax savings @ 30%	0.041	0.031	0.023	0.017	0.036
Net cash flows*	(0.459)	0.031	0.023	0.017	0.036
PV @ 5%	(0.459)	0.030	0.021	0.015	0.030

PV of cash flow $= (0.363)$, i.e. (£0.363 m)

Hence, leasing is the preferable (least-cost) method of financing (£0.174 m compared to £0.363 m), i.e. the lease has a net advantage of £0.189 m.

(iii) The third stage in the evaluation is now a formality – the optimal decision is to acquire the equipment using lease finance.

CHAPTER 16

1 (i)

	Pence
3 old shares prior to rights issue at 320p	960
1 new share at £2	200
4 shares worth	1,160
1 share therefore worth 1,160/4	290

(ii) Value of the rights is the difference between the offer price and the ex-rights price: $(290p - 200p) = 90p$ for every new share issued.

2 Proposed preference share issue

(a) Benefits to the company:
- Dividends are only paid if funds are available.
- No asset security required as with some loans.
- Lower risk than for ordinary share capital, giving a cheaper source of finance.
- Suitable when a company does not want to increase the number of ordinary shares but is concerned that its gearing is already high.

Drawbacks to the company:
- Cost of preference shares is usually higher than for debentures because the risks are greater.
- No tax relief on the dividends (unlike loan interest).

(b) Benefits to the investor:
- Should produce a higher yield than fixed-interest securities.
- Lower risk than ordinary shares.

– Redeemable preference shares provide a means of liquidating the investment where markets are thin or non-existent.

Drawbacks to the investor:
– Unable to participate fully in the profits.
– Not usually secured.
– No guaranteed dividend.

Many of the drawbacks can be overcome where preference shares are cumulative, participating, redeemable and convertible into equity if desired.

4 (a) Burnsall needs additional finance to fund both working and fixed capital needs. As sales are expected to increase by 20 per cent, and since working capital needs are expected to rise in line with sales, the predicted working capital needs will be 20 per cent above the existing working capital level, i.e.

$1.2 \times$ [stock + debtors + cash − trade creditors]

$= 1.2 \times$ [£16 m + £23 m + £6 m − £18 m]

$= £32.4$ m, increase of £5.4 m.

Together with the additional capital expenditures of £20 m, the total funding requirement = (£5.4 m + £20 m) = £25.4 m.

This funding requirement can be met partly by internal finance and partly by new external capital. The internal finance available will derive from depreciation provisions and retained earnings, after accounting for anticipated liabilities, such as taxation, that is, from cash flow.

Note that the profit margin on sales of £100 m (£10 × 10 m units) before interest and tax was 16 per cent in 1994–95. If depreciation of £5 m for 1994–95 is added back, this yields a 'cash flow margin' of 21 per cent (ignoring movements in current assets/liabilities).

Using the same margin, and making a simple operating cash flow projection based on the accounts and other information provided:

Inflows	£m
Sales in 1995–96 = (£10 × 10 m) + 20%	120.00
Cost of sales before depreciation @ 79%	(94.80)
Operating cash flow	25.20
Outflows	
Tax liability for 1994–95	(5.00)
Interest payments: 12% × £20 m	(2.40)
Dividends: 1.1 × £5 m	(5.50)
	(12.90)
Net internal finance generated	10.30
Funding requirements	(25.40)
Net additional external finance required	(13.10)

(b) (i) Some new equipment could be leased, via a long-term capital lease. Tax relief is available on rental payments, lowering the effective cost of using the equipment. Lessors may 'tailor' a leasing package to suit Burnsall's specific needs regarding timing of payments and provision of ancillary services. Alternatively, good quality property assets at present owned by Burnsall could be sold to a financial institution and their continued use secured via a leaseback arrangement, although this arrangement involves losing any capital appreciation of the assets.

(ii) If Burnsall's assets are of sufficient quality, i.e. easily saleable, it may be possible to raise a mortgage secured on them. This enables retention of ownership.

(iii) Burnsall could make a debenture issue, interest on which would be tax-allowable. The present level of gearing (long-term debt-to-equity) is relatively low at £20 m/£120 m = 16%, Burnsall has no short-term debt apart from trade creditors and the Inland Revenue, and its interest cover is healthy with profit before tax and interest divided by interest charges = [16% × £100 m]/[12% × £20 m] = 6.6 times. It is likely that Burnsall could make a sizeable debt issue without unnerving the market. Any new debenture would be subordinate to the existing long-term debt and probably carry a higher interest rate.

(iv) An alternative to a debt issue is a rights issue of ordinary shares. Because rights issues are made at a discount to the existing market price, they result in lower EPS and thus market price, although if existing shareholders take up their allocations, neither their wealth nor control is diluted. If the market approves of the intended use of funds, a capital gain may ensue, although the company and its advisers must carefully manage the issue regarding the declared reasons and its timing in order to avoid unsettling the market.

(v) Burnsall could approach a venture capitalist such as 3i, which specialises in extending development capital to small-to-medium-sized firms. However, 3i may require an equity stake, and possibly insist on placing an appointee on the Board to monitor its interests.

(vi) Burnsall could utilise official sources of aid, such as a regional development agency depending on its location, or perhaps the European Investment Bank.

CHAPTER 17

1 With a residual dividend policy:

Dividends = [Distributable Earnings − Capital Expenditure]

Tom has issued 10 million shares with par value 50p each, giving total book value of £5 million. Total dividends are (20p × 10 m) = £2 m. Hence, capex = (£10 m − £2 m) = £8 m.

2 As all projects are indivisible, only whole projects with IRR > 20 per cent can be selected, i.e. B + C + E, with total expenditure of £14 million. As this infringes the capital availability constraint of £9 million, Dick is restricted to only B + C, with joint outlay of £7 million, leaving £2 million as a residual dividend.

3 According to Lintner's target adjustment theory, companies only partially adjust their dividends in line with earnings changes. The change in dividend will be half of the difference between current EPS and the last dividend, i.e.

$$
\begin{aligned}
\text{Dividend change} &= 0.5 \times [0.5\ \text{EPS}_t - \text{Div}_{t-1}] \\
&= 0.5 \times [0.5 \times (\text{£}3.0) - \text{£}1] \\
&= 0.5 \times \text{£}0.5 = \text{£}0.25
\end{aligned}
$$

so that the new dividend = (£1.0 + £0.25) = £1.25.

4 Tamas' tax charge = (30% × £30 m) = £9 m

Profit after tax = (£30 − £9 m) = £21 m

Dividend = (50% × £21 m) = £10.5 m

i.e. (£10.5 m/100 m) = 10.5p per share

Net (i.e. post-tax) cash flow = [Pre-tax profit + depreciation − replacement investment − tax]

= [£30 m + £2 m − £2 m − £9 m] = £21 m

Value cum dividend = (£21 m/15%) + £10.5 m

= (£140 m + £10.5 m) = £150.5 m

Per share, this is (£150.5 m/100 m) = £1.505 (cum dividend).

The ex-dividend share price is £1.505 reduced by the dividend per share of 10.5p, i.e. (£1.505 − £0.105) = £1.40.

5 (a) The price per share is given by:

$$
P_o = \frac{D_1}{(k_e - g)}
$$

where D_1 is next year's dividend, k_e is the shareholder's required return and g is the expected rate of growth in dividends.

The growth rate can be found from the expression:

$$5.0p(1 + g)^4 = 7.3p$$

where g is the past (compound) growth rate.

$$(1 + g)^4 = \frac{7.3p}{5.0p} = 1.46$$

or

$$\frac{1}{(1 + g)^4} = 0.6849$$

From the present value tables, $g = 10\%$, whence:

$$P_o = \frac{7.3p(1.1)}{(16\% - 10\%)} = \frac{8.03p}{0.06} = £1.34$$

(b) With D_1 at just 5.0p, using managerial expectations for the investment:

$$P_o = \frac{5.0p}{(16\% - 14\%)} = \frac{5.0p}{0.02} = £2.50$$

(c) To break even, share price must not fall below £1.34, i.e.

$$£1.34 = \frac{5.0p}{(16\% - g)}$$

Solving for g, we find $g = 12.3\%$, marginally above the assessment of the more pessimistic managers.

(d) Until 1990, Galahad pursued a policy of distributing 40–50 per cent of profit after tax as dividend. Each year, it has offered a steady dividend increase, even in 1989, when its earnings actually fell. This was presumably out of reluctance to lower the dividend, fearing an adverse market reaction, and reflecting a belief that the earnings shortfall was a temporary phenomenon. In 1990 it offered a 12 per cent dividend increase, the highest percentage increase in the time series, possibly to compensate shareholders for the relatively small increase (only 8 per cent) in 1989. It would appear that Galahad has either already built up a clientèle of investors whose interests it is trying to safeguard, or that it is trying to do so.

The proposed dividend cut to 5.0p per share would represent a sharply increased dividend cover of 3.5, on the assumption that EPS also grows at 10 per cent p.a. Such a sharp rise in the dividend safety margin is likely to be construed by the market as implying that Galahad's managers expect earnings to be depressed in the future, especially as it follows a year of record dividend increase. Such an abrupt change in dividend policy is thus likely to offend its clientèle of shareholders at best, and at worst, to alarm the market as to the reliability of future earnings.

In an efficient capital market, with homogeneous investor expectations, the share price would increase by the amount calculated in (b), at least, if the market agreed with the managers' views about the attractions of the projected expenditure. However, in view of the information content of dividends, Galahad's board will have to be very confident of its ability to persuade the market of the inherent desirability of the proposed investment programme. This may well be a difficult task, especially given the stated doubts of some of its managers. The board will have to explain why they feel internal financing is preferable to raising capital externally, either by a rights issue, or by raising further debt finance. While the level of indebtedness of Galahad is not given, the implication is that it is unacceptably high, so as to obviate the issue of additional borrowing instruments. If this is the case, then it seems doubly risky to propose a dividend cut, as it may signal fears regarding Galahad's ability to service a high level of debt.

If the dividend cut is greeted adversely, then the ability of the shareholder clientèle to home-make dividends will be impaired, since, apart from the transactions costs involved, there will perhaps be no capital gain to realise. Any significant selling to convert capital into income will further depress share price.

If the investment programme is truly worthwhile, Galahad's managers perhaps should not shrink from offering a rights issue, since, despite the costs of such issues, shareholders will eventually reap the benefits in the form of higher future earnings and dividends. However, this might suggest a short-term reduction in share price, which may penalise short-term investors, but who still have the option of protecting their interests by selling their rights.

CHAPTER 18

1 LTD/Equity = (£200 m/£500 m) = 40%
LTD/(LTD + Equity) = £200 m/(£200 m + £500 m) = 28%
Total Debt/Equity = (£200 m + £50 m)/£500 m = 50%
Net Debt = (£200 m + £50 m) − (£20 m + £40 m) = £190 m.
Net Debt/Equity = (£190 m/£500 m) = 38%
Interest Cover = (£120 m/£25 m) = 4.8 times.
Income Gearing = (1/4.8) × 100 = 21%
Total Liabilities/Total Assets = (£300 m/£800 m) = 37.5%

2 Market value of debt = £100 × (£45 m/£50 m) = £90

(i) cost of debt = [8% × £100]/£90 = (£8/£90) = 8.9%
(ii) cost of debt = £8(1 − 30%)/£90 = (£5.6/£90) = 6.2%
(iii) cost of debt is the solution R to:

$$£90 = \frac{£8}{(1 + R)} + \frac{£8}{(1 + R)^2} + \cdots + \frac{£8 + £100}{(1 + R)^6}$$

Solution value for R = 10.3%
(iv) The cost of debt is the solution R to:

$$£90 = \frac{£8(1 − 30\%)}{(1 + R)} + \frac{£5.6}{(1 + R)^2} + \cdots + \frac{(£5.6 + £100)}{(1 + R)^6}$$

Solution value for R = 8.25%

3 (a) £m

	prob. = 0.3	prob. = 0.5	prob. = 0.2
EBIT	5	50	150

Current Structure – £200 m equity.

Tax @ 30%	(1.5)	(15)	(45)
PAT	3.5	35	105
ROE	1.8%	17.5%	52.5%

Programme (i) – £40 m debt/£160 m equity.

Interest @ 8%	(3.2)	(3.2)	(3.2)
Taxable profit	1.8	46.8	146.8
Tax @ 30%	(0.5)	(14.0)	(44.0)
PAT	1.26	32.8	102.8
ROE	0.8%	20.5%	64.3%

Programme (ii) – £80 m debt/£120 m equity.

Interest @ 8%	(6.8)	(6.8)	(6.8)
Taxable profit	–	43.2	143.2
Tax @ 30%	–	(14.0)	(44.0)
PAT	–	30.2	100.2
ROE	(1.5%)	25.2%	83.5%

(b) Risk is measured by the standard deviation of the returns on equity (ROE) around the respective expected values (EVs).

Current Capital Structure.
EV of ROE = (0.3 × 1.8) + (0.5 × 17.5) + (0.2 × 52.5) = 19.8%

Outcome	EV	Deviation	Squared	× probability
1.8	19.8	−18.0	324	97.2
17.5	19.8	−2.3	5.3	2.65
52.5	19.8	32.7	1069.3	213.86
			Total = variance =	313.71

Standard deviation = $\sqrt{313.71}$ = 17.7, i.e. 17.7%

Programme (i)

EV of ROE = $(0.3 \times 0.8) + (0.5 \times 20.5) + (0.2 \times 64.3) = 23.4\%$

Outcome	EV	Deviation	Squared	× probability
0.8	23.4	−22.6	510.76	153.228
20.5	23.4	−2.9	8.41	4.205
64.3	23.4	40.9	1672.81	334.56
			Total = variance =	491.993

Standard deviation = $\sqrt{491.993}$ = 22.18, i.e. 22.2%

Programme (ii)

EV of ROE = $(0.3 \times -1.5) + (0.5 \times 25.2) + (0.2 \times 83.5) = 28.9\%$

Outcome	EV	Deviation	Squared	× probability
−1.5	28.9	−30.4	924.16	277.248
25.2	28.9	−3.7	13.69	6.845
83.5	28.9	54.6	2981.16	596.230
			Total variance =	880.323

Standard deviation = $\sqrt{880.323}$ = 29.7, i.e. 29.7%

(c) Business risk, exposure to the risk of fluctuations in business activity, can be measured in different ways, e.g. the variability in sales or in some measure of profitability. This example shows how ROE varies under three scenarios. By abstracting from financial risk, the ROE in the all-equity case shows inherent business risk – the ROE varies from 1.8 per cent to 52.5 per cent, with a standard deviation of 17.7 per cent.

Gearing adds a second layer of risk because it imposes an extra fixed cost, i.e. the prior interest charge – the higher the level of gearing, the higher the financial risk. Consequently, we see that the standard deviation of the ROE rises to 28.9 per cent under the relatively modest level of gearing, programme (i), while programme (ii) raises it to 22.2 per cent.

4 (a) (i) Perpetual debt:

$$\text{PV of Tax Savings} = \frac{(T \times i \times \text{Nominal Debt Value})}{i} = \frac{TiB}{i}$$

$$= (30\% \times 10\% \times £100\text{ m})/10\%$$

$$= (£3\text{ m}/0.1) = £30\text{ m}$$

(ii) When debt is repaid in full after 5 years:
PV = £3 m p.a. over 5 years @ 10% discount rate = (£3 m × 3.7908) = £11.37 m

(iii) Debt repaid in equal tranches:

Year	Start-year debt	Interest	Tax saving	PV	Repayment
1	100	10	3	2.73	20
2	80	8	2.4	1.98	20
3	60	6	1.8	1.35	20
4	40	4	1.2	0.82	20
5	20	2	0.6	0.37	20
				Total = £7.25 m	

(b) The value of the tax shield is higher:
  – the higher the interest rate
  – the higher the tax rate
  – the higher the amount of debt
  – the longer the term of the loan
  – the slower that debt is repaid
  – the greater the firm's taxable capacity

5 Book value weights:

			weight	cost
Equity [£10 m + £20 m] =	£30 m	i.e.	66.7%	20%
Debt	£15 m		33.3%	10% pre-tax
	£45 m		100%	

WACC = (20% × 66.7%) + (10%[1 − 30%] × 33.3%)
   = (13.3% + 2.3%) = 16.6%

Market value of equity = (20 m shares × £4.50) = £90 m. The weights become:

Equity: £90 m/£105 m = 85.7%

Debt: £15 m/£105 m = 14.3%

WACC = (20% × 85.7%) + (10%[1 − 30%] × 14.3%)
   = (17.1% + 1.0%) = 18.1%

CHAPTER 19

1 (i) From MM's Proposition 1, the cost of equity is:

$$k_{eg} = k_{eu} + (k_{eu} - k_d)\frac{V_B}{V_S}$$

If the proportion of debt to total finance is 20%, the ratio of debt to equity must be 1:4, i.e. 0.25. Hence,

$$k_{eg} = 20\% + (20\% - 8\%)(0.25) = 23\%$$

2 (i) Because the WACC is constant at 20% at all gearing levels, the figures correspond to the MM-no tax theory.
   (ii) This now illustrates the MM-with tax theory, wherein the WACC falls continuously as gearing increases. The cost of equity becomes:

$$k_{eg} = k_{eu} + (k_{eu} - k_d)(1 - T)\frac{V_B}{V_S}$$

The amended table is:

% Debt	% Equity	k_d	k_e	WACC
–	100	–	20%	20%
25	75	5.6%	22.8%	18.5%
50	50	5.6%	28.4%	17.2%
75	25	5.6%	45.2%	15.5%

3 Assume the investor holds 10 per cent of Geared's equity.

Value of stake = 10% × £900 = £90

Personal income initially = 10% × [£100 − (5% × £200)]

= 10% × £90 = £9

Geared's gearing = £200/£900 = 22.2%

Assume the investor sells the stake in Geared for £90, borrows £20 at 5 per cent to duplicate Geared's gearing and invests the whole stake in Ungeared's equity.

Now entitled to earnings of (£110/£950) × £100 = £11.58

Personal interest liability = (5% × £20) = (£1.00)

Net income = £10.58

The investor is better off by (£10.58 − £9.00) = £1.58

9 (a) CAPM

The stated Beta is an equity Beta so that the required return on equity is:

$ER_j = R_f + \beta_j (ER_m − R_f)$

$= 12\% + 1.4(18\% − 12\%) = 20.4\%$

This is unsuitable as a discount rate because:

(i) It is the required return on equity rather than the required return on the overall company.

(ii) The equity Beta of 1.4 reflects the financial risk of Folten's equity. Wemere's gearing differs from that of Folten, hence their equity Betas will differ.

The inflation adjustment is unnecessary since ER_m and R_f already incorporate the expected impact of inflation.

The equity Beta for Wemere can be estimated by ungearing Folten's equity Beta and regearing to reflect the financial risk of Wemere.

The market value-weighted gearing figures are:

Folten

Equity (138p × 7.2 m shares) = £9.936 m, i.e. 69.3% of total

Debt = £4,400 m, i.e. 30.7% of total

Total = £14,336 m

Wemere

Equity (using the takeover bid offer) = £10.6 m, i.e. 81.5% of total

Debt = £2.4 m , i.e. 18.5% of total

Total = £13.0 m

Assuming corporate debt is risk-free, the ungeared equity Beta is

$$\beta_u = \beta_g \times \cfrac{1}{\left[1 + \dfrac{V_B}{V_S} \times (1 − T) \right]} = 1.4 \times \cfrac{1}{1 + (0.44)(1 − 35\%)}$$

$= 1.089$

Regearing Beta for Wemere,

$$\beta_g = \beta_u\left[1 + \frac{V_B}{V_S} \times (1 - T)\right]$$
$$= 1.089\ [1 + 0.23(1 - 35\%)] = 1.25$$

The cost of equity for Wemere is thus:

$12\% + 1.25[18\% - 12\%] = 19.5\%$

Given the cost of debt is 13%:

WACC $= [13\%(1 - 35\%) \times 18.5\%] + [19.5\% \times 81.5\%]$
$= 1.56\% + 15.89\% = 17.5\%$

However, the WACC is only suitable as a discount rate if the systematic risk of the new investment is similar to that of the company as a whole.

Dividend Growth Model
The expression for this model relates to the cost of equity, not the overall cost of capital, i.e.

$$k_e = \frac{D_1}{P_o} + g = \frac{14.20p}{138p} + 9\% = 19.3\%$$

No inflation adjustment is required.
The WACC is: $[13\%(1 - 35\%) \times 18.5\%] + [19.3\% \times 81.5\%]$
$= 1.56\% + 15.73\% = 17.3\%$

(b) Neither method is problem-free. The surrogate company is unlikely to have identical characterisitics, either at an operating level or in terms of financial characteristics. For example, the cost of equity in the Dividend Growth Model is derived from a different set of data regarding dividend policy, growth and share prices.

Folten's managers may have different capabilities, and the company may face different growth opportunities. Before using the estimated WACC, Wemere must be confident that the two companies are a sufficiently close fit.

Even so, the calculated WACC is inappropriate if the systematic risk of any new project differs from that of the company as a whole, and/or if project financing involves moving to a new capital structure.

CHAPTER 20

1 (a) Value of target now:

$$P_o = \frac{D_1}{k_e - g}$$

$$20 = \frac{80p}{k_e - 6\%}$$

whence $k_e = 10\%$.

Value of target would become

$$P = \frac{80p}{10\% - 8\%} = £40\ \text{per share}$$

Value of equity $= £40 \times 0.6\ \text{m} = £24\ \text{m}$, i.e. an increase of £12 m.

(b) Cost of acquisition: $(£25 - £20) = £5$ per share. In total, $£5 \times 0.6\ \text{m} = £3\ \text{m}$. NPV of acquisition: $(£12\ \text{m} - £3\ \text{m}) = £9\ \text{m}$. Advise to proceed.

(c) Number of new shares required: 0.6 m/3 = 0.2 m; new total = 1.2 m.

$$\text{Value of new company} = \frac{£90\text{ m} + £12\text{ m} + £12\text{ m}}{1.2\text{ m}}$$
$$= £114/1.2\text{ m} = £95\text{ per share}$$

Cost of acquisition = (0.2 m × £95 m) − £12 m = £7 m. NPV of acquisition = (£12 m − £7 m) = £5 m. Again, advise to proceed.

(d) (i) Cost of cash bid unchanged. Pointless to proceed as there are no gains.
(ii) With the share exchange:

$$\text{Value of new company} = \frac{£90\text{ m} + £12\text{ m}}{1.2\text{ m}} = £85\text{ per share}$$

NPV of acquisition: (0.2 m × £85) − £12 m = (£5 m). Advise not to proceed on this basis.

7 (a) The Balance Sheet net asset value is total assets minus total liabilities, i.e. £620 m. Land and buildings have an estimated value of £150 m × $(1.25)^4$ = £366 m, i.e. £216 m higher than the book value. Hence, the adjusted NAV is £836 m.
Applying Grapper's 12 per cent growth rate, estimated PAT for the coming five years is:

£151(1.12) + £151$(1.12)^2$, etc. = £1,074 m

This yields total value of £836 m + £1,074 m = £1,910 m. Grapper's market value is currently (400 m shares × share price 470p) = £1,880 m. The premium is thus £30 m or 7.5p per share.
  This is not a sound basis for valuation as it ignores the time-value of money. The premium of 1.6 per cent above the current market price is very small compared with those achieved in many 'real' bids.
  Using the Dividend Valuation Model:

$$P_o = \frac{D_1}{k_e - g} = \frac{D_o(1 + g)}{k_e - g}$$

$$\text{Current dividend per share} = \frac{£76\text{ m}}{400\text{ m}} = 19\text{p}$$

Hence D_1 = 19p(1.12) = 21.3p.
From the CAPM:
$k_e = ER_j = R_f + \beta_j(ER_m - R_f) = 10\% + 1.05(16\% - 10\%) = 16.3\%$

Thus:

$$P_o = \frac{21.3\text{p}}{16.3\% - 12\%} = 495\text{p}$$

i.e. 5.3 per cent above the market price.
  Restrictive assumptions underlying such a valuation include a constant growth rate, and an unchanged dividend policy. It is more rational to assess the value of Grapper incorporating post-merger rationalisation.
(b) The post-merger sales revenue of Woppit will be over £5,000 million, a size which could deter other takeover raiders, at least from the United Kingdom. However, bids from US and other European sources should not be ruled out. In addition, debt-financed bids from consortia like Hoylake (which bid for BAT) show that size alone is not an adequate protection against a takeover bid.
(c) An indication of the scope for improving Grapper's efficiency can be obtained by examination of key financial ratios.

	Woppit	**Grapper**
Operating profit margin (PBIT/sales)	20%	16.6%
Asset turnover (sales/total assets)	1.80	1.36
Debtors' collection period	31 days	50 days
Stock turnover	10.3	6.4
Current ratio	1.65:1	2.45:1

There are clear opportunities to improve Grapper's performance by rationalisation and restructuring of activities. For example:

- Grapper's operating profit margin could be brought into line with Woppit's by a price increase and/or cost reduction.
- Grapper's stock level looks high by comparison. There could well be stockholding economies in an expanded operation.
- Grapper's cash holdings look excessive – again, centralised cash management may generate economies.
- Grapper's asset turnover is relatively low. Some assets could well be sold and others worked more intensively.
- Grapper seems to have scope for reducing its investment in debtors.
- Introduction of such economies may well close the gap between Woppit's return on assets of 36 per cent and Grapper's present 22.5 per cent.

CHAPTER 21

1 Remember the bank always wins, so it sells euros at 1.6296, and buys at 1.6320.
   (a) Selling £10 million, its receipts are $(10 \times 1.6296) = €16.296$ million.
   (b) Selling €10 million, its receipts are $(10 \text{ m}/1.6320) = £6.127 \text{ m}$.

2 (a) Amount invoiced $= (£10 \text{ m} \times €1.6) = €16 \text{ m}$.
   (b) With spot at €1.7 vs. £1, proceeds $= (€16 \text{ m}/1.7) = £9.412$. Hence, the loss compared to the current spot rate $= (£10 \text{ m} - £9.412) = £0.588 \text{ m}$.
   (c) With spot at €1.5 vs. £1, the sale proceeds are $(€16 \text{ m}/1.5 \text{ m}) = £10.667$. In this case, the exporter gains £0.667 m from the exchange rate change.
   (d) If the exporter sells forward, the contracted proceeds are $€16 \text{ m}/1.62 = £9.877 \text{ m}$.
      (i) The cost of the hedge is thus £0.123 m, i.e. 1.2% of the sterling value of the deal.
      (ii) If sterling falls to €1.5 vs. £1, the forward contract guarantees the exporter £9.877 million, but it could have received £10.667 million had it not hedged. There is thus an opportunity cost of $(£10.667 \text{ m} - £9.877 \text{ m}) = £0.790 \text{ m}$.
      (iii) If sterling rises to €1.7 vs. £1, the forward contract still guarantees the exporter £9.877 m, but it would have received £9.412 m had it not hedged. The exporter is thus better off by $(£9.877 \text{ m} - £9.412 \text{ m}) = £0.465 \text{ m}$.

   If the exporter thinks there is an equal chance of a ten per cent variation in the €/£ exchange rate, it must balance an opportunity cost from hedging of £0.75 million if sterling falls against being better off by £0.465 million if sterling rises.

3 The forward outrights are:

KRONA	$10.960 - 10.967$
Less forward premium	$(25 - 11)$
Forward outright	$10.935 - 10.956$
YEN	$230.11 - 230.16$
Less forward premium	$(1.10 - 97)$
Forward outright	$229.01 - 229.19$
KRONOR	$11.717 - 11.723$
Less forward premium	$(21 - 4)$
Forward outright	$11.696 - 11.719$

5 According to Interest Rate Parity,

$$\text{Forward rate} = \text{Spot rate} \times \frac{(1 + \text{US interest rate})}{(1 + \text{UK interest rate})}$$

$$= 1.6000 \times \frac{(1.09)}{(1.10)} = 1.5834, \quad \text{i.e. USD stronger on forward market}$$

The currency of the country in which interest rates are lower (presumably due to lower expected inflation) would be traded at a premium.

6 It is assumed it is desired to hold US$ at the year-end. By lending US$ at 7.625 per cent, the end-year balance will be £1 m(1.07625) = $1,076,250.

If wishing to lend in Swiss francs, the treasurer would convert from dollars at spot of 1.3125 to obtain CHF of $1 m × 1.3125 = SFr1,312,500. Over one year, invested at 4.5625 per cent, this would accumulate to CHF1,372,383.

To cover the risk of adverse exchange rate movements, he will then sell CHF forward at the ruling rate of 1.275 to guarantee US$ delivery in one year's time of CHF1,372,383/1.275 = $1,076,379.

The minimal difference of $129 can be attributed to the operation of IRP.

Transactions costs would wipe out any gain from arbitrage.

11 With £1 = S$2.80:

Value of deal at today's spot = S$28 m/2.80 = £10.00 m

Hedged using a forward contract:

Value = S$28 m/2.79 = £10.04 m

Money market hedge:

Ashton borrows S$28 m at 2%/4 = 0.5%
Borrowing = S$28 m/1.005 = S$27.86 m
Converts at spot (S$2.8 = £1) to yield S$27.86 m/2.80 = £9.95 m
Invested at 3%/4 (i.e. 0.0075%) this generates £9.95(1.0075) = £10.03 m

The forward hedge is superior.

(a) If future spot is £2.78 = £1

Value of deal at spot = S28 m/2.78 = **£10.07 m**

Not hedging would have been preferable (with hindsight!)
The OTC option is out of the money and the premium is lost:

Net value = [£10.07 m − premium of £0.2 m] = £9.87 m

(b) If future spot is £1 = S2.82:

Value of deal at spot = S$28 m/2.82 = £9.93 m

Hedging would have been preferable (using the forward contract)
The OTC option is in the money.
Net of the premium, Ashton receives:

[S$28 m/2.785] − £0.2 m = £10.05 m − £0.2 m = £9.85 m

But the forward contract would have been superior to this.

CHAPTER 22

1 In this question, we need to show that the receipts in sterling after adjusting for inflation at both locations remains unchanged.
Current exchange = $US 1.50 against £1.
Sterling equivalent of US revenue = $150 m/1.50 = 100 m

Exchange rate	$ revenue	£ revenue	£ revenue (real terms)
(i) 1.05/1.05 × 1.50 = $1.50 : £1	$157.5 m	£105 m	£105 m/1.05 = £100 m
(ii) 1.02/1.05 × 1.50 = £1.457 : £1	$153 m	£105 m	£105 m/1.05 = £100 m
(iii) 1.05/1.02 × 1.50 = £1.544 : £1	$157.5 m	£102 m	£102 m/1.02 = £100 m

2 (a) (i) With exchange controls:

Year	PAT (SA$000)	OJ share (SA$000)	50% div (SA$000)	Sterling (£000)	PV @ 16% (£000)
0	–	–	–	(450)	(450)
1	4,250	2,125	1,062	106	91
2	6,500	3,250	1,625	108	80
3	8,350	4,175	2,088	100	64
			4,775	277	146
			(balance)		NPV = (69)

In this scenario, OJ should reject the project.

(ii) No exchange control:

Year	PAT (SA$000)	OJ share (SA$000)	Sterling (£000)	PV @ 16% (£000)
0	–	–	(450)	(450)
1	4,250	2,125	212	183
2	6,500	3,250	217	161
3	8,350	4,175	199	127
				NPV = +21

In this scenario, the positive NPV indicates acceptance, but the project is marginal e.g. the Profitability Index (NPV/Outlay = 21/450) is only 0.047. Given the risk of exchange controls being imposed, this suggests that OJ should treat the project with great caution.

3 PG plc

Year	0	1	2	3	4
Method 1					
C$ Initial investment	(150,000)				50,000
Other cash flows		60,000	60,000	60,000	45,000
Net cash flows	(150,000)	60,000	60,000	60,000	95,000
C$ per £1	1.700	1.785	1.874	1.968	2.066
Sterling	(88,235)	33,613	32,017	30,488	45,983
DF @ 14%	1.000	0.877	0.769	0.675	0.592
PV	(88,235)	29,479	24,621	20,579	27,222
NPV = £13,666					
Method 2					
C$ net cash flows	(150,000)	60,000	60,000	60,000	95,000
DF @ 19.7%	1.000	0.835	0.698	0.583	0.487
C$	(150,000)	50,100	41,888	34,980	46,265
£ PV @ 1.7	(88,235)	29,479	24,621	20,579	27,222
NPV = £13,666					

For the two approaches to generate the same answer, the discount rate applied to the C$ cash flows must be the combination of the sterling discount rate (14 per cent) and the expected strengthening of sterling, according to PPP. This yields:

$$(1.14 \times 1.05) - 1 = (1.197 - 1) = 0.197, \quad \text{i.e. } 19.7\%$$

A forecast 5 per cent appreciation of sterling against the C$ will be associated with UK inflation rates being 5 per cent less than the rate experienced in Canada. In practice, one might inflate the cash flows in C$ to reflect inflation internal to the Canadian economy.

6 Palmerston plc

Forecast Cash Flow Statement (£m)

Euros per £1	1.65	1.60	1.55
Sales			
UK	200	210	220
Germany	**182**	**188**	**194**
Total	382	398	414
Cost of Goods Sold:			
UK	(120)	(120)	(120)
Germany	(121)	(125)	(129)
Total	(241)	(245)	(249)
Gross Profit	141	153	165
Operating Expenses:			
UK – fixed	(50)	(50)	(50)
UK – variable (20% of total sales)	(76)	(80)	(83)
Total	(126)	(130)	(133)
Net Cash Flow	15	23	32
Firm value @ 15%:	$(15 \times 5.019^*)$	(23×5.019)	(32×5.019)
*Annuity factor	= £75.3 m	= £115.4 m	= £160.61

The analysis suggests that Palmerston benefits from a strong euro and vice versa. It could further reduce its exposure by shifting its cost base to Germany, or elsewhere in the euro area, preferably to a low-cost location, say, Greece or Portugal.

Present value interest factor (PVIF)

per £1.00 due at the end of *n* years for interest rate of:

n	1%	2%	3%	4%	5%	6%	7%	8%	9%	10%	n
1	0.99010	0.98039	0.97007	0.96154	0.95238	0.94340	0.93458	0.92593	0.91743	0.90909	1
2	0.98030	0.96117	0.94260	0.92456	0.90703	0.89000	0.87344	0.85734	0.84168	0.82645	2
3	0.97059	0.94232	0.91514	0.88900	0.86384	0.83962	0.81630	0.79383	0.77218	0.75131	3
4	0.96098	0.92385	0.88849	0.85480	0.82270	0.79209	0.76290	0.73503	0.70843	0.68301	4
5	0.95147	0.90573	0.86261	0.82193	0.78353	0.74726	0.71299	0.68058	0.64993	0.62092	5
6	0.94204	0.88797	0.83748	0.79031	0.74622	0.70496	0.66634	0.63017	0.59627	0.56447	6
7	0.93272	0.87056	0.81309	0.75992	0.71068	0.66506	0.62275	0.58349	0.54703	0.51316	7
8	0.92348	0.85349	0.78941	0.73069	0.67684	0.62741	0.58201	0.54027	0.50187	0.46651	8
9	0.91434	0.83675	0.76642	0.70259	0.64461	0.59190	0.54393	0.50025	0.46043	0.42410	9
10	0.90529	0.82035	0.74409	0.67556	0.61391	0.55839	0.50835	0.46319	0.42241	0.38554	10
11	0.89632	0.80426	0.72242	0.64958	0.58468	0.52679	0.47509	0.42888	0.38753	0.35049	11
12	0.88745	0.78849	0.70138	0.62460	0.55684	0.49697	0.44401	0.39711	0.35553	0.31863	12
13	0.87866	0.77303	0.68095	0.60057	0.53032	0.46884	0.41496	0.36770	0.32618	0.28966	13
14	0.86996	0.75787	0.66112	0.57747	0.50507	0.44230	0.38782	0.34046	0.29925	0.26333	14
15	0.86135	0.74301	0.64186	0.55526	0.48102	0.41726	0.36245	0.31524	0.27454	0.23939	15
16	0.85282	0.72845	0.62317	0.53391	0.45811	0.39365	0.33873	0.29189	0.25187	0.21763	16
17	0.84438	0.71416	0.60502	0.51337	0.43630	0.37136	0.31657	0.27027	0.23107	0.19784	17
18	0.83602	0.70016	0.58739	0.49363	0.41552	0.35034	0.29586	0.25025	0.21199	0.17986	18
19	0.82774	0.68643	0.57029	0.47464	0.39573	0.33051	0.27651	0.23171	0.19449	0.16351	19
20	0.81954	0.67297	0.55367	0.45639	0.37689	0.31180	0.25842	0.21455	0.17843	0.14864	20
21	0.81143	0.65978	0.53755	0.43883	0.35894	0.29415	0.24151	0.19866	0.16370	0.13513	21
22	0.80340	0.64684	0.52189	0.42195	0.34185	0.27750	0.22571	0.18394	0.15018	0.12285	22
23	0.79544	0.63414	0.50669	0.40573	0.32557	0.26180	0.21095	0.17031	0.13778	0.11168	23
24	0.78757	0.62172	0.49193	0.39012	0.31007	0.24698	0.19715	0.15770	0.12640	0.10153	24
25	0.77977	0.60953	0.47760	0.37512	0.29530	0.23300	0.18425	0.14602	0.11597	0.09230	25

n	11%	12%	13%	14%	15%	16%	17%	18%	19%	20%	n
1	0.90090	0.89286	0.88496	0.87719	0.86957	0.86207	0.85470	0.84746	0.84034	0.83333	1
2	0.81162	0.79719	0.78315	0.76947	0.75614	0.74316	0.73051	0.71818	0.70616	0.69444	2
3	0.73119	0.71178	0.69305	0.67497	0.65752	0.64066	0.62437	0.60863	0.59342	0.57870	3
4	0.65873	0.63552	0.61332	0.59208	0.57175	0.55229	0.53365	0.51579	0.49867	0.48225	4
5	0.59345	0.56743	0.54276	0.51937	0.49718	0.47611	0.45611	0.43711	0.41905	0.40188	5
6	0.53464	0.50663	0.48032	0.45559	0.43233	0.41044	0.38984	0.37043	0.35214	0.33490	6
7	0.48166	0.45235	0.42506	0.39964	0.37594	0.35383	0.33320	0.31392	0.29592	0.27908	7
8	0.43393	0.40388	0.37616	0.35056	0.32690	0.30503	0.28487	0.26604	0.24867	0.23257	8
9	0.39092	0.36061	0.33288	0.30751	0.28426	0.26295	0.24340	0.22546	0.20897	0.19381	9
10	0.35218	0.32197	0.29459	0.26974	0.24718	0.22668	0.20804	0.19106	0.17560	0.16151	10
11	0.31728	0.28748	0.26070	0.23662	0.21494	0.19542	0.17781	0.16192	0.14756	0.13459	11
12	0.28584	0.25667	0.23071	0.20756	0.18691	0.16846	0.15197	0.13722	0.12400	0.11216	12
13	0.25751	0.22917	0.20416	0.18207	0.16253	0.14523	0.12989	0.11629	0.10420	0.09346	13
14	0.23199	0.20462	0.18068	0.15971	0.14133	0.12520	0.11102	0.09855	0.08757	0.07789	14
15	0.20900	0.18270	0.15989	0.14010	0.12289	0.10793	0.09489	0.08352	0.07359	0.06491	15
16	0.18829	0.16312	0.14150	0.12289	0.10686	0.09304	0.08110	0.07078	0.06184	0.05409	16
17	0.16963	0.14564	0.12522	0.10780	0.09293	0.08021	0.06932	0.05998	0.05196	0.04507	17
18	0.15282	0.13004	0.11081	0.09456	0.08080	0.06914	0.05925	0.05083	0.04367	0.03756	18
19	0.13768	0.11611	0.09806	0.08295	0.07026	0.05961	0.05064	0.04308	0.03669	0.03130	19
20	0.12403	0.10367	0.08678	0.07276	0.06110	0.05139	0.04328	0.03651	0.03084	0.02608	20
21	0.11174	0.09256	0.07680	0.06383	0.05313	0.04430	0.03699	0.03094	0.02591	0.02174	21
22	0.10067	0.08264	0.06796	0.05599	0.04620	0.03819	0.03162	0.02622	0.02178	0.01811	22
23	0.09069	0.07379	0.06014	0.04911	0.04017	0.03292	0.02702	0.02222	0.01830	0.01509	23
24	0.08170	0.06588	0.05322	0.04308	0.03493	0.02838	0.02310	0.01883	0.01538	0.01258	24
25	0.07361	0.05882	0.04710	0.03779	0.03038	0.02447	0.01974	0.01596	0.01292	0.01048	25

continued

n	21%	22%	23%	24%	25%	26%	27%	28%	29%	30%	n
1	0.82645	0.81967	0.81301	0.80645	0.80000	0.79365	0.78740	0.78125	0.77519	0.76923	1
2	0.68301	0.67186	0.66098	0.65036	0.64000	0.62988	0.62000	0.61035	0.60093	0.59172	2
3	0.56447	0.55071	0.53738	0.52449	0.51200	0.49991	0.48819	0.47684	0.46583	0.45517	3
4	0.46651	0.45140	0.43690	0.42297	0.40960	0.39675	0.38440	0.37253	0.36111	0.35013	4
5	0.38554	0.37000	0.35520	0.34111	0.32768	0.31488	0.30268	0.29104	0.27993	0.26933	5
6	0.31863	0.30328	0.28878	0.27509	0.26214	0.24991	0.23833	0.22737	0.21700	0.20718	6
7	0.26333	0.24859	0.23478	0.22184	0.20972	0.19834	0.18766	0.17764	0.16822	0.15937	7
8	0.21763	0.20376	0.19088	0.17891	0.16777	0.15741	0.14776	0.13878	0.13040	0.12259	8
9	0.17986	0.16702	0.15519	0.14428	0.13422	0.12493	0.11635	0.10842	0.10109	0.09430	9
10	0.14864	0.13690	0.12617	0.11635	0.10737	0.09915	0.09161	0.08470	0.07836	0.07254	10
11	0.12285	0.11221	0.10258	0.09383	0.08590	0.07869	0.07214	0.06617	0.06075	0.05580	11
12	0.10153	0.09198	0.08339	0.07567	0.06872	0.06245	0.05680	0.05170	0.04709	0.04292	12
13	0.08391	0.07539	0.06780	0.06103	0.05498	0.04957	0.04472	0.04039	0.03650	0.03302	13
14	0.06934	0.06180	0.05512	0.04921	0.04398	0.03934	0.03522	0.03155	0.02830	0.02540	14
15	0.05731	0.05065	0.04481	0.03969	0.03518	0.03122	0.02773	0.02465	0.02194	0.01954	15
16	0.04736	0.04152	0.03643	0.03201	0.02815	0.02478	0.02183	0.01926	0.01700	0.01503	16
17	0.03914	0.03403	0.02962	0.02581	0.02252	0.01967	0.01719	0.01505	0.01318	0.01156	17
18	0.03235	0.02789	0.02408	0.02082	0.01801	0.01561	0.01354	0.01175	0.01022	0.00889	18
19	0.02673	0.02286	0.01958	0.01679	0.01441	0.01239	0.01066	0.00918	0.00792	0.00684	19
20	0.02209	0.01874	0.01592	0.01354	0.01153	0.00983	0.00839	0.00717	0.00614	0.00526	20
21	0.01826	0.01536	0.01294	0.01092	0.00922	0.00780	0.00661	0.00561	0.00476	0.00405	21
22	0.01509	0.01259	0.01052	0.00880	0.00738	0.00619	0.00520	0.00438	0.00369	0.00311	22
23	0.01247	0.01032	0.00855	0.00710	0.00590	0.00491	0.00410	0.00342	0.00286	0.00239	23
24	0.01031	0.00846	0.00695	0.00573	0.00472	0.00390	0.00323	0.00267	0.00222	0.00184	24
25	0.00852	0.00693	0.00565	0.00462	0.00378	0.00310	0.00254	0.00209	0.00172	0.00152	25

n	31%	32%	33%	34%	35%	36%	37%	38%	39%	40%	n
1	0.76336	0.75758	0.75188	0.74627	0.74074	0.73529	0.72993	0.72464	0.71942	0.71429	1
2	0.58272	0.57392	0.56532	0.55692	0.54870	0.54066	0.53279	0.52510	0.51757	0.51020	2
3	0.44482	0.43479	0.42505	0.41561	0.40644	0.39754	0.38890	0.38051	0.37235	0.36443	3
4	0.33956	0.32939	0.31959	0.31016	0.30107	0.29231	0.28387	0.27573	0.26788	0.26031	4
5	0.25920	0.24953	0.24029	0.23146	0.22301	0.21493	0.20720	0.19980	0.19272	0.18593	5
6	0.19787	0.18904	0.18067	0.17273	0.16520	0.15804	0.15124	0.14479	0.13865	0.13281	6
7	0.15104	0.14321	0.13584	0.12890	0.12237	0.11621	0.11040	0.10492	0.09975	0.09486	7
8	0.11530	0.10849	0.10214	0.09620	0.09064	0.08545	0.08058	0.07603	0.07176	0.06776	8
9	0.08802	0.08219	0.07680	0.07179	0.06714	0.06283	0.05882	0.05509	0.05163	0.04840	9
10	0.06719	0.06227	0.05774	0.05357	0.04973	0.04620	0.04293	0.03992	0.03714	0.03457	10
11	0.05129	0.04717	0.04341	0.03998	0.03684	0.03397	0.03134	0.02893	0.02672	0.02469	11
12	0.03915	0.03574	0.03264	0.02984	0.02729	0.02498	0.02287	0.02096	0.01922	0.01764	12
13	0.02989	0.02707	0.02454	0.02227	0.02021	0.01837	0.01670	0.01519	0.01383	0.01260	13
14	0.02281	0.02051	0.01845	0.01662	0.01497	0.01350	0.01219	0.01101	0.00995	0.00900	14
15	0.01742	0.01554	0.01387	0.01240	0.01109	0.00993	0.00890	0.00798	0.00716	0.00643	15
16	0.01329	0.01177	0.01043	0.00925	0.00822	0.00730	0.00649	0.00578	0.00515	0.00459	16
17	0.01015	0.00892	0.00784	0.00691	0.00609	0.00537	0.00474	0.00419	0.00370	0.00328	17
18	0.00775	0.00676	0.00590	0.00515	0.00451	0.00395	0.00346	0.00304	0.00267	0.00234	18
19	0.00591	0.00512	0.00443	0.00385	0.00334	0.00290	0.00253	0.00220	0.00192	0.00167	19
20	0.00451	0.00388	0.00333	0.00287	0.00247	0.00213	0.00184	0.00159	0.00138	0.00120	20
21	0.00345	0.00294	0.00251	0.00214	0.00183	0.00157	0.00135	0.00115	0.00099	0.00085	21
22	0.00263	0.00223	0.00188	0.00160	0.00136	0.00115	0.00098	0.00084	0.00071	0.00061	22
23	0.00201	0.00169	0.00142	0.00119	0.00101	0.00085	0.00072	0.00061	0.00051	0.00044	23
24	0.00153	0.00128	0.00107	0.00089	0.00074	0.00062	0.00052	0.00044	0.00037	0.00031	24
25	0.00117	0.00097	0.00080	0.00066	0.00055	0.00046	0.00038	0.00032	0.00027	0.00022	25

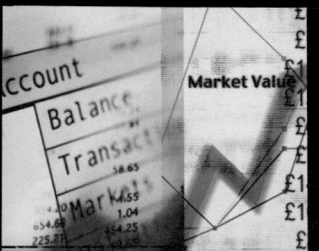

Present value interest factor for an annuity (PVIFA)

of £1.00 for a series of *n* years for interest rate of:

n	1%	2%	3%	4%	5%	6%	7%	8%	9%	10%	n
1	0.9901	0.9804	0.9709	0.9615	0.9524	0.9434	0.9346	0.9259	0.9174	0.9091	1
2	1.9704	1.9416	1.9135	1.8861	1.8594	1.8334	1.8080	1.7833	1.7591	1.7355	2
3	2.9410	2.8839	2.8286	2.7751	2.7232	2.6730	2.6243	2.5771	2.5313	2.4868	3
4	3.9020	3.8077	3.7171	3.6299	3.5459	3.4651	3.3872	3.3121	3.2397	3.1699	4
5	4.8535	4.7134	4.5797	4.4518	4.3295	4.2123	4.1002	3.9927	3.8896	3.7908	5
6	5.7955	5.6014	5.4172	5.2421	5.0757	4.9173	4.7665	4.6229	4.4859	4.3553	6
7	6.7282	6.4720	6.2302	6.0020	5.7863	5.5824	5.3893	5.2064	5.0329	4.8684	7
8	7.6517	7.3254	7.0196	6.7327	6.4632	6.2098	5.9713	5.7466	5.5348	5.3349	8
9	8.5661	8.1622	7.7861	7.4353	7.1078	6.8017	6.5152	6.2469	5.9852	5.7590	9
10	9.4714	8.9825	8.5302	8.1109	7.7217	7.3601	7.0236	6.7101	6.4176	6.1446	10
11	10.3677	9.7868	9.2526	8.7604	8.3064	7.8868	7.4987	7.1389	6.8052	6.4951	11
12	11.2552	10.5753	9.9539	9.3850	8.8632	8.3838	7.9427	7.5361	7.1607	6.8137	12
13	12.1338	11.3483	10.6349	9.9856	9.3925	8.8527	8.3576	7.9038	7.4869	7.1034	13
14	13.0038	12.1062	11.2960	10.5631	9.8986	9.2950	8.7454	8.2442	7.7861	7.3667	14
15	13.8651	12.8492	11.9379	11.1183	10.3796	9.7122	9.1079	8.5595	8.0607	7.6061	15
16	14.7180	13.5777	12.5610	11.6522	10.8377	10.1059	9.4466	8.8514	8.3125	7.8237	16
17	15.5624	14.2918	13.1660	12.1656	11.2740	10.4772	9.7632	9.1216	8.5436	8.0215	17
18	16.3984	14.9920	13.7534	12.6592	11.6895	10.8276	10.0591	9.3819	8.7556	8.2014	18
19	17.2261	15.6784	14.3237	13.1339	12.0853	11.1581	10.3356	9.6036	8.9501	8.3649	19
20	17.8571	16.3514	14.8774	13.5903	12.4622	11.4699	10.5940	9.8181	9.1285	8.5136	20
21	18.0457	17.0111	15.4149	14.0291	12.8211	11.7640	10.8355	10.0168	9.2922	8.6487	21
22	19.6605	17.6580	15.9368	14.4511	13.1630	12.0416	11.0612	10.2007	9.4424	8.7715	22
23	20.4559	18.2921	16.4435	14.8568	13.4885	12.3033	11.2722	10.3710	9.5802	8.8832	23
24	21.2435	18.9139	16.9355	15.2469	13.7986	12.5503	11.4693	10.5287	9.7066	8.9847	24
25	22.0233	19.5234	17.4131	15.6220	14.0939	12.7833	11.6536	10.6748	9.8226	9.0770	25

n	11%	12%	13%	14%	15%	16%	17%	18%	19%	20%	n
1	0.9009	0.8929	0.8850	0.8772	0.8696	0.8621	0.8547	0.8475	0.8403	0.8333	1
2	1.7125	1.6901	1.6681	1.6467	1.6257	1.6052	1.5852	1.5656	1.5465	1.5278	2
3	2.4437	2.4018	2.3612	2.3216	2.2832	2.2459	2.2096	2.1743	2.1399	2.1065	3
4	3.1024	3.0373	2.9745	2.9137	2.8550	2.7982	2.7432	2.6901	2.6486	2.5887	4
5	3.6959	3.6048	3.5172	3.4331	3.3522	3.2743	3.1993	3.1272	3.0576	2.9906	5
6	4.2305	4.1114	3.9976	3.8887	3.7845	3.6847	3.5892	3.4976	3.4098	3.3255	6
7	4.7122	4.5638	4.4226	4.2883	4.1604	4.0386	3.9224	3.8115	3.7057	3.6046	7
8	5.1461	4.9676	4.7988	4.6389	4.4873	4.3436	4.2072	4.0776	3.9544	3.8372	8
9	5.5370	5.3282	5.1317	4.9464	4.7716	4.6065	4.4506	4.3030	4.1633	4.0310	9
10	5.8892	5.6502	5.4262	5.2161	5.0188	4.8332	4.6586	4.4941	4.3389	4.1925	10
11	6.2065	5.9377	5.6869	5.4527	5.2337	5.0286	4.8364	4.6560	4.4865	4.3271	11
12	6.4924	6.1944	5.9176	5.6603	5.4206	5.1971	4.9884	4.7932	4.6105	4.4392	12
13	6.7499	6.4235	6.1218	5.8424	5.5931	5.3423	5.1183	4.9095	4.7147	4.5327	13
14	6.9819	6.6282	6.3025	6.0021	5.7245	5.4675	5.2293	5.0081	4.8023	4.6106	14
15	7.1909	6.8109	6.4624	6.1422	5.8474	5.5755	5.3242	5.0916	4.8759	4.6755	15
16	7.3792	6.9740	6.6039	6.2651	5.9542	5.6685	5.4053	5.1624	4.9377	4.7296	16
17	7.5488	7.1196	6.7291	6.3729	6.0472	5.7487	5.4746	5.2223	4.9897	4.7746	17
18	7.7016	7.2497	6.8399	6.4674	6.1280	5.8178	5.5339	5.2732	5.0333	4.8122	18
19	7.8393	7.3658	6.9380	6.5504	6.1982	5.8575	5.5845	5.3162	5.0700	4.8435	19
20	7.9633	7.4694	7.0248	6.6231	6.2593	5.9288	5.6278	5.3527	5.1009	4.8696	20
21	8.0751	7.5620	7.1016	6.6870	6.3125	5.9731	5.6648	5.3837	5.1268	4.8913	21
22	8.1757	7.6446	7.1695	6.7429	6.3587	6.0113	5.6964	5.4099	5.1486	4.9094	22
23	8.2664	7.7184	7.2297	6.7921	6.3988	6.0442	5.7234	5.4321	5.1668	4.9245	23
24	8.3481	7.7843	7.2829	6.8351	6.4338	6.0726	5.7465	5.4509	5.1822	4.9371	24
25	8.4217	7.8431	7.3300	6.8729	6.4641	6.0971	5.7662	5.4669	5.1951	4.9476	25

continued

n	21%	22%	23%	24%	25%	26%	27%	28%	29%	30%	n
1	0.8264	0.8197	0.8130	0.8065	0.8000	0.7937	0.7874	0.7813	0.7752	0.7692	1
2	1.5095	1.4915	1.4740	1.4568	1.4400	1.4235	1.4074	1.3916	1.3761	1.3609	2
3	2.0739	2.0422	2.0114	1.9813	1.9520	1.9234	1.8956	1.8684	1.8420	1.8161	3
4	2.5404	2.4936	2.4483	2.4043	2.3616	2.3202	2.2800	2.2410	2.2031	2.1662	4
5	2.9260	2.8636	2.8035	2.7454	2.6893	2.6351	2.5827	2.5320	2.4830	2.4356	5
6	3.2446	3.1669	3.0923	3.0205	2.9514	2.8850	2.8210	2.7594	2.7000	2.6427	6
7	3.5079	3.4155	3.3270	3.2423	3.1611	3.0833	3.0087	2.9370	2.8682	2.8021	7
8	3.7256	3.6193	3.5179	3.4212	3.3289	3.2407	3.1564	3.0758	2.9986	2.9247	8
9	3.9054	3.7863	3.6731	3.5655	3.4631	3.3657	3.2728	3.1842	3.0997	3.0915	9
10	4.0541	3.9232	3.7993	3.6819	3.5705	3.4648	3.3644	3.2689	3.1781	3.1090	10
11	4.1769	4.0354	3.9018	3.7757	3.6564	3.5435	3.4365	3.3351	3.2388	3.1473	11
12	4.2785	4.1274	3.9852	3.8514	3.7251	3.6060	3.4933	3.3868	3.2859	3.1903	12
13	4.3624	4.2028	4.0530	3.9124	3.7801	3.6555	3.6381	3.4272	3.3224	3.2233	13
14	4.4317	4.2646	4.1082	3.9616	3.8241	3.6949	3.5733	3.4587	3.3507	3.2487	14
15	4.4890	4.3152	4.1530	4.0013	3.8593	3.7261	3.6010	3.4834	3.3726	3.2682	15
16	4.5364	4.3567	4.1894	4.0333	3.8874	3.7509	3.6228	3.5026	3.3896	3.2832	16
17	4.5755	4.3908	4.2190	4.0591	3.9099	3.7705	3.6400	3.5177	3.4028	3.2948	17
18	4.6079	4.4187	4.2431	4.0799	3.9279	3.7861	3.6536	3.5294	3.4130	3.3037	18
19	4.6346	4.4415	4.2627	4.0967	3.9424	3.7985	3.6642	3.5386	3.4210	3.3105	19
20	4.6567	4.4603	4.2786	4.1103	3.9539	3.8083	3.6726	3.5458	3.4271	3.3158	20
21	4.6750	4.4756	4.2916	4.1212	3.9631	3.8161	3.6792	3.5514	3.4319	3.3198	21
22	4.6900	4.4882	4.3021	4.1300	3.9705	3.8223	3.6844	3.5558	3.4356	3.3230	22
23	4.7025	4.4985	4.3106	4.1371	3.9764	3.8273	3.6885	3.5592	3.4384	3.3254	23
24	4.7128	4.5070	4.3176	4.1428	3.9811	3.8312	3.6918	3.5619	3.4406	3.3272	24
25	4.7213	4.5139	4.3232	4.1474	3.9849	3.8342	3.6943	3.5640	3.4423	3.3286	25

n	31%	32%	33%	34%	35%	36%	37%	38%	39%	40%	n
1	0.7634	0.7576	0.7519	0.7463	0.7407	0.7353	0.7299	0.7246	0.7194	0.7143	1
2	1.3461	1.3315	1.3172	1.3032	1.2894	1.2760	1.2627	1.2497	1.2370	1.2245	2
3	1.7909	1.7663	1.7423	1.7188	1.6959	1.6735	1.6516	1.6302	1.6093	1.5889	3
4	2.1305	2.0957	2.0618	2.0290	1.9969	1.9658	1.9355	1.9060	1.8772	1.8492	4
5	2.3897	2.3452	2.3021	2.2604	2.2200	2.1807	2.1427	2.1058	2.0699	1.9352	5
6	2.5875	2.5342	2.4828	2.4331	2.3852	2.3388	2.2936	2.2506	2.2086	2.1680	6
7	2.7386	2.6775	2.6187	2.5620	2.5075	2.4550	2.4043	2.3555	2.3083	2.2628	7
8	2.8539	2.7860	2.7208	2.6582	2.5982	2.5404	2.4849	2.4315	2.3801	2.3306	8
9	2.9419	2.8681	2.7976	2.7300	2.6653	2.6033	2.5437	2.4866	2.4317	2.3790	9
10	3.0091	2.9304	2.8553	2.7836	2.7150	2.6495	2.5867	2.5265	2.4689	2.4136	10
11	3.0604	2.9776	2.8987	2.8236	2.7519	2.6834	2.6180	2.5555	2.4956	2.4383	11
12	3.0995	3.0133	2.9314	2.8534	2.7792	2.7084	2.6409	2.5764	2.5148	2.4559	12
13	3.1294	3.0404	2.9559	2.8757	2.7994	2.7268	2.6576	2.5916	2.5286	2.4685	13
14	3.1522	3.0609	2.9744	2.8923	2.8144	2.7403	2.6698	2.6026	2.5386	2.4775	14
15	3.1696	3.0764	2.9883	2.9047	2.8255	2.7502	2.6787	2.6106	2.5457	2.4839	15
16	3.1829	3.0882	2.9987	2.9140	2.8337	2.7575	2.6852	2.6164	2.5509	2.4885	16
17	3.1931	3.0971	3.0065	2.9209	2.8398	2.7629	2.6899	2.6202	2.5546	2.4918	17
18	3.2008	3.1039	3.0124	2.9260	2.8443	2.7668	2.6934	2.6236	2.5573	2.4941	18
19	3.2067	3.1090	3.0169	2.9299	2.8476	2.7697	2.6959	2.6288	2.5592	2.4958	19
20	3.2112	3.1129	3.0202	2.9327	2.8501	2.7718	2.6977	2.6274	2.5606	2.4970	20
21	3.2154	3.1158	3.0227	2.9349	2.8519	2.7734	2.6991	2.6285	2.5616	2.4979	21
22	3.2173	3.1180	3.0246	2.9365	2.8533	2.7746	2.7000	2.6294	2.5623	2.4985	22
23	3.2193	3.1197	3.0260	2.9377	2.8543	2.7754	2.7008	2.6300	2.5628	2.4989	23
24	3.2209	3.1210	3.0271	2.9386	2.8550	2.7760	2.7013	2.6304	2.5632	2.4992	24
25	3.2220	3.1220	3.0279	2.9392	2.8556	2.7765	2.7017	2.6307	2.5634	2.4994	25

Glossary

Acceptance credit: a facility to issue bank-guaranteed bills (**bank bills**) by a firm wanting to raise short-term finance. They can be sold on the money market, but are unrelated to specific trading transactions. The bank accepts the liability to exchange cash for bills when presented at the due date.

Acquirees: taken-over firms. Also, 'targets' or 'victims'.

Acquirers: firms that make takeovers. Also, 'predators'.

Adjusted NAV: the NAV as per the accounts, adjusted for any known or suspected deviations between book values and market, or realisable values.

Adjusted present value (APV): the basic NPV of an activity adjusted for 'bolt-on extras' like financing costs and benefits, e.g. the tax shield, costs of issuing new finance.

Agency costs: costs that owners (principals) have to incur in order to ensure that their agents (managers) make financial decisions consistent with their best interests.

Aggressive stocks: generate returns that vary by a larger proportion than overall market returns. Their Betas exceed 1.0.

Alternative Investment Market (AIM): where smaller, younger companies can acquire a stock market listing.

American options: can be exercised at any time up to the maturity date.

Amortisation: repayment of debt by a series of instalments. Also used as a term for depreciation of intangible assets.

Annual percentage rate (APR): the true annualised cost of finance.

Annuity: a finite series of cash flows.

Arbitrage: the profitable exploitation of divergences between the prices of goods (or between interest rates), that violate the Law of One Price. Also applied in MM's capital structure analysis to refer to the process of equalising the values of geared and ungeared firms. Hence, **arbitageur**.

The **Arbitrage Pricing Model (APT):** an extension of the CAPM to include more than one factor (hence, an example of a **multi-factor model**) used to explain the returns on securities. Each factor has its own Beta coefficient.

Articles of Association: a document drawn up at the formation of an enterprise, detailing the rights and obligations of shareholders and directors.

Asset or **activity Beta:** the inherent systematic riskiness of a firm's operations, before allowing for gearing. Also known as **firm Beta**, **company Beta**, or **ungeared Beta**.

Asset-backed securities are bonds issued on the security of a stream of highly reliable income flows, e.g. mortgage payments to a bank, out of which interest payments are made.

Asset stripping: selling off the assets of a taken-over firm, often in order to recoup the initial outlay.

Asymmetric information: one party to a contract is in possession of more information than the other.

Balloon repayment: most of the loan repayment is made on maturity.

Bancassurance: a term coined to denote the combination of banking and insurance business within the same organisation.

A **bank loan** is usually extended for a fixed term with a pre-agreed schedule of interest and capital repayments. Interest is usually payable on the initial amount borrowed, regardless of the falling balance as repayments are made.

Barter: the simplest form of counter-trade, involving direct exchange of goods with no money being exchanged.

Betas (or Beta coefficients) relate the responsiveness of the returns on individual securities to variations in the return on the overall market portfolio.

Beta geared: the Beta attaching to the ordinary shares of a geared firm. These bear a risk higher than the firm's basic activity.

Beta ungeared: the geared Beta stripped of the effect of gearing. Corresponds to the activity Beta in an equivalent ungeared firm.

Bilateral netting is operated by pairs of firms in the same group netting off their respective positions regarding payables and receivables.

Bill of Exchange: a promise to pay at a specific time, issued to suppliers by purchasers in exchange for goods. Bills may be held to maturity or sold at a discount on the money market if cash is required sooner.

Bill of Lading: a document that transfers title to exported goods to the bank that finances the deal when the goods are shipped.

Bird-in-the-hand fallacy: the mistaken belief that dividends paid early in the future are worth more than dividends expected in later time periods, simply because they are nearer in time and viewed as less risky.

Bonds: any form of borrowing that firms can undertake in the form of a medium- or long-term security, that commits them to specific repayment dates, at fixed or variable interest.

Bonus or **scrip issues:** issues of free shares to existing shareholders *in lieu* of, or in addition to, cash dividends. Reflected in lower reserves (hence the alternative label, **capitalisation issue**).

Book-to-market ratio: ratio of the book value of equity to the market value of the shares.

Break-up value (BUV): the value that can be obtained by selling off the firm's assets piecemeal to the highest bidders.

Bullet repayment: where a loan is repaid wholly at the maturity date.

Business angels: wealthy private investors who take equity stakes in small, high-risk firms.

The **Business Expansion Scheme** was established to enable investors to obtain tax relief when purchasing ordinary shares in unquoted firms seeking 'seed-corn' funds for development (now defunct).

Buy-back: a method of obtaining payment for building a manufacturing unit overseas by taking the future physical product of the plant in return.

Capital: strictly, the funds invested in a firm by shareholders when they purchase ordinary shares, but often used to indicate all forms of equity, and often to refer to any form of finance, whether equity or debt.

Capital allowances: tax allowances for capital expenditure.

Capital asset: any investment that offers a prospective return, with or without risk. However, in finance, the term is usually applied to securities and ordinary shares in particular.

Capital Asset Pricing Model (CAPM): a theory used to explain how efficient capital markets value securities, i.e. capital assets, by discounting future expected returns at risk-adjusted discount rates.

Capital gains tax is paid on realising an increase in share value. Capital gains are currently treated as income in the UK at the investor's marginal tax rate.

Capital gearing: the mixture of debt and equity in a firm's capital structure, which influences variations in shareholders' profits in response to sales and EBIT variations.

Capitalisation: the procedure of converting (by discounting) a series of future cash flows into a single capital sum.

Capitalisation rate: a discount rate used to convert a series of future cash flows into a single capital sum.

The **capital market line (CML)** traces out the efficient combinations of risk and return available to investors when combining a risk-free asset with the market portfolio.

Capital structure: the mixture of debt and equity resulting from decisions on financing operations.

Cash operating cycle: length of time between cash payment to suppliers and cash received from customers.

The **characteristics line (CL)** relates the periodic returns on a security to the returns on the market portfolio. Its slope is the Beta of the security. The regression model used to estimate Betas is called the **market model**.

Chartist: analyst who relies on charts of past share movements to predict future movements.

City Code: the non-statutory rules laid down by the Take-Over Panel to guide the conduct of participants in the take-over process.

A **classical tax system** initially taxes company profits, and then also taxes any dividend income. This double taxation of dividends thus provides an incentive to retain profits.

Clientèle effect: the notion that a firm attracts investors by establishing a set dividend policy that suits a particular group of investors.

Commercial paper: a short-term promissory note or IOU, issued by a highly credit-worthy corporate borrower to financial institutions and other cash-rich corporates.

co-movement or **co-variability:** the tendency for two variables, e.g. the returns from two investments, to move in parallel. It can be measured using either:

(i) the **correlation coefficient:** a relative measure of co-movement that locates assets on a scale between -1 and $+1$. Where returns move exactly in unison, perfect positive correlation exists, and where exactly opposite movements occur, perfect negative correlation exists. Most investments fall in between, mainly, with positive correlation.

(ii) the **covariance:** an absolute measure of co-movement with no upper or lower limits.

A **concentric acquisition** is undertaken to exploit synergies in marketing of two firms' products, without production economies.

Conglomerate takeover: the acquisition of a target firm in a field apparently unrelated to the acquiror's existing activities.

Contra-cyclical: a term applied to an investment whose returns fluctuate in opposite ways to general trends in business activity, i.e. contrary to the cycle.

Convertible loan stock: a debenture that can be converted into ordinary shares, often on attractive terms, usually at the option of the holder. Some preference shares are convertible.

Cost of debt: the yield a firm would have to offer if undertaking further borrowing at current market rates.

Cost of equity: the minimum rate of return a firm must offer owners to compensate for waiting for their returns, and also for bearing risk.

Counter-party risk: the risk that the opposite party to a contract defaults on its obligations.

Counter-trade: a form of trade involving reciprocal obligations with a trading partner, or counter-party, e.g. a commitment to buy from a firm or country that the firm sells to.

Country risk: the risk of adverse effects on the net cash flows of a MNC due to political and economic factors peculiar to the country of location of FDI.

Coupon rate of interest: the fixed rate of interest, as printed on the debt security, that a firm must pay to lenders.

Crest: an electronic mechanism for settling and registering shares sold on the London Stock Exchange.

Critical mass: the minimum size of firm thought necessary to compete effectively, e.g. to finance R&D.

Currency futures contract: a commitment to deliver a specific amount of foreign exchange at a specified future date at an agreed price incorporated in the contract. Contracts can be traded on an exchange in standard sizes.

Currency option: the right, but not the obligation, to buy or sell a fixed amount of currency at a predetermined rate at a specified future date.

Currency swap: a transfer of cash payment obligations denominated in foreign currencies. The two parties initially exchange the principal of their respective borrowings, plus the interest commitments in the currencies over an agreed period, and re-exchange the principal at the end of this period.

Currency switching: where a firm uses foreign exchange received in the course of operations to settle obligations to a third party, often located in a third country.

Current cost accounting (CCA) attempts to capture the effect of inflation on asset values (and liabilities) by recording them at their current replacement cost, i.e. the cost of obtaining an identical replacement.

Current ratio: ratio of current assets to current liabilities.

Debentures: in law, any form of borrowing that commits a firm to pay interest and repay capital. In practice, usually applied to long-term loans that are secured on a firm's assets.

The **debt capacity** of an investment or a whole firm is the maximum amount of debt finance, and hence interest payments that it can support without incurring financial distress.

Default: the failure by a borrower to adhere to a pre-agreed schedule of interest and/or capital payments on a loan.

A **defensive stock** generates returns that vary by a smaller proportion than overall market returns. Its Beta is less than one.

Derivative: financial instrument whose value derives from an underlying asset.

Discount rate: any percentage required return used to convert future expected cash flows into their equivalent present values.

Discounted cash flow: future cash flows adjusted for the time-value of money.

Disintermediation: the process whereby firms borrow and lend funds directly without going through a bank or other intermediary.

Diversifiable risk can be removed by efficient portfolio diversification.

Diversification: extension of a firm's activities into new and unrelated fields. Although this may generate cost savings, e.g. via shared distribution systems, as a by-product, the fundamental motive for diversification is to reduce exposure to fluctuations in economic activity.

Dividend irrelevance: the theory that, when firms have access to external finance, it is irrelevant to firm value whether they pay a dividend or not.

Dividend Valuation Model: a way of assessing the value of shares by capitalising the future dividends. With growing dividend payments, it becomes the **Dividend Growth Model**.

Dividend yield: gross dividend per ordinary share (including both interim and final payments) divided by current share price.

Double tax agreements (DTAs): reciprocal arrangements between countries whereby tax paid in one location is credited in the second, thus avoiding doubling up the firm's tax bill. Hence, Double Tax Relief (DTR).

Earnings before interest, tax, depreciation & amortisation (EBITDA): a rough measure of operating cash flow, effectively, operating profit with depreciation added back. It differs from the 'Net Cash Inflow from Operating Activities' shown in cash flow statements due to working capital movements.

Earnings dilution: the dampening effect on EPS of issuing further shares at a discount as in a rights issue.

Earnings yield: EPS divided by current share price. Sometimes, it refers to expected or 'prospective' EPS, becoming the '**prospective earnings yield**'. It is a simple way of expressing the investor's Return on Investment on the share.

EBIT: earnings (i.e. profits) before interest and taxation.

Economic order quantity (EOQ): the most economic quantity to be ordered that minimises holding and ordering costs.

Economic value added (EVA): post-tax accounting profit generated by a firm reduced by a charge for using the equity (usually, cost of equity times book value of equity).

The **efficient frontier** traces out all the available portfolio combinations that either minimise risk for a stated expected return or maximise expected return for a specified measure of risk.

Efficient markets: where current share prices fully reflect the information available.

Enhanced scrip dividends: scrip alternatives offered to investors that are worth more than the alternative cash payment.

The **Enterprise Investment Scheme** replaced the BES in 1994, incorporating less generous tax breaks.

Enterprise value: the value of the whole firm.

Entrepreneurial companies: are driven by the growth ambitions and desire of the owners to create significant wealth.

Equity, or **equity value:** the value of the owners' stake in a firm, however calculated.

The **equity Beta** indicates the systematic riskiness attaching to the returns on ordinary shares. It equates to the asset Beta for an ungeared firm, or is adjusted upwards to reflect the extra riskiness of shares in a geared firm, to become '**Beta geared**'.

Equivalent loan: the loan that would involve the same schedule of interest and loan repayments as the profile of rentals required by an equipment lessor.

Equivalent risk class: a concept used by MM to include all firms subject to the same business risks (i.e. all having the same Activity Betas).

Eurobonds(or international bonds): securities issued by borrowers in a market outside that of their domestic currency.

European options: can only be exercised at the specified maturity date.

Exchange agio: the percentage difference between the spot and forward rates of exchange between two currencies.

A share is quoted **ex-dividend (ex-div.** or **xd)** when subsequent purchasers no longer qualify for the forthcoming dividend payment. Until this point, the shares are quoted **cum-dividend**.

Exercise (strike) price: the price at which the option to buy or sell can be transacted.

Expectations (or unbiased forward predictor) theory: the postulate that the expected change in the spot rate of exchange is equal to the difference between the current spot rate and the current forward rate for the relevant period.

Externalisation: the transfer of key functions and expertise to an overseas strategic partner.

Factoring: a means of obtaining faster cash inflow, and thus increased funds. A firm appoints the factor to collect outstanding accounts payable and to administer debtors' accounts. It also lends money to the client based on the value of the firm's sales.

Finance lease: a method of acquiring an asset that involves a series of rental payments extending over the whole expected life-time of the asset.

Financial distress: in narrow terms, the difficulty that a firm encounters in meeting obligations to creditors. More broadly, it refers to the adverse consequences, e.g. restrictions on behaviour that result, usually from excessive borrowing by a firm.

Financial gearing includes both capital gearing and income gearing.

Financial intermediaries: specialist financial institutions which collect funds from savers and lend to corporate and other borrowers.

Financial Services Authority (FSA): a regulatory body for maintaining confidence in the financial markets.

A **fixed charge** applies when a lender can force the sale of pre-specified company's assets in order to recover debts in the event of default on interest and/or capital payments.

The **flat yield (or running yield)** on a bond is the ratio of the fixed interest payment to the current market price of the bond.

A **floating charge** applies when a lender can force the sale of any (i.e. unspecified) of a company's assets in order to recover debts in the event of default on interest and/or capital payments. (Ranks behind a fixed charge)

Floating rate note (FRN): a bond issue where interest is paid at a variable rate (often a Eurobond).

Foreign bonds: Loan stock issued on the domestic market by non-resident firms or organisations. In London, called 'bulldogs', in New York 'Yankees'.

Foreign currency swap: a way of extending the delivery date incorporated in a forward contract. A spot/forward swap involves completing the original contract by a spot transaction and entering a new forward contract for the additional of time.

Foreign direct investment (FDI): investment in fixed assets located abroad for operating distribution and/or production facilities.

Foreign exchange exposure: the risk of loss stemming from exposure to adverse foreign exchange rate movements.

Forfaiting: the practice whereby a bank purchases an exporter's sales invoices or promissory notes, that usually carry the guarantee of the importer's bank.

Forward contract: a legal obligation to deliver a specified amount of currency at some specified future date. The rate of exchange is fixed at the date of the contract.

Forward option: a forward currency contract that incorporates a flexible settlement date between two fixed dates.

Forward rate of exchange: the rate fixed for transactions that involve delivery and settlement at some specified future date.

Free cash flow (FCF): a firm's cash flow free of obligatory payments. Strictly, it is cash flow after interest, tax and replacement investment, although it is measured in many other ways in practice, e.g. after all investment.

Fundamental analysis: estimation of the 'true' value of a share based on expected future returns.

FX: abbreviation for foreign exchange.

Generally Accepted Accounting Principles (GAAP): the set of legal regulations and accounting standards that dictate 'best practice' in constructing company accounts.

Global companies serve a range of overseas markets both by exporting and direct investment.

Going concern value (GCV): the value of the assets as stated in the accounts that assume that the firm will continue as a viable entity as it stands, i.e. as an ongoing activity.

Hedging: attempting to minimise the risk of loss stemming from exposure to adverse foreign exchange rate movements.

Hire purchase: a means of obtaining the use of an asset before payment is completed. An HP contract involves an initial, or 'down payment', followed by a series of hire charges at the end of which ownership passes to the user.

Home-made dividends: cash released when an investor realises part of his/her investment in a firm in order to supplement his/her income.

Home-made gearing: personal borrowing undertaken in the process of arbitraging between the ordinary shares of geared and ungeared firms.

Horizontal integration: the acquisition of a competitor in pursuit of market power and/or scale economies.

Hybrid: a security that embodies features of both equity and debt, and is thus difficult to classify under either category.

Imputation systems of taxation offer shareholders tax credits (fully or partially) in respect of company tax already paid when assessing their income tax liability on dividends paid out.

Income gearing: the proportion of EBIT pre-empted by prior interest commitments, i.e. the inverse of interest cover.

The **incremental hypothesis** suggests that firms tend to gradually build their degree of involvement in foreign markets, beginning with exporting and culminating in FDI.

Information asymmetry: the imbalance in access to information about a firm's affairs as between directors and owners.

Information content: the extra, unstated intelligence that investors deduce from the formal announcement by a firm of any financial news, i.e. what people read 'between the lines', or 'financial body language'.

Initial public offering (IPO): the first issue of shares by an existing or a newly-formed firm to the general public.

Insider trading: dealing in shares using information not publicly available.

Interest agio: the percentage difference between interest rates prevailing in the money markets for lending/borrowing in two currencies.

Interest cover: the number of times the profit before interest exceeds loan interest.

Interest rate parity (IRP) asserts that the difference between the spot and forward exchanges is equal to the differential between interest rates prevailing in the money markets for lending/borrowing in the respective currencies.

Internal rate of return the discount rate that equates the present value of future cash flows with initial investment cost.

Internalisation: the retention by the MNC of key management functions and technology.

The **International** (or **Open**) **Fisher Theory:** the notion that, because real rates of interest are equalised throughout the world, given freedom of capital mobility, any observed differences in nominal rates between different locations must be due to different expectations of inflation between those locations.

Invoice discounting: a service less comprehensive than factoring, involving the sale of approved invoices to a financial institution.

Joint venture: a strategic alliance involving the formal establishment of a new marketing and/or production operation involving two or more partners.

Junk bonds: low-quality, risky bonds with no credit rating.

Lagging: settling as late as possible a payable (receivable) denominated in a currency expected to weaken (strengthen).

Law of One Price: the proposition that any good or service will sell for the same price, adjusting for the relevant exchange rate, throughout the world.

Leading: advancing before the due date a payable denominated in a foreign currency that is expected to strengthen, or advancing a receivable in a currency expected to weaken.

Letter of credit: a credit drawn up by an importer in favour of an exporter. It is endorsed by a bank that guarantees payment provided the beneficiary delivers the Bill of Lading proving that goods have been shipped.

Licensing involves the assignment of production and selling rights to producers located in foreign locations in return for royalty payments.

Listed companies: firms whose shares are quoted on the Main List of the Stock Exchange.

Loan Guarantee Scheme: a facility whereby banks are able to lend to firms that would not otherwise qualify for bank finance due to lack of track record, the loan being guaranteed by the Department of Trade and Industry.

Main list: daily list of securities and prices traded on the London Stock Exchange.

Management buy-in (MBI): acquisition of an equity stake in an existing firm by new management that injects expertise as well as capital into the enterprise.

Management buy-out (MBO): acquisition of an existing firm by its existing management usually involving substantial amounts of straight debt and mezzanine finance.

Marginal efficiency of investment (MEI): a schedule listing available investments, in declining order of attractiveness.

Market capitalisation: the market value of a firm's equity, i.e. number of ordinary shares issued times market price.

The **market portfolio** includes all securities traded on the stock market weighted by their respective capitalisations. Usually, a more limited portfolio such as the FT All Share Index is used as a proxy.

Matching: offsetting a currency inflow in one currency, e.g. a stream of revenues, by a corresponding stream of costs, thus leaving only the profit element unmatched. Firms may also match operating cash flows against financial flows, e.g. a stream of interest and capital payments resulting from overseas borrowing in the same currency.

Mergers: pooling by firms of their separate interests into newly-constituted business, each party participating on roughly equal terms.

Mezzanine finance covers hybrids such as convertibles that embody both debt and equity features.

Modified internal rate of return (MIRR): the internal rate of return modified for the reinvestment assumption.

Modigliani & Miller's (MM) Capital Structure Theories are:

(i) MM-no tax, which 'proves' that no optimal capital structure exists, and that the WACC is invariant to debt/equity ratio.

(ii) MM-with tax which suggests that the tax shield should be exploited up to the point of almost 100 per cent debt financing.

Monetary Policy Committee: a body whose members are appointed by the Bank of England, responsible for setting UK interest rates at monthly meetings.

Money market cover: involves an exporter borrowing on the money market (i.e. creating a liability) in the same currency in which it expects to receive a payment.

Moral hazard: the temptation facing managers to engage in risky activities when they are protected from the consequences of failure, e.g. by guaranteed severance payments.

Multilateral netting: a central Treasury department operation to minimise net flows of currency throughout an organisation.

Multi-national company (MNC): one that conducts a significant proportion of its operations abroad.

Natural hedge: where the adverse impact of FX rate variations on cash inflows are offset by the effect on cash outflows, or *vice versa*.

Net advantage of a lease (NAL): the NPV of the acquisition of an asset adjusted for financing benefits.

Net asset value (NAV): the value of owners' stake in a firm, found by deducting total liabilities (i.e. debts) from total assets.

Net debt: a firm's net borrowing including both long-term and also short-term debt, offset by cash holdings. Expressed either in absolute terms, or in relation to owner's equity.

Net present value: the value of a stream of cash flows adjusted for the time-value of money. A positive NPV adds value.

Netting: offsetting a firm's internal currency inflows and outflows in the same currency to minimise the net flow in either direction.

Neutral stocks generate returns that vary by the same proportion as overall market returns. Their Betas equal 1.0. Also called 'market-tracking' investments.

New Issue Market: the market for selling and buying newly-issued securities. It has no physical existence.

Niche companies serve a limited segment of their markets, usually offering high-quality, differentiated products at a high margin.

Non-recourse (as distinct from recourse) factoring operates where factors are unable to reclaim bad debts from a client's accounts.

Operating gearing is the importance of fixed expenses within a firm's overall cost structure. It can be measured in various ways, for example, by looking at the responsiveness of operating profit to sales variations.

Operating gearing factor: a ratio that compares the operating gearing of a particular activity, e.g. a product division within a larger firm to that of a larger entity such as the whole firm.

Operating/strategic exposure: the risk that adverse foreign exchange rate movements will affect the present value of the firm's future cash flows (effectively, long-term transactions exposure).

Operating lease: a method of hiring assets over periods less than the expected lifetime of those assets.

Opportunity cost: the value forgone by opting for a particular course of action.

Optimal capital structure: the financing mix that minimises the overall cost of finance and maximises market value.

Optimal portfolio: the one chosen by an investor to achieve his/her most desired combination of risk and return. This choice depends on the investor's attitude to risk, or risk–return preference, i.e. how he/she rates different combinations of risk and return. If a risk-free asset is available, the optimal portfolio of risky assets is the market portfolio.

Option: the right but not the obligation to buy or sell something at some time in the future at a given price.

Overdraft: short-term finance extended by banks subject to instant recall. A maximum deficit balance is pre-agreed and interest is paid on the actual daily balance outstanding.

Overtrading: where a firm has insufficient long-term capital to finance business growth.

Owner's equity: in accounting terms, simply the NAV, but can also be expressed in market value terms, i.e. share price times number of ordinary shares issued, or 'capitalisation'.

Parallel matching applies where a firm offsets inflows in one currency with outflows denominated in a closely correlated currency.

Perpetuity: an infinite series of cash flows.

Poison pill: a provision designed to damage the interests of a takeover bidder, e.g. handsome severance terms for departing managers, activated on completion of the bid.

Political risk: the risk of politically-motivated interference by a foreign government in the affairs of a MNC, that adversely affects its net cash flows.

Portfolio: a combination of investments – securities or physical assets – into a single 'bundled' investment. A well-diversified portfolio has the potential capacity to lower the investor's exposure to the risk of fluctuations in the overall economy.

Portfolio effect: the tendency for the risk on a well-diversified holding of investments to fall below the risk of most and sometimes, all of its individual components.

Portfolio investment: investment in paper claims such as ordinary shares, without obtaining a voice in management.

Post-completion audit: audit of a capital project at an agreed time following implementation.

Preference shares: hybrid securities that rank ahead of ordinary shares for dividend payment, usually at a fixed rate, and also in distributing the proceeds of a liquidation. Normally, they carry no voting rights.

Price: Earnings ratio (PER): the current share price divided by the latest reported earnings (i.e. profits after tax) per share.

Profitability index: ratio of the present value of benefits to costs.

Project risk factor: the product of the Revenue Sensitivity Factor and the Operating Gearing Factor multiplied together.

Proprietorial companies are run by founders and their heirs to provide a livelihood for their families. They usually have limited growth aims.

Provision: a notional deduction from profits to allow for some highly likely future financial contingency. In accounting terms, an appropriation of profit after taxation.

A proxy Beta is used when the firm has no market listing and thus no Beta of its own. It is taken from a comparable listed firm, and adjusted as necessary for relative financial gearing levels. Hence, **proxy discount rate**.

Purchasing power parity (PPP): the theory that foreign exchange rates are in equilibrium when a currency can purchase the same amount of goods at the prevailing exchange rate.

Random walk theory: share price movements are independent of each other so that tomorrow's share price cannot be predicted by looking at today's.

Real assets: assets in the business (tangible or intangible).

Real options: capital investment options rather than financial options.

Record day: the cut-off date beyond which further entrants to the shareholder register do not qualify for the next dividend.

Relevant risk: the component of total risk taken into account by the stock market when assessing the appropriate risk premium for determining capital asset values.

Reserves: the funds that shareholders invest in a firm in addition to their initial subscription of capital.

The **residual theory of dividends** asserts that firms should only pay cash dividends when they have financed new investments. It assumes no access to external finance.

Retained earnings: reserves represented by retention of profits. Sometimes, labelled 'profit & loss account' on the balance sheet. Also called **revenue reserves**.

Revenue sensitivity: the extent to which revenue of an activity varies in response to general economic fluctuations.

Revenue sensitivity factor: the revenue sensitivity of a particular activity, e.g. a product division, relative to that of a larger entity, such as the whole firm.

A **revolving credit facility** enables a firm to borrow up to a pre-specified amount usually over 1–5 years. As repayments of outstanding balances are made, the loan facility is replenished.

Rights issues: sales of further ordinary shares at less than market price to existing shareholders who are usually able to sell the rights on the market should they not wish to purchase additional shares.

Risk-free assets: securities with zero variation in overall returns.

Risk premium: the additional return demanded by investors above the risk-free rate to compensate for exposure to systematic risk.

Scale economies: cost efficiencies, e.g. bulk-buying, due to increasing a firm's size of operation.

A **scrip dividend** is offered to investors in lieu of the equivalent cash payment. Also called a **scrip alternative**.

SEAQ: a computer-based quotation system on the London Stock Exchange where market-makers report bid and offer prices and trading volumes.

Securitisation: the technique of packaging non-tradable claims into a traded security backed by an asset such as a flow of low risk income payments.

Security market line (SML): an upward-sloping relationship tracing out all combinations of expected return and systematic risk, available in an efficient market. All traded securities locate on this schedule. In effect, the capital market line adjusted for systematic risk.

Sensitivity analysis: analysis of the impact of changes in assumptions on investment returns.

Share buy-back: repurchase by a firm of its existing shares, either via the market or by a tender to all shareholders.

Shareholder value analysis (SVA): a way of assessing the inherent value of the equity in a company, taking into account the sources of value creation and the time horizon over which the firm enjoys competitive advantages over its rivals.

Share premium account: a reserve set up to account for the issue of new shares at a price above their par value.

Share splits (USA: **stock splits**): a way of reducing the share price of 'heavyweight' shares (prices above £10). Achieved by reducing the par value of issued shares, e.g. two shares of par value 50p to replace one share at £1 is a one-for-one split, halving the share price.

Short selling: selling securities not yet owned in the expectation of being able to buy them later at a lower price.

Signalling: using financial announcements to deliver more information than is actually spelt out in detail.

Specific risk: the variability in the return on a security due to exposure to risks relating to that security in isolation, e.g. risk of losing market share due to poor marketing decisions.

Spot rate: the rate of exchange quoted for transactions involving immediate settlement. Hence, **spot market**.

Spread: the difference between the exchange rates (interest rates) at which banks buy and sell foreign exchange (lend and borrow).

Straight, or **plain vanilla, debt:** fixed rate borrowing with no additional features such as convertibility rights or warrants.

Sunk cost: a cost already incurred, or committed to.

Synergies: gains in revenues or cost savings resulting from takeovers and mergers, not resulting from firm size, i.e. stemming from a 'natural match' between two sets of assets.

Systematic risk: variability in a security's return due to exposure to risks affecting all firms traded in the market (hence, **market risk**), e.g. the impact of exchange rate changes.

Takeover: acquisition of the share capital of another firm, resulting in its identity being absorbed into that of the acquiror.

Takeover Panel: a non-statutory body set up by, and with the participation of, leading financial organisations to oversee the conduct of takeover bids.

Tax breaks: tax concessions, e.g. relief of interest payments against profits tax.

Tax credit: see **imputation system**.

Tax shield: a method of sheltering profits from corporation tax. It is measured by the discounted value of future tax savings generated by the available tax reliefs.

Theoretical ex-rights price (TERP): the market price to which the ordinary shares should gravitate following the completion of a rights issue.

Time-value of money: the notion that money received in the future is worth less than the same amount received today.

Total shareholder return (TSR): the overall return enjoyed by investors, including dividend and capital appreciation, expressed as a percentage of their initial investment. Related to individual years, or to a lengthier time period, and then converted into an annualised, or equivalent annual return.

Trade credit: temporary financing extended by suppliers of goods and services pending the customer's settlement.

Traditional theory of capital structure: the theory that an optimal capital structure exists, where the WACC is minimised and market value is maximised.

Transaction exposure: the risk of loss due to adverse foreign exchange rate movements that affect the home currency value of import and export contracts denominated in a foreign currency.

A **transfer price:** the cost applied to goods transferred between operating units owned by the same firm.

Translation exposure: the risk of loss from adverse foreign exchange movements that affect sterling values of balance sheet items held overseas and past transactions in foreign currency.

Treasury Bills: short-dated (up to three months) securities issued by the Bank of England on behalf of the UK government to cover short-term financing needs.

Unit Trust: investment business attracting funds from investors by issuing units of shares or bonds to invest in.

Value-based management: a managerial approach where the whole aim, strategies and actions are linked to shareholder value creation.

Venture capital: finance, usually equity, offered by specialist merchant banks wanting to take a stake in firms with high growth potential, but involving a high risk of loss.

Vertical integration: extension of a firm's activities further back, or forward, along the supply chain from existing activities.

Warrants: options to buy ordinary shares at a predetermined 'exercise price'. Usually attached to issues of loan stock.

Weighted average cost of capital (WACC): the overall return a firm must achieve in order to meet the requirements of all its investors.

White knight: a takeover bidder emerging after a hostile bid has been made, usually offering alternative bid terms that are more favourable to the defending management.

Working capital: current assets less current liabilities.

Yield: income from a security as a percentage of market price.

Yield curve: a graph of the relationship between the yield on bonds and their current length of time to maturity.

Z-score: a mathematically-derived critical value below which firms are associated with failure.

Zero coupon bond: a bond that does not pay interest but is issued at a discount and redeemed at par (full) value.

References

3i *Making an Acquisition.*

3i (1993), *Dividend Policy,* April.

Ainley, M., A. Mashayekhi, R. Hicks, Rahman and A. Ravalia (2007) *Islamic Finance in the United Kingdom: Regulation and Challenges* (London: Financial Services Authority).

Alkaraan, F. and D. Northcott (2006) 'Strategic capital investment decision-making: A role for emergent analysis tools? A study of practice in large UK manufacturing companies', *British Accounting Review,* Vol. 38, No. 12, [June], pp. 149–173.

Altman, E.I. (1968) 'Financial Ratios, Discriminant Analysis and the Prediction of Corporate Bankruptcy', *Journal of Finance,* Vol. 23, No. 4, [September], pp. 589–609.

Andersen, J.A. (1987) *Currency and Interest Rate Hedging* (Prentice Hall).

Andrade, G. and S.N. Kaplan (1998) 'How Costly is Financial (not Economic) Distress? Evidence from Highly-Leveraged Transactions that became Distressed', *Journal of Finance,* Vol. 53, [October], pp. 1443–1493.

Andrews, G.S. and C. Firer (1987) 'Why Different Divisions Require Different Hurdle Rates', *Long Range Planning,* Vol. 20, No. 5, [October], pp. 62–68.

Ang, J. and P.P. Peterson (1984) 'The Leasing Puzzle', *Journal of Finance,* Vol. 39, No. 4, pp. 1055–1065.

Angwin, D. (2000) *Managing Successful Post-Acquisition Integration* (FT/Prentice Hall).

Angwin, D. (2004) 'Speed in M&A integration: The first 100 days', *European Management Journal,* Vol. 22, No. 4, pp. 418–430.

Antill, N. and K. Lee (2005) *Company Valuation under IFRSs* (Harrison House Publishing).

Arnold, G. and P. Hatzopoulos (2000) 'The theory-practice gap in capital budgeting: evidence from the United Kingdom', *Journal of Business Finance & Accounting,* Vol. 27, Issue 5/6, [June], pp. 603–626.

Arzac, E. (2004) *Valuation for Mergers, Buyouts and Restructuring* (Wiley).

Ashkenas, R., L.J. De Monaco and S.C. Francis (1998) 'Making the Deal Real: How GE Capital Integrates Acquisitions', *Harvard Business Review,* Vol. 76, No. 1, [January–February], pp. 165–178.

Ashton, D. and D. Acker (2003) 'Establishing Bounds on the Tax Advantage of Debt', *British Accounting Review* Vol. 35, No. 4, [December], pp. 385–399.

Asquith, P. and D. Mullins (1986) 'Signalling with dividends, stock purchases and equity issues', *Financial Management,* Vol. 15, Issue 3, [Autumn], pp. 27–44.

Ball, J. (1991) 'Short Termism—Myth or Reality', *National Westminster Bank Quarterly Review,* August.

Ball, R. and P. Brown (1968) 'An empirical evaluation of accounting income numbers', *Journal of Accounting Research,* Vol. 6, No. 2, [Autumn], pp. 159–178.

Bank for International Settlements (2007) 'Triennial Central Bank Survey of Foreign Exchange and Derivatives Market Activity in April 2007' (www.bis.org).

Bank of England (1988) 'Share Repurchase by Quoted Companies', *Quarterly Bulletin,* Vol. 28, No. 3, [August], pp. 382–390.

Barber, B. and T. Odean (2000) 'Trading is Hazardous to Your Wealth: The Common Stock Investment Performance of Individual Investors', *Journal of Finance,* Vol. 55, No. 2, pp. 773–806.

Barclay, M.J., C.W. Smith and R.L. Watts (1995) 'The Determinants of Corporate Leverage and Dividend Policies', *Journal of Applied Corporate Finance,* Vol. 7, No. 4, [Winter], pp. 4–19.

Barclay, M. and C. Smith, (2006) 'The capital structure puzzle: Another look at the evidence', in Rutterford, J., Upton, M. and Kodwani, D. (eds) *Financial Strategy,* 2nd edn (Wiley).

Bartram, S., G. Brown and F. Fehle (2006) 'International evidence of financial derivative usage', *Social Science Research Network,* October.

Barwise, P., P. Marsh, and R. Wensley (1989) 'Must Finance and Strategy Clash?' *Harvard Business Review,* Vol. 67, No. 5, [September–October], pp. 85–90.

Baumol, W. (1952) 'The Transactions Demand for Cash: An Inventory Theoretic Approach', *Quarterly Journal of Economics,* Vol. 66, No. 4, [November], pp. 545–556.

Beenstock, M. and K. Chan (1986) 'Testing the Arbitrage Pricing Theory in the UK', *Oxford Bulletin of Economics and Statistics,* Vol. 48, No. 2, [May], pp. 121–141.

Benito and Young (2001) *Hard Times or Great Expectations: Dividend Omissions and Dividend Cuts by UK Firms,* Bank of England Working Paper 147.

Bennet, N. (2001) 'One day we will go out of business', *Sunday Telegraph,* 9 December.

Bhattacharya, S. (1979) 'Imperfect Information, Dividend Policy and the Bird-in-the-Hand Fallacy', *Bell Journal of*

Economics and Management Science, Vol. 10, No. 1, [Spring], pp. 259–270.

Bierman Jr, H. and J.E. Hass (1973) 'Capital Budgeting under Uncertainty: A Reformulation', *Journal of Finance,* Vol. 28, No. 1, [March], pp. 1119–1129.

Black, F. and M. Scholes (1973) 'The Pricing of Options and Corporate Liabilities', *Journal of Political Economy,* Vol. 81, No. 3, [May–June], pp. 637–654.

Black, F., M.C. Jensen and M. Scholes (1972) 'The Capital Asset Pricing Model: Some Empirical Tests', in M. Jensen (ed.) *Studies in the Theory of Capital Markets* (Praeger).

Black, F. (1976) 'The Dividend Puzzle', *Journal of Portfolio Management,* Vol. 2, No. 2, [Winter], pp. 5–8.

Black, F. (1993a) 'Estimating Expected Return', *Financial Analysts Journal,* Vol. 43, No. 2, pp. 507–528.

Black, F. (1993b) 'Return and Beta', *Journal of Portfolio Management,* Vol. 20, No. 1, pp. 8–18.

Block, S. and G. Hirt (1994) *Foundations of Financial Management* (Irwin).

Bodie, Z. and R. Merton (2000) *Finance* (Prentice-Hall).

Bowman, R.G. (1980) 'The Debt Equivalence of Leases: An Empirical Investigation', *Accounting Review,* Vol. 55, No. 2, pp. 237–253.

Bradley, M., G. Jarrell and E. Kim (1984) 'The existence of an optimal capital structure: Theory and evidence, *Journal of Finance,* Vol. 39, No. 3, pp. 857–878.

Branson, R. (1998) *Losing my Virginity* (Virgin Publishing).

Brealey, R.A., S.C. Myers and F. Allen (2005) *Principles of Corporate Finance* (McGraw-Hill).

Brennan, M. (1971) 'A Note on Dividend Irrelevance and the Gordon Valuation Model', *Journal of Finance,* Vol. 26, No. 5, [December], pp. 1115–1121.

Brennan, M. and L. Trigeorgis (eds) (2000) *Project Flexibility, Agency and Competition: New Developments in the Theory and Application of Real Options* (Oxford, Oxford University Press).

Brett, M. (2003) *How to Read the Financial Papers* (4th edn) (Random House).

Brickley, J., C. Smith and J. Zimmerman (1994) 'Ethics, Incentives, and Organisational Design', *Journal of Applied Corporate Finance,* Vol. 7, Issue 2, [Summer], pp. 20–30.

Brickley, J., C. Smith and L. Zimmerman (2003) 'Corporate Governance, Ethics and Organisational architecture', *Journal of Applied Corporate Finance,* Vol. 15, [Spring], pp. 34–45.

Brigham, E.F. and L.C. Gapenski (1996) *Financial Management Theory and Practice* (Dryden).

Bromwich, M. and A. Bhimani (1991) 'Strategic Investment Appraisal', *Management Accounting,* Vol. 69, [March], pp. 45–48.

Brooks, R. and L. Catao (2000) 'The New Economy and Global Stock Returns', *IMF Working Paper 216,* December.

Brown, P. (2006) *An Introduction to the Bond Markets* (Wiley).

Bruner, R. and J. Perella (2004) *Applied Mergers and Acquisitions* (Wiley).

Buckley, A. (1995) *International Capital Budgeting* (Prentice Hall).

Buckley, A. (2003) *Multinational Finance* (Prentice Hall).

Buckley, P.J. and M. Casson (1981) 'The Optimal Timing of a Foreign Direct Investment', *Economic Journal,* Vol. 92, No. 361, [March], pp. 75–87.

Butler, R., L. Davies, R. Pike and J. Sharp (1993) *Strategic Investment Decisions* (Routledge).

BVCA (2008) *The Economic Impact of Private Equity in the UK 2007* (The British Private Equity and Venture Capital Association).

BZW (2008) *Equity–Gilt Study,* London.

Campbell, J. and T. Vuolteenako (2004) 'Bad Beta, Good Beta', *American Economic Review,* Vol. 94, No. 5, [December], pp. 1249–1275.

Chang, K. and C. Osler (1999) 'Methodical Madness: Technical Analysis and the Irrationality of Exchange Rate Forecasts', *Economic Journal,* Vol. 109, Issue 458, [October], pp. 636–661.

Chesley, G.R. (1975) 'Elicitation of Subjective Probabilities: A Review', *Accounting Review,* Vol. 50, No. 2, [April], pp. 325–337.

Clare, A.D. and S.II. Thomas (1994) 'Macroeconomic Factors, the Arbitrage Pricing Theory and the UK Stockmarket', *Journal of Business Finance and Accounting,* Vol. 21, Issue 3, [April], pp. 309–330.

Clark, T.M. (1978) *Leasing* (McGraw-Hill).

Collier, P. and E.W. Davies (1985) 'The Management of Currency Transaction Risk by UK Multinational Companies', *Accounting and Business Research,* Vol. 15, Issue 6, [Autumn], pp. 327–334.

Collier, P., T. Cooke and J. Glynn (1988) *Financial and Treasury Management* (Heinemann).

Cooke, T.E. (1986) *Mergers and Acquisitions* (Blackwell).

Cooper, D.J. (1975) 'Rationality and Investment Appraisal', *Accounting and Business Research,* Vol. 5, No. 19, [Summer], pp. 198–202.

Copeland, T., T. Koller and J. Murrin (2000) *Valuation* (New York: Wiley).

Copeland, T.E., J.F. Weston and K. Shastri (2004) *Financial Theory and Corporate Policy,* (4th edn) (Addison-Wesley).

Cox, J., S. Ross and M. Rubinstein (1979) 'Option Pricing: A Simplified Approach', *Journal of Financial Economics,* Vol. 7, Issue 3, [September], pp. 229–263.

Czinkota, M., T. Ronkainen and M. Moffet (1994) International Business, in Dunning, J. and Lundan, S. *Multinational Enterprises and the Global Economy* (Edward Elgar).

Damodoran, A. (2002) *Investment Valuation* (Wiley).

Daniels, J.D. and L.H. Radebaugh (2004) *International Business: Environments and Operations* (10th edn), (Reading, Mass., Addison-Wesley).

Day, R., D. Allen, I. Hirst and J. Kwiatkowski (1987) 'Equity, Gilts, Treasury Bills and Inflation', *The Investment Analyst*, No. 83, [January], pp. 11–18.

DeAngelo, H. and R. Masulis (1980) 'Optimal capital structure under corporate and personal taxation', *Journal of Financial Economics*, Vol. 8, Issue 1, pp. 3–30.

Dean, J. (1951) *Capital Budgeting* (New York, Columbia University Press).

Department of Trade and Industry (1988) *Mergers Policy* (HMSO).

De Pamphilis, D. (2001) *Mergers, Acquisitions and Other Restructuring Activities* (Academic Press).

Devine, M. (2002) *Successful Mergers: Getting the People Issues Right* (The Economist/Profile Books).

Dimson, E. (1993) 'Appraisal Techniques', Proceedings of the Capital Projects Conference, May, Institute of Actuaries and Faculty of Actuaries.

Dimson, E. and R.A. Brealey (1978) 'The Risk Premium on UK Equities', *The Investment Analyst*, No. 52, [December], pp. 14–18.

Dimson, E. and P. Marsh (1982) 'Calculating the Cost of Capital', *Long Range Planning*, Vol. 15, No. 2, [April], pp. 112–120.

Dimson, E. and P. Marsh (1986) 'Event Study Methodologies and the Size Effect: The Case of UK Press Recommendations', *Journal of Financial Economics*, September.

Dimson, E., P. Marsh and M. Staunton (2002) *Triumph of the Optimists* (Princeton University Press).

Dion, C., D. Allay, D. Derain and G. Lahiri (2007) *Dangerous Liaisons: The Integration Game* (The Hay Group).

Dissanaike, G. (1997) 'Do stock market investors overreact?' *Journal of Business Finance and Accounting*, Vol. 24, No. 1, pp. 27–49.

Dixit, A. and R. Pindyck (1995) 'The Options Approach to Capital Investment', *Harvard Business Review*, Vol. 73, No. 3, [May–June], pp. 105–115.

Dobbs, R. and W. Rehm (2006) 'The value of share buybacks', in J. Rutterford, M. Upton and D. Kodwani (eds) *Financial Strategy* (Wiley).

Doyle, P. (1994) 'Setting Business Objectives and Measuring Performance', *Journal of General Management*, Vol. 20, No. 2, [Winter], pp. 1–19.

Drucker, P.F. (1981) 'Five Rules for Successful Acquisition', *Wall Street Journal*, 15 October.

Drury, J.C. and S. Braund (1990) 'The Leasing Decision: A Comparison of Theory and Practice', *Accounting and Business Research*, Vol. 20, Issue 79, [Summer], pp. 179–191.

Economist (2000) 'Making Mergers Work' (The Economist Newspaper Ltd, London).

Eiteman, D.K., A.I. Stonehill and M.H. Moffet (2003) *Multinational Business Finance* (10th edn), (Reading, Mass., Addison-Wesley).

Elton, E.J. (1970) 'Capital Rationing and External Discount Rates', *Journal of Finance*, Vol. 25, No. 3, [June], pp. 573–584.

Elton, E.J. and M. Gruber (1970) 'Marginal Stockholder Tax Rates and the Clientele Effect', *Review of Economics and Statistics*, Vol. 52, No. 1, [February], pp. 68–74.

Elton, E.J., M.J. Gruber, S.J. Brown and W.N. Goetzman (2007) *Modern Portfolio Theory and Investment Analysis* (Wiley).

Emery, D. and J. Finnerty (1997) *Corporate Financial Management* (Prentice Hall).

Evans, C. (2005) 'Private Lessons', *Accountancy*, Vol. 135, No. 1337, [January], pp. 46–47.

Fama, E. (1980) 'Agency Problems and the Theory of the Firm', *Journal of Political Economy*, Vol. 88, No. 2. [April], pp. 288–307.

Fama, E. (1998) 'Market Efficiency, Long-term Returns, and Behavioural Finance', *Journal of Financial Economics*, Vol. 49, Issue 3, [September], pp. 283–306.

Fama, E.F. (1970) 'Efficient Capital Markets: A Review of Theory and Empirical Work', *Journal of Finance*, Vol. 25, No. 2, [May], pp. 383–417.

Fama, E.F. (1991) 'Efficient Capital Markets', *Journal of Finance*, Vol. 46, No. 5, [December], pp. 1575–1617.

Fama, E. and K. French (1993) 'Common risk factors in the returns on stocks and bonds', *Journal of Financial Economics*, Vol. 33, Issue 1, pp. 3–56.

Fama, E. and K. French (1995) 'Size and book-to-market factors in earnings and returns', *Journal of Finance*, Vol. 50, No. 1, [March], pp. 131–155.

Fama, E. and K. French (1996) 'Multifactor explanations of asset pricing anomalies', *Journal of Finance*, Vol. 51, No. 1, [March], pp. 55–84.

Fama, E. and K. French (2001) 'Disappearing dividends: Changing firm characteristics or lower propensity to pay?' *Journal of Financial Economics*, Vol. 60, Issue 1, [April], pp. 3–43.

Fama, E. and K. French (2002) 'The Equity Premium', *Journal of Finance*, Vol. 57, No. 2, pp. 637–659.

Fama, E. and K. French (2004) 'The CAPM – Theory and Evidence', *Journal of Economic Perspectives*, Vol. 18, No. 3, [Summer], pp. 25–46.

Fama, E.F. and K.R. French (1992) 'The Cross-Section of Expected Stock Returns', *Journal of Finance,* Vol. 47, No. 2, [June], pp. 427–465.

Fama, E.F. and J. McBeth (1973) 'Risk, Return and Equilibrium: Empirical Tests', *Journal of Political Economy,* Vol. 81, No. 3, [May/June], pp. 607–636.

Fama, E.F. and M.H. Miller (1972) *The Theory of Finance* (Holt, Rinehart and Winston).

Feldman, S. (2005) *Principles of Private Firm Valuation* (Wiley).

Ferguson, A. (1989) 'Hostage to the Short Term', *Management Today,* March.

Finnie, J. (1988) 'The Role of Financial Appraisal in Decisions to Acquire Advanced Manufacturing Technology', *Accounting and Business Research,* Vol. 18, No. 70, [Spring], pp. 133–139.

Firth, M. and S. Keane (1986) *Issues in Finance* (Philip Allan).

Fisher, I. (1930) *The Theory of Interest*, reprinted in 1977 by Porcupine Press.

Foley, B.J. (1991) *Capital Markets* (Macmillan).

Fosback, N. (1985) *Stock Market Logic* (The Institute for Economic Research, Fort Lauderdale).

Franks, J. and J. Broyles (1979) *Modern Managerial Finance* (Wiley).

Franks, J.R. and R.S. Harris (1989) 'Shareholder Wealth Effects of Corporate Takeovers: The UK Experience 1955–85', *Journal of Financial Economics,* Vol. 23, Issue 2, pp. 225–249.

Franks, J. and C. Mayer (1996a) 'Do Hostile Take-overs Improve Performance?', *Business Strategy Review,* Vol. 7, No. 4, pp. 1–6.

Franks, J. and C. Mayer (1996b) 'Hostile Take-overs and the Correction of Managerial Failure', *Journal of Financial Economics,* Vol. 40, No. 1, pp. 163–181.

Friedman, M. (1953) 'The Methodology of Positive Economics', in *Essays in Positive Economics* (The University of Chicago Press).

Frykman, D. and J. Tolleryd (2003) *Corporate Valuation: An Easy Guide to Measuring Value* (Pearson Education).

Galati, G., A. Heath and P. McGuire (2007) 'Evidence of Carry Trade Activity', *Bank of International Settlements Quarterly Review,* [September], pp. 27–42.

Gaughan, P. (2007) *Mergers, Acquisitions and Corporate Restructuring* (Wiley).

Gentry, J. (1988) 'State of the Art of Short-Run Financial Management', *Financial Management,* Vol. 17, No. 2, [Summer], pp. 41–57.

Ghosh, C. and J. Woolridge (1989) 'Stock Market Reaction to Growth – Induced Dividend Cuts: Are Investors Myopic?', *Managerial & Decision Economics,* Vol. 10, No. 1, [March], pp. 25–35.

Giddy, I.H. (1994) *Global Financial Markets* (D.C. Heath).

Gluck, F.W. (1988) 'The Real Takeover Defense', *The McKinsey Quarterly,* Issue 1, [Winter], pp. 2–16.

Gordon, M. (1959) 'Dividends, Earnings and Stock Prices', *Review of Economics and Statistics,*Vol. 41 [May], pp. 99–105.

Gordon, M. (1963) 'Optimal Investment and Financing Policy', *Journal of Finance,* Vol. 18, No. 2, pp. 264–272.

Graham, B., D. Dodd and S. Cottle (1962) *Security Analysis: Principles and Techniques* (4th edn) (McGraw-Hill).

Grant, R.M. (2004) *Contemporary Strategy Analysis* (5th edn, Blackwell Publishers).

Grant, R.M. (2005) *Contemporary Strategic Analysis* (Basil Blackwell).

Grant, R. and L. Soenen (2004) 'Strategic Management of Operating Exposure', *European Management Journal,* Vol. 22, No. 1, [February], pp. 353–362.

Graves, S.B. (1988) 'Institutional Ownership and Corporate R&D in the Computer Industry', *Academy of Management Journal,* Vol. 31, Issue 2, [June], pp. 417–428.

Gregoriu, G. and K. Neuhauser (2007) *Mergers and Acquisitions* (Palgrave).

Gregory, A. (1997) 'An Examination of the Long-Run Performance of UK Acquiring Firms', *Journal of Business Finance and Accounting,* Vol. 24, No. 7–8, [September], pp. 971–1002.

Gregory, A. and S. McCorriston (2004) 'Foreign Acquisitions by UK Limited Companies: Short and Long-Run Performance', University of Exeter Centre for Finance and Investment Working Paper 04/01.

Grice, J.S. and R.W. Ingram (2001) 'Tests of the Generalisability of Altman's Bankruptcy Prediction Model', *Journal of Business Research,* Vol. 54, Issue 1, [October], pp. 53–61.

Grinyer, J.R. (1986) 'An Alternative to Maximisation of Shareholders' Wealth in Capital Budgeting Decisions', *Accounting and Business Research,* Vol. 16, Issue 64, [Autumn], pp. 319–326.

Grubb, M. (1993/4) 'A Second Generation of Low Inflation', *Professional Investor,* [December/January], pp. 53–57.

Gup, B.E. and S.W. Norwood (1982) 'Divisional Cost of Capital: A Practical Approach', *Financial Management,* Vol. 11, Issue 1, [Spring], pp. 20–24.

Habeck, M., F. Kroger and M.R. Traem (2000) *After the Merger* (Pearson Education).

Hamada, R.S. (1969) 'Portfolio Analysis: Market Equilibrium and Corporate Finance', *Journal of Finance*, Vol. 24, No. 1, [March], pp. 13–31.

Harar, S. (1998) 'Islamic Banking: An Overview', in J. Rutterford (ed.) *Financial Strategy: Adding Stakeholder Value* (Open Business School/Wiley).

Harrington, D.R. (1983) 'Stock Prices, Beta and Strategic Planning', *Harvard Business Review*, Vol. 6, No. 3, [May–June], pp. 157–164.

Harrington, D. (1987) *Modern Portfolio Theory, The Capital Asset Pricing Model and Arbitrage Pricing Theory: A User's Guide* (Prentice Hall).

Harris, M. and A. Raviv (1990) 'Capital structure and the informational role of debt', *Journal of Finance*, Vol. 45, No. 2, pp. 321–350.

Harris, M. and A. Raviv (1991) 'The theory of capital structure', *Journal of Finance*, Vol. 46, No. 1, pp. 297–356.

Harrison, J.S. (1987) 'Alternatives to Merger – Joint Ventures and Other Strategies', *Long Range Planning*, Vol. 20, No. 6, [December], pp. 78–83.

Haspelagh, P. and D. Jemison (1991) *Managing Acquisitions: Creating Value Through Corporate Renewal* (The Free Press).

Hassan, M.K. and M.K. Lewis (eds) (2005) *Handbook of Islamic Financing* (Edward Elgar).

Healy, P. and G. Palepu (1988) 'Earnings information conveyed by dividend initiations and omissions', *Journal of Financial Economics*, Vol. 21, Issue 2, [September], pp. 149–175.

Hertz, D.B. (1964) 'Risk Analysis in Capital Investment', *Harvard Business Review*, Vol. 42, Issue 1, [January–February], pp. 95–106.

Hexter, O. (2007) 'Positive Feeling', *Euromoney*, [September], pp. 200–206.

Hirshleifer, J. (1958) 'On the Theory of Optimal Investment Decision', *Journal of Political Economy*, Vol. 66. No. 4, [August], pp. 329–352.

Hodgkinson, L. (1989) *Taxation and Corporate Investment* (CIMA).

Horngren, C.T., A. Bhimani, G. Foster and S.M. Datar (1998) *Management and Cost Accounting* (Prentice Hall Europe).

Hunt, J., J. Grumber, S. Lees and P. Vivien (1987) *Acquisition – the Human Factors* (London Business School and Egon Zehnder Associates).

ICAEW (1989) *Accounting for Brands*, P. Barwise, C. Higson, A. Likierman and P. Marsh (Institute of Chartered Accountants in England and Wales, London Business School).

Iqbal, Z. (1999) 'Financial engineering in Islamic finance', *Thunderbird International Business Review*, Vol. 41, No. 4/5, [July–October], pp. 541–560.

Jarvis, R., J. Collis and P. Bainbridge (2000) 'The Finance Leasing Market in the 1990s: a Chronological Review', Association of Certified and Corporate Accountants Occasional Paper No. 28.

Jensen, M.C. (1984) 'Takeovers: Folklore and Science', *Harvard Business Review*, Vol. 62, No. 6, [November–December], pp. 109–121.

Jensen, M.C. (2001) 'Value maximisation, stakeholder theory, and the corporate objective function', *Journal of Applied Corporate Finance*, Vol. 14, [Fall], p. 8.

Jensen, M.C. and W.H. Meckling (1976) 'Theory of the Firm: Managerial Behaviour, Agency Costs and Ownership Structure', *Journal of Financial Economics*, Vol. 3, No. 4, [October], pp. 305–360.

Jensen, M.C. and W.H. Meckling (1994) 'The Nature of Man', *Journal of Applied Corporate Finance*, Vol. 7, Issue 2, [Summer], pp. 4–19.

Jensen, M.C. and R.S. Ruback (1983) 'The Market for Corporate Control: The Scientific Evidence', *Journal of Financial Economics*, Vol. 11, Issue 1/4, [April], pp. 5–50.

Johanson, J. and F. Wiedersheim-Paul (1975) 'The Internationalisation of the Firm – Four Swedish Cases', *Journal of Management Studies*, Vol. 12, Issue 3, [October], pp. 305–323.

Jones, C.S. (1982) *Successful Management of Acquisitions* (Derek Beattie Publishing).

Jones, C.S. (1983) *The Control of Acquired Companies* (Chartered Institute of Cost and Management Accountants).

Jones, C.S. (1986) 'Integrating Acquired Companies', *Management Accounting*, April.

JPMorgan/Cazenove (2008) European Listed Private Equity Bulletin.

Junankar, S. (1994) 'Realistic Returns: How do Manufacturers Assess New Investment?', Confederation of British Industry, July.

Jupe, R.E. and B.A. Rutherford (1997) 'The Disclosure of "Free Cash Flow" in Published Financial Statements: a Research Note', *British Accounting Review*, Vol. 29, No. 3, [September], pp. 231–243.

Kahneman, D. and A. Tversky (1979) 'Prospect Theory: An Analysis of Decisions Under Risk', *Econometrica*, Vol. 47, [March], pp. 263–291.

Kahneman, D. and A. Tversky (1982) *Judgement Under Uncertainty: Heuristics and Biases* (Cambridge University Press).

Kanodia, C., R. Bushman and J. Dickart (1989) 'Escalation Errors and the Sunk Cost Effect: An Explanation Based on Reputation and Information Asymmetries', *Journal of Accounting Research*, Vol. 27, No. 1, [Spring], pp. 59–77.

Kaplan, R.S. (1986) 'Must CIM be Justified by Faith Alone?', *Harvard Business Review,* Vol. 64, No. 2, [March–April], pp. 87–97.

Kaplanis, E. (1997) 'Benefits and Costs of International Portfolio Investments' in *Financial Times Mastering Finance* (FT/Pitman Publishing, London).

Keane, S. (1974) 'Dividends and the Resolution of Uncertainty', *Journal of Business Finance and Accountancy,* Vol. 1, Issue 3, [September], pp. 389–393.

Keane, S.M. (1983) *Stock Market Efficiency: Theory, Evidence, Implications* (Philip Allan).

Kerr, H.S. and R.J. Fuller (1981) 'Estimating the Divisional Cost of Capital: An Analysis of the Pure-Play Technique', *Journal of Finance,* Vol. 36, No. 5, [December], pp. 997–1009.

Kester, W.C. (1984) 'Today's Options for Tomorrow's Growth', *Harvard Business Review,* Vol. 62, No. 2, [March–April], pp. 153–160.

Kindleberger, C.P. (1978) *International Economics* (Irwin).

King, P. (1975) 'Is the Emphasis of Capital Budgeting Misplaced?', *Journal of Business Finance and Accounting,* Vol. 2, No. 1, [Spring], pp. 69–82.

Klemm, M., S. Sanderson and G. Luffman (1991) 'Mission Statements: Selling Corporate Values to Employees', *Long-Range Planning,* Vol. 24, No. 3, [June], pp. 73–78.

Klieman, R. (1999) 'Some new evidence on EVA companies', *Journal of Applied Corporate Finance,* Vol. 12, Issue 2, [Summer], pp. 80–91.

Koh, P. (2006) 'Leasing gives loans a run for their money', *Euromoney,* October 2006, pp. 90–92.

Koller, T., M. Goedhart and D. Wessels (2005) *Valuation: Measuring and Managing the Value of Companies* (Wiley).

KPMG (2006) *The Morning After* (KPMG Group).

Lambert, R.A. and D.F. Larcker (1985) 'Executive Compensation, Corporate Decision-making and Shareholder Wealth: A Review of the Evidence', *Midland Corporate Finance Journal,* Vol. 2, [Winter], pp. 6–22.

Larcker, D.F. (1983) 'Association between Performance Plan Adoption and Capital Investment', *Journal of Accounting and Economics,* Vol. 5, No. 1, [April], pp. 3–30.

Lee, E., M. Walker and H. Christensen (2006) *The cost of capital in Europe,* Certified Accountants Educational Trust.

Lees, S. (1992) 'Auditing Mergers and Acquisitions – Caveat Emptor', *Managerial Auditing Journal,* Vol. 7, No. 4, pp. 6–11.

Lerner, J. (2008) *The Global Economic Impact of Private Equity Report 2008, Globalization of Alternative Investments,* Volume 1 (World Economic Forum).

Lessard, D. (1985) 'Evaluating foreign projects: An adjusted present value approach', in D. Lessard (ed.) *International Financial Management* (Wiley).

Lessard, D.R. and J.B. Lightstone (2006) 'Operating Exposure', in J. Rutherford, M. Upton and D. Kodwani, *Financial Strategy* (John Wiley & Sons).

Levich, R.M. (1989) 'Is the Foreign Exchange Market Efficient?', *Oxford Review of Economic Policy,* Vol. 5, No. 3, pp. 40–60.

Levinson, M. (2002) *Guide to Financial Markets,* 3rd edn (London: Economist Books).

Levis, M. (1985) 'Are Small Firms Big Performers?', *The Investment Analyst,* No. 76, [April], pp. 21–27.

Levy, H. and M. Sarnat (1994) *Portfolio and Investment Selection: Theory and Practice* (Prentice Hall).

Levy, H. and M. Sarnat (1994) *Capital Investment and Financial Decisions* (Prentice Hall).

Limmack, R.J. (1991) 'Corporate Mergers and Shareholder Wealth Effects: 1977–1986', *Accounting and Business Research,* Vol. 21, Issue 83, [Summer], pp. 239–251.

Lintner, J. (1956) 'The Distribution of Incomes of Corporations among Dividends, Retained Earnings and Taxes', *American Economic Review,* Vol. 46, [May], pp. 97–113.

Lorie, J.H. and L.J. Savage (1955) 'Three Problems in Capital Rationing', *Journal of Business,* Vol. 28, No. 4, [October], pp. 229–239.

Luehrman, T.A. (1997a) 'What's it worth? A General Manager's Guide to Valuation', *Harvard Business Review,* Vol. 75, No. 3, [May–June], pp. 132–142.

Luehrman, T.A. (1997b) 'Using APV: A Better Tool for Valuing Operations', *Harvard Business Review,* Vol. 75, No. 3, [May–June], pp. 145–154.

Madura, J. (2006) *International Financial Management* (West Publishing Co.).

Madura, J. and A.M. Whyte (1990) 'Diversification Benefits of Direct Foreign Investment', *Management International Review,* Vol. 30, No. 1, pp. 73–85.

Manson, S., A. Stark and M. Thomas (1994) 'A Cash Flow Analysis of the Operational Gains from Takeovers', *ACCA Certified Research Report 35.*

Mao, J.C.T. and J.F. Helliwell (1969) 'Investment Decisions under Uncertainty: Theory and Practice', *Journal of Finance,* Vol. 24, No. 2, [May], pp. 323–338.

Marais, D. (1982) 'Corporate Financial Strength', *Bank of England Quarterly Bulletin,* June.

Markowitz, H.M. (1952) 'Portfolio Selection', *Journal of Finance,* Vol. 7, No. 1, [March], pp. 77–91.

Markowitz, H.M. (1991) 'Foundations of Portfolio Theory', *Journal of Finance,* Vol. 46, No. 2, [June], pp. 469–477.

Marsh, P. (1982) 'The choice between debt and equity: An empirical study', *Journal of Finance*, Vol. 37, No. 1, [March], pp. 121–144.

Marsh, P. (1990) *Short-termism on Trial* (International Fund Managers Association).

Mason, C. (2006) 'Informal Sources of Venture Finance', in S.C. Parker (ed.) *The Life Cycle of Entrepreneurial Ventures* (New York: Springer), pp. 259–299.

Mason, C. and R. Harrison (2002) 'Is it Worth it? The Rates Return from Informal Venture Capital Investments', *Journal of Business Venturing*, Vol. 17, Issue 3, [May], pp. 211–236.

Mason, C.M. and R.T. Harrison (2004) 'Improving Access to Early Stage Venture Capital in Regional Economies', *Local Economy*, Vol. 19, No. 2, [May], pp. 159–173.

Mastering Finance (1998) (Pitman).

Mathur, I. and De, S. (1989) 'A Review of the Theories of and Evidence on Returns Related to Mergers and Takeovers', *Managerial Finance*, Vol. 15, No. 4, pp. 1–11.

McDaniel, W.R., D.E. McCarty and K.A. Jessell (1988) 'Discounted Cash Flow with Explicit Reinvestment Rates: Tutorial and Extension', *The Financial Review*, Vol. 23, Issue 3, [August], pp. 369–385.

McGowan, C.B. and J.C. Francis (1991) 'Arbitrage Pricing Theory Factors and their Relationship to Macro-economic Variables', in C.F. Lee, T.J. Frecka and L.O. Scott (eds.), *Advances in Quantitative Analysis of Finance and Accounting* (JAI Press).

McIntyre, A.D. and N.J. Coulthurst (1985) 'Theory and Practice in Capital Budgeting', *British Accounting Review*, Vol. 17, No. 2, [Autumn], pp. 24–70.

McRae, T.W. (1996) *International Business Finance* (John Wiley and Sons).

Meall, L. (2001) 'Dot.com dot.gone', *Accountancy*, Vol. 128, No. 1296, [August], p. 70.

Mehra, R. and E.C. Prescott (1985) 'The Equity Premium: A Puzzle', *Journal of Monetary Economics*, Vol. 15, Issue 2, [March], pp. 145–161.

Merton, R. (1998) 'Applications of option pricing theory: twenty five years later', *American Economic Review*, Vol. 88, Issue 3, [June], pp. 323–349.

Miller, M. (1977) 'Debt and Taxes', *American Economic Review*, Vol. 32, No. 2, [May], pp. 261–275.

Miller, M. (1986) Behavioural Rationality in finance: The Case of Dividends', *Journal of Business*, Vol. 59, [October], pp. 451–468.

Miller, M. (1999) 'The history of finance: An eyewitness account', in Stern and Chew (2003) *The Revolution in Corporate Finance* (Blackwell Publishing).

Miller, M. (2000) 'The History of Finance: An eyewitness account,' *Journal of Applied Corporate Finance*, Vol. 13, [Summer], pp. 8–14.

Miller, M.H. (1991) 'Leverage', *Journal of Finance*, Vol. 46, No. 2, [June], pp. 479–488.

Miller, M.H. and F. Modigliani (1961) 'Dividend Policy, Growth and the Valuation of Shares', *Journal of Business*, Vol. 34, No. 4, [October], pp. 411–433.

Miller, M. and D. Orr (1966) 'A Model of the Demand for Money by Firms', *Quarterly Journal of Economics*, Vol. 80, Issue 2, [August], pp. 413–435.

Mills, R.W. (1988) 'Capital Budgeting Techniques Used in the UK and USA', *Management Accounting*, Vol. 61, [January], pp. 26–27.

Modigliani, F. and M. Miller (1958) 'The Cost of Capital, Corporation Finance and the Theory of Investment', *American Economic Review*, Vol. 48, No. 3, [June], pp. 261–297.

Modigliani, F. and M.H. Miller (1963) 'Corporate Income Taxes and the Cost of Capital: A Correction', *American Economic Review*, Vol. 53, No. 3, [June], pp. 433–443.

Mossin, J. (1966) 'Equilibrium in a Capital Assets Market', *Econometrica*, Vol. 34, No. 4, [October], pp. 768–783.

Murphy, J. (1989) *Brand Valuation: A True and Fair View* (Hutchinson).

Myers, S. and N. Majluf (1984) 'Corporate financing and decisions when firms have information that investors do not have', *Journal of Financial Economics*, Vol. 13, No. 2, pp. 187–221.

Myers, S.C. (1974) 'Interactions of Corporate Financing and Investment Decisions—Implications for Capital Budgeting', *Journal of Finance*, Vol. 29, No. 1, [March], pp. 1–25.

Myers, S.C., D.A. Dill and A.J. Bautista (1976) 'Valuation of Lease Contracts', *Journal of Finance*, Vol. 31, No. 3, [June], pp. 799–820.

Myers, S.C. (1984) 'The Capital Structure Puzzle', *Journal of Finance*, Vol. 39, No. 3, [July], pp. 575–592.

Narayanaswamy, V.J. (1994) 'The Debt Equivalence of Leases in the UK: An Empirical Investigation, *British Accounting Review*, Vol. 26, No. 1, pp. 337–351.

Neale, B., A. Milsom, C. Hills and J. Sharples (1998) 'The Hostile Takeover Process: A Case Study of Granada Versus Forte', *European Management Journal*, Vol. 16, No. 2, [April], pp. 230–241.

Neale, C.W. and D.E.A. Holmes (1988) 'Post-Completion Audits: The Costs and Benefits', *Management Accounting*, Vol. 66, No. 3, [March], pp. 27–30.

Neale, C.W. and D.E.A. Holmes (1990) 'Post-auditing Capital Investment Projects', *Long Range Planning*, Vol. 23, No. 4, [August], pp. 88–96.

Neale, C.W. and P.J. Buckley (1992) 'Differential British and US Adoption Rates of Investment Project Post-Completion Auditing', *Journal of International*

Business Studies, Third Quarter, Vol. 23, No. 3, pp. 419–442.

Neale, C.W. and D.E.A. Holmes (1991) *Post-Completion Auditing* (Pitman).

O'Shea, D. (1986) *Investing for Beginners*, Financial Times Business Information.

Owen, G., J. Black and S. Arcot (2007) *From Local to Global – The Rise of the AIM* (London Stock Exchange Publications).

Payne, A.F. (1987) 'Approaching Acquisitions Strategically', *Journal of General Management*, Vol. 13, Issue 2, [Winter], pp. 5–27.

Peacock, A. and G. Bannock (1991) *Corporate Takeovers and the Public Interest* (David Hume Institute).

Pearce, R. and S. Barnes (2002) *Raising Venture Capital* (Wiley).

Peel, M.J. (1995) 'The Impact of Corporate Re-structuring: Mergers, Divestments and MBOs', *Long Range Planning*, Vol. 28, No. 2, [April], pp. 92–101.

Peters, E.E. (1991) *Chaos and Order in the Capital Markets* (Wiley).

Peters, E.E. (1993) *Fractal Market Analysis* (Wiley).

Pettit, J. (2001) 'Is a Share Buyback Right for Your Company?', *Harvard Business Review*, Vol. 79, No. 4, [October], pp. 141–147.

Pettit, R. (1984) 'Dividend announcements, security performance, and capital market efficiency, *Journal of Finance*, Vol. 27, No. 4, [September], pp. 993–1007.

Pike, R. (1996) 'A Longitudinal Survey on Capital Budgeting Practices', *Journal of Business Finance and Accounting*, Vol. 23, No. 1, [January], pp. 79–92.

Pike, R.H. (1982) *Capital Budgeting in the 1980s* (Chartered Institute of Management Accountants).

Pike, R.H. (1983) 'The Capital Budgeting Behaviour and Corporate Characteristics of Capital-Constrained Firms', *Journal of Business Finance and Accounting*, Vol. 10, No. 4, [Winter], pp. 663–671.

Pike, R.H. (1988) 'An Empirical Study of the Adoption of Sophisticated Capital Budgeting Practices and Decision-Making Effectiveness', *Accounting and Business Research*, Vol. 18, No. 2, [Autumn], pp. 341–351.

Pike, R.H. and S.M. Ho (1991) 'Risk Analysis Techniques in Capital Budgeting Contexts', *Accounting and Business Research*, Vol. 21, No. 83, pp. 227–238.

Pike, R.H. and M. Wolfe (1988) *Capital Budgeting in the 1990s* (Chartered Institute of Management Accountants).

Pike, R., J. Sharp and D. Price (1989) 'AMT Investment in the Larger UK Firm', *International Journal of Operations and Production Management*, Vol. 9, No. 2, pp. 13–26.

Pike, R., N. Cheng and L. Chadwick (1998) *Managing Trade Credit for Competitive Advantage* (CIMA).

Pilbeam, K. (2006) *International Finance* (Palgrave Macmillan).

Pinches, G. (1982) 'Myopic Capital Budgeting and Decision-Making', *Financial Management*, Vol. 11, No. 3, [Autumn], pp. 6–19.

Pohlman, R.A., E.S. Santiago and F.L. Markel (1988) 'Cash Flow Estimation Practices of Larger Firms', *Financial Management*, Vol. 17, No. 2, [Summer], pp. 71–79.

Pointon, J. (1980) 'Investment and Risk: The Effect of Capital Allowances', *Accounting and Business Research*, Vol. 10, Issue 40, [Autumn].

Poon, S. and S.J. Taylor (1991) 'Macroeconomic Factors and the UK Stock Market', *Journal of Business Finance and Accounting*, Vol. 18, Issue 5, [September], pp. 619–636.

Porter, M.E. (1985) *Competitive Advantage* (Free Press).

Porter, M.E. (1987) 'From Competitive Advantage to Corporate Strategy', *Harvard Business Review*, Vol. 65, No. 3, [May–June], pp. 43–59.

Porter, M.E. (1992) 'Capital Disadvantage – America's Failing Capital Investment System', *Harvard Business Review*, Vol. 70, No. 4, [September–October], pp. 65–82.

Prasad, S.B. (1987) 'American and European Investment Motives in Ireland', *Management International Review* (Third quarter).

Price, J. and S.K. Henderson (1988) *Currency and Interest Rate Swaps* (Butterworths).

Prindl, A. (1978) *Currency Management* (John Wiley).

Pritchett, P., D. Robinson and R. Clarkson (1997) *After the Merger* (McGraw-Hill).

Pruitt, S.W. and L.J. Gitman (1987) 'Capital Budgeting Forecast Biases: Evidence from the Fortune 500', *Financial Management*, Vol. 16, No. 1, [Spring], pp. 46–51.

Quah, P. and S. Young (2005) 'Post-merger integration: A phases approach for cross-border M&As', *European Management Journal*, Vol. 23, No. 1, pp. 65–75.

Rajan, R. and L. Zingales (1995) 'What do we know about capital structure? Some evidence from international data', *Journal of Finance*, Vol. 11, No. 3, pp. 1421–1460.

Rankine, D., G. Steadman and M. Borner (2003) *Due Diligence* (Financial Times/Prentice Hall).

Rappaport, A. (1986) *Creating Shareholder Value: The New Standard for Business Performance* (Macmillan).

Rappaport, A. (1987) 'Stock Market Signals to Managers', *Harvard Business Review*, Vol. 65, No. 6, [November–December], pp. 57–62.

Redhead, K. (1990) *Introduction to Financial Futures and Options* (Woodhead-Faulkner).

Reimann, B.C. (1990) 'Why Bother with Risk-Adjusted Hurdle Rates?' *Long Range Planning*, Vol. 23, Issue 3, [June], pp. 57–65.

Risk Measurement Service, London Business School.

Ritter, J. (2006) 'Initial Public Offerings', in J. Rutterford, M. Upton and D. Kodwani (eds) *Financial Strategy*, 2nd edn (Wiley).

Robbins, S. and R. Stobaugh (1973) 'The Bent Measuring Stick for Foreign Subsidiaries', *Harvard Business Review*, Vol. 51, No. 5, [September–October], pp. 80–88.

Rodriguez, R.M. (1981) 'Corporate Exchange Risk Management: Theme and Aberrations', *Journal of Finance*, Vol. 36, No. 2, [May], pp. 427–444.

Roll, R. (1977) 'A Critique of the Asset Pricing Theory's Tests; Part I: On Past and Potential Testability of the Theory', *Journal of Financial Economics*, Vol. 4, No. 2, [March], pp. 129–176.

Ross, S.A. (1976) 'The Arbitrage Theory of Capital Asset Pricing', *Journal of Economic Theory*, Vol. 13, No. 3, pp. 341–360.

Ross, S.A. (1977) 'The Determination of Financial Structure: the Incentive Signalling Approach', *Bell Journal of Economics*, Vol. 8, No. 1, [Spring], pp. 23–40.

Ross, S., R. Westerfield and J. Jaffe (2005) *Corporate Finance* (McGraw-Hill).

Ross, S., R. Westerfield and J. Jaffe (2002) *Fundamentals of Corporate Finance* (McGraw-Hill).

Ross, S., R. Westerfield and B. Jordan (2008) *Corporate Finance Fundamentals*, 8th edn (New York: McGraw-Hill).

Ruback, R.S. (1988) 'An Overview of Takeover Defenses', in A.J. Auerbach (ed.), *Mergers and Acquisitions* (University of Chicago Press).

Rubinstein, M. (2002) 'Markowitz's "Portfolio Selection": A fifty-year retrospective', *Journal of Finance*, Vol. 57, No. 3, pp. 1041–1045.

Rugman, A.M., D.J. Lecraw and L.D. Booth (1985) *International Business, Firm and Environment* (McGraw-Hill).

Rutterford, J. (1992) *Handbook of UK Corporate Finance* (Butterworths).

Sartoris, W. and N. Hill (1981) 'Evaluating Credit Policy Alternatives: A Present Value Framework', *Journal of Financial Research*, Vol. 4, Issue 1, [Spring], pp. 81–89.

Shanken, J., R. Sloan and S. Kothari (1995) 'Another look at the cross-section of expected stock returns', *Journal of Finance*, Vol. 50, No. 1, [March], pp. 185–224.

Shao, L.P. (1996) (ed) 'Capital Budgeting for the Multi-national Enterprise', *Managerial Finance*, Vol. 22, No. 1.

Shapiro, A.C. (2006) *Multinational Financial Management* (Wiley).

Sharpe, P. and T. Keelin (1998) 'How SmithKline Beecham Makes Better Resource-Allocation Decisions', *Harvard Business Review*, Vol. 76, No. 3, [March/April], pp. 45–57.

Sharpe, W. (1981) *Investments* (Prentice Hall).

Sharpe, W.F. (1963) 'A Simplified Model for Portfolio Analysis', *Management Science*, Vol. 9, Issue 2, [January], pp. 277–293.

Sharpe, W.F. (1964) 'Capital Asset Prices – A Theory of Market Equilibrium under Conditions of Risk', *Journal of Finance*, Vol. 19, No. 3, [September], pp. 425–442.

Sharpe, W.F. and G.J. Alexander and Bailey (1999) *Investments* (Prentice Hall).

Shefrin, H. (1999) *Beyond Greed and Fear: Understanding Behavioural Finance and the Psychology of Investing* (Harvard Business School Press).

Short, T. (2000) 'Should Foreign Investors Buy Polish Shares?' in T. Kowalski and S. Letza (eds) *Financial Reform and Institutions* (Poznan University of Economics).

Smith, B.M. (2001) *Toward Rational Exuberance: The evolution of the Modern Stock Market* (Farrar, Strauss & Giroux).

Smith, K.V. (1988) *Readings in Short-term Financial Management* (West Publishing).

Solnik, B.H. (1974) 'Why not Diversify Internationally Rather than Domestically?', *Financial Analysts Journal*, Vol. 30, No. 4, [July/August], pp. 48–54.

Staw, B.M. (1976) 'Knee-deep in the Big Muddy: A Study of Escalating Commitment to a Chosen Course of Action', *Organisational Behaviour and Human Performance*, Vol. 16, No. 1, [June], pp. 27–44.

Staw, B.M. (1981) 'The Escalation of Commitment to a Chosen Course of Action', *Academy of Management Review*, Vol. 6, No. 4, [October], pp. 577–587.

Staw, J. and S. Ross (1987) 'Knowing When to Pull the Plug', *Harvard Business Review*, Vol. 65, No. 2, [March/April], pp. 68–74.

Stonham, P. (1995) 'Reuter's Share Re-purchase', *European Management Journal*, Vol. 13, No. 1, pp. 99–109.

Strong, N. and X.G. Xu (1997) 'Explaining the Cross-Section of UK Expected Stock Returns', *British Accounting Review*, Vol. 29, No. 1, [March], pp. 1–23.

Sudarsanam, P.S. (1995) *The Essence of Mergers and Acquisitions* (Prentice Hall).

Sudarsanam, P., P. Holl and A. Salami (1996) 'Shareholder Wealth Gains in Mergers: Effect of Synergy and Ownership Structure', *Journal of Business Finance and Accounting*, Vol. 23, No. 5/6, [July], pp. 673–698.

Sudarsanam, S. (2003) *Creating Value Through Mergers and Acquisitions: The Challenges* (FT/Prentice Hall).

Swalm, R.O. (1966) 'Utility Theory – Insights into Risk-taking', *Harvard Business Review,* Vol. 44, Issue 6, [November–December], pp. 123–136.

Taffler, R. (1991) 'Z-scores: An Approach to the Recession', *Accountancy,* July.

Thaler, R., R. Michaely and S. Bernatzi (1997) 'Do changes in dividend policy signal the past or the future?' *Journal of Finance,* Vol. 52, Issue 3, [July], pp. 1007–1034.

Tobin, J. (1958) 'Liquidity Preference as Behaviour Towards Risk', *Review of Economic Studies,* Vol. 25, [February], pp. 65–86.

Tomkins, C. (1991) *Corporate Resource Allocation* (Basil Blackwell).

Tomkins, C.R., J.F. Lowe and E.J. Morgan (1979) *An Economic Analysis of the Financial Leasing Industry* (Saxon House).

UNCTAD (2007) *World Investment Report,* United Nations Conference on Trade and Development **(www.unctad.org).**

Vaga, T. (1990) 'The Coherent Market Hypothesis', *Financial Analysts Journal,* Vol. 46, No. 6, [November/December], pp. 36–49.

van Horne, J. (1975) 'Corporate Liquidity and Bankruptcy Costs', Research Paper 205, Stanford University.

van Horne, J. (2001) *Financial Management and Policy* (Prentice Hall).

Vermaelen, T. (1981) 'Common stock repurchases and market signalling: An empirical study', *Journal of Financial Economics,* Vol. 9, Issue 2, pp. 139–183.

Very, P. (2004) *Management of Mergers and Acquisitions* (Wiley).

Wallace, J.S. (2003) 'Value maximization and stakeholder theory: compatible or not?' *Journal of Applied Corporate Finance,* 15, [Spring], pp. 120–127.

Walters, A. (1991) *Corporate Credit Analysis* (Euromoney Publications).

Wardlow, A. (1994) 'Investment Appraisal Criteria and the Impact of Low Inflation', Bank of England *Quarterly Bulletin,* Vol. 34, No. 3, [August], pp. 250–254.

Warner, J. (1977) 'Bankruptcy costs: Some evidence', *Journal of Finance,* Vol. 32, No. 2, pp. 337–347.

Watts, R. (1973) 'The information content of dividends', *Journal of Business,* Vol. 46, No. 2, [April], pp. 191–211.

Wearing, R.T. (1989) 'Cash Flow and the Eurotunnel', *Accounting & Business Research,* Vol. 20, Issue 77, [Winter], pp. 13–24.

Weaver, S.C., D. Peters, R. Cason and J. Daleiden (1989) 'Capital Budgeting', *Financial Management,* Vol. 18, No. 1, [Spring], pp. 10–17.

Weaver, S. (1989) 'Divisional Hurdle Rates and the Cost of Capital', *Financial Management,* Vol. 18, [Spring], pp. 18–25.

Weingartner, H. (1977) 'Capital Rationing: Authors in Search of a Plot', *Journal of Finance,* Vol. 32, No. 5, [December], pp. 1403–1431.

Weston, J.F. and T.E. Copeland (1992) *Managerial Finance* (Cassell).

Wilkie, A.D. (1994) 'The Risk Premium on Ordinary Shares', Institute of Actuaries and Faculty of Actuaries, November.

Wilson Committee (1980) 'Report of the Committee to Review the Functioning of Financial Institutions', Cmnd. 7937, HMSO.

Wilson, M. (1990) (ed) 'Capital Budgeting for Foreign Direct Investments', *Managerial Finance,* Vol. 16, No. 2.

Wood, A. (2006) 'Death of the Dividend?' in J. Rutterford, M. Upton and D. Kodwani (eds) *Financial Strategy* (Wiley).

Wright, M. and K. Robbie (1991) 'Corporate Restructuring, Buy-Outs and Managerial Equity: The European Dimension', *Journal of Applied Corporate Finance,* Vol. 3, Issue 4, [Winter], pp. 47–58.

Young, D. (1997) 'Economic Value Added: A Primer for European Managers', *European Management Journal,* Vol. 15, No. 4, [August].

Young, D. and B. Sutcliffe (1990) 'Value Gaps – Who is Right? The Raiders, the Market or the Manager?', *Long Range Planning,* Vol. 23, No. 4, [August], pp. 20–34.

Young, S. and S. O'Byrne (2001) *EVA and Value-Based Management: A Practical Guide to Implementation* (McGraw-Hill).

Index

Note: Page entries in **bold** refer to terms in the Glossary